Foundations of Molecular Pharmacology

Volume 1

Medicinal and Pharmaceutical Chemistry

# Foundations of Molecular Pharmacology

## Volume 1

# Medicinal and Pharmaceutical Chemistry

J. B. STENLAKE

PH.D., D.SC., F.P.S., C. CHEM., F.R.I.C., F.R.S.E.

*Professor of Pharmacy and Pharmaceutical Chemistry*
*University of Strathclyde, Glasgow*

THE ATHLONE PRESS *of the University of London* 1979

Carnegie Library
Livingstone College
28144

*Published by*
THE ATHLONE PRESS
UNIVERSITY OF LONDON
4 *Gower Street, London* WC1

*Distributed by Tiptree Book Services Ltd*
*Tiptree, Essex*

*USA and Canada*
*Humanities Press Inc*
*New Jersey*

© *J. B. Stenlake* 1979

*British Library Cataloguing in Publication Data*
Stenlake, John Bedford
Foundations of molecular pharmacology
Vol 1: Medicinal and pharmaceutical chemistry
1. Pharmacology
I. Title
615'.1 RM300
ISBN 0 485 11171 3

*Set in Monophoto Times by*
COMPOSITION HOUSE LTD, SALISBURY

*Printed by photolithography in Great Britain by*
WHITSTABLE LITHO LTD, WHITSTABLE, KENT

615.1
St825

# Preface

This text has emerged from some thirty years of teaching undergraduate courses and conducting research in medicinal and pharmaceutical chemistry. It is conceived essentially as a foundation course in the basic principles of organic chemistry applied to the study of medicinal agents and the formulations in which they are used. It is intended primarily to cater for the needs of undergraduate students of pharmacy and medicinal chemistry up to Honours level. References to original papers, however, should extend its use to postgraduate students and others engaged in the search for new drugs.

My intention was to contain the text within the covers of a single volume, concentrating essentially on the fundamental groundwork chemistry which must of necessity be taught in any undergraduate course. Experience, however, has shown the value of more general discussion of certain selected topics of wide general applicability in the study of drug action, and it was always my objective to conclude the book in this way. In the event, I have been defeated, partly by the ramifications of the subject, but mainly by my enthusiasm and attempts to achieve a realistic degree of coverage. Publishing costs, too, have risen enormously in the ten years of writing. Attempts to overcome the twin difficulties of coverage and cost, therefore, left no alternative other than to divide the book between two volumes.

It is just possible that some readers may find virtue in the necessity which has forced this publication of the Foundations of Molecular Pharmacology in two separate volumes. I hope, nonetheless, that serious students will not be deterred by this somewhat artificial division from pursuing the broader approach to the subject contained in Volume 2. In order, therefore, to reinforce the continuity of the subject, I have provided a system of cross-referencing between chapters, both within and between the two volumes. Such cross-references are denoted by two numbers, the first indicating volume, and the second chapter; thus, for example, (**1**, 13) indicates Volume 1, Chapter 13, and (**2**, 5) Volume 2, Chapter 5.

The basic philosophy underlying the text is that those concerned with the design and use of drugs and medicines are interested fundamentally in properties rather than in methods of manufacture. Accordingly, the chemistry in this book almost entirely ignores the synthesis of medicinal agents. Instead, attention is focused, in Volume 1, on the physical and chemical properties of medicinal agents, pharmaceutical additives and cellular components, that determine the way in which they interact with each other. To achieve this end, substantial accounts of relevant intermediary tissue metabolism, drug transport and metabolism, and other factors affecting both stability and availability of drugs from

#90.00 Ballen Direct 10-31-79
108670

dosage forms have been brought together in the general body of the text. This approach emphasises the close similarity between chemical and biochemical transformations, and should help to give students and others engaged in the design of new drugs a better understanding of the fundamental mechanisms which control interactions between drugs and body chemistry.

The more general, but essentially similar approach to the Chemical Basis of Drug Action adopted in Volume 2, which reinforces the basic principles for the specialist, should also appeal in its own right to clinical pharmacologists and others whose interests lie rather more in the action and use of drugs than in their design.

Since this book is designed to assist in the education of students, many of whom will be engaged in later life in the handling and use of drugs in practice, I have deliberately chosen to draw my examples from drugs in current use in western medicine. My text, however, is essentially British, and British Approved Names, denoted by italics, are used throughout, notwithstanding the difficulties that this may make for North American readers. Fortunately, British and American drug nomenclature is convergent, but where important and confusing differences still exist, I have endeavoured to overcome them by also giving the United States Adopted Name.

It is an unfortunate fact of life that the vast majority of modern drugs have chemical structures which are infinitely more complex than those of the simple examples commonly used in most textbooks of organic chemistry. Indeed, their very complexity frequently presents an educational hurdle, so that students of medicinal and pharmaceutical chemistry often fail to grasp the essential simplicity of drug action mechanisms and transformations. I am, therefore, most grateful to the publishers for their help and co-operation in the use of printing devices involving bold type and colour to focus attention on the simple stepwise transformations of otherwise complex compounds.

I am very much indebted to my colleagues, past and present, and friends, who between them provided the stimulus to write this book, and all those who, once I was embarked upon it, so patiently answered my questions, and helped to resolve the many problems I inevitably encountered. I am especially grateful to Dr G. A. Smail and Dr R. E. Bowman, both of whom read the entire original draft and commented so helpfully upon it. I am sure others will still find errors, oversights and misconceptions, but there would have been many more without the help of these two colleagues. For similar reasons, I am also grateful for all the many valuable comments and criticisms I received from the Athlone Press's own anonymous referees. My most grateful thanks are also due to Tom Moody for help with the preparation of diagrams, to Dr N. C. Dhar for assistance in locating and checking references, and especially to my ever willing Secretary, Mrs Sylvia Cohen, for her invaluable help in typing the manuscript, for countless hours devoted to the dull routine of checking text and references at every stage right through to the final proofs, and for her help in compiling the index.

The time I have taken to write this book has been taken away from many things I might otherwise have done, and most of all, taken from my wife, Anne, and our family. Their tolerance and support made it possible. I have tried to make this book one that they, too, can be proud of, and worthy of the hours of pleasure in their company which I have sacrificed.

1978 John B. Stenlake

# Contents

# Contents of Volume 2

# 1 Introduction

Pharmaceutical Chemistry is the study of the chemistry of drugs. It is concerned primarily with the chemical and physico-chemical properties of drugs insofar as these are relevant to an understanding of their action on living tissue. Drug action, also, can only be explained in its ultimate analysis in terms of chemical and physico-chemical reactions of the drug with the chemical constituents of living matter. Of necessity, therefore, the subject also embraces the study of the so-called receptor molecules of living tissue with which drugs react, and in this sense Pharmaceutical Chemistry can be considered to bear the same relationship to Biochemistry as Pharmacology does to Physiology. It is based, however, on the study of chemical properties and chemical reactions rather than physiological reactions, and employs the art, technique and method of the chemist, with all the refinements of modern physico-chemical methods of analysis to study both drug and drug–tissue–receptor interactions.

The very complexity of living matter even in the simplest organism presents a phenomenal problem to the would-be-investigator of the chemical reactions of drugs with the molecules of living tissues. Often all too little is known of the precise structure of receptor molecules to permit analysis of the exact mechanism of their reaction with drugs. This is especially true of the natural biopolymers such as peptides, proteins, nucleic acids and carbohydrates. For this reason, more than any other, Pharmaceutical Chemistry was for many years concerned almost entirely with the study of the properties of the drug molecules themselves. This aspect of the subject, limited though it is, is no less essential than it ever was, but the refinements which are now possible with modern separation and analytical methods are slowly allowing extension of the subject to include interpretation of the all-important drug–receptor interactions in the basically chemical, rather than merely descriptive terms, which is so essential to the full development of the subject.

Pharmaceutical Chemistry must, therefore, eventually aim to explain both the symptomatic and curative properties of drugs in man and domestic animals, and in the same way also to provide an understanding of the toxic reactions of drugs which have selective host-parasite action, and which are used to control and eradicate infection by pathogenic bacteria, viruses, and multi-cellular parasites. The complex structural and tissue organisation of higher animals requires consideration of the chemical and physico-chemical properties of drugs responsible for the various processes at tissue level, which together lead

Table 1. The Periodic Table of Elements

| | METALS | | | | | | | | | | | NON-METALS | | | | | |
|---|---|---|---|---|---|---|---|---|---|---|---|---|---|---|---|---|---|
| Group | IA | IIA | IIIB | IVB | VB | VIB | VIIB | VIII | | | IB | IIB | IIIA | IVA | VA | VIA | VIIA | 0 |
| Period | | | | | | | | | | | | | | | | | | |
| 1 | H | | | | | | | | | | | | | | | | | He |
| 2 | Li | Be | | | | | | | | | | | B | C | N | O | F | Ne |
| 3 | Na | Mg | Transition metals | | | | | | | | | | Al | Si | P | S | Cl | Ar |
| 4 | K | Ca | Sc | Ti | V | Cr | Mn | Fe | Co | Ni | Cu | Zn | Ga | Ge | As | Se | Br | Kr |
| 5 | Rb | Sr | Y | Zr | Nb | Mo | Tc | Ru | Rh | Pd | Ag | Cd | In | Sn | Sb | Te | I | Xe |
| 6 | Cs | Ba | La | Hf | Ta | W | Re | Os | Ir | Pt | Au | Hg | Tl | Pb | Bi | Po | At | Rn |
| 7 | Fr | Ra | Ac | | | | | | | | | | | | | | | |
| Lanthanides | | | | | Ce | Pr | Nd | Pm | Sm | Eu | Gd | Tb | Dy | Ho | Er | Tm | Yb | Lu |
| Actinides | | | | | Th | Pa | U | Np | Pu | Am | Cm | Bk | Cf | Es | Fm | Md | No | Lw |

to the observed response. Consideration must, therefore, be given to each of the following gross factors which can influence response to the drug:

(a) absorption and transport,
(b) selectivity of location,
(c) selectivity of action,
(d) bio-transformation,
(e) excretion.

Appreciation of the rôle of these factors in drug–receptor interactions at molecular level rests in the first instance on an understanding of the forces leading to bonding between atoms and molecules, and of other relevant forms of non-bonded interaction between molecular species. Secondly, it requires an understanding of the influence which small modifications in chemical structure of drugs can have on their physical properties and chemical reactivity. The modifying influence of the additives and processes used in pharmaceutical formulation must also be considered as factors affecting the mechanism of drug–receptor interactions.

Pharmaceutical chemistry is concerned largely with organic compounds. A number of inorganic compounds, including water, various elements, their ions and certain complex ions are also of considerable importance, and require consideration for a full understanding of the action of drugs. The **Periodic Table** of elements is set out (Table 1) for reference.

## ATOMIC STRUCTURE

### The Atom

Atoms consist essentially of a positively-charged **nucleus** made up of a number of **protons** and **neutrons** surrounded by associated negatively-charged **electrons**. Since protons are positively charged and neutrons have no charge, the overall charge of the nucleus is determined by the number of protons it contains. Each element has a different number of protons in its nucleus and this number is known as the **Atomic Number** of the element. The mass of the atom is concentrated almost entirely in the nucleus since the mass of an electron is only about 1/1840th of that of a proton or neutron. The nett charge on any atom is zero, hence the number of nuclear protons must equal the number of extra-nuclear electrons.

Many elements have several different **isotopes**. Hydrogen, for example, has an Atomic Number of 1. There is one proton in the nucleus, hence one extra-nuclear electron. The hydrogen nucleus can, however, contain either none, one or two *neutrons* and so there are three forms of the element. All have one proton and one electron and thus identical chemical properties. The isotope with one neutron and mass number, two, is the isotope, deuterium. That with two neutrons and mass number, three, is tritium.

**Extra-nuclear Electrons**

The extra-nuclear electrons can be considered either as particles rotating in orbits about the nucleus, or as stationary waves having a maximum amplitude at a surface or **orbital** about the nucleus. According to the classical theory of Bohr and Sommerfeld, based on the observation of the emission spectrum of hydrogen, the extra-nuclear electron of hydrogen can be considered as a particle orbiting about the nucleus in a defined 'shell' with a fixed radius determined by the electrostatic force between the electron and the nucleus. The wave-like properties of electrons were proposed by De Broglie, who drew an analogy between the behaviour of light and other forms of radiation involving electrons. The problem of measuring the precise location and momentum of an electron gave rise to the **Heisenberg uncertainty principle**. This stated that both the precise position and the momentum of an electron cannot ever be known at the same time. Application of this concept in wave mechanics has therefore led to a modification of the classical concept of electron orbits, in which the *probability* of finding an electron at various distances from the nucleus is defined by a three-dimensional **orbital**. Thus, the orbital delineates a three-dimensional surface in relation to the nucleus of the atom, in which there is the highest probability of finding a particular electron.

The energy level of each orbital is defined by principal, $n$, orbital, $l$, and magnetic, $m$, quantum numbers, with restrictions on the possible values of $l$ and $m$ as in the Bohr–Sommerfeld concept. Each orbital, however, may accommodate only two electrons, which by the **Pauli exclusion principle** must have opposed spins, since according to this principle no two electrons in the same atom can have the same values for the four quantum numbers, $n$, $l$, $m$ and $s$. It is this principle which establishes that each electron in an atom is unique in its total energy content, and which therefore determines the total number of electrons which can be accommodated at each of the principal energy levels (or shells in the classical concept). Thus, when $n = 1$, the only possible value of $l$ is zero. Therefore, there is only one orbital at this energy level, which at most can accommodate two electrons with opposed spins. This is the so-called 1*s* orbital, which is present in hydrogen and helium, containing one and two electrons respectively (p. 5). The terminology, 1*s*, which is used to describe this orbital, defines the principal quantum number, one; the letter *s*, however, is derived from a reference to the associated line spectrum and not to the spin quantum number and refers to the fact that $l = 0$.

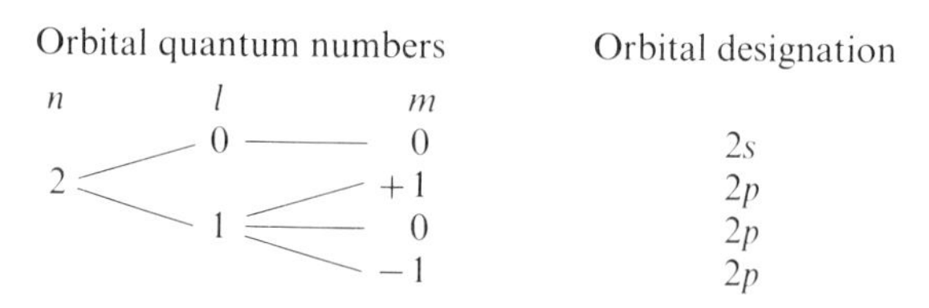


Fig. 1 Permissible values of azimuthal and magnetic quantum numbers when $n = 2$

Similarly, when $n = 2$, possible values of $l$ are 0 and 1, and possible values of $m$ are $+1$, 0, and $-1$. Figure 1 shows the combinations of these permissible values for $n$, $l$ and $m$ and defines four orbitals, each of which is capable of accommodating a maximum of two electrons with opposed spins ($s = +\frac{1}{2}$ or $-\frac{1}{2}$). These four orbitals are the 2*s* and three 2*p* orbitals which are progressively filled in the elements lithium, beryllium, boron, carbon, nitrogen, oxygen, fluorine and neon in the first short period of the Periodic Table (Table 1).

Table 2 shows the possible combinations of values attributable to the principal orbital and magnetic quantum numbers, which define the various atomic orbitals and determine the maximum possible number of electrons which can be accommodated at each of the principal energy levels from one to five. The order in which individual orbitals are filled is shown in Table 3. This depends, however, on their relative energy levels (Fig. 2) so that they are not filled exactly in sequence.

The shapes of atomic orbitals are important, and are determined by the energy state of their electrons. The two 1*s* electrons occupy a single three-dimensional orbital (1*s* orbital) which is spherically symmetrical about the nucleus, but separated from it by a region or spherical nodal surface in which the probability of finding an electron approaches zero. Figure 3 is a planar (two-dimensional) representation showing the relationship of the 1*s* orbital to the nucleus in the atoms of hydrogen and helium as might be seen in sectional view. The only differences between the orbitals of the two atoms are that of size, and that in hydrogen, the orbital contains a single unpaired electron, whilst in helium, there are two electrons with opposite spins in the one orbital.

Table 2. Disposition of Electrons in Shells and Energy States

| Quantum number | | | Orbital designation | Number of electrons |
|---|---|---|---|---|
| $n$ | $l$ | $m$ | | |
| 1 | 0 | 0 | 1*s* | 2 |
| 2 | 0 | 0 | 2*s* | 2 |
| | 1 | 0, $\pm 1$ | 2*p* | 6 |
| 3 | 0 | 0 | 3*s* | 2 |
| | 1 | 0, $\pm 1$ | 3*p* | 6 |
| | 2 | 0, $\pm 1$, $\pm 2$ | 3*d* | 10 |
| 4 | 0 | 0 | 4*s* | 2 |
| | 1 | 0, $\pm 1$ | 4*p* | 6 |
| | 2 | 0, $\pm 1$, $\pm 2$ | 4*d* | 10 |
| | 3 | 0, $\pm 1$, $\pm 2$, $\pm 3$ | 4*f* | 14 |
| 5 | 0 | 0 | 5*s* | 2 |
| | 1 | 0, $\pm 1$ | 5*p* | 6 |
| | 2 | 0, $\pm 1$, $\pm 2$ | 5*d* | 10 |
| | 3 | 0, $\pm 1$, $\pm 2$, $\pm 3$ | 5*f* | 14 |
| | 4 | 0, $\pm 1$, $\pm 2$, $\pm 3$, $\pm 4$ | 5*g* | 18 |

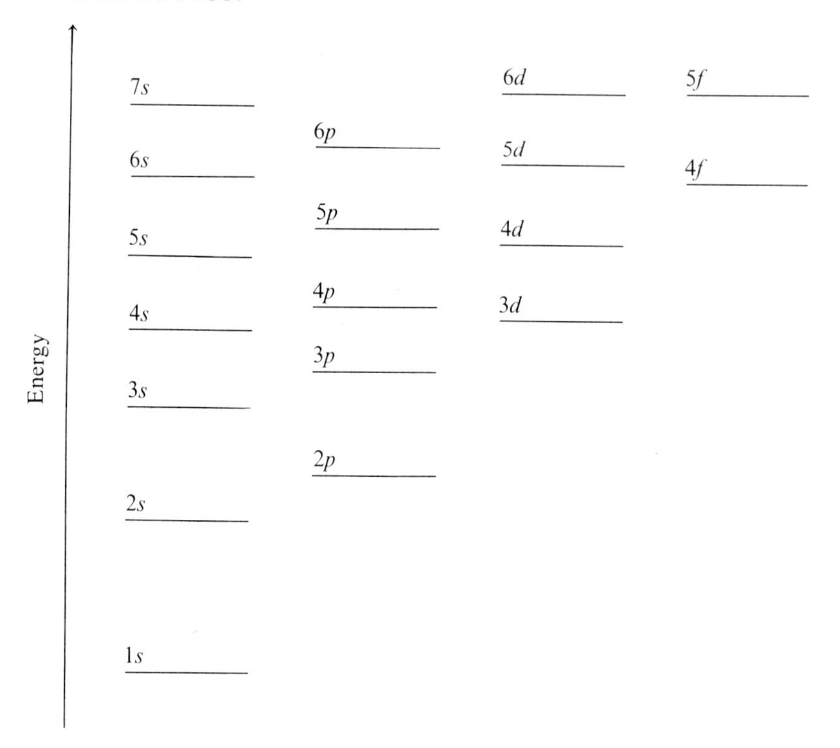


Fig. 2 Relative energy levels of atomic orbitals

It must be remembered, however, that Fig. 3 is merely a sectional two-dimensional representation of a three-dimensional structure. The spatial relationship between the nucleus and the orbital could perhaps be imagined as similar to that of the pip at the centre of an orange relative to the skin of the orange.

Similarly, the 2*s* valency electrons of lithium and beryllium, elements three and four in the Periodic Table, occupy a 2*s* orbital, which is spherically symmetrical about both the nucleus and the 1*s* orbital, and separated from the latter by a spherical nodal plane in which the probability of finding an electron also approaches zero (Fig. 4). Like the 1*s* orbital, the 2*s* orbital may contain a single (unpaired) electron as in lithium, or two electrons with opposite (paired) spins as in beryllium. Both 1*s* and 2*s* orbitals being spherically symmetrical, however, are not directionally orientated, and the completion of such orbitals by the receipt of electrons in chemical combination gives rise to bonds which have no element of directional orientation.

The three 2*p* orbitals are directionally orientated, so that the probability of finding an electron is greater in certain directions than in others. They are dumb-bell shaped, and directed at right angles to each other along notional axes.

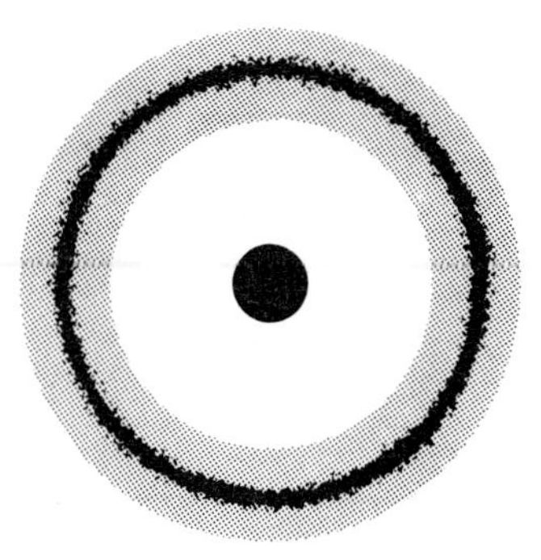

Fig. 3 Planar representation of the relationship between the 1s orbital and the nucleus in the hydrogen and helium atoms

These axes are designated *x*, *y* and *z* axes, and the orbitals corresponding to them are described as $2p_x$, $2p_y$ and $2p_z$ orbitals respectively. Each orbital exhibits a **nodal plane**, passing through the nucleus of the atom at right angles to the axis, in which the probability of finding an electron is zero, thereby creating the 'dumb-bell' shape (Fig. 5).

The entry of electrons into the three equivalent 2*p* orbitals is governed by **Hund's rules**, which are based on the fact that mutual repulsion energy will be less if the electrons have unpaired spins. For this reason, when electrons are added successively as in the elements boron, carbon, nitrogen, oxygen, fluorine

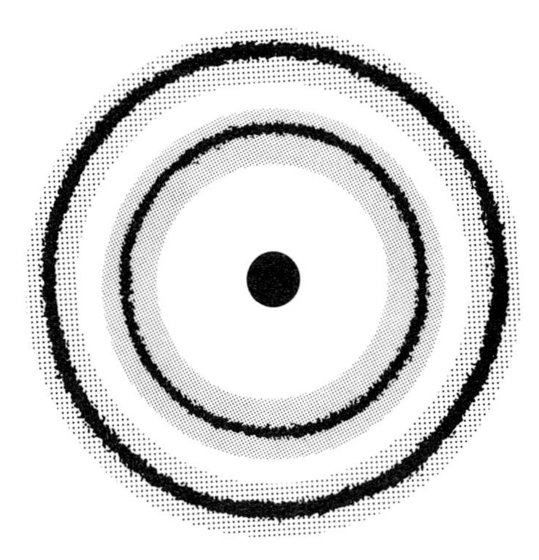

Fig. 4 Planar representation of 1*s* and 2*s* orbitals in lithium and beryllium

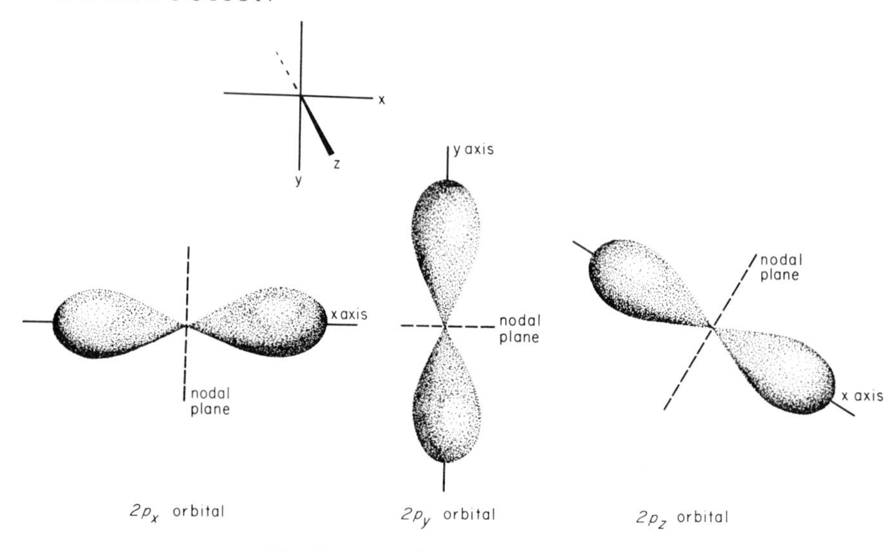


Fig. 5 Dumb-bell shaped $2p$ orbitals

and neon, as many orbitals as possible are singly occupied before any pairing occurs. It follows also that in the ground state, two electrons each occupying a pair of equivalent orbitals, tend to have parallel spins.

## Valency

The electronic theory of valency is based on the assumption that normally only the electrons in the highest energy state (i.e. the outermost orbital) participate in the formation of chemical bonds. In some cases, only electrons in the highest sub-levels are involved; in others, lower sub-levels in either the highest principal energy state or next highest principal energy state are also implicated. The theory, due to Kossel (1916) and Lewis (1916), relates the chemical stability of the rare gases to the possession of a complete set of electrons in the highest energy state. In all but helium, which has only two extra-nuclear electrons in the $1s$ level, this means an **octet** consisting of two $s$ and six $p$ electrons (Table 3).

Atoms combine by the gain or loss or sharing of valency electrons in an attempt to achieve the stable octet configuration of the inert gases by the formation of chemical bonds. Three principal types of chemical bond are recognised in the formation of chemical compounds:

(a) electrovalent,
(b) covalent,
(c) co-ordinate-covalent.

All three, together with a number of low energy bonding forces, such as hydrogen bonding, van der Waals bonding, and charge-transfer complexing, are also important in the interaction of drugs with biological receptor molecules (**2**, 2).

Table 3. Electron Configuration of the Elements

| Atomic Number | Element | Symbol | k $n = 1$ | l 2 | | m 3 | | | n 4 | | | | o 5 | | | | p 6 | | | q 7 |
|---|---|---|---|---|---|---|---|---|---|---|---|---|---|---|---|---|---|---|---|---|
| | | | *s* | *s* | *p* | *s* | *p* | *d* | *s* | *p* | *d* | *f* | *s* | *p* | *d* | *f* | *s* | *p* | *d* | *s* |
| 1 | Hydrogen | H | 1 | | | | | | | | | | | | | | | | | |
| **2** | **Helium** | **He** | **2** | | | | | | | | | | | | | | | | | |
| 3 | Lithium | Li | 2 | 1 | | | | | | | | | | | | | | | | |
| 4 | Beryllium | Be | 2 | 2 | | | | | | | | | | | | | | | | |
| 5 | Boron | B | 2 | 2 | 1 | | | | | | | | | | | | | | | |
| 6 | Carbon | C | 2 | 2 | 2 | | | | | | | | | | | | | | | |
| 7 | Nitrogen | N | 2 | 2 | 3 | | | | | | | | | | | | | | | |
| 8 | Oxygen | O | 2 | 2 | 4 | | | | | | | | | | | | | | | |
| 9 | Fluorine | F | 2 | 2 | 5 | | | | | | | | | | | | | | | |
| **10** | **Neon** | **Ne** | **2** | **2** | **6** | | | | | | | | | | | | | | | |
| 11 | Sodium | Na | 2 | 2 | 6 | 1 | | | | | | | | | | | | | | |
| 12 | Magnesium | Mg | 2 | 2 | 6 | 2 | | | | | | | | | | | | | | |
| 13 | Aluminium | Al | 2 | 2 | 6 | 2 | 1 | | | | | | | | | | | | | |
| 14 | Silicon | Si | 2 | 2 | 6 | 2 | 2 | | | | | | | | | | | | | |
| 15 | Phosphorus | P | 2 | 2 | 6 | 2 | 3 | | | | | | | | | | | | | |
| 16 | Sulphur | S | 2 | 2 | 6 | 2 | 4 | | | | | | | | | | | | | |
| 17 | Chlorine | Cl | 2 | 2 | 6 | 2 | 5 | | | | | | | | | | | | | |
| **18** | **Argon** | **A** | **2** | **2** | **6** | **2** | **6** | | | | | | | | | | | | | |
| 19 | Potassium | K | 2 | 2 | 6 | 2 | 6 | | 1 | | | | | | | | | | | |
| 20 | Calcium | Ca | 2 | 2 | 6 | 2 | 6 | | 2 | | | | | | | | | | | |
| 21 | Scandium | Sc | 2 | 2 | 6 | 2 | 6 | 1 | 2 | | | | | | | | | | | |
| 22 | Titanium | Ti | 2 | 2 | 6 | 2 | 6 | 2 | 2 | | | | | | | | | | | |
| 23 | Vanadium | V | 2 | 2 | 6 | 2 | 6 | 3 | 2 | | | | | | | | | | | |
| 24 | Chromium | Cr | 2 | 2 | 6 | 2 | 6 | 5 | 1 | | | | | | | | | | | |
| 25 | Manganese | Mn | 2 | 2 | 6 | 2 | 6 | 5 | 2 | | | | | | | | | | | |
| 26 | Iron | Fe | 2 | 2 | 6 | 2 | 6 | 6 | 2 | | | | | | | | | | | |
| 27 | Cobalt | Co | 2 | 2 | 6 | 2 | 6 | 7 | 2 | | | | | | | | | | | |

Table 3. (*continued*)

| Atomic Number | Element | Symbol | k $n = 1$ | l 2 | | m 3 | | | n 4 | | | | o 5 | | | | p 6 | | | q 7 |
|---|---|---|---|---|---|---|---|---|---|---|---|---|---|---|---|---|---|---|---|---|
| | | | *s* | *s* | *p* | *s* | *p* | *d* | *s* | *p* | *d* | *f* | *s* | *p* | *d* | *f* | *s* | *p* | *d* | *s* |
| 28 | Nickel | Ni | 2 | 2 | 6 | 2 | 6 | 8 | 2 | | | | | | | | | | | |
| 29 | Copper | Cu | 2 | 2 | 6 | 2 | 6 | 10 | 1 | | | | | | | | | | | |
| 30 | Zinc | Zn | 2 | 2 | 6 | 2 | 6 | 10 | 2 | | | | | | | | | | | |
| 31 | Gallium | Ga | 2 | 2 | 6 | 2 | 6 | 10 | 2 | 1 | | | | | | | | | | |
| 32 | Germanium | Ge | 2 | 2 | 6 | 2 | 6 | 10 | 2 | 2 | | | | | | | | | | |
| 33 | Arsenic | As | 2 | 2 | 6 | 2 | 6 | 10 | 2 | 3 | | | | | | | | | | |
| 34 | Selenium | Se | 2 | 2 | 6 | 2 | 6 | 10 | 2 | 4 | | | | | | | | | | |
| 35 | Bromine | Br | 2 | 2 | 6 | 2 | 6 | 10 | 2 | 5 | | | | | | | | | | |
| **36** | **Krypton** | **Kr** | **2** | **2** | **6** | **2** | **6** | **10** | **2** | **6** | | | | | | | | | | |
| 37 | Rubidium | Rb | 2 | 2 | 6 | 2 | 6 | 10 | 2 | 6 | | | 1 | | | | | | | |
| 38 | Strontium | Sr | 2 | 2 | 6 | 2 | 6 | 10 | 2 | 6 | | | 2 | | | | | | | |
| 39 | Yttrium | Y | 2 | 2 | 6 | 2 | 6 | 10 | 2 | 6 | 1 | | 2 | | | | | | | |
| 40 | Zirconium | Zr | 2 | 2 | 6 | 2 | 6 | 10 | 2 | 6 | 2 | | 2 | | | | | | | |
| 41 | Niobium | Nb | 2 | 2 | 6 | 2 | 6 | 10 | 2 | 6 | 4 | | 1 | | | | | | | |
| 42 | Molybdenum | Mo | 2 | 2 | 6 | 2 | 6 | 10 | 2 | 6 | 5 | | 1 | | | | | | | |
| 43 | Technetium | Tc | 2 | 2 | 6 | 2 | 6 | 10 | 2 | 6 | 6 | | 1 | | | | | | | |
| 44 | Ruthenium | Ru | 2 | 2 | 6 | 2 | 6 | 10 | 2 | 6 | 7 | | 1 | | | | | | | |
| 45 | Rhodium | Rh | 2 | 2 | 6 | 2 | 6 | 10 | 2 | 6 | 8 | | 1 | | | | | | | |
| 46 | Palladium | Pd | 2 | 2 | 6 | 2 | 6 | 10 | 2 | 6 | 10 | | | | | | | | | |
| 47 | Silver | Ag | 2 | 2 | 6 | 2 | 6 | 10 | 2 | 6 | 10 | | 1 | | | | | | | |
| 48 | Cadmium | Cd | 2 | 2 | 6 | 2 | 6 | 10 | 2 | 6 | 10 | | 2 | | | | | | | |
| 49 | Indium | In | 2 | 2 | 6 | 2 | 6 | 10 | 2 | 6 | 10 | | 2 | 1 | | | | | | |
| 50 | Tin | Sn | 2 | 2 | 6 | 2 | 6 | 10 | 2 | 6 | 10 | | 2 | 2 | | | | | | |
| 51 | Antimony | Sb | 2 | 2 | 6 | 2 | 6 | 10 | 2 | 6 | 10 | | 2 | 3 | | | | | | |
| 52 | Tellurium | Te | 2 | 2 | 6 | 2 | 6 | 10 | 2 | 6 | 10 | | 2 | 4 | | | | | | |
| 53 | Iodine | I | 2 | 2 | 6 | 2 | 6 | 10 | 2 | 6 | 10 | | 2 | 5 | | | | | | |

| | | | | | | | | | | | | | | | | | | | | |
|---|---|---|---|---|---|---|---|---|---|---|---|---|---|---|---|---|---|---|---|---|
| **54** | **Xenon** | **Xe** | **2** | **2** | **6** | **2** | **6** | **10** | **2** | **6** | **10** | | **2** | **6** | | | | | | |
| 55 | Caesium | Cs | 2 | 2 | 6 | 2 | 6 | 10 | 2 | 6 | 10 | | 2 | 6 | | | 1 | | | |
| 56 | Barium | Ba | 2 | 2 | 6 | 2 | 6 | 10 | 2 | 6 | 10 | | 2 | 6 | | | 2 | | | |
| 57 | Lanthanum | La | 2 | 2 | 6 | 2 | 6 | 10 | 2 | 6 | 10 | | 2 | 6 | 1 | | 2 | | | |
| 58 | Cerium | Ce | 2 | 2 | 6 | 2 | 6 | 10 | 2 | 6 | 10 | 1 | 2 | 6 | 1 | | 2 | | | |
| ⋮ | ⋮ | | | | | | | | | | | ⋮ | | | | | | | | |
| 71 | Lutecium | Lu | 2 | 2 | 6 | 2 | 6 | 10 | 2 | 6 | 10 | 14 | 2 | 6 | 1 | | 2 | | | |
| 72 | Hafnium | Hf | 2 | 2 | 6 | 2 | 6 | 10 | 2 | 6 | 10 | 14 | 2 | 6 | 2 | | 2 | | | |
| 73 | Tantalum | Ta | 2 | 2 | 6 | 2 | 6 | 10 | 2 | 6 | 10 | 14 | 2 | 6 | 3 | | 2 | | | |
| 74 | Tungsten | W | 2 | 2 | 6 | 2 | 6 | 10 | 2 | 6 | 10 | 14 | 2 | 6 | 4 | | 2 | | | |
| 75 | Rhenium | Re | 2 | 2 | 6 | 2 | 6 | 10 | 2 | 6 | 10 | 14 | 2 | 6 | 5 | | 2 | | | |
| 76 | Osmium | Os | 2 | 2 | 6 | 2 | 6 | 10 | 2 | 6 | 10 | 14 | 2 | 6 | 6 | | 2 | | | |
| 77 | Iridium | Ir | 2 | 2 | 6 | 2 | 6 | 10 | 2 | 6 | 10 | 14 | 2 | 6 | 7 | | 2 | | | |
| 78 | Platinum | Pt | 2 | 2 | 6 | 2 | 6 | 10 | 2 | 6 | 10 | 14 | 2 | 6 | 9 | | 1 | | | |
| 79 | Gold | Au | 2 | 2 | 6 | 2 | 6 | 10 | 2 | 6 | 10 | 14 | 2 | 6 | 10 | | 1 | | | |
| 80 | Mercury | Hg | 2 | 2 | 6 | 2 | 6 | 10 | 2 | 6 | 10 | 14 | 2 | 6 | 10 | | 2 | | | |
| 81 | Thallium | Tl | 2 | 2 | 6 | 2 | 6 | 10 | 2 | 6 | 10 | 14 | 2 | 6 | 10 | | 2 | 1 | | |
| 82 | Lead | Pb | 2 | 2 | 6 | 2 | 6 | 10 | 2 | 6 | 10 | 14 | 2 | 6 | 10 | | 2 | 2 | | |
| 83 | Bismuth | Bi | 2 | 2 | 6 | 2 | 6 | 10 | 2 | 6 | 10 | 14 | 2 | 6 | 10 | | 2 | 3 | | |
| 84 | Polonium | Po | 2 | 2 | 6 | 2 | 6 | 10 | 2 | 6 | 10 | 14 | 2 | 6 | 10 | | 2 | 4 | | |
| 85 | Astatine | At | 2 | 2 | 6 | 2 | 6 | 10 | 2 | 6 | 10 | 14 | 2 | 6 | 10 | | 2 | 5 | | |
| **86** | **Radon** | **Rn** | **2** | **2** | **6** | **2** | **6** | **10** | **2** | **6** | **10** | **14** | **2** | **6** | **10** | | **2** | **6** | | |
| 87 | Francium | Fr | 2 | 2 | 6 | 2 | 6 | 10 | 2 | 6 | 10 | 14 | 2 | 6 | 10 | | 2 | 6 | | 1 |
| 88 | Radium | Ra | 2 | 2 | 6 | 2 | 6 | 10 | 2 | 6 | 10 | 14 | 2 | 6 | 10 | | 2 | 6 | | 2 |
| 89 | Actinium | Ac | 2 | 2 | 6 | 2 | 6 | 10 | 2 | 6 | 10 | 14 | 2 | 6 | 10 | | 2 | 6 | 1 | 2 |
| 90 | Thorium | Th | 2 | 2 | 6 | 2 | 6 | 10 | 2 | 6 | 10 | 14 | 2 | 6 | 10 | | 2 | 6 | 2 | 2 |
| 91 | Protoactinium | Pa | 2 | 2 | 6 | 2 | 6 | 10 | 2 | 6 | 10 | 14 | 2 | 6 | 10 | 2 | 2 | 6 | 1 | 2 |
| 92 | Uranium | U | 2 | 2 | 6 | 2 | 6 | 10 | 2 | 6 | 10 | 14 | 2 | 6 | 10 | 3 | 2 | 6 | 1 | 2 |
| 93 | Neptunium | Np | 2 | 2 | 6 | 2 | 6 | 10 | 2 | 6 | 10 | 14 | 2 | 6 | 10 | 4 | 2 | 6 | 1 | 2 |
| 94 | Plutonium | Pu | 2 | 2 | 6 | 2 | 6 | 10 | 2 | 6 | 10 | 14 | 2 | 6 | 10 | 6 | 2 | 6 | | 2 |
| 95 | Americium | Am | 2 | 2 | 6 | 2 | 6 | 10 | 2 | 6 | 10 | 14 | 2 | 6 | 10 | 7 | 2 | 6 | | 2 |
| 96 | Curium | Cm | 2 | 2 | 6 | 2 | 6 | 10 | 2 | 6 | 10 | 14 | 2 | 6 | 10 | 7 | 2 | 6 | 1 | 2 |

## ELECTROVALENT BONDS

The term **electrovalence** is due to Langmuir (1919). It describes the characteristic bond of metallic salts in which two **ions** of opposite charge are held together in a crystal lattice by electrostatic forces. Metallic elements with one, two or three valency electrons are able to achieve a stable inert gas electron configuration by loss of an electron (or electrons) to a non-metallic element or radical which similarly achieves stability by the gain of an electron (or electrons) to form a stable octet. Thus, sodium chloride (NaCl) is formed by the transfer of the single 3*s* valency electron of a sodium atom which has the electron configuration $1s^2 2s^2 2p^6 3s$ completely into the sphere of influence of a chlorine atom which has the electron configuration $1s^2 2s^2 2p^6 3s^2 3p^5$, Thus, the molecule is stabilised by the loss of an electron from sodium which achieves the stable configuration of neon, $1s^2 2s^2 2p^6$ and the gain of an electron by chlorine to achieve the stable configuration of argon, $1s^2 2s^2 2p^6 3s^2 3p^6$. Considered in terms of the gain and loss of valency electrons, however, the process can be expressed simply.

$$\mathrm{Na}\cdot \curvearrowright \cdot\ddot{\underset{\cdot\cdot}{\mathrm{Cl}}}: \longrightarrow \mathrm{Na}^+\mathrm{Cl}^-$$

Since the two atoms combining in the formation of the electrovalent bond were electrically neutral initially, transfer of the electron leaves the sodium positively charged and creates a negative charge on the chlorine. The two **ions** thus produced are held together in the solid state crystal lattice of sodium chloride by electrostatic attraction, so that the molecule as a whole is still electrically neutral. The existence of this ionic state, however, confers certain specific properties on compounds which exhibit electrovalency, which are characteristics of metallic salts in general. These include:

(a) high melting and boiling points, due to the lattice structure of such compounds in the crystalline state,
(b) ability to conduct electricity when fused,
(c) solubility in water, to give solutions of the component ions which are capable of conducting an electric current,
(d) insolubility in non-polar organic solvents.

Not all metallic ions have outer shells with inert gas octet configurations. Thus, the elements copper and zinc, silver and cadmium, and gold and mercury, which complete the various transition groups, give rise to ions with eighteen electrons in their outer shell (Table 3). These ions of the IB and 2B sub-group elements are stable, but somewhat less so than those with typical inert gas structure in the corresponding 1A and 2A sub-groups of the Periodic Table (Table 1). B Sub-group elements are characterised by multiple valency, and like the remaining transition elements, they exhibit strong tendencies to form complex ions in their attempts to achieve a more stable configuration.

Only the halogens and the elements oxygen, hydrogen, nitrogen, sulphur, selenium and tellurium form simple anions. Most anions are complex, such as the nitrate ($NO_3^-$), sulphate ($SO_4^{2-}$) and phosphate ($PO_4^{3-}$) anions. Salts formed between such inorganic anions, whether simple or complex, and metallic ions are for the most part salts of strong acids and strong bases, and are completely ionised in solution. Only a few organic compounds, notably quaternary ammonium compounds, such as tetramethylammonium chloride and the detergent, *Cetylpyridinium Chloride*, and metal salts of organic sulphates and sulphonates, such as *Sodium Lauryl Sulphate* and sodium benzenesulphonate, exhibit bonds which are strongly electrovalent in character leading to almost if not complete ionisation in solution.

$(CH_3)_4\overset{+}{N}Cl^-$

Tetramethylammonium chloride

$[C_5H_5\overset{+}{N}\cdot(CH_2)_{15}\cdot CH_3]Cl^-\cdot H_2O$

*Cetylpyridinium Chloride*

$CH_3\cdot(CH_2)_{10}\cdot CH_2\cdot O\cdot SO_2\cdot O^-Na^+$

*Sodium Lauryl Sulphate*

$C_6H_5SO_2\cdot O^-Na^+$

Sodium benzenesulphonate

Most other organic salts are salts of comparatively weak organic acids and bases, and hence, whatever the origin of the bond, are only partially dissociated into ions in solution to an extent determined by the dissociation constant ($pK_a$) of the acid, or base-conjugate acid, and the pH of the solution (Chapters 11 and 16). Such compounds usually react in biological systems in the ionic form, and this is undoubtedly true of many important reactions between basic

$$C_6H_4(CO\cdot OH)(O\cdot CO\cdot CH_3) \rightleftharpoons C_6H_4(CO\cdot O^-)(O\cdot CO\cdot CH_3) + \mathbf{H^+}$$

Reversible dissociation of *Aspirin*

$$H_2N\cdot C_6H_4\cdot CO\cdot O\cdot CH_2\cdot CH_2\cdot \overset{+}{N}H(Et)_2\ Cl^- \rightleftharpoons H_2N\cdot C_6H_4\cdot CO\cdot O\cdot CH_2\cdot CH_2\cdot \ddot{N}(Et)_2 + \mathbf{H^+} + Cl^-$$

Reversible dissociation of *Procaine Hydrochloride*

and acidic drugs and the free acidic and basic groups of receptor proteins. Thus, the electrovalent bond forms one important mechanism whereby drugs are bound to receptor or carrier peptides and proteins as, for example, the binding of acidic anti-rheumatic drugs, such as *Aspirin* (acetylsalicylic acid), *Phenylbutazone*, *Indomethacin* and *Flufenamic Acid* to lysine and other basic residues of serum proteins.

Similarly, the anionic site of acetylcholinesterase, which is responsible for binding the cationic head of acetylcholine prior to the initiation of its hydrolysis, appears to be the carboxylate anion of a glutamate or aspartate residue in the enzymic protein, so that enzyme and substrate are linked initially by this electrovalent bond (**2**, 2).

## COVALENT BONDS

Covalent bond formation is typical of the elements of Groups III, IV and V of the Periodic Table (Table 1). The elements of these groups have either three, four or five electrons in the outermost orbital. They enter into electrovalent combination less readily than the elements of Groups I, II and VII, because the gain or loss of three or four electrons required to form an electrovalent bond is less readily achieved than the gain or loss of one or two electrons. Elements in these groups of the Periodic Table, therefore, more often complete their valency shells by the mutual sharing of electrons with the formation of covalent bonds. This property reaches a peak in Group IV with the elements carbon and silicon, and to a lesser extent, tin and lead, which also typically form covalent bonds.

Most covalent compounds exhibit properties which contrast with those of typical ionic compounds. These include:

(a) low melting and boiling points,
(b) poor electrical conductivity,
(c) low solubility in water,
(d) good solubility in non-polar organic solvents, such as hexane.

### Molecular Orbitals

The sharing of electrons in the formation of covalent bonds is achieved by orbital overlap and in some cases hybridisation of the contributing atomic orbitals. The degree of orbital overlap determines the strength of the bond. The two overlapping atomic orbitals can be considered to give rise to two **molecular orbitals**, one with less energy and the other with more than the sum of the energies of the contributing atomic orbitals. The lower energy molecular orbital, known as the **bonding** or $\sigma$-orbital, contains the two paired electrons which form the bond. The higher energy molecular orbital, called the **antibonding** or $\sigma^*$-orbital, is not involved in the formation of stable bonds.

### The Covalent Bond in the Hydrogen Molecule

The covalent bond is formed by the **sharing** of a pair of electrons between atoms, one electron being contributed by each atom. The simplest type of covalent bond

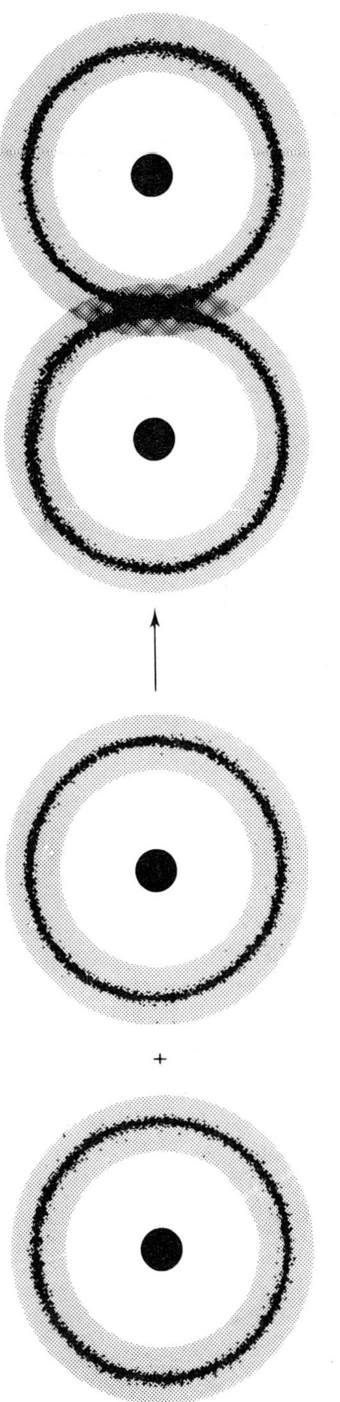

Fig. 6 Formation of the hydrogen molecule by overlap of *s* orbitals

is that formed between the constituent atoms of a symmetrical diatomic molecule, such as hydrogen, chlorine or oxygen. In the hydrogen molecule, each atom contributes its single valency to the covalent bond.

H:H

For simplicity, the bond is usually represented by a line which is taken to imply a pair of shared electrons.

H—H

In terms of orbital theory, the covalent bond is considered to be formed by the overlap of the two contributing atomic 1*s* orbitals to form a **molecular orbital** occupied by two electrons with paired spins and surrounding the two nuclei (Fig. 6).

The molecular orbital is radially symmetrical about the axis joining the two nuclei, and in this respect resembles the constituent atomic *s* orbitals; it is known as a **sigma** (Gk $\sigma$) molecular orbital. The **area of overlap** defines the region in which there is a relatively high probability of finding both electrons. The **degree of overlap**, provides a measure of the strength of the bond; the greater the overlap, the stronger the bond. In the hydrogen molecule, overlap is appreciable and the sigma bond formed is a strong one.

**Electron Distribution in the Carbon Atom**

The carbon atom has six orbital electrons, which in the ground (unexcited) state are distributed in the 1*s*, 2*s* and 2*p* orbitals in accordance with Hund's rules (p. 7).

| $1s$ | $2s$ | $2p_x$ | $2p_y$ | $2p_z$ |
|---|---|---|---|---|
| ↑↓ | ↑↓ | ↑ | ↑ | |

Both the 1*s* and 2*s* orbitals are complete; the $2p_x$ and $2p_y$ orbitals each contain one electron and the $2p_z$ orbital is vacant; only the two unpaired electrons in the $2p_x$ and $2p_y$ orbitals are available for bonding, so the element is apparently divalent.

One of the 2*s* electrons, however, is readily promoted in an excited state to the $2p_z$ orbital, so that the electrons are now distributed between all five orbitals.

| $1s$ | $2s$ | $2p_x$ | $2p_y$ | $2p_z$ |
|---|---|---|---|---|
| ↑↓ | ↑ | ↑ | ↑ | ↑ |

In this way, four unpaired electrons are available for bond formation in the orbitals 2*s*, $2p_x$, $2p_y$, and $2p_z$, of which only three are directionally orientated. All four bonds formed of the saturated carbon atom are, however, directionally orientated. These are formed by **hybridisation** of the one 2*s* and three 2*p* orbitals to form four new $sp^3$ **hybrid orbitals** (Fig. 7). These four hybrid orbitals are **identical**, and all are **directionally orientated**, being mutually inclined at an angle

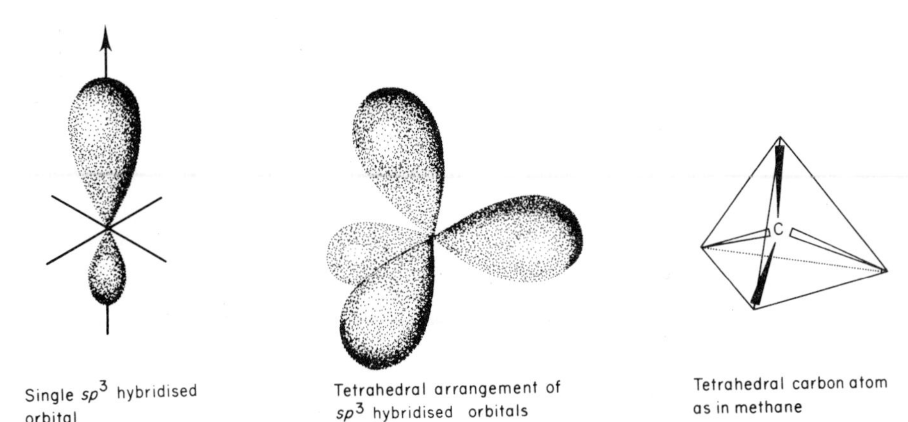


Fig. 7 $sp^3$ Hybridisation and the tetrahedral carbon atom

of 109° 28′. This is in agreement with the long-accepted concept of the tetrahedral carbon atom, in which the atom is regarded as being placed at the centre of a regular tetrahedron, with its four valencies directed towards the corners of the tetrahedron.

The tetrahedral disposition of carbon valencies is consistent with the fact that methylene dichloride, for example, exists in only one form, although different planar representations of the structure might at first sight suggest otherwise.

| Cl (top); Cl, H, H | Cl (top); H, H, Cl | H (top); H, Cl, Cl | H (top); Cl, H, Cl |
|---|---|---|---|

All these forms are superposable by suitable rotation of the entire molecule, whereas a planar distribution of carbon valencies would provide for two distinct non-superposable methylene dichlorides.

$$\begin{array}{c} \text{H} \\ | \\ \text{H—C—}\mathbf{Cl} \\ | \\ \mathbf{Cl} \end{array} \quad \text{and} \quad \begin{array}{c} \text{H} \\ | \\ \mathbf{Cl}\text{—C—}\mathbf{Cl} \\ | \\ \text{H} \end{array}$$

The concepts of the asymmetric carbon atom and of optical isomerism (Chapter 13) are also dependent upon the tetrahedral disposition of carbon valencies. The **tetrahedral angle** of carbon valencies determines the typical disposition of carbon atoms in a carbon chain such as that of a paraffin hydrocarbon.

C C C C C C

The zig-zag conformation of the chain is determined by non-bonded interactions between hydrogen atoms (Chapter 3).

### The Carbon–Carbon Single Bond

The carbon–carbon single bond in saturated hydrocarbons such as ethane similarly results from the equal sharing of two electrons, one supplied by each carbon atom. The bond is formed by the formation and coalescence of two $sp^3$ hybridised orbitals, which overlap endways to give a molecular orbital occupied by the two bonding electrons (Fig. 8).

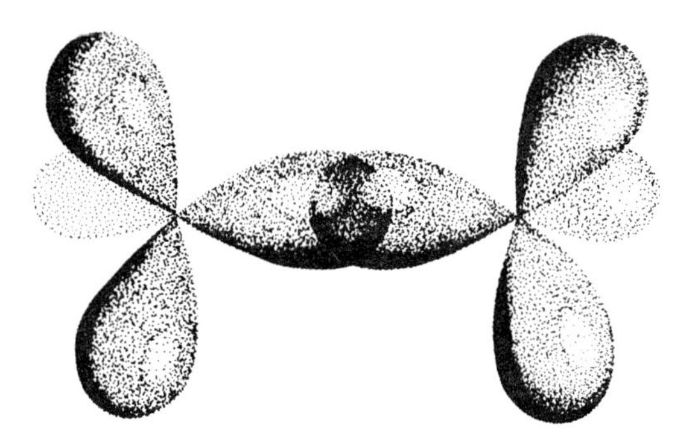

Fig. 8 Formation of the C—C single bond by overlap of $sp^3$ orbitals

As in the hydrogen molecule, the orbital is symmetrical about the axis joining the two nuclei. The area of overlap of the constituent orbitals is considerable, so that a stable sigma bond results. The bond retains the directional orientation of the tetrahedrally disposed consitituent $sp^3$ hybridised orbitals. This means that the remaining bonds on one carbon atom have a defined relationship to the remaining bonds on the other carbon atom in the sense that each group of three bonds is directed outwards from the carbon–carbon bond towards the apices of two opposed regular tetrahedrons.

Because of the radial symmetry of the molecular orbital comprising the carbon–carbon single bond, there is said to be free-rotation about the bond so that the remaining three bonds on one carbon atom may take up any orientation with respect to those on the others ranging from the fully **staggered** to the **eclipsed** forms (conformation) shown in the Newman projections below.

Fully staggered

Fully eclipsed

Newman projections

**The Carbon–Hydrogen Bond**

Methane, the simplest hydrocarbon, is formed from four hydrogen atoms each sharing its valence electron with one of the four valence electrons of carbon, and may be represented:

```
   H                 H
   ..                |
H:C:H      or    H—C—H
   ..                |
   H                 H
```

The C—H bond in methane, and other hydrocarbons, consists of a molecular orbital formed by end-on fusion of an $sp^3$ carbon orbital with a 1*s* hydrogen orbital (Fig. 9).

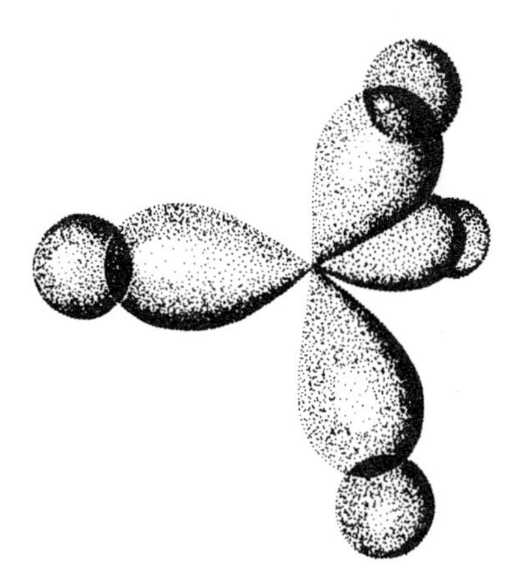

Fig. 9 Orbital overlap in the molecule of methane

As in the hydrogen molecule, and the carbon–carbon single bond, the orbital is symmetrical about the axis joining the two nuclei. **Endways overlap** of the two contributing orbitals ensures that the C—H bond is a stable **sigma** bond. The directional component of the hybridised $sp^3$ carbon orbitals determines the tetrahedral disposition of the four C—H bonds in methane.

The C—H bond, however, differs from the H—H and C—C covalent bonds, in that it is formed between two different elements. In general, such bonds

(C—X) exhibit unequal sharing of electrons, due to the differing electronegativities of the elements concerned. Such differences result in permanent polarisation of the bond with the more electronegative element having the greater share of the electrons, X : Y where Y is more electronegative than X.

**The Electronegativity Scale**

The electronegativity scale provides a measure of the ability with which an element is able to gain electrons. Clearly, this is greatest for elements which only require to gain one or two electrons to complete their valency shell, and least for metallic elements which would require to gain seven, six or five electrons to achieve the same result. **Electronegativity**, therefore, **increases from left to right across the Periodic Table**.

Similarly, elements with only a small number of orbital electrons attract additional electrons much more readily than the larger elements with many electrons, since in the former the incoming electrons experience a greater attraction to the positively charged nucleus by virtue of being able to approach it more closely. **Electronegativity**, therefore, **increases from the bottom to the top of the Periodic Table**.

A scale of electronegativities was drawn up by Pauling (Fig. 10). It is expressed in arbitrary units from 0 to 4, based on bond energy measurements. It can be used to obtain a rough assessment of the relative polarity of bonds by taking

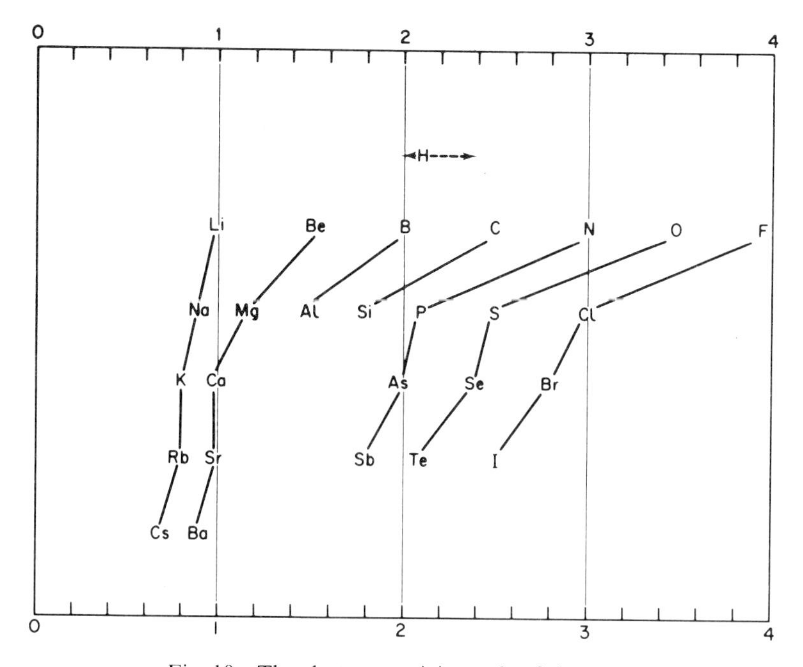


Fig. 10 The electronegativity scale of elements

the difference ($\Delta E$) of the scale electronegativity values for the two elements composing the bond. Bonds formed between elements with electronegativities differing by more than three scale units (e.g. NaF; $\Delta E = 3.1$) are purely ionic, and conversely, bonds with electronegativity differences of less than 0.5 units are purely covalent (e.g. C—C; $\Delta E = 0$). Between these two extremes, however, lie bonds between pairs of dissimilar elements, with varying degrees of polarity, depending upon the electronegativity difference between the two elements as, for example, in the molecules of hydrogen chloride ($\Delta E = 1.0$) and methyl chloride ($\Delta E_{C-Cl} = ca\ 0.5$).

Unequal sharing of electrons can be indicated in a number of ways.

H :Cl $\quad\quad$ $\overset{\delta+}{H}$—$\overset{\delta-}{Cl}$ $\quad\quad$ H→Cl

Hydrogen chloride

```
   H                H                 H
   ..               |δ+  δ-           |
H:C :Cl          H—C—Cl           H—C→Cl
   ..               |                 |
   H                H                 H
```

Methyl chloride

It gives rise to permanent polarisation of the bond. As a result, the molecule possesses a permanent **Dipole Moment**, i.e. it exhibits a turning moment when placed in an electric field, a property which is capable of measurement. It is the unequal sharing of electrons, and consequent bond polarity, which allows chemical attack of covalent bonds by ionic reagents, such as $OH^-$, $H^+$, $NO_2^+$ and so on.

The effect of polarisation, resulting from the electronegativity differences is transmissible to adjacent bonds. Such influence is known as the **Inductive Effect (I)**. Inductive effects may be positive (+I) or negative (−I) depending upon whether the effect is one of electron repulsion (+I) or electron attraction (−I).

## Electronegativity and the C—H Bond

The electronegativity difference ($\Delta E = 0.4$) between hydrogen and carbon does accurately represent the properties of the C—H bond, which in fact is almost 100% covalent in character. None-the-less, carbon is slightly more electronegative than hydrogen, and summation of this small contribution from each of the H→C bonds in the methyl group leads to a positive repulsion of electrons in the remaining bond. As a result, the methyl group is electron repelling (+I effect) even when attached through a C—C single bond to another carbon atom.

```
    H
    ↓  |
H→C→C—
    ↑  |
    H
```

The inductive effect, whether repelling or attracting electrons, can be transmitted along adjoined covalent bonds up to a limit of two saturated carbon atoms.

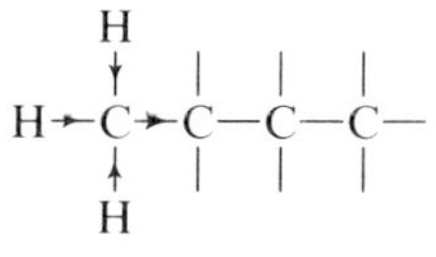

```
      H
      ↓   |   |   |
  H→ C → C — C — C —
      ↑   |   |   |
      H
```

Primary inductive effect

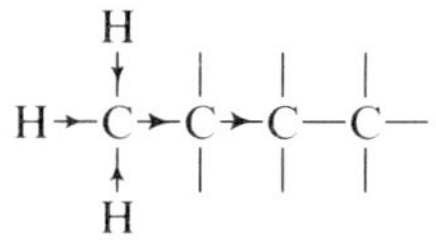

```
      H
      ↓   |   |   |
  H→ C → C → C — C —
      ↑   |   |   |
      H
```

Primary and secondary inductive effects

The low polarity of the H—C bond compared with that of the H—N ($\Delta E$ = 0.9), H—O ($\Delta E$ = 1.4) and H—F ($\Delta E$ = 1.9) bonds explains why paraffin hydrocarbons do not participate in hydrogen bond formation (Chapter 3), and hence also explains at least in part their lack of affinity for water.

### The Covalent Bond in Drug Action

The majority of drugs are organic and hence covalent compounds. As already explained, many are organic bases or acids and enter into biological reactions at least in part by electrovalent bonding. A few, however, give rise to the formation of covalent bonds in biochemical reactions which are irreversible. Most covalent bonds have bond energies (Chapter 2) ranging from about 140–800 kJ $mol^{-1}$. They are much stronger than individual ionic bonds, which for the most part are of the order of 4–8 kJ $mol^{-1}$, and hence have much higher stability. Many of the reactions of intermediary metabolism involve the making and breaking of covalent bonds, but with the few exceptions briefly mentioned below, the great majority of drugs do not initiate their pharmacological response in this way.

Typical examples include the selective trypanocidal action of organic arsenicals which combine with sulphydryl groups of protozoal respiratory enzymes, the interruption of bacterial cell-wall synthesis through the acylation of enzymic $\beta$-mercaptoethylamine units by penicillin, and the inactivation of acetylcholinesterase by phosphorus insecticides which acylate serine residues essential to the enzymic hydrolysis of acetylcholine. These and further examples of covalent bond formation in drug action are examined in detail elsewhere (**2**, 2).

## CO-ORDINATE-COVALENT BONDS

Co-ordinate-covalent bonds, like covalent bonds, are formed by the sharing of electrons between the elements forming the bond. They differ from the normal covalent bond, however, in that both electrons are derived from one of the two

Table 4. Electron Distribution in Covalent and Co-ordinate-Covalent Bond Formation

| | | | | | | |
|---|---|---|---|---|---|---|
| Nitrogen | 1*s* | 2*s* | $2p_x$ | $2p_y$ | $2p_z$ | |
| | ↑↓ | ↑↓ | ↑ | ↑ | ↑ | |
| $NH_3$ | 1*s* | | $2sp^3$ | | | |
| | ↑↓ | ↑↓ | ↑↓ | ↑↓ | (↑↓) | lone pair |
| Boron | 1*s* | 2*s* | $2p_x$ | $2p_y$ | $2p_z$ | $H_3N \longrightarrow BF_3$ |
| | ↑↓ | ↑↓ | ↑ | | | |
| $BF_3$ | 1*s* | | $2sp^2$ | | $2p_z$ | |
| | ↑↓ | ↑↓ | ↑↓ | ↑↓ | | |

elements forming the bond; electrons are donated from one element to the bond, so that the donor element acquires a partial positive charge and the acceptor element a partial negative charge. A typical example of co-ordinate-covalent bond formation occurs in the reaction between boron trifluoride and ammonia.

Bonding with formation of the complex is favoured by the electron-deficiency of the boron atom in boron trifluoride. This arises because, although the electron structure of boron is promoted from $1s^2 2s^2 2p$ in the ground (unexcited) state to $1s^2 2s 2p^2$ when it combines with fluorine, the six electrons comprising the three trigonal ($sp^2$) B—F bonds of boron trifluoride occupy only three of the four available bonding orbitals (Table 4). Boron trifluoride is, therefore, able to accommodate a further electron pair in the $2p_z$ orbital from any base, such

$$H_3N + BF_3 \longrightarrow H_3N \longrightarrow BF_3$$

as ammonia, free to donate an unshared (non-bonded) electron pair. It will be evident from Table 4 that all four $sp^3$ hybridised bonding orbitals of nitrogen are filled in the ammonia molecule. Only three of these orbitals, however, are involved in N—H bond formation. The fourth, the unshared electron pair, is available for donation to an electron-deficient atom to form a co-ordinate-covalent bond.

Donation of the electron pair from nitrogen completes the boron octet, but leaves the nitrogen deficient in charge. The resulting bond may be represented as either $H_3N \longrightarrow BF_3$ or $H_3\overset{+}{N}—\overset{-}{B}F_3$ to indicate the source and direction of electron movement, and additionally in the latter the electrical dipole which is created thereby. Completion of the boron octet is accompanied by $sp^3$ orbital hybridisation so that boron is tetrahedral in the complex.

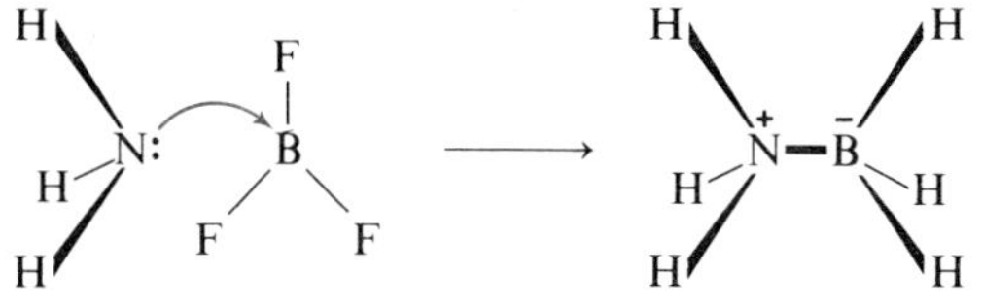

Similarly, the new N—H bond, formed when ammonia reacts with a proton in acid solution to give an ammonium ion, is a co-ordinate-covalent bond since both electrons are donated from the nitrogen atom. All four N—H bonds of the resulting ammonium ion are equivalent and tetrahedrally disposed about the positively charged nitrogen atom. In this case, however, the charge on the nitrogen is counterbalanced by the negative charge on the anion originally associated with the proton.

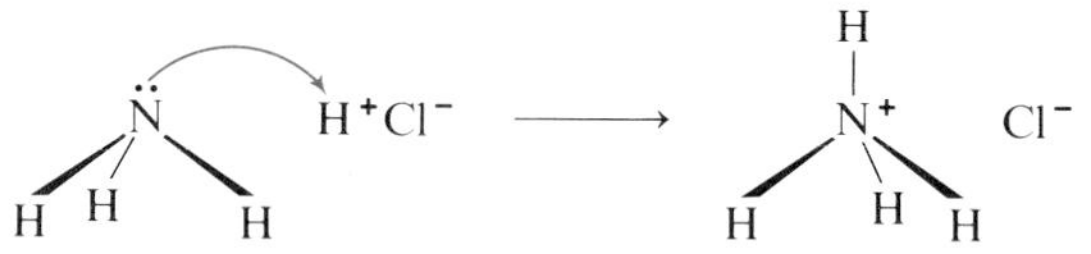

$$\ddot{N}H_3 + H^+Cl^- \longrightarrow NH_4^+ \quad Cl^-$$

Organic bases, amines (Chapter 16), which are substituted ammonias, similarly form base conjugate acids by co-ordinate-covalent bond formation with a proton in acid solution. The resulting ammonium salt, being ionic in character, confers the property of water solubility on the product.

$$R \cdot NH_2 + H^+Cl^- \rightleftharpoons R \cdot \overset{+}{N}H_3Cl^-$$

The formation of amine oxides, which like the boron trifluoride-ammonia complex may be represented either as $R_3N \rightarrow O$ or $R_3\overset{+}{N}-\overset{-}{O}$, from tertiary amines, and sulphoxides from organic sulphides, similarly involves co-ordinate-covalent bond formation.

$$R_3\ddot{N} \quad [\ddot{O}:] \longrightarrow R_3\overset{+}{N}-\overset{-}{O}$$

Tertiary amine → Amine oxide

$$R_2\ddot{S} \quad [\ddot{O}:] \longrightarrow R_2\overset{+}{S}-\overset{-}{O}$$

Sulphide → Sulphoxide

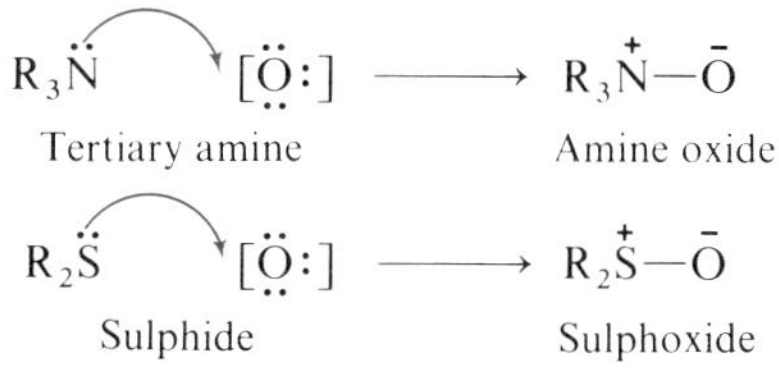

Amine oxides and sulphoxides are often formed as products of drug metabolism (**2**, 5). The dipolar character of $\overset{+}{N}-\overset{-}{O}$ and $\overset{+}{S}-\overset{-}{O}$ bonds favours solubility of the product in polar solvents such as water, and reduces lipid solubility. This factor favours urinary excretion of the product.

Co-ordinate-covalency occurs widely in the formation of complex ions involving neutral molecules. Metal ion complexes feature in many enzyme systems. Thus, iron-porphyrin complexes (Chapter 23) are essential to the cytochrome enzymes, peroxidases, and to haemoglobin, the oxygen-carrying pigment of red blood cells. Co-ordinate-covalent complexes of cobalt, copper, manganese, molybdenum and zinc are also essential to the activity of a number of enzymes.

Complexation of toxic metals with ethylenediaminetetra-acetic acid, administered as *Sodium Calcium Edetate* and *Trisodium Edetate*, forms the basis of a useful method for the treatment of lead and vanadium poisoning. The antibacterial action of 8-hydroxyquinoline (oxine) has been ascribed to the formation of its 1:1 iron(III)—oxine complex (Albert, Rubbo, Goldacre and Balfour, 1947). The actions of a number of other antibacterial and antiviral drugs including *Isoniazid*, *Thiacetazone* (Amithiazone), *Methisazone*, and the tetracycline antibiotics are also believed to be enhanced by metal complex formation (**2**,2).

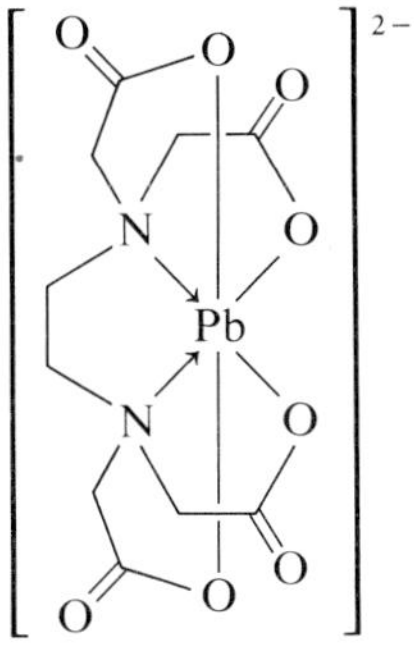

Lead edetate

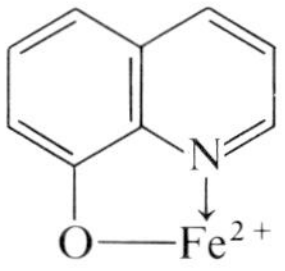

1:1 Iron—oxine complex

## HYDROGEN BONDS

Although hydrogen has only one valency electron, certain molecules containing hydrogen appear to function as if the hydrogen were divalent. Typically, hydrogen bonds are formed between such molecules by the attraction which unshared electron pairs of fluorine, oxygen or nitrogen have for the positively charged nucleus of a hydrogen atom of a second molecule. Thus, hydrogen bonds are formed between water molecules in the liquid state.

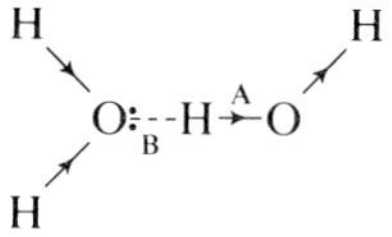

The hydrogen bond is unique. Such bonds are feasible only with hydrogen, since it is the only element with a single orbital electron. It is, therefore, the only element which, whilst already in molecular combination, is small enough to allow another (electron donor) molecule (or group) close enough for interaction with its positively charged nucleus leading to intermolecular (or intramolecular) association. A second important contributory factor in hydrogen bond formation is the unequal sharing of electrons in the H—O covalent bond (A) which reduces what would otherwise be a competing repulsive force on the unshared electron pair (B) of the adjacent water molecule, which forms the bond. Thus, hydrogen bond formation is facilitated when hydrogen is attached to the strongly electronegative elements, fluorine, oxygen and nitrogen, and to a lesser extent, chlorine and sulphur. Carbon, on the other hand, is not sufficiently electronegative to engender hydrogen bond formation. Even with strongly electronegative elements, the hydrogen bond is weak compared with typical covalent bonds. Thus, bond energies (Chapter 2) of most hydrogen bonds lie between

5 and 40 kJ mol$^{-1}$ compared with that of a typical covalent bond, such as the C—H bond, which is 412 kJ mol$^{-1}$. Examples of typical hydrogen bonds and their bond energies expressed in kJ mol$^{-1}$ are:

| | |
|---|---|
| O—H --- O | 13–35 |
| O—H --- X (halogen) | 7–15 |
| O—H --- N | 25–35 |
| N—H --- O | 12–27 |
| N—H --- N | 7–15 |

Hydrogen bonds may be either **inter**molecular or **intra**molecular. **Inter**molecular hydrogen bonded compounds show a marked lowering of volatility compared with related molecules which are not intermolecularly hydrogen-bonded. Thus, hydrogen sulphide (non-hydrogen-bonded) is a gas at room temperature, compared with water, a liquid which boils at 100°C. Carboxylic acids show intermolecular hydrogen bonding. This leads to complete dimerisation in the solid and liquid state and in solution in non-polar solvents, such as benzene. Intermolecular hydrogen bonding of carboxylic acids in the liquid state accounts for their boiling points being higher than those of the corresponding esters, which are incapable of similar intermolecular hydrogen bonding.

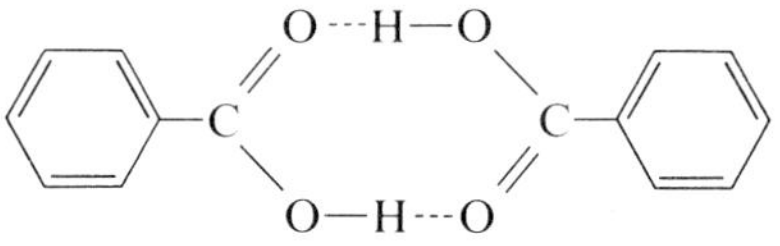

**Intra**molecular hydrogen bonds are formed by *o*-nitrophenol; in consequence, it is more volatile than the corresponding *para* isomer, which can only associate by intermolecular hydrogen bonding.

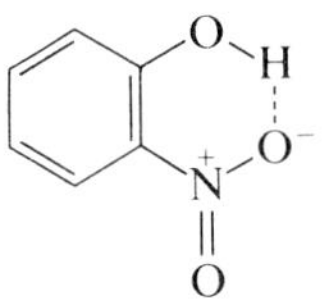

*o*-nitrophenol

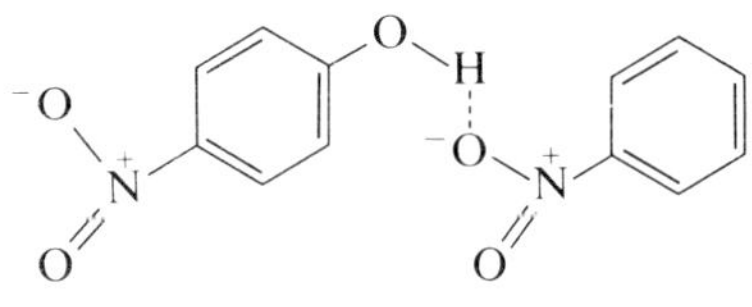

*p*-nitrophenol

Physical evidence of hydrogen bond formation is available from infrared (IR) and proton magnetic resonance (PMR) spectra. The former show characteristic frequency shifts of the hydroxyl absorption bands, which normally appear in solution at 3625 cm$^{-1}$, to around 3300 cm$^{-1}$ when hydrogen-bonded. Intermolecular hydrogen bonding is concentration-dependent; intramolecular hydrogen bonding, on the other hand, is unaffected by dilution. PMR signals for hydroxyl protons, which normally appear in the 5$\tau$ region, experience characteristic downfield shifts due to deshielding when hydrogen-bonded.

The rôle of both intramolecular and intermolecular hydrogen bonding in biological reactions is discussed in detail in Volume 2 (Chapter 2).

# 2 The Making and Breaking of Bonds

## SOME PHYSICAL ATTRIBUTES OF THE COVALENT BOND

### Bond Lengths

The size of a molecule is determined by the distances and angles between its constituent atoms. These inter-aromic distances are expressed in terms of **bond lengths** and **bond angles**. Precise bond lengths are measured by X-ray crystallography of pure materials. Thus, the carbon–carbon single bond length is 1.54 Å (1 Å = $10^{-10}$ m) in diamond, a crystalline form of pure carbon. Precise bond lengths, so measured, are average internuclear distances when the molecule is in the ground state. They are determined partly by the degree of orbital overlap, but more particularly by the nature of the molecular orbitals composing the bond, as illustrated in Table 5. Bond lengths and bond angles, however, can be affected by structural features of a particular molecule such as ring strain or repulsive steric effects.

Calculation of molecular dimensions is facilitated by the use of covalent radii, rather than bond lengths. Covalent radii are derived from actual measured bond distances. Thus, the covalent single bond radius of carbon is taken as 0.77 Å, which is half the carbon–carbon single bond length in diamond. Tables

Table 5. Bond Lengths

| Bond type | Example | Bond length (Å) |
|---|---|---|
| C—H($sp^3$—$s$) | $CH_3$—H | 1.12 |
| C—C($sp^3$—$sp^3$) | $CH_3$—$CH_2$—$CH_3$ | 1.54 |
| C—C($sp^3$—$sp^2$) | $CH_3$—CH=CH—$CH_3$ | 1.51 |
| C—C($sp^3$—$sp$) | $CH_3$—C≡CH | 1.46 |
| C—C($sp^2$—$sp^2$) | $CH_2$=CH—CH=$CH_2$ | 1.48 |
| C—C($sp^2$—$sp$) | $CH_2$=CH—C≡CH | 1.45 |
| C—C($sp$—$sp$) | HC≡C—C≡CH | 1.38 |
| C—N | $CH_3$—$NH_2$ | 1.47 |
| C—O | $CH_3$—OH | 1.43 |
| C—S | $CH_3$—$CH_2$—SH | 1.81 |
| O—H | $CH_3$—O—H | 0.96 |
| C=C | $CH_2$=$CH_2$ | 1.33 |
| C=O | $CH_2$=O | 1.22 |
| C≡C | HC≡CH | 1.20 |
| C≡N | $CH_3$C≡N | 1.17 |

Table 6. Covalent Single Bond and van der Waals Radii

| Element | Covalent Single Bond Radii (Å) | van der Waals Radii (Å) |
|---|---|---|
| Hydrogen | 0.3 | 1.2 |
| Carbon | 0.77 | 2.0 |
| Nitrogen | 0.70 | 1.5 |
| Phosphorus | 1.10 | 1.9 |
| Oxygen | 0.66 | 1.4 |
| Sulphur | 1.04 | 1.85 |
| Fluorine | 0.64 | 1.33 |
| Chlorine | 0.98 | 1.80 |
| Bromine | 1.14 | 1.95 |
| Iodine | 1.33 | 2.15 |

of covalent radii (Table 6) permit calculations of molecular dimensions in compounds of novel atomic composition and molecular structure.

## Bond Energy

When energy is absorbed by a molecule, its potential energy is raised above the ground state. Vibrational energy is increased in this excited state, so that the average distance separating its constituent nuclei is also increased. The potential energy–nuclear separation curve for a diatomic molecule is illustrated in Fig. 11. At infinite separation, the force between the nuclei is zero, but as they move toward each other they experience attraction forces, and potential energy decreases. When, however, the nuclei approach each other very closely, they begin to experience repulsive forces, and the potential energy increases once again. As a result, the curve passes through the characteristic minimum shown. In the excited state, the lip of the curve is reached at a lower vibrational level than in the ground state. Suitable excitation from the ground to the excited state can, therefore, promote the vibrational energy of the molecule to a point beyond the critical level for dissociation, when the bond will break.

The energy required to break a bond is called the **bond dissociation energy**. This is a measure of the strength of the bond (Cottrell, 1958). Some typical bond dissociation energies are given in Table 7. Bond dissociation energy is equal to the energy released when the bond is formed, and it is this release of energy which permits a reaction to proceed to form a product of lower total energy content. Bond energies are derived from thermochemical measurements of **heats of combustion** or **heats of formation** of appropriate compounds.

The total energy released in a chemical reaction is known as the **free energy** ($\Delta G$); by convention, large negative values of $\Delta G$ are favourable to the reaction. The free energy of a reaction is not to be confused with the **heat of reaction** or enthalpy ($\Delta H$), since some reactions release energy as heat and are described

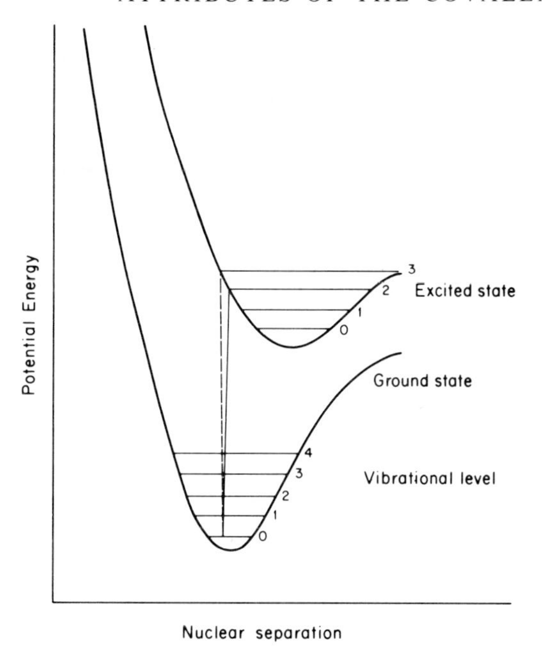


Fig. 11 Potential energy—nuclear separation curve for a diatomic molecule

as **exothermic** ($\Delta H$ negative), while others absorb energy as heat and are described as **endothermic** ($\Delta H$ positive). The difference between the free energy of a reaction and the enthalpy is accounted for by **entropy** ($\Delta S$), which is essentially a measure of the degree of disorder of the system. The entropy of a system depends on the degree of translational, vibrational or rotational disorder, which are temperature related. In general the greater the entropy, the greater the stability of the system. For example, cyclohexane is constrained by its ring structure in a way that the open-chain compound, n-hexane, is not. The latter may, therefore, adopt any one of a large number of conformations (Chapter 3), whereas cyclohexane is restricted to either a chair or boat structure. As a result, the degree of disorder and hence the entropy of the system is much greater in n-hexane than in cyclohexane.

The relationship between free energy, enthalpy and entropy is given by the expression:

$$\Delta G = \Delta H - T\Delta S$$

Entropy changes, therefore, provide a useful measure of the degree of structural ordering or disorganisation achieved in a particular chemical reaction. Thus, the exceptionally large entropy changes accompanying denaturation of proteins, provide evidence of the very high degree of structural organisation in the undenatured (native) protein molecules.

Table 7. Bond Dissociation Energies

| Type of bond | Bond dissociation energy ($kJ\ mol^{-1}$) |
|---|---|
| $C—C(CH_3—CH_3)$ | 347 |
| $C—C(PhCH_2—CH_3)$ | 264 |
| $C—C(PhCH_2—CH_2Ph)$ | 197 |
| C=C | 611 |
| C≡C | 803 |
| $C—H(CH_3—H)$ | 427 |
| $C—H(CH_3 \cdot CH_2—H)$ | 402 |
| $C—H[(CH_3)_2 \cdot CH—H]$ | 393 |
| $C—H[(CH_3)_3 \cdot C—H]$ | 372 |
| C—O | 356 |
| $C=O(H_2C=O)$ | 695 |
| $C=O(CH_3 \cdot CH=O)$ | 736 |
| $C=O[(CH_3)_2 \cdot C=O]$ | 749 |
| C—N | 305 |
| C=N | 615 |
| C≡N | 891 |
| C—S | 272 |
| C=S | 536 |
| C—F | 485 |
| C—Cl | 339 |
| C—Br | 285 |
| C—I | 213 |
| H—H | 435 |
| H—O | 464 |
| H—N | 389 |
| H—S | 347 |
| H—F | 565 |
| H—Cl | 431 |
| H—Br | 368 |
| H—I | 297 |
| O—O | 146 |
| O—N | 222 |
| O=N | 607 |
| N—N | 163 |
| N=N | 418 |
| S—S | 226 |
| F—F | 155 |
| Cl—Cl | 243 |
| Br—Br | 193 |
| I—I | 151 |

Values from T. L. Cottrell (1958) and L. Pauling (1960)

**Resonance Energy**

The electronic structures of some molecules can be depicted in more than one way, although the relative positions of the elements comprising the molecule are unchanged. Thus, the molecule of acetic acid may be represented by the following non-equivalent structures.

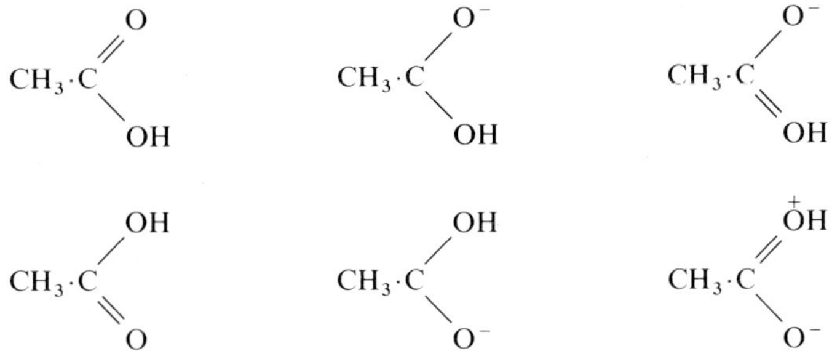

$$CH_3{\cdot}C{\Large(}\begin{matrix}=O\\ -OH\end{matrix} \qquad CH_3{\cdot}C{\Large(}\begin{matrix}-O^-\\ -OH\end{matrix} \qquad CH_3{\cdot}C{\Large(}\begin{matrix}-O^-\\ =OH\end{matrix}$$

$$CH_3{\cdot}C{\Large(}\begin{matrix}-OH\\ =O\end{matrix} \qquad CH_3{\cdot}C{\Large(}\begin{matrix}-OH\\ -O^-\end{matrix} \qquad CH_3{\cdot}C{\Large(}\begin{matrix}=\overset{+}{O}H\\ -O^-\end{matrix}$$

Provided none of the possible structures differ too greatly in energy from the remainder, the molecule will exist as a resonance hybrid to which each of the individual, so-called **canonical**, forms contribute in varying degree. Such hybridised structures always possess a greater degree of stability, and hence a lower energy content than that calculated for the individual contributing forms. Thus, the heat of combustion is less and the heat of formation more for the resonance hybrid than that calculated for the most stable of the canonical forms. This energy difference is known as the **resonance energy**.

Resonance interactions between canonical forms are shown by the symbol, ↔. Various forms of resonance interaction are recognisable in arrangements of multiple unsaturated systems including ethylenic bonds (Chapter 4) and aromatic rings (Chapter 5), which contain mobile electrons known as $\pi$-electrons. The following combinations of such groups with or without other unsaturated groups or groups with non-bonded electron pairs ($n$-electrons) show resonance.

(a) Conjugation. Interactions between two groups of $\pi$-electrons ($\pi$-$\pi$ interactions).

(b) Extended conjugation. Interactions between $\pi$-electrons and non-bonded $p$-electrons ($n$-$\pi$ interactions).

(c) Hyperconjugation. Interactions between $\pi$-electrons and $\sigma$-electrons.

(d) Interactions between $\pi$-electrons and vacant $p$-orbitals.

**Conjugation** arises in a variety of systems, in all of which $\pi$-bonds alternate with single bonds. Characteristic systems include conjugated dienes, C=C—C=C, conjugated enones, C=C—C=O, and the rather special case of benzene (Chapter 5). For complete conjugation, the entire system of alternating double and single bonds must be co-planar, though partial conjugation can be achieved where this requirement is not fully satisfied.

The possible canonical forms for butadiene may be represented as follows.

$$H_2C{=}CH{-}\overset{+}{C}H{-}\overset{-}{C}H_2 \longleftrightarrow H_2\overset{+}{C}{-}CH{=}CH{-}\overset{-}{C}H_2$$

a

a

$$H_2C{=}CH{-}CH{=}CH_2$$

b

b

$$H_2\overset{-}{C}{-}\overset{+}{C}H_2{-}CH{=}CH_2 \longleftrightarrow H_2\overset{-}{C}{-}CH{=}CH{-}\overset{+}{C}H_2$$

The symbol ⌒ denotes movement of an electron pair (Budziewicz, Djerassi and Williams, 1964).

They indicate the extent to which the charge arising from the $\pi$-electrons is distributed over the molecule. The electrons are said to be **delocalised** and the resonance hybrid may be represented as follows to show this delocalisation.

$$CH_2{\cdots}CH{\cdots}CH{\cdots}CH_2$$

**Extended conjugation.** This type of resonance involves interactions between $\pi$-electrons and non-bonded $p$-electron pairs (lone pairs) on neutral atoms such as nitrogen, halogen, or oxygen. This type of interaction is seen in the resonance of carboxylic acids (Chapter 11). It is also characteristic of phenols, phenolic ethers, aromatic amines and aromatic halides (*q.v.*), as for example:

$$>C{=}C{-}\ddot{N}H_2 \longleftrightarrow -\overset{-}{C}{-}C{=}\overset{+}{N}H_2$$

Resonance also occurs as a result of interactions between $\pi$-electrons and the electron pair forming the negative charge on a carbanion or oxanion as, for example, in the carboxylate ion and the nitro group.

$$-C(=O)O^- \longleftrightarrow -C(O^-)=O \qquad -\overset{+}{N}(=O)O^- \longleftrightarrow -\overset{+}{N}(O^-)=O$$

Carboxylate ion

Nitro group

Other examples include the enolate and phenolate ions, and ozone.

$$>C{=}C{-}O^- \longleftrightarrow -\overset{-}{C}{-}C{=}O$$

Enolate ion

Phenolate ion

$$O{=}\overset{+}{O}{-}\overset{-}{O} \longleftrightarrow \overset{-}{O}{-}\overset{+}{O}{=}O$$

Ozone

Extended conjugation also occurs in the interaction of a $\pi$-electron system with the unpaired $p$-electron of a free radical as, for example, in the phenoxy radical ($C_6H_5O\cdot$). However, in all these interactions, complete conjugation is, again, only achieved if the interacting $\pi$- and $p$-electrons are in a co-planar arrangement.

**Hyperconjugation**, or no-bond resonance, results from interaction between $\pi$-electrons and $\sigma$-bonds in allylic and benzylic systems which comply with the co-planarity requirement. Hyperconjugation in general reinforces the inductive effect as, for example, the +I effect of the methyl group in propylene ($CH_3CH{=}CH_2$) for which the following canonical forms can be written.

$$H_3C{-}CH{=}CH_2 \longleftrightarrow H^+\ H_2C{=}CH{-}\overset{-}{C}H_2 \longleftrightarrow H^+\ H_2C{=}CH{-}\overset{-}{C}H_2 \longleftrightarrow H^+\ H_2C{=}CH{-}\overset{-}{C}H_2$$

The rôle of hyperconjugation in resonance stabilisation is discussed further on p. 41.

**Interactions between $\pi$-electrons and vacant $p$-orbitals.** This type of interaction is analogous to that which exists in extended conjugation, giving rise to de-localisation of positive charge through the movement of $\pi$-electrons. The vacant $p$-orbital and interacting $\pi$-electrons must similarly be co-planar, a requirement which is met by allylic and benzylic carbonium ions, and in other carbonium ions (p. 42).

$$>C{=}C{-}\overset{+}{C}< \longleftrightarrow >\overset{+}{C}{-}C{=}C<$$

Allylic carbonium ion

An extension of this effect occurs in the interaction between $\pi$-electrons and vacant $d$-orbitals in elements of the second and higher periods of the Periodic Table, such as sulphur, which in loose terminology is sometimes said to be able to **expand its octet**. Interactions, such as the following, may be encountered.

$$\gt C{=}C{-}S{-}R \longleftrightarrow \gt \overset{+}{C}{-}C{=}\overset{-}{S}{-}R$$

The gain in resonance energy in products resulting from hydrolysis is an important driving force in many biological reactions involving so-called energy-rich compounds (p. 35).

### Activation Energy

In considering the energy requirements of a reaction, it is not sufficient to consider the total free energy change, since this merely represents the balance sheet when the reaction is complete. Energy released by the reaction may not be available at the beginning of the reaction and in this event must be supplied to the system. Also, repulsive forces between interacting species must be overcome so that the reactants can approach each other sufficiently closely for the reaction to proceed. The energy required to overcome these repulsive interactions is known as the **activation energy**. This is frequently, but not always, supplied in the form of heat and in such cases the relationship between the enthalpy of reactants and products ($\Delta H$) and the heat of activation ($\Delta H^*$) can be illustrated diagrammatically (Figs. 12A and B).

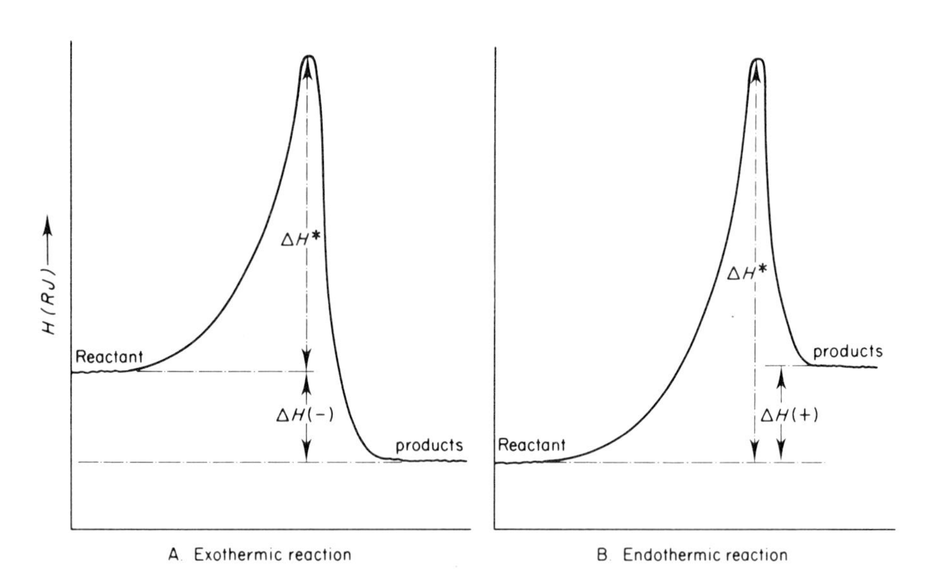


Fig. 12 Relationship between heat of reaction and heat of activation

**Energy-rich Bonds in Biological Systems**

Whilst most chemical reactions in the laboratory can be induced and accelerated by externally applied heat (p. 38), the nature of most biological reactions in animal systems is such that they must necessarily proceed within extremely narrow limits of temperature. Accordingly, their energy requirements are met from other than thermal sources.

The energy for biological reactions is ultimately derived either from light, as in photosynthesis by green plants, or by oxidative processes relying on the supply of molecular oxygen. The complexity of most biological systems calls for the existence of reaction sequences in which the energy released in one reaction is consumed in another. Compounds, capable of releasing energy in this way, are described as **energy-rich** compounds. The term is relative, but an acceptable definition of an energy-rich compound is one whose reaction with a substance commonly present in the environment is accompanied by a large negative free energy change at physiological pH (Jencks, 1962). In practice in biochemical systems, this is interpreted quantitatively as including any reaction in which the standard free energy (at pH7) is greater than $-30$ kJ mol$^{-1}$.

Many of the so-called energy-rich compounds are esters or anhydrides, and the release of energy occurs on hydrolysis (Table 8). Large amounts of energy are also released in oxidation/reduction processes. The energy released in such processes can be calculated from the relationship:

$$\Delta G^\circ = -nF\Delta E^\circ$$

From this it can be deduced that compounds with oxidation—reduction potentials less than that of the oxygen/water system ($E^\circ$ at pH 7.0, +0.85 V) by 0.5 V or more may be classed as biochemically energy-rich.

The most commonly encountered energy-rich compounds are phosphate esters and anhydrides, such as adenosine triphosphate (ATP). Hydrolysis of ATP

Table 8. Energy-rich Compounds

| Class of Compound | Example | Free energy of hydrolysis kJ mol$^{-1}$ at pH 7.0 |
|---|---|---|
| Phosphoric anhydrides | Adenosine triphosphate (ATP) → ADP | 31 |
| | Adenosine triphosphate (ATP) → AMP | 32 |
| Phosphoric-carboxylic anhydrides | Acetyl phosphate | 44 |
| | Acetyl adenylate | 56 |
| Enol phosphates | Phospho-enol pyruvate | 54 |
| Aminoacid esters | Methyl glycinate | 35 |
| Thiol esters | Acetyl-$\overline{\text{SCoA}}$ | 32 |
| | Acetoacetyl-$\overline{\text{SCoA}}$ | 44 |
| Guanidine phosphates | Phosphocreatine | 38 |

to adenosine diphosphate (ADP) and phosphate at pH 7.0 leads to a free energy change of about 32 kJ $mol^{-1}$ at 37°C. Two factors are important in these hydrolyses. One is the gain in resonance energy in the products, as the result of removal on hydrolysis of competing factors which arise in the anhydride links, as shown.

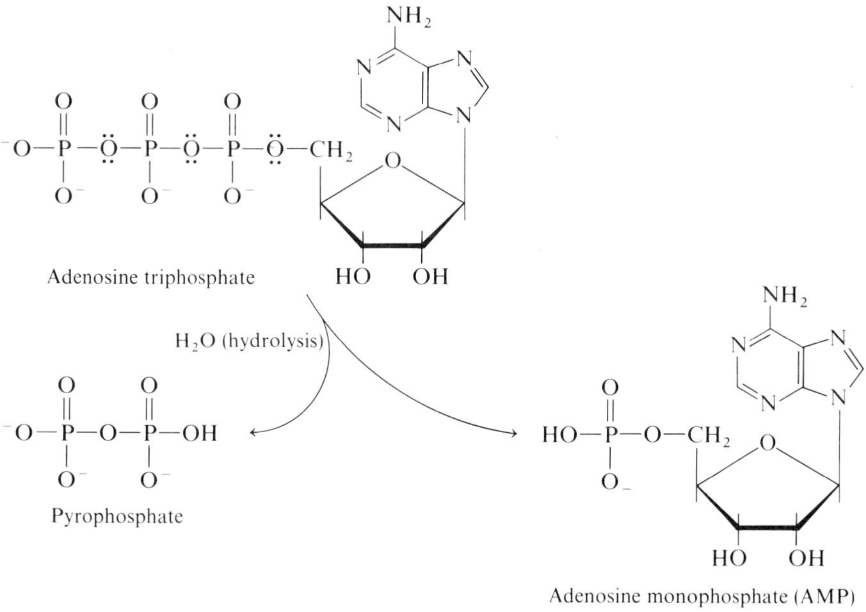

$NH_2$

N N N N

O O O

$^{-}O-P-\ddot{O}-P-\ddot{O}-P-\ddot{O}-CH_2$ O

$O^-$ $O^-$ $O^-$

Adenosine triphosphate HO OH

$H_2O$ (hydrolysis)

$NH_2$

N N N N

O O O

$^{-}O-P-O-P-OH$ $HO-P-O-CH_2$ O

$O^-$ $O^-$ $O_-$

Pyrophosphate

HO OH

Adenosine monophosphate (AMP)

The second factor is that hydrolysis to the diphosphate (ADP) or monophosphate (AMP) relieves the electrostatic repulsion between the anionic centres in the polyanion triphosphate, with a corresponding gain in energy. Relief from electrostatic interaction is also important in the hydrolysis of phosphosulphates, and sulphate transfer in drug metabolism (**2**, 5).

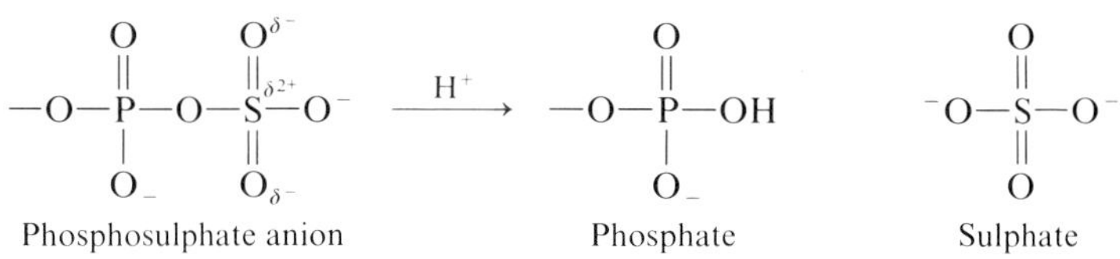

O $O^{\delta-}$ O O

$-O-P-O-\overset{\delta 2+}{S}-O^-$ $\xrightarrow{H^+}$ $-O-P-OH$ $^{-}O-S-O^-$

$O_-$ $O_{\delta-}$ $O_-$ O

Phosphosulphate anion Phosphate Sulphate

The major source of free energy in this and most other energy-rich compounds, however, is the gain in resonance energy which is achieved in the products of hydrolysis. Thus, the normal resonance of the guanidinium ion of creatine is suppressed both by the positively charged phosphorus of the attached phosphate group in phosphocreatine, and by the loss of symmetry of the molecule.

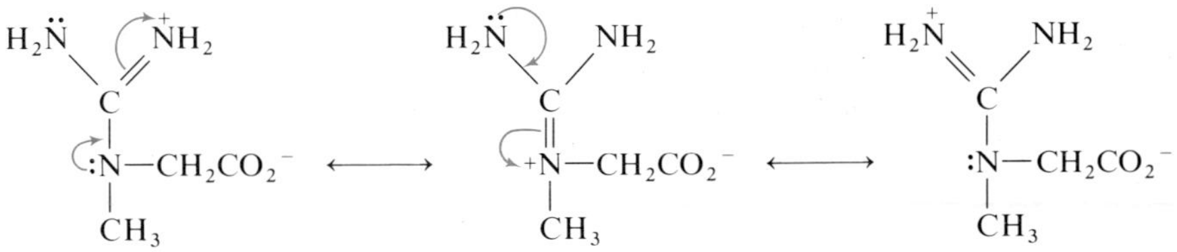

Creatine

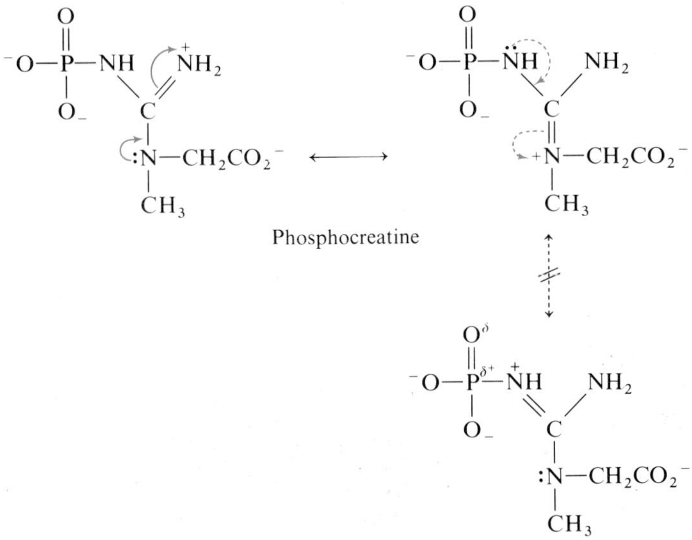

Phosphocreatine

Similarly, thiol esters such as acetyl-$\overline{\text{SCoA}}$ have lower resonance energy than their oxygen analogues, owing to the lower ability of sulphur to donate electrons and form double bonds (Pitzer, 1948). The normal carboxylic ester resonance, therefore, tends to be suppressed in the thiol esters, and a considerable gain of free energy occurs on hydrolysis (attack by water) with formation of the parent carboxylic acid group which is capable of resonance (p. 31).

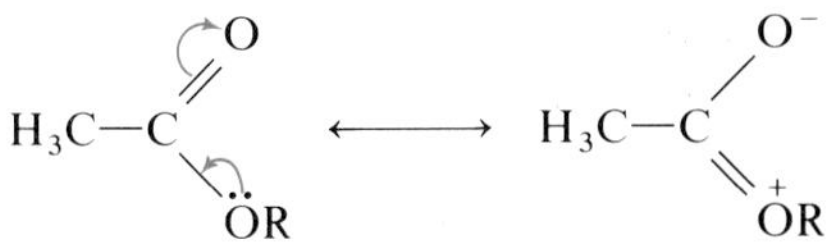

Carboxylic ester

Thiol ester

Enol phosphates, such as phospho-enol pyruvate, have an even larger free energy of hydrolysis (*ca* -54 kJ $mol^{-1}$). They are, therefore, extremely powerful phosphorylating agents, and capable of bringing about the transformation of ADP to ATP. The energy of this bond arises very largely because the enol structure present in the phosphate ester is much less stable than its tautomeric keto form. This energy is released with the formation of pyruvate (keto form) on hydrolysis.

$$H_2C{=}C(OPO_3^{2-})\cdot CO_2^- \xrightarrow{\text{hydrolysis}} H_2C{=}C(OH)H\cdot CO_2^- + PO_4^{3-} \rightleftharpoons CH_3\cdot CO\cdot CO_2^-$$

Phospho-enol pyruvate — Phosphate — Pyruvate

## HETEROLYTIC BOND CLEAVAGE

### Factors Affecting Bond Cleavage

The energy required in a chemical reaction may be derived from a number of sources. In a few cases, the reacting molecules possess sufficient energy at room temperature to overcome the activation energy requirements, and the reaction is able to proceed spontaneously. Examples of such reactions include the formation of some quaternary salts, and the reaction between hydrazine and hydrogen peroxide, which is used for rocket propulsion in space-craft.

$$H_2N\cdot NH_2 + H_2O_2 \longrightarrow 2NH_3 + O_2 + \text{Energy}$$

Thermal, or some other form of energy is frequently required to initiate reaction between organic molecules. Thus, the application of heat will often promote reaction at an elevated temperature which either would not proceed at all at room temperature or, alternatively, would only proceed so slowly as to be totally uneconomic. The application of thermal energy will increase both vibrational, rotational and translational energy of the molecules, leading to an increased prospect of both bond dislocation and molecular collision.

Thermal energy alone, however, is unable to provide sufficient energy at temperatures up to about 100°C to break bonds of more than a few kJ mol$^{-1}$ in strength. Reference to Table 7, showing a number of bond dissociation energies, would suggest that few organic reactions would be feasible under such conditions. However, other factors such as the existence of bond dipoles (inductive effect), mesomeric (Chapter 5) and other polarisability effects (Chapter 9) promote electron drifts which weaken covalent bonds sufficiently to permit the making and breaking of bonds under moderate reaction conditions, when appropriately substituted molecules are brought into contact with each other.

Thus, whilst the C—H bond of a paraffin hydrocarbon is insufficiently polarised to permit reaction with the hydroxyl ion, the withdrawal of electrons from carbon as a result of the electronegativity of iodine in methyl iodide allows a reaction to proceed, which results in the displacement of iodine and its substitution by hydroxyl.

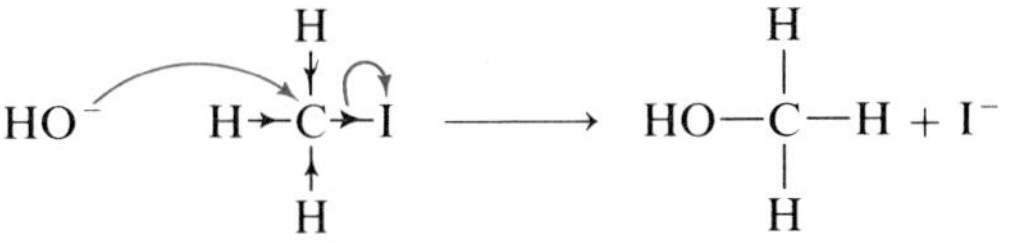

$$HO^- \quad H \rightarrow \underset{\uparrow \atop H}{\overset{H \atop \downarrow}{C}} \rightarrow I \longrightarrow HO{-}\underset{H}{\overset{H}{C}}{-}H + I^-$$

**Ionic Reactions**

Reactions which are promoted by bond polarisation and which result in the breaking of the bond by the gain (and loss) of both bonding electrons are known as **heterolytic** or **ionic** reactions. Such reactions rarely occur in the gaseous state, but are favoured in the molten state and in solution in polar solvents. In solution, the polarity of the solvent exerts a marked influence on the rate of the reaction. Ionic reactions are favoured with charged reactants such as $HO^-$, $RO^-$, $H_3O^+$, $H^-$, and are often catalysed by the presence of inorganic salts or Lewis acids ($AlCl_3$, $SbCl_3$, $FeCl_3$, $BF_3$), i.e. covalent compounds which can accept electron pairs into their incomplete penultimate shells.

In the above example, the essential feature in the breaking of the carbon–iodine bond is the transfer of both electrons to iodine with the release of the iodide ion. In some reactions, as, for example, in that between t-butyl iodide and sodium hydroxide, the loss of the electron pair occurs as a distinct step with the formation of a **carbonium** ion as a discrete positively charged species.

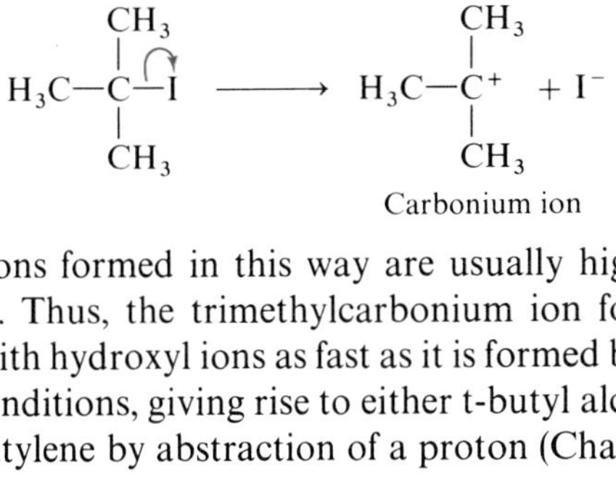

$$H_3C{-}\underset{CH_3}{\overset{CH_3}{C}}{-}I \longrightarrow H_3C{-}\underset{CH_3}{\overset{CH_3}{C^+}} + I^-$$

Carbonium ion

The carbonium ions formed in this way are usually highly reactive and of extremely short life. Thus, the trimethylcarbonium ion formed in the above reaction will react with hydroxyl ions as fast as it is formed by one of two routes, depending on the conditions, giving rise to either t-butyl alcohol by direct combination or to isobutylene by abstraction of a proton (Chapter 9).

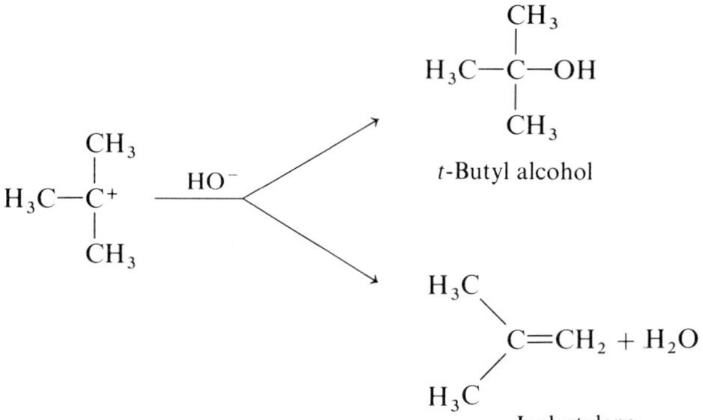

$$H_3C{-}\underset{CH_3}{\overset{CH_3}{C^+}} \xrightarrow{HO^-} H_3C{-}\underset{CH_3}{\overset{CH_3}{C}}{-}OH$$

*t*-Butyl alcohol

$$H_3C{-}\underset{CH_3}{\overset{CH_3}{C^+}} \xrightarrow{HO^-} (H_3C)_2C{=}CH_2 + H_2O$$

Isobutylene

In other reactions, bond breaking does not occur as a distinct step with the formation of a carbonium ion. Instead, formation of a new bond to the attacking reagent occurs simultaneously with the breaking of the existing bond, and the existence of an intermediate transition state may be postulated.

$$HO^- + R{-}\overset{H}{\underset{H}{C}}{\rightarrow}I \longrightarrow HO{\cdots}\overset{H}{\underset{R\ \ H}{C}}{\cdots}I \longrightarrow HO{-}\overset{H}{\underset{R}{C}}{-}H + I^-$$

Transition state

Other types of heterolytic reaction occur in which the movement of electrons in bond breaking (shown by a curved double-headed arrow ⌒) is towards carbon giving rise to a **carbanion**.

$$EtO^- \curvearrowright H{-}\overset{H}{\underset{H}{C}}{-}CHO \longrightarrow {}^{-}\overset{H}{\underset{H}{C}}{-}CHO \quad (+\ EtOH)$$

## CARBONIUM IONS

### Formation of Carbonium Ion Intermediates

Carbonium ions are formed by heterolytic bond cleavage typically in the following general reactions.

(a) *Hydrolysis of tertiary alkyl halides*

$$(CH_3)_3C{-}Br \xrightarrow{HO^-} (CH_3)_3C^+ + Br^-$$

(b) *Protonation of alcohols with mineral acid*

$$R\ddot{O}H + H^+ \longrightarrow R{-}\underset{H}{\overset{+}{O}}{-}H \longrightarrow R^+ + H_2O$$

(c) *The action of a Lewis acid on an alkyl halide* in the Friedel Crafts alkylation of aromatic compounds.

$$R{\cdot}\ddot{C}l + AlCl_3 \longrightarrow R{-}ClAlCl_3 \longrightarrow R^+ + AlCl_4^-$$

(d) *Thermal decomposition of aryldiazonium salts*

$$C_6H_5{-}\overset{+}{N_2}Cl^- \longrightarrow C_6H_5^+ + N_2 + Cl^-$$

(e) *Acid-catalysed additions to ethylenic compounds*

$$CH_3{\cdot}CH{=}CH_2 + H^+ \longrightarrow CH_3{\cdot}\overset{+}{C}H{\cdot}CH_3$$

(f) *Acid-catalysed additions to aldehydes and ketones*

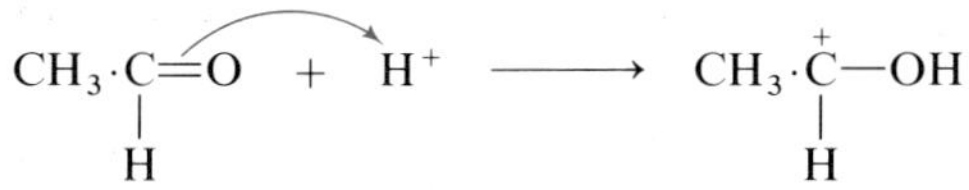

$$CH_3{\cdot}\underset{\displaystyle H}{\underset{|}{C}}{=}O \; + \; H^+ \longrightarrow CH_3{\cdot}\underset{\displaystyle H}{\underset{|}{\overset{+}{C}}}{-}OH$$

**Relative Stability**

Since a positively charged carbon atom is much more electronegative than a fully substituted saturated carbon atom, which is electrically neutral, the carbonium carbon will exert an inductive effect (−I) attracting electrons from attached alkyl groups. This inductive effect reduces the full positive charge on the carbon and at the same time weakens the C—H bonds of the alkyl substituents. The result is a form of resonance known as **hyperconjugation** (no-bond resonance; p. 33), the number of extreme (canonical) forms which contribute to the resonance hybrid increasing with the number of C—H bonds on carbon atoms α to the carbonium carbon. Since the stability of any resonance hybrid is directly related to the number of contributing forms, the relative stability of carbonium ions is also directly related to the number of C—H bonds, as this is a measure of the extent to which the charge can be spread over the carbonium ion as a whole.

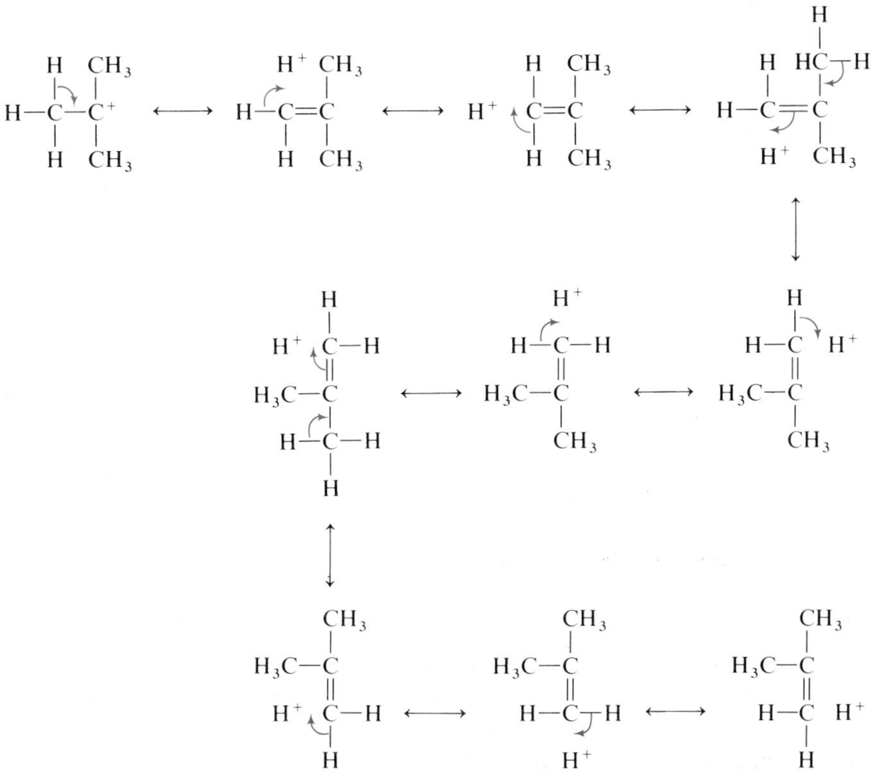

```
                                                                         H
                                                                         |
   H  CH3              H+ CH3               H  CH3                  H  HC—H
   |  |                   |                 |  |                    |  |
H—C—C+     <——>     H—C=C      <——>   H+  C=C       <——>       H—C=C
   |  |                |  |                 |  |                    |  |
   H  CH3              H  CH3               H  CH3                  H+ CH3
                                                                       ↕
              H                 H+                  H
              |                                     |
         H+  C—H             H—C—H             H—C   H+
              ||                ||                  ||
         H3C—C     <——>    H3C—C      <——>    H3C—C
              |                 |                   |
           H—C—H               CH3                 CH3
              |
              H
              ↕
             CH3               CH3                 CH3
              |                 |                   |
         H3C—C             H3C—C               H3C—C
              ||                ||                  ||
         H+  C—H   <——>      H—C—H      <——>     H—C  H+
              |                 H+                  |
              H                                     H
```

The relative order of stability is, therefore, 3° > 2° > 1°, whilst the actual stabilisation in the t-butyl carbonium ion has been calculated as 335 kJ $mol^{-1}$ (Müller and Mulliken, 1958).

The terms primary (1°), secondary (2°) and tertiary (3°) are used to describe carbon atoms, ions or radicals which are bonded to one, two or three other carbon atoms respectively.

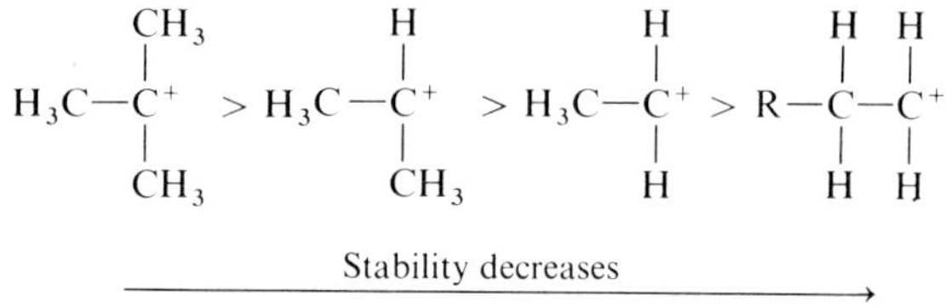

$$H_3C-C^+(CH_3)_2 > H_3C-C^+H(CH_3) > H_3C-C^+H_2 > R-CH_2-C^+H_2$$

Stability decreases →

| | Tertiary (3°) | Secondary (2°) | Primary (1°) | Primary (1°) |
|---|---|---|---|---|
| No of α C—H bonds | 9 | 6 | 3 | 2 |

Benzylic and allylic carbonium ions are also stabilised by resonance, and to a greater degree than simple primary carbonium ions.

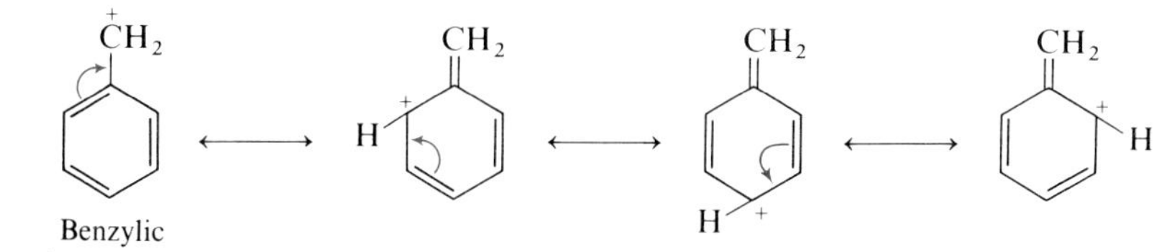

Benzylic carbonium ion

$$\overset{+}{C}H_2{=}CH{-}\overset{+}{C}H_2 \longleftrightarrow \overset{+}{C}H_2-CH=CH_2$$

Allylic carbonium ion

Although most carbonium ions are very reactive, a few carbonium salts, such as triphenylcarbonium perchlorate, are sufficiently stable to permit isolation, whilst tropylium bromide is even stable in aqueous solution.

$Ph_3\overset{+}{C}ClO_4^-$ Triphenylcarbonium perchlorate

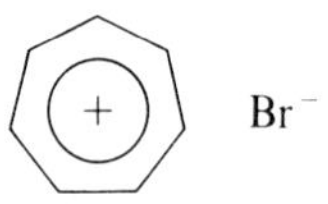

$Br^-$ Tropylium bromide

## The Configuration of Carbonium Ions

Carbonium ions are planar, with their three C—C bonds mutually inclined at an angle of 120°, and may be considered as $sp^2$ hybridised carbon atoms each with a vacant *p*-orbital. As a result, carbonium ions formed as intermediates in the reaction of optically active compounds give rise to racemic products

(Chapter 13), since there is an equal chance of attack from either above or below the plane of the ion. As a further consequence of their planar configuration, tertiary carbonium ions are formed most readily from compounds with bulky substituents, due to the relief from steric interactions which is achieved in changing from a tetrahedral to a planar structure.

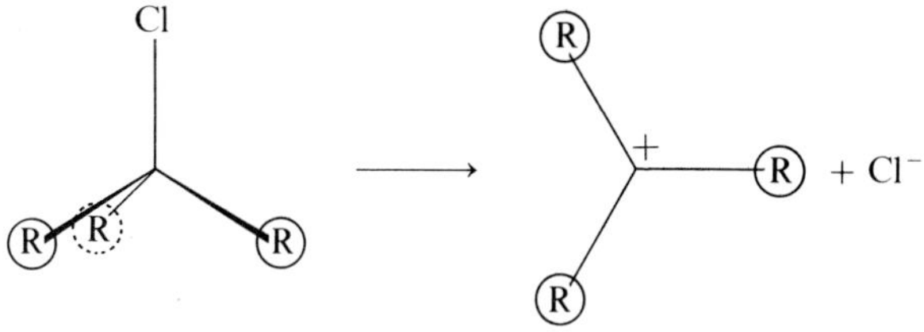

$$R_3C\text{—}Cl \longrightarrow R_3C^+ + Cl^-$$

Bridgehead carbon atoms are sterically incapable of adopting a planar configuration. For this reason, the following compound cannot form a carbonium ion, and shows none of the usual substitution reactions of alkyl halides.

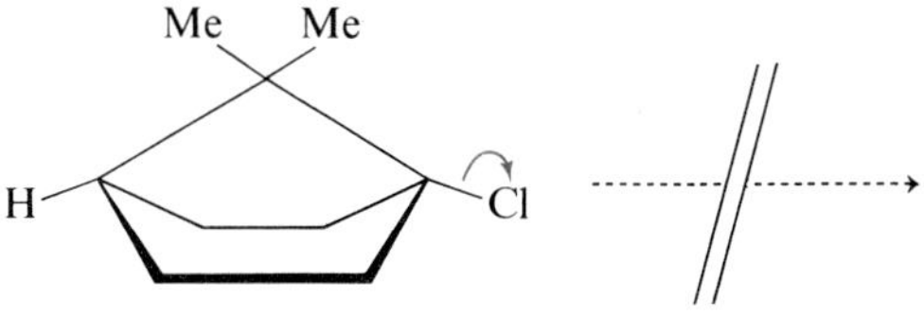

### Some Reactions of Carbonium Ions

(a) *Reaction with nucleophilic reagents.* 1. The term **nucleophile** means 'positive centre-loving'. Nucleophilic reagents, therefore, include negative ions (inorganic anions and carbanions) or other reactants with available electrons, such as the lone pairs on oxygen, nitrogen and sulphur.

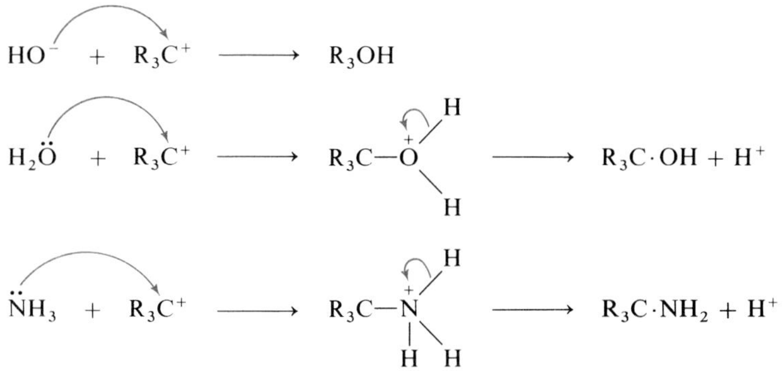

$$HO^- + R_3C^+ \longrightarrow R_3OH$$

$$H_2\ddot{O} + R_3C^+ \longrightarrow R_3C\text{—}\overset{+}{O}H_2 \longrightarrow R_3C\cdot OH + H^+$$

$$\ddot{N}H_3 + R_3C^+ \longrightarrow R_3C\text{—}\overset{+}{N}H_3 \longrightarrow R_3C\cdot NH_2 + H^+$$

2. The coenzyme nicotinamide adenine dinucleotide (NAD) functions as a carbonium ion in the hydride ion transfer catalysed by liver-alcohol dehydrogenase during the oxidation of ethanol. The coenzyme NAD may be represented

by the resonance hybrid, in which the carbonium ion form acts as the hydride ion acceptor.

$$H_3C-CH(H)-O-H \;+\; \text{NAD} \longrightarrow H_3C-CH{=}O \;+\; \text{NADH}$$

NAD NADH

(b) *Elimination of a β-proton to form an olefine*

$$H_3C-CH_2-\overset{+}{C}H_2 \longrightarrow H_3C-CH{=}CH_2 + H^+$$

(c) *Rearrangement to a more stable carbonium ion*, by means of a hydride shift from a β-carbon atom. This implies the shift of a hydrogen atom accompanied by the electrons of the β-carbon—hydrogen bond, i.e. the shift of a hydride ion ($H^-$).

$$H_3C-CH_2-\overset{+}{C}H_2 \longrightarrow H_3C-\overset{+}{C}H-CH_3$$

Primary (1°) carbonium ion — Secondary (2°) carbonium ion

(d) *Carbon–carbon rearrangement* (*Wagner–Meerwein rearrangement*). Such rearrangements are in general promoted in the interests of achieving greater stability. An alkyl group migrates from a β-carbon atom, accompanied by the bond electrons.

$$H_3C-C(CH_3)_2-\overset{+}{C}H_2 \longrightarrow H_3C-\overset{+}{C}(CH_3)-CH_2-CH_3$$

Primary (1°) carbonium ion — Tertiary (3°) carbonium ion

## CARBANIONS

### Formation of Carbanion Intermediates

Carbanions (p. 40) are formed by heterolytic cleavage, generally of C—H bonds in which the carbon atom is linked to a powerful electron-withdrawing group.

(a) *The reaction of active methylene compounds with strong bases.* Carbanions are the characteristic intermediates of reactions involving active methylene groups, and are readily formed from compounds containing the following, in which R is an electron-withdrawing group.

$$R{\cdot}CH_2{\cdot}\underset{|}{C}{=}O \qquad R{\cdot}CH_2{\cdot}\underset{OEt}{\underset{|}{C}}{=}O \qquad R{\cdot}CH_2{\cdot}NO_2 \qquad R{\cdot}CH_2{\cdot}CN$$

Carbanion formation occurs under the influence of strong bases ($EtO^-$, $R_4\overset{+}{N}OH^-$) which aid the removal of a proton from the active methylene group.

$$EtO^- \;\; CH_3{\cdot}\underset{O}{\underset{\|}{C}}{-}\overset{H}{\overset{|}{C}}H{\cdot}COOEt \xrightarrow{-EtOH} CH_3{\cdot}\underset{O^-}{\underset{|}{C}}{=}CH{\cdot}COOEt \longleftrightarrow CH_3{\cdot}\underset{O}{\underset{\|}{C}}{-}\bar{C}H{\cdot}COOEt$$

Carbanion formation can also occur in attack on active methyl groups by strong bases.

$$EtO^- \;\; H_2\overset{H}{\overset{|}{C}}{-}\underset{O}{\underset{\|}{C}}{-}CH_3 \xrightarrow{-EtOH} H_2C{=}\underset{O^-}{\underset{|}{C}}{-}CH_3 \longleftrightarrow {}^-H_2C{-}\underset{O}{\underset{\|}{C}}{-}CH_3$$

$$EtO^- \;\; H_2\overset{H}{\overset{|}{C}}{-}N\begin{smallmatrix}\nearrow O\\ \searrow\!\!\!\backslash O\end{smallmatrix} \xrightarrow{-EtOH} H_2C{=}N\begin{smallmatrix}\nearrow O\\ \searrow O^-\end{smallmatrix} \longleftrightarrow H_2\bar{C}{-}N\begin{smallmatrix}\nearrow O\\ \searrow\!\!\!\backslash O\end{smallmatrix}$$

The carbanion, once formed, is stabilised by resonance, and this is an important factor in promoting its formation. The ease with which the proton is removed by base depends on its acidity, which is a function of resonance in the parent compound. Thus, acetone will form a carbanion more readily than ethyl acetate in which there is a competing resonance lowering the acidity of the proton.

$$EtO^- + H_2C(H)-C(=O)-\ddot{O}Et \xrightarrow{a} H_2C=C(O^-)-OEt \longleftrightarrow {}^-CH_2 \cdot CO \cdot OEt \quad (+\ EtOH)$$

$$\updownarrow b$$

$$CH_3 \cdot C(O^-)=\overset{+}{O}Et$$

(b) *Heterolytic cleavage of carbon–metal bonds.* (1) Acetylenes in liquid ammonia react with alkali metals to give metal acetylides, in which the acetylide ion can function as a carbanion.

$$HC{\equiv}CH \xrightarrow[-35^\circ]{Na/Liq.NH_3} HC{\equiv}C^- Na^+$$

(2) Reactive species generated in reactions with Grignard reagents (RMgX) and metal alkyls (BuLi) are usually regarded formally as carbanions.

$$R \cdot H_2C-MgBr \longrightarrow R \cdot CH_2^- + {}^+MgBr$$

## Stability

It is doubtful whether carbanions can be regarded as having the same discrete existence comparable with that observed for some carbonium ions. Once formed, they experience a driving force to lose, or alternatively, spread the negative charge which they have acquired. Electron-withdrawing groups will, therefore, tend to stabilise the carbanion by reducing the electron density on the carbanion carbon atom itself. The contribution of resonance to stabilisation is evident in the examples of carbanions from active methylene compounds such as diethyl malonate and ethyl acetoacetate, and also in the benzylic carbanion.

$$C_6H_5CH_2^- \longleftrightarrow CH_2{=}C_6H_5^- \longleftrightarrow \ldots \longleftrightarrow \ldots$$

Some lesser degree of stabilisation is also achieved in the presence of elements which confer a strong inductive effect away from carbon ($-I$ effect). The trifluoromethyl and thio-ether groups are effective in this way.

```
   F  H              H
   ↑  |              |
F←-C←-C⁻        R·S←-C⁻
   ↓  |              ↓
   F  R             R·S
```

## The Configuration of Carbanions

The precise configuration of carbanions is a matter of some doubt. Whilst it may be argued that these anions have a pyramidal arrangement, as in tertiary amines with which they are isoelectronic, effective stabilisation by charge delocalisation demands co-planarity of the interacting groups. This can only apply if the three bonds of the carbanion are planar. As a result, molecular asymmetry, if present on the carbanion carbon atom, is destroyed with carbanion formation, and such compounds are racemised very readily in the presence of base (Chapter 13). Carbanions are, therefore, comparable with tertiary amines, which similarly lack optical asymmetry, though this is due to continuous and rapid inversion of a pyramidal structure.

R, R″, R′

Planar carbanion

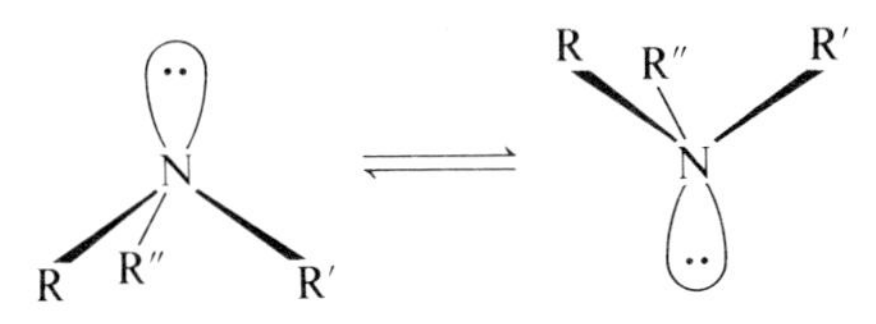


Nitrogen inversion in tertiary amine

## Some Reactions of Carbanions

Carbanions are powerful nucleophiles, and hence readily react by addition to carbonyl compounds.

(a) *Benzoin condensation*, between two molecules of aromatic aldehyde to give an $\alpha$-hydroxyketone.

$$\underset{\text{Benzaldehyde}}{2\,\mathrm{Ph{\cdot}CHO}} \longrightarrow \underset{\text{Benzoin}}{\mathrm{Ph{\cdot}CO{\cdot}CHOH{\cdot}Ph}}$$

This reaction occurs in the presence of cyanide ions ($CN^-$), and is initiated by addition of the ion to the carbonyl group of benzaldehyde to form a cyanohydrin.

The latter undergoes a prototropic shift to give a **carbanion**, which in turn adds to the carbonyl group of a second molecule of aldehyde.

$$CN^- \quad Ph-\underset{\displaystyle H}{\underset{|}{C}}=O \rightleftharpoons Ph-\overset{\displaystyle CN}{\overset{|}{\underset{\displaystyle H}{\underset{|}{C}}}}-O^- \rightleftharpoons Ph-\overset{\displaystyle CN}{\overset{|}{\underset{\displaystyle OH}{\underset{|}{C}}}}{}^-$$

Cyanhydrin Carbanion

$$Ph-\overset{\displaystyle CN}{\overset{|}{\underset{\displaystyle OH}{\underset{|}{C}}}}{}^- \quad \overset{\displaystyle H}{\overset{|}{\underset{\displaystyle Ph}{\underset{|}{C}}}}=O \rightleftharpoons Ph-\overset{\displaystyle CN}{\overset{|}{\underset{\displaystyle OH}{\underset{|}{C}}}}-\overset{\displaystyle H}{\overset{|}{\underset{\displaystyle Ph}{\underset{|}{C}}}}-O^- \underset{}{\overset{H^+}{\rightleftharpoons}} Ph-\underset{\displaystyle O}{\underset{\|}{C}}-\overset{\displaystyle H}{\overset{|}{\underset{\displaystyle Ph}{\underset{|}{C}}}}-OH$$

(b) *Addition of thiamine pyrophosphate to pyruvate*, in the enzymic formation of acetoin. This addition is catalysed by α-carboxylase for which thiamine pyrophosphate is the coenzyme. The hydrogen at C-2 in the latter is readily exchanged for deuterium in $D_2O$ (Breslow, 1958; Breslow and McNelis, 1959) and it is deduced that the corresponding anion is the key intermediate in the addition to the carbonyl group of pyruvate.

NH₂ · Me · N · N · S · $CH_2 \cdot CH_2 \cdot O \cdot \overset{O}{\overset{\|}{\underset{O^-}{\underset{|}{P}}}} \cdot O \cdot \overset{O}{\overset{\|}{\underset{O^-}{\underset{|}{P}}}} \cdot O^-$

$$\mathbf{Me-\underset{\displaystyle CO\cdot O^-}{\underset{|}{C}}=O}$$

$\searrow$

NH₂ · Me · N · N · S · $\mathbf{Me}$ · C · $\mathbf{O^-}$ · $\mathbf{CO\cdot O^-}$ · $CH_2 \cdot CH_2 \cdot O \cdot \overset{O}{\overset{\|}{\underset{O^-}{\underset{|}{P}}}} \cdot O \cdot \overset{O}{\overset{\|}{\underset{O^-}{\underset{|}{P}}}} \cdot O^-$

(c) *The addition of a Grignard reagent to the carbonyl group* of aldehydes and ketones. The actual mechanism of addition is complex, but may be represented in a simplified manner.

$$Me-\overset{\displaystyle H}{\overset{|}{C}}=O \quad R\cdot CH_2^- \xrightarrow[{}^+MgBr]{} Me-\overset{\displaystyle H}{\overset{|}{\underset{\displaystyle R\cdot CH_2}{\underset{|}{C}}}}-O^-\overset{+}{M}gBr \longrightarrow Me-\overset{\displaystyle H}{\overset{|}{\underset{\displaystyle R\cdot CH_2}{\underset{|}{C}}}}-OH$$

$$R\cdot CH_2-MgBr$$

## HOMOLYTIC BOND CLEAVAGE

Ionic reactions are normally feasible in organic solvents at temperatures within the boiling points of the common laboratory solvents. At much higher temperatures of the order of 400°C or more, gas phase pyrolyses are feasible which lead to a breaking of covalent bonds by a **homolytic** (or **free radical**) mechanism. In this, the bond is broken with the movement of one electron (shown by a single-headed, fish-hook, curved arrow ⌒) from the pair forming the covalent bond to each of the separating atoms.

$$-\overset{|}{\underset{|}{C}}-\overset{|}{\underset{|}{C}}- \longrightarrow -\overset{|}{\underset{|}{C}}\cdot + \cdot\overset{|}{\underset{|}{C}}-$$

Homolytic cleavage gives rise to two free radicals, which are electrically neutral. Each radical has an unpaired electron, and accordingly is highly reactive and generally of short life.

Typical examples of pyrolyses leading to homolytic bond cleavage occur in the high temperature cracking processes used to degrade high molecular weight paraffin hydrocarbons to lower molecular weight fractions for use as motor fuel. The process is conducted under pressure in either the liquid or vapour phase in the presence of metal oxide catalysts and usually at temperatures between 450° and 600°. Ethylenic hydrocarbons are formed as by-products in a process which may be depicted as follows.

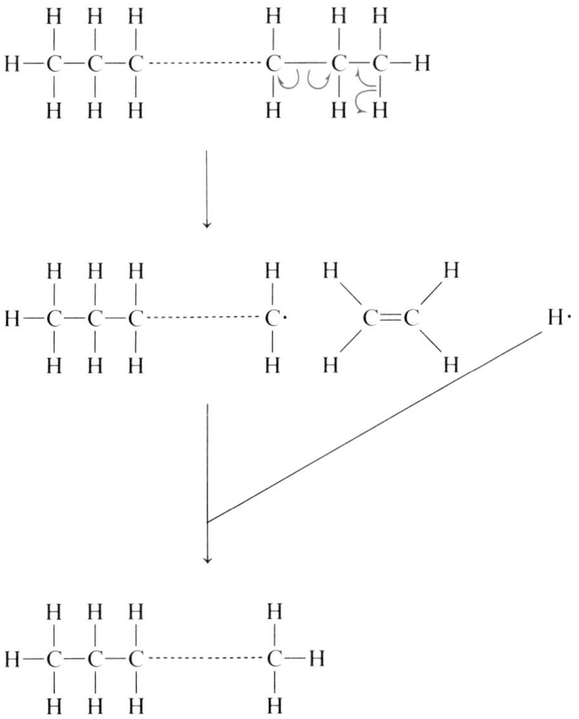

$$H-\overset{H}{\underset{H}{C}}-\overset{H}{\underset{H}{C}}-\overset{H}{\underset{H}{C}}\cdots\cdots\overset{H}{\underset{H}{C}}-\overset{H}{\underset{H}{C}}-\overset{H}{\underset{H}{C}}-H$$

$$\downarrow$$

$$H-\overset{H}{\underset{H}{C}}-\overset{H}{\underset{H}{C}}-\overset{H}{\underset{H}{C}}\cdots\cdots\overset{H}{\underset{H}{C}}\cdot \quad \overset{H}{\underset{H}{>}}C=C\overset{H}{\underset{H}{<}} \quad H\cdot$$

$$\downarrow$$

$$H-\overset{H}{\underset{H}{C}}-\overset{H}{\underset{H}{C}}-\overset{H}{\underset{H}{C}}\cdots\cdots\overset{H}{\underset{H}{C}}-H$$

Homolytic reactions also occur in solution in non-polar solvents. Such reactions are initiated by light and/or the presence of free radicals. In some cases, light is the prime source of energy, as for example in the breaking of the Cl—Cl bond; this is demonstrated by the spontaneous chlorination of methane at room temperature in diffused daylight, as opposed to the almost complete lack of reaction in the dark.

$$Cl{-}Cl \longrightarrow 2Cl\cdot$$

$$H_3C{-}H \;\; \cdot Cl \longrightarrow H_3C\cdot + H{-}Cl$$

$$H_3C\cdot \;\; Cl_2 \longrightarrow H_3C{-}Cl + Cl\cdot$$

Homolytic cleavages are often initiated by other free radicals. This is particularly true of oxidative decomposition by peroxides, which readily generate peroxide radicals (RO·), and molecular oxygen which has a considerable element of di-radical character.

$$\cdot O{-}O\cdot$$

Oxidative decomposition by molecular oxygen and peroxides, and its prevention by anti-oxidants, is an important consideration in the stability and stabilisation of many pharmaceutical chemicals.

Kinetically, homolytic reactions are often anti-catalytic, so that they frequently show a slow induction period, followed by a fast chain reaction which is only limited by the availability of the reactants. Alternatively, they may be inhibited, or if established, halted by free radical trapping agents, such as phenols (Chapter 8).

## FREE RADICALS

### Formation of Radical Intermediates

(a) *Atomic hydrogen* (H·) is also a radical, and is formed transiently from molecular hydrogen at the catalyst surface during catalytic hydrogenation. Atomic hydrogen may also be generated by dissolving metals, which are themselves free radicals in acid, for example Zn/HCl, Mg/HCl, Na-Hg/EtOH.

Produced in this way, in the presence of a reactant, it is sometimes described as **nascent** hydrogen.

$$\text{Zn:} + 2\,\text{H}^+\text{Cl}^- \longrightarrow \text{Zn}^{2+} + 2\text{H}\cdot + 2\text{Cl}^-$$

$$\text{Et·O—H} + \text{Na}\cdot \longrightarrow \text{EtO}^-\text{Na}^+ + \text{H}\cdot$$

(b) *Pyrolysis of metal alkyls* (Paneth, 1929). Certain metal alkyls, such as lead tetramethyl are thermally unstable, and hence readily pyrolysed to give alkyl radicals.

$$\text{Pb(Me)}_4 \xrightarrow{\text{Pyrolysis}} \cdot\ddot{\text{Pb}}\cdot + 4\text{Me}\cdot$$

Lead tetramethyl　　　　Methyl radical

The methyl radical is typical of alkyl radicals in general, having a short life; its half-life is about $8 \times 10^{-3}$ s.

(c) *Thermal decomposition of acyl derivatives.* (1) Mercuric acetate and lead acetate undergo thermal decomposition to yield the acetoxy radical.

$$CH_3\cdot CO\cdot O—Hg—O\cdot CO\cdot CH_3 \longrightarrow \cdot Hg\cdot + 2CH_3\cdot CO\cdot O\cdot$$

The acetoxy radical is unstable and readily decomposes to give a methyl radical and carbon dioxide.

$$H_3C—C(=O)—O\cdot \longrightarrow CH_3\cdot + CO_2$$

(2) Acyl peroxides, such as benzoyl peroxide, also readily undergo homolytic cleavage at temperatures of 30° and above.

$$Ph—C(=O)—O—O—C(=O)—Ph \longrightarrow 2Ph—C(=O)—O\cdot$$

CARNEGIE LIBRARY
LIVINGSTONE COLLEGE
SALISBURY, N. C. 28144

The benzoyl radical, like the acetyl radical, decomposes further to yield the phenyl radical and carbon dioxide.

$$Ph-C(=O)O\cdot \longrightarrow Ph\cdot + CO_2$$

(d) *Thermal decomposition of N-Bromosuccinimide.* N-Bromosuccinimide undergoes thermal decomposition to yield bromine and succinimide radicals.

$$\text{(succinimide)N}-Br \longrightarrow \text{(succinimide)N}\cdot + Br\cdot$$

The decomposition is used as a source of bromine radicals (in allylic bromination) since the succinimide radical which is also formed in the decomposition is stabilised by resonance, making the formation of a new N—C bond unfavourable.

(e) *Photochemical decomposition.* (1) Formation of chlorine radicals by homolytic cleavage of the chlorine—chlorine bond occurs at room temperature when chlorine gas is exposed to diffused sunlight.

$$Cl-Cl \longrightarrow 2Cl\cdot$$

(2) Acetone is typical of many carbonyl compounds which undergo photochemical decomposition on exposure to ultraviolet light.

$$H_3C-C(CH_3)=O \longrightarrow CH_3\cdot + CH_3-\dot{C}=O$$

$$H_3C-\dot{C}=O \longrightarrow CH_3\cdot + CO$$

(f) *Oxidation of phenols by one-electron oxidising agents*, such as potassium ferricyanide, gives rise to the reactive phenyl radical.

$$C_6H_5O-H \xrightarrow{e} C_6H_5O\cdot + H^+$$

$$[Fe(CN)_6]^{3-} \qquad [Fe(CN)_6]^{4-}$$

(g) *Electrolysis of sodium salts of carboxylic acids* (Kolbé electrolysis). The acyl ion loses an electron at the anode forming an unstable acyl radical, which immediately decomposes to give an alkyl radical. Combination of alkyl radicals so formed gives a paraffin hydrocarbon, containing an even number of carbon atoms. The reaction provides a useful method for synthesising pure hydrocarbons.

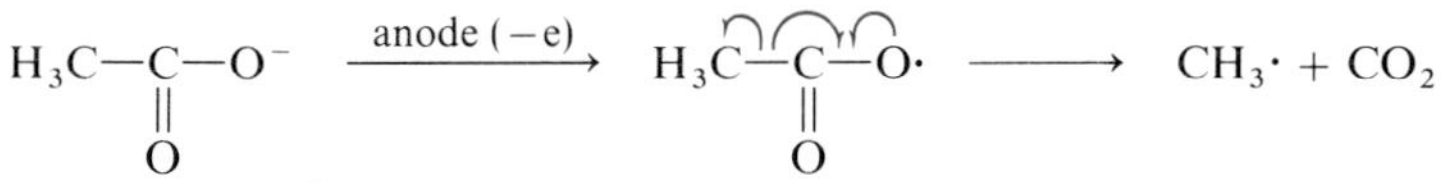

$$H_3C{-}C({=}O){-}O^- \xrightarrow{\text{anode }(-e)} H_3C{-}C({=}O){-}O\cdot \longrightarrow CH_3\cdot + CO_2$$

Acetate ion — Acetyl radical — Methyl radical

$$2CH_3\cdot \longrightarrow CH_3{\cdot}CH_3$$

Ethane

The method is also applicable to the electrolysis of dibasic acids, and also to mixtures of acids, when hydrocarbons with an odd number of carbon atoms can be obtained.

**Stability**

The majority of free radicals are of comparatively short-life. The half-life of the methyl radical is only about $8 \times 10^{-3}$ s. Chain-branching favours stability, and the relative order of stability is:

$$3^\circ > 2^\circ > 1^\circ$$

Allylic and benzylic free radicals are stabilised by resonance.

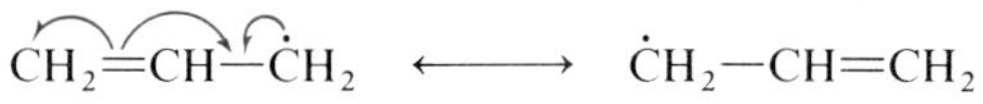

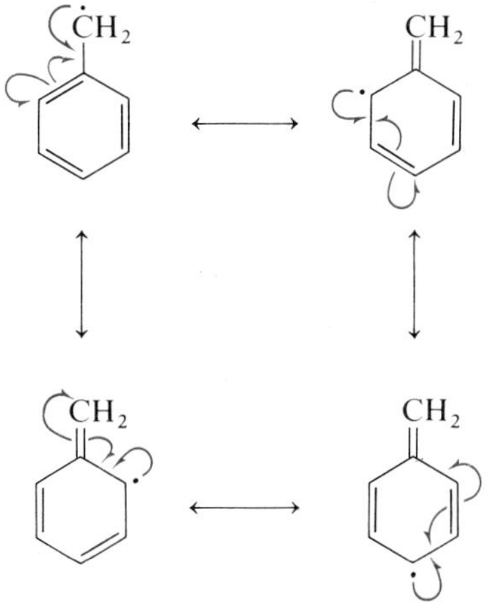

$$\dot{C}H_2{=}CH{-}\dot{C}H_2 \longleftrightarrow \dot{C}H_2{-}CH{=}CH_2$$

$\dot{C}H_2$ $CH_2$ $CH_2$ $CH_2$

Triphenylmethyl has a very high degree of resonance, and hence has much greater stability. Hexaphenylethane dissociates in solution to exist very largely as the radical (Gomberg, 1900).

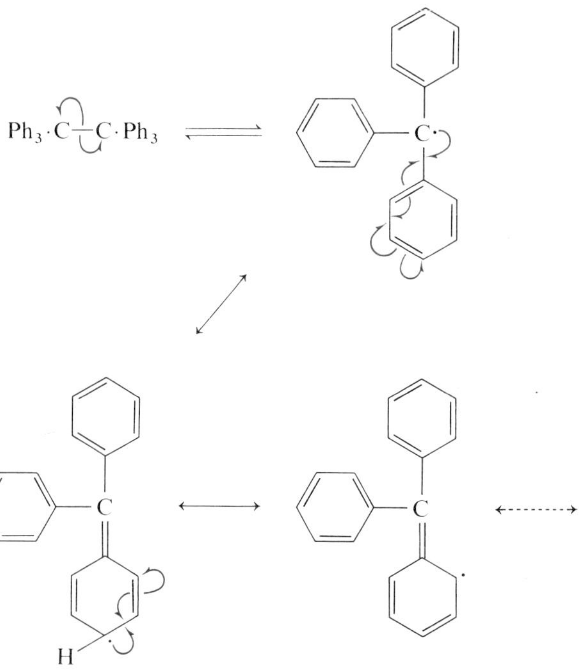

$Ph_3 \cdot C-C \cdot Ph_3$ ⇌

H

Other free radicals are stabilised due to the release of steric compression achieved when the radical is formed, as in tri-t-butyl phenol. The hydroxyl group is sterically hindered by the adjacent ortho t-butyl groups; this hindrance is relieved with the breaking of the O—H bond when the radical is formed.

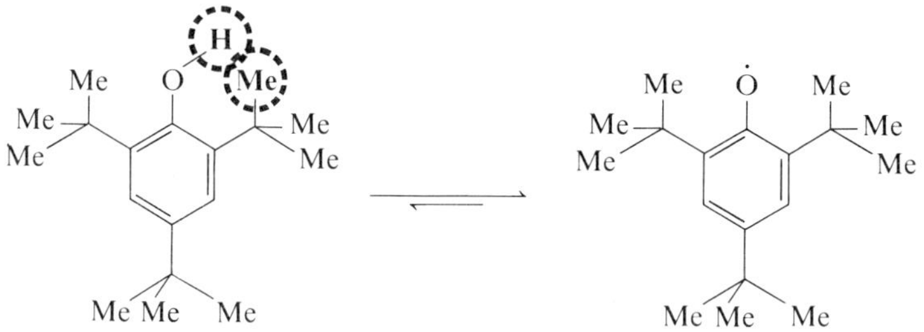

H
Me O Me
Me Me
Me Me
Me Me Me

Tri-t-butylphenol

Me O Me
Me Me
Me Me
Me Me Me

Tri-t-butylphenyl

Stabilised free radicals of this type are useful as anti-oxidants owing to their ability to act as radical scavengers (Chapter 8).

## The Configuration of Free Radicals

Free radicals are intrinsically tetrahedral and capable of easy inversion to the mirror image form.

In practice they are, however, only tetrahedral when their configuration is so constrained, as for example in the apocamphyl radical where the radical bearing atom is in a bridgehead position.

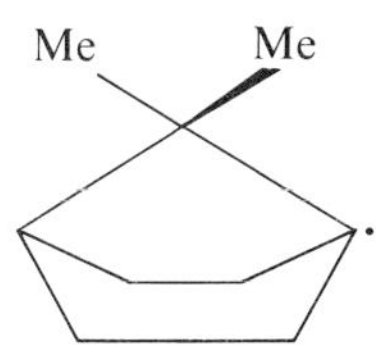


In contrast, wherever there is no such constraint, and particularly where there is appreciable delocalisation, stabilisation favours a planar structure. In general, however, the degree of stabilisation achieved by radicals is less than that for ions.

## Some Reactions of Free Radicals

(a) *Combination with other free radicals* to form a covalent bond. An example occurs in the Kolbé (p. 53) reaction.

$$CH_3{\cdot}CH_2{\cdot}\ {\cdot}CH_2{\cdot}CH_3 \longrightarrow CH_3{\cdot}CH_2{\cdot}CH_2{\cdot}CH_3$$

(b) *Addition to olefines through homolytic cleavage* of the $\pi$-bond, as in catalytic hydrogenation.

(c) *Polymerisation of olefines*, such as styrene to give polystyrene. This is an example of a chain reaction, in which the three main steps are chain initiation, chain propagation and chain termination. The reaction is initiated through the thermal decomposition of benzoyl peroxide (p. 70), and subsequent formation of styrene radicals.

$$Ph{\cdot}CO{\cdot}O{-}O{\cdot}CO{\cdot}Ph \longrightarrow Ph{\cdot}CO{\cdot}O{\cdot} + Ph{\cdot} + CO_2$$

$$Ph{\cdot}\ \ Ph{\cdot}CH{=}CH_2 \longrightarrow Ph{\cdot}\underset{\displaystyle Ph}{\underset{|}{CH}}{\cdot}CH_2{\cdot}$$

Chain propagation then occurs by reaction with further styrene molecules.

$$Ph{\cdot}CO{\cdot}O{\cdot}\underset{\displaystyle Ph}{\underset{|}{CH}}{\cdot}CH_2{\cdot}\ \ \underset{\displaystyle Ph}{\underset{|}{CH}}{=}CH_2 \longrightarrow Ph{\cdot}CO{\cdot}O{\cdot}\underset{\displaystyle Ph}{\underset{|}{CH}}{\cdot}CH_2{\cdot}\underset{\displaystyle Ph}{\underset{|}{CH}}{\cdot}CH_2{\cdot}$$

$$\xrightarrow{Ph{\cdot}CH{=}CH_2\ (n\ \text{moles})} Ph{\cdot}CO{\cdot}O{\cdot}\underset{\displaystyle Ph}{\underset{|}{CH}}{\cdot}CH_2(\underset{\displaystyle Ph}{\underset{|}{C}}HCH_2)_n\underset{\displaystyle Ph}{\underset{|}{CH}}{\cdot}CH_2{\cdot}$$

Theoretically, the reaction is terminated by the complete reaction of all styrene present. In practice, the reaction ceases before this point is reached by the intervention of a number of events, including collision between two growing polymer chains, by collision between the polymer chain and initiator radicals, and by disproportionation.

(d) *Abstraction of a hydrogen radical from a* C—H *bond*

$$Cl{\cdot}\ \ H{-}CH_3 \longrightarrow HCl + CH_3{\cdot}$$

(e) *Disproportionation* arises as a result of abstraction of a hydrogen radical from the $\beta$-carbon of a second alkyl radical.

$$CH_3{\cdot}CH_2{\cdot}\ \ H{-}CH_2{-}\dot{C}H_2 \longrightarrow \underset{\text{Ethane}}{CH_3{\cdot}CH_3} + \dot{C}H_2{-}\dot{C}H_2$$

$$\dot{C}H_2{-}\dot{C}H_2 \longrightarrow \underset{\text{Ethylene}}{CH_2{=}CH_2}$$

## CARBENES

Carbenes are sometimes considered to be carbon di-radicals, arising from the simultaneous homolytic cleavage of two separate bonds to the same carbon atom. It is clear, however, from many of their reactions that they do not always function as di-radicals, but merely as carbon derivatives deficient in two electrons. The simplest compound of this type, carbene, $CH_2$, is formed by photolysis or pyrolysis of diazomethane.

$$H_2C{=}\overset{+}{N}{=}\overset{-}{N} \longrightarrow \dot{\underset{\cdot}{C}}H_2 + N_2$$

Two types of reaction are typical of carbene, insertion in C—H, O—H or C—Cl bonds, and addition across double bonds to form cyclopropanes.

$$CH_3{\cdot}CH_2{\cdot}CH_2{\cdot}Cl + \dot{\underset{\cdot}{C}}H_2 \longrightarrow CH_3{\cdot}CH_2{\cdot}CH_2{\cdot}CH_2{\cdot}Cl$$

$$CH_3{\cdot}CH{=}CH{\cdot}CH_3 + \dot{\underset{\cdot}{C}}H_2 \longrightarrow CH_3{\cdot}\underset{\diagdown\ \diagup}{CH{-}CH}{\cdot}CH_3 \quad (\text{ring closed by } CH_2)$$

Dichlorocarbene is formed under ionic conditions, in the alkaline decomposition of chloroform.

$$Cl_2CHCl \xrightarrow{HO^-} Cl_2C\colon + HCl$$

Dichlorocarbene is considered to be the reactive intermediate in the Reimer–Tiemann reaction in which phenol reacts with chloroform and base to give salicylaldehyde.

## RADICAL-IONS

A few compounds, such as the semi-quinones, which are intermediates in the reduction of quinones to hydroquinones (Chapter 10) under strongly alkaline conditions (pH 13), exist as radical-ions, i.e. they are at one and the same time both radicals and ions. The formation of semi-quinones plays an important part in the oxidation–reduction of riboflavine and flavoprotein enzymes (Chapter 23).

Benzoquinone $\xrightarrow{+e}$ Semiquinone $\xrightarrow{+e}$ Hydroquinone (di-anion)

Similarly, oxidation of the bisulphite anion gives rise to a radical-ion. Its formation is, therefore, important in the use of sodium metabisulphite ($Na_2S_2O_5$)

as an anti-oxidant. Sodium metabisulphite functions in aqueous solution as sodium bisulphite.

$$O{=}S(O^-)(O{-}H) \longrightarrow O{=}S(O^-)(O\cdot)$$

Radical ions are also formed from organic compounds by electron impact in the mass spectrometer. This may be represented as follows, where M represents the molecule under bombardment.

$$M + e \longrightarrow [M]^{+\cdot} + 2e$$

## THE CLASSIFICATION OF CHEMICAL REACTIONS

Heterolytic and homolytic bond breaking together give rise to three classes of chemical reaction.

| | |
|---|---|
| Heterolytic cleavage | (a) Electrophilic reactions |
| | (b) Nucleophilic reactions |
| Homolytic cleavage | (c) Free radical reactions |

Heterolytic reactions always involve the reaction of an electrophile with a nucleophile, but it is usual to classify the reaction from the point of view of the reagent, rather than the compound under attack. Thus, an electrophilic reagent (electrophile) is one which is electron-deficient, and is seeking electrons. A nucleophilic reagent nucleophile) is one with an excess of electrons, which it is seeking to lose in reaction with an electron-deficient centre.

Irrespective of whether reactions are electrophilic, nucleophilic or free radical, all can be classified alternatively into one of the following.

| | |
|---|---|
| (a) Addition | (1) Electrophilic |
| | (2) Nucleophilic |
| | (3) Frcc radical |
| (b) Substitution | (1) Electrophilic |
| | (2) Nucleophilic |
| | (3) Free radical |
| (c) Elimination | (1) Ionic |
| | (2) Free radical |
| (d) Rearrangement | |

## CATALYSIS

Catalysts are substances which increase the rate at which a chemical reaction proceeds. In general, they do so both by lowering the activation energy and by increasing the frequency of molecular collisions leading to the reaction. The former is achieved, however, not by reduction of the activation energy of the uncatalysed reaction, but by the creation of a new reaction mechanism with a lower activation energy in which the catalyst participates. The principal criteria of catalysis are:

(a) the catalyst may be recovered chemically unchanged at the end of the reaction,

(b) the catalytic effect is often out of all proportion to the amount of catalyst present. The molar ratio of reactant to catalyst may be of the order of 1 000 000 to 1 or even higher,

(c) the catalyst does not affect the position of the equilibrium in a reversible reaction. It follows that both the forward and reverse reactions in an equilibrium system are similarly affected. Thus, mineral acid catalyses both the esterification of alcohols by carboxylic acids, and the hydrolysis of the resulting esters.

**Acid-base Catalysis**

Catalytic processes may be classified as homogeneous or heterogeneous depending on the number of phases involved in the process.

The type of homogeneous catalysis most frequently encountered in ionic organic reactions involves the intervention of protons, and is known as acid-base catalysis. The proton is small and possesses no electrons, hence the work required to move a proton close to another atom or group is relatively small. The addition or removal of a proton is, therefore, facilitated and in most cases easily and rapidly accomplished. The function of acid catalysis in the esterification of acetic acid by methanol is demonstrated by the following mechanism.

$$H_3C{-}C(=\ddot{O})OH \xrightarrow{H^+} H_3C{-}C(=\overset{+}{O}H)OH \longleftrightarrow H_3C{-}\overset{+}{C}(OH)OH$$

$$CH_3\ddot{O}H + H_3C{-}\overset{+}{C}(OH)OH \longrightarrow H_3C{-}C(H\overset{+}{O}CH_3)(OH){-}OH \rightleftharpoons H_3C{-}C(OCH_3)(\overset{+}{O}H_2){-}O{-}H$$

$$\xrightarrow{-H_2O,\ -H^+} H_3C{-}C(OCH_3)=O$$

Many such reactions, which are catalysed by protons, are also catalysed by Lewis acids generally. Similarly, reactions which are catalysed by hydroxide ions are also often catalysed equally effectively by organic bases, such as amines and the anions of weak acids. This is known as **general acid catalysis** and **general base catalysis** respectively.

*Salt effects.* Catalysis of reactions by acids and bases is influenced by the presence of salts in two ways; the primary salt effect and the secondary kinetic salt effect. The primary salt effect is due to the total ionic strength of the solution; the observed rate constant is a function of the ionic strength. The secondary kinetic salt effect is a mass action effect of the salt concentration of the catalytically effective ion; this will, therefore, only be appreciable when the latter is formed by dissociation of the salt of a weak acid or base.

*Metal ion catalysts.* Specific metallic ions, such as $Zn^{2+}$, $Cu^{2+}$, $Ni^{2+}$ and $Mg^{2+}$, are able to act as specific catalysts in reactions which are facilitated by the presence of a charged moiety. According to Orgel (1958), the effectiveness of such ions increases with the charge on the ion and decreases as the radius of the ion increases, i.e. it depends on the charge density of the ion. Manganese (II) and iron (III) ions effectively catalyse the decarboxylation of $\beta$-ketodicarboxylic acids, such as oxaloacetic acid and dimethyloxaloacetic acid (Gelles and Hay, 1958).

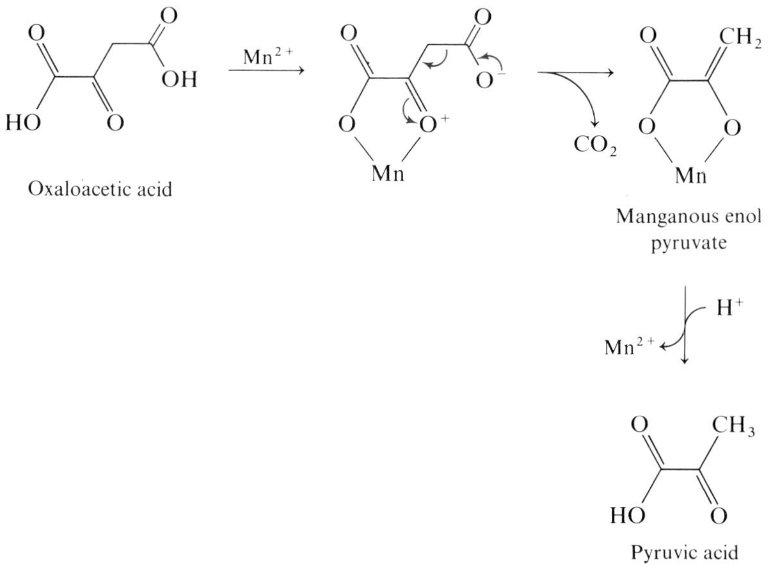

Oxaloacetic acid $\xrightarrow{Mn^{2+}}$ (Mn chelate) $\rightarrow$ $CO_2$ + Manganous enol pyruvate

Manganous enol pyruvate $\xrightarrow{H^+}$ $Mn^{2+}$ + Pyruvic acid

Similarly, the enzyme-catalysed decarboxylation of oxaloacetic acid requires the presence of metal ions. The precise mechanism of the two reactions, however, appears to be different, as the enzymic reaction is some $10^8$ times faster than the $Mn^{2+}$ catalysed reaction (Seltzer, Hamilton and Westheimer, 1959).

**Heterogeneous Catalysis** (Surface catalysis)

The interaction between gases on the surface of solid catalysts is dependent on the nature and properties of the catalyst surface. Adsorption of the reacting molecules at the catalyst surface to form a monomolecular film, provided both reactants are absorbed appreciably but not to their mutual exclusion, will promote their interaction merely by increasing the probability of contact between the reacting species. Reaction will then occur provided the contacting molecules possess the necessary activation energy.

The hydrogenation of olefines at a metal catalyst surface, such as nickel, platinum or palladium, involves the formation of metal-olefine bonds between the $\pi$-electrons of the olefine and the unsatisfied valencies of the metal atoms at the surface of the metal (Fig. 13). Hydrogen, similarly, forms a weak complex at the metal surface and the formation of such a complex at a point on the surface adjacent to the metal olefine complex leads to the alignment, and combination through a partially bonded structure.

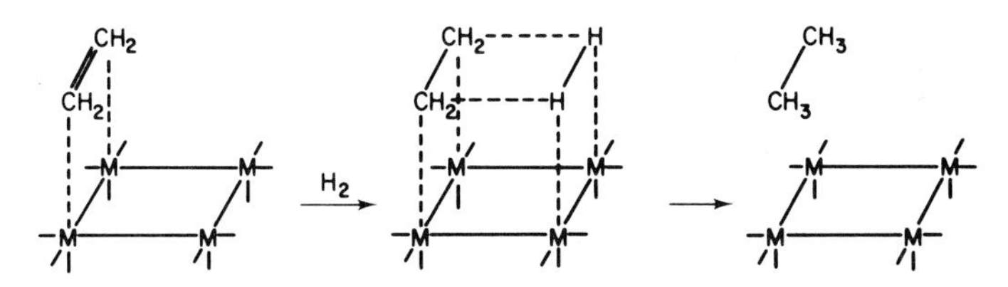


Fig. 13 Metal-olefine complex in catalytic hydrogenation

This mechanism accounts for the *cis* addition of hydrogen to olefines, and for the fact that catalytic ability resides only in certain metals, with structures where the inter-atomic distance is appropriate for reaction. In practice, the metals are prepared in a finely divided form, either by reduction (with hydrogen) of the metal oxide (platinum and palladium catalysts) or in the case of nickel catalysts by dissolving out the aluminium from a nickel-aluminium alloy with sodium hydroxide.

The specific influence of the catalyst composition on the course of a reaction is illustrated by the catalytic reaction of ethanol which can be brought about with ethanol using a variety of catalysts.

$$CH_3CH_2OH \xrightarrow{Al_2O_3/300^\circ} CH_3CH_2{\cdot}O{\cdot}CH_2CH_3$$

$$CH_3CH_2OH \xrightarrow{Al_2O_3/375^\circ} H_2C{=}CH_2 + H_2O$$

$$CH_3CH_2OH \xrightarrow{Cu/400^\circ} CH_3CHO + H_2$$

# 3 Alkanes (Paraffinic Hydrocarbons)

## CLASSIFICATION OF HYDROCARBONS

Compounds which contain carbon and hydrogen only are classed as hydrocarbons. They are further classified according to the relative proportion of carbon and hydrogen in the structure; those containing the maximum proportion of hydrogen are said to be **saturated**; all other hydrocarbons are said to be **unsaturated**. Saturated hydrocarbons, whether open-chain or cyclic, are also known as paraffinic hydrocarbons from the Latin **parum affinis**, meaning slight affinity, on account of their chemical inertness. Unsaturated hydrocarbons are classified as:

alkenes (olefinic or ethylenic hydrocarbons),
alkynes (acetylenes),
aromatic (benzenoid hydrocarbons),

according to the particular form of unsaturation present in the molecule.

## THE HOMOLOGOUS SERIES OF ALKANES

There exists a series of paraffin hydrocarbons related to the simplest saturated hydrocarbon, methane, $CH_4$, and having the same general properties. The molecular formula of each member of the series differs from that of the previous member by an increment of $CH_2$, so that all members of the series can be expressed by a general formula, which in the paraffin series is $C_nH_{2n+2}$. Physical properties such as melting point, boiling point, and specific gravity show a regular gradation throughout the series in parallel with increasing molecular weight. Such a series of compounds, known as an **homologous series**, may be defined as 'a series of uniform chemical type, capable of being represented by a general formula, and exhibiting a regular gradation of physical properties' (Table 9).

### Structure and Isomerism of Alkanes

Since all four hydrogen atoms of methane are equivalent, replacement of any one by a methyl group ($CH_3$) gives ethane for which only one structure is possible.

All six hydrogen atoms of ethane are identically bound to carbon and are identically situated with respect to each other. Only one monosubstitution product is possible. There is, therefore, only one structure possible for the next higher homologue, propane, which may be regarded as being formed by replacement of a hydrogen atom in ethane with a methyl group.

Table 9. Straight-chain Saturated Hydrocarbons

| Hydrocarbon | Formula | b.p. (°C) | m.p. (°C) |
|---|---|---|---|
| Methane | $CH_4$ | −162 | −188 |
| Ethane | $CH_3{\cdot}CH_3$ | −87 | −172 |
| Propane | $CH_3{\cdot}CH_2{\cdot}CH_3$ | −42 | −187 |
| n-Butane | $CH_3{\cdot}(CH_2)_2{\cdot}CH_3$ | −0.5 | −135 |
| n-Pentane | $CH_3{\cdot}(CH_2)_3{\cdot}CH_3$ | +36 | −130 |
| n-Hexane | $CH_3{\cdot}(CH_2)_4{\cdot}CH_3$ | 69 | −94 |
| n-Heptane | $CH_3{\cdot}(CH_2)_5{\cdot}CH_3$ | 98.4 | −90 |
| n-Octane | $CH_3{\cdot}(CH_2)_6{\cdot}CH_3$ | 125.6 | −57 |
| n-Nonane | $CH_3{\cdot}(CH_2)_7{\cdot}CH_3$ | 149.5 | −51 |
| n-Decane | $CH_3{\cdot}(CH_2)_8{\cdot}CH_3$ | 173 | −32 |
| n-Undecane | $CH_3{\cdot}(CH_2)_9{\cdot}CH_3$ | 195 | −26.5 |
| n-Dodecane | $CH_3{\cdot}(CH_2)_{10}{\cdot}CH_3$ | 215 | −12 |
| n-Tridecane | $CH_3{\cdot}(CH_2)_{11}{\cdot}CH_3$ | 234 | −6.2 |
| n-Tetradecane | $CH_3{\cdot}(CH_2)_{12}{\cdot}CH_3$ | 252 | +5 |
| n-Pentadecane | $CH_3{\cdot}(CH_2)_{13}{\cdot}CH_3$ | 270 | 10 |
| n-Hexadecane | $CH_3{\cdot}(CH_2)_{14}{\cdot}CH_3$ | 287 | 18 |
| n-Heptadecane | $CH_3{\cdot}(CH_2)_{15}{\cdot}CH_3$ | 303 | 22.5 |
| n-Octadecane | $CH_3{\cdot}(CH_2)_{16}{\cdot}CH_3$ | 317 | 28 |
| n-Nonadecane | $CH_3{\cdot}(CH_2)_{17}{\cdot}CH_3$ | 330 | 32 |
| n-Eicosane | $CH_3{\cdot}(CH_2)_{18}{\cdot}CH_3$ | 208/15 mm | 37 |
| n-Heneicosane | $CH_3{\cdot}(CH_2)_{19}{\cdot}CH_3$ | 219/15 | 40.4 |
| n-Docosane | $CH_3{\cdot}(CH_2)_{20}{\cdot}CH_3$ | 230/15 | 44.4 |
| n-Tricosane | $CH_3{\cdot}(CH_2)_{21}{\cdot}CH_3$ | 240/15 | 47.7 |
| n-Tetracosane | $CH_3{\cdot}(CH_2)_{22}{\cdot}CH_3$ | 250/15 | 51.1 |

*Butanes.* Substitution of hydrogen by methyl in propane can, however, give rise to two products depending upon whether substitution occurs on the terminal or on the central carbon atom. Substitution on either of the terminal carbon atoms gives rise to a straight chain or normal butane (n-butane), whereas substitution on the central atom leads to a branched chain isobutane.

$$CH_3{\cdot}CH_2{\cdot}CH_2{\cdot}CH_3 \qquad\qquad CH_3{\cdot}\underset{\displaystyle CH_3}{\underset{|}{CH}}{\cdot}CH_3$$

n-Butane — Isobutane

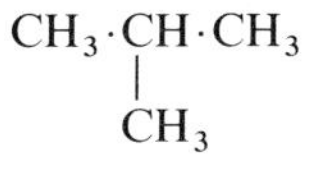

The two butanes, which are structural **isomers**, have different physical properties, the boiling point of the branched chain isomer, isobutane (b.p. −12°C) being lower than that of n-butane (b.p. −0.5°C). Examination of the two structures shows that the various carbon atoms are not equivalent. Four different types of carbon atom may be distinguished in saturated hydrocarbons.

(a) Primary—joined to only one other carbon atom

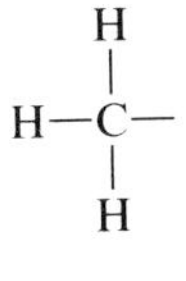

$$\mathrm{H}-\overset{\displaystyle \mathrm{H}}{\overset{|}{\underset{\displaystyle \mathrm{H}}{\underset{|}{\mathrm{C}}}}}-$$

(b) Secondary—joined to two other carbon atoms

```
   H
   |
H—C—
   |
```

(c) Tertiary—joined to three other carbon atoms

```
   |
H—C—
   |
```

(d) Quaternary—joined to four other carbon atoms

```
  |
—C—
  |
```

Thus, n-butane has two primary and two secondary carbons, whilst isobutane has one tertiary and three primary carbons.

*Pentanes.* Three structures are possible for pentane, $C_5H_{12}$, and in fact three pentanes are known:

n-Pentane $CH_3 \cdot CH_2 \cdot CH_2 \cdot CH_2 \cdot CH_3$ b.p. 30°C

Isopentane b.p. 28°C

```
CH3·CH·CH2·CH3
    |
    CH3
```

Neopentane b.p. 9.5°C

```
        CH3
        |
  H3C—C—CH3
        |
        CH3
```

Once again, physical properties can be related to chain-branching, the most highly branched isomer having the lowest boiling point. A method for quantitating the effects of chain-branching on the physical properties of hydrocarbons by means of a **molecular connectivity index**, $\chi$, is described elsewhere (**2**, 1).

## Nomenclature of Alkanes

As the paraffin series is ascended, the number of isomers increases rapidly with each additional carbon atom. There are five isomeric hexanes, nine heptanes, eighteen octanes and thirty-five nonanes. A systematic method of naming compounds is, therefore, important. The method adopted was first worked out in Geneva in 1892 and subsequently taken over by the International Union of Pure and Applied Chemistry in 1922. The rules have been modified on several occasions and those now used are known as the IUPAC Rules.

Under these rules, the generic name of saturated acyclic hydrocarbons is **alkane**. The trivial names methane, ethane, propane and butane are retained for lower members of the series, but higher members of the series are named by combining a numerical prefix with the suffix-**ane** (Table 9).

Univalent radicals derived from the alkanes by removal of hydrogen from a terminal carbon atom (straight chain alkyls) are named by replacing the termination **-ane** by **-yl**, and numbered from the carbon atom with the free valency.

| | |
|---|---|
| Methyl | $CH_3\cdot$ |
| Ethyl | $\overset{2}{C}H_3\cdot\overset{1}{C}H_2\cdot$ |
| n-Propyl | $\overset{3}{C}H_3\cdot\overset{2}{C}H_2\cdot\overset{1}{C}H_2\cdot$ |
| n-Butyl | $\overset{4}{C}H_3\cdot\overset{3}{C}H_2\cdot\overset{2}{C}H_2\cdot\overset{1}{C}H_2\cdot$ |
| n-Pentyl | $\overset{5}{C}H_3\cdot\overset{4}{C}H_2\cdot\overset{3}{C}H_2\cdot\overset{2}{C}H_2\cdot\overset{1}{C}H_2\cdot$ |
| n-Hexyl | $\overset{6}{C}H_3\cdot\overset{5}{C}H_2\cdot\overset{4}{C}H_2\cdot\overset{3}{C}H_2\cdot\overset{2}{C}H_2\cdot\overset{1}{C}H_2.$ |

Complex hydrocarbons are named as derivatives of the longest straight-chain, which is numbered in such a way as to give the lowest numbers possible to the side-chains.

3-Methylpentane

$$\overset{1}{C}H_3\cdot\overset{2}{C}H_2\cdot\underset{\substack{|\\ CH_3}}{\overset{3}{C}H}\cdot CH_2\cdot CH_3$$

2-Methylpentane (Isopentane)

$$\overset{1}{C}H_3\cdot\underset{\substack{|\\ CH_3}}{\overset{2}{C}H}\cdot\overset{3}{C}H_2\cdot\overset{4}{C}H_2\cdot\overset{5}{C}H_3$$

**not**

$$\overset{5}{C}H_3\cdot\underset{\substack{|\\ CH_3}}{\overset{4}{C}H}\cdot\overset{3}{C}H_2\cdot\overset{2}{C}H_2\cdot\overset{1}{C}H_3$$

2,3,5-Trimethylhexane

$$\overset{1}{C}H_3\cdot\underset{\substack{|\\ CH_3}}{\overset{2}{C}H}\cdot\underset{\substack{|\\ CH_3}}{\overset{3}{C}H}\cdot\overset{4}{C}H_2\cdot\underset{\substack{|\\ CH_3}}{\overset{5}{C}H}\cdot\overset{6}{C}H_3$$

## PHYSICAL PROPERTIES

The physical, chemical and biological properties of the alkanes can be related to the fundamental characteristics of C—H and C—C single bonds. The almost complete lack of polarity of these bonds explains their typical behaviour as non-polar substances. Thus, they are not merely insoluble (if gaseous) or immiscible (if liquid) with polar solvents such as water and ethanol, but neutral in character, and stable to the action of polar reagents such as acids and bases.

Although the polarity of the C—H bond is towards carbon, C←H (Chapter 2), it is so slight that when a hydrocarbon is mixed with water, adjacent water molecules are unable to solvate it by hydrogen bond formation. In consequence, adjacent solvent molecules arrange themselves into a quasi-crystalline structure, forming cavities to accommodate the hydrocarbon chains within the body of the solvent. The presence of the hydrocarbon, therefore, introduces a degree of order in neighbouring solvent molecules, which was not previously present and which also is absent from the bulk of the liquid. This results in a loss of entropy, which the system resists by forcing the hydrocarbon out of the water, thus accounting for its insolubility.

It is also evident from the examination of flexible molecules more amenable to experiments which depend on water solubility, that where a large hydrocarbon fragment is present in the molecule, this will tend to orientate itself in aqueous solution to reduce the area of the total hydrocarbon–water interface. Thus, conductimetric studies (Elworthy, 1963b) have shown that the extended conformation of decamethonium is not favoured in aqueous or ethanolic solution, and that in an endeavour to reduce the lipid–aqueous interface between the hydrocarbon chain and the solvent, the chain spirals, contracting the molecule and reducing the inter-onium distance from 14 Å in the fully staggered form to approximately 9.5 Å (**2**, 2).

## The Conformation of Saturated Hydrocarbons

The symmetry of the sigma orbital which comprises the C—C single bond accounts for the classical concept of free rotation about such bonds. Non-bonded interactions between hydrogen atoms or between hydrogen and other atoms or groups attached to carbon arise as a result of the repulsive interactions of their respective electron orbitals. These interactions increase in intensity with the size of the interacting atoms or groups creating energy barriers, which can be partly or fully relieved if certain staggered conformations about the C—C bond are adopted. The various conformations which a four-carbon hydrocarbon may adopt range from **fully staggered** in which the energy barriers are least, through **partially staggered** and **partially eclipsed** to **fully eclipsed**, in which the potential energy is at a maximum (Fig. 24; **2**, 3).

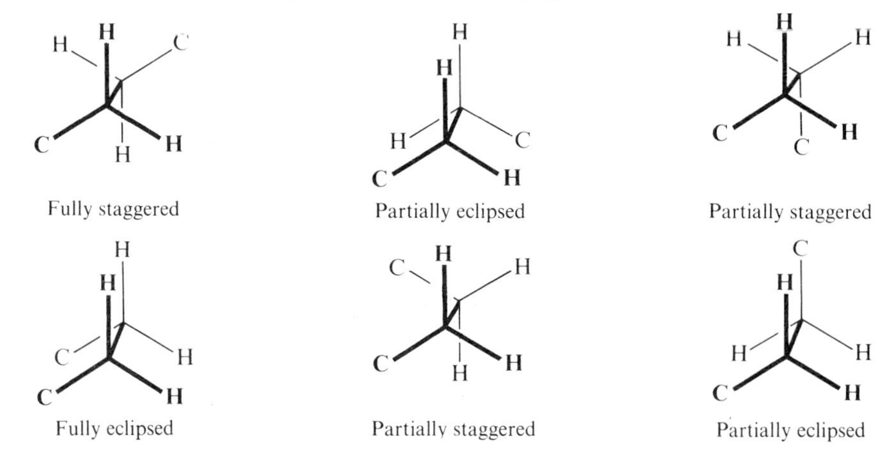

Adoption of the strainless fully staggered conformation in saturated hydrocarbons leads to the classical concept of the unstrained fully extended, zig-zag, carbon chain conformation (Chapter 1).

## CHEMICAL PROPERTIES

The chemical properties of alkanes similarly reflect the lack of polarity of C—H and C—C bonds. Thus, they are resistant to attack by strong mineral acids and caustic alkalies so much so that both concentrated sulphuric acid and sodium hydroxide solution are used in the purification of mineral oil fractions for medicinal purposes (p. 74). They are also resistant to attack by most chemical oxidising agents, and their chemical reactions are limited to oxidation by combustion and other reactions, mainly substitutions, involving free radical intermediates.

### Catalytic Cracking

Cracking is the term used to denote the thermal decomposition of paraffins, a process which has been developed on a large scale to convert some of the less valuable higher-boiling fractions into petrol. Paraffins are decomposed into smaller molecules when they are heated to between 500 and 600°C. The products obtained from a given paraffin depend on a number of factors, including the structure of the paraffin, the pressure under which the cracking is carried out, and the presence or absence of catalysts. The latter are mixed oxides, and all contain silica and alumina, with the possible addition of zirconium or thorium oxides. Besides petrol, large quantities of gas are produced in these processes containing both saturated and unsaturated hydrocarbons with up to four carbon atoms. The process involves homolysis of C—H and C—C bonds.

$$R{\cdot}CH_2{\cdot}CH_2{-}\overset{H}{\underset{H}{C}}{-}H \longrightarrow R{\cdot}CH_2{\cdot}CH_2{\cdot}CH_2{\cdot} + H{\cdot}$$

$$R{\cdot}CH_2{-}CH_2{-}\dot{C}H_2 \longrightarrow R{\cdot}CH_2{\cdot} + H_2C{=}CH_2$$

$$R{\cdot}CH_2{\cdot} + H{\cdot} \longrightarrow R{\cdot}CH_3$$

Two main types of cracking process are used.

(a) *Liquid phase cracking*. Lubricating oil fractions are heated to a temperature of 475–530°C under pressures of up to 70 atmospheres (i.e. *ca* 7000 kN $m^{-2}$)*. The use of pressure ensures that the cracked material stays in the liquid state. About 60–65% of the oil is converted into petrol by this method.

* 1 atmosphere = 101.325 kN $m^{-2}$

(b) *Vapour phase cracking.* Much lower pressures up to about 10 atmospheres are used, but in consequence a somewhat higher cracking temperature is necessary (600°C). Lower-boiling petroleum fractions may be cracked by this method, which is used primarily for the production of ethylene, propylene and butylene.

Special cracking processes have been designed specifically for the production of ethylene, which is in considerable demand. The most widely used method is the steam-cracking process in which hydrocarbon vapours are mixed with steam and pyrolysed at about 750°C.

Other cracking processes employing direct heat transfer from strongly heated refractory materials operate at temperatures as high as 1400°C. Hydrocarbons, particularly methane, exposed to these temperatures for only very short periods of time, and then rapidly cooled, produce high yields of acetylene(s).

### Combustion

All paraffin hydrocarbons are highly flammable; additionally, the lower gaseous and liquid alkanes have low flash points. Typically, they burn with a non-luminous flame in an adequate supply of air or oxygen forming carbon dioxide and water.

$$\underset{\text{Propane}}{CH_3{\cdot}CH_2{\cdot}CH_3} + 5O_2 \longrightarrow 3CO_2 + 4H_2O$$

Combustion is a strongly exothermic process, and natural gas consisting of lower members of the series is now used for fuel. It is fed into national grid lines

$$R{\cdot}CH_2{\cdot}CH_2{\cdot}CH_2{-}H \longrightarrow R{\cdot}CH_2{\cdot}CH_2{\cdot}CH_2{\cdot} + H{\cdot}$$

$$R{\cdot}CH_2{\cdot}CH_2{\cdot}CH_2{\cdot} + {\cdot}O{-}O{\cdot} \longrightarrow R{\cdot}CH_2{\cdot}CH_2{\cdot}CH_2{-}O{-}O{\cdot}$$

$$R{\cdot}CH_2{\cdot}CH_2{\cdot}CH_2{-}O{-}O{\cdot} \longrightarrow R{\cdot}CH_2{\cdot}CH_2{\cdot}CH{=}O + HO{\cdot}$$

$$R{\cdot}CH_2{\cdot}CH_2{\cdot}CH{=}O + HO{\cdot} \longrightarrow R{\cdot}CH_2{\cdot}CH_2{\cdot}\dot{C}{=}O + H_2O$$

$$R{\cdot}CH_2{\cdot}CH_2{\cdot}\dot{C}{=}O \longrightarrow R{\cdot}CH_2{\cdot}CH_2{\cdot} + CO$$

$$R{\cdot}CH_2{\cdot}CH_2{\cdot} + {\cdot}O{-}O{\cdot} \longrightarrow \cdots$$

in Britain and other parts of Europe and America where suitable natural sources exist close at hand. Combustion of the more volatile liquid hydrocarbons obtained by fractionation of petroleum is also used as a source of energy in the internal combustion engine and in jet engines. Certain mixtures of these gaseous hydrocarbons with air and oxygen are explosive when ignited, and methane represents a serious explosion hazard in coal mines.

It is clear from consideration of the equation for the complete combustion of propane that the burning process is complex, since the probability of five oxygen molecules simultaneously coming together with a single propane molecule is negligible. There is evidence that burning involves both formation of hydrocarbon free radicals by homolytic cleavage of C—H bonds at the high burning temperature and interaction with oxygen which is itself a di-radical (·O—O·).

**Catalytic Oxidation**

The catalytic oxidation of methane provides an important route for the commercial production of methanol. A methane–oxygen mixture (9:1) is passed under pressure (*ca* 100 atmospheres) through copper tubes heated to about 200°C. Higher temperatures give rise to further oxidation to formaldehyde.

$$CH_4 + O_2 \longrightarrow 2CH_3OH$$

**Chlorination**

Chlorine has no action on methane in the dark. Dissociation of molecular chlorine into chlorine radicals is photo-initiated, and chlorine–methane mixtures explode in bright sunlight with the formation of hydrogen chloride and carbon.

$$CH_4 + 2Cl_2 \xrightarrow{h\nu} C + 4HCl$$

In diffused daylight, no explosion occurs and the four hydrogen atoms of methane are successively replaced to yield chloromethane (methyl chloride), dichloromethane (methylene dichloride), trichloromethane (chloroform) and finally tetrachloromethane (carbon tetrachloride).

$$CH_4 \xrightarrow[-HCl]{Cl_2} CH_3{\cdot}Cl \xrightarrow[-HCl]{Cl_2} CH_2{\cdot}Cl_2 \xrightarrow[-HCl]{Cl_2} CH{\cdot}Cl_3 \xrightarrow[-HCl]{Cl_2} CCl_4$$

Chloro-methane; Dichloro-methane; Trichloro-methane; Tetrachloro-methane

Unless modified, the reaction invariably gives complex mixtures of the chlorinated hydrocarbons, though chloromethane can be obtained by this method in 90% yield using the reactants mixed with nitrogen in the ratio methane–chlorine–nitrogen 8:1:80 in the presence of a cupric chloride catalyst.

The reaction is initiated by light, and is therefore said to be photochemically induced. The light initially provides the energy for homolytic cleavage of the chlorine molecule into two chlorine free radicals (atomic chlorine).

$$Cl{-}Cl \xrightarrow{h\nu} 2Cl\cdot$$

The chlorine radicals are highly reactive and attack one of the C—H bonds to form an alkyl radical.

$$H_3C{-}H \; \cdot Cl \longrightarrow H_3C\cdot$$

The alkyl radical so formed in turn attacks a further chlorine molecule to give the alkyl halide, and regenerates a further chlorine radical, so sustaining the reaction process.

$$CH_3\cdot \; Cl{-}Cl \longrightarrow CH_3{\cdot}Cl + Cl\cdot$$

In more complex hydrocarbons, hydrogen atoms are substituted at rates in the order tertiary (3°) > secondary (2°) > primary (1°). Thus, propane ($CH_3{\cdot}CH_2{\cdot}CH_3$) chlorinates faster at C-2 than at C-1.

A free radical induced chlorination of alkanes can also be brought about under comparatively mild conditions by the action of sulphuryl chloride in the presence of benzoyl peroxide as catalyst (Kharasch and Mayo, 1933). Benzoyl peroxide is readily decomposed at about 30°C to give the phenyl radical, and thus acts as the reaction initiator (Chapter 2). Phenyl radicals so formed attack sulphuryl chloride to give the $\cdot SO_2Cl$ radical which in turn initiates a self-propagating chain reaction leading to the formation of the alkyl chloride.

$$C_6H_5\cdot \; + \; Cl{-}SO_2{-}Cl \longrightarrow C_6H_5{\cdot}Cl + \cdot SO_2Cl$$

Chlorobenzene

$$Cl{-}\dot{S}O_2 \longrightarrow Cl\cdot + SO_2$$

$$R{\cdot}CH_3 + Cl\cdot \longrightarrow R{\cdot}CH_2\cdot + HCl$$

$$R{\cdot}CH_2\cdot + SO_2{\cdot}Cl_2 \longrightarrow R{\cdot}CH_2{\cdot}Cl + \cdot SO_2Cl$$

The extent of the chlorination process is controlled by the amount of sulphuryl chloride present in the reaction mixture.

**Bromination and Iodination**

Substitution of alkanes by bromine is similarly induced photochemically. The reaction is less vigorous than chlorination, and moreover does not proceed beyond monosubstitution. This difference results from the relative stability of the various carbon–halogen bonds in relation to that of the C—H bond with respect to attack by their respective halogen free radicals. The stability order is:

$$C{-}F > C{-}Cl > C{-}H > C{-}Br > C{-}I$$

It follows that treatment of methane with bromine under free radical conditions gives rise solely to methyl bromide, since further free radical attack leads to cleavage of the C—Br bond rather than of one of the C—H bonds.

Similar considerations apply to reaction with iodine, but additionally the reaction is reversible owing to the powerful reducing action of hydriodic acid which is also formed. At very high temperatures (685°C), iodine abstracts hydrogen from ethane to yield a mixture of ethylene (72%) and acetylene (10%) (Raley, Mullineaux and Bittner, 1963).

**Fluorination**

Most organic compounds react violently with fluorine resulting in combustion and fragmentation of the carbon skeleton.

$$CH_3{\cdot}CH_3 + 3F_2 \longrightarrow 2C + 6HF$$

Perfluorocarbons can be prepared from hydrocarbons by exhaustive fluorination in the vapour-phase at 150°C with a fluorine–nitrogen mixture in a closed vessel packed with silver fluoride.

$$CH_3{\cdot}(CH_2)_5{\cdot}CH_3 \xrightarrow[150^\circ]{AgF} CF_3{\cdot}(CF_2)_5{\cdot}CF_3$$

The use of copper catalysts gives perfluorocarbon mixtures as a result of chain fragmentation.

Alkyl fluorides are readily prepared by substitution on heating other alkyl halides with a metal fluoride.

$$R{\cdot}CH_2{\cdot}I + AgF \longrightarrow R{\cdot}CH_2{\cdot}F + AgI$$

## THE BIOLOGICAL REACTIVITY OF ALKANES

The resistance of paraffin hydrocarbons to attack at ambient temperatures by all but free radical reagents, together with their water-repellent properties, probably accounts for their apparent stability in biological systems. Thus, it is widely held that medicinal *Liquid Paraffin* (Mineral Oil)* administered orally

* British Approved Names in italics. United States Adopted Names in brackets.

in high doses as a laxative is not absorbed, and is excreted from the bowel unchanged. Similarly, *Soft Paraffins* (White Petrolatum) applied to the skin as greasy ointment bases are not appreciably absorbed. The fate of *Hard Paraffin* (Paraffin) used to coat granules and so delay drug absorption from the intestines has, however, not been established, though it is assumed to be excreted unchanged. On the other hand, it is clear that volatile low molecular weight hydrocarbons are readily absorbed from the lungs into the blood stream, and, owing to their high lipid solubility, concentrated in the central nervous system. Natural petroleum gas will stupify, and in sufficient concentration render the individual unconscious. This absorption process is essentially reversible, and consciousness returns as the concentration of the gas in the body falls. The low molecular weight hydrocarbon, *Cyclopropane*, is used as a volatile anaesthetic in this way (p. 79).

Although paraffin hydrocarbons are generally considered to be resistant to biological oxidation, McCarthy (1964) has shown that when $^{14}$[C]-octadecane and $^{14}$[C]-hexadecane are administered orally to rats, chickens or goats, the radioactivity is recovered in the body fat as fatty acids of the same chain length. It has also been demonstrated that small quantities of paraffin hydrocarbons with both unbranched and branched chains are normally associated with human plasma proteins in amounts corresponding to some 4–5% of the total serum lipids (Skipski, *et al.*, 1967). It appears, therefore, that paraffin hydrocarbons may have some function as intermediates in fatty acid metabolism.

### Microsomal Oxidation

There is ample evidence that hydrocarbons are oxidised in the liver. Oxidation is brought about by enzymes associated with the microsomal fraction of the endoplasmic reticulum, which have been described as mixed function oxidases (Mason, 1957). The enzymes are associated with the lipid membrane, and treatment with desoxycholic acid which solubilises the membrane destroys activity. Nicotinamide adenine dinucleotide is a co-factor for the enzymes, and the oxidations also require molecular oxygen and not water as the source of oxygen, as shown in experiments with $^{18}O_2$ and $H_2{}^{18}O$. Only one oxygen atom of each molecule, however, is transferred to the substrate, the remaining atom being reduced to water. A particular cytochrome enzyme known as P450 (peak at 450 nm when in the reduced form) plays a key rôle in the oxidation (**2**, 5).

$$>CH_2 + {}^{18}O_2 + NADH + H^+ \longrightarrow >CHOH + NAD^+ + H_2{}^{18}O$$

It has been shown that methylcyclohexane is readily hydroxylated in the rabbit giving a mixture of 3- and 4-hydroxymethylcyclohexanes (Elliott, Tao and Williams, 1965).

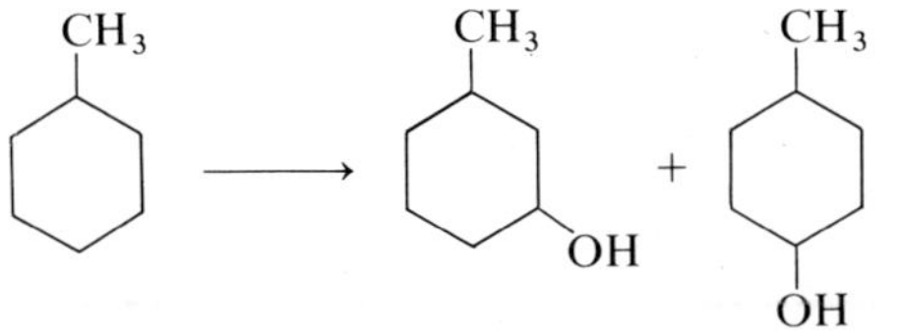

The hydroxylations show a high degree of stereospecificity, as shown by the oxidation of *cis*-decalin to *cis-cis*-2-decalol (Elliott, Robertson and Williams, 1966). The product was recovered as the $\beta$-glucuronide in 67% of the administered dose.

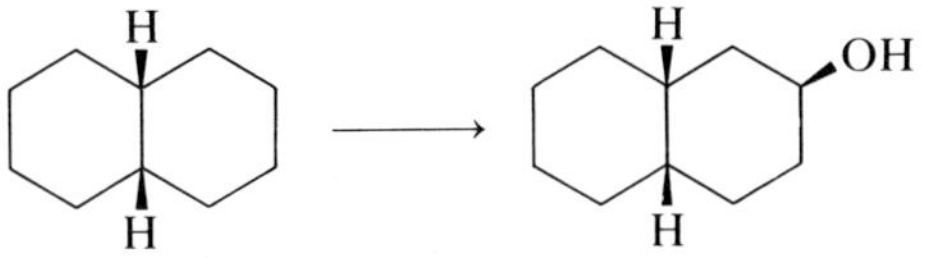

*cis*-Decalin *cis-cis*-2-Decalol

Hydrocarbon residues in a wide variety of drugs undergo similar metabolic oxidations (**2**, 5), as for example in the anti-diabetic agent, *Tolbutamide*, (Louis *et al.*, 1956) and the alkyl chains of barbiturates such as *Pentobarbitone* (Pentobarbital) *Sodium* (Cooper and Brodie, 1955).

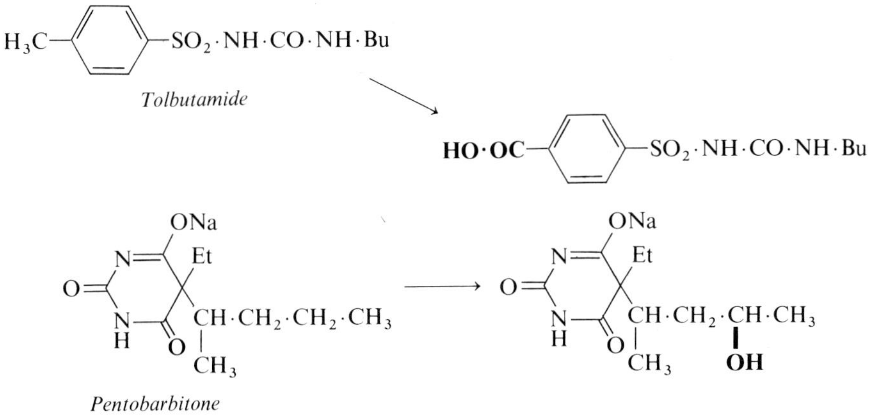

*Tolbutamide*

*Pentobarbitone*

## Steroid Hydroxylation

Specific steroid hydroxylases are present in the adrenal cortex, each specific for oxidation at a particular position in the steroid nucleus, and together capable of achieving the conversion of progesterone to a variety of corticosteroids (Chapter 22; Usui and Yamasaki, 1960; Halkerston, Eichhorn and Hechter, 1961). The mould *Rhizopus nigricans* is also capable of hydroxylating steroids specifically in the 11$\alpha$-position (Hayano, Gut, Dorfman, Sebek and Peterson, 1958), a property which has been utilised in a number of commercial syntheses of hydrocortisone and its derivatives.

## MEDICINAL PARAFFINS

Paraffin fractions used medicinally are derived from the lubricating oil fraction of natural petroleum, as outlined in Fig. 14. Fractionation into liquid, hard and soft paraffin fractions is preceded by purification to remove the more objectionable impurities. Unsaturated hydrocarbons are seldom natural constituents of petroleum, but may arise due to incipient cracking in the preliminary distillations. Washing with concentrated sulphuric acid removes any such unsaturated compounds by addition (Chapter 4) and also sulphur compounds such as thiophene which are soluble. Thiols are oxidised to disulphides with chlorine and aqueous sodium hydroxide.

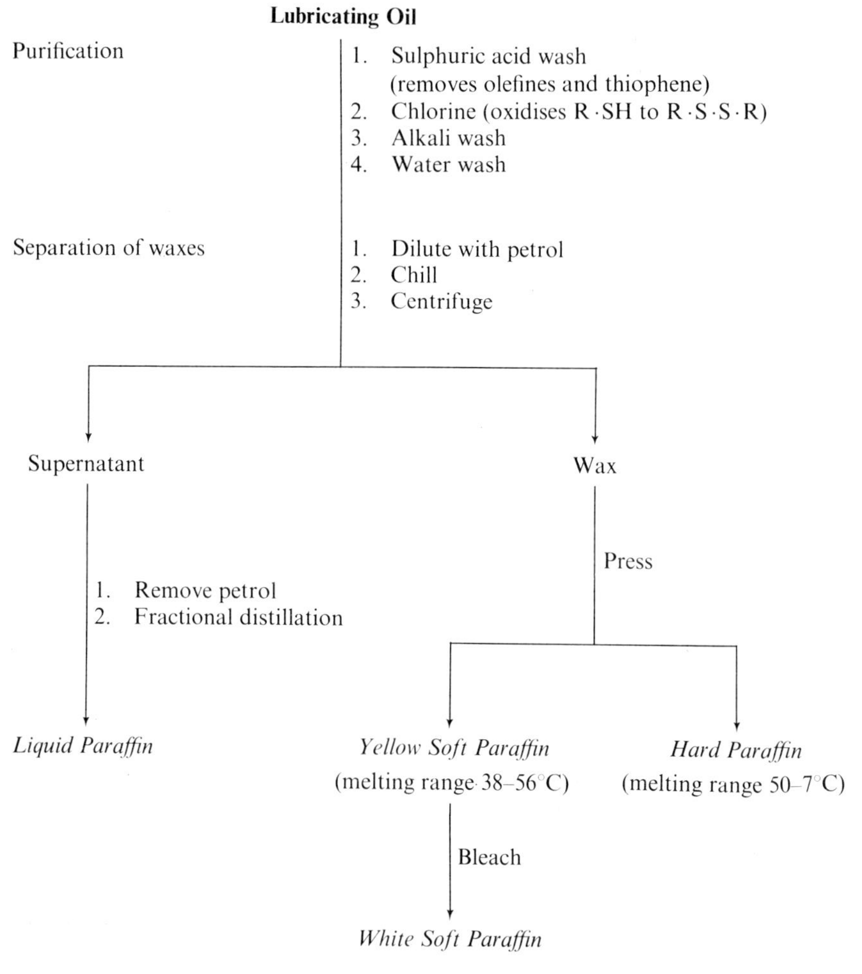


Fig. 14 Purification and fractionation of lubricating oil to yield pharmaceutical quality paraffin fractions

Some soft paraffins commercially available, prepared by judicious admixture of liquid and hard paraffin fractions, show a tendency to separate. The addition of anti-oxidants such as *Butylated Hydroxytoluene* is permitted up to 10 ppm in medicinal *Liquid Paraffin* as a stabiliser. The amount of such addition is controlled by imposing a limit on the ultraviolet absorption between 240 and 280 nm, a region in which the paraffin hydrocarbons themselves are transparent. A similar limit in ultraviolet absorption at 290 nm in the *Soft Paraffins* provides a means of excluding undesirable amounts of polycyclic aromatic hydrocarbons (Chapter 22) which may be carcinogenic.

## CYCLOALKANES

### Nomenclature

Cycloalkanes are cyclic saturated hydrocarbons of general formula $C_nH_{2n}$ corresponding to the open-chain saturated hydrocarbons ($C_nH_{2n+2}$). Their nomenclature follows that of the open-chain compound containing the same number of carbon atoms, but prefixed by, **cyclo-**.

Cyclopropane $(CH_2)_3$ b.p. −34.5°C

Cyclobutane $(CH_2)_4$ b.p. 11–12°C

Cyclopentane $(CH_2)_5$ b.p. 49°C

Cyclohexane $(CH_2)_6$ b.p. 81°C

### Physical Properties

Cycloalkanes resemble the alkanes closely in physical properties. They are neutral, oily liquids or gases showing a gradation of boiling point throughout the series. Boiling points and densities are higher in each case than those of the corresponding alkanes.

### Chemical Properties

Chemical properties of cyclopentane and cyclohexane closely resemble those of n-pentane and n-hexane. Cyclopropane and cyclobutane, however, show special properties, which result from ring strain. This arises in two ways. The first, known as Baeyer strain, results from the distortion of the normal C—C bond angle when three- and four-membered rings are formed. The second, is a consequence of high repulsive energy arising from the presence of fully eclipsed C—H bonds.

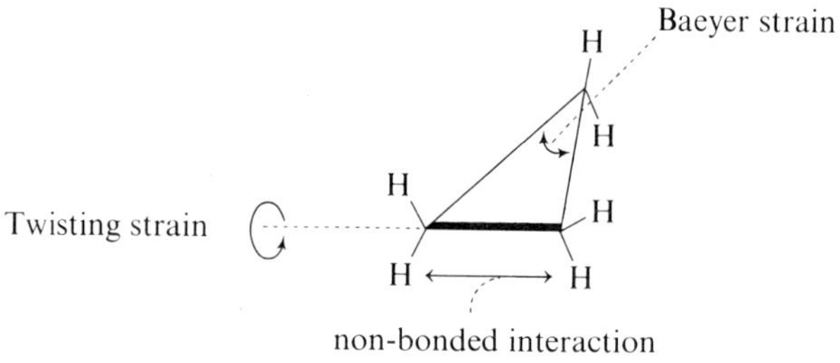

Baeyer strain

Twisting strain

non-bonded interaction

Cyclopropane and cyclobutane necessarily have planar rings, and considerable deviation from the normal tetrahedral bond angle occurs in their formation. The calculated deviations are based upon the assumption that the bonds in these particular cyclic structures remain straight. It is now assumed that this is not so, and that the bonds are in fact bent (banana bonds). Bond bending releases some of the strain on the ring, which is therefore not as great as it would appear at first sight. The rings, however, are still planar, so that all C—H bonds are fully eclipsed (Chapter 1), and their repulsive forces place a twisting strain on the C—C bonds.

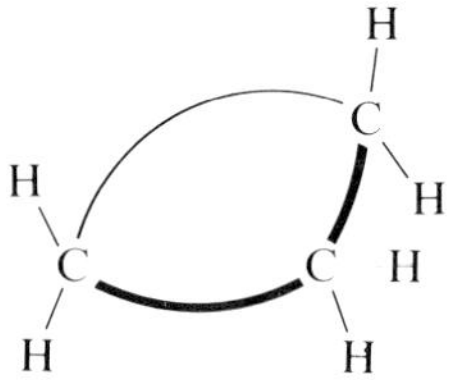

The combination of these two forms of strain is reflected in their significantly higher heats of combustion compared with cyclohexane and open-chain n-alkanes which are almost if not completely strain-free, and also in their enhanced chemical reactivity (Table 10). Thus, whereas cyclopentane and cyclohexane are stable, cyclopropane and cyclobutane undergo ring cleavage when heated with hydrogen and a nickel catalyst. The lower temperature for the reaction with cyclopropane reflects the greater instability of the three-membered ring.

*Cyclopropane*, but not cyclobutane, undergoes ring scission with bromine, hydrobromic acid or sulphuric acid at room temperature. In this respect, *Cyclopropane* resembles propylene (Chapter 4). All cycloalkanes, however, including *Cyclopropane* and cyclobutane are stable to oxidation by ozone and potassium permanganate. The reaction with sulphuric acid is used in control

Table 10. Bond Angle Deviations in Planar Rings

Tetrahedral bond angle

Planar bond angle

Deviation = $\frac{1}{2}$(Tetrahedral bond angle − planar ring bond angle)

| Cycloalkane | Planar ring bond angle | Deviation (assuming straight bonds) | Heat of Combustion (kJ/$CH_2$ group) |
|---|---|---|---|
| Cyclopropane | 60° | 24° 44′ | 697 |
| Cyclobutane | 90° | 9° 44′ | 686 |
| Cyclopentane | 108° | 0° 44′ | 664 |
| Cyclohexane | 120° | −5° 16′ | 659 |
| Cycloheptane | 128° 34′ | −9° 33′ | 662 |

$$\text{(cyclo-}C_3H_6\text{)}\ \xrightarrow[120^\circ]{H_2-Ni}\ CH_3{\cdot}CH_2{\cdot}CH_3$$

$$\text{(cyclo-}C_4H_8\text{)}\ \xrightarrow[200^\circ]{H_2-Ni}\ CH_3CH_2{\cdot}CH_2{\cdot}CH_3$$

of the quality of *Cyclopropane* for use as an anaesthetic, ≮99% of the sample being required to be absorbed by the reagent.

$$\text{(cyclo-}C_3H_6\text{)}\ \xrightarrow{Br_2}\ Br{\cdot}CH_2{\cdot}CH_2{\cdot}CH_2{\cdot}Br \quad \text{1,3-Dibromopropane}$$

$$\text{(cyclo-}C_3H_6\text{)}\ \xrightarrow{HBr}\ CH_3{\cdot}CH_2{\cdot}CH_2{\cdot}Br \quad \text{n-Propyl bromide}$$

$$\text{(cyclo-}C_3H_6\text{)}\ \xrightarrow{H_2SO_4}\ CH_3{\cdot}CH_2{\cdot}CH_2{\cdot}O{\cdot}SO_2{\cdot}OH \quad \text{n-Propyl hydrogen sulphate}$$

**Strainless Rings**

Six-membered and larger carbocyclic rings do not show the instability which would be expected of strained planar rings, as postulated by Baeyer (1885). The heat of combustion per methylene group for cyclohexane is identical with that for open-chain n-alkanes, and those for larger rings are not appreciably greater (Table 10). They owe their stability to the adoption of non-planar puckered conformations, which preserve the normal tetrahedral carbon valency angle, and hence are strainless in accord with the predictions of Sachse (1890) and

Mohr (1918). According to this concept, two conformations are possible for cyclohexane, the so-called **boat** and **chair** forms, though cyclohexane exists only in one physical form.

Chair form

Boat form

The energy difference between the two forms is insufficient for separate existence, but at ambient temperature the chair form has greater stability, since all its C—H bonds are fully staggered, whereas in the boat conformation C—H bonds at the adjacent C-2 and C-3, and C-3 and C-4 positions are fully eclipsed.

Two types of C—H bond can be distinguished in the chair form of cyclohexane. Those which are directed vertically above and below the ring are known as **axial** bonds, and those which lie approximately in the equatorial plane of the carbon ring are known as **equatorial** bonds.

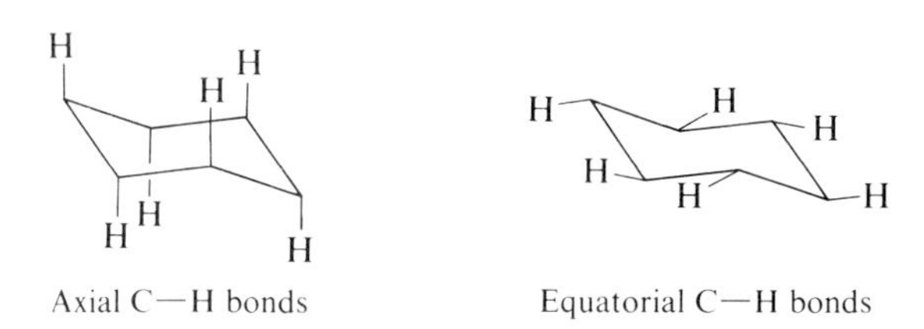


Axial C—H bonds

Equatorial C—H bonds

The flexibility of the ring calls for two inter-convertible chair forms of cyclohexane, in which the axial and equatorial C—H bonds of one form become equatorial and axial respectively in the other.

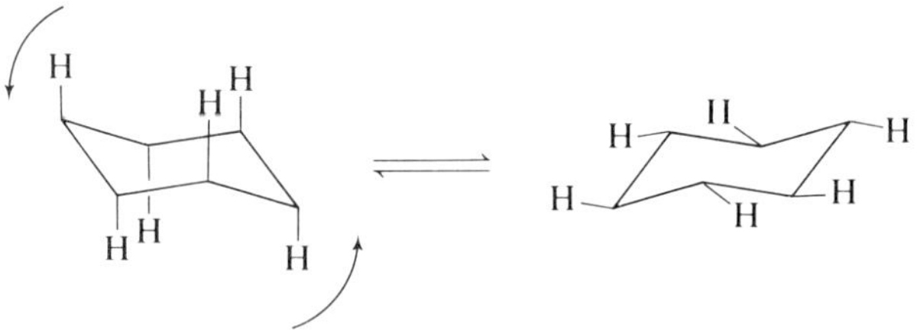

These two forms are physically indistinguishable in cyclohexane itself, but Hassel and Viervoll (1943) showed by electron diffraction studies that two distinct conformations are possible in substituted cyclohexanes. The greater stability of the equatorially mono-substituted cyclohexanes has been rationalised by Barton (1950) as a consequence of unfavourable 1,3-non-bonded interactions in the axially substituted conformation, and forms the basis of what has become known as **conformational analysis** (Barton and Cookson, 1956; Hassel, 1953).

Unfavourable non-bonded interactions

Favoured conformation

The conformation of saturated six-membered rings represents an important aspect of their three-dimensional structure, which is relevant to their biological action. For example, only one of the isomeric benzene hexachlorides, the gamma isomer (Chapter 5), is an effective pesticide. The flexibility of these six-membered rings is, however, a factor which may override the otherwise more stable conformation of particular compounds when they react with biological intermediates, and must always be borne in mind.

The heat of combustion for cyclopentane shows that the molecule is slightly strained. The ring, however, is not planar, as this would not only distort the bond angles, but also place the C—H bonds in eclipsed positions. Some deformation of the ring occurs, and it is considered to adopt alternative puckered forms, known as the **envelope** and **twist** forms respectively, in which one or two atoms respectively are out-of-plane.

Envelope(E) form

Twist(T) form

Cyclopentane conformations

## Cyclopropane as a Structurally Non-specific Anaesthetic

*Cyclopropane* used in concentrations ranging from 4 to 25% in admixture with oxygen is a potent anaesthetic. Like most other general anaesthetics, it owes its anaesthetic activity to its physical properties rather than to its chemical reactivity. Thus, we have an example of structurally non-specific drug action, the common factor between a wide variety of chemical types capable of producing general anaesthesia, *Nitrous Oxide*, *Cyclopropane*, *Chloroform*, *Ether* and *Halothane*, being their lipid solubility. This property leads to a high affinity for the lipid-based tissues of the nervous system. Dependence on a physical rather than a chemical property for biological activity is emphasized by the complete reversibility of controlled anaesthesia, and the relationship between lipid solubility and anaesthetic potency. Anaesthetic action may, therefore, be classified as a structurally non-specific biological effect (**2**, 1). The ready solubility of cyclopropane in fatty tissues has been used as the basis of a method for measuring total body fat by absorption of cyclopropane (Lesser, Blumberg, Steele, Reiter and Porosowska, 1952).

# 4 Alkenes (Olefines)

## THE ETHYLENIC BOND

Alkenes or olefines are unsaturated hydrocarbons having the general formula, $C_nH_{2n}$, and characterised by the presence of a carbon–carbon double bond. The first member of the series is ethylene, $CH_2{=}CH_2$, and the double bond is often referred to as the ethylenic bond.

A mode of hybridisation of 2*s* and 2*p* atomic orbitals, alternative to the $sp^3$ hybridisation of the saturated carbon atom, occurs in the formation of carbon–carbon double bonds. Hybridisation of one 2*s* with two 2*p* orbitals leads to the formation of three new ($sp^2$) hybrid orbitals, which are co-planar and mutually inclined at an angle of 120° (trigonal hybridisation). The remaining 2*p* orbital is orientated at right angles to the plane occupied by the trigonal hybridised orbitals. The carbon–carbon double bond, therefore, results in part from the overlap (endways) of two $sp^2$ hybrid orbitals and in part from the overlap (sideways) of the two unhybridised 2*p* orbitals (Fig. 15A). Thus, in ethylene, strong sigma ($\sigma$) C—H bonds are formed by the overlapping of two of the $sp^2$ orbitals of each carbon atom with the 1*s* orbitals of each of two hydrogen atoms, and in the same plane, a C—C sigma bond is formed by the overlapping of the remaining $sp^2$ atomic orbitals. The overlap of the two unhybridised 2*p* orbitals (Fig. 15B) forms a molecular orbital, known as a $\pi$-orbital (Fig. 15C), which because of the orientation of the constituent 2*p* orbitals, lies above and below the plane of the trigonal bonds.

The combination of $\sigma$- and $\pi$-bonds in the carbon–carbon double bond increases the strength of bond and shortens the bond distance between the two nuclei. The carbon–carbon double bond (C=C) distance is 1.34 Å compared with the carbon–carbon single bond (C—C) distance of 1.54 Å. Thus, the C=C bond energy in ethylene is 598 kJ mol$^{-1}$ compared with 347 kJ mol$^{-1}$ for the C—C bond energy in ethane. Additional energy contributed by the $\pi$-bond is less than that of a $\sigma$-bond, because of sideways rather than endways overlapping of orbitals in the former. Sideways overlap of *p*-electrons also places a restriction on rotation about the carbon–carbon double bond; overlap is at a maximum when the two carbons and four hydrogens (in ethylene) are coplanar, and out-of-plane twisting would tend to reduce the overlap and hence

H H / C=C / HO·OC CO·OH

Maleic acid (*cis*)

H CO·OH / C=C / HO·OC H

Fumaric acid (*trans*)

the strength of the bond. This explains the existence of *cis* and *trans* isomers, which are stabilised by resistance of the double bond to rotation, as for example in maleic and fumaric acids.

### Nomenclature

The simpler members of the series are known by trivial names, but systematic names are used for more complicated molecules. In general, alkenes are named as derivatives of the corresponding saturated hydrocarbon, and distinguished by the suffix **ene**. The position of the double bond in the hydrocarbon chain is indicated by a numeral. The principles are illustrated in Table 11.

### Physical Properties

The alkenes possess physical properties which in many respects closely resemble those of the corresponding paraffins. For example, they are immiscible with, and float on water, and show a similar gradation of boiling point. Ethylene, propylene and butylene are all gaseous at room temperature; those containing five to fifteen carbon atoms are liquids, and the higher members solids. They are flammable, and burn in air with a luminous and smoky flame, owing to the high proportion of carbon to hydrogen.

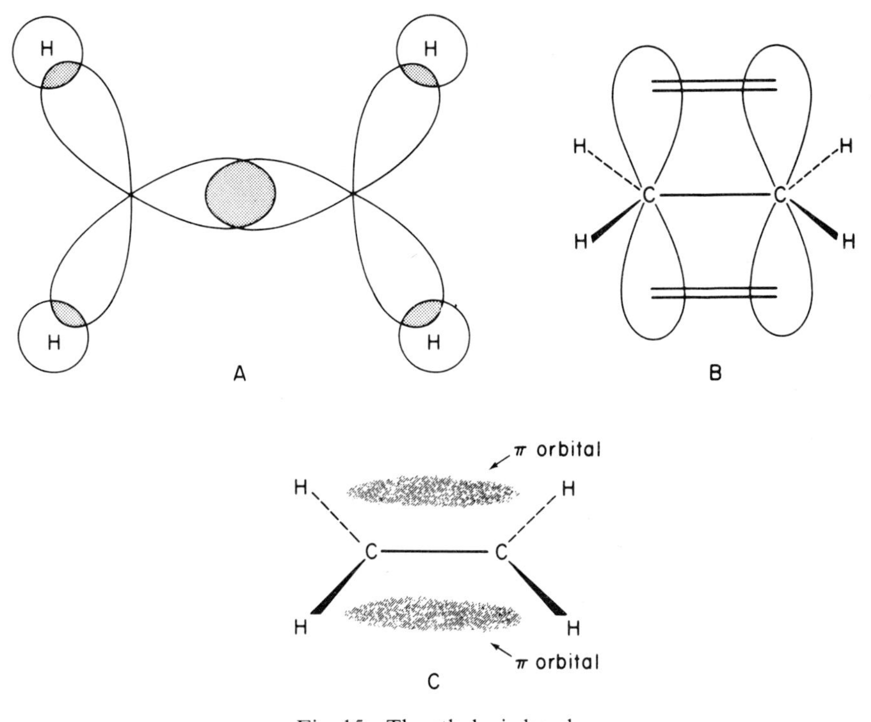


Fig. 15 The ethylenic bond

Table 11. Nomenclature of Alkenes

| Alkene | Trivial Name | Systematic Name |
|---|---|---|
| $H_2C{=}CH_2$ | Ethylene | Ethylene |
| $CH_3{\cdot}CH{=}CH_2$ | Propylene | Propene |
| $CH_3{\cdot}CH_2{\cdot}CH{=}CH_2$ | Butylene | But-1-ene |
| $CH_3{\cdot}CH{=}CH{\cdot}CH_3$ | $\beta$-Butylene | But-2-ene |
| $(CH_3)_2{\cdot}C{=}CH{\cdot}CH_3$ | $\beta$-iso-Amylene | 2-Methylbut-2-ene |
| $\overset{4}{C}H_3{\cdot}\overset{3}{C}H(CH_3){\cdot}\overset{2}{C}(CH_2{\cdot}CH_3){=}\overset{1}{C}H_2$ | | 2-Ethyl-3-methylbut-1-ene |

## ADDITION REACTIONS

In contrast to the inertness of the paraffins, olefines are characterised by great chemical reactivity, forming products by addition rather than substitution. This reactivity of the carbon–carbon double bond is due to the polarisability of the electrons in the $\pi$-orbital, lying above and below the plane of the trigonal $sp^2$ bonds. The $\pi$-electrons are less firmly bound, and more readily polarisable than those forming the $\sigma$-bonds. They can respond to the approach of an attacking reagent in one of two ways depending on the reaction conditions, and whether the reagent is ionic or free radical. Non-polar solvents and light, conditions which favour the production of free radicals, favour free radical addition to the double bond which proceeds through the following mechanism.

$$H_2C{=}CH_2 \longrightarrow H_2\dot{C}{-}\dot{C}H_2$$

Attack by electrophilic reagents, on the other hand, is favoured by reaction in polar solvents, when polarisation of the bond occurs.

$$H_2C{=}CH_2 \longrightarrow H_2\overset{+}{C}{-}\overset{-}{C}H_2$$

### Ionic Addition of Halogens

Addition of halogens to the ethylenic bond occurs under both ionic and free radical conditions. Ionic addition occurs readily with bromine in carbon tetrachloride, and can be followed by the discharge of the red colour of the bromine as the reaction proceeds.

$$\underset{\text{Ethylene}}{H_2C{=}CH_2} + Br_2 \longrightarrow \underset{\text{1,2-Dibromoethane}}{BrCH_2{\cdot}CH_2Br}$$

Chlorine adds readily to olefines, but iodine is much less reactive. The reaction occurs stepwise; approach of the halogen causes polarisation of the bromine molecule under the influence of the $\pi$-electrons. Interaction of the two polar forms then proceeds through the formation of the so-called $\pi$-complex to give a carbonium ion, which in turn is attacked by a bromide ion ($Br^-$) to give 1,2-dibromoethane.

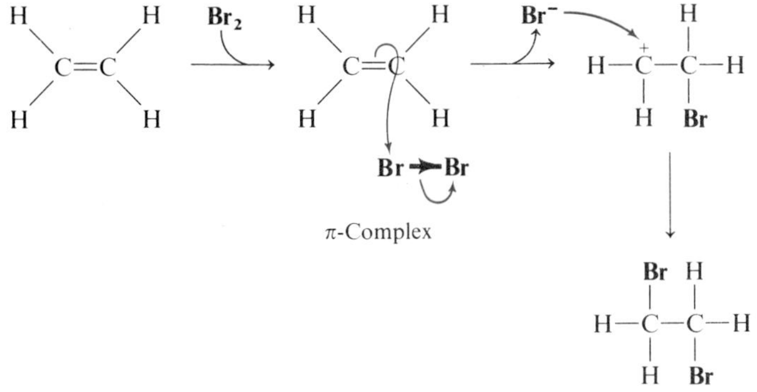

$\pi$-Complex

The stepwise addition is evident from the product of addition in the presence of chloride ions ($Cl^-$), when a mixture of 1,2-dibromoethane and 1-bromo-2-chloroethane is obtained.

$$H_2C{=}CH_2 + Br_2 \longrightarrow BrCH_2\cdot\overset{+}{C}H_2 + Br^-$$

$$BrCH_2\cdot\overset{+}{C}H_2 + Br^- \longrightarrow BrCH_2\cdot CH_2\cdot \mathbf{Br}$$

$$BrCH_2\cdot\overset{+}{C}H_2 + Cl^- \longrightarrow BrCH_2\cdot CH_2\cdot \mathbf{Cl}$$

In order to account for the stereochemistry of the addition which occurs **trans** to the double bond, it is necessary to postulate the formation of a bromonium ion intermediate which, because of the bulk of the ion, favours attack from the opposite side.

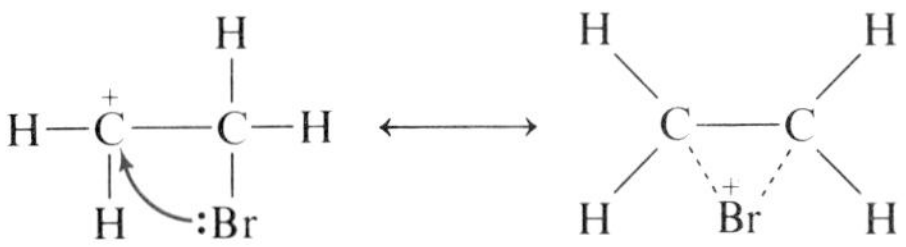

The stereochemistry of addition is not obvious in reactions with simple open-chain olefines, but is readily apparent in the formation of *trans*-1,2-dibromocyclohexane from cyclohexene.

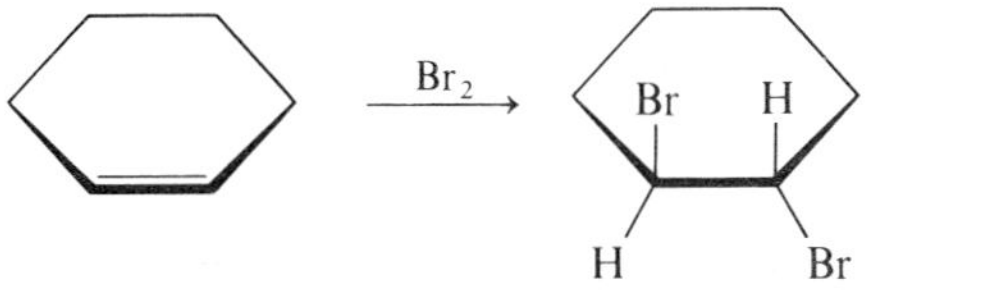

Cyclohexene *trans*-1,2-Dibromocyclohexane

Addition of iodine monochloride (ICl) is used in the quantitative determination of olefines. The reagent is used in excess, unused iodine monochloride being converted to iodine with potassium iodide, and titrated with sodium thiosulphate.

$$>C{=}C< + I{\rightarrow}Cl \longrightarrow -\underset{I}{\overset{|}{\underset{|}{C}}}-\underset{Cl}{\overset{|}{\underset{|}{C}}}-$$

$$I^- \quad I{-}Cl \longrightarrow I_2 + Cl^-$$

Addition of iodine monochloride is used in the determination of Iodine Value of fixed oils (Chapter 11) and also as a test to limit traces of unsaturated hydrocarbons in the anaesthetic, *Cyclopropane*.

**The effect of substituents on the rate of addition.** The rate of addition is influenced by substituents attached to the ethylenic bond. Methyl substituents, which are electron-repelling, promote the reaction by increasing the availability of electrons. This favours $\pi$-complex formation.

$$H_3C{\rightarrow}CH{=}CH_2 \longrightarrow H_3C{\rightarrow}CH{=}CH_2 \quad Br{\leftarrow}Br$$

As a result, propylene reacts twice as fast as ethylene, and additional methyl substituents still further increase the rate at which the reaction proceeds. The rate of addition relative to ethylene increases with methyl substitution as follows.

| $H_2C{=}CH_2$ | $Me{\rightarrow}CH{=}CH_2$ | $Me_2{\cdot}C{=}CH_2$ | $Me_2{\cdot}C{=}CHMe$ | $Me_2{\cdot}C{=}C{\cdot}Me_2$ |
|---|---|---|---|---|
| 1 | 2 | 5 | 10 | 15 |

Relative rate

In contrast, electron-attracting substituents such as carboxyl (COOH) and bromine impede the reaction, since they lead to a deficiency of electrons on carbon.

$$H_2C{=}CH{\rightarrow}Br \equiv H_2\overset{\delta^+}{C}{=}CH{-}\overset{\delta^-}{Br}$$

The following relative rates of addition have been observed.

| $H_2C{=}CH_2$ | $H_2C{=}CH{\cdot}CO{\cdot}OH$ | $H_2C{=}CH{\cdot}Br$ |
|---|---|---|
| 1 | 0.03 | 0.03 |

Relative rate

Phenyl substituents, however, speed the reaction owing to resonance stabilisation of the carbonium ion intermediate, which facilitates its formation.

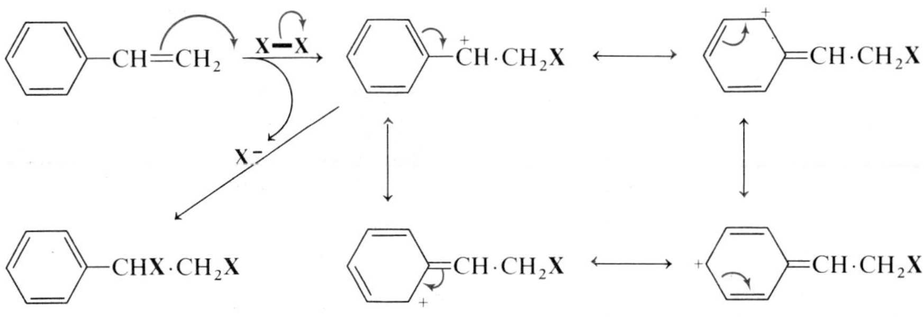

## Ionic Addition of Halogen Acids

Halogen acids, like halogens, readily add to the carbon–carbon double bond by an ionic mechanism. Because of electronegativity differences between hydrogen and halogens, the halogen forms the negative end of the dipole in the halogen acids. Polarisation is increased by approach of the double bond. Stepwise addition occurs as with halogens. The π-complex, formed by interaction of the π-electron cloud with the electron-deficient hydrogen of the halogen acid, leads to a carbonium ion intermediate. Addition is completed by the attack of $X^-$ on the carbonium ion at the opposite side of the molecule (**trans** addition).

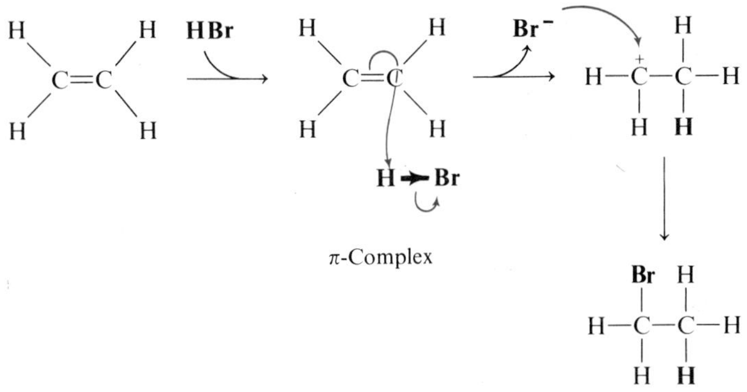

π-Complex

The reactivity of the halogen acids in additions to olefines increases in order of increasing acid strength. This increases in parallel with increasing polarisability and contrary to the permanent polarisation of the hydrogen–halogen bond. Thus, the strength of the acid is determined by the ease with which the halogen can accept the electrons released to it upon ionisation. Charge-density decreases and polarisability increases, as the size of the ion increases, hence the observed gradation of acid strength among the halogen acids.

H—F    H—Cl    H—Br    H—I

Acid strength increases →

Polarisability increases →

← Permanent polarisation increases

**Addition to unsymmetrical olefines—Markownikoff's rule.** Addition of a halogen acid to an unsymmetrical olefine, such as propylene, can lead to two possible products, n-propyl bromide ($CH_3 \cdot CH_2 \cdot CH_2Br$) and isopropyl bromide ($CH_3 \cdot CHBr \cdot CH_3$). The direction of addition is governed by an empirical rule, the Markownikoff rule, which requires that the negative centre of the attacking reagent (i.e. the halogen) adds to the more highly substituted ethylenic carbon atom. Isopropyl bromide is therefore the major product.

The rule is capable of theoretical interpretation, since a secondary carbonium ion intermediate will be formed in preference to the alternative less stable primary carbonium ion.

$$H_3C \rightarrow CH{=}CH_2 \xrightarrow{H-Br} \begin{cases} CH_3 \cdot CH_2 \cdot \overset{+}{C}H_2 \ \text{(1° Carbonium ion)} \dashrightarrow CH_3 \cdot CH_2 \cdot CH_2 \cdot Br \quad \text{(not favoured)} \\ CH_3 \cdot \overset{+}{C}H \cdot CH_3 \ \text{(2° Carbonium ion)} \longrightarrow CH_3 \cdot CHBr \cdot CH_3 \end{cases}$$

The stability of carbonium ions is promoted by hyperconjugation (no bond resonance) which spreads the charge by interaction between the carbonium carbon and adjacent (α) C—H bonds. Thus, the secondary carbonium ion is a resonance hybrid of seven contributing forms.

$$H_3C{-}CH_2{-}\overset{+}{C}H{-}CH_3 \ \text{form:}\ H{-}\underset{H}{\overset{H}{C}}{-}\overset{+}{C}H{-}CH_3$$

(a) $H^+\ HC(H){=}CH{-}CH_3 \longleftrightarrow H^+\ C(H)(H){=}C(H)CH_3 \longleftrightarrow H{-}C(H){=}C(H)\ H^+\ H_2C{-}H$

(b) $H_3C{-}CH{=}CH_2\ H^+ \longleftrightarrow H_3C{-}CH{=}C(H){-}H\ H^+ \longleftrightarrow H_3C{-}CH{=}C(H)H\ H^+$

In contrast, only three forms contribute to the resonance hybrid of the primary carbonium ion. A much lower degree of stabilisation is achieved, and the pathway involving the secondary carbonium ion is favoured.

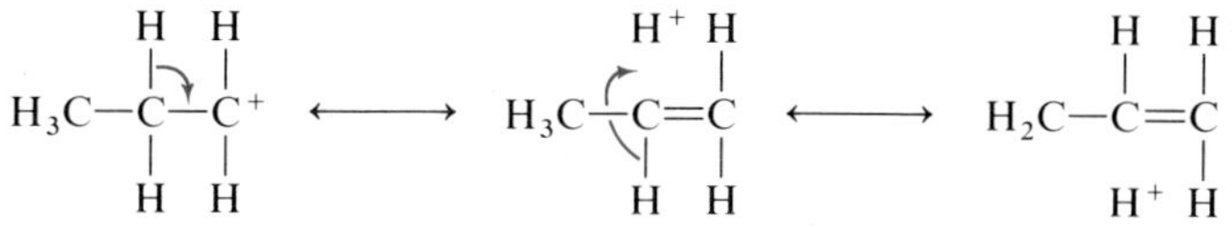

$$H_3C-\overset{H}{\underset{H}{C}}-\overset{H}{\underset{H}{C^+}} \longleftrightarrow H_3C-\overset{H^+}{\underset{H}{C}}=\overset{H}{\underset{H}{C}} \longleftrightarrow H_2C-\overset{H}{C}=\overset{H}{\underset{H}{C}}\;\;(H^+)$$

In general, the number of C—H bonds α to the carbonium carbon provides a measure of the stability of the ion. We thus have the relative stability order.

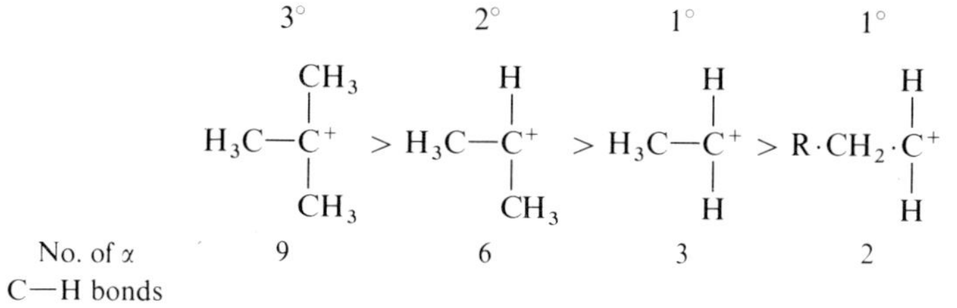

$$\underset{3^\circ}{H_3C-\overset{CH_3}{\underset{CH_3}{C^+}}} > \underset{2^\circ}{H_3C-\overset{H}{\underset{CH_3}{C^+}}} > \underset{1^\circ}{H_3C-\overset{H}{\underset{H}{C^+}}} > \underset{1^\circ}{R\cdot CH_2\cdot\overset{H}{\underset{H}{C^+}}}$$

No. of α C—H bonds: 9, 6, 3, 2

### Addition of Hypohalous Acids

Reaction of ethylene with hypochlorous acid (chlorine in water) gives ethylene chlorhydrin.

$$H_2C{=}CH_2 + HOCl \longrightarrow HO\cdot CH_2\cdot CH_2\cdot Cl$$

Ethylene chlorhydrin

Reaction proceeds by an ionic mechanism similar to that in the addition of halogen acids probably via the formation of a chlorinium ion. The Markownikoff rule applies, with addition of hydroxyl to the more highly substituted ethylenic carbon.

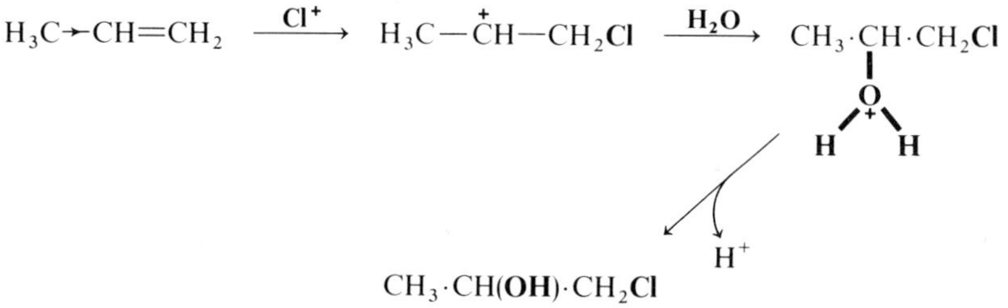

$$H_3C\rightarrow CH{=}CH_2 \xrightarrow{Cl^+} H_3C-\overset{+}{C}H-CH_2\mathbf{Cl} \xrightarrow{H_2O} CH_3\cdot \underset{\overset{+}{O}H_2}{CH}\cdot CH_2\mathbf{Cl} \xrightarrow{-H^+} CH_3\cdot CH(\mathbf{OH})\cdot CH_2\mathbf{Cl}$$

Propylene chlorhydrin

As in the other ionic reactions of alkenes, addition occurs **trans** to the double bond. Hypobromous acid reacts similarly.

### Addition of Sulphuric Acid

Absorption of olefines into cold concentrated sulphuric acid at 0° leads to a similar addition, to which the Markownikoff rule again applies, with formation of the corresponding alkyl hydrogen sulphate.

In contrast, dilute (aqueous) sulphuric acid catalyses the hydration of olefines to form alcohols. Alkyl hydrogen sulphates are also readily hydrolysed in warm aqueous solution to the corresponding alcohol (Chapter 7).

$$H_3C \rightarrow CH{=}CH_2 \xrightarrow{H-O\cdot SO_2\cdot OH} H_3C-\overset{+}{C}H-CH_3 \xrightarrow{HO\cdot SO_2\cdot O^-} CH_3\cdot CH(O\cdot SO_2\cdot OH)\cdot CH_3$$

Isopropyl hydrogen sulphate

### Acid-catalysed Hydration

The acid-catalysed hydration of low molecular weight olefines is of considerable economic importance for the large-scale production of alcohols. Thus, ethylene is readily hydrated in 10% sulphuric or phosphoric acid at a temperature of about 240°C. Substituted olefines are much more readily hydrated, isobutylene being readily converted to t-butyl alcohol in 10% sulphuric acid at 25°C. Hydration rather than addition occurs in dilute acid solution because $HO\cdot SO_2\cdot O^-$ is too weak a **nucleophile** to compete with water in attack on the intermediate carbonium ion. The reaction is reversible, so that alcohols are dehydrated to olefines at the same rate and through the same intermediates, the nature of the product being dependent on the supply and withdrawal of reactants and products from the system.

$$CH_3\cdot CH{=}CH_2 \underset{}{\overset{H^+}{\rightleftharpoons}} CH_3\cdot\overset{+}{C}H\cdot CH_3 \overset{H_2O}{\rightleftharpoons} CH_3\cdot CH(\overset{+}{O}H_2)\cdot CH_3 \overset{-H^+}{\rightleftharpoons} CH_3\cdot CH(OH)\cdot CH_3$$

### Biological Counterparts of Acid-catalysed Hydration

A number of highly specific enzymes (hydratases) are concerned with the biological hydration of olefines and the dehydration of alcohols. Two such enzymes, fumarase and aconitase, are involved in the citric acid cycle, which is central to the metabolism of most living cells. Fumarase catalyses the conversion of fumarate to L-malate.

Fumarate + $H_2O$ $\xrightleftharpoons{\text{Fumarase}}$ L-Malate

Aconitase catalyses the interconversion of citrate and isocitrate through *cis*-aconitate.

Citrate $\xrightleftharpoons{H_2O}$ *cis*-Aconitate $\xrightleftharpoons{H_2O}$ Isocitrate

pH–Rate studies of these enzymes provide evidence of two reactive centres, one basic and the other acidic, one responsible for transfer of a proton at the beginning of the reaction and the other responsible for removal of the proton at the end of the reaction.

The interconversion of $\alpha\beta$-unsaturated acids and $\beta$-hydroxy-acids in fatty acid metabolism similarly requires the addition or subtraction of the elements of water under the influence of the enzyme enoyl-$\overline{\text{CoA}}$ hydratase.

$$R{\cdot}CH_2{\cdot}CH{=}CH_2{\cdot}CO{\cdot}S\overline{CoA} \xrightleftharpoons{H_2O} R{\cdot}CH_2{\cdot}CH(OH){\cdot}CH_2{\cdot}CO{\cdot}S\overline{CoA}$$

$\overline{\text{CoA}}$ = Coenzyme A

**Free Radical Addition of Halogens**

A photochemically-induced addition of chlorine or bromine also occurs readily with olefines in the gas phase on exposure to sunlight. The reaction occurs by a free radical mechanism, initiated by the photochemical decomposition of a chlorine molecule to give chlorine free radicals. The reaction possesses the advantage that it will occur readily at double bonds which react only slowly by ionic addition, as for example in tetrachloroethylene.

$$Cl{-}Cl \longrightarrow 2Cl\cdot$$

$$Cl\cdot + Cl_2C{=}CCl_2 \longrightarrow Cl_3C{-}\dot{C}Cl_2$$

$$Cl_3C{-}\dot{C}Cl_2 + Cl{-}Cl \longrightarrow Cl_3C{-}CCl_3 + Cl\cdot$$

Hexachloroethane

In contrast, alkyl-substituted olefines undergo substitution in the allylic position. This pathway is favoured because of the resonance stabilisation of the allyl radical which is formed as an intermediate.

$$Cl\cdot + H{-}CH_2{-}CH{=}CH_2 \xrightarrow{-HCl} \dot{C}H_2{-}CH{=}CH_2 \longleftrightarrow CH_2{=}CH{-}\dot{C}H_2$$

$$\xrightarrow{Cl_2,\ -Cl\cdot} CH_2{=}CH{\cdot}CH_2\mathbf{Cl}$$

Allyl chloride

**Free Radical Addition of Halogen Acids**

The direction of addition of hydrogen bromide to olefines is reversed in the presence of peroxides (Kharasch and Mayo, 1933). Thus, treatment of propylene with hydrogen bromide in the presence of catalytic amounts of benzoyl peroxide gives n-propyl bromide (anti-Markownikoff addition).

The reaction proceeds by a free radical mechanism, and is initiated by decomposition of the benzoyl peroxide to give benzoyl and phenyl radicals.

$$C_6H_5{\cdot}CO{\cdot}O{\cdot}O{\cdot}CO{\cdot}C_6H_5 \longrightarrow C_6H_5{\cdot}CO{\cdot}O\cdot + C_6H_5\cdot + CO_2$$

These attack hydrogen bromide to form bromine radicals.

$$C_6H_5\cdot + HBr \longrightarrow C_6H_6 + Br\cdot$$

Radical attack at the double bond follows only one of the two possible pathways.

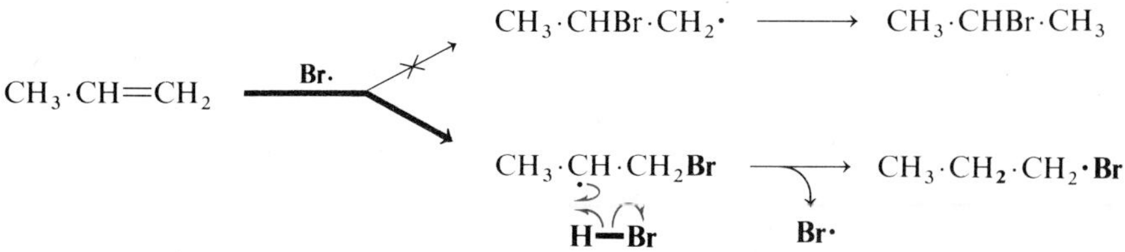

$$CH_3{\cdot}CH{=}CH_2 \xrightarrow{Br\cdot} \begin{cases} CH_3{\cdot}CHBr{\cdot}CH_2\cdot \;\not\rightarrow\; CH_3{\cdot}CHBr{\cdot}CH_3 \\ CH_3{\cdot}\dot{C}H{\cdot}CH_2Br \xrightarrow{H-Br} CH_3{\cdot}CH_2{\cdot}CH_2{\cdot}Br + Br\cdot \end{cases}$$

The orientation of addition follows from the relative ease of formation and stability of the two possible free radical intermediates. The ease of formation of the radicals is determined by the relative bond dissociation energies of the C—H bonds present (Table 7), and is in the order tertiary (3°) > secondary (2°) > primary (1°). Resonance stabilisation of the radical intermediates due to hyperconjugation also follows the same order, since, like that of carbonium ions (p. 86), it depends upon the number of C—H bonds in the α-position to the radical bearing carbon. Thus, the secondary radical intermediate is stabilised as a result of hyperconjugation involving six contributing forms.

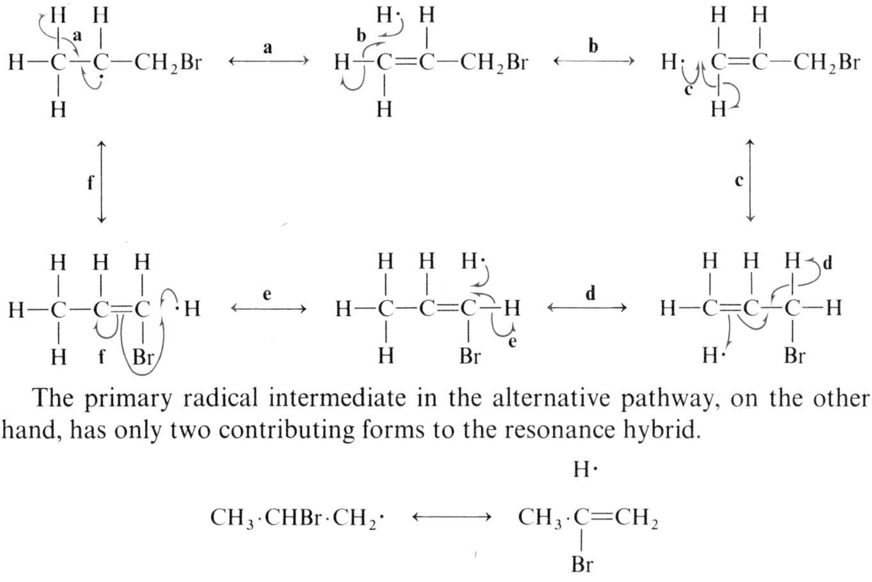

$$H_3C{-}\dot{C}H{-}CH_2Br \overset{a}{\longleftrightarrow} H\cdot\;H_2C{=}CH{-}CH_2Br \overset{b}{\longleftrightarrow} H\cdot\;H_2C{=}CH{-}CH_2Br \overset{c}{\longleftrightarrow} \cdots$$

$$H_3C{-}CH{=}CHBr\;\cdot H \overset{e}{\longleftrightarrow} H_3C{-}CH{=}CBr{-}H\;\;H\cdot \overset{d}{\longleftrightarrow} H_2C{=}CH{-}CH_2Br\;\;H\cdot \quad (f)$$

The primary radical intermediate in the alternative pathway, on the other hand, has only two contributing forms to the resonance hybrid.

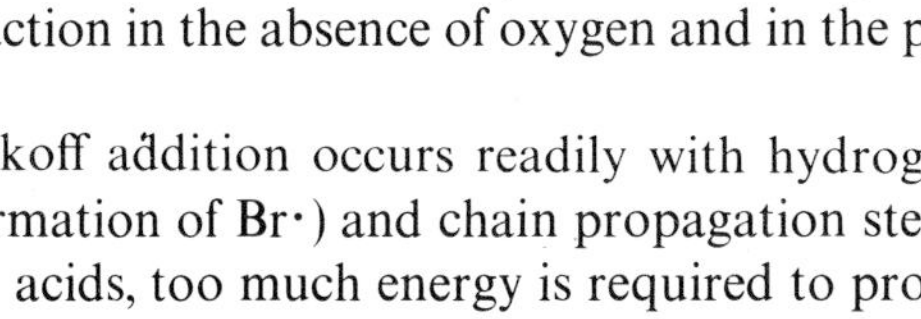

$$CH_3{\cdot}CHBr{\cdot}CH_2\cdot \longleftrightarrow CH_3{\cdot}C(Br){=}CH_2 \quad H\cdot$$

Alkenes frequently form peroxides on exposure to air, so that the normal Markownikoff ionic addition occurs only if these are rigorously excluded by performing the reaction in the absence of oxygen and in the presence of an anti-oxidant.

Anti-Markownikoff addition occurs readily with hydrogen bromide, since both initiation (formation of Br·) and chain propagation steps are exothermic. For other halogen acids, too much energy is required to produce F· from HF; I· is formed, but is not sufficiently reactive to attack the olefine; Cl· reacts, but the reaction is slow at room temperature and Markownikoff addition occurs more readily.

Other substances capable of adding to olefines under free radical conditions include thiols (RSH), chloroform ($CHCl_3$), bromoform ($CHBr_3$) and trifluoroiodomethane ($CF_3I$). The nature of the addition product is in part dictated by the relative stability of carbon–halogen and carbon–hydrogen bonds (Chapter 3) to homolytic cleavage.

**Polymerisation of Alkenes**

Ethylene undergoes radical-induced polymerisation at high temperature (100°C) and under high pressure (1000 atmospheres) in the presence of peroxide catalysts to give polyethylene. The catalysts function as free radical initiators for chain propagation, the reaction being terminated eventually by either radical combination, or **disproportionation**, in which one molecule acquires hydrogen at the expense of another in an intermolecular hydrogenation (p. 56).

$$R\cdot \; H_2C{=}CH_2 \longrightarrow R\cdot CH_2\cdot CH_2\cdot \xrightarrow{n(H_2C=CH_2)} R(CH_2\cdot CH_2)_n\cdot CH_2\cdot CH_2\cdot$$

$$2R(CH_2\cdot CH_2)_n\cdot CH_2\cdot CH_2\cdot \xrightarrow{\text{Radical combination}} R(CH_2\cdot CH_2)_{2n+2}\cdot R$$

$$2R(CH_2\cdot CH_2)_n\cdot CH_2\cdot CH_2\cdot \xrightarrow{\text{Disproportionation}} R(CH_2\cdot CH_2)_nCH{=}CH_2 + R(CH_2\cdot CH_2)_n\cdot CH_2\cdot CH_3$$

Propylene and styrene ($C_6H_5CH{=}CH_2$), which also undergo similar radical-induced polymerisation, give rise to **atactic** polymers by this process. In these, the alkyl (or aryl) substituents are sterically disposed in a random manner with respect to the backbone hydrocarbon chain. A **low density** polypropylene results, which being non-crystalline, is capable of being produced in sheet or film, of being moulded, or extruded into tubing. The product is resistant to acids, bases and most organic solvents, and is used extensively for the manufacture of plastic containers and packing materials.

An alternative method of polymerisation, known as co-ordination polymerisation, using a titanium–aluminium catalyst suspension in a hydrocarbon solvent (Ziegler catalyst) is effective at atmospheric pressure and temperature.

$$n\,H_2C{=}CH_2 \longrightarrow -CH_2\cdot CH_2(CH_2\cdot CH_2)_{n-2}\cdot CH_2\cdot CH_2-$$

Polypropylene and polystyrene prepared by this process are stereoregular, **isotactic**, polymers, in which the substituent alkyl (or aryl) groups are sterically disposed on the same side of the backbone hydrocarbon chain. These **high density** polymers are more highly crystalline and require higher moulding temperatures than the corresponding atactic polymers. A second type of stereoregular polymer is the **syndiotactic** polymer, in which the substituent groups are regularly arranged alternately on each side of the polymer chain.

## OXIDATION

### Oxidative Scission

Oxidation of olefines with powerful oxidising agents such as potassium permanganate in aqueous acid or with chromic acid in acetic acid fragments the carbon chain, initially at the double bond.

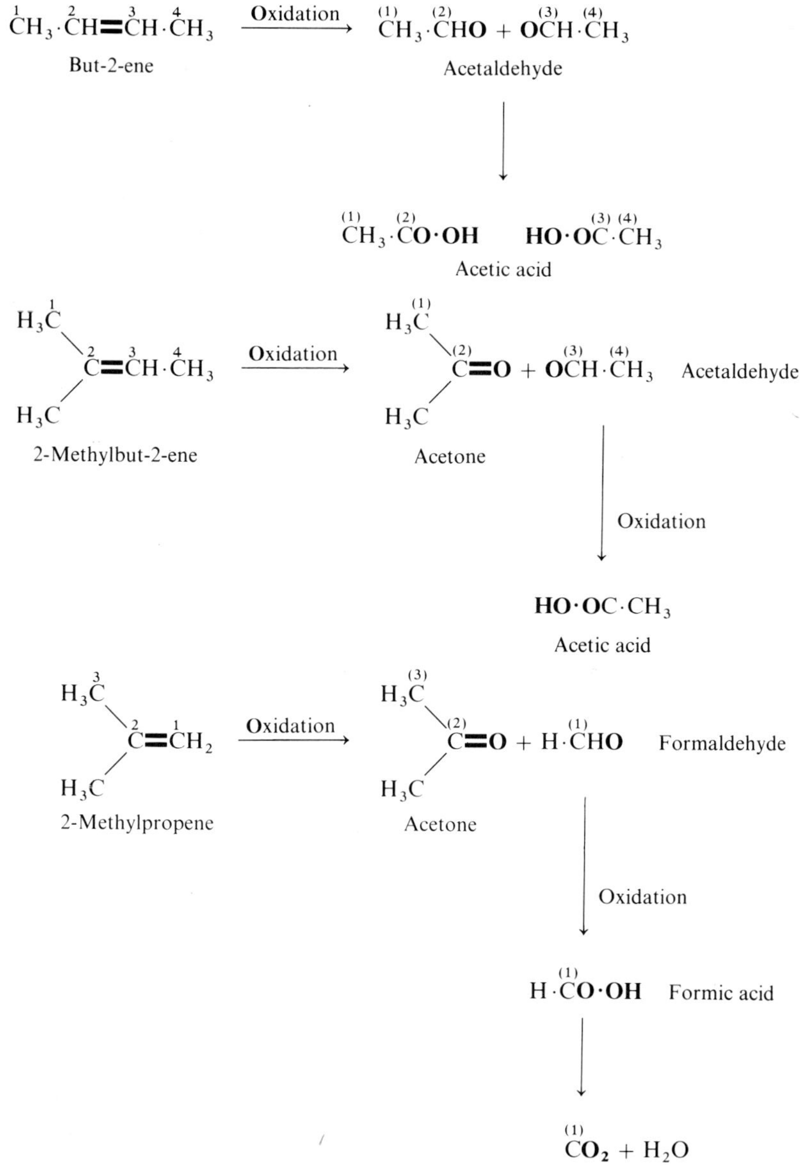

$$\overset{1}{C}H_3\cdot\overset{2}{C}H{=}\overset{3}{C}H\cdot\overset{4}{C}H_3 \xrightarrow{\text{Oxidation}} \overset{(1)}{C}H_3\cdot\overset{(2)}{C}\mathbf{HO} + \mathbf{O}\overset{(3)}{C}H\cdot\overset{(4)}{C}H_3$$

But-2-ene → Acetaldehyde

↓

$$\overset{(1)}{C}H_3\cdot\overset{(2)}{C}\mathbf{O\cdot OH} \quad \mathbf{HO\cdot O}\overset{(3)}{C}\cdot\overset{(4)}{C}H_3$$

Acetic acid

$$(\overset{1}{C}H_3)(CH_3)\overset{2}{C}{=}\overset{3}{C}H\cdot\overset{4}{C}H_3 \xrightarrow{\text{Oxidation}} (\overset{(1)}{C}H_3)(CH_3)\overset{(2)}{C}{=}\mathbf{O} + \mathbf{O}\overset{(3)}{C}H\cdot\overset{(4)}{C}H_3$$

2-Methylbut-2-ene → Acetone + Acetaldehyde

↓ Oxidation

$$\mathbf{HO\cdot O}C\cdot CH_3$$

Acetic acid

$$(\overset{3}{C}H_3)(CH_3)\overset{2}{C}{=}\overset{1}{C}H_2 \xrightarrow{\text{Oxidation}} (\overset{(3)}{C}H_3)(CH_3)\overset{(2)}{C}{=}\mathbf{O} + H\cdot\overset{(1)}{C}\mathbf{HO}$$

2-Methylpropene → Acetone + Formaldehyde

↓ Oxidation

$$H\cdot\overset{(1)}{C}\mathbf{O\cdot OH}$$ Formic acid

↓

$$\overset{(1)}{\mathbf{CO_2}} + H_2O$$

Figures in brackets indicate the position of the carbon atom in the original ethylenic hydrocarbon

The reaction can be used to identify the position of the double bond in the chain by characterisation of the products. Yields, however, are not always good, particularly from complex olefines, as the initial products frequently undergo further fragmentation.

Oxidation by aqueous potassium permanganate, with discharge of the purple colour, forms a useful identity test for the characterisation of unsaturated substances, such as *Undecenoic Acid* (Undecylenic Acid), $CH_2{=}CH \cdot (CH_2)_8 \cdot CO \cdot OH$. Oxidation is also useful as a test to limit unsaturated impurities in drugs, as for example cinnamylcocaine in *Cocaine*; the cinnamic acid derivative is oxidised with potassium permanganate in acidic solution at the carbon–carbon double bond to give benzaldehyde and ultimately benzoic acid.

$$C_6H_5{-}CH{=}CH \cdot CO \cdot OH \xrightarrow{\text{Oxidation}} C_6H_5{-}CHO \xrightarrow{\text{Oxidation}} C_6H_5{-}CO \cdot OH$$

Cinnamic acid — Benzaldehyde — Benzoic acid

## Ozonolysis

Ozonolysis generally provides a milder and more satisfactory method for oxidative cleavage of double bonds. Ozone is formed in about 5% concentration when oxygen is passed through a high voltage electrical discharge.

$$\ddot{O}{-}\ddot{O}: \;\; \ddot{O}: \longrightarrow O{=}\overset{+}{O}{-}\overset{-}{O} \longleftrightarrow \overset{-}{O}{-}\overset{+}{O}{=}O \longleftrightarrow \overset{-}{O}{-}O{-}\overset{+}{O}$$

Ozonolysis occurs when ozonised oxygen is passed through a solution of the olefine in a suitable solvent, e.g. chloroform. The ozonides are seldom isolated, since they are generally oily liquids or glassy solids, which do not crystallise. Moreover some are explosive. Decomposition of the ozonide by water or by hydrogenolysis gives two carbonyl compounds, which are structurally related to the ethylenic compound undergoing ozonolysis.

The reaction proceeds as a cyclo-addition with ozone reacting in its 1,3-dipolar forms to form the molozonide; the latter then undergoes rearrangement to the ozonide. The rearrangement of molozonide to ozonide is supported by ozonolysis experiments which have been conducted in the presence of an identifiable carbonyl compound, when the latter is incorporated in the product ozonide.

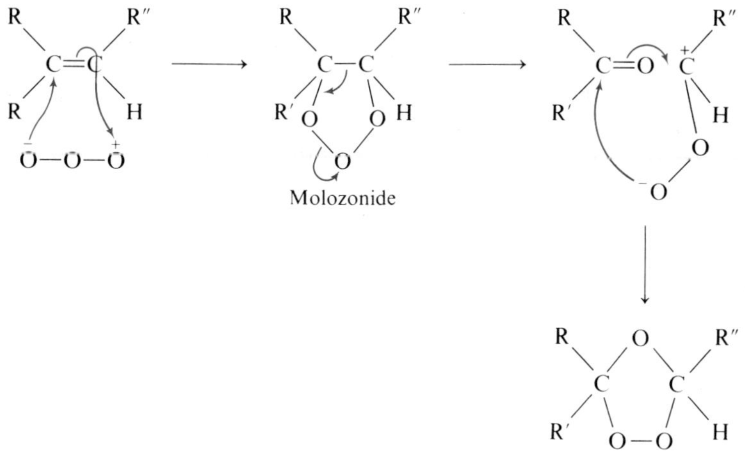

Molozonide

Ozonide

The ozonide is usually decomposed either with water or by hydrogenolysis at a palladium catalyst. Decomposition with water suffers from the disadvantage that aldehydes so formed may become oxidised to the corresponding carboxylic acid by the hydrogen peroxide so released. This can be suppressed by the addition of titanous chloride, or avoided in the alternative decomposition with palladium and hydrogen.

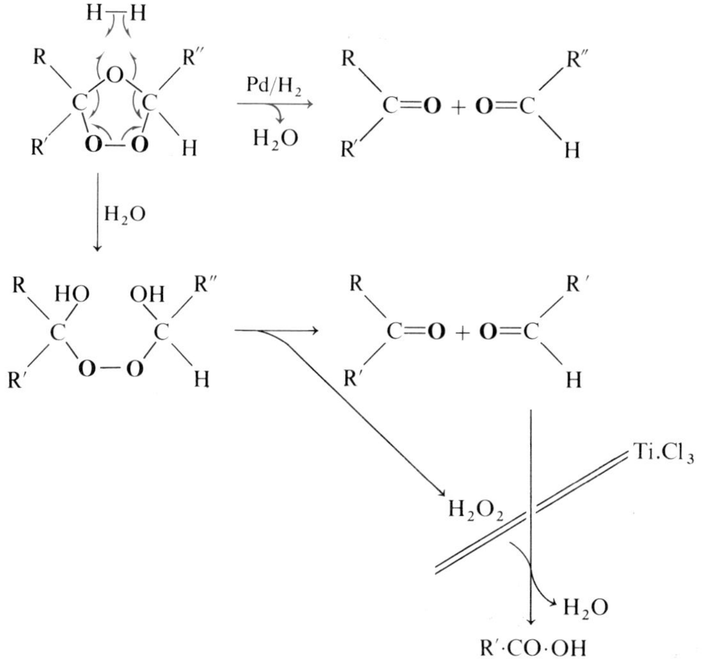

Pd/$H_2$

$H_2O$

$H_2O$

Ti.$Cl_3$

$H_2O_2$

$H_2O$

R′·CO·OH

**Oxidative Addition to Olefines**

Treatment of olefines with osmium tetroxide in ethereal solution leads to precipitation of a black cyclic osmate ester. Addition of water or a mild reducing agent ($Na_2SO_3$) gives the corresponding *cis*-glycol in which the hydroxyl groups have been added to the same side of the olefinic carbon atoms.

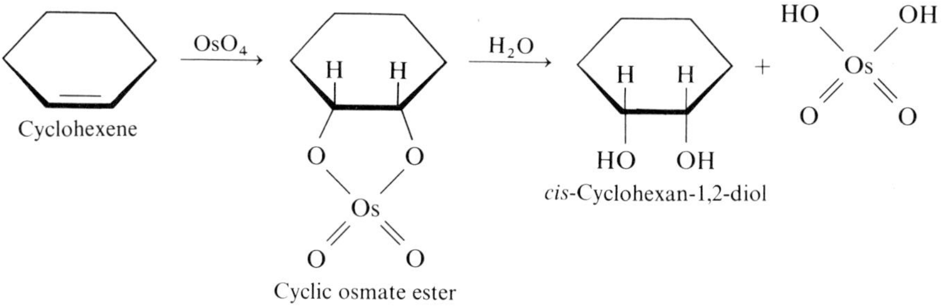

Cyclohexene

$OsO_4$

Cyclic osmate ester

$H_2O$

*cis*-Cyclohexan-1,2-diol

The *cis*-glycol is also obtained by oxidation with potassium permanganate, in cold aqueous alkaline solution with a miscible co-solvent such as t-butanol.

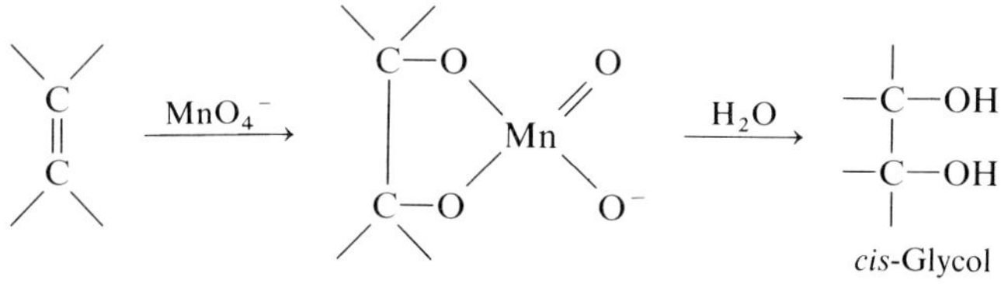

$MnO_4^-$

$H_2O$

*cis*-Glycol

Oxidation with peracids, such as peracetic or perbenzoic acid, leads to formation of an epoxide.

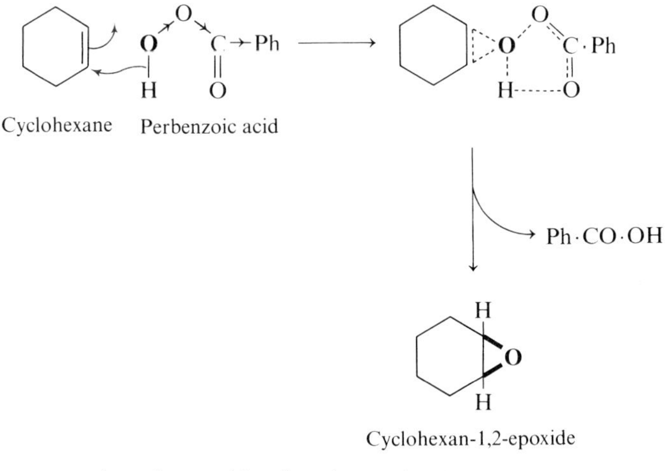

Cyclohexane Perbenzoic acid

Ph·CO·OH

Cyclohexan-1,2-epoxide

The three-membered epoxide ring is strained, and hence highly reactive. Epoxides are readily converted both by acid and by alkali to the corresponding *trans*-glycol (Chapter 7).

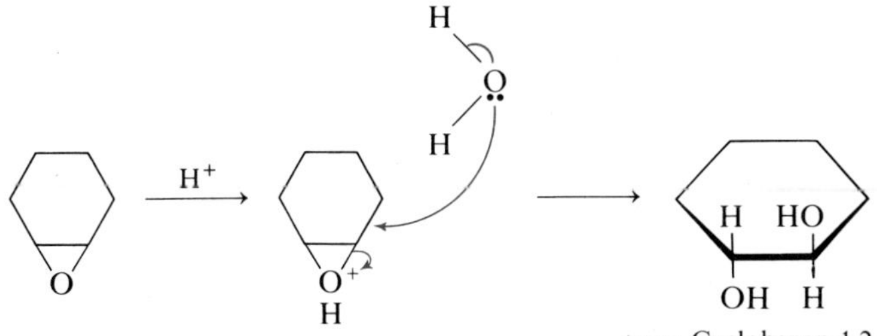

H⁺

trans-Cyclohexan-1,2-diol

## Biological Oxidation

Biological oxidation occurs with some olefines, as for example styrene, which in the rat is metabolised mainly to benzoic acid via phenylglycol. Some mandelic acid is also formed.

$$\underset{\text{Styrene}}{Ph{\cdot}CH{=}CH_2} \xrightarrow[\text{Oxidation}]{\text{Metabolic}} \underset{\text{Phenylglycol}}{Ph{\cdot}CHOH{\cdot}CH_2OH} \longrightarrow \underset{\text{Mandelic acid}}{Ph{\cdot}CHOH{\cdot}CO{\cdot}OH}$$

$$\xrightarrow{-CO_2} \underset{\text{Benzoic acid}}{Ph{\cdot}CO{\cdot}OH}$$

Other unsaturated compounds are converted metabolically to epoxides. Thus, metabolism of the insecticide, *Aldrin*, in flies and mammals gives rise to the corresponding epoxide, *Dieldrin*, also an insecticide, which is metabolised in turn by hydrolysis to the glycol, 6,7-*trans*-dihydroxydihydroaldrin.

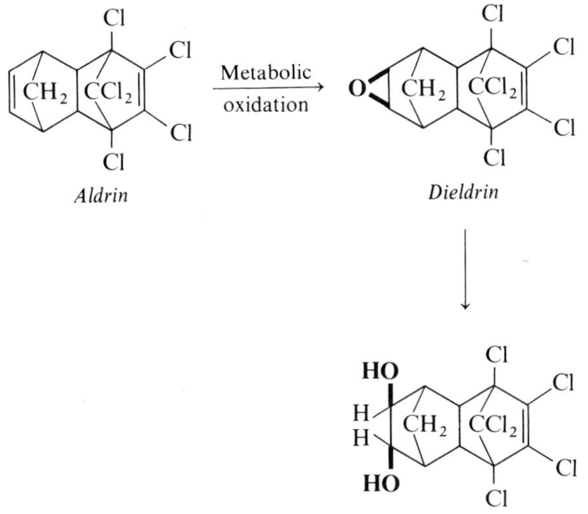

*Aldrin* $\xrightarrow[\text{oxidation}]{\text{Metabolic}}$ *Dieldrin* $\longrightarrow$ 6,7-*trans*-Dihydroxydihydroaldrin

Both *Aldrin* and *Dieldrin* suffer from the very serious disadvantage that because of their high halogen content they accumulate in the liver causing serious damage to it.

### Allylic Oxidation

Unsaturated compounds readily undergo autoxidation at the allylic position, i.e. on the saturated carbon atom immediately adjacent to the ethylenic bond. The oxidations, which occur on exposure to air, are light-catalysed, as for example by exposure to sunlight and proceed through the formation of peroxide intermediates. Allylic oxidation also proceeds readily at elevated temperatures which catalyse the reaction, as in the catalytic oxidation of propylene to acraldehyde. Attack occurs initially at the double bond (Gunstone and Hilditch, 1946), followed by attack at the allylic position of a second molecule by the radical first formed.

$$H_2C{=}CH{\cdot}CH_3 \xrightarrow{O_2} \underset{\displaystyle O{\cdot}O{\cdot}}{CH}{\cdot}\dot{C}H{\cdot}CH_3$$

$$H_2C{=}CH{\cdot}CH_2{\cdot} \qquad \underset{\displaystyle O{\cdot}OH}{CH}{\cdot}\dot{C}H{\cdot}CH_3$$

$$\downarrow O_2$$

$$H_2C{=}CH{\cdot}\overset{H}{C}{-}O \quad (H,\ O) \longrightarrow H_2C{=}CH{\cdot}\overset{H}{C}{=}O \quad + \quad HO{\cdot}$$

Acraldehyde

Peroxidation of unsaturated esters and acids, such as *Ethyl Oleate* which is used as solvent in the preparation of oily injections, and *Undecenoic Acid* occurs in this way. Thus, there is evidence that all four 8-, 9-, 10-, and 11-hydroperoxides are formed in equal amount from ethyl oleate (Swern, 1961) as a result of autoxidation. These and other unsaturated substances should be stored in well-filled, sealed containers, in the absence of air (displaced by nitrogen) and protected from light, in order to limit oxidative decomposition (Chapter 11). The accumulation of peroxides in *Ethyl Oleate* could, for example, have a deleterious effect on any medicament dissolved in it.

$$CH_3(CH_2)_7{\cdot}CH{=}CH(CH_2)_7{\cdot}CO{\cdot}OEt \qquad H_2C{=}CH(CH_2)_8{\cdot}CO{\cdot}OH$$

*Undecenoic Acid*
(Undecylenic Acid)

# REDUCTION

## Hydrogenation

Catalytic hydrogenation of ethylenes yields the corresponding saturated hydrocarbon.

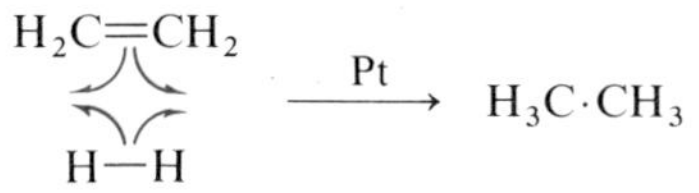

$$H_2C{=}CH_2 + H{-}H \xrightarrow{Pt} H_3C{\cdot}CH_3$$

The alkene and hydrogen are both adsorbed at the catalyst surface; alignment of reactants on the catalyst surface leads to **cis** addition (Chapter 2).

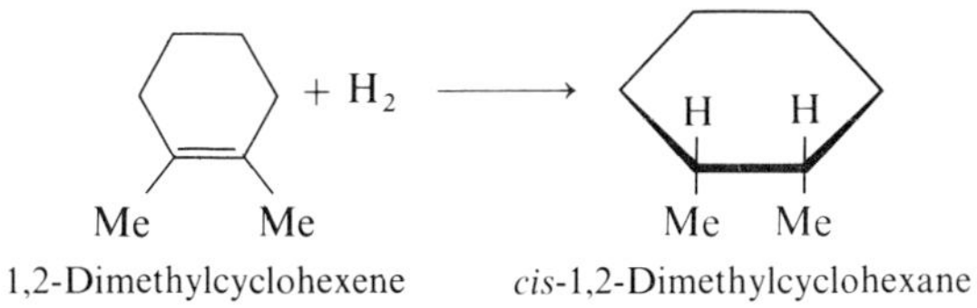

1,2-Dimethylcyclohexene $+ H_2 \longrightarrow$ *cis*-1,2-Dimethylcyclohexane

Reaction occurs at room temperature with the most active catalysts, platinum, palladium and Raney nickel. Less active catalysts require elevated temperatures and higher pressures.

## Metabolic Reduction

Metabolic reduction of unsaturated compounds occurs in mammalian liver under the influence of a group of enzymes, known as reductases. Saturation of the double bond in $\Delta^4$-3-ketosteroids is a feature of the deactivation of many progestational and corticosteroid hormones prior to excretion (Atherden, 1959; Wettstein, Neher and Urech, 1959; Stenlake, Templeton and Taylor, 1968).

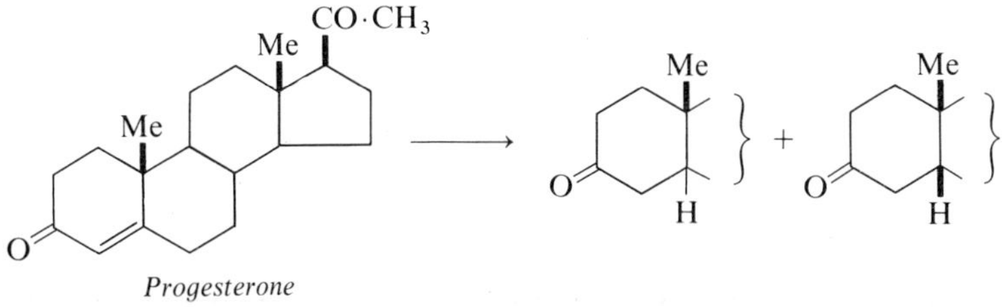

*Progesterone*

Reduction of the ethylenic bond in the $C_{17}$ side-chain of desmosterol, the precursor of cholesterol, has been shown to proceed by a Markownikoff type electrophilic addition of an enzyme-bound proton to the more electron-rich carbon of the double bond to give a carbonium ion, followed by addition of a hydride ion ($H^-$) from the 4-position of NADP$^3$[H] (Akhtar, Munday, Rahimhula, Watkinson and Wilton, 1969).

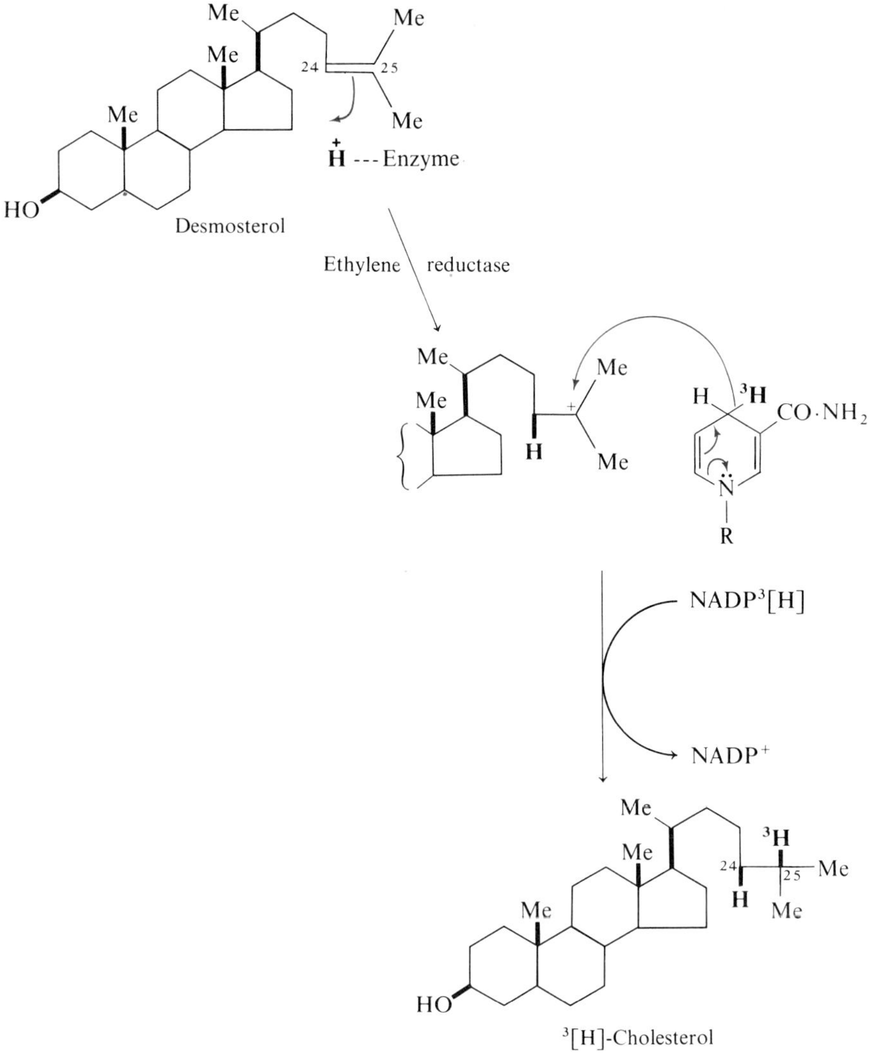

Desmosterol

Ethylene reductase

$NADP^3[H]$

$NADP^+$

$^3$[H]-Cholesterol

## CONJUGATED DIENES

Multiple unsaturation in hydrocarbons gives rise to compounds in which the double bonds are either isolated as in penta-1,4-diene or, alternatively, conjugated as in butadiene and penta-1,3-diene.

$$\overset{1}{H_2C}{=}\overset{2}{C}H{\cdot}\overset{3}{C}H_2{\cdot}\overset{4}{C}H{=}\overset{5}{C}H_2 \qquad H_2C{=}CH{\cdot}CH{=}CH_2 \qquad \overset{1}{H_2C}{=}\overset{2}{C}H{\cdot}\overset{3}{C}H{=}\overset{4}{C}H{\cdot}\overset{5}{C}H_3$$

Penta-1,4-diene    Butadiene    Penta-1,3-diene

Compounds, such as penta-1,4-diene, behave in every respect as molecules with two independent double bonds. The C=C and C—C bond lengths are the same as in simpler systems, i.e. 1.34 and 1.54 Å respectively. The alternating system of single and double bonds in butadiene and penta-1,3-diene, known as **conjugation**, permits a considerable measure of orbital overlap, with a consequent modification of properties. $sp^2$ Hybridisation leads to the formation of a stable planar arrangement of C—C and C—H bonds.

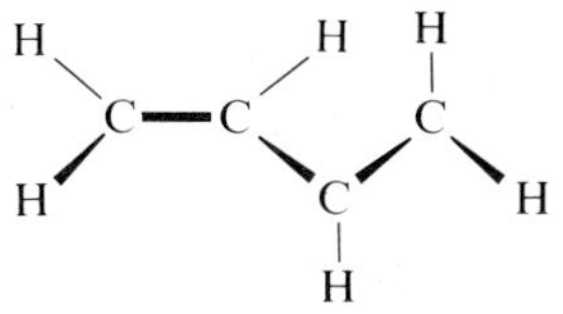

Sideways overlap of the four unhybridised $2p$ orbitals (one derived from each of the participating carbon atoms), which lie at right angles to the main axis of the molecule, gives rise to a new $\pi$-orbital.

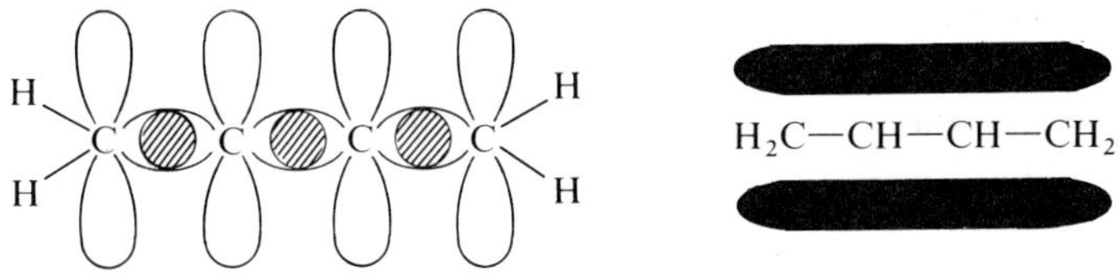

$H_2C—CH—CH—CH_2$

$$H_2C \underset{1.38\,Å}{=} CH \overset{1.48\,Å}{—} CH{=}CH_2$$

This orbital delocalises the four $p$-electrons, spreading their influence, though still unevenly, across all four carbon atoms in the form of two electron clouds, one above and the other below the plane of the molecule. This effect is reflected in the carbon–carbon bond distances in butadiene. Thus, the two terminal double bonds are somewhat longer (1.38 Å) than a normal double bond and the single bond is somewhat shorter (1.48 Å) than a normal single bond. Also, the heat of formation of butadiene is marginally greater by 15.5 kJ mol$^{-1}$ than that calculated, a difference which represents a measure of the increased resonance stabilisation achieved by conjugation.

### Addition of Bromine

Conjugated dienes are characterised by a readiness to participate in reactions which lead to 1,4-addition products. In some reactions 1,4-addition occurs exclusively; in others, 1,2-addition may also occur to an extent determined by the actual reaction conditions. 1,4-Addition of bromine to butadiene gives mainly *trans*-1,4-dibromobut-2-ene together with some 3,4-dibromobut-2-ene. At low temperature (−15°C), both products are produced in approximately equal amounts, as a consequence of a reaction mechanism which provides for two possible carbonium ion intermediates.

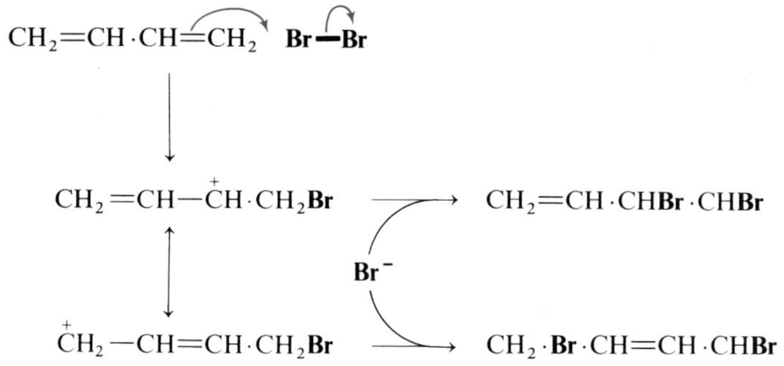

$$CH_2{=}CH{\cdot}CH{=}CH_2 \quad Br{-}Br$$

$$\downarrow$$

$$CH_2{=}CH{-}\overset{+}{C}H{\cdot}CH_2Br \xrightarrow{Br^-} CH_2{=}CH{\cdot}CHBr{\cdot}CHBr$$

$$\updownarrow$$

$$\overset{+}{C}H_2{-}CH{=}CH{\cdot}CH_2Br \xrightarrow{Br^-} CH_2{\cdot}Br{\cdot}CH{=}CH{\cdot}CHBr$$

The two products are interconvertible at higher temperatures (200°C), and at equilibrium, the 1,4-addition product predominates (70%).

**Addition of Halogen Acids**

A similar mechanism of predominantly 1,4-addition applies with halogen acids and conjugated dienes. The reaction is initiated by addition of a proton to the terminal double bond, followed by 2- or 4-addition of the halide ion.

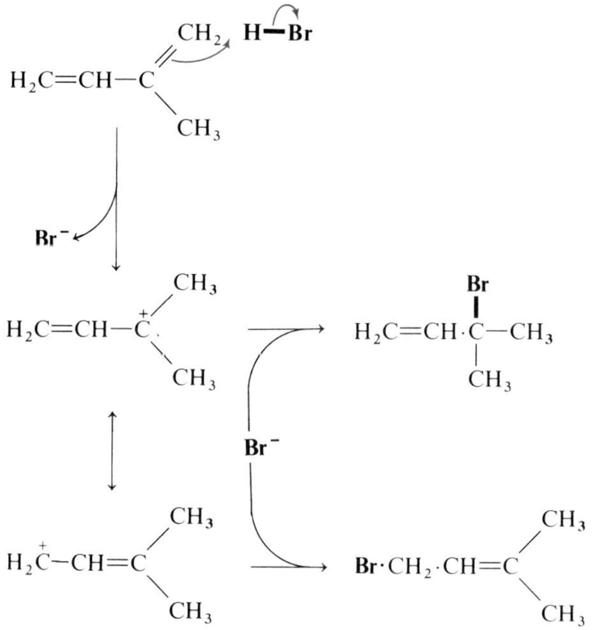

$$H_2C{=}CH{-}C({=}CH_2)(CH_3) \quad H{-}Br$$

$$\downarrow \; (Br^-)$$

$$H_2C{=}CH{-}\overset{+}{C}(CH_3)_2 \xrightarrow{Br^-} H_2C{=}CH{\cdot}CBr(CH_3){-}CH_3$$

$$\updownarrow$$

$$H_2\overset{+}{C}{-}CH{=}C(CH_3)_2 \xrightarrow{Br^-} Br{\cdot}CH_2{\cdot}CH{=}C(CH_3)_2$$

**Diels–Alder Reaction**

Conjugated dienes react by cyclo-addition with suitably substituted ethylenic compounds to form carbocyclic adducts.

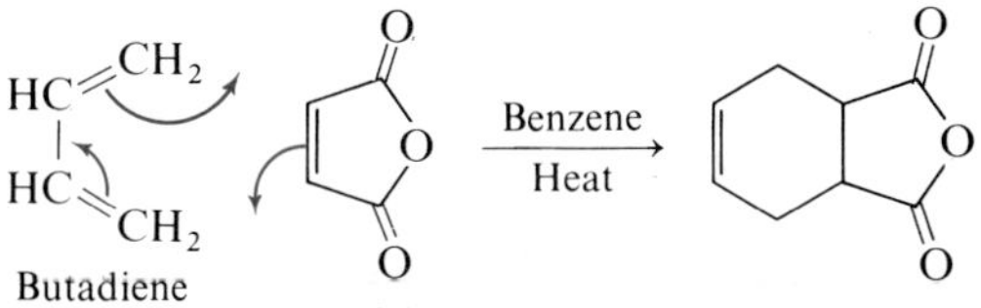

Butadiene — Maleic anhydride —(Benzene, Heat)→ Adduct

DIENE DIENOPHILE ADDUCT

The reaction is promoted by electron donating substituents (e.g. $CH_3$) in the diene, and requires that the dienophile be bonded to an electron attracting substituent (e.g. COOH, COOR, CHO, COR, CN). The reaction is useful in the preparation of bridged ring systems, and is stereospecific for the formation of addition products with the *endo* (as opposed to the *exo*) configuration.

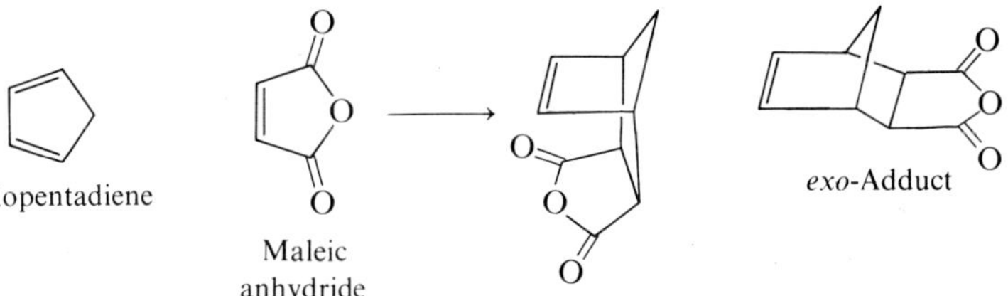

Cyclopentadiene — Maleic anhydride → *endo*-Adduct; *exo*-Adduct

**Radical-induced Rearrangements**

A number of pharmaceutically important dienes and polyenes including *Calciferol*, *Dihydrotachysterol*, *Retinol* and *Dienoestrol* are particularly sensitive to radical-induced reactions which include oxidation, polymerisation and rearrangement. They are all required to be stored in the absence of light, and the majority also in sealed containers under nitrogen, in order to inhibit such changes. Thus, over-exposure of *Calciferol* to ultraviolet light leads to the formation of suprasterol-II (Dauben, Bell, Hutton, Laws, Rheiner and Urscheler, 1958), which has no antirachitic activity.

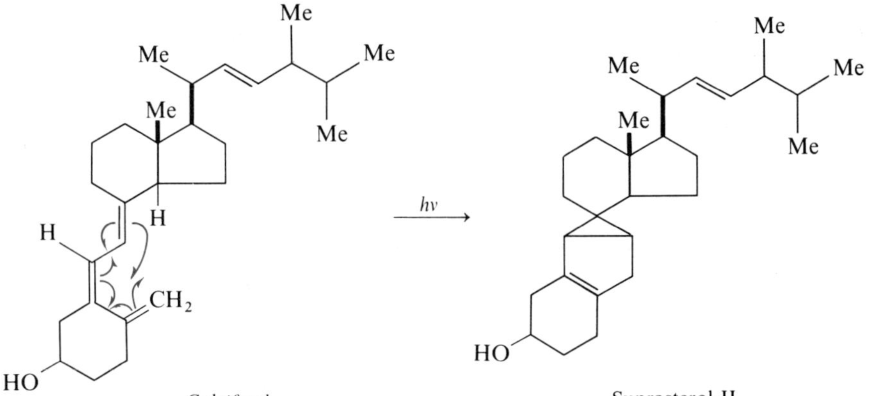

*Calciferol* —$h\nu$→ Suprasterol-II

**Polymerisation**

Butadiene and isoprene undergo self-polymerisation when heated with sodium at 60°C to form a synthetic rubber.

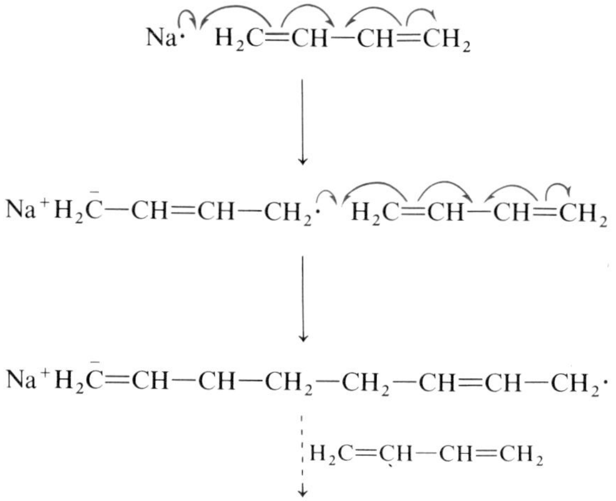

$$Na\cdot \quad H_2C{=}CH{-}CH{=}CH_2$$

$$\downarrow$$

$$Na^+H_2\bar{C}{-}CH{=}CH{-}CH_2\cdot \quad H_2C{=}CH{-}CH{=}CH_2$$

$$\downarrow$$

$$Na^+H_2\bar{C}{=}CH{-}CH{-}CH_2{-}CH_2{-}CH{=}CH{-}CH_2\cdot$$

$$\downarrow H_2C{=}CH{-}CH{=}CH_2$$

Natural rubber is similarly considered to be formed by 1,4-polymerisation of isoprene. The natural biological precursor, mevalonic acid, is present in the latex of the rubber tree, and in the conversion of latex to rubber undergoes decarboxylation, dehydration and head-to-tail polymerisation (Park and Bonner, 1958; Keckwick, Archer, Barnard, Higgins, McSweeney and Moore, 1959).

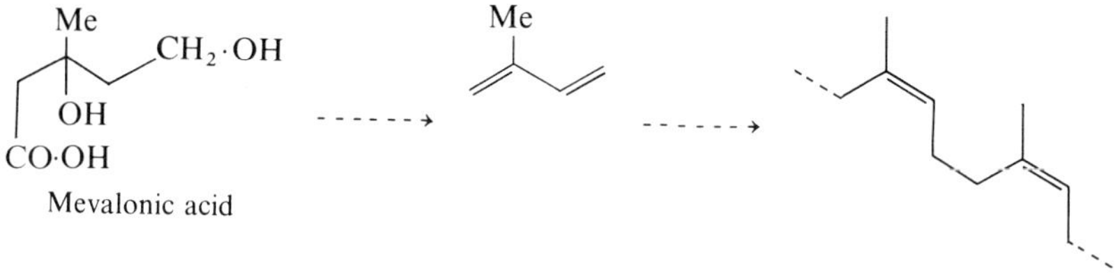

Me, $CH_2\cdot OH$, OH, CO·OH ------→ Me ------→

Mevalonic acid

Mevalonic acid (Wolf, Hoffman, Aldrich, Skeggs, Wright and Folkers, 1956, 1957), the biological equivalent of the isoprene unit, is the essential biosynthetic intermediate between acetate and both terpenes and steroids (Chapter 22). It is formed biosynthetically from acetate and acetoacetate via $\beta$-hydroxygluturate.

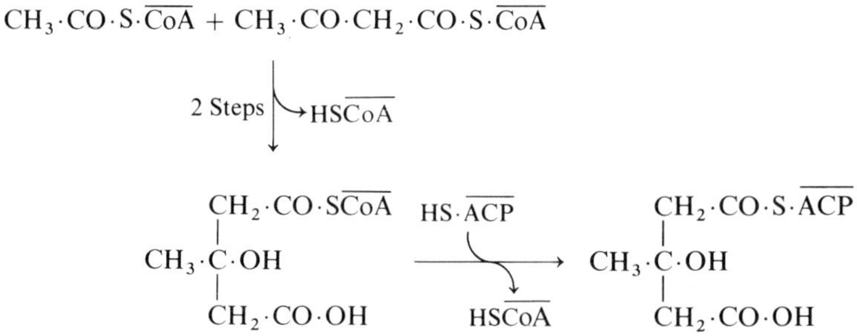

$$CH_3\cdot CO\cdot S\cdot \overline{CoA} + CH_3\cdot CO\cdot CH_2\cdot CO\cdot S\cdot \overline{CoA}$$

$$\downarrow \text{2 Steps} \rightarrow HS\overline{CoA}$$

$$CH_3\cdot \overset{\displaystyle CH_2\cdot CO\cdot S\overline{CoA}}{\underset{\displaystyle CH_2\cdot CO\cdot OH}{C\cdot OH}} \xrightarrow[HS\overline{CoA}]{HS\cdot \overline{ACP}} CH_3\cdot \overset{\displaystyle CH_2\cdot CO\cdot S\cdot \overline{ACP}}{\underset{\displaystyle CH_2\cdot CO\cdot OH}{C\cdot OH}}$$

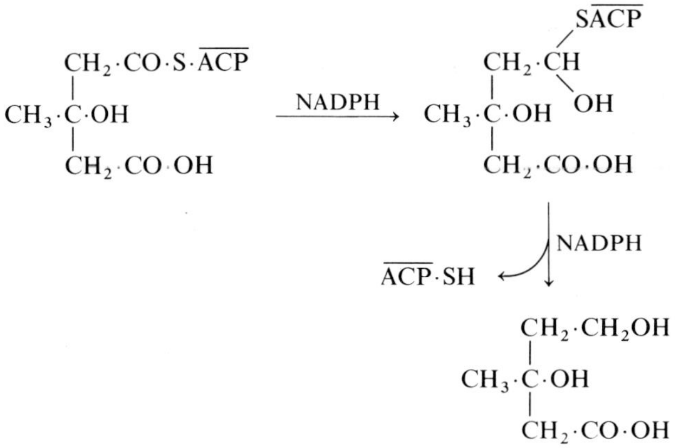

$$\begin{array}{l} CH_2\cdot CO\cdot S\cdot \overline{ACP} \\ | \\ CH_3\cdot C\cdot OH \\ | \\ CH_2\cdot CO\cdot OH \end{array} \xrightarrow{NADPH} \begin{array}{l} CH_2\cdot CH(S\overline{ACP})(OH) \\ | \\ CH_3\cdot C\cdot OH \\ | \\ CH_2\cdot CO\cdot OH \end{array}$$

$$\xrightarrow[-\ \overline{ACP}\cdot SH]{NADPH} \begin{array}{l} CH_2\cdot CH_2OH \\ | \\ CH_3\cdot C\cdot OH \\ | \\ CH_2\cdot CO\cdot OH \end{array}$$

$\overline{ACP}$ = Acyl carrier protein

Mevalonic acid

Naturally occurring terpene hydrocarbons and their derivatives, such as geraneol, are formed biosynthetically from mevalonic acid via isopentenyl pyrophosphate and its isomer, αα-dimethylallyl pyrophosphate.

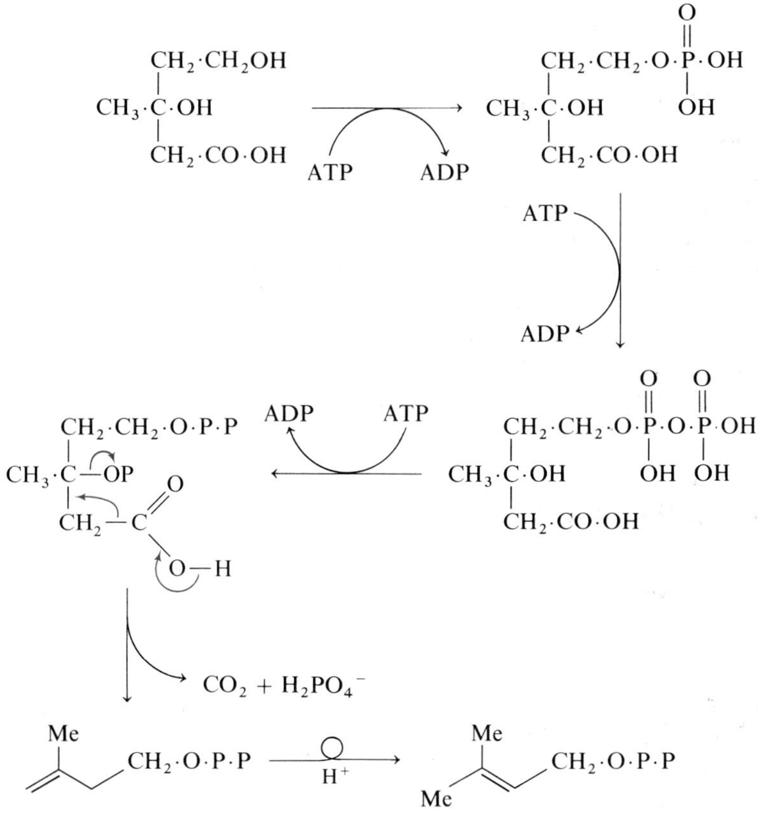

$$\begin{array}{l} CH_2\cdot CH_2OH \\ | \\ CH_3\cdot C\cdot OH \\ | \\ CH_2\cdot CO\cdot OH \end{array} \xrightarrow[ATP \rightarrow ADP]{} \begin{array}{l} CH_2\cdot CH_2\cdot O\cdot P(=O)(OH)\cdot OH \\ | \\ CH_3\cdot C\cdot OH \\ | \\ CH_2\cdot CO\cdot OH \end{array}$$

$$\xrightarrow[ATP \rightarrow ADP]{} \begin{array}{l} CH_2\cdot CH_2\cdot O\cdot P(=O)(OH)\cdot O\cdot P(=O)(OH)\cdot OH \\ | \\ CH_3\cdot C\cdot OH \\ | \\ CH_2\cdot CO\cdot OH \end{array}$$

$$\xrightarrow[ATP \rightarrow ADP]{} \begin{array}{l} CH_2\cdot CH_2\cdot O\cdot P\cdot P \\ | \\ CH_3\cdot C - OP \\ | \\ CH_2 - C(=O) - O - H \end{array} \longrightarrow CO_2 + H_2PO_4^-$$

$$CH_2{=}C(Me)CH_2\cdot CH_2\cdot O\cdot P\cdot P \xrightarrow[H^+]{\ominus} Me_2C{=}CH\cdot CH_2\cdot O\cdot P\cdot P$$

Isopentenyl pyrophosphate

α,α-Dimethylallyl pyrophosphate

Geranyl pyrophosphate is formed from isopentenyl pyrophosphate and dimethylallyl pyrophosphate with elimination of pyrophosphate, possibly via a carbonium ion intermediate.

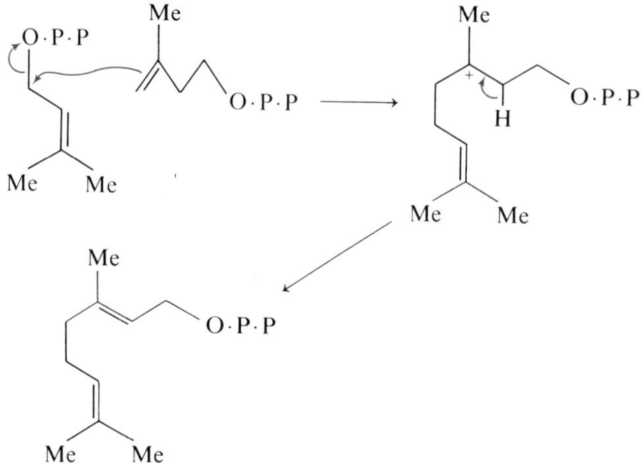

Geranyl pyrophosphate

## CAROTENOIDS AND VITAMINS A

### Carotenes

The carotenoids form a group of fat-soluble pigments occurring naturally in both plant and animal kingdoms. Carotenes are polyethylenic, polyisoprenoid hydrocarbons of which *β*-carotene is an important precursor of vitamin A (Moore, 1930) and the visual pigments. The majority of carotenes are $C_{40}$ structures, consisting of a central fully conjugated $C_{22}$ branched nonaene attached to two $C_9$ unsaturated terminal structures, which may be either cyclic or acyclic. Only those with the *β*-ionone type (5,6-double bond) rings as in *β*-carotene are able to act as precursors of vitamin A.

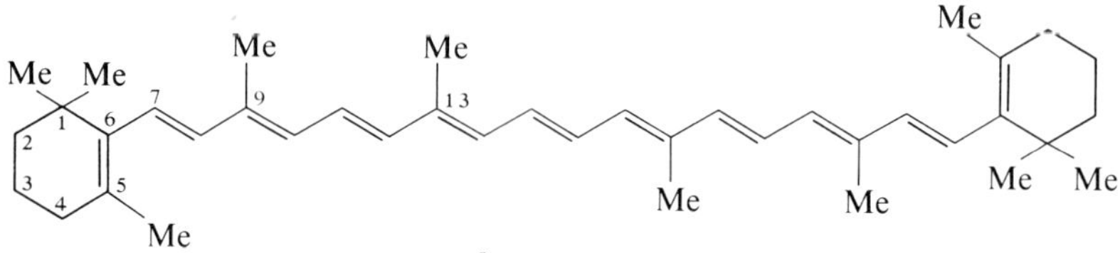

*β*-carotene

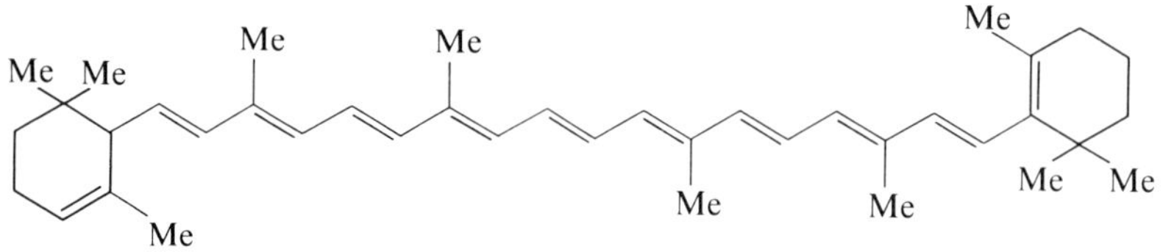

*α*-carotene

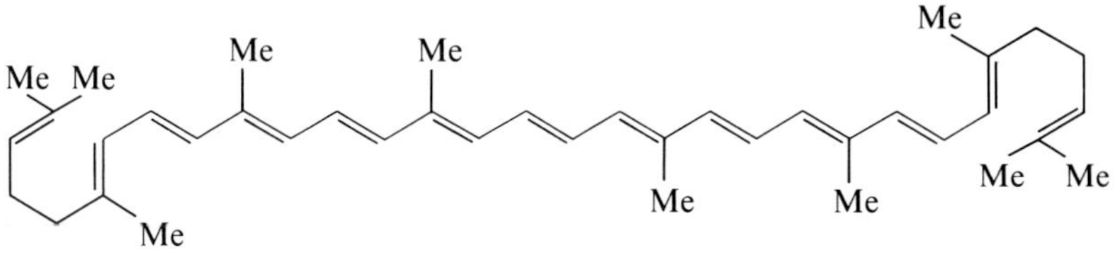

Lycopene

Both light absorption and NMR experiments indicate that the ring 5,6-double bond of β-carotene and vitamin $A_1$ is not co-planar with the polyene chain. Two conformations about the 6,7-single bond are possible *s-cis* and *s-trans*. X-Ray studies also suggest that *all-trans*-vitamin $A_1$ acid (retinoic acid) adopts the *s-cis*-conformation in the crystalline state (Stam and MacGillvary, 1963).

**Retinol** (Vitamin $A_1$)

Vitamin $A_1$ occurs naturally both as the free alcohol (*Retinol*) or as its esters. It is present in most dairy products, and in much greater concentration in certain fish-liver oils, notably cod, halibut and tuna. In man, vitamin A is acquired as such or esterified from fat sources in the diet, and additionally from dietary carotenoid precursors. Conversion of carotenes to vitamin A occurs in the intestinal wall in mammals (Glover, Goodwin and Morton, 1948), from whence it passes to the liver, where it is stored mainly as the palmitate. Molecular oxygen is essential for the oxidation which proceeds with the quantitative formation of retinal (Goodman and Huang, 1965), and subsequent reduction of the latter to *Retinol*.

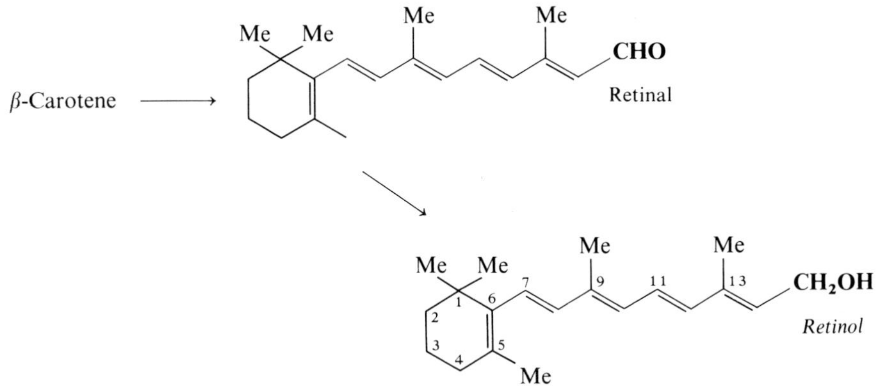

β-Carotene → Retinal → *Retinol*

Vitamin A is mainly available for medicinal purposes as a constituent of fish-liver oils, or in the form of a vitamin A ester concentrate consisting of natural or synthetic esters in solution in *Arachis Oil* (Peanut Oil). Crystalline *all-trans*-vitamin $A_1$ and vitamin $A_1$ acetate are now available by synthesis

(Isler, Huber, Ronco and Kofler, 1947). Vitamin $A_2$ (3,4-dehydroretinol) and 3,4-dehydroretinal, both naturally occurring vitamins, retinoic acid (vitamin $A_1$ acid) and various *cis*-isomers of *Retinol*, including the hindered 11-*cis*- and 11,13-di-*cis*- and the unhindered 9-*cis*-,13-*cis*- and 9,13-di-*cis*-isomers of retinol, have all been synthesised. Retinoic acid has an activity comparable with *Retinol* as a growth vitamin in the rat, but is not reduced *in vivo* to either retinal or *Retinol* (Arens and van Dorp, 1946), and consequently does not alleviate eye disorders resulting from retinal deprivation. Vitamin $A_2$ has only about 30–40% of the vitamin activity of *Retinol*, and like the latter is very sensitive to atmospheric oxidation.

**Properties of Retinol**

Pure *all-trans* vitamin $A_1$ (*Retinol*) is a crystalline solid, which is readily soluble in fats and oils. As an allylic alcohol, it is sensitive to both chemical and enzymic oxidation. Thus, it is readily converted to retinal by such chemical oxidants as manganese dioxide (Ball, Goodwin and Morton, 1948) and *in vivo* by alcohol dehydrogenase with $NA\overset{+}{D}$ as coenzyme. Vitamin $A_1$ acetate is more resistant to oxidation. *Retinol* is also sensitive to oxidation by air with epoxide formation. For this reason, vitamin A preparations should be stored under nitrogen and in hermetically-sealed containers. Epoxidation occurs predominantly, though not exclusively, at the 5,6-double bond. The epoxide is particularly sensitive to even traces of acid, undergoing rearrangement to a furanoid oxide. (See scheme on opposite page.)

Even in the total absence of air, vitamin $A_1$ forms an inactive dimer, kitol, on exposure to sunlight (Embree and Shantz, 1943). Kitol is only reconverted to the monomer by heat.

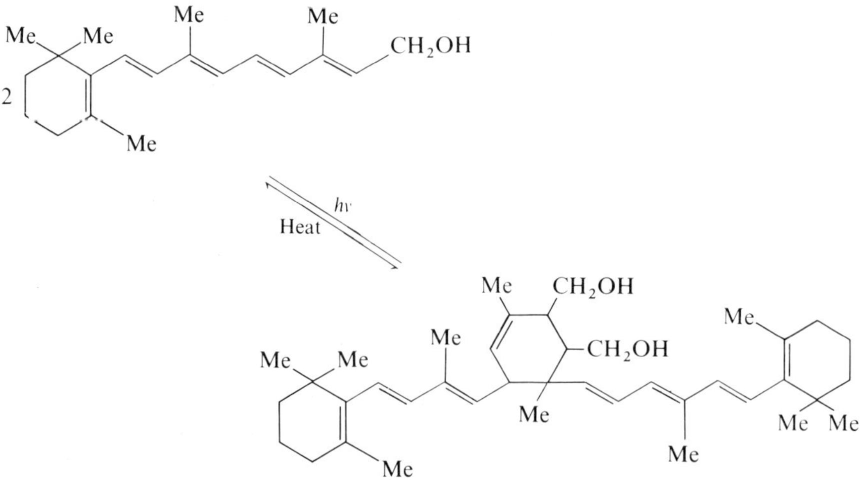

Me Me Me Me $CH_2OH$ 2 Me

$h\nu$ Heat

Me $CH_2OH$ Me Me $CH_2OH$ Me Me Me Me Me Me Me Me

Kitol

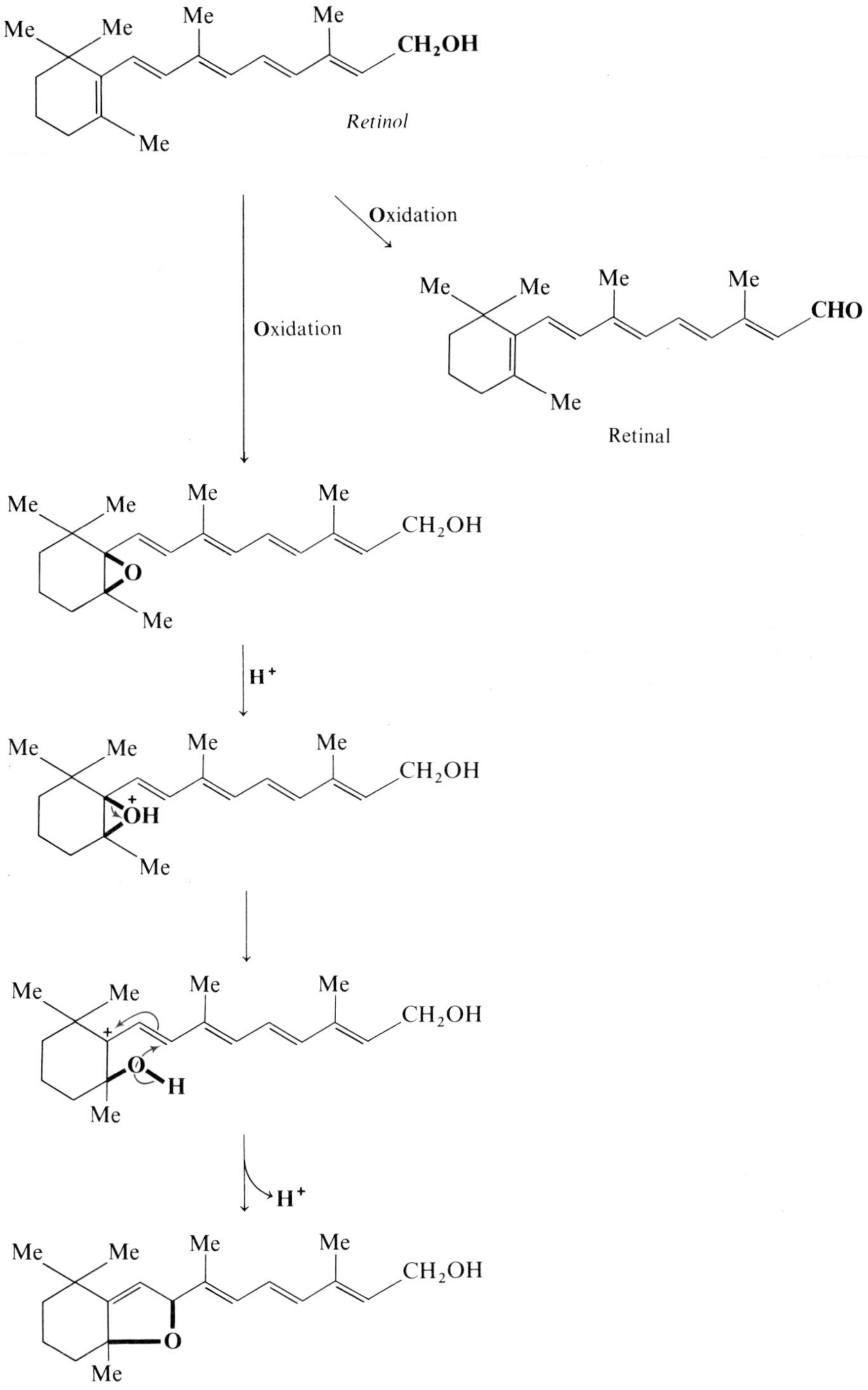

Me Me Me Me $CH_2OH$

Me

*Retinol*

**O**xidation

**O**xidation

Me Me Me Me CHO

Me

Retinal

Me Me Me Me $CH_2OH$

O

Me

$H^+$

Me Me Me Me $CH_2OH$

$\overset{+}{O}H$

Me

Me Me Me Me $CH_2OH$

+

O H

Me

$H^+$

Me Me Me Me $CH_2OH$

O

Me

Kitol is clearly analogous to the Diels–Alder adducts which *all-trans*-vitamin $A_1$ acetate readily forms with suitably activated ethylenic compounds (maleic anhydride in ether) even at ambient temperatures.

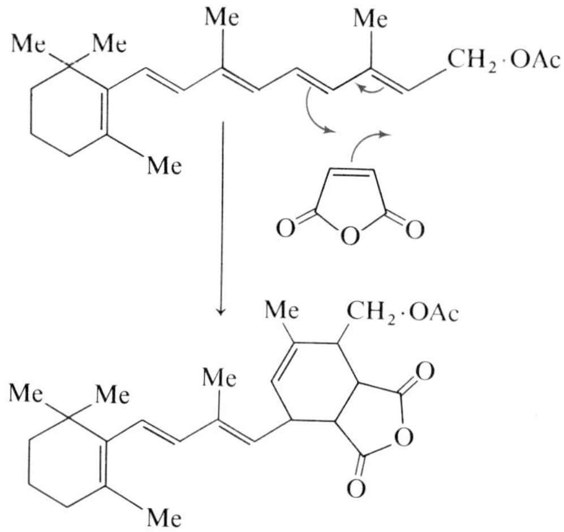

Vitamin A also undergoes spontaneous *trans-cis* isomerism with loss of vitamin A activity on prolonged storage even in the dark and under nitrogen. Two *cis*-isomers are formed with unhindered and hindered double bonds respectively (Zechmeister and Polgár, 1943; Zechmeister, 1960, 1962).

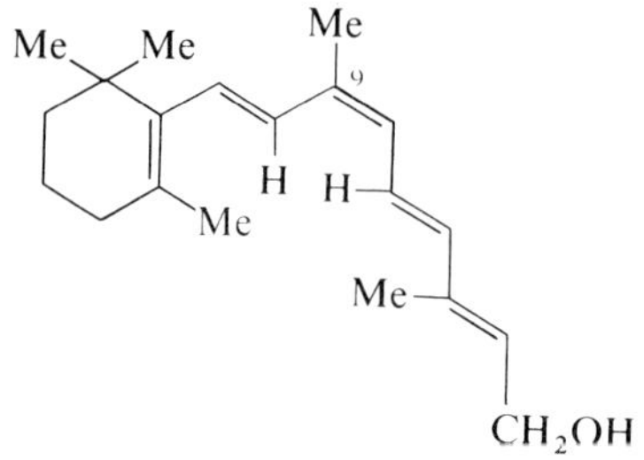

9-*cis*-Retinol
(unhindered)

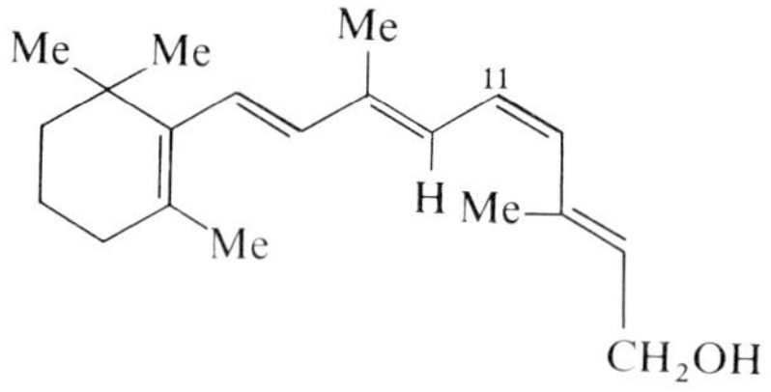

11-*cis*-Retinol
(hindered)

A similar *all-trans* → 11-*cis*-isomerisation is almost certainly involved in the regeneration of 11-*cis*-retinal from *all-trans*-retinal (Futterman, 1963) which is formed in the bleaching of the visual pigments (**2**, 3).

Although less sensitive to acid than retinol epoxide, *Retinol*, as an allylic alcohol, is unstable in the presence of acid. Brief contact with acid (anhydrous hydrogen chloride in ethanol) or Lewis acids (antimony trichloride in chloroform) causes dehydration with the formation of anhydrovitamin $A_1$, a polyene with only about 0.4% of vitamin A activity (Hawkins and Hunter, 1944). More prolonged acid treatment with ethanolic hydrogen chloride gives rise to 5-ethoxy-anhydrovitamin $A_1$, with only weak vitamin A activity.

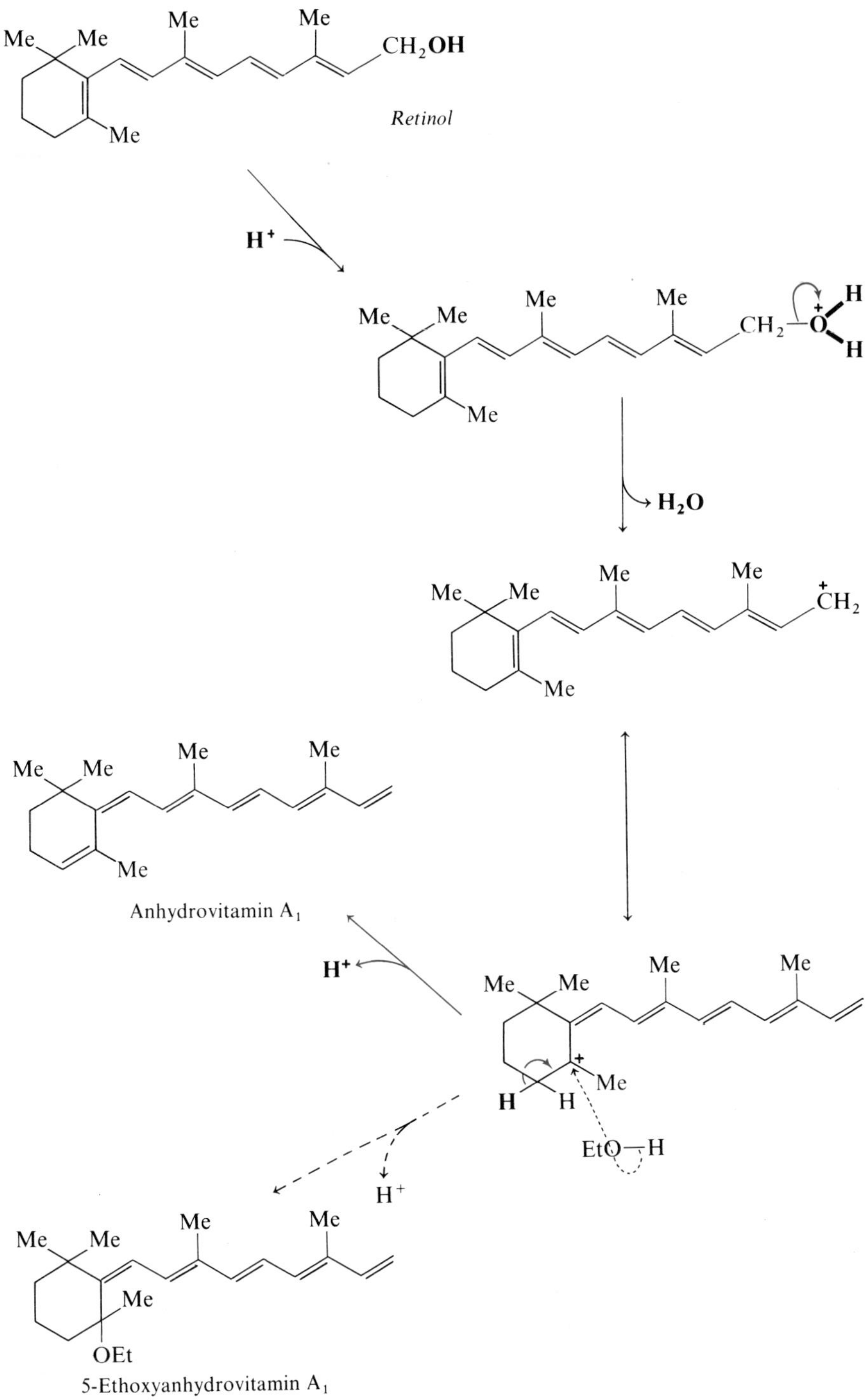

Me Me Me Me $CH_2OH$ Me

*Retinol*

$H^+$

Me Me Me Me $CH_2-\overset{+}{O}H_2$ Me

$H_2O$

Me Me Me Me $\overset{+}{C}H_2$ Me

Me Me Me Me Me

Anhydrovitamin $A_1$

$H^+$

Me Me Me Me + Me H H

EtO—H

$H^+$

Me Me Me Me Me OEt

5-Ethoxyanhydrovitamin $A_1$

# 5 Benzenoid Aromatic Hydrocarbons

## BENZENE

### Structure

Benzene is the parent hydrocarbon of the aromatic series. The term **aromatic** is derived from the description of some of the first compounds of this type to be isolated, which had fragrant or aromatic odours. Benzene has the molecular formula, $C_6H_6$, and hence possesses a carbon/hydrogen ratio which implies a high degree of unsaturation compared with other hydrocarbons. In many respects, however, benzene has much greater stability than the alkenes, although it does show evidence of unsaturation in the uptake of 3 moles of hydrogen under catalytic conditions to yield cyclohexane.

$$\text{Benzene} + 3H_2 \xrightarrow{\text{Ni}} \text{Cyclohexane}$$

Benzene Cyclohexane

The nature of the unsaturation in benzene is quite different from that of the alkenes and dienes, since benzene reacts characteristically by substitution rather than by addition, as for example in the formation of bromobenzene.

$$C_6H_6 + Br_2 \longrightarrow C_6H_5Br + HBr$$

The classical structure of benzene incorporating three double bonds was proposed by Kekulé in 1865. This structure, however, was not readily accepted, since although it satisfied the valency requirements of a compound, $C_6H_6$, and was consistent with the lack of isomerism in monosubstitution compounds, it failed to rationalise a number of other important properties. It did not explain the lack of extreme reactivity which might be expected of a substance with three conjugated double bonds. Thus, benzene does not readily enter into the characteristic addition reactions of the olefines, and is for the most part unaffected by oxidising and reducing agents. More specifically, Ladenburg (1869) argued that Kekulé's formula should give rise to four isomeric disubstitution products (two *ortho*, one *meta* and one *para*), and not three as was generally found.

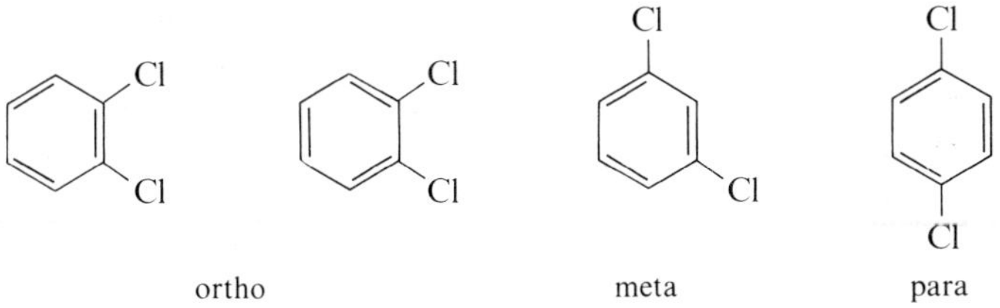

Cl Cl Cl Cl Cl Cl Cl Cl

ortho meta para

This objection was overcome in a revised structure by Kekulé (1872) who postulated a tautomeric equilibrium between two equivalent forms.

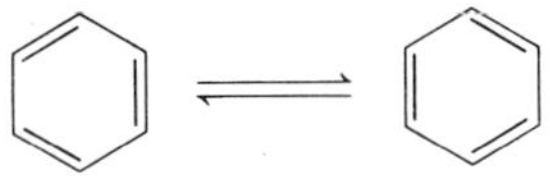

The objection to this formulation, and other alternative proposals such as that of Dewar, remains that they do not account for its hexagonal symmetry, revealed by X-ray analysis, nor for the stability of the system. These difficulties are to a large extent overcome by the concept of resonance which provides for the contribution of a number of extreme Kekulé and Dewar structures in a single resonance hybrid.

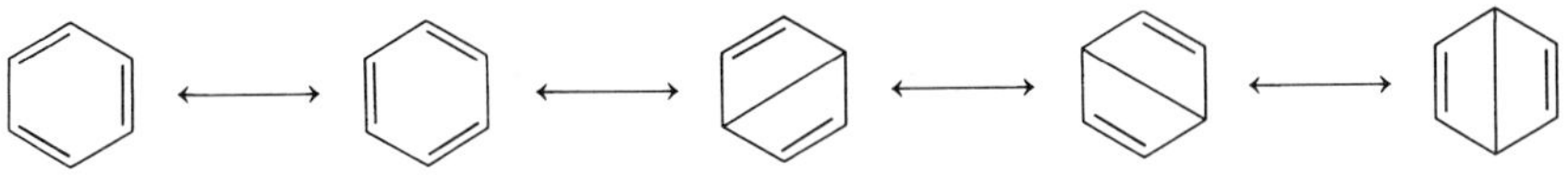

Contributing resonance forms

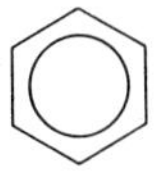

Resonance hybrid structure

The resonance hybrid is not to be regarded as a dynamic equilibrium between the contributing forms, but rather as a single structure formed by hybridisation of all the contributing forms. This concept is supported by evidence from X-ray analysis. This shows that the six carbon atoms are linked in a regular hexagonal manner, that they and the six hydrogen atoms all lie in the same plane, and that the carbon–carbon bonds are all of equal length. The actual carbon–carbon bond length in benzene is 1.40 Å, which is less than that of the carbon–carbon single bond (1.54 Å) and greater than that of the carbon–carbon double bond (1.34 Å). The aromatic carbon–carbon bond distance is, therefore, less than that of the central bond of butadiene (1.48 Å), which, as already discussed, has some double bond character.

The stabilisation achieved in the benzene molecule by resonance is also apparent from thermochemical data. The calculated energy of formation of benzene obtained by summation of the separate bond energies based on the

Kekulé structure is less than the experimentally determined figure by 163 kJ mol$^{-1}$. This is the resonance energy of the molecule. Comparison with the figure of 15.5 kJ mol$^{-1}$ for the resonance energy of butadiene gives an indication of a marked increase in stabilisation arising from the symmetry conferred by the cyclic structure.

The concept of resonance in the benzene molecule is further clarified by the application of orbital theory. The linking of six carbon atoms in a regular and planar cyclic structure provides an almost ideal arrangement for the overlap of the respective atomic orbitals, leading to delocalisation of bonding electrons and hence conferring stability. The essential planar bonding of the ring carbons with each other and hydrogen arises as a result of endways overlap of trigonal $sp^2$ hybrid orbitals which are planar and have a bond angle of 120°. Since the hexagonal angle is also 120°, perfect overlap is achieved and strong $\sigma$-bonds are formed between each carbon atom and the next. The planarity of the $sp^2$ orbital ensures that the C—H bonds are also co-planar with the ring, so that the arrangement of orbital overlap shown in Fig. 16A results.

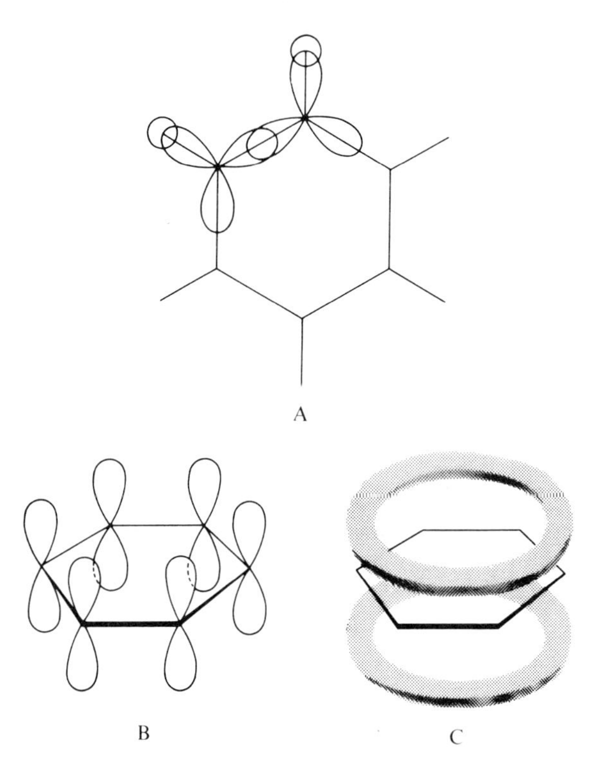


Fig. 16 Orbital overlap and hybridisation in benzene

A Overlap of $sp^2$ orbitals to give $\sigma$ bonds C—C and C—H bonds

B Participating $p$-orbitals

C Hybridisation of $\pi$-orbitals to form the aromatic sextet

The remaining *p* orbitals on each carbon atom, which are situated in a plane at right angles to that of the $sp^2$ orbitals (Fig. 16B), overlap each other sideways to form $\pi$-bonds (Fig. 16C). In so doing, any one electron on a single carbon atom may be considered to come partly under the influence of each of the adjacent carbon atoms, resulting in a large element of delocalisation. Six $\pi$-orbitals are, therefore, formed, the $\pi$-electrons forming two doughnut-shaped clouds, one above and one below the ring. These six electrons, known as the **aromatic sextet** (Robinson), account for the characteristic and so-called aromatic properties of benzene and benzenoid derivatives.

### General Properties

Benzene shows many of the general properties of hydrocarbons. It is a colourless, flammable liquid, which boils at 80.4°C and crystallises at 4.5°C. It has a characteristic odour, and is insoluble in water, a characteristic of aromatic compounds in general. It is miscible in all proportions with ether and light petroleum.

Benzene is highly toxic when absorbed as liquid through the skin and from the gastrointestinal tract, or as vapour via the lungs. It has a high affinity for nervous tissue on account of its lipophilic properties, and owes its toxicity to oxidation in the liver to phenol (*q.v.*), which is a protoplasmic poison. There is also evidence that benzene can give rise to occupational leukaemogenesis in those exposed to its effects over long periods (Vigliani and Forni, 1966). Toluene, which is much less toxic, should therefore be used in preference to benzene as a solvent, wherever this is feasible.

## ELECTROPHILIC SUBSTITUTION OF BENZENE

Delocalisation of the aromatic sextet of electrons provides a rational explanation of the characteristic reaction of aromatic compounds by **electrophilic substitution** rather than by addition. The concentration of electrons above and below the plane of the ring invites attack by cationic reagents ($X^+$) and other electron-deficient species which function as electrophilic reagents. At the same time, the $\pi$-electron clouds above and below the plane of the ring shield the molecule from attack by nucleophilic reagents ($X^-$).

Consideration of the following electrophilic substitutions shows how substitution preserves the delocalisation of the aromatic sextet in contrast to addition, which would limit delocalisation of the remaining $\pi$-electrons to a *cis*-diene system, with consequent loss of resonance stabilisation.

### Nitration

The nitration of benzene is carried out most efficiently with a mixture of concentrated nitric and sulphuric acids. The reaction is slow in the absence of sulphuric acid, which is essential primarily to promote formation from nitric acid of the nitronium ion $\overset{+}{N}O_2$, the effective reagent.

$$HO\cdot NO_2 + H_2SO_4 \rightleftharpoons HSO_4^- + H_2\overset{+}{O}{-}NO_2$$

$$H_2\overset{+}{O}{-}NO_2 + H_2SO_4 \rightleftharpoons HSO_4^- + H_3\overset{+}{O} + \overset{+}{N}O_2$$

Nitration can, therefore, be regarded as an electrophilic attack by the nitronium ion $\overset{+}{N}O_2$ on the aromatic sextet. There is evidence that it occurs stepwise.

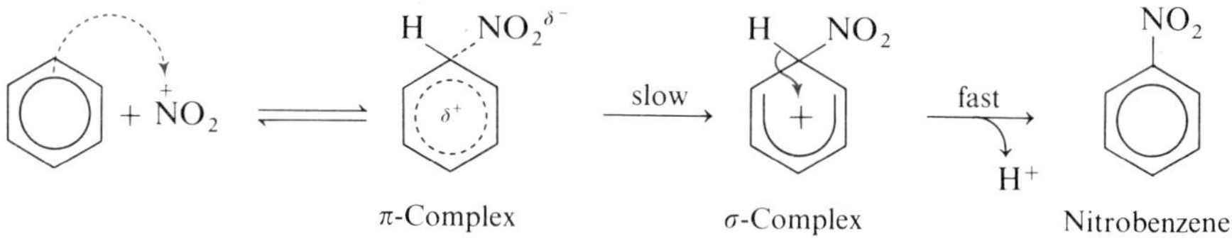

+ $\overset{+}{N}O_2$ ⇌ H $NO_2^{\delta-}$ $\delta^+$ —slow→ H $NO_2$ + —fast→ $H^+$ $NO_2$

π-Complex σ-Complex Nitrobenzene

Interaction between the nitronium ion and the π-electrons leads to the formation of a π-complex. At this stage, the reaction is reversible. As the nitronium ion actually becomes bonded to one of the ring carbon atoms, the so-called σ-complex is formed. Formation of this new bond requires the deployment of two electrons from the aromatic sextet, and the ring is therefore left with a deficiency of charge. The σ-complex as represented is a resonance hybrid which distributes the positive charge over the remaining five carbon atoms by hybridisation of the three contributing forms.

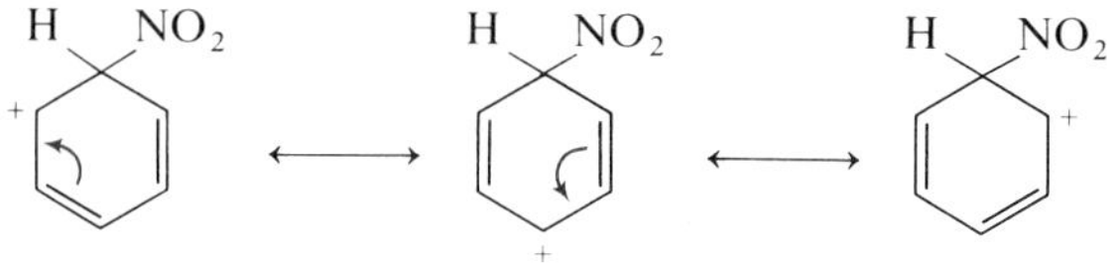

H $NO_2$ + ⟷ H $NO_2$ + ⟷ H $NO_2$ +

This hybrid is best represented by a single structure.

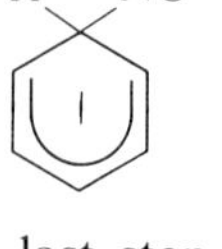

H $NO^2$

Expulsion of the proton, in the last step, restores the fully delocalised π-orbital of the aromatic system, but with substitution of H by $NO_2$. The stability which is achieved when the σ-complex reverts to an aromatic system is a powerful driving force for the expulsion of the proton and completion of the reaction. The proton is removed as $H_2SO_4$ through association with $HSO_4^-$.

**Sulphonation**

Treatment of benzene with hot concentrated sulphuric acid leads to the formation of benzenesulphonic acid by substitution of $-SO_2OH$ for H. This reaction is known as **sulphonation**. The reaction occurs more readily with fuming sulphuric acid, which contains dissolved sulphur trioxide, considered to be the reactive species. It is also present in concentrated sulphuric acid as a result of dissociation.

$$2H_2SO_4 \rightleftharpoons H_3\overset{+}{O} + HSO_4^- + SO_3$$

Sulphur trioxide functions as an electrophilic reagent by virtue of the imbalance of charge in the molecule. Oxygen is more electronegative than sulphur so that bonding electrons are more closely associated with oxygen than sulphur, leaving the sulphur atom electron-deficient.

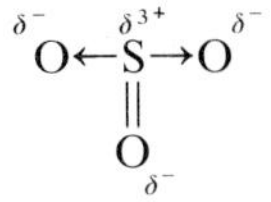

$$\overset{\delta^-}{O}\leftarrow\overset{\delta^{3+}}{S}\rightarrow\overset{\delta^-}{O}, \quad S=\overset{\delta^-}{O}$$

Sulphonation, therefore, results from electrophilic attack by the electron-deficient sulphur atom (Brand, Jarvie and Horning, 1959).

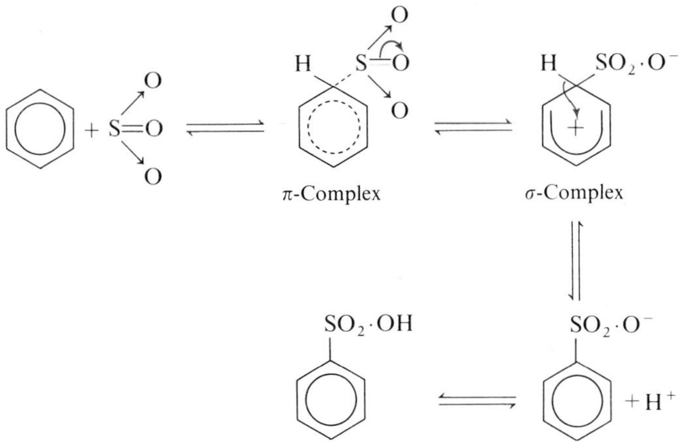

π-Complex

σ-Complex

$SO_2 \cdot OH$

$SO_2 \cdot O^-$

$+H^+$

Benzenesulphonic acid

In contrast to nitration, sulphonation is reversible and benzenesulphonic acid can be converted to benzene on treatment with superheated steam. Use is made of this reversibility to achieve a separation of the xylenes. *m*-Xylene, the most reactive of the three xylenes, dissolves slowly in 80% sulphuric acid at room temperature to yield *m*-xylene-4-sulphonic acid. *o*- and *p*-Xylenes are not attacked under these conditions, and can be separated from the sulphuric acid phase. *m*-Xylene can then be recovered by treatment with dilute sulphuric acid and superheated steam.

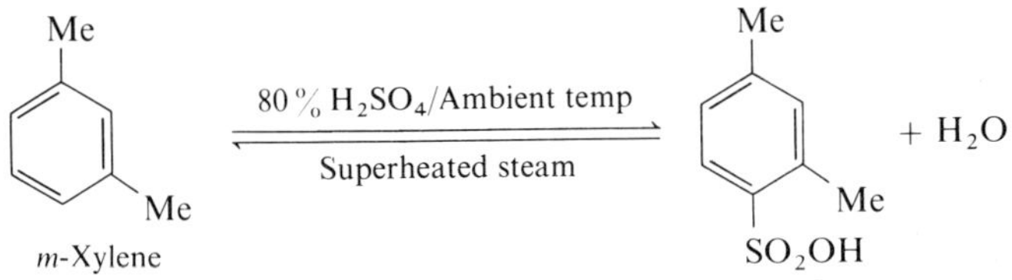

80% $H_2SO_4$/Ambient temp

Superheated steam

$+ H_2O$

Me

Me

*m*-Xylene

Me

Me

$SO_2OH$

*m*-Xylene-4-sulphonic acid

*o*- and *p*-Xylenes are sulphonated in 98% sulphuric acid. The resulting sulphonic acids can be separated, and the parent hydrocarbons recovered from them by steam decomposition as for *m*-xylene.

### Halogenation

Depending upon the reaction conditions, chlorine can be made to undergo either addition to, or substitution in, the benzene nucleus. The electrophilic substitution of aromatic compounds by chlorine and bromine only takes place in the presence of an appropriate catalyst, such as anhydrous zinc chloride, aluminium chloride, aluminium bromide or ferric bromide. The function of the catalyst is to induce polarisation of the halogen, and thereby promote attack by the aromatic $\pi$-electron system at the positive end of the dipole.

$Cl—Cl + AlCl_3$

H Cl→Cl---$AlCl_3$

$\pi$-Complex

$AlCl_4^-$

H Cl

$\sigma$-Complex

$H^+$

Cl

Chlorobenzene

$AlCl_3 + HCl$

Bromination proceeds similarly in the presence of an appropriate catalyst such as $AlBr_3$ or $FeBr_3$. Iodination, however, is only possible in the presence of iodic acid ($HIO_3$) or some other oxidising agent ($HNO_3$ or HgO) capable of destroying the hydriodic acid formed as by-product. If the ring is sufficiently activated by an appropriate substituent (OH or $NH_2$), then iodination is also possible with the inter-halogen compound, iodine monochloride (ICl), the ring being attacked by the less electronegative halogen, i.e. iodine.

### **Acylation** (Friedel–Crafts Reaction)

Benzene reacts with acid chlorides and acid anhydrides in the presence of anhydrous aluminium chloride to form aromatic ketones. The function of the aluminium chloride is to complex with the acid chloride and promote, in effect, formation of an **acylium ion**, $R\cdot\overset{+}{C}{=}O$, which then attacks the benzene ring.

$$CH_3\cdot CO\cdot Cl + AlCl_3 \rightleftharpoons CH_3\cdot\overset{+}{C}O\cdots Cl\cdots\overset{-}{Al}Cl_3$$

$$C_6H_6 + CH_3\cdot\overset{+}{C}O\cdots Cl\cdots\overset{-}{Al}Cl_3 \longrightarrow [\sigma\text{-Complex}] + AlCl_4^- \longrightarrow \text{Acetophenone } (C_6H_5\cdot CO\cdot CH_3) + H^+;\quad AlCl_4^- + H^+ \longrightarrow AlCl_3 + HCl$$

σ-Complex

Acetophenone

Although the function of the aluminium chloride is catalytic, equimolar quantities of aluminium chloride, benzene and acid chloride are required, since the catalyst is removed by complex formation with the product.

$$C_6H_5\cdot CO\cdot CH_3 + AlCl_3 \rightleftharpoons (C_6H_5)(CH_3)C{=}O\cdots AlCl_3$$

Ketone–aluminium chloride 1:1-complex

On completion of the reaction, the ketone is usually isolated by pouring the reaction mixture including solvent (usually carbon disulphide or nitrobenzene) onto a mixture of ice and dilute hydrochloric acid. This destroys the ketone–aluminium chloride complex. The aluminium chloride dissolves in the aqueous phase, and the ketone in the organic phase.

Acylation of benzene occurs equally readily with acid anhydrides, except that a further mole of catalyst is consumed in the formation of the acylium ion.

$$(CH_3\cdot CO)_2O + AlCl_3 \longrightarrow CH_3\cdot\overset{+}{C}O\cdots\overset{-}{Al}Cl_4 + CH_3\cdot CO\cdot O\cdot AlCl_2$$

Introduction of the acyl substituent deactivates the ring (*q.v.*), so that, unlike the comparable Friedel–Crafts alkylation (p. 123), the reaction ceases with the formation of the monosubstitution product. Thus, reaction of phthalic anhydride with benzene gives *o*-benzoylbenzoic acid, and does not proceed to the anthraquinone (Chapter 22).

$$\text{Phthalic anhydride} + C_6H_6 \longrightarrow o\text{-}C_6H_5\cdot CO\cdot C_6H_4\cdot CO\cdot OH$$

Phthalic anhydride

*o*-Benzoylbenzoic acid

## FREE RADICAL REACTIONS OF BENZENE

### Free Radical Addition of Halogens

The $\pi$-electron cloud above and below the plane of the aromatic ring shields the ring from attack by nucleophilic reagents. High energy particles, such as free-radicals, however, are able to penetrate this barrier, and both radical additions and radical substitutions are known.

Addition of chlorine to benzene (as opposed to substitution) can be brought about, in the absence of the usual halogen carriers and oxygen, by passing chlorine into benzene irradiated with ultraviolet light.

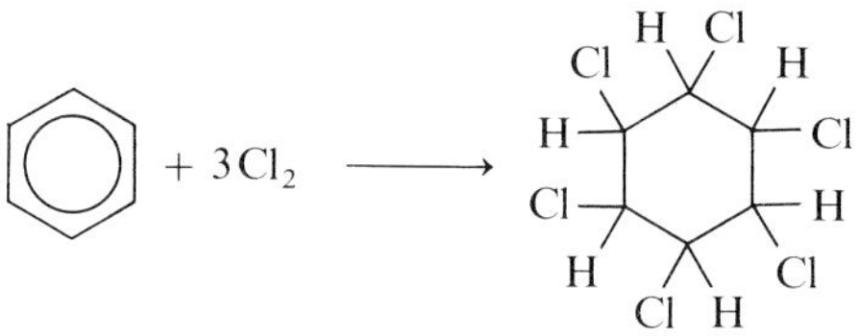

$$C_6H_6 + 3Cl_2 \longrightarrow C_6H_6Cl_6$$

$$Cl_2 \xrightarrow{h\nu} 2Cl\cdot$$

Benzene hexachloride

The reaction is catalysed by peroxides and proceeds through the formation of chlorine free radicals.

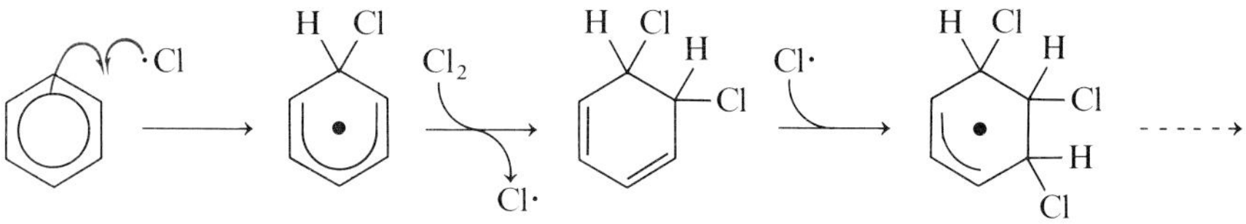

The reaction gives rise to a mixture of six of the nine possible stereoisomers. One, the $\gamma$-isomer, *Gamma Benzene Hexachloride*, which has powerful insecticidal properties, forms about 10% of the product. The stereochemistry of these isomers is discussed elsewhere (**2**, 3).

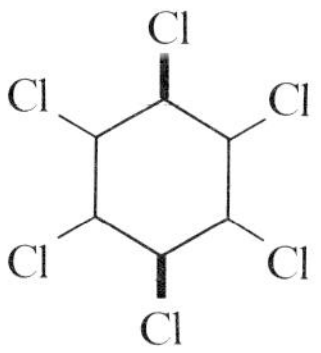

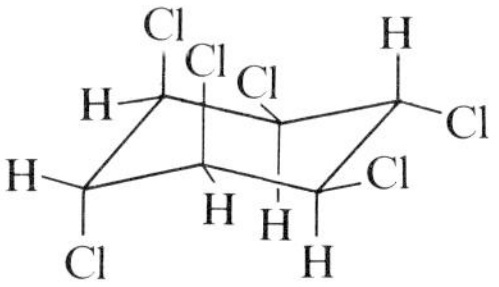

*Gamma Benzene Hexachloride*

### Free Radical Hydroxylation

Hydroxylation of aromatic compounds can be brought about with Fenton's reagent (Dermer and Edmison, 1957), hydrogen peroxide and ferrous salts, which generates hydroxyl radicals (Lindsay-Smith and Norman, 1963).

$$Fe^{2+} \; HO{-}OH \longrightarrow Fe^{3+} + HO^{-} + HO\cdot$$

The hydroxyl radicals attack benzene to give a resonance-stabilised radical intermediate, which expels a hydrogen radical to reform the aromatic sextet so giving rise to phenol.

$$C_6H_6 + \cdot OH \xrightarrow{\text{slow}} [C_6H_6OH]\cdot \xrightarrow[-H_2O]{HO\cdot} C_6H_5OH$$

Phenol

Expulsion of a hydrogen radical from the intermediate leading to aromatisation and ring substitution contrasts with the radical coupling and addition which occurs in the corresponding reaction with chlorine (p. 120). This difference may be explained by the much greater energy of chlorine radicals compared with that of hydroxyl radicals.

The hydroxyl radicals also attack benzene to give phenyl radicals leading to the formation of biphenyl.

$$C_6H_5{-}H + \cdot OH \longrightarrow C_6H_5\cdot + H_2O$$

$$C_6H_6 + C_6H_5\cdot \longrightarrow [C_6H_6C_6H_5]\cdot \xrightarrow[-H_2O]{HO\cdot} C_6H_5{-}C_6H_5$$

Biphenyl

Hydroxylation of benzene can also be brought about in the presence of water by ionising radiation ($\gamma$-rays) from a suitable radioactive source, or on exposure to intense sources of ultraviolet light.

$$H{-}OH \longrightarrow H\cdot + \cdot OH$$

Hydroxyl radicals so formed lead to the hydroxylation of benzene, as above. Phenyl radicals may also react either with hydroxyl or hydrogen radicals. The latter reaction, however, is inhibited in the presence of oxygen.

$$C_6H_5\cdot \xrightarrow{HO\cdot} C_6H_5OH$$

$$C_6H_5\cdot \xrightarrow{H\cdot} C_6H_6 \qquad H\cdot + O_2 \longrightarrow H_2O$$

Hydroxylation reactions of this sort may occur in the use of ionising radiations for the sterilisation of aqueous injections containing drugs which are aromatic in character; they also provide a likely pathway for the decomposition of such solutions on storage, if unnecessary exposure to ultraviolet light occurs.

### Enzymatic Hydroxylation

Enzymatic hydroxylation of aromatic molecules forms an important step in their detoxication by the body. The reaction which is catalysed by oxidases, situated primarily in the liver in mammals, requires molecular oxygen (demonstrated by experiments with $^{18}O$). Only one atom of oxygen is incorporated in the aromatic substituent, the remaining atom being reduced to water. The oxidases require reduced nicotinamide adenine dinucleotide phosphate (NADPH) as coenzyme, and also are pH-dependent.

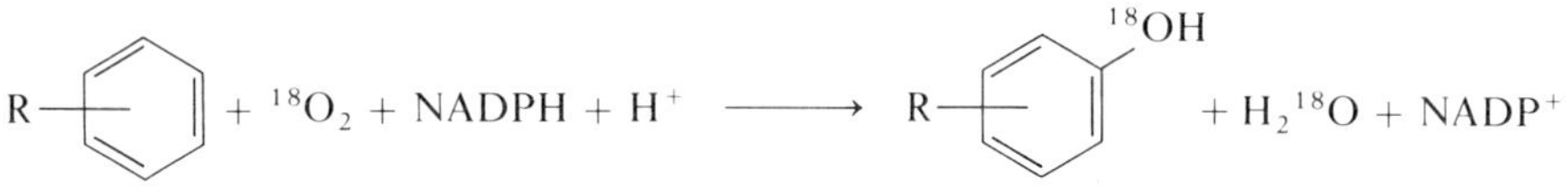

$$R\text{-}C_6H_5 + {}^{18}O_2 + NADPH + H^+ \longrightarrow R\text{-}C_6H_4\text{-}{}^{18}OH + H_2{}^{18}O + NADP^+$$

Benzene is metabolised principally to phenol and catechol by this route via the 1,2-epoxide (Sato, Fukuyama, Suzuki and Yoshikawa, 1963). The mechanism of the NIH shift provides evidence for the 1,2-epoxide intermediate, and the pathway followed in its conversion to both phenol and catechol (**2**, 5).

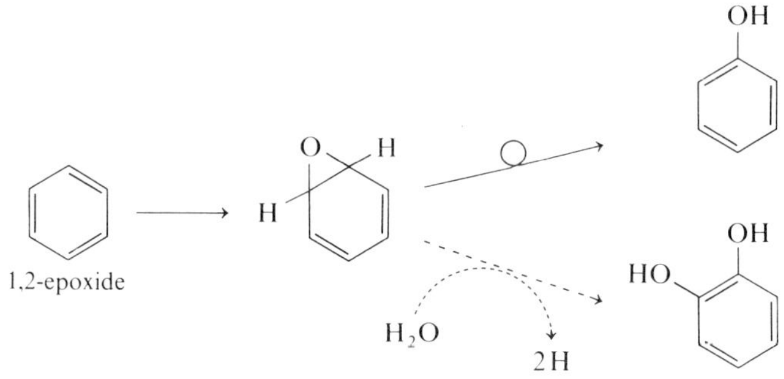

1,2-epoxide

O H H OH HO OH $H_2O$ 2H

## SOURCES OF AROMATIC HYDROCARBONS

### Coal Tar

Aromatic hydrocarbons are formed by destructive distillation of bituminous coal in the absence of air at temperatures between 1000 and 1300°C. The lower temperature is used for coke production; somewhat higher temperatures are used in gas works retorts. The main products of the distillation are:

(a) coke,
(b) coal gas,
(c) aqueous ammoniacal liquors,
(d) coal tar.

Coal gas is purified by passage through scrubbers to remove tar and ammonia, and then through absorption tanks containing high boiling oil for recovery of crude naphtha, a hydrocarbon fraction consisting mainly of benzene and toluene. The crude coal tar is re-distilled and collected in fractions (Table 12). Further fractionation and refinement permits separation of the individual components.

Table 12. Coal Tar Fractions

| Boiling range (°C) | Description | Constituents |
|---|---|---|
| up to 100 | Crude naphtha | Benzene, toluene and xylenes<br>Pyridine, picolines, collidene<br>Thiophene |
| 180–210 | Carbolic acid | Phenol, cresols, xylenols |
| 210–240 | Naphthalene oil | Naphthalene |
| 240–290 | Wash oil | |
| 290–360 | Anthracene oil | Anthracene, phenanthrene, carbazole |
| Residue | Pitch | |

**Petroleum**

Aromatic hydrocarbons are also present in most natural petroleum fractions, but the isolation of benzene and its homologues is seldom undertaken. Large quantities of benzene, toluene and xylenes are now recovered from the catalytic reforming processes, which are used to improve the octane number of petroleum fractions. These aromatic hydrocarbons are formed as a result of the cyclisation of alkanes when they are passed over heated chromium or molybdenum oxide catalysts around 500°C.

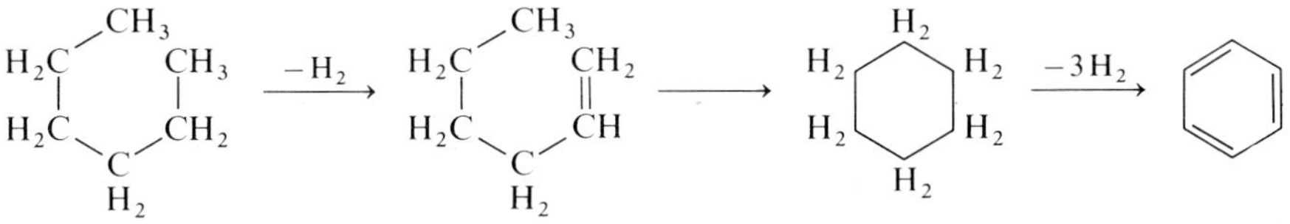

$$\text{H}_2\text{C(CH}_3\text{)—CH}_3 \ldots \xrightarrow{-\text{H}_2} \ldots \longrightarrow \text{cyclohexane} \xrightarrow{-3\text{H}_2} \text{benzene}$$

**Friedel–Crafts Alkylation of Benzene**

Treatment of benzene with an alkyl halide in the presence of anhydrous aluminium chloride leads to alkylation with the formation of a homologue.

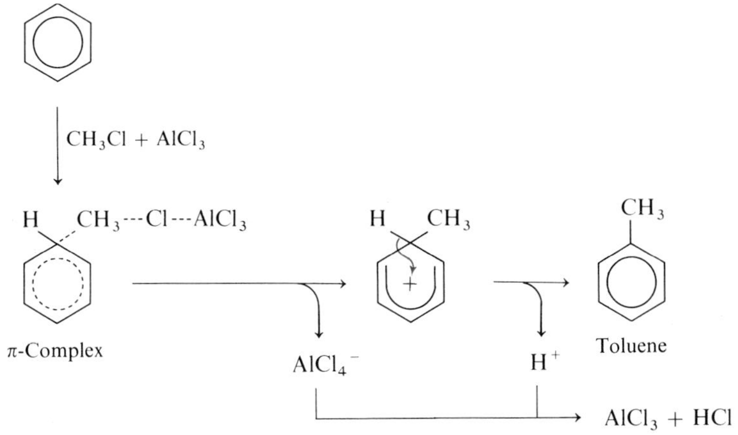

$$\text{C}_6\text{H}_6 \xrightarrow{\text{CH}_3\text{Cl} + \text{AlCl}_3} \text{H} \;\; \text{CH}_3\text{---Cl---AlCl}_3$$

$\pi$-Complex $\longrightarrow$ ($AlCl_4^-$) H CH$_3$ (+) $\longrightarrow$ ($H^+$) Toluene

$AlCl_4^-$ + $H^+$ $\longrightarrow$ $AlCl_3 + HCl$

Nitrobenzene, which is inert in this reaction because of deactivation of the ring (p. 130), is often used as a solvent for alkylation of solid aromatic compounds. The reaction, however, applies generally to aromatic compounds which are not deactivated. The course of the reaction can be influenced by the choice of catalyst. Thus, n-propyl chloride and benzene yield n-propylbenzene with aluminium chloride, but n-propyl bromide in the presence of gallium bromide yields mainly isopropylbenzene. Rearrangements such as the latter reflect the extent to which the initial complex formed by the alkyl halide and the catalyst lead to actual carbonium ion formation.

$$CH_3{\cdot}CH_2{\cdot}CH_2{\cdot}Cl \xrightarrow{AlCl_3} CH_3{\cdot}CH_2{\cdot}CH_2{\cdots}Cl{\cdots}AlCl_3$$

$$CH_3{\cdot}CH_2{\cdot}CH_2{\cdot}Br \xrightarrow{GaBr_3} CH_3{\cdot}CH_2{\cdot}\overset{+}{C}H_2\overset{-}{Ga}Br_4 \longrightarrow CH_3{\cdot}\overset{+}{C}H{\cdot}CH_3\overset{-}{Ga}Br_4$$

Increased electron availability in the ring of the product alkylbenzene (p. 126) means that the product is alkylated more readily than the initial reactant. The reaction is, therefore, difficult to stop at the monosubstitution stage, and polyalkylation often occurs.

Olefines will also function as alkylating agents in the presence of Lewis acids. An important industrial application is the preparation of cumene (isopropylbenzene) from propylene and benzene. Cumene is an important industrial product as it can be readily oxidised to yield phenol and acetone (p. 129), which are required on a large scale in the plastics industry. Alkylation of benzene with ethylene similarly yields ethylbenzene, which is readily dehydrogenated to styrene, another important intermediate for the production of plastics.

$$CH_3{\cdot}CH{=}CH_2 \xrightarrow{AlCl_3} CH_3{\cdot}\overset{+}{C}H{\cdot}CH_2{\cdots}\overset{-}{Al}Cl_3 \xrightarrow{C_6H_6} [\text{arenium ion: H and } CH(CH_3){\cdot}CH_2{\cdots}\overset{-}{Al}Cl_3 \text{ on ring}^{+}] \longrightarrow C_6H_5{\cdot}CH(CH_3)_2 + AlCl_3$$

Cumene

## Biosynthesis of Aromatic Compounds

Biosynthesis of aromatics occurs in many plants by head-to-tail condensation of acetate units, as shown by feeding experiments with 1-$^{14}$[C] acetate.

Another biosynthetic route to aromatic compounds is via shikimic acid, which is formed from carbon dioxide by photosynthesis and carbohydrate metabolism. In man, aromatisation of steroidal androgens to oestrogens occurs in the placenta, and elsewhere. The metabolic pathway is through the 19-hydroxymethyl and 19-formyl derivatives (Longchampt, Gual, Ehrenstein and

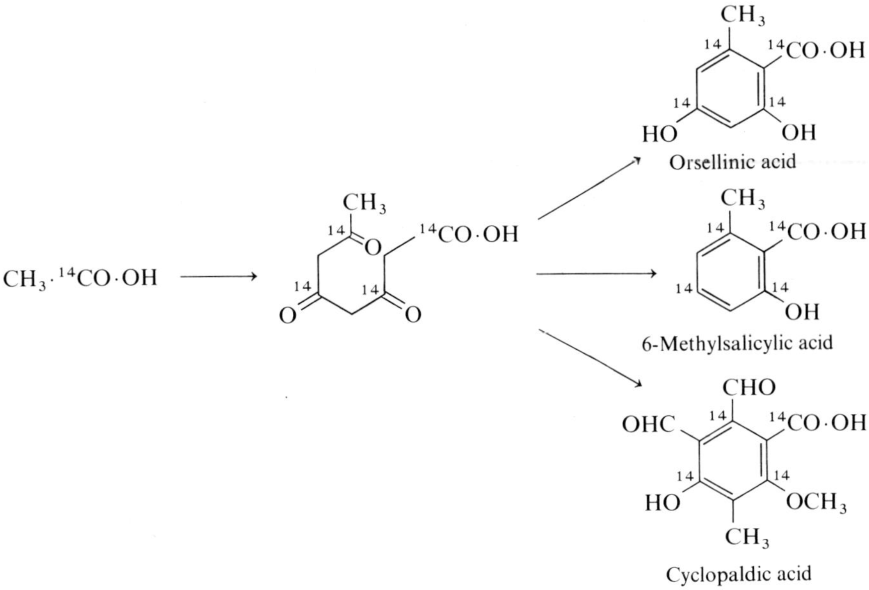

$CH_3 \cdot {}^{14}CO \cdot OH$ ⟶ [triketo acid intermediate: $CH_3$, $^{14}CO \cdot OH$, O, O, O; 14, 14, 14]

Orsellinic acid [$CH_3$, $^{14}CO \cdot OH$, HO, OH; 14, 14, 14]

6-Methylsalicylic acid [$CH_3$, $^{14}CO \cdot OH$, OH; 14, 14, 14]

Cyclopaldic acid [CHO, OHC, $^{14}CO \cdot OH$, HO, $OCH_3$, $CH_3$; 14, 14, 14]

Dorfman, 1960; Morato, Hayano, Dorfman and Axelrod, 1961) since 19-nor-compounds are not readily aromatised.

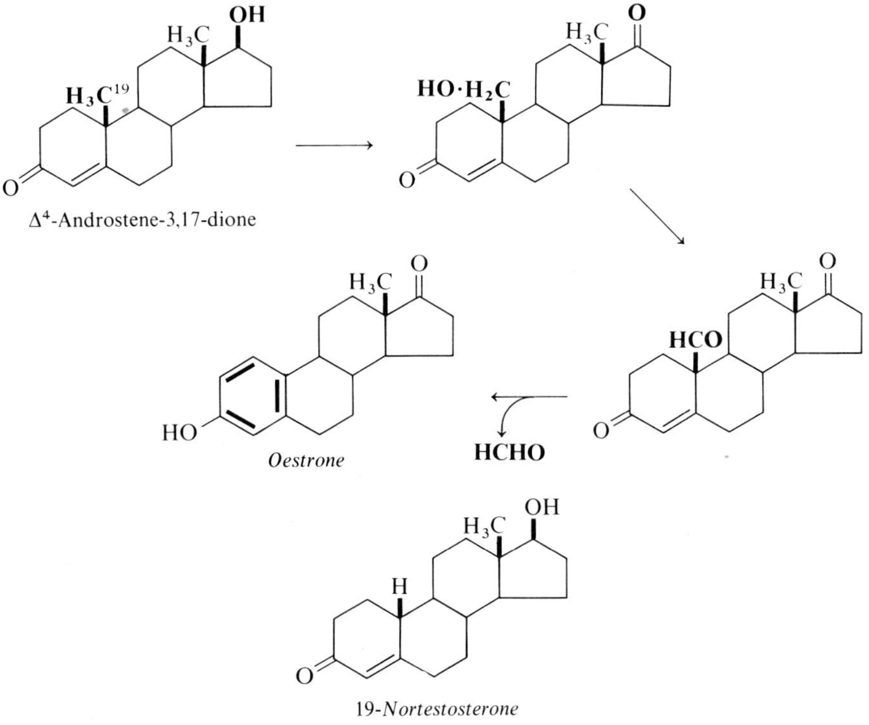

$\Delta^4$-Androstene-3,17-dione [OH, $H_3C$, $H_3C^{19}$, O] ⟶ [O, $H_3C$, $HO \cdot H_2C$, O] ⟶ [O, $H_3C$, HCO, O] ⟶ (HCHO) *Oestrone* [O, $H_3C$, HO]

19-*Nortestosterone* [OH, $H_3C$, H, O]

## ALKYLBENZENES

### Ring Substitution

The presence of an alkyl substituent in the benzene ring increases reactivity of the ring to attack by electrophilic reagents. This is to be expected because alkyl groups are electron-repelling and hence increase electron availability at certain points in the ring by hyperconjugation. Toluene can be considered to be a resonance hybrid of a number of contributing forms.

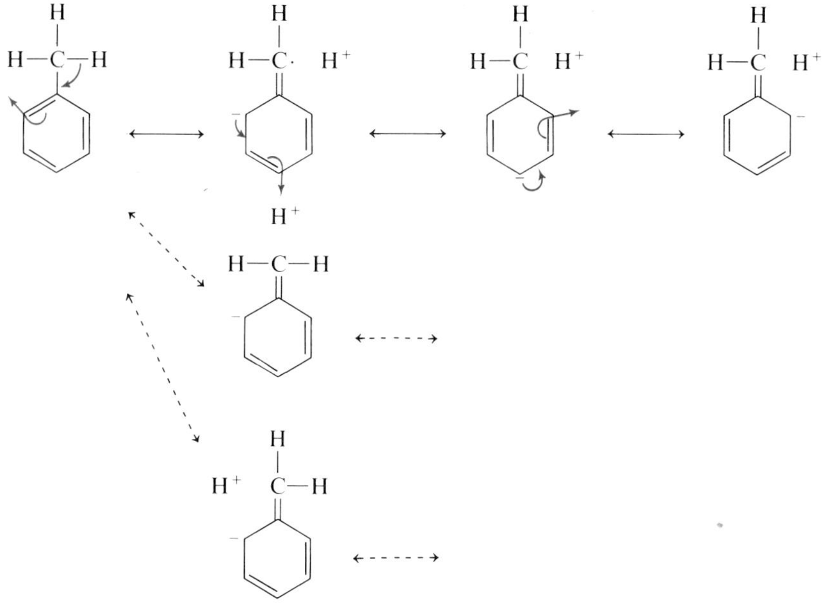

The overall effect of this resonance is that the molecule of toluene has a dipole with the negative end of the dipole centred on the ring, and will experience a turning moment, the dipole moment, of strength $0.3\mu$, when placed in an electric field. This permanent increase of electron density on the ring carbon atoms known as the **resonance** or **mesomeric** effect increases the ease of electrophilic substitution. Hence, toluene nitrates some 25 times faster than benzene.

Since electron availability is greatest in the *ortho* and *para* positions, electrophilic substitution will occur predominantly in these two positions. Theoretically, the *ortho* position is doubly favoured on grounds of electron availability, and yields of isomers should be in the ratio 67% *ortho* to 33% *para*.

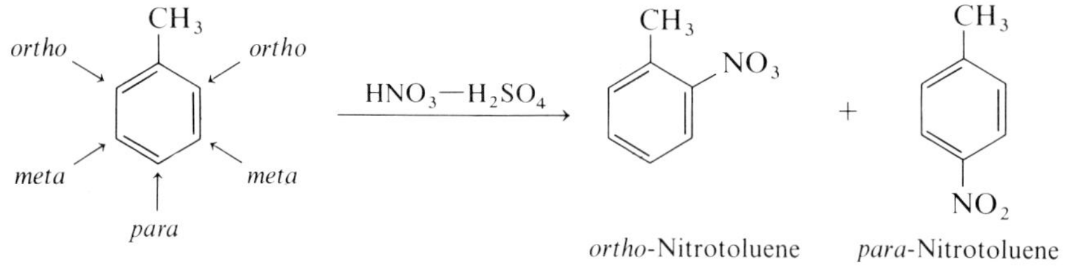

$$\xrightarrow{HNO_3-H_2SO_4}$$

*ortho*-Nitrotoluene + *para*-Nitrotoluene

Steric effects, however, play an important part in fixing the actual ratio of isomers produced, steric hindrance and hence low yields of *ortho*-isomers becoming quite marked with more bulky alkyl substituents. This is clearly seen in the percentage yields of isomers formed by nitration of a series of alkylbenzenes under equivalent conditions.

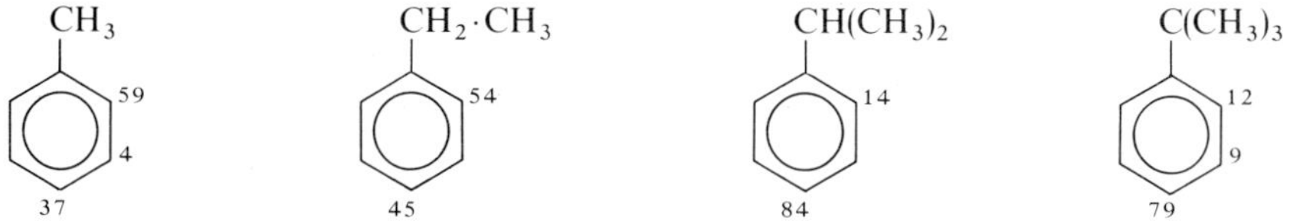

$CH_3$ (ortho 59, meta 4, para 37) $CH_2{\cdot}CH_3$ (ortho 54, para 45) $CH(CH_3)_2$ (ortho 14, para 84) $C(CH_3)_3$ (ortho 12, meta 9, para 79)

## Side-Chain Substitution

Toluene is susceptible to attack by chlorine. Substitution can occur in either the ring or in the side-chain. Moderate temperatures and the use of an appropriate catalysts, as with benzene (p. 118), lead to substitution in the nucleus in the *ortho* and *para* positions. Free radical conditions, on the other hand, lead to substitution in the side-chain, and are favoured at high temperature (i.e. boiling point) and on exposure to ultraviolet light. Attack occurs at the $\alpha$C—H bonds, which are activated by the bond weakening effect of hyperconjugation. Side-chain attack is favoured because the electrons of the chlorine radical are repulsed by the $\pi$-electron cloud of the ring. Radical attack is also favoured by resonance stabilisation in the benzyl radical formed as intermediate.

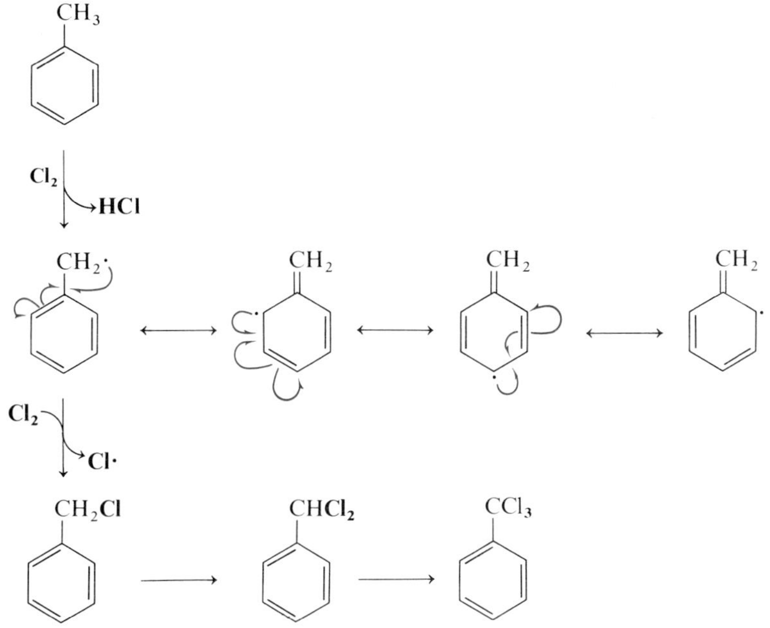

$CH_3$

$Cl_2$ → HCl

$CH_2\cdot$ ⟷ $CH_2$ ⟷ $CH_2$ ⟷ $CH_2$

$Cl_2$ → Cl·

$CH_2Cl$ ⟶ $CHCl_2$ ⟶ $CCl_3$

Benzyl chloride Benzylidene dichloride Benzotrichloride

Direct bromination can be similarly achieved, or alternatively, by the use of *N*-bromosuccinimide. The precise mechanism of action of *N*-bromosuccinimide is uncertain, but radical attack is involved.

$$\text{succinimide-NBr} \xrightarrow{h\nu/\text{Radical initiator}} \text{succinimide-N}\cdot + Br\cdot$$

$$C_6H_5CH_3 + \text{succinimide-N}\cdot \rightarrow C_6H_5CH_2\cdot + \text{succinimide-NH}$$

$$C_6H_5CH_2\cdot + \text{succinimide-NBr} \rightarrow C_6H_5CH_2\cdot\mathbf{Br} + \text{succinimide-N}\cdot$$

## Side-Chain Oxidation

The side-chain of alkylbenzenes is much more readily susceptible to oxidation than the ring. Toluene is readily oxidised to benzoic acid with potassium permanganate and sodium carbonate, potassium dichromate and sulphuric acid, or dilute nitric acid under pressure.

$$\underset{\text{Toluene}}{C_6H_5CH_3} \xrightarrow{KMnO_4} \underset{\text{Benzoic acid}}{C_6H_5CO\cdot OH}$$

The preferred oxidation of the side-chain reflects the comparative stability of the aromatic ring. Attack in the side-chain is favoured by the activation of the hydrogens on the $\alpha$-carbon atom of the alkyl group, which results from delocalisation of the bonding electrons. This is evident from the fact that ethyl-, propyl- and isopropyl-benzenes all yield benzoic acid on oxidation with potassium permanganate, whereas t-butylbenzene which has no $\alpha$C—H bond is stable to oxidation.

$\overset{\alpha}{C}H_2 \cdot CH_3$ (Ethylbenzene) $\overset{\alpha}{C}H_2 \cdot CH_2CH_3$ (Propylbenzene) $\overset{\alpha}{C}H(CH_3)_2$ (Isopropylbenzene) $\overset{\alpha}{C}(CH_3)_3$ (t-Butylbenzene)

Ethylbenzene Propylbenzene Isopropylbenzene t-Butylbenzene

In recent years, use has been made of the vulnerability of alkylbenzenes to attack at the α-carbon atom to devise a large-scale method for the manufacture of phenol and acetone from isopropylbenzene (cumene). Catalytic oxidation with air in the presence of sodium hydroxide yields cumene hydroperoxide, which undergoes an acid-catalysed rearrangement to phenol and acetone.

$$C_6H_5CH(CH_3)_2 \xrightarrow{O_2/NaOH} C_6H_5C(CH_3)_2 \cdot O \cdot OH \xrightarrow{\text{Ⓗ}} C_6H_5OH + CH_3 \cdot CO \cdot CH_3$$

Cumene Cumene hydroperoxide Phenol Acetone

Diphenylmethane and triphenylmethane are both readily oxidised at the alkyl carbon to benzophenone and triphenylcarbinol respectively. Ease of oxidation at the α-carbon atom increases with multiplicity of aryl substituents as activation of αC—H bonds is increased by the increasing possibilities for delocalisation of bonding electrons. Thus, triphenylmethane is oxidised by shaking with oxygen in the presence of trace amounts of aluminium chloride.

$$C_6H_5CH_2 \cdot C_6H_5 \xrightarrow{KMnO_4} C_6H_5 \cdot CO \cdot C_6H_5$$

Diphenylmethane Benzophenone

$$(C_6H_5)_3 \cdot CH \xrightarrow{O_2/AlCl_3} (C_6H_5)_3 \cdot C \cdot OH$$

Triphenylmethane Triphenylcarbinol

## REACTIVITY AND ORIENTATION OF ENTERING SUBSTITUENTS IN ELECTROPHILIC SUBSTITUTION OF BENZENE DERIVATIVES

Nitration of toluene (p. 126) shows that the presence of an alkyl substituent influences the reactivity of the ring in further substitution, and determines the position of the entering substituent. In this case, reactivity is increased, and the ring is said to be **activated** for substitution because of increased availability of π-electrons. Also, entering substituents are directed mainly into the *ortho* and *para* positions by the combined influence of **mesomeric** (+M) and steric effects. The same effect, i.e. *ortho* and *para* substitution with activation, is seen with certain oxygenated and nitrogenous substituents, such as OH, $O^-$, $OCH_3$ and $NH_2$. These substituents although electron-withdrawing (−I effect) as a result of the electronegativity difference between carbon and oxygen (or nitrogen), are able to make electrons available to the ring by an **electromeric** effect (+E effect) in response to the approach of a positively charged (electrophilic) reagent. The electromeric effect is due to the interaction (orbital overlap) of unshared electron pairs with delocalised $\pi$-electrons of the ring, and opposes and overrides

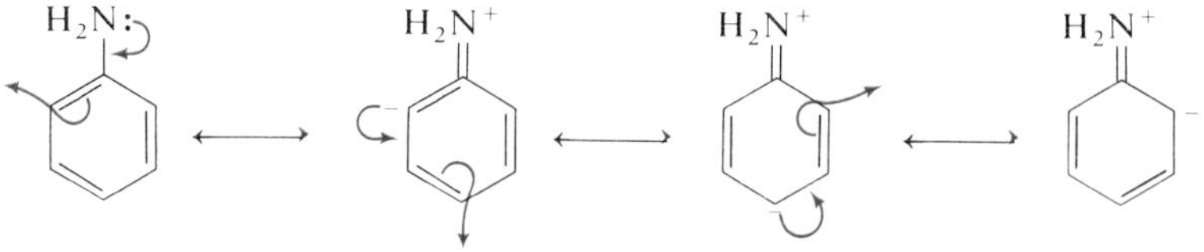

$H_2N$: ⟷ $H_2N^+$ ⟷ $H_2N^+$ ⟷ $H_2N^+$

the inductive effect. The ring is thus again activated for reaction, and electron density is concentrated at the *ortho* and *para* positions, as for example in aniline.

The very much more strongly electronegative halogens are also able to promote **ortho/para** electrophilic substitution by a similar electromeric effect through the lone pair of electrons on the halogen, which is brought into play on demand by positively charged (electrophilic) attacking reagents. The greater electronegativity of the halogen compared with nitrogen and oxygen, and the greater C—Cl bond length, which attenuates orbital overlap with the aromatic sextet, is such that the ring as a whole is **deactivated**, though less so in the *ortho* and *para* positions than the *meta* position. Thus, although chlorobenzene nitrates in the *ortho* and *para* positions, it is more difficult to nitrate than benzene.

Other substituents, which are electron-attracting, withdraw electrons from the ring, so **deactivating** the ring for substitution. Such electron withdrawal is mainly from positions *ortho* and *para* to the substituents, so that the *meta* position is left as the position of greater electron density, and substitution occurs in this position. Thus, the nitro group decreases electron density over the ring

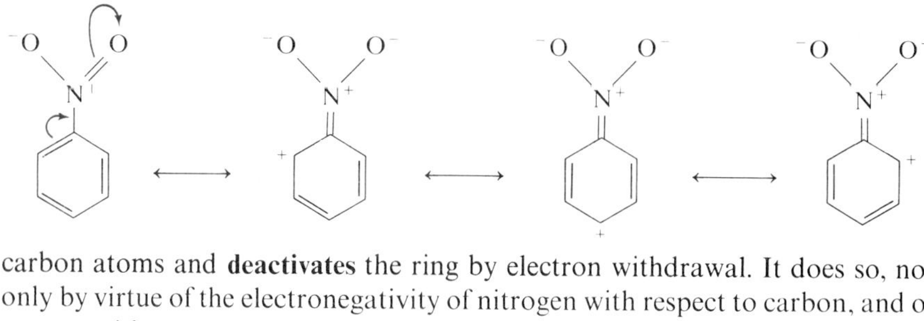

⟷ ⟷ ⟷

carbon atoms and **deactivates** the ring by electron withdrawal. It does so, not only by virtue of the electronegativity of nitrogen with respect to carbon, and of oxygen with respect to nitrogen (−I effect), but also by virtue of a mesomeric effect, which arises through the interaction of its unshared electron pairs with the delocalised $\pi$-electrons of the ring and operates in the same direction (−M effect) as the inductive effect. Electron deficiency is, therefore, localised at the *ortho* and *para* positions, and electrophilic attack occurs preferentially in the *meta* position.

Table 13. Orientation and Reactivity Effects of Ring Substituents in Electrophilic Substitution of Benzene Derivatives

| **Ortho/para** orientation with activation | **Ortho/para** orientation with deactivation | **Meta** orientation with deactivation |
|---|---|---|
| $—NMe_2$ | —F | $—\overset{+}{N}Me_3$ |
| $—NH_2$ | —Cl | $—NO_2$ |
| —OH | —Br | $—CF_3$ |
| $—O^-$ | —I | —CN |
| —OR | | $—SO_2·OH$ |
| $—NH·CO·CH_3$ | | —CHO |
| $—O·CO·CH_3$ | | —CO·R |
| —alkyl | | —CO·OH |
| —phenyl | | —CO·OR |
| | | $—CO·NH_2$ |
| | | $—\overset{+}{N}H_3$ |

Even greater deactivation occurs when the electron-withdrawing group is also positively charged, as in the quaternary ammonium compound $Ph\overset{+}{N}Me_3Cl^-$, since attack by a cationic (electrophilic) reagent will also be retarded by the mutual repulsion of like charges.

In summary, there are three classes of substituent:
(a) **ortho/para** directing, and activating,
(b) **ortho/para** directing, and deactivating,
(c) **meta** directing, and deactivating.

The substituents in each class are shown in Table 13, and placed approximately in order (from top to bottom of the table) of reactivity. This order provides a means of assessing competing effects if these are present, when two or more substituents are already present in the ring undergoing substitution.

Whilst the balance between the additive or competitive influence of inductive, mesomeric and electronic effects provides a simple working basis for determining the orientation of incoming substituents in electrophilic substitution, consideration of resonance stabilisation in $\sigma$-complex transition states provides a much more satisfactory solution. Thus, substitution is favoured at those positions where resonance stabilisation of $\sigma$-complex transition states is greatest. This is evident from the degree of stabilisation achieved in transition states relating to the nitration of anisole in the *ortho* and *para* positions, which is not attained in *meta* substitution. Ignoring resonance stabilisation within the rings, which is the same irrespective of the position of the incoming substituents, stabilisation of the *o*- and *p*-transition states involves two canonical forms compared with only one for the *meta* intermediate. In consequence, *o*- and *p*-substitution predominates.

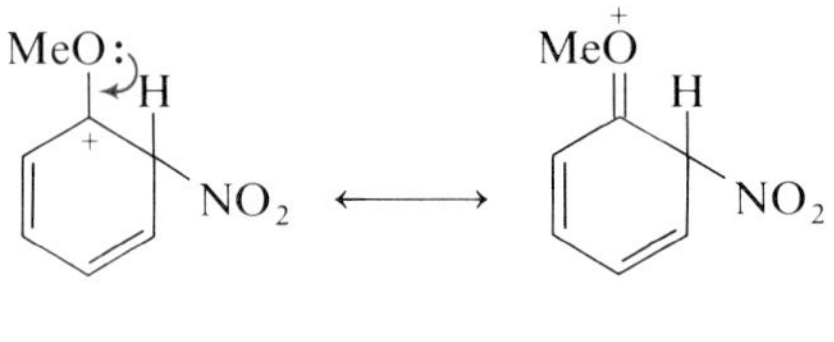

*o*-Substitution

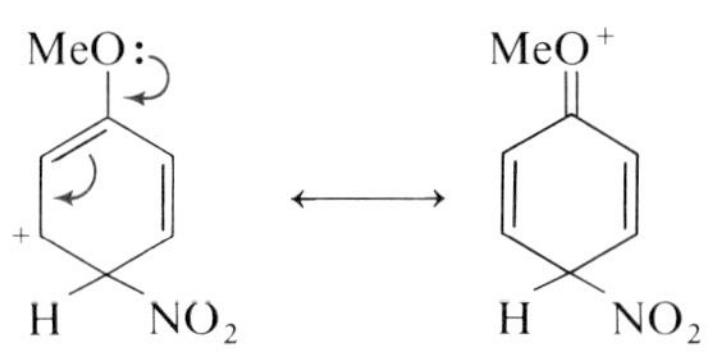

*p*-Substitution

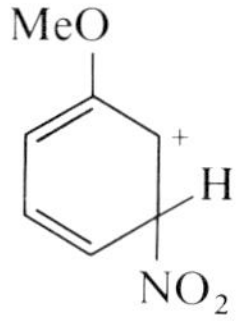

*m*-Substitution

In the further nitration of nitrobenzene, charge delocalisation in the transition state leading to *m*-substitution is stabilised by resonance between three forms, in which the positive charge on the ring is clearly separated from that on nitrogen. In contrast, one resonance form in each of the transition states leading to *o*- and *p*-substitution carries a positive charge on the ring carbon immediately adjacent to the positively charged nitrogen. These latter transition states are, therefore, less stable and in consequence *m*-substitution is favoured.

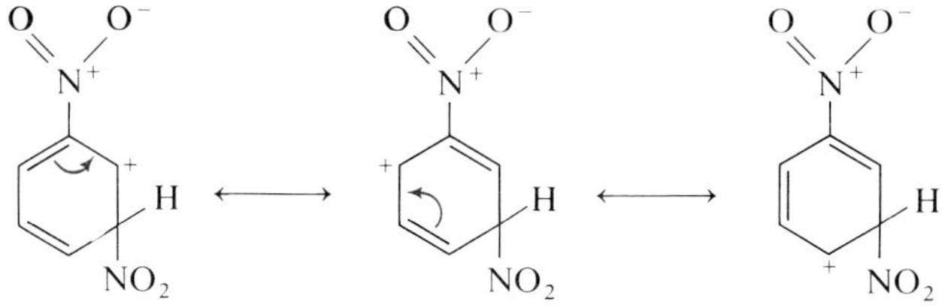

*m*-Substitution

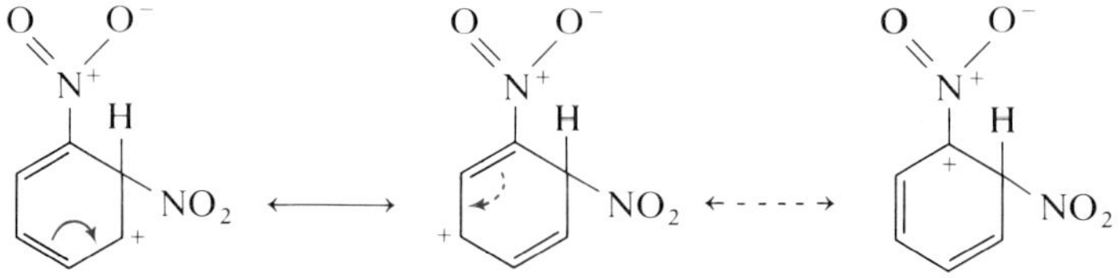

*o*-Substitution

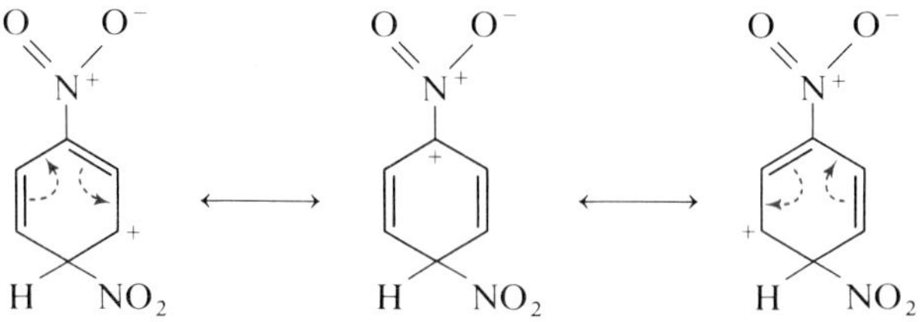

*p*-Substitution

# 6 Alkynes

## THE ACETYLENIC BOND

Alkynes, or acetylenes, are unsaturated hydrocarbons having the general formula $C_nH_{2n-2}$ and characterised by the presence of a carbon–carbon triple bond. The first member of the series is acetylene, HC≡CH, and the triple bond is often referred to as the **acetylenic bond**.

The acetylenic bond embodies a mode of hybridisation of carbon orbitals involving a single $2p$ orbital and one $2s$ orbital into two hybrid ($sp$) orbitals, the axes of which lie at an angle of 180° with respect to each other (Fig. 17A). The remaining two unhybridised $2p$ orbitals are situated at right angles to each other and to the axis of the $sp$ hybridised orbitals (Fig. 17B). The carbon–carbon triple bond, therefore, results in part from the endways overlap of two $sp$ hybrid orbitals, and in part from the sideways overlap of the four unhybridised $2p$ orbitals. Thus, in acetylene, we have strong sigma ($\sigma$) C—H bonds formed by overlapping of one of the $sp$ hybrid orbitals of each carbon atom with the $1s$ orbital of each of two hydrogen atoms, and in the same digonal axis, a C—C sigma bond formed by the overlapping of the remaining pair of $sp$ atomic orbitals (Fig. 17C).

Overlap of the two pairs of unhybridised $2p$ orbitals generates two new $\pi$-orbitals which because of the orientation of the participating $2p$ orbitals concentrate the electron density above and below, and in front and behind the axis of the digonal sigma bonds (Figs. 17D and E). As in the ethylenic bond, the combination of $\sigma$ and $\pi$-bonds increases the strength of the bond and shortens the bond distance between the two nuclei. The C≡C bond distance is 1.20 Å (C=C 1.34 Å) and the bond energy 803 kJ $mol^{-1}$.

### Nomenclature

Simple monoacetylenes are named as derivatives of acetylene. The IUPAC system relates the compound to the parent saturated hydrocarbon, employing the suffix **-yne** in place of **-ane** for the saturated compounds (Table 14). The position of the acetylenic group in the chain is indicated by numbering the chain from the end which gives the lowest number to the first of the two acetylenic carbon atoms.

### General Properties

Acetylene is a highly flammable gas, which normally burns with a luminous flame due to the formation of carbon particles in the combustion process. Thus,

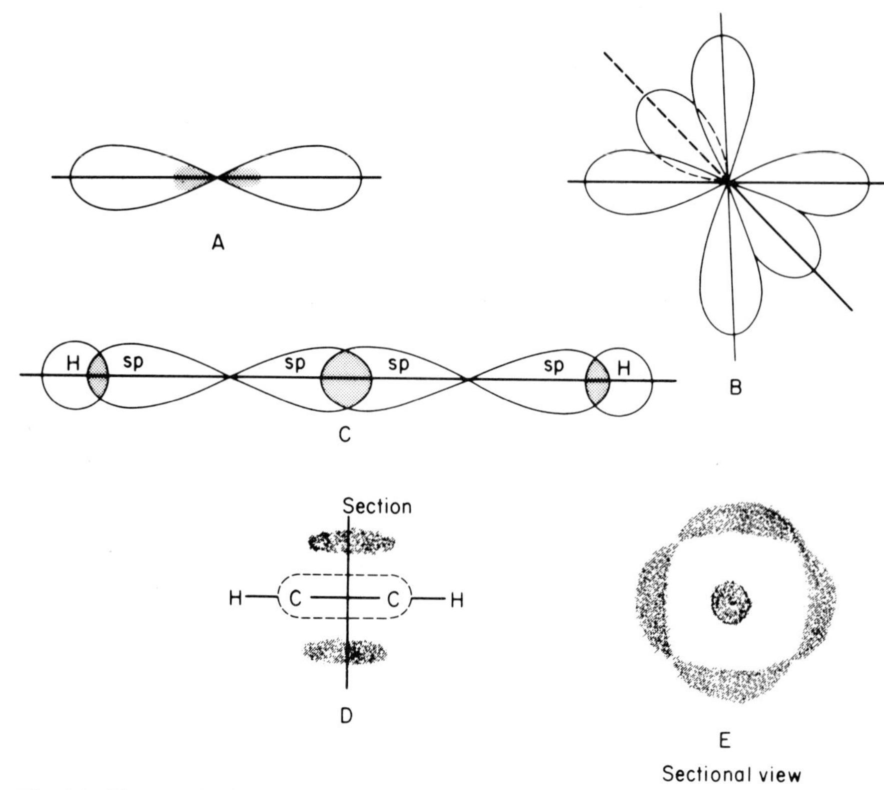


Fig. 17 The acetylenic bond
A Digonal *sp* hybridised orbitals
B Arrangement of unhybridised $2p$ orbitals mutually inclined at right angles to each other and to the axis of the *sp* hybridised orbitals
C Orbital overlap in C—C and C—H bond formation
D $\pi$-orbitals
E Sectional view showing electron density

incomplete combustion in air arises as a result of the high carbon to hydrogen ratio in acetylene; complete combustion is only obtained in the oxy-acetylene flame. The latter reaches very high temperatures around 2700°C due to the release of the high bond energy of the acetylenic bond. The oxy-acetylene flame is used because of its high temperature for the cutting and welding of metals. Certain mixtures of air and acetylene are explosive.

Acetylene is gaseous at room temperatures (b.p. $-84$°C), and is usually transported as a solution in acetone, adsorbed under pressure onto kieselguhr in cylinders. This method of storage is safe and free from the explosion hazard which is always present if acetylene is merely liquefied under pressure. This instability reflects the high energy content of the triple bond (high negative heat of formation).

Table 14. Nomenclature of Alkynes

| Alkyne | Systematic Name | IUPAC Name |
|---|---|---|
| $HC{\equiv}CH$ | Acetylene | Ethyne |
| $CH_3{\cdot}C{\equiv}CH$ | Methylacetylene | Prop-1-yne |
| $CH_3{\cdot}CH_2{\cdot}C{\equiv}CH$ | Ethylacetylene | But-1-yne |
| $CH_3{\cdot}CH_2{\cdot}CH_2{\cdot}C{\equiv}CH$ | Propylacetylene | Pent-1-yne |
| $(CH_3)_2CH{\cdot}C{\equiv}CH$ | Isopropylacetylene | 3-Methylbut-1-yne |
| $CH_3{\cdot}C{\equiv}C{\cdot}CH_3$ | Dimethylacetylene | But-2-yne |
| $CH_3{\cdot}C{\equiv}C{\cdot}CH_2{\cdot}CH_3$ | Methylethylacetylene | Pent-2-yne |

As with alkenes, the high electron density in the $\pi$-orbitals of the acetylenic bond promotes reactivity in electrophilic addition, and acetylenes undergo a series of addition reactions closely analogous to those of the olefines. Acetylenes are also able to function as weak acids, forming metal acetylides, the acetylenic anion being useful in certain nucleophilic substitution and addition reactions.

## ELECTROPHILIC ADDITION REACTIONS

### Addition of Halogens

Addition of chlorine or bromine to acetylene occurs readily in the gas phase, and is catalysed by the presence of metal halides. With chlorine, the initial product is *trans*-1,2-dichloroethylene; further reaction occurs less readily owing to the electron-withdrawing effect of the halogen substituents, but with excess of halogen gives rise to tetrachloroethane.

Iodine adds to acetylene less readily, but 1,2-di-iodothylene is formed in ethanolic solution. Bromine water similarly gives the mono addition product; liquid bromine, however, readily forms tetrabromoethane.

In the presence of metal halides, or in aqueous solution, the addition is ionic, and analogous to that occurring in additions to olefines.

$$HC{\equiv}CH \;\; Cl{-}Cl \xrightarrow{Cl^-} H{-}\overset{+}{C}{=}C(H)(Cl) \longrightarrow \underset{trans\text{-1,2-Dichloro ethylene}}{(Cl)(H)C{=}C(H)(Cl)} \longrightarrow \underset{\text{Tetrachloroethane}}{CHCl_2{\cdot}CHCl_2}$$

The acetylenic bond is less reactive than the ethylenic bond in the addition of halogens, and selective addition to the ethylenic bond is possible in molecules containing both types of unsaturation.

**Addition of Halogen Acids**

Addition of halogen acids is also similar to that in olefines, but occurs stepwise, giving with hydrogen chloride first vinyl chloride and then ethylidene chloride (1,1-dichloroethane). The reactivity of the halogen acids is in the order HI > HBr > HCl > HF, and the Markownikoff rule (the negative dipole of the attacking reagent attacks the more highly substituted carbon atom) applies both in the initial addition and at the second stage.

$$HC{\equiv}CH \xrightarrow{H-Cl} H\overset{+}{C}{=}CH_2 \; (Cl^-) \longrightarrow \underset{Cl}{\overset{H}{>}}C{=}CH_2$$

Vinyl chloride

$$\underset{Cl}{\overset{H}{>}}C{=}CH_2 \xrightarrow{H-Cl} \underset{Cl}{\overset{H}{>}}\overset{+}{C}-CH_3 \; (Cl^-) \longrightarrow Cl_2{\cdot}CH{\cdot}CH_3$$

Ethylidene chloride

The addition of chlorine to vinyl chloride is determined by the polarisability of the molecule, which overrides the competing inductive effect of the C→Cl bond. Resonance stabilisation of vinyl chloride favours the formation of ethylidene chloride, as shown, rather than the alternative 1,2-dichloroethane, which would result if the alternative carbonium ion intermediate were formed as a result of the inductive effect.

$$\ddot{C}l-CH{=}CH_2 \longleftrightarrow \overset{+}{C}l{=}CH-CH_2^-$$

The isolated acetylenic bond is less reactive than the ethylenic bond in the addition of halogen acids. The ready conversion of the ene-yne, vinylacetylene, to chloroprene is probably the result of an initial 1,4-addition and rearrangement, rather than direct 1,2-addition to the acetylenic bond.

$$H_2C{=}CH-C{\equiv}CH \xrightarrow{HCl} H_2C{=}CH-CCl{=}CH_2$$

Chloroprene

Chloroprene is used in the manufacture of neoprene, a synthetic rubber, formed from it by polymerisation.

**Hydration**

Acid-catalysed hydration of the acetylenic bond occurs in the presence of dilute sulphuric acid (40%) and mercuric sulphate. The function of the mercuric ions is to promote the formation of water-soluble π-complexes, which increase the solubility, and reactivity, of the hydrocarbon.

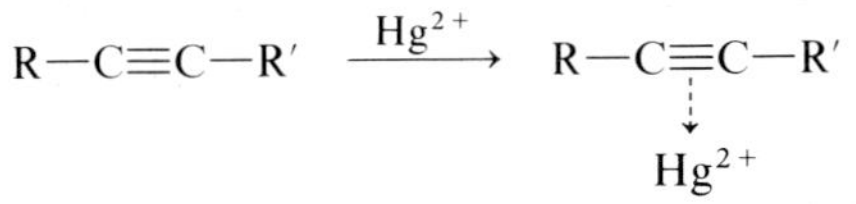

$$R{-}C{\equiv}C{-}R' \xrightarrow{Hg^{2+}} R{-}C{\equiv}C{-}R' \;(\downarrow Hg^{2+})$$

The hydration of acetylene gives rise to an unstable product, vinyl alcohol, which immediately undergoes a 1,3-prototropic rearrangement to yield acetaldehyde. The product, produced on a commercial scale by this method, is used mainly as a source of acetic acid, which is obtained from it by catalytic oxidation with air over a heated manganese catalyst (Chapter 10).

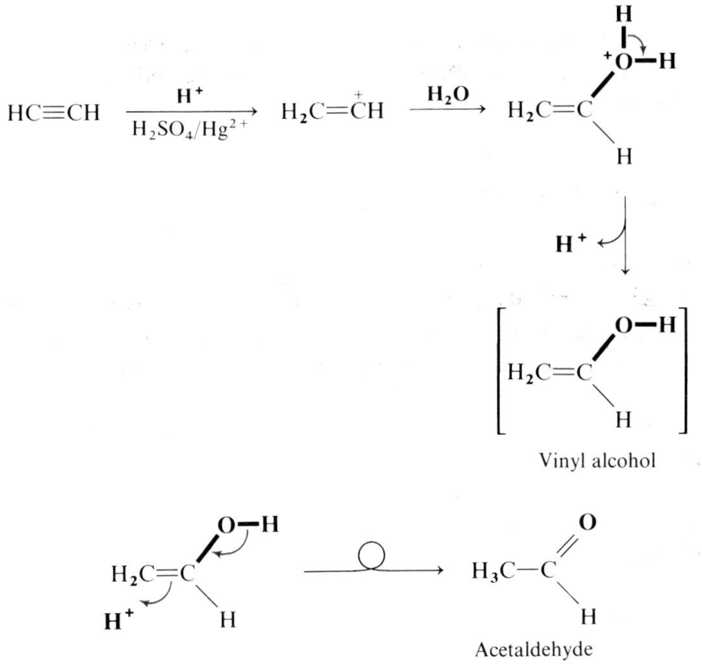

$$HC{\equiv}CH \xrightarrow[H_2SO_4/Hg^{2+}]{H^+} H_2C{=}\overset{+}{C}H \xrightarrow{H_2O} H_2C{=}C(H){-}\overset{+}{O}H_2 \xrightarrow{-H^+} \left[H_2C{=}C(H){-}O{-}H\right]$$

Vinyl alcohol

$$H_2C{=}C(H){-}O{-}H \longrightarrow H_3C{-}C(H){=}O$$

Acetaldehyde

The hydration of acetylenes is a general route to substituted acetylenes. Addition occurs according to the Markownikoff rule; thus, methylacetylene yields acetone.

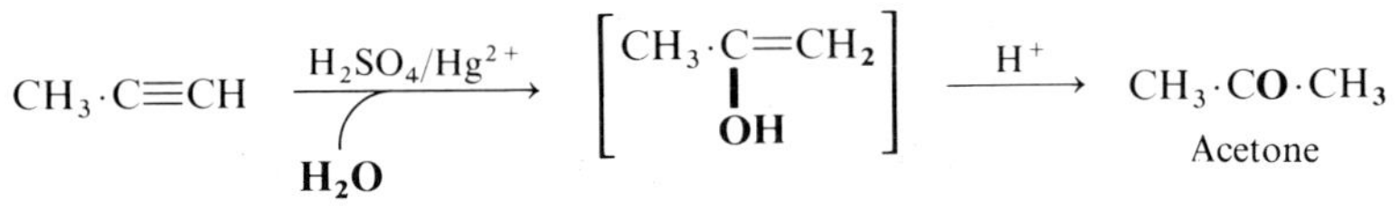

$$CH_3{\cdot}C{\equiv}CH \xrightarrow[H_2O]{H_2SO_4/Hg^{2+}} \left[CH_3{\cdot}C(OH){=}CH_2\right] \xrightarrow{H^+} CH_3{\cdot}CO{\cdot}CH_3$$

Acetone

### Addition of Alcohols

Alcohols add across the acetylenic bond in the presence of mercuric ions to yield vinyl ethers, which in contrast to the parent alcohols are isolable without rearrangement.

$$HC{\equiv}CH + \mathbf{CH_3{\cdot}OH} \xrightarrow{Hg^{2+}} H_2C{=}CH{\cdot}\mathbf{O{\cdot}CH_3}$$

Methyl vinyl ether

### Addition of Acetic Acid

Addition of acetic acid to acetylene in the presence of mercuric salts is carried out on an industrial scale in the production of vinyl acetate.

$$HC{\equiv}CH + \mathbf{CH_3{\cdot}CO{\cdot}OH} \xrightarrow{Hg^{2+}} H_2C{=}CH{\cdot}\mathbf{O{\cdot}CO{\cdot}CH_3}$$

Vinyl acetate

Vinyl acetate is stable, but readily polymerised under free radical conditions to give polyvinyl acetate, which is useful as an adhesive and as a source of polyvinyl alcohols (Chapter 20).

## METAL ACETYLIDES

### Acidic Properties of Alkynes

The *sp* hybridised orbitals of the acetylenic C—H bond have a proportionally higher *s* orbital content than the $sp^2$ and $sp^3$ hybridised orbitals of ethylenic and paraffinic C—H bonds. In consequence, the electron pairs of acetylenic C—H bonds are drawn in closer to the carbon nucleus than in ethylenes and paraffins. It follows that the acetylenic C—H bonds are more readily broken to yield a proton under ionic conditions than their saturated and ethylenic counterparts. Acetylene is, therefore, able to function as a weak acid ($K_a$ *ca* $10^{-22}$), and acetylenic hydrogens are replaceable by metals such as sodium, potassium, lithium, silver and copper.

### Metal Acetylides

Alkali metal acetylides are formed readily by passing acetylene into a solution of metal in liquid ammonia. Sodium dissolves in liquid ammonia to give a bright blue solution of sodium metal. This is converted to sodamide in the presence of a catalyst ($Fe^{3+}$), which readily reacts with acetylene to yield the acetylide.

$$2Na + 2NH_3 \xrightarrow[Fe^{3+}/-35^\circ]{\text{Liquid } NH_3} 2NaNH_2 + H_2$$

$$HC{\equiv}CH + NaNH_2 \longrightarrow HC{\equiv}C^-Na^+ + NH_3$$

The formation of an insoluble silver acetylide by the addition of silver nitrate is used both as a test for identity and in the assay of a number of pharmaceutical products including *Ethchlorvynol* and *Methylpentynol.*

Alkali metal acetylides are decomposed by water ($K_a$ *ca* $10^{-16}$) or ethanol ($K_a$ *ca* $10^{-18}$), with re-formation of acetylene.

$$HC{\equiv}C^-Na^+ + CH_3{\cdot}CH_2{\cdot}OH \longrightarrow HC{\equiv}CH + CH_3{\cdot}CH_2ONa$$

**Formation of Substituted Acetylenes**

Metal acetylides are very reactive in nucleophilic displacement and addition reactions. Thus, sodium acetylide reacts with alkyl halides in liquid ammonia to form alkyl-substituted acetylenes.

$$HC{\equiv}C^- \; CH_3{-}I \longrightarrow HC{\equiv}C{\cdot}CH_3 + I^-$$

Methylacetylene

Long-chain alkyl halides are much less reactive because of low solubility in liquid ammonia, and effective conversion to alkylacetylene is only accomplished by the addition of ether or by heating the metal acetylide (first formed in liquid ammonia) with the alkyl halide in dimethylformamide.

Alkylacetylenes are also acidic, and behave similarly.

$$R{\cdot}C{\equiv}CH \xrightarrow{\text{Na/Liquid } NH_3} R{\cdot}C{\equiv}C^-Na^+$$

$$R{\cdot}C{\equiv}C^- + CH_3I \longrightarrow R{\cdot}C{\equiv}C{\cdot}CH_3 + I^-$$

Dialkylacetylene

**Self-addition**

Absorption of acetylene into a solution of cuprous chloride and ammonium chloride in dilute hydrochloric acid leads first to the formation of cuprous acetylide, and then to further reaction in which the acetylenic carbanion attacks acetylene to yield vinylacetylene.

$$HC{\equiv}CH \xrightarrow{[Cu(NH_3)_2]^+Cl^-} HC{\equiv}C^-Cu^+$$

$$HC{\equiv}C^- \; HC{\equiv}CH \longrightarrow HC{\equiv}C{-}CH{=}\bar{C}H_2 \longrightarrow HC{\equiv}C{-}CH{=}CH_2$$

Vinylacetylene

**Addition to Carbonyl Compounds**

Acetylene and monosubstituted acetylenes, as their alkali metal acetylides, react by nucleophilic addition to aldehydes and ketones. Formaldehyde yields propargyl alcohol, and with the aldehyde in excess, but-2-yne-1,4-diol. The reaction is carried out in liquid ammonia. Volatilisation of the solvent at the end of the reaction and acidification yields the product.

$$HC{\equiv}C^- \; H_2C{=}O \longrightarrow HC{\equiv}C{\cdot}CH_2O^- \xrightarrow{H^+} HC{\equiv}C{\cdot}CH_2OH$$

Propargyl alcohol

$$HO{\cdot}CH_2{\cdot}C{\equiv}CH \xrightarrow{\text{Na/Liquid NH}_3} HOCH_2{\cdot}C{\equiv}C^-$$

$$\downarrow \text{1. } CH_2O \;\; \text{2. } H^+$$

$$HOCH_2{\cdot}C{\equiv}C{\cdot}CH_2OH$$

But-2-yne-1,4-diol

*Methylpentynol* is prepared similarly by the addition of sodium acetylide to methyl ethyl ketone in liquid ammonia.

$$CH_3{\cdot}CH_2{\cdot}C(CH_3){=}O + HC{\equiv}C^- \longrightarrow CH_3{\cdot}CH_2{\cdot}C(CH_3)(C{\equiv}CH){-}O^- \xrightarrow{H^+} CH_3{\cdot}CH_2{\cdot}C(CH_3)(C{\equiv}CH){\cdot}OH$$

*Methylpentynol*

The product is a colourless, flammable liquid, b.p. 104°C, with a sharp characteristic geranium-like odour. It is soluble (1 in 10) in water, and miscible with organic solvents and fixed oils. Its high lipid solubility explains its hypnotic and anaesthetic effects on the central nervous system. (**2**, 1). As a monosubstituted acetylenic compound it forms metal derivatives. The formation from silver nitrate of its silver acetylide with release of nitric acid, which can be titrated with standard alkali solution, is used as the basis of a method of assay.

$$CH_3{\cdot}CH_2{\cdot}C(CH_3)(C{\equiv}CH){\cdot}OH \xrightarrow[HNO_3]{AgNO_3} CH_3{\cdot}CH_2{\cdot}C(CH_3)(C{\equiv}C{\cdot}Ag){\cdot}OH$$

## OXIDATION AND REDUCTION

### Reduction

Catalytic hydrogenation of acetylenes occurs readily with the formation of the corresponding olefines. The latter in turn are also subject to reduction, but this reaction can be suppressed by the use of a partially poisoned palladium catalyst when the *cis* addition product can be isolated.

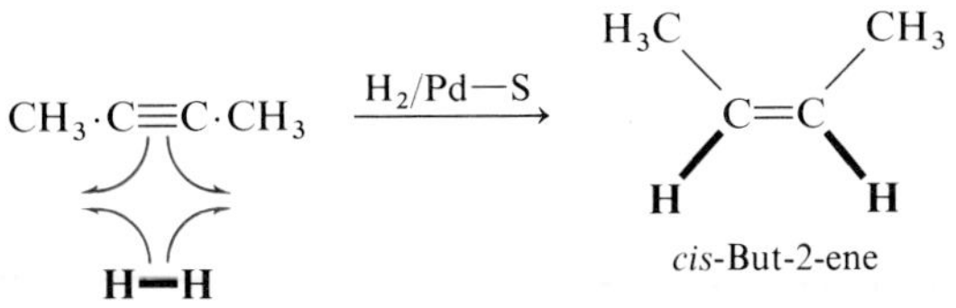

$$CH_3\cdot C{\equiv}C\cdot CH_3 \xrightarrow{H_2/Pd{-}S} \text{cis-But-2-ene}$$

H—H

*cis*-But-2-ene

In contrast, chemical reduction with sodium or lithium in liquid ammonia yields the *trans* addition product.

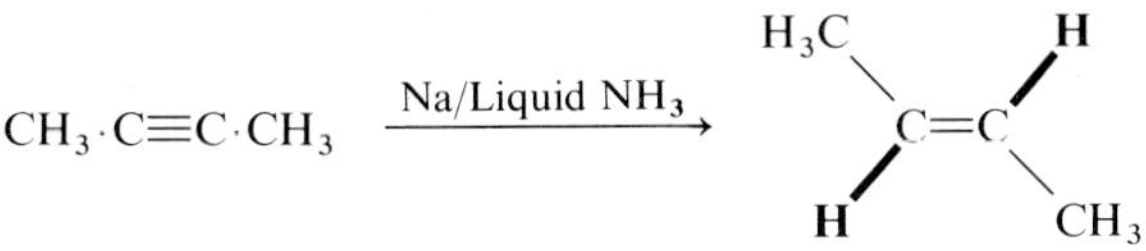

$$CH_3\cdot C{\equiv}C\cdot CH_3 \xrightarrow{Na/Liquid\ NH_3} \text{trans-But-2-ene}$$

*trans*-But-2-ene

**Oxidation**

The acetylenic bond is more stable than the olefinic bond to oxidation both by peracids and chromic acid, which react by ionic mechanisms, as shown by the oxidation of en-ynes to ynoic acids.

$$HC{\equiv}C(CH_2)_7\cdot CH{=}CMe_2 \xrightarrow{CrO_3(40\%)/HAc} HC{\equiv}C(CH_2)_7\cdot CO\cdot OH$$

11-Methyldodec-10-en-1-yne — Dec-9-ynoic acid

There is little detailed information on the bio-stability of acetylenes, though 5,6-$^{14}$[C]-eicosa-5,8,11,14-tetraynoic acid is significantly metabolised in the rat and man with release of $^{14}CO_2$ (Stenlake, Taylor and Templeton, 1971). It is possible that breaking of the chain is the result of both reduction (to the corresponding 5,6-ethylenic compound) and oxidation (as found with unsaturated acids).

**Autoxidation**

Mono- and poly-acetylenes all show considerable evidence of oxidative instability on exposure to light and air. Whether this is due to radical attack by oxygen on the allynic position (cf ethylenes) is not clear, though it seems likely that radical mechanisms are involved. Many acetylenic pharmaceutical chemicals, such as *Methylpentynol*, *Ethchlorvynol* and *Ethisterone*, which have no allynic hydrogen (i.e. R—C≡C · C · **H**), however, show oxidative instability, and require protection by storage in the absence of air.

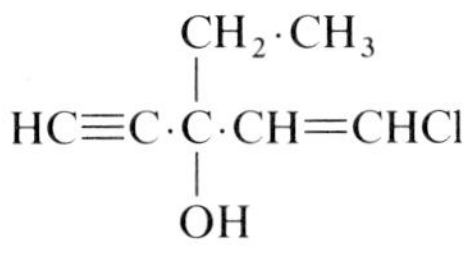

$$HC{\equiv}C\cdot C(CH_2\cdot CH_3)(OH)\cdot CH{=}CHCl$$

*Ethchlorvynol*

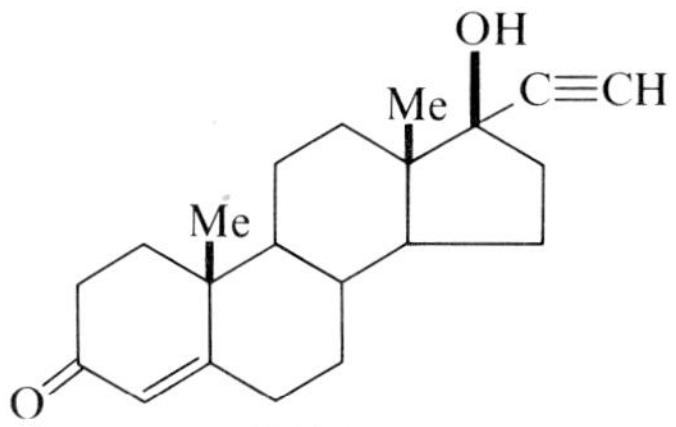

*Ethisterone*

## POLYACETYLENES

Considerable interest has been aroused in polyacetylenes following the discovery of their widespread occurrence naturally in micro-organisms, fungi and higher plants. The fungal antibiotic, mycomycin, is a typical example. It is very unstable and readily undergoes rearrangement in the presence of alkali to the more stable, but inactive, isomycomycin.

$$HC{\equiv}C{\cdot}C{\equiv}C{\cdot}CH{=}C{=}CH(CH{=}CH)_2{\cdot}CH_2{\cdot}CO{\cdot}OH \quad \text{Mycomycin}$$

$$\downarrow HO^-$$

$$CH_3(C{\equiv}C)_3{\cdot}(CH{=}CH)_2{\cdot}CH_2{\cdot}CO{\cdot}OH \quad \text{Isomycomycin}$$

The acetylenic chains are biosynthesised from the same precursors as those required for the biosynthesis of saturated fatty acids (Chapter 11), i.e. from acetyl coenzyme A and malonyl coenzyme A. The mechanism whereby the acetylenic unsaturation is achieved is not clear, but may include progressive dehydrogenation or, alternatively, elimination of enol esters of acylmalonic acids.

$$-C(OX){=}C(CO{\cdot}OH){-}C({=}O){-}O^- \longrightarrow -C{\equiv}C{\cdot}CO{\cdot}OH + XO^- + CO_2$$

The synthesis of di- and poly-acetylenes has been achieved by oxidative coupling.

$$2\ R{\cdot}C{\equiv}CH \xrightarrow[-H_2O]{O_2/NH_4Cl/Cu} R{\cdot}C{\equiv}C{\cdot}C{\equiv}C{\cdot}R$$

# 7 Monohydric Alcohols

## GENERAL PROPERTIES

Alcohols may be regarded as substitution products of the paraffins in which one or more hydrogens have been replaced by hydroxyl (OH) groups. They are classified as monohydric, dihydric, trihydric or polyhydric alcohols according to whether there are one, two, three or many such hydroxyl groups in the molecule.

Monohydric alcohols, which form an homologous series of general formula $C_nH_{2n+1}OH$ are further classified as primary ($—CH_2OH$), secondary ($>CHOH$) or tertiary ($\geqslant C \cdot OH$) according to whether the hydroxyl group is attached to a primary, secondary or tertiary carbon atom. Methanol ($CH_3OH$), ethanol ($CH_3 \cdot CH_2OH$) and propanol ($CH_3 \cdot CH_2 \cdot CH_2OH$) are examples of primary alcohols, and are characterised by the presence in the molecule of the primary alcohol group, $—CH_2OH$. Secondary alcohols, such as isopropanol ($CH_3 \cdot CHOH \cdot CH_3$), secondary butyl alcohol ($CH_3 \cdot CH_2 \cdot CHOH \cdot CH_3$) contain the characteristic group, $>CHOH$, whilst tertiary alcohols, such as t-butyl alcohol ($(CH_3)_3 \cdot C \cdot OH$) all contain the $\geqslant C \cdot OH$ group.

### Nomenclature

Trivial names, in which the alcohol is named as the derivative of an alkyl radical, e.g. methyl alcohol, n-propyl alcohol, are commonly adopted for the simple alcohols. Methyl alcohol is also known as carbinol, and some more complex alcohols are given trivial names in which they are named as derivatives of carbinol. Systematic names based on the IUPAC system of nomenclature are nearly always used for the higher alcohols. According to this system, the alcohol is named, with the suffix **-ol** as a derivative of the paraffin hydrocarbon which has the same number and arrangement of carbon atoms. The position of the hydroxyl (and of other substituents) is indicated by a numeral, the chain being numbered from the end nearest to which substitution occurs. Reference to Table 15 will indicate the application of this system of nomenclature.

### Physical Properties

The lower alcohols are colourless low boiling liquids, possessing distinctive spiritous odours, a burning taste and exhibiting pronounced physiological effects. As in other homologous series, the boiling points increase as the series is ascended. Chain-branching, however, tends to lower the boiling point progressively with increasing molecular complexity, so that secondary and tertiary alcohols in general have lower boiling points than the corresponding primary alcohols. This is a reflection of **molecular connectivity** (**2**, 1).

Table 15. Nomenclature of Alcohols

| Formula | Trivial Name | Systematic Name | b.p. (°C) |
|---|---|---|---|
| $CH_3OH$ | Methyl alcohol (Carbinol) | Methanol | 64.5 |
| $CH_3{\cdot}CH_2OH$ | Ethyl alcohol | Ethanol | 78.5 |
| $CH_3{\cdot}CH_2{\cdot}CH_2OH$ | n-Propyl alcohol | Propan-1-ol | 97.8 |
| $CH_3{\cdot}CHOH{\cdot}CH_3$ | Isopropyl alcohol | Propan-2-ol | 82.3 |
| $CH_3{\cdot}CH_2{\cdot}CH_2{\cdot}CH_2OH$ | n-Butyl alcohol | Butan-1-ol | 118 |
| $CH_3{\cdot}CH(CH_3){\cdot}CH_2OH$ | Isobutyl alcohol | 2-Methylpropan-1-ol | 107 |
| $CH_3{\cdot}CHOH{\cdot}CH_2{\cdot}CH_3$ | s-Butyl alcohol | Butan-2-ol | 99.5 |
| $(CH_3)_3{\cdot}C{\cdot}OH$ | t-Butyl alcohol | 1,1-Dimethylethanol | 83 |
| $CH_3(CH_2)_3{\cdot}CH_2OH$ | n-Amyl alcohol | Pentan-1-ol | 138 |
| $(CH_3)_2{\cdot}CH{\cdot}CH_2{\cdot}CH_2OH$ | Isopentyl alcohol | 3-Methylbutan-1-ol | 131 |
| $CH_3{\cdot}CH_2{\cdot}CH(CH_3){\cdot}CH_2OH$ | s-Butylcarbinol | 2-Methylbutan-1-ol | 128 |
| $(CH_3)_3C{\cdot}CH_2OH$ | Neopentyl alcohol | 2,2-Dimethylpropanol | 113 |
| $CH_3{\cdot}CH_2{\cdot}CH_2{\cdot}CHOH{\cdot}CH_3$ | Methylpropylcarbinol | Pentan-2-ol | 119 |
| $(CH_3)_2{\cdot}CH{\cdot}CHOH{\cdot}CH_3$ | Methylisopropylcarbinol | 3-Methylbutan-2-ol | 112.5 |
| $CH_3{\cdot}CH_2{\cdot}CHOH{\cdot}CH_2{\cdot}CH_3$ | Diethylcarbinol | Pentan-3-ol | 117 |
| $(CH_3)_2{\cdot}COH{\cdot}CH_2{\cdot}CH_3$ | t-Pentyl alcohol | 2-Methylbutan-2-ol | 102 |
| $CH_3(CH_2)_4{\cdot}CH_2OH$ | n-Hexyl alcohol | Hexan-1-ol | 156 |
| $CH_3(CH_2)_5{\cdot}CH_2OH$ | n-Heptyl alcohol | Heptan-1-ol | 176 |
| $CH_3(CH_2)_6{\cdot}CH_2OH$ | n-Octyl alcohol | Octan-1-ol | 194 |
| $CH_3(CH_2)_7{\cdot}CH_2OH$ | n-Nonyl alcohol | Nonan-1-ol | 215 |
| $CH_3(CH_2)_8{\cdot}CH_2OH$ | n-Decyl alcohol | Decan-1-ol | 231 |
| $CH_3(CH_2)_{10}{\cdot}CH_2OH$ | Lauryl alcohol | Dodecan-1-ol | 259 |
| $CH_3(CH_2)_{14}{\cdot}CH_2OH$ | Cetyl alcohol | Hexadecan-1-ol | m.p. (°C) 49 |
| $CH_3(CH_2)_{16}{\cdot}CH_2OH$ | Stearyl alcohol | Octadecan-1-ol | 59.5 |
| $CH_3(CH_2)_{18}{\cdot}CH_2OH$ | Arachidyl alcohol | Eicosan-1-ol | 65.5 |
| $CH_3(CH_2)_{24}{\cdot}CH_2OH$ | Ceryl alcohol | Hexacosan-1-ol | 79 |

The lower alcohols are volatile and flammable, and burn with a non-luminous flame; methanol when vaporised forms explosive mixtures with air. The first members of the series are mobile liquids, but viscosity increases as the molecular weight and size of the hydrocarbon fragment increases. The higher alcohols, e.g. cetyl and stearyl alcohols, are low melting solids. Solubility in water decreases rapidly with increasing molecular weight; methanol, ethanol and the propanols are completely miscible with water. Hydrogen bond formation between alcohol and water molecules is an important solubilising factor, but the insolubility of the hydrocarbon moiety rapidly becomes the major solubility-determining factor as molecular weight increases. Thus, n-butanol is only partially miscible with water and the amyl alcohols are almost completely immiscible. Chain-branching favours water solubility, and t-butanol, unlike n-butanol, is completely miscible with water.

The polarity of long-chain alcohols engendered by the terminal water-miscible hydroxyl group and the long water-insoluble hydrocarbon chain, causes them to concentrate at oil–water interfaces and for this reason, alcohols such as cetyl, stearyl, *Cetostearyl Alcohol* (a commercial mixture of cetyl and stearyl alcohols) and *Wool Alcohols* (Chapter 22) are used extensively as emulsifying agents.

**Solvent Properties of Alcohols**

Methanol is a useful solvent for a wide variety of organic compounds, but its use pharmaceutically is restricted by its extreme toxicity. The latter arises, not from the toxicity of methanol, *per se*, but because of the toxicity and slow elimination of the formaldehyde and formic acid which are formed by oxidation in the liver. It can, however, be used quite safely as solvent where there is no danger of contaminating a final product designed for internal administration. Removal of the last traces of solvent from complex organic molecules, however, is a tiresome and troublesome process, and many substances long believed to be free from contamination by traces of methanol used as solvent have in recent years been shown by gas-liquid chromatography to retain significant amounts. A number of antibiotics, including *Streptomycin Sulphate*, have been shown to be so contaminated, and limit tests to control solvent contamination are now imposed.

In contrast, methanol is also used on account of its toxic properties for the denaturation of ethanol in the preparation of *Industrial Methylated Spirits* (Denaturated Alcohol), a procedure which is designed to render the product unfit for drinking.

Ethanol also possesses important solvent and preservative properties. It is a good solvent for crystallisation for a wide range of organic chemicals. It is used extensively in the extraction and purification of alkaloids, steroids, essential oils and resins, and in the preparation of extracts, tinctures, and flavouring essences. It is also used occasionally in the preparation of solutions of medicinal compounds which have low water solubilities. Compounds, such as *Sulphathiazole* (Sulfathiazole) and *Phenobarbitone* (Phenobarbital), show increased solubilities in aqueous alcoholic solutions. The logarithm of the solubility is in an almost linear relationship with alcohol concentration (Higuchi, Gupta and Busse, 1953).

The main disadvantage of alcohols as solvents is their flammability, but in aqueous solution, they possess the advantage of functioning as preservatives. Dilute aqueous solutions (20% and over) inhibit the action of most enzymes and suppress the growth of moulds, fungi and bacteria; as a bacteriostatic, ethanol is most effective at concentrations around 70%.

A number of pharmaceutical and cosmetic preparations depend for their effectiveness not only on the solvent properties of ethanol, but also on its volatility which permits the product to dry out rapidly when applied to the skin. Similar considerations also apply in perfumery, though isopropanol is often preferred because of freedom in the United Kingdom from Excise control. Isopropanol, like methanol, has the advantage over ethanol that it is readily obtained anhydrous by simple distillation. It is, however, more toxic than ethanol, and as with methanol, its use must be restricted to preparations for external use, such as lotions, cosmetics and skin antiseptics.

Butanol and amyl alcohol are also widely used as industrial solvents. They are especially useful for the extraction of steroidal saponins from aqueous solution. They are also used in metabolic studies for the extraction of highly polar organic substances such as glucuronides and guanidine bases from samples of urine.

## O—H BOND CLEAVAGE

### Acidic Properties

Although alcohols are generally regarded as neutral liquids, they are capable of functioning as very weak acids. Separation of a proton is facilitated by the polarity of the O—H bond, and alcohols react with strongly electropositive elements such as alkali and alkaline earth metals with displacement of the hydroxylic hydrogen.

$$2CH_3{\cdot}CH_2OH + 2Na \longrightarrow 2CH_3{\cdot}CH_2{\cdot}O^-Na^+ + H_2\uparrow$$

Alkali metal alkoxides are usually soluble in excess of the parent alcohol. Magnesium and alkaline earth metal alkoxides are often somewhat less soluble. On evaporation of their solutions, metal alkoxides remain as non-volatile white deliquescent solids. They are readily soluble in, and decomposed by, water to the parent alcohol and alkali metal hydroxide.

$$CH_3{\cdot}CH_2{\cdot}ONa + H_2O \longrightarrow CH_3{\cdot}CH_2{\cdot}OH + NaOH$$

Decomposition occurs because alcohols ($K_a$ *ca* $10^{-18}$) are weaker acids than water ($K_a$ *ca* $10^{-16}$). The electron-releasing tendencies of the alkyl group tend to increase electron density on the oxygen of alcohols ($R{\cdot}CH_2{\rightarrow}CH_2{\rightarrow}O{\leftarrow}H$) compared with water. The oxygen function of alcohols is, therefore, less able than that of water to accommodate the negative charge formed on loss of a proton; consequently, alcohols are weaker acids than water.

The reactivity of alcohols towards alkali metals, and hence their acidity, decreases with alkyl substitution adjacent to the hydroxyl group corresponding to the extent of electron release towards the α-carbon atom.

$$\underset{\text{methanol}}{H{-}\overset{H}{\underset{H}{C}}{-}OH} > \underset{\text{other primary alcohols}}{Me{\rightarrow}\overset{H}{\underset{H}{C}}{-}OH} > \underset{\text{secondary alcohols}}{Me{\rightarrow}\overset{Me\downarrow}{\underset{H}{C}}{-}OH} > \underset{\text{tertiary alcohols}}{Me{\rightarrow}\overset{Me\downarrow}{\underset{Me}{C}}{-}OH}$$

In accordance with these differences in reactivity, t-butyl alcohol reacts only slowly with sodium, but much more readily with potassium, which is more electropositive.

Alcohols are more strongly acidic than either acetylene or ammonia, and will displace them from sodium acetylide and sodamide respectively. The following order of relative acidities therefore applies.

$$\begin{array}{ccccc} H_2O > & ROH > & HC{\equiv}CH > & NH_3 > & CH_4 \\ K_a\ ca & 10^{-16} & 10^{-18} & 10^{-22} & 10^{-35} \end{array}$$

The corresponding cations lie in the reverse order of basicity.

$$HO^- < RO^- < HC{\equiv}C^- < NH_2^- < CH_3^-$$

Alkoxide ions ($RO^-$) function as strong bases in alcoholic solution, and are useful as basic catalysts in a number of synthetic reactions (Chapter 13). The basic strength of the ions varies inversely with acidic strength of the parent alcohol, so that basicity increases with increasing alkyl substitution, in the order

tertiary alkoxide > secondary alkoxide > primary alkoxide

**Ether Formation** (Williamson's method)

Metal alkoxides have important synthetic use in the formation of ethers when heated under reflux with an alkyl halide.

$$CH_3{\cdot}CH_2{\cdot}O^- \ \mathbf{CH_3}{-}I \longrightarrow CH_3{\cdot}CH_2{\cdot}O{\cdot}\mathbf{CH_3} + I^-$$

Ethyl methyl ether

This method has the advantage over the alternative acid-catalysed etherification of alcohols (p. 152) in that it can be used to prepare mixed alkyl ethers. Since also secondary (2°) and tertiary (3°) halides readily form olefines under alkaline conditions (Chapter 9), branched-chain ethers are best prepared by reaction of a 2° or 3° alkoxide with a primary halide.

$$(CH_3)_2{\cdot}CHO^- \ \mathbf{CH_3}{-}I \longrightarrow (CH_3)_2{\cdot}CH{\cdot}O{\cdot}\mathbf{CH_3} + I^-$$

Isopropyl methyl ether

**Esterification**

Alcohols combine with both organic and inorganic acids with elimination of water. The products of the reaction with carboxylic acids are known as esters. They are formed when an alcohol and acid are heated together under reflux. The reaction is reversible and, in the absence of factors which would disturb the equilibrium, gives rise to a mixture of reactants and products, though in some cases the reaction may proceed only very slowly.

$$\underset{\text{Acetic acid}}{CH_3{\cdot}CO{\cdot}OH} + \underset{\text{Ethanol}}{CH_3{\cdot}CH_2{\cdot}OH} \rightleftharpoons \underset{\text{Ethyl acetate}}{CH_3{\cdot}CO{\cdot}O{\cdot}CH_2{\cdot}CH_3} + \underset{\text{Water}}{H_2O}$$

The forward reaction is favoured by the use of either one of the reactants in excess and is catalysed by small amounts (*ca* 1%) of anhydrous mineral acids, such as concentrated sulphuric acid or dry hydrogen chloride. Under these conditions, both acid and alcohol are protonated. The normal polarisation of the carboxyl C=O due to the difference in electronegativity between carbon and oxygen is enhanced by protonation. On the other hand, protonation of the alcohol reduces its reactivity, due to the repulsive influence of like charges. Protonation of the acid, however, compensates for this lowered reactivity.

$$CH_3{\cdot}C(=O)OH \overset{H^+}{\rightleftharpoons} CH_3{\cdot}C(=\overset{+}{O}H)OH + CH_3{\cdot}CH_2{\cdot}\ddot{O}H \rightleftharpoons CH_3{\cdot}C(OH)_2{\cdot}\overset{+}{O}(H){\cdot}CH_2{\cdot}CH_3$$

$$\xrightarrow{-H_2O} CH_3{\cdot}C(=O){\cdot}\overset{+}{O}(H)CH_2{\cdot}CH_3 \overset{-H^+}{\rightleftharpoons} CH_3{\cdot}C(=O){\cdot}O{\cdot}CH_2{\cdot}CH_3$$

The product esters are oily liquids, insoluble in water, mostly with pleasant fruity odours.

## C—O BOND CLEAVAGE

### Basic Properties

Alcohols act as proton acceptors and as such function as weak bases.

$$CH_3\ddot{O}H + H^+ \rightleftharpoons CH_3\overset{+}{O}H_2$$

The positively charged oxygen in the resulting protonated molecule promotes electron withdrawal from the adjacent carbon atom, and facilitates C—O bond cleavage in reactions with strong mineral acids.

### Sulphate Ester Formation

Alcohols react with cold concentrated sulphuric acid to form alkyl hydrogen sulphates (sulphate esters). The mechanism of the reaction, which results in C—O bond cleavage, differs according to the nature of the alcohol. Primary alcohols most probably react by an $S_N2$ mechanism (Chapter 9) involving nucleophilic attack of bisulphate ($HO{\cdot}SO_2{\cdot}O^-$) on the hydroxonium ion. Tertiary alcohols probably react by an $S_N1$ process.

$$CH_3{\cdot}CH_2{\cdot}OH \overset{H^+}{\rightleftharpoons} CH_3{\cdot}CH_2{-}\overset{+}{O}H_2 \; (+\; HO{\cdot}SO_2{\cdot}O^-) \overset{-H_2O}{\rightleftharpoons} CH_3{\cdot}CH_2{\cdot}O{\cdot}SO_2{\cdot}OH$$

Excess sulphuric acid not only displaces the equilibrium to the right, but also acts as a dehydrating agent removing water as it is formed. Ethyl hydrogen sulphate is acidic and forms a water-soluble barium salt $(CH_3CH_2 \cdot O \cdot SO_2 \cdot O)_2Ba$ on neutralisation of the reaction mixture with barium carbonate. The water-soluble barium salt is thus readily separated from barium sulphate and excess carbonate; it can be reconverted to ethyl hydrogen sulphate by titration with the equivalent amount of sulphuric acid.

### Metabolic Formation of Sulphate Esters

Certain high molecular weight alcohols, such as sterols, are metabolised to water-soluble steroid sulphates in the liver. Suphate is transferred from 3′-phosphoadenosine-5′-phosphosulphate (PAPS) by substrate specific sulpho-kinases in the liver.

$$R-\overset{+}{O}H_2 \quad + \quad {}^{-}O \cdot SO_2 \cdot O-\underset{\underset{O^-}{|}}{\overset{\overset{O}{\|}}{P}}-O-R \cdots \text{Sulphokinase}$$

3′-Phosphoadenosine-5′-phosphosulphate (PAPS)

$$\downarrow$$

$$R \cdot O \cdot SO_2 \cdot OH + HO \cdot \underset{\underset{O^-}{|}}{\overset{\overset{O}{\|}}{P}} \cdot O-R \cdots \text{Sulphokinase}$$

3′-Phosphoadenosine-5′-phosphate (PAP)

### Sodium Lauryl Sulphate

*Sodium Lauryl Sulphate*, a powerful anionic detergent, consists of the sodium salts of sulphated primary alcohols. The alcohols, obtained by hydrogenation of coconut oil, consisting mainly of lauryl alcohol, are treated with concentrated sulphuric acid, and the product neutralised with sodium hydroxide.

$$\underset{\text{Lauryl alcohol}}{CH_3(CH_2)_{10} \cdot CH_2OH} \xrightarrow{H_2SO_4} \underset{\text{Lauryl hydrogen sulphate}}{CH_3(CH_2)_{10} \cdot CH_2 \cdot O \cdot SO_2 \cdot OH}$$

$$\downarrow \text{NaOH}$$

$$\underset{\textit{Sodium Lauryl Sulphate}}{CH_3(CH_2)_{10} \cdot CH_2 \cdot O \cdot SO_2 \cdot O^- Na^+}$$

*Sodium Lauryl Sulphate* has powerful detergent properties in water. It is the salt of a strong acid and strong base and is completely ionised in aqueous solution. Solutions of the pure substance are neutral, but in practice *Sodium Lauryl Sulphate* solutions are often alkaline owing to the use of excess sodium hydroxide in the final neutralisation process of manufacture. Such solutions are stable to small additions of acid or alkali.

The detergent action of *Sodium Lauryl Sulphate* arises from the combination of the long fat-soluble alkyl chain, and the terminal water-soluble sulphate anion, which assists concentration of the molecule at the oil–water interface and hence aids their mutual solubility. Its detergent properties, unlike those of soaps, are retained in hard water, since both calcium and magnesium salts are water-soluble. *Sodium Lauryl Sulphate* is bacteriostatic for Gram-positive organisms, and is used for cleansing the skin prior to operations; both its detergent and bacteriostatic action are antagonised by cationic detergents (Chapter 16). It is used to assist the solution of compounds of low water solubility from solid dose formulations such as tablets and capsules.

*Sodium Lauryl Sulphate* is hydrolytically decomposed by hot aqueous hydrochloric acid. The reaction forms the basis of an assay, in which the resulting alcohols are extracted with ether and weighed.

$$CH_3(CH_2)_{10}\cdot CH_2\cdot O\cdot SO_2\cdot ONa + H_2O \xrightarrow{H^+} CH_3(CH_2)_{10}\cdot CH_2OH + NaHSO_4$$

## Phosphate Esters

Neutral trialkylphosphates are readily formed by reaction of phosphorus oxychloride with an excess of alcohol.

$$3\,R\cdot OH + POCl_3 \longrightarrow PO(OR)_3 + 3\,HCl$$

Monophosphate esters are widespread in biological systems where they are formed directly by the action of adenosine triphosphate (ATP) and the appropriate enzyme (Chapters 2 and 23).

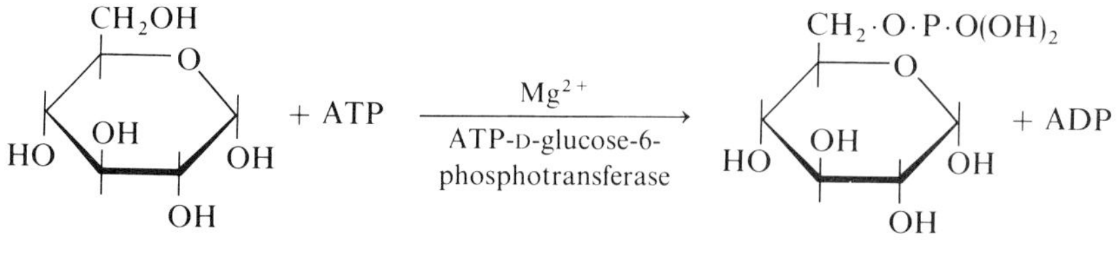

In the laboratory, however, monophosphate esters of alcohols can only be prepared satisfactorily by indirect methods, such as that based on the use of *p*-nitrophenyl phosphodichloridate.

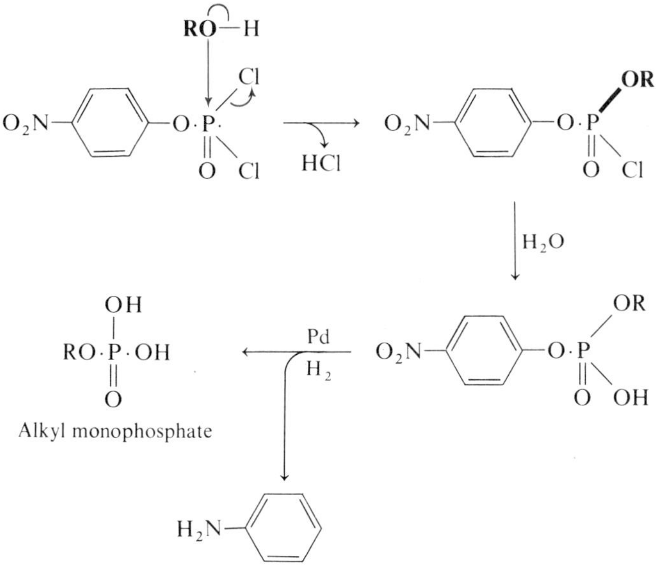

Alkyl monophosphate

Monophosphate esters of sterols and other water-insoluble alcohols form water-soluble sodium salts, which are useful in the preparation of stable aqueous injections, e.g. *Triclofos Sodium* and *Dexamethasone Sodium Phosphate*.

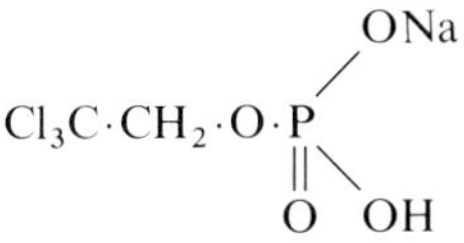

*Triclofos Sodium*

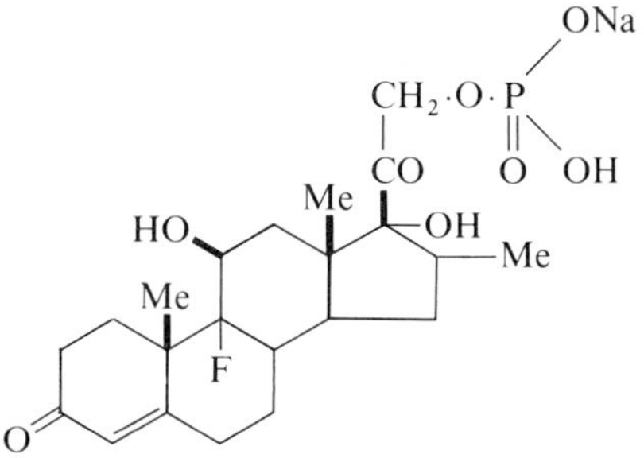

*Dexamethasone Sodium Phosphate*

**Dyflos** (DFP). *Dyflos* (di-isopropylfluorophosphonate) is a powerful inhibitor of the enzyme acetylcholinesterase (**2**, 2). It blocks enzymic action by phosphorylation of serine residues which have a key rôle in the hydrolysis of acetylcholine (Schaffer, May and Summerson, 1954; Cohen *et al.*, 1959; Sanger, 1963).

*Dyflos* is hydrolysed in aqueous solution at room temperature with the liberation of hydrogen fluoride; for this reason it is usually administered as an injection in *Arachis Oil* (Peanut Oil). The vapour is extremely toxic, and contaminated materials should be immersed in sodium hydroxide solution for several hours. It can be safely removed from the skin by washing with soap and water.

$$(Me_2CH{\cdot}O)_2P(=O)F + H{-}O{\cdot}CH_2{\cdot}CH(NH{-}CO{-}){\cdot}CO{\cdot}NH{-} \xrightarrow{-HF} (Me_2CH{\cdot}O)_2P(=O){-}O{\cdot}CH_2{\cdot}CH(NH{-}CO{-}){\cdot}CO{\cdot}NH{-}$$

Dyflos

## Acid-catalysed Dehydration

Alcohols are readily dehydrated to olefines by strong acids at elevated temperatures. Thus, ethanol is catalytically dehydrated to ethylene by concentrated sulphuric acid at 160–70°C.

$$CH_3{\cdot}CH_2{\cdot}\mathbf{OH} \underset{}{\overset{\mathbf{H^+}}{\rightleftharpoons}} H{-}CH_2{-}CH_2{-}\overset{+}{\mathbf{O}}\mathbf{H_2} \underset{}{\overset{\mathbf{H_2O}}{\rightleftharpoons}} H{-}CH_2{-}CH_2^{+} \overset{-\mathbf{H^+}}{\rightleftharpoons} H_2C{=}CH_2\uparrow$$

The reaction sequence is reversible and the same intermediates apply to the reverse reaction, the acid-catalysed hydration of olefines (Chapter 4). The dehydration is favoured by the removal of volatile olefines from the reaction medium.

The relative ease with which 1°, 2° and 3° alcohols dehydrate accords with the relative stabilisation of the intermediate carbonium ion, and hence is in the order 3° > 2° > 1°. The reaction may also be accompanied by rearrangement, depending on the relative stabilisation of the intermediate carbonium ion first formed.

## Acid-catalysed Etherification

Ethers are formed from primary alcohols in the presence of acid catalysts when the alcohol is maintained in excess. Thus, ethanol yields diethyl ether when heated at about 140°C with concentrated sulphuric acid, and the distillate collected.

The reaction is probably an $S_N2$ displacement of water from the hydroxonium ion in which ethanol acts as the nucleophile.

$$CH_3{\cdot}CH_2{\cdot}\mathbf{OH} \underset{}{\overset{H^+}{\rightleftharpoons}} CH_3{\cdot}CH_2{-}\overset{+}{\mathbf{O}}\mathbf{H_2} \xrightarrow[H_2O\,+\,H^+]{} CH_3{\cdot}CH_2{\cdot}\mathbf{O}{\cdot}CH_2{\cdot}CH_3$$

$CH_3{\cdot}CH_2{-}O{-}H$ (attacking nucleophile)

Diethyl ether

## Alkyl Halide Formation

Alcohols react with hydrogen halides (HX) under reflux to form alkyl halides.

$$CH_3{\cdot}CH_2{\cdot}\mathbf{OH} \overset{H^+}{\rightleftharpoons} CH_3{\cdot}CH_2{-}\overset{+}{\mathbf{O}}\mathbf{H_2} \xrightarrow[H_2O]{} CH_3{\cdot}\overset{+}{C}H_2 \xrightarrow{X^-} CH_3{\cdot}CH_2{\cdot}\mathbf{X}$$

The reactivity of the acid depends upon acidic strength, and is in the order HCl < HBr < HI. Reaction of primary and secondary alcohols with hydrogen chloride requires the presence of a catalyst (anhydrous zinc chloride or aluminium chloride) to promote breaking of the C—O bond.

$$CH_3{\cdot}CH_2{\cdot}OH + ZnCl_2 \rightleftharpoons CH_3{\cdot}CH_2{-}\underset{\displaystyle H}{\overset{}{O}}\cdots ZnCl_2$$

$$CH_3{\cdot}CH_2{-}\underset{\displaystyle H}{O}\cdots ZnCl_2 \longrightarrow CH_3{\cdot}\overset{+}{C}H_2 + [ZnCl_2(OH)]^-$$

$$CH_3{\cdot}\overset{+}{C}H_2 \xrightarrow{Cl^-} CH_3{\cdot}CH_2{\cdot}\mathbf{Cl} \qquad [ZnCl_2(OH)]^- \xrightarrow{H^+} ZnCl_2 + H_2O$$

Tertiary alcohols, such as t-butanol, which are more reactive, yield the corresponding halide at room temperature with concentrated hydrochloric acid alone.

The usual order of reactivity of alcohols in accord with the decreasing stability of the intermediate carbonium ion is:

$$\text{benzyl} > \text{allyl} > 3° > 2° > 1°$$

The Lucas test, which is based on this reactivity order, provides a useful means of distinguishing the lower primary, secondary and tertiary alcohols. The test, which is applicable to alcohols up to the hexanols only, depends upon the fact that the alcohols are soluble in a mixutre of concentrated hydrochloric acid and zinc chloride, whilst the corresponding alkyl halides are insoluble. The speed with which a cloudiness appears in the solution provides a means of distinguishing the various classes of alcohol. Tertiary alcohols react immediately, secondary alcohols within approximately five minutes, and primary alcohols very much more slowly, if at all.

Rearrangements are not uncommon in the course of alkyl halide formation from alcohols due to the rearrangement of intermediate carbonium ions to more stable ion intermediates.

Alcohols also react with phosphorus tri- and penta-halides to yield alkyl halides. The reaction is initiated by nucleophilic attack of the oxygen lone pairs on the electron-deficient phosphorus atom of the phosphorus halide. The choice of reagent is usually dictated by the volatility of the product halides. Phosphorus trichloride is particularly useful for the conversion of low molecular weight alcohols to alkyl halides, since the volatile product is readily separated from the by-product, phosphorous acid, which is non-volatile.

$$3\,CH_3{\cdot}CH_2{\cdot}OH + PCl_3 \longrightarrow \underset{\text{Ethyl chloride}}{3CH_3{\cdot}CH_2Cl\uparrow} + H_3PO_3$$

Phosphorus pentachloride is more useful for higher molecular weight alcohols, where the corresponding alkyl halide is less volatile, and can be isolated by volatilisation of the phosphorus oxychloride, which is formed at the same time.

$$CH_3(CH_2)_{10}{\cdot}CH_2OH + PCl_5 \longrightarrow CH_3(CH_2)_{10}{\cdot}CH_2Cl + POCl_3\uparrow + HCl\uparrow$$

Thionyl chloride is a more generally useful reagent since the two by-products, sulphur dioxide and hydrogen chloride, are both readily volatile at room temperature.

$$R{\cdot}CH_2{\cdot}OH + Cl_2S{=}O \longrightarrow R{\cdot}CH_2{-}O{-}S(Cl){=}O + HCl \longrightarrow R{\cdot}CH_2{\cdot}Cl + SO_2\uparrow$$

# OXIDATION OF ALCOHOLS

## Dehydrogenation

Primary and secondary alcohols are readily oxidised to aldehydes and ketones respectively. The products, which result from the cleavage of O—H and C—H bonds, are formed without disruption of the carbon chain or of the C—O bond. Catalytic oxidation occurs in the vapour phase when the alcohol vapour is passed over heated copper turnings at temperatures of 200–300°C. The oxidation product, which is formed by dehydrogenation in a thermally-inspired radical decomposition, is separated from unreacted alcohol by distillation.

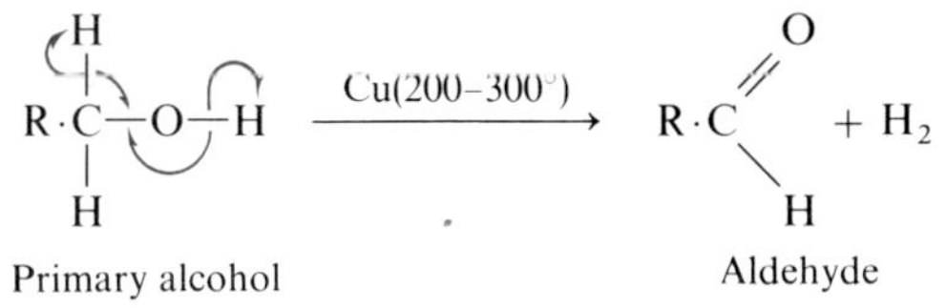

$$R{\cdot}CH_2{-}O{-}H \xrightarrow{Cu(200-300^\circ)} R{\cdot}C(=O)H + H_2$$

Primary alcohol → Aldehyde

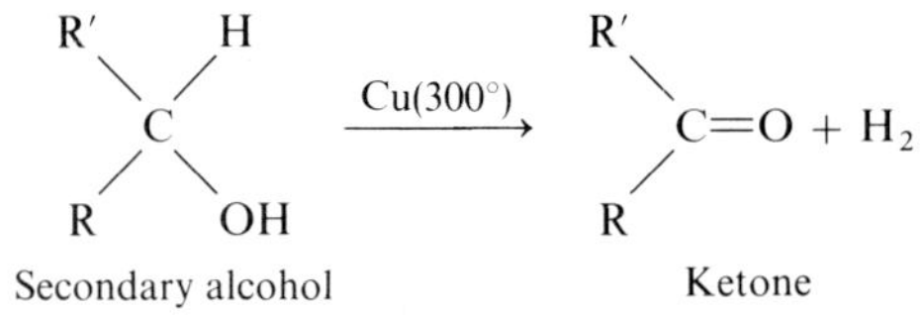

$$R'R\,CH{-}OH \xrightarrow{Cu(300^\circ)} R'R\,C{=}O + H_2$$

Secondary alcohol → Ketone

In contrast, tertiary alcohols, which lack an αC—H bond, merely dehydrate to olefines when heated under the same conditions.

**Chemical Oxidation**

Primary and secondary alcohols are also readily oxidised by chemical oxidants, such as potassium dichromate and sulphuric acid, or potassium permanganate in acid solution. Oxidation with potassium dichromate can be followed by the disappearance of the orange colour of the dichromate ion ($Cr_2O_7^{2-}$) and its replacement by the chromic ion ($Cr^{3+}$). The effective oxidising agent is chromic acid (chromium trioxide, $CrO_3$) which is formed in acid solution.

$$K_2Cr_2O_7 + H_2SO_4 \longrightarrow K_2SO_4 + H_2O + 2CrO_3$$

Oxidation results in the reduction of hexavalent chromium ($Cr^{6+}$) to tetravalent chromium ($Cr^{4+}$). The latter, however, is unstable and disproportionates in acid solution to a mixture of hexavalent ($CrO_3$) and trivalent ($Cr^{3+}$) states.

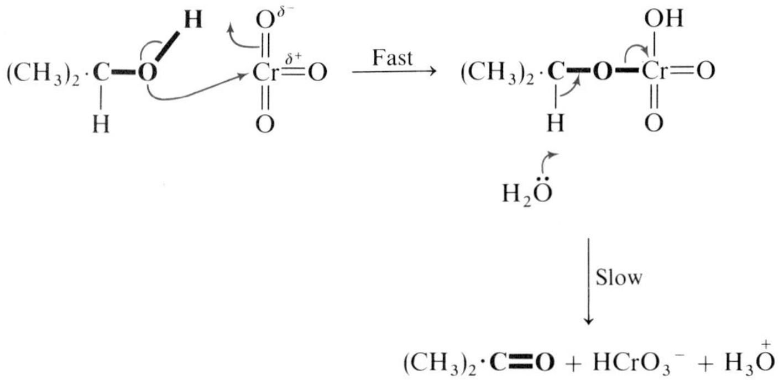

$$(CH_3)_2{\cdot}CH{-}O{-}H + O^{\delta-}{=}Cr^{\delta+}({=}O)_2 \xrightarrow{Fast} (CH_3)_2{\cdot}CH{-}O{-}Cr(OH)({=}O)_2 \quad (+\ H_2\ddot{O})$$

$$\xrightarrow{Slow} (CH_3)_2{\cdot}C{=}O + HCrO_3^- + H_3\overset{+}{O}$$

$$3HCrO_3^- + 9H^+ \longrightarrow CrO_3 + 2Cr^{3+} + 6H_2O$$

The rate-determining step is that involving the breaking of the C—H bond. This is established by the kinetic isotope effect observed in the comparable oxidation of $(CH_3)_2CDOH$, which proceeds at a rate of 0.6 compared with that of 1.0 for the oxidation of isopropanol (Westheimer and Nicolaides, 1949).

The participation of C—H bond breaking in the oxidation mechanism accounts for the further oxidation of aldehydes to the corresponding carboxylic acid (Chapter 10) by the same reagents. It also explains the termination of the oxidation of secondary alcohols at the ketone stage. Similarly, tertiary alcohols resist oxidation because of the lack of an αC—H bond. In strongly acidic conditions, however, they are dehydrated to the corresponding alkene, which is then cleaved oxidatively at the double bond (Chapter 4).

### The Haloform Reaction

Certain alcohols when treated with a hypohalous acid (HOX) in sodium hydroxide solution, undergo oxidation and subsequent halogenation and hydrolysis. The reaction is characteristic of all alcohols of structure $R{\cdot}CHOH{\cdot}CH_3$, where R is H, alkyl or aryl. Reaction is almost instantaneous when aqueous iodine is added dropwise to a solution of the alcohol in aqueous sodium hydroxide, and results in the formation of a yellow precipitate of iodoform. The initial reaction consists of oxidation to the corresponding aldehyde or ketone. Halogenation of the carbonyl compound and hydrolysis of the product yield iodoform (Chapter 10).

$$R{\cdot}CHOH{\cdot}CH_3 \xrightarrow{NaIO} R{\cdot}CO{\cdot}CH_3 \xrightarrow{NaIO} R{\cdot}CO{\cdot}I_3$$

$$R{\cdot}CO{\cdot}I_3 \xrightarrow{NaOH} R{\cdot}CO{\cdot}ONa + \underset{\text{Iodoform}}{CHI_3\downarrow}$$

It is evident from the above series of reactions that methyl ketones also undergo the iodoform reaction. The reaction affords a means of distinguishing ethanol, which gives a positive response to the test, from methanol and propanol which do not.

### Metabolic Oxidation

Primary alcohols are rapidly metabolised in the liver, and when administered in small quantities are almost completely oxidised by liver alcohol dehydrogenase by pathways similar to those of chemical oxidation. Thus, ethanol is oxidised first to acetaldehyde and then to acetic acid. Further oxidation of acetic acid to carbon dioxide and water occurs mainly in muscle tissue.

Liver alcohol dehydrogenase requires the presence of nicotinamide adenine dinucleotide ($NA\overset{+}{D}$) as co-factor. Deuterium studies have shown that oxidation

$$CH_3{\cdot}CH_2OH \underset{\text{dehydrogenase}}{\overset{\text{Liver alcohol}}{\rightleftharpoons}} CH_3{\cdot}CHO \xrightarrow[\text{dehydrogenase}]{\text{Liver alcohol}} CH_3{\cdot}CO{\cdot}OH \longrightarrow CO_2 + H_2O$$

requires the breaking of an $\alpha$C—H(D) bond and transfer of a hydride ion to $\overset{+}{\text{NAD}}$. The hydroxylic hydrogen is removed as a proton by a basic group on the enzymic surface.

CH$_3$·C(D)(D)—O—H (B̈) Enzyme ⇌ CH$_3$·C(D)=O H(B)$^+$ Enzyme

(NAD$^+$) (NAD$^2$[H])

Secondary alcohols are similarly oxidised by liver alcohol dehydrogenase to the corresponding ketone. Oxidation of isopropanol is extremely rapid in man, acetone appearing in the breath and urine within 15 min. More complex secondary alcohols, however, are usually less readily oxidised than the corresponding primary alcohols, and more usually are excreted as the *O*-glucuronide or sulphate ester (p. 144 and **2**, 5). Tertiary alcohols are not oxidised by liver alcohol dehydrogenase, but are metabolised to the sulphate or glucuronide.

## UNSATURATED ALCOHOLS

The simplest unsaturated alcohol, vinyl alcohol, $CH_2{=}CHOH$, is unknown, since all attempts to prepare it lead to the isomeric acetaldehyde. Several of its derivatives, however, are well known, e.g. vinyl bromide, vinyl acetate and vinyl cyanide (acrylonitrile) and are available synthetically from acetylene (Chapter 6).

### Allyl Alcohol

Allyl alcohol is manufactured from propylene via allyl chloride.

$$H_2C{=}CH{-}CH_3 \xrightarrow{Cl_2(600^\circ)} H_2C{=}CH{\cdot}CH{\cdot}CH_2Cl \xrightarrow{HO^-} H_2C{=}CH{\cdot}CH_2OH$$

It can also be obtained by heating glycerol with formic or oxalic acid at 260°C.

$$CH_2OH\cdot CHOH\cdot CH_2OH \xrightarrow{H\cdot CO\cdot OH} CH_2OH\cdot CHOH\cdot CH_2\cdot O\cdot CO\cdot H$$

$$\downarrow 260^\circ \quad (\rightarrow CO_2 + H_2O)$$

$$H_2C{=}CH\cdot CH_2OH$$

A number of allyl derivatives occur naturally in plants; for example, allyl isothiocyanate in mustard oils, and allyl disulphide in onions and garlic.

Allyl alcohol is a colourless, mobile liquid (b.p. 97°C), with an acrid odour. It is miscible with water in all proportions, toxic and irritating to the skin. Chemically, allyl alcohol behaves both as a typical olefine and as an alcohol. Halogens and halogen acids add readily to the double bond. The addition of bromine gives 2,3-dibromopropanol, an intermediate in the preparation of *Dimercaprol* (British Anti-Lewisite; Chapter 14).

$$H_2C{=}CH\cdot CH_2OH + Br_2 \longrightarrow \underset{\text{2,3-Dibromopropanol}}{CH_2Br\cdot CHBr\cdot CH_2OH}$$

$$\downarrow NaSH$$

$$\underset{\textit{Dimercaprol}}{CH_2SH\cdot CHSH\cdot CH_2OH}$$

Preparation of *Dimercaprol* by this route may lead to contamination of the product by 2,3-dibromopropanol. A limit test for bromo compounds is, therefore, included in the monograph specification for *Dimercaprol*.

Allyl alcohol is more reactive than saturated aliphatic alcohols in reactions, such as alkyl halide formation and acid-catalysed dehydration, which involve C—O bond cleavage. Such reactions proceed via an intermediate carbonium ion, which is a resonance hybrid. The reaction is, therefore, facilitated by the enhanced stability which resonance gives to the carbonium ion.

$$H_2C{=}CH{-}\overset{+}{C}H_2 \longleftrightarrow H_2\overset{+}{C}{-}CH{=}CH_2$$

## ARYL ALCOHOLS

### Benzyl Alcohol

*Benzyl Alcohol* is usually manufactured from toluene by chlorination (Chapter 5) and subsequent hydrolysis with calcium hydroxide of the benzyl chloride produced.

$$C_6H_5CH_3 \xrightarrow[h\nu/\text{heat}]{Cl_2} C_6H_5CH_2Cl \xrightarrow[Cl^-]{HO^-} C_6H_5CH_2OH$$

*Benzyl Alcohol* combines well-defined antiseptic properties with mild local anaesthetic activity. It is often used on this account in lotions and ointments for topical application, and occasionally as a combined anaesthetic and preservative in injection solutions. *Benzyl Alcohol* is sensitive to oxidation due to the reactivity of the benzylic position (Chapter 5), and official monograph specifications usually include a limit test for benzaldehyde, the immediate oxidation product. There is also some evidence that the antiseptic action of *Benzyl Alcohol* is due to its ability to enter into radical reactions. Thus, the toxicity of a series of ring-substituted benzyl alcohols to *Aspergillus penicillium* is a function of the resonance substituent constant, $E_R$, in the modified Hammett equation (Hansch, Kutter and Leo, 1969).

$$\frac{\log k}{k_0} = \rho\sigma + \gamma E_R$$

in which $E_R$ = resonance substituent constant
$\sigma$ = polar substituent constant, and $\rho$ and $\gamma$ are reaction constants

Since the resonance substituent constant has been shown to have a linear relationship with radical reaction parameters (Price, 1948; Yamamoto and Otsu, 1967), it would appear that the antiseptic action of benzyl alcohols is due to interference with a radical mechanism.

**Di- and Tri-phenylcarbinols**

Diphenylcarbinol (benzhydrol) is also obtained from diphenylmethane by chlorination (or bromination) and hydrolysis.

$$C_6H_5{\cdot}CH_2{\cdot}C_6H_5 \xrightarrow{Cl_2/h\nu} C_6H_5{\cdot}CHCl{\cdot}C_6H_5 \xrightarrow{HO^-} C_6H_5{\cdot}CHOH{\cdot}C_6H_5$$

Triphenylcarbinol can be obtained readily by direct air oxidation of triphenylmethane or by oxidation with manganese dioxide.

$$(C_6H_5)_3{\cdot}C{-}H \xrightarrow{O_2} (C_6H_5)_3{\cdot}C{-}OH$$

The aryl alcohols, such as benzyl alcohol, are even more reactive than allyl alcohol (*q.v.*) in reactions involving C—O bond fission, owing to the increased ease of formation and enhanced stability of the intermediate carbonium ion as a result of charge delocalisation. This property increases with increasing aryl substitution, as in diphenylcarbinol and triphenylcarbinol.

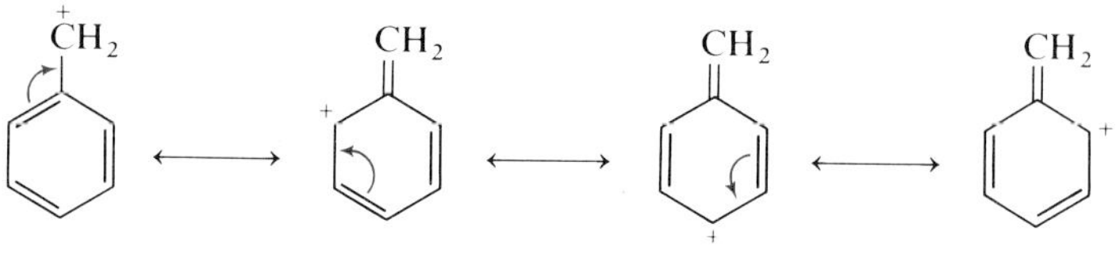

$$\overset{+}{C}H_2\text{-}C_6H_5 \longleftrightarrow CH_2{=}C_6H_5^{+}\,(ortho) \longleftrightarrow CH_2{=}C_6H_5^{+}\,(para) \longleftrightarrow CH_2{=}C_6H_5^{+}\,(ortho')$$

## ETHERS

### General Properties

Simple dialkyl ethers are generally mobile, low boiling, flammable liquids, which are usually immiscible with water. The high volatility of ethers and immiscibility with water reflect their hydrocarbon characteristics, and the masking of the hydroxyl group, which in the parent alcohols leads to hydrogen bonding. Diethyl ether has b.p. 35°C, and a low flash point; it is immiscible with water, but soluble (1 in 10) in water, whilst water is soluble (*ca* 1%) in ether. Ethers have good solvent properties for a wide range of organic compounds, and are widely used as extraction solvents for the isolation of such materials from aqueous systems. In mammals, they have a high affinity for fatty tissue and hence diethyl ether and divinyl ether are used as inhalation anaesthetics. Their anaesthetic activity like that of other general anaesthetics, such as *Cyclopropane* and *Chloroform*, is structurally non-specific, and depends on their ability to achieve an adequate degree of saturation of lipid components of the central nervous system (**2**, 1). *Anaesthetic Ether* (Ether) is not an ideal anaesthetic because of its flammability, and proneness to radical oxidation with the formation of peroxides (p. 162).

### Basic Function

Ethers function as weak bases and act as proton acceptors in acid solution. Thus, they are soluble in concentrated inorganic acids with the formation of oxonium salts.

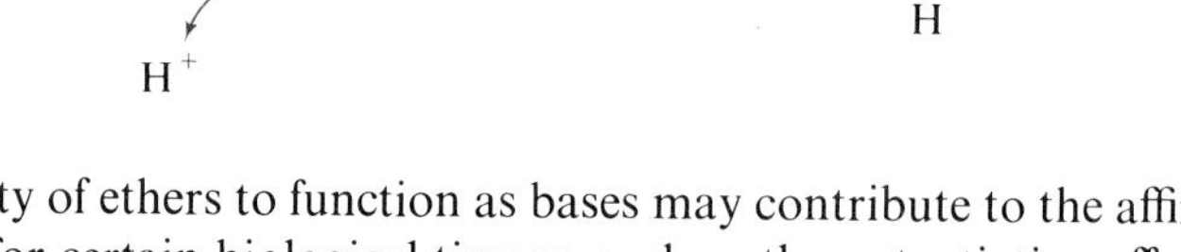

$$CH_3{\cdot}CH_2{\cdot}\ddot{O}{\cdot}CH_2{\cdot}CH_3 \xrightarrow{H^+} CH_3{\cdot}CH_2{\cdot}\overset{+}{O}(H){\cdot}CH_2{\cdot}CH_3$$

The ability of ethers to function as bases may contribute to the affinity which they show for certain biological tissues, such as the potentiating effect they are known to exhibit on neuromuscular blockade by curare alkaloids and similar neuromuscular blocking agents (Stenlake, 1963).

Table 16. Nomenclature of Ethers

| Formula | Name | b.p. (°C) |
|---|---|---|
| $CH_3 \cdot O \cdot CH_3$ | Dimethyl ether | −23.6 |
| $CH_3 \cdot O \cdot CH_2 \cdot CH_3$ | Ethyl methyl ether | +10.8 |
| $CH_3 \cdot CH_2 \cdot O \cdot CH_2 \cdot CH_3$ | Diethyl ether | 34.5 |
| $CH_3 \cdot CH_2 \cdot O(CH_2)_2 \cdot CH_3$ | Ethyl propyl ether | 63.6 |
| $CH_3(CH_2)_2 \cdot O(CH_2)_2 \cdot CH_3$ | Dipropyl ether | 91 |
| $CH_3(CH_2)_3 \cdot O(CH_2)_3 \cdot CH_3$ | Dibutyl ether | 142 |
| $CH_2{=}CH \cdot O \cdot CH{=}CH_2$ | Divinyl ether | 39 |

**Acid-catalysed Decomposition**

Strong halogen acids (HBr and HI) promote the decomposition of ethers at room temperature by protonation and subsequent attack at the α-carbon atom by the halide ion. The ion generally attacks the smaller of the two alkyl groups, except where the larger group has a tertiary α-carbon atom.

$$\mathbf{CH_3 \cdot O \cdot CH_2 \cdot CH_3} \underset{I^-}{\overset{HI}{\rightleftharpoons}} CH_3{-}\overset{+}{\underset{H}{O}} \cdot CH_2CH_3 \longrightarrow \mathbf{CH_3I} + \mathbf{HO \cdot CH_2 \cdot CH_3}$$

If the ether is heated with the halogen acid, then both alkyl groups give an alkyl halide, any alcohol initially formed being converted immediately to alkyl halide.

$$CH_3 \cdot O \cdot CH_2 \cdot CH_3 + 2HI \xrightarrow{\text{Reflux}} CH_3I + CH_3 \cdot CH_2I + H_2O$$

This reaction forms the basis for the Zeisel determination of methoxyl ($OCH_3$) in phenolic ethers, the methyl iodide released being absorbed into aqueous ethanolic silver nitrate solution to give a precipitate of silver iodide which can be collected and weighed.

$$C_6H_5\text{-}\mathbf{OCH_3} + \mathbf{HI} \longrightarrow C_6H_5\text{-}\mathbf{OH} + \mathbf{CH_3I}$$

$$CH_3I \xrightarrow{AgNO_3} AgI\downarrow$$

$$CH_3O{-} \equiv CH_3I \equiv AgI$$

*O*-Dealkylation of ethers to the corresponding alcohols forms the main pathway for their metabolism in mammals (**2**, 5).

**Halogenation**

Ethers are attacked by chlorine and bromine in the dark to give halogenated derivatives.

$$CH_3 \cdot CH_2 \cdot O \cdot CH_2 \cdot CH_3 \xrightarrow{Cl_2} CH_3 \cdot CHCl \cdot O \cdot CH_2 \cdot CH_3$$

$$\downarrow Cl_2$$

$$CH_3 \cdot CHCl \cdot O \cdot CHCl \cdot CH_3$$

1,1′-Dichlorodiethyl ether

**Ether Peroxides**

Ethers readily form peroxides by light-catalysed air oxidation. According to Rieche and Meister (1936), that from diethyl ether has the structure $CH_3 \cdot CH(O \cdot OH) \cdot O \cdot CH_2 \cdot CH_3$. Decomposition of this peroxide leads to the formation of acetaldehyde and acetic acid, which are also possible contaminants of *Anaesthetic Ether*. A more serious hazard, however, is the liability of ether peroxide to explode on distillation of the parent ether. The peroxides are non-volatile and concentrate in the non-volatile fraction as the distillation proceeds increasing the explosion risk. This explosive decomposition of ether peroxides can be prevented by the addition of ferrous sulphate crystals prior to distillation.

The explosion hazard due to peroxides represents a particularly serious risk to the patient anaesthetised with *Anaesthetic Ether*, and particular care is necessary to ensure that the anaesthetic is as near as possible free from contamination by peroxides. The official monograph specification, therefore, includes a sensitive test based on the use of potassium iodide and starch solution. Peroxides liberate iodine from aqueous potassium iodide, and the solution is required to give no brown or reddish colour due to starch–iodine complex formation after shaking with a sample of the *Anaesthetic Ether* and allowing the mixture to stand in the dark for thirty minutes.

**Vinyl Ethers**

*Vinyl Ether* is used as an anaesthetic. Such compounds are readily susceptible to hydrolysis and oxidation, and require protection from light, and the addition of a suitable stabiliser (*N*-phenyl-1-naphthylamine) to prevent excessive decomposition on storage. Hydrolytic decomposition of vinyl ethers gives rise to the corresponding aldehyde.

The detection of aldehydogenic material in phospholipid preparations (Rapport, Lerner, Alonzo and Franzl, 1957) is due to hydrolytic decomposition of plasmalogens (Chapter 20).

$H_2C{=}CH{\cdot}O{\cdot}CH{=}CH_2$

*Vinyl Ether*

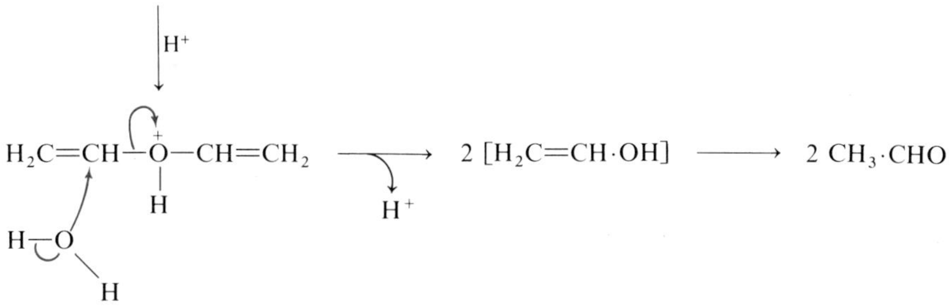

$$H_2C{=}CH{-}\overset{+}{O}(H){-}CH{=}CH_2 \xrightarrow[-H^+]{} 2\,[H_2C{=}CH{\cdot}OH] \longrightarrow 2\,CH_3{\cdot}CHO$$

(H⁺ adds to the ether oxygen; $H_2O$ attacks.)

## OXIRANES (EPOXIDES)

Cyclic ethers, such as tetrahydrofuran and tetrahydropyran, exhibit properties which are essentially the same as those of the dialkyl ethers. The cyclic ether, ethylene oxide, however, is characteristic of a very reactive group of compounds, known as oxiranes or epoxides, which owe their reactivity to the highly strained three-membered ring.

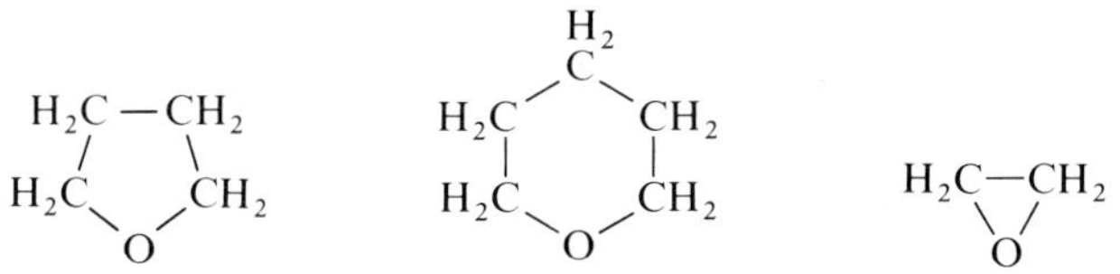

Tetrahydrofuran   Tetrahydropyran   Ethylene oxide

Oxiranes are formed in the laboratory by direct oxidation of alkenes with peracids (Chapter 4), or by cyclisation of vicinal halohydrins with calcium hydroxide solution.

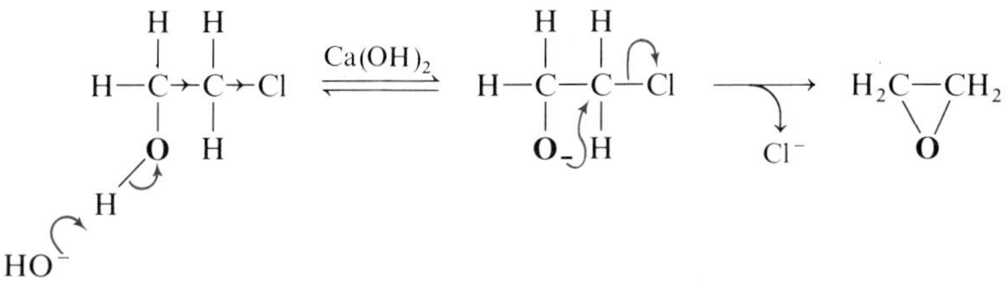

$$H{-}CH(OH){-}CH_2{-}Cl \underset{}{\overset{Ca(OH)_2}{\rightleftharpoons}} H{-}CH(O^-){-}CH_2{-}Cl \xrightarrow[-Cl^-]{} H_2C{-}CH_2\ (O)$$

HO⁻

On a manufacturing scale, ethylene and propylene oxides are produced by passing a mixture of air (or oxygen) and olefine over a silver catalyst at 250°C.

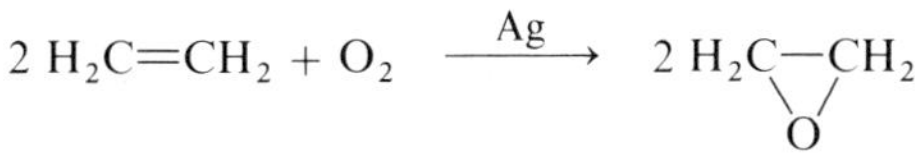

$$2\,H_2C{=}CH_2 + O_2 \xrightarrow{Ag} 2\,H_2C{-}CH_2\ (O)$$

Air is kept in excess to reduce the risk of explosion. About 50–60% conversion of ethylene to ethylene oxide is achieved, the product being removed from the effluent gas by absorption into water under pressure, from which it is subsequently distilled (b.p. 10°C).

Ethylene oxide is widely used as a cold gaseous sterilising agent for use with non-reactive compounds where heat sterilisation or aqueous filtration procedures through bacterial filters are considered unsuitable. Before implementing sterilisation processes using ethylene oxide, it is essential that the stability of the compound and container materials to ethylene oxide be confirmed. Also there must be no retention of traces of ethylene oxide or chlorocarbons arising from the sterilisation process. The process suffers from the further limitation that the gas is unable to penetrate solid materials, and hence only surface sterilisation results.

## Acid-catalysed Hydrolysis

Oxiranes are readily hydrolysed by aqueous mineral acid to yield the corresponding glycol. Protonation of the ether oxygen atom increases the electron deficiency of the adjacent carbon atoms, and promotes nucleophilic attack of a water molecule. Ring strain in the three-membered ring facilitates breaking of the C—O bond.

$$H_2C\text{—}CH_2\ (\text{—O—}) \xrightleftharpoons{H^+} H_2C\text{—}CH_2\ (\text{—}\overset{+}{O}H\text{—}) \xrightarrow{H_2\ddot{O}} H_2\overset{+}{O}\cdot CH_2\cdot CH_2\cdot OH \xrightarrow{-H^+} HO\cdot CH_2\cdot CH_2\cdot OH$$

Ethyleneglycol

$$CH_3\cdot CH\text{—}CH_2\ (\text{—O—}) \xrightleftharpoons{H^+} CH_3\text{—}CH\text{—}CH_2\ (\text{—}\overset{+}{O}H\text{—}) \xrightarrow{H_2\ddot{O}} CH_3\cdot CHOH\cdot CH_2\overset{+}{O}H_2 \xrightarrow{-H^+} CH_3\cdot CHOH\cdot CH_2\cdot OH$$

Propyleneglycol

Hydrolysis yields the corresponding *trans*-glycol with compounds such as cyclohexane epoxide, where the geometry of the oxirane and product glycol is fixed.

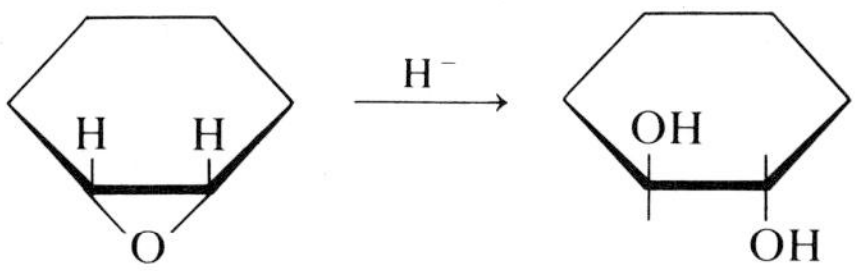

$$\text{Cyclohexane-1,2-epoxide} \xrightarrow{H^-} \textit{trans}\text{-Cyclohexane-1,2-diol}$$

Cyclohexane-1,2-epoxide *trans*-Cyclohexane-1,2-diol

**Base-catalysed Hydrolysis**

Oxiranes are also readily hydrolysed by aqueous sodium hydroxide to form the corresponding (*trans*) diol.

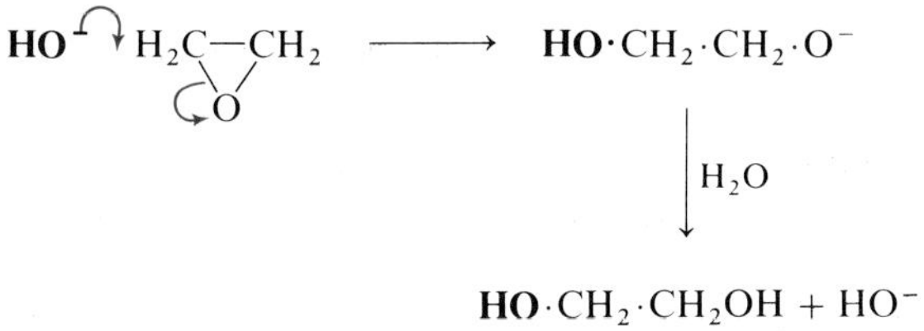

$$HO^- + H_2C{-}CH_2(O) \longrightarrow HO{\cdot}CH_2{\cdot}CH_2{\cdot}O^- \xrightarrow{H_2O} HO{\cdot}CH_2{\cdot}CH_2OH + HO^-$$

**Epoxides as Metabolic Intermediates**

There is some indirect evidence that epoxides are formed as unstable intermediates in the metabolism of aromatic hydrocarbons such as benzene, naphthalene and anthracene (**2**, 5), which are usually excreted as the corresponding mercapturic acids (Boyland and Williams, 1965). It is thought that epoxides formed metabolically from carcinogenic polycyclic aromatic hydrocarbons may be the active proximate carcinogens, since the hydroxy-compounds formed via the epoxides are non-carcinogenic.

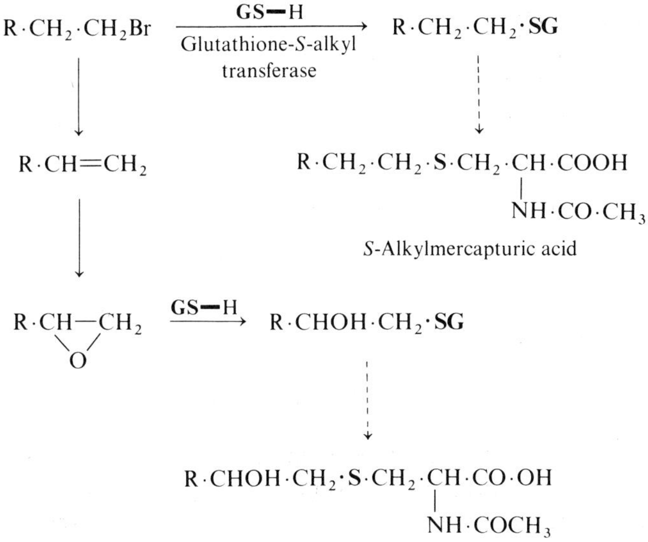

$$R{\cdot}CH_2{\cdot}CH_2Br \xrightarrow[\text{Glutathione-}S\text{-alkyl transferase}]{GS{-}H} R{\cdot}CH_2{\cdot}CH_2{\cdot}SG \dashrightarrow R{\cdot}CH_2{\cdot}CH_2{\cdot}S{\cdot}CH_2{\cdot}CH(NH{\cdot}CO{\cdot}CH_3){\cdot}COOH$$

*S*-Alkylmercapturic acid

$$R{\cdot}CH_2{\cdot}CH_2Br \longrightarrow R{\cdot}CH{=}CH_2 \longrightarrow R{\cdot}CH{-}CH_2(O) \xrightarrow{GS{-}H} R{\cdot}CHOH{\cdot}CH_2{\cdot}SG \dashrightarrow R{\cdot}CHOH{\cdot}CH_2{\cdot}S{\cdot}CH_2{\cdot}CH(NH{\cdot}COCH_3){\cdot}CO{\cdot}OH$$

G—SH = Glutathione (Chapter 14)

Epoxides have similarly been postulated as intermediates in the metabolism of bromoalkanes which form both alkyl and hydroxyalkylmercapturic acids (Barnsley, 1964). The latter are believed to be formed via the corresponding alkene and epoxide (James and Jeffery, 1964).

## Glycol Ether Formation

Oxiranes react under acidic conditions with alcohols such as methanol and ethanol to yield glycol monoalkyl ethers.

$$\mathbf{CH_3\ddot{O}H} \quad H_2C{-}CH_2 \text{ (ring } \overset{+}{O}H) \longrightarrow \mathbf{CH_3{\cdot}O}CH_2{\cdot}CH_2OH$$

Ethyleneglycol monomethyl ether

Ethyleneglycol monomethyl ether (methyl cellosolve) is used as a solvent for nitrocellulose in the preparation of cellulose lacquers. It also has good solvent properties for a wide range of organic compounds. Ethyl cellosolve is ethyleneglycol monoethyl ether.

Further reaction of the cellosolves with ethylene oxide gives the carbitols (diethyleneglycol monoalkyl ethers) which also have valuable solvent properties.

$$CH_3O{\cdot}CH_2{\cdot}CH_2{\cdot}\mathbf{\ddot{O}H} \quad H_2C{-}CH_2 \text{ (ring } \overset{+}{O}H) \longrightarrow CH_3O{\cdot}CH_2{\cdot}CH_2{\cdot}\mathbf{O}{\cdot}CH_2{\cdot}CH_2OH$$

Methyl carbitol

## Polyglycol Formation

Condensation of ethylene or propylene oxides with either water, ethylene glycol or preferably, diethylene glycol, at high temperature under pressure, and in the presence of sodium hydroxide, yields polyglycols. Diethylene glycol forms an alkoxide with sodium hydroxide which facilitates reaction with the alkylene oxide.

$$HO{\cdot}CH_2{\cdot}CH_2{\cdot}O{\cdot}CH_2{\cdot}CH_2{\cdot}O^- \quad (H_2C{-}CH_2 \text{ (ring O)})_{n-1}$$

$$\downarrow$$

$$HO{\cdot}CH_2{\cdot}CH_2(O{\cdot}CH_2{\cdot}CH_2)_n{\cdot}O^-$$

The alkaline catalyst is neutralised with acetic or phosphoric acid during the cooling process. According to the supply of alkylene oxide, polyglycols of any required molecular weight can be formed. The polyglycols and their derivatives are used as dispersing and emulsifying agents (Chapter 20).

**Addition to Amines**

Epoxides react readily with primary and secondary amines to form alkanolamines (Chapter 16).

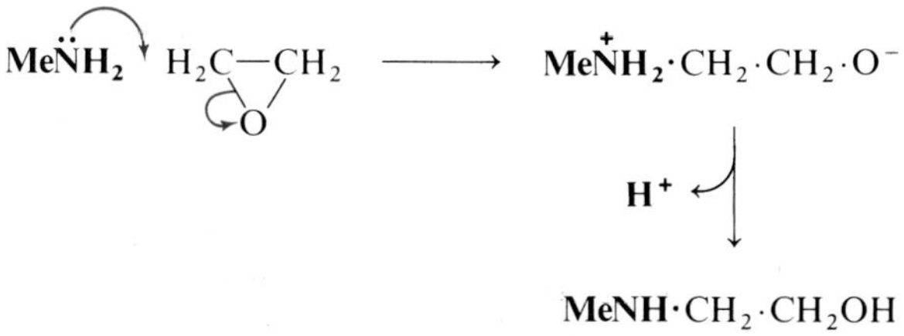

$$\mathbf{Me\ddot{N}H_2} \; H_2C{-}CH_2 \text{ (epoxide, O bridging)} \longrightarrow \mathbf{Me\overset{+}{N}H_2}{\cdot}CH_2{\cdot}CH_2{\cdot}O^-$$

$$\xrightarrow{H^+} \mathbf{MeNH}{\cdot}CH_2{\cdot}CH_2OH$$

*N*-Methylethanolamine

# 8 Phenols

### Structure and Classification

Phenols are hydroxy-aromatic compounds, in which the hydroxyl group is attached directly to the aromatic ring. The simplest example is phenol itself, which lends its name to the whole class of compounds. Its structure, like that of other benzene derivatives, can be represented as a hybrid of various resonance forms, which arise by interaction of unshared electron pairs with the $\pi$-electrons of the aromatic ring.

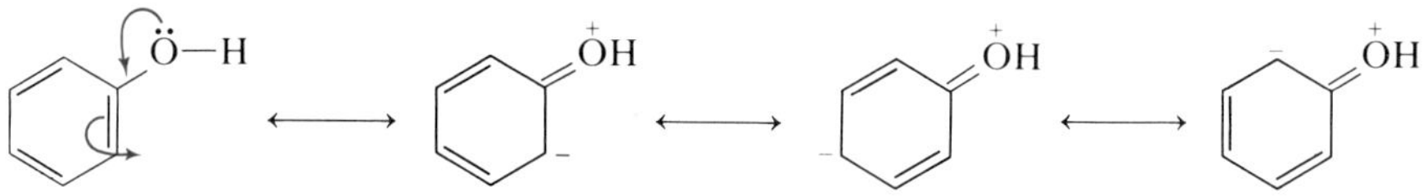

Phenols, like alcohols, can be classified as monohydric (Chapter 7), dihydric, trihydric (Chapter 20) and polyhydric according to whether they possess one, two, three or more hydroxyl groups.

## MONOHYDRIC PHENOLS

### General Properties

Many, though not all, phenols possess a distinctive and characteristic odour. Most are insoluble or only slightly soluble in water. *Phenol* is hygroscopic and soluble (1 in 20) in water at 20°C. At higher concentrations, phenol and water form a two-phase system consisting of two saturated solutions, one of water in phenol and the other of phenol in water. Mutual solubility increases as the critical solution temperature (65.85°C) is approached; above this temperature, phenol and water are miscible in all proportions.

*Liquefied Phenol*, which is phenol (80% w/w) in water, is used as an antiseptic and preservative, and also as an extraction solvent. *Phenol* is caustic when applied to the skin, and in high concentration acts as a general protoplasmic poison (Hugo, 1957). Lower concentrations disrupt the cytoplasmic membrane of bacteria, causing leakage of cellular constituents (Gale and Taylor, 1947). There is evidence that these actions in part relate to the ability of phenols to function as weak acids capable of ionisation, but the antibacterial activity is also more importantly related to their surface activity. Thus, bacterial efficiency of lower n-alkyl phenols increases with increasing chain length in parallel with their increasing surface activity. Solubility in water, on the other hand, also decreases sharply with increasing molecular weight, and in consequence alkyl phenols with more than five or six carbon alkyl chains show a marked fall in antibacterial activity compared with their lower homologues (Table 17). This is expressed in terms of phenol coefficients, which show the antibacterial activity against a particular organism (*B. typhosus*) relative to that of phenol.

**Acidity**

Phenols are acidic and ionise in aqueous media to form phenoxide ions.

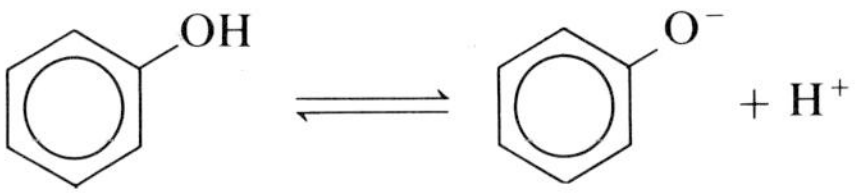

$$\text{OH} \rightleftharpoons \text{O}^- + \text{H}^+$$

Phenols are more strongly acidic than alcohols (p$K_a$ *ca* 16–18), as a result of resonance, which is only possible when the hydroxyl group is directly attached to an aromatic ring system. This effect operates by increasing the stability of the phenoxide ion.

Two opposing factors operate in determining the extent to which phenols are ionised as distinct from alcohols. Oxygen, being more electronegative than carbon, tends to withdraw electrons from the ring, and thus inhibits the loss of a proton. This is the inductive effect, which depresses ionisation, and is the sole effect operating in alcohols.

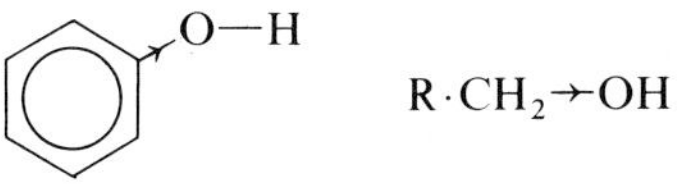

$$\text{O—H} \qquad R\cdot CH_2 \rightarrow OH$$

This electron withdrawal, however, is opposed and outweighed in phenols by the much more powerful mesomeric effect, which arises by interaction of unshared electron pairs on the oxygen atom with the $\pi$-electrons of the aromatic ring (p. 129). The positively charged oxygen atom in the dipolar ion forms

Table 17. Phenol Coefficients of *p*-Alkylphenols (Suter, 1941)

| Alkyl group (R) | Phenol coefficient against *B. typhosus* (20°C) |
|---|---|
| — | 1 |
| Methyl | 2.5 |
| Ethyl | 7.5 |
| n-Propyl | 20 |
| n-Butyl | 70 |
| n-Amyl | 104 |
| n-Hexyl | 90 |
| n-Heptyl | 20 |

weakens the O—H bond and facilitates separation of the proton. Ionisation is, therefore, promoted. The ion, once formed, is also stabilised by resonance.

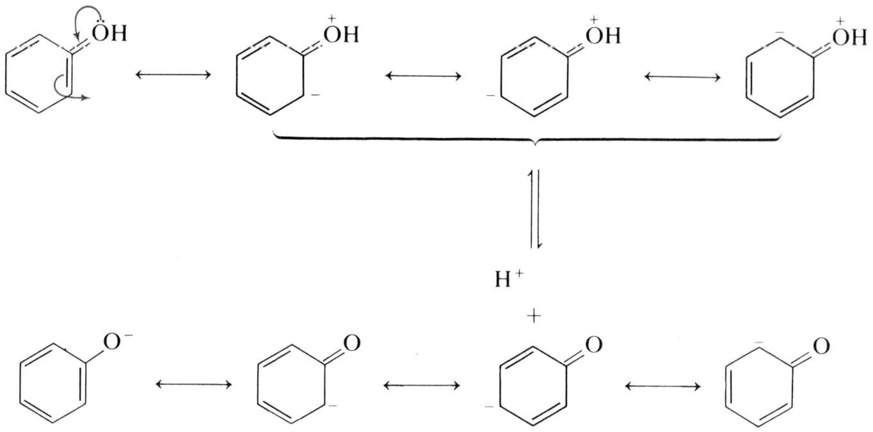

It will be apparent that whereas the resonance forms of the phenoxide ion merely carry a negative charge, those of phenol carry both positive and negative charges. Energy is required to promote the separation of charge, hence the greater energy content and lower stability of phenol relative to the phenoxide ion. Ionisation is thus promoted by the greater stability of the phenoxide ion.

The interaction of phenols with synthetic peptides such as nylon, and natural proteins such as silk, keratin and bacterial cell-wall proteins can be related to the acidic function of phenols in interaction with the peptide bond.

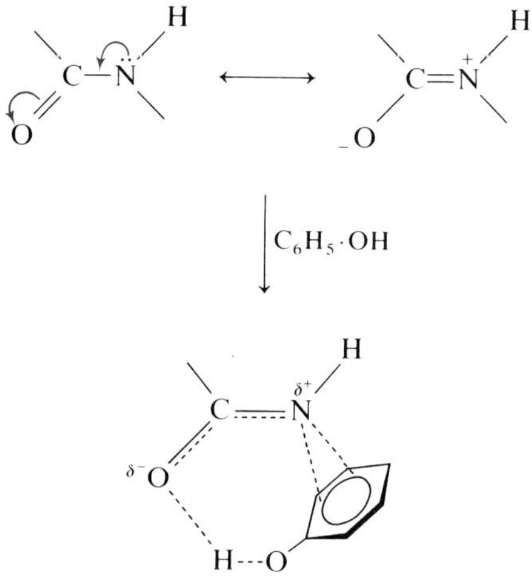

$C_6H_5 \cdot OH$

This type of interaction is probably important in the binding of oestrogens to albumin and specific oestrogen-binding globulins (**2**, 4).

**Salt Formation**

Phenols are acidic and dissolve in solutions of alkali metal hydroxides to form water-soluble salts, known as phenoxides or phenates.

$$C_6H_5OH + NaOH \longrightarrow C_6H_5ONa + H_2O$$

In this respect, they differ from alcohols which form salts only on treatment with alkali metals.

$$CH_3 \cdot CH_2 \cdot OH \xrightarrow{NaOH} \text{No reaction}$$

$$CH_3 \cdot CH_2 \cdot OH \xrightarrow{Na} CH_3 \cdot CH_2 \cdot ONa$$

The parent phenols are liberated from their alkali metal salts by the action of dilute mineral acids, sulphonic acids, carboxylic acids or carbon dioxide. Phenols are, therefore, weaker acids than carbonic, carboxylic or sulphonic acids, and in the absence of modifying substituents have a $pK_a$ of about 10. Thus, unlike carboxylic or sulphonic acids, phenols do not react with sodium carbonate or sodium bicarbonate. This difference in acidity, and hence reactivity, forms the basis of a convenient method for the separation of mixtures of phenols and carboxylic or sulphonic acids, in which an ethereal solution of the mixture is shaken with aqueous sodium carbonate solution. The acid component forms a water-soluble sodium salt and passes into the aqueous phase, whilst the phenol remains in the ethereal layer.

Similarly, alkali metal phenoxides, unlike the corresponding salts of carboxylic acids, react with carbon dioxide to liberate the parent phenol.

$$CO_2 + H_2O \rightleftharpoons H_2CO_3$$

$$2\ C_6H_5 \cdot ONa + H_2CO_3 \longrightarrow 2\ C_6H_5 \cdot OH + Na_2CO_3$$

Reactions such as this form the basis of the use of resins containing free phenolic hydroxyl groups as weak cation-exchange resins.

**The Effect of Nuclear Substitution on the Ionisation of Phenols**

Electron-repelling substituents, such as $CH_3$ or $C_2H_5$, oppose resonance stabilisation of the phenoxide ion, and hence decrease acidic strength and depress ionisation. In consequence, all three cresols are weaker acids than phenol (Table 18).

$$R\text{-}C_6H_4\text{-}OH \rightleftharpoons R\text{-}C_6H_4\text{-}O^- + H^+$$

Table 18. Dissociation Constants of Substituted Phenols (25°C)

| Phenol | p$K_a$ | Phenol | p$K_a$ |
|---|---|---|---|
| Phenol | 9.98 | *o*-Chlorophenol | 8.48 |
| *o*-Cresol | 10.28 | *m*-Chlorophenol | 9.02 |
| *m*-Cresol | 10.08 | *p*-Chlorophenol | 9.38 |
| *p*-Cresol | 10.14 | *o*-Methoxyphenol | 9.98 |
| *o*-Nitrophenol | 7.23 | *m*-Methoxyphenol | 9.65 |
| *m*-Nitrophenol | 8.40 | *p*-Methoxyphenol | 10.21 |
| *p*-Nitrophenol | 7.15 | *o*-Aminophenol | 9.71 |
| 2,4-Dinitrophenol | 4.00 | *m*-Aminophenol | 9.87 |
| 2,4,6-Trinitrophenol | 0.71 | *p*-Aminophenol | 10.30 |

Conversely, electron-attracting substituents ($NO_2$, $\overset{+}{N}(CH_3)_3$, CHO, CO·R, CO·OR, CN, $CF_3$) promote the withdrawal of electrons from the phenolic oxygen into the ring, and hence increase acidic strength and ionisation.

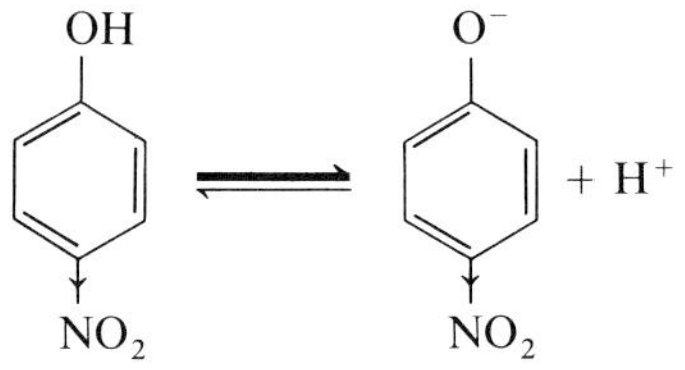

OH O$^-$

$\rightleftharpoons$ + $H^+$

$NO_2$ $NO_2$

Since, however, electron withdrawal is greatest from the *o*- and *p*-positions, *o*- and *p*-nitrophenols are stronger acids than the corresponding *m*-isomer. Di- and tri-substitution increases acidity still further. Thus, both 2,4-dinitrophenol and 2,4,6-trinitrophenol (picric acid) are much more readily soluble in aqueous sodium bicarbonate than the mononitrophenols.

Substituents such as hydroxyl, methoxyl, halogen and amino with opposing inductive and mesomeric effects will either increase or decrease ionisation depending on the balance between the two effects and the position of the substituent in the ring. Thus, +M, −I substituents, such as $NH_2$ and $CH_3O$, in which the +M effect predominates decrease acidic strength in the *p*-position, but in the *m*-position where only the −I effect is operative, increase acidic strength. In contrast, substituents such as halogens in which the −I effect predominates, increase acidic strength irrespective of the position of the substituent. It should be noted, however, that the electronic influence of *o*-substituents is also modified by other factors such as hydrogen bonding and steric effects.

The enhanced antibacterial activity of chlorophenols, such as *Chlorocresol* and *Chloroxylenol*, compared with unsubstituted phenols relates in part to their greater acidic strength and hence their enhanced capability for ionisation at physiological pH.

**Free Radical Formation**

Apart from their readiness to lose a proton and form a phenoxide ion, phenols readily undergo homolytic cleavage of the O—H bond to form a phenoxy radical. Free radical formation occurs readily under the influence of one-electron oxidising agents, such as ferric chloride, potassium ferricyanide, lead and manganese dioxides and lead tetra-acetate, in either neutral or alkaline solution.

$$\mathrm{Ph{\cdot}O{-}H} \quad \mathrm{Fe^{3+}} \longrightarrow \mathrm{PhO{\cdot}} + \mathrm{H^{+}} + \mathrm{Fe^{2+}}$$

$$\mathrm{Ph{\cdot}O^{-}} + [\mathrm{Fe(CN)_6}]^{3-} \longrightarrow \mathrm{PhO{\cdot}} + [\mathrm{Fe(CN)_6}]^{4-}$$

These phenoxy radicals are stabilised to a greater degree than simple alkoxy radicals because the charge due to the odd electron is spread by resonance over the aromatic ring, mainly into the *ortho* and *para* positions.

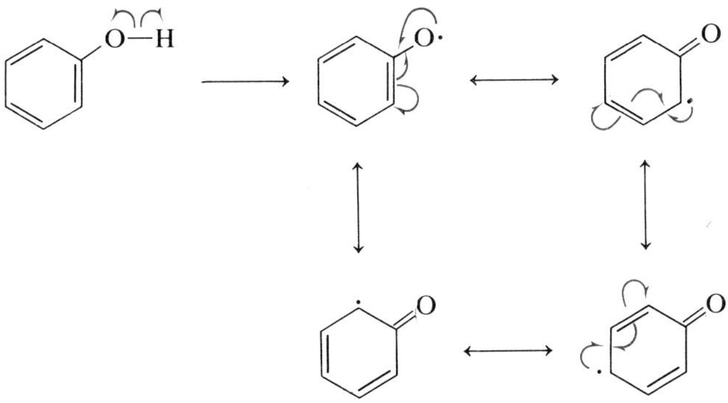

Free radicals of this type will react by coupling, as for example in the case of 2,6-di-t-butylphenol (Kharasch and Joshi, 1957).

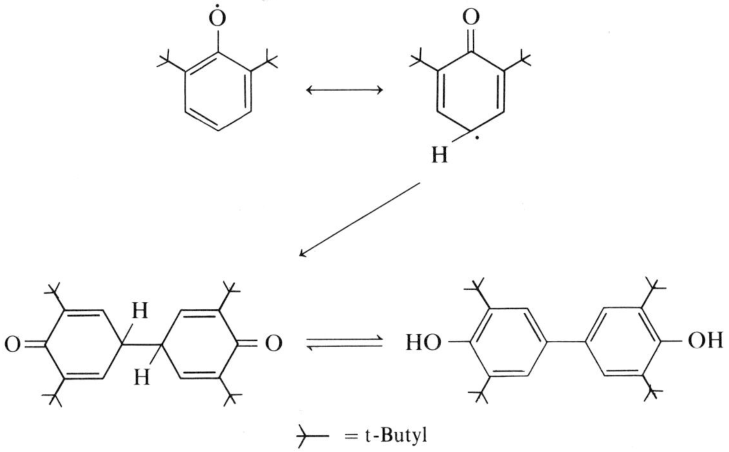

$+$ = t-Butyl

*Para-para* and similar *ortho-ortho* coupling no doubt accounts for the action of substances such as *Thymol*, *Octyl Gallate* and *Butylated Hydroxyanisole* (BHA), which are used extensively as antioxidants.

## Antioxidants

Many pharmaceutical products are susceptible to deterioration as a result of autoxidation. This can often be retarded, though seldom prevented by the presence of an antioxidant. Autoxidations, however, are usually radical-induced chain reactions, so that once initiated they cannot be slowed or reversed by the subsequent addition of an antioxidant.

Most antioxidants are themselves highly susceptible to radical-induced oxidation, and function as preferentially oxidisable free-radical scavengers. Many are phenolic in character and presumably owe their effectiveness to the ease with which this type of compound produces radicals capable of dimerisation by radical coupling, a process which favours termination rather than initiation or propagation of radical-induced chain reactions. Thus, the typical autoxidation of unsaturated oils and fats (Chapter 4) is interrupted at the initiation stage by radical attack on the phenolic antioxidant in preference to the allylic position of the unsaturated hydrocarbon chain.

$-CH_2\cdot\dot{C}H\cdot CH(OO\cdot)-$  HO–C₆H₄–R

$O_2$

$-CH_2\cdot CH{=}CH-$

$-\dot{C}H\cdot CH{=}CH-$

$-CH_2\cdot CH_2\cdot CH(OOH)-$  ·O–C₆H₄–R

Typical antioxidants used in the preservation of oils and fats from rancidity include *Thymol*, *Butylated Hydroxyanisole*, *Octyl Gallate* and *Dodecyl Gallate*. They are also used for the stabilisation of volatile anaesthetics and other pharmaceutical products which are susceptible to oxidation, causing loss of potency and formation of toxic or otherwise undesirable breakdown products. *Thymol* is also used as an antibacterial agent (phenol coefficient *ca* 37). It occurs as the principal constituent of thyme and ajowan oils, and is also prepared by rearrangement of piperitone, the principal constituent of Japanese peppermint oil.

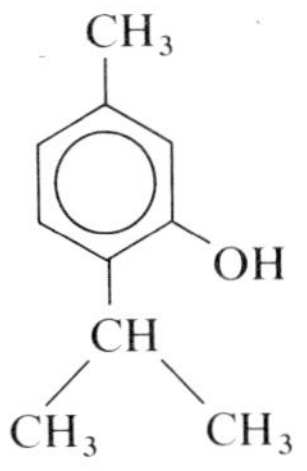

*Thymol*

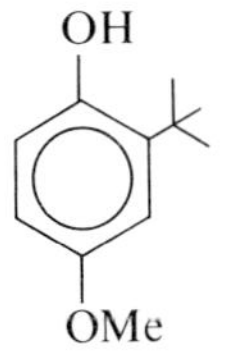

*Butylated Hydroxyanisole* (BHA)

*Butylated Hydroxytoluene* (BHT)

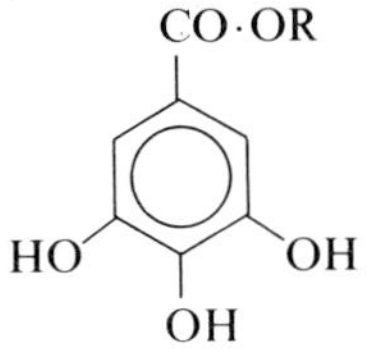

*Propyl Gallate* (R = $CH_3(CH_2)_2—$)
*Octyl Gallate* (R = $CH_3(CH_2)_7—$)
*Dodecyl Gallate* (R = $CH_3(CH_2)_{11}—$)

= t-Butyl

*Propyl Gallate*, although only slightly soluble in water (1 in 1000), is more soluble in water but much less soluble in vegetable oils than *Octyl* and *Dodecyl Gallates*, and is therefore useful as an antioxidant in aqueous creams and similar products. *Butylated Hydroxyanisole* (BHA) has the particular advantages that it is heat resistant and non-volatile in steam. *Para-para*-coupling is blocked by the methoxyl substituent, and autoxidation probably proceeds by coupling through the vacant *ortho* position. *Butylated Hydroxyanisole* is metabolised in mammalian systems by conjugation, and is excreted as the *O*-glucuronide and sulphate ester (Astill *et al.*, 1960, 1962).

The presence of αC—H groups, particularly in the *ortho* or *para* positions to the phenolic hydroxyl, as in *Butylated Hydroxytoluene* (BHT), leads to coupling with the formation of methylenequinones by a process of autoxidation. This reaction has been shown with BHT to proceed by rearrangement of the resonance-stabilised phenoxy radical to the benzyl radical, and thence by coupling to a diphenylethane. Oxidation of the latter to a stable quinone methine then proceeds at the expense of the benzyl radical which is itself converted to BHT (Cook, Nash and Flanagan, 1955).

Electron-spin resonance provides evidence of the resonance stabilisation of the primary phenoxy radical. Similar studies also provide evidence of hyperconjugation involving αC—H bonds in the *ortho* and *para* positions to the phenoxy group. Rearrangement to the benzyl radical, and rearrangements resulting from attack on the *ortho*-t-butyl C—H bonds by the phenoxy radical are probably also significant in mammalian metabolism of BHT, as shown by the end-products which are excreted (Ladomery, Ryan and Wright, 1967).

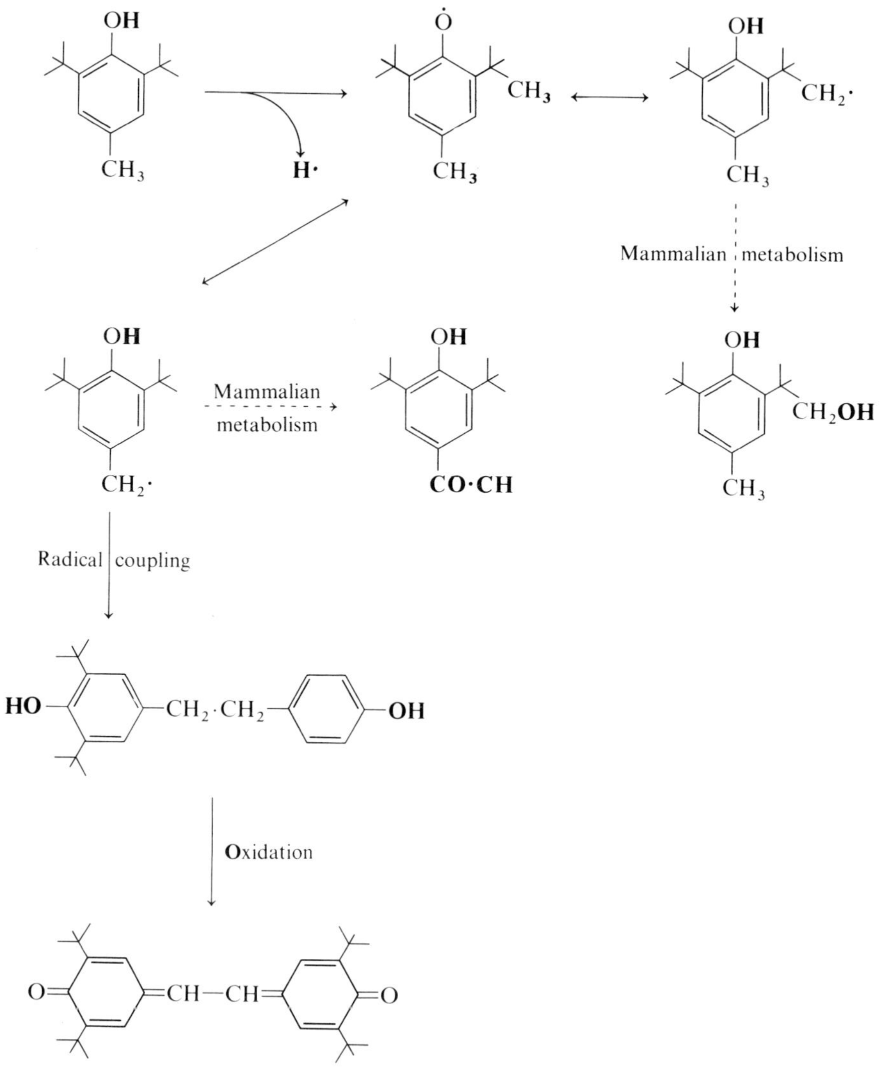

OH CH$_3$ H· Ȯ CH$_3$ CH$_3$ OH CH$_2$· CH$_3$

Mammalian metabolism

OH CH$_2$· Mammalian metabolism OH CO·CH OH CH$_2$OH CH$_3$

Radical coupling

HO CH$_2$·CH$_2$ OH

Oxidation

O =CH—CH= O

The latter reaction, however, is evidently not important in physico-chemical systems *in vitro*, since *ortho*-disubstituted phenols, such as 2,4,6-tri-t-butylphenol, give rise to long-lived 'stable' free radicals as a result of steric hindrance.

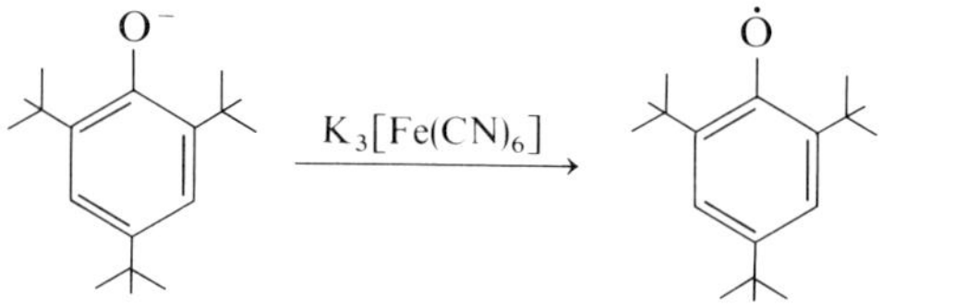

O$^-$ $K_3[Fe(CN)_6]$ Ȯ

Deep blue 'stable' radical

The tocopherols (Vitamins E), which occur naturally in a number of vegetable oils including *Maize Oil* (Corn Oil), wheat germ, soya bean and cottonseed oils, are also used as antioxidants, particularly in the food industry. The tocopherols are all closely related chemically, differing only in the extent of methyl substitution in the chroman ring.

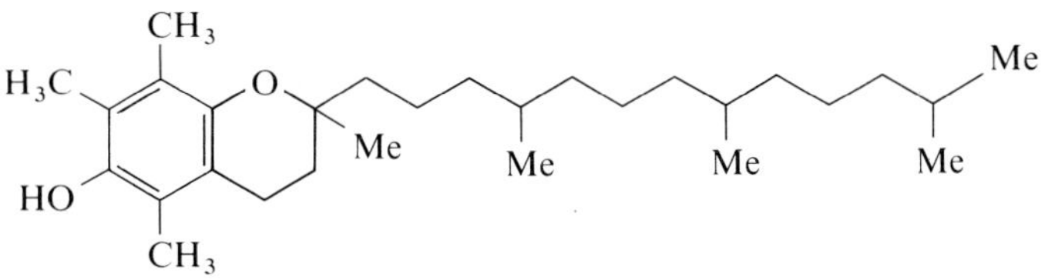

$\alpha$-Tocopherol

They are readily susceptible to oxidation in air and light as well as by chemical agents, the first step in the oxidation being the formation of $\alpha$-tocopheroxide. The latter undergoes hydrolysis, dehydration and ring cleavage to yield the corresponding quinone.

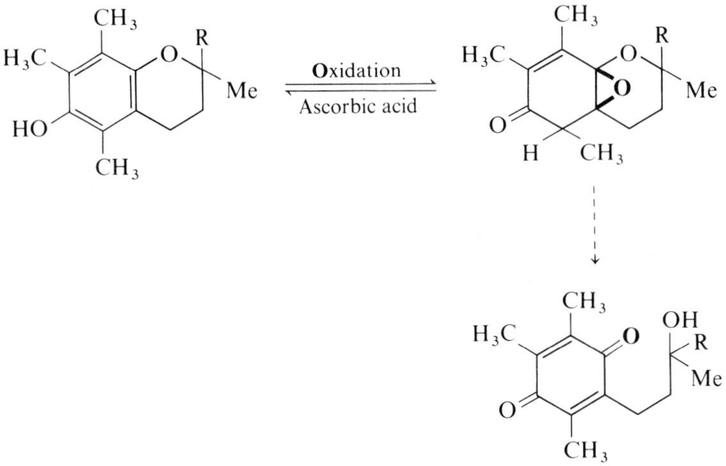

Oxidation

Ascorbic acid

The initial epoxidation is readily reversed by reagents such as ascorbic acid, and this has led to suggestions that the Vitamins E act as an integral part of certain natural oxidation–reduction systems. The tocopherols affect reproductive activity and the course of pregnancy in the female rat, but there is no real evidence that they are important in human reproduction.

**Colour Reactions and Discolouration of Phenols**

Phenols give characteristic colour reactions with ferric chloride, which are thought to be due to free radical formation and autoxidation to form coloured

Table 19. Colour Reactions of Phenols with Ferric Chloride

| Trivial Name | Systematic Name | Colour[1] |
|---|---|---|
| Phenol | Phenol | Violet |
| Salicylic acid | *o*-Hydroxybenzoic acid | Violet[1] |
| Catechol | *o*-Dihydroxybenzene | Green |
| Resorcinol | *m*-Dihydroxybenzene | Violet |
| Hexylresorcinol | 4-Hexyl-1,2-dihydroxybenzene | Green |
| Quinol | *p*-Dihydroxybenzene | Violet |
| Pyrogallol | 1,2,3-Trihydroxybenzene | Red |
| Phloroglucinol | 1,3,5-Trihydroxybenzene | Bluish-violet |
| Hydroxyquinol | 1,2,4-Trihydroxybenzene | Bluish-violet |

[1] All ferric chloride colours, except that of salicylic acid, are discharged by acetic acid.

quinonoid products. The colours produced vary according to the number and position of phenolic groups in the aromatic ring (Table 19) and are useful for characterisation purposes. Coloured products of a similar nature may also be formed causing discolouration of phenolic pharmaceutical chemicals, by oxidative radical decomposition through air oxidation and contact with heavy metal one-electron oxidising agents (salts of metals such as iron, copper, manganese, cobalt, cerium and vanadium capable of existence in more than one oxidation state). In consequence, avoidance of metal containers or the use of plastic-lacquered metal containers is essential for the proper packaging of products such as *Salicylic Acid Ointment*. Phenoxy radicals are formed particularly readily under alkaline conditions from the phenoxide ion, so that soft glass containers capable of yielding alkali should also be avoided.

Radical-producing chemicals form an alternative source of reactants capable of triggering off radical decomposition of phenolic drugs. The Gibbs' reagent (2,6-dichlorobenzoquinonechlorimide), which gives a characteristic blue colour with phenols unsubstituted in the *para* position, is an example of such a reagent. It is unlikely that such reactive radical-forming reagents would be included in a pharmaceutical formulation, but the possibility of photochemically- or oxidatively-induced decomposition of pharmaceutical additives must always be borne in mind.

## Radical Combination Reactions

Phenoxy radicals react readily *in vitro* by radical combination, as for example in the reaction of phenol with hydrogen peroxide and ferrous iron (Fenton's reagent; Haber and Weiss, 1934; Weiss, 1951; Stein and Weiss, 1950).

Similar radical addition of acetoxy-radicals to give the *ortho*- and *para*-acetoxy phenols occurs on reaction with lead tetra-acetate (Cavill, Cole, Gilham and McHugh, 1954).

$$Fe^{2+} + H_2O_2 \longrightarrow Fe^{3+} + HO^- + HO\cdot$$

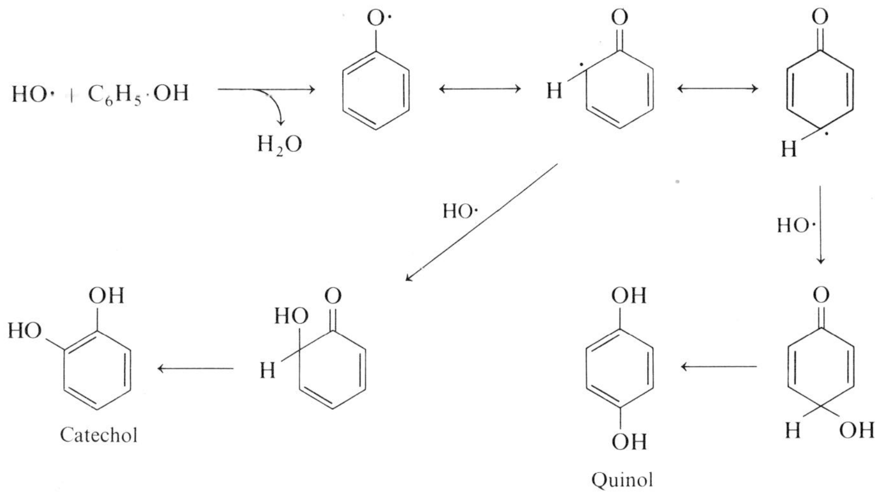

HO· + $C_6H_5\cdot OH$ → (− $H_2O$) phenoxy radical ⟷ ortho radical ⟷ para radical; ortho radical + HO· → ortho adduct → Catechol; para radical + HO· → para adduct → Quinol

Catechol

Quinol

# PHENOLIC ETHERS

## Williamson's Etherification

Phenolic ethers are formed by the reaction of phenols in alkaline solution (NaOH) with alkyl halides or alkyl sulphates. The reactive species is the phenoxide ion, a nucleophilic reagent, which attacks and displaces the halide or sulphate.

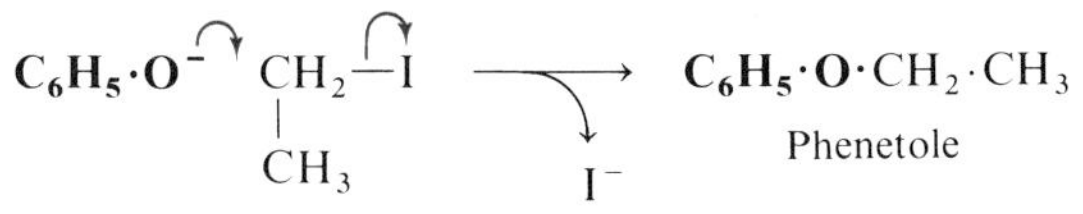

$$\mathbf{C_6H_5\cdot O^-}\ CH_2(CH_3)\text{—}I \longrightarrow \mathbf{C_6H_5\cdot O}\cdot CH_2\cdot CH_3 + I^-$$

Phenetole

Methyl ethers are frequently prepared by shaking the phenol in sodium hydroxide solution with dimethyl sulphate or methyl *p*-toluenesulphonate.

$$\mathbf{C_6H_5\cdot O^-}\ CH_3\text{—}O\cdot SO_2\cdot CH_3 \longrightarrow \mathbf{C_6H_5\cdot O}\cdot CH_3 + CH_3\cdot O\cdot SO_2\cdot O^-$$

$$\mathbf{C_6H_5O^-}\ CH_3\text{—}O\cdot SO_2\cdot C_6H_4\cdot CH_3 \longrightarrow \mathbf{C_6H_5\cdot O}\cdot CH_3 + CH_3\cdot C_6H_4\cdot SO_2\cdot O^-$$

The reaction is not generally applicable to aryl halides because of their limited reactivity towards nucleophilic reagents, and hence cannot be used for the preparation of diphenyl ethers, unless a catalyst such as copper powder is present.

Activation of the halogen by electron-attracting groups (e.g. $NO_2$) in the *ortho* or *para* positions, however, promotes reactivity, as for example in 2,4-dinitrofluorobenzene, which can be used in the preparation of diphenyl ethers.

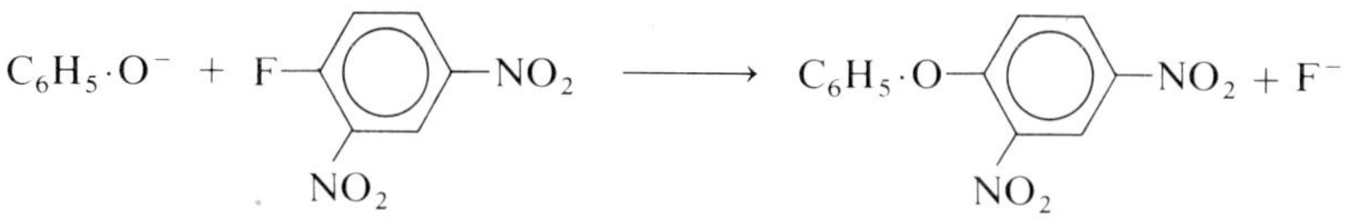

$$C_6H_5 \cdot O^- + F\text{—}C_6H_3(NO_2)\text{—}NO_2 \longrightarrow C_6H_5 \cdot O\text{—}C_6H_3(NO_2)\text{—}NO_2 + F^-$$

### Methylation with Diazomethane

Phenols are sufficiently acidic to permit formation of ethers by the action of diazomethane (in ether).

$$C_6H_5 \cdot OH + \mathbf{CH_2}N_2 \longrightarrow C_6H_5 \cdot O \cdot \mathbf{CH_3} + N_2$$

### Naturally Occurring Phenolic Ethers

A number of naturally occurring phenolic ethers are important constituents of volatile oils from various plants. They are used as food and pharmaceutical flavours, and in perfumery. They include anethole, the principal constituent of *Anise Oil* and *Fennel Oil*, eugenol from *Clove Oil*, safrole from *Sassafras Oil* and *Vanillin* from vanilla beans. Guaiacol is present in high proportion in beechwood tar.

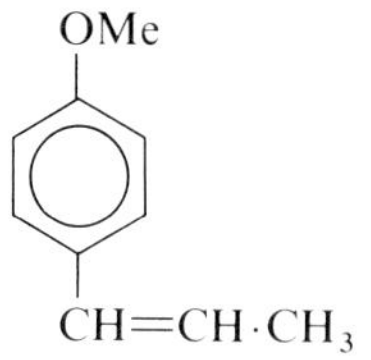

OMe / $CH{=}CH \cdot CH_3$

Anethole

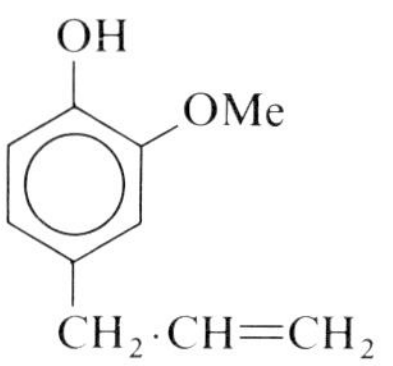

OH / OMe / $CH_2 \cdot CH{=}CH_2$

Eugenol

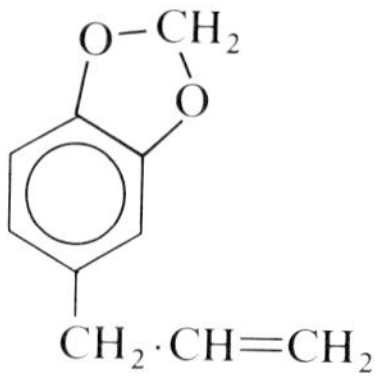

$O{-}CH_2{-}O$ / $CH_2 \cdot CH{=}CH_2$

Safrole

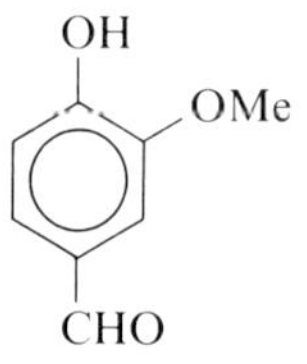

OH / OMe / CHO

Vanillin

OMe / OH

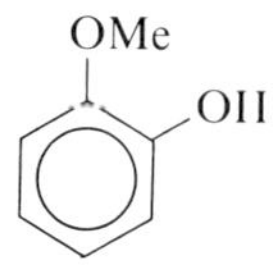

Guaiacol

A number of important naturally occurring bases, including codeine, papaverine and tubocurarine, may also be regarded as complex examples of phenolic ethers.

### General Properties

Substitution of alkyl for hydrogen in the phenolic hydroxyl group masks the acidic function of the latter, destroys reactivity towards free radical producing reagents, and in general reduces toxicity to biological systems. Phenolic ethers resemble their aliphatic analogues in that they are relatively stable neutral compounds with increased affinity for lipid media. As a result, they are more readily transported across cell and tissue boundaries in mammalian systems and, being neutral, transport is not influenced by pH gradients across such boundaries. Ethers are still capable of hydrogen bonding to biological acceptor molecules through the lone pair of electrons on oxygen. This type of interaction appears to play an important part in the pharmacological activity of the curare alkaloids and their derivatives (Stenlake, 1968a; Marshall, Murray, Smail and Stenlake, 1967).

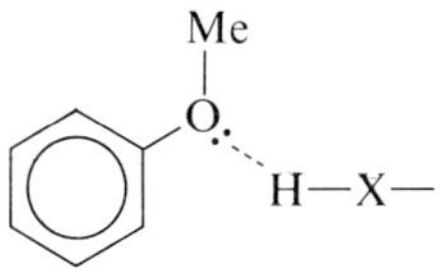

Me

O

H—X—

X = O or N

### Acid Hydrolysis

Phenolic ethers are readily cleaved by concentrated hydrobromic and hydriodic acids to the parent phenol and the alkyl halide. The reaction with hydriodic acid forms the basis of the Zeisel method for the determination of alkoxyl groups (Chapter 7); the alkyl halide is absorbed in ethanolic silver nitrate when silver iodide is precipitated quantitatively.

$$C_6H_5{\cdot}O{\cdot}CH_3 + HI \longrightarrow C_6H_5{\cdot}OH + CH_3I$$

### Metabolic *O*-Dealkylation

Phenolic ethers are metabolised in mammalian liver by oxidative dealkylation to the corresponding phenol and aliphatic aldehyde (Axelrod, 1956). Thus, *Codeine* is metabolised by *O*-demethylation to the analgesic *Morphine*, with release of formaldehyde (Adler, 1963).

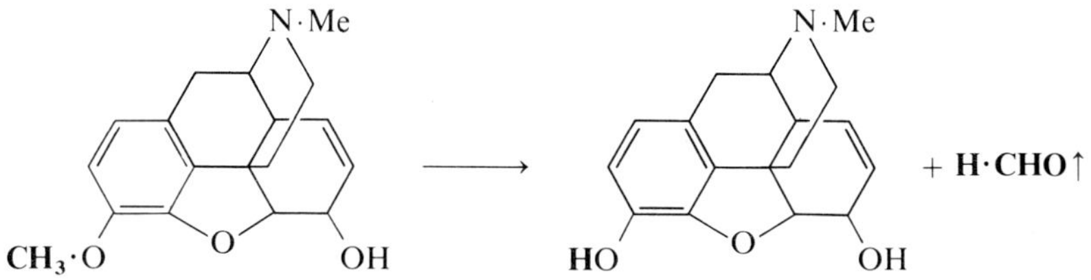

N·Me — N·Me

$CH_3{\cdot}O$ O OH → HO O OH + **H·CHO↑**

Codeine

Unlike *Morphine*, which behaves as if it were zwitterionic and is lipid-insoluble, *Codeine* is appreciably lipid-soluble and hence readily absorbed into the blood and circulation when administered orally. *Codeine* is, therefore, able to function as a mild oral analgesic. Similarly, the analgesic activity of *Phenacetin* is attributable to the formation *in vivo* of *Paracetamol* by *O*-de-ethylation.

*Paracetamol* (Acetaminophen), which is used in its own right as a mild pain killer, is also formed *in vivo* from *Acetanilide* (Acetanilid) by ring hydroxylation.

$O{\cdot}CH_2{\cdot}CH_3$ / $NH{\cdot}CO{\cdot}CH_3$ (Phenacetin) —Oxidative dealkylation→ $OH$ / $NH{\cdot}CO{\cdot}CH_3$ (Paracetamol) ←Ring hydroxylation— $NH{\cdot}CO{\cdot}CH_3$ (Acetanilide)

Phenacetin Paracetamol Acetanilide

The oxidative enzymes involved are NADPH-dependent, and molecular oxygen is consumed in the oxidation. A study of the oxidative demethylation of *p*-methoxyacetanilide by liver microsomes in the presence of $^{18}O_2$ has shown that heavy oxygen is not incorporated into the product *p*-hydroxyacetanilide (Renson, Weissbach and Udenfriend, 1965). It is concluded that oxidation proceeds by attack on the methyl C—H bonds (**2**, 5).

$O{\cdot}CH_3$ / $NH{\cdot}CO{\cdot}CH_3$ —$^{18}O_2$, $NADPH_2$→ $O{\cdot}CH_2{}^{18}OH$ / $NH{\cdot}CO{\cdot}CH_3$ —($HCH^{18}O$)→ $OH$ / $NH{\cdot}CO{\cdot}CH_3$

## PHENOLIC ESTERS

Phenols, like alcohols, are readily esterified by the action of acid chlorides and acid anhydrides.

$$C_6H_5CO{\cdot}Cl + C_6H_5{\cdot}OH \longrightarrow C_6H_5{\cdot}CO{\cdot}O{\cdot}C_6H_5 + HCl$$

Benzoyl chloride Phenyl benzoate

$$(CH_3{\cdot}CO)_2O + C_6H_5{\cdot}OH \longrightarrow CH_3{\cdot}CO{\cdot}O{\cdot}C_6H_5 + CH_3{\cdot}CO{\cdot}OH$$

Acetic anhydride Phenyl acetate

Phenolic esters are neutral substances, capable of being hydrolysed chemically or enzymically to the parent acid and phenol, but more readily so than aliphatic esters. Like phenolic ethers, they also have greater affinity than the parent phenol for lipid media, and are readily transported across tissue boundaries at rates which are independent of the pH gradient. Thus, 17$\beta$-oestradiol is usually administered intramuscularly as *Oestradiol Benzoate* (Estradiol Monobenzoate) or as *Oestradiol Dipropionate* (Estradiol Dipropionate) in oily solution or in aqueous suspension, to provide a depôt from which the drug can leach into the system and release the parent steroid hormone after hydrolysis by serum and other esterases.

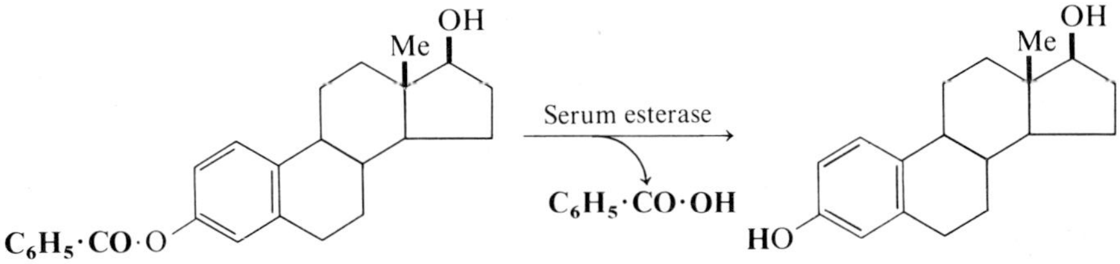

$C_6H_5 \cdot CO \cdot O$– (Oestradiol Benzoate) $\xrightarrow[C_6H_5 \cdot CO \cdot OH]{\text{Serum esterase}}$ HO– (17-β-Oestradiol)

*Oestradiol Benzoate* *17-β-Oestradiol*

### Phenol Sulphates

Phenols are metabolised and excreted in man and other mammals partly as their glucuronides and partly as sulphate esters. Sulphation occurs mainly in the liver, but also to a lesser extent in the kidney and the intestinal mucosa by substrate specific sulphokinases in the presence of 3′-phosphoadenosine-5′-phosphosulphate (PAPS), by a mechanism similar to that described for the sulphation of alcohols (Chapter 7).

$$C_6H_5 \cdot OH + PAPS \xrightarrow{\text{Phenol sulphokinase}} C_6H_5 \cdot O \cdot SO_2 \cdot OH + PAP$$

### Phenol Phosphates

The synthetic oestrogen, *Stilboestrol* (Diethylstilbestrol), is used for the treatment of prostatic hypertrophy and prostrate carcinoma in the form of *Stilboestrol Diphosphate* (Diethylstilbestrol Diphosphate). The concentration of alkaline phosphatase in the prostate gland facilitates the release of the parent phenol in high concentration at the site of action.

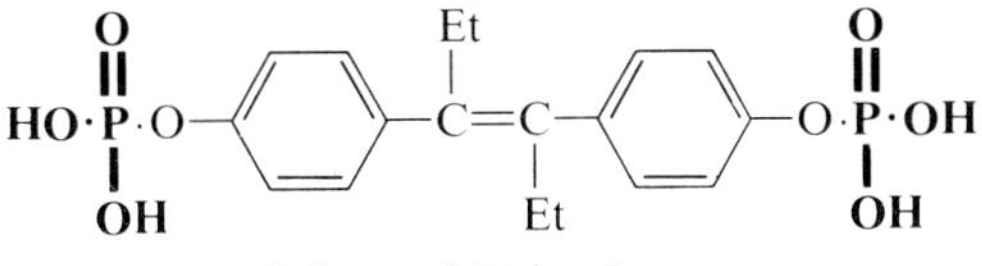

$HO \cdot P(=O)(OH) \cdot O-C_6H_4-C(Et)=C(Et)-C_6H_4-O \cdot P(=O)(OH) \cdot OH$

*Stilboestrol Diphosphate*

↓ Alkaline phosphatase

$HO-C_6H_4-C(Et)=C(Et)-C_6H_4-OH$

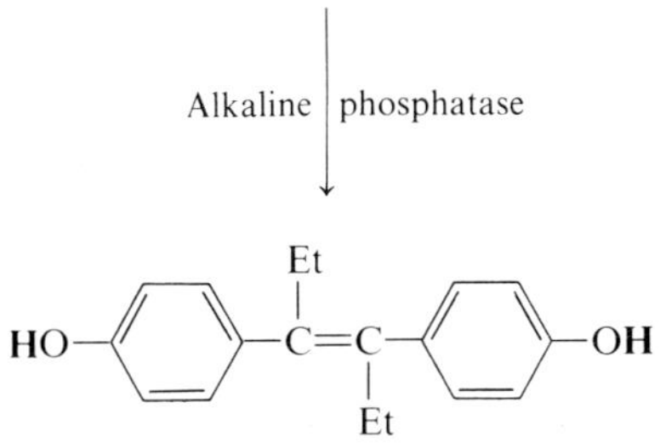

*Stilboestrol*

## REACTIONS OF PHENOLS IN THE AROMATIC RING

Both undissociated phenols and phenoxides are resonance hybrids such that electron density is high in the *ortho* and *para* positions with respect to oxygen. This concentration of electron density leads to considerable reactivity in electrophilic substitution reactions, such as nitration, sulphonation, halogenation and diazo coupling; moreover, this influence is so strong that it outweighs the influence of virtually any other substituent that may also be present in the ring.

### Nitration and Nitrosation

Nitration of phenols occurs more readily than with benzene due, in part, to increased availability of electrons in the ring at the *ortho* and *para* positions (Chapter 5). Thus, phenol is nitrated with cold dilute nitric acid to give a mixture of *o*- and *p*-nitrophenols.

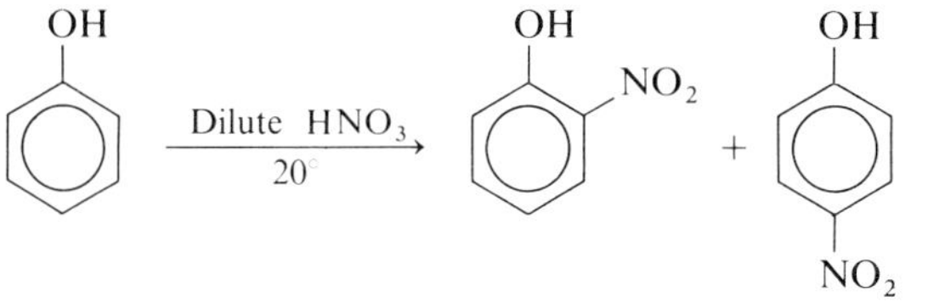

OH Dilute $HNO_3$, 20° → OH $NO_2$ + OH $NO_2$

*o*-Nitrophenol *p*-Nitrophenol

The isomers can be separated by steam distillation, since only the *ortho* isomer is steam volatile. This is due to intramolecular hydrogen bonding in *o*-nitrophenol, which prevents intermolecular hydrogen bonds from being formed with water molecules. Such intermolecular hydrogen bonds reduce the vapour pressure of *p*-nitrophenol and hence its volatility at the boiling point of water.

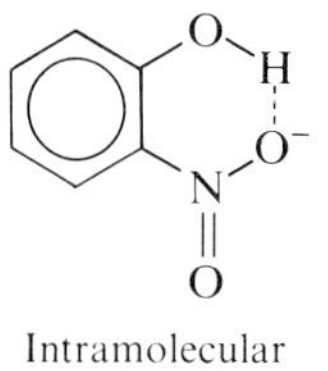

Intramolecular hydrogen bond

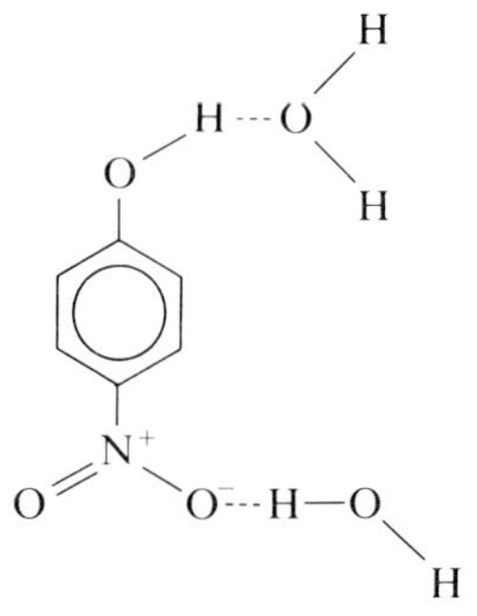

Intermolecular hydrogen bonds

The readiness with which phenols undergo nitration is only partly explained by increased electron availability in the ring. The high rate of reaction is also due to the presence in nitric acid of nitrous acid and hence nitrosonium ions ($\overset{+}{N}O$); these attack the ring giving the nitrosophenol, which is then rapidly oxidised by nitric acid to the nitrophenol.

$$HNO_2 + 2\,HNO_2 \rightleftharpoons H_3\overset{+}{O} + 2\,NO_3^- + \overset{+}{N}O$$

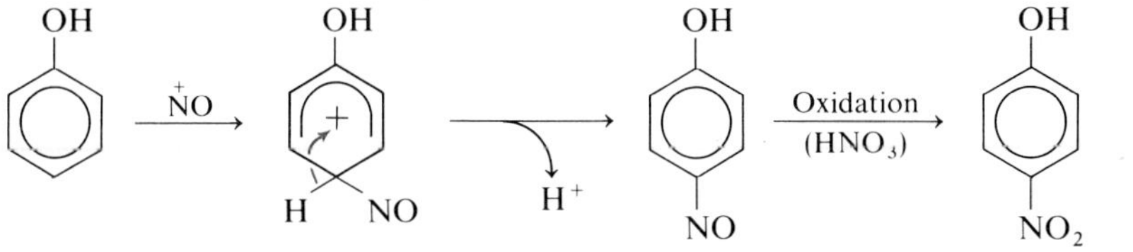

OH, $\overset{+}{N}O$ →, OH (H, NO) → $H^+$ → OH (NO) Oxidation ($HNO_3$) → OH ($NO_2$)

*p*-Nitrosophenol *p*-Nitrophenol

Direct nitrosation of phenol with nitrous acid, generated *in situ* from sodium nitrite and dilute sulphuric or acetic acid at 0°, gives solely *p*-nitrosophenol, which is tautomeric with *p*-quinone monoxime.

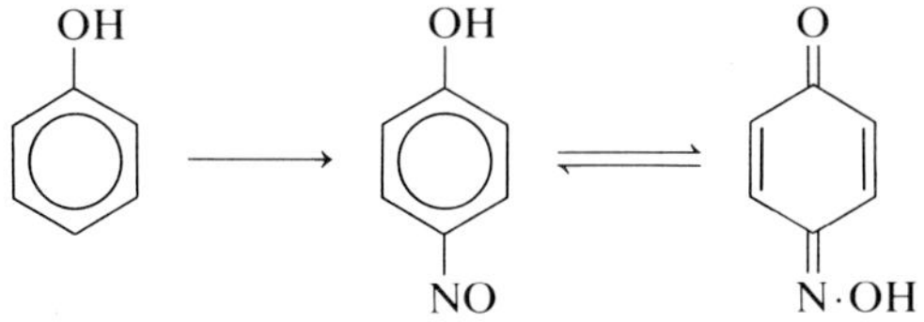

OH → OH (NO) ⇌ O (N·OH)

When the *para* position is blocked, the nitroso group will substitute in the *ortho* position.

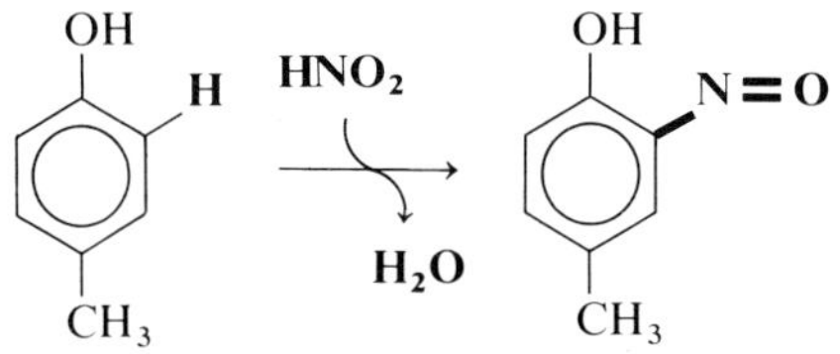

OH, H, $CH_3$ —$HNO_2$, $H_2O$→ OH, N=O, $CH_3$

Further nitration of either *o*- or *p*-nitrophenol in sulphuric acid finally yields picric acid (2,4,6-trinitrophenol).

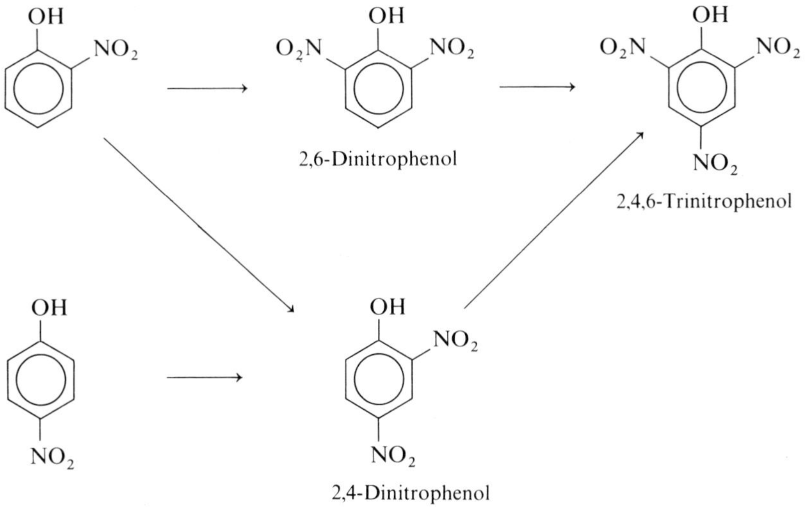

OH, $NO_2$ → $O_2N$, OH, $NO_2$ → $O_2N$, OH, $NO_2$, $NO_2$

2,6-Dinitrophenol

2,4,6-Trinitrophenol

OH, $NO_2$ → OH, $NO_2$, $NO_2$

2,4-Dinitrophenol

These reaction conditions, however, give rise to oxidative processes yielding by-products, since nitric acid is also an oxidising agent as well as a nitrating agent. Picric acid is, therefore, usually manufactured by one of the following routes, both of which give high yields.

(a)

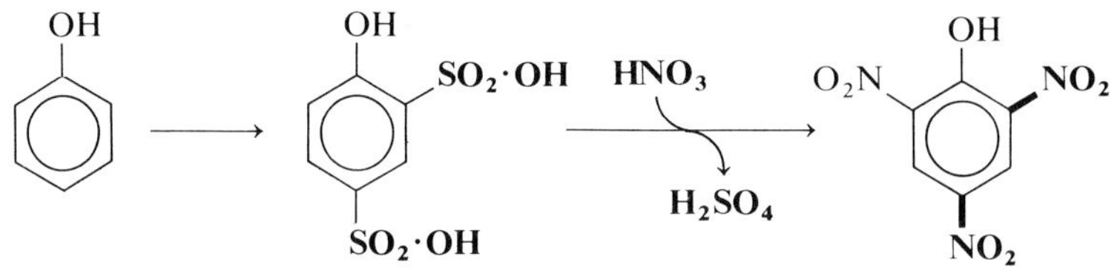

OH → OH, $SO_2 \cdot OH$, $SO_2 \cdot OH$ $\xrightarrow{HNO_3 \quad H_2SO_4}$ OH, $O_2N$, $NO_2$, $NO_2$

Phenol-2,4-disulphonic acid

This method ensures a low electron density on the benzene ring during the nitration due to the electron-withdrawing properties of the sulphonic acid groups. It is dependent on the reversibility of the sulphonation reaction, which permits displacement of the sulphonic acid groups.

(b)

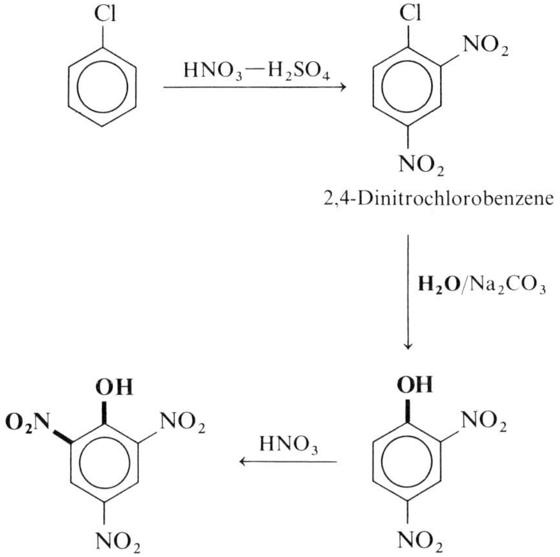

Cl $\xrightarrow{HNO_3—H_2SO_4}$ Cl, $NO_2$, $NO_2$

2,4-Dinitrochlorobenzene

$\downarrow H_2O/Na_2CO_3$

OH, $NO_2$, $NO_2$ $\xrightarrow{HNO_3}$ OH, $O_2N$, $NO_2$, $NO_2$

2,4-Dinitrophenol → 2,4,6-Trinitrophenol

Picric acid, a yellow solid, m.p. 112°C, is strongly acidic in reaction, and explosive. Its metallic salts are even more explosive than the parent substance. As a safety precaution picric acid is stored under water. Picric acid forms crystalline salts with organic bases of structure [Base],$C_6H_2(NO_2)_3OH$; it also forms crystalline addition compounds with polycyclic aromatic hydrocarbons such as anthracene. Base picrates are readily formed when ethanolic solutions of picric acid and the organic base are mixed and set aside to crystallise.

### Sulphonation

Sulphonation of phenol occurs readily to give a mixture of *o*- and *p*-phenolsulphonic acids. Higher temperatures favour production of the *p*-isomer. The phenolsulphonic acids and their salts have antiseptic properties and are less toxic than phenol.

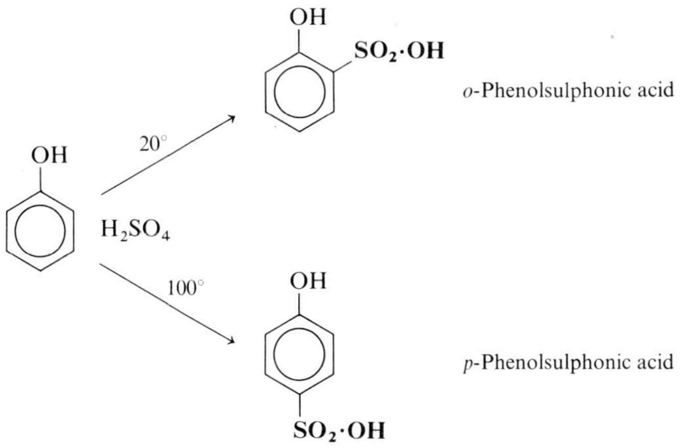

OH, SO$_2$·OH — *o*-Phenolsulphonic acid

OH — 20° / $H_2SO_4$ / 100°

OH, SO$_2$·OH — *p*-Phenolsulphonic acid

### Diazo Coupling

Phenols yield bright red azo dyes with diazonium salts in alkaline solution, a reaction which is frequently employed for the characterisation of phenols. Coupling occurs preferentially in a position *para* to the phenolic hydroxyl, but if this is blocked, *ortho* coupling results.

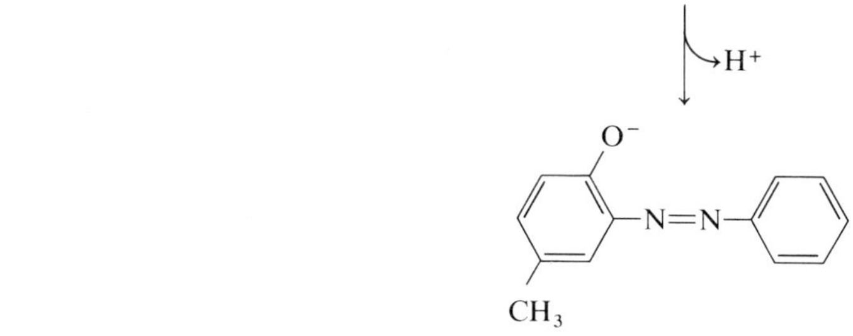

$O^-$, $CH_3$, $N\equiv \overset{+}{N}$—, $Cl^-$ → $Cl^-$; O, H, N=N—, $CH_3$

→ $H^+$

$O^-$, —N=N—, $CH_3$

**C-Alkylation**

Phenol condenses with aldehydes to form *ortho*- and *para*-*C*-alkyl derivatives. Aqueous formaldehyde (40 %) reacts slowly with phenol at room temperature under acidic conditions, but more rapidly in alkali, to give a mixture of *o*- and *p*-hydroxybenzyl alcohol.

$O^-$ $H_2C{=}O$ $\xrightarrow{HO^-}$ O H $CH_2{\cdot}O^-$

$\downarrow$

$O^-$ $CH_2OH$

At high temperatures and with excess formaldehyde, phenol under alkaline conditions forms phenol–formaldehyde resins by a series of polycondensations, which proceed through the formation of the 4,4′- and 2,2′-dihydroxydiphenylmethanes.

$^-O$–$C_6H_4$–$CH_2$–OH $\rightleftharpoons$ O=$C_6H_4$=$CH_2$ + $C_6H_5O^-$

$\downarrow$

HO–$C_6H_4$–$CH_2$–$C_6H_4$–OH $\xleftarrow{H^+}$ $^-O$–$C_6H_4$–$CH_2$–(H)$C_6H_4$=O

Chain polymerisation and cross-linking occurs through the available *ortho* and *para* positions to give Bakelite, a three-dimensional polymer. In practice, the polymerisation is carried to an incomplete pre-polymer stage, at which the product is low melting and capable of being moulded. Polymerisation of the moulded polymer is completed by heating to form an insoluble non-melting polymer.

Acid-catalysed *C*-alkylation with formaldehyde is used in the preparation of the antiseptic chlorophenols *Dichlorophen* and *Hexachlorophane*. *Dichlorophen* is also used extensively in the treatment of tapeworm infection.

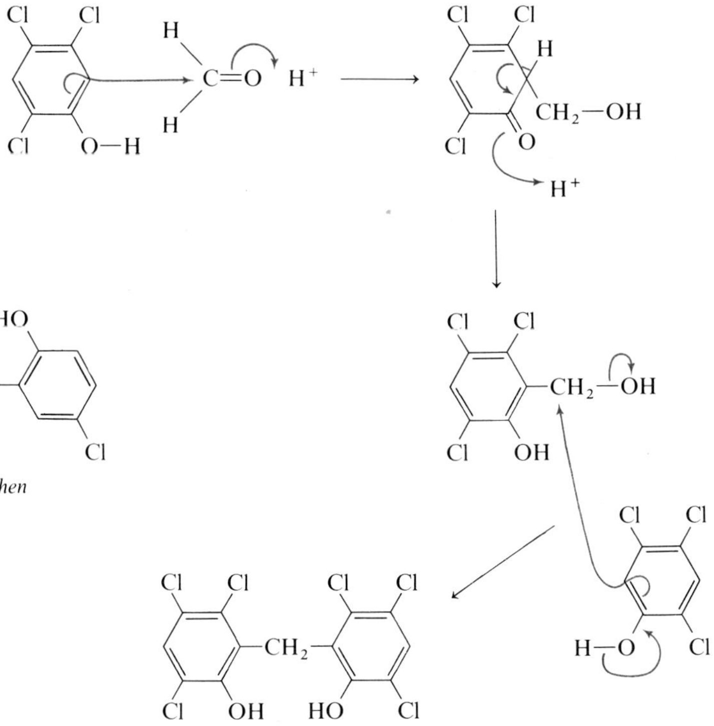

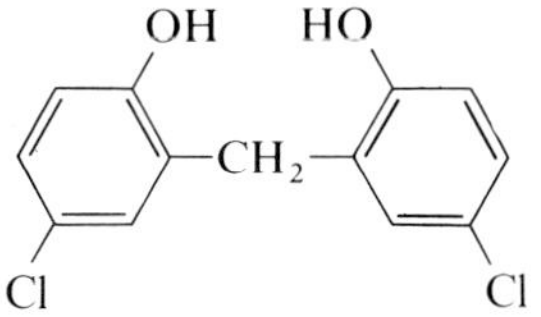

Dichlorophen

Hexachlorophane

**Carboxylation** (Kolbé-Schmitt Reaction)

Sodium phenoxide reacts with carbon dioxide under pressure at temperatures around 125°C to form sodium salicylate. At higher temperatures (200–50°C) and more effectively with potassium phenate, the principal product is *p*-hydroxybenzoic acid.

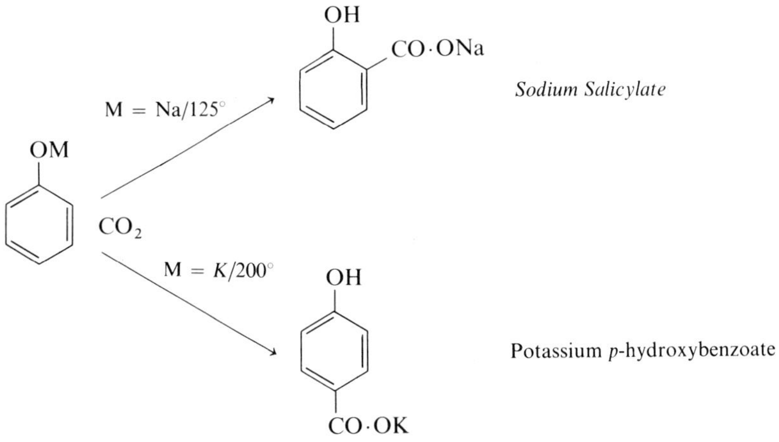

Sodium Salicylate

Potassium *p*-hydroxybenzoate

The mechanism of the reaction is uncertain, but probably involves a cyclic intermediate in the formation of sodium salicylate. It is possible that the larger potassium ion inhibits formation of the cyclic intermediate, thus favouring *p*-carboxylation.

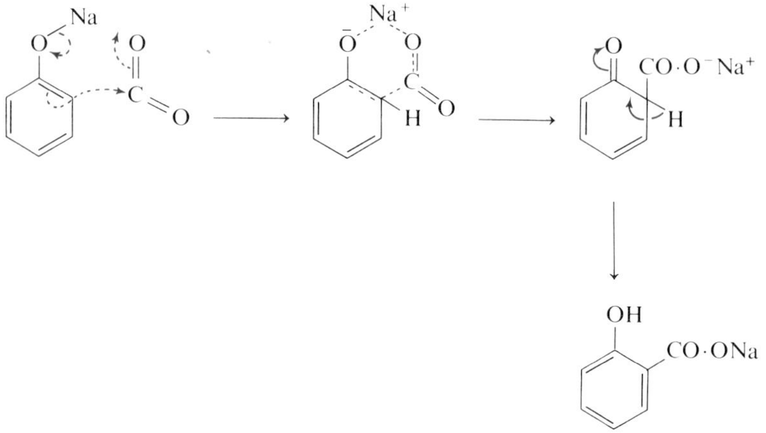

Na O C O O → Na$^+$ O O C O H → O CO·O$^-$Na$^+$ H → OH CO·ONa

## HALOPHENOLS

Both chlorine and bromine react very much more readily with phenol than with benzene, attack being mainly in the *ortho* and *para* positions as dictated by the high electron density at these positions. When chlorine is passed into phenol, a mixture of *o*- and *p*-chlorophenols is first formed. Excess chlorine gives 2,4,6-trichlorophenol.

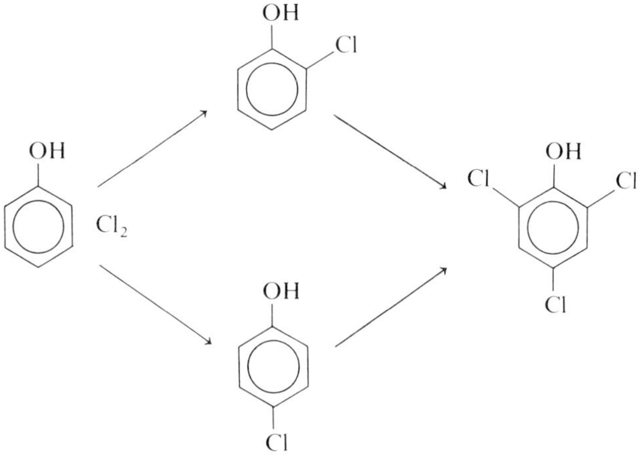

OH $Cl_2$ → OH Cl ; OH Cl → OH Cl Cl Cl

The isomeric monochlorophenols are separable by fractional distillation owing to the considerable differences in boiling point. This is attributed to intramolecular hydrogen bonding in the *ortho* isomer, which exists in discrete monomolecular units. The *para* isomer, in contrast, exhibits intermolecular hydrogen bonding, so that additional energy must be applied to break up the liquid into single molecules before the liquid can boil. The *para* isomer, therefore, has a higher boiling point than the *ortho* isomer.

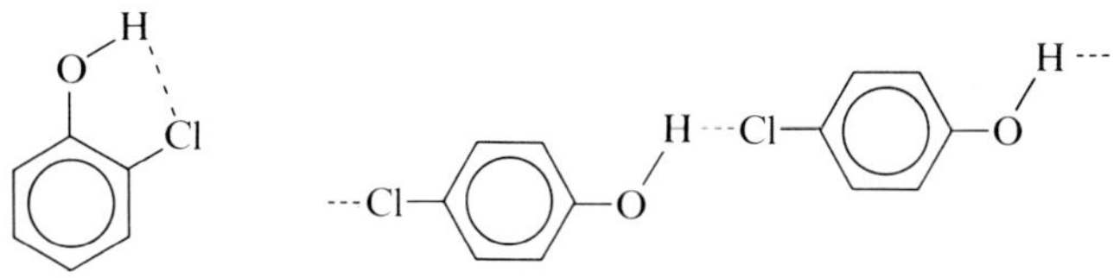

*o*-Chlorophenol

*p*-Chlorophenol

Bromine, which is more reactive than chlorine in electrophilic addition, yields 2,4,6-tribromophenol as the sole product when phenol is treated in aqueous solution with bromine water. The reaction in polar solvents such as water is promoted by the enhanced ionisation of phenol which occurs under these conditions. In contrast, monobromination to yield mainly *p*-bromophenol is favoured in non-polar solvents such as carbon disulphide.

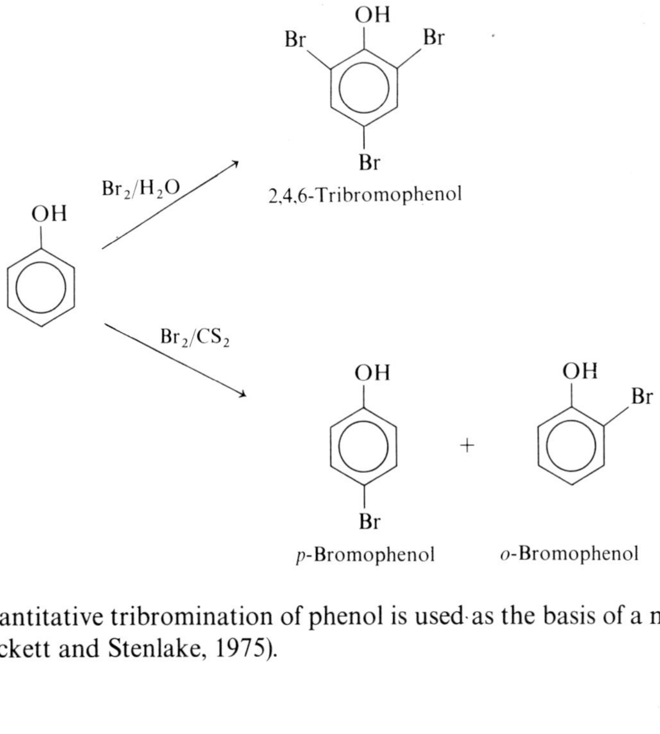

$Br_2/H_2O$

2,4,6-Tribromophenol

$Br_2/CS_2$

*p*-Bromophenol

*o*-Bromophenol

The quantitative tribromination of phenol is used as the basis of a method of assay (Beckett and Stenlake, 1975).

**Enzymic Iodination**

Enzymic iodination of tyrosine and other phenolic thyroid hormone intermediates is catalysed by an enzyme present in the thyroid gland. Mechanistically, the iodination seems likely to be analogous to chemical halogenation.

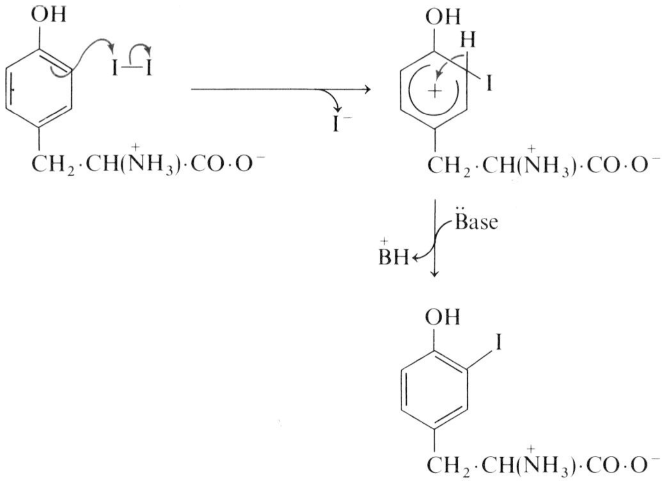

**Chlorocresol; Chloroxylenol**

*Chlorocresol* (*p*-chloro-*m*-cresol) and *Chloroxylenol* (*p*-chloro-*m*-xylenol) are prepared from *m*-cresol and *m*-xylenol by chlorination with sulphuryl chloride.

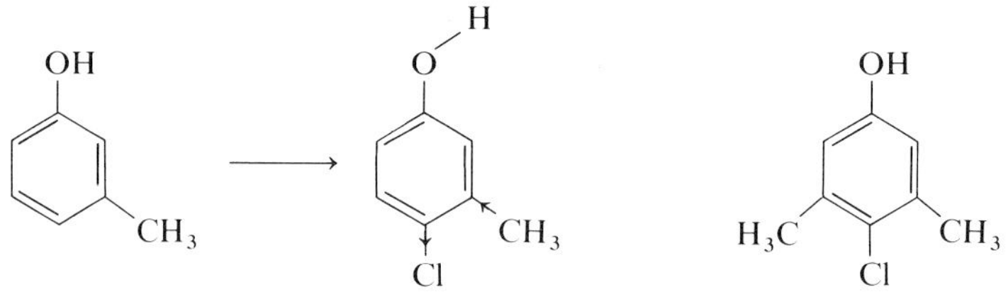

*Chlorocresol* *Chloroxylenol*

Both products have lower water solubility, but are readily soluble in non-polar solvents, particularly terpenes and oils. The overall electron-attracting properties of the halogen substituent not only increase the acidity of the phenolic hydroxyl compared with that of phenol itself, but also increase the dipolar character of the molecule. This is only partly opposed by the electron-repelling influence of the methyl substituent(s), which is outweighed by the more powerful effect due to the halogen. The greater dipolar character compared with the unhalogenated phenol enhances the ability to become orientated and absorbed at the surface of lipid biological membranes, and at least in part accounts for their enhanced antibacterial properties. The use of terpineol and soap solutions to formulate these substances for use as surface antiseptics is also based on their dipolar and solubility characteristics.

## MANUFACTURE OF PHENOLS

### Extraction from Coal Tar

Phenol and its simple homologues are obtained commercially from the carbolic acid fraction (b.p. 180–210°C) from the coal tar distillation process. The crude distillate is pressed free of naphthalene and extracted by agitation with aqueous sodium hydroxide solution, when the phenolic substances dissolve as their water-soluble sodium salts. Traces of hydrocarbons (naphthalene) and water-soluble bases (pyridine and quinoline) are displaced by heating and blowing a current of air through the solution. On cooling, the sodium phenates are decomposed with carbon dioxide.

$$2\ Ph{\cdot}ONa + CO_2 + H_2O \longrightarrow 2\ Ph{\cdot}OH + Na_2CO_3$$

The oily product is fractionally distilled to give phenol (b.p. 182°C), a mixture of cresols (b.p. 190–205°C) and a mixture of xylenols (b.p. 210–255°C). Careful fractionation of the cresols gives a partial separation into *o*-cresol (b.p. 188°C) and a mixture of *m*- and *p*-cresols (b.p. 200–2°C).

Although large quantities of phenol are still obtained from coal tar, production from this source represents only a small fraction of the total commercial requirement. A number of important synthetic processes based on benzene as the basic raw material have been developed.

### Sulphonation of Benzene

Phenol can be synthesised from benzene by sulphonation, and fusion of the resulting benzenesulphonic acid (as its sodium salt) with sodium hydroxide.

$$C_6H_6 + H_2SO_4 \longrightarrow C_6H_5{\cdot}O{\cdot}SO_2{\cdot}OH$$

$$C_6H_5{\cdot}O{\cdot}SO_2{\cdot}ONa + 2NaOH \xrightarrow{350^\circ} C_6H_5{\cdot}ONa + Na_2SO_3 + H_2O$$

This process was used extensively in the First World War to produce phenol required for the production of picric acid.

### Chlorination of Benzene

The Dow process introduced in 1928 is based on the hydrolysis of chlorobenzene by aqueous sodium hydroxide at high temperature (350°C) and under high pressure (200 atmospheres).

$$C_6H_5{\cdot}Cl + 2NaOH \longrightarrow C_6H_5ONa + NaCl + H_2O$$

The reaction, which is run as a continuous process, gives rise to diphenyl ether as a by-product. This side-reaction is, fortunately, reversible and its formation is kept to a minimal amount by re-cycling sufficient diphenyl ether to suppress the reaction.

$$C_6H_5ONa + C_6H_5OH \rightleftharpoons \underset{\text{Diphenyl ether}}{C_6H_5{\cdot}O{\cdot}C_6H_5} + NaOH$$

**Oxidation of Cumene**

A recent development has been the synthesis of phenol (and acetone) from cumene, which is obtained by Friedel–Crafts alkylation of benzene with propylene. Cumene is oxidised to cumene hydroperoxide by blowing air through it in the presence of a trace of sodium hydroxide; cumene hydroperoxide is then rearranged with acid to yield phenol and acetone.

$$C_6H_5CH(CH_3)_2 \xrightarrow{O_2} C_6H_5C(CH_3)_2\cdot O\cdot OH \xrightarrow{H_3\overset{+}{O}} C_6H_5OH + (CH_3)_2\cdot CO$$

The decomposition of cumene hydroperoxide involves a 1,2-shift probably by the following mechanism.

$$C_6H_5C(CH_3)_2\text{—}O\cdot OH \xrightarrow{H^+} C_6H_5C(CH_3)_2\text{—}O\text{—}\overset{+}{O}H_2 \xrightarrow{-H_2O} C_6H_5C(CH_3)_2\text{—}\overset{+}{O} \longrightarrow (CH_3)_2\overset{+}{C}\text{—}O\text{—}C_6H_5$$

$$(CH_3)_2\overset{+}{C}\text{—}O\text{—}C_6H_5 \xrightarrow{H_2O} (CH_3)_2C(\overset{+}{O}H_2)\text{—}O\text{—}C_6H_5 \rightleftharpoons (CH_3)_2C(OH)\text{—}O\text{—}C_6H_5 + H^+ \longrightarrow C_6H_5OH + (CH_3)_2\cdot CO$$

## BIOSYNTHESIS OF PHENOLS

Biosynthesis of phenols occurs in plants by two main pathways. Of these, the simplest is the acetate pathway in which acetic acid, activated by combination with the thiol, coenzyme A, undergoes a head-to-tail polymerisation to a poly-keto acid intermediate.

$$CH_3{\cdot}CO{\cdot}OH + \overline{CoA}SH \longrightarrow CH_3{\cdot}CO{\cdot}S{\cdot}\overline{CoA}$$

Acetyl coenzyme A

$$\xrightarrow{CO_2} HO{\cdot}OC{\cdot}CH_2{\cdot}CO{\cdot}S\overline{CoA}$$

Malonylcoenzyme A

$$HO{\cdot}OC{\cdot}CH_2{\cdot}CO{\cdot}S\overline{CoA} + CH_3{\cdot}CO{\cdot}S{\cdot}\overline{CoA} \longrightarrow CH_3{\cdot}CO{\cdot}CH_2{\cdot}CO{\cdot}S\overline{CoA} + CO_2$$

$$\downarrow$$

$$R{\cdot}CO{\cdot}CH_2{\cdot}CO{\cdot}CH_2{\cdot}CO{\cdot}CH_2{\cdot}CO{\cdot}S\overline{CoA}$$

A similar head-to-tail acetate polymerisation occurs in mammals, and intermediates of this type are involved in fatty acid and steroid biosynthesis in both plant and animal tissue. Cyclisation to phenolic products appears to be characteristic mainly of plant rather than animal metabolism, and can follow a variety of pathways (Birch and Donovan, 1953).

$$R{\cdot}\overset{7}{C}O{\cdot}\overset{6}{C}H_2{\cdot}\overset{5}{C}O{\cdot}\overset{4}{C}H_2{\cdot}\overset{3}{C}O{\cdot}\overset{2}{C}H_2{\cdot}\overset{1}{C}O{\cdot}S\overline{CoA}$$

2,7-cyclisation → (ring with R, CO·OH, OH, OH)

1,6-cyclisation → Trihydric phenol (ring with R·OC, OH, HO, OH)

Reduction → Alkylsalicylic acid (ring with R, CO·OH, OH)

→ $CO_2$ → Dihydric phenol (ring with R, OH, OH)

Birch, Massy-Westropp and Moye (1955) have shown that 6-methylsalicylic acid formed by *Penicillium griseofulvum* from 1-$^{14}$[C]-acetic acid carries the isotopic label in positions which are wholly consistent with the above biosynthetic pathway.

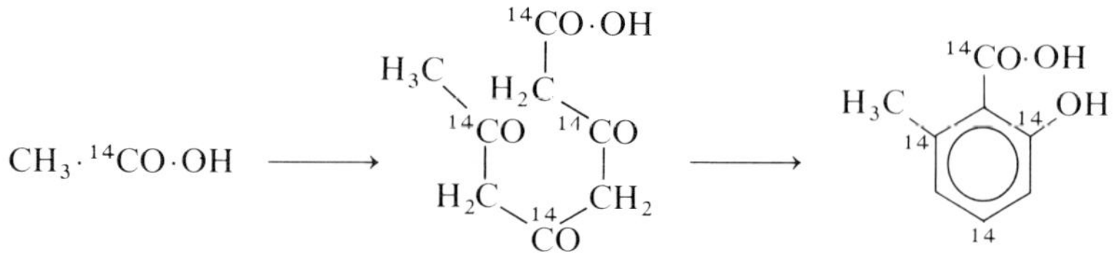

$$CH_3\cdot{}^{14}CO\cdot OH \longrightarrow \text{(}H_3C\cdot{}^{14}CO\cdot CH_2\cdot{}^{14}CO\cdot CH_2\cdot{}^{14}CO\cdot CH_2\cdot{}^{14}CO\cdot OH\text{)} \longrightarrow \text{(ring-labelled } H_3C\text{, }{}^{14}CO\cdot OH\text{, OH aromatic acid)}$$

The phenolic amino acid, tyrosine, and its precursor, phenylalanine, are synthesised by the shikimic acid path (Chapters 18 and 22) in plant and micro-organisms, but not in mammals. In man, tyrosine is formed by hydroxylation of phenylalanine.

## DIHYDRIC AND TRIHYDRIC PHENOLS

All three dihydroxybenzenes, catechol (*ortho*), resorcinol (*meta*) and quinol (*para*) are known.

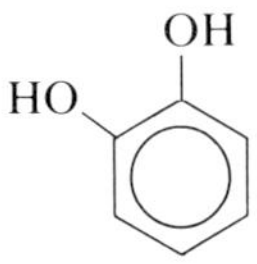

Catechol 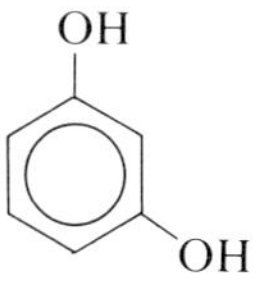    Resorcinol 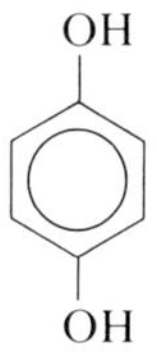    Hydroquinone (Quinol)

Catechol itself and various catechol derivatives occur naturally. Thus, natural catechol derivatives include guaiacol and creosol from beechwood creosote, *Eugenol* from *Clove Oil*, and the sympathomimetric amines of mammalian systems, *Adrenaline* (Epinephrine) and *Noradrenaline* (Norepinephrine). Hydroquinone occurs in bearberry leaves as the aglycone of the glycoside, arbutin.

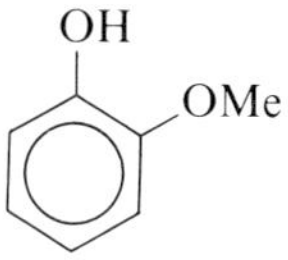

*Guaiacol* 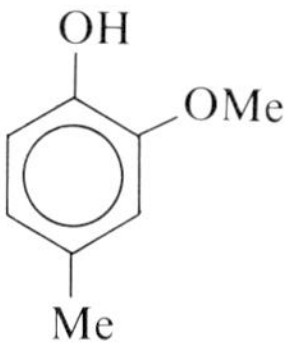    *Creosol* (Me)    *Eugenol* ($CH_2\cdot CH{=}CH_2$)    $CHOH\cdot CH_2\cdot NHR$ *Noradrenaline* (R = H) *Adrenaline* (R = Me)

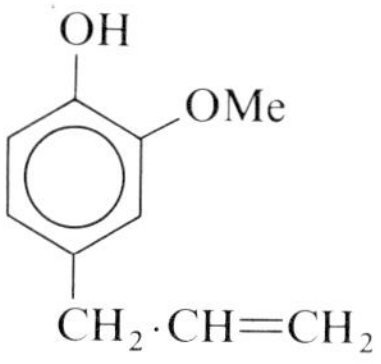

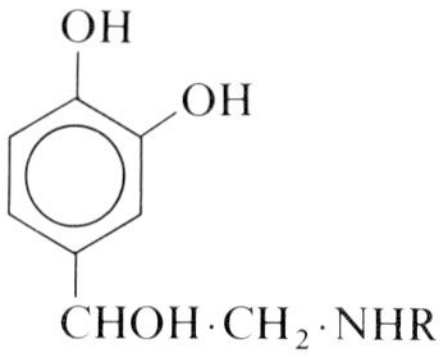

### Oxidation

The dihydroxybenzenes possess typical phenolic properties, but compared with phenol they have much greater water solubility, and markedly increased susceptibility to oxidation. Catechol is readily oxidised to *o*-quinone, and for this reason is a powerful reducing agent, reducing Fehling's solution and silver nitrate solution at room temperature.

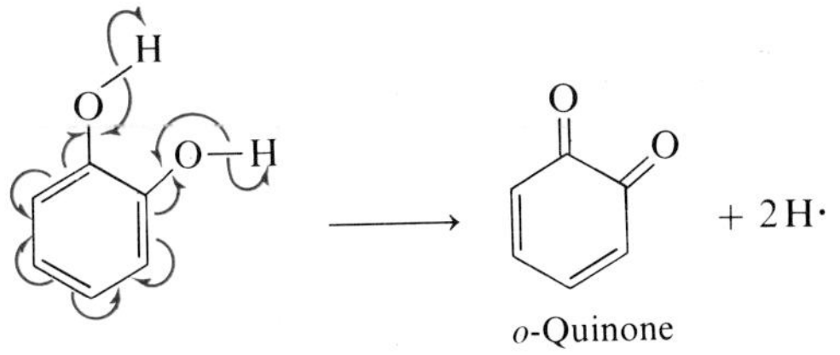

$\longrightarrow$ + 2H·

*o*-Quinone

*Adrenaline* (Epinephrine), similarly, is very susceptible to oxidation, and aqueous solutions rapidly discolour on exposure to air unless protected by an antioxidant such as sodium metabisulphite ($NaHSO_3$). The oxidation product, an *o*-quinone, is further oxidised to adrenochrome, which ultimately polymerises to a dark-brown or black insoluble polymer.

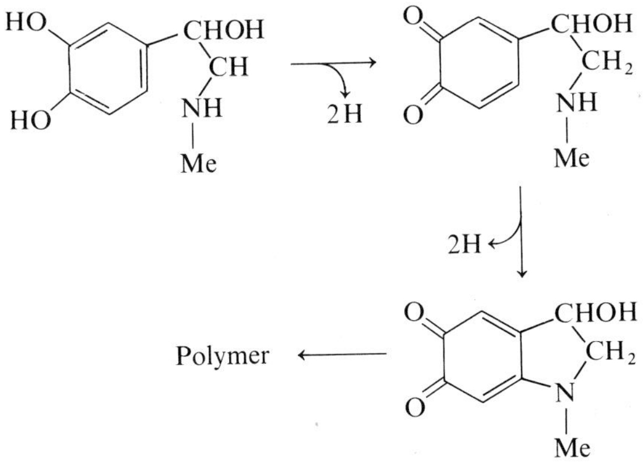

Polymer

The oxidation products are physiologically inactive. Oxidation occurs much less readily in acidic solution, and solutions at pH 4 can be heat-sterilised by autoclaving without serious decomposition.

Resorcinol and resorcinol derivatives, such as hexylresorcinol, are also oxidised on exposure to air especially in alkaline solution, though somewhat less readily than catechol. Hexylresorcinol is a much more effective antiseptic

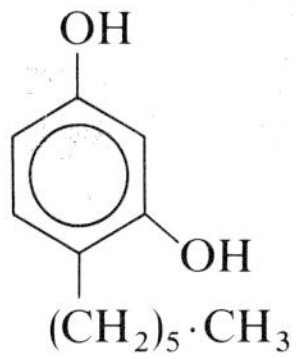

$(CH_2)_5 \cdot CH_3$

Hexylresorcinol

than resorcinol, and provides a further example of the favourable influence of alkyl substitution on antibacterial activity. It is, however, used mainly as an anthelmintic.

Quinol is a powerful reducing agent, being readily converted by chemical oxidants to benzoquinone. The oxidation is reversible and many important biological oxidations and reductions are based on quinol-quinone systems (Chapter 10).

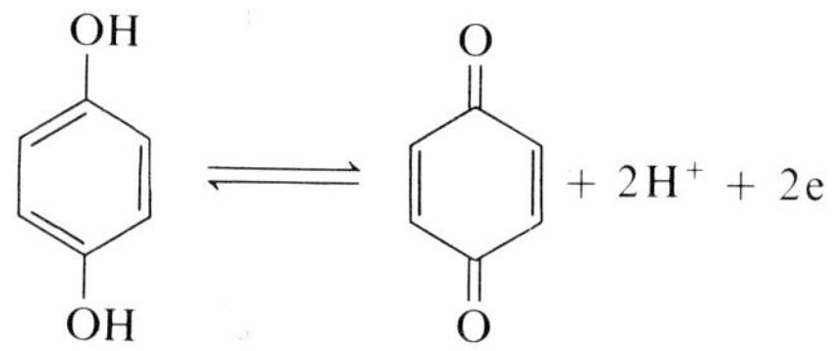

$$\rightleftharpoons \quad + 2H^+ + 2e$$

Chemical oxidation of quinol with ferric chloride leads to the formation of a 1:1-molecular complex of quinol and benzoquinone, known as quinhydrone, which only yields benzoquinone on further prolonged oxidation.

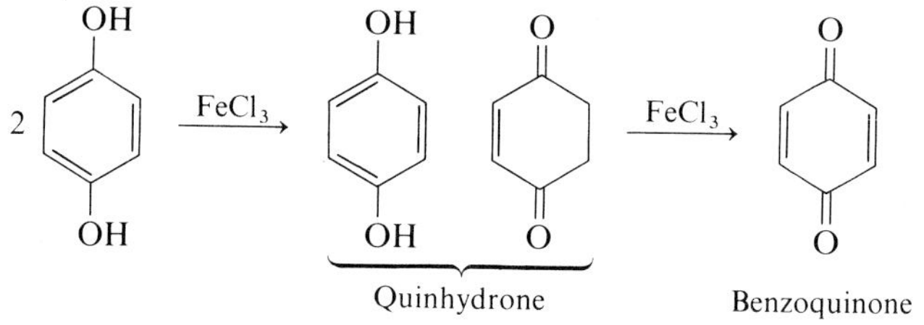

$$2 \xrightarrow{FeCl_3} \text{Quinhydrone} \xrightarrow{FeCl_3} \text{Benzoquinone}$$

### Complex Formation

*o*-Dihydric phenols such as *Adrenaline* and its analogues catalyse the cyclisation of adenosine triphosphate by the enzyme adenylcyclase in a reaction which is known to require magnesium ions (Rall and Sutherland, 1959). It is suggested that the 3-phenolic hydroxyl complexes through a water cluster with the magnesium ion, which in turn is associated with the two terminal phosphate residues of adenosine triphosphate (Belleau, 1960; Bloom, Goldman and Belleau, 1969; Belleau, 1967; **2**, 3).

### Enzymic *O*-Methylation of Catecholamines

*O*-Methylation forms an important metabolic pathway for catecholamines such as *Adrenaline* (Axelrod, Senoh and Witkop, 1958). Methylation occurs by transfer of a methyl substituent from *S*-adenosylmethionine and is catalysed by substrate specific *O*-methyltransferases (Senoh, Daly, Axelrod and Witkop, 1959). The reaction is assisted by the weakening of the H—O bond as a result of hydrogen bonding through the *ortho*-hydroxyl group.

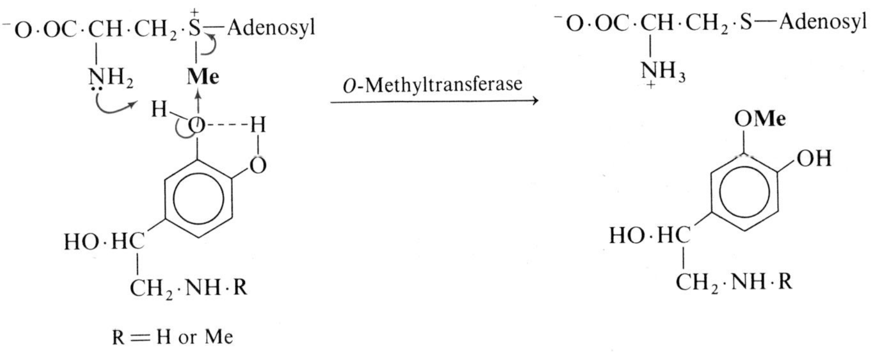

$^{-}O \cdot OC \cdot CH(NH_2) \cdot CH_2 \cdot \overset{+}{S}(Me)$—Adenosyl + HO·HC(CH₂·NH·R)-catechol (R = H or Me) —*O*-Methyltransferase→ $^{-}O \cdot OC \cdot CH(\overset{+}{N}H_3) \cdot CH_2 \cdot S$—Adenosyl + HO·HC(CH₂·NH·R)-C₆H₃(OMe)(OH)

R = H or Me

**Clathrates**

Quinol when crystallised from water containing sulphur dioxide, hydrogen sulphide or methanol forms inclusion complexes $3C_6H_4(OH)_2,Z$, in which the quinol molecules are multiply-linked by hydrogen bonds to form large polymeric structures with cavities containing the complexed molecules. The complexes, known as clathrates, are stable in the crystalline state, but the trapped molecules are released by melting or solution of the quinol in a suitable solvent.

**Trihydric Phenols**

The trihydric phenols, pyrogallol, phloroglucinol and hydroxyquinol, are similarly characterised by appreciable water solubility and susceptibility to oxidation.

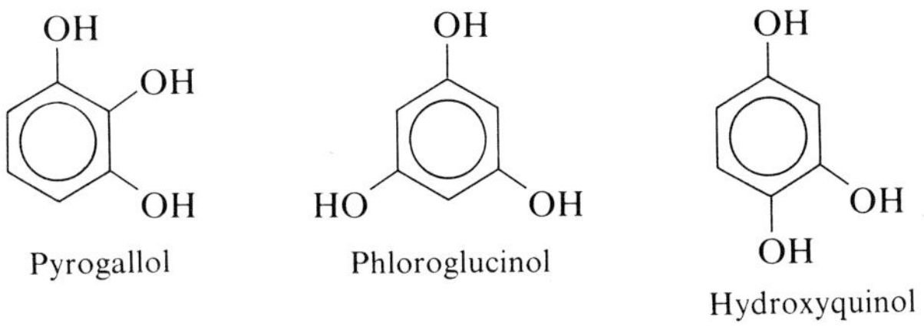

Pyrogallol Phloroglucinol Hydroxyquinol

In some respects, phloroglucinol is unusual in that it is capable of reacting in a tautomeric form, cyclohexane-1,3-5-trione, with hydroxylamine to give the corresponding tri-oxime.

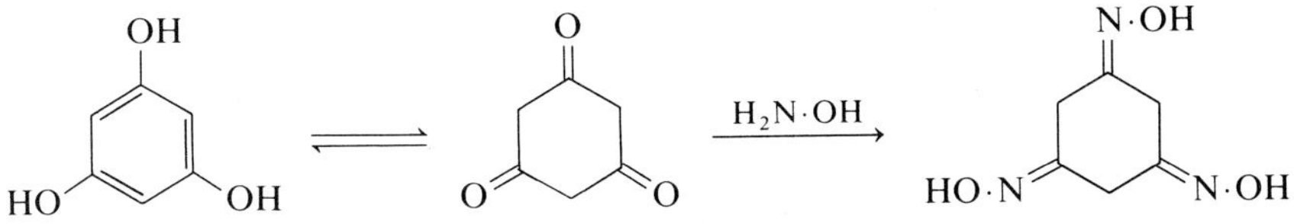

Phloroglucinol ⇌ cyclohexane-1,3,5-trione —$H_2N \cdot OH$→ tri-oxime (HO·N, N·OH, N·OH)

# 9 Halogenated Hydrocarbons and Esters of Inorganic Acids

Halogen-substituted hydrocarbons rarely occur naturally, but many synthetic compounds of this type are of considerable importance on account of their use as anaesthetics, and as insecticides and pesticides.

Alkyl halides may be considered not merely as halogenated hydrocarbons, but also as esters derived from alcohols and inorganic acids. In consequence, they show certain properties in common with esters of inorganic oxyacids, such as those of phosphoric and sulphuric acids, which are frequently found in biological systems. These same properties also have considerable bearing on the wide synthetic utility of alkyl halides in the laboratory. Their study, therefore, provides a useful foundation for consideration of a number of important biosynthetic and metabolic processes.

## ALKYL HALIDES

### Nomenclature

The simplest alkyl halides are monohalogenated hydrocarbons. They are usually named by reference to the corresponding alkyl radical and the substituting halogen, as for example, ethyl bromide. Alternatively, and systematically, they are named as halogen substitution products of the parent hydrocarbon. Thus, ethyl bromide is 1-bromoethane.

## General Properties

### Physical Properties

Apart from the lowest members of the series, which are gases, alkyl halides are mobile volatile liquids. They show a typical homologous gradation of physical properties. Volatility and density decrease with increasing molecular weight for a given halogen. Ethyl chloride is used as a local anaesthetic on account of its high volatility (b.p. 12.5°C); its rapid evaporation when sprayed on the skin produces intense cooling and hence numbness and temporary localised anaesthesia. Stable low boiling fluorinated hydrocarbons (Arctons and Freons), such as *Dichlorodifluoromethane* and *Trichlorofluoromethane* (Trichloromonofluoromethane; $CCl_3F$) are also used as refrigerants and propellants in aerosol packs, similarly on account of their high volatility.

The highly electronegative halogen atom induces a strong bond dipole in the carbon–halogen bond (C—X), and simple alkyl halides have pronounced dipole moments (i.e. turning moments) when placed in an electric field. The compounds, however, are not ionic and they do not give an immediate precipitate with cold aqueous ethanolic silver nitrate solution. Despite this high polarity, they are almost insoluble in water because of their inability to compete with water molecules in the formation of hydrogen bonds. The formation of C—X---H hydrogen bonds with biological receptors situated in a hydrophobic environment is, none-the-less, a distinct possibility, and may account in part for the prolonged retention of some halogen compounds in most animal species. Such interactions may also partially explain the rôle of the C-9 α-fluorosubstituents in such anti-inflammatory steroids as *Betamethasone Valerate*, and the retention of anti-inflammatory activity in the C-21 chloro-steroid, *Clobetasol* 17-Propionate (Chapter 22).

High polarity favours solubility in hydrophobic polar solvents such as chloroform and fatty acid esters (natural oils and fats), as a result of dipole–dipole interaction. The ready absorption of alkyl halides through the skin (Gemmell and Morrison, 1957), passage across lipid membranes and tissue boundaries, and their facility for accumulation in fatty tissues is largely due to this affinity. The accumulation in natural oils and animal fats of metabolically-stable chlorinated hydrocarbons, such as *Dicophane* (Chlorphenothane; DDT), *Chlordane*, *Aldrin* and *Dieldrin*, which have been used as insecticides and pesticides, presents a world problem owing to their inherent toxicity to man and many birds and animals (Carson, 1963).

### Solvent Properties

Most halogentated hydrocarbons have good solvent properties for a wide variety of organic compounds. Solvent power increases with the extent of halogenation, and low boiling polyhalogen compounds are used extensively as industrial and pharmaceutical solvents with the particular advantage over alternative solvents of being non-flammable (p. 219). These include *Chloroform* (b.p. 60–2°C), carbon tetrachloride (b.p. 76–8°C) and tetrachloroethane (b.p. 146°C), and the unsaturated chlorohydrocarbon, *Trichloroethylene* (b.p. 88–90°C).

## Nucleophilic Substitution and Elimination

Alkyl halides are hydrolysed to the corresponding alcohol and halogen acid. The reaction with water is slow, but conversion to alcohol occurs much more rapidly in hot aqueous sodium or potassium hydroxide solution.

$$R{\cdot}X + KOH \longrightarrow R{\cdot}OH + KX$$

The hydrolysis requires the breaking of the carbon—halogen bond. This is facilitated by the electronegativity of the halogen which varies with the nature of the halogen and decreases in the order:

$$R{\cdot}F > R{\cdot}Cl > R{\cdot}Br > R{\cdot}I$$

This permanent polarisation of the bond leads to unequal sharing of the electrons constituting the bond and creates a deficiency of charge on carbon, thus laying it open to attack by any electron-rich (nucleophilic) reagent. In the alkaline hydrolysis of an alkyl halide, the attacking reagent, a hydroxide ion ($HO^-$) displaces the halogen, so that the reaction is an example of **nucleophilic substitution.**

Whilst the halogen facilitates polarisation of the bond, the ease with which it is able to function as a **leaving group** depends not so much on the extent to which the bond is permanently polarised, but rather more on the ability of the halogen to accept the electron pair from the bond when it is broken under the influence of an attacking reagent. This property, known as **polarisability**, increases with increasing atomic number and hence size of the halogen; it is a function of the ability of the halogen to reduce the charge per unit volume. Hence the order of polarisability and reactivity is:

$$R{\cdot}I > R{\cdot}Br > R{\cdot}Cl > R{\cdot}F$$

Two types of mechanism are distinguished in nucleophilic substitutions according to the kinetics of the reaction:

(a) bimolecular ($S_N2$),
(b) unimolecular ($S_N1$).

**Bimolecular Substitution ($S_N2$)**

This is a single stage mechanism, involving both reactant and reagent, proceeding via a transition state in which the new bond to the attacking nucleophile is being formed as the bond to the departing halide ion is being broken.

$$HO^- + H_3C{-}Br \xrightarrow{KOHaq} \overset{\delta^-}{HO}\cdots CH_3 \cdots \overset{\delta^-}{Br} \longrightarrow HO{-}CH_3 + Br^-$$

The rate of the reaction is dependent on the concentrations of both reacting species, alkyl halide and nucleophile, and hence, mechanistically, the reaction is a bimolecular process. This mechanism is designated, according to Ingold, as $S_N2$ (substitution, nucleophilic, bimolecular).

The geometry of the transition state is also important. The attacking group ($HO^-$) approaches at an angle of 180° to the carbon—halogen bond. The new bond is, therefore, on the opposite side of the molecule to the bond which is broken and a **Walden inversion** of configuration is said to occur. This, however, is only evident where the central carbon atom is asymmetric, i.e. attached to four distinctly different groups (Chapter 13). Such molecules may be resolved into optically active forms, and the change in configuration which occurs with the Walden inversion may be detected by changes in the rotation of plane- or circularly-polarised light when passed through solutions of the compound.

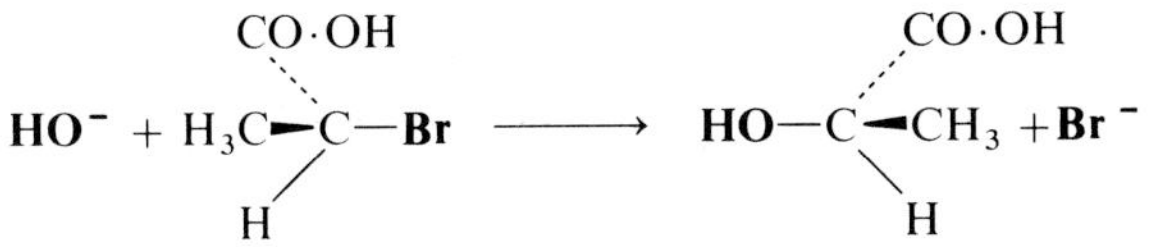

$$HO^- + H_3C{-}C(CO{\cdot}OH)(H){-}Br \longrightarrow HO{-}C(CO{\cdot}OH)(H){-}CH_3 + Br^-$$

The central carbon atom in the transition state has trigonal symmetry with respect to the three atoms or groups not involved in the reaction. These are centred on a plane which is at right angles to the line joining the central carbon atom and the entering and leaving groups. The central carbon, however, is penta-co-ordinate, and crowding is least in the transition state of primary halides ($R{\cdot}CH_2X$) and greatest in that of tertiary halides ($R_3C{\cdot}X$). The bimolecular mechanism, therefore, is most favoured with primary halides and least favoured with tertiary halides.

## $S_N2$ Substitution versus E2 Elimination

Solvent is also important. In aqueous solution, solvation of the bipolar transition stage overcomes what would otherwise be an unfavourable separation of charges. In ethanolic solution, however, the much less polar ethanol molecules are unable to achieve the same degree of solvation, so that an alternative transition state with greater separation of charge is favoured. This leads to an alternative reaction pathway in which an olefine is formed by elimination of hydrogen and halide ion from the alkyl halide. This reaction, which is also bimolecular, is therefore described as an E2 elimination (elimination, bimolecular). Since, also, the departing proton and halogen are derived from adjacent (α and β) carbon atoms, the reaction is sometimes described as a β-elimination.

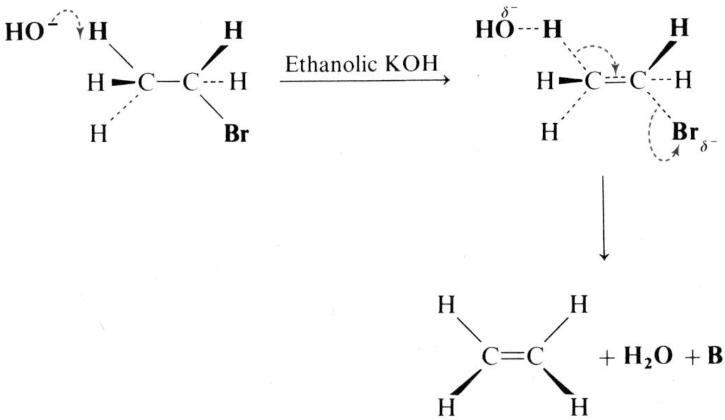

$$HO^- + H{-}CH_2{-}CH_2{-}Br \xrightarrow{\text{Ethanolic KOH}} \overset{\delta^-}{HO}{\cdots}H{\cdots}CH_2{\cdots}CH_2{\cdots}Br^{\delta^-}$$

$$\downarrow$$

$$H_2C{=}CH_2 + H_2O + Br^-$$

Solvent, however, is only one of several factors which help to determine whether nucleophilic attack on an alkyl halide favours substitution or elimination. These include temperature, the relative basicity and nucleophilicity of the attacking nucleophile, and steric factors. Thus, whilst bulky substituents on the α-C atom inhibit $S_N2$ substitution, E2 elimination, which is initiated by attack on a βC—H bond, is unaffected by them. Similarly, breaking of the βC—H bond required for elimination is favoured by strong bases. Hence, treatment of 2-chlorohexane by a strong base ($HO^-$) favours elimination with formation of hex-1-ene, in preference to substitution and formation of hexan-2-ol.

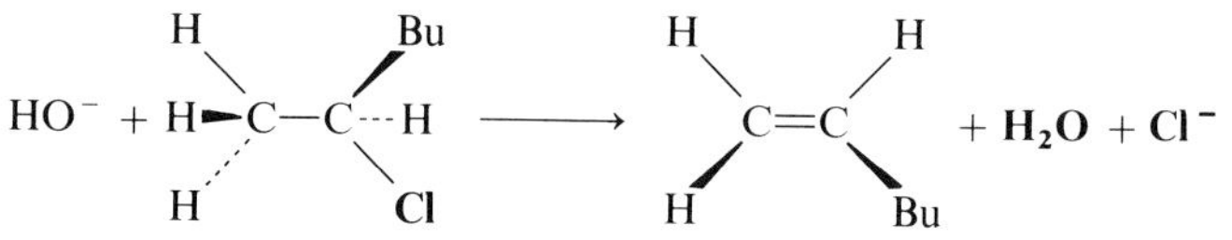

$$HO^- + H\text{—}CH_2\text{—}CH(Bu)Cl \longrightarrow H_2C\text{=}CHBu + \mathbf{H_2O} + \mathbf{Cl^-}$$

In contrast, reaction with acetate ($CH_3CO \cdot O^-$), which is a weak base, favours nucleophilic substitution, though by a unimolecular ($S_N1$) mechanism (p. 205).

The usual stereochemical requirement of E2 eliminations is that all four atoms involved (H—C—C—X) should lie in the same plane with the βC—H and αC—X bonds in a **trans** and **antiparallel** arrangement (i.e. parallel, but opposite in direction). This is easily achieved in alkyl halides where there is free rotation about the C—C sigma bonds of the chain. Certain cyclic structures, such as that of *Gamma Benzene Hexachloride* (Gammexane), which have preferred conformations, demonstrate the significance of this steric requirement, and dehydrohalogenation occurs readily at those (two) positions in the molecule where it is met. The corresponding β-isomer is only able to undergo *cis*-elimination and undergoes dehydrohalogenation by sodium ethoxide at about 1/5000th of the rate of the γ-compound.

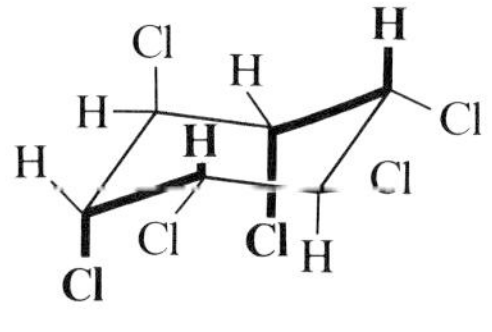

γ-isomer

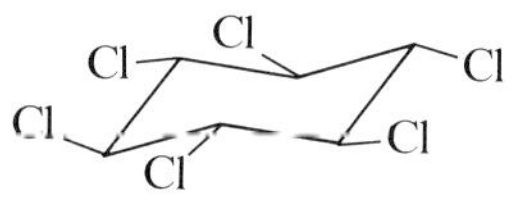

β-isomer

It should be noted, however, that the official assay of *Gamma Benzene Hexachloride* is based on the elimination of three $Cl^-$ ions from the molecule by ethanolic potassium hydroxide. Loss of one mole of HCl by elimination leads to a pentachlorohexene of a totally different conformation, but with two allylic C—Cl bonds readily capable of further dehydrohalogenation to 1,2,4-trichlorobenzene. Displaced $Cl^-$ ions are titrated with silver nitrate. *Gamma Benzene Hexachloride* undergoes similar stepwise dehydrohalogenation in mammals and insects (**2**, 5).

$$\therefore C_6H_6Cl_6 \equiv 3Cl^-$$

The potential for an analogous dehydrohalogenation of the unsaturated chlorocarbon anaesthetic, *Trichloroethylene*, is of special interest, since this gives rise to 1,2-dichloroacetylene which is toxic. For this reason, *Trichloroethylene* must not be used in a closed-circuit anaesthetic apparatus, as the heat produced in the removal of carbon dioxide and water vapour by the soda lime is sufficient to initiate elimination.

$$Cl_2C{=}CHCl \xrightarrow{HO^-} Cl{\cdot}C{\equiv}C{\cdot}Cl + HCl$$

*Trichloroethylene* Dichloroacetylene

**Unimolecular Substitution ($S_N1$)**

The hydrolysis of tertiary halides in alkali proceeds at a rate which is directly proportional to the concentration of the alkyl halide and independent of the hydroxyl ion concentration. That is, the reaction has first order kinetics and accordingly is described as **unimolecular substitution** ($S_N1$). Since the overall reaction rate depends solely on the concentration of the alkyl halide, it means that the hydroxyl ion concentration plays no part in the rate-determining step. This is interpreted as implying that the rate-determining step consists of a **slow** heterolytic cleavage of the already polarised carbon—halogen bond, followed by a fast stage in which the carbonium ion combines with a nucleophile to form the product.

$$Me_3C{-}Cl \xrightarrow[-Cl^-]{Slow} Me_3C^+ \xrightarrow[Fast]{HO^-} Me_3C{-}OH$$

t-Butyl chloride t-Butanol

This mechanism is favoured in tertiary halides where the bulk of the three alkyl groups on the α-carbon atom would hinder formation of a transition state in the alternative bimolecular process. Furthermore, the intermediate carbonium ion is also trigonal and planar, so that steric compression resulting from the presence of large bulky groups in the alkyl halide is relieved either partially or completely in the carbonium ion. Since the carbonium ion is planar, nucleophilic attack can occur from either side. Where the central carbon atom is asymmetric, substitution is, therefore, accompanied by racemisation (Chapter 13).

Other factors which assist ionisation also assist the $S_N1$ process. These include reaction in solvents of high dipole moment, and catalysis by heavy metal salts which promote ionisation through the halogen lone electron pairs.

$$R-\ddot{X}\ Ag^+ \rightleftharpoons \overset{\delta^+}{R}\cdots X \cdots \overset{\delta^+}{Ag} \longrightarrow R^+ + AgX$$

The effect of solvent polarity, which favours ionisation, is seen in the vast increase in the rate of the hydrolysis of t-butyl chloride in aqueous ethanol compared with that in absolute ethanol. Unimolecular heterolysis is also favoured as the number of electron-repelling alkyl substituents increases. It is, therefore, increasingly favoured in secondary and tertiary halides, both as a result of the increased electron density on the central carbon atom, which opposes attack by nucleophilic reagents, and by the increased transmission of charge to the carbon—halogen bond, which promotes separation of the halide ion. Stabilisation of the intermediate carbonium ions by hyperconjugation (Chapters 2 and 7) also favours unimolecular substitution and elimination reactions in secondary and, particularly, in tertiary halides.

$$(H_3C)_3C \rightarrow Cl \rightleftharpoons (H_3C)_3C^+ + Cl^-$$

Trigonal carbonium ion

## $S_N1$ Substitution versus E1 Elimination

The carbonium ion, once formed, can interact in a fast stage reaction with the attacking nucleophile, either at the α-carbon leading to nucleophilic substitution or at a βC—H bond leading to elimination. Of the two alternatives, elimination is favoured as the size of the alkyl group increases, and it becomes increasingly difficult for the nucleophile to attack the carbonium carbon. Elimination is also favoured by the attack of nucleophiles which are also strong bases and hence readily capable of removing a proton from the β-carbon atom. In contrast, substitution is favoured by the nucleophilic attack of groups which are both strong nucleophiles and weak bases, e.g. $CH_3COO^-$, or $RS^-$.

$(CH_3)\cdot C{=}CH_2$ ← $CH_3\cdot CO\cdot O^-$ — $(CH_3)_3C^+$ — $HO^-$ → $(CH_3)_2\cdot C{=}CH_2$

$(CH_3)_3\cdot C\cdot O\cdot CO\cdot CH_3$ — t-Butyl acetate

$(CH_3)_3\cdot C\cdot OH$ — t-Butanol

The extent of competition between substitution and elimination reactions is also influenced by reaction temperature. Elimination almost invariably requires a higher activation energy than substitution. Higher reaction temperatures, therefore, favour elimination.

**Saytzeff Rule** (Direction of elimination)

When alternative products can arise in an elimination of halogen halide depending upon which $\beta$-hydrogen is eliminated, the Saytzeff rule applies. In this, **the predominant product is formed by elimination of hydrogen from the more highly alkylated β-carbon atom**.

$$CH_3 \cdot \overset{\beta}{C}H_2 \cdot \underset{X}{\overset{\alpha}{C}H} \cdot \overset{\beta}{C}H_3 \longrightarrow CH_3 \cdot CH{=}CH \cdot CH_3$$

But-2-ene

$$CH_3 \cdot \overset{\beta}{C}H_2 \cdot \underset{X}{\overset{\alpha}{C}H} \cdot \overset{\beta}{C}H_3 \not\longrightarrow CH_3 \cdot CH_2 \cdot CH{=}CH_2$$

But-1-ene

The major product (but-2-ene) with two alkyl substituents attached directly to olefinic carbons is thus the more highly substituted of the two possible product olefines. Several factors influence the direction of elimination, but the most important is the relative stabilities of the two product olefines. In the example, the relative stabilities are determined by hyperconjugation. Thus, but-2-ene has six $\alpha$C—H bonds, whereas but-1-ene has only two such bonds; in consequence, the former offers more opportunity for charge delocalisation by hyperconjugation.

$$CH_3 \cdot CH{=}CH{-}\underset{H}{\overset{H}{C}}{-}H \longleftrightarrow CH_3 \cdot \bar{C}H{-}CH{=}\underset{H^+}{\overset{H}{C}}{-}H$$

$$\updownarrow$$

$$CH_3 \cdot \bar{C}H{-}CH{=}\underset{H}{\overset{H}{C}} \; H^+ \longleftrightarrow$$

The Saytzeff rule applies in both E1 and E2 eliminations.

Table 20. Nucleophiles and Leaving Groups in Descending Order of Reactivity

| Nucleophiles | Leaving Groups |
|---|---|
| $HS^-$ | $-\overset{+}{N}\equiv N$ |
| $CN^-$ | $-O\cdot SO_2^-$ |
| $SCN^-$ | $-O\cdot SO_2\cdot O\cdot CH_3$ |
| $NH_2$ | $-I$ |
| $HO^-$ | $-Br$ |
| $RO^-$ | $-ONO_2$ |
| $N_3^-$ | $-Cl$ |
| $CH_3\cdot COO^-$ | $-\overset{+}{O}H_2$ |
| $H_2O$ | $-\overset{+}{N}R_3$ |
| | $-OR$ |

## Synthetic Applications of Nucleophilic Substitution

Nucleophilic substitution reactions have widespread applications. The facility with which such reactions occur depends both on the nucleophilicity of the attacking reagent, and the ease with which the displaced radical will act as a leaving group. Table 20 lists a number of nucleophiles and leaving groups in descending order of reactivity.

Nucleophilic substitution of alkyl halides provides a useful method of synthesising a wide variety of substituted alkyl compounds.

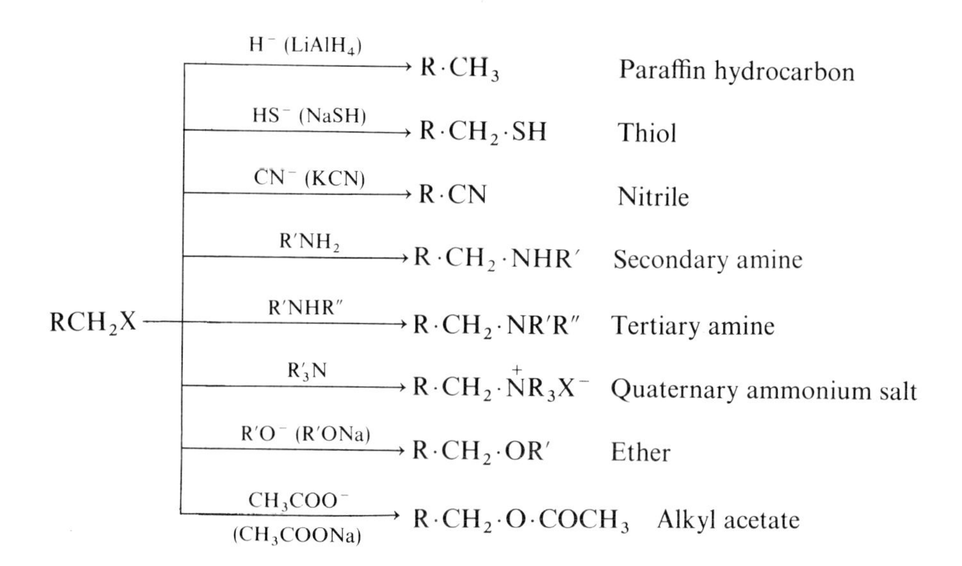

**Substitution of One Halogen by Another**

One halogen is able to displace another, more polarisable halogen from the carbon—halogen bond. Thus, fluoride normally displaces chloride; chloride displaces bromide, and bromide displaces iodide. Certain exceptions to this rule are observed as in the Finkelstein displacement of chloride or bromide by iodide in acetone, which is possible because of the greater solubility of sodium iodide in acetone compared with that of sodium chloride or bromide.

The normal displacement, however, is especially useful in the preparation of fluorocarbon compounds (Barbour, 1969), which are not available by direct fluorination owing to the violence and uncontrollable nature of the latter reaction.

$$2CCl_4 + 3HF \xrightarrow[\text{Pressure}]{SbCl_3/50\text{–}150^\circ} \underset{\substack{\textit{Dichlorodifluoromethane} \\ \text{(Propellant 12)}}}{CCl_2F_2} + \underset{\substack{\textit{Trichlorofluoromethane} \\ \text{(Propellant 11)}}}{CCl_3F} \quad (-\,3HCl)$$

$$CHCl_3 + 2HF \xrightarrow{SbF_3} \underset{\text{Chlorodifluoromethane}}{CHClF_2} \quad (-\,2HCl)$$

The chlorofluoromethanes are used as refrigerants and as propellants in aerosol packs. Chlorodifluoromethane is used as a source of tetrafluoroethylene, which on peroxide-catalysed radical polymerisation yields polytetrafluoroethylene (Teflon), a tough and chemically inert plastic.

$$2CHClF_2 \xrightarrow{800^\circ} F_2C{=}CF_2 \xrightarrow{\text{Polymerisation}} \text{Polytetrafluoroethylene}$$

Polytetrafluoroethylene is reputedly non-toxic and chemically inert below about 250°. It has a very low coefficient of friction and has been used successfully in the surgical repair of arthritic human limb joint sockets. Its potential use as a tablet lubricant (Hotko, 1967) has been investigated (Alpar, Deer, Hersey and Shotton, 1969). It has been shown to be about as effective for this purpose as *Magnesium Stearate* with some advantages.

*Tetrafluoroethylene* is also used as a source of *Dichlorotetrafluoroethane* (Propellant 114; $CClF_2{\cdot}CClF_2$), which because of its low boiling point and stability is used as an aerosol propellant often in combination with *Dichlorodifluoromethane* (Propellant 12).

The displacement of chlorine by fluorine is also an important step in the following synthesis of *Halothane* which is but one of about a dozen different routes to this anaesthetic.

$$CCl_3{\cdot}CH_2{\cdot}Cl \xrightarrow{HF} CF_3{\cdot}CH_2Cl \xrightarrow{Br_2/500^\circ} \underset{\textit{Halothane}}{CF_3{\cdot}CHBrCl}$$

**$S_N$i Substitutions**

A number of reactions which follow second order kinetics have been shown to proceed with retention of configuration (in contrast to the Walden inversion of configuration which is the normal outcome of $S_N2$ substitution). One such reaction is the displacement of OH by Cl in the conversion of alcohols to alkyl halides with thionyl chloride (Chapter 7).

$$\begin{matrix} & & Cl & & \\ -C-O & & S{=}O & \longrightarrow & -C-O-\overset{Cl}{S}{=}O \xrightarrow[SO_2]{} -C-Cl \\ & H & Cl & & \end{matrix}$$

No change of configuration occurs, since it is only in the second stage of the reaction that the C—O bond is broken, and because of the structure of the intermediate, the new C—Cl bond is formed on the same side of the molecule. The reaction is, therefore, an internal bimolecular nucleophilic substitution, and is designated $S_N$i (substitution, nucleophilic, internal).

## Nucleophilic Substitutions of Alkyl Halides in Biological Systems

**Detoxification of Halogenated Hydrocarbons**

Alkyl (and aryl) halides are intrinsically toxic to mammalian systems and, where the reactivity of the halogen is sufficiently high, detoxification occurs through a nucleophilic substitution by the sulphydryl compound, glutathione (glutamylcysteinylglycine). Thus, alkyl halides are converted to *S*-alkylglutathiones by enzymes known as glutathione-*S*-alkyl transferases. The resulting *S*-alkyl glutathione is then degraded stepwise to an *S*-alkylcysteine, which is acetylated and oxidised prior to excretion as the corresponding mercapturic acid and mercapturic acid *S*-oxide (reaction scheme p. 211 and **2**, 5).

Mercapturic acid conjugation also occurs through the medium of glutathione-*S*-aryl transferases, with suitably activated aryl halides, as for example with 2,4-dinitrochlorobenzene. The detoxification mechanism fails, however, for the unactivated aryl halogens in *Dicophane* (Chlorophenothane; DDT) and for the deactivated vinylic halogens in *Trichloroethylene*, *Chlordane*, *Dieldrin* and *Aldrin*. Mercapturation is also impeded where the halogen is sterically hindered as in the trichloromethyl group of *Dicophane*, and the dichloromethyl groups of *Chlordane*, *Dieldrin* and *Aldrin*. It is also inhibited where the halogen occupies a bridgehead position in *Chlordane*, *Dieldrin* and *Aldrin* from which halogen displacement is resisted since formation of a planar carbonium ion is not possible.

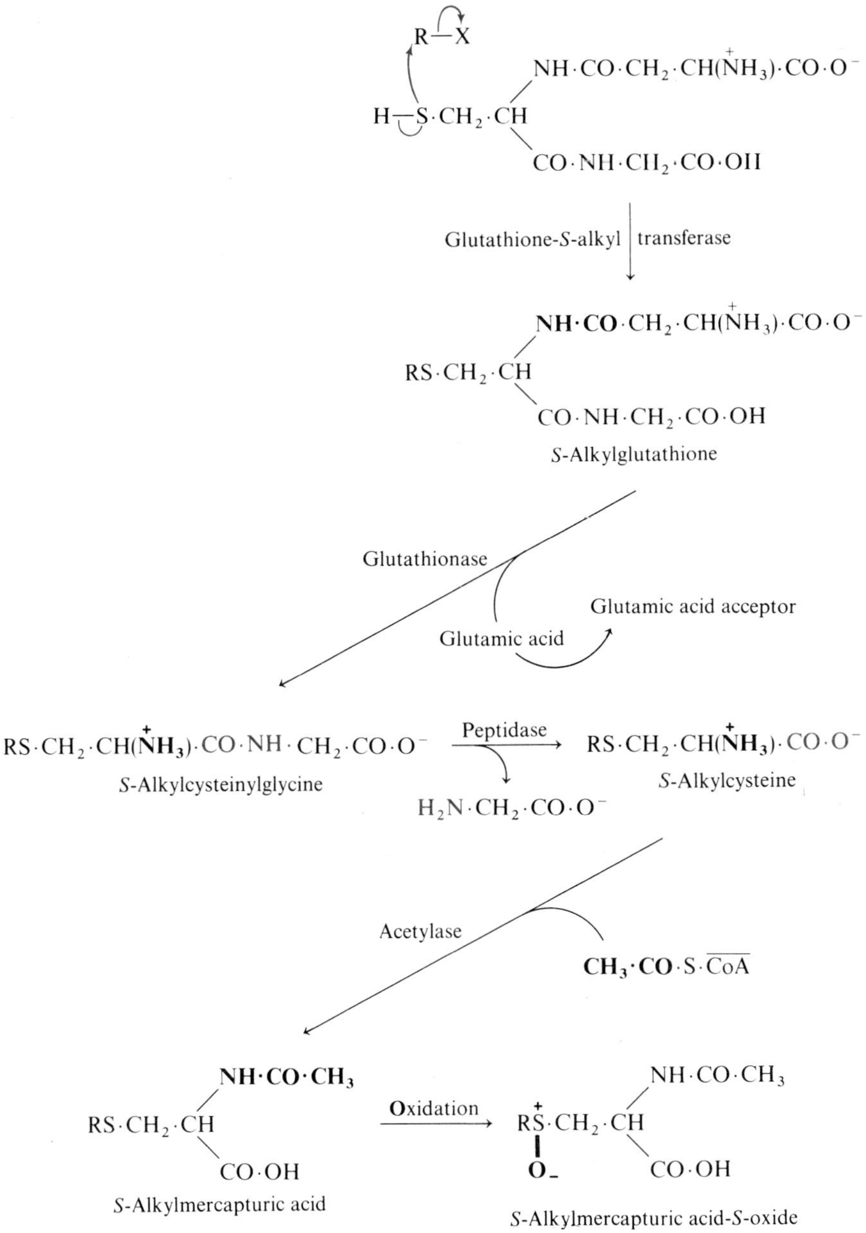

R—X

$NH \cdot CO \cdot CH_2 \cdot CH(\overset{+}{N}H_3) \cdot CO \cdot O^-$

$H—S \cdot CH_2 \cdot CH$

$CO \cdot NH \cdot CH_2 \cdot CO \cdot OH$

Glutathione-*S*-alkyl transferase

$\mathbf{NH \cdot CO} \cdot CH_2 \cdot CH(\overset{+}{N}H_3) \cdot CO \cdot O^-$

$RS \cdot CH_2 \cdot CH$

$CO \cdot NH \cdot CH_2 \cdot CO \cdot OH$

*S*-Alkylglutathione

Glutathionase

Glutamic acid acceptor

Glutamic acid

$RS \cdot CH_2 \cdot CH(\overset{+}{\mathbf{N}}\mathbf{H_3}) \cdot CO \cdot NH \cdot CH_2 \cdot CO \cdot O^-$ —Peptidase→ $RS \cdot CH_2 \cdot CH(\overset{+}{\mathbf{N}}\mathbf{H_3}) \cdot CO \cdot O^-$

*S*-Alkylcysteinylglycine

*S*-Alkylcysteine

$H_2N \cdot CH_2 \cdot CO \cdot O^-$

Acetylase

$\mathbf{CH_3 \cdot CO} \cdot S \cdot \overline{CoA}$

$\mathbf{NH \cdot CO \cdot CH_3}$

$RS \cdot CH_2 \cdot CH$

$CO \cdot OH$

*S*-Alkylmercapturic acid

Oxidation

$NH \cdot CO \cdot CH_3$

$R\overset{+}{S} \cdot CH_2 \cdot CH$

$\mathbf{O}_-$

$CO \cdot OH$

*S*-Alkylmercapturic acid-*S*-oxide

*Dicophane* (Chlorophenothane) is, however, metabolised in the rat mainly to DDD (Peterson and Robison, 1964). *Dieldrin* undergoes metabolic epoxidation to *Aldrin*, but the latter resists further metabolic degradation (Nakatsugawa, Ishida and Dahm, 1965).

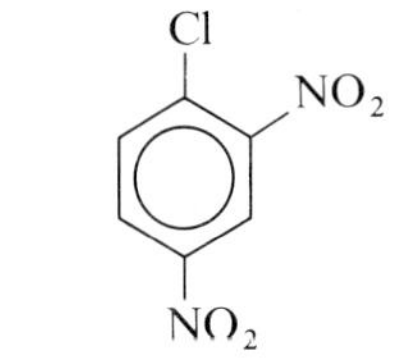

2,4-Dinitrochlorobenzene

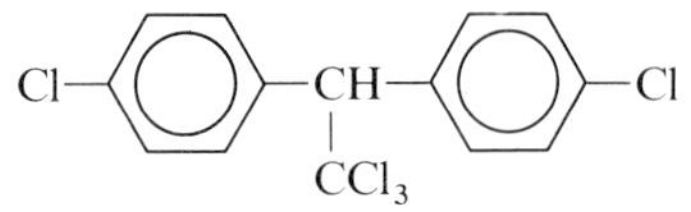

*Dicophane* (DDT)
(Chlorophenothane)

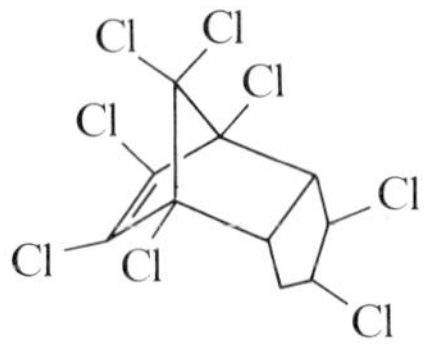

*Chlordane*

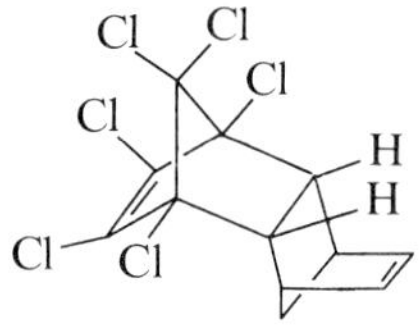

*Aldrin*

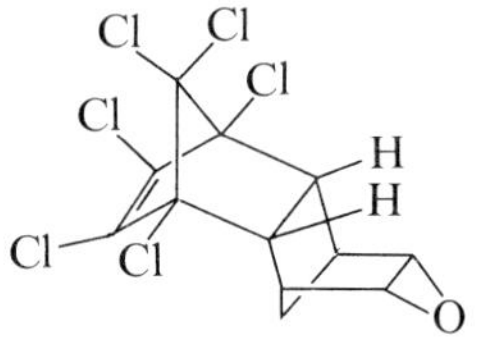

*Dieldrin*

**Nitrogen Mustards**

Nitrogen mustards are bis-dialkylaminoethyl chlorides, and are used as biological alkylating agents in the control of certain types of cancer.

$\beta$-Dialkylaminoethyl chlorides are typical of a wide range of compounds which show enhanced reactivity through **neighbouring group participation** in $S_Ni$ substitutions. Thus, $\beta$-dimethylaminoethyl chloride is hydrolysed many times more rapidly than ethyl chloride. This is due to the close proximity of the dimethylamino group on the $\beta$-carbon atom to the C—Cl bond, which leads to the formation of a highly strained ethyleneimmonium ion in the rate determining step.

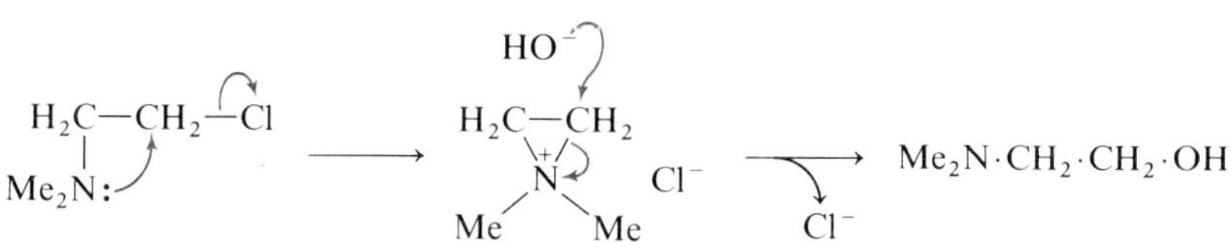

$$Me_2N{:}\,CH_2{-}CH_2{-}Cl \longrightarrow \underset{Me\quad Me}{\overset{+}{N}}(CH_2{-}CH_2)\ Cl^- \xrightarrow[-Cl^-]{HO^-} Me_2N{\cdot}CH_2{\cdot}CH_2{\cdot}OH$$

The action of nitrogen mustards such as *Melphalan* depends on a partially selective alkylation of purine residues in the double helix of DNA in such a way that it inhibits the replication of nucleic acid in the rapidly dividing cancerous cells. Alkylation proceeds through an internal nucleophilic attack on carbon ($S_Ni$ reaction) to form an ethyleneimmonium ion, which is then open to nucleophilic attack by the DNA purine moiety (**2**, 2).

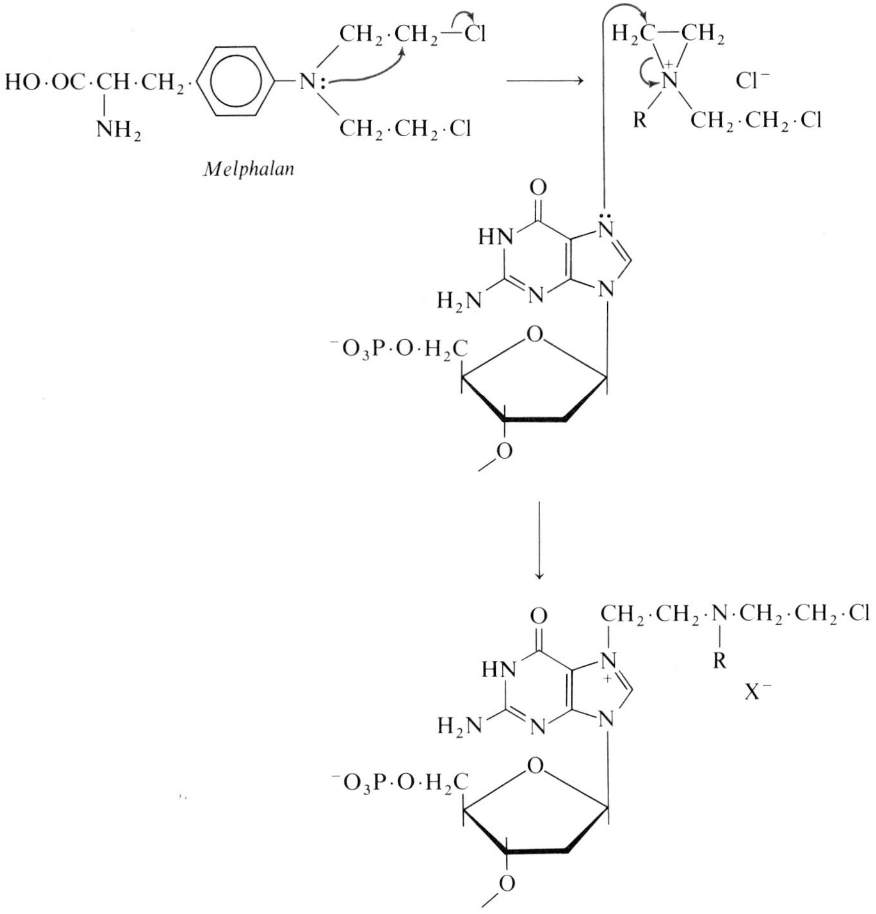

$CH_2{\cdot}CH_2{-}Cl$

$HO{\cdot}OC{\cdot}CH{\cdot}CH_2{\cdot}C_6H_4{-}N{:}$

$NH_2$

$CH_2{\cdot}CH_2{\cdot}Cl$

*Melphalan*

$\longrightarrow$

$H_2C{-}CH_2$

$\overset{+}{N}$ $Cl^-$

$R$ $CH_2{\cdot}CH_2{\cdot}Cl$

O

HN

$\ddot{N}$

$H_2N$ N N

$^-O_3P{\cdot}O{\cdot}H_2C$ O

O

$\downarrow$

O $CH_2{\cdot}CH_2{\cdot}N{\cdot}CH_2{\cdot}CH_2{\cdot}Cl$

HN $\overset{+}{N}$ R

$X^-$

$H_2N$ N N

$^-O_3P{\cdot}O{\cdot}H_2C$ O

O

## Homolytic Cleavage of the Carbon—Halogen Bond

### Light Sensitivity

The carbon—halogen bond may be broken homolytically in light-catalysed, thermal or radical-induced reactions. Sensitivity to light increases with polarisability of the halogen, and discolouration of alkyl iodides is often noticeable as a result of iodine formation.

$$R{-}I \longrightarrow R\cdot + I\cdot$$

$$2R\cdot \longrightarrow R{\cdot}R$$

$$2I\cdot \longrightarrow I_2$$

**Pyrolysis**

Pyrolytic (thermal) decomposition of alkyl halides also involves a radical mechanism of *cis*-elimination.

$$R\text{—}CH(H)\text{—}CH(H)(Br)\text{—}R' \longrightarrow (H)(R)C{=}C(H)(R') + HBr$$

**Reduction**

Radical-induced homolysis of the carbon—halogen bond occurs in reduction by nascent hydrogen, produced for example with constant-boiling hydriodic acid and red phosphorus, sodium and ethanol, or zinc and acetic acid.

$$R\cdot CH_2\text{—}Br \ \cdot H \longrightarrow R\cdot CH_2\cdot + HBr$$
$$R\cdot CH_2\cdot + H\cdot \longrightarrow R\cdot CH_3$$

**Formation of Metal Alkyls and Grignard Reagents**

Halogen is also displaced by radical attack of reactive metals, such as sodium, zinc and magnesium in the formation of metal alkyls and Grignard reagents (Musgrave, 1964) and in the Reformatsky reaction. With the more electropositive metals, such as sodium, the metal alkyl reacts as fast as it is formed with a second molecule of alkyl halide in a nucleophilic substitution to yield a hydrocarbon (Wurtz reaction).

$$CH_3\text{—}I \ \cdot Na \longrightarrow CH_3\cdot + NaI$$
$$CH_3\cdot + Na\cdot \longrightarrow CH_3^-Na^+$$
$$CH_3^-Na^+ + CH_3I \longrightarrow CH_3\cdot CH_3 + NaI$$

Lithium, however, which is less electropositive, gives rise to less reactive lithium alkyls, which can in consequence be obtained directly from lithium and the alkyl halide.

$$R\cdot Br + Li\cdot \longrightarrow R\cdot \ (+\ LiBr) \xrightarrow{Li\cdot} R\cdot Li$$

Magnesium reacts similarly with alkyl halides in dry ether to give solutions of alkylmagnesium halides or Grignard reagents.

$$R\cdot CH_2\text{—}I \ \cdot Mg\cdot \longrightarrow R\cdot CH_2\cdot + I\text{—}Mg\cdot \longrightarrow R\cdot CH_2\cdot Mg\cdot I$$

Both halide and ether must be completely dry; the magnesium also should be completely dry and free from contamination with traces of grease. The reaction vessel must also be thoroughly dried and protected from access of moisture, carbon dioxide and oxygen, which would otherwise decompose the reagent, as formed. The reactivity of halides in the reaction decreases in the order RI > RBr > RCl in accord with their decreasing susceptibility to radical attack; chlorides are often slow to react and fluorides do not react at all. Ease of formation of Grignard reagents also decreases with increasing molecular weight of the alkyl halide. Notwithstanding this factor, tertiary chlorides are more effective precursors than tertiary iodides, since the latter more readily undergo elimination to form olefines. Vinyl and aryl chlorides form Grignard reagents at the higher reaction temperatures which are achieved when tetrahydrofuran is used as a solvent in place of ether, though the corresponding bromides react more readily.

### Radical Oxidation

The stability order of C—X and C—H bonds in free radical reactions is:

$$\underset{\text{Reactivity decreases}}{\xrightarrow[\hspace{20em}]{\text{Stability increases}}}$$

C—I  C—Br  C—H  C—Cl  C—F

Stability increases →

Reactivity decreases

Radical oxidation of bromides and iodides, therefore, leads to breaking of the carbon—halogen bond. In contrast, alkyl chlorides and fluorides are subject to radical oxidation of the $\alpha$C—H bond. With chloroform, this gives rise to carbonyl chloride (phosgene), possibly via a peroxide intermediate, $CCl_3 \cdot O \cdot OH$ (Chapman, 1935).

$$2CHCl_3 + O_2 \longrightarrow 2CO \cdot Cl_2 + 2HCl$$

The susceptibility to radical oxidation of the anaesthetics, *Halothane* ($CHClBr \cdot CF_3$), *Methoxyfluorane* ($CHCl_2 \cdot CF_2 \cdot O \cdot CH_3$) and *Trichloroethylene* ($CHCl{=}CCl_2$), on exposure to air in bright light leads to the stringent storage requirements for these products in airtight containers, protected from light and in a cool place.

## Grignard Reagents

Grignard reagents are extremely reactive, providing a source of carbanions in a wide range of synthetic reactions. The reagents themselves, although highly polarised, are not completely ionic, and reaction only occurs with compounds which are themselves sufficiently polarised to promote attack. Thus, the polarity of the C—O bond of ether, the normal reaction solvent, is insufficient to initiate carbanion attack. The reagent, however, is associated in a 2:2-co-ordination complex with solvent molecules which are not readily removed under reduced pressure at ambient temperatures.

**Reactions with Acidic Hydrogen**

Grignard reagents react readily with acids, even such weakly acidic substances as water, alcohols, primary and secondary amines and acetylenes, to yield hydrocarbons. The use of methyl magnesium iodide, and measurement of the volume of methane released forms the basis of the Zerewitinoff method for the determination of active hydrogen in such compounds.

$$CH_3{\cdot}Mg{\cdot}I + H_2O \longrightarrow CH_4 + HO{\cdot}Mg{\cdot}I$$

$$CH_3{\cdot}Mg{\cdot}I + CH_3{\cdot}CH_2{\cdot}OH \longrightarrow CH_4 + CH_3{\cdot}CH_2{\cdot}O{\cdot}Mg{\cdot}I$$

$$CH_3{\cdot}Mg{\cdot}I + R{\cdot}NH_2 \longrightarrow CH_4 + R{\cdot}NH{\cdot}Mg{\cdot}I$$

By using deuterium oxide ($D_2O$) in place of water, the Grignard reaction can be used to introduce deuterium as a substituent in place of a halogen.

$$H_3C{-}C_6H_4{-}Br \xrightarrow{Mg|Et_2O} H_3C{-}C_6H_4{-}MgBr \xrightarrow{D_2O} H_3C{-}C_6H_4{-}D$$

*p*-Bromotoluene — *p*-Deuterotoluene

**Additions to Carbonyl Compounds**

One of the most important reactions of Grignard reagents is with carbonyl groups ($>C{=}O$) in aldehydes and ketones (Chapter 10), esters (Chapter 11) and carbon dioxide. Grignard reagents also react with compounds containing other polar multiple bonds, including $>C{=}S$, $-C{\equiv}N$, $-N{=}O$, $-N{\equiv}C$, and $>S{=}O$.

The addition of Grignard reagents to carbonyl compounds is thought to be mediated through a cyclic intermediate involving two molecules of the reagent, one of which increases the polarity of the carbonyl group, thereby promoting attack by the second molecule at the electron-deficient carbonyl carbon.

$$(CH_3CH_2)(H)C{=}O + \overset{\delta^-}{CH_3}{-}\overset{\delta^+}{Mg}{\cdot}I \longrightarrow \text{cyclic intermediate: } CH_3{\cdot}CH_2{\cdot}\overset{+}{C}H{-}O^-{\cdots}(H_3C{-}Mg{\cdot}I)(CH_3{-}Mg{\cdot}I)$$

Propionaldehyde

$$\longrightarrow CH_3{\cdot}CH_2{\cdot}CH(CH_3){\cdot}OMgI + CH_3{\cdot}Mg{\cdot}I$$

$$CH_3{\cdot}CH_2{\cdot}CH(CH_3){\cdot}OMgI \longrightarrow CH_3{\cdot}CH_2{\cdot}CH(OH){\cdot}CH_3$$

Butan-2-ol

The reaction with carbon dioxide is particularly useful for the preparation of carboxylic acids of which the corresponding cyanide is not readily accessible.

$$(CH_3)_3\cdot C\cdot Cl \xrightarrow{Mg/Et_2O} (CH_3)_3\cdot \overset{\delta-}{C}-\overset{\delta+}{MgCl}$$

$$\downarrow CO_2\text{(solid)}$$

$$(CH_3)_3C\cdot CO\cdot OH \xleftarrow{H^+/H_2O} (CH_3)_3\cdot C\cdot CO\cdot OMgCl$$

**Relative Reactivity of Functional Groups to Grignard Reagents**

Reactions involving reactants with more than one Grignard reactive group pose problems of relative reactivity, unless excess reagent is used. In general, however, the most reactive groups are active hydrogen in descending order of acidic strength. Carbonyl groups are less reactive, but aldehydes are more reactive than ketones, and both more reactive than esters according to the reactivity order:

aldehydes > ketones > esters

## Polyhalogen Compounds

### Trihalogen Compounds

Chloroform and bromoform exhibit some acidic properties. Thus, both undergo deuterium exchange in alkaline solution.

$$CHCl_3 + HO^- \rightleftharpoons H_2O + {}^-CCl_3 \overset{D_2O}{\rightleftharpoons} CDCl_3 + DO^-$$

Chloroform also adds to the carbonyl group of acetone in the presence of potassium hydroxide to form *Chlorbutol* (Chlorobutanol), a mild hypnotic, now used mainly as a preservative.

$$(CH_3)_2\cdot C{=}O \;(\uparrow {}^-CCl_3) \longrightarrow (CH_3)_2\cdot C(CCl_3){-}O^- \xrightarrow{H_2O} (CH_3)_2\cdot C(OH)(CCl_3)$$

*Chlorbutol*
(Chlorobutanol)

Formation of the $^-CCl_3$ carbanion appears to be facilitated by the stability which it derives from resonance stabilisation. This requires expansion of the chlorine valency shells to form decets, through participation of otherwise vacant *d*-orbitals.

$$Cl{-}\overset{Cl}{\underset{Cl}{C^-}} \longleftrightarrow Cl{-}\overset{Cl^-}{\overset{\|}{\underset{Cl}{C}}} \longleftrightarrow {}^-Cl{=}\overset{Cl}{\underset{Cl}{C}} \longleftrightarrow Cl{-}\overset{Cl}{\underset{Cl^-}{\underset{\|}{C}}}$$

This type of resonance is not feasible where two or more of the halogens are fluorine, since fluorine lacks the required *d*-orbitals. This explains the inability of such fluorocarbons as $CHF_3$ and $CHClF_2$ to ionise readily under the same conditions. The marked stability of such hydrocarbons as *Halothane* ($CHClBr \cdot CF_3$) to both acid and alkaline hydrolysis is also derived in this way.

### Carbenes

In the presence of strong bases, such as potassium t-butoxide or butyl lithium, the $^-CCl_3$ carbanion, which is first formed from chloroform, loses $Cl^-$ to form an unstable intermediate **carbene**. In effect, this in an $\alpha$-elimination since both $H^+$ and $Cl^-$ are eliminated from the same carbon atom.

$$CHCl_3 \xrightarrow{\text{t-BuOK}} {}^-C \cdot Cl_3 \longrightarrow \underset{\text{Carbene}}{:C \cdot Cl_2} + Cl^-$$

Methylene dichloride similarly eliminates HCl to give a carbene, whilst inter-halogen compounds eliminate the more polarisable halogen.

$$CH_2Cl_2 \longrightarrow {}^-CHCl_2 \longrightarrow :CH \cdot Cl + Cl^-$$

$$CHBrCl_2 \longrightarrow {}^-CBrCl_2 \longrightarrow :C \cdot Cl_2 + Br^-$$

Carbenes are highly reactive in addition reactions with ethylenic compounds.

### Carbon Tetrachloride

Carbon tetrachloride is much less reactive than chloroform to alkali. The stability of carbon tetrachloride arises from the tetrahedral symmetry of the molecule which leads to it having a net zero dipole moment. It is stable to heat at temperatures up to 500°C, and because of its low chemical reactivity is sometimes used as a fire extinguisher. It is, however, decomposed by water at high temperature to carbonyl chloride, which is toxic. Rapid dispersal of fumes is, therefore, important where carbon tetrachloride fire extinguishers are used.

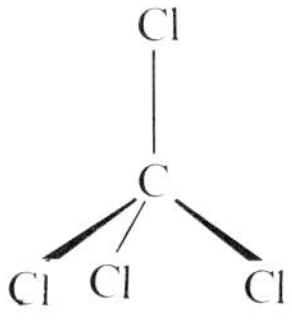

Cl
|
C
Cl Cl Cl

### Solvent Properties

Polyhalogen compounds in general are good solvents for a wide variety of non-polar organic compounds. The polarity of chloroform, however, is such that it is also an effective solvent for a substantial number of polar compounds. Thus, it is miscible with ethanol and is a useful solvent for carboxylic acids, phenols and organic bases. Certain more ionic compounds including some higher molecular weight amine hydrochlorides and even quaternary salts are also significantly soluble in chloroform.

Both chloroform and carbon tetrachloride are non-flammable, and also immiscible with and heavier than water, properties which combine to make them valuable extraction solvents for the separation of solvent-soluble compounds from water-soluble materials in aqueous systems. Hence, their widespread use in the isolation and purification of soluble reaction products from synthetic processes and natural products from plant sources; in laboratory analysis of pharmaceutical products and their dosage forms; and in the examination of blood, urine and other biological materials for drug substances and their metabolites. The unusual ability of chloroform to dissolve complex organic ion pairs forms the basis of its use in combination with various indicator dyestuffs in so-called indicator extraction methods of analysis (Beckett and Stenlake, 1975).

### Polyhalogen Aerosol Propellants

The combination of good solubility properties, non-flammability, low boiling point, and low chemical reactivity in certain low molecular weight polyhalogen fluorocarbon derivatives of methane and ethane has led to their extensive use as aerosol propellants. The very low boiling *Dichlorodifluoromethane* (b.p. $-29.8°C$), $CCl_2F_2$, which is gaseous at room temperature, but readily compressible to a liquid, is used as the main propellant (Propellant 12) in aerosol packs. Its volatility is often reduced by the active ingredients and additives with which it is formulated, but in preparations where it is not so affected, it is often used in combination with other less volatile fluorocarbon propellants, such as *Dichlorotetrafluoroethane* (Propellant 114; b.p. 3.5°C), $CClF_2 \cdot CClF_2$, and *Trichlorofluoromethane* (Propellant 11; b.p. 23.7°C), $CCl_3F$.

Pressurised aerosols are used as media for the administration of medicaments directly into the lungs and bronchi, particularly for rapid control of bronchospasm in asthmatics. The medicament is released as fine powder or droplets from solutions and suspensions in the aerosol propellant. The use of special metering devices in combination with suitable mouth adaptors permits release of measured doses of the medicament in finely divided solid or liquid form directly into the airway passages.

### Polyhalogen Anaesthetics

The combination of high volatility, high affinity for lipid tissue and non-flammability found in alkyl polyhalides provide the essential requirement for their use as general anaesthetics. They are administered by inhalation and absorbed via the lungs. High volatility is essential to achieve an effective concentration in the inspired air. High lipid solubility is necessary to ensure rapid absorption across the alveolar membrane, high blood carrier capacity and effective distribution from blood to the central nervous system. These properties and that of non-flammability are found to an exceptionally high degree in the saturated polyhalogenated hydrocarbons, such as *Chloroform* ($CHCl_3$), *Halothane* ($CHClBr\cdot CF_3$), *Methoxyfluorane* ($CHCl_2\cdot CF_2\cdot O\cdot CH_3$), and the unsaturated compound, *Trichloroethylene* ($CHCl{=}CCl_2$).

An exceptionally high standard of purity is essential in all inhalent anaesthetics, and specific limit tests must be imposed to ensure virtual freedom from toxic impurities arising either as a result of the method of manufacture or through decomposition on storage. Thus, the official monograph for *Halothane* imposes a limit of 50 parts per million of higher boiling and more toxic volatile related compounds, which if present in significant amounts would be unduly hazardous to the patient unfortunate enough to be anaesthetised from the residues rather than from a full container.

**Retention and Metabolism of Halothane and Methoxyfluorane**

Volatile polyhalogen anaesthetics and aerosol propellants are largely excreted unchanged via the lungs. In common with all other polyhalogen compounds, however, there is some evidence that repeated exposure to such compounds can occasionally result in liver damage and jaundice (Inman and Mushin, 1974), and subcellular binding of 1-$^{14}$C-Halothane has been observed in mouse liver and brain tissue (Howard, Brum and Blake, 1973). In contrast to simple alkyl halides, however, metabolism of *Halothane* and *Methoxyfluorane* proceeds by reductive dehalogenation and not by mercapturation (Chapter 28).

$$CHClBr\cdot CF_3 \longrightarrow CH_3\cdot CF_3$$

$$CHCl_2\cdot CF_2\cdot O\cdot CH_3 \longrightarrow CH_3\cdot CF_2\cdot OCH_3$$

**Non-specific Action of Anaesthetics**

The tissues of the central nervous system have a high content of lecithins and cephalins (Chapter 18) and in consequence have a predominantly lipid character with a high affinity for lipid-soluble compounds. It is this special affinity for lipid tissue which the polyhalogenated hydrocarbons share with a diverse variety of other chemical types having general anaesthetic properties, including *Cyclopropane*, *Diethyl Ether* and *Nitrous Oxide*. General anaesthesia is, therefore, the result of a structurally non-specific drug action, depending essentially on physical rather than chemical properties (**2**, 1).

Meyer (1899) and Overton (1901) working independently developed a lipid theory of narcosis, based on observations that the depressant properties of lipid-soluble substances were greatest in cells which themselves have a high content of lipid tissue, and that the depressant effects increased in parallel with the oil/water partition coefficients of the substances. It was left, however, to Ferguson (1939) to produce a rational explanation of non-specific drug action based on thermodynamic principles.

The choice of olive oil or some similar system as the lipid phase in partition experiments relating to general anaesthesia has given rise to objections on the grounds that it is unrepresentative of the lipid components of nervous tissue.

These objections, however, are completely overcome in Ferguson's rationalisation of the equilibrium achieved in anaesthesia as a steady state between the internal drug-saturated biophase (central nervous tissue) and the external vapour phase in the lungs. Once equilibrium has been reached between external and internal phases, the thermodynamic activity is the same in both phases. Thus, the thermodynamic activity of the drug in the lungs, which is a purely physical concept, provides a direct measure of the thermodynamic activity of the drug in the biophase. Hence, the physical effect of the drug in the biophase is capable of measurement irrespective of whether or not its precise location can be defined.

The relative thermodynamic activity of any substance in equilibrium with the anaesthetised tissue is, therefore, calculable from the degree of saturation of its vapour in the lungs. The degree of biological response is thus related, not to actual concentration on a w/w or w/v basis, but to the relative thermodynamic activity of the systems. Substances with the same relative activity will thus show approximately the same degree of biological response. In practice, relative thermodynamic activity is given by the relationship:

$$\text{Relative Activity} = \frac{P_t}{P_s}$$

where $P_t$ is the partial pressure of the vapour in the lungs at equilibrium in the anaesthetised animal,
$P_s$ is the saturation vapour pressure of the substance.

Table 21 (Ferguson, 1939) shows that, whereas the isonarcotic concentrations of a group of non-specific anaesthetics in mice vary from 0.5 to 100%, the relative activities ($P_t/P_s$) all lie within relatively narrow limits.

The validity of this concept was borne out by the discovery that nitrogen (under pressure) and the rare gas, xenon, cause loss of reflex nervous action in mice (Lawrence, Loomis, Tobias and Turpin, 1946). This led to a demonstration of the clinical effectiveness of xenon as a general anaesthetic (Cullen and Gross, 1951).

There is little factual evidence on which to base any theory of anaesthetic action. General anaesthetics, however, do depress the excitability of isolated nerve fibres, mainly by reducing acetylcholine release at synapses. Certain anaesthetics, notably *Halothane* and *Ether*, depress the sensitivity of motor-end-plates to acetylcholine. The central action of general anaesthetics is, therefore, thought to lie in a similar depression of synaptic transmission and excitability of post-synaptic membranes in the brain. According to Pauling (1961), this depression of central nervous activity is brought about by changes in the structural organisation of adjacent water molecules, as a result of hydrophobic forces (p. **2**, 2). Thus, the hydrophobic anaesthetic agents form clathrate hydrates, which entrap inorganic ions and ionised protein side-chains, that have an essential rôle in the electrical changes responsible for nervous transmission.

Table 21. Isonarcotic Concentrations of Anaesthetics for Mice at 37°C (Ferguson, 1939)

| Anaesthetics | Saturation Vapour Pressure at 37°C (mm Hg) | Narcotic Concentration (% by volume) | Activity ($P_t/P_s$) |
|---|---|---|---|
| Nitrous oxide | 59 300 | 100 | 0.01 |
| Acetylene | 51 700 | 65 | 0.01 |
| Methyl ether | 6100 | 12 | 0.02 |
| Methyl chloride | 5900 | 14 | 0.01 |
| Ethylene oxide | 1900 | 5.8 | 0.02 |
| Ethyl chloride | 1780 | 5.0 | 0.02 |
| Diethyl ether | 830 | 3.4 | 0.03 |
| Methylal | 630 | 2.8 | 0.03 |
| Ethyl bromide | 725 | 1.9 | 0.02 |
| Dimethyl acetal | 288 | 1.9 | 0.05 |
| Diethyl formal | 110 | 1.0 | 0.07 |
| Dichloroethylene | 450 | 0.95 | 0.02 |
| Carbon disulphide | 560 | 1.1 | 0.02 |
| Chloroform | 324 | 0.5 | 0.01 |

## ALKYL PHOSPHATES

### The Analogy to Alkyl Halides

Alkyl phosphates and sulphates function as important biological analogues of alkyl halides, but differ markedly from the latter in several important respects. As oxyacid derivatives they are characterised by possession of oxygen and hydroxyl functions which facilitate hydrogen bond formation. In consequence, alkyl phosphates and sulphates contrast with the alkyl halides in that they are highly water-soluble. Hence their ability to function in the aqueous media of biological systems.

The sodium salts of phosphate esters also have high water solubility. Advantage is taken of this fact pharmaceutically to prepare water-soluble salts of otherwise insoluble alcohols and sterols in forms which are suitable for the preparation of stable injectable aqueous solutions. Examples of this type of chemical modification include *Triclofos Sodium* and *Betamethasone Sodium Phosphate.*

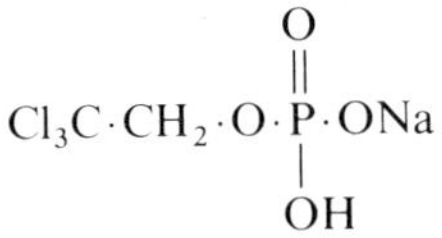

$Cl_3C{\cdot}CH_2{\cdot}O{\cdot}P(=O)(OH){\cdot}ONa$

*Triclofos Sodium*

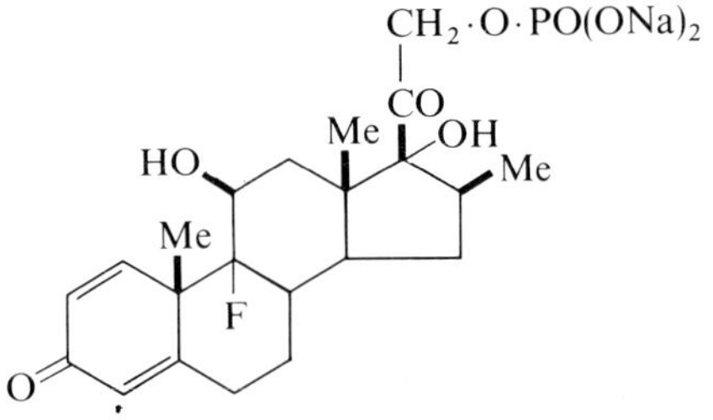

$CH_2{\cdot}O{\cdot}PO(ONa)_2$ — CO (steroid skeleton: HO, Me, Me, OH, Me, F, O)

*Betamethasone Sodium Phosphate*

Table 22. Dissociation Constants of Phosphoric Acids (Pitzer, 1937)

| Acid | $pK_1$ | $pK_2$ | $pK_3$ | $pK_4$ |
|---|---|---|---|---|
| Orthophosphoric ($H_3PO_4$) | 2.13 | 7.21 | 13.0 | — |
| Pyrophosphoric ($H_4P_2O_7$) | 0.85 | 1.96 | 6.68 | 9.39 |
| Triphosphoric ($H_5P_3O_{10}$) | — | 1.06 | 2.30 | 6.26 |

**Monoalkyl Phosphates**

Natural phosphate esters are derived from orthophosphoric acid ($H_3PO_4$), pyrophosphoric acid ($H_4P_2O_7$) or triphosphoric acid ($H_5P_3O_{10}$). All three are strong acids (Table 22), and all alkyl phosphates (Table 23), irrespective of the acid from which they are derived, contain at least one ionisable hydroxyl group. In consequence, alkyl phosphates are all appreciably ionised at physiological pH.

$$R\cdot O\cdot P(=O)(OH)-OH \underset{}{\overset{H^+}{\rightleftharpoons}} RO\cdot P(=O)(OH)-O^- \overset{H^+}{\rightleftharpoons} RO\cdot P(=O)(O^-)-O^-$$

These ionised forms are probably important in determining the association of coenzymes and co-factors, such as pyridoxal phosphate, $NADP^+$, AMP, ADP and ATP, by bonding to the cationic centres of enzymic proteins. The monoanion appears to be especially important in the hydrolysis of monoalkylorthophosphates. Kinetic studies show a maximum in the pH–rate profile at pH 4.0 (Bunton, Llewellyn, Oldham and Vernon, 1958), due to the greater reactivity of the monoanion compared with that of the neutral molecule. Studies with $H_2{}^{18}O$, however, reveal that whereas the neutral molecule is hydrolysed by C—O bond fission, the anion is cleaved by P—O bond fission. It has been suggested (Vernon, 1959) that the unionised hydroxyl serves to supply a proton to the departing alkoxyl group, while the ionised hydroxyl accepts a proton from the attacking water molecule.

$$O{=}P(OR)(O^-)(O{-}H) + H_2{}^{18}O \longrightarrow H^{18}O{-}P(=O)(OH){-}O^- + R\cdot OH$$

The much slower hydrolysis of the neutral molecule leading to C—O bond fission may be regarded as nucleophilic substitution.

$$H{-}{}^{18}O{-}H \quad R{-}O{-}P(=O)(OH){-}OH \longrightarrow H^+ + R\cdot{}^{18}OH + {}^-O{-}P(=O)(OH){-}OH$$

Table 23. Classes of Natural Phosphate Esters

| Class | Structure | Examples |
|---|---|---|
| Monoalkylorthophosphates | $RO\cdot PO_3\cdot H_2$ | D-Glucose-1-phosphate<br>D-Glucose-6-phosphate<br>D-Fructose-1,6-diphosphate<br>D-Glyceraldehyde-3-phosphate<br>2-Phospho-D-glyceric acid<br>Adenosine monophosphate<br>Pyridoxal phosphate |
| Dialkylorthophosphates | O<br>‖<br>RO·P·OR′<br>\|<br>OH | Adenosine-2′,3′-cyclic phosphate<br>Ribonucleic acids (RNA)<br>Deoxyribonucleic acids (DNA)<br>Phosphorylcholines (lecithins) |
| Alkylpyrophosphates | O O<br>‖ ‖<br>RO·P·O·P·OH<br>\| \|<br>OH OH | Adenosine diphosphate (ADP) |
| Dialkylpyrophosphates | O O<br>‖ ‖<br>RO·P·O·P·OR′<br>\| \|<br>OH OH | Nicotinamide Adenine<br>Dinucleotide; Coenzyme A |
| Alkyltriphosphates | O O O<br>‖ ‖ ‖<br>RO·P·O·P·O·P·OH<br>\| \| \|<br>OH OH OH | Adenosine triphosphate (ATP) |

**Nucleophilic Substitution of Monoalkyl Phosphates with P—O Bond Fission**

Whilst the reaction pathway in chemical systems *in vitro* is controlled by pH, enzymic hydrolysis of alkyl phosphates is largely a matter of enzyme specificity. Thus, phosphate cleavage by acid or alkaline phosphatase provides a source of phosphate and a mechanism for phosphorylation through P—O bond fission. This has been demonstrated for the hydrolysis of D-glucose-1-phosphate (Cohn, 1949; Bunton, Silver and Vernon, 1957) by prostatic acid phosphatase. The use of stilboestrol diphosphate as a transport form for the localisation of stilboestrol in the treatment of prostatic cancer rests on the acid phosphatase activity of the cancerous tissue (Druckrey and Raabe, 1952; Flocker, Marberger, Begley and Prendergist, 1955). The diphosphate is non-oestrogenic and has negligible cytostatic activity. Acid phosphatase activity in the cancerous tissue of the prostate is much higher than in most other tissues, and hence leads to the release of the cytostatic, stilboestrol, in high concentration at the required site of action.

The phosphorylation of glucose by adenosine triphosphate (ATP), which is catalysed by hexokinase, may similarly be formulated as a nucleophilic attack on the terminal phosphorus atom resulting in P—O bond cleavage, since in the presence of $H_2{}^{18}O$, there is no incorporation of $^{18}O$ in either product (Cohn, 1956).

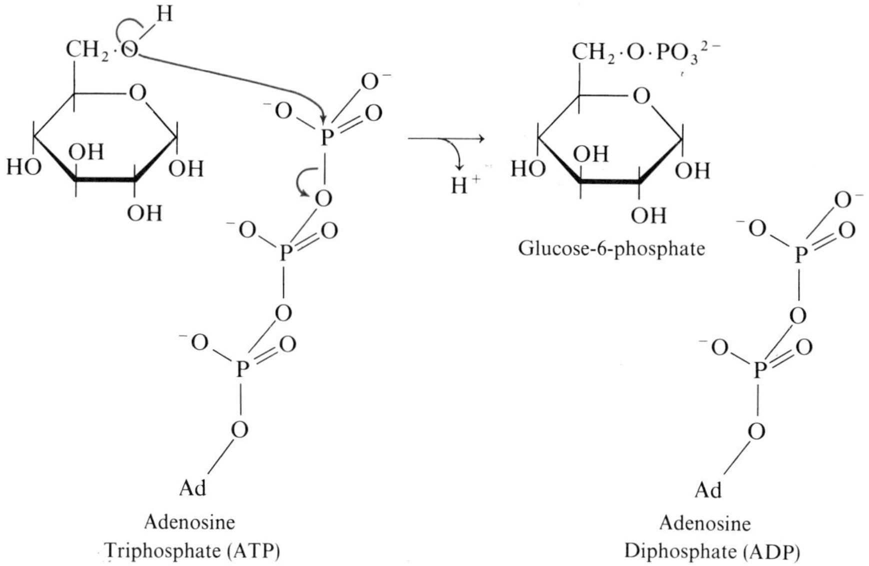

$CH_2\cdot O\cdot PO_3{}^{2-}$

Glucose-6-phosphate

Adenosine Triphosphate (ATP)

Adenosine Diphosphate (ADP)

Adenosine triphosphate can also undergo an alternative nucleophilic displacement of pyrophosphate, $P_2O_7{}^{4-}$, with P—O bond fission, as for example in the formation of acetyl adenylate in the presence of acetic thiokinase.

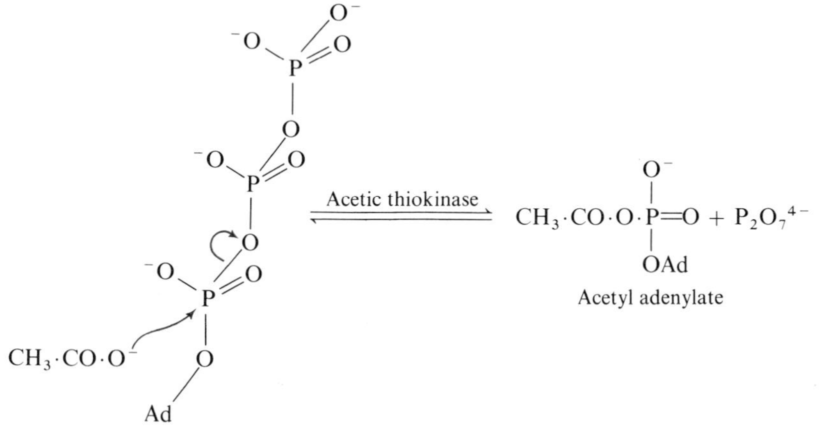

$$CH_3\cdot CO\cdot O^- + \text{ATP} \xrightleftharpoons{\text{Acetic thiokinase}} CH_3\cdot CO\cdot O\cdot P(O^-)(=O)(OAd) + P_2O_7{}^{4-}$$

Acetyl adenylate

Acetyl adenylate, a mixed phosphate carboxylic anhydride, is a key intermediate in the biosynthesis of acetylcoenzyme A.

Some of the enzymic reactions of adenosine triphosphate are catalysed by divalent metal ions, as for example the formation of ADP by ATP-ase.

$$ATP + HPO_4^{2-} \xrightleftharpoons{Mn^{2+}} ADP + P_2O_7^{4-} + H^+$$

The metal ions form ionic bonds and hence neutralise some of the charge on the triphosphate ion, which would otherwise inhibit the attack by $HPO_4^{2-}$. The presence of the positively charged metal ions also promotes electron withdrawal from phosphorus towards oxygen in the P—O bond under attack.

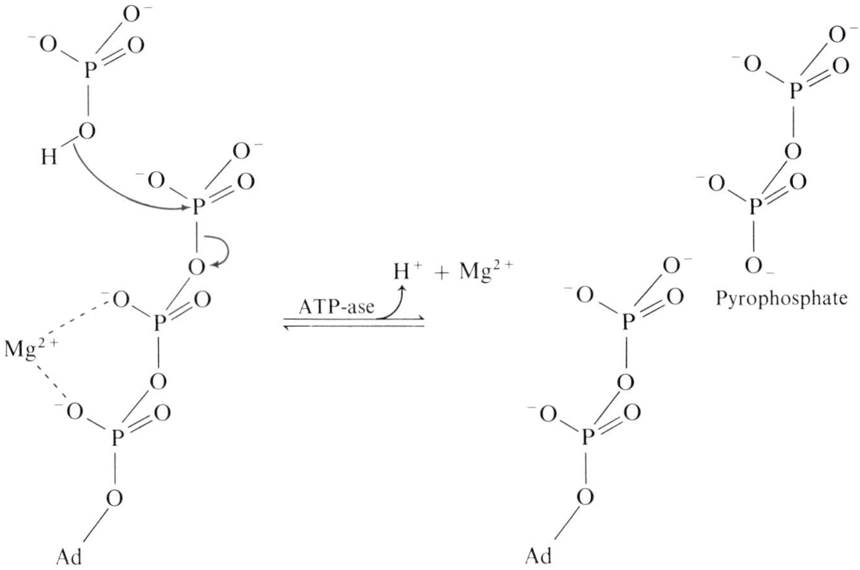

Adenosine triphosphate (ATP) $\xrightleftharpoons{\text{ATP-ase}}$ ($H^+ + Mg^{2+}$) Adenosine diphosphate (ADP) + Pyrophosphate

## Nucleophilic Substitution of Monoalkyl Phosphates with C—O Bond Fission

Phosphorylases catalyse C—O bond fission of alkyl phosphates leading to alkylation reactions, as in the glucosylation of fructose by glucose-1-phosphate (Doudoroff, Barker and Hassid, 1947) in the presence of sucrose transglucosylase.

$$\text{Glucose-1-phosphate} + \text{fructose} \xrightleftharpoons{\text{sucrose transglucosylase}} \text{sucrose} + H_2PO_4^-$$

Polysaccharide phosphorylase similarly controls the interconversion of glycogen and glucose-1-phosphate.

$$(\text{glucose})_n + H_3PO_4 \rightleftharpoons \text{glucose-1-phosphate} + (\text{glucose})_{n-1}$$

Both these reactions proceed with retention of stereochemical configuration at C-1 of glucose, and hence are assumed to proceed by a double displacement (each with inversion) via an intermediate enzyme–substrate complex (Chapter 21).

**Dialkyl Phosphates and Cyclic Alkyl Phosphates**

α-Hydroxymonophosphates readily undergo an internal nucleophilic displacement with the formation of highly reactive cyclic esters, as in the conversion of α- to β-glycerophosphoric acid.

α-Glycerophosphoric acid

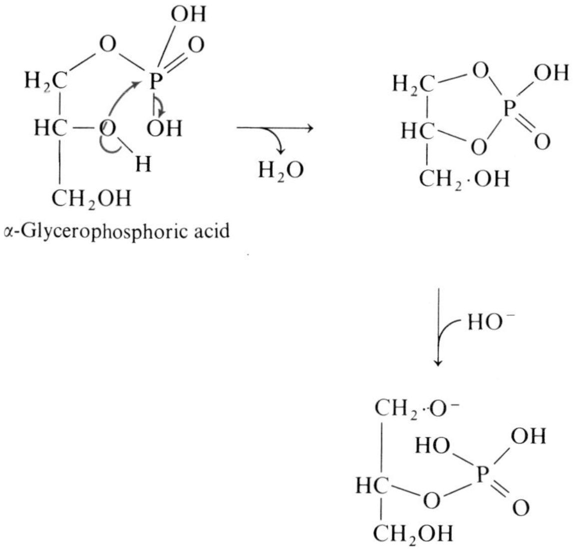

β-Glycerophosphoric acid

Dialkyl phosphates, of which the cyclic glyceryl-1,2-diphosphate is a special example, differ from monoalkyl phosphates in that they do not show a maximum in the pH–rate profile for hydrolysis (Bunton, Mhala, Oldham and Vernon, 1960). The neutral diphosphate rather than its anion is, therefore, the reactive species in the hydrolysis. The intermediate glyceryl-1,2-cyclomonophosphate, however, is typical of many similar cyclic phosphates (Kumamoto, Cox and Westheimer, 1956) and is hydrolysed with extreme ease. Hence, conversion of α- to β-glycerophosphoric acid is readily achieved. α-Hydroxydialkyl phosphates as found in the ribonucleic acids similarly form cyclic phosphates under alkaline conditions, or enzymatically with ribonuclease leading to disruption of the ribose phosphate backbone of the nucleic acid (reaction scheme p. 228).

Further attack on the cyclic ester by alkali or water does not discriminate between the 2′ and 3′-positions and a mixture of both 2′- and 3′-phosphates is formed.

Adenyl cyclase similarly catalyses the formation of cyclic-AMP (adenosine-3′,5′-phosphate) from ATP in the presence of $Mg^{2+}$ ions. It has been shown that *Isoprenaline*, the most effective β-adrenergic receptor agonist, is also the most

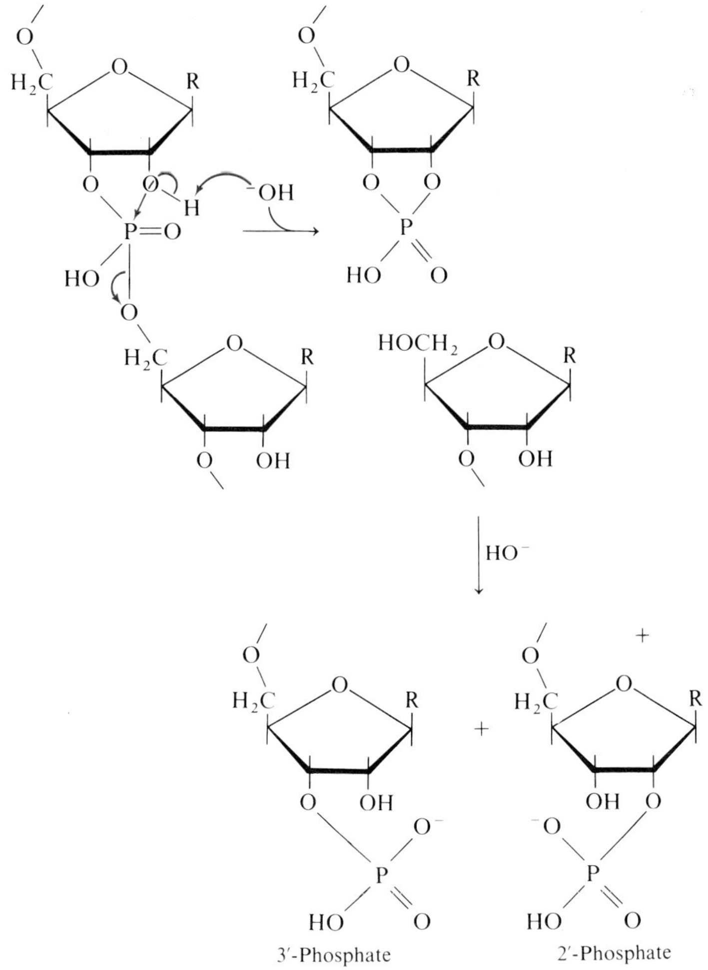

3′-Phosphate 2′-Phosphate

effective stimulator of cyclic-AMP formation, and it appears that cyclic-AMP is capable of mimicking the pharmacological actions of $\beta$-agonists.

## ALKYL SULPHATES

Monoalkylsulphates, such as ethyl hydrogen sulphate ($CH_3 \cdot CH_2 \cdot O \cdot SO_2 \cdot OH$), have many properties in common with alkyl phosphates. They are highly water-soluble, fully ionised in aqueous solution, strongly acidic and readily form water-soluble sodium salts when neutralised with sodium hydroxide. For this reason, the sodium salts of long-chain alkyl sulphates, such as *Sodium Lauryl Sulphate*, are useful as detergents and as antibacterial agents for cleansing surgical instruments and skin prior to operations (Chapter 7). They are hydrolysed by hot aqueous mineral acid to the parent alcohol.

Dialkyl sulphates, such as dimethyl sulphate, are powerful alkylating agents for phenols, amines and sulphydryl compounds, particularly in alkaline solution, when they undergo nucleophilic displacement.

$$C_6H_5{\cdot}O^- \; CH_3{-}O{\cdot}SO_2{\cdot}O{\cdot}CH_3 \longrightarrow C_6H_5{\cdot}O{\cdot}CH_3 + {}^-O{\cdot}SO_2{\cdot}O{\cdot}CH_3$$

The vapour of dimethyl sulphate is extremely toxic, doubtless due to its efficiency as a biological alkylating agent.

The formation of sulphate esters by sulphokinases is an important route for mammalian metabolism of alcohols, sterols and phenols (Chapters 7 and **2**, 5). The co-factor in these reactions, 3′-phosphoadenosine-5′-phosphosulphate (PAPS), is formed from ATP via adenosine-5′-phosphosulphate (APS). It is interesting to note that APS is formed from ATP by ATP-sulphurylase in an overall displacement of pyrophosphate by sulphate. The sulphate ion ($SO_4^{2-}$), however, is generally regarded as a weak nucleophile, and it has been suggested that at least two steps are involved with participation of an intermediate enzyme–substrate complex (Kosower, 1962). The participation of $Mg^{2+}$, however, reduces charge repulsion, whilst the nucleophilicity of the sulphate ion may be enhanced by association of the sulphate ion in an enzyme complex, with a concomitant reduction in the resonance stabilisation of the ion.



$$\text{APS} + \text{ATP} \xrightarrow{\text{APS-Kinase}} \text{PAPS} + \text{ADP}$$

## ALKYL NITRITES

Volatile alkyl nitrites are used in medicine in the control of angina pectoris. *Amyl Nitrite* first introduced for this purpose in 1897 is now little used and has been largely replaced by *Octyl Nitrite* (2-ethylhexyl nitrite; $CH_3(CH_2)_3 \cdot CH(C_2H_5) \cdot CH_2 \cdot O \cdot NO$. *Octyl Nitrite*, despite the fact that its boiling point is as high as 169°C, is administered by inhalation. It is less volatile than its lower molecular weight analogues, is less toxic, and has a longer duration of action. Alkyl nitrites appear to act as specific inhibitors of an ATP-ase which is essential for the supply of energy in smooth muscle (Krantz, Carr and Knapp, 1951).

*Octyl Nitrite* is usually packed for use in vitrellae, which are small crushable glass capsules wrapped in gauze. It is, however, somewhat unstable and liable to decompose with evolution of nitric oxide, particularly if it is allowed to become acid in reaction prior to filling the containers.

$$CH_3(CH_2)_3 \cdot CH(Et) \cdot CH_2 \cdot O \cdot NO \xrightarrow{H^+} CH_3(CH_2)_3 \cdot CH(Et) \cdot \overset{+}{C}H_2 + HNO_2$$

$$3HNO_2 \longrightarrow HNO_3 + H_2O + 2NO\uparrow$$

Most nitrite esters are readily subject to photochemical decomposition and should be protected from light.

$$R{-}O{-}N{=}O \xrightarrow{h\nu} RO\cdot + \cdot NO$$

This instability has been put to use in the synthesis of *Aldosterone* (Barton and Beaton, 1960) in which the breaking and substitution of a C—H bond of the C-18 angular methyl group in the steroid skeleton was effected by the radical generated in the photochemical decomposition of the sterically adjacent 11$\beta$-nitrite ester (Chapter 22).

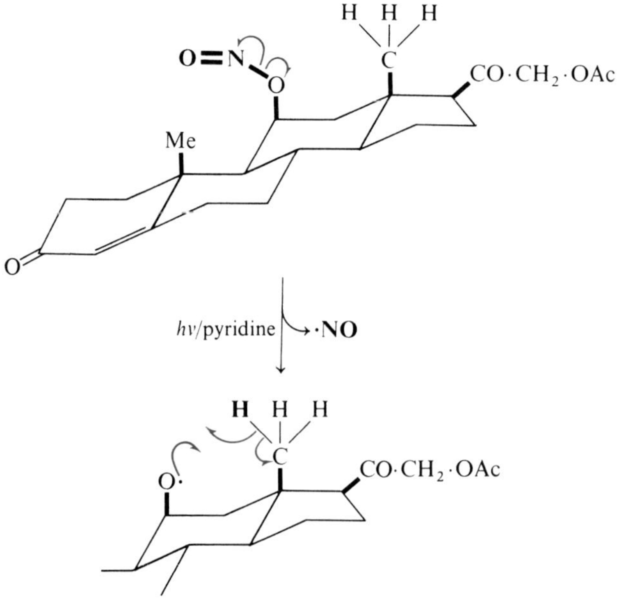

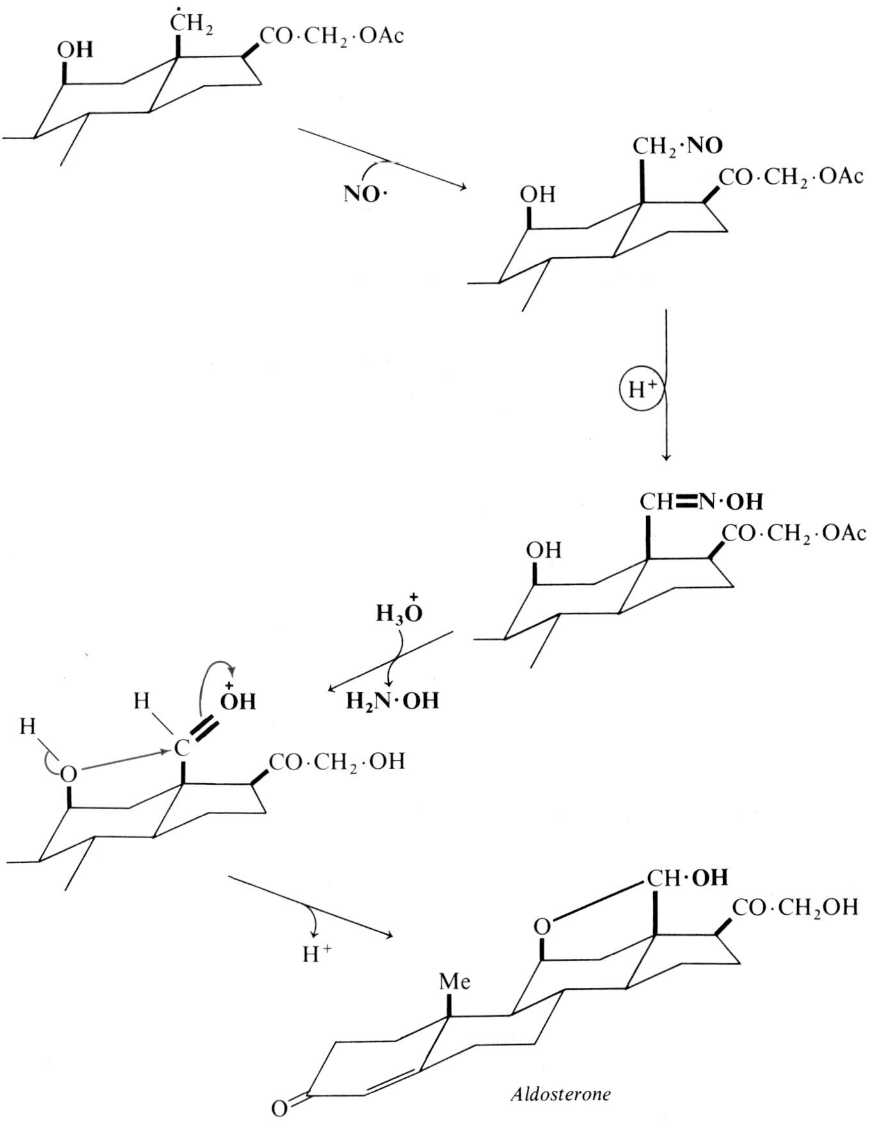

$\dot{C}H_2$ OH $CO{\cdot}CH_2{\cdot}OAc$

$NO{\cdot}$

$CH_2{\cdot}NO$ OH $CO{\cdot}CH_2{\cdot}OAc$

$H^+$

$CH{=}N{\cdot}OH$ OH $CO{\cdot}CH_2{\cdot}OAc$

$H_3\overset{+}{O}$

$H_2N{\cdot}OH$

H H $\overset{+}{O}H$ C O $CO{\cdot}CH_2{\cdot}OH$

$H^+$

$CH{\cdot}OH$ O $CO{\cdot}CH_2OH$ Me O

*Aldosterone*

## ALKYL NITRATES

Certain alkyl polynitrates, such as *Glyceryl Trinitrate* (Nitroglycerin) and *Erythrityl Tetranitrate*, are used in the control of angina pectoris. They are less volatile and somewhat more stable than alkyl nitrites which are used for the same purpose. Nitrates, however, are explosive and need to be handled with care. They are also partially hydrolysed in mammalian systems; thus *Glyceryl Trinitrate* is metabolised to glyceryl 1,2- and 1,3-dinitrates (Needleman and

Krantz, 1965) in the liver. For this reason, they are active when absorbed through the buccal mucosa, but ineffective when absorbed from the gastrointestinal tract. It seems likely that the nitrate ion so released is probably the effective therapeutic agent, since it is known that sodium nitrite has similar actions to organic nitrites.

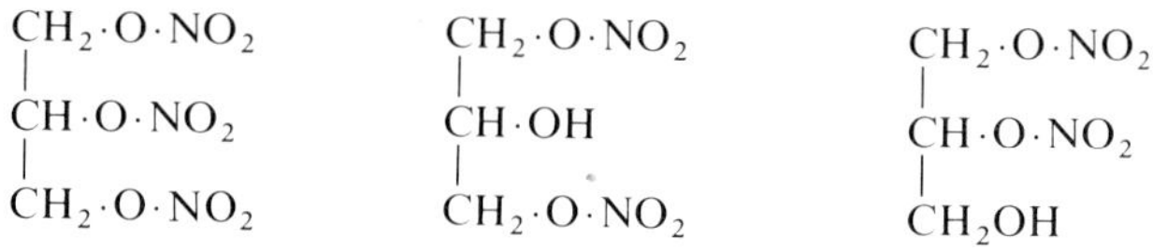

$$\begin{array}{l} CH_2 \cdot O \cdot NO_2 \\ | \\ CH \cdot O \cdot NO_2 \\ | \\ CH_2 \cdot O \cdot NO_2 \end{array} \qquad \begin{array}{l} CH_2 \cdot O \cdot NO_2 \\ | \\ CH \cdot OH \\ | \\ CH_2 \cdot O \cdot NO_2 \end{array} \qquad \begin{array}{l} CH_2 \cdot O \cdot NO_2 \\ | \\ CH \cdot O \cdot NO_2 \\ | \\ CH_2OH \end{array}$$

*Glyceryl Trinitrate* (Nitroglycerin) — Glyceryl-1,3-dinitrate — Glyceryl-1,2-dinitrate

# NITRILES

## General Properties

Nitriles may be considered to be alkyl or aryl esters of hydrogen cyanide and are sometimes named as such. As nitriles, however, they are considered to be substitution derivatives of the carboxylic acids containing the same number of carbon atoms.

| | |
|---|---|
| $CH_3 \cdot CN$ | Acetonitrile |
| $CH_3 \cdot CH_2 \cdot CN$ | Propionitrile |
| $C_6H_5 \cdot CN$ | Benzonitrile |

Nitriles are stable neutral compounds, which unlike hydrogen cyanide are not highly toxic. They readily undergo hydrolysis, first to the corresponding acid amide and then to the corresponding carboxylic acid. Hydrolysis to the amide can usually be accomplished by dissolving the nitrile in concentrated sulphuric acid and pouring the solution into cold water.

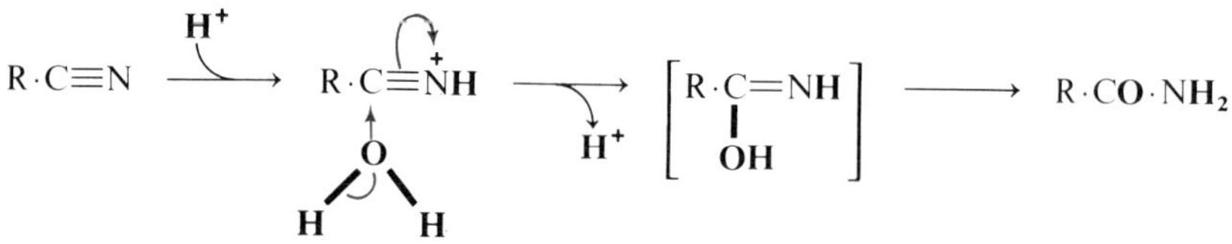

$$R \cdot C \equiv N \xrightarrow{H^+} R \cdot C \equiv \overset{+}{N}H \;(\leftarrow H_2O) \xrightarrow{-H^+} \left[ \begin{array}{c} R \cdot C = NH \\ | \\ OH \end{array} \right] \longrightarrow R \cdot CO \cdot NH_2$$

Further hydrolysis of the amide so formed by either acid or alkali gives the corresponding carboxylic acid (Chapter 11). Aromatic nitriles are metabolised in part by hydrolysis to the corresponding acid, but alkyl nitriles undergo a more complex oxidative hydrolysis, which results in the release of cyanide ions (**2**, 5). Sterically hindered nitriles like *Diphenoxalate*, which is used in the treatment of diarrhoea, are resistant to hydrolysis of the nitrile group, and are preferentially metabolised by ester hydrolysis.

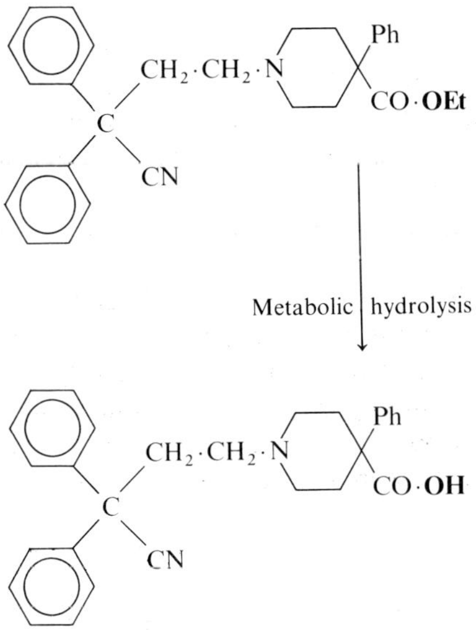

Ph

CH$_2$·CH$_2$·N

CO·**OEt**

*Diphenoxalate*

C

CN

Metabolic hydrolysis

Ph

CH$_2$·CH$_2$·N

CO·**OH**

Principal metabolite

C

CN

Nitriles also undergo alcoholysis in solution in alcohols in the presence of strong mineral acids ($H_2SO_4$ and HCl) to yield the corresponding ester.

$$R \cdot CN + EtOH + H_2O \xrightarrow{H^+} R \cdot CO \cdot OEt + NH_3$$

Nitriles also undergo a number of addition reactions which are of preparative importance, including addition of hydrogen (reduction) to yield primary amines, addition of ammonia to form amidines, and addition of Grignard reagents to form ketones.

## ARYL HALIDES

### Nucleophilic and Electrophilic Substitution

Aryl halides, in which the halogen is directly attached to the aromatic nucleus, have properties which in some respects resemble those of aliphatic halides, but are modified in other ways by the properties of the aromatic ring. Thus, in contrast to alkyl halides, aryl halides are normally resistant to nucleophilic substitution. Thus, the release of non-bonded electrons from the halogen to the ring in the mesomeric effect creates a degree of double bond character in the carbon—halogen bond.

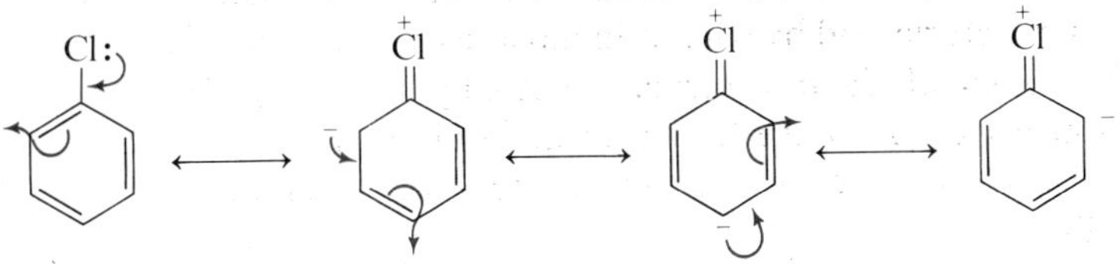

Cl: ⟷ $\overset{+}{Cl}$ ⟷ $\overset{+}{Cl}$ ⟷ $\overset{+}{Cl}$

This opposes formation of the carbonium ion which is a requirement of $S_N1$ type nucleophilic substitution. Similarly, the broadside attack by the nucleophile required for substitution by the $S_N2$ mechanism is precluded by the $\pi$-electron cloud of the aromatic system. This latter impediment is, however, removed when the ring becomes activated by the presence of sufficiently powerful electron-withdrawing substituents. These distort the $\pi$-electron cloud sufficiently to permit attack at the centres of low electron density created in the ring.

Activation of the carbon—halogen bond for nucleophilic substitution occurs when powerful electron-attracting substituents are present in the *ortho* and *para* positions in the ring. Thus, 2,4-dinitrochlorobenzene is readily converted to 2,4-dinitrophenol by aqueous sodium hydroxide. This is due to weakening of the carbon—halogen bond by the powerful mesomeric (and inductive) effects of the nitro substituents, and by the resonance stabilisation of the anion intermediate.

Cl: NO$_2$ N$^+$ O O$^-$ ⟷ HO$^-$ $\overset{+}{Cl}$ NO$_2$ N$^+$ $^-$O O$^-$ ⟶ HO Cl NO$_2$ N$^+$ $^-$O O$^-$ → Cl$^-$ → OH NO$_2$ N$^+$ O O$^-$

Mesomeric forms of
2,4-dinitrochlorobenzene

2,4-Dinitrophenol

A similar nucleophilic substitution occurs in the mercapturation of 2,4-dinitrochlorobenzene by glutathione in mammalian metabolism, which leads to an *S*-arylmercapturic acid by what is in most respects exactly the same pathway as that for the metabolic transformation of alkyl halides (p. 211). Nucleophilic substitution also occurs in the Dow process for the conversion of chlorobenzene to phenol by the action of sodium hydroxide at high temperature and pressure (Chapter 8).

Electrophilic substitution is also feasible due to the mesomeric effect which increases electron availability at the *ortho* and *para* positions, thus activating them for attack by electrophilic reagents. The halogens, however, are more electronegative than carbon, so that in the ground state, the ring is deactivated relative to benzene itself by the $-I$ effect of the halogen. As a result of this almost equal competition between the deactivating influence of the inductive effect and activation due to the mesomeric effect, a balance is achieved which makes electrophilic substitution in the *ortho* and *para* positions possible, but at a rate which is slower than that for the parent hydrocarbon, benzene.

### Radical Reactions

The carbon—halogen bond of aryl halides is cleaved in certain radical-induced reactions. Thus, aryl halides readily form Grignard reagents (p. 214) with magnesium in dry ether.

$$C_6H_5Cl \xrightarrow{Mg/Et_2O} C_6H_5MgCl$$

Like alkyl halides, they are also reactive toward alkali metals as in the Fittig synthesis of alkylbenzenes, though the mechanism of this reaction is uncertain.

$$C_6H_5 \cdot Br + CH_3 \cdot Br \xrightarrow{2Na} C_6H_5 \cdot CH_3 + 2NaBr$$

The Ullman reaction, which is used in the synthesis of biphenyls, also involves a radical reaction mechanism. In this, an aryl iodide is heated with copper powder.

$$2C_6H_5 \cdot I + Cu \longrightarrow C_6H_5\text{—}C_6H_5 + CuI_2$$

Biphenyl

# 10 Aldehydes and Ketones

Aldehydes and ketones are characterised by the presence of the carbonyl group ($>C=O$). They are distinguished by the nature of the groups to which the carbonyl group is attached. The carbonyl group of aldehydes is always linked to at least one hydrogen (two in formaldehyde), whereas both groups linked to ketone carbonyls are either alkyl or aryl.

$$R\cdot C\begin{matrix} \diagup\!\!\!\diagup O \\ \diagdown H \end{matrix}$$

Aldehyde (R = H, alkyl or aryl)

$$R\cdot C\begin{matrix} \diagup\!\!\!\diagup O \\ \diagdown R' \end{matrix}$$

Ketone (R, R′ = alkyl or aryl)

## NOMENCLATURE

### Aldehydes

The term aldehyde arises because of the relationship to alcohols (**al**cohol **dehyd**rogenated). The characteristic group, —CHO, is known as the formyl group, but this is only used in substitutive nomenclature of more complex aldehydes (formylcyclohexane, $C_6H_{11}\cdot CHO$). Simple aldehydes are named in relation to the corresponding carboxylic acids, which are formed from them by oxidation. In the IUPAC system, aldehydes are named as derivatives of the parent saturated hydrocarbon with the suffix **-al**.

### Ketones

The term ketone is derived from that of acetone, the first member of the series. Ketones are named by reference to the two substituting alkyl or aryl groups. In

Table 24. Nomenclature of Aldehydes

| Trivial name | Systematic name | Structural formula | b.p. (°C) |
|---|---|---|---|
| Formaldehyde | Methanal | $H\cdot CHO$ | −21 |
| Acetaldehyde | Ethanal | $CH_3\cdot CHO$ | +21 |
| Propionaldehyde | Propanal | $CH_3\cdot CH_2\cdot CHO$ | 49 |
| n-Butyraldehyde | Butanal | $CH_3(CH_2)_2\cdot CHO$ | 73 |
| n-Valeraldehyde | Pentanal | $CH_3(CH_2)_3\cdot CHO$ | 102 |
| iso-Valeraldehyde | 3-Methylbutanal | $(CH_3)_2CH\cdot CH_2\cdot CHO$ | 92.5 |
| Benzaldehyde | — | $C_6H_5\cdot CHO$ | 179 |

Table 25. Nomenclature of Ketones

| Ketone | Systematic name | Structural formula | b.p. (°C) |
|---|---|---|---|
| Dimethyl ketone (Acetone) | Propan-2-one | $CH_3 \cdot CO \cdot CH_3$ | 56 |
| Ethyl methyl ketone | Butan-2-one | $CH_3 \cdot CO \cdot CH_2 \cdot CH_3$ | 79.6 |
| Diethyl ketone | Pentan-3-one | $CH_3 \cdot CH_2 \cdot CO \cdot CH_2 \cdot CH_3$ | 101.5 |
| Methyl phenyl ketone (Acetophenone) | — | $C_6H_5 \cdot CO \cdot CH_3$ | 202 |
| Ethyl phenyl ketone (Propiophenone) | — | $C_6H_5 \cdot CO \cdot CH_2 \cdot CH_3$ | 218 |
| Diphenyl ketone (Benzophenone) | — | $C_6H_5 \cdot CO \cdot C_6H_5$ | m.p. 49 |

the IUPAC system, they are named as derivatives of the saturated hydrocarbon corresponding to the longest carbon chain (including the carbonyl carbon) with the suffix **-one**.

## GENERAL PROPERTIES

### The Structure of the Carbonyl Group

The carbon—oxygen double bond is a combination of a $\sigma$- and a $\pi$-bond. The $\sigma$-bond is formed by orbital overlap of $sp^2$ hybridised oxygen and carbon, and the $\pi$-bond by sideways overlap of the unhybridised $2p$- orbital from each atom. This leaves two $sp^2$ orbitals on carbon available for bond formation and two $sp^2$ orbitals on oxygen, each with a lone pair of electrons. Fusion of $sp^2$ hybridised orbitals, which closely resembles that of the ethylenic bond, leads to a trigonal and planar arrangement in which the non-bonding electron pairs on oxygen lie in the same plane as the carbon and oxygen nuclei and the two substituents attached to carbon (Fig. 18).

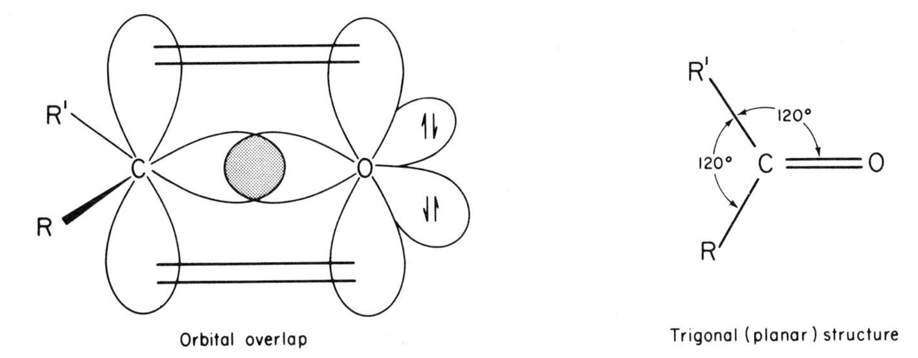


Fig. 18 The carbonyl group

Since oxygen is more electronegative than carbon, the electrons of both $\sigma$- and $\pi$-bonds are not equally shared, and the carbonyl group experiences a permanent dipole, which favours attack by nucleophilic species at the electron-deficient carbon atom.

$$X^- + \rangle C{=}O \longrightarrow X{-}\overset{|}{\underset{|}{C}}{-}O^-$$

Attack by an electrophile at the electron-rich oxygen atom is equally possible. This occurs most readily in attack on the non-bonded electron pairs when the electrophile is $H^+$, a Lewis acid, or a metallic cation ($M^+$). Such electrophilic attack promotes withdrawal of the $\pi$-electron onto oxygen, thereby catalysing nucleophilic attack at the carbonyl carbon atom.

$$\rangle C{=}\ddot{O} \; H^+ \rightleftharpoons \rangle C{=}\overset{+}{O}H \longleftrightarrow \rangle\overset{+}{C}{-}OH \xrightarrow{X^-} {-}\underset{X}{\overset{|}{C}}{-}OH$$

## Physical Properties

Formaldehyde is gaseous at room temperature (b.p. $-21°C$), but most other simple aldehydes and ketones are low boiling liquids. Higher members of both series are either liquids or crystalline solids.

Aldehydes and ketones in general are characterised by solubility properties primarily derived from the constituent alkyl and aryl substituents, but modified by the special characteristics of the carbonyl group. Thus, they are usually readily soluble in organic solvents, whilst the lower members are also appreciably soluble in hydroxylic solvents (water, alcohols) as a result of hydrogen bonding between the carbonyl group and the hydroxyl groups (Chapters 1 and **2**, 2). Thus, many of the lower aldehydes, but not ketones, are actually hydrated in aqueous solution.

$$\rangle C{=}O\cdots H{-}O{\diagdown}R$$

Hydrogen bonding is also clearly important for the solubilisation of otherwise comparatively insoluble ketonic steroid hormones and drugs in the aqueous media of biological systems. Steroid ketones also show enhanced solubility in the presence of plasma proteins, particularly albumin, as a result of hydrogen bonding (and possibly other interactions involving charge-transfer complexing) with the phenolic hydroxyl of tyrosine units in the protein (Oyakawa and Levedahl, 1958).

## NUCLEOPHILIC ADDITIONS TO CARBONYL COMPOUNDS

### Hydration

In contrast to ketones, most lower aldehydes are hydrated in aqueous solution. Thus, formaldehyde solution contains about 37% of formaldehyde in water, and exists wholly in the hydrated form, as shown by the almost complete absence of carbonyl absorption ($n \rightarrow \pi^*$ transition) in its ultraviolet absorption spectrum at room temperature.

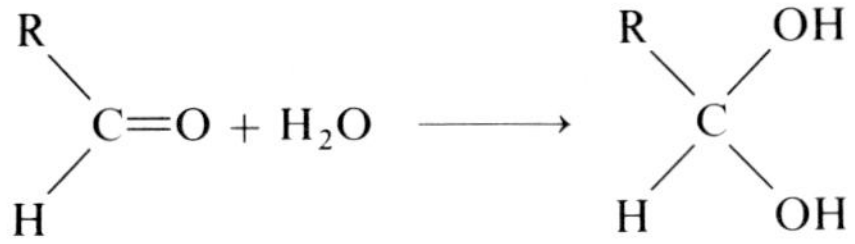

$$\text{R(H)C=O} + H_2O \longrightarrow \text{R(H)C(OH)}_2$$

The extent of hydration accords with the known effects of substituents. Thus, hydration is depressed by electron-repelling alkyl groups; acetaldehyde is incompletely hydrated, and ketones not at all. In contrast, electron-attracting groups favour hydration, so much so that chloral is able to exist in a stable (but hygroscopic) crystalline hydrate.

$Cl_3C \cdot CH(OH)_2$

*Chloral Hydrate*

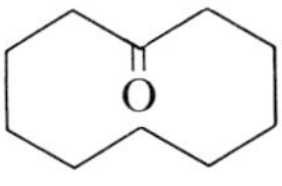

Cyclodecanone

The hydration of aldehydes is characteristic of a wide range of nucleophilic addition reactions to the carbonyl group. The gradation of reactivity, HCHO > R·CHO > R·CO·R′, observed in hydration is typical of other nucleophilic additions, with reactivity decreasing as alkyl substituents increase in number (from zero in formaldehyde to two in ketones) and size of group. This reflects both the +I inductive effect of the alkyl substituents and steric effects with larger groups which inhibit access of the reagent to the carbonyl group. The latter effect is seen in its most extreme form in certain medium size ring ketones, such as cyclodecanone, in which the carbonyl group is orientated such that it faces in towards the centre of the ring. In contrast, five- and six-membered ring ketones are generally more reactive than aliphatic ketones, since in the ring structure both alkyl substituents are constrained to be on one side of the carbonyl group.

Since oxygen is more electronegative than carbon, the carbonyl carbon—oxygen double bond is polarised, and exists as a resonance hybrid of two forms, with a net bond dipole moment of 2.75D.

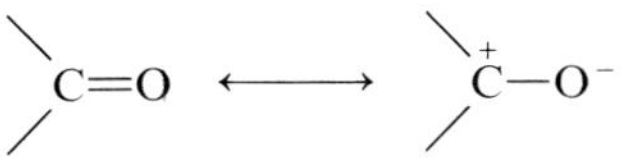

$$\rangle C{=}O \longleftrightarrow \rangle \overset{+}{C}{-}O^{-}$$

Nucleophilic addition is initiated by attack at the electron-deficient carbon atom, and hydration of aldehydes can be illustrated as follows.

$$\mathrm{R(H)C{=}O} + \mathrm{H_2O} \rightleftharpoons \mathrm{R(H)C(O^-)(OH)} + \mathrm{H^+} \rightleftharpoons \mathrm{R(H)C(OH)_2}$$

## Acetals and Ketals

Aldehydes react similarly with alcohols to form hemi-acetals. This reaction is normally slow in neutral solution, but is catalysed by traces of acid, which promote polarisation of the C=O bond, through formation of a resonance-stabilised carbonium ion.

$$\mathrm{CH_3(H)C{=}O} \overset{H^+}{\rightleftharpoons} \mathrm{CH_3(H)C{=}\overset{+}{O}H} \longleftrightarrow \mathrm{CH_3(H)\overset{+}{C}{-}OH} \xrightleftharpoons[]{EtOH,\ -H^+} \underset{\text{Hemi-acetal}}{\mathrm{CH_3(H)C(OH)(OEt)}}$$

In the presence of excess alcohol, the product hemi-acetal, however, usually undergoes a further acid-catalysed nucleophilic substitution to form an acetal.

$$\mathrm{CH_3(H)C(OH)(OEt)} \overset{H^+}{\rightleftharpoons} \mathrm{CH_3(H)C(\overset{+}{O}H_2)(OEt)} \rightleftharpoons \mathrm{CH_3(H)\overset{+}{C}(OEt)} \xrightleftharpoons[]{EtOH,\ -H^+} \underset{\text{Acetal}}{\mathrm{CH_3(H)C(OEt)_2}}$$

$$\mathrm{CH_3(H)\overset{+}{C}(OEt)} \longleftrightarrow \mathrm{CH_3(H)C{=}\overset{+}{O}Et}$$

Ketones react similarly to form ketals, as for example in *Triamcinolone Acetonide*. Both acetals and ketals are stable to alkali, but are readily hydrolysed to the parent carbonyl compound by dilute aqueous mineral acid. As derivatives, they form a simple means of protecting reactive carbonyl groups or, conversely, as with *Triamcinolone Acetonide*, of converting otherwise hydrophilic alcohols to more readily fat-soluble compounds to increase absorption and transport in the body. Ketones also react with glycol and ethylenedithiol to form cyclic ketals and thioketals.

Suitable juxtaposition of hydroxyl and carbonyl functions within one and the same molecule leads to stabilisation as intramolecular cyclic hemi-acetals or -ketals, even without catalytic assistance. Cyclic acetals and ketals are formed typically from $\gamma$- and $\delta$-hydroxy-aldehydes and -ketones, in which formation of a stable five- or six-membered ring provides a powerful driving force for the reaction. Examples are seen in furanose and pyranose sugars (Chapter 21) and the mineralocorticosteroid, *Aldosterone* (Chapters 9 and 22) which regulates sodium ion excretion and potassium retention in the body.

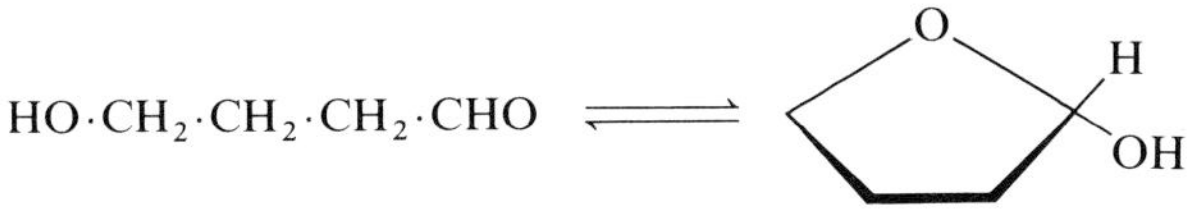

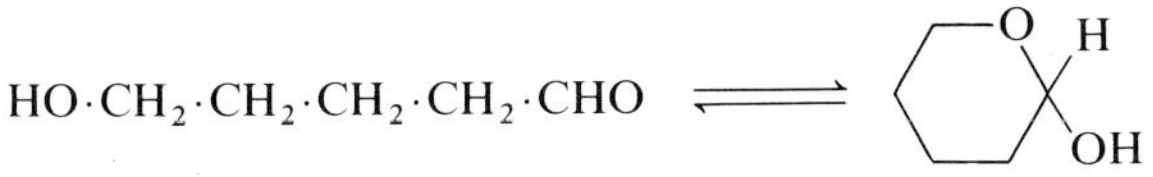

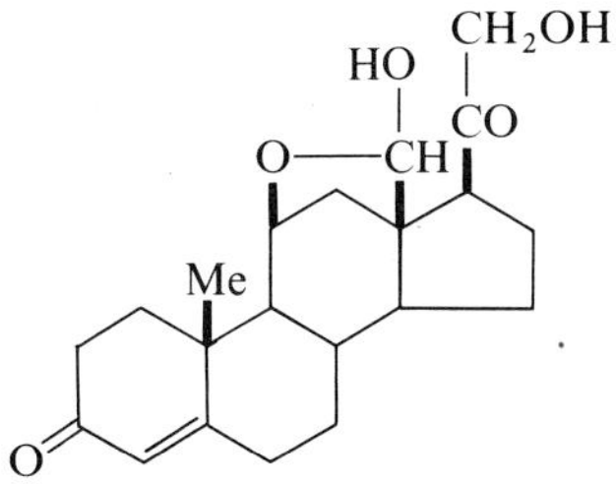

$$HO{\cdot}CH_2{\cdot}CH_2{\cdot}CH_2{\cdot}CHO \rightleftharpoons$$

$$HO{\cdot}CH_2{\cdot}CH_2{\cdot}CH_2{\cdot}CH_2{\cdot}CHO \rightleftharpoons$$

*Aldosterone*

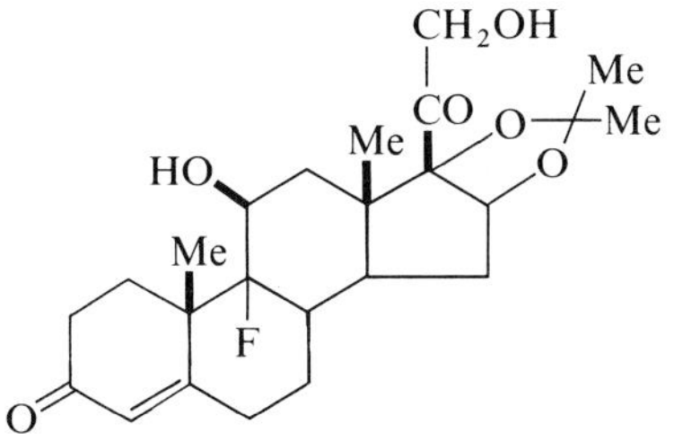

*Triamcinolone Acetonide*

**Cyanohydrins**

Addition of hydrogen cyanide to aliphatic (but not aromatic) aldehydes and ketones to form cyanohydrins is of considerable preparative importance in the preparation of $\alpha$-hydroxy-acids. Potassium cyanide, slowly treated with acid, is used as the source of cyanide ion ($CN^-$), which is the effective nucleophile.

$$HCN + HO^- \rightleftharpoons CN^- + H_2O$$

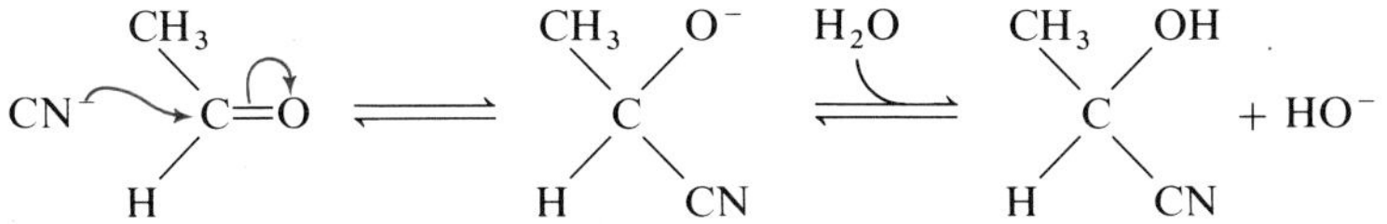

Cyanohydrin

**Sodium Bisulphite Addition Compounds**

Lower molecular weight aldehydes and ketones react with saturated (40%) aqueous sodium bisulphite solutions to form crystalline addition compounds which separate from solution. The salt-like product, although water-soluble and insoluble in organic solvents, is salted out of solution in the presence of excess sodium bisulphite. Formation of such compounds can be used to purify the lower aldehydes and ketones, the carbonyl compound being recoverable by heating the compound with sodium carbonate solution or hydrochloric acid.

$$(CH_3)_2C{=}O + NaO{-}\ddot{S}({=}O){-}O{-}H \rightleftharpoons (CH_3)_2C(O^-)SO_2\cdot ONa + H^+ \underset{HO^-}{\rightleftharpoons} (CH_3)_2C(OH)SO_2\cdot ONa$$

Bisulphite addition is not confined to aliphatic aldehydes and ketones. Thus, benzaldehyde readily forms a sodium bisulphite addition complex.

**Polymerisation of Aldehydes**

Acetaldehyde is polymerised on treatment with cold concentrated sulphuric acid to yield paraldehyde, a mobile, neutral, volatile liquid, with a characteristic and unpleasant smell, which is used as a mild hypnotic.

$$CH_3CH{=}O \xrightarrow{H^+} CH_3CH{=}\overset{+}{O}H + 2\,CH_3\cdot CH{=}O \longrightarrow \text{cyclic } \overset{+}{O}H \text{ intermediate} \xrightarrow{-H^+} \text{Paraldehyde}$$

Paraldehyde

The cyclic trioxy-ether structure is evident from its inability to enter into addition reactions, and from its stability, relative to acetaldehyde, to oxidation.

Light-catalysed oxidations to peroxidic compounds with decomposition to acetaldehyde can, however, occur on storage if this is not carefully controlled.

Evaporation of aqueous formaldehyde gives rise to chain polymerisation by loss of water from the hydrate which occurs in aqueous solution. The product, paraformaldehyde, is a solid polymer, insoluble in water, which decomposes when heated at 180–200°C to generate anhydrous gaseous formaldehyde.

$$n\text{-HO}\cdot\text{CH}_2\cdot\text{OH} \xrightarrow[\text{H}_2\text{O}]{} \underset{\text{Paraformaldehyde}}{\text{HO}\cdot\text{CH}_2(\text{OCH}_2)_{n-2}\cdot\text{CH}_2\text{OH}} \xrightarrow{180\text{–}200^\circ} n\text{-HCHO}\uparrow$$

Solutions of formaldehyde in sulphuric acid undergo an alternative polymerisation to a trimer, trioxymethylene, which is analogous to paraldehyde.

$$3\text{HCHO} \xrightarrow{\text{H}^+} \text{(CH}_2\text{O)}_3 \text{ (cyclic: O–CH}_2\text{–O–CH}_2\text{–O–CH}_2\text{)}$$

**Addition of Hydride Ion**

Aldehydes and ketones are reduced to the corresponding alcohols by complex metal hydrides, lithium aluminium hydride ($LiAlH_4$) and sodium borohydride ($NaBH_4$), and also by certain coenzymes, such as NADH and NADPH, all of which act as hydride ion ($H^-$) sources. Aldehydes yield primary alcohols; ketones yield secondary alcohols. Hydride reduction is generally characteristic of enzyme-catalysed reductions of aldehydes and ketones.

$$\text{H}^- + \text{R(H)C=O} \xrightarrow{\text{LiAlH}_4} \text{R}\cdot\text{CH}_2\cdot\text{O}\cdot\bar{\text{A}}\text{lH}_3\text{Li}^+ \xrightarrow{\text{H}^+/\text{H}_2\text{O}} \text{R}\cdot\text{CH}_2\cdot\text{OH} \quad \text{Primary alcohol}$$

$$\text{R}'\text{RC=O} \longrightarrow \text{R}'\text{RC(H)}(\text{O}\bar{\text{A}}\text{lH}_3\text{Li}^+) \xrightarrow{\text{H}^+/\text{H}_2\text{O}} \text{R}'\text{RC(H)OH} \quad \text{Secondary alcohol}$$

**Enzymic Reduction**

Enzymic reduction of aldehydes and ketones to alcohols is favoured in mammalian systems, despite the inherent reversibility of the process, since liver NADPH is mainly in the reduced form. Thus, *Chloral Hydrate* is reduced by liver alcohol dehydrogenase to trichloroethanol ($Cl_3C\cdot CH_2OH$), which is the effective hypnotic agent. The enzyme is pH-dependent, and reduction requires both the provision of a hydride ion from the coenzyme (NADPH) and of a proton from the enzyme protein. Only one of the two hydrogen atoms in the 4-position of the coenzyme is available for release as a hydride ion, and dehydrogenase enzymes are stereospecific for either one or the other hydrogens (**2**, 5). This has been elegantly demonstrated in experiments with yeast alcohol dehydrogenase using deuterated coenzyme, which reduces acetaldehyde to 1-[$^2$H]-ethanol (deuteroethanol).

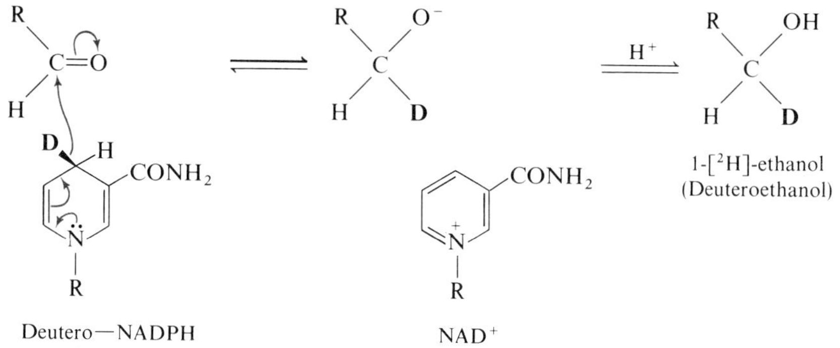

Deutero—NADPH

$NAD^+$

1-[$^2$H]-ethanol
(Deuteroethanol)

### The Cannizzaro Reaction

The Cannizzaro reaction, a disproportionation which takes place in strong alkali, is also a hydride ion transfer (p. 259). The reaction takes place between any two aldehydes which lack α-hydrogen atoms (aromatic aldehydes and formaldehyde).

### Addition of Carbanions

The addition of carbanions, generated *in situ*, to aldehydes and ketones has considerable preparative importance. The reaction of chloroform with acetone in the presence of potassium hydroxide to give the preservative, *Chlorbutol* (Chlorobutanol), is an interesting example.

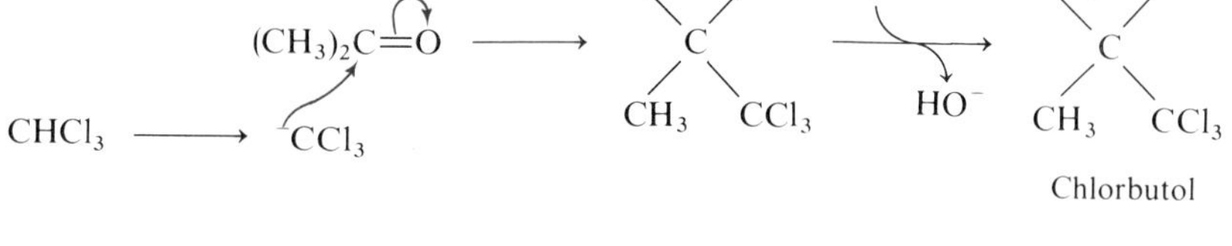

$$CHCl_3 \longrightarrow {}^-CCl_3 + (CH_3)_2C{=}O \longrightarrow (CH_3)_2C(O^-)CCl_3 \xrightarrow{H_2O} (CH_3)_2C(OH)CCl_3 + HO^-$$

Chlorbutol

Grignard reagents (Chapter 9), which function as a source of carbanions, react with formaldehyde to form primary alcohols, with other aldehydes to form secondary alcohols, and with ketones to give tertiary alcohols.

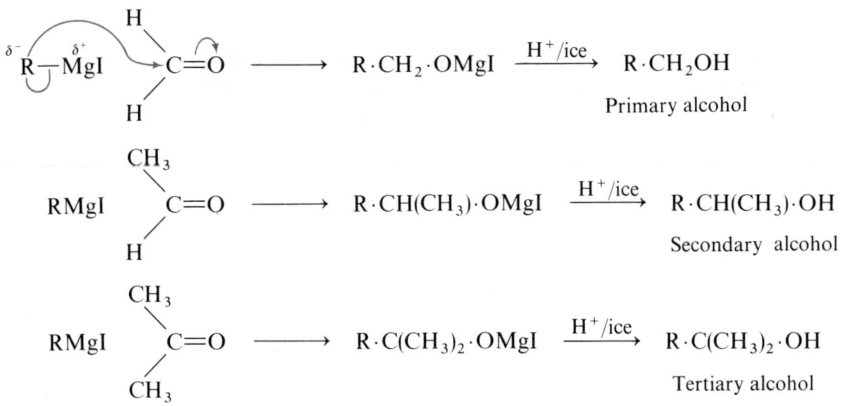

$$\overset{\delta^-}{R}-\overset{\delta^+}{MgI} + H_2C{=}O \longrightarrow R\cdot CH_2\cdot OMgI \xrightarrow{H^+/ice} R\cdot CH_2OH$$

Primary alcohol

$$RMgI + CH_3(H)C{=}O \longrightarrow R\cdot CH(CH_3)\cdot OMgI \xrightarrow{H^+/ice} R\cdot CH(CH_3)\cdot OH$$

Secondary alcohol

$$RMgI + (CH_3)_2C{=}O \longrightarrow R\cdot C(CH_3)_2\cdot OMgI \xrightarrow{H^+/ice} R\cdot C(CH_3)_2\cdot OH$$

Tertiary alcohol

**Phenol–Formaldehyde Resins**

The formation of phenol–formaldehyde (bakelite) resins from phenol and formaldehyde in aqueous solution proceeds through nucleophilic addition of the phenolate ion to the carbonyl group to yield *ortho*- and *para*-hydroxybenzyl alcohols.

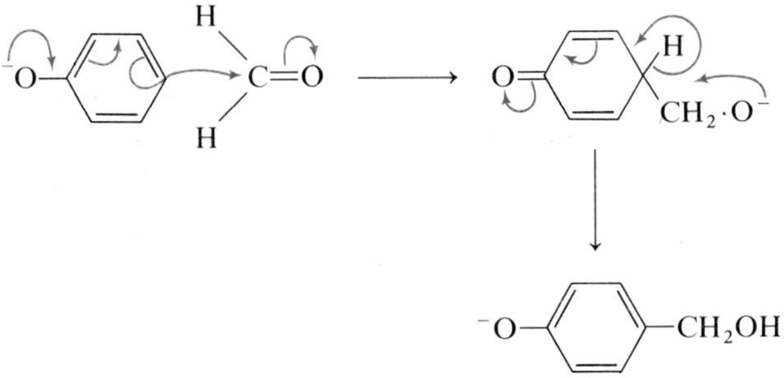

$$^-O-C_6H_5 + H_2C{=}O \longrightarrow O{=}C_6H_5(H)(CH_2\cdot O^-) \longrightarrow {}^-O-C_6H_4-CH_2OH$$

The latter then undergo further condensation in both the *ortho* and *para* positions with formaldehyde to form a cross-linked polymer resin (bakelite).

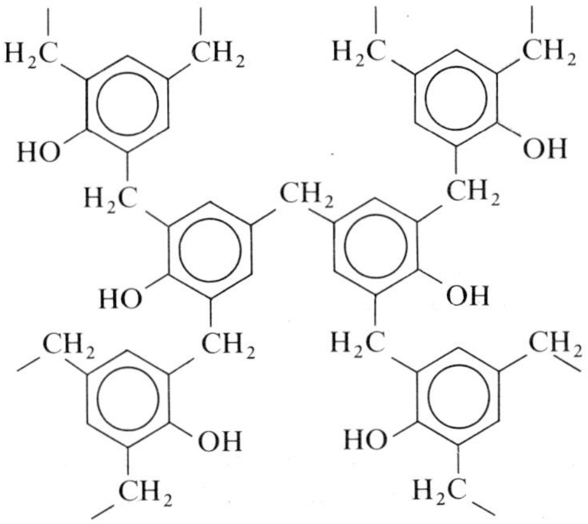

H$_2$C CH$_2$ H$_2$C CH$_2$
HO OH
H$_2$C CH$_2$ CH$_2$
HO OH
CH$_2$ CH$_2$ H$_2$C CH$_2$
OH HO
CH$_2$ H$_2$C

**Chloromethylation**

In the presence of concentrated hydrochloric acid and zinc chloride, formaldehyde reacts with aromatic compounds to give chloromethyl substitution products.

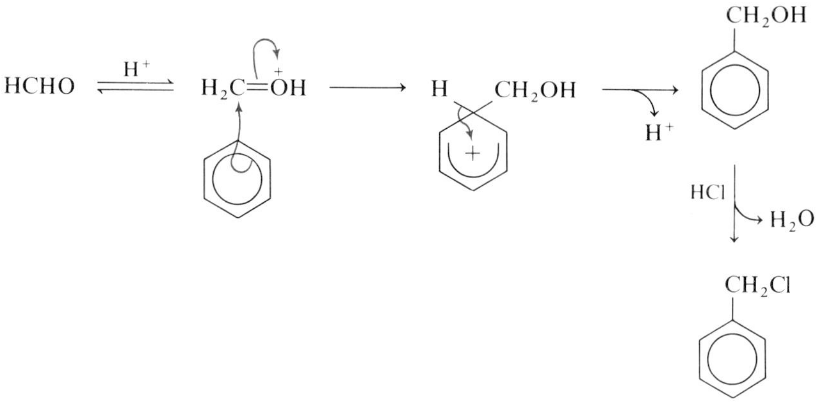

$$HCHO \underset{}{\overset{H^+}{\rightleftharpoons}} H_2C{=}\overset{+}{O}H \longrightarrow \text{H} \quad CH_2OH \xrightarrow{-H^+} C_6H_5CH_2OH \xrightarrow[-H_2O]{HCl} C_6H_5CH_2Cl$$

Chloromethyl benzene

It has been shown, however, that mixtures of concentrated hydrochloric acid and formaldehyde also react rapidly to produce bis-chloromethyl ether ($ClCH_2 \cdot O \cdot CH_2Cl$) within less than a minute of mixing (Keene, 1973), and that in common with a number of other bis-alkylating agents capable of effecting DNA cross-linking, bis-chloromethyl ether is highly carcinogenic (American Industrial Hygiene Journal, 1972).

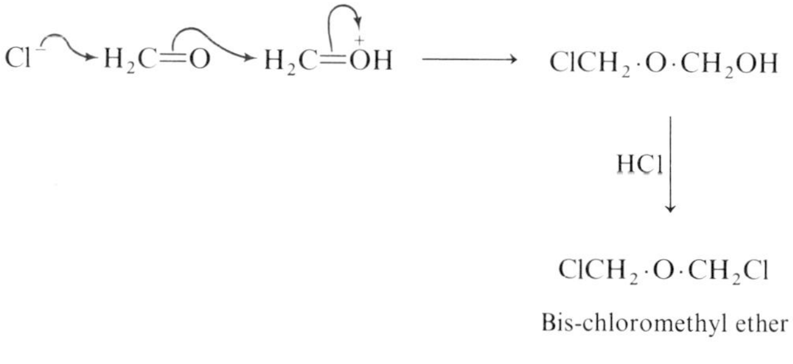

$$Cl^- \; H_2C{=}O \; H_2C{=}\overset{+}{O}H \longrightarrow ClCH_2 \cdot O \cdot CH_2OH \xrightarrow{HCl} ClCH_2 \cdot O \cdot CH_2Cl$$

Bis-chloromethyl ether

## NUCLEOPHILIC ADDITIONS WITH ELIMINATION OF WATER

**Carbinolamines**

Ammonia adds readily to aldehydes (except formaldehyde) to form carbinolamines. The products are not particularly stable and readily undergo dehydration and polymerisation. Formaldehyde also reacts with ammonia, but to give a tricyclic addition product, hexamine (hexamethylenetetramine). Hexamine is

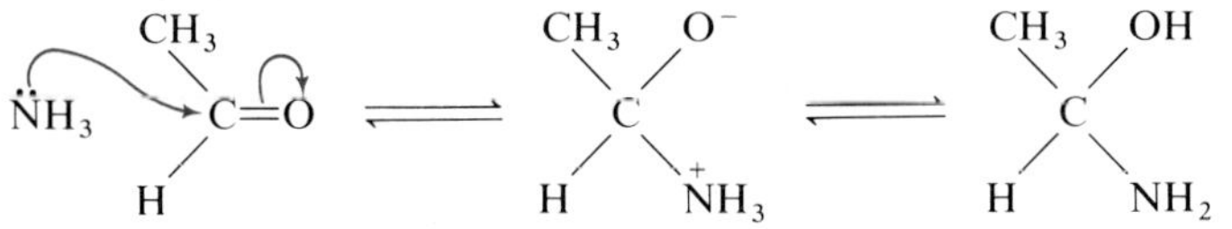

$$\ddot{N}H_3 + (CH_3)(H)C{=}O \rightleftharpoons (CH_3)(H)C(O^-)(\overset{+}{N}H_3) \rightleftharpoons (CH_3)(H)C(OH)(NH_2)$$

Carbinol amine

decomposed by acid with release of formaldehyde, and was formerly used on this account as an antiseptic for the treatment of urinary tract infections, since it is readily execreted in the urine after oral administration.

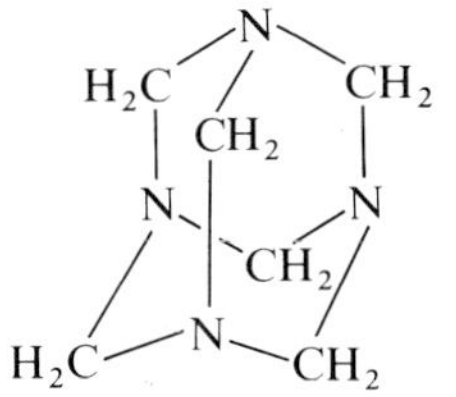

Hexamine

A number of suitably substituted cyclic aminoketones show evidence of transannular intramolecular interaction arising from attack by the lone pair of electrons on the electron-deficient carbonyl carbon. Thus, the infrared absorption spectrum of the cyclic aminoketone, 1-methyl-1-azacyclo-octan-5-one, shows an absorption band at 1683 $cm^{-1}$ (Leonard, Oki and Chiavarelli, 1955). This abnormally low carbonyl absorption frequency (aliphatic ketones absorb in the region 1700–1725 $cm^{-1}$) is largely due to depletion of double bond character due to transannular interaction. This reaches an extreme with the formation of a transannular bond in the perchlorate salt, which is transparent in the infrared at 1683 $cm^{-1}$, but shows characteristic OH group absorption at 3380 $cm^{-1}$.

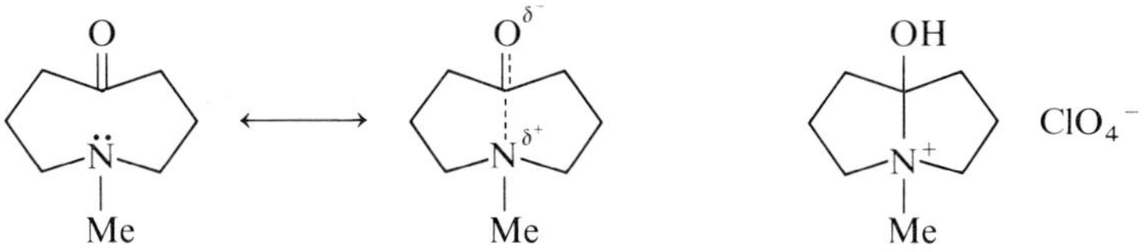

1-Methyl-1-azacyclo-octan-5-one

1-Methyl-1-azacyclo-octan-5-one perchlorate

Similar infrared data and dissociation constants have been cited by Beckett (1956) as evidence of intramolecular interaction between amino and ketone groups in the analgesic aminoketones, *Methadone Hydrochloride* and *Dipipanone Hydrochloride*. Such interaction leads to the adoption of molecular conformations which relate to those of such more highly structured analgesics as *Pethidine*

(Meperidine) and *Betaprodine*.

$CH_3 \cdot CH_2 \cdot CO \cdot CPh_2 \cdot CH_2 \cdot CHMe \cdot NMe_2 \cdot HCl$

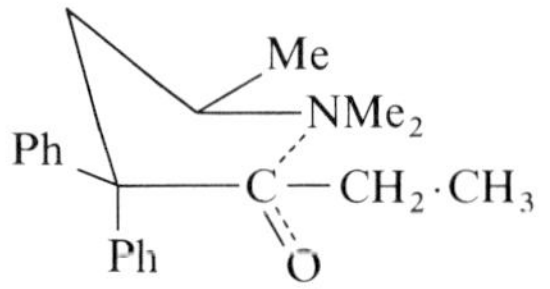

*Methadone Hydrochloride*

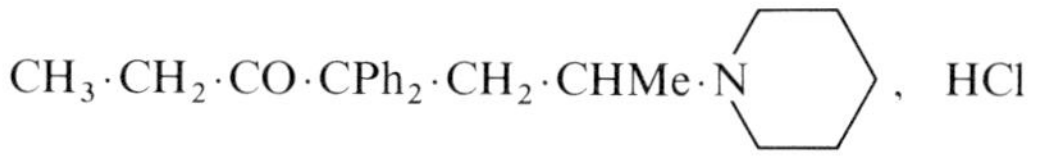

$CH_3 \cdot CH_2 \cdot CO \cdot CPh_2 \cdot CH_2 \cdot CHMe \cdot N(C_5H_{10})$, HCl

*Dipipanone Hydrochloride*

**Imines**

Primary amines and a variety of ammonia derivatives including hydroxylamine, hydrazine and substituted hydrazines, react with aldehydes and less readily with ketones by nucleophilic addition and subsequent elimination of water. The reactions are acid-catalysed, but reaction rates are pH-dependent, with the optimum pH such that maximal protonation of the carbonyl group is achieved with minimal protonation of the base.

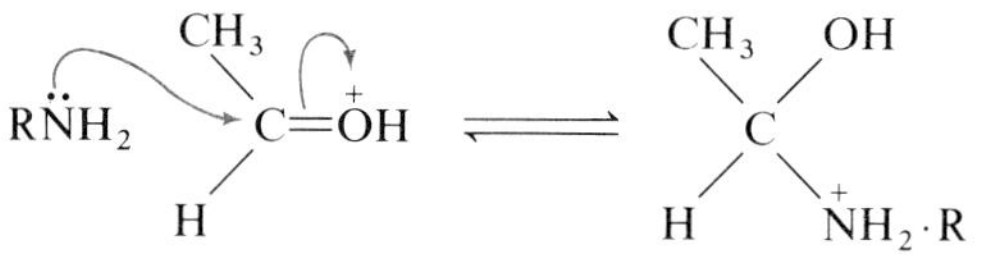

$$R\ddot{N}H_2 + CH_3(H)C{=}\overset{+}{O}H \rightleftharpoons CH_3(H)C(OH)\overset{+}{N}H_2 \cdot R$$

This slow (rate-determining) addition step is followed by a fast acid-catalysed dehydration to give the product, which in the case of a primary amine, is a Schiff's base, or imine. Reactions such as these are liable to lead to interactions between the active ingredients with loss of activity, when for example corticosteroids are formulated with amino type antibiotics such as *Neomycin* for topical application.

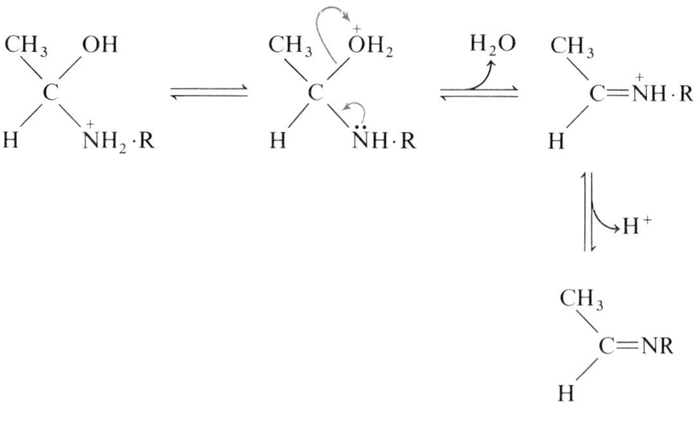

$$CH_3(H)C(OH)\overset{+}{N}H_2 \cdot R \rightleftharpoons CH_3(H)C(\overset{+}{O}H_2)\ddot{N}H \cdot R \xrightleftharpoons{-H_2O} CH_3(H)C{=}\overset{+}{N}H \cdot R \xrightleftharpoons{-H^+} CH_3(H)C{=}NR$$

Schiff's base

The sterilising action of formaldehyde is believed to be brought about by the combination of formaldehyde with the free amino groups of proteins to form methyleneimines. Transaminations requiring the participation of the coenzyme pyridoxal phosphate have also been shown to proceed via an imine intermediate. Studies on model systems (Metzler, Ikawa and Snell, 1954) have demonstrated the rôle of metal ion catalysts and the formation of an imine-ion chelate in enzymic transamination (Snell, 1958; Braunshtein and Shemyakin, 1953).

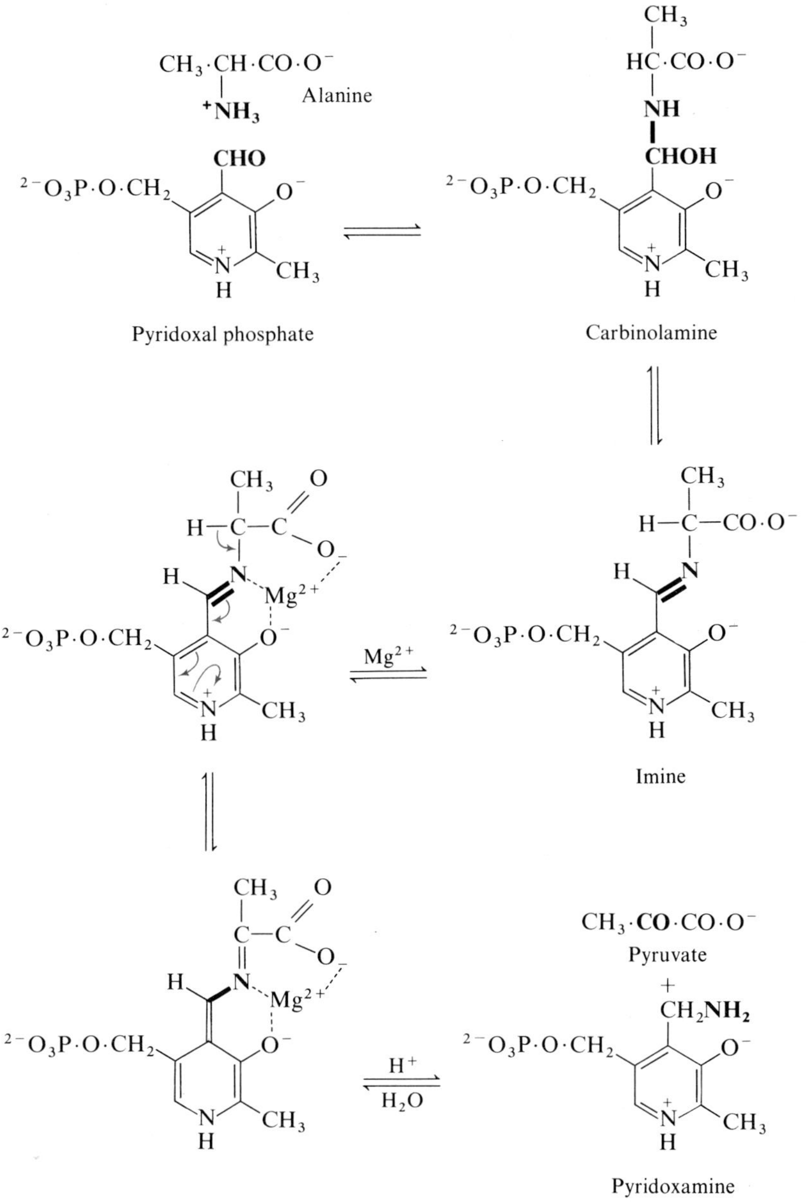

$CH_3{\cdot}CH(NH_3^+){\cdot}CO{\cdot}O^-$ Alanine

Pyridoxal phosphate ⇌ Carbinolamine

⇅

Imine ⇌ ($Mg^{2+}$) Mg chelate

⇅

Chelate ⇌ ($H^+$, $H_2O$) $CH_3{\cdot}\mathbf{CO}{\cdot}CO{\cdot}O^-$ Pyruvate + Pyridoxamine ($CH_2\mathbf{NH_2}$)

Imines formed by retinal and ethanolamine phosphoglycerides are believed to play a key rôle in the functioning of the visual pigments (Daemen and Bonting, 1969). Proximity of the phosphate group to the imine function in the retinylidene phosphoethanolamine derivative provides a source of protons for the important protonation of the imine group.

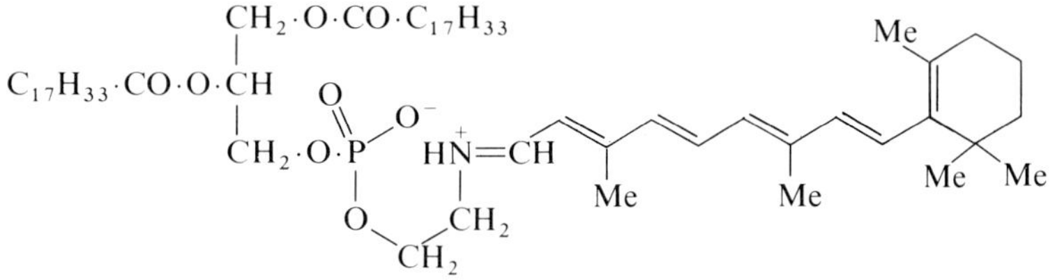

$$\begin{array}{l} CH_2\cdot O\cdot CO\cdot C_{17}H_{33} \\ C_{17}H_{33}\cdot CO\cdot O\cdot CH \\ CH_2\cdot O\cdot P(=O)(O^-)\cdots H\overset{+}{N}=CH\text{-(retinylidene: Me, Me, Me, Me, Me)} \\ \quad O\text{—}CH_2\text{—}CH_2 \end{array}$$

## Oximes

Condensation of aldehydes and ketones with hydroxylamine to form aldoximes and ketoximes follows a similar sequence of addition and loss of water. The products are usually highly crystalline and, therefore, useful in the characterisation of liquid carbonyl compounds. The reaction with hydroxylamine hydrochloride can be conducted in such a manner as to be quantitative. Titration of the hydrochloric acid released in this way forms the basis of a method for the determination of aldehydes, such as citral in volatile oils.

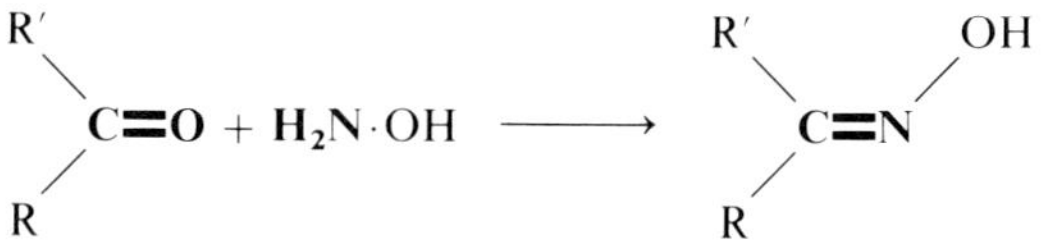

$$\begin{matrix} R' \\ R \end{matrix}\!\!>C=O + H_2N\cdot OH \longrightarrow \begin{matrix} R' \\ R \end{matrix}\!\!>C=N\text{—}OH$$

(R, R′ = H, alkyl or aryl) Oxime

Oximes may function as nucleophiles, as for example when *Pralidoxime Iodide* (2-Hydroxy iminomethyl-1-methyl pyridinium iodide) is used to reverse

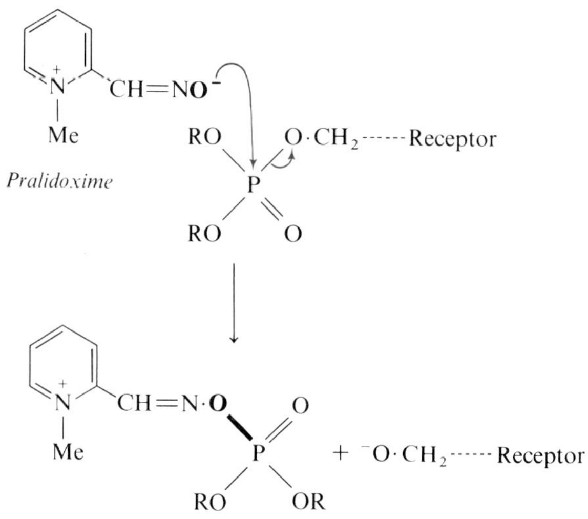

$$\text{Pralidoxime: } (\text{pyridinium-}\overset{+}{N}\text{-Me})\text{-}CH=NO^- + (RO)_2P(=O)\text{—}O\cdot CH_2\text{------Receptor}$$

$$\downarrow$$

$$(\text{pyridinium-}\overset{+}{N}\text{-Me})\text{-}CH=N\cdot O\text{—}P(=O)(RO)(OR) + {}^-O\cdot CH_2\text{------Receptor}$$

*Pralidoxime*

the anticholinesterase activity of phosphorus insecticides. The latter phosphorylate serine hydroxyl groups at the esteratic site of the enzyme forming enzymatically inactive phosphorus esters. *Pralidoxime Iodide* is able to bring about nucleophilic displacement of hydroxyl (**2**, 2). The high nucleophilicity of oximates in such reactions is due to the α-effect (Filippini and Hudson, 1972). This arises from electron pair repulsion by the lone pair of electrons adjacent to the nucleophilic atom ($—CH{=}\ddot{N}—O^-$).

**Hydrazones, Phenylhydrazones, Semicarbazones and Thiosemicarbazones**

Analogous additions with dehydration occur in reactions of aldehydes and ketones with hydrazine, phenylhydrazine, 2,4-dinitrophenylhydrazine, semicarbazide and thiosemicarbazide to form crystalline derivatives, which are extensively used for characterisation purposes.

$R'RC{=}O \xrightarrow{H_2N\cdot NH_2} R'RC{=}N\cdot NH_2$ — Hydrazone

$R'RC{=}O \xrightarrow{H_2N\cdot NH\cdot Ph} R'RC{=}N\cdot NHC_6H_5$ — Phenylhydrazone

$R'RC{=}O \xrightarrow{H_2NNH\cdot C_6H_3(NO_2)_2} R'RC{=}N\cdot NH{-}C_6H_3(NO_2)_2$ — 2,4-Dinitrophenyl hydrazone

$R'RC{=}O \xrightarrow{H_2N\cdot NH\cdot CO\cdot NH_2} R'RC{=}N\cdot NH\cdot CO\cdot NH_2$ — Semicarbazone

$R'RC{=}O \xrightarrow{H_2NNH\cdot CS\cdot NH_2} R'RC{=}N\cdot NH\cdot CS\cdot NH_2$ — Thiosemicarbazone

The acetohydrazide derivatives, known as Girard reagents, are of special utility in the isolation of small quantities of aldehydes and ketones from aqueous fluids, and have proved of particular value in the isolation of urinary steroidal ketones. The quaternary ammonium semicarbazones are highly water-soluble; non-carbonyl compounds can be extracted in their presence by organic solvents and the parent carbonyl compounds liberated by acid hydrolysis.

$(CH_3)_3\overset{+}{N}\cdot CH_2\cdot CO\cdot NH\cdot NH_2\ \ Cl^-$

Girard reagent-T

$C_5H_5\overset{+}{N}-CH_2\cdot CO\cdot NH\cdot NH_2\ \ Cl^-$

Girard reagent-P

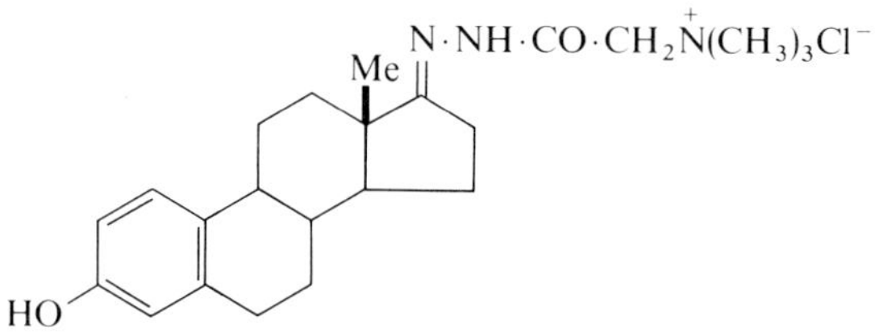

$N\cdot NH\cdot CO\cdot CH_2\overset{+}{N}(CH_3)_3Cl^-$ Me HO

Oestrone—Girard-T derivative

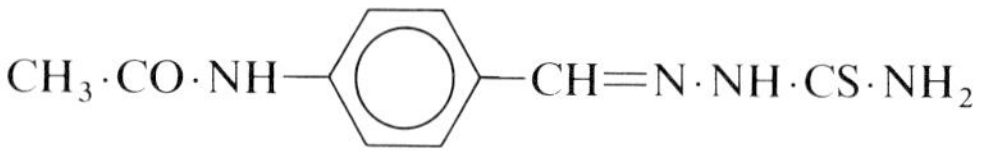

$CH_3\cdot CO\cdot NH-C_6H_4-CH{=}N\cdot NH\cdot CS\cdot NH_2$

Thiacetazone

Thiacetazone (*p*-acetylaminobenzaldehyde thiosemicarbazone) is effective in the treatment of tuberculosis (Domagk, Behnische, Mietzsch and Schmidt, 1946; Behnische, Mietzsch and Schmidt, 1950) and is especially useful in the treatment in combination with *Isoniazid* (East African/MRC Thiacetazone Investigations 1959, 1963) of infections which are resistant to other anti-tubercular agents.

## ENOLS AND ENOLATE IONS

The phenomenon of enolisation is characteristic of all compounds with either a methylene ($-CH_2-$) or methine ($>CH-$) group adjacent to the carbonyl group. It is, therefore, characteristic of most aliphatic aldehydes and ketones, but specifically not feasible with either formaldehyde or aromatic aldehydes. Such juxtaposition of carbonyl and α-methylene or methine groups permits migration of an α-hydrogen to the carbonyl oxygen, and the existence of a tautomeric equilibrium between ketonic and enolic forms.

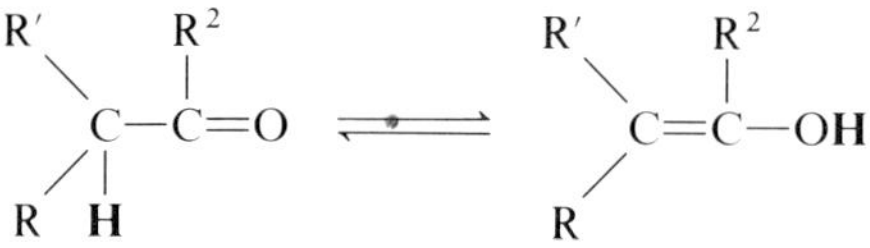

$$RR'CH-C(R^2){=}O \rightleftharpoons RR'C{=}C(R^2)-OH$$

Keto form Enol form

Comparison of the bond energy of keto and enol groups suggests that the keto form should be more stable than the enol by 50 kJ mol$^{-1}$. The equilibrium, however, may favour one or other of the two forms, depending on the nature of the substituents R, $R^1$ and $R^2$ and their ability to stabilise the enol form. It is also influenced by the pH of the medium in which the ketone is examined, since enols are weak acids and, like phenols, form alkali metal salts.

**Acid-catalysed Enolisation**

Enolisation can be catalysed by both acids and bases. Acid-catalysed enolisation involves protonation of the carbonyl group. This weakens the $\alpha$ C—H bond for removal of a proton by water, as seen for example in the total exchange of $\alpha$-hydrogens for deuterium which is possible with deuterium oxide ($D_2O$).

$$H_3C-C(=O)-CH_3 \underset{}{\overset{H^+}{\rightleftharpoons}} H_3C-C(=\overset{+}{O}H)-CH_2H \xrightleftharpoons{H_2\ddot{O}} H_3C-C(:OH)=CH_2 + H_3\overset{+}{O}$$

Acetone

$$H_3C-C(=O)-CH_2D \rightleftharpoons H_3C-C(=\overset{+}{O}H)-CH_2D \xrightleftharpoons{HD_2\overset{+}{O}} H_3C-C(:OH)=CH_2$$

Monodeuteroacetone

Bromination, in which the $\alpha$-hydrogens are successively replaced, occurs by a similar mechanism involving reaction of bromine with the carbon—carbon double bond of the enol.

$$(CH_3)_2C=O + H^+ \rightleftharpoons (CH_3)_2C=\overset{+}{O}H \rightleftharpoons CH_3(CH_2=)C-OH + H^+$$

$$CH_3(CH_2=)C-OH + Br-Br \longrightarrow CH_3(BrCH_2)C=\overset{+}{O}H$$

$$CH_3(BrCH_2)C=O + H^+ \rightleftharpoons CH_3(BrCH_2)C=\overset{+}{O}H$$

$\alpha$-Bromoacetone

Successive bromine atoms will predominantly substitute on the same carbon atom as the first, since the electron-attracting effect of the first bromine substituent promotes the release of adjacent protons, and enolisation in its direction.

### Base-catalysed Enolisation

Base catalysis promotes removal of an α-proton, the driving force for removal being achieved by resonance stabilisation of the enolate anion which is formed. In consequence, the reverse addition of a proton to the enolate can theoretically occur on acidification at either the carbanion or the oxanion. The former is by far the stronger base of the two, so that reformation of the ketone is favoured.

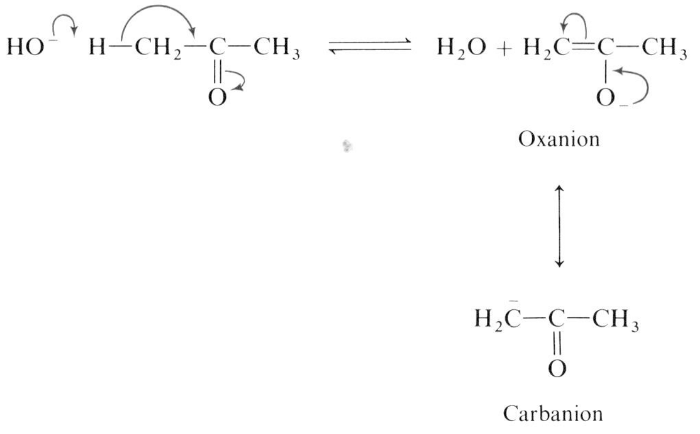

$$HO^- \; H{-}CH_2{-}\underset{\underset{O}{\|}}{C}{-}CH_3 \rightleftharpoons H_2O + H_2C{=}\underset{\underset{O^-}{|}}{C}{-}CH_3$$

Oxanion

$$\updownarrow$$

$$H_2\bar{C}{-}\underset{\underset{O}{\|}}{C}{-}CH_3$$

Carbanion

### The Haloform Reaction

The formation of iodoform from methyl ketones in the haloform reaction (Chapter 7) with iodine is base-catalysed and proceeds through the enolate ion.

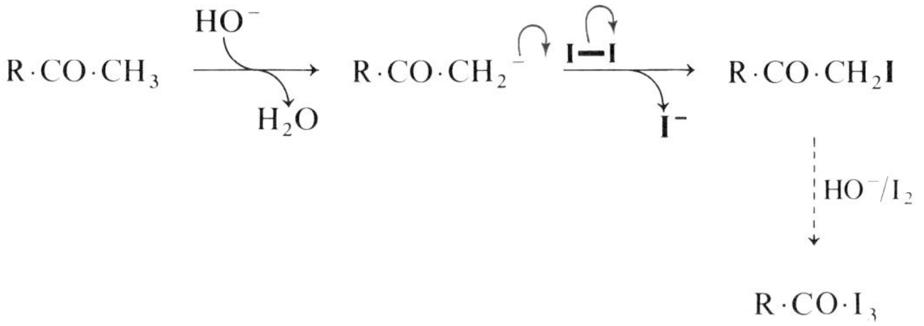

$$R{\cdot}CO{\cdot}CH_3 \xrightarrow[H_2O]{HO^-} R{\cdot}CO{\cdot}CH_2^- \xrightarrow[I^-]{I{-}I} R{\cdot}CO{\cdot}CH_2I \overset{HO^-/I_2}{\dashrightarrow} R{\cdot}CO{\cdot}I_3$$

The resulting trihalo-carbonyl compound is hydrolysed as it is formed to yield a yellow crystalline precipitate of iodoform.

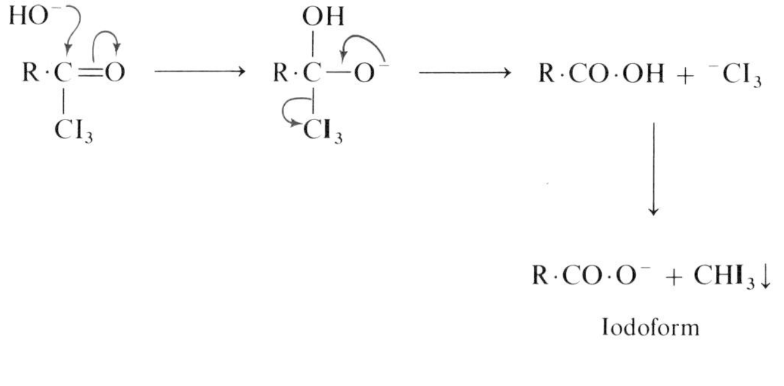

$$\underset{\underset{CI_3}{|}}{R{\cdot}C{=}O} \;(HO^-) \longrightarrow R{\cdot}\underset{\underset{CI_3}{|}}{\overset{\overset{OH}{|}}{C}}{-}O^- \longrightarrow R{\cdot}CO{\cdot}OH + {}^-CI_3$$

$$\downarrow$$

$$R{\cdot}CO{\cdot}O^- + CHI_3\downarrow$$

Iodoform

### Aldol Condensation

The aldol condensation occurs when the carbanion is formed in the presence of the parent aldehyde and attacks the latter. A weak base (to avoid polymerisation), such as potassium carbonate, with the aldehyde in moist ether, often provides the necessary slow release of carbanion which favours the reaction.

$$\mathrm{CH_3(H)C{=}O} + {}^{-}\mathrm{CH_2{\cdot}CHO} \rightleftharpoons \mathrm{CH_3(H)C(O^-)CH_2{\cdot}CHO} \rightleftharpoons \mathrm{CH_3(H)C(OH)CH_2{\cdot}CHO} + \mathrm{HO^-}$$

Aldol

The product aldols readily dehydrate to form $\alpha\beta$-unsaturated aldehydes which polymerise in the presence of strong aqueous alkali to resinous products. This polymerisation in strong alkali is characteristic of all aliphatic aldehydes, except formaldehyde, and forms a useful distinction from ketones.

Aldol-type condensations are also possible with ketones and between aldehydes and ketones, provided one member of each pair has at least one $\alpha$ C—H bond from which to form the enolate ion. The most useful reactions are those in which only one of the reacting carbonyl compounds can form an enolate, as for example when this is formaldehyde or benzaldehyde; otherwise complex mixtures of products may be formed. In the absence of an enolisable carbonyl compound, however, formaldehyde and benzaldehyde undergo an alternative reaction in the presence of base, the Cannizzaro reaction (p. 259).

In mixed aldol condensations between aldehydes and ketones, it is the ketone which functions preferentially as the carbanion. This is seen in the enzymic aldol reaction catalysed by aldolase in which dihydroxyacetone phosphate condenses with D-glyceraldehyde-3-phosphate to give fructose-1,6-diphosphate.

$$\mathrm{CH_2{\cdot}O{\cdot}PO_3H_2{-}CO{-}CH_2OH} + \mathrm{CHO{-}CH(OH){-}CH_2{\cdot}O{\cdot}PO_3H_2} \xrightleftharpoons{\text{Aldolase}} \mathrm{CH_2{\cdot}O{\cdot}PO_3H_2{-}CO{-}C(HO)H{-}C(H)OH{-}C(H)OH{-}CH_2{\cdot}O{\cdot}PO_3H_2}$$

Fructose-1,6-diphosphate

The stereospecificity of the reaction is evident from the formation of fructose-1,6-diphosphate as the sole product. It has also been demonstrated in reactions using stereospecific tritium labelling of dihydroxyacetone phosphate, in which either one or the other of the two hydrogen atoms on C-3 are replaced by tritium

(**2**, 3). In one case, aldolase gives 3-[$^3$H]-fructose-1,6-diphosphate, and in the other a tritium-free product (Rose and Rieder, 1958).

$$
\begin{array}{c}
CH_2{\cdot}O{\cdot}PO_3H_2 \\
| \\
CO \\
| \\
HO{-}C{-}H \\
| \\
H
\end{array}
$$

Aldolase + $\mathbf{T_2O}$ (left arrow); Triosephosphate isomerase + $\mathbf{T_2O}$ (right arrow)

$$
\begin{array}{c}
CH_2{\cdot}O{\cdot}PO_3H_2 \\
| \\
CO \\
| \\
HO{-}C{-}H \\
| \\
\mathbf{T}
\end{array}
\qquad\qquad
\begin{array}{c}
CH_2{\cdot}O{\cdot}PO_3H_2 \\
| \\
CO \\
| \\
HO{-}C{-}\mathbf{T} \\
| \\
H
\end{array}
$$

Aldolase + D-Glyceraldehyde-3-phosphate

$$
\begin{array}{c}
CH_2{\cdot}OPO_3H_2 \\
| \\
CO \\
| \\
HO{-}C{-}H \\
| \\
H{-}C{-}OH \\
| \\
H{-}C{-}OH \\
| \\
CH_2{\cdot}OPO_3H_2
\end{array}
\qquad\qquad
\begin{array}{c}
CH_2{\cdot}O{\cdot}PO_3H_2 \\
| \\
CO \\
| \\
HO{-}C{-}\mathbf{T} \\
| \\
H{-}C{-}OH \\
| \\
H{-}C{-}OH \\
| \\
CH_2OH
\end{array}
$$

Fructose-1,6-diphosphate — 3-[$^3$H]-fructose-1,6-diphosphate

## FREE RADICAL REACTIONS

### Additions to the Carbonyl Group

Both aldehydes and ketones undergo addition of hydrogen in the presence of a suitable metal catalyst (Pt, Pd or Ni) and molecular hydrogen. Aldehydes form primary alcohols; ketones give secondary alcohols.

$$R'R\,C{=}O + H{-}H \xrightarrow{Ni} R'R\,CHOH$$

Similar reductions occur with nascent hydrogen generated by dissolving metals (Na and ethanol; Zn and hydrochloric acid). Dimerisation of free radicals generated in the reduction of acetone with magnesium in moist ether results in the formation of pinacol.

$$2(CH_3)_2C{=}O + Mg \longrightarrow \begin{matrix}(CH_3)_2\cdot C{-}O\\ |\qquad\quad Mg\\ (CH_3)_2\cdot C{-}O\end{matrix} \longrightarrow \begin{matrix}(CH_3)_2\cdot C\cdot OH\\ |\\ (CH_3)_2\cdot C\cdot OH\end{matrix}$$

Pinacol

### Carbon—Carbon Bond Cleavage

Cleavage of the α carbon—carbonyl carbon bond sometimes occurs on irradiation with light. This, however, is more likely to occur in the vapour phase than in solid state or solution, since the heat required to vaporise also predisposes towards a thermally-induced radical decomposition.

$$R'R\,C{=}O \longrightarrow R\cdot + R'\cdot\dot{C}{=}O$$

Radicals so formed may become stabilised by further radical decomposition, and/or recombination as seen, for example, in the formation of methane and ketene from acetone.

$$CH_3\cdot CO\cdot CH_3 \longrightarrow CH_3\cdot + H{-}CH_2{-}\dot{C}{=}O$$

$$CH_3\cdot + H{-}CH_2{-}\dot{C}{=}O \longrightarrow CH_4 \qquad H{-}CH_2{-}\dot{C}{=}O \longrightarrow H\cdot + H_2C{=}C{=}O$$

Ketene

Storage requirements for steroid ketones and their preparations such as, for example, those for *Progesterone* and *Progesterone Injection* which specify protection from light are designed to prevent this type of decomposition.

The light-catalysed epimerisation of C-17 steroid ketones which gives rise to their C-13 epimers is almost certainly the result of radical-induced cleavage of the C-13—C-17 carbon—carbon bond followed by recombination with the epimeric stereochemistry.

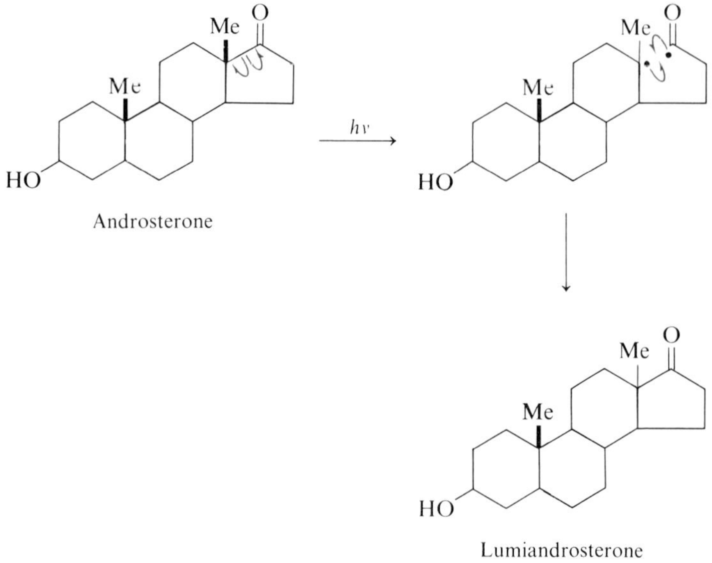

Androsterone $\xrightarrow{h\nu}$ → Lumiandrosterone

## OXIDATION OF ALDEHYDES

Aldehydes are readily oxidised by a variety of reagents (such as moist silver oxide, potassium permanganate, chromic acid) to the corresponding carboxylic acids. Metabolic oxidation proceeds by a similar pathway (**2**, 5).

$$R{\cdot}CHO \xrightarrow{\text{Oxidation}} R{\cdot}CO{\cdot}OH$$

In consequence, aliphatic aldehydes are powerful reducing agents and readily reduce Tollen's reagent (ammoniacal silver nitrate solution—formed by the addition of strong ammonia solution to silver nitrate solution, dropwise until the precipitate first formed just redissolves) and also Fehling's solution (alkaline potassium cupri-tartrate). Tollen's reagent is reduced to metallic silver, usually seen as a silver mirror, whilst Fehling's solution, which is deep blue in colour, is decolourised with formation of a red precipitate of cuprous oxide.

Aldehydes are very susceptible to autoxidation in air (Phillips, Frostick and Starcher, 1957), in a process which is catalysed by light and metallic ions, notably the cobaltous ion, $Co^{2+}$. The reaction pathway and products are temperature-dependent. At 20°C, the main product in the autoxidation of acetaldehyde is acetic acid, but at lower temperatures a hydroxy-peroxide is formed which dissociates at somewhat higher temperature into peracetic acid and acetaldehyde.

$$CH_3{\cdot}C(=O){-}H + {\cdot}O{-}O{\cdot} \longrightarrow CH_3{\cdot}C(=O){\cdot} + {\cdot}O{-}OH \longrightarrow CH_3{\cdot}C(=O){-}O{-}OH$$

$$CH_3{\cdot}C(=O){-}O{-}O{-}H + H_3C(H)C{=}O \underset{>20^\circ C}{\overset{0^\circ C}{\rightleftharpoons}} CH_3{\cdot}C(=O{\cdots}H{-}O){-}O{-}O{-}C(CH_3)(H)$$

The reactivity of peracetic acid in nucleophilic attack on a second molecule of aldehyde depends on the $\alpha$-effect (Fillipini and Hudson, 1972) as in the case of the oximate ion (p. 251).

Ketones are much more resistant to oxidation, but can be oxidised with cleavage of the enolate $\alpha$ carbon—carbonyl carbon double bond in either acid or alkaline solution.

Aldehydes and ketones can also be distinguished by the characteristic reaction of aldehydes with Schiff's reagent, which is prepared by bleaching a solution of the red dye, rosaniline, with sulphur dioxide. Most aldehydes restore the colour to Schiff's reagent.

## AROMATIC ALDEHYDES

Certain reactions are peculiarly characteristic of aromatic aldehydes. Thus, they fail to reduce Fehling's solution. Also, the absence of $\alpha$ C—H bonds makes it impossible for benzaldehyde to function as the carbanion source in an aldol condensation in the presence of mild alkali. As a result, the sequence of condensation, dehydration, and polymerisation to resinous products, which is a feature of aliphatic aldehydes (except formaldehyde), does not occur with aromatic aldehydes. Instead, aromatic aldehydes disproportionate to the corresponding alcohol and carboxylic acid in the Cannizzaro reaction.

### Cannizzaro Reaction

The Cannizzaro reaction is a disproportionation reaction between two aromatic aldehydes in strong alkali to give the corresponding carboxylic acid and alcohol.

$$\underset{\text{Benzaldehyde}}{2C_6H_5{\cdot}CHO} + HO^- \longrightarrow \underset{\text{Benzoate}}{C_6H_5{\cdot}CO{\cdot}O^-} + \underset{\text{Benzyl alcohol}}{C_6H_5{\cdot}CH_2{\cdot}OH}$$

The reaction also occurs with formaldehyde and other aliphatic aldehydes which have no $\alpha$ C—H bonds. The mechanism consists of a nucleophilic addition of $HO^-$ to the carbonyl group of the aldehyde, followed by a hydride ion shift and attack at the carbonyl carbon of a second aldehyde molecule. The reaction is trimolecular in terms of aldehyde and hydroxyl ion concentration.

The reaction is believed to proceed by the following pathway.

$$\text{Ph—CH=O} + \text{HO}^- \rightleftharpoons \text{Ph—CH(OH)—O}^- \xrightarrow{\text{Ph·CHO}} \text{Ph·CO·OH} + \text{Ph·CH}_2\text{O}^-$$

$$\downarrow$$

$$\text{Ph·CO·O}^- + \text{Ph·CH}_2\text{OH}$$

**Benzoin Condensation**

Aromatic aldehydes form cyanhydrins with hydrogen cyanide, but with alkali metal cyanides, the reaction proceeds a stage further. The electron-attracting properties of both phenyl and nitrile substituents induce formation of a carbanion, which then adds to a second molecule of aldehyde to give a benzoin.

$$\text{Ph(H)C=O} \underset{}{\overset{\text{CN}^-}{\rightleftharpoons}} \text{Ph(H)C(O}^-\text{)CN} \rightleftharpoons \text{Ph}\overset{-}{\text{C}}\text{(CN)OH}$$

$$\text{Ph}\overset{-}{\text{C}}\text{(CN)OH} + \text{Ph(H)C=O} \rightleftharpoons \text{Ph·C(OH)(CN)—C(O}^-\text{)(H)·Ph}$$

$$\rightleftharpoons$$

$$\text{Ph·CO·CHOH·Ph} + \text{CN}^-$$

Benzoin

## QUINONES

Quinones are unsaturated cyclic dicarbonyl compounds derived from aromatic systems. The simplest and best known examples are *p*-benzoquinone and *o*-benzoquinone.

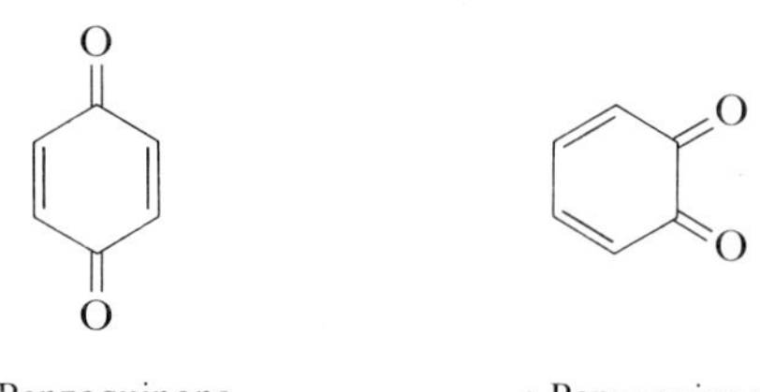

*p*-Benzoquinone *o*-Benzoquinone

Their reactions and properties are those of $\alpha\beta$-unsaturated ketones. Both structures incorporate non-linear arrangements of ethylenic and carbonyl groups, and hence are examples of cross-conjugation. They may be distinguished by colour and volatility. *p*-Benzoquinone is yellow, has a characteristic odour and is volatile in steam. *o*-Benzoquinone is light-red in colour, but can also be obtained in an unstable green form, and is non-volatile in steam.

**Reduction**

The reduction of *p*-benzoquinone to quinol is a rapid and reversible process. The overall equation (Chapter 8) shows that the position of the equilibrium varies as the square of the hydrogen ion concentration, so that measurement of electrical potential in such a system provides a useful basis for pH measurement.

The reduction of quinones, which is of considerable biological significance, is a more complex process than appears from the overall equation. Thus, chemical reduction of benzoquinone with sulphurous acid or hydrogen sulphide leads to an intermediate, quinhydrone. This is also formed as a greenish-black solid on admixture of equal parts of benzoquinone and quinol in ethanol. Quinhydrone is a 1 : 1-charge-transfer complex, stabilised by resonance and hydrogen bonding, with the quinol acting as electron donor and the quinone as electron acceptor.

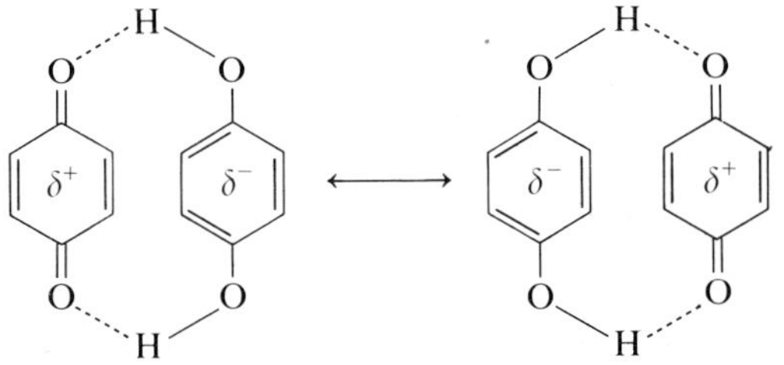

The formation of this crystalline dimeric complex represents a half-way stage in the complete reduction of quinones to quinols, which is a two-electron process. In recent years, it has been established that electrolytic reduction of quinones can lead to the transfer of a single electron with the formation of a radical ion or semiquinone in solution. Such radical ions are in the same oxidation state as quinhydrones. They are extremely reactive and probably play an important part as intermediates in various biological oxidation–reduction systems based on quinone–quinol systems.

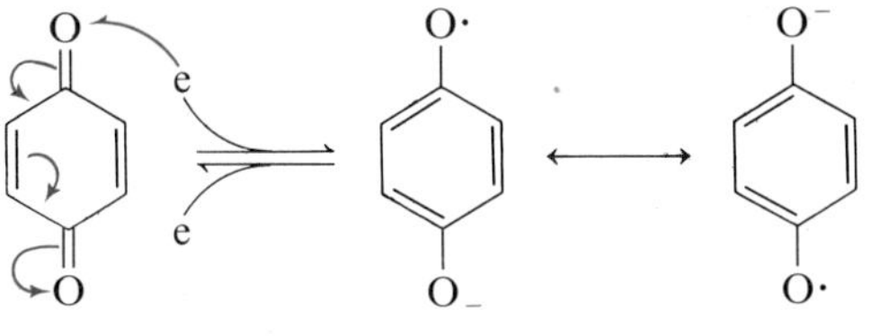

Benzoquinone is also capable of dehydrogenating hydroaromatic compounds, with transfer of hydride ions and concomitant reduction of benzoquinone to quinol.

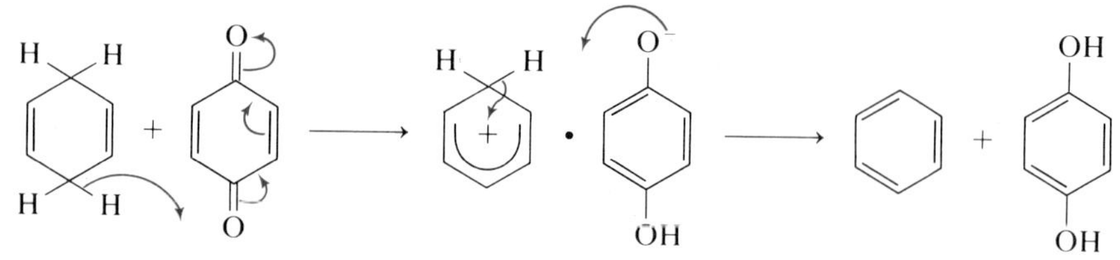

### The Rôle of Quinones in Oxidative Phosphorylation

It has been suggested that various naturally-occurring quinones present in mitochondria may function as both phosphate and electron carriers in oxidative

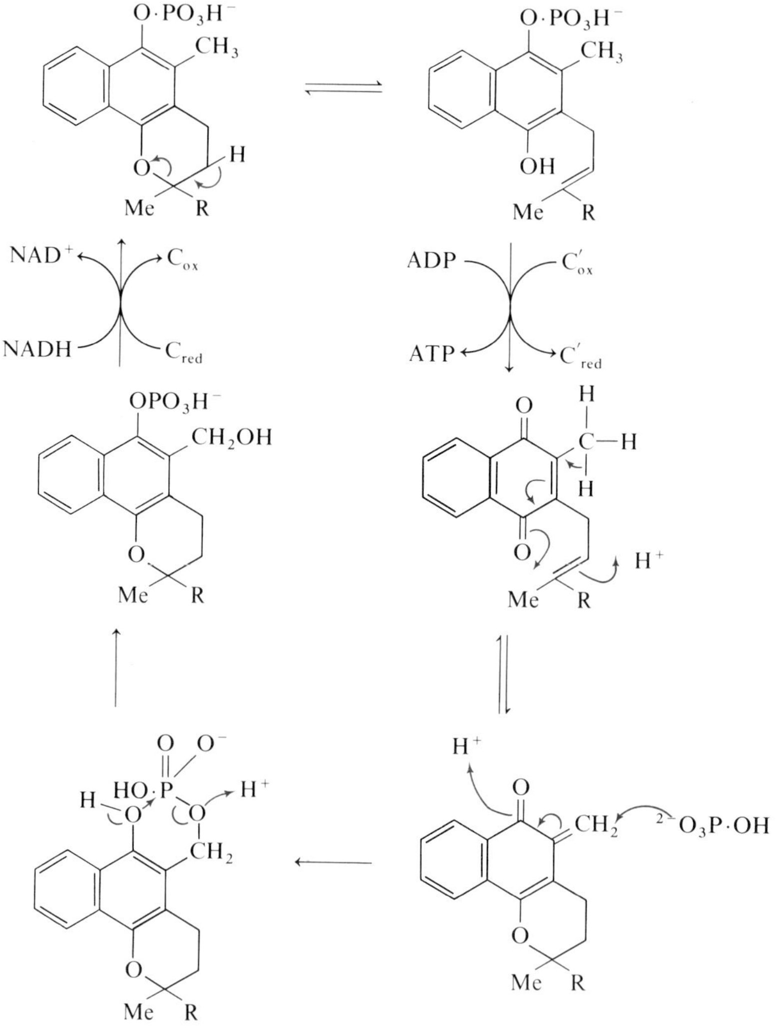

$C_{ox}$, $C'_{ox}$, $C_{red}$, $C'_{red}$, are oxidised and reduced cytochromes.

phosphorylation (Vilkas and Lederer, 1962). Thus, there is evidence that vitamin $K_2$ is reductively cyclised to the corresponding 6-chromanyl phosphate in *Mycobacterium phlei* in a cycle coupling the electron transport chain with phosphorylation, which is visualised as set out on page 262. (Asano, Brodie, Wagner, Wittreich and Folkers, 1962).

### Addition Reactions

As $\alpha\beta$-unsaturated ketones, quinones undergo a number of addition reactions. Bromine adds directly to the ethylenic bonds, but hydrogen halides, primary amines and primary alcohols react by 1,4-addition.

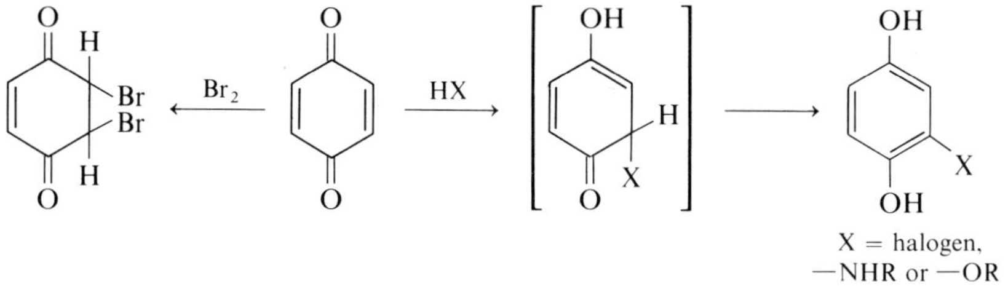

X = halogen, —NHR or —OR

The double bonds are also activated for addition as dienophiles in the Diels–Alder reaction (Chapter 4) leading to the formation of fused carbocyclic compounds.

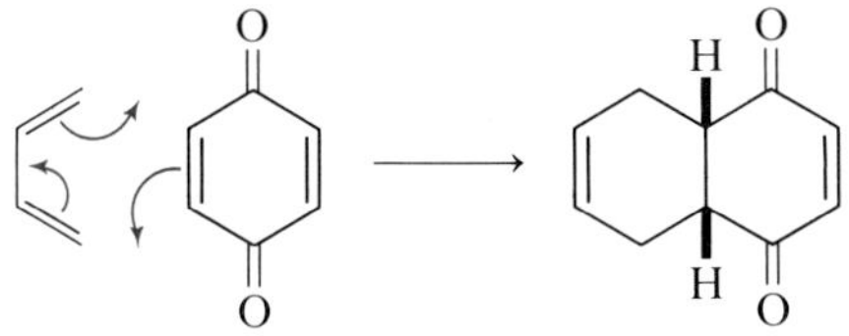

# 11 Monocarboxylic Acids and Esters

The saturated aliphatic monocarboxylic acids form a homologous series of general formula $C_nH_{2n+1}CO_2H$. They are also known as fatty acids, a term which derives from the fact that higher members of the series, particularly palmitic and stearic acids, occur in combined form as constituents of natural fats. The characteristic property of carboxylic acids, namely the ability to react as an acid in water, arises in the carboxylic acid group $-\mathrm{C}(=\mathrm{O})\mathrm{OH}$. The electron-withdrawing properties of the carbonyl oxygen promotes the separation of hydrogen as a proton, whilst the ionised state is still further favoured by resonance stabilisation of the carboxylate ion which is also formed.

$$-\mathrm{C}(=\mathrm{O})-\mathrm{O}-\mathrm{H} \longleftrightarrow -\mathrm{C}(-\mathrm{O}^-)=\overset{+}{\mathrm{O}}-\mathrm{H} \rightleftharpoons \mathrm{H}^+ + -\mathrm{C}(-\mathrm{O}^-)=\mathrm{O} \longleftrightarrow -\mathrm{C}(=\mathrm{O})-\mathrm{O}^-$$

In consequence, the carbonyl group of carboxylic acids has properties which, in comparison with the carbonyl groups in aldehydes and ketones, are significantly modified. The C=O bond length in the undissociated acid is 1.23Å and almost identical with that (1.22Å) in aldehydes and ketones. The C—O bond length of 1.36Å, however, is significantly shorter than that of alcohols and ethers (1.43Å), implying that it has some element of double bond character in keeping with the resonance picture. This is also reflected in the C=O infrared stretching vibrations, which in undissociated carboxylic acids appear at 1725–1700 $cm^{-1}$ (Cross, 1964) compared with that of saturated aldehydes at 1740–1720 $cm^{-1}$. In contrast, carboxylic acid esters, in which the alkyl substituent is joined to one particular oxygen atom $-\mathrm{C}(=\mathrm{O})\mathrm{OR}$, possess properties more closely resembling those of typical carbonyl compounds, exhibiting infrared carbonyl stretching frequencies in the 1750–1735 $cm^{-1}$ region.

Table 26. Nomenclature of Saturated Aliphatic Monocarboxylic Acids

| Formula | Trivial Name | IUPAC Name | b.p. (°C) |
|---|---|---|---|
| $HCOOH$ | Formic acid | Methanoic acid | 101 |
| $CH_3 \cdot COOH$ | Acetic acid | Ethanoic acid | 118 |
| $CH_3 \cdot CH_2 \cdot COOH$ | Propionic acid | Propanoic acid | 141 |
| $CH_3(CH_2)_2 \cdot COOH$ | Butyric acid | Butanoic acid | 163 |
| $CH_3(CH_2)_3 \cdot COOH$ | Valeric acid | Pentanoic acid | 186 |
| $CH_3(CH_2)_4 \cdot COOH$ | Caproic acid | Hexanoic acid | 205 |
| $CH_3(CH_2)_5 \cdot COOH$ | Oenanthic acid | Heptanoic acid | 224 |
| $CH_3(CH_2)_6 \cdot COOH$ | Caprylic acid | Octanoic acid | 237 |
| $CH_3(CH_2)_7 \cdot COOH$ | Pelargonic acid | Nonanoic acid | 255 |
| $CH_3(CH_2)_8 \cdot COOH$ | Capric acid | Decanoic acid | m.p. 31.5 |
| $CH_3(CH_2)_9 \cdot COOH$ | Undecylic acid | Undecanoic acid | 28.6 |
| $CH_3(CH_2)_{10} \cdot COOH$ | Lauric acid | Dodecanoic acid | 44 |
| $CH_3(CH_2)_{11} \cdot COOH$ | Tridecylic acid | Tridecanoic acid | 45.5 |
| $CH_3(CH_2)_{12} \cdot COOH$ | Myristic acid | Tetradecanoic acid | 58 |
| $CH_3(CH_2)_{13} \cdot COOH$ | Pentadecylic acid | Pentadecanoic acid | 53 |
| $CH_3(CH_2)_{14} \cdot COOH$ | Palmitic acid | Hexadecanoic acid | 64 |
| $CH_3(CH_2)_{15} \cdot COOH$ | Margaric acid | Heptadecanoic acid | 61 |
| $CH_3(CH_2)_{16} \cdot COOH$ | Stearic acid | Octadecanoic acid | 72 |
| $CH_3(CH_2)_{17} \cdot COOH$ | Nondecylic acid | Nonadecanoic acid | 69 |
| $CH_3(CH_2)_{18} \cdot COOH$ | Arachidic acid | Eicosanoic acid | 77 |
| $CH_3(CH_2)_{19} \cdot COOH$ | — | Heneicosanoic acid | 75 |
| $CH_3(CH_2)_{20} \cdot COOH$ | Behenic acid | Docosanoic acid | 82 |
| $CH_3(CH_2)_{22} \cdot COOH$ | Lignoceric acid | Tetracosanoic acid | 88 |
| $CH_3(CH_2)_{24} \cdot COOH$ | Cerotic acid | Hexacosanoic acid | 88 |

**Nomenclature**

Many of the saturated aliphatic monocarboxylic acids are known by trivial names. In the IUPAC system of nomenclature, acids are named as derivatives of the corresponding saturated hydrocarbon with the same number of carbon atoms, by applying the suffix **oic**. Thus, formic acid is methanoic acid, and acetic acid, ethanoic, and so on (Table 26). The position of side-chains is indicated by numbering the main carbon chain commencing from the carboxyl group, as for example in 3-methylhexanoic acid.

$$\overset{6}{C}H_3 \cdot \overset{5}{C}H_2 \cdot \overset{4}{C}H_2 \cdot \underset{\displaystyle CH_3}{\underset{|}{\overset{3}{C}H}} \cdot \overset{2}{C}H_2 \cdot \overset{1}{C}O \cdot OH$$

## GENERAL PROPERTIES OF CARBOXYLIC ACIDS

**Physical Properties**

The lower carboxylic acids are colourless liquids with characteristic pungent odours. Decanoic and the higher fatty acids are low melting, odourless waxy solids. The waxy texture of stearic acid provides the basis for its use as a lubricant in tabletting. Their melting and boiling points are significantly higher than those of the corresponding alcohols, reflecting the existence of the acids as internally hydrogen-bonded dimers.

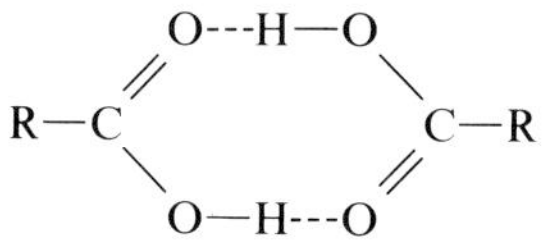

```
      O---H—O                         O···H—O···H—O
     //       \                      //       |       \
R—C             C—R             R—C           H         C—R
     \        //                     \      H   H      //
      O—H---O                         O—H···O—H···O
```

Dimer

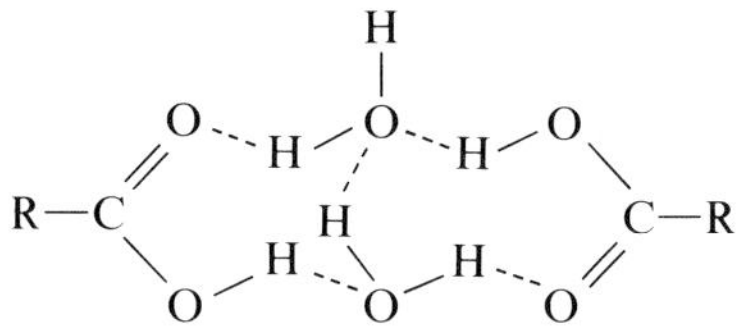

Hydrated dimer

The lower members are completely miscible with water, as a result of hydrogen bonding to water molecules, possibly as hydrated dimers. Solubility in water, however, decreases rapidly in the higher members of the series with increase in length of the hydrocarbon chain.

**Acidity of Aliphatic Acids**

Carboxylic acids are much stronger acids than alcohols or phenols as a result of the electron-withdrawing properties of the carbonyl group (Chapter 10). With the exception of formic acid (p$K_a$ 3.77), all the aliphatic monocarboxylic acids have p$K_a$ values within the range 4.75–5.0, and readily form alkali metal salts in reaction with sodium carbonate (distinction from alcohols and phenols *q.v.*) or sodium bicarbonate.

$$R{\cdot}CO{\cdot}OH + NaHCO_3 \longrightarrow R{\cdot}CO{\cdot}ONa + H_2O + CO_2\uparrow$$

The strength of carboxylic acids is affected by substituents. Alkyl groups, which are electron-releasing, inhibit the separation of the carboxylic proton, and furthermore, tend to destabilise the resulting carboxylate ion. As a result, ionisation is inhibited. This is seen in the marked shift of p$K_a$ between formic and acetic acids.

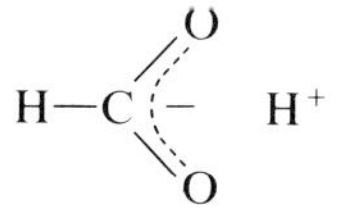

```
     O                      O                         O
    //                     //                        //
H—C ⁝–   H⁺       H₃C→C ⁝–   H⁺       ClCH₂←C ⁝–   H⁺
    \\                     \\                        \\
     O                      O                         O
```

Formic acid (p$K_a$ 3.77) 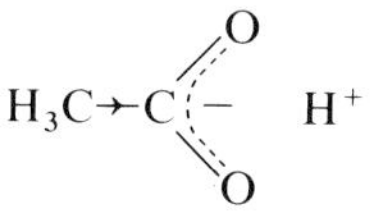 Acetic acid (p$K_a$ 4.76) 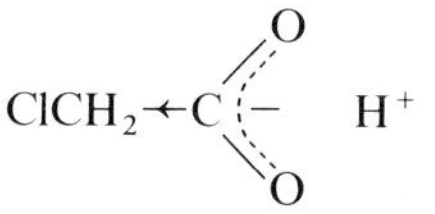 Monochloroacetic acid (p$K_a$ 2.86)

Electron-withdrawing substituents, on the other hand, promote the separation of the carboxylic acid proton, and stabilise the resulting carboxylate ion. Thus, monochloroacetic acid is a much stronger acid than acetic acid. The effect diminishes as the distance between the electron-withdrawing substituent and carboxyl group is increased (Table 27). It increases, however, with increasing electronegativity of the substituent (I < Br < Cl < F) and also with multiple substitution of electron-withdrawing substituents.

Table 27. Effect of Substituents on the Dissociation of Aliphatic Carboxylic Acids

| Acid | p$K_a$ | Acid | p$K_a$ |
|---|---|---|---|
| $H \cdot COOH$ | 3.77 | $F \cdot CH_2 \cdot COOH$ | 2.66 |
| $CH_3 \cdot COOH$ | 4.76 | $Br \cdot CH_2 \cdot COOH$ | 2.86 |
| $CH_3 \cdot CH_2 \cdot COOH$ | 4.88 | $I \cdot CH_2 \cdot COOH$ | 3.12 |
| $Cl \cdot CH_2 \cdot COOH$ | 2.86 | $Cl_2 \cdot CH \cdot COOH$ | 1.29 |
| $Cl \cdot CH_2 \cdot CH_2 \cdot COOH$ | 4.08 | $Cl_3 \cdot C \cdot COOH$ | 0.65 |
| $Cl \cdot CH_2 \cdot CH_2 \cdot CH_2 \cdot COOH$ | 4.52 | $C_6H_5 \cdot COOH$ | 4.19 |
| $Cl \cdot CH_2 \cdot CH_2 \cdot CH_2 \cdot CH_2 \cdot COOH$ | 4.70 | $C_6H_5 \cdot CH_2 \cdot COOH$ | 4.25 |

**Acidity of Aromatic Acids**

Phenylacetic acid (p$K_a$ 4.25) is a stronger acid than acetic acid (p$K_a$ 4.76), due to the electron-attracting effect of the aromatic ring. When, however, the carboxyl group is attached directly to the aromatic ring as in typical aromatic acids, such as benzoic acid, the electron-withdrawing effect is opposed by a resonance effect which results from conjugation of the carbonyl group with the ring. As a result, the double bond character of the carbonyl group is reduced, as is evident from the shift in the infrared carbonyl stretching frequency to 1700–1680 $cm^{-1}$ from 1725–1700 $cm^{-1}$ in typical aliphatic acids.

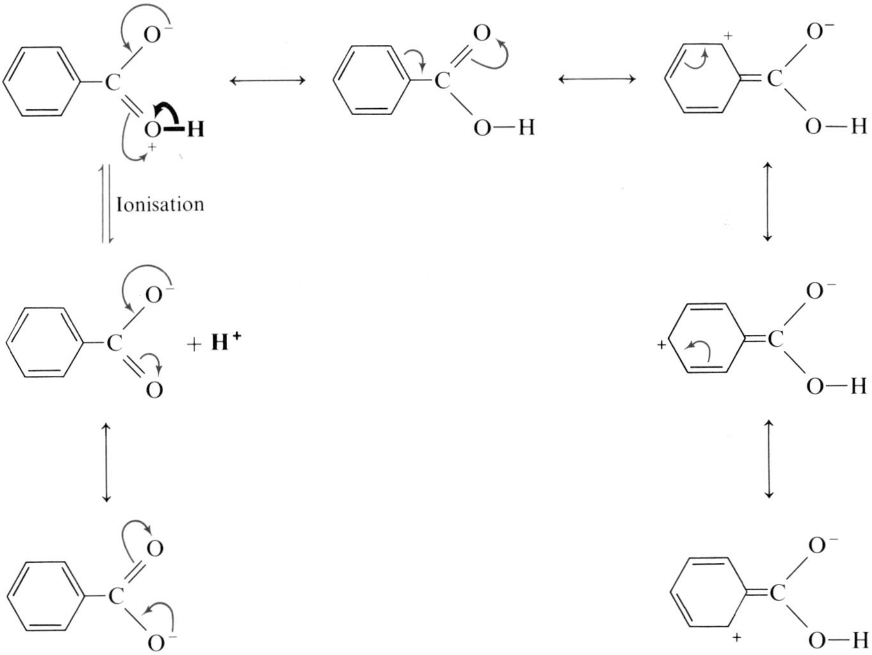

Ionisation

+ $H^+$

Carboxylate ion resonance

Carboxylic acid resonance

Since it is the electron-withdrawing properties of the carbonyl group which are primarily responsible for the separation of the proton, this reduction of double bond character in the carbonyl group impedes ionisation of the acid. In consequence, benzoic acid (p$K_a$ 4.19) is a weaker acid than formic acid (p$K_a$ 3.77). It is, however, a slightly stronger acid than acetic acid, so that the combination of inductive and resonance effects of the aromatic nucleus is less inhibitory of ionisation than the electron-repelling (+I) effect of the methyl group.

Ring substitution in benzoic acid modifies the acidic strength. The overall electron-withdrawing or electron-repelling influence of the substituent is primarily responsible for determining the strength of *m*- and *p*-substituted acids, but steric effects and hydrogen bonding have an overriding effect when the substituent is in the *ortho* position.

Hammett substituent constants ($\sigma$) provide a quantitative measure of the overall electron-withdrawing (+) or electron-releasing (−) power of particular substituents relative to hydrogen (zero). They relate to polar effects of *meta*- and *para*-substituted compounds, and therefore take no account of steric and hydrogen bonding effects encountered in the *ortho* position.

Substituent constants derive from the Hammett equation, which is based on a series of observations concerning the effect of substitution on the physical properties and reactivity of substituted benzoic acids. Thus, the effect of *m*- and *p*-substituents on the dissociation constants of benzoic acid is closely paralleled by the effects of the same substituents on other physical properties, such as for example, the rate constants for the hydrolysis of alkyl benzoates. Thus, a plot of p$K_a$($-\log K_a$) for a series of either *m*- or *p*-substituted benzoic acids against $\log k$, the log of the rate constant for the alkaline hydrolysis of the corresponding esters, exhibits a straight line relationship (linear free energy relationship), which can be expressed by the general equation

$$\log k = \rho \log K + c \tag{1}$$

where $\rho$ is the slope and $c$ the intercept on the ordinate.

If $K_0$ is the dissociation constant for benzoic acid and $k_0$ the rate constant for the hydrolysis of ethyl benzoate, then the following equation relates to the special case of the unsubstituted acid and ester.

$$\log k_0 = \rho \log K_0 + c \tag{2}$$

Subtracting (2) from (1)

$$\log \frac{k}{k_0} = \rho \log \frac{K}{K_0}$$

and

$$\log \frac{k}{k_0} = \rho\sigma$$

where

$$\sigma = \log \frac{K}{K_0}$$

The **substituent constant**, $\sigma$, measures the **polar effect** relative to hydrogen of the substituent in a given position (Jaffé, 1953) and in general is substantially independent of the nature of the reaction. Despite one or two exceptions, substituent constants reflect the fact that, in general, the inductive effect (+I or −I) only is operative for *meta*-substituted compounds, whereas both the inductive effect and the mesomeric effect (+M or −M) affect the reactivity of *para*-substituted compounds. It should be noted, however, that the Ingold sign convention for I and M effects (Chapter 1), which is based on the use of negative symbols with electron-attracting and electron-withdrawing effects, runs counter to the sign conventions which have become accepted for use with polar substituent constants (see below).

The constant $\rho$ in the Hammett equation represents the slope of the line obtained when $\log k/k_0$ is plotted against the substituent constant ($\sigma$). The numerical value of $\rho$ varies with the reaction under consideration, hence $\rho$ is known as the **reaction constant**. It is dependent upon the nature of the reaction, and the actual reaction conditions, and is a measure of the influence of polar effects on a particular reaction under defined conditions. Large values imply that the reaction rate is sensitive to substituent effects. Low values ($<0.5$) mean relative insensitivity to substituent effects. The reaction constant may additionally be either negative or positive. Reactions which are promoted by high electron density at the point of attack have negative values. Reactions which are promoted by electron-deficiency at the reaction centre have positive values.

Table 28 shows that *m*- and *p*-substituted benzoic acids with negative substituent constants (*m*- and *p*-alkyl-; *m*- and *p*-amino-; *p*-hydroxy- and *p*-methoxy-) are all weaker acids than benzoic, whereas *m*- and *p*-substituted acids with positive substituent constants are stronger acids than benzoic acid. It is noteworthy, too, that whereas both *m*- and *p*-aminobenzoic acids are weaker

Table 28. Substituent Constants ($\sigma$) and Dissociation Constants ($pK_a$) for Substituted Aromatic Acids

| Substituent | *meta* | | *para* | | *ortho* |
|---|---|---|---|---|---|
| | $\sigma_m$ | $pK_a$ | $\sigma_p$ | $pK_a$ | $pK_a$ |
| $NH_2$ | −0.160 | 4.79 | −0.660 | 4.92 | 4.98 |
| OH | — | — | −0.367 | 4.58 | 2.98 |
| OMe | — | — | −0.268 | 4.47 | 4.09 |
| Me | −0.069 | 4.27 | −0.170 | 4.37 | 3.91 |
| H | 0.000 | 4.19 | 0.000 | 4.19 | 4.19 |
| OMe | +0.115 | 4.09 | — | — | — |
| OH | +0.121 | 4.08 | — | — | — |
| F | +0.337 | 3.87 | +0.062 | 4.14 | 3.27 |
| Cl | +0.373 | 3.83 | +0.227 | 3.98 | 2.94 |
| $CF_3$ | +0.415 | 5.11 | +0.551 | 4.96 | — |
| $NO_2$ | +0.710 | 3.49 | +0.778 | 3.43 | — |

than benzoic, only the *para*-substituted hydroxy- and methoxy-benzoic acids are similarly weaker than benzoic, whereas the corresponding *m*-hydroxy- and *m*-methoxy-benzoic acids are stronger acids, thereby reflecting the greater electronegativity of the C—O bond compared with the C—N bond of the amino compounds.

Substituents *ortho* to the carbonyl group impede the co-planarity of the carbonyl group and the ring which is essential for maximum orbital overlap. Any departure from co-planarity, therefore, reduces the resonance contribution from the ring and its substituents, and thus increases the electron-withdrawing influence of the carbonyl group. This favours ionisation, so that acidic strength is increased.

The effect of increasing steric hindrance on acidic strength is seen in the two series:

| | Benzoic acid | 2-Methylbenzoic acid | 2,6-Dimethylbenzoic acid |
|---|---|---|---|
| $pK_a$ | 4.19 | 3.91 | 3.21 |

| | Benzoic acid | 2-Nitrobenzoic acid | 2,6-Dinitrobenzoic acid |
|---|---|---|---|
| $pK_a$ | 4.19 | 2.17 | 1.89 |

The marked effect of the *ortho*-hydroxyl group on acidic strength of salicylic acid ($pK_a$ 2.98) is due to hydrogen bonding between the phenoxyl hydrogen and the carbonyl oxygen, which assists co-planarity with the ring, but promotes release of the acidic proton. In the corresponding *o*-aminobenzoic acid (anthranilic acid; $pK_a$ 4.98), however, release of the acidic proton is inhibited by O—H---N hydrogen bonding; co-planarity is also enhanced and consequently acidic strength is decreased.

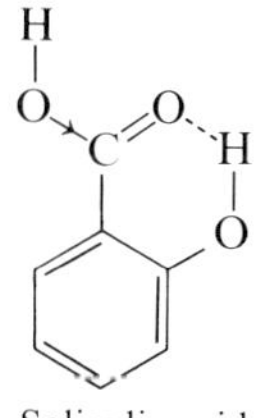

Salicylic acid

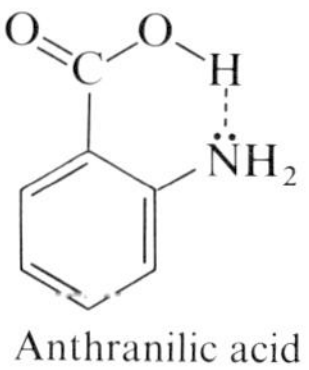

Anthranilic acid

**Salt Formation**

Neutralisation of carboxylic acids by inorganic and organic bases gives rise to salts. Alkali metal and alkaline earth metal salts of the lower aliphatic acids are soluble in water, but because they are salts of weak acids and strong bases they are not completely ionised in solution and are subject to hydrolysis. Aqueous solutions, therefore, have an alkaline reaction.

$$CH_3{\cdot}CO{\cdot}ONa + H_2O \rightleftharpoons CH_3{\cdot}CO{\cdot}OH + Na^+ + HO^-$$

None-the-less, water-soluble alkali metal salts provide a convenient method for obtaining appreciable concentrations of otherwise water-insoluble acids in aqueous media. Alkali metal salts of aromatic acids are also readily water-soluble. Thus, *Sodium Benzoate*, which is used as a preservative, is soluble 1 in 2 in water. *Potassium Benzoate* is also sufficiently soluble in water to form a useful means of administering potassium in a water-soluble form.

The carboxylate ion is stabilised as the resonance hybrid of the two extreme carbonyl forms. In accordance with the resonance concept, the two C—O bonds are of equal length (1.27Å) in the carboxylate ion.

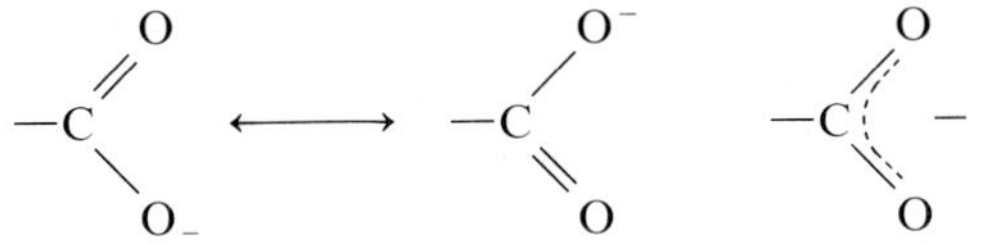

$$-C(=O)O^- \longleftrightarrow -C(O^-)=O \qquad -C(\overset{\cdot\cdot}{=}O)(\overset{\cdot\cdot}{=}O)\;^-$$

Canonical forms — Hybrid

Carboxylate ion

Carboxylic acids also form water-soluble salts with ammonia and organic bases.

$CH_3(CH_2)_2CO \cdot O^- \; \overset{+}{N}H_4$ — Ammonium butyrate

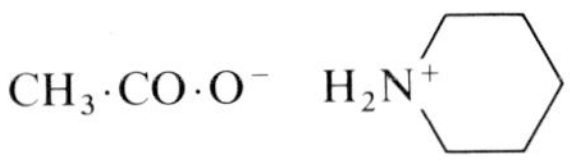

$CH_3 \cdot CO \cdot O^- \; H_2\overset{+}{N}C_5H_{10}$ (piperidinium) — Piperidine acetate

Ammonium butyrate is used as a constituent of butter flavouring in some pharmaceutical products. Because of the ease with which it is hydrolysed, however, care must be taken not to formulate it with inorganic bases such as *Aluminium Hydroxide Gel* or *Magnesium Carbonate*.

Such salts, particularly those with the more highly water-soluble and more strongly acidic hydroxyacids (citric and tartaric acids) provide a means of solubilising organic bases for injection. The use of di- and tri-basic acids also has the added advantage that acid salts, e.g. *Adrenaline Acid Tartrate* (Epinephrine Acid Tartrate) are not only readily water-soluble, but give rise to acidic solutions which may, as in this example, assist the maintenance of stability.

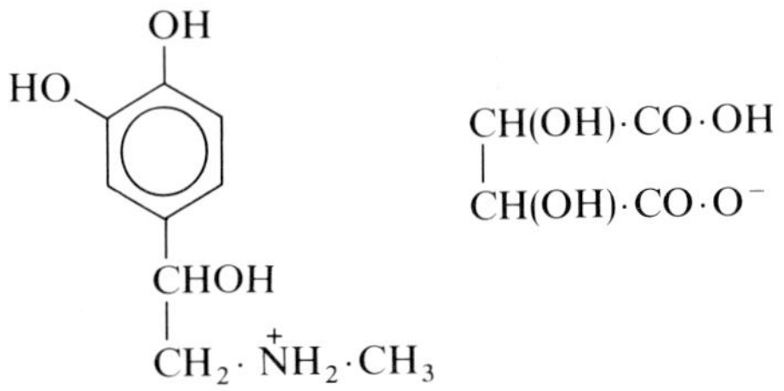

$$(HO)_2C_6H_3 \cdot CHOH \cdot CH_2 \cdot \overset{+}{N}H_2 \cdot CH_3 \qquad CH(OH) \cdot CO \cdot OH \,|\, CH(OH) \cdot CO \cdot O^-$$

*Adrenaline Acid Tartrate*

$$Ph \cdot CH_2 \cdot CO \cdot NH\text{-(penam: S, Me, Me, N, O, } CO \cdot O^-) \qquad H_2N\text{—}C_6H_4\text{—}CO \cdot O \cdot CH_2 \cdot CH_2 \cdot \overset{+}{N}HEt_2$$

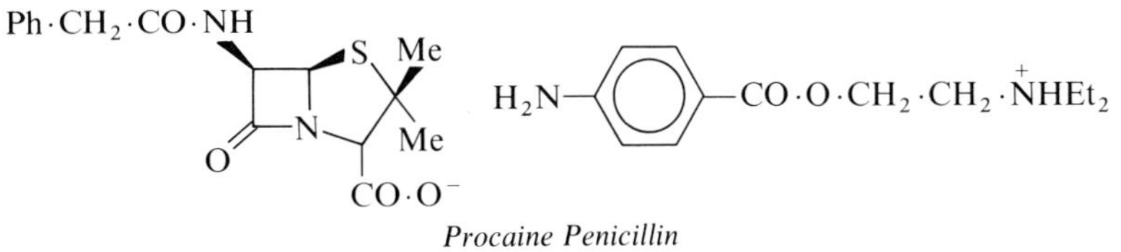

*Procaine Penicillin*

Other salts with organic bases, such as *Procaine Penicillin* (Penicillin G Procaine) and *Benzathine Penicillin* (Penicillin G Benzathine) owe their specific utility to their very low water solubility. When injected in aqueous suspension, they provide a depôt from which the active drug (which may be either the acid or base, or both) can slowly leach away to maintain therapeutic concentrations for considerably longer than a single injection of a more soluble salt.

*Chlorpromazine Embonate*, which is water-insoluble and almost tasteless, forms a useful alternative to the water-soluble, very bitter *Hydrochloride* in preparations for oral administration (in suspension).

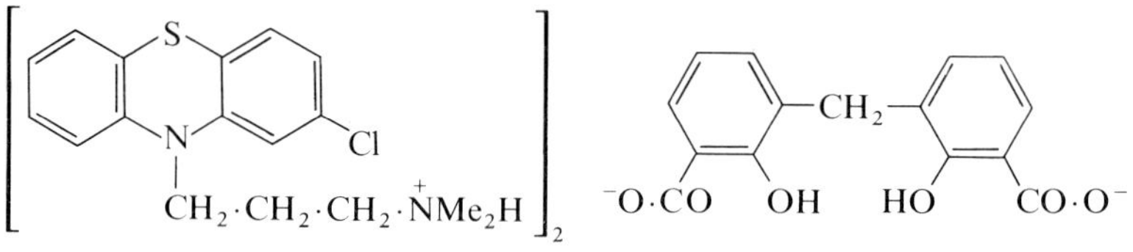

$$\left[ \text{(chlorophenothiazine)}\text{-N-}CH_2{\cdot}CH_2{\cdot}CH_2{\cdot}\overset{+}{N}Me_2H \right]_2 \quad {}^{-}O{\cdot}CO\text{-}C_{10}H_5(OH)\text{-}CH_2\text{-}C_{10}H_5(OH)\text{-}CO{\cdot}O^{-}$$

*Chlorpromazine Embonate*

Ammonium salts and salts of primary and secondary amines (but not tertiary amines) lose the elements of water when heated to form amides.

$$R{\cdot}CO{\cdot}ONH_4 \xrightarrow{\text{Heat}} R{\cdot}CO{\cdot}NH_2 + H_2O$$

Salt formation also plays an important part in the binding of carboxylic acids to carrier and receptor proteins (**2**, 4). Carboxylic acids, however, are only capable of diffusing across biological membranes in their undissociated forms. The dissociation constant of the acid, and pH of the environment on either side of the membrane are, therefore, important factors in determining the distribution and excretion of acids in animal systems (**2**, 4).

## Soaps

The alkali metal salts of long-chain fatty acids have rather poor water solubility. The long fatty hydrocarbon chain attached to the hydrophilic carboxylate ion, however, enables them to concentrate at oil–water interfaces, and so reduce interfacial tension. In consequence, they are widely used as soaps for cleansing. Soaps are prepared from glycerol esters of fatty acids by saponification of natural oils and fats with either sodium or potassium hydroxide. When saponification is complete, the soap is salted out from the resulting solution with common salt, washed free from salt, pressed free from water to form a solid cake, and dried.

$$\begin{array}{l} CH_2{\cdot}O{\cdot}CO{\cdot}R^1 \\ | \\ CH{\cdot}O{\cdot}CO{\cdot}R^2 \\ | \\ CH_2{\cdot}O{\cdot}CO{\cdot}R^3 \end{array} + 3NaOH \longrightarrow \begin{array}{l} CH_2OH \\ | \\ CHOH \\ | \\ CH_2OH \end{array} + \begin{array}{l} R^1{\cdot}CO{\cdot}ONa \\ R^2{\cdot}CO{\cdot}ONa \\ R^3{\cdot}CO{\cdot}ONa \end{array}$$

In the solid state, soaps have valuable lubricating properties. Thus, calcium and magnesium stearates are waxy solids; both are used extensively as lubricants for addition to tablet granules prior to compression to aid ejection of the tablet from the machine. In cases where an alkaline lubricant would react with the medicament, stearic acid itself is used as the lubricant.

The characteristic properties of soaps in solution are shown by the alkali metal salts of fatty acids with chain lengths from 6–22 carbon atoms. The salts of longer chain acids are so sparingly soluble in water that little or no detergent action can be observed. Aqueous solutions of soaps differ from those of typical electrolytes in a number of important respects. They reduce the surface tension from about 0.072 N $m^{-1}$ (72 dyne $cm^{-1}$) of pure water down to values of the order of 0.025 N $m^{-1}$, and it is this property which is mainly responsible for the lathering power of the solution. The viscosity of soap solutions increases rapidly with concentration, so that solutions varying in viscosity from watery fluids to stiff gels are readily obtained. Their osmotic properties are anomalous, and the electrical conductivity is greater than that expected of simple ionised molecules. These phenomena are explained in terms of the formation of **ionic micelles**, in which large numbers of fatty acid anions are aggregated together (Fig. 19). Such micelles are extensively hydrated, and have high viscosities and high mobilities (high conductivity), the latter on account of the large number of electrical charges which they carry. Micellation of soaps and detergents provides a means of effectively solubilising a number of otherwise water-insoluble substances. Soaps, however, suffer from the disadvantage that their water solubility is comparatively low (*ca* 1 in 20). Also, they are decomposed by acid, which precipitates

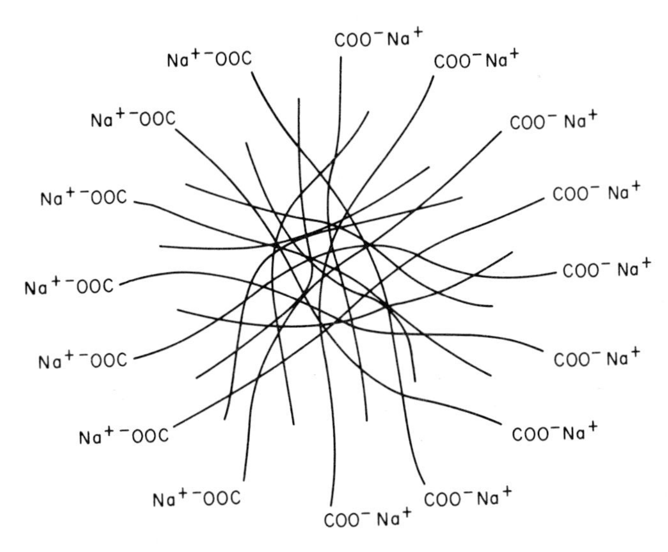


Fig. 19 Cross-sectional diagram showing the arrangement of soap molecules in a typical micelle

the free fatty acids, and form insoluble calcium and magnesium salts in hard water.

$$CH_3(CH_2)_n\cdot CO\cdot O^-Na^+ + HX \longrightarrow CH_3(CH_2)_n\cdot CO\cdot OH\downarrow + NaX$$

$$2CH_3(CH_2)_n\cdot CO\cdot O^- Na^+ + CaCl_2 \longrightarrow [CH_3(CH_2)_n\cdot CO\cdot O]_2Ca\downarrow + 2NaCl$$

Soft soap is a jelly-like product obtained by concentrating the total hydrolysate from the saponification of natural oils with potassium hydroxide. Glycerol, which is also formed in the saponification, is retained in the product, which in consequence is soft and unctuous. Chlorophyll or other green colouring matter is often added to improve the appearance. Soft soap is quite soluble in water (*ca* 1 in 4) and solutions in water and surgical spirit are used both for cleansing the skin, and also either alone or in admixture with certain medicaments, in liniments.

## NUCLEOPHILIC ATTACK AT THE CARBOXYLIC CARBONYL

Carboxylic acids and esters undergo nucleophilic attack at the carbonyl carbon atom with a wide variety of reagents. The properties of the carboxyl carbonyl group, however, are modified by the resonance which is responsible for acidity (p. 264). In consequence, carboxylic acids are much less reactive in nucleophilic additions to carbonyl than aldehydes and ketones. The majority of these reactions result in displacement of hydroxyl; these include the formation of esters, acid chlorides and acid anhydrides, and also hydride ion reductions.

### Ester Formation

Carboxylic acids react with alcohols when they are heated together to form esters.

$$\underset{\text{Acetic acid}}{CH_3\cdot CO\cdot OH} + \underset{\text{Ethanol}}{CH_3\cdot CH_2\cdot OH} \rightleftharpoons \underset{\text{Ethyl Acetate}}{CH_3\cdot CO\cdot O\cdot CH_2} + H_2O$$

The reaction is an equilibrium one, and does not normally proceed to completion unless one of the reactants is used in excess. Alternatively, water may be removed as the reaction proceeds, by azeotropic distillation with benzene, by the addition of a dehydrating agent such as sulphuric acid, or by the use of molecular sieves. Sulphuric acid and other acids, such as hydrogen chloride and boron trifluoride, have the added advantage that they catalyse the reaction by protonating the carbonyl group, so assisting nucleophilic attack and displacement of HO by the attacking alcohol.

$$CH_3\cdot C(=O)OH \underset{}{\overset{H^+}{\rightleftharpoons}} CH_3\cdot \overset{+}{C}(OH)OH \;\; + \;\; H\text{–}\mathbf{O}\text{–}CH_2\cdot CH_3$$

$$CH_3\cdot C(OH)(\ddot{O}H)\text{–}\mathbf{O}^{+}(H)CH_2\cdot CH_3 \;\; H^+ \longrightarrow CH_3\cdot C(OH)(\overset{+}{O}H_2)\text{–}\mathbf{O}\text{–}CH_2\cdot CH_3$$

$$\xrightarrow{-H_2O} CH_3\cdot \overset{+}{C}(O\text{–}H)\text{–}\mathbf{O}\cdot CH_2\cdot CH_3 \xrightarrow{-H^+} CH_3\cdot C(=O)\text{–}\mathbf{O}\cdot CH_2\cdot CH_3$$

It has been shown by experiment with an $^{18}O$-labelled alcohol that it is the acyl—oxygen bond (and not the alkyl—oxygen bond of the alcohol) which is broken in esterification, since the isotope becomes incorporated in the product ester.

$$R^1\cdot CO\cdot OH + R\cdot {}^{18}OH \rightleftharpoons R^1\cdot CO\cdot {}^{18}OR + H_2O$$

The carbonyl carbon, which is trigonal in the carboxylic acid ($sp^2$ hybridisation) and the corresponding ester, becomes tetrahedral ($sp^3$ hybridisation) in the intermediate addition complex. Substituents which impede approach of the esterifying alcohol inhibit the reaction, and in consequence the rate of esterification is sharply decreased by alkyl substituents adjacent to the carbonyl group (Loening, Garrett and Newman, 1952; Table 29). This effect, however, is greater in the acid-catalysed esterification of $\beta$-substituted aliphatic acids than in that of their $\alpha$-substituted analogues.

Esterification of aromatic acids is similarly hindered sterically by the presence of bulky substituents, irrespective of whether they are electron-donating or electron-repelling. Thus, both 2,6-dimethoxybenzoic acid and 2,6-dibromobenzoic acid are more difficult than benzoic acid to esterify directly with methanolic hydrogen chloride, a reagent which reacts readily with benzoic acid. Both

Table 29. Relative Rates of Esterification of Aliphatic Acids by Methanol

| Acid | Rate of Esterification |
|---|---|
| $CH_3 \cdot COOH$ | 1.0 |
| $CH_3 \cdot CH_2 \cdot COOH$ | 0.51 |
| $(CH_3)_3C \cdot COOH$ | 0.037 |

acids, however, are readily esterified with diazomethane ($CH_2N_2$) in ether, since this reagent is less bulky than methanol, and it is easier for it to approach the reactive centre, which in this case is the much less hindered acidic hydroxyl and not the carbonyl carbon as in the esterification with methanol/HCl.

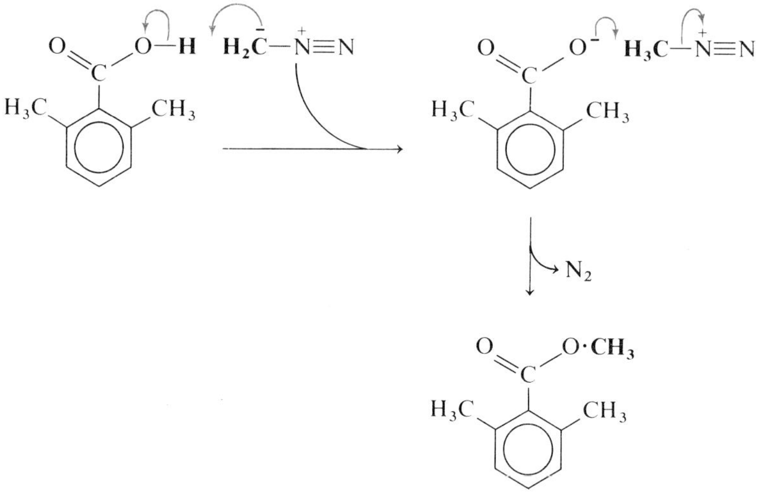

Although esterification of acids with diazomethane is effected in ether, it is the anion which is reactive, and the reaction may on occasion be promoted by the addition of a drop of water to assist ionisation which would otherwise be severely repressed.

**Acid Chloride Formation**

Carboxylic acids react with phosphorus halides, thionyl chloride or oxalyl chloride to form acid chlorides by an overall displacement of hydroxyl. The choice of reagent depends to some extent on the volatility of the acid chloride, since the only satisfactory method of isolating the product acid chloride is by distillation. Phosphorus trichloride is used in the preparation of low boiling acid chlorides, such as acetyl chloride (b.p. 52°C), since the by-product, phosphorous acid, is non-volatile.

$$3CH_3 \cdot CO \cdot OH + PCl_3 \longrightarrow 3CH_3 \cdot CO \cdot Cl\uparrow + H_3PO_3\downarrow$$

High boiling acid chlorides, such as benzoyl chloride (b.p. 197°C), are best prepared using phosphorus pentachloride, since the lower boiling phosphorus oxychloride (b.p. 107°C) is readily separated by fractional distillation.

$$C_6H_5 \cdot CO \cdot OH + PCl_5 \longrightarrow C_6H_5 \cdot CO \cdot Cl + POCl_3 + HCl$$

The reactions proceed in two stages, first by formation of a mixed anhydride, which then undergoes nucleophilic attack at the carbonyl carbon by chloride ions formed in the initial displacement.

$$C_6H_5 \cdot C(=O)-O-H + PCl_5 \longrightarrow H^+ + Cl^- + C_6H_5 \cdot C(=O)-O-PCl_4$$

$$C_6H_5 \cdot C(O^-)(Cl)-O-PCl_4 \longrightarrow Cl^- + POCl_3 + C_6H_5 \cdot C(=O)Cl$$

The advantage of thionyl chloride as a reagent for the preparation of acid chlorides is that both by-products, sulphur dioxide and hydrogen chloride, are volatile, and hence readily separate from the product acid chloride.

$$R \cdot CO \cdot OH + SOCl_2 \longrightarrow R \cdot CO \cdot Cl + SO_2\uparrow + HCl\uparrow$$

The mechanism of the reaction is similar to that with the phosphorus halides, with the initial formation of a mixed anhydride and subsequent attack and displacement by chloride ions.

$$R \cdot C(=O)-O-H + Cl_2S=O \longrightarrow H^+ + Cl^- + R \cdot C(=O)-O-S(=O)Cl$$

$$R \cdot C(O^-)(Cl)-O-S(=O)Cl \longrightarrow Cl^- + SO_2 + R \cdot C(=O)Cl$$

## Acid Anhydride Formation

Formic acid forms neither acid chloride nor acid anhydride. Dehydration leads to formation of carbon monoxide.

$$H{\cdot}CO{\cdot}OH \longrightarrow CO + H_2O$$

Other acid anhydrides are formed by nucleophilic attack of the carboxylate ion at the carbonyl carbon of an acid chloride, when the sodium salt of the acid is heated with an acid chloride.

$$CH_3{\cdot}CO{\cdot}\mathbf{O}^- + R{\cdot}C(Cl){=}O \xrightarrow[Cl^-]{} CH_3{\cdot}CO{\cdot}\mathbf{O}{\cdot}\mathbf{CO}{\cdot}R$$

The reaction is facilitated by the presence of pyridine, quinoline or other tertiary base. A salt is formed in which the electron-deficiency of the carbonyl carbon is increased, thereby promoting attack by the carboxylate ion.

$$C_5H_5N: + R{\cdot}C(Cl){=}O \longrightarrow \left[ C_5H_5\overset{+}{N}{\leftarrow}C(R){=}O \right] Cl^- \quad (\text{attacked by } R'{\cdot}CO{\cdot}\mathbf{O}^-)$$

$$\xrightarrow[Cl^-]{} C_5H_5\overset{+}{N}{\leftarrow}C(R)(O{\cdot}CO{\cdot}R'){-}O^- \xrightarrow[C_5H_5N]{} R'{\cdot}CO{\cdot}\mathbf{O}{\cdot}\mathbf{CO}{\cdot}R$$

Higher molecular weight anhydrides can be prepared by refluxing the carboxylic acid in acetic anhydride or acetyl chloride, when the acetic acid which is formed functions as a solvent from which the anhydride crystallises.

$$\begin{array}{l} CH_2{\cdot}CO{\cdot}\mathbf{OH} \\ | \\ CH_2{\cdot}CO{\cdot}\mathbf{OH} \end{array} + CH_3{\cdot}CO{\cdot}Cl \longrightarrow \begin{array}{l} CH_2{\cdot}CO \\ | \qquad\quad \diagdown \\ | \qquad\qquad \mathbf{O} \\ | \qquad\quad \diagup \\ CH_2{\cdot}CO \end{array} + CH_3{\cdot}CO{\cdot}\mathbf{OH} + \mathbf{H}Cl$$

Succinic anhydride

## Mixed Anhydrides

Mixed (acetic) anhydrides are formed by the action of ketene (*see* pyrolysis of acetone, Chapter 10) on the carboxylic acid. Formic–acetic anhydride can be prepared by this method, despite the fact that the simple formic anhydride is unknown.

$$\mathrm{R{\cdot}CO{\cdot}OH + H_2C{=}C{=}O \longrightarrow R{\cdot}CO{\cdot}O{\cdot}CO{\cdot}CH_3}$$

These mixed anhydrides disproportionate when heated to the two simple anhydrides which can be separated by distillation.

$$\mathrm{R{\cdot}CO{\cdot}O{\cdot}CO{\cdot}CH_3 \longrightarrow (R{\cdot}CO)_2O + (CH_3{\cdot}CO)_2O}$$

Certain mixed carboxylic phosphoric anhydrides, such as acetyl phosphate and acetyl adenylate, are important energy transducers in intermediary metabolism. Acetyl phosphate is formed enzymically by transfer of an acetyl residue from hydroxyethylthiaminepyrophosphate (hydroxyethyl TPP), an intermediate in the oxidative decarboxylation of pyruvate, to inorganic phosphate, either directly or via acetyl-$\overline{\mathrm{CoA}}$ ($\overline{\mathrm{CoA}}\cdot$SH).

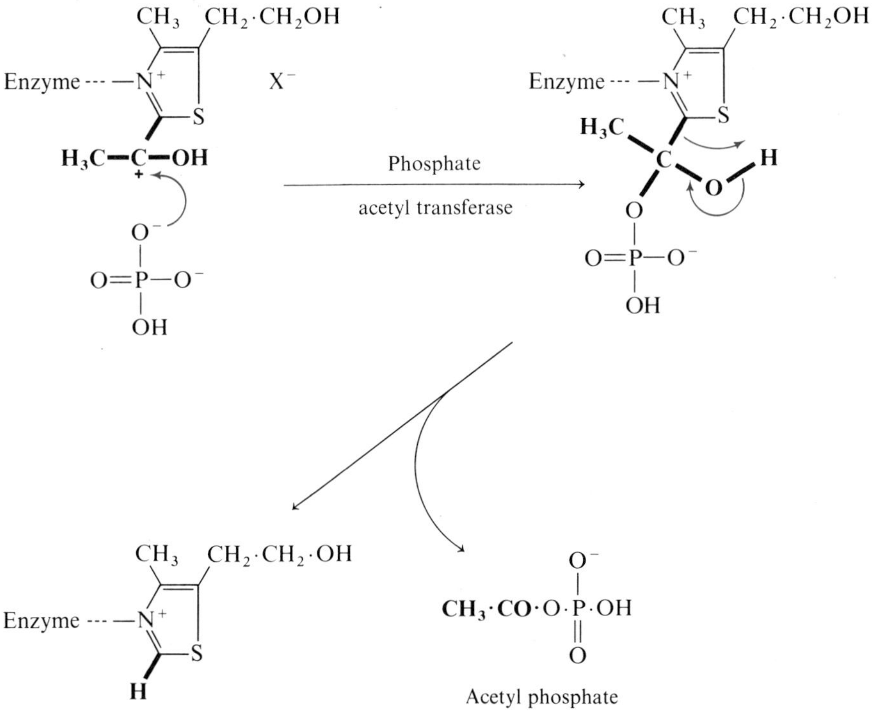

$CH_3$ $CH_2{\cdot}CH_2OH$
Enzyme --- $N^+$ S $X^-$
$H_3C-\overset{+}{C}-OH$
$O^-$
$O{=}P{-}O^-$
OH

Phosphate
acetyl transferase

$CH_3$ $CH_2{\cdot}CH_2OH$
Enzyme --- $N^+$ S
$H_3C$ C O H
O
$O{=}P{-}O^-$
OH

$CH_3$ $CH_2{\cdot}CH_2{\cdot}OH$
Enzyme --- $N^+$ S
H

$$\mathrm{CH_3{\cdot}CO{\cdot}O{\cdot}\overset{\overset{\Large O^-}{|}}{\underset{\underset{\Large O}{||}}{P}}{\cdot}OH}$$

Acetyl phosphate

## Hydride Ion Reduction of Acids

The carboxylic acid group is, in general, resistant to oxidation and reduction. Catalytic reduction is ineffective, but carboxylic acids can be reduced to the corresponding primary alcohol by lithium aluminium hydride. The reagent acts as a source of hydride ions, and the reaction appears to proceed essentially by nucleophilic displacement of the alkoxide function to give the corresponding aldehyde, which is then further reduced to the primary alcohol (Chapter 10).

$$R\cdot C(=O)-O-H \xrightarrow{Li^+(AlH_4)H^-} H_2 + R\cdot C(=O)OLi(AlH_3) \xrightarrow{Li^+(AlH_3)H^-} R\cdot C(=O)H + {}^-OLi(AlH_3)$$

The reaction is sometimes slow, since the ease with which it takes place depends upon the solubility of the lithium salt of the acid in the reaction solvent (dry ether, dioxan or tetrahydrofuran). It is, therefore, usually preferable to carry out the reduction by first converting the acid to its ester.

## RADICAL REACTIONS OF CARBOXYLIC ACIDS

### Hydrogen Abstraction

Carboxylic acids are subject to radical attack leading to abstraction of hydrogen from the hydroxyl group, followed by reaction of the radical so formed. A radical reaction is involved when metals, such as zinc, dissolve in glacial acetic acid in the presence of a suitable hydrogen acceptor.

$$2CH_3\cdot CO\cdot O-H + \dot{Z}n \longrightarrow 2CH_3\cdot CO\cdot O^- + Zn^{2+} + 2H\cdot$$

Hydrogen abstraction from the α-carbon also occurs readily with radical generating reagents, such as acetyl peroxide, leading to dimerisation (Kharasch and Gladstone, 1943).

$$R\cdot CH_2\cdot CO\cdot OH \xrightarrow{CH_3\cdot CO\cdot O\cdot \;\to\; CH_3\cdot CO\cdot OH} R\cdot \dot{C}H\cdot CO\cdot OH \longrightarrow \begin{matrix} R\cdot CH\cdot CO\cdot OH \\ | \\ R\cdot CH\cdot CO\cdot OH \end{matrix}$$

α,α′-Dialkylsuccinic acid

### Decarboxylation

The Kolbé electrolysis of alkali metal salts of carboxylic acids already discussed (Chapter 3) is an example of an electrolytically-induced radical reaction leading to decarboxylation and hydrocarbon formation. The silver salts of carboxylic acids also decarboxylate in the presence of bromine to give an alkyl bromide in the Hunsdiecker reaction, for which a radical mechanism is favoured.

$$R{\cdot}CO{\cdot}O{-}Ag \quad Br{-}Br \longrightarrow R{\cdot}C(=O)O{\cdot} + AgBr + Br{\cdot}$$

$$R{-}C(=O)O{\cdot} \longrightarrow R{\cdot} + CO_2$$

$$R{\cdot} + Br{\cdot} \longrightarrow R{-}Br$$

## IONIC DECARBOXYLATION

### Thermal Decarboxylation

Thermal decarboxylation of carboxylic acids can generally be achieved by heating the acid strongly with copper powder and quinoline (or aniline). The mechanism is uncertain, but probably ionic. Quinoline acts both as a high boiling solvent, and also assists ionisation of the acid. The rôle of the copper is also uncertain, but probably also induces ionisation by complex formation.

$$R{-}C(=O){-}O{-}H \quad C_7H_7\ddot{N} \xrightarrow{Cu} R{-}C(=O)O^- \quad C_7H_7\overset{+}{N}{-}H \longrightarrow RH + CO_2 \quad (C_7H_7N)$$

Thermal decarboxylation of carboxylic acids occurs more readily in acids with strong electron-attracting $\alpha$-substituents, such as cyanacetic acid and trichloroacetic acid, both of which decarboxylate in the presence of base at temperatures around 100–150°C. Malonic acid decarboxylates at this temperature even in the presence of base.

$$Cl_3C{-}C(=O)O^- \xrightarrow{-CO_2} Cl_3C^- \xrightarrow{H^+} Cl_3{\cdot}CH$$

### Enzymic Decarboxylation

There is evidence also that the enzymic decarboxylation of mevalonic acid pyrophosphate occurs in an analogous manner, possibly through an intermediate *O*-phosphate (Bloch, Chaykin, Phillips and De Waard, 1959).

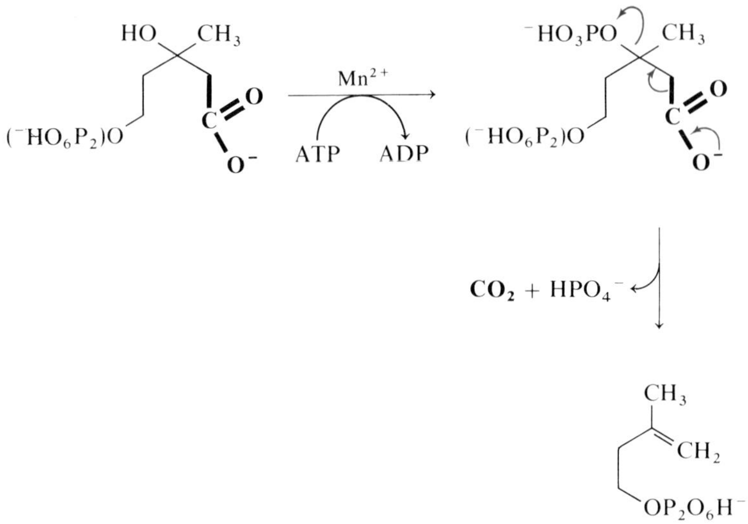

HO CH$_3$ ($^-$HO$_6$P$_2$)O C=O O$^-$ —$Mn^{2+}$ (ATP → ADP)→ $^-$HO$_3$PO CH$_3$ ($^-$HO$_6$P$_2$)O C=O O$^-$

$CO_2$ + $HPO_4^-$

CH$_3$ CH$_2$ OP$_2$O$_6$H$^-$

Isopentyl pyrophosphate

## ESTERS

### General Properties

Most esters are mobile, oily liquids, neutral in reaction, and almost insoluble in water. They are miscible with ethanol, ether and chloroform, and are good solvents for a wide variety of organic compounds. In general, esters are more lipophilic than the acids and alcohols from which they are derived. Esterification is, therefore, often used as a means of converting high molecular weight alcohols and phenols to derivatives capable of accumulation in, and delayed release from, body fats. The lower members are readily volatile; they have distinctive fruity odours, and are used as constituents of artificial flavours. Similarly, the taste of high molecular weight esters is frequently quite bland compared with the parent acid or alcohol. Thus, *Chloramphenicol Palmitate* is virtually free from the persistent bitter taste which characterises the parent antibiotic.

### Fixed Oils, Fats and Waxes

Chemically, oils and fats are related, being naturally occurring glycerol esters of the higher saturated and unsaturated aliphatic monocarboxylic acids. There is no clear-cut distinction between oils and fats, but oils are fluid at room temperature, whereas fats are solid or semi-solid. Waxes differ chemically from oils and fats in that the acids are esterified with high molecular weight aliphatic alcohols or sterols. In appearance they are hard, brittle, non-greasy solids.

All natural oils and fats consist of complex mixtures of triglycerides, though on hydrolysis (saponification; p. 272), they may be shown to be derived from a comparatively small number of fatty acids. The physical properties of the oil or

$$\begin{array}{l} CH_2O\cdot CO\cdot R^1 \\ | \\ CHO\cdot CO\cdot R^2 \\ \\ CH_2O\cdot CO\cdot R^3 \end{array}$$

Oils and fats (glycerol esters)

$$CH_3(CH_2)_nO\cdot CO(CH_2)_m\cdot CH_3$$

Waxes

fat depends largely on the composition of the glyceride mixture, and two substances (e.g. cocoa butter and mutton tallow) may have similar fatty acid content, but widely differing physical properties. Most fats, however, contain a much higher proportion of saturated fatty acids, and are frequently of animal origin. Oils, on the other hand, are largely of vegetable origin and contain a higher proportion of esters of unsaturated acids.

Oils and fats form useful vehicles for a wide variety of organic compounds. Oils provide useful non-aqueous media for the injection of oil-soluble steroids, such as *Oestradiol Benzoate* (Estradiol Monobenzoate), an ester which is administered intramuscularly to provide depôt medication, from which the medicament is slowly released. The general stability of oils to moderate heat permits heat sterilisation of the injection solution at a temperature of 150°C. At this temperature, there is little evidence of the pyrolytic decomposition to olefines which is liable to occur at higher temperatures (500–600°C). The oil, being an ester, is dispersed and enzymically hydrolysed by the normal fat-metabolising enzymes of the body, and hence to all intents and purposes is innocuous. Animal fats also have similar advantages over semi-solid paraffin hydrocarbons as vehicles for ointments, since they are readily miscible with the natural oils of the skin. They are, therefore, readily absorbed, so aiding absorption of admixed medicaments if this is required. The simple ester, isopropyl myristate, also exhibits a balance of hydrophilic and hydrophobic properties which resemble those of skin itself, and for this reason is a useful constituent of creams for topical application.

**Phospholipids**

Many fatty acids are widely distributed in mammalian tissue as phospholipids. These consist of the phosphatidylcholines, also termed lecithins, and the phosphatidylethanolamines and phosphatidylserines or cephalins. The parent α-glycerophosphoric acid is asymmetric at the β-carbon atom, and naturally occurring phospholipids are derived from α-L-glycerophosphoric acid. Most of the phosphatidylcholines are esterified with a saturated fatty acid in the α′-position and an unsaturated fatty acid in the β-position, but some contain two saturated or two unsaturated acids. They form important constituents of fatty tissue and are key structural components of biological membranes.

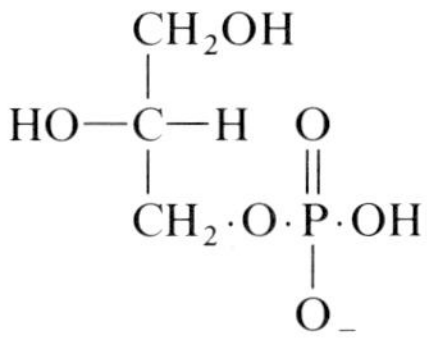

$CH_2OH \cdot CH(OH) \cdot CH_2 \cdot O \cdot PO(OH) \cdot O^-$

α-L-Glycerophosphoric acid

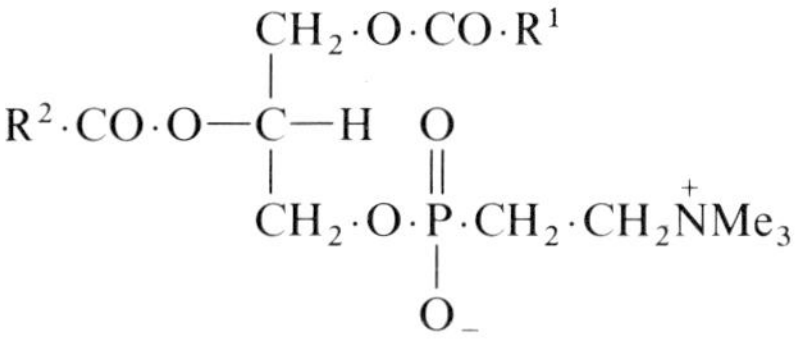

$R^1 \cdot CO \cdot O \cdot CH_2 \cdot CH(O \cdot CO \cdot R^2) \cdot CH_2 \cdot O \cdot PO(O^-) \cdot CH_2 \cdot CH_2\overset{+}{N}Me_3$

Phosphatidylcholine

$R^1 \cdot CO \cdot O \cdot CH_2 \cdot CH(O \cdot CO \cdot R^2) \cdot CH_2 \cdot O \cdot PO(O^-) \cdot O \cdot CH_2 \cdot CH_2\overset{+}{N}H_3$

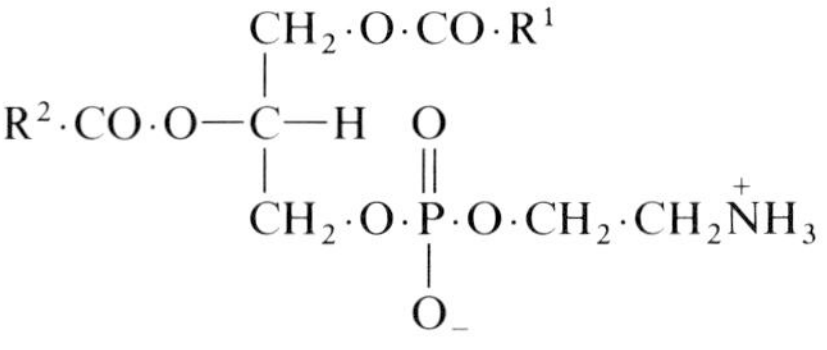

Phosphatidylethanolamine

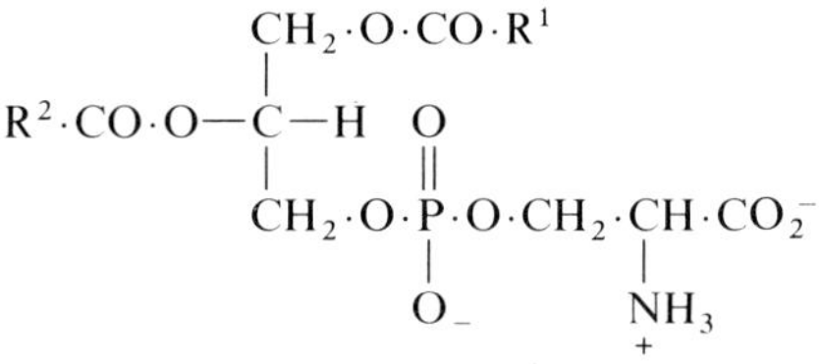

$R^1 \cdot CO \cdot O \cdot CH_2 \cdot CH(O \cdot CO \cdot R^2) \cdot CH_2 \cdot O \cdot PO(O^-) \cdot O \cdot CH_2 \cdot CH(\overset{+}{N}H_3) \cdot CO_2^-$

Phosphatidylserine

**Acid-catalysed Hydrolysis**

Esters may be cleaved hydrolytically into their constituent acids and alcohols. This is a reversal of the equilibrium reaction whereby they are formed (p. 274). Just as the forward reaction can be favoured by using either acid or alcohol in

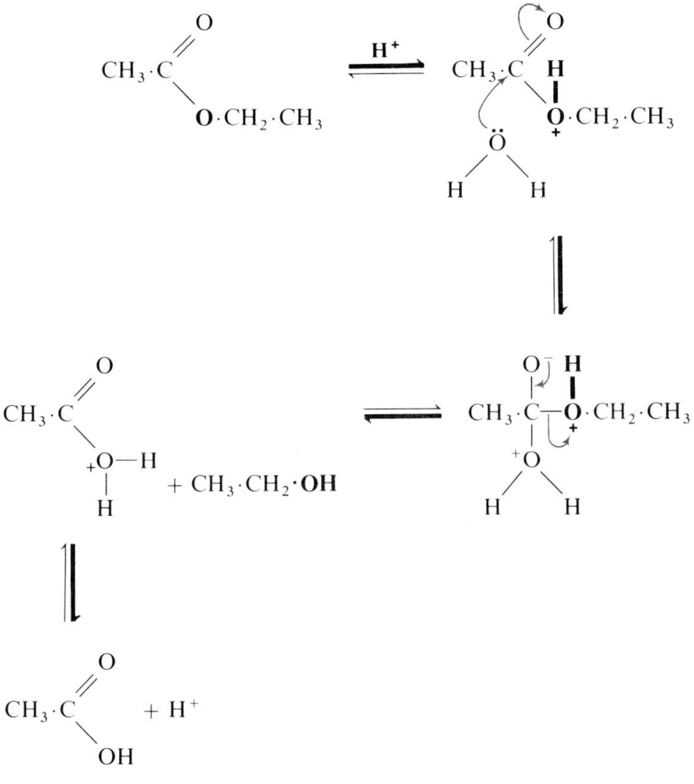

$$CH_3 \cdot CO \cdot O \cdot CH_2 \cdot CH_3 \overset{H^+}{\rightleftharpoons} CH_3 \cdot CO \cdot \overset{+}{O}H \cdot CH_2 \cdot CH_3 \; (+ H_2\ddot{O})$$

$$\rightleftharpoons CH_3 \cdot C(O^-)(\overset{+}{O}H_2) \cdot \overset{+}{O}H \cdot CH_2 \cdot CH_3$$

$$\rightleftharpoons CH_3 \cdot CO \cdot \overset{+}{O}H_2 + CH_3 \cdot CH_2 \cdot \mathbf{OH}$$

$$\rightleftharpoons CH_3 \cdot CO \cdot OH + H^+$$

excess, so hydrolysis is favoured when water is present in excess. The acid-catalysed aqueous hydrolysis of esters is, therefore, the reverse of acid-catalysed esterification. The first step is protonation on oxygen. The positively charged oxygen acts as an electron sink, and hence increases the normal electron-deficiency of the carbonyl carbon, thereby promoting nucleophilic attack by the water molecules which are present in excess.

### Base-catalysed Hydrolysis

Hydrolysis also occurs readily in alkaline solution, with sodium or potassium hydroxide. The base removes the acid from the reaction by salt formation, so that in contrast to acid-catalysed hydrolysis, the base-catalysed reaction is irreversible. The reaction is usually carried out in methanol or ethanol under reflux, since ethanol is a good solvent for both the ester and potassium hydroxide.

$$CH_3{\cdot}CO{\cdot}O{\cdot}CH_2{\cdot}CH_3 + KOH \longrightarrow CH_3{\cdot}CO{\cdot}OK + CH_2{\cdot}CH_2OH$$

The hydrolysis is initiated by nucleophilic attack of $HO^-$ on the electron-deficient carbonyl carbon atom. The intermediate so formed may then either revert to the original ester by the loss of $HO^-$ or alternatively lose $CH_3CH_2O^-$ to form the carboxylic acid. Since the latter is immediately removed from the reaction by salt formation, the equilibrium is disturbed and reaction then proceeds to completion in favour of hydrolysis.

$$CH_3{\cdot}C(=O){\cdot}O{\cdot}CH_2{\cdot}CH_3 + HO^- \rightleftharpoons CH_3{\cdot}C(O^-)(OH){-}O{\cdot}CH_2{\cdot}CH_3$$

$$CH_3{\cdot}C(=O){\cdot}OH + CH_3{\cdot}CH_2O^- \longrightarrow CH_3{\cdot}C(=O){\cdot}O^- + CH_3{\cdot}CH_2{\cdot}OH$$

### Enzymic Hydrolysis

The hydrolysis of esters in the animal body is brought about by enzymes known as esterases. These include the pancreatic lipases, which are relatively specific for natural fats, acetylcholinesterase of nervous tissue and red blood cells, which destroys acetylcholine, and the relatively non-specific esterases of plasma, liver, skin and intestinal tissue. Proteolytic enzymes, such as chymotrypsin, also catalyse the hydrolysis of esters as well as peptides.

Certain esterases, notably acetylcholinesterase, are able to function by a combination of acid and base catalysis in a double displacement reaction. The primary nucleophile in the enzymatic hydrolysis of acetylcholine has been identified as a serine hydroxyl group (**2**, 3). This is acetylated in the first stage of the reaction with release of the alcohol, choline. The acetylated enzyme is then hydrolysed in a second displacement reaction with liberation of the enzyme (E—OH) and acetic acid.

$$CH_3{\cdot}CO{\cdot}O{\cdot}CH_2{\cdot}CH_2{\cdot}\overset{+}{N}Me_3 + E{-}OH$$

Acetylcholine Enzyme

$$\rightleftharpoons$$

$$E{-}O{\cdot}CO{\cdot}CH_3 + HO{\cdot}CH_2{\cdot}CH_2{\cdot}\overset{+}{N}Me_3$$

Acetylated enzyme Choline

$$E{-}O{\cdot}CO{\cdot}CH_3 + H_2O \rightleftharpoons E{-}OH + CH_3CO{\cdot}OH$$

pH-dependence studies (Krupka, 1966a and b; 1967) of the hydrolysis of acetylcholine by bovine erythrocyte cholinesterase have established that two

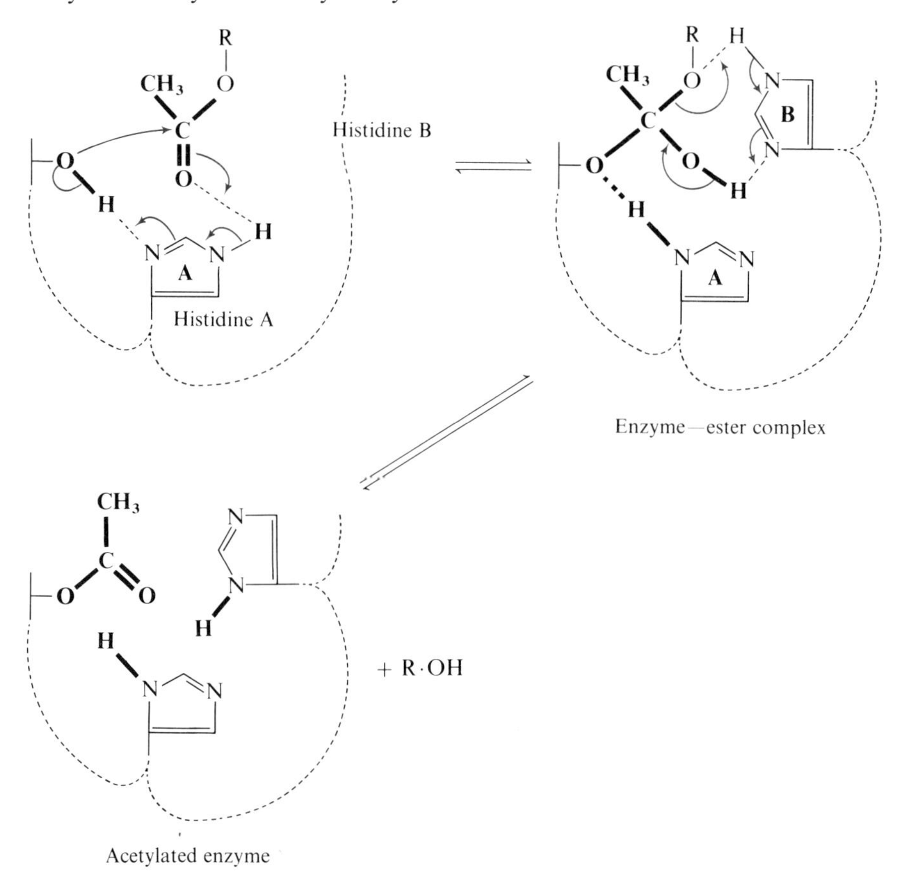

basic groups of p$K_a$ 5.5 and 6.3, present at the active centre of the enzyme, catalyse these displacements when in the unprotonated form. Both are considered to be histidine residues, and both act by the simultaneous acceptance and donation of a proton (Brestkin and Rozengart, 1965; Bender and Kézdy, 1965). One histidine group is adjacent to the serine residue and initiates the reaction. The second group is brought into the reaction by a conformational change in the shape of the enzyme–ester addition complex, and completes the acetylation stage of the reaction.

Hydrolysis of the acetylated enzyme probably proceeds by the reverse of this process, with attack by a water molecule (ROH = $H_2O$).

**Plasma Esterases and Biological Stability of Ester-type Medicaments**

The enzymatic hydrolysis of esters, particularly by plasma esterases, plays an important part in determining the biological stability and biological half-life of a number of ester-based drugs. It provides the basis for the limited duration of action of the muscle relaxant, *Suxamethonium Bromide* (Succinylcholine Bromide), which is enzymatically hydrolysed in two stages to products with negligible motor-end-plate activity (Whittaker and Wijesundera, 1952; Tsuji and Foldes, 1953). Metabolism of ester-based local anaesthetics such as *Procaine* is also effected in human subjects primarily by plasma esterases (Kalow, 1952; **2**, 5).

$$Me_3\overset{+}{N}\cdot CH_2\cdot CH_2\cdot O\cdot CO\cdot CH_2\cdot CH_2\cdot CO\cdot O\cdot CH_2\cdot CH_2\cdot \overset{+}{N}Me_3 \quad 2Br^-$$

$$\downarrow \rightarrow HO\cdot CH_2\cdot CH_2\cdot \overset{+}{N}Me_3 \quad Br^-$$

$$Me_3\overset{+}{N}\cdot CH_2\cdot CH_2\cdot O\cdot CO\cdot CH_2\cdot CH_2\cdot CO\cdot OH \quad Br^-$$

$$\downarrow$$

$$Me_3\overset{+}{N}\cdot CH_2\cdot CH_2OH \quad Br^- + HO\cdot OC\cdot CH_2\cdot CH_2\cdot CO\cdot OH$$

Delayed release of oestradiol is achieved by depôt administration of *Oestradiol Benzoate* (Estradiol Monobenzoate) which is subject to enzymatic hydrolysis to the parent phenol. The fat solubility of the hydroxylated steroid *Hydrocortisone* is markedly increased in *Hydrocortisone Acetate*. This affects tissue distribution after injection and permits slow release of the parent hormone by enzymatic hydrolysis.

*Cellulose Acetate Phthalate* (Cellacephate) is a partially-acetylated cellulose derivative, obtained by treating a partly-acetylated cellulose with phthalic anhydride. It is used for the enteric film coating of tablets. The ester is relatively impervious to acid, but soluble in alkali due to the free carboxyl groups and hydrolysed by pancreatic enzymes. Tablets film-coated with solutions of the ester, therefore, pass through the stomach unchanged, but disintegrate with hydrolysis of the film in the alkaline conditions of the intestines.

$R\cdot CH_2$ … $]_n$ R=HO—, $CH_3\cdot CO\cdot O$— or $C_6H_4(CO\cdot O—)(CO\cdot OH)$

## Electronic and Steric Factors Affecting the Hydrolysis of Esters

(a) **Base-catalysed Hydrolysis.** The rate of base-catalysed hydrolysis of esters ($R\cdot COOR'$) is influenced by substituents in both the acid and the alcohol moiety (Kindler, 1926; 1927; 1936; Tommila and Hinshelwood, 1938; Palomaa, 1938). Electron-repelling groups in either group (R or R′) reduce electron-deficiency on the carbonyl carbon impeding attack by $HO^-$ and depressing the rate of hydrolysis (Table 30). Electron-withdrawing substituents have the reverse effect, increasing susceptibility to attack by $HO^-$. Such substituents, therefore, decrease the stability of the ester to alkali, and hence increase the rate of hydrolysis.

Similar effects, but less pronounced are seen with electron-donating and electron-withdrawing substituents in the *para* position of esters of aromatic acids and phenolic esters (Table 31).

Steric effects also influence the rate of base-catalysed hydrolysis. Thus, the rate of hydrolysis of *ortho*-substituted aromatic acid esters is depressed relative

Table 30. Relative Rates of Base-catalysed Hydrolysis of Aliphatic Esters

| | | | | |
|---|---|---|---|---|
| | $CH_3\cdot CO\cdot OR'$ | | | |
| R′ | Me | Et | i-Pr | t-Bu |
| Relative rate ($H_2O$/25°C) | 1 | 0.601 | 0.146 | 0.0084 |
| | $R\cdot CO\cdot OEt$ | | | |
| R | Me | Et | i-Pr | t-Bu |
| Relative rate ($H_2O$/25°C) | 1 | 0.47 | 0.1 | 0.105 |
| | $R\cdot CO\cdot OMe$ | | | |
| R | Me | H | $CH_2Cl$ | $CHCl_2$ |
| Relative rate (85% aq EtOH/30°C) | 1 | 223 | 761 | 16 000 |

Table 31. Relative Rates of Base-catalysed Hydrolysis of Aromatic Acid Esters and Phenolic Esters

R—$C_6H_4$—CO·OEt

| R | $NH_2$ | OMe | Me | H | F | Cl | Br | I | $NO_2$ |
|---|---|---|---|---|---|---|---|---|---|
| Relative rate (85% aq EtOH/25°C) | 0.023 | 0.209 | 0.456 | 1 | 2.03 | 4.31 | 5.25 | 5.05 | 110 |

$CH_3$·CO·O—$C_6H_4$—R

| R | $NH_2$ | Me | H | $NO_2$ |
|---|---|---|---|---|
| Relative rate (60% aq acetone/0°C) | 0.51 | 0.602 | 1 | 18.8 |

to the corresponding *para*-substituted esters with all substituents except fluorine (Table 32).

Table 32. Relative rates of Base-catalysed Hydrolysis of *o*- and *p*-Substituted Aromatic Acid Esters

| R | *o*-R—$C_6H_4$—CO·OEt | R—$C_6H_4$—CO·OEt (*p*) |
|---|---|---|
| | Relative rates (85% aq EtOH/25°C) | |
| Me | 0.125 | 0.456 |
| H | 1 | 1 |
| F | 3.74 | 2.03 |
| Cl | 2.24 | 4.31 |
| $NO_2$ | 8.71 | 110 |

(b) **Acid-catalysed Hydrolysis.** In contrast to their effect on base-catalysed hydrolysis, polar (electronic) effects have only marginal influence on the rate of hydrolysis of esters irrespective of the type of ester (Table 33). This is due to the positive charge of the protonated acyl/carbonyl oxygen, which because of its proximity has an overriding influence on the carbonyl carbon, swamping out other effects. Steric effects, which depress the rate of hydrolysis even with electron-withdrawing substituents, are clearly apparent in esters of both substituted aliphatic acids (Table 33) and *o*-nitro-substituted aromatic acids (Table 34). Similar steric effects are also seen in the acid-catalysed esterification of aromatic acids.

Table 33. Relative Rates of Acid-catalysed Hydrolysis of Esters

$CH_3 \cdot CO \cdot OR'$

| R′ | Me | Et | iPr | t-Bu | $CH_2Ph$ | Ph |
|---|---|---|---|---|---|---|
| Relative rate (aq HCl/25°C) | 1 | 0.97 | 0.53 | 1.15 | 0.96 | 0.69 |

$R \cdot CO \cdot OEt$

| R | Me | $CH_2Cl$ | $CHCl_2$ | $CCl_3$ |
|---|---|---|---|---|
| Relative rate ($Ph \cdot SO_2 \cdot OH$/60% aq EtOH/60°C) | 1 | 0.91 | 0.41 | 0.11 |

R—$C_6H_4$—CO·OEt

| R | H | Me | OMe | Br | $NO_2$ |
|---|---|---|---|---|---|
| Relative rate ($Ph \cdot SO_2 \cdot OH$/60% aq EtOH/60°C) | 1 | 0.97 | 0.92 | 0.98 | 1.03 |

(c) **Enzymic Hydrolysis.** It has been suggested that enzymic hydrolyses of esters appear to follow the pattern of base-catalysed hydrolysis, and are subject to both electronic and steric influences. Examples certainly exist where electronic and/or steric effects are significant, but in the present state of knowledge it is doubtful whether any such precise generalisation can be made. Thus, the studies of Levine and Clark (1955) on the enzymic hydrolysis of ring-substituted diethylaminoethyl benzoates by human serum esterases shows that the rate of hydrolysis of the *p*-fluoro ester (electron-withdrawing) is six times that of the *p*-amino ester (electron-repelling), whilst that of the *p*-ethoxy ester is intermediate between the two (Table 35). Their studies have also shown that substitution on the α-C of either the acid or alcohol moiety of the ester impedes hydrolysis, indicating that it is also subject to steric effects.

Table 34. Relative Rates of Acid-catalysed Hydrolysis of Ethyl Nitrobenzoates

$O_2N$—$C_6H_4$—CO·OEt

| Position of Substituent | Rate of Hydrolysis (60% aq EtOH/25°C) relative to Ethyl Benzoate (1.00) |
|---|---|
| *para* | 1.03 |
| *meta* | 0.99 |
| *ortho* | 0.36 |

Table 35. Hydrolysis of Ring-substituted Diethylaminoethyl Benzoates by Human Plasma Esterases (Levine and Clark, 1955)

| | Substituent | μg/ml Serum Hydrolysed/hr |
|---|---|---|
| R–$C_6H_4$–$CO{\cdot}O{\cdot}CH_2{\cdot}CH_2{\cdot}NEt_2$ | F | 3000 |
| | $CH_3{\cdot}CH_2{\cdot}O$ | 1000 |
| | $NH_2$ | 500 |

The influence of steric factors is similarly apparent in enzymic hydrolysis of acetylcholine derivatives (Table 36). Hydrolysis of the acetyl-β-methylcholines by acetylcholinesterase is significantly inhibited compared with that of acetylcholine, whereas hydrolysis of the acetyl-α-methylcholines in which the methyl substituent is further removed from the ester link, is not nearly so markedly affected (Beckett, 1962; Beckett, Harper, Clitherow and Lesser, 1961).

$$[Me_3\overset{+}{N}{\cdot}CH_2{\cdot}CH_2{\cdot}O{\cdot}CO{\cdot}CH_3]I^-$$

Acetylcholine iodide

$$[Me_3\overset{+}{N}{\cdot}\underset{\displaystyle Me}{\underset{|}{CH}}{\cdot}CH_2{\cdot}O{\cdot}CO{\cdot}CH_3]I^-$$

Acetyl-α-methylcholine iodide

$$[Me_3\overset{+}{N}{\cdot}CH_2{\cdot}\underset{\displaystyle Me}{\underset{|}{CH}}{\cdot}O{\cdot}CO{\cdot}CH_3]I^-$$

Acetyl-β-methylcholine iodide

The differences in rates of hydrolysis and activity on cat blood pressure (Table 36) between the members of each isomeric pair reflect the stereoselectivity of the biological receptors for the individual esters. Reference to the mechanism of acetylcholinesterase hydrolysis (p. 286) in which histidine acts as a hydrogen donor to the acyl oxygen, indicates a certain parallel with acid-catalysed ester hydrolysis, which as already described, is similarly influenced by steric parameters.

Table 36. Relative Rates of Hydrolysis by Acetylcholinesterase and Activity of Acetyl α- and β-Methylcholine Iodides

| Ester | Relative Hydrolysis by Acetylcholinesterase (%) | Activity on Cat Blood Pressure (mols equivalent to 1 mol of acetylcholine) |
|---|---|---|
| Acetylcholine iodide | 100 | 1 |
| *S*-(−)-Acetyl-α-methylcholine iodide | 97 | 144 |
| *R*-(+)-Acetyl-α-methylcholine iodide | 78 | 13 |
| *S*-(+)-Acetyl-β-methylcholine iodide | 55 | 0.75 |
| *R*-(−)-Acetyl-β-methylcholine iodide | 0 | 202 |

**Transesterification**

Transesterification can be effected by alcoholysis in which the ester is refluxed with excess alcohol in the presence of an acid catalyst or catalytic amounts of the corresponding sodium alkoxide. It is usually most effective in replacing a higher alcohol by a more volatile lower one, but the reverse is also possible by using an excess of the replacing alcohol.

$$CH_3{\cdot}CO{\cdot}OR + CH_3{\cdot}OH \xrightleftharpoons{CH_3{\cdot}ONa} CH_3{\cdot}CO{\cdot}OCH_3 + ROH$$

**Acidolysis**

Acidolysis is the replacement of the acidic moiety of an ester by another acid. Its effectiveness is probably dependent on the relative volatility of the two acids concerned.

$$CH_3(CH_2)_8{\cdot}CO{\cdot}OH + CH_3{\cdot}CO{\cdot}O{\cdot}CH{=}CH_2 \rightleftharpoons CH_3(CH_2)_8{\cdot}CO{\cdot}O{\cdot}CH{=}CH_2 + CH_3{\cdot}CO{\cdot}OH$$

**Ammonolysis**

Esters react with ammonia and amines, with displacement of the alcohol to form amides. The nucleophilic addition of amine to the ester carbonyl carbon is reversible, but the completion step is base-catalysed, with the overall kinetics of the reaction in accord with a combination of unimolecular and bimolecular decomposition of the addition complex.

$$CH_3{\cdot}C(OR')({=}O) + R{\cdot}\ddot{N}H_2\ (R{=}H \text{ or alkyl}) \rightleftharpoons CH_3{\cdot}C(OR')(O^-){\cdot}\overset{+}{N}H_2R$$

$$\xrightarrow{\text{Unimolecular}} CH_3{\cdot}CO{\cdot}NHR + R'OH$$

$$\xrightarrow[+\,R{\cdot}NH_2]{\text{Bimolecular}} CH_3{\cdot}CO{\cdot}NHR + R\overset{+}{N}H_3 + R'O^-$$

Esters also react similarly with hydrazine to form acid hydrazides and with hydroxylamine to form hydroxamic acids.

$$R{\cdot}CO{\cdot}OCH_3 + H_2N{\cdot}NH_2 \longrightarrow \underset{\text{Acid hydrazide}}{R{\cdot}CO{\cdot}NH{\cdot}NH_2} + CH_3OH$$

$$R{\cdot}CO{\cdot}OCH_3 + H_2N{\cdot}OH \longrightarrow \underset{\text{Hydroxamic acid}}{R{\cdot}CO{\cdot}NHOH} + CH_3OH$$

**Reduction**

Reduction of esters to primary alcohols can be brought about either catalytically at high temperature under pressure (copper chromite catalyst; 200–300°C; 200 atm.) or at room temperature by hydride ion reduction.

Catalytic reduction is used in the commercial production of lauryl alcohol from coconut oil, which consists largely of glyceryl esters of lauric acid.

$$\begin{array}{l} CH_2{\cdot}O{\cdot}CO(CH_2)_{11}{\cdot}CH_3 \\ | \\ CH{\cdot}O{\cdot}CO(CH_2)_{11}{\cdot}CH_3 \\ | \\ CH_2{\cdot}O{\cdot}CO(CH_2)_{11}{\cdot}CH_3 \end{array} \xrightarrow[200°/200\ atm]{H_2/\text{Copper chromite}} \begin{array}{l} CH_2OH \\ | \\ CHOH \\ | \\ CH_2OH \end{array} + \underset{\text{Lauryl alcohol}}{3CH_3(CH_2)_{11}{\cdot}CH_2OH}$$

The latter is used without further purification in the production of the anionic detergent *Sodium Lauryl Sulphate* (Chapter 7).

Hydride ion reduction with lithium aluminium hydride proceeds via the intermediate formation of the aldehyde, which is further reduced to the primary alcohol.

Esters are also reduced to the corresponding primary alcohol by the action of sodium in ethanol (Bouveault-Blanc reduction). The reaction is capable of being rationalised by the following mechanism, which is in part analogous to the acyloin condensation (below).

$$R{\cdot}C(=O){\cdot}OEt \xrightarrow{Na\cdot} R{\cdot}\overset{-}{C}(OEt){-}O\cdot\ Na^+ \xrightarrow{H{-}OEt,\ -EtO^-Na^+} R{\cdot}CH(OEt){-}O\cdot \xrightarrow{Na\cdot} R{\cdot}CH(OEt){-}O^-Na^+ \xrightarrow{-EtO^-Na^+} R{\cdot}CH{=}O \xrightarrow{Na/EtOH} R{\cdot}CH_2OH$$

**Radical Decomposition**

Esters are relatively stable, but exposure to free radicals or radical generators, such as acetyl peroxide, leads to radical attack at the α-carbon atom and binary combination (Kharasch, McBay and Urry, 1945; Ansell, Hickinbottom and Holton, 1955).

$$R{\cdot}CH_2{\cdot}CO{\cdot}OCH_3 \xrightarrow{R\cdot \ \to\ RH} R{\cdot}\dot{C}H{\cdot}CO{\cdot}OCH_3 \longrightarrow \begin{array}{l} R{\cdot}CH{\cdot}CO{\cdot}OCH_3 \\ | \\ R{\cdot}CH{\cdot}CO{\cdot}OCH_3 \end{array}$$

**Acyloin Formation**

A somewhat similar radical-induced dimerisation occurs when esters of low molecular weight alcohols are treated in an inert solvent (ether or benzene) with metallic sodium, and subsequently with acid to form acyloins (α-ketoalcohols).

$$CH_3{\cdot}CH_2{\cdot}C(=O)OEt \xrightarrow{Na\cdot} CH_3{\cdot}CH_2{\cdot}\dot{C}(O^-Na^+)OEt \longrightarrow \begin{matrix} CH_3{\cdot}CH_2{\cdot}C(O^-Na^+)OEt \\ | \\ CH_3{\cdot}CH_2{\cdot}C(O^-Na^+)OEt \end{matrix}$$

$$\xrightarrow{-2EtO^-Na^+} \begin{matrix} CH_3{\cdot}CH_2{\cdot}C{=}O \\ | \\ CH_3{\cdot}CH_2{\cdot}C{=}O \end{matrix} \xrightarrow{2Na\cdot} \begin{matrix} CH_3{\cdot}CH_2{\cdot}CO^-Na^+ \\ \| \\ CH_3{\cdot}CH_2{\cdot}CO^-Na^+ \end{matrix} \xrightarrow{H^+} \left[\begin{matrix} CH_3{\cdot}CH_2{\cdot}C{\cdot}OH \\ \| \\ CH_3{\cdot}CH_2{\cdot}C{\cdot}OH \end{matrix}\right] \longrightarrow \begin{matrix} CH_3{\cdot}CH_2{\cdot}C{=}O \\ | \\ CH_3{\cdot}CH_2{\cdot}CH{\cdot}OH \end{matrix}$$

**Claisen Condensation**

The self-condensation of ethyl acetate to give the β-keto ester, ethyl acetoacetate, is typical of a group of reactions between esters and carbanions, known as the Claisen condensation. The reaction is base-catalysed usually with sodium ethoxide, which initiates the reaction by formation of a carbanion.

$$EtO^-(Na^+) + \mathbf{CH_3}{\cdot}CO{\cdot}OEt \rightleftharpoons {}^-\mathbf{CH_2}{\cdot}CO{\cdot}OEt + EtOH$$

The reaction continues with addition of the carbanion to the carbonyl carbon of a second ester molecule (see facing page).

The reaction is reversible at each stage, but the product ethyl acetoacetate is a stronger acid ($pK_a$ 10.7) than ethanol ($pK_a \approx 18$) so that formation of the product enolate ion is favoured and the reverse reaction suppressed.

Mixed Claisen condensations between two different esters both with α-C—H bonds give rise to mixtures consisting of all four possible products. If one of the

$$CH_3{\cdot}C(=O)OEt + (Na^+)^-CH_2{\cdot}CO{\cdot}OEt \rightleftharpoons CH_3{\cdot}C(O^-)(OEt){\cdot}CH_2{\cdot}CO{\cdot}OEt$$

$$\rightleftharpoons CH_3{\cdot}\mathbf{CO{\cdot}CH_2}CO{\cdot}OEt + EtO^-(Na^+)$$

$$CH_3{\cdot}C(O^-(Na^+))\!=\!CH{\cdot}CO{\cdot}OEt + EtOH \rightleftharpoons CH_3{\cdot}\mathbf{CO{\cdot}CH_2}CO{\cdot}OEt + EtO^-(Na^+)$$

$$\xrightarrow{H^+} CH_3{\cdot}\mathbf{CO{\cdot}CH_2}{\cdot}CO{\cdot}OEt$$

esters is without $\alpha$-C—H bonds (formate, oxalate, carbonate, benzoate), the direction in which the reaction can proceed becomes more limited.

$$R{\cdot}C(=O)OEt \xrightarrow{CH_3{\cdot}CO{\cdot}OEt/EtO^-} R{\cdot}CO{\cdot}CH_2{\cdot}CO{\cdot}OEt$$

(R = H; —CO·OEt; —O·CO·OEt; Ph)

## ACID HALIDES

### Halogen Displacement

Acid chlorides are extremely reactive. The lower members are fuming liquids, extremely lachrymatory, and fume in moist air owing to instantaneous hydrolysis.

$$CH_3{\cdot}CO{\cdot}Cl + H_2O \longrightarrow CH_3{\cdot}CO{\cdot}OH + HCl\uparrow$$

Higher molecular weight and aromatic acid chlorides are more resistant to hydrolysis owing to their much lower solubility in water.

Acid chlorides react readily with alcohols and phenols to form esters; also with ammonia, and primary and secondary amines to form amides. As with hydrolysis, these reactions result in the displacement of the halogen which shows a high degree of reactivity in contrast to the otherwise structurally comparable, but unreactive vinylic halogen ($CH_2$=CH—Cl). The reactivity of the halogen in these reactions is, therefore, considered to lie not in the acid chloride itself, but in the intermediate formed by the initial addition of the reactant to the carbonyl group.

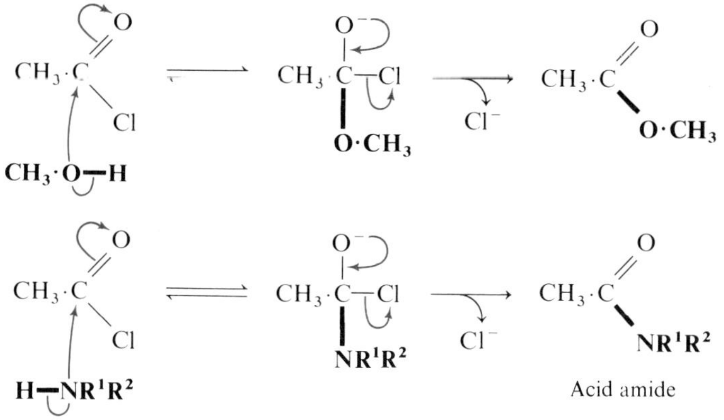

$$CH_3\cdot C(=O)Cl + CH_3\cdot O{-}H \longrightarrow CH_3\cdot C(O^-)(Cl)(O\cdot CH_3) \xrightarrow{-Cl^-} CH_3\cdot C(=O)O\cdot CH_3$$

$$CH_3\cdot C(=O)Cl + H{-}NR^1R^2 \rightleftharpoons CH_3\cdot C(O^-)(Cl)(NR^1R^2) \xrightarrow{-Cl^-} CH_3\cdot C(=O)NR^1R^2$$

Acid amide

Ammonia $R^1 = R^2 = H$
Primary amine $R^1 = H$, $R^2 =$ alkyl
Secondary amine $R^1$, $R^2 =$ alkyl

Acid chlorides also react with hydrogen sulphide (excess) in pyridine to form thio-acids.

**Reactions with Organo–metallic Compounds**

Acid chlorides react with Grignard reagents to form ketones, and, in the presence of excess reagent, tertiary alcohols. There is evidence that the reaction proceeds by addition of the reagent to the carbonyl carbon followed by elimination of magnesium halide.

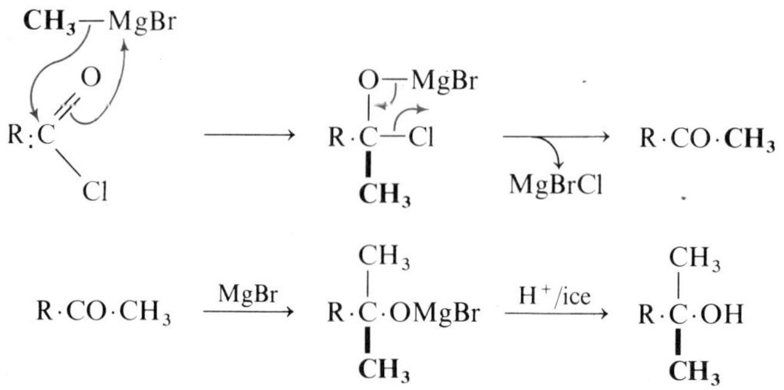

$$R\cdot C(=O)Cl + CH_3{-}MgBr \longrightarrow R\cdot C(O{-}MgBr)(Cl)(CH_3) \xrightarrow{-MgBrCl} R\cdot CO\cdot CH_3$$

$$R\cdot CO\cdot CH_3 \xrightarrow{MgBr} R\cdot C(CH_3)_2\cdot OMgBr \xrightarrow{H^+/ice} R\cdot C(CH_3)_2\cdot OH$$

**Arndt–Eistert Reaction**

Diazomethane reacts with acid chlorides to form diazoketones, which readily lose nitrogen and rearrange in the presence of moist silver oxide to form the homologous carboxylic acid.

$$R\cdot CO\cdot Cl + CH_2N_2 \xrightarrow[-HCl]{} R\cdot CO\cdot CH\cdot N_2 \text{ (Diazoketone)} \xrightarrow[-N_2]{H_2O,\ AgOH} R\cdot CH_2\cdot CO\cdot OH$$

Diazoketones also react with alcohols and ammonia in the presence of silver oxide to form the corresponding esters and amides.

$$R\cdot CO\cdot CHN_2 \xrightarrow{Ag_2O,\ R^1\cdot OH} R\cdot CH_2\cdot CO\cdot OR^1$$

$$R\cdot CO\cdot CHN_2 \xrightarrow{Ag_2O,\ NH_3} R\cdot CH_2\cdot CO\cdot NH_2$$

The mechanism of the Wolff rearrangement as it is known, which gives rise to a keten intermediate, is uncertain, but is thought to involve a 1,2-shift.

$$R\cdot C(=O)\cdot CH{=}\overset{+}{N}{=}\overset{-}{N} \xrightarrow[-N_2]{} R{-}C(=O){-}\overset{-}{CH} \longrightarrow R\cdot CH{=}C{=}O \text{ (Ketene)}$$

$$R\cdot CH{=}C{=}O \longleftrightarrow R\cdot \overset{-}{C}H{=}\overset{+}{C}{=}O$$

$$R\cdot \overset{-}{C}H{=}\overset{+}{C}{=}O + HO^-(Ag^+) \longrightarrow R\cdot \overset{-}{C}H\cdot C(=O)OH\ (Ag^+) \xrightarrow{H^+} R\cdot CH_2\cdot CO\cdot OH$$

## Hell-Volhard-Zelinsky Reaction

The Hell-Volhard-Zelinsky reaction is a reaction of carboxylic acids with bromine in the presence of small amounts of red phosphorus and gives rise to $\alpha$-bromocarboxylic acids. The reaction, however, is slow in the absence of

$$CH_3\cdot C(=O)Br \overset{H^+}{\rightleftharpoons} H{-}CH_2{-}C(=\overset{+}{O}H)Br \xrightarrow[-H^+]{} H_2C{=}C(O{-}H)Br \xrightarrow[-HBr]{Br{-}Br} Br\cdot CH_2\cdot CO\cdot Br$$

phosphorus, which forms phosphorus tribromide, which in turn gives the acid bromide.

Acid-catalysed enolisation of the acid bromide occurs more readily than with the acid, thereby promoting the reaction with bromine.

The resulting bromo-acid bromide exchanges with a second molecule of carboxylic acid, releasing a further mole of acid bromide for continued reaction.

$$Br{\cdot}CH_2{\cdot}CO{\cdot}Br + CH_3{\cdot}CO{\cdot}OH \longrightarrow Br{\cdot}CH_2{\cdot}CO{\cdot}OH + CH_3{\cdot}CO{\cdot}Br$$

## ACID ANHYDRIDES

Acid anhydrides are neutral liquids or solids, which are soluble in organic solvents and generally immiscible with water. They are, however, reactive to a greater or less extent with amines, alcohols and water.

### Hydrolysis

The molecules of acid anhydrides exhibit the phenomenon of competing resonance, in which the two carbonyl groups are competing simultaneously for the electrons on the central oxygen atom.

$$CH_3{\cdot}\overset{\overset{\displaystyle O}{\|}}{C}-\ddot{\underset{\cdot\cdot}{O}}-\overset{\overset{\displaystyle O}{\|}}{C}-CH_3$$

This competition is relieved by hydrolysis, since the resonance energy of acetic anhydride is less than that of the two separate acetate ions which are formed. In consequence, hydrolysis is readily promoted. Acid anhydrides, however, are less readily hydrolysed than acid chlorides. Thus, acetic anhydride is soluble in cold water and only slowly hydrolysed at room temperature over a period of an hour or so. It can, therefore, be used as an acetylating agent for more reactive substrates in cold aqueous solution. It is, however, hydrolysed to acetic acid in boiling water.

$$(CH_3{\cdot}CO)_2O + H_2O \longrightarrow 2CH_3{\cdot}CO{\cdot}OH$$

Acetic anhydride similarly reacts with hydrogen sulphide in the presence of a small quantity of potassium hydroxide, to form thioacetic acid.

$$(CH_3{\cdot}CO)_2O + H_2S \longrightarrow CH_3{\cdot}CO{\cdot}SH + CH_3{\cdot}CO{\cdot}OH$$

The relative stability of anhydrides in aqueous media accounts for the importance of mixed anhydrides, such as acetyl phosphate, which act as important energy transducers in certain biological systems. Acetyl phosphate has a standard free energy of hydrolysis at pH 7.0 of 44 kJ. This is considerably greater than that of acetic anhydride and reflects the greater electron-withdrawing powers and, hence, destabilising influence of the phosphate group compared with that of the carbonyl group. This energy is made available to assist the enzyme-catalysed transfer of phosphate from acetyl phosphate to ADP in the synthesis of ATP (Chapter 2).

### Acylation of Alcohols and Phenols

Acid anhydrides react with primary and secondary alcohols, and with phenols to form esters, with displacement of the carboxylate ion.

$$(CH_3\cdot CO)_2O + CH_3\cdot OH \xrightarrow{H^+} CH_3\cdot C(=O)-O-C(CH_3)(OCH_3)O^- \longrightarrow CH_3\cdot CO\cdot O^- + CH_3\cdot \mathbf{CO\cdot OCH_3}$$

$$CH_3\cdot CO\cdot O^- \xrightarrow{H^+} CH_3\cdot CO\cdot OH$$

Methyl acetate

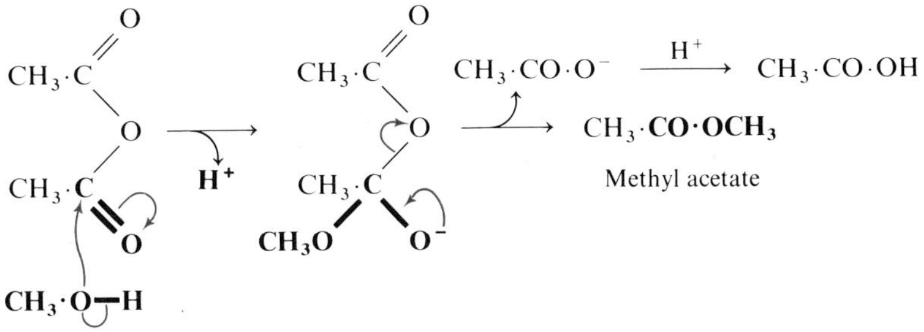

The reaction is facilitated by acidic and basic catalysts. The latter may promote formation of alkoxide and phenoxide ions, which are more reactive than the parent alcohol in nucleophilic addition to the anhydride carbonyl group. Weak organic bases, such as pyridine, however, are more likely to form acyl pyridinium complexes, analogous to those formed with acid chlorides.

### Acylation of Amines

Reaction of anhydrides with ammonia, primary and secondary amines occurs readily with the displacement of the carboxylate ion to form amides.

$$(CH_3\cdot CO)_2O + 2NH_3 \longrightarrow CH_3\cdot CO\cdot NH_2 + CH_3\cdot CO\cdot ONH_4$$

Acetamide

The mechanism of the reaction is analogous to the formation of esters, and consists of nucleophilic attack by the base at the anhydride carbonyl carbon.

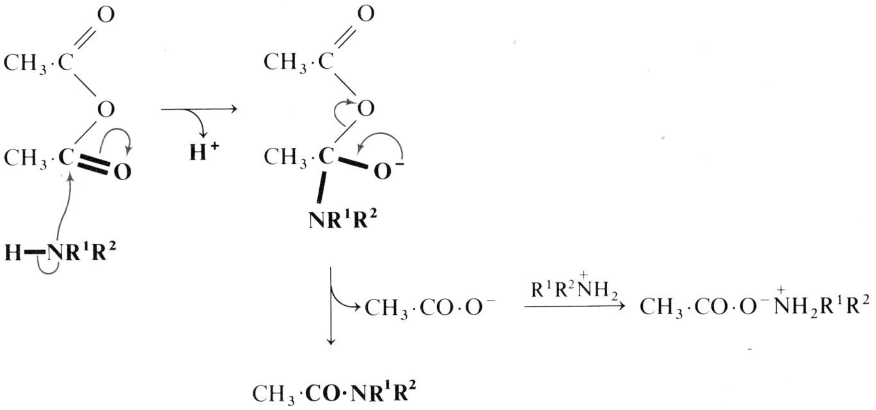

$$(CH_3\cdot CO)_2O + H-NR^1R^2 \xrightarrow{H^+} CH_3\cdot C(=O)-O-C(CH_3)(NR^1R^2)O^- \longrightarrow CH_3\cdot \mathbf{CO\cdot NR^1R^2} + CH_3\cdot CO\cdot O^-$$

$$CH_3\cdot CO\cdot O^- \xrightarrow{R^1R^2\overset{+}{N}H_2} CH_3\cdot CO\cdot O^-\overset{+}{N}H_2R^1R^2$$

### Acylation by Mixed Anhydrides

Rates of hydrolysis, alcoholysis and aminolysis of acid anhydrides are increased by substituents which decrease electron density on the carbonyl carbon. Thus, electron-withdrawing groups in substituted benzoic anhydrides increase the rate of hydrolysis. Similarly, trifluoroacetic anhydride ($CF_3CO{\cdot}O{\cdot}COCF_3$) is a much more powerful acylating agent than acetic anhydride.

By analogy, mixed anhydrides, such as formic-acetic anhydride, undergo nucleophilic attack in acylation reactions at the more electron-deficient of the two carbon groups. In the case of formic-acetic anhydride, this will be the formyl carbonyl group, since the methyl group is electron-repelling. Formic-acetic anhydride is, therefore, a formylating and not an acetylating agent.

$CH_3{\cdot}CO{\cdot}O{\cdot}CHO + CH_3{\cdot}OH \xrightarrow{H^+} CH_3{\cdot}C(=O){\cdot}O{\cdot}CH(O^-)(O{\cdot}CH_3) \longrightarrow CH_3{\cdot}CO{\cdot}O^- + H{\cdot}CO{\cdot}OCH_3$

$CH_3{\cdot}CO{\cdot}O^- \xrightarrow{H^+} CH_3{\cdot}CO{\cdot}OH$

H·CO·OCH₃ Methyl formate

Steric factors, however, seem to be paramount as in the acylation of amines with 2-cyanoacetyl 2′-trichloroacetyl anhydride which yield the 2-cyanoacetylamine.

### Acetate Transfer in Intermediary Metabolism

Biological acetylations involving the transfer of acetate from mixed anhydrides, such as acetyl phosphate and acetyl adenylate, to coenzyme A is analogous to formylation by formic-acetic anhydride. The carbonyl carbon is less electronegative and less hindered than phosphorus so that it invites preferential nucleophilic attack by the thiolate anion.

$\overline{CoA}{\cdot}S^- + CH_3{\cdot}C(=O){\cdot}O{\cdot}P(O^-)(=O){\cdot}OAd \longrightarrow CH_3{\cdot}C(O^-)(S{\cdot}\overline{CoA}){\cdot}O{\cdot}P(O^-)(=O){\cdot}OAd \longrightarrow CH_3{\cdot}CO{\cdot}S{\cdot}\overline{CoA} + Ad{\cdot}O{\cdot}P(O^-)_2{=}O$

Coenzyme A; Acetyl adenylate; Acetyl coenzyme A

# AMIDES

## General Properties

Amides are usually crystalline solids which are soluble in organic solvents; only the lower members, however, are soluble in water. Unsubstituted and mono-substituted amides exhibit weakly acidic and basic properties due to resonance involving the following forms.

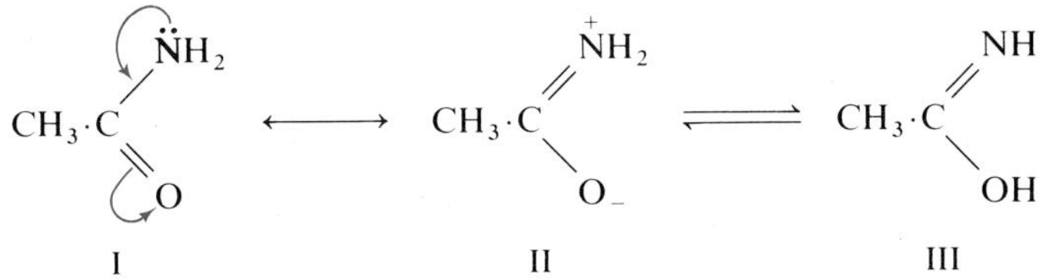

I $\longleftrightarrow$ II $\rightleftharpoons$ III

Thus, amides form sodium salts with sodamide in ether, which are unstable in water and apparently derived from II. Rather more stable mercury salts, apparently derived from I, in which the metal is believed to be covalently bound to nitrogen, are formed in aqueous solution from water-soluble amides and mercuric oxide.

Certain amides, such as ergometrine and other lysergic acid amides, are soluble in aqueous sodium hydroxide. It is interesting to note that their stereo-isomers of the isolysergic acid series, which differ only in the orientation of the carboxyamide link, are insoluble in alkali, since ionisation of the amide proton is inhibited by hydrogen bonding to the adjacent ring nitrogen (Craig, Shedlovsky, Gould and Jacobs, 1938; Stenlake, 1953; 1955).

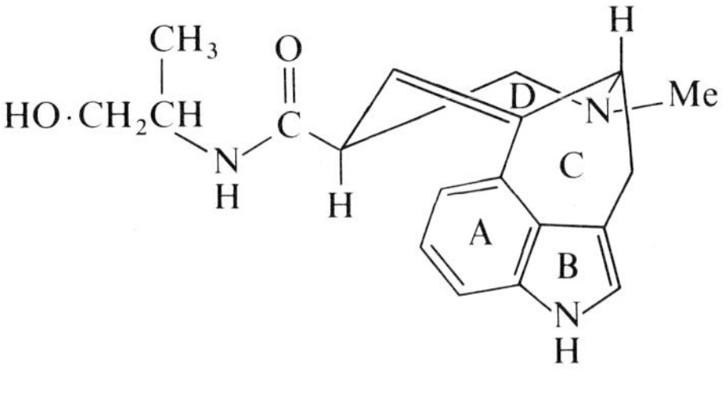

Ergometrine (p$K_a$ 6.70)

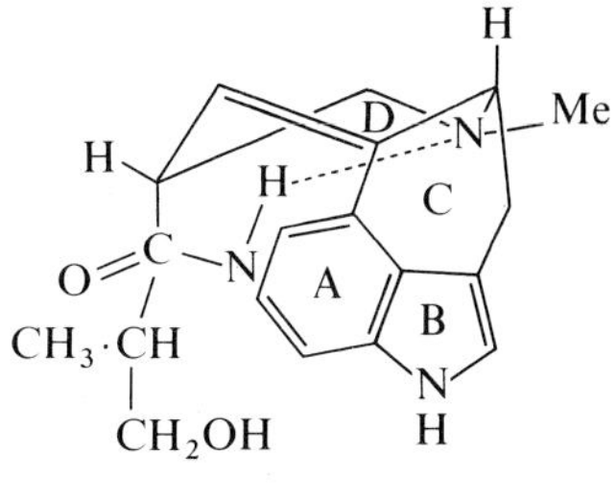

Ergometrinine (p$K_a$ 7.30)

The difference in alkali solubility of the two stereoisomers is reflected in the different dissociation constants of the basic group in ring D. It also forms the basis of a limit test to control contamination of ergometrine, which is pharmacologically active, by ergometrinine, its inactive isomer.

Amides also form distinct, but readily hydrolysable salts with acids, such as nitric acid ($CH_3 \cdot CO \cdot NH_2 \cdot NHO_3$) and oxalic acid.

These weakly acidic and basic properties account in part for some of the weak interactions between proteins and acidic and basic drugs, though clearly much more powerful interactions occur in the free acidic and basic groups of proteins (Chapter 19). The same properties also contribute to the unusually wide solvent properties which dimethylformamide ($H \cdot CO \cdot NMe_2$) and *NN*-dimethylacetamide ($CH_3 \cdot CO \cdot NMe_2$) have for many polar organic compounds. A number of acid amides, including *NN*-dimethylacetamide, have been shown to be useful for increasing the solubility of tetracycline antibiotics in aqueous solution (Gans and Higuchi, 1957).

### Stereochemistry

Resonance accounts for a certain degree of double bond character in both C—O and C—N bonds, and in consequence mono-substituted amides are planar and adopt a semi-rigid configuration isosteric with esters in which the amide hydrogen is orientated *trans* to the carbonyl oxygen (**2**, 3).

Secondary amide

Ester

Isosterism of secondary amides and esters has been used to produce related compounds with similar biological activities, e.g. *Procaine* and *Procainamide* (**2**, 3). In other cases, it has been used to produce inhibitors of physiological transmitter substances.

### Hydrolysis

Amides, like esters, are hydrolysed by aqueous acid, and more readily by alkali. Acid hydrolysis requires nucleophilic attack by water following protonation.

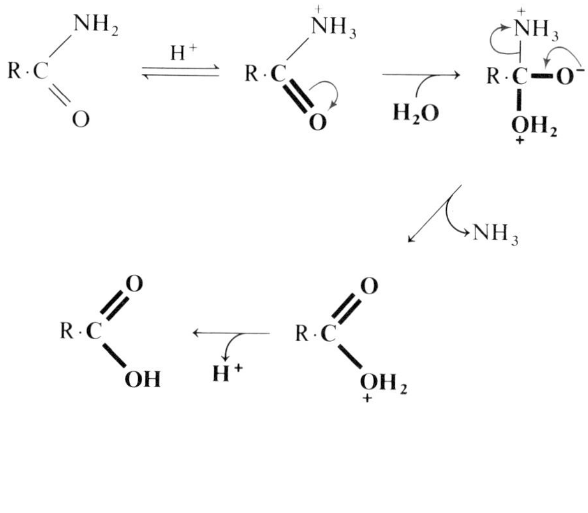

Alkaline hydrolysis probably occurs as follows by attack of $HO^-$.

$$R{\cdot}C(=O)NH_2 + HO^- \rightleftharpoons R{\cdot}C(NH_2)(OH){-}O^- \rightleftharpoons R{\cdot}C(=O)OH + {}^-NH_2 \xrightarrow{NH_3} R{\cdot}CO{\cdot}O^-$$

The enzymatic hydrolysis of *Procainamide* by serum amidases is slower than the corresponding esterase hydrolysis of *Procaine*. In consequence, the former has more prolonged local anaesthetic activity (Bowman, Rand and West, 1968). Hydrolysis of both drugs yields *p*-aminobenzoic acid, which antagonises the antibacterial action of sulphonamides (Chapter 14). Steric inhibition of amide

$H_2N{-}C_6H_4{-}CO{\cdot}NH{\cdot}CH_2{\cdot}CH_2{\cdot}NEt_2$

*Procainamide*

$H_2N{-}C_6H_4{-}CO{\cdot}O{\cdot}CH_2{\cdot}CH_2NEt_2$

*Procaine*

hydrolysis is demonstrated in *Lignocaine* and its derivatives, in which it has been shown that resistance to acid hydrolysis and local anaesthetic activity increase with increasing number and bulk of the *ortho* substituents (Sekera, Sova and Vrba, 1955; Table 37).

Enzymatic hydrolysis of aromatic amides in mammalian systems is also influenced by the electronic effects of ring substituents. The extent of hydrolysis is roughly in accord with the substituent effect (Tables 28 and 38), but deviations are apparent in these *in vivo* experiments due to the intervention of alternative metabolic pathways (Parke, 1968).

Table 37. Hydrolysis Rates and Local Anaesthetic Activity of Lignocaine and its Derivatives

$2\text{-}R^1\text{-}6\text{-}R^2\text{-}C_6H_3{-}NH{\cdot}CO{\cdot}CH_2NEt_2$

| $R^1$ | $R^2$ | Rate Constant ($s^{-1}$) in 5N HCl at 99.5°C | Local Anaesthetic Activity Relative to Procaine (1.0) (Guinea pig) | Duration (min) (Rabbit cornea) |
|---|---|---|---|---|
| H | H | $8.15 \times 10^{-5}$ | 0.12 | Incomplete |
| H | $CH_3$ | $2.64 \times 10^{-5}$ | 0.24 | $4.5 \pm 2$ |
| $CH_3$ | $CH_3$ | $7.27 \times 10^{-7}$ | 1.4 | $52.5 \pm 2$ |

Table 38. Metabolic Hydrolysis of *p*-Substituted Benzamides

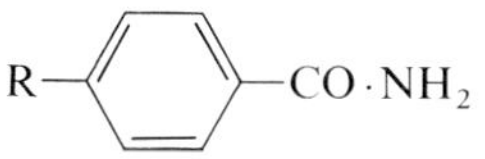

R—⟨benzene ring⟩—$CO \cdot NH_2$

| Substituent | Amide Hydrolysis (%) |
|---|---|
| H | 100 |
| Cl | 95 |
| $NH_2$ | 20 |
| OH | 4 |

**Dehydration**

Amides lose the elements of water to form nitriles with dehydrating agents, such as phosphorus pentoxide.

$$R \cdot CO \cdot NH_2 \longrightarrow R \cdot CN + H_2O$$

**Hofmann Reaction**

Amides are converted to amines with loss of a carbon atom when treated with bromine in aqueous alkali.

$$R \cdot CO \cdot NH_2 \dashrightarrow R \cdot NH_2$$

The first step in the reaction is *N*-bromination of the amide. Loss of a proton and expulsion of the halogen from the *N*-bromoamide gives a nitrene. The latter is electron-deficient (*cf* carbene, Chapters 2 and 9), and the nitrogen octet is completed by a 1,2-alkyl shift from carbon to nitrogen yielding the isocyanate.

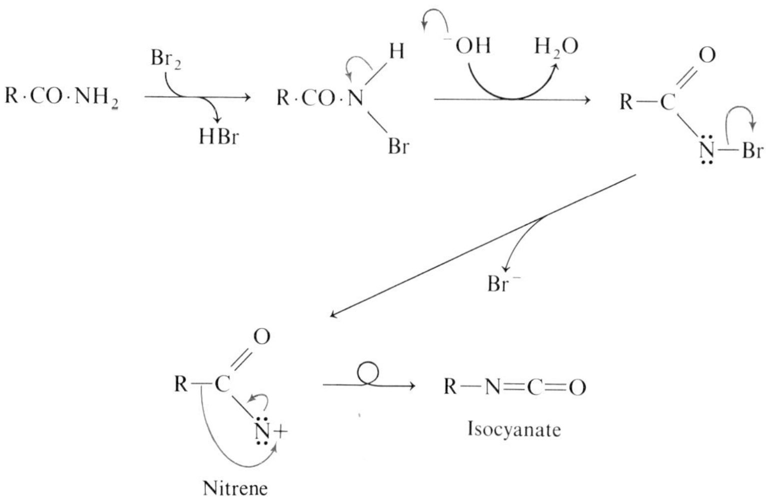

$R \cdot CO \cdot NH_2$ —($Br_2$; $-HBr$)→ $R \cdot CO \cdot N(H)(Br)$ —(${}^-OH$; $H_2O$)→ $R—C(=O)—\ddot{N}—Br$ —($-Br^-$)→ $R—C(=O)—\ddot{N}+$ (Nitrene) ⟶ $R—N{=}C{=}O$ (Isocyanate)

The isocyanate is rapidly hydrolysed to the carbamic acid (RNH·COOH) which decarboxylates spontaneously to the amine.

$$R{\cdot}N{=}C{=}O + H_2O \xrightarrow{H^+} R{\cdot}NH{\cdot}CO{\cdot}OH \xrightarrow{-CO_2} R{\cdot}NH_2$$

### Hydride Ion Reduction

Hydride ion reduction of amides occurs readily with lithium aluminium hydride to yield the corresponding amine.

$$R{\cdot}CO{\cdot}NH_2 \longrightarrow R{\cdot}CH_2{\cdot}NH_2$$

### Thioamide Formation

Treatment of amides with phosphorus pentasulphide gives thioamides.

$$R{\cdot}CO{\cdot}NH_2 \longrightarrow R{\cdot}CS{\cdot}NH_2$$

## ACID HYDRAZIDES

### Hydrolysis

Acid hydrazides in many respects resemble acid amides, but are less readily hydrolysed by acid and alkali than amides.

$$\text{Isonicotinic acid hydrazide (4-pyridyl}{\cdot}CO{\cdot}NH{\cdot}NH_2) \xrightarrow[H_2O \quad -H_2N{\cdot}NH_2]{} \text{Isonicotinic acid (4-pyridyl}{\cdot}CO{\cdot}OH)$$

Isonicotinic acid hydrazide (*Isoniazid*) — Isonicotinic acid

Isonicotinic acid hydrazide is extensively hydrolysed in the dog, but much less so in man, where the main metabolite is the *N*-acetyl derivative.

$$\text{4-pyridyl}{\cdot}CO{\cdot}NH{\cdot}NH{\cdot}CO{\cdot}CH_3$$

*N*-Acetylisonicotinic acid hydrazide

**Acylhydrazones**

Acid hydrazides condense readily with aldehydes and ketones to form acyl hydrazones. This reaction forms the basis of the use of Girard's reagents for the isolation of small amounts of urinary steroids.

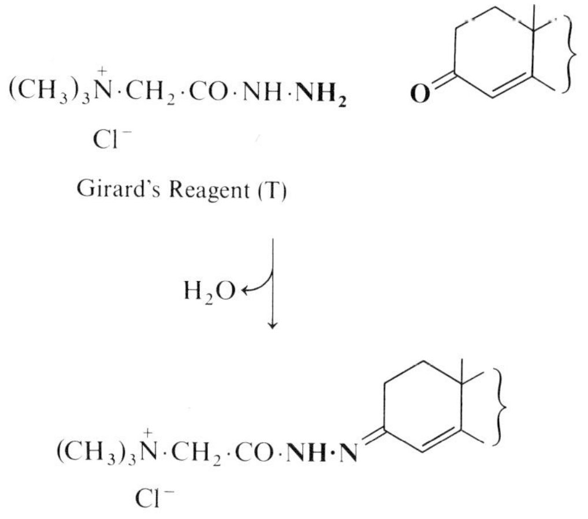

$$(CH_3)_3\overset{+}{N}\cdot CH_2\cdot CO\cdot NH\cdot \mathbf{NH_2} \quad Cl^-$$

Girard's Reagent (T)

$\downarrow H_2O$

$$(CH_3)_3\overset{+}{N}\cdot CH_2\cdot CO\cdot \mathbf{NH\cdot N}= \quad Cl^-$$

The steroid-Girard hydrazones, being quaternary ammonium compounds, are highly water-soluble and remain in aqueous solution during the removal of the neutral organic components when the product solution is extracted with organic solvents. Subsequent hydrolysis of the acyl hydrazone releases the steroid which can then be extracted with an immiscible organic solvent.

**Reducing Properties**

Acid hydrazides are powerful reducing agents. They are readily oxidised by iodine and other oxidising agents to diacylhydrazones.

$$2R\cdot CO\cdot NH\cdot \mathbf{NH_2} + \mathbf{I_2} \longrightarrow (R\cdot CO\cdot NH)_2 + \mathbf{N_2} + \mathbf{2HI}$$

Nitrous acid is reduced with the concurrent formation of an acid azide.

$$R\cdot CO\cdot NH\cdot NH_2 + HO\cdot NO \longrightarrow R\cdot CO\cdot N_3 + 2H_2O$$

The acid azides undergo the Curtius rearrangement in boiling alcohols to yield *N*-substituted urethanes. The rearrangement may be formulated in a similar manner to the Hofmann rearrangement of amides.

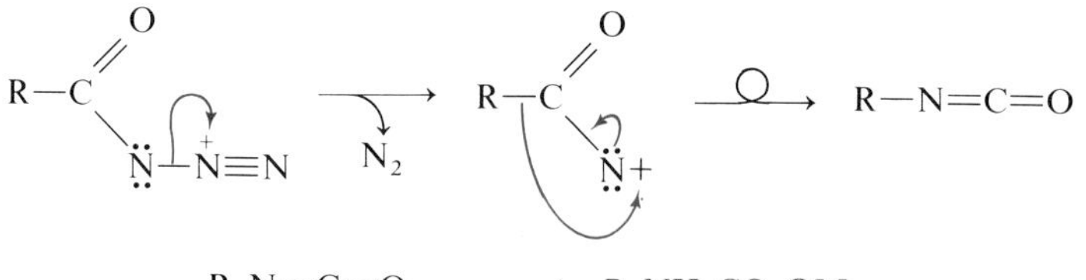

$$R-C(=O)-\ddot{N}-\overset{+}{N}\equiv N \xrightarrow{-N_2} R-C(=O)-\ddot{N}^+ \longrightarrow R-N=C=O$$

$$R\cdot N=C=O \xrightarrow{\mathbf{MeOH}} R\cdot NH\cdot CO\cdot \mathbf{OMe}$$

### Metal Complexation

Acid hydrazides form metal chelate complexes with copper and other heavy metals. For example, *Isoniazid* readily forms metal complexes (Albert, 1953), but these do not account for the tuberculostatic action of the drug. Thus, although *N*-methylisonicotinic acid hydrazide, which is tuberculostatically-inactive, is unable to form similar complexes (Cymerman-Craig, Rubbo, Willis and Edgar, 1955), nicotinic acid and picolinic acid hydrazides, which do form metal chelates, are also inactive as tuberculostatics (Albert, 1956).

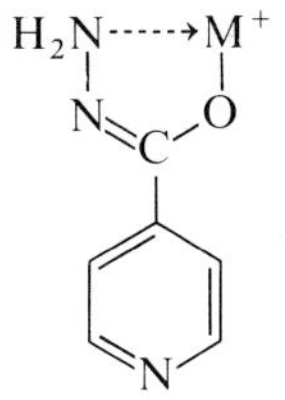

$H_2N \cdots\rightarrow M^+$ ; N=C–O

1:1-Isoniazid metal complex

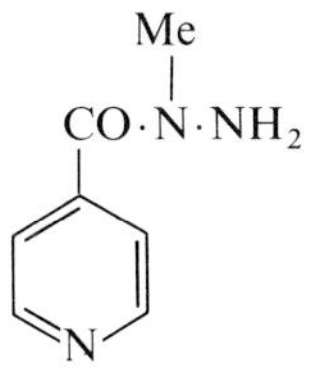

Me ; $CO \cdot N \cdot NH_2$

*N*-Methylisoniazid

## HYDROXAMIC ACIDS

Hydroxamic acids exhibit tautomerism, but spectroscopic evidence indicates that the hydroxy-amide structure normally predominates.

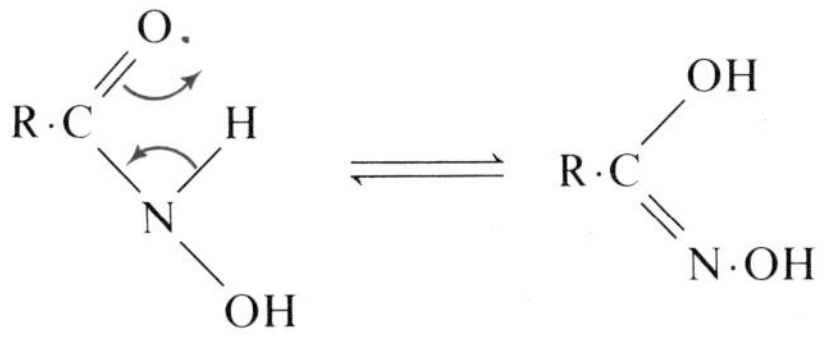

$$R \cdot C(=O)\cdot NH \cdot OH \rightleftharpoons R \cdot C(OH){=}N \cdot OH$$

Hydroxy-amide　　Hydroxy-imide

Hydroxamic acids are weak acids forming water-soluble salts, which are ionised largely in the oximido form (Exner and Kakac, 1963).

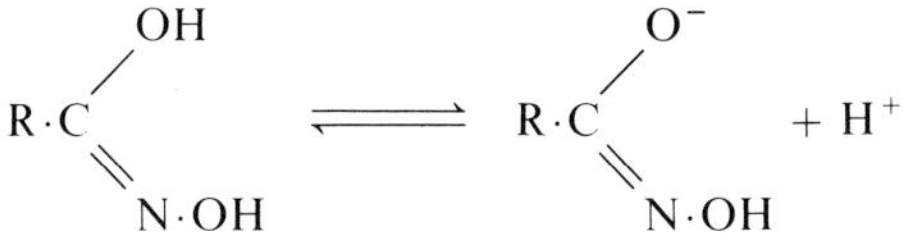

$$R \cdot C(OH){=}N \cdot OH \rightleftharpoons R \cdot C(O^-){=}N \cdot OH + H^+$$

*N*-Substituted hydroxamic acids, which cannot tautomerise, ionise in the normal way (Aubort and Hudson, 1970).

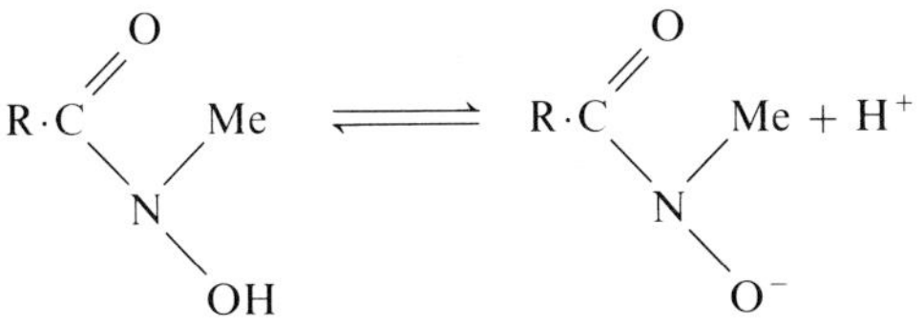

$$R \cdot C(=O) \cdot N(Me) \cdot OH \rightleftharpoons R \cdot C(=O) \cdot N(Me) \cdot O^- + H^+$$

The sodium salt of formamidohydroxamic acid slowly decomposes to give sodium bicarbonate and ammonia. Metal salts of homologous hydroxamic acids are considerably more stable and only decompose when strongly heated. Salts with polyvalent metals, such as ferric iron, form coloured complexes, which provide the basis for the colorimetric determination of esters.

$$\left[ \begin{array}{c} \text{R·C} \overset{\text{O}}{\diagup} \\ \diagdown\!\!\diagdown \text{N—ÖH} \end{array} \right]_3 \rightarrow \text{Fe}$$

Hydroxamic acids react readily with esters to form *O*-acyl hydroxamic acids (Aubort and Hudson, 1970).

$$\text{R·C(=N—OH)O}^- + \text{R}^1\text{·CO·OEt} \rightleftharpoons \text{R·C(=O)—NH—O—C(O}^-\text{)(R}^1\text{)OEt}$$

$$\xrightarrow{-\text{EtO}^-} \text{R·CO·NH·}\mathbf{O·CO}\text{·R}^1$$

A similar mechanism almost certainly applies in the ring closure which follows nucleophilic displacement of halogen by the hydroxamate anion in the synthesis of the antitubercular antibiotic, *Cycloserine*, which is a cyclic *O*-alkylhydroxamic acid.

$$\text{Cl—CH}_2\text{—CH(NH}_2\text{)·CO·NHOH} \xrightarrow[\text{resin}]{\text{Base exchange}} \text{Cl—CH}_2\text{—CH(NH}_2\text{)—C(O}^-\text{)=N—O—H}$$

$$\downarrow$$

$$\text{H}_2\text{C—CH·NH}_2 \text{ (ring: } \text{H}_2\text{C—O—NH—C(=O)—CH·NH}_2\text{)}$$

*Cycloserine*

### Hydrolysis

Hydroxamic acids are hydrolysed to the parent acid and hydroxylamine by aqueous acid. Hydrolytic enzymes in rabbit liver similarly convert benzohydroxamic acid to benzoic acid and hydroxylamine, the benzoic acid being excreted as the glycine conjugate, hippuric acid ($C_6H_5 \cdot CO \cdot NH \cdot CH_2 \cdot COOH$).

$$C_6H_5 \cdot CO \cdot NHOH \xrightarrow[H_2O]{H^+ \text{ (or enzyme)}} C_6H_5 \cdot CO \cdot OH + H_2N \cdot OH$$

*Cycloserine* is likewise unstable to aqueous acid.

$$\text{Cycloserine} \xrightarrow[H_2O]{H^+} \left[ \begin{array}{l} H_2C - CH \cdot NH_2 \\ \;\; | \quad\;\; | \\ \;\; O \quad CO \cdot OH \\ \;\; | \\ \;\; NH_2 \end{array} \right] \xrightarrow[H_2O \quad NH_2 \cdot OH]{H^+} \begin{array}{l} HO \cdot CH_2 \cdot CH \cdot NH_2 \\ \qquad\qquad\;\; | \\ \qquad\qquad CO \cdot OH \end{array}$$

Serine

### Lossen Rearrangement

Hydroxamic acids undergo the Lossen rearrangement when heated to yield isocyanates, which in the presence of strong mineral acids are converted to primary amines. In practice, the Lossen rearrangement is normally carried out on the *O*-acyl (e.g. benzoyl) derivatives in the presence of a basic catalyst.

$$R \cdot CO \cdot N(H) \cdot OBz \xrightarrow[OH^- \quad H_2O]{} R{-}C({=}O){-}\ddot{N}{-}OBz \xrightarrow[BzO^-]{} R{-}C({=}O){-}\ddot{N}^{+} \longrightarrow R \cdot N{=}C{=}O$$

In accordance with the proposed mechanism, the reaction is accelerated by *m*- and *p*-electron-withdrawing substituents in the *O*-benzoyl ring and decelerated by electron-repelling substituents.

## UNSATURATED MONOCARBOXYLIC ACIDS

The properties of unsaturated acids of structure $R \cdot CH{=}CH(CH_2)_n \cdot COOH$ are essentially those of carboxylic acids and olefines where $n$ is 3 or more. Unusual properties due to the proximity and interaction of the two groups appear where n is 0, 1 or 2.

### Acidity

Proximity of the two groups increases the acidic strength of the acid by electron-withdrawal, so that $\alpha\beta$-, $\beta\gamma$- and $\gamma\delta$-unsaturated acids are stronger than the corresponding saturated acids. The acidic strength decreases as the distance between the double bond and the carboxyl group increases (Table 39), as shown by a comparison of acrylic, vinylacetic and allylacetic acids.

| $H_2C{=}CH{\cdot}CO{\cdot}OH$ | $H_2C{=}CH{\cdot}CH_2{\cdot}CO{\cdot}OH$ | $H_2C{=}CH{\cdot}CH_2{\cdot}CH_2{\cdot}CO{\cdot}OH$ |
|---|---|---|
| Acrylic acid ($pK_a$ 4.25) | Vinylacetic acid ($pK_a$ 4.34) | Allylacetic acid ($pK_a$ 4.68) |

### Isomerism

Unsaturated acids other than those with terminal double bonds, such as acrylic and methacrylic acids, are subject to geometrical isomerism about the double bond. Most naturally occurring long-chain fatty acids are found in the *cis*-form. They may, however, be isomerised to the higher melting *trans*-form, as for example in the conversion of oleic acid to elaidic acid on treatment with nitrous acid.

$$\underset{\text{Oleic acid (m.p. 16°C)}}{\begin{matrix} H & & H \\ & C{=}C & \\ CH_3(CH_2)_7 & & (CH_2)_7{\cdot}CO{\cdot}OH \end{matrix}} \longrightarrow \underset{\text{Elaidic acid (m.p. 45°C)}}{\begin{matrix} CH_3(CH_2)_7 & & H \\ & C{=}C & \\ H & & (CH_2)_7{\cdot}CO{\cdot}OH \end{matrix}}$$

*Oleic Acid* is used as a constituent of ointments to aid absorption of metal oxides and organic bases with which it forms salts capable of passing into the lower layers of the skin. Elaidic acid, which has similar properties, gives a firmer product owing to its high melting point. *Ethyl Oleate* is used as a solvent in the preparation of intramuscular injections for slow release of injectable products.

### Double Bond Migration

Treatment of $\beta\gamma$- and $\gamma\delta$-unsaturated acids with alkali metal hydroxides leads to isomerisation by double bond migration.

$$R{\cdot}CH{=}CH{\cdot}CH_2{\cdot}CO{\cdot}O^- \underset{}{\overset{HO^-}{\rightleftharpoons}} R{\cdot}CH_2{\cdot}CH{=}CH{\cdot}CO{\cdot}O^-$$

The formation of the $\alpha\beta$-unsaturated acid is greatly favoured, since a proton is readily removed from the activated 2-methylene group of the $\beta\gamma$-unsaturated

$$\underset{\phantom{x}H}{R{\cdot}CH{=}CH{\cdot}CH{\cdot}CO{\cdot}O^-} \xrightarrow[{}^-OH \quad H_2O]{} R{\cdot}CH{=}CH{-}\bar{C}H{\cdot}CO{\cdot}O^-$$

$$\updownarrow$$

$$R{\cdot}CH_2{\cdot}CH{=}CH{\cdot}CO{\cdot}OH \xleftarrow{H^+} R{\cdot}\bar{C}H{\cdot}CH{=}CH{\cdot}CO{\cdot}O^-$$

acid and, additionally, $\alpha\beta$-unsaturation is stabilised by the gain in resonance energy with conjugation to the acid carbonyl group.

Polyene fatty acids are isomerised to the corresponding fully conjugated system when heated at 140°C with potassium t-butoxide in t-butanol. The characteristic ultraviolet absorption maximum of the conjugated acid so formed can be used to determine the unsaturation of natural oils.

Fusion of unsaturated acids with potassium hydroxide similarly causes double bond migration, but this is followed by cleavage of the two terminal carbon atoms.

$$CH_3(CH_2)_7\cdot CH{=}CH(CH_2)_7\cdot CO\cdot OH \longrightarrow CH_3(CH_2)_{14}\cdot CH{=}CH\cdot CO\cdot OH$$

Oleic acid

$$CH_3(CH_2)_{14}\cdot CH{=}CH\cdot CO\cdot OH \xrightarrow{-\,CH_3\cdot COOH} CH_3(CH_2)_{14}\cdot COOH$$

Palmitic acid

**Addition of Hydrogen Halides**

Halogen acids add to unsaturated acids to give a halogen-substituted saturated acid. Addition to $\alpha\beta$- and $\beta\gamma$-unsaturated acids is anti-Markownikoff (Chapter 4) due to the inductive effect of the carboxyl group. $\gamma\delta$-Unsaturated and other unsaturated acids, however, undergo addition in accordance with the Markownikoff rule.

Additions to $\alpha\beta$-unsaturated acids follow a different pathway from that which applies to olefines, the first step being the addition of a proton to the carbonyl oxygen.

$$H_2C{=}CH{-}C(=\ddot{O})OH \xrightarrow[-\,Br^-]{H{-}Br} H_2C{=}CH{-}C(=\overset{+}{O}H)OH \longleftrightarrow H_2\overset{+}{C}{-}CH{=}C(OH)OH$$

$$\xrightarrow{Br^-} Br\cdot CH_2\cdot CH{=}C(O{-}H)OH \xrightarrow{-\,H^+} Br\cdot CH_2\cdot CH_2\cdot C(=O)OH$$

$\beta$-Bromopropionic acid

Table 39 Nomenclature of Unsaturated Monocarboxylic Acids

| Structure | Trivial Name | Systematic Name | m.p. (°C) | p$K_a$ |
|---|---|---|---|---|
| $CH_2{=}CH\cdot CO\cdot OH$ | Acrylic acid | Propenoic acid | 13 | 4.25 |
| $CH_2{=}C(Me)\cdot CO\cdot OH$ | Methacrylic acid | 2-Methylpropenoic acid | 16 | |
| $H_3C$   $CO\cdot OH$<br>$C{=}C$<br>$H$   $H$ | *cis*-Crotonic acid | *cis*-But-2-enoic acid | 15.5 | |
| $H$   $CO\cdot OH$<br>$C{=}C$<br>$H_3C$   $H$ | *trans*-Crotonic acid | *trans*-But-2-enoic acid | 72 | 4.69 |
| $CH_2{=}CH\cdot CH_2\cdot CO\cdot OH$ | Vinylacetic acid | But-3-enoic acid | | 4.34 |
| $H_3C$   $CO\cdot OH$<br>$C{=}C$<br>$H$   $CH_3$ | Angelic acid | 2-Methyl-*cis*-crotonic acid | 46 | 4.30 |
| $H$   $CO\cdot OH$<br>$C{=}C$<br>$H_3C$   $CH_3$ | Tiglic acid | 2-Methyl-*trans*-crotonic acid<br>*trans*-Pent-2-enoic acid<br>*trans*-Pent-3-enoic acid | 64 | 5.02<br>4.83<br>4.47 |
| $CH_3\cdot CH_2\cdot CH{=}CH\cdot CO\cdot OH$ | Allylacetic acid | Pent-4-enoic acid | | 4.68 |
| $CH_3\cdot CH{=}CH\cdot CH_2\cdot CO\cdot OH$ | Undecenoic acid | Undec-10-enoic acid | 28.5 | |
| $CH_2{=}CH\cdot CH_2\cdot CH_2\cdot CO\cdot OH$ | Myristoleic acid | Tetradec-*cis*-9-enoic acid | | |
| $CH_2{=}CH(CH_2)_8\cdot CO\cdot OH$ | Palmitoleic acid | Hexadec-*cis*-9-enoic acid | | |

| | | | | |
|---|---|---|---|---|
| $CH_3(CH_2)_3 \cdot CH{=}CH(CH_2)_7 \cdot CO \cdot OH$ | Petroselenic acid | Octadec-*cis*-6-enoic acid | | |
| $CH_3(CH_2)_5CH{=}CH(CH_2)_7 \cdot CO \cdot OH$ | Oleic acid | Octadec-*cis*-9-enoic acid } | 16 | |
| $CH_3(CH_2)_{10}CH{=}CH(CH_2)_4 \cdot CO \cdot OH$ | Elaidic acid | Octadec-*trans*-9-enoic acid } | 45 | |
| $CH_3(CH_2)_7 \cdot CH{=}CH(CH_2)_7 \cdot CO \cdot OH$ | Linoleic acid | Octadec-*cis*-9,*cis*-12-dienoic acid | | |
| $CH_3(CH_2)_4(CH{=}CH \cdot CH_2)_2 \cdot (CH_2)_6 \cdot CO \cdot OH$ | α-Linolenic acid | Octadec-*cis*-9,*cis*-12,*cis*-15-trienoic acid | | |
| $CH_3 \cdot CH_2(CH{=}CH \cdot CH_2)_3 \cdot (CH_2)_6 \cdot CO \cdot OH$ | Arachidonic acid | Eicosa-*cis*-5,*cis*-8,*cis*-11,*cis*-14-tetraenoic acid | | |
| $CH_3 \cdot CH_2(CH{=}CH \cdot CH_2)_4 \cdot (CH_2)_2 \cdot CO \cdot OH$ | | | | |

**Aromatic-substituted Unsaturated Monocarboxylic Acids**

| Structure | Trivial Name | Systematic Name | m.p. (°C) | p$K_a$ |
|---|---|---|---|---|
| $C_6H_5$(H)C=C(H)$CO \cdot OH$ ($C_6H_5$ and $CO \cdot OH$ on the same side) | Allocinnamic acid | *cis*-3-Phenylacrylic acid | 68 | 3.85 |
| (H)($C_6H_5$)C=C($CO \cdot OH$)(H) ($C_6H_5$ and $CO \cdot OH$ on opposite sides) | Cinnamic acid | *trans*-3-Phenylacrylic acid | 133 | 4.45 |
| (H)(o-HO$C_6H_4$)C=C($CO \cdot OH$)(H) (trans) | *o*-Coumaric acid | *trans*-3-*o*-Hydroxyphenylacrylic acid | 208 | 4.67 |

### Hydration

Acid-catalysed hydration occurs by a similar mechanism to hydrogen halide addition, giving the corresponding hydroxyacid.

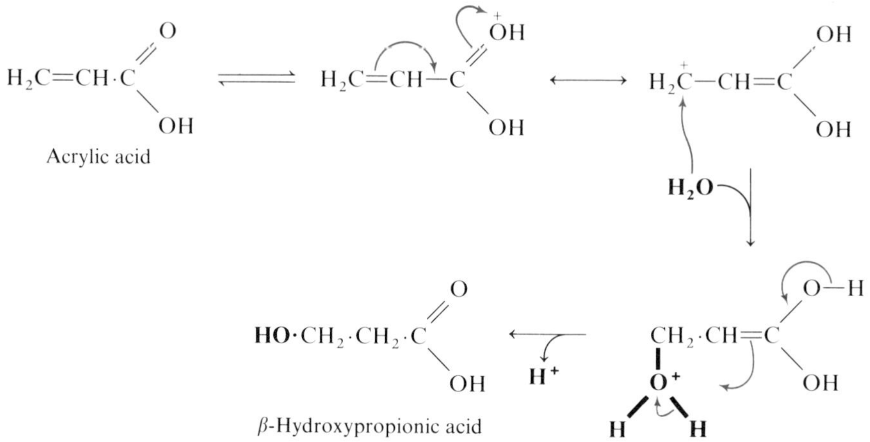

$$H_2C{=}CH{\cdot}C(=O)OH \rightleftharpoons H_2C{=}CH{-}C(={\overset{+}{O}}H)OH \longleftrightarrow H_2\overset{+}{C}{-}CH{=}C(OH)OH$$

Acrylic acid

$$\xrightarrow{H_2O} CH_2(\overset{+}{O}H_2){\cdot}CH{=}C(OH)OH \xrightarrow{-H^+} HO{\cdot}CH_2{\cdot}CH_2{\cdot}C(=O)OH$$

$\beta$-Hydroxypropionic acid

The analogous enzyme-catalysed hydration of $\alpha\beta$-unsaturated acids by enol hydratase is an important step in the metabolic oxidation of fatty acids.

### Lactone Formation

$\gamma\delta$- and $\delta\varepsilon$-unsaturated acids yield $\gamma$- and $\delta$-lactones when heated with dilute sulphuric acid. Other unsaturated acids form saturated hydrogen sulphates by addition. The lactones are formed by protonation of the double bond and cyclisation through intramolecular nucleophilic attack by the carboxyl group at the carbonium ion.

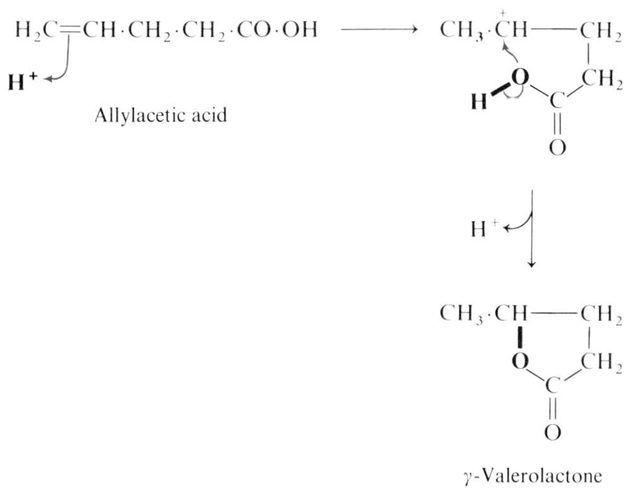

$$H_2C{=}CH{\cdot}CH_2{\cdot}CH_2{\cdot}CO{\cdot}OH \xrightarrow{H^+} CH_3{\cdot}\overset{+}{C}H{-}CH_2{-}CH_2{-}C(=O){-}OH \xrightarrow{-H^+} CH_3{\cdot}CH{-}CH_2{-}CH_2{-}C(=O){-}O \text{ (ring)}$$

Allylacetic acid

$\gamma$-Valerolactone

$\beta\gamma$-Unsaturated acids undergo decarboxylation in hot aqueous acid.

$$H_2C{=}CH{\cdot}CH_2{\cdot}CO{\cdot}OH \xrightarrow[-CO_2]{H^+} CH_3{\cdot}CH{=}CH_2$$

### Pyrazoline Formation

Diazomethane adds to $\alpha\beta$-unsaturated esters to form pyrazolines. Diazomethane is a resonance hybrid, and reaction presumably occurs with the carbonium form.

$$H_2\bar{C}{-}\overset{+}{N}{\equiv}N \longleftrightarrow H_2C{=}\overset{+}{N}{=}\bar{N} \longleftrightarrow H_2\overset{+}{C}{-}N{=}\bar{N}$$

$$H_2C{=}CH{\cdot}CO{\cdot}OEt + H_2\overset{+}{C}{-}N{=}\bar{N} \longrightarrow \text{(pyrazoline with N=N)}{\cdot}CO{\cdot}OEt \longrightarrow \text{(pyrazoline with N–NH)}{\cdot}COOEt$$

### Addition of Carbanions (Michael reaction)

$\alpha\beta$-Unsaturated esters are activated for addition of carbanions. The reaction, which is base-catalysed, is not restricted to $\alpha\beta$-unsaturated esters, but is characteristic of a variety of unsaturated compounds in which the double bond is activated by carbonyl, nitro or cyano groups.

The addendum is usually derived from an active methylene compound, such as diethyl malonate, ethyl acetoacetate, or ethyl cyanoacetate, and the reaction conducted in the presence of sodium ethoxide. The function of the base is two-fold, to generate the carbanion, and to activate the unsaturated carbonyl compound for addition.

$$CH_2(CO{\cdot}OEt)_2 + EtO^- \longrightarrow \bar{C}H(CO{\cdot}OEt)_2 + EtOH{\cdot}$$

$$R{\cdot}CH{=}CH{-}C({=}O)OEt + {}^-CH(CO{\cdot}OEt)_2 \longrightarrow R{\cdot}CH(CH(CO{\cdot}OEt)_2){-}CH{=}C(O^-)OEt \xrightarrow{H^+} R{\cdot}CH(CH(CO{\cdot}OEt)_2){\cdot}CH_2{\cdot}CO{\cdot}OEt$$

### Addition of Amines

Ammonia, primary and secondary amines also undergo nucleophilic addition to $\alpha\beta$-unsaturated esters, such as methyl acrylate. Multiple additions are apt to occur with ammonia and primary amines, and the reaction is simplest with secondary amines, which having only a single N—H group react with only one mole of unsaturated ester.

$$\mathrm{R{\cdot}CH{=}CH{-}C({=}O)OEt} + \mathrm{R'_2N{-}H} \xrightarrow{} \underset{\mathrm{R'_2N}}{\mathrm{R{\cdot}CH}}\mathrm{{-}CH{=}C(O^-)OEt} \xrightarrow{H^+} \underset{\mathrm{R'_2N}}{\mathrm{R{\cdot}CH}}\mathrm{{\cdot}CH_2{\cdot}CO{\cdot}OEt}$$

## Polymerisation

All unsaturated acids and their derivatives are prone to both ion- and radical-induced polymerisation whatever the position of the double bond. Polymerisation of the lower molecular weight $\alpha\beta$-unsaturated acids and their derivatives, which are readily available synthetically, is carried out on a production scale. Polymerisation of methyl methacrylate gives the transparent glass-like plastic, Perspex.

$$\mathrm{R\cdot} \quad \mathrm{H_2C{=}C(Me){\cdot}CO{\cdot}OMe}$$

$$\downarrow$$

$$\mathrm{R{\cdot}CH_2{\cdot}\dot{C}(Me){\cdot}CO{\cdot}OMe} \quad + \quad \mathrm{H_2C{=}C(Me){\cdot}CO{\cdot}OMe} \longrightarrow \mathrm{R{\cdot}CH_2{\cdot}C(Me){\cdot}CO{\cdot}OMe} \atop {|\atop \mathrm{CH_2{\cdot}\dot{C}(Me){\cdot}CO{\cdot}OMe}}$$

$$\downarrow \text{1. } \mathrm{n(H_2C{=}C(Me){\cdot}CO{\cdot}OMe)} \quad \text{2. } \mathrm{H^+}$$

$$\mathrm{R{\cdot}CH_2{\cdot}C(Me){\cdot}CO{\cdot}OMe}{-}\left[\mathrm{CH_2{\cdot}C(Me){\cdot}CO{\cdot}OMe}\right]_n{-}\mathrm{CH_2{\cdot}CH(Me){\cdot}CO{\cdot}OMe}$$

Similarly, ion-induced polymerisation of alkyl 2-cyanoacrylate monomers forms the basis of their action as wound adhesives (von Schmeissner, 1970). Typical monomers, such as butyl 2-cyanoacrylate, which can be applied in liquid form, are polymerised in the presence of water and proteinous wound exudates to form pliable biodegradable plastic films linked to the tissue. They are stabilised for application by small amounts of dehydrating agents, and are considered to polymerise by an anionic mechanism initiated either by hydroxyl ions (from water) or glutamate and aspartate ions (from protein).

$A^- \; H_2C{=}C(CN)\cdot CO\cdot O\cdot Bu$

↓

$A{-}CH_2\cdot C(CN)\cdot CO\cdot OBu \longrightarrow A\cdot CH_2\cdot C(CN)\cdot CO\cdot OBu$ (bonded through C to $CH_2\cdot C(CN)\cdot CO\cdot OBu$)

$H_2C{=}C(CN)\cdot CO\cdot OBu$

1. $n(H_2C{=}C(CN)\cdot CO\cdot OBu)$
2. $H^+$

$A = HO^-$

or

$-CO\cdot CH\cdot NH-$ with side chain $(CH_2)_n\cdot CO\cdot O^-$

$A\cdot CH_2\cdot C(CN)\cdot CO\cdot OBu$ — $[CH_2\cdot C(CN)\cdot CO\cdot OBu]_n$ — $CH_2\cdot C(CN)\cdot CO\cdot OBu$

**Oxidation**

Unsaturated acids, esters and natural oils containing unsaturated acids are subject to air oxidation, and eventually develop a rancid odour due to oxidative decomposition. The initial step appears to be the formation of a hydroperoxide by radical attack at the double bond (Gunstone and Hilditch, 1946). Oxidation then proceeds with attack at the allyl position of a second molecule by the radical first formed (Chapter 4). Skellon and Wharry (1963), however, point out that other types of complex peroxides and non-peroxidic oxygen compounds are also formed. The mechanism of the initial peroxide formation is also influenced by traces of metal complexes, such as metallo-porphyrins, which are often present in oils and promote autoxidation.

Despite these uncertainties, there is evidence that autoxidation of ethyl oleate is promoted by hydrogen abstraction at the $C_8$ and $C_{11}$ allylic carbon atoms, and that all four $C_8$, $C_9$, $C_{10}$ and $C_{11}$ hydroperoxides are formed (Swern, 1961).

Subsequent decomposition of the hydroperoxides leads to a mixture of ketones, alcohols, aldehydes and acids, resulting from chain cleavage in the 8, 9, 10 and 11-positions. To counter this type of oxidative degradation, *Ethyl Oleate* should be stored in well-filled, well-closed containers of limited capacity, under nitrogen and protected from light.

$$CH_3(CH_2)_6\cdot\overset{11}{C}H_2\cdot\overset{10}{C}H{=}\overset{9}{C}H\cdot\overset{8}{C}H_2(CH_2)_6\cdot CO\cdot OEt$$

$$\cdot O{-}O$$

$$\longrightarrow -CH_2\cdot\overset{10}{C}H(O\cdot O\cdot){-}\dot{C}H\cdot CH_2{-} \dashrightarrow$$

$$\longrightarrow -CH_2\cdot\dot{C}H{-}\overset{9}{C}H(O\cdot O\cdot)\cdot CH_2{-}$$

$$CH_3(CH_2)_6\cdot\overset{11}{C}H_2\cdot\overset{10}{C}H{=}\overset{9}{C}H\cdot\overset{8}{C}H_2\cdot(CH_2)_6\cdot CO\cdot OEt$$

$$-\underset{11}{\dot{C}}H\cdot CH{=}CH\cdot\overset{8}{C}H_2{-} \qquad -\overset{11}{C}H_2\cdot CH{=}CH\cdot\underset{8}{\dot{C}}H{-}$$

$$CH_3(CH_2)_6\cdot CH_2\cdot CH_2\cdot\overset{9}{C}H(O\cdot O\cdot)\cdot CH_2(CH_2)_6\cdot CO\cdot OEt$$

## BIOSYNTHESIS AND METABOLISM OF FATTY ACIDS

Fatty acids play a central rôle in the metabolism of a wide variety of living matter, and both the biosynthetic and metabolic pathways are similar if not identical in quite unrelated species.

## BIOSYNTHESIS

The key substance in the biosynthesis of fatty acids is acetyl coenzyme A ($CH_3\cdot CO\cdot S\cdot\overline{CoA}$), and a fundamental step in the process is its conversion to malonyl-S$\cdot\overline{CoA}$ by the enzyme acetyl-$\overline{CoA}$ carboxylase.

$$CH_3\cdot CO\cdot S\cdot\overline{CoA} + \mathbf{CO_2} \xrightarrow{ATP/Mg^{2+}} \mathbf{HO\cdot OC}\cdot CH_2\cdot CO\cdot S\cdot\overline{CoA} \quad (1)$$

Malonyl-S$\cdot\overline{CoA}$

Studies of fatty acid biosynthesis in *E. coli* (Alberts, Majerus, Talamo and Vagelos, 1964) have shown that acetyl and malonyl groups are then transferred to an acyl carrier protein (ACP) which acts as a carrier in the build up of the fatty acid from the separate component units (Goldman, Alberts and Vagelos, 1963). The acyl carrier protein has been shown to have a single —SH group, and transfer of the acetyl and malonyl groups is effected by specific acetyl and malonyl transacylases.

$$CH_3\cdot CO\cdot S\cdot \overline{CoA} + \mathbf{HS\cdot \overline{ACP}} \xrightarrow{\text{Acetyltransacylase}} CH_3\cdot CO\cdot \mathbf{S\cdot \overline{ACP}} + \mathbf{HS}\cdot \overline{CoA} \quad (2)$$

$$\mathbf{HO\cdot OC}\cdot CH_2\cdot CO\cdot S\cdot \overline{CoA} + HS\cdot \mathbf{\overline{ACP}}$$

$$\Big\downarrow \text{Malonyltransacylase}$$

$$\mathbf{HO\cdot CO}\cdot CH_2\cdot CO\mathbf{\cdot \overline{ACP}} + HS\cdot \overline{CoA}$$

Acetyl-ACP and malonyl-ACP react in the presence of acetoacetyl synthetase to give acetoacetyl-ACP.

$$\mathbf{CH_3\cdot CO}\cdot S\cdot \overline{ACP} + \mathbf{HO\cdot OC}\cdot CH_2\cdot CO\cdot S\cdot \overline{ACP} \quad (3)$$

$$\Big\downarrow\Big\uparrow \text{Acetoacetylsynthetase} \quad (4)$$

$$\mathbf{CH_3\cdot CO}\cdot CH_2\cdot CO\cdot S\cdot \overline{ACP} + HS\cdot \overline{ACP} + \mathbf{CO_2}$$

Reduction by $\beta$-ketoacetyl-ACP-reductase, dehydration by enoyl-ACP-hydratase, reduction by flavine mononucleotide (FMN) and finally displacement of the carrier protein complete the process of chain extension.

A number of important observations support this proposed pathway. Thus, in the presence of excess malonyl-S$\cdot\overline{CoA}$, labelled acetyl-S$\cdot\overline{CoA}$ is incorporated only into the two terminal positions of the fatty acid chain, e.g. $C_{15}$ and $C_{16}$ in palmitic acid; all the remaining carbon atoms of the fatty acid are derived from malonate. Labelled carbon dioxide is not incorporated; this would not be expected, as the carbon dioxide incorporated in malonate synthesis (1) is lost in the formation of the $\beta$-keto ester (4). Similarly, studies with tritiated NADPH lead to tritium incorporation on carbon atom 3 and higher odd numbered carbon atoms (5); it is not incoporated on carbon atom 2, since reduction at step (7) involves hydrogen transfer from the flavine mononucleotide and not directly from NADPH (Brady, Bradley and Trams, 1960).

$$CH_3 \cdot \mathbf{CO} \cdot CH_2 \cdot CO \cdot S \cdot \overline{ACP} + NADPH + H^+$$

$$\rightleftharpoons \quad \beta\text{-Ketoacetyl-}\overline{ACP}\text{-reductase} \qquad (5)$$

$$CH_3 \cdot \mathbf{CH(OH)} \cdot CH_2 \cdot CO \cdot S \cdot \overline{ACP} + NADP^+$$

D-(−)-β-Hydroxybutyryl-$\overline{ACP}$

$$CH_3 \cdot CH\mathbf{(OH)} \cdot CH_2 \cdot CO \cdot S \cdot \overline{ACP}$$

$$\rightleftharpoons \quad \text{Enoyl-}\overline{ACP}\text{ hydratase} \qquad (6)$$

$$CH_3 \cdot CH{=}CH \cdot CO \cdot S \cdot \overline{ACP} + H_2O$$

Crotonyl-$\overline{ACP}$

$$CH_3 \cdot CH{=}CH \cdot CO \cdot S \cdot \overline{ACP} + NADPH + \mathbf{H^+}$$

$$\rightleftharpoons \quad \text{FMN} \qquad (7)$$

$$CH_3 \cdot CH_2 \cdot CH_2 \cdot CO \cdot S \cdot \overline{ACP} + NADP^+$$

Butyryl-$\overline{ACP}$

$$CH_3 \cdot CH_2 \cdot CH_2 \cdot CO \cdot S \cdot \overline{ACP} + \mathbf{HS \cdot \overline{CoA}}$$

$$\rightleftharpoons \qquad (8)$$

$$CH_3 \cdot CH_2 \cdot CH_2 \cdot CO \cdot \mathbf{S \cdot \overline{CoA}} + \mathbf{HS} \cdot \overline{ACP}$$

## METABOLISM

The metabolism of fatty acids by β-oxidation was established by Knoop as early as 1905, from feeding experiments with acids labelled on the terminal carbon atom with a phenyl group, and identification of the end-products in the urine of the experimental animals. When ω-phenyl fatty acids with an even number of carbon atoms in the chain were fed, phenylacetylglycine ($C_6H_5 \cdot CH_2 \cdot CO \cdot NH \cdot CH_2 \cdot COOH$) was excreted in the urine. Fatty acids with an odd number of carbon atoms and an ω-phenyl substituent, on the other hand, led to the excretion of benzoylglycine ($C_6H_5 \cdot CO \cdot NH \cdot CH_2 \cdot COOH$). The pattern of metabolism was consistent with the idea that the carbon chain was degraded by the successive cleavage of the two carbon units, by a series of $\beta$-oxidations.

$$C_6H_5\cdot CH_2\cdot CH_2\cdot CO\cdot OH \longrightarrow R\cdot CHOH\cdot CH_2\cdot CO\cdot OH \rightleftharpoons R\cdot CO\cdot CH_2\cdot CO\cdot OH \longrightarrow R\cdot CO\cdot OH + CH_3\cdot CO\cdot OH$$

One major difficulty in acceptance of the $\beta$-oxidation theory was the co-production of acetoacetate. This was overcome by the work of MacKay, Wick and Barnum (1940), which established that the acetoacetate is formed by self-condensation of acetate. It is now established that the fatty acid undergoing degradation is activated by conversion to the corresponding acyl-S·$\overline{\text{CoA}}$, before undergoing oxidation according to the following sequence.

$$R\cdot CH_2\cdot CH_2\cdot CO\cdot \mathbf{OH} \xrightarrow[CH_3\cdot CO\cdot \mathbf{S}\cdot \overline{\mathbf{CoA}} \;\to\; CH_3\cdot CO\cdot \mathbf{OH}]{\text{Thiokinase}} R\cdot CH_2\cdot CH_2\cdot CO\cdot \mathbf{S}\cdot \overline{\mathbf{CoA}}$$

$$R\cdot CH_2\cdot CH_2\cdot CO\cdot S\cdot \overline{CoA} \xrightarrow[FAD \;\to\; FADH_2]{\text{Acyl dehydrogenase}} R\cdot CH\mathbf{=}CH\cdot CO\cdot S\cdot \overline{CoA}$$

$$R\cdot CH\mathbf{=}CH\cdot CO\cdot S\cdot \overline{CoA} \xrightarrow[\text{Enol hydratase}]{\mathbf{H_2O}} R\cdot \mathbf{CH(OH)}\cdot CH_2\cdot CO\cdot S\cdot \overline{CoA}$$

$$R\cdot \mathbf{CH(OH)}\cdot CH_2\cdot CO\cdot S\cdot \overline{CoA} \xrightarrow[NAD^+ \;\to\; \mathbf{NADH + H^+}]{\beta\text{-Hydroxyacyl dehydrogenase}} R\cdot \mathbf{CO}\cdot CH_2\cdot CO\cdot S\cdot \overline{CoA}$$

$$R\cdot \mathbf{CO}\cdot CH_2\cdot CO\cdot S\cdot \overline{CoA} \xrightarrow[\mathbf{HS}\cdot \overline{\mathbf{CoA}} \;\to\; CH_3\cdot CO\cdot S\cdot \overline{CoA}]{\beta\text{-Ketothiolase}} R\cdot \mathbf{CO}\cdot S\cdot \overline{\mathbf{CoA}}$$

## BIOLOGICAL FUNCTION

Fatty acids in the form of natural fats form an important article of diet for many animals because of their high calorific value. When oxidised, they provide a large amount of energy, and they are readily stored as components of fatty tissue.

Certain of the long-chain unsaturated acids are essential dietary constituents. The uptake of linoleic, linolenic and arachidonic acids appears to be important in controlling the levels of blood cholesterol and hence in determining the incidence of atherosclerosis (Jolliffe and Goodhart, 1960). Rats on a diet low in unsaturated fatty acids show impaired growth and reproduction, and develop scaliness of the skin (Sinclair, 1964). The synthetic polyacetylenic analogue of arachidonic acid, eicosa-5,8,11,14-tetraynoic acid, has been shown to be active in the suppression of sebaceous gland activity, possibly through inhibition of cholesterol synthesis (Strauss, Pochi and Whitman, 1967).

$$CH_3(CH_2)_4 \cdot (C{\equiv}C \cdot CH_2)_4 \cdot (CH_2)_3 \cdot CO \cdot OH$$

Eicosa-5,8,11,14-tetraynoic acid

## Prostaglandins

The prostaglandins are a group of naturally occurring fatty acid derivatives related to a parent saturated $C_{20}$ structure, prostanoic acid. Individual prostaglandins which are found in prostate extracts and in other tissues are classified according to the substituents in the cyclopentane ring, and designated by letters (E and F). Subscript numerals indicate the number of double bonds. Additional subscripts, $\alpha$ and $\beta$, indicate stereoisomers ($PGF_{2\alpha}$ and $PGF_{2\beta}$). Only the $\alpha$-isomers occur naturally; both enantiomers can be formed by reduction of the carbonyl group in the corresponding PGE.

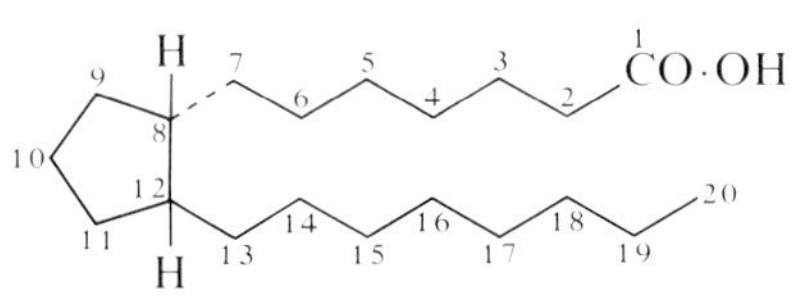

Prostanoic acid

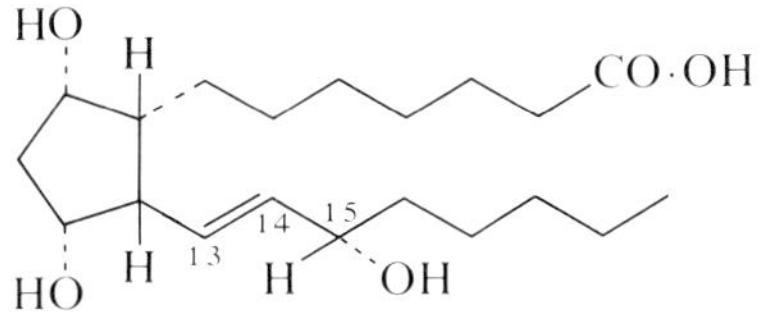

$PGF_{1\alpha}$

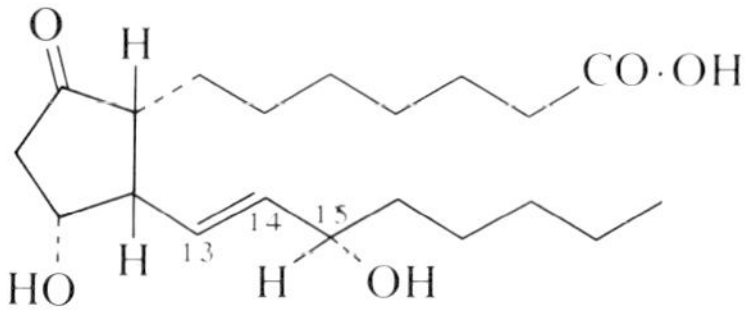

$PGE_1$

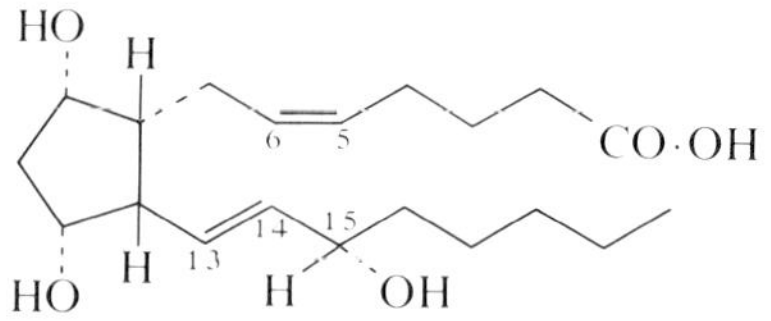

$PGF_{2\alpha}$

Most mammalian cells are capable of synthesising prostaglandins, and their biosynthesis can be initiated or enhanced by a wide variety of physiological, pharmacological and pathological stimuli (Bergström, Carlson and Weeks, 1968). Prostaglandin, $PGE_1$, lowers blood pressure in man by producing peripheral vasodilation. The prostaglandins also appear to have a physiological rôle in uterine function, and both $PGE_2$ and $PGF_{2\alpha}$ are effective in the induction of labour and as abortifacients. $PGE_1$ also greatly reduces the formation of free fatty acids from adipose tissue, possibly by blocking the formation of cyclic AMP, required for activation of lipolysis (Horton, 1969; Pickles, 1969).

Prostaglandins, which are widely distributed in mammalian tissue, are formed biosynthetically from dietary linoleic acid. This is converted into arachidonic acid, which is stored as phospholipid until required for oxidation and conversion to prostaglandin. Peroxidation occurs first at $C_{11}$ to form a hydroperoxide which undergoes further oxidation and stereospecific cyclisation.

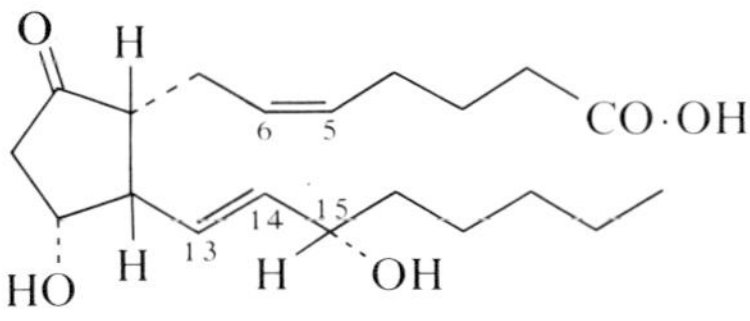

$PGE_2$ 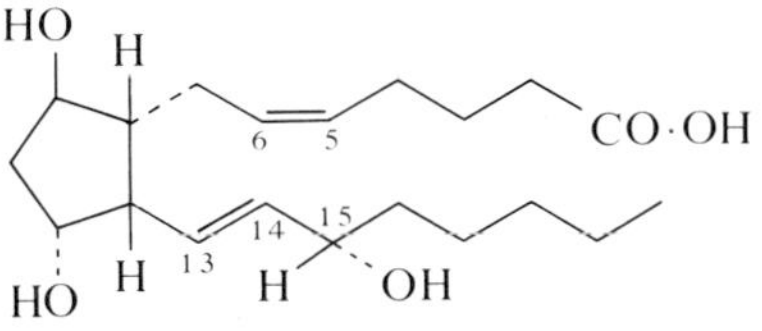 $PGF_{2\beta}$

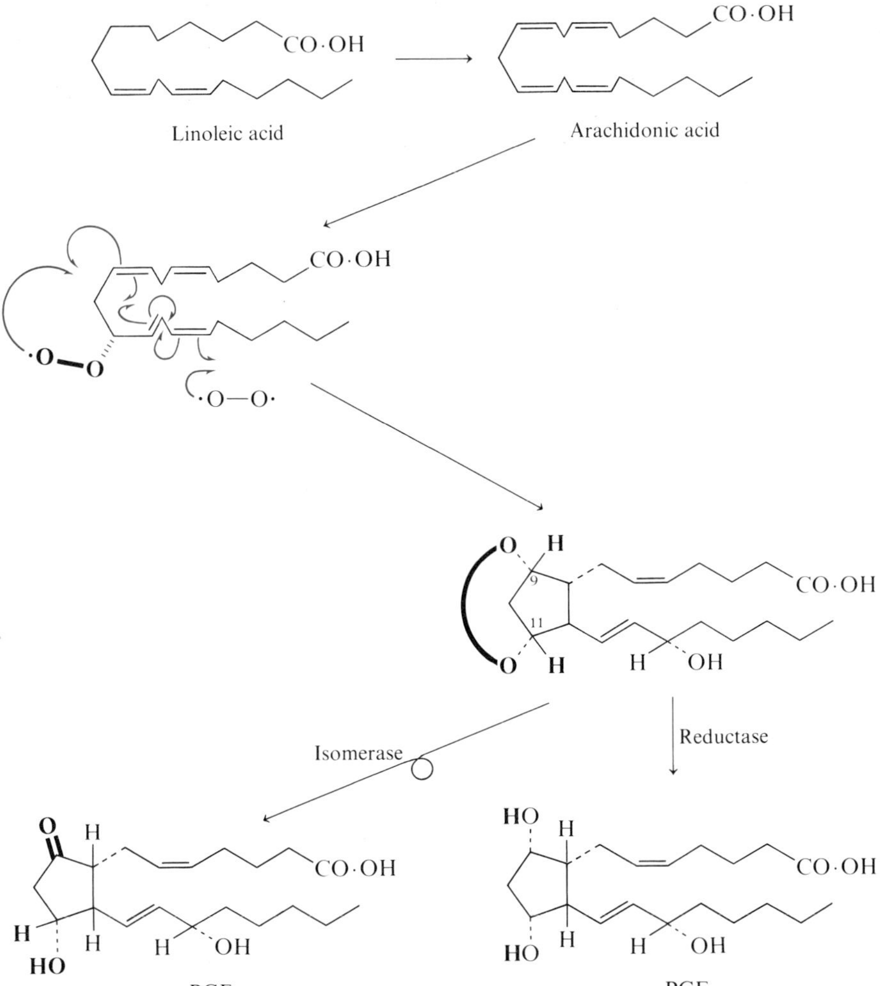

Linoleic acid → Arachidonic acid

Isomerase Reductase

$PGE_2$ $PGF_{2\alpha}$

Prostaglandin endoperoxides are also transformed by microsomal enzymes in arterial walls to an unstable derivative, 9-deoxy-6,9α-epoxy-$\Delta^5$-$PGF_{1\alpha}$, designated PGX (Moncada, Gryglewski, Bunting and Vane, 1976). PGX is a powerful inhibitor of human platelet aggregation, some thirty times more potent than $PGE_1$, and relaxes blood vessels. The formation of an epoxy ring in PGX leads to a molecular structure which, in contrast to the parent prostaglandin, is virtually linear in shape. Prostaglandin endoperoxides are also converted by a microsomal enzyme system present in blood platelets to a non-prostaglandin, thromboxane $A_2$, $TXA_2$ (Needleman, Moncada, Bunting, Vane, Hamberg and Samuelsson, 1976). In contrast to PGX, and in common with the prostaglandin endoperoxides, $TXA_2$ causes platelet aggregation and smooth muscle contraction.

It appears that the prostaglandins are biosynthesised on demand, and are metabolised rapidly to limit their physiological action. Metabolic deactivation occurs primarily by oxidation of the allylic hydroxyl at $C_{15}$.

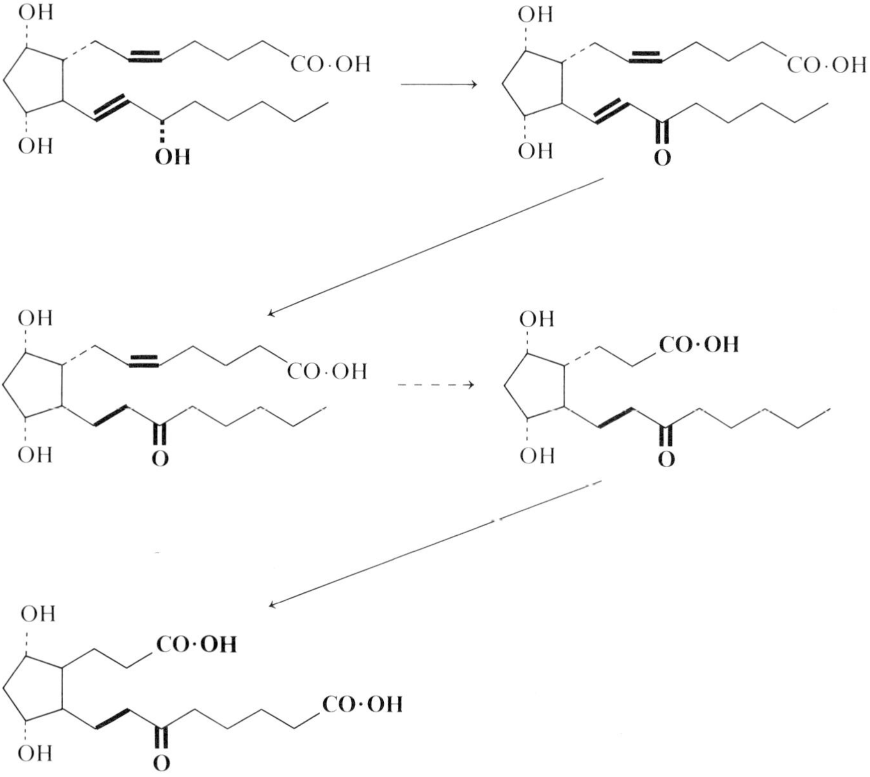

OH CO·OH OH OH → OH CO·OH OH O

OH CO·OH OH O - - - → OH CO·OH OH O

OH CO·OH CO·OH OH O

Biosynthesis of prostaglandins is inhibited by *Aspirin* and other non-steroid anti-inflammatory agents such as *Indomethacin* (Vane, 1971). It is considered that the inhibition of prostaglandin synthesis may explain some of the therapeutic effects of this group of drugs. Thus prostaglandin $PGE_1$ induces fever, and prostaglandin infusions have been shown to induce headache (Bergström, Carlson and Weeks, 1968). Antipyretic and analgesic effects could, therefore, be attributed to reduction of prostaglandin synthesis in the brain. Some types of inflammation have similarly been ascribed to prostaglandin release. *Aspirin* also reduces blood platelet aggregation (Smith and Willis, 1971) by blocking the formation of a labile prostaglandin derivative, which appears to be a 9,11-cyclic endoperoxide (Flower, 1975).

# 12 Dibasic Acids

## SATURATED DIBASIC ACIDS

### Structure and Nomenclature

Saturated dibasic acids form part of a homologous series of general formula $C_nH_{2n}(CO \cdot OH)_2$. In oxalic acid, the first member of the series, $n = 0$. Commonly-available higher members of the series have the carboxyl groups at the two terminal positions of a polymethylene chain $HOOC(CH_2)_n \cdot COOH$. Most of the lower members have long-established trivial names. Under the IUPAC system of nomenclature, they are named as *dioic* acids derived from the parent saturated hydrocarbon with the same total number of carbon atoms (Table 40).

### General Physical Properties

The saturated dibasic acids are crystalline solids, and in accordance with the presence of two lipophilic functional groups (COOH) relatively more soluble in water than the corresponding monocarboxylic acids. Their physical properties, notably melting point and solubility, show decided alternation as the series is ascended, due to intermolecular hydrogen bonding of one carboxyl group to another. This effect, which results in dimerisation of monocarboxylic acids, leads dicarboxylic acids to form polymer chains in the crystal structure (Allen and Caldin, 1953).

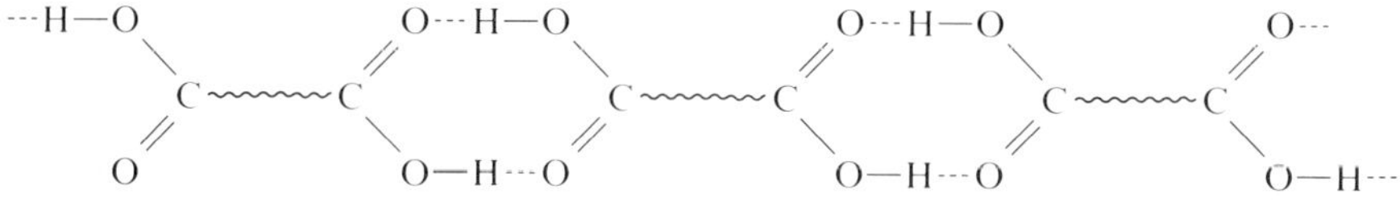

```
---H—O              O---H—O              O---H—O              O---
      \            //      \            //      \            //
       C~~~~~~~~~C           C~~~~~~~~~C           C~~~~~~~~~C
      //           \        //          \         //          \
     O              O—H---O              O—H---O               O—H---
```

The relative orientation of the two carboxyl groups is different in odd and even numbered dicarboxylic acids due to the zig-zag arrangement of the polymethylene chain separating the two groups.

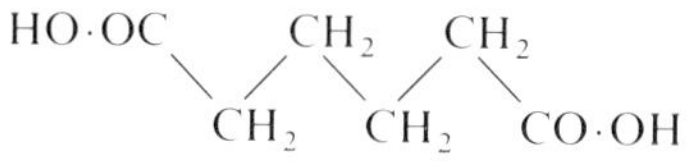

```
HO·OC       CH2     CH2                 HO·OC       CH2     CO·OH
     \     /   \   /   \                     \     /   \   /
      CH2       CH2     CO·OH                 CH2       CH2
          even-chain                              odd-chain
```

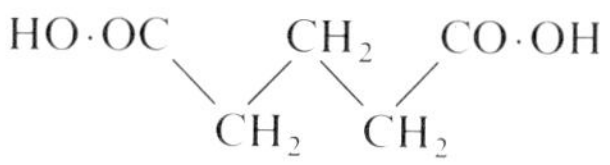

In the **even**-chain acids, chain polymerisation is achieved with the entire carbon chain lying naturally without strain in a single plane. In the **odd**-chain acids, however, chain polymerisation can only be achieved with twisting of the polymethylene chain, and in consequence these acids have a higher energy content. This shows in their lower melting points and other unusual properties.

Table 40. Nomenclature of Saturated Dibasic Acids

| Structure | Trivial Name | Systematic Name | m.p. (°C) | p$K_{a_1}$ | p$K_{a_2}$ |
|---|---|---|---|---|---|
| HOOC·COOH | Oxalic acid | Ethanedioic acid | 189.5[1] | 1.27 | 4.27 |
| HOOC·$CH_2$·COOH | Malonic acid | Propanedioic acid | 135.5 | 2.86 | 5.70 |
| HOOC$(CH_2)_2$·COOH | Succinic acid | Butanedioic acid | 185 | 4.21 | 5.64 |
| HOOC$(CH_2)_3$·COOH | Glutaric acid | Pentanedioic acid | 99 | 4.34 | 5.27 |
| HOOC$(CH_2)_4$·COOH | Adipic acid | Hexanedioic acid | 153 | 4.41 | 5.28 |
| HOOC$(CH_2)_5$·COOH | Pimelic acid | Heptanedioic acid | 105 | | |
| HOOC$(CH_2)_6$·COOH | Suberic acid | Octanedioic acid | 144 | | |
| HOOC$(CH_2)_7$·COOH | Azelaic acid | Nonanedioic acid | 107 | | |
| HOOC$(CH_2)_8$·COOH | Sebacic acid | Decanedioic acid | 134.5 | | |

[1] anhydrous oxalic acid

Oxalic acid crystallises readily as a dihydrate, but the anhydrous form is readily formed on drying at 70–80°C or on boiling with benzene and distilling off the benzene–water azeotrope. The remaining dibasic acids are anhydrous.

**Acidity**

Dibasic acids ionise in two stages.

$$\mathrm{HO\cdot OC\cdot(CH_2)_n\cdot CO\cdot OH} \underset{}{\overset{K_{a_1}}{\rightleftharpoons}} \mathrm{HO\cdot OC\cdot(CH_2)_n\cdot CO\cdot O^-} + \mathbf{H^+}$$

$$\mathrm{HO\cdot OC\cdot(CH_2)_n\cdot CO\cdot O^-} \underset{}{\overset{K_{a_2}}{\rightleftharpoons}} \mathrm{^-O\cdot OC\cdot(CH_2)_n\cdot CO\cdot O^-} + \mathbf{H^+}$$

In the first stage, the acid can lose a proton from either carboxyl, but mono-anions can accept a proton in only one position. Ionisation is, therefore, kinetically more favoured in dibasic than in monobasic acids. Additionally, ionisation is favoured by the electron-withdrawing influence of each carboxyl group on the other. This effect, however, diminishes with the number of methylene groups

Table 41. Phthalic Acids

| | Phthalic acid | Isophthalic acid | Terephthalic acid |
|---|---|---|---|
| Structure | benzene-1,2-di(CO·OH) | benzene-1,3-di(CO·OH) | benzene-1,4-di(CO·OH) |
| m.p. (°C) | 231 | 345–347 | sublimes *ca* 300 |
| p$K_{a_1}$ | 2.89 | 3.54 | 3.51 |
| p$K_{a_2}$ | 5.41 | 4.60 | 4.82 |

separating the two carboxyl groups, so that the first dissociation constant shifts from a p$K_a$ of 1.27 in oxalic acid through 2.86 in malonic acid to a near constant value in succinic (4.21), glutaric (4.34) and adipic (4.14) acids.

In the second stage of ionisation, dissociation of the acidic proton is inhibited by hydrogen bonding to the carboxylate ion (CO·O$^-$). This effect is greatest in the monomalonate ion (p$K_{a_2}$ 5.70), which is able to form a six-membered resonance-stabilised chelate ring.

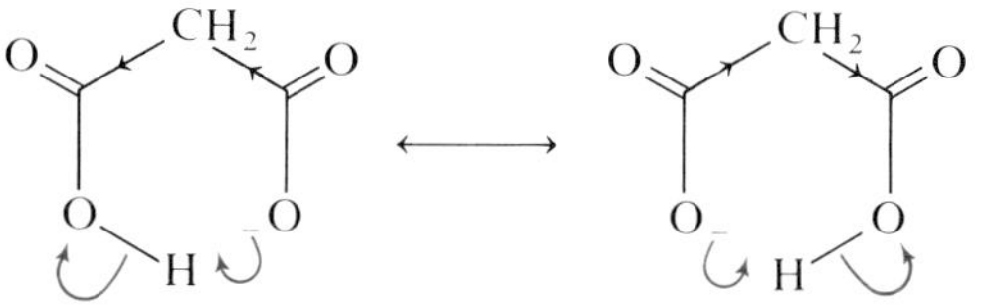

Intramolecular chelates are less feasible in succinate and higher dibasic acid monoanions. The $-$I effect of the carboxyl group is also reduced with the introduction of additional methylene groups. Acidic strength at the second dissociation stage, therefore, increases marginally with chain length beyond malonic acid, as in succinic acid (p$K_{a_2}$ 5.64) and glutaric acid (p$K_{a_2}$ 5.27). There is, however, no really adequate explanation for the apparently anomalous second acid dissociation constant of oxalic acid (p$K_{a_2}$ 4.27), which is the strongest acid of the series at this stage of dissociation.

Kinetically, the second dissociation equilibrium is also influenced by the fact that di-anions can add a proton to either one of the two available carboxylate ions. Association is, therefore, more favoured than in the monocarboxylic acids. In consequence, dibasic acids are weaker acids at the second dissociation stage than monocarboxylic acids.

**Unsaturated dibasic acids.** Maleic and fumaric acids are vinylogues of oxalic acid (i.e. with a vinyl group, >C=C<, interposed between the two carboxyl groups). The *cis*-monomaleate ion is stabilised by internal hydrogen bonding,

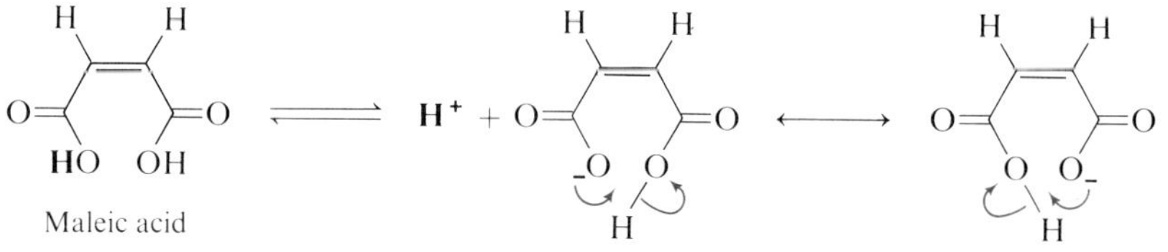

Maleic acid

Monomaleate ion

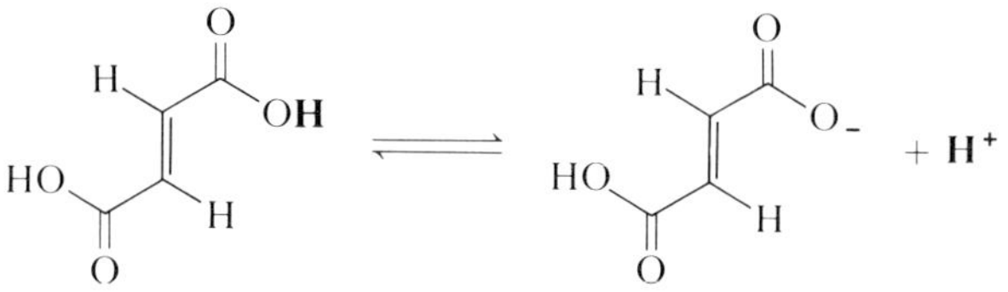

Fumaric acid

Monofumarate ion

but it is geometrically impossible for the *trans*-monofumarate ion to be so stabilised. In consequence, the monomaleate ion is stabilised with respect to the unionised acid, and maleic acid is, therefore, a stronger acid (p$K_{a_1}$ 1.93) than fumaric acid (p$K_{a_1}$ 3.03) at the first dissociation.

At the second stage of dissociation, the position is reversed and maleic acid (p$K_{a_2}$ 6.58) is a weaker acid than fumaric acid (p$K_{a_2}$ 4.54), since the adjacent negatively-charged carboxylate ion hinders removal of the proton in the *cis*-acid (Stenlake, 1953).

**Aromatic acids.** The dissociation of aromatic dibasic acids is similarly dictated by a combination of electronic and steric effects. In phthalic acid, however, the ring competes with the adjacent carboxyl group for electrons from the first carboxyl group. Phthalic acid (p$K_{a_1}$ 2.89) is, therefore, a somewhat weaker acid than oxalic acid (p$K_{a_1}$ 1.27). In isophthalic acid, however, the electron-attracting effect of the carboxyl groups is not transmitted to the *meta* position, so that it is a weaker acid (p$K_{a_1}$ 3.54) than phthalic acid. The order of acidic strengths is reversed at the second dissociation stage, because the chelate effect inhibits ionisation of the second acidic proton only in the *ortho*-disubstituted phthalic acid.

## Salt Formation

Oxalic acid is present in large amounts as calcium oxalate in rhubarb, and is widely distributed as such throughout the plant kingdom. Oxalic acid is toxic in man only in large doses. Small amounts are normally ingested in foodstuffs and pass through the human body to be excreted unchanged in the urine. The bulk of the normal daily urinary output of oxalic acid (6–45 mg/24 hr in normal subjects) is derived from the metabolism of ascorbic acid (Ch. 21), glycollic acid and glyoxalic acid. Excessive urinary excretion of oxalic acid (100–400 mg/24 hr) occurs in primary hyperoxaluria, a hereditary disease due to impairment of glycollate and glyoxylate metabolism, and can lead to deposition of calcium oxalate crystals in the urine. Excessive formation and excretion of oxalic acid also occurs in ethyleneglycol poisoning.

Heavy metal salts of dibasic acids are relatively insoluble in water. Blood coagulation can be prevented by small concentrations (0.1%) of oxalic acid (or citric acid) which acts as a de-calcifier, since this anticoagulant effect can be inhibited by the addition of calcium salts. The ferrous salts of the much less toxic succinic and fumaric acids are used for the oral administration of iron in the treatment of iron-deficiency anaemias.

Dibasic acids, being stronger acids than the monocarboxylic acids, in general form good crystalline salts with organic bases. Oxalic acid is particularly useful in this respect and oxalates are used in the isolation of naturally occurring and synthetic organic bases (e.g. ephedrine from plant extracts). In such cases, there is always a danger that oxalate contamination may be carried through as a toxic hazard in the final product; suitable tests to guard against this hazard must be applied to final products resulting from such procedures. Other less toxic dibasic acids form a useful means of producing stable crystalline water-soluble salts of organic bases for oral administration, e.g. *Piperazine Adipate*.

*Ergometrine Maleate* (Ergonovine Maleate) is an example of the use of the acid salt of a dibasic acid to provide pH stability for the base in solutions used for injection, since ergometrine is readily equilibrated to its pharmacologically inert epimer, ergometrinine, in neutral or alkaline solution.

Certain complex aromatic dibasic acids, such as embonic acid, are used in the preparation of water-insoluble salts, which are poorly absorbed from the gut, as for example *Pyrantel Embonate* (Pyrantel Pamoate), which is administered orally in the treatment of hookworm and pinworm infestation of the gastro-intestinal tract, and the anthelmintic, *Viprynium Embonate* (Pyrvinium Pamoate).

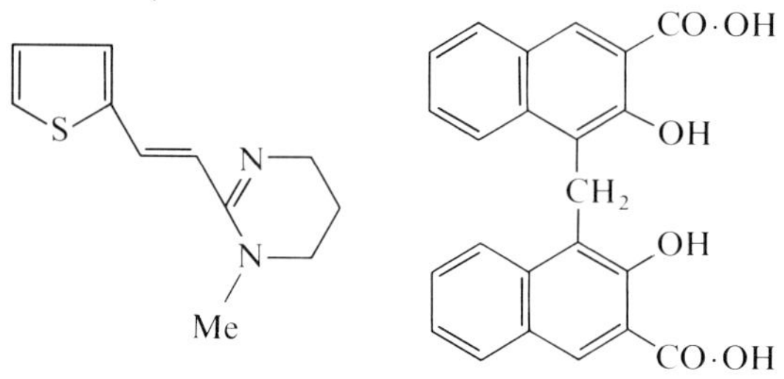

S N N Me CO·OH OH $CH_2$ OH CO·OH

*Pyrantel Embonate*

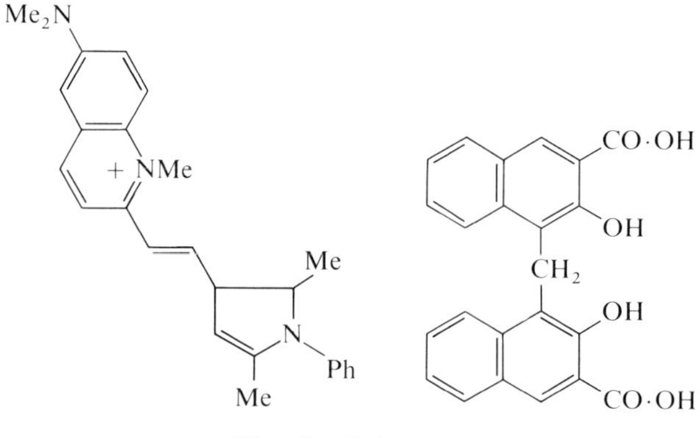

$Me_2N$ + NMe Me N Ph Me CO·OH OH $CH_2$ OH CO·OH

*Viprynium Embonate*

## Thermal and Light Stability

Oxalic acid decarboxylates when heated strongly to give carbon dioxide and formic acid; decomposition of the latter will also occur unless precautions are taken to prevent it, as in its manufacture from oxalic acid and glycerol.

$$(CO{\cdot}OH)_2 \xrightarrow[CO_2]{} H{\cdot}CO{\cdot}OH \longrightarrow H_2O + CO_2$$

Malonic acid, similarly, undergoes thermal decarboxylation at temperatures above 130°C, a property which is important in the use of malonic esters in synthesis (p. 339).

$$HO\cdot OC\cdot CH_2\cdot CO\cdot OH \longrightarrow CH_3CO\cdot OH + CO_2$$

In contrast, succinic and glutaric acids, and the *ortho* dibasic aromatic acid, phthalic acid, undergo cyclodehydration on heating alone, or in a high boiling solvent (acetic anhydride is frequently used) to yield the corresponding anhydrides.

$$\begin{array}{l} CH_2\cdot CO\cdot OH \\ | \\ CH_2\cdot CO\cdot OH \end{array} \longrightarrow \begin{array}{l} CH_2\cdot CO \\ | \qquad\quad \rangle O \\ CH_2\cdot CO \end{array} + H_2O$$

Succinic anhydride

$$C_6H_4(CO\cdot OH)_2 \longrightarrow C_6H_4(CO)_2O + H_2O$$

Phthalic anhydride

Unsaturated dibasic acids, maleic and fumaric acid, differ in that whereas the *cis*-isomer, maleic acid, readily forms an anhydride on heating, fumaric acid does not, though when strongly heated the latter also forms maleic anhydride.

Adipic and all higher saturated dibasic acids form cyclic ketones on pyrolysis in yields which depend on the ring size of the product; optimum yields are obtained in the formation of cyclopentanone and cyclohexanone from the calcium salts (Chapter 10).

$$(CH_2)_4(CO\cdot OH)_2 \xrightarrow{300^\circ} C_5H_8{=}O + CO_2 + H_2O$$

Adipic acid — Cyclopentanone

Dibasic acids undergo photodecarboxylation in the presence of iron (III) chloride (Kuhnle, Lunden and Waiss, 1972).

### Anhydrides, Hemi-esters and -amides

Dibasic acid anhydrides have properties similar to those of the monocarboxylic acids. They are, however, especially useful in the preparation of hemi-esters and -amides, and in the preparation of cyclic imides.

Hemi-esters, such as *Hydrocortisone Hydrogen Succinate*; readily form water-soluble sodium salts for use in the preparation of aqueous intravenous injections, which are readily degraded by metabolic hydrolysis to the parent steriod.

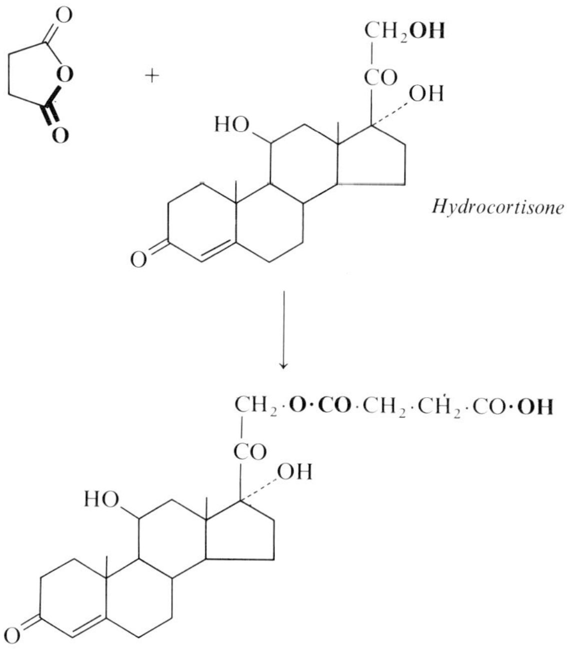

$CH_2OH$ CO OH HO O

*Hydrocortisone*

$CH_2$·**O·CO**·$CH_2$·$CH_2$·CO·**OH** CO OH HO O

*Hydrocortisone Hydrogen Succinate*

The parent, unesterified steroid, is almost insoluble in water. Hemi-esters of succinic acid and phthalic acid are also used as intermediates in the resolution of optically active alcohols (Chapter 13).

Analogous reactions of cyclic anhydrides with ammonia, primary and secondary amines similarly give rise to hemi-amides, via the ammonium or amine salt.

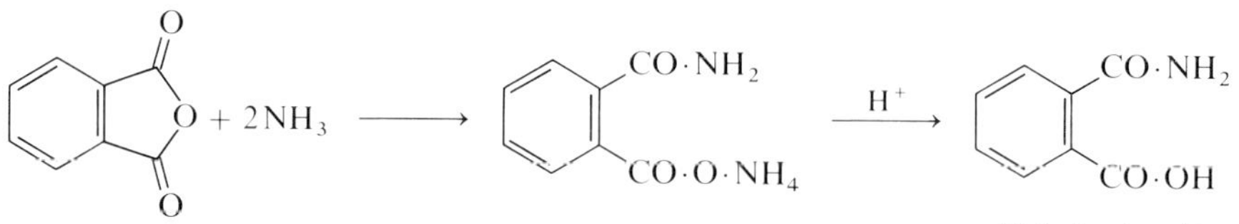

$$\text{O} + 2NH_3 \longrightarrow \text{CO·NH}_2,\ \text{CO·O·NH}_4 \xrightarrow{H^+} \text{CO·NH}_2,\ \text{CO·OH}$$

Phthalamic acid

Oxalic and malonic acids do not form anhydrides, but they do yield hemi-acids and hemi-amides by alternative reactions.

*Succinylsulphathiazole* (Succinylsulfathiazole) and *Phthalylsulphathiazole* (Phthalylsulfathiazole) are typical examples of hemi-amides. They are almost completely insoluble in water, and virtually insoluble in lipid solvents. They are fully ionised under alkaline conditions, and hence when administered orally are largely retained in the gut. The amide link, however, is susceptible to enzymatic hydrolysis by bacterial proteinases in the lower part of the alimentary tract with the release of the parent sulphonamide. They are, therefore, useful for reducing the bacterial content of the large intestine prior to operation, and also in the treatment of bacillary dysentery.

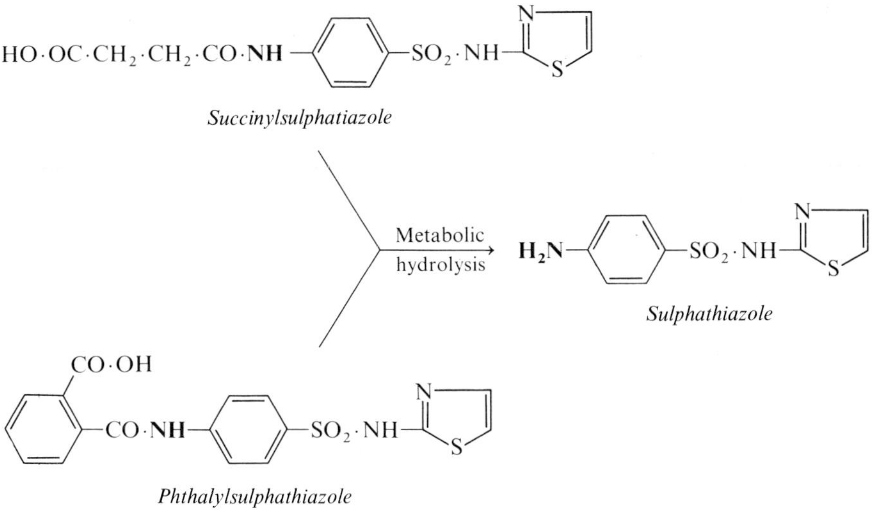

$HO \cdot OC \cdot CH_2 \cdot CH_2 \cdot CO \cdot \mathbf{NH}$—$C_6H_4$—$SO_2 \cdot NH$—(thiazolyl)

*Succinylsulphatiazole*

Metabolic hydrolysis →

$\mathbf{H_2N}$—$C_6H_4$—$SO_2 \cdot NH$—(thiazolyl)

*Sulphathiazole*

CO·OH

$C_6H_4$—$CO \cdot \mathbf{NH}$—$C_6H_4$—$SO_2 \cdot NH$—(thiazolyl)

*Phthalylsulphathiazole*

**Imides**

Succinamic, glutaramic and phthalamic acids cyclise when heated strongly to give the corresponding imide.

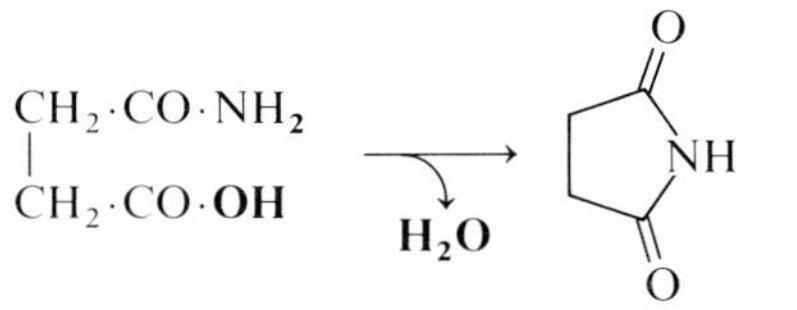

$$\begin{matrix} CH_2 \cdot CO \cdot NH_2 \\ | \\ CH_2 \cdot CO \cdot \mathbf{OH} \end{matrix} \xrightarrow[\mathbf{H_2O}]{} \text{Succinimide}$$

Succinamic acid Succinimide

These cyclic amides are weakly acidic, and lose a proton from the imide nitrogen to form a resonance-stabilised anion in strong alkali (potassium hydroxide).

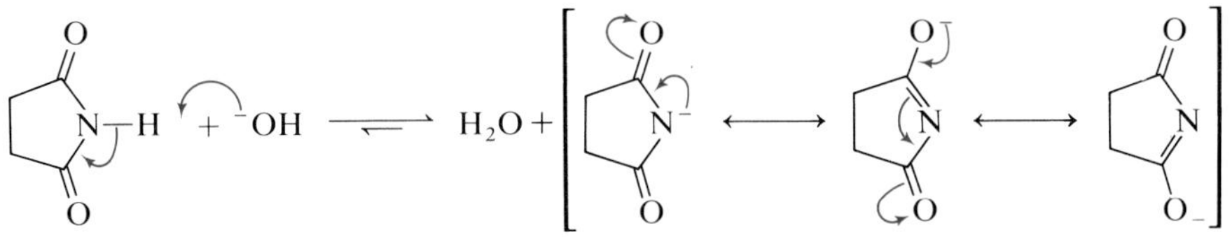

$$\text{Succinimide (N—H)} + {}^-OH \rightleftharpoons H_2O + [\text{N}^- \leftrightarrow \text{O}^- \leftrightarrow \text{O}^-]$$

The alkali metal salts, so formed, are decomposed by carbon dioxide. They also react with alkyl halides to form *N*-alkylimides.

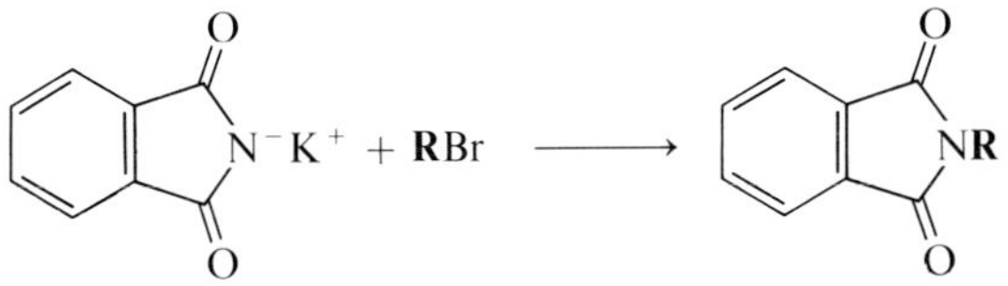

$$\text{Phthalimide } N^-K^+ + \mathbf{R}Br \longrightarrow \text{N}\mathbf{R}$$

*N*-Alkylphthalimide

A number of cyclic imides, such as *Glutethimide* and *Ethosuximide*, are important hypnotics and anticonvulsants.

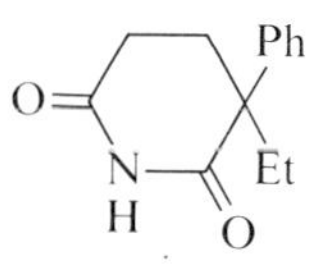

*Glutethimide*

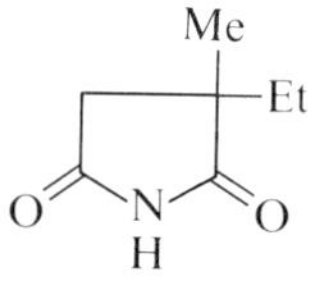

*Ethosuximide*

*Thalidomide* also has similar properties, but is no longer considered safe because of its frightful teratogenic properties, which appear if taken during the first three months of pregnancy. The study of a number of simpler compounds has established that it is the *N*-substituted, phthalimido group which is responsible for this effect (Fabro, Schumacher, Smith and Williams, 1964), and it has been suggested that this may be due to the reactivity of the phthalimido ring as a biological acylating agent (Fabro Smith and Williams, 1965). Hydrolytic stability and mammalian metabolic studies (Schumacher, Smith and Williams, 1965) have established that the phthalimido ring is cleaved in both acid (pH 6.7) and alkaline conditions, whereas the glutarimide ring is cleaved only at pH 7.0 and above. *Thalidomide* has a half-life of 5 hr at pH 7.4. In agreement with this, Fabro, Smith and Williams (1965) have shown spectrophotometrically that the phthalimido ring of *Thalidomide* is cleaved in the presence of certain natural diamines, such as putrescine ($H_2N(CH_2)_4NH_2$), cadaverine ($H_2N(CH_2)_5NH_2$) and spermidine ($H_2N(CH_2)_4NH(CH_2)_3NH_2$).

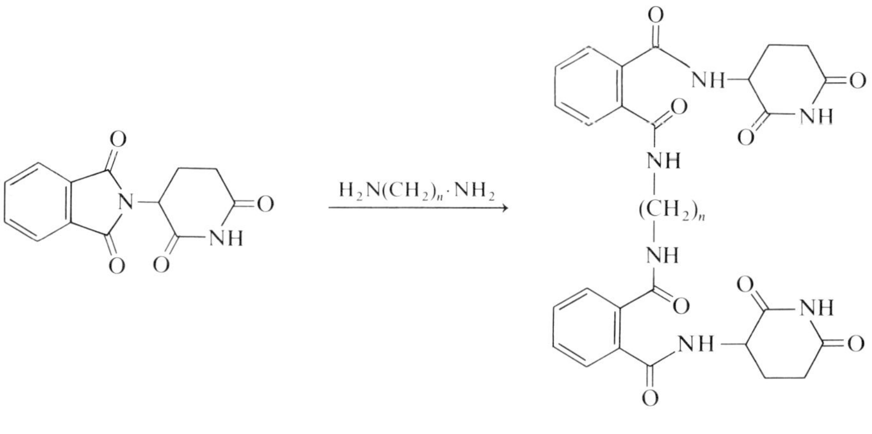

$$\xrightarrow{H_2N(CH_2)_n \cdot NH_2}$$

Metabolism studies show that *Thalidomide* is hydrolysed mainly to 4-phthalimidoglutaramic acid in man, but to α-(*o*-carboxybenzamidoglutarimide) in the dog and rat; some twelve hydrolytic products have been identified in all.

Hydroxylation in the 3- and 4-positions of the aromatic ring has also been shown to occur in the rabbit.

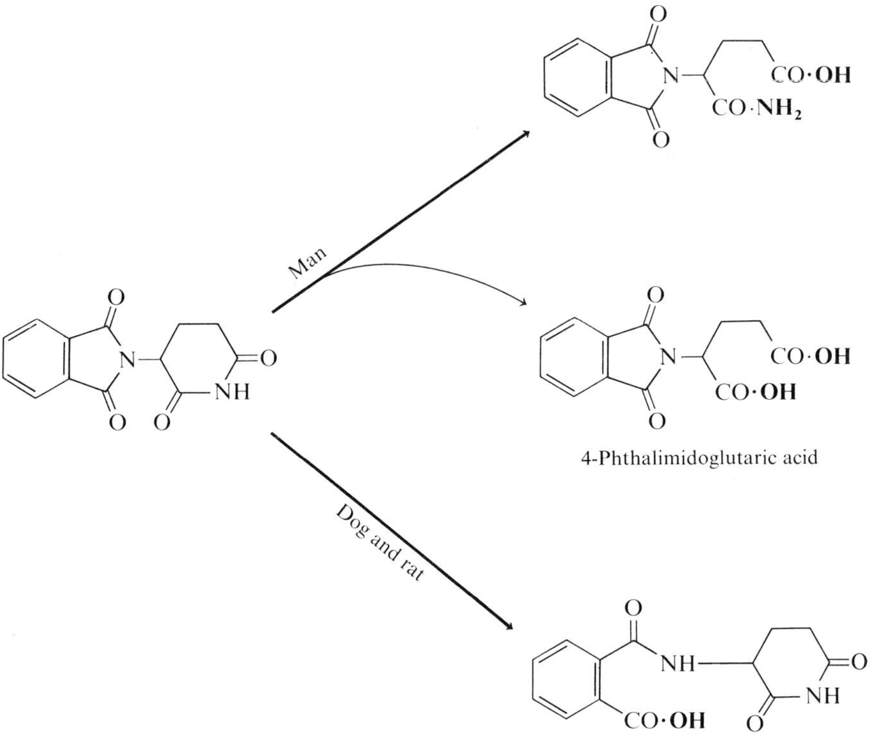

4-Phthalimidoglutaric acid

Both *Glutethimide* and *Ethosuximide* are extremely soluble in lipid solvents ($<1$ in 1 in $CHCl_3$), a property which favours partition into the lipid tissues of the central nervous system. Ring size clearly affects the pharmacological properties of these compounds, but these must also be influenced by the nature, position and stereochemistry of the ring substituents. The latter may possibly also influence the ease and direction of hydrolysis, or biological acylation reactions (*cf Thalidomide*).

The influence of stereochemical configuration on biological response in this series is illustrated by the metabolism of ($\pm$)-*Glutethimide* in the dog (Keberle, Riess and Hoffmann, 1963). The two optical isomers are metabolised by distinctly different pathways (see page 336).

Significantly, 96% of the administered dose was accounted for, and as with *Thalidomide* (in the dog), there was no evidence of cleavage of the glutarimide ring. However, cyclic imides in general are readily hydrolysed in neutral or alkaline solution. Also, *Glutethimide* reacts readily with hydroxylamine in alkaline solution to form a hydroxamic acid. Since also *Thalidomide* is metabolised in man primarily by cleavage of the glutarimide ring, there are good grounds for

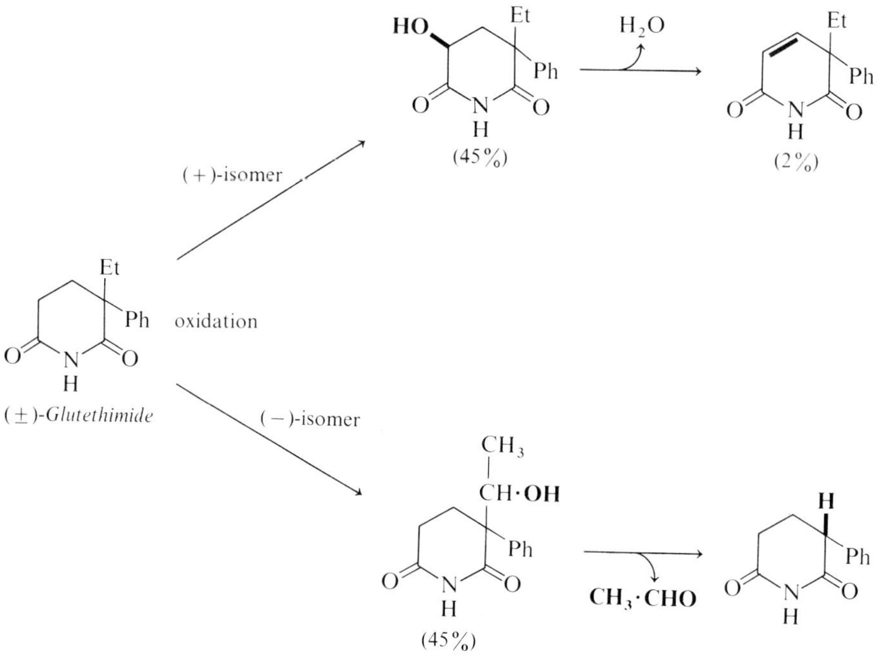

(+)-isomer

(45%)

$H_2O$

(2%)

(±)-*Glutethimide* oxidation

(−)-isomer

(45%)

$CH_3 \cdot CHO$

assuming that a similar pathway may obtain in the case of *Glutethimide* (and possibly also *Ethosuximide*).

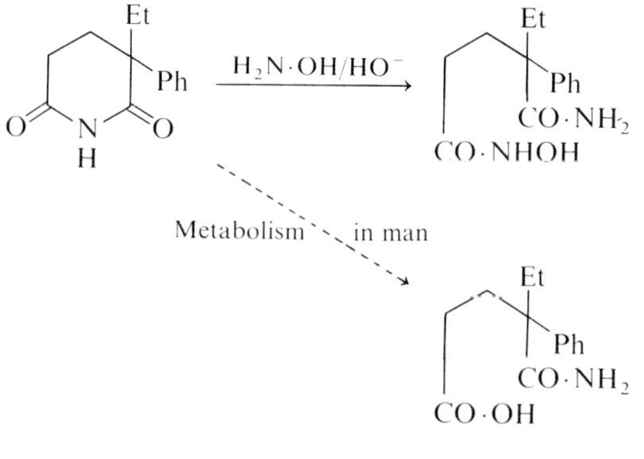

$H_2N \cdot OH/HO^-$

Metabolism in man

Although central nervous system (CNS) depressants are generally regarded as structurally non-specific drugs, the close similarity in the three-dimensional structure of *Phenobarbitone* (Phenobarbital) and *Glutethimide* seems significant. Moreover, in contrast, the glutarimide derivative, *Bemegride*, in which the ring substituents are at C-4, is an analeptic. It is used in respiratory failure and collapse during anaesthesia, and in the treatment of barbiturate poisoning. The molecular mechanism of its pharmacological action is unknown, but the structure of *Bemegride* is sufficiently similar to *Glutethimide* and *Phenobarbitone* that it is feasible for it to approach at its upper ($\beta$) face the same bio-receptor molecule, though with significantly less steric interference from substituents than would obtain with the depressant drugs.

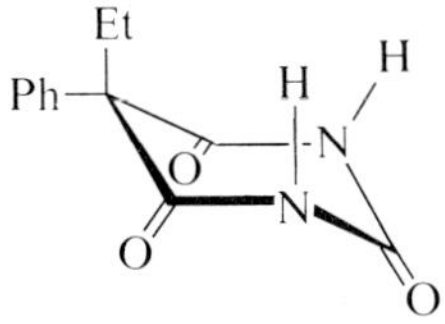

Phenobarbitone

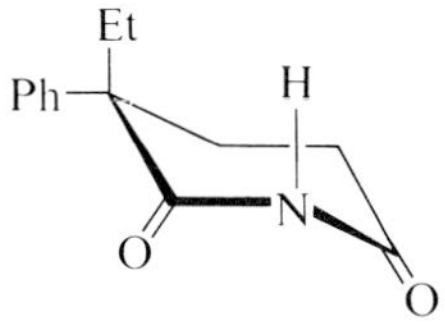

Glutethimide

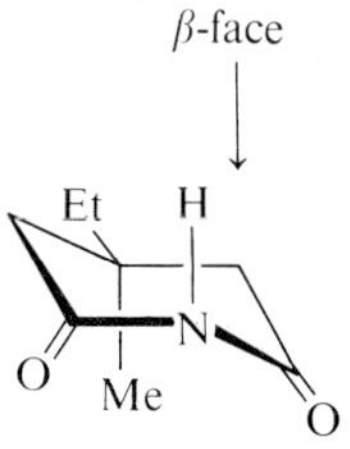

Bemegride

**N-Bromosuccinimide.** Treatment of succinimide with bromine in the presence of sodium hydroxide at 0°C yields *N*-bromosuccinimide.

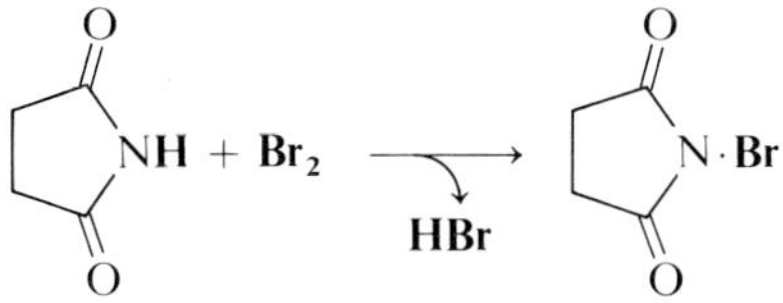

$$\text{succinimide (NH)} + Br_2 \xrightarrow{-HBr} \text{N·Br}$$

*N*-Bromosuccinimide is especially reactive in radical-induced substitutions of allylic compounds (Ziegler, Späth, Schaaf, Schumann and Winkelmann, 1942), which are catalysed by peroxides and light. The *N*-bromosuccinimide functions as a low concentration source of molecular bromine by interaction with hydrogen bromide (Adam, Gosselain and Goldfinger, 1953; McGrath and Tedder, 1961).

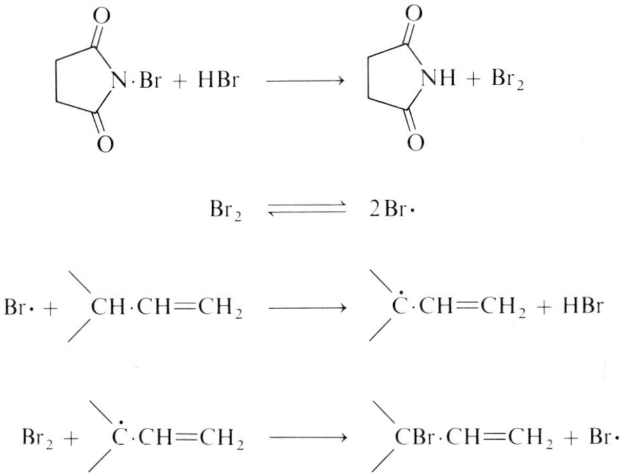

$$\text{N·Br} + HBr \longrightarrow \text{NH} + Br_2$$

$$Br_2 \rightleftharpoons 2Br\cdot$$

$$Br\cdot + \; >CH\cdot CH{=}CH_2 \longrightarrow \; >\dot{C}\cdot CH{=}CH_2 + HBr$$

$$Br_2 + \; >\dot{C}\cdot CH{=}CH_2 \longrightarrow \; >CBr\cdot CH{=}CH_2 + Br\cdot$$

## Di- and Poly-Esters and -Amides

Di-esters, such as *Dimethyl Phthalate*, *Ditophal* (Chapter 14) and *Suxamethonium Bromide* (Succinylcholine Bromide), and di-amides, such as the cholecystographic agent *Iodipamide*, have properties which are typical of esters and amides. In *Suxamethonium Bromide*, the ester links are weakened by the I effect of the quaternary nitrogens. It is readily hydrolysed in aqueous solution, and insufficiently stable for aqueous injections to be sterilised by autoclaving. *Suxamethonium Bromide* owes its short action as a neuromuscular blocking agent to enzymic hydrolysis to yield inactive products.

$$[Me_3\overset{+}{N}\cdot CH_2\cdot CH_2\cdot \mathbf{O\cdot CO}\cdot CH_2\cdot CH_2\cdot \mathbf{CO\cdot O}\cdot CH_2\cdot CH_2\cdot \overset{+}{N}Me_3]Br^-$$

*Suxamethonium Bromide*

$$\downarrow \; \mathbf{H_2O}, \text{ Pseudochloinesterase} \; \rightarrow Me_3\overset{+}{N}\cdot CH_2\cdot CH_2\cdot \mathbf{OH}$$

$$Me_3\overset{+}{N}\cdot CH_2\cdot CH_2\cdot \mathbf{O\cdot CO}\cdot CH_2\cdot CH_2\cdot \mathbf{CO\cdot OH}$$

Monosuccinylcholine

$$\downarrow \; \mathbf{H_2O}, \text{ Pseudocholinesterase}$$

$$Me_3N\cdot CH_2\cdot CH_2\cdot \mathbf{OH} + \mathbf{HO\cdot OC}\cdot CH_2\cdot CH_2\cdot \mathbf{CO\cdot OH}$$

$C_6H_4(CO\cdot OMe)_2$

*Dimethyl Phthalate*

HO·OC, I, I, I (ring)—NH·CO(CH$_2$)$_4$·CO·NH—(ring) I, I, I, CO·OH

*Iodipamide*

Dibasic acids also react with glycols and diamines to yield linear poly-esters and poly-amides, which are capable of being spun into fibres. Terylene is a typical poly-ester formed by chain-polymerisation of dimethyl terephthalate and ethylene glycol. Nylon is a poly-amide formed by polymerisation of adipic acid and hexamethylene diamine.

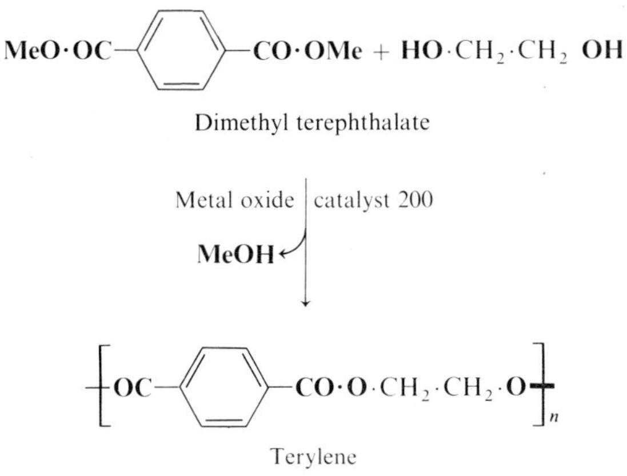

$$MeO\cdot OC-C_6H_4-CO\cdot OMe + HO\cdot CH_2\cdot CH_2\cdot OH$$

Dimethyl terephthalate

$$\xrightarrow[\text{Metal oxide catalyst } 200]{-MeOH}$$

$$\left[-OC-C_6H_4-CO\cdot O\cdot CH_2\cdot CH_2\cdot O-\right]_n$$

Terylene

$$HO\cdot OC(CH_2)_4\cdot CO\cdot OH + H_2N(CH_2)_6\cdot NH_2$$

$$\xrightarrow[280]{-H_2O}$$

$$\left[-OC(CH_2)_4\cdot CO\cdot NH(CH_2)_6\cdot NH-\right]_n$$

Nylon 66

## Malonic Esters

The methylene group hydrogens of malonic esters are ionisable in the presence of a suitable base (sodium ethoxide). The resulting carbanion, which is resonance-stabilised, is reactive towards alkyl halides to form mono- and dialkylmalonic esters.

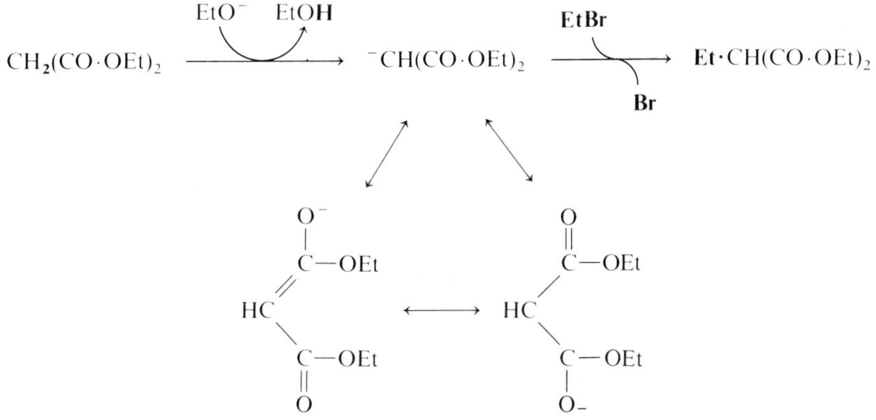

$$CH_2(CO\cdot OEt)_2 \xrightarrow{EtO^- \quad EtOH} {}^-CH(CO\cdot OEt)_2 \xrightarrow[Br^-]{EtBr} Et\cdot CH(CO\cdot OEt)_2$$

$$HC(=C(O^-)-OEt)-C(=O)-OEt \longleftrightarrow HC(-C(=O)-OEt)=C(O^-)-OEt$$

The ionisation of the second proton occurs less readily (Pearson, 1949), but dialkylmalonic esters can none-the-less be readily obtained.

$$EtCH(CO\cdot OEt)_2 \xrightarrow{EtO^-} Et\bar{C}(CO\cdot OEt)_2 \xrightarrow{EtBr} Et_2C(CO\cdot OEt)_2$$

Alkaline hydrolysis of the resulting esters followed by acid-catalysed decarboxylation of the resulting substituted malonic acid provides a convenient route to the synthesis of straight chain and branched (2-alkyl) monobasic acids.

$$EtCH(CO\cdot \mathbf{OEt})_2 \xrightarrow[\mathbf{EtOH}]{\mathbf{HO^-}} EtCH(CO\cdot \mathbf{O^-})_2 \xrightarrow{\mathbf{H^+}} Et\text{—}CH(C(=O)\text{—}O\text{—}\mathbf{H})\text{—}C(=O)\text{—}OH$$

$$\xrightarrow{H^+} CH_3\cdot CH_2\cdot CH{=}C(OH)\text{—}O\text{—}H \xrightarrow{H^+} CH_3\cdot CH_2\cdot CH_2\cdot CO\cdot OH$$

$$Et_2C(CO\cdot OEt)_2 \xrightarrow{HO^-} Et_2C(CO\cdot O^-)_2 \xrightarrow{H^+} Et_2C(CO\cdot OH)_2$$

$$Et_2C(CO\cdot OH)_2 \xrightarrow{-CO_2} CH_3\cdot CH_2\cdot CH(CH_2\cdot CH_3)\cdot CO\cdot OH$$

Condensation of dialkyl substituted malonic esters with urea leads to formation of barbiturates (Ch. 17). As already described (Ch. 12), malonyl coenzyme A, which is similarly reactive at the $\alpha$-methylene group, is a key intermediate in the biosynthesis of fatty acids.

Similarly, the reaction between diethyl malonate and halogenated aliphatic acids leads through similar reactions to the formation of dibasic acids.

$$(CH_2)_2(COOEt)_2 \xrightarrow{Br(CH_2)_n\cdot CO\cdot OEt} (EtO\cdot OC)_2\cdot CH(CH_2)_n\cdot CO\cdot OEt$$

$$\xrightarrow{1.\ HO^-;\ 2.\ H^+} HO\cdot OC(CH_2)_{n+1}\cdot CO\cdot OH$$

**Oxidation and Dehydrogenation**

Oxalic acid is readily and quantitatively oxidised with potassium permanganate in acid solution, a reaction which is widely used in analytical work.

$$(CO{\cdot}OH)_2 \longrightarrow 2CO_2 + H_2O$$

Mild oxidation of higher dibasic acids (and their esters) gives rise to α-keto acids (and esters).

$$CH_2(COOEt)_2 \xrightarrow{HNO_3} \underset{\text{Ethyl mesoxalate}}{EtO{\cdot}OC{\cdot}CO{\cdot}CO{\cdot}OEt}$$

The enzymic conversion of succinate to fumarate by succinate dehydrogenase is a key step in the citric acid cycle, which is central to the metabolism of almost all living cells. The enzyme is present in the mitochondria of plant and animal cells, and in the bacterial cell membranes of aerobic bacteria. The constitution of the enzyme is not fully known, but it is established that flavine adenine dinucleotide (FAD), which is firmly bound, is the hydrogen acceptor.

$$^{-}O{\cdot}OC{\cdot}CH_2{\cdot}CH_2{\cdot}CO{\cdot}O^{-} + E{-}FAD \rightleftharpoons {}^{-}O{\cdot}OC{\cdot}CH{=}CH{\cdot}CO{\cdot}O^{-} + E{-}FADH_2$$

## UNSATURATED DIBASIC ACIDS

Unsaturated dibasic acids resemble both the corresponding saturated dibasic acids in their acidic properties (p. 327), and the analogous unsaturated monobasic acids in their olefinic properties. They are, therefore, susceptible to both ionic and radical addition, and exhibit geometrical isomerism as exemplified by maleic and fumaric acids

$$\underset{\text{Fumaric acid}}{(H)(CO{\cdot}OH)C{=}C(H)(HO{\cdot}OC)} \xrightarrow{\text{Oxidation}} \underset{\text{Mesotartaric acid}}{CO{\cdot}OH{-}CH(OH){-}CH(OH){-}CO{\cdot}OH}$$

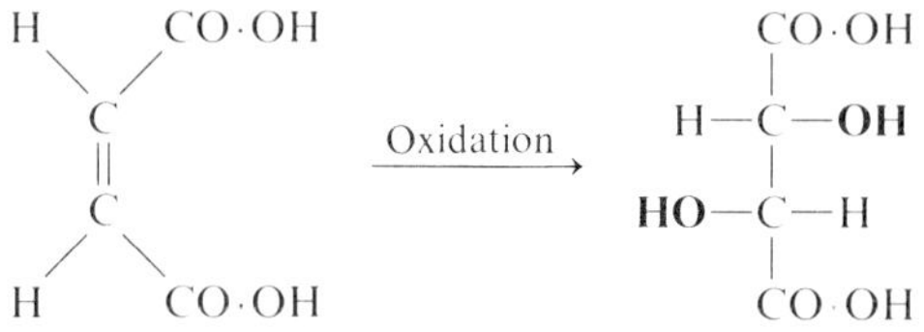

H CO·OH / C=C / H CO·OH — Oxidation → CO·OH / H—C—**OH** / **HO**—C—H / CO·OH

Maleic acid    Racemic acid

They are less stable than their saturated counterparts, and are readily oxidised to the corresponding dihydroxy acids by peroxides. Addition of HO radicals is **trans** to the double bond, so that fumaric acid gives mesotartaric acid, whilst maleic acid forms racemic acid (Church and Blumberg, 1951).

Exposure to ultraviolet light causes dimerisation, as in the case of dimethyl fumarate, which yields one isomer of methyl cyclobutane tetracarboxylate (Griffin, Basinski and Vallturo, 1960).

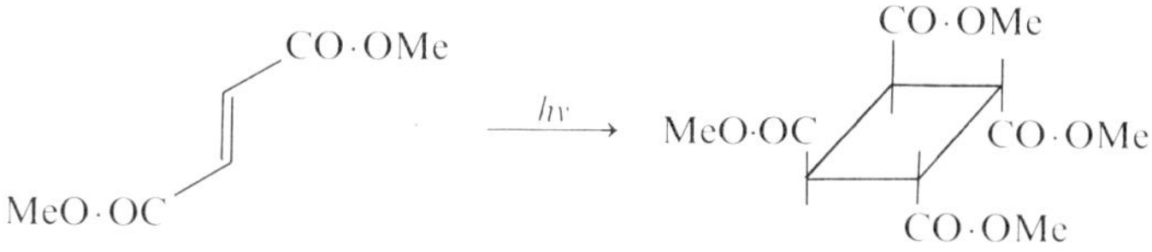

MeO·OC—CH=CH—CO·OMe —$h\nu$→ MeO·OC, CO·OMe, CO·OMe, CO·OMe (cyclobutane)

The reversible hydration of fumarate to L-malate, which is catalysed by the enzyme fumarase, forms part of the citric acid cycle. The hydration involves participation of both an acid and a basic centre on the enzyme surface which appear to be histidine residues. The reaction has been shown by NMR spectroscopy and ORD studies in $D_2O$ (Eggerer, Remberger and Gruenewaelder, 1964) to be stereospecific.

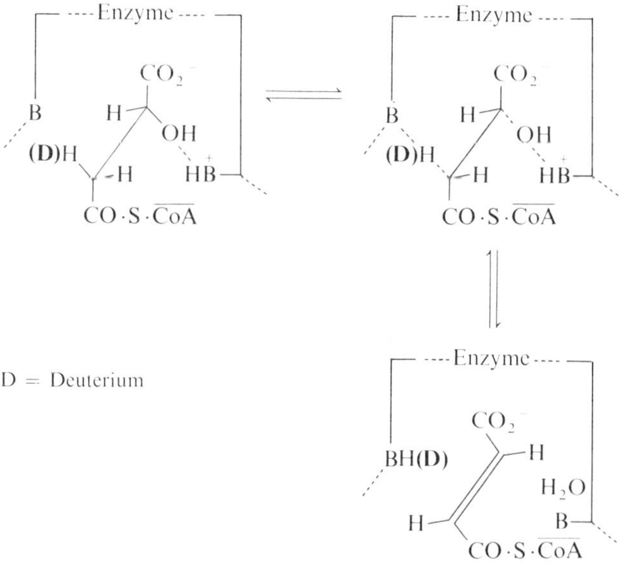

----Enzyme---- B (**D**)H H $CO_2^-$ OH H $\overset{+}{H}B$ CO·S·$\overline{CoA}$ ⇌ ----Enzyme---- B (**D**)H H $CO_2^-$ OH H $\overset{+}{H}B$ CO·S·$\overline{CoA}$

⇅

----Enzyme---- BH(**D**) $CO_2^-$ H H $H_2O$ B H CO·S·$\overline{CoA}$

D = Deuterium

# 13 Oxo- and Hydroxy-Carboxylic Acids

Carbonyl derivatives of carboxylic acids fall into two classes. In the aldehydo-acids, the carbonyl group is in a chain terminal position, whereas in the keto-acids it occupies a non-terminal position.

## ALDEHYDO-ACIDS

### Glyoxylic Acid

Glyoxylic acid, the only α-aldehydo acid, shows properties which are typical of both aldehydes and carboxylic acids. It is a viscous liquid, but normally is obtained as a crystalline monohydrate [$(HO)_2CH \cdot CO \cdot OH$]. As an aldehyde, it is unstable in alkali and undergoes the Cannizzaro reaction.

$$2\ OCH \cdot CO \cdot OH \xrightarrow{HO^-} HO \cdot CH_2 \cdot CO \cdot OH + HO \cdot OC \cdot CO \cdot OH$$

Glyoxylic acid is also a typical acid. It cannot, however, be esterified directly, since with ethanolic hydrogen chloride it gives the ethyl diethoxyacetate.

$$OCH \cdot CO \cdot OH \xrightarrow{EtOH/HCl} (EtO)_2CH \cdot CO \cdot OEt$$

Ethyl glyoxalate can be obtained by oxidation of diethyl tartrate with lead tetra-acetate in benzene.

$$\begin{array}{l} CHOH \cdot CO \cdot OEt \\ | \\ CHOH \cdot CO \cdot OEt \end{array} \xrightarrow{Pb(OAc)_4} 2\ OCH \cdot CO \cdot OEt$$

Glyoxylic acid is formed in mammalian intermediary metabolism from serine, proline and glycine, and is further metabolised by a number of alternative routes, including decarboxylation (Crawhall and Watts, 1962), oxidation to oxalic acid (Richardson and Tolbert, 1961), and transamination to glycine (Meister, Sober, Tice and Fraser, 1952). The latter requires pyridoxal phosphate as co-factor, though a similar non-enzymatic transamination does not require the co-factor (Nakada and Weinhouse, 1953).

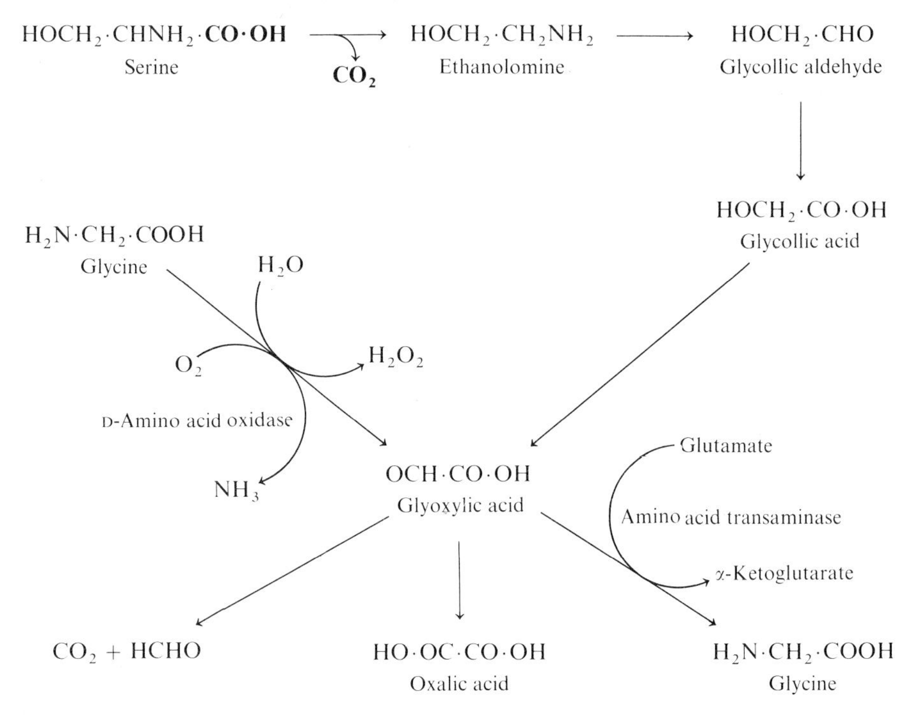


Glyoxylic acid is also incorporated biosynthetically by aldol-type reactions into isocitrate and malate in certain micro-organisms.

$$\begin{array}{l} CO{\cdot}O^- \\ | \\ CHO \\ \\ CH_2{\cdot}CO{\cdot}O^- \\ | \\ CH_2{\cdot}CO{\cdot}O^- \\ \text{Succinate} \end{array} \quad \underset{}{\overset{\text{Isocitrase}}{\rightleftharpoons}} \quad \begin{array}{l} CO{\cdot}O^- \\ | \\ CHOH \\ | \\ CH{\cdot}CO{\cdot}O^- \\ | \\ CH_2{\cdot}CO{\cdot}O^- \\ \text{Isocitrate} \end{array}$$

$$\begin{array}{l} CO{\cdot}O^- \\ | \\ CHO \\ \\ CH_3 \\ | \\ CO{\cdot}S{\cdot}\overline{CoA} \end{array} \quad \longrightarrow \quad \begin{array}{l} \quad\;\; CO{\cdot}O^- \\ \quad\quad\; | \\ HO{-}C{-}H \\ \quad\quad\; | \\ \quad\;\; CH_2 \\ \quad\quad\; | \\ \quad\;\; CO{\cdot}S{\cdot}\overline{CoA} \\ \quad\;\; \text{L-Malate} \end{array}$$

**β- and γ-Aldehydo-acids**

*β*-Aldehydo-carboxylic acids are unstable and have not been isolated as such. Synthetic methods which might be expected to lead to *β*-aldehydo-acids lead only to decomposition products, which appear to result from their decarboxylation. Certain of their derivatives such as formylacetic acid acetal and ethyl formylacetate are known. The latter, however, is stable only as the sodium salt of the **aci**-tautomer of the ester, which is formed by condensation of formic and acetic esters (Chapter 11).

$(Et_2O)_2CH{\cdot}CH_2{\cdot}COOH$ — Formylacetic acid acetal

$OCH{\cdot}CH_2{\cdot}CO{\cdot}OEt \rightleftharpoons HO{\cdot}CH{=}CH{\cdot}CO{\cdot}OEt$
Ethyl formylacetate — *aci*-form

$\xrightarrow{NaOH} NaOCH{=}CH{\cdot}CO{\cdot}OEt$

Ethyl formylacetate itself readily undergoes self-condensation.

$\beta$-Formylpropionic acid and $\gamma$-formyl-$\alpha$-aminobutyric acid are important in mammalian intermediary metabolism. The former arises from $\gamma$-aminobutyric acid and the latter from arginine and proline by multi-stage deamination. They enter the citric acid cycle via succinic acid and $\alpha$-keto-glutaric acids respectively.

$H_2N{\cdot}CH_2{\cdot}CH_2{\cdot}CH_2{\cdot}CO{\cdot}OH \dashrightarrow OCH{\cdot}CH_2{\cdot}CH_2{\cdot}CO{\cdot}OH$
$\gamma$-Aminobutyric acid — $\beta$-Formylpropionic acid

$\downarrow$

$HO{\cdot}OC{\cdot}CH_2{\cdot}CH_2{\cdot}COOH$
Succinic acid

Proline (pyrrolidine ring, N–H, CO·OH)

$\downarrow$ (dashed)

$OCH(CH_2)_2{\cdot}CH(NH_2){\cdot}CO{\cdot}OH \longrightarrow HO{\cdot}OC(CH_2)_2{\cdot}CH(NH_2){\cdot}CO{\cdot}OH$
$\gamma$-Formyl-$\alpha$-aminobutyric acid — Glutamic acid

$\uparrow$ (dashed) — $\downarrow$

$(HN{=})(H_2N)C{\cdot}NH(CH_2)_3{\cdot}CH(NH_2){\cdot}CO{\cdot}OH$ — $HO{\cdot}OC(CH_2)_2{\cdot}CO{\cdot}CO{\cdot}OH$
Arginine — $\alpha$-Ketoglutaric acid

## KETO-ACIDS

Most keto-carboxylic acids possess properties which are characteristic of both ketones and carboxylic acids. Certain keto-acids, however, possess additional properties which result from the position of the keto group in relation to the carboxylic acid group. Systematically, the position of the keto group is defined numerically by IUPAC nomenclature as an oxo-derivative of the corresponding saturated carboxylic acid (Table 42). Designation of the acids generically as $\alpha$-, $\beta$-, $\gamma$- and $\delta$-keto acids, signifying the relative positions of ketone and carboxyl groups, is also in widespread use.

### $\alpha$-Keto-acids

**General Properties**

Pyruvic acid, typically, is a somewhat stronger acid (p$K_a$ 2.49) than propionic acid (p$K_a$ 4.88). It forms water-soluble salts with alkali metals, insoluble salts with alkaline earth metals, and a coloured complex with ferric ion.

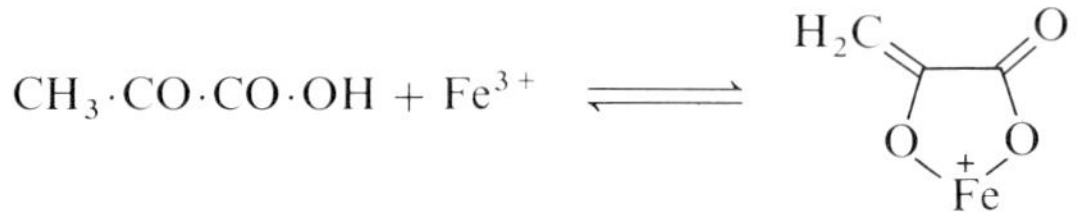

$$CH_3{\cdot}CO{\cdot}CO{\cdot}OH + Fe^{3+} \rightleftharpoons$$

H$_2$C, O, O, O, $\overset{+}{Fe}$

Pyruvic acid, the simplest $\alpha$-keto-acid, is of central importance in intermediary carbohydrate and protein metabolism. It is formed as the end-product of the oxidative metabolism of glucose (Chapter 21). Pyruvic acid and the $\alpha$-keto-acids of the tricarboxylic acid cycle also represent the link between carbohydrate and protein metabolism through reversible pathways for interconversion with the corresponding $\alpha$-keto-acids.

Table 42. Nomenclature of Keto-acids

| Structure | Systematic Name | Trivial Name |
|---|---|---|
| $CH_3{\cdot}CO{\cdot}CO{\cdot}OH$ | 2-Oxopropionic acid | Pyruvic acid |
| $CH_3{\cdot}CO{\cdot}CH_2{\cdot}CO{\cdot}OH$ | 3-Oxobutanoic acid | Acetoacetic acid |
| $CH_3{\cdot}CO{\cdot}CH_2{\cdot}CH_2{\cdot}CO{\cdot}OH$ | 4-Oxopentanoic acid | Laevulinic acid |
| $HO{\cdot}OC{\cdot}CO{\cdot}CO{\cdot}OH$ | 2-Oxomalonic acid | Mesoxalic acid |
| $HO{\cdot}OC{\cdot}CO{\cdot}CH_2{\cdot}CO{\cdot}OH$ | 2-Oxosuccinic acid | Oxaloacetic acid |
| $HO{\cdot}OC{\cdot}CO{\cdot}CH(CH_2{\cdot}CO{\cdot}OH){\cdot}CO{\cdot}OH$ | 2-Carboxymethyl-3-oxosuccinic acid | Oxalosuccinic acid |
| $HO{\cdot}OC{\cdot}CH{=}CH{\cdot}(CO{\cdot}CH_2)_2{\cdot}CO{\cdot}OH$ | 4,6-Dioxo-oct-2-ene-1,8-dioic acid | Fumarylaceto-acetic acid |

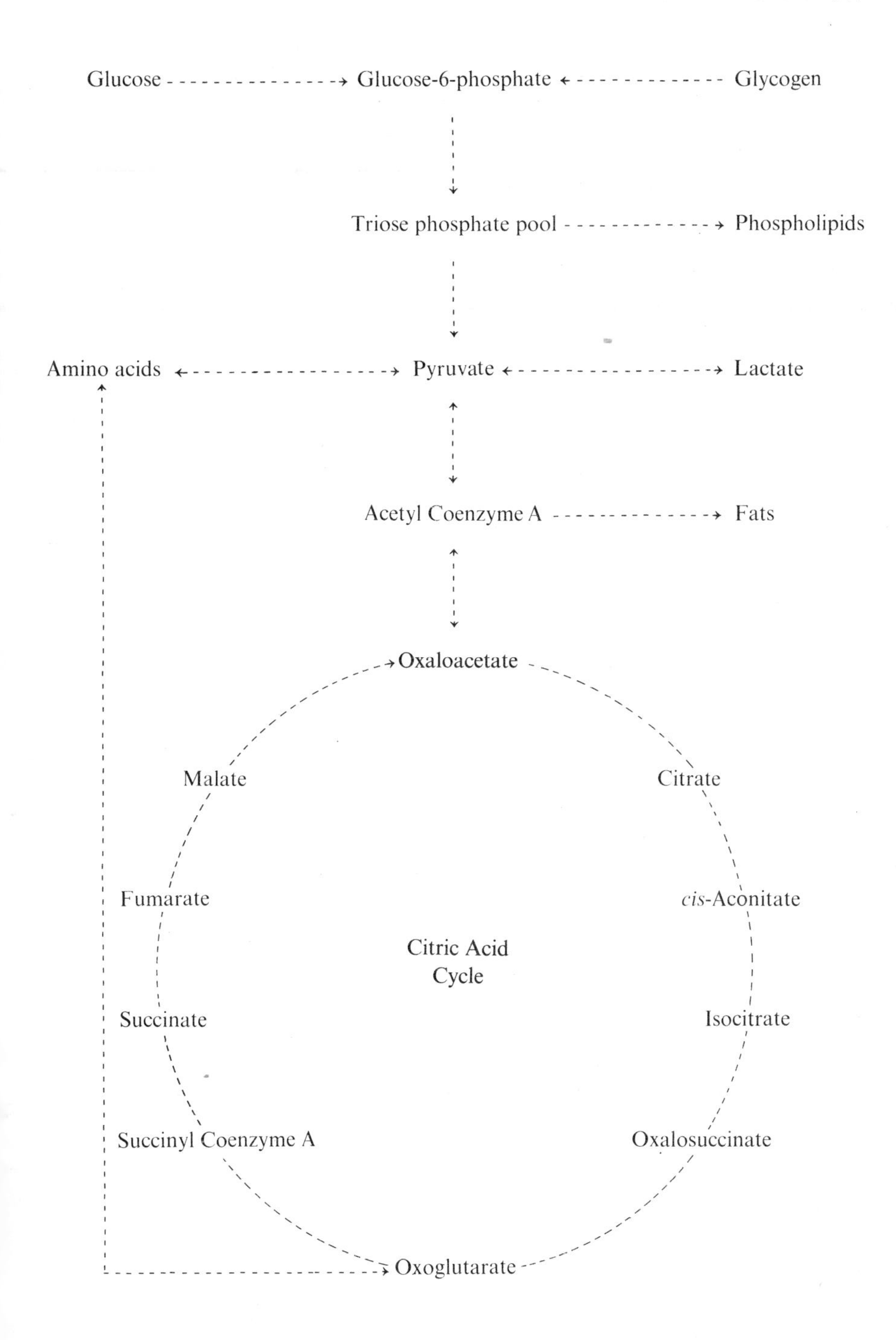
Glucose
Glucose-6-phosphate
Glycogen
Triose phosphate pool
Phospholipids
Amino acids
Pyruvate
Lactate
Acetyl Coenzyme A
Fats
Oxaloacetate
Malate
Citrate
Fumarate
cis-Aconitate
Citric Acid
Cycle
Succinate
Isocitrate
Succinyl Coenzyme A
Oxalosuccinate
Oxoglutarate

**Decarboxylation**

Unlike ketones, α-keto-acids show reducing properties (reduction of ammoniacal silver nitrate to metallic silver), presumably as a result of decarboxylation to the corresponding aldehyde, and oxidation of the latter.

$$CH_3{\cdot}CO{\cdot}CO{\cdot}OH \xrightarrow{-CO_2} CH_3{\cdot}CHO \xrightarrow[-Ag]{Ag_2O} CH_3{\cdot}CO{\cdot}OH$$

Decarboxylation of pyruvic acid occurs readily in warm dilute sulphuric acid to yield acetaldehyde and carbon dioxide.

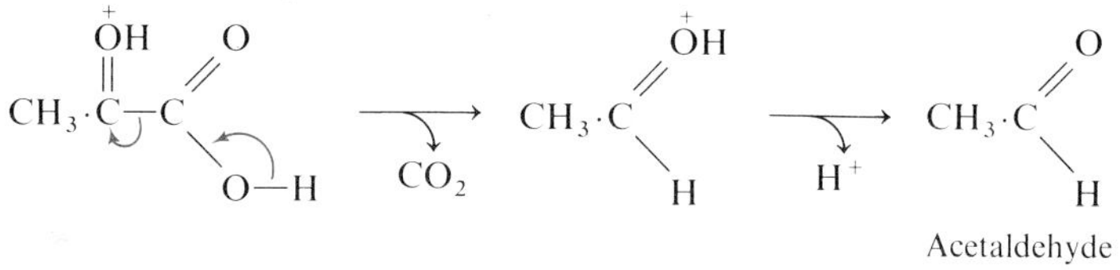

$$CH_3{\cdot}C(=\overset{+}{O}H){-}C(=O){-}O{-}H \xrightarrow{-CO_2} CH_3{\cdot}C(=\overset{+}{O}H){-}H \xrightarrow{-H^+} CH_3{\cdot}C(=O){-}H$$

Acetaldehyde

The decarboxylation of pyruvate to yield acetaldehyde by yeast **pyruvate decarboxylase** represents one of the terminal steps in the alcoholic fermentation

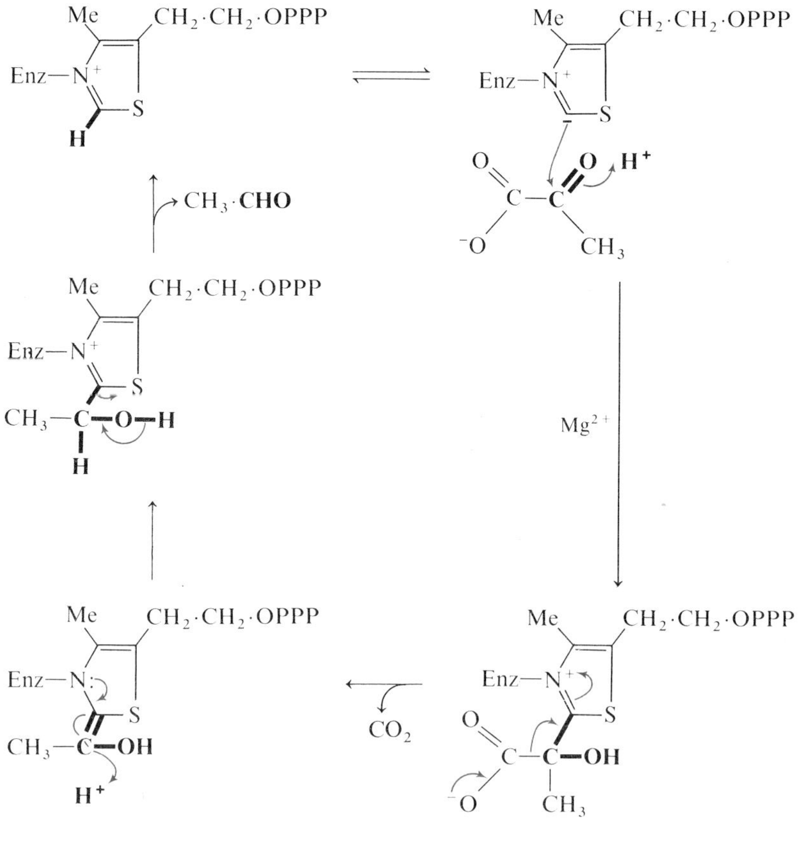

Me CH$_2$·CH$_2$·OPPP

Enz—N$^+$ S

H

⇌

Me CH$_2$·CH$_2$·OPPP

Enz—N$^+$ S

O O H$^+$

C—C

$^-$O CH$_3$

→ CH$_3$·**CHO**

Me CH$_2$·CH$_2$·OPPP

Enz—N$^+$ S

CH$_3$—**C—O—H**

**H**

Mg$^{2+}$

Me CH$_2$·CH$_2$·OPPP

Enz—N: S

CH$_3$—**C—OH**

**H$^+$**

← CO$_2$

Me CH$_2$·CH$_2$·OPPP

Enz—N$^+$ S

O

C—**C—OH**

$^-$O CH$_3$

of sugars. The decarboxylation requires the intervention of thiamine pyrophosphate and magnesium ions.

**Decarbonylation**

α-Keto-acids and their esters exhibit thermal instability (Hurd and Raterink, 1934). Decomposition proceeds by at least two pathways; both carbon monoxide and carbon dioxide are formed in the decomposition of phenylpyruvic acid.

$$Ph{\cdot}CH_2{\cdot}CO{\cdot}CO{\cdot}OH \xrightarrow{250°} \begin{cases} [Ph{\cdot}CH_2{\cdot}CHO] + CO_2 \quad \text{not identified} \\ Ph{\cdot}CH_2{\cdot}CO{\cdot}OH (35\%) + CO \quad \text{Phenylacetic acid} \end{cases}$$

Thermal decarbonylation of the α-keto-ester formed from the condensation of ethyl phenylacetate and ethyl oxalate is an important step in the synthesis of diethyl phenylmalonate which is used in the production of phenobarbitone. The thermal decomposition is catalysed in soda glass vessels.

$$Ph{\cdot}CH_2{\cdot}CO{\cdot}OEt \xrightarrow{EtO{\cdot}OC{\cdot}CO{\cdot}OEt} Ph{\cdot}CH(CO{\cdot}OEt){\cdot}CO{\cdot}CO{\cdot}OEt \xrightarrow[-CO]{180°} Ph{\cdot}CH(CO{\cdot}OEt)_2$$

Decarbonylation of pyruvic and other α-keto-acids also occurs in hot concentrated sulphuric acid. Calvin and Lemmon (1947) have shown that the reaction leads to the expulsion of the carboxylic carbonyl group.

$$CH_3{\cdot}CO{\cdot}\mathbf{CO}{\cdot}OEt \xrightarrow{conc\ H_2SO_4} CH_3{\cdot}{}^{14}C(=O){\cdot}OEt + \mathbf{CO}\uparrow$$

Exposure to ultraviolet light also leads to the decomposition of α-keto-esters (Leermakers, Warren and Vesley, 1964) with concurrent decarbonylation and fragmentation.

$$CH_3 \cdot \underset{\underset{O}{\|}}{C} - \underset{\underset{O}{\|}}{C} - O - \underset{\underset{CH_3}{|}}{\overset{\overset{D}{|}}{C}} - CH_3 \xrightarrow[CO]{} CH_3 \cdot \dot{C}O \quad \dot{O} - \underset{\underset{CH_3}{|}}{\overset{\overset{D}{|}}{C}} - CH_3$$

$$\downarrow$$

$$CH_3 \cdot CDO + (CH_3)_2CO$$

**Carboxylation**

The carboxylation of pyruvate to oxaloacetate by **pyruvate carboxylase** is an important route to the biosynthesis of dibasic acids. The enzyme, which is located in the mitochondria, is ATP-dependent, requires the intervention of biotin, and is activated by acetyl-coenzyme A (Scrutton, Keech and Utter, 1965).

$$O{=}C{=}O + \overset{H}{CH_2} \cdot CO \cdot CO \cdot O^- \ (\text{Pyruvate}) \xrightarrow[H_2O + ATP \ \rightarrow \ ADP + P_i + H^+]{Mg^{2+}/\text{Biotin}/\text{Acetyl CoA}} {}^-\mathbf{O \cdot OC \cdot CH_2}CO \cdot CO \cdot O^- \ (\text{Oxaloacetate})$$

**Reduction**

Pyruvic acid is readily reduced catalytically to lactic acid.

$$CH_3 \cdot \mathbf{CO} \cdot CO \cdot OH \xrightarrow{H_2} CH_3 \cdot \mathbf{CH(OH)} \cdot COOH$$

The corresponding enzymic reduction by **lactate dehydrogenases** is reversible. The enzymes present in skeletal and heart muscle are stereospecific giving rise to only L-(+)-lactic acid. NADH is the effective reducing agent.

$$CH_3 \cdot \mathbf{CO} \cdot CO \cdot O^- + \mathbf{NADH} + \mathbf{H^+} \rightleftharpoons CH_3 \cdot \mathbf{CH(OH)} \cdot CO \cdot O^- + NAD^+$$

**Reductive Amination and Transamination**

α-Keto-acids are readily converted chemically to α-amino-acids by reductive amination, i.e. catalytic reduction in the presence of ammonia, primary or secondary amines.

$$CH_3 \cdot \mathbf{CO} \cdot CO \cdot OH + \mathbf{NH_3} \xrightarrow{H_2} CH_3 \cdot \mathbf{CH(NH_2)} \cdot CO \cdot OH + \mathbf{H_2O}$$

The direct reductive amination of pyruvate to alanine has been observed in at least one micro-organism, *Bacillus subtilis*. In most mammals, however, the reaction is one of transamination analogous to that of glyoxylic acid to glycine, though unlike the latter, transamination of α-keto-acids is reversible. The reversible transamination of α-ketoglutarate and glutamate plays a key rôle in the entry of the latter amino acid into the citric acid cycle, and in the biosynthesis of amino acids from the former. Pyridoxal and pyridoxamine are important coenzymes in these reactions (Chapter 18).

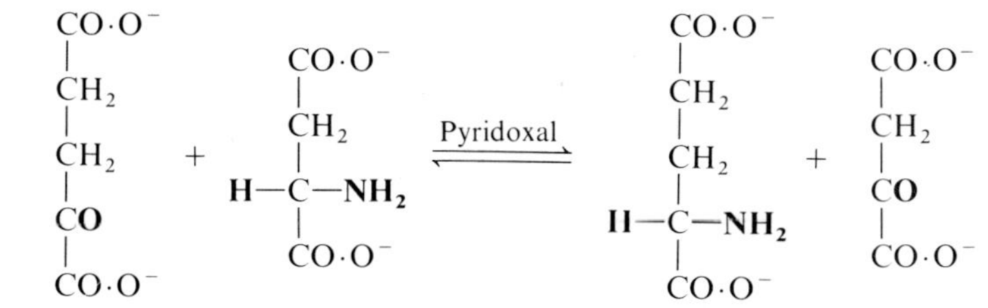

$$\underset{\alpha\text{-Ketoglutarate}}{{}^{-}O\cdot OC\cdot CH_2\cdot CH_2\cdot CO\cdot CO\cdot O^-} + \underset{\text{Aspartate}}{{}^{-}O\cdot OC\cdot CH_2\cdot CH(NH_2)\cdot CO\cdot O^-} \underset{}{\overset{\text{Pyridoxal}}{\rightleftharpoons}} \underset{\text{Glutamate}}{{}^{-}O\cdot OC\cdot CH_2\cdot CH_2\cdot CH(NH_2)\cdot CO\cdot O^-} + \underset{\text{Oxaloacetate}}{{}^{-}O\cdot OC\cdot CH_2\cdot CO\cdot CO\cdot O^-}$$

**Aldol Condensation**

Oxaloacetate undergoes an enzymically-controlled aldol-type condensation with thiol esters such as acetyl-coenzyme A.

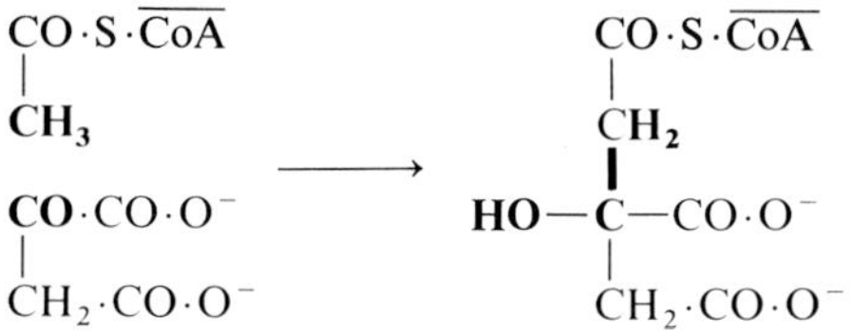

$$CH_3\cdot CO\cdot S\cdot \overline{CoA} + CO(CO\cdot O^-)\cdot CH_2\cdot CO\cdot O^- \longrightarrow \underset{\text{Citrate}}{\overline{CoA}\cdot S\cdot CO\cdot CH_2\cdot C(OH)(CO\cdot O^-)\cdot CH_2\cdot CO\cdot O^-}$$

## β-Keto-acids

**Enolisation**

A characteristic feature of *β*-keto-acids and their derivatives (*q.v.*) is their ability to exhibit keto-enol tautomerism.

$$\underset{\text{keto}}{CH_3\cdot \mathbf{CO\cdot CH_2}\cdot CO\cdot OH} \rightleftharpoons \underset{\text{enol}}{CH_3\cdot \mathbf{C(OH){=}CH}\cdot CO\cdot OH}$$

Acetoacetic acid

Keto-enol tautomerism is a general characteristic of all 1,3-dicarbonyl compounds, including, for example, cyclohexane-1,3-dione. The anti-inflammatory compound *Phenylbutazone* exhibits a similar type of tautomerism.

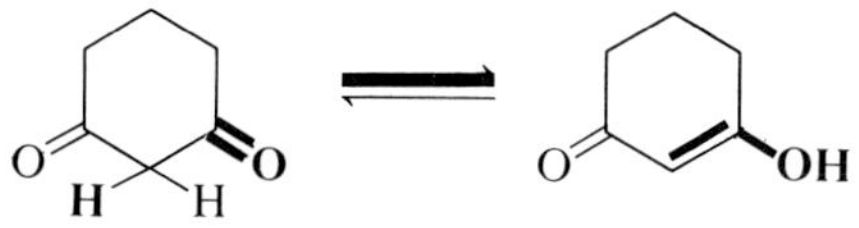

Cyclohexane-1,3-dione

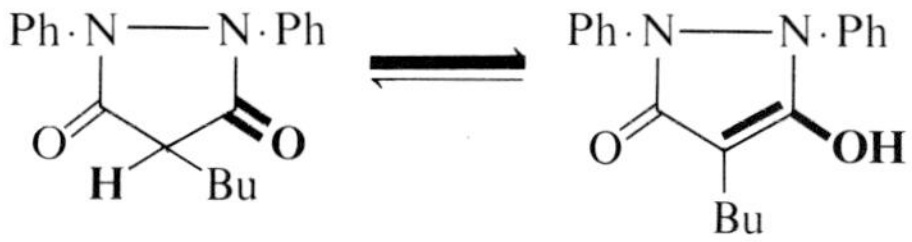

*Phenylbutazone*

The position of the equilibrium varies enormously from compound to compound. According to Hughes and Ingold, enolisation involves the transfer of proton, irrespective of whether the reaction is acid- or base-catalysed. The acidic strengths of enols in proton-accepting solvents will, therefore, roughly parallel the proportion of enol present at equilibrium.

Enols also resemble phenols in that they react with ferric chloride to give a red coloured complex.

### Decarboxylation

$\beta$-Keto-acids and their salts are unstable. They readily undergo thermal decarboxylation with evolution of carbon dioxide and formation of a ketone.

$$CH_3{\cdot}CO{\cdot}CH_2{\cdot}\mathbf{CO{\cdot}OH} \longrightarrow CH_3{\cdot}CO{\cdot}CH_3 + \mathbf{CO_2}$$

The decarboxylation involves both the undissociated acid and the carboxylate ion.

$CH_3$, $H_2C$, C, O, O, O, **H** $\rightleftharpoons$ $CH_3$, $H_2C$, C, O, O, $O^-$ + $\mathbf{H^+}$

$\downarrow$ $CO_2$ $\qquad$ $\downarrow$ $CO_2$

$CH_3$, $H_2C$, C, O, H $\qquad$ $CH_3$, $H_2C$, C, $O^-$ ($H^+$)

$\searrow$ $CH_3$, $H_3C$, C, O $\swarrow$

The decarboxylation is catalysed by primary amines (Pedersen, 1938). Similarly, the formation of acetone by **acetoacetate decarboxylase** in diabetes involves a Schiff's base intermediate (Fridovich and Westheimer, 1962; see p. 353).

## β-Keto-esters

### Tautomerism

Unlike the acids, $\beta$-keto-esters obtained by Claisen condensation of saturated monocarboxylic acid esters are thermally stable and can be distilled without

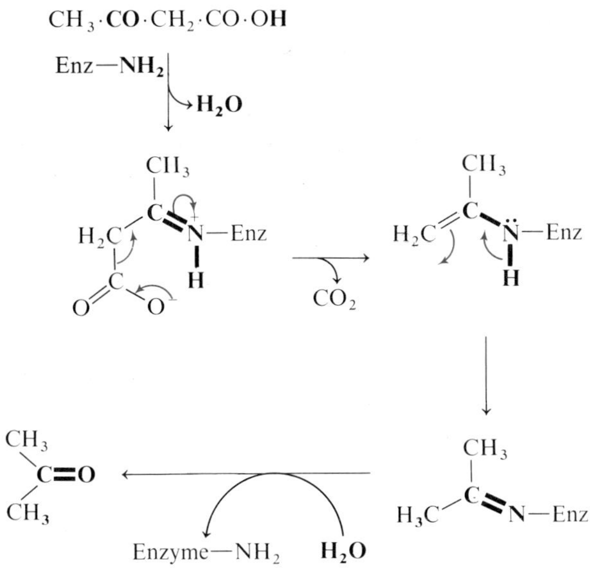

$CH_3 \cdot \mathbf{CO} \cdot CH_2 \cdot CO \cdot OH$

Enz—$\mathbf{NH_2}$

$\mathbf{H_2O}$

$CH_3$ $H_2C$ C N—Enz H C O O

$CO_2$

$CH_3$ $H_2C$ C N—Enz H

$CH_3$ C=O $CH_3$

Enzyme—$NH_2$ $\mathbf{H_2O}$

$CH_3$ $H_3C$ C N—Enz

decomposition. Like the acids, however, they are capable of enolisation and exist as tautomeric mixtures of keto- and enol-forms.

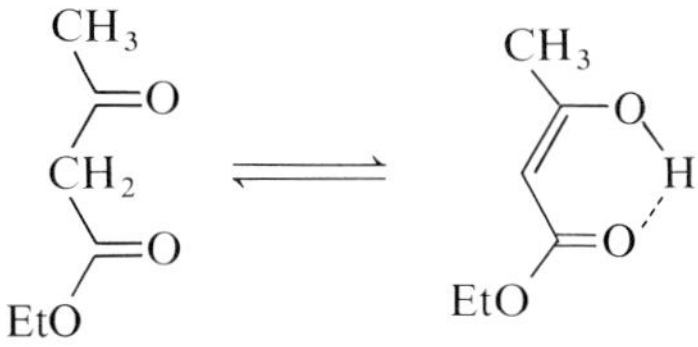

$CH_3$ =O $CH_2$ =O EtO

Keto

$CH_3$ O H O EtO

Enol

The composition of these mixtures is both temperature- and solvent-dependent. The equilibrium in ethyl acetoacetate favours the keto-tautomer at room temperature, 92.3% being in this form (Meyer, 1911); cooling promotes formation of the keto-form, which can be crystallised (m.p. $-39°C$) from ethanol at $-78°C$. Internal hydrogen bonding (Sidgwick, 1925) makes the enol more volatile, and pure enol-form has been obtained by fractional distillation under reduced pressure.

**Ketonic Properties**

As a result of enolisation, $\beta$-keto-esters exhibit properties of both keto- and enol-forms. The carbonyl group is readily reduced to the corresponding secondary alcohol, both chemically, and enzymatically in the course of fatty acid biosynthesis (Chapter 11). Thus, acetoacetyl-S$\cdot\overline{ACP}$ ($\overline{ACP}$ = acyl carrier protein) and other $\beta$-keto-ester intermediates are readily reduced to $\beta$-hydroxy-esters.

$$CH_3 \cdot \mathbf{CO} \cdot CH_2 \cdot CO \cdot S \cdot \overline{ACP} \longrightarrow CH_3 \cdot \mathbf{CH(OH)} \cdot CH_2 \cdot CO \cdot S \cdot \overline{ACP}$$

Diabetes, in which carbohydrate metabolism is deranged, is characterised by release of free acetoacetic acid, which is partly decarboxylated to acetone (see above) and partly reduced by a similar stereospecific reduction to D-(−)-β-hydroxybutyric acid.

As ketones, β-keto-esters also condense readily with ammonia, primary amines, hydrazines, hydroxylamine and semicarbazide to form typical derivatives (Chapter 10).

$$CH_3{\cdot}\mathbf{CO}{\cdot}CH_2{\cdot}CO{\cdot}OEt + R{\cdot}\mathbf{NH_2} \longrightarrow CH_3{\cdot}\underset{\mathbf{NR}}{\underset{\|}{\mathbf{C}}}{\cdot}CH_2{\cdot}CO{\cdot}OEt$$

$R = —H, —alkyl, —aryl, —NH_2, —NHPh, —OH, —NH{\cdot}CO{\cdot}NH_2$

The hydrazones and oximes cyclise readily with elimination of ethanol to form substituted pyrazolones and isoxazolones.

$CH_3{\cdot}C(=N{-}\mathbf{NH}{-}Ph){-}CH_2{-}CO{\cdot}\mathbf{OEt} \xrightarrow{-\mathbf{EtOH}}$ 3-Methyl-1-phenylpyrazolone

$CH_3{\cdot}C(=N{-}\mathbf{OH}){-}CH_2{-}CO{\cdot}\mathbf{OEt} \xrightarrow{-\mathbf{EtOH}}$ 3-Methylisoxazalone

## Enolic Properties

The enol is characterised by the ease with which bromine adds to the double bond. It is also a weak acid (p$K_a$ 10.5) and readily forms a sodium salt with metallic sodium or sodium ethoxide.

$$CH_3{\cdot}CO{\cdot}CH_2{\cdot}CO{\cdot}OEt \rightleftharpoons CH_3{\cdot}C(OH){=}CH{\cdot}CO{\cdot}OEt$$

$$CH_3{\cdot}C(OH){=}CH{\cdot}CO{\cdot}OEt \xrightarrow[\text{NaOEt} \to \text{EtOH}]{} CH_3{\cdot}C(ONa){=}CH{\cdot}CO{\cdot}OEt$$

The enolate ion is resonance-stabilised.

$$CH_3{\cdot}C(O^-){=}CH{\cdot}CO{\cdot}OEt \longleftrightarrow CH_3{\cdot}C({=}O){-}\overset{-}{C}H{\cdot}CO{\cdot}OEt$$

The carbanion readily displaces halogen from alkyl halides to form an α-alkyl-β-keto-ester. The latter also forms a sodio-derivative which is capable of reaction with a second molecule of alkyl halide to give an αα-dialkyl-β-keto-ester.

$$[CH_3\cdot CO\cdot \bar{C}H\cdot CO\cdot OEt]Na^+ \xrightarrow[\mathbf{R}-Br]{NaBr} CH_3\cdot CO\cdot CH\mathbf{R}\cdot CO\cdot OEt$$

$$\downarrow NaOEt$$

$$CH_3\cdot CO\cdot C\mathbf{RR'}\cdot CO\cdot OEt \xleftarrow[NaBr]{\mathbf{R'}Br} [CH_3\cdot CO\cdot \bar{C}\mathbf{R}\cdot CO\cdot OEt]Na^+$$

**Hydrolysis**

β-Keto-esters are hydrolysed with cold aqueous sodium hydroxide (5%) to the sodium salt of the acid. Acidification gives the free acid, which on warming decarboxylates to form a ketone. The hydrolysis and decarboxylation of ethyl acetoacetate derivatives provides a useful method for the preparation of substituted methyl ketones.

$$CH_3\cdot CO\cdot CRR'\cdot CO\cdot OEt \xrightarrow{NaOHaq} CH_3\cdot CO\cdot CRR'\cdot CO\cdot ONa$$

$$\downarrow H^+/Heat \quad (-CO_2)$$

$$CH_3\cdot CO\cdot CHRR'$$

**Reverse-Claisen Condensation**

In hot concentrated (20%) ethanolic potassium hydroxide, the ethoxide ion attacks the keto-carbonyl carbon of β-keto-esters leading to cleavage of the carbon chain. The product esters are simultaneously hydrolysed to carboxylic acids.

$$CH_3\cdot C(=O)-CRR'\cdot CO\cdot OEt \xrightarrow{EtO^-} CH_3\cdot C(O^-)(OEt)-CRR'\cdot CO\cdot OEt \xrightarrow{H^+}$$

$$CH_3\cdot CO\cdot OEt + R'\cdot RCH\cdot CO\cdot OEt$$

The breakdown of $\beta$-ketoadipate ($^{-}O \cdot OC \cdot CH_2 \cdot CH_2 \cdot CO \cdot CH_2 \cdot CO \cdot O^{-}$) to succinate and acetate in **Pseudomonas** and the cleavage of 4-fumarylacetoacetate to fumarate and acetoacetate in the terminal stages of mammalian phenylalanine and tyrosine metabolism are clearly analogous reactions. Similarly, $\beta$-**ketothiolases** present in beef mitochondria and heart muscle, which catalyse analogous cleavages of $\beta$-keto-acyl-S·$\overline{CoA}$ compounds, also appear to act by a corresponding mechanism.

$$R \cdot \overset{O}{\overset{\|}{C}}-CH_2 \cdot CO \cdot S \cdot \overline{CoA} \;\; (+\; ^{-}S-\overline{CoA}) \longrightarrow R \cdot \underset{S \cdot \overline{CoA}}{\overset{O^{-}}{\overset{|}{\underset{|}{C}}}}-CH_2 \cdot CO \cdot S \cdot \overline{CoA} \xrightarrow{H^+} R \cdot CO \cdot S \cdot \overline{CoA} + CH_3 \cdot CO \cdot S \cdot \overline{CoA}$$

## γ-and δ-Keto-acids

$\gamma$- and $\delta$-keto-acids are characterised by the ease with which they are dehydrated on distillation either alone or in the presence of acid catalysts (phosphoric acid) to form the corresponding enol-lactone. Thus, laevulinic acid gives $\alpha$-angelicalactone; the latter is readily isomerised to the corresponding $\alpha\beta$-unsaturated lactone ($\beta$-angelicalactone), and the product usually contains both isomers.

$$\underset{\text{Laevulinic acid}}{CH_3 \cdot CO \cdot CH_2 \cdot CH_2 \cdot CO \cdot OH} \xrightarrow{-H_2O} \alpha\text{-} \longrightarrow \beta\text{-}$$

Angelicalactones

# ALIPHATIC HYDROXY-ACIDS

Most hydroxy-acids possess properties which are typical of both alcohols and carboxylic acids. Their properties, however, reflect the precise nature of the hydroxyl group depending upon whether it is primary, secondary or tertiary alcoholic. They also reflect the special relationships of the hydroxyl and carboxylic acid groups in the molecule as a whole.

### Nomenclature

Certain of the lower aliphatic hydroxy-acids have well-established trivial names. Alternatively, they are frequently named as derivatives of the corresponding saturated carboxylic acid with the position of the hydroxyl group denoted by use of a Greek letter ($\alpha$-, $\beta$-, $\gamma$-, etc.) symbolising the position of the attached carbon atom with respect to the carboxyl carbon. The IUPAC system is similar, but with the chain numbered from the carboxyl group (1) as in the saturated acids, and the position of the hydroxyl also denoted by a number (Table 43).

## General Properties

Aliphatic hydroxy-acids are usually much more soluble in water than their parent saturated carboxylic acids due to the intramolecular hydrogen bonding affinity of the hydroxyl group which stabilises the anion. They are also significantly stronger acids than their substituted counterparts (acetic acid $pK_a$ 4.76, glycollic acid $pK_a$ 3.83; propionic acid $pK_a$ 4.9, lactic acid $pK_a$ 3.86), reflecting the electronegativity of the hydroxylic oxygen. Dissociation of the α-hydroxyacids is probably also affected by intermolecular hydrogen bonding, since glycollic acid is a weaker acid than α-methoxyacetic acid ($pK_a$ 3.53). β-Hydroxypropionic acid, however, is a somewhat stronger acid ($pK_a$ 2.51) than the α-hydroxy-acids, but the corresponding alkoxy-acid, β-ethoxypropionic acid ($pK_a$ 2.49) is of almost identical acidic strength.

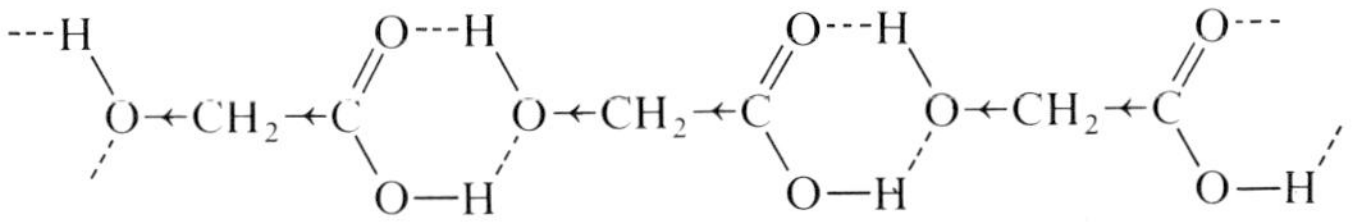

```
---H             O---H               O---H               O---
    \           //     \            //     \            //
     O←CH2←C            O←CH2←C             O←CH2←C
    /           \      /            \      /            \     /
                 O—H                 O—H                 O—H
```

Possible association of α-hydroxy-acids

Most aliphatic hydroxy-acids form water-soluble alkali metal salts. *Sodium Lactate* solutions are used to provide an injectable lactide-free (p. 360) form of lactic acid of near physiological pH. *Calcium Lactate* provides an acceptable method for the administration of a water-soluble calcium salt in the treatment of calcium deficiency, whilst the water-soluble *Antimony Sodium Tartrate* is used as a readily assimilable source of trivalent antimony for the treatment of schistosomiasis (**2**, 2).

Low mammalian toxicity and high water solubility are typical of most hydroxy-acids. Tartaric acid which shows these properties is often the acid of choice for the preparation of water-soluble salts of organic bases required for internal administration. As a dibasic acid, tartaric acid can also be used to prepare stable acid salts, which give immediate pH stabilisation in aqueous solution on the acid side of neutrality where this is essential to ensure stability of the product solution, e.g. *Adrenaline Acid Tartrate* (Epinephrine Bitartrate).

## Oxidation

Aliphatic hydroxy-acids are subject to oxidation as for alcohols (Chapter 7) with the production of aldehydo- and keto-acids from primary and secondary hydroxy-acids respectively; tertiary hydroxy-acids, however, undergo chain disruption on oxidation. The corresponding enzymic oxidation of lactate to pyruvate by **lactic dehydrogenase** is specific in mammalian heart and skeletal

Table 43. Nomenclature of Aliphatic Hydroxy-acids

| Structure | Systematic Name | Trivial Name(s) |
|---|---|---|
| $HO \cdot CH_2 \cdot CO \cdot OH$ | 2-Hydroxyethanoic acid | 2-Hydroxyacetic acid<br>Glycollic acid |
| $CH_3 \cdot CHOH \cdot CO \cdot OH$ | 2-Hydroxypropanoic acid | $\alpha$-Hydroxypropionic acid<br>Lactic acid |
| $HO \cdot CH_2 \cdot CH_2 \cdot CO \cdot OH$ | 3-Hydroxypropanoic acid | $\beta$-Hydroxypropionic acid |
| $CH_3 \cdot CH_2 \cdot CHOH \cdot CO \cdot OH$ | 2-Hydroxybutanoic acid | $\alpha$-Hydroxybutyric acid |
| $CH_3 \cdot CHOH \cdot CH_2 \cdot CO \cdot OH$ | 3-Hydroxybutanoic acid | $\beta$-Hydroxybutyric acid |
| $HO \cdot CH_2 \cdot CH_2 \cdot CH_2 \cdot CO \cdot OH$ | 4-Hydroxybutanoic acid | $\gamma$-Hydroxybutyric acid |
| $HO \cdot OC \cdot CH_2 \cdot CHOH \cdot CO \cdot OH$ | 2-Hydroxybutanedioic acid | $\alpha$-Hydroxysuccinic acid<br>Malic acid |
| $HO \cdot OC \cdot CHOH \cdot CHOH \cdot CO \cdot OH$ | 2,3-Dihydroxybutanedioic acid | $\alpha\beta$-Dihydroxysuccinic acid<br>Tartaric acid |
| $HO \cdot OC \cdot CH_2 \cdot C(OH)(COOH) \cdot CH_2 \cdot CO \cdot OH$ | 3-Hydroxypentane-1,3,5-trioic acid | Citric acid |
| $HO \cdot OC \cdot CHOH \cdot CH(CO \cdot OH) \cdot CH_2 \cdot CO \cdot OH$ | 2-Hydroxypentane-1,3,5-trioic acid | Isocitric acid |
| $C_6H_5 \cdot CHOH \cdot CO \cdot OH$ | 2-Hydroxy-2-phenylacetic acid | Mandelic acid |
| $(C_6H_5)_2 \cdot C(OH) \cdot CO \cdot OH$ | 2,2′-Diphenyl-2-hydroxyacetic acid | Benzilic acid |

muscle for the optically active form L-(+)-lactate. The oxidation requires the participation of nicotinamide adenine dinucleotide (NAD), and can be reversed by the same enzyme under appropriate conditions.

$$CH_3\cdot\mathbf{CH(OH)}\cdot CO\cdot O^- + NAD^+ \rightleftharpoons CH_3\cdot\mathbf{CO}\cdot CO\cdot O^- + NAD\mathbf{H} + \mathbf{H^+}$$

α-Hydroxy-acids are also smoothly and quantitatively oxidised chemically with loss of the carboxyl carbon in a similar manner to glycols (Chapter 20) by lead tetra-acetate and sodium bismuthate.

$$CH_3\cdot CH(OH)\cdot CO\cdot OH \xrightarrow{\text{Oxidation}} CH_3\cdot CHO + H_2O + CO_2$$

## Hydroxylic Properties

The alcoholic hydroxyl groups behave typically in the displacement of hydrogen by alkali metals, and in the formation of *O*-acyl derivatives, as for example with acetic anhydride.

$$\underset{\text{Lactic acid}}{CH_3\cdot CH(\mathbf{OH})\cdot CO\cdot OH} + \mathbf{(CH_3\cdot CO)_2\cdot O}$$

$$\downarrow$$

$$\underset{\text{Lactyl acetate}}{CH_3\cdot CH(\mathbf{O\cdot CO\cdot CH_3})\cdot CO\cdot OH} + \mathbf{CH_3\cdot CO\cdot OH}$$

The *O*-acyl acids are thermally-unstable and pyrolyse readily to give the corresponding unsaturated acids.

$$CH_3\cdot CH(\mathbf{O\cdot CO\cdot CH_3})\cdot CO\cdot OH \xrightarrow{\text{Heat}} \underset{\text{acrylic acid}}{H_2C{=}CH\cdot COOH} + \mathbf{CH_3\cdot CO\cdot OH}$$

β-Hydroxy-acids and esters also decompose thermally with loss of water to form αβ-unsaturated (and possibly βγ-unsaturated) acids.

$$\mathbf{HO}\cdot CH_2\cdot \mathbf{CH_2}\cdot CO\cdot OH \longrightarrow H_2C{=}CH\cdot CO\cdot OH + \mathbf{H_2O}$$

These dehydrations are possibly self (acid) catalysed, as they bear close analogy to the acid-catalysed hydrations–dehydrations which form the basis of the enzymically-controlled interconversion of citrate and (+)-isocitrate via *cis*-aconitate. Aconitase, the enzyme responsible for these interconversions, appears to be dependent on the presence of ferrous iron (Dickman, 1961). There is some doubt, however, whether *cis*-aconitate is a true intermediate in the interconversion of citrate and isocitrate, and on the basis of studies with deuterium oxide (Englard and Colowick, 1957), the related carbonium ion-ferrous ion hydrate chelate now appears to be more likely.

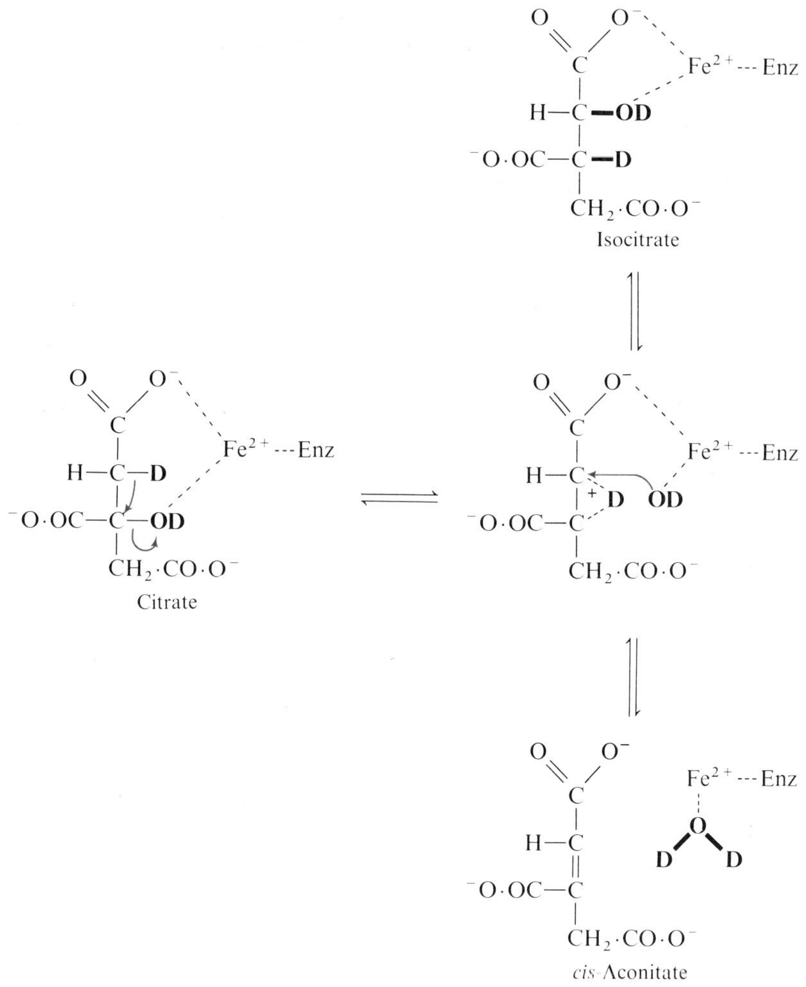

O O⁻ C $Fe^{2+}$ --- Enz
H—C—**OD**
$^{-}O \cdot OC$—C—**D**
$CH_2 \cdot CO \cdot O^-$

Isocitrate

⇌

O O⁻ C $Fe^{2+}$ --- Enz
H—C—**D**
$^{-}O \cdot OC$—C—**OD**
$CH_2 \cdot CO \cdot O^-$

Citrate

⇌

O O⁻ C $Fe^{2+}$ --- Enz
H—C + **D** **OD**
$^{-}O \cdot OC$—C
$CH_2 \cdot CO \cdot O^-$

⇌

O O⁻ C $Fe^{2+}$ --- Enz
H—C **O** **D** **D**
$^{-}O \cdot OC$—C
$CH_2 \cdot CO \cdot O^-$

*cis*-Aconitate

## Esterification

Aliphatic hydroxy-acids readily form esters. α-Hydroxy-acids, however, also undergo a bimolecular cyclodehydration equally readily with the formation of a cyclic di-ester or lactide. The reaction occurs slowly at ambient temperatures and rapidly on heating, so that lactic acid, for example, prepared by distillation consists largely of lactide; it also contains some lactoyl-lactic acid.

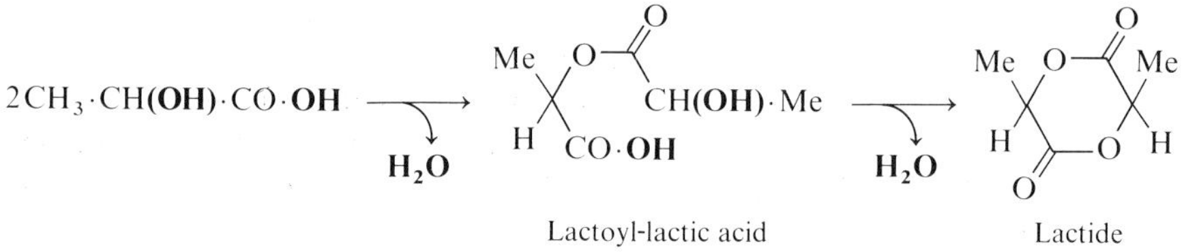

$2CH_3 \cdot CH(\mathbf{OH}) \cdot CO \cdot \mathbf{OH} \xrightarrow{-\mathbf{H_2O}}$ Me, O—C(=O)—CH(**OH**)·Me, H, CO·**OH** $\xrightarrow{-\mathbf{H_2O}}$ Me, O, O, Me, H, O, H, O

Lactoyl-lactic acid

Lactide

Both lactoyl-lactic acid and lactide are hydrolysed to the monomeric acid by alkali. Dilute mineral acid, however, leads to further decomposition of lactic acid (and other α-hydroxy-acids) with the formation of formic acid and acetaldehyde, possibly by the following pathway.

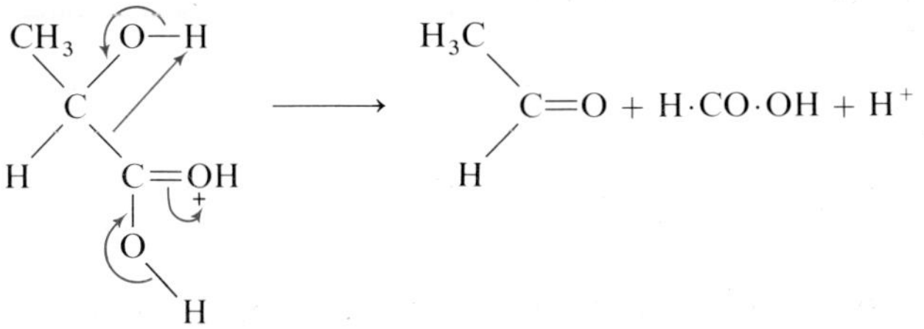

$$CH_3(H)C(OH)\cdot C(=\overset{+}{O}H)OH \longrightarrow H_3C(H)C{=}O + H\cdot CO\cdot OH + H^+$$

## LACTONES

γ- and δ- and some higher hydroxy-acids also undergo cyclodehydration on heating, but intramolecularly with the formation of cyclic esters known as lactones.

$$\underset{\gamma\text{-Hydroxybutyric acid}}{HO\cdot CH_2\cdot CH_2\cdot CH_2\cdot \mathbf{CO\cdot OH}} \longrightarrow \gamma\text{-Butyrolactone (1,4: butanolide)}$$

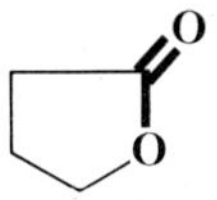

$$\underset{\delta\text{-Hydroxyvaleric acid}}{HO\cdot CH_2\cdot CH_2\cdot CH_2\cdot CH_2\cdot \mathbf{CO\cdot OH}} \longrightarrow \delta\text{-Valerolactone(1,5-pentanolide)}$$

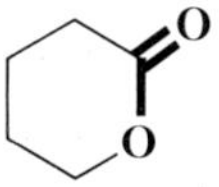

In contrast, β-hydroxy-acids do not cyclodehydrate. β-Lactones, however, are readily obtainable by the reaction of ketenes with carbonyl compounds. Thus, ketene and formaldehyde give β-propiolactone.

$$\underset{\text{Ketene}}{H_2C{=}C{=}O} + HCHO \longrightarrow \beta\text{-Propiolactone (1,3-propanolide)}$$

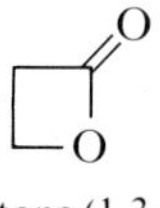

Lactones are often given trivial names which in the case of the lower molecular weight members relate the structure to the parent hydroxy-acid. According to the IUPAC system of nomenclature, however, lactones are known as **-olides**.

Lactones are in effect internal esters, and readily hydrolysable in alkaline solution to give the corresponding hydroxy-acids as their alkali metal salts.

$$\gamma\text{-butyrolactone} + \mathbf{NaOH} \longrightarrow HO\cdot CH_2\cdot CH_2\cdot CH_2\cdot \mathbf{CO\cdot ONa}$$

Lactones are soluble in acid solution. Thus, *Digoxin*, which is an $\alpha\beta$-unsaturated lactone derivative (Chapter 22) appears to be absorbed intact from the acid media of the stomach, and although it is strongly bound to plasma skeletal muscle and heart muscle proteins, there is no evidence as to whether or not this is hydroxyamide formation, analogous to the reaction *in vitro* of simple lactones with amines.

$$\text{(γ-lactone)} + R \cdot NH_2 \longrightarrow \mathbf{HO \cdot CH_2 \cdot CH_2 \cdot CH_2 \cdot CO \cdot NHR} \pm \text{(lactam, NHR)}$$

Reactivity of lactones, generally, depends to some extent on both ring size and substitution. The strained four-membered $\beta$-lactones are particularly reactive both chemically and biologically. $\beta$-Propiolactone itself is reported to be carcinogenic (Walpole, Roberts, Rose, Hendry and Homer, 1954; Dickens, 1964). Its hydrolysis product, $\beta$-hydroxypropionic acid, however, is free from carcinogenicity (Dickens and Jones, 1963), showing that tumorogenic activity resides in the alkylating capability of the $\beta$-lactone ring.

$\beta$-Propiolactone is very reactive and readily hydrolysed in aqueous solution with a half-life of about 3 hr at 25°C. It is generally susceptible to nucleophilic attack in increasing order of activity by chloride, acetate, and hydroxyl ions in aqueous solution (Bartlett and Small, 1950). According to Zaugg (1954), such ionic nucleophiles cause $H_2C—O$ bond cleavage, whereas non-ionic nucleophiles cause O—CO ring cleavage. The reaction pathway, however, is dependent on reagent and lactone, since whilst $\beta$-propiolactone reacts preferentially with cysteine at the sulphydryl rather than the amino group to give *S*-2-carboxyethylcysteine (Dickens and Jones, 1961), the homologous $\beta$-butyrolactone reacts with coenzyme A to give the thiol ester (Decker and Lynen, 1955).

$$HO \cdot CO \cdot CH(NH_2) \cdot CH_2 \cdot SH + \beta\text{-Propiolactone} \longrightarrow HO \cdot OC \cdot CH(NH_2) \cdot CH_2 \cdot S \cdot CH_2 \cdot CH_2 \cdot CO \cdot OH$$

Cysteine; $\beta$-Propiolactone; *S*-2-Carboxyethylcysteine

$$R \cdot CH_2 \cdot SH + \beta\text{-Butyrolactone (Me)} \longrightarrow HO \cdot CH(Me) \cdot CH_2 \cdot CO \cdot S \cdot CH_2 \cdot R$$

$\beta$-Butyrolactone

$\beta$-Propiolactone has also been reported to react with guanosine to give the corresponding *N*-carboxyethylguanosine, isolated after acid hydrolysis as 7-(2′-carboxyethyl)guanine (Roberts and Warwick, 1962). This reaction may have some bearing on its carcinogenic activity.

A number of unsaturated lactones have also been shown to be potent hepatic carcinogens. The most potent of these are aflatoxin-B and the related aflatoxin-G produced by the common mould, *Aspergillus flavus*, which has been known to be present in substantial amount in mould-contaminated groundnuts (peanuts) and arachis oil (Sargeant, Sheridan, O'Kelly and Carnaghan, 1961).

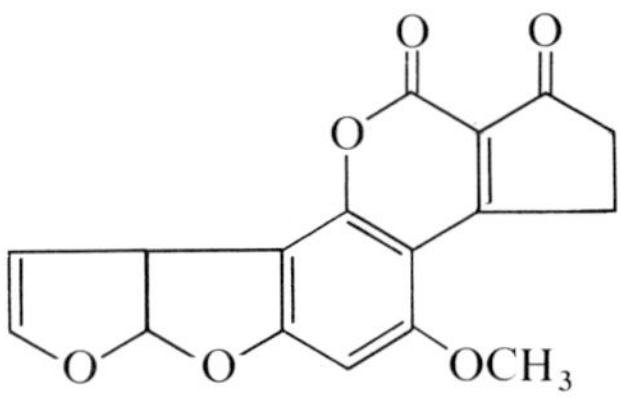

Aflatoxin B

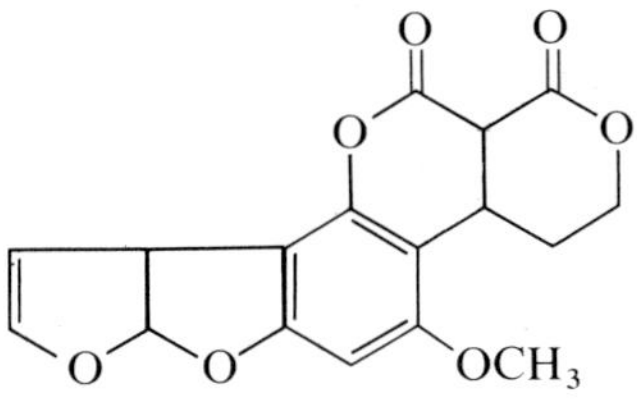

Aflatoxin G

## MACROLIDES

The macrolide group of antibiotics is so named because each member is characterised by the presence of a large lactone ring containing 12 or more carbon atoms. The lactone ring is linked glycosidically in every case to an amino sugar, and in some, either the lactone ring or the amino sugar is joined by a glycosidic link to a second sugar moiety.

### Erythromycin

*Erythromycin* is a metabolic product of *Streptomyces erythreus* (McGuire, Bunch, Anderson, Boaz, Flynn, Powell and Smith, 1952). It is a monoacidic base (p$K_a$ 8.6 in 66% dimethylformamide), and only slightly soluble in water, though it forms crystalline water-soluble salts with mineral acids.

*Erythromycin* is sensitive to both acid and alkali. It is readily inactivated by dilute acid, which cleaves the glycosidic link to cladinose, and catalyses dehydration and intramolecular ketal formation to give erythralosamine. For this reason, *Erythromycin* is administered orally in tablets with an acid-resistant coating (Smith, Dyke and Griffith, 1953).

The lactone ring is also extremely sensitive to alkali, even in very dilute solution (*0.005 N*) when the corresponding hydroxy-acid is formed. Stronger solutions of sodium hydroxide (*0.01N*) also cause dehydration. The antibiotic would, therefore, be expected to be present in one or other of these forms at physiological pH(7.4). There is, however, no evidence of their lipid solubility relative to that of the parent antibiotic (*Erythromycin* is chloroform-soluble), or whether or not the antibiotic is absorbed from the gastro-intestinal tract and transported within the body to act intact at the site of infection.

The bitter taste of *Erythromycin* is lost in its ethylsuccinate form (2′-succinoyl erythromycin). The analogous *Erythromycin Stearate*, which is formed by fermentation of *Streptomyces erythreus* in the presence of stearic acid and sodium stearate, is actually a mixture of *ca* 77% stearate together with sodium stearate and free stearic acid. In consequence, it retains a slightly bitter taste and also is incompletely soluble in chloroform. *Erythromycin Estolate* (2′-propionylerythromycin laurylsulphate) which is both tasteless and chloroform-soluble is now used extensively. It is said to be more stable to acid than the free base, presumably due to steric hindrance. It is readily absorbed from the gastro-intestinal tract, and has been identified in blood and urine as erythromycin base.

As aliphatic organic bases, erythromycin and its derivatives are all sensitive to light-catalysed oxidation, and should be stored in well-closed containers protected from light. *Erythromycin* is effective as an antibacterial agent against

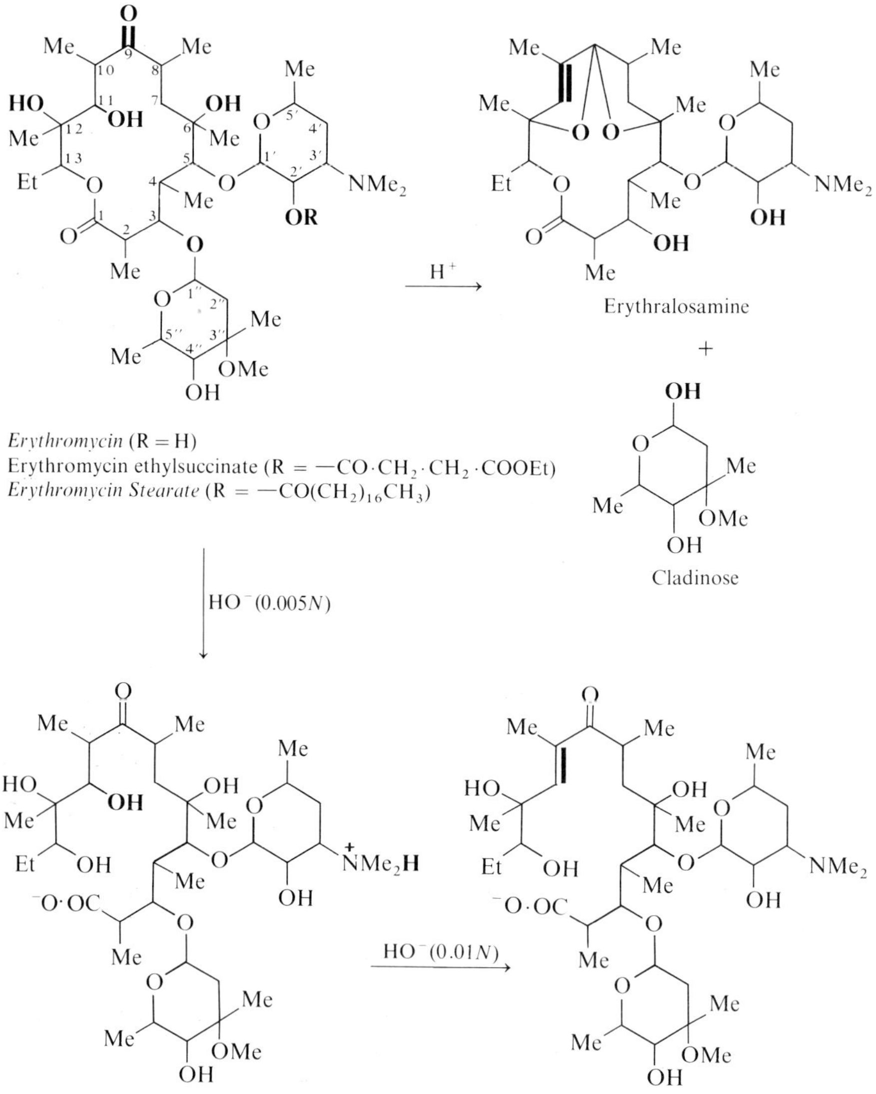

*Erythromycin* (R = H)
Erythromycin ethylsuccinate (R = $—CO\cdot CH_2\cdot CH_2\cdot COOEt$)
*Erythromycin Stearate* (R = $—CO(CH_2)_{16}CH_3$)

$\xrightarrow{H^+}$ Erythralosamine + Cladinose

$\downarrow$ $HO^-$ (0.005*N*)

$\xrightarrow{HO^-(0.01N)}$

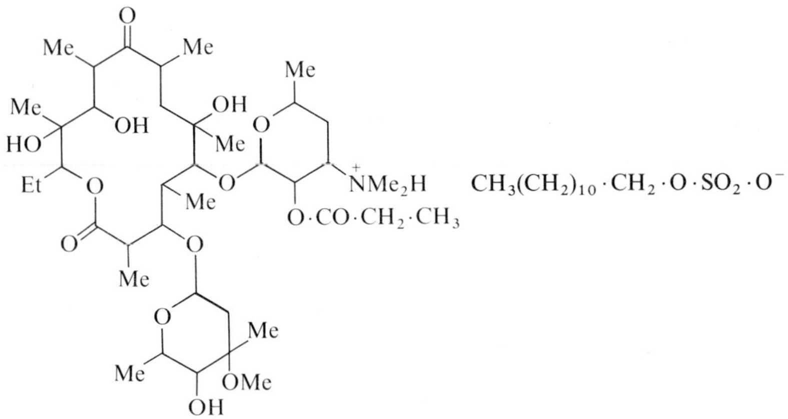

*Erythromycin Estolate*

most Gram-positive and some Gram-negative organisms, and is most active in the growth phase. It is useful in the treatment of acute streptococcal infections, pneumonia and staphylococcal enteritis.

## Polyene Macrolides

A group of polyunsaturated macrolide antibiotics are characterised by their antifungal properties. Two important members of this group are the hexaene macrolide, *Nystatin*, and the heptaene macrolide, *Amphotericin B*. As high molecular weight polyhydroxylic amino carboxylic acids, they both have low water and lipid solubilities, which are not enhanced in the alkaline media of the intestines. In consequence, they are poorly absorbed from the gastro-intestinal tract, skin and mucous membranes, and, therefore, are particularly effective in the treatment of intestinal, vulvovaginal and cutaneous moniliasis due to *Candida albicans*.

Because of the high level of unsaturation, polyene antibiotics are particularly sensitive to light-catalysed oxidation. They should, therefore, be stored in well-sealed containers, protected from light and temperatures not greater than 5°C.

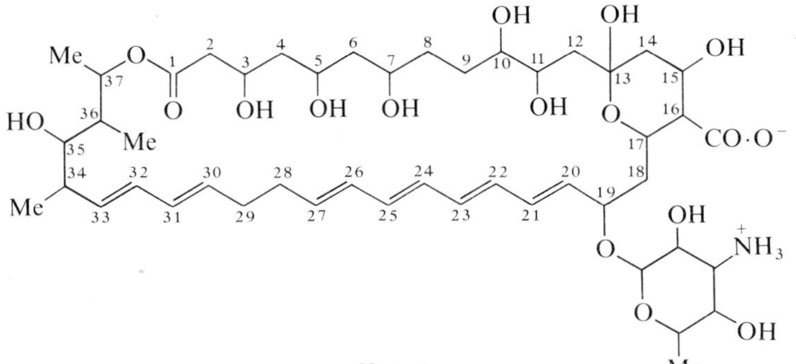

*Nystatin*

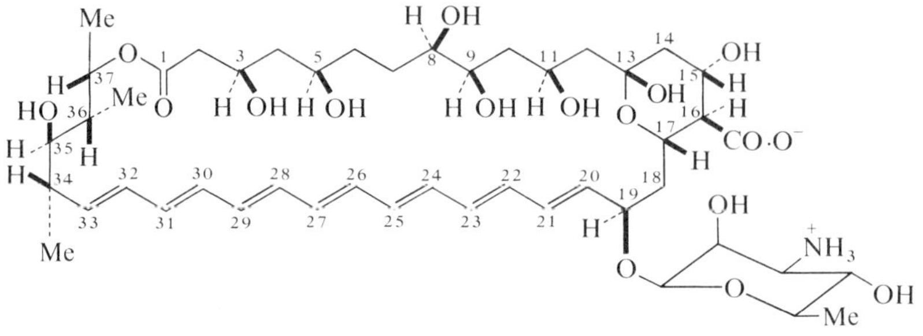

*Amphotericin B*

Polyene antifungal antibiotics appear to act by inhibiting the incorporation of steriods into cell membranes (Chapter 22).

## AROMATIC HYDROXY-ACIDS

The three isomeric monohydroxy aromatic acids, salicylic, *m*- and *p*-hydroxybenzoic acids, show properties which are characteristic both of phenols and carboxylic acids. As acids, they react with sodium bicarbonate to form the water-soluble salt of the carboxylic acid; with aqueous sodium hydroxide, they form the di-anion salts. Salicylic acid and *p*-hydroxybenzoic acid also show the characteristic ferric chloride colour reaction of phenols giving violet and red colours respectively. *m*-Hydroxybenzoic acid does not react with ferric chloride.

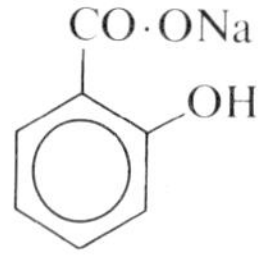

*Sodium Salicylate* 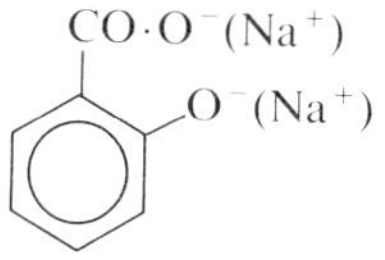 — Salicylate dianion 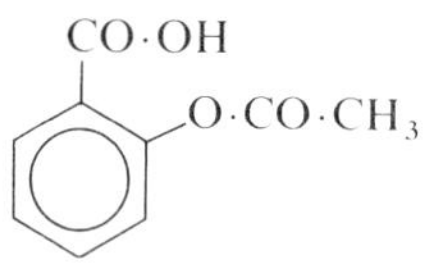 — Acetylsalicylic acid (*Aspirin*)

Salicylic acid is a much stronger acid than either benzoic acid or its iso-isomeric *m*- and *p*-hydroxybenzoic acids; the *o*-hydroxy group of salicylic acid is internally hydrogen-bonded to the carbonyl group and thus promotes dissociation of the acidic proton.

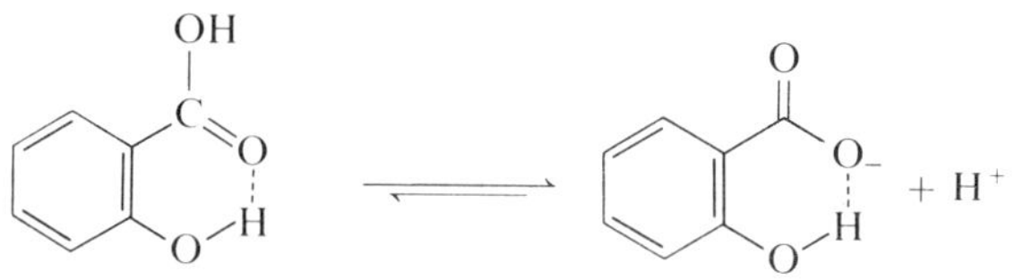

Salicylic acid (p$K_a$ 2.79)

Salicylic acid is used on account of its strongly acidic properties as a keratolytic agent for topical application to corns and warts.

The difference in strength between *m*-hydroxybenzoic acid (p$K_a$ 4.08) and *p*-hydroxybenzoic acid (p$K_a$ 4.54) can be attributed to the inhibiting influence on ionisation of the hydroxyl +M effect, which is transmitted to the carboxyl group only in the *para* isomer, with destabilisation of the anion.

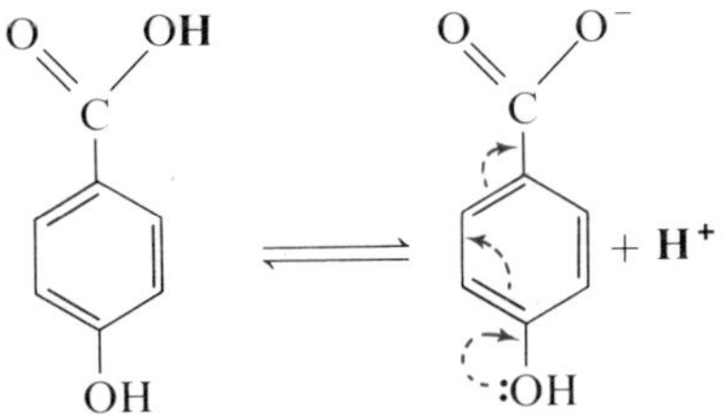

O OH C OH ⇌ O O⁻ C :OH + $H^+$

*p*-Hydroxybenzoic acid (p$K_a$ 4.54)

Salicylic acid is administered orally as sodium salicylate and as its *O*-acetyl derivative, *Aspirin*, for its antipyretic and analgesic effect. Acetylsalicylic acid is a slightly weaker acid (p$K_a$ 3.5) than salicylic acid. Both are rapidly absorbed into the blood stream from the stomach, but acetylsalicylic acid as the weaker acid is less damaging to the gastro-intestinal tract. Both acids are also strongly bound to plasma proteins (**2**, 4). Acetylsalicylic acid has been reported, however, to acetylate human plasma proteins (Hawkins, Pinckard and Farr, 1968; Pinckard, Hawkins and Farr, 1968). Acetylsalicylic acid has also been shown to acetylate amines, alcohols and phenols at significant rates even in solid state mixtures at temperatures of 60°C and over (Shami, Dudzinski, Lachman and Tingstad, 1973). Acetylsalicylic acid is hydrolysed rapidly to salicylic acid in the liver. *Aspirin* is excreted mainly as salicylic acid (*ca* 60 %), and the remainder after conjugation with glycine (15 %) and glucuronic acid (20 %); a small amount (1–5%) is further hydroxylated to gentisic acid (2,5-dihydroxybenzoic acid). Salicylic acid is conjugated and hydroxylated to a similar extent.

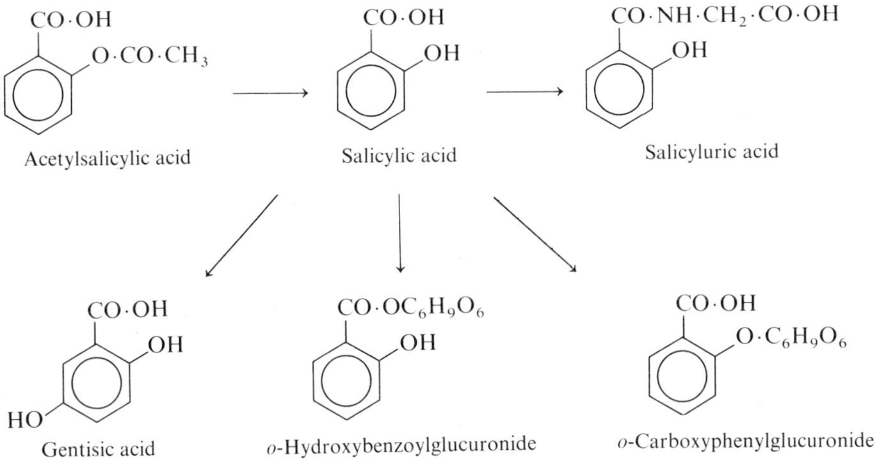

CO·OH, O·CO·CH$_3$ — Acetylsalicylic acid → CO·OH, OH — Salicylic acid → CO·NH·CH$_2$·CO·OH, OH — Salicyluric acid

CO·OH, OH, HO — Gentisic acid

CO·OC$_6$H$_9$O$_6$, OH — *o*-Hydroxybenzoylglucuronide

CO·OH, O·C$_6$H$_9$O$_6$ — *o*-Carboxyphenylglucuronide

Both *o*- and *p*-hydroxybenzoic acids react readily with halogens, with displacement of the carboxyl group.

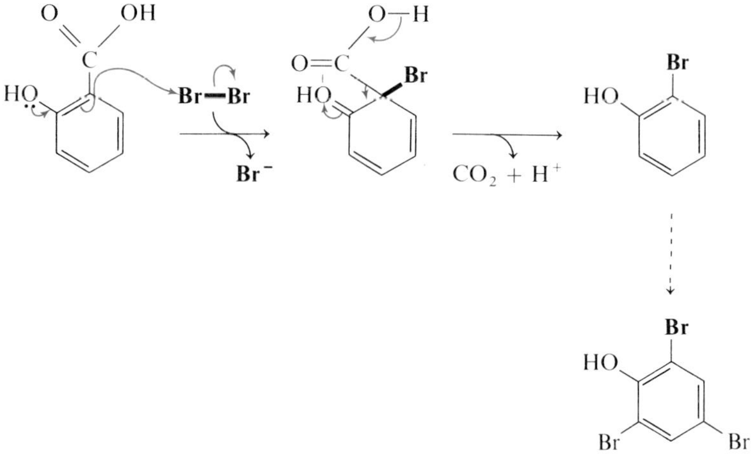

Gallic acid (3,4,5-trihydroxybenzoic acid) decarboxylates spontaneously when heated at its melting point.

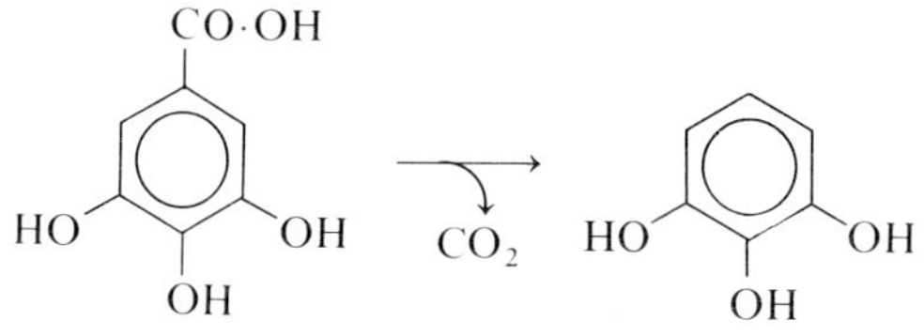

Gallic acid

Pyrogallol

**Esters**

Methyl salicylate (oil of wintergreen) occurs naturally. It is used for external application to the skin as a counter-irritant. *Methyl* and *Propyl Hydroxybenzoates* (Methylparaben and Propylparaben), and *Propyl*, *Octyl* and *Dodecyl*

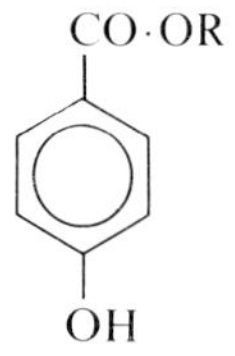

*Methyl Hydroxybenzoate* (R = Me)
*Propyl Hydroxybenzoate*
(R = $-CH_2 \cdot CH_2 \cdot CH_3$)

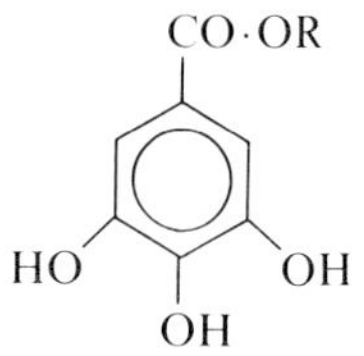

*Propyl Gallate* (R = $-CH_2 \cdot CH_2 \cdot CH_3$)
*Octyl Gallate* (R = $-CH_2(CH_2)_6 \cdot CH_3$)
*Dodecyl Gallate* (R = $-CH_2(CH_2)_{10} \cdot CH_3$)

*Gallates* are used as preservatives and antioxidants respectively. The former are particularly useful as antifungal agents in creams and emulsions, and are often used in combination to provide protection from a wide spectrum of mould and fungal contamination. *Dodecyl* and *Octyl Gallates*, which are soluble 1 in 30 and 1 in 33 in arachis oil respectively, are used as antioxidants for oils and fats. The propyl ester, which is much less soluble in oils and fats (solubility 1 in 2000 in arachis oil), is more suitable for protecting ether, volatile oils and paraldehyde from oxidation. As polyhydric phenols, gallates are highly susceptible to radical oxidation (Chapter 8). Their antioxidant activity is due to preferential oxidation, and in consequence they require careful storage in airtight containers, protected from light and contact with metal surfaces, all of which accelerate oxidative decomposition.

**Coumarin Derivatives**

Coumarin is the lactone of *o*-hydroxyallocinnamic acid. It is a natural perfume, occurring in certain grasses, but is normally prepared synthetically. It is the parent of a number of important 4-hydroxycoumarins which include the synthetic blood anticoagulants, *Warfarin Sodium* and *Nicoumalone* (Acenocoumarol), and the antibiotic, *Novobiocin*.

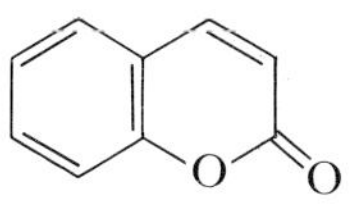

Coumarin

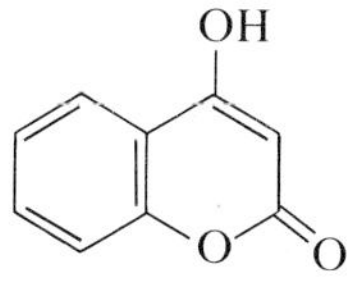

4-Hydroxycoumarin

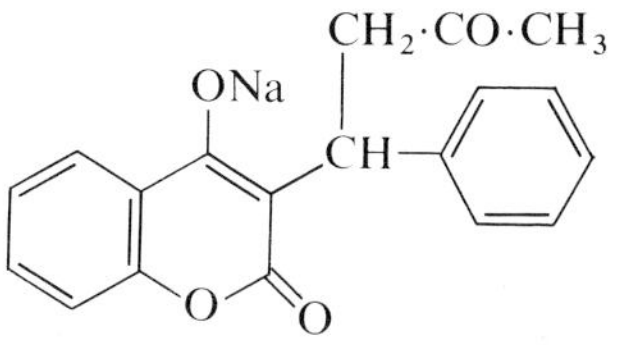

*Warfarin Sodium*

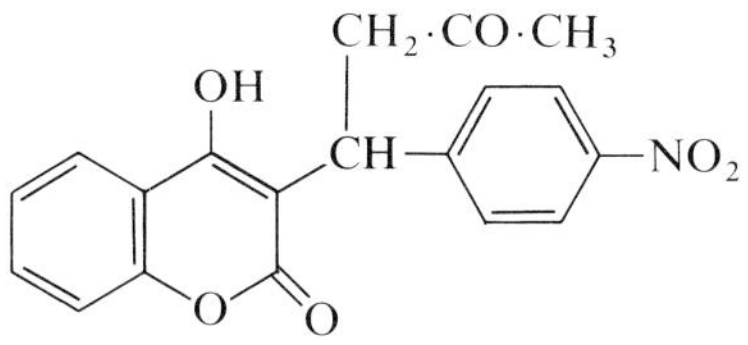

*Nicoumalone*

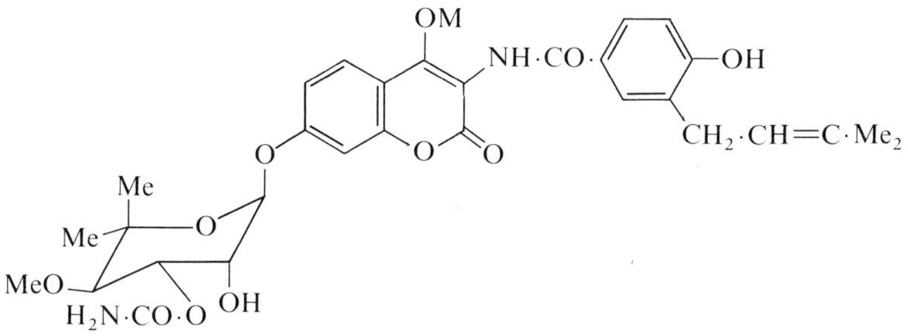

*Novobiocin* (M=H)
*Novobiocin Sodium* (M=Na)
*Novobiocin Calcium* (M=$\frac{1}{2}$Ca)

4-Hydroxycoumarins are strongly acidic due to the enolic hydroxyl which in *Novobiocin* has a p$K_a$ of 4.3. As a result, they are strongly protein-bound, but displaceable from protein-binding sites by other acidic drugs, such as *Aspirin*, *Phenylbutazone* and most sulphonamides (**2**, 4).

The parent 4-hydroxycoumarins are poorly soluble in water, but form water soluble sodium and calcium salts, as for example *Warfarin Sodium* and *Novobiocin Sodium*. *Novobiocin*, although poorly soluble, is available in an amorphous form, which is readily absorbed from the gastro-intestinal tract. This form, however, is metastable and in aqueous suspensions reverts to a crystalline form from which it is less readily available. *Novobiocin Sodium* and *Novobiocin Calcium*, on the other hand, are stable and hence preferable in the formulation of solid dosage products (Mullins and Macek, 1960).

The mode of action of *Novobiocin* is not fully established. It has, however, a pronounced effect on DNA synthesis and on the integrity of bacterial cell membranes (Morris and Russell, 1971).

### Sodium Cromoglycate

*Sodium Cromoglycate* is an oxochromene derivative, Such compounds are functionally vinylogous lactones. *Sodium Cromoglycate* is used in the treatment of asthma by direct insufflation into the lungs. It has a strong affinity for water, and normally crystallises as a hydrate, the water-content of which is an

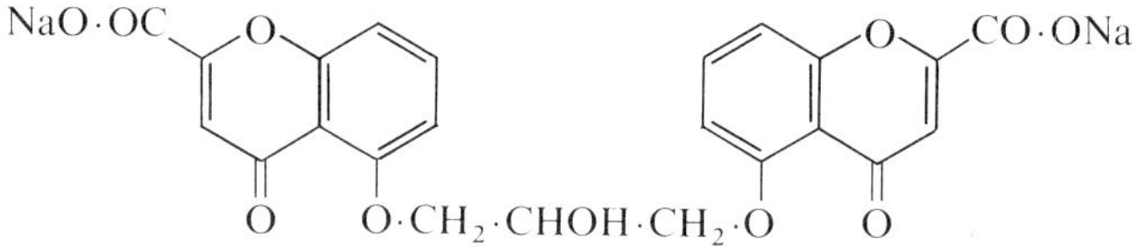

NaO·OC O O CO·ONa

O $O \cdot CH_2 \cdot CHOH \cdot CH_2 \cdot O$ O

important determinant of its flow properties. It is soluble in water, but is generally administered as a micronised powder of particle size in the 5–20$\mu$m range to ensure adequate penetration into the lungs and rapid absorption (Cox, Beach, Blair, Clarke, King, Lee, Loveday, Moss, Orr, Ritchie and Sheard, 1970).

## OPTICAL ISOMERISM

### Molecular Asymmetry

The carbon atom at C-2 in lactic acid is bonded to four completely different atoms or groups of atoms. The disposition of these groups about the central carbon atom is tetrahedral (Chapter 2). This confers an element of asymmetry on the molecule as a whole, which permits the existence of two different compounds each with the same molecular and structural formula. These two compounds are, therefore, identical in all respects except the relative orientation in space of the four groups about the central carbon atom. The two compounds are mirror images each of the other, and may be compared as left hand and right hand. They are not, therefore, superposable one upon the other. Such compounds are called **enantiomers**. The carbon atom at C-2 is the centre of asymmetry, and the molecule is said to be asymmetric. The two compounds are **stereoisomers**.

CO·OH H $CH_3$ OH      CO·OH HO $H_3C$ H

Some compounds do not possess an asymmetric carbon atom (or other asymmetric atom), but none-the-less show molecular asymmetry such that the molecule and its mirror image are non-superposable. Molecules which lack all elements of symmetry (except axes of rotation) are said to exhibit **chirality**. Such centres of asymmetry are termed **chiral centres**.

### Optical Isomers

Both enantiomeric (mirror image) forms of lactic acid are known, and they may be distinguished by their optical properties. When placed in a beam of plane-polarised light, one form of lactic acid rotates the plane in a clockwise and the other in an anti-clockwise direction. The compounds are **optical isomers**. The former is said to be **dextrorotatory** and is designated as (+)-lactic acid, whilst the latter is said to be **laevorotatory** and is designated as (−)-lactic acid. Lactic acid isolated from mammalian tissues (sarcolactic acid) is (+)-lactic acid; that obtained by the fermentation of sucrose with *Bacillus acidi laevolactii* is laevorotatory. In contrast, lactic acid prepared by chemical synthesis in which the asymmetric centre is newly created, e.g. by catalytic reduction of pyruvic acid, does not rotate the plane of polarised light, and consists of equal amounts of (+)- and (−)-forms. This optically inactive product is described as **racemic** lactic acid, and is designated (±)-lactic acid. Such compounds, which are optically inactive by virtue of consisting of equal parts of optically active forms of opposite configuration, are said to be optically inactive by **external compensation**.

Racemates are mixtures consisting of equal numbers of molecules of each enantiomer. It is necessary, however, to distinguish between racemic mixtures and racemic compounds. A **racemic mixture** consists of a mechanical mixture or conglomerate of individual crystals of the (+)- and (−)-forms. Such a mixture, therefore, consists of two solid phases, and if the crystal habit of the two forms is sufficiently distinctive, the two isomers can be separated mechanically, if laboriously. The mixture will have the same density, solubilities and refractive index as the component optical isomers, but will show zero optical rotation. Such mixtures are, however, eutectic mixtures. Consequently, the melting point of the conglomerate will be lower than that of the individual isomers (Fig. 20A),

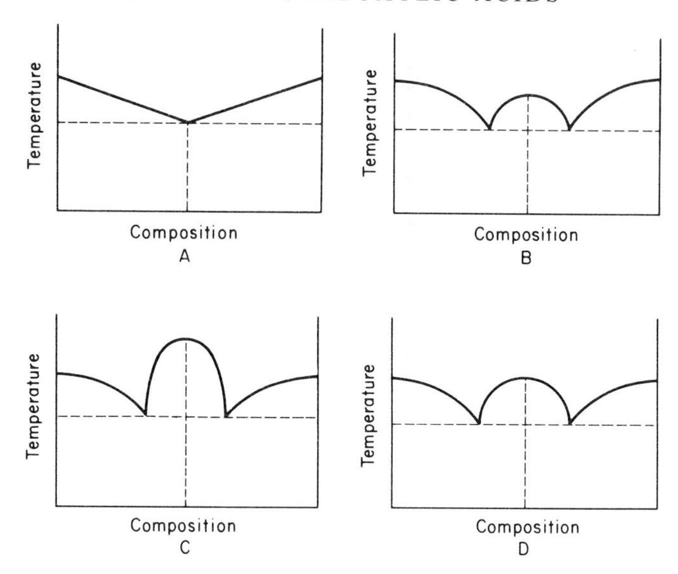


Fig. 20 Melting point composition diagrams for racemic mixtures and compounds
A Racemic mixture (conglomerate)
B Racemic compound (mp less than enantiomers)
C Racemic compound (mp greater than enantiomers)
D Racemic compound (mp the same as enantiomers)

and the melting point of the eutectic conglomerate will be raised by the addition of either enantiomer.

Alternatively, the racemate may crystallise as a **racemic compound**. In this, equal numbers of molecules of each enantiomer are crystallised together in crystals of identical form. There is, therefore, only one solid phase, and mechanical separation of the enantiomers is impossible. The physical properties of the racemic compound, such as density, solubilities, refractive index, and rotation (zero), will all be different from those of the component isomers. The melting point of a racemic compound may be above (Fig. 20C), below (Fig. 20B), or at the same temperature (Fig. 20D) as that of the enantiomers, but in contrast to that of the racemic mixture, will be depressed by small additions of either optical isomer.

Differences in physical properties between a racemic compound and its enantiomers can sometimes be used to achieve a particular therapeutic effect, even when they do not differ in biological activity. Thus, both enantiomers of the cytotoxic agent, (±)-1,2-di(2,6-dioxo-4-piperazinyl)-propane exhibit the same biological activity as the racemate. The latter, however, is too insoluble to achieve therapeutic blood levels by intravenous administration. This solubility problem can be overcome by use of one of the equally active enantiomers which are some five times more soluble (Repta, Baltezor and Bansal, 1976).

**Essential Criteria of Molecular Asymmetry**

Whilst the presence of optical asymmetry (chirality) is evident from the actual measured rotation of optical isomers, the identity of a racemate capable of resolution can only be determined by consideration of its molecular structure. The essential criterion is molecular asymmetry, established by the non-superposability of the three-dimensional molecular structure and its mirror image. Thus, it is seen that a three-dimensional representation of the molecular structure of carbon tetrachloride is identical with and superposable upon its mirror image, merely by rotation of the molecule. Carbon tetrachloride, therefore, in contrast to lactic acid, is not capable of resolution into optical isomers.

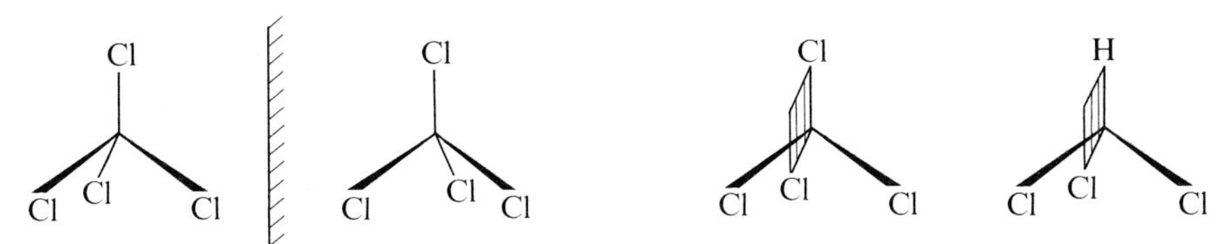


Carbon tetrachloride
Superposable mirror image forms

Planes of symmetry in carbon tetrachloride and chloroform

It is not always essential to examine the superposability of a molecule and its mirror image to decide whether or not a centre of symmetry is present. Other geometrical criteria may be adopted, such as the presence or absence of planes, centres and axes of symmetry. Thus, it is apparent that the molecules of carbon tetrachloride and chloroform are each bisected by a plane of symmetry, so that half the molecule on one side of the plane is the mirror image of the other half on the other side of the plane. Such molecules will be optically inactive. No such plane, however, can be drawn through the molecule of lactic acid, or any other molecule which has a similar centre of asymmetry.

Absence of a plane of symmetry is usually, but not always, a criterion for optical isomerism. Some molecules which lack a plane of symmetry have nonethe-less either a centre or axis of symmetry. As a consequence, they are superposable upon their mirror image, though this may not be immediately obvious. Such molecules will not be capable of resolution into optical isomers. A centre of symmetry is defined as any point within the molecule such that the straight line drawn from any group through the centre of symmetry and continued for an equal distance in the same direction will meet an identical group.

The particular 1,3-dibromo-2,4-dimethylcyclobutane shown below has a centre of symmetry. The two carbon atoms to which the bromine substituents are attached lie on an axis of symmetry AA equidistant from the centre of symmetry. Similarly, the two carbon atoms to which methyl substituents are attached also lie on an axis of symmetry BB equidistant from the centre of symmetry. Rotation of the molecule through 180° about axis AA followed by rotation of the molecule through 90° about axis CC gives the mirror image form. The molecule and its mirror image are, therefore, identical and superposable, and the compound in consequence is not resolvable into optical isomers.

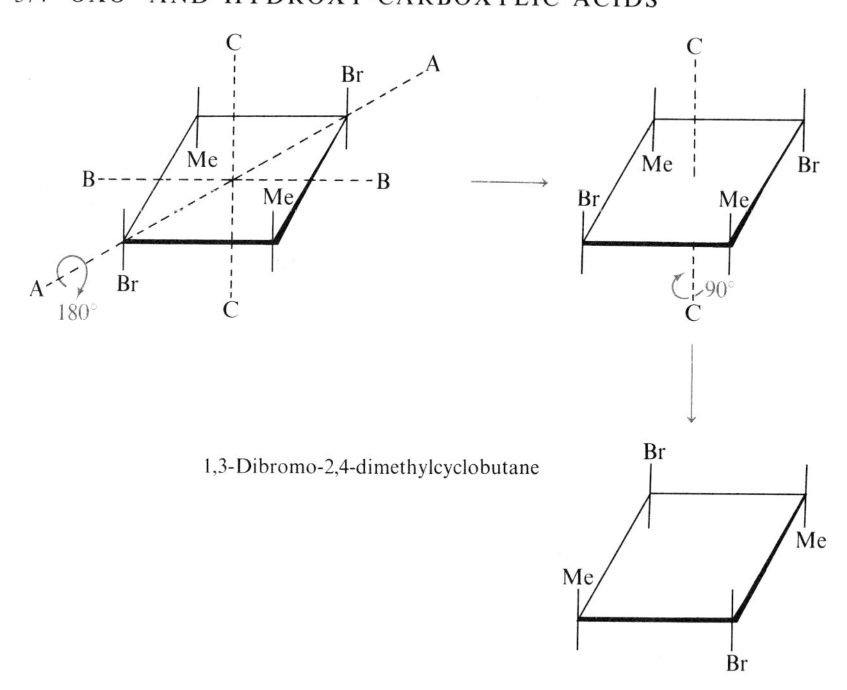


1,3-Dibromo-2,4-dimethylcyclobutane

## Measurement of Optical Rotation

Unpolarised light consists of planar electromagnetic vibrations perpendicular to the direction of the light ray. Unpolarised light is multiplanar. Plane-polarised light is composed of vibrations in a single plane.

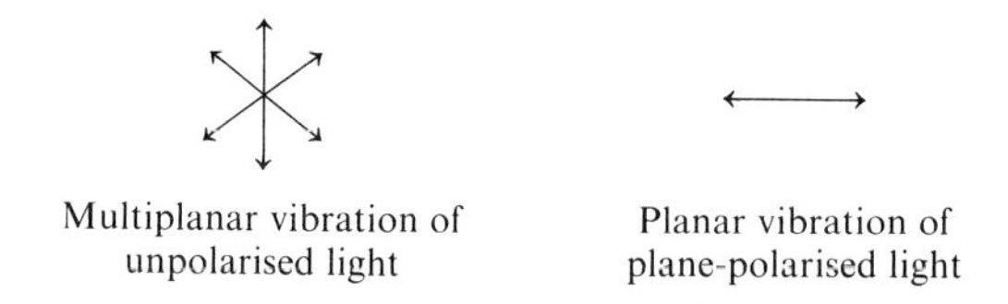

Multiplanar vibration of unpolarised light

Planar vibration of plane-polarised light

Plane-polarised light can be generated by special arrangement of prisms of the mineral Iceland spar, known as Nicol prisms. These make use of the fact that Iceland spar has differing refractive indices for light vibrating parallel and perpendicular to the principal plane of the crystal. Rotation of plane-polarised light by a compound can be measured in a polarimeter which consists of an arrangement whereby the compound usually in solution is placed in a cell between two such Nicol prisms. The first Nicol prism, the polariser, generates the beam of plane-polarised light; the second, the analyser, is attached to a circular scale marked out in degrees, and is used to measure the angle through which the plane-polarised beam has been rotated by the compound under examination.

The **optical rotation,** $[\alpha]$, of a pure liquid is the angle through which the plane of polarisation is rotated by a layer of substance one decimeter in length. Most measurements are made at ambient temperature, but the measured rotation can vary with temperature and the precise temperature should be recorded. Optical rotation also varies, often markedly, with the wavelength of light, and simple measurements are made either at the wavelength of the sodium D line (589.3 nm) or of the mercury green line at 546.1 nm. The optical rotation of a liquid expressed as $[\alpha]_{D}^{22°} = -8.3°$ implies that a layer of the liquid one decimeter in length has an anti-clockwise rotation of 8.3° when measured at 22°C at the sodium D line.

The **specific rotation**, $[\alpha]$, of a pure liquid is the optical rotation of a layer one decimeter in length calculated on the assumption that it contains 1 g of the substance per ml of liquid. It, therefore, contains a correction for the density ($\rho$) of the liquid, and specific rotation of a liquid is given by the expression

$$[\alpha]_{D}^{t} = \frac{\alpha}{l\rho}$$

where $\alpha$ = observed rotation
$t$ = temperature in °C
$l$ = path length of cell in decimetres
$\rho$ = density of the liquid

The specific rotation of a solid in solution is given by the expression

$$[\alpha]_{D}^{t} = \frac{100\alpha}{lc}$$

where $\alpha$ = observed rotation
$l$ = path length of cell in decimetres
$c$ = concentration of the solid in g/100 ml of solution.

The specific rotation is both solvent- and concentration-dependent and both solvent and concentration should be stated.

Comparisons of optical rotatory power between molecules are often valuable in making structural correlations. Such comparisons are more meaningful if made on a molecular basis as molecular rotation, $[M]$. This is given by the expression.

$$[M]_{D}^{t} = \frac{M[\alpha]_{D}^{t}}{100} = \frac{M\alpha}{lc}$$

where $M$ is the molecular weight of the compound.

**Graphical Representation of Stereochemical Configurations on Plane Surfaces**

Stereochemical structures are three-dimensional. Certain conventions are, therefore, essential if they are to be represented graphically without ambiguity on two dimensional surfaces. The Fischer convention illustrated below implies that bonds (and attached atoms and groups) to left (H) and right (OH) of the asymmetric carbon (C-2) project in front of the plane of the paper. Bonds (and attached atoms and groups) above (COOH) and below ($CH_3$) of the asymmetric carbon atom project behind the plane of the paper.

CO·OH | H—C—OH | $CH_3$ implies H◄C◄OH (CO·OH above, $CH_3$ below, dashed bonds)

For comparison purposes, rotation of structures in the plane of the paper is only permissible if this is through an angle of 180° (A $\xrightarrow{90^\circ}$ B $\xrightarrow{90^\circ}$ C). Thus, A and B are mirror images, and have opposite configurations; A and C are identical, and have the same configuration.

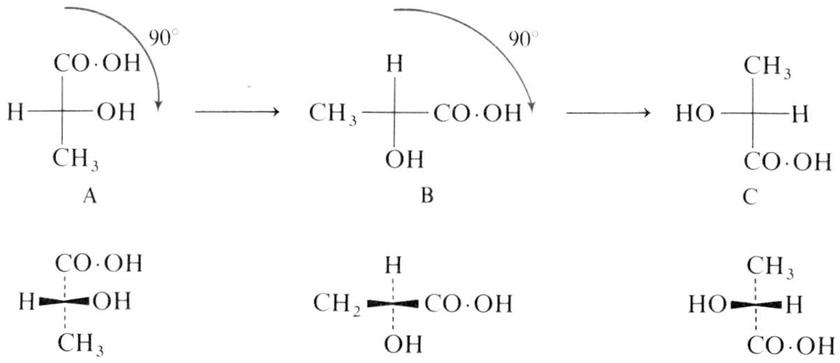

A: CO·OH above; H—OH; $CH_3$ below $\xrightarrow{90^\circ}$ B: H above; $CH_3$—CO·OH; OH below $\xrightarrow{90^\circ}$ C: $CH_3$ above; HO—H; CO·OH below

A: CO·OH; H◄OH; $CH_3$ — B: H; $CH_2$◄CO·OH; OH — C: $CH_3$; HO◄H; CO·OH

**Compounds with Two Identical Asymmetric Centres**

The molecule of the dihydroxy-dibasic acid, tartaric acid, possesses two identical asymmetric carbon atoms, each being attached to four distinct atoms or groups (—H, —OH, —COOH and —CHOH·COOH). Examination of its molecular structure shows that it is possible for the molecule to exist in four different forms. Two of these are mirror image forms and are optically active; a third is the racemic form derived from the two optically active isomers. Natural tartaric acid is $L_g$-(+)-tartaric acid (page 381).

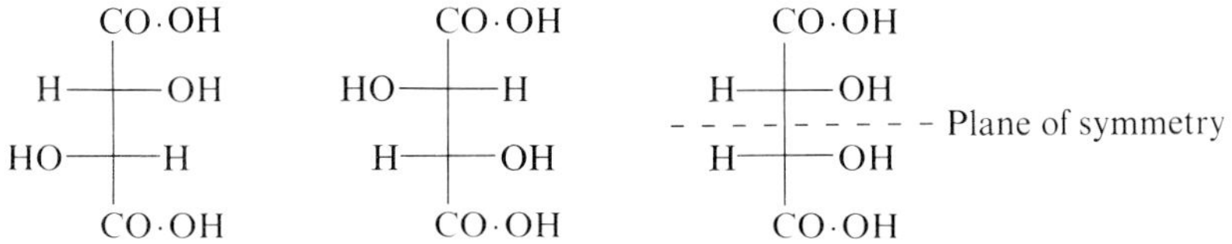

| | | |
|---|---|---|
| CO·OH | CO·OH | CO·OH |
| H—OH | HO—H | H—OH |
| HO—H | H—OH | - - - - - - Plane of symmetry<br>H—OH |
| CO·OH | CO·OH | CO·OH |
| $L_g$-(+)-Tartaric acid | $D_g$-(−)-Tartaric acid | *meso*-Tartaric acid |

The fourth isomer, meso-tartaric acid, is also optically inactive, since it has a plane of symmetry which bisects the molecule, and in consequence the molecule is identical with and superposable upon its mirror image.

An alternative way of examining the three isomeric forms shown above is to consider that each optical centre makes either a positive or negative rotatory contribution to that of the molecule as a whole. In this way, there are contributions as follows:

| + | − | + |
|---|---|---|
| + | − | − |
| $\text{L}_g$-(+)-tartaric acid | $\text{D}_g$-(−)-tartaric acid | *meso*-tartaric acid |

It is then apparent that the rotatory contributions of the two centres in *meso*-tartaric acid are self-cancelling. Such compounds are said to be optically inactive by **internal compensation**.

Stereoisomers, which are not related as object and mirror image, are known generally as **diastereoisomers**. *meso*-Tartaric and $\text{L}_g$-(+)-tartaric acids are diastereoisomers. The term **epimer** is reserved for those diastereoisomers which differ in configuration at only one optical centre. *meso*-Tartaric and $\text{L}_g$-(+)-tartaric acids, therefore, are also epimers.

**Compounds with Two or more Non-identical Asymmetric Centres**

Ephedrine and its stereoisomers possess two asymmetric carbon atoms which are non-identical. Such a compound will give rise to $2^2$ (i.e. four) optically active isomers. In general, compounds with n asymmetric centres will give rise to $2^n$ optically active isomers. In the case of ephedrine, all four isomers are known; none of the isomers has a plane of symmetry. The four isomers consist of two mirror image pairs, and can be represented graphically using the Fischer convention as follows.

| | | | |
|---|---|---|---|
| $CH_3$ | $CH_3$ | $CH_3$ | $CH_3$ |
| $CH_3NH$—+—H | H—+—$NHCH_3$ | H—+—$NHCH_3$ | $CH_3NH$—+—H |
| H—+—OH | HO—+—H | H—+—OH | HO—+—H |
| Ph | Ph | Ph | Ph |
| D-(−)-ψ-Ephedrine | L-(+)-ψ-Ephedrine | D-(−)-Ephedrine | L-(+)-Ephedrine |

The mirror image pair, D-(−)-ψ-ephedrine and L-(+)-ψ-ephedrine are **enantiomers**; similarly, D-(−)-ephedrine and L-(+)-ephedrine are **enantiomers**. Pairs of isomers, such as D-(−)-ψ-ephedrine and D-(−)-ephedrine, which differ in the configuration of only one of two optical centres (i.e. that at $C_2$), are known as **epimers**. Interconversion of two such epimers, which in this case can be brought about by refluxing with 25% hydrochloric acid, is described as **epimerisation**. This particular epimerisation is an equilibrium reaction which favours the formation of D-(−)-ephedrine.

$$\text{D-(−)-}\psi\text{-Ephedrine} \xrightleftharpoons{25\%\ HCl} \text{D-(−)-Ephedrine}$$

## Other Forms of Molecular Asymmetry

Molecular asymmetry can arise about other atoms than carbon. Some typical examples include sulphoxides, sulphonium salts, and quaternary ammonium salts.

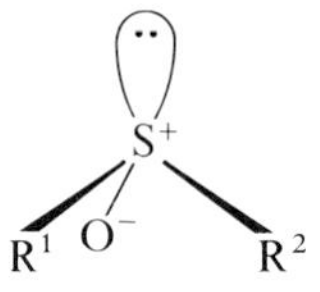

Sulphoxide 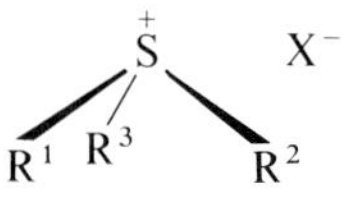 Sulphonium salt 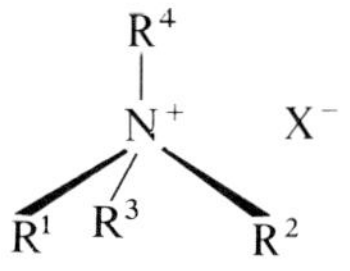 Ammonium salt

Molecular asymmetry can also arise in compounds which do not have asymmetric atoms as in certain allenes, spiranes and diphenyls.

Allene derivative 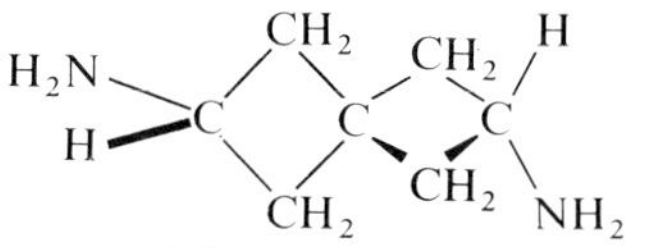 Spirane derivative 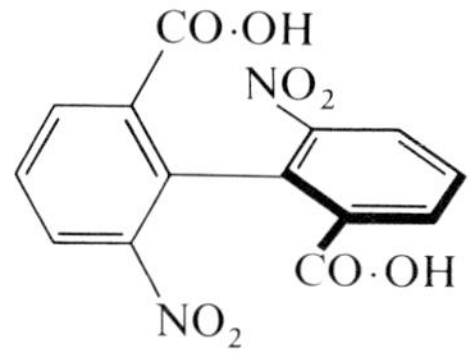 Diphenyl derivative

## Optical Resolution

Racemic forms of optically active compounds may be resolved into their optically active components. Some crystalline racemic mixtures have sufficiently distinctive and sufficiently well developed crystal forms to permit mechanical separation by hand sorting of the two crystalline forms. The method has been used to resolve lactic acid as zinc ammonium lactate, and tartaric acid as sodium ammonium tartrate.

A more suitable and more usual method for the resolution of racemic acids is the fractional crystallisation of salts with an optically active base, such as quinine or brucine. Salt formation from a racemic acid and an optically active base, which is for example dextrorotatory, gives rise to two compounds which are not enantiomorphs, but **diastereoisomers**.

$$2(\pm)\text{-}CH_3\cdot CHOH\cdot COOH + 2(+)\text{-base}$$

$$\downarrow$$

$$[(+)\text{-}CH_3\cdot CHOH\cdot CO\cdot OH(+)\text{-base}] + [(-)\text{-}CH_3\cdot CHOH\cdot CO\cdot OH(+)\text{-base}]$$

Diastereoisomers have different physical properties, and can usually be readily separated by fractional crystallisation. Decomposition of the pure diastereoisomers with an equivalent of sodium hydroxide gives the sodium salt of the optically active acid and the base. The latter can be separated from the sodium salt of the acid by solvent extraction, and the free optically active acid isolated by acid decomposition of the sodium salt.

$$[(+)\text{-}CH_3\cdot CHOH\cdot COOH(+)\text{-base}] \xrightarrow[\searrow (+)\text{-base}]{NaOH} (+)\text{-}CH_3\cdot CHOH\cdot CO\cdot ONa \xrightarrow[\swarrow NaCl]{HCl} (+)\text{-}CH_3\cdot CHOH\cdot COOH$$

(±)-Lactic acid has been resolved by this method using the strychnine salt. Racemic bases may be resolved similarly by salt formation with an optically active acid, such as (+)-tartaric acid. Racemic neutral compounds, such as alcohols, usually require the preparation of a suitable derivative which is itself acidic or basic, and capable of forming separable diastereoisomers.

**Racemisation**

Racemisation is the conversion of optically active forms of a compound into the optically inactive racemate. In effect, this is the interconversion of one optically active form with the other. The method of racemisation depends very much on the nature of the compound, and a wide variety of chemical reagents, and also thermal and light energy may bring about racemisation. Thus, (+)- or (−)-lactic acids are racemised to (±)-lactic acid on heating in sodium hydroxide solution. Racemisation of some compounds can also occur spontaneously under ambient conditions.

**Relative and Absolute Configuration**

It is often useful to know the relative orientation in space of like substituents in similar molecules. Fischer (1891) pointed out that the (+)- and (−)-sign of the actual measured rotation direction is not a guide to the relative configuration of chemically-related molecules. Thus, esterification of (+)-lactic acid with methanol gives (−)-methyl lactate, although the reaction does not affect the optical centre, and both molecules have the same relative configuration about the asymmetric carbon atom.

$$\begin{array}{c} CO\cdot OH \\ HO-\!\!\!+\!\!\!-H \\ CH_3 \end{array} \xrightarrow{MeOH} \begin{array}{c} CO\cdot OMe \\ HO-\!\!\!+\!\!\!-H \\ CH_3 \end{array}$$

L-(+)-Lactic acid　　　　L-(−)-Methyl lactate

The difficulty of assigning relative configurations was overcome by Fischer by the use of the plane projection diagrams, already described (p. 375) and related to an arbitrarily-assigned configuration for *dextro*-saccharic acid. From this, he deduced that *dextro*-glucose had the same relative configuration as *dextro*-saccharic acid. The prefixes *d* and *l*, which were formerly used, are now replaced by the symbols D and L to indicate the relative configuration of like molecules, and avoid confusion with the terms dextro (*d*) and laevo (*l*), which refer to actual observed rotations. Actual rotation is now designated as (+) or

(−) to indicate the direction of the observed optical rotation when this is **dextro** and **laevo** respectively.

```
      CO·OH                 ¹CHO
H ─────┼───── OH      H ──2──┼───── OH
HO ────┼───── H       HO ─3──┼───── H
H ─────┼───── OH      H ──4──┼───── OH               ¹CHO
H ─────┼───── OH      H ──5──┼───── OH          H ──2──┼───── OH
      CO·OH                 ⁶CH₂OH                    ³CH₂OH
```

D-Saccharic acid — D-Glucose — D-(+)-Glyceraldehyde

D-(+)-Glyceraldehyde, which has been shown chemically to have the same configuration at C-2 as the centre at C-5 in D-glucose (Wohl and Momber, 1971), has been adopted as a standard. All substances which can be shown to have the same relative arrangement of substituents as D-(+)-glyceraldehyde, and hence written with the carbon chain vertical in the projection formula with the lowest numbered carbon at the top, and a specified substituent to the right (hydrogen to the left), have the D-configuration. Thus, (−)-lactic acid has been shown to have the D-configuration by the following series of reactions, which do not involve the asymmetric carbon atom.

```
     CHO                    CO·OH                    COOH
H ───┼─── OH   ──HgO──→  H ───┼─── OH  ←──HNO₂──  H ───┼─── OH
    CH₂OH                   CH₂·OH                  CH₂NH₂
D-(+)-Glyceraldehyde    (−)-Glyceric acid        (+)-Isoserine
                                                      │ NaOBr
                                                      ↓
                            CO·OH                    CO·OH
                         H ───┼─── OH  ←──Na—Hg──  H ───┼─── OH
                             CH₃                     CH₂Br
                        D-(−)-Lactic acid
```

Similarly (+)-alanine has been shown to belong to the L-series by the reactions shown on the opposite page.

In this sequence, both the reaction of hydroxyls ($HO^-$) and azide ($N_3^-$) on the D-(+)-α-bromopropionic acid are bimolecular nuclear substitution reactions ($S_N2$), in which the entering and departing groups are on opposite sides of the molecule, thereby effecting an inversion at the optical centre. The reduction of the α-azidopropionic acid to L-(+)-alanine does not affect the optical centre.

The original assignment of the standard configuration to D-(+)-glyceraldehyde was arbitrary, and at the time there was no means of knowing whether or not this correctly represented the actual configuration of the molecule. D-(+)-Glyceraldehyde has since been converted to (−)-tartaric acid. The latter, therefore, can be assigned the $D_g$ configuration (the subscript indicates the

CO·OH
HO—H
$CH_3$
L-(+)-Lactic acid

$\xleftarrow{HO^-}$

CO·OH
H—Br
$CH_3$
D-(+)-α-Bromo-propionic acid

$\xrightarrow{N_3^-}$

CO·OH
$N_3$—H
$CH_3$

↓ Pt—$H_2$

CO·OH
$H_2N$—H
$CH_3$
L-(+)-Alanine

relationship to glyceraldehyde). Natural (+)-tartaric acid, therefore, has the $L_g$-configuration. The absolute configuration of the tartaric acids has now been shown by X-ray crystallography of sodium rubidium-(+)-tartrate (Bijvoet, Peerdeman and van Bommel, 1951) to be the same as that which was arbitrarily assigned to it, thus confirming that the arbitrarily-assigned configuration of the standard, D-(+)-glyceraldehyde, is also the absolute configuration in fact.

**Absolute Configuration**

Certain disadvantages attach to the use of the D,L-convention in assigning absolute configurations, where there is more than one optical centre in the molecule. Thus, L-threonine is sterically related to L-alanine at $C_2$ and to D-glyceraldehyde at $C_3$.

$^1$CO·OH
$H_2N$—$^2$—H
H—$^3$—OH
$^4CH_3$
L-Threonine

CO·OH
$H_2N$—H
$CH_3$
L-Alanine

CHO
H—OH
$CH_2OH$
D-Glyceraldehyde

Similarly, since D-(+)-glyceraldehyde can be converted to (−)-tartaric acid, natural (+)-tartaric acid can be assigned the $L_g$-configuration. It is, however, also possible to degrade (+)-tartaric acid to D-(−)-glyceric acid, which leads to the converse assignment of the $D_g$-configuration to (+)-tartaric acid. Such conflicts are overcome by the Cahn–Ingold–Prelog (1956) **sequence rule**, which provides an unambiguous method of assigning absolute stereochemistry (*see* also Cahn, 1964).

**The Sequence Rule (R and S convention)**

The assignment of configuration is based on the use of a sequence rule, which is applied to the four substituents, designated A, B, C and D, attached to the asymmetric atom as follows.

(a) The four substituents are in the priority order A, B, C, D. (*See* below for Priority Rules.)

(b) The asymmetric centre and its four substituents are viewed from the side opposite the substituent with the lowest priority (substituent D).

(c) The substance has the **R(rectus)** configuration if the substituents in the descending order of priority have a clockwise sequence; conversely, the substance has the **S(sinister)** configuration if the substituents in descending order of priority have an anti-clockwise sequence.

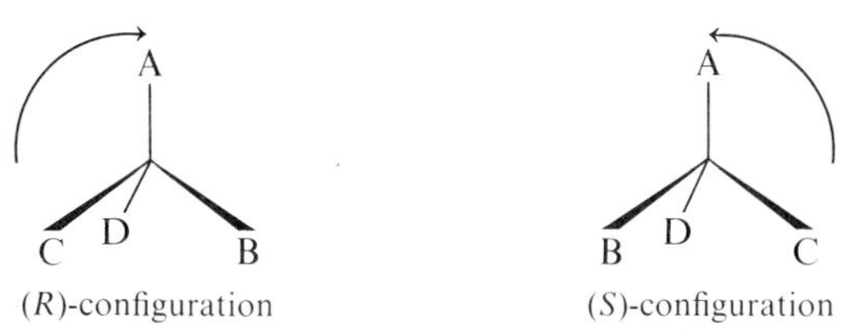


($R$)-configuration    ($S$)-configuration

*Priority Rules*

(a) Priority is assigned according to descending order of atomic number.

I, Br, Cl, S, F, O, N, C, H

(b) Where two or more substituent atoms are the same, priority is assigned according to the second atom which has the higher atomic number, and so on.

$\therefore$ $-CH_2 \cdot CH_3$ takes precedence over $-CH_3$,

and

$-CH_2 \cdot CH_2 \cdot OH$ takes precedence over $-CH_2 \cdot CH_3$

*Examples*

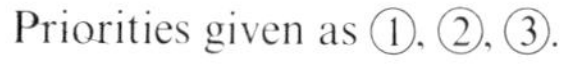


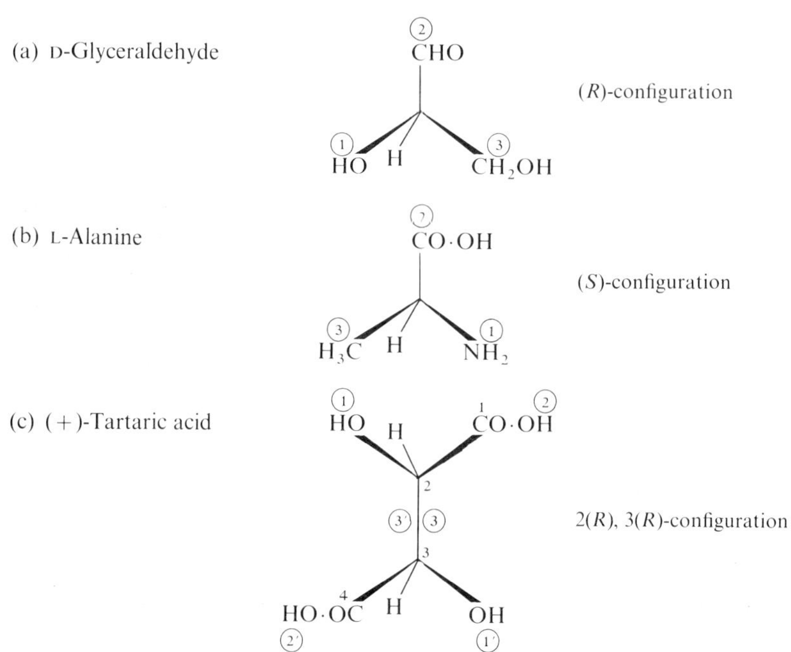

# 14 Organo-Sulphur Compounds

## THIOLS AND THIOPHENOLS

Thiols and thiophenols are the sulphur analogues of alcohols and phenols. Low molecular weight thiols occur in natural petroleum. Thiols were originally named mercaptans because of the ease with which they react with mercuric oxide to give crystalline mercury di-thio compounds (see below). Under the IUPAC system of nomenclature, they are known as thiols. The term **thiol** is used as a suffix in naming the compounds as derivatives of the corresponding saturated hydrocarbon (Table 44). The —SH group is often described as the sulphydryl group.

### General Properties

Apart from methanethiol, which is gaseous at room temperature, lower molecular weight thiols are low boiling liquids. Most simple thiols have intense and highly unpleasant odours, and are generally capable of noticeably polluting the atmosphere of any enclosed space at extremely low concentration. As little as 1 in $5 \times 10^9$ parts of ethanethiol can be detected by smell. Smell decreases markedly with increasing molecular weight and molecular complexity.

Sulphur is less electronegative than oxygen. Intermolecular hydrogen bonding between thiols is, therefore, less than in the corresponding alcohols. As a result, they have lower boiling points and are less water-soluble than alcohols. Methanethiol is soluble in about 4 parts of water (methanol is completely miscible at all concentrations) and ethanethiol only about 1 in 67.

### Acidity and Salt Formation

Thiols and thiophenols are more strongly acidic than alcohols. This is due to the greater polarisability of sulphur compared with oxygen, and consequent greater ease with which sulphur can accept and spread the negative charge in the anion. Ethanethiol (p$K_a \approx 11$) is, therefore, a stronger acid than water (p$K_a \approx 16$) or ethanol (p$K_a \approx 17$). In consequence, thiols are soluble in aqueous sodium hydroxide to form solutions of alkali metal mercaptides.

$$CH_3{\cdot}CH_2{\cdot}SH + NaOH \rightleftharpoons CH_3{\cdot}CH_2{\cdot}S^-Na^+ + H_2O$$

It follows from the greater acidity of thiols compared with alcohols that they are also weaker bases and are less ready to accept a proton. Unlike alcohols, thiols, therefore, do not react readily with hydrogen bromide to form alkyl bromides.

Table 44. Nomenclature of Thiols and Thiophenols

| Structure | Systematic Name | Trivial Name | b.p. (°C) |
|---|---|---|---|
| $CH_3 \cdot SH$ | Methanethiol | Methyl mercaptan | 6 |
| $CH_3 \cdot CH_2 \cdot SH$ | Ethanethiol | Ethyl mercaptan | 35 |
| $CH_3 \cdot CH_2 \cdot CH_2 \cdot SH$ | Propane-1-thiol | n-Propyl mercaptan | 68 |
| $(CH_3)_2 \cdot CH \cdot SH$ | Propane-2-thiol | Isopropyl mercaptan | 58 |
| $CH_3 \cdot CH_2 \cdot CH_2 \cdot CH_2 \cdot SH$ | Butane-1-thiol | n-Butyl mercaptan | 99 |
| $C_6H_5 \cdot SH$ | Thiophenol | | 168 |

Thiols also react with heavy metal oxides (HgO; PbO) and soluble heavy metal salts (lead acetate; mercuric chloride) to form insoluble metal mercaptides which, not infrequently, are crystallisable.

$$2 \cdot EtSH + HgO \longrightarrow (EtS)_2Hg\downarrow + H_2O$$

Mercuric diethylmercaptide

$$2\ EtSH + (CH_3 \cdot CO \cdot O)_2Pb \longrightarrow (EtS)_2Pb\downarrow + CH_3 \cdot CO \cdot OH$$

Lead diethylmercaptide

*Propylthiouracil* may be assayed titrimetrically with mercuric acetate in neutral solution with formation of the mercuric dimercaptide.

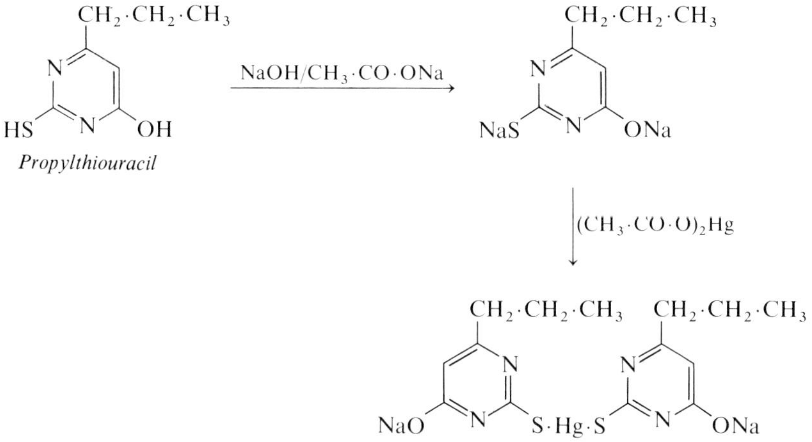

$CH_2 \cdot CH_2 \cdot CH_3$ / HS / N / OH — $\xrightarrow{NaOH/CH_3 \cdot CO \cdot ONa}$ — $CH_2 \cdot CH_2 \cdot CH_3$ / NaS / N / ONa

*Propylthiouracil*

$\downarrow (CH_3 \cdot CO \cdot O)_2Hg$

$CH_2 \cdot CH_2 \cdot CH_3$ $CH_2 \cdot CH_2 \cdot CH_3$ / NaO / N / $S \cdot Hg \cdot S$ / N / ONa

These heavy metal dimercaptides decompose when heated to yield dialkyl sulphides.

$$(CH_3 \cdot CH_2 \cdot S)_2Hg \longrightarrow CH_3 \cdot CH_2 \cdot S \cdot CH_2 \cdot CH_3 + HgS$$

Mercury mercaptides, such as the antiseptic, *Thiomersal* (Thimerosal), are also light-sensitive, and must be protected from light-induced radical decomposition.

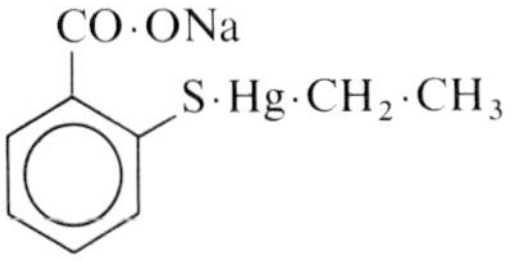

CO·ONa
S·Hg·$CH_2$·$CH_3$

*Thiomersal*

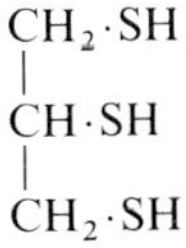

$CH_2$·SH
|
CH·SH
|
$CH_2$·SH

*Dimercaprol*

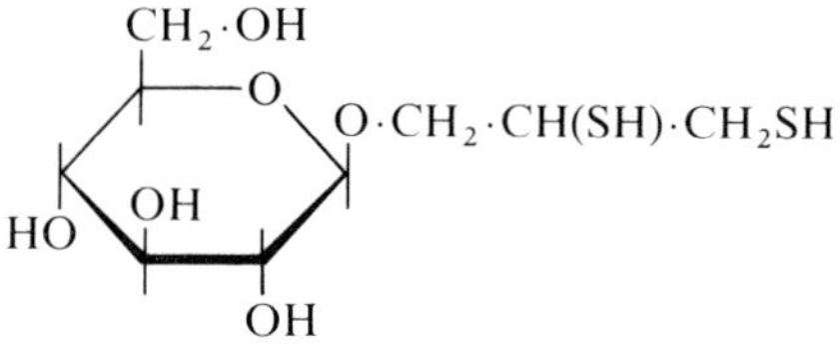

$CH_2$·OH
O
O·$CH_2$·CH(SH)·$CH_2$SH
OH
HO
OH

Dimercaprol-*O*-*β*-glucoside

The dithiol, *Dimercaprol* (British Anti-Lewisite; BAL) is used to combat lead, arsenic, mercury, zinc and cadmium poisoning in man and higher animals. The cyclic dimercaptides which it forms with these toxic metals are more stable to hydrolysis than acyclic monothiol mercaptides (Stocken and Thompson, 1946); they are also sufficiently water-soluble to be readily excreted from the body (**2**, 2). BAL-*O*-glucoside is itself sufficiently water-soluble to be administered intravenously (Danielli, Danielli, Fraser, Mitchell, Owen and Shaw, 1947; Weatherall, 1949). It forms metal dimercaptides which are highly water-soluble, the zinc complex being excreted about 20 times as fast as that with BAL itself (McCance and Widdowson, 1946).

The organo-arsenical drugs, such as *Tryparsamide* and *Acetarsol* (Acetarsone), and the analogous organo-antimonial drugs, *Stibophen* and *Antimony Sodium Tartrate*, act as acceptable and assimilable sources of these otherwise toxic metals, and combine selectively in their trivalent state with sulphydryl (—SH) groups of key protozoal enzymes (**2**, 2).

## Nucleophilic Displacements

The mercaptide and thiophenoxide anions have much higher nucleophilic reactivity than the corresponding alkoxides and phenoxides. Streitweiser (1956) gives their average relative displacement rates as butyl mercaptide ($BuS^-$) 5.8, thiophenoxide ($PhS^-$) 5.7, ethoxide ($EtO^-$) 3.0 and phenoxide ($PhO^-$) 2.6. Accordingly, alkali metal mercaptides react readily with alkyl halides providing a useful route to dialkyl sulphides in a Williamson type thio-etherification.

$$CH_3{\cdot}CH_2{\cdot}S^- \;\mathbf{CH_3}\text{—}I \longrightarrow CH_3{\cdot}CH_2{\cdot}S\textbf{—}\mathbf{CH_3} + I^-$$

Sodium thiophenoxide has also been used as an efficient reagent for the *N*-demethylation of quaternary ammonium salts.

$$C_6H_5{\cdot}S^- \; CH_3{-}\overset{+}{N}R_3 \longrightarrow C_6H_5{\cdot}S{-}CH_3 + NR_3$$

**Metabolic Detoxification of Labile Organo-halogen Compounds**

The metabolic detoxification of labile organo-halogen compounds to *S*-alkyl- and *S*-aryl-mercapturic acid derivatives requires an initial attack by the sulphydryl group of glutathione. The attack and displacement is catalysed by *S*-alkyl- and *S*-aryl-transferases, which are present in liver, kidney and heart tissue, and which, presumably, are capable of accepting the sulphydryl proton (Johnson, 1963). Mercapturation of benzyl chloride, 2,4-dinitrochlorobenzene, 2- and 4-chloropyridines and 3-chlorocrotonic acid occurs by this process. That with 2,4-dinitrochlorobenzene proceeds by the following mechanism.

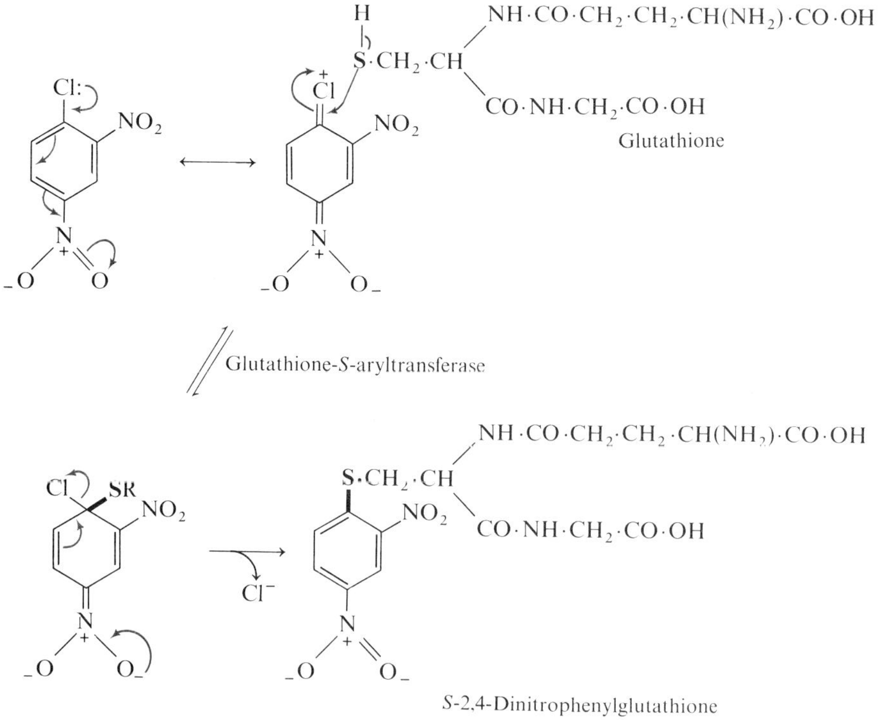

$H{-}S{\cdot}CH_2{\cdot}CH(NH{\cdot}CO{\cdot}CH_2{\cdot}CH_2{\cdot}CH(NH_2){\cdot}CO{\cdot}OH)(CO{\cdot}NH{\cdot}CH_2{\cdot}CO{\cdot}OH)$

Glutathione

Glutathione-*S*-aryltransferase

Cl⁻

*S*-2,4-Dinitrophenylglutathione

The product of this displacement reaction, *S*-2,4-dinitrophenylglutathione, is metabolically degraded by the following pathway to yield the corresponding mercapturic acid. Mercapturic acid conjugates are readily excreted from mammalian systems in the urine.

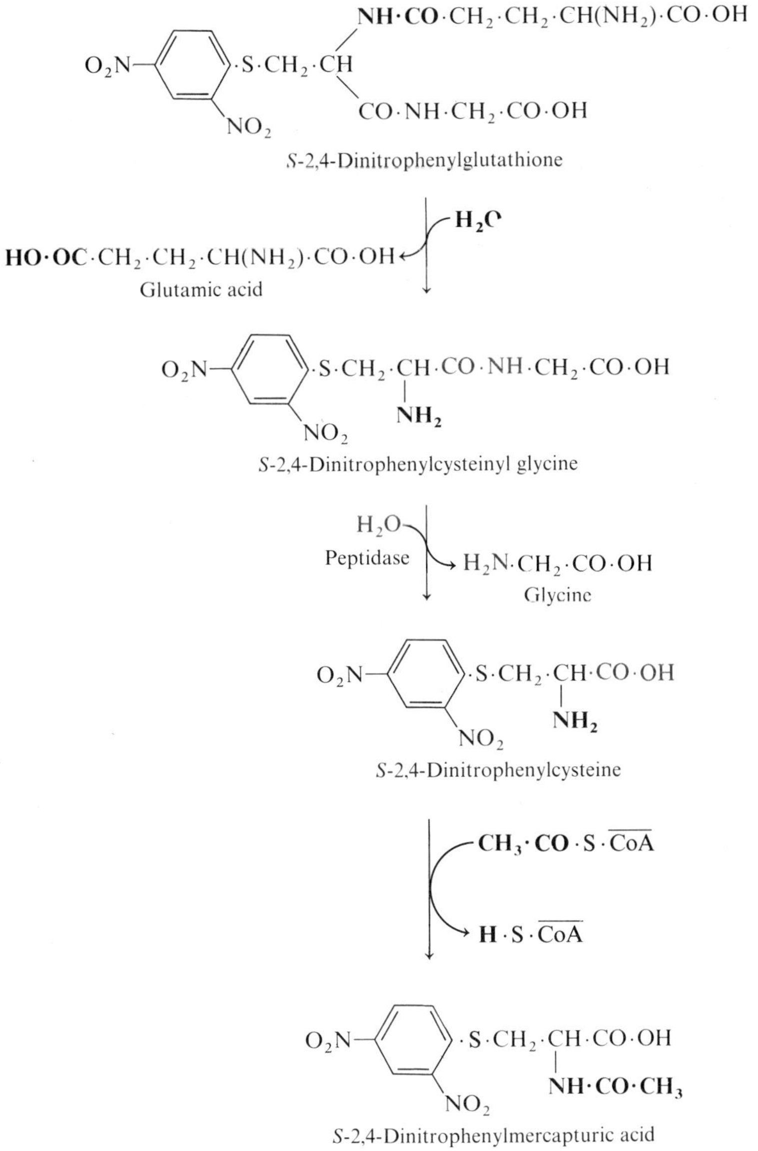

$$O_2N-C_6H_3(NO_2)\cdot S\cdot CH_2\cdot CH(\mathbf{NH\cdot CO}\cdot CH_2\cdot CH_2\cdot CH(NH_2)\cdot CO\cdot OH)(CO\cdot NH\cdot CH_2\cdot CO\cdot OH)$$

*S*-2,4-Dinitrophenylglutathione

$\downarrow$ $H_2O$ → $\mathbf{HO\cdot OC}\cdot CH_2\cdot CH_2\cdot CH(NH_2)\cdot CO\cdot OH$
Glutamic acid

$$O_2N-C_6H_3(NO_2)\cdot S\cdot CH_2\cdot CH(\mathbf{NH_2})\cdot CO\cdot NH\cdot CH_2\cdot CO\cdot OH$$

*S*-2,4-Dinitrophenylcysteinyl glycine

$\downarrow$ $H_2O$, Peptidase → $H_2N\cdot CH_2\cdot CO\cdot OH$
Glycine

$$O_2N-C_6H_3(NO_2)\cdot S\cdot CH_2\cdot CH(\mathbf{NH_2})\cdot CO\cdot OH$$

*S*-2,4-Dinitrophenylcysteine

$\downarrow$ $\mathbf{CH_3\cdot CO}\cdot S\cdot \overline{CoA}$ → $\mathbf{H}\cdot S\cdot \overline{CoA}$

$$O_2N-C_6H_3(NO_2)\cdot S\cdot CH_2\cdot CH(\mathbf{NH\cdot CO\cdot CH_3})\cdot CO\cdot OH$$

*S*-2,4-Dinitrophenylmercapturic acid

**Esterification**

Thiols, like alcohols, can be esterified directly by reaction with a carboxylic acid. Yields, however, are unsatisfactory since the equilibrium favours hydrolysis rather than esterification (p. 433).

$$CH_3\cdot CH_2\cdot \mathbf{SH} + CH_3\cdot CO\cdot \mathbf{OH} \rightleftharpoons CH_3\cdot CH_2\cdot \mathbf{S\cdot CO}\cdot CH_3 + H_2O$$

Thiol esters can, however, be readily obtained by the action of acid chlorides on thiols.

$$CH_3 \cdot CH_2 \cdot \mathbf{SH} + CH_3 \cdot \mathbf{CO \cdot Cl} \longrightarrow CH_3 \cdot CH_2 \cdot \mathbf{S \cdot CO} \cdot CH_3 + HCl\uparrow$$

The direct enzyme-catalysed esterification of thiols by acetate is of widespread biosynthetic importance. The acetylation of coenzyme A, which plays a vital intermediary rôle in mammalian metabolism, and the properties of some of the biologically more important thiol esters are described in Chapter 11.

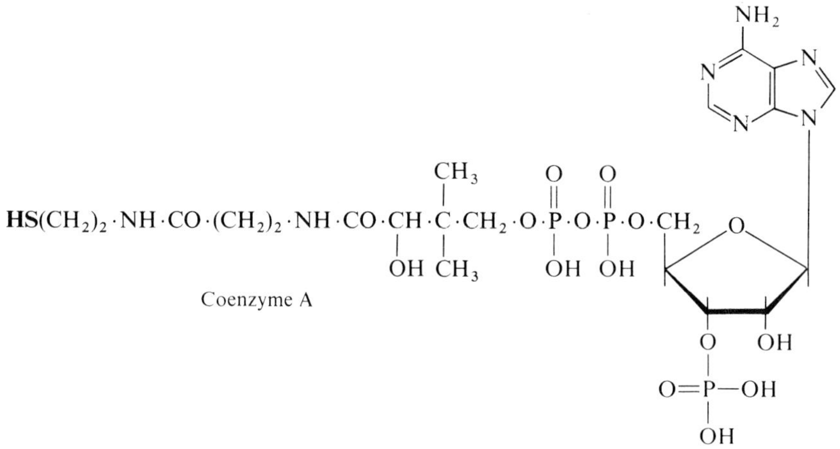

$$\mathbf{HS}(CH_2)_2 \cdot NH \cdot CO \cdot (CH_2)_2 \cdot NH \cdot CO \cdot CH(OH) \cdot C(CH_3)_2 \cdot CH_2 \cdot O \cdot P(=O)(OH) \cdot O \cdot P(=O)(OH) \cdot O \cdot CH_2\text{-(ribose-3'-phosphate)-adenine}$$

Coenzyme A

## Mercaptals

Thiols condense with both aldehydes and ketones to form mercaptals. The condensations with aldehydes, which are acid-catalysed, give rise to either hemi-mercaptals or, more usually, mercaptals.

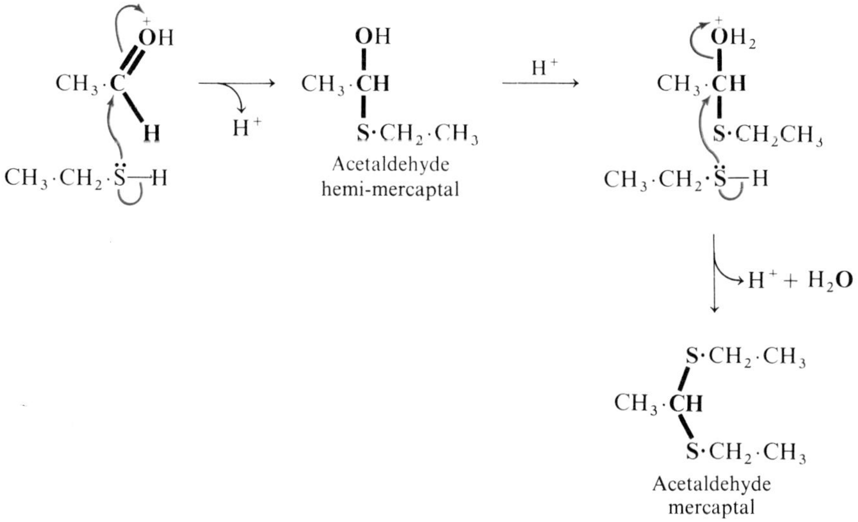

$$CH_3 \cdot CH{=}\overset{+}{O}H + CH_3 \cdot CH_2 \cdot \ddot{S}H \xrightarrow{-H^+} CH_3 \cdot CH(OH) \cdot S \cdot CH_2 \cdot CH_3$$

Acetaldehyde hemi-mercaptal

$$CH_3 \cdot CH(OH) \cdot S \cdot CH_2 \cdot CH_3 \xrightarrow{H^+} CH_3 \cdot CH(\overset{+}{O}H_2) \cdot S \cdot CH_2CH_3 \xrightarrow[-H^+ + H_2O]{CH_3 \cdot CH_2 \cdot \ddot{S}H} CH_3 \cdot CH(S \cdot CH_2 \cdot CH_3)_2$$

Acetaldehyde mercaptal

Glutathione forms a hemi-mercaptal with formaldehyde, which functions as an intermediate in its oxidation by the NAD-linked liver formaldehyde dehydrogenase (Strittmatter and Ball, 1955).

$$G{\cdot}SH + HCHO \longrightarrow G{\cdot}S{\cdot}CH_2{\cdot}OH$$

Glutathione hemi-thioacetal

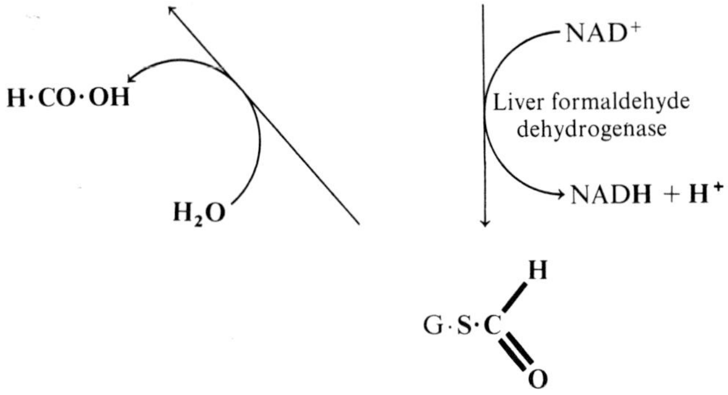

H·CO·OH ← $H_2O$ ; $NAD^+$ → Liver formaldehyde dehydrogenase → $NADH + H^+$

G·S·C(H)=O

Formylglutathione

Similar condensations occur between aldehydes and ketones, and cysteine and cysteine derivatives to form thiazolidines, as, for example, in the condensation of penicillamine with acetone.

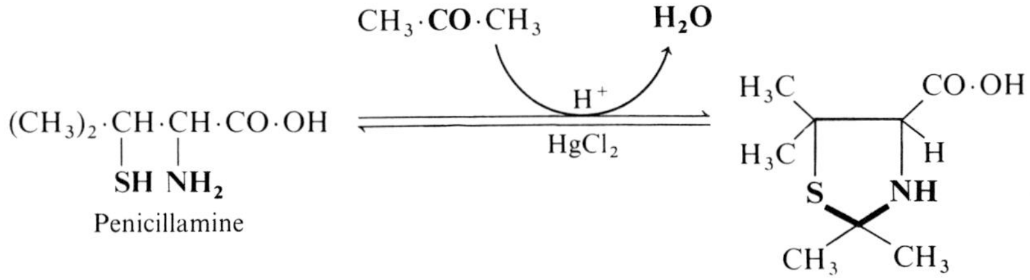

$$(CH_3)_2{\cdot}CH(SH){\cdot}CH(NH_2){\cdot}CO{\cdot}OH \xrightarrow[HgCl_2]{CH_3{\cdot}CO{\cdot}CH_3,\ H^+,\ -H_2O}$$

Penicillamine

(thiazolidine: $H_3C$, $H_3C$, CO·OH, H, S, NH, $CH_3$, $CH_3$)

Mercaptals are stable to alkali, but susceptible to hydrolysis in dilute aqueous mineral acid. Thus, the antibiotic *Lincomycin* (Hoeksema, Argoudelis and Wiley, 1962; Hoeksema *et al.*, 1964) and *Clindamycin* (Magerlein, Birkenmeyer and

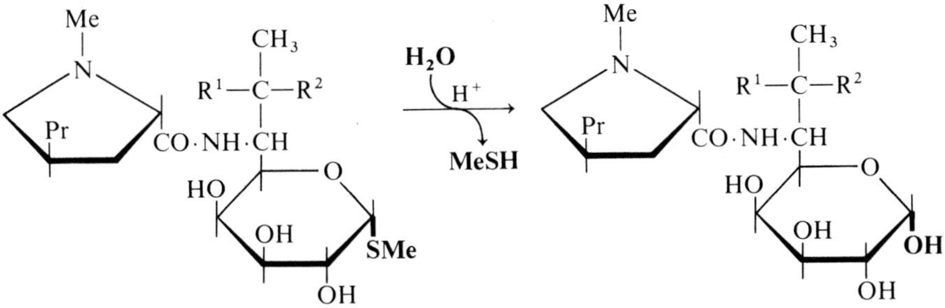

(Me, N, Pr, $CH_3$, $R^1$—C—$R^2$, CO·NH·CH, HO, O, OH, SMe, OH) $\xrightarrow{H_2O,\ H^+}$ MeSH + (Me, N, Pr, $CH_3$, $R^1$—C—$R^2$, CO·NH·CH, HO, O, OH, OH, OH)

*Lincomycin* $R^1 = OH$, $R^2 = H$
*Clindamycin* $R^1 = H$, $R^2 = Cl$

Kagan, 1966), both of which incorporate thioacetal links, are so susceptible to hydrolysis that injection solutions are required to be sterilised by cold filtration rather than by autoclaving to minimise decomposition to inactive derivatives.

Thiazolidines are also readily decomposed to the constituent carbonyl compound and thiol by mercuric chloride. Thus, benzylpenicilloic acid (the product of alkaline hydrolysis of benzylpenicillin) is degraded by mercuric chloride to penicillamine and benzylpenaldic acid.

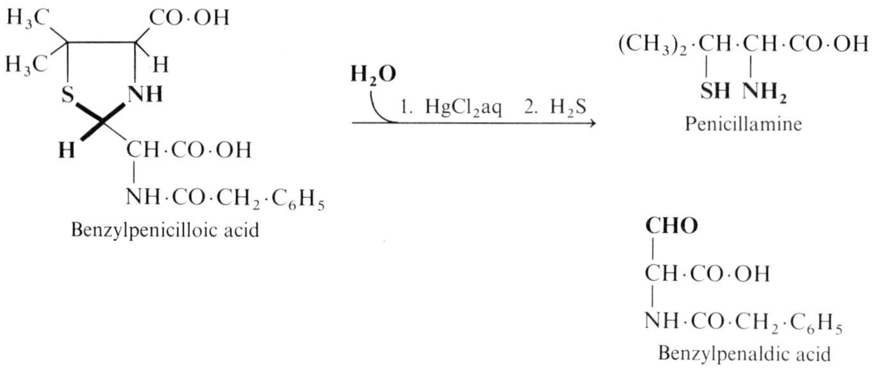

$H_3C$, $H_3C$, CO·OH, H, S, NH, H, CH·CO·OH, $NH·CO·CH_2·C_6H_5$

Benzylpenicilloic acid

$H_2O$ — 1. $HgCl_2$aq 2. $H_2S$ ⟶

$(CH_3)_2·CH·CH(SH)·CH(NH_2)·CO·OH$

Penicillamine

CHO–CH(CO·OH)–$NH·CO·CH_2·C_6H_5$

Benzylpenaldic acid

Cyclic dithioketals are often difficult to hydrolyse due to steric effects. In such cases, conversion to the corresponding mono-sulphonium salt facilitates acid hydrolysis (Fetizon and Jurion, 1972).

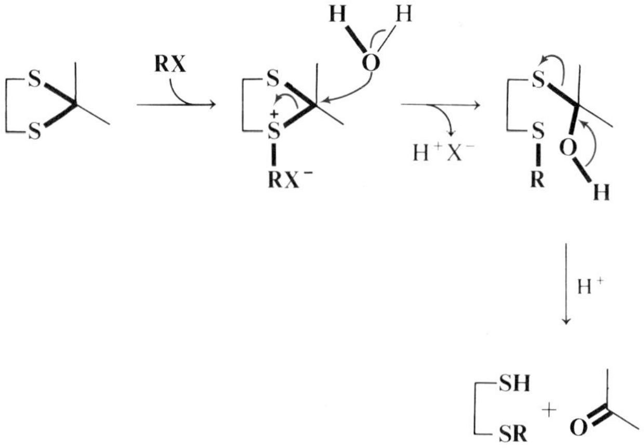

RX ⟶ [S-alkylsulphonium salt, $RX^-$] + $H_2O$ ⟶ ($H^+X^-$) [S, S–R, O–H] ⟶ ($H^+$) [–SH, –SR] + O=C<

## Radical Reactions

Thiols readily undergo addition to olefines to form dialkyl sulphides. The reaction, however, is slow in the absence of light and peroxides. Under these conditions, ionic addition occurs according to the Markownikoff rule.

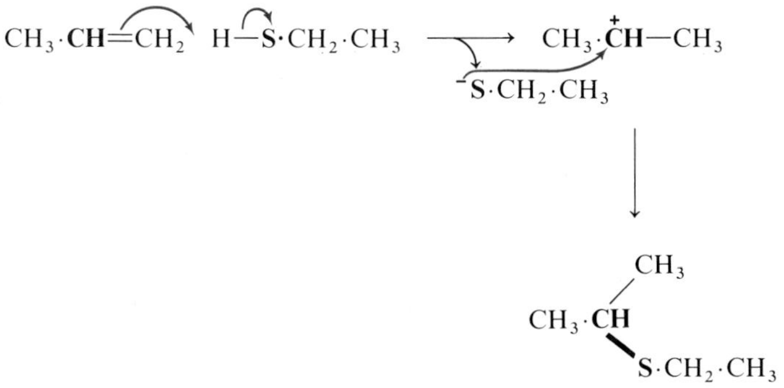

$$CH_3\cdot CH{=}CH_2 \quad H{-}S\cdot CH_2\cdot CH_3 \longrightarrow CH_3\cdot \overset{+}{C}H{-}CH_3 \quad {}^{-}S\cdot CH_2\cdot CH_3$$

$$\downarrow$$

$$CH_3\cdot CH(CH_3)\cdot S\cdot CH_2\cdot CH_3$$

Ethyl isopropyl sulphide

A much more rapid radical-induced anti-Markownikoff addition occurs in the presence of peroxide or on exposure to ultraviolet light.

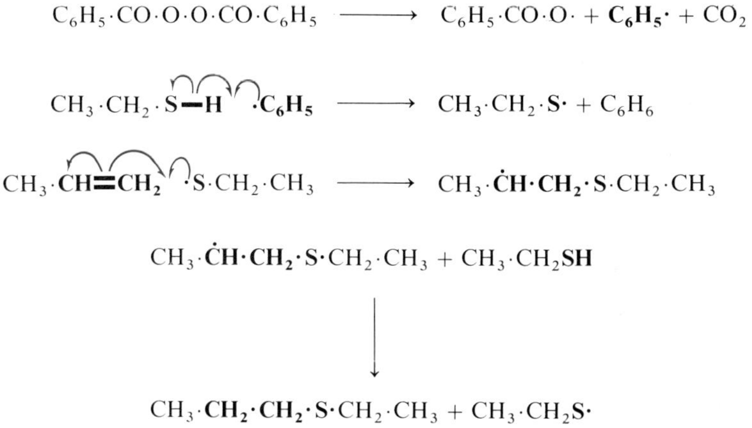

$$C_6H_5\cdot CO\cdot O\cdot O\cdot CO\cdot C_6H_5 \longrightarrow C_6H_5\cdot CO\cdot O\cdot + C_6H_5\cdot + CO_2$$

$$CH_3\cdot CH_2\cdot S{-}H \quad \cdot C_6H_5 \longrightarrow CH_3\cdot CH_2\cdot S\cdot + C_6H_6$$

$$CH_3\cdot CH{=}CH_2 \quad \cdot S\cdot CH_2\cdot CH_3 \longrightarrow CH_3\cdot \dot{C}H\cdot CH_2\cdot S\cdot CH_2\cdot CH_3$$

$$CH_3\cdot \dot{C}H\cdot CH_2\cdot S\cdot CH_2\cdot CH_3 + CH_3\cdot CH_2SH$$

$$\downarrow$$

$$CH_3\cdot CH_2\cdot CH_2\cdot S\cdot CH_2\cdot CH_3 + CH_3\cdot CH_2S\cdot$$

The inhibitory action of the diuretic, *Ethacrynic Acid*, on the renal enzymes involved in the reabsorption of $Na^+$ ions, is probably an attack on enzymatic sulphydryl groups by the highly activated double bond of the drug.

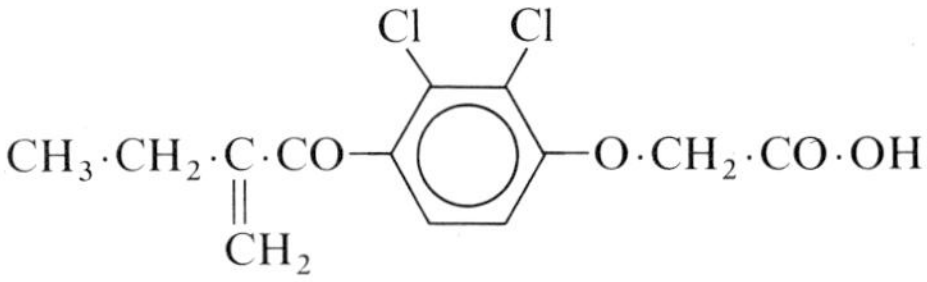

$$CH_3\cdot CH_2\cdot C({=}CH_2)\cdot CO{-}C_6H_2Cl_2{-}O\cdot CH_2\cdot CO\cdot OH$$

*Ethacrynic Acid*

Thiyl radicals also catalyse *cis-trans*-isomerism of olefines. Isomerisation proceeds through the same radical intermediate as addition; the inversion of configuration and elimination leading to isomerism, however, occurs some 80 times faster than the alternative attack on thiol to form sulphide (Walling and Helmreich, 1959).

*cis*-Butene $\xrightarrow{MeS\cdot}$ $H-\dot{C}-CH_3$ / $H-C(SMe)-CH_3$ $\underset{\text{equilibrium}}{\overset{\text{Fast}}{\rightleftharpoons}}$ $H_3C-\dot{C}-H$ / $H-C(SMe)-CH_3$ → *trans*-Butene

$\xrightarrow{MeSH \rightarrow MeS\cdot}$ $H-C(H)-CH_3$ / $H-C(SMe)-CH_3$

2-Butyl methyl sulphide

It has been established that the conversion of maleyl pyruvate to fumaryl pyruvate by maleyl pyruvate isomerase is similarly mediated by glutathione (Lack, 1961).

## Oxidation

Thiols are readily oxidised quantitatively to the corresponding disulphides. The ease with which oxidation occurs decreases in the following order.

$$Ar{\cdot}SH > R{\cdot}CH_2{\cdot}SH > R_2{\cdot}CH{\cdot}SH > R_3C{\cdot}SH$$

Thiophenols and primary alkylthiols are susceptible to air oxidation, particularly in alkaline solution (Cecil and McPhee, 1959). Thiyl radicals, formed by radical decomposition of the S—H link, couple to form the disulphide.

$$\mathrm{R{\cdot}SH} \xrightarrow{\cdot O{-}O\cdot \;\rightarrow\; HO{\cdot}O\cdot} \mathrm{R{\cdot}S\cdot}$$

$$\mathrm{R{\cdot}SH} \xrightarrow{HO{\cdot}O\cdot \;\rightarrow\; HO{\cdot}OH} \mathrm{R{\cdot}S\cdot}$$

$$2\,\mathrm{R{\cdot}S\cdot} \longrightarrow \mathrm{R{\cdot}S{\cdot}S{\cdot}R}$$

$$2\mathrm{R{\cdot}SH} \xrightarrow{HO{\cdot}OH} \mathrm{R{\cdot}S{\cdot}S{\cdot}R} + 2H_2O$$

Dialkyl disulphide

Oxidation of thiols to disulphides by air, oxygen or hydrogen peroxide is catalysed by polyvalent metal ions, particularly $Fe^{3+}$ and $Cu^{2+}$ (Lamfrom and Nielsen, 1957).

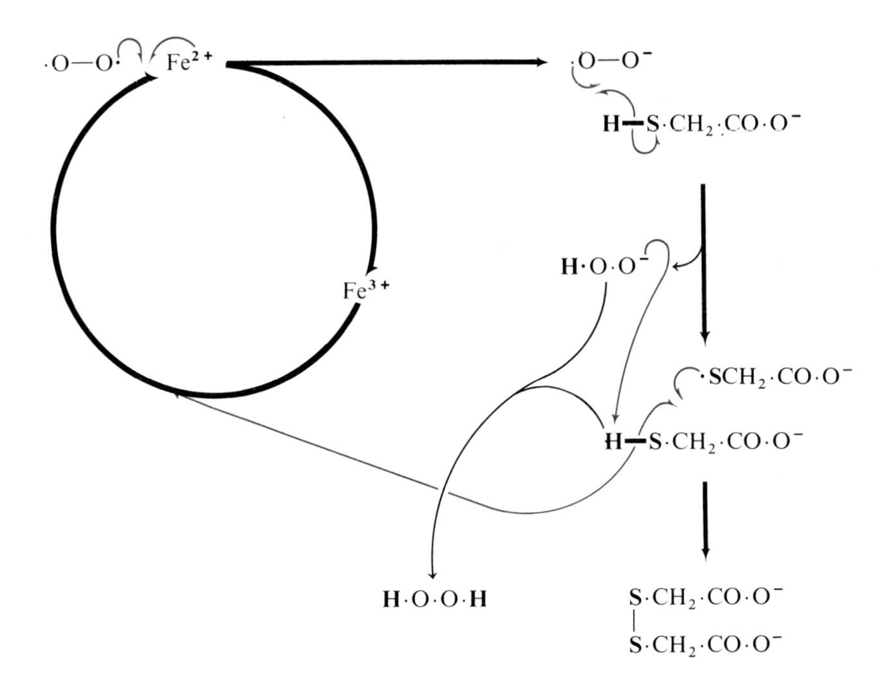


Oxidation to disulphides is also effected by a wide variety of mild chemical oxidants, such as iodine in alkaline solution. The reaction probably occurs by a mechanism similar to that which obtains for the oxidation of thiosulphate by iodine (Raschig, 1915).

$$\mathrm{CH_3{\cdot}CH_2{\cdot}S^-\;\; I{-}I} \xrightarrow[\;I^-]{} \mathrm{CH_3{\cdot}CH_2{\cdot}S{-}I} \;\; (+\,\mathrm{CH_3{\cdot}CH_2{\cdot}S^-}) \xrightarrow[\;I^-]{} \mathrm{CH_3{\cdot}CH_2{\cdot}S{\cdot}S{\cdot}CH_2{\cdot}CH_3}$$

It has been suggested (Miller, Roblin and Astwood, 1945; Campbell, Landgrebe and Morgan, 1944) that the action of sulphydryl compounds in the treatment of thyrotoxicosis is due to reduction of molecular iodine to iodide, thereby inhibiting its incorporation into thyronine.

It has been shown (Field and White, 1973) that stable peptide sulphenyl iodides (RSI) can be prepared, and it is considered (Maloof and Soodak, 1963; Cunningham, 1964) that a sulphenyl iodide derived from thyroglobulin is central to the iodination of tyrosine in the thyroid gland. According to Cunningham (1964), antithyroid sulphydryl compounds, such as *Methimazole*, *Carbimazole* and *Propylthiouracil* interfere with this iodination process by displacing iodide from the sulphenyl iodide. There is evidence (Marchant and Alexander, 1972) that these drugs are protein-bound by a disulphide link since they are displaceable by sulphydryl compounds.

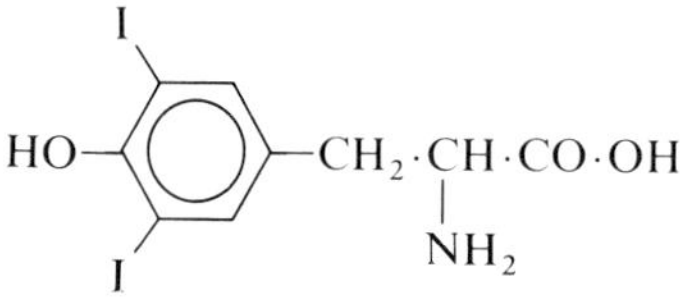

$HO-C_6H_2I_2-CH_2\cdot CH(NH_2)\cdot CO\cdot OH$

3,5-Diiodotyrosine

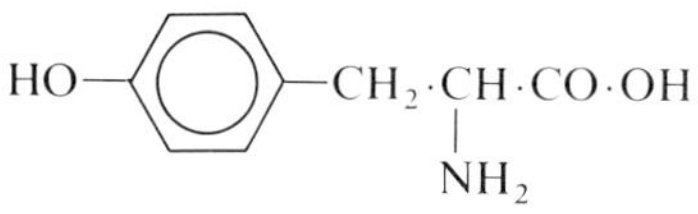

$HO-C_6H_4-CH_2\cdot CH(NH_2)\cdot CO\cdot OH$

Tyrosine

THYROGLOBULIN—S—H ⇌ THYROGLOBULIN—S—I

I—I, I⁻

Thyroid peroxidase

MeN NH (C=S) ⇌ MeN N (C—S—H)

$H^+I^-$

Me·N N (C—S—S—THYROGLOBULIN)

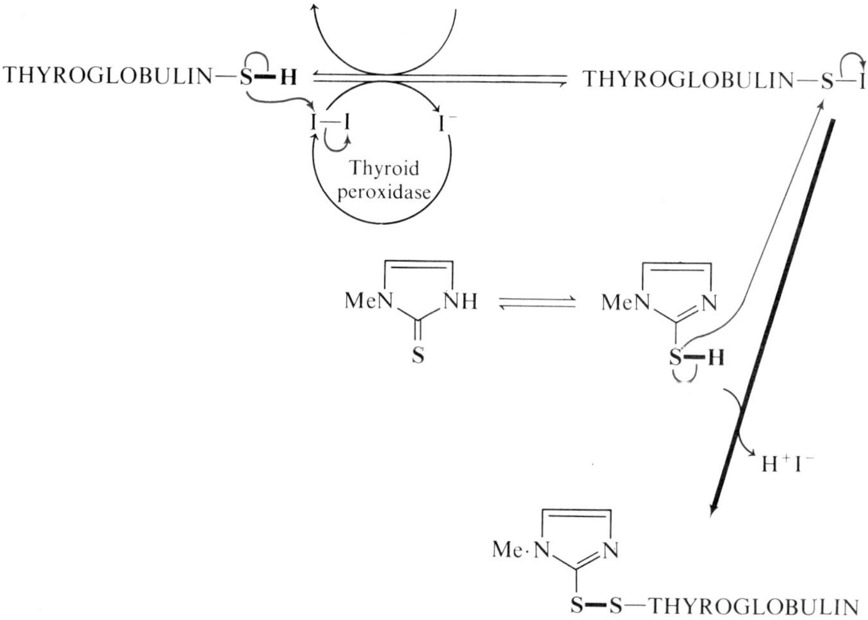

More vigorous oxidation of thiols and thiophenols with potassium permanganate or nitric acid leads to the formation of sulphonic acids.

$$C_6H_5\cdot \mathbf{SH} + \mathbf{3[O]} \longrightarrow C_6H_5\cdot \mathbf{SO_2\cdot OH}$$

Benzenesulphonic acid

## THIOETHERS

(Dialkyl and Diaryl Sulphides)

### General Properties

Thioethers or sulphides are the sulphur analogues of ethers. They are usually named as sulphides (dimethyl sulphide; diphenyl sulphide), but more complex aliphatic compounds are named systematically in the IUPAC system as alkylthioalkanes (methyl propyl sulphide = methylthiopropane).

Thioethers are usually colourless or pale-yellow liquids and solids. They are mostly insoluble in water. The lower aliphatic compounds have distinctive sulphurous, but not unpleasant, odours. Thioethers are weakly basic and like ethers are soluble in strong mineral acids as a result of protonation. Thioethers also readily form addition complexes with halogens ($R_2S$, $Br_2$) and heavy metal salts ($R_2S$, $HgCl_2$). As with *O*-ethers, prolonged treatment with strong acids causes slow decomposition to the corresponding thiol and alcohol.

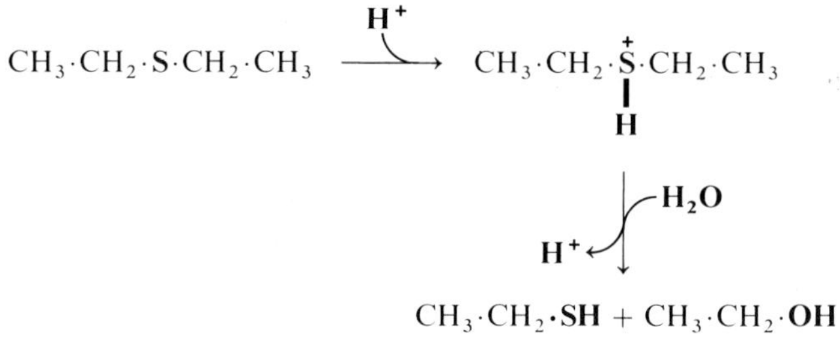

$$CH_3{\cdot}CH_2{\cdot}S{\cdot}CH_2{\cdot}CH_3 \xrightarrow{H^+} CH_3{\cdot}CH_2{\cdot}\overset{+}{S}(H){\cdot}CH_2{\cdot}CH_3$$

$$\xrightarrow[-H^+]{H_2O} CH_3{\cdot}CH_2{\cdot}\mathbf{SH} + CH_3{\cdot}CH_2{\cdot}\mathbf{OH}$$

Cyanogen bromide, similarly, cleaves the sulphide link, giving rise to a mixture of isothiocyanate and bromide (Tarbell and Harnish, 1951).

$$CH_3{\cdot}CH_2{\cdot}S{\cdot}CH_2CH_3 + \mathbf{CNBr} \longrightarrow CH_3{\cdot}CH_2{\cdot}\mathbf{SCN} + CH_3{\cdot}CH_2{\cdot}\mathbf{Br}$$

Unsymmetrical sulphides usually give a mixture of all four possible products, but methyl alkyl sulphides usually form methyl bromide and alkyl isothiocyanate.

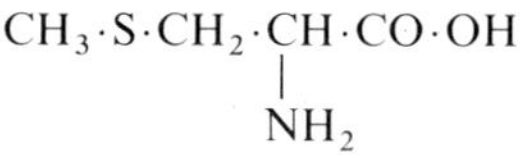

$CH_3{\cdot}S{\cdot}CH_2{\cdot}CH(NH_2){\cdot}CO{\cdot}OH$

*S*-Methylcysteine

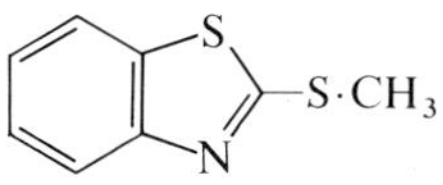

2-*S*-Methylthiobenzthiazole

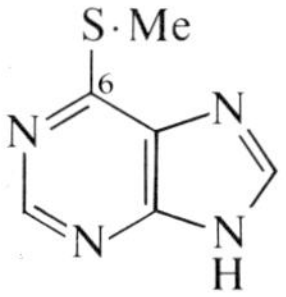

*Mercaptopurine*

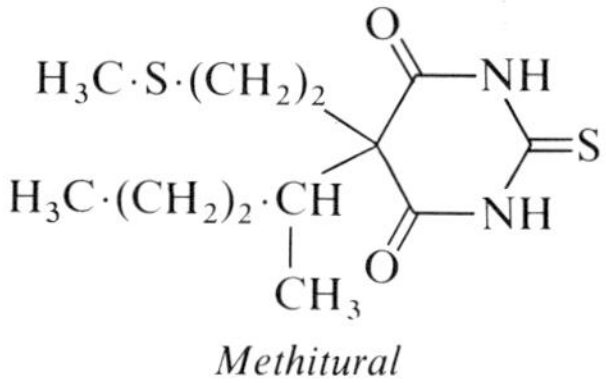

*Methitural*

Oxidative *S*-demethylation of *S*-methylcysteine, *S*-methylthiobenzothiazole, *Mercaptopurine*, and the *S*-methylthiobarbiturate, *Methitural*, to the corresponding thiol with release of formaldehyde is catalysed by microsomal enzymes (Mazel, Henderson and Axelrod, 1964). The enzymes, which require NADPH and molecular oxygen, differ from those involved in *O*- and *N*-demethylations in that they are neither inhibited by SKF-525A nor stimulated by pretreatment with phenobarbitone.

### Oxidation

Sulphides are oxidised chemically at moderate temperatures to sulphoxides and sulphones. Hydrogen peroxide and organic hydroperoxides normally yield sulphoxides only.

$$(CH_3 \cdot CH_2)_2S \xrightarrow{RO \cdot OH} (CH_3 \cdot CH_2)_2\overset{+}{S}{-}O^-$$

Diethyl sulphoxide

The oxidation, which is first order with respect to both sulphide and peroxide is acid-catalysed, and thought to occur by the following mechanism.

$$R_2S: \; \underset{H}{O}{-}\underset{H}{OR} \; (X) \longrightarrow R_2\overset{+}{S}{-}O^- + HX + R \cdot OH$$

Uncatalysed air oxidation of sulphides only occurs appreciably at temperatures over 100°C. Syrups containing *Chlorpromazine*, however, can suffer significant losses in potency if stored for prolonged periods in containers which are only partially filled.

$$\text{Chlorpromazine} \xrightarrow{O_2} \text{Chlorpromazine-}S\text{-oxide}$$

S, N, Cl, $CH_2(CH_2)_2 \cdot NMe_2$ → $O^-$, $\overset{+}{S}$, N, Cl, $CH_2(CH_2)_2 \cdot NMe_2$

*Chlorpromazine* — Chlorpromazine-*S*-oxide

Oxidation to sulphoxide is catalysed by ultraviolet radiation, and chemically as in the commercial synthesis of dimethyl sulphoxide, which is mediated by nitrogen peroxide.

$$(CH_3)_2S \xrightarrow{NO_2 \rightarrow NO} (CH_3)_2\overset{+}{S}{-}O^-$$

$$NO + O_2 \rightarrow NO_2$$

Peracids oxidise sulphides to sulphoxides, and the latter to sulphones.

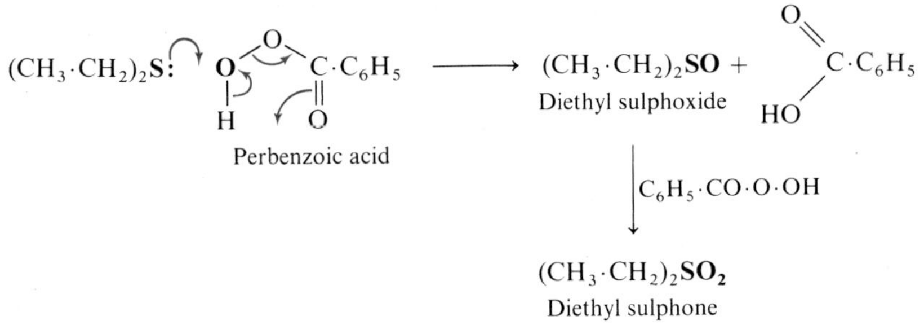

$$(CH_3 \cdot CH_2)_2\mathbf{S}: \;\; O(H)\text{–}O\text{–}C(=O) \cdot C_6H_5 \longrightarrow (CH_3 \cdot CH_2)_2\mathbf{SO} + HO\text{–}C(=O) \cdot C_6H_5$$

Perbenzoic acid

Diethyl sulphoxide

$$\downarrow C_6H_5 \cdot CO \cdot O \cdot OH$$

$$(CH_3 \cdot CH_2)_2\mathbf{SO_2}$$

Diethyl sulphone

Enzymatically, sulphides are oxidised in the liver to both sulphoxides and sulphones. Thus, *Chlorpromazine* is excreted partly as its *S*-oxide, though the

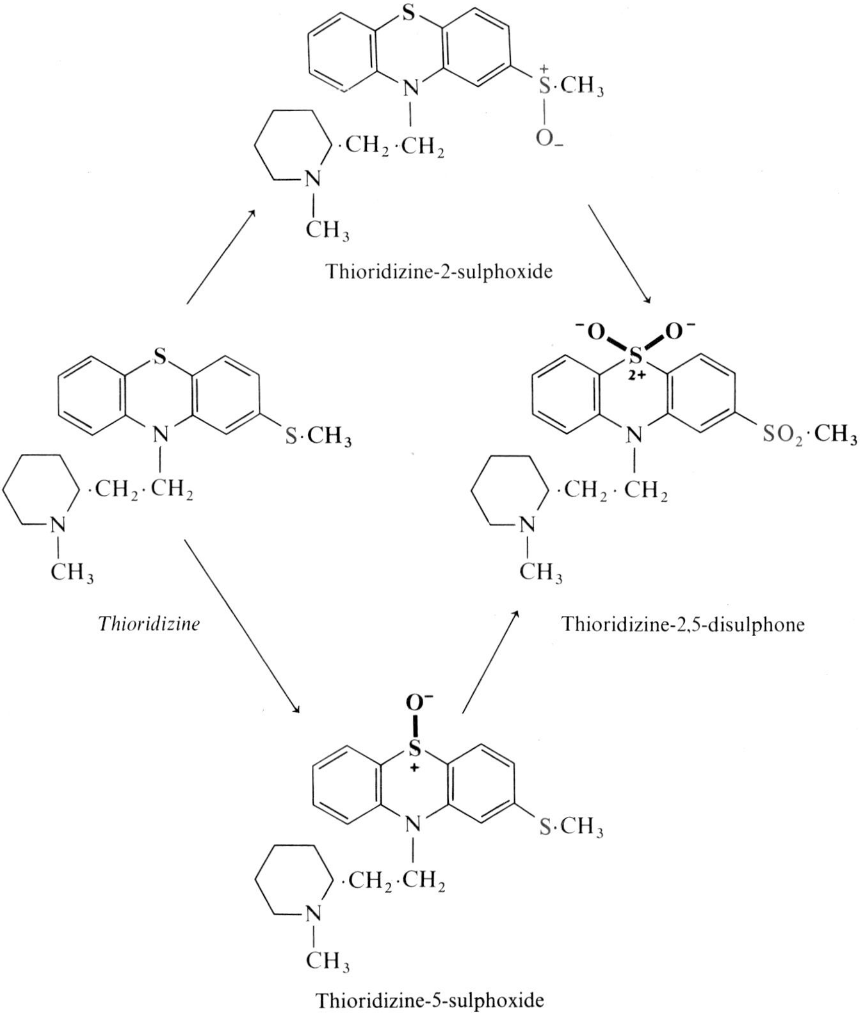

S, N, $\overset{+}{S} \cdot CH_3$, $O_-$, $\cdot CH_2 \cdot CH_2$, N, $CH_3$

Thioridizine-2-sulphoxide

S, N, $S \cdot CH_3$, $\cdot CH_2 \cdot CH_2$, N, $CH_3$

*Thioridizine*

$^-O$, $O^-$, S, 2+, N, $SO_2 \cdot CH_3$, $\cdot CH_2 \cdot CH_2$, N, $CH_3$

Thioridizine-2,5-disulphone

$O^-$, S, +, N, $S \cdot CH_3$, $\cdot CH_2 \cdot CH_2$, N, $CH_3$

Thioridizine-5-sulphoxide

*N*-oxide is the major metabolite (Beckett and Hewick, 1967), whilst *Thioridazine* is metabolised to both monosulphoxides, the disulphoxide and disulphone.

In contrast, *Methylene Blue* appears to be metabolised solely to the corresponding sulphone.

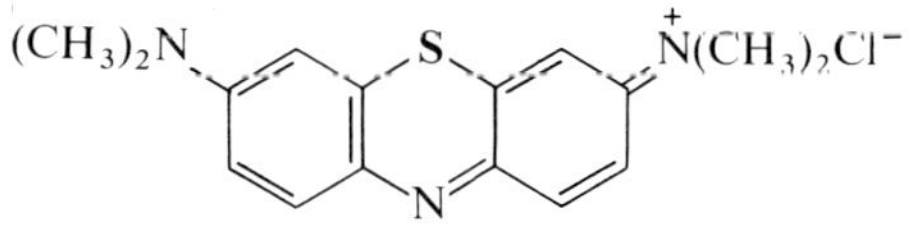

*Methylene Blue*

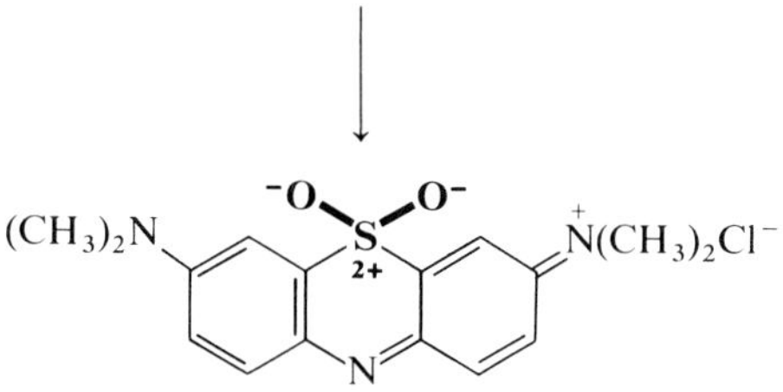

Oxidation of the thioether link in the antibacterial and trichomonicidal drug, *Nifuratel*, markedly reduces its activity; replacement of S—$CH_3$ by other *S*-alkyl groups also markedly decreases antibacterial activity and destroys trichomonicidal activity (Launchbury, 1970).

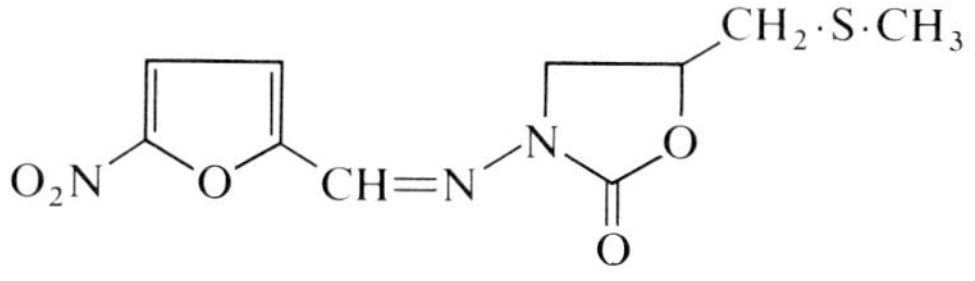

*Nifuratel*

**Reductive Desulphurisation**

Treatment of dialkyl sulphides with hydrogen in the presence of Raney nickel leads to reductive desulphurisation, with complete cleavage of both carbon–sulphur bonds.

$$R{\cdot}S{\cdot}R' + \mathbf{2H_2} \xrightarrow{\text{Raney Ni}} R{\cdot}\mathbf{H} + R'{\cdot}\mathbf{H} + \mathbf{H_2S}$$

Such reductive desulphurisation has proved useful in structural studies of cyclic sulphur compounds, as in the conversion of *Benzylpenicillin* (Penicillin G) to desthiobenzylpenicillin.

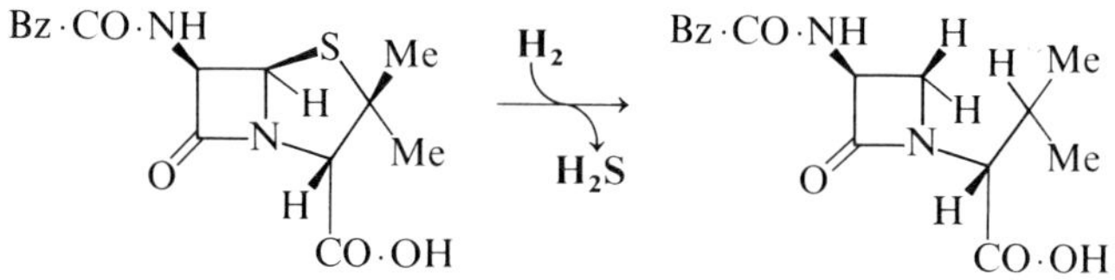

*Benzylpenicillin* Desthiobenzylpenicillin

**Sulphonium Salts**

The much greater reactivity of thioethers compared with their oxygen analogues is demonstrated by the ease with which they react with alkyl halides to form stable sulphonium salts.

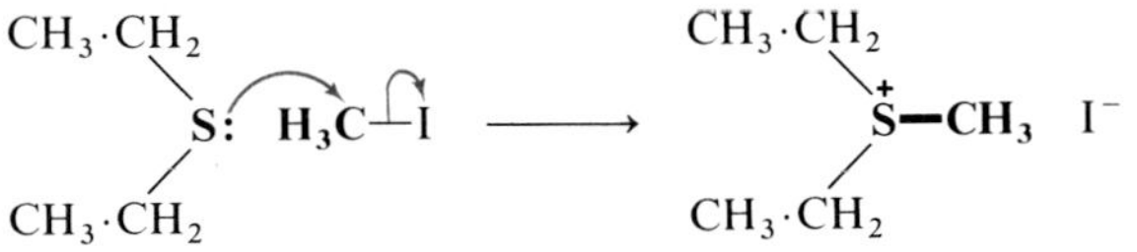

$$(CH_3{\cdot}CH_2)_2S\colon + H_3C{-}I \longrightarrow (CH_3{\cdot}CH_2)_2\overset{+}{S}{-}CH_3 \; I^-$$

Diethylmethylsulphonium iodide

The metabolic sulphonium salt intermediate, *S*-adenosylmethionine is formed, similarly, from methionine and adenosine triphosphate (ATP).

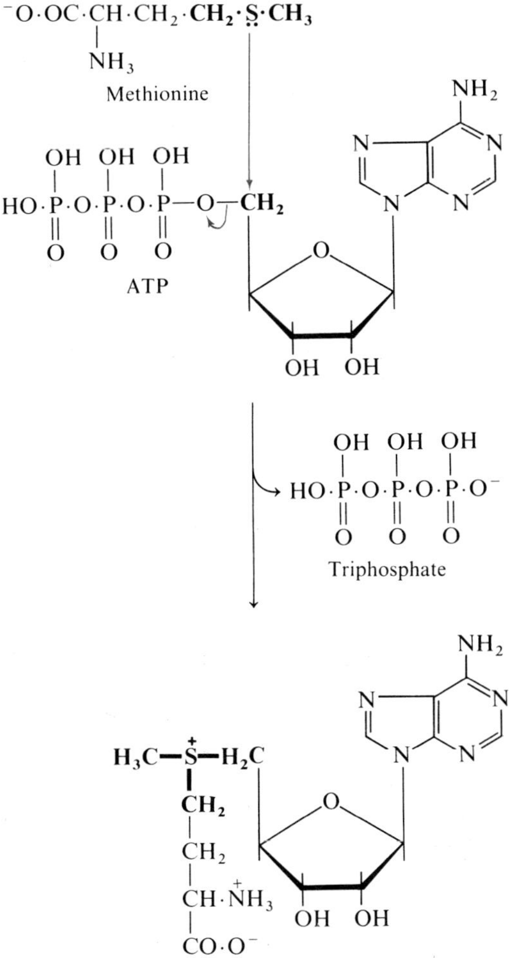

$$^-O{\cdot}OC{\cdot}CH(NH_3){\cdot}CH_2{\cdot}CH_2{\cdot}\ddot{S}{\cdot}CH_3$$

Methionine

ATP

Triphosphate

*S*-Adenosylmethionine

Sulphonium salts are acetylcholine antagonists; they inhibit ganglionic and neuromuscular transmission of nervous impulses. They are somewhat less stable than the corresponding quaternary ammonium salts and decompose on heating to give a sulphide and alkyl halide.

$$(CH_3{\cdot}CH_2)_2{\cdot}\overset{+}{S}{-}\mathbf{CH_3} \quad I^- \longrightarrow (CH_3{\cdot}CH_2)_2S + \mathbf{CH_3I}$$

Similarly, *S*-adenosylmethionine is able to act as a methyl donor in a number of biologically important methylations (**2**, 5).

Certain appropriately substituted sulphonium salts, like the corresponding quaternary ammonium salts (Chapter 16), undergo elimination to form olefines when treated with bases.

Sulphur is $sp^3$ hybridised in sulphonium compounds. Sulphonium salts are, therefore, tetrahedral, and when asymmetrically substituted as in ethylmethylpropylsulphonium iodide or *S*-adenosylmethionine, are capable of being resolved into optically active enantiomeric (non-superposable mirror image) forms.

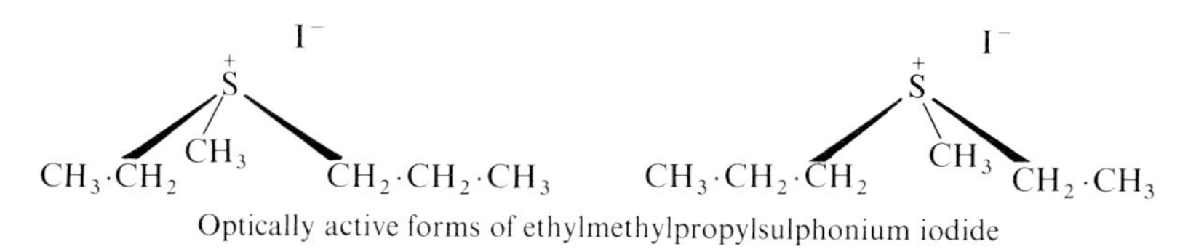

Optically active forms of ethylmethylpropylsulphonium iodide

## DISULPHIDES

### General Properties

Disulphides (RS·SR) are usually named as dialkyl or diaryl disulphides, e.g. dimethyl disulphide. The IUPAC system describes **dithio**-compounds as substitutive derivatives of hydrocarbons, e.g. ethyldithioethane ($CH_3{\cdot}CH_2{\cdot}S{\cdot}S{\cdot}CH_2{\cdot}CH_2$).

The bond strength of the S—S bond is usually about 210–290 kJ mol$^{-1}$, and is somewhat greater than that of the O—O bond in peroxides (*ca* 170 kJ mol$^{-1}$). The ease with which the S—S bond is cleaved is due to the ability of sulphur to make use of its 3*d* orbitals in the transition state. Both radical and ionic scissions of the S—S bond are known. Radical decomposition can be induced photochemically, by radical reagents such as alkali metals, and by nascent hydrogen produced either catalytically or by electro-reduction.

### Reductive Cleavage

Disulphides are readily reduced catalytically and enzymatically to the corresponding thiols:

$$\underset{\text{Cystine}}{\begin{matrix} CH_2-S-S-CH_2 \\ | \quad\quad\quad | \\ CH\overset{+}{N}H_3 \quad CH\overset{+}{N}H_3 \\ | \quad\quad\quad | \\ CO\cdot O^- \quad CO\cdot O^- \end{matrix}} \underset{H_2O \quad O_2}{\overset{H_2,\ \text{Catalyst}}{\rightleftharpoons}} 2\ \underset{\text{Cysteine}}{\begin{matrix} CH_2\cdot \mathbf{SH} \\ | \\ CH\cdot\overset{+}{N}H_3 \\ | \\ CO\cdot O^- \end{matrix}}$$

The ease with which the interconversion of disulphides and thiols is accomplished by oxidation and reduction accounts for the considerable biological significance of the amino acid pair, cystine and cysteine. Disulphide bridges in a number of important peptides and proteins including, insulin, serum albumin and keratin, are formed *in situ* by oxidation of appropriately situated peptide links involving cysteine (thiol) residues. Permanent-waving of hair, which is composed largely of keratin, is accomplished by breaking the disulphide bonds by chemical reduction using sodium thioglycollate. The hair, 'softened' in this way, can be curled on rollers, and then set by oxidation with dilute hydrogen peroxide to re-form the disulphide links.

The disulphide, (+)-lipoic acid, thioctic acid, plays a key rôle in the oxidative decarboxylation of α-keto acids. In particular, the conversion of pyruvate to acetyl-S. $\overline{\text{CoA}}$ is mediated by multi-enzyme complexes associated with the mitochondria in mammalian systems, and with the cell membranes of bacteria. These complexes are composed of sub-units with three distinct types of enzyme activity, pyruvate decarboxylase, lipoate-reductase-transacetylase and dihydrolipoate dehydrogenase (Koike, Reed and Carroll, 1963). Lipoic acid, linked as its amide to the ε-amino group of a lysine residue into the enzymic protein, is reductively acetylated by enzyme-linked hydroxyethylthiamine pyrophosphate (formed by the pyruvate decarboxylase; Chapter 13), as shown on page 402.

The reactivity of dithiolanes, such as lipoic acid, has been ascribed to ring strain. The dihedral angle between the two S—S—C planes, which is about 90° in open-chain disulphides, is significantly reduced due to orbital overlap of *p*-electron pairs on S (Foss and Tjomsland, 1958). This also accounts for the ready polymerisation of lipoic acid when heated or exposed to ultraviolet light, with the formation of an open-chain polymer. Lipoic acid, however, does undergo mild oxidation with peroxides to give the 8-*S*-oxide.

$$\text{Lipoic acid: dithiolane ring (S—S) bearing } H \text{ and } (CH_2)_4\cdot CO\cdot OH \xrightarrow[R\cdot OH]{R\cdot O\cdot \mathbf{OH}} \text{8-}S\text{-oxide: } {}^-O-\overset{+}{S}-S \text{ ring, } H,\ (CH_2)_4\cdot \overset{1}{C}O\cdot OH \ (\text{positions } 8,\ 6)$$

Reductive cleavage forms the normal metabolic pathway for some disulphide drugs such as *Antabuse* (Tetraethylthiuram Disulphide) which is used in the treatment of alcoholism (**2**, 5).

$$\underset{\textit{Antabuse}}{(CH_3\cdot CH_2)_2N\cdot CS\cdot \mathbf{S\cdot S}\cdot CS\cdot N(CH_2\cdot CH_3)_2} \xrightarrow{H_2} \underset{\text{Diethyldithiocarbamic acid}}{2(CH_3\cdot CH_2)_2N\cdot CS\cdot \mathbf{SH}}$$

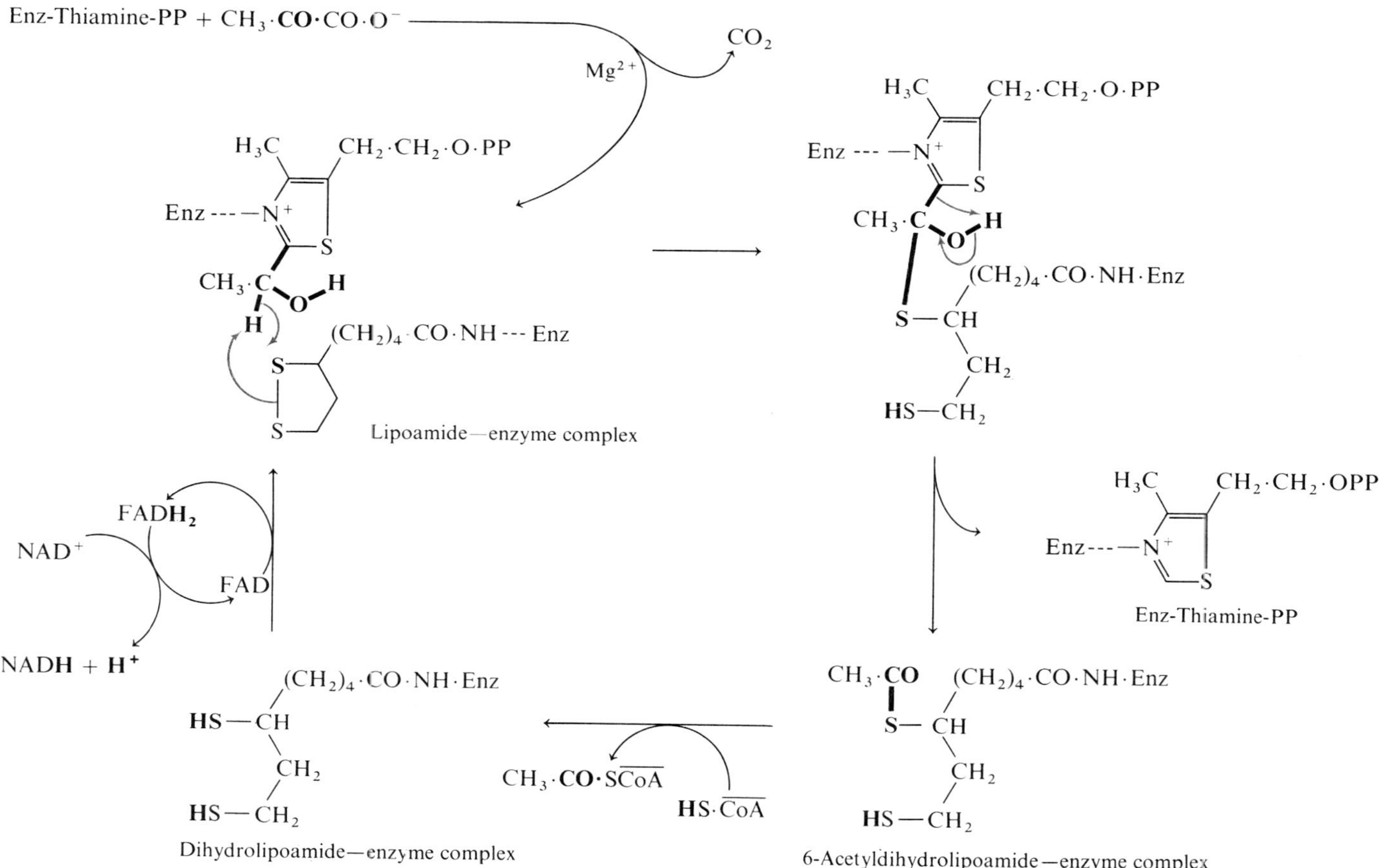
Enz-Thiamine-PP + CH₃·CO·CO·O⁻
Mg²⁺
CO₂
Lipoamide—enzyme complex
Enz-Thiamine-PP
FADH₂
NAD⁺
FAD
NADH + H⁺
CH₃·CO·SCoA
HS·CoA
Dihydrolipoamide—enzyme complex
6-Acetyldihydrolipoamide—enzyme complex

**Ionic Cleavage**

Diaryl disulphides are readily cleaved by alkaline hydrolysis; dialkyl disulphides are much more stable.

$$C_6H_5{\cdot}S{-}S{\cdot}C_6H_5 + HO^- \longrightarrow C_6H_5{\cdot}\mathbf{S{\cdot}OH} + C_6H_5{\cdot}\mathbf{S^-}$$

$$\downarrow\uparrow$$

$$C_6H_5{\cdot}\mathbf{SO^-} + C_6H_5{\cdot}\mathbf{SH}$$

Disulphides are similarly cleaved by halogens to form sulphenyl halides and by ammonia to sulphenamides.

$$CH_3{\cdot}CH_2{\cdot}S{\cdot}S{\cdot}CH_2{\cdot}CH_3 + \mathbf{Br_2} \longrightarrow 2CH_3{\cdot}CH_2{\cdot}\mathbf{SBr}$$

Ethylsulphenyl bromide

$$CH_3{\cdot}CH_2{\cdot}S{\cdot}S{\cdot}CH_2{\cdot}CH_3 + \mathbf{2NH_3}$$

$$\downarrow$$

$$CH_3{\cdot}CH_2{\cdot}S{\cdot}\mathbf{NH_2} + CH_3{\cdot}CH_2{\cdot}S^- + \overset{+}{N}\mathbf{H_4}$$

Ethylsulphenamide

A further important reaction of this type is that of thiol-disulphide exchange.

$$C_6H_5{\cdot}\mathbf{SH} + CH_3{\cdot}CH_2{\cdot}S{\cdot}S{\cdot}CH_2CH_3 \longrightarrow C_6H_5{\cdot}\mathbf{S}{\cdot}S{\cdot}CH_2{\cdot}CH_3 + CH_3{\cdot}CH_2{\cdot}S\mathbf{H}$$

$$C_6H_5{\cdot}\mathbf{SH} + C_6H_5{\cdot}S{\cdot}S{\cdot}CH_2CH_3 \longrightarrow C_6H_5{\cdot}\mathbf{S}{\cdot}S{\cdot}C_6H_5 + CH_3{\cdot}CH_2{\cdot}S\mathbf{H}$$

This is particularly well exemplified by the alkaline depolymerisation of lipoic acid and similar dithiolane polymers.

$$\text{Lipoic acid polymer} \longrightarrow (n+1)\ \text{Lipoic acid}$$

Lipoic acid polymer (R = $(CH_2)_4{\cdot}CO{\cdot}OH$)

Lipoic acid (R = $(CH_2)_4{\cdot}CO{\cdot}OH$)

## SULPHOXIDES

The balance of physical evidence leads to the conclusion that sulphoxides have essentially co-ordinate covalent bond structures, $>\overset{+}{S}{-}O^-$. The sulphur, however, is able to expand its octet of electrons to a decet by accepting *p*-electrons from oxygen into its *d*-orbitals, giving rise to some resonance involving the >S=O structure.

Double bond character of the S—O group is favoured by the presence of electron-withdrawing substituents on sulphur. Even so, the S—O stretching frequency varies only within quite narrow limits (1030–1055 $cm^{-1}$) for a wide range of compounds. Similarly, the R—S—O bond angle is remarkably constant at 107°. The polarity of the $\overset{+}{S}$—$\overset{-}{O}$ bond accounts for its weak electron-withdrawing properties. Aromatic sulphoxides deactivate the ring and give rise to *meta*-substitution.

A further consequence of $\overset{+}{S}$—$\overset{-}{O}$ bond formation by overlap of *p*- and *d*-orbitals is that unlike the *p*-$\pi$-orbital overlap of the carbon—oxygen double bond, there is no requirement for the sulphur and oxygen atoms to be in the same plane. Sulphoxides are, therefore, non-planar, and unsymmetrically alkyl- or aryl-substituted sulphoxides are capable of resolution into optical enantiomers (mirror image, non-superposable forms).

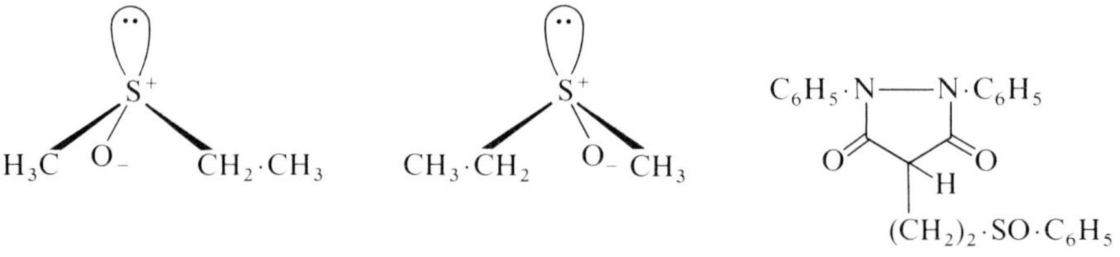

$S^+$ — $H_3C$, $O^-$, $CH_2 \cdot CH_3$ $\qquad$ $S^+$ — $CH_3 \cdot CH_2$, $O^-$, $CH_3$

$C_6H_5 \cdot N$—$N \cdot C_6H_5$, O, O, H, $(CH_2)_2 \cdot SO \cdot C_6H_5$

*Sulphinpyrazone*

The sulphoxide uricosuric agent, *Sulphinpyrazone*, has been prepared in its (+)- and (−)-forms; it is the (+)-enantiomer only which is excreted when the corresponding sulphide is metabolised in man, indicating that metabolic sulphoxidation is stereospecific.

**General Properties**

Dimethyl sulphoxide, a typical aliphatic sulphoxide, is a colourless, odourless liquid with a slightly bitter taste. Most aromatic sulphoxides are colourless or very pale yellow solids. The sulphoxide oxygen is capable of hydrogen bonding. Thus, most sulphoxides are readily soluble in chloroform by C—H···O bonding, and much less readily soluble in ether. Low molecular weight aliphatic sulphoxides are hygroscopic (dimethyl sulphoxide takes up about 70% of its weight of water at 20°C and 65% relative humidity). The affinity, which all sulphoxides have for water, accounts for the importance of sulphoxide formation in the metabolism of organic sulphides (e.g. *Chloropromazine* and related phenothiazines), since this clearly aids urinary excretion.

Dimethyl sulphoxide reacts with sodium hydroxide under strongly alkaline conditions to form a carbanion, which has important synthetic applications.

$$\mathbf{HCH_2 \cdot SO \cdot CH_3 + NaOH} \xrightarrow[\mathbf{H_2O}]{} \mathbf{Na^+}\ {}^-CH_2 \cdot SO \cdot CH_3$$

### Basic Properties

Sulphoxides are able to accept protons and Lewis acids, and hence are capable of functioning as bases.

$$(CH_3)_2\cdot\overset{+}{S}—O^- + \mathbf{HX} \rightleftharpoons (CH_3)_2\cdot\overset{+}{S}\cdot OH \quad X^-$$

Typically, sulphoxides are more soluble in aqueous acid due to protonation than they are in water. The acid pH of the tubular lumen in the kidney, therefore, aids the excretion of sulphoxide metabolites; excretion will also be inhibited under conditions which alkalise the urine. The sulphoxide, *Sulphinpyrazone*, acts as a uricosuric agent, promoting the excretion of uric acid in the treatment of gout. It is not clear whether the sulphoxide group exerts any special rôle in the action of this compound, though it probably assists excretion by reducing urinary pH.

The sulphoxide oxygen is also capable of donating electrons into vacant orbitals of transition metals to form salts and complexes. Such interactions undoubtedly account for some of the unusual solvent properties of dimethyl sulphoxide, which is an excellent solvent for both inorganic salts and aromatic (and unsaturated) hydrocarbons. Dimethyl sulphoxide is also completely miscible with many common organic solvents including alcohols, aldehydes, ketones, acids, amines and ethers. It should be noted that dimethyl sulphoxide reacts explosively with perchloric acid.

### Thermal Stability

Di-t-butyl sulphoxide has been shown to undergo thermal decomposition in both polar and non-polar solvents to form a sulphenic acid (Shelton and Davis, 1967).

$$(CH_3)_3\cdot C\cdot \overset{+}{S}(—\overset{-}{O})—C(CH_3)_2—CH_2—H \xrightarrow{\text{Benzene}/80^\circ} (CH_3)_3\cdot C\cdot \mathbf{S\cdot OH} + CH_3\cdot C(CH_3){=}CH_2$$

*t*-Butylsulphenic acid

The rearrangement of phenoxymethylpenicillin sulphoxide methyl ester to a cephalosporin derivative by acid catalysts in xylene (Morin, Jackson, Mueller, Lavagnino, Scanlon and Andrews, 1963) has been shown to take place by way of such sulphenic acid intermediates (Barton, Comer, Greig, Sammes, Cooper, Hewitt and Underwood, 1971).

### Nucleophilic Properties

The sulphoxide oxygen is also capable of acting as a nucleophile in reactions with activated halogen compounds, such as ethyl bromoacetate and phenacyl bromide. Dissociation of the adduct leads to formation of sulphide, aldehyde and halogen acid.

$$(CH_3)_2\cdot\overset{+}{S}{-}O^- + CH_2\cdot CO\cdot C_6H_5 \text{ (Br)} \longrightarrow (CH_3)_2\cdot\overset{+}{S}{-}O{-}CH_2\cdot CO\cdot C_6H_5 + Br^-$$

Phenacyl bromide

$$\xrightarrow{-HBr} (CH_3)_2\cdot S + H\cdot C(=O)\cdot CO\cdot C_6H_5$$

The reaction of dimethyl sulphoxide with acyl halides is particularly violent. The formation of sulphoxonium compounds, as for example with dimethyl sulphoxide and methyl iodide, is probably initiated in the same way. The products appear, however, to be stabilised in the *S*-trialkyl form.

$$(CH_3)_2\cdot\overset{+}{S}{-}O^- \;\; H_3C{-}I \longrightarrow (CH_3)_2\cdot\overset{+}{S}{-}O(CH_3)\;\; I^- \longrightarrow \left[(CH_3)_3\cdot\overset{2+}{S}{-}O^-\right]I^-$$

**Carbonyl-like Properties**

The sulphoxide group having only marginal double bond character is only weakly electron-withdrawing, which in aromatic systems is of the same order as the carboxyl group. Nonetheless, sulphoxides exhibit some properties which are analogous to those of carbonyl compounds. The formation of sulphilimines from sulphonamides under dehydrating conditions is typical.

$$R_2\cdot SO + H_2N\cdot SO_2\cdot Ar \xrightarrow[-H_2O]{P_2O_5/CHCl_3} R_2\cdot S{=}N\cdot SO_2\cdot Ar$$

Sulphilimine

**Reduction**

Sulphoxides are reduced quite readily to the corresponding sulphide by a variety of chemical reducing agents, such as zinc and acetic acid and lithium aluminium hydride. Dimethyl sulphoxide is reduced in cats to dimethyl sulphide, which is then excreted largely via the lungs.

**Oxidation**

Sulphoxides are not appreciably oxidised by peroxides, but they are readily oxidised by peracids and potassium permanganate to sulphones (p. 397). Dimethyl sulphoxide is oxidised metabolically in the rat, rabbit and guinea pig to dimethyl sulphone (Hucker, Ahmad and Miller, 1966).

## SULPHONES

### General Properties

As with sulphoxides, there is some physical evidence of polarity in the sulphur—oxygen bond of sulphones. There is also evidence of $\pi_{d-p}$ bond formation giving rise to resonance involving S=O bonds as in the structures:

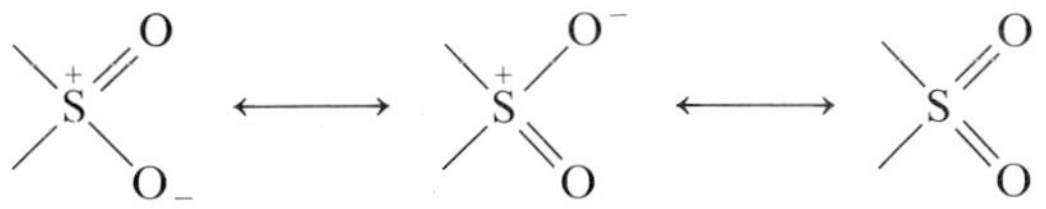

$$>\overset{+}{S}(=O)-O^- \longleftrightarrow >\overset{+}{S}(-O^-)=O \longleftrightarrow >S(=O)=O$$

It is evident, too, from the ultraviolet absorption spectra of methylene sulphones in alkaline solution that the electron configuration about sulphur can be expanded still further to a duodecet. Enol formation, however, does not occur.

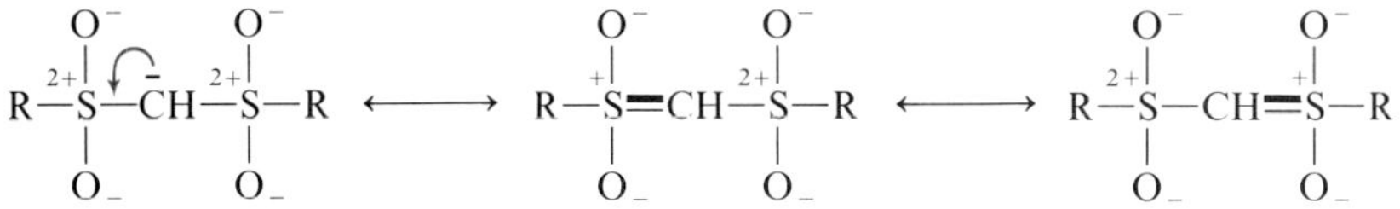

$$R-\overset{2+}{S}(O^-)_2-\overset{-}{C}H-\overset{2+}{S}(O^-)_2-R \longleftrightarrow R-\overset{+}{S}(O^-)_2=CH-\overset{2+}{S}(O^-)_2-R \longleftrightarrow R-\overset{2+}{S}(O^-)_2-CH=\overset{+}{S}(O^-)_2-R$$

The polarity of the sulphone group accounts for its electron-withdrawing properties, so that in aromatic systems it gives rise to *meta*-substitution. This is the result purely of its inductive effect. There is no mesomeric effect, as in the case of the nitro group, and in consequence it is less deactivating than the nitro group.

Electron-attraction gives rise to activation of adjacent (α)-methylene groups. α-Methylenesulphones are soluble in alkali, and also readily undergo deuterium exchange. The sulphone group also activates adjacent ethylenic compounds in Diels Alder additions.

$$\text{butadiene} + CH_2=CH\cdot SO_2\cdot C_6H_5 \longrightarrow \text{cyclohexenyl}-SO_2\cdot C_6H_5$$

Electron-withdrawal also deactivates the halogen in α-halogeno-sulphones.

### Chemical Stability of the Sulphone Group

The sulphone group itself is characterised by great chemical stability. It is capable of becoming protonated, but less readily so than the sulphoxide group. Thus, sulphones are usually soluble in concentrated sulphuric acid and the solutions show a molecular depression of freezing point compatible with protonation. Most sulphones are resistant to chemical reduction, though some, such as thiophene-1,1-dioxide, can be reduced with zinc and hydrochloric acid or the prolonged action of lithium aluminium hydride. There is no evidence that the sulphone group of the anti-leprotic sulphones, *Dapsone* and *Solapsone*, is reduced *in vivo*.

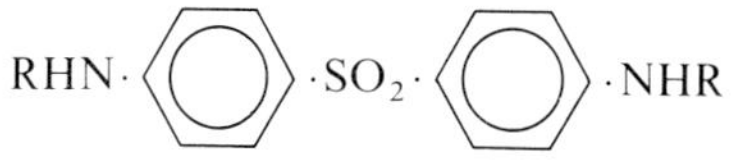

RHN·⟨C₆H₄⟩·SO₂·⟨C₆H₄⟩·NHR

*Dapsone* (R = H)
*Solapsone* (R = $C_6H_5 \cdot CH(SO_3Na) \cdot CH_2 \cdot CH(SO_3Na)$—)

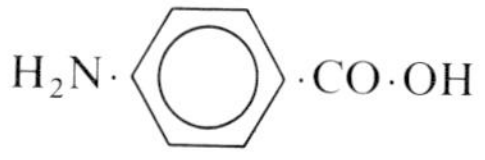

$H_2N \cdot C_6H_4 \cdot CO \cdot OH$

*p*-Aminobenzoic acid (PABA)

*Dapsone*, like the analogous sulphonamides (p. 412) interferes with folic acid metabolism by competing with PABA (*p*-aminobenzoic acid) for an essential enzyme. It is an effective antibacterial agent for species which require to synthesise their own folic acid, but is considerably more toxic than most sulphonamides.

### Cleavage of the C—S Bond in Sulphones

The C—S bond of sulphones is broken only with difficulty. Cleavage occurs when aryl sulphones are treated with strong bases to give a sulphinic acid, a process which may be rationalised as follows.

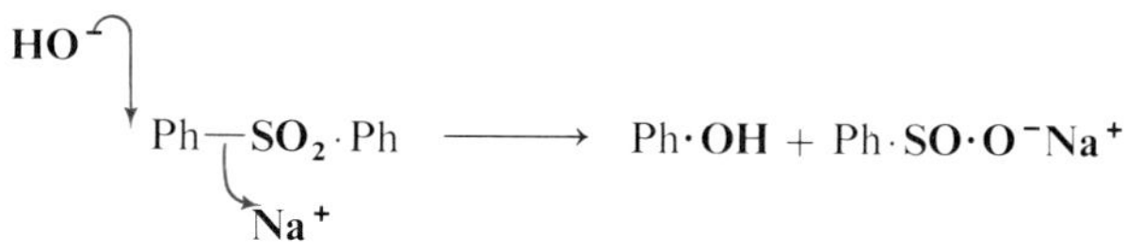

$$HO^- + Ph—SO_2 \cdot Ph \;(Na^+) \longrightarrow Ph \cdot OH + Ph \cdot SO \cdot O^-Na^+$$

Diaryl and aralkyl sulphones are similarly cleaved by refluxing with sodamide in pyridine to yield an *N*-arylpiperidine and a sodium aryl- or aralkyl-sulphinate.

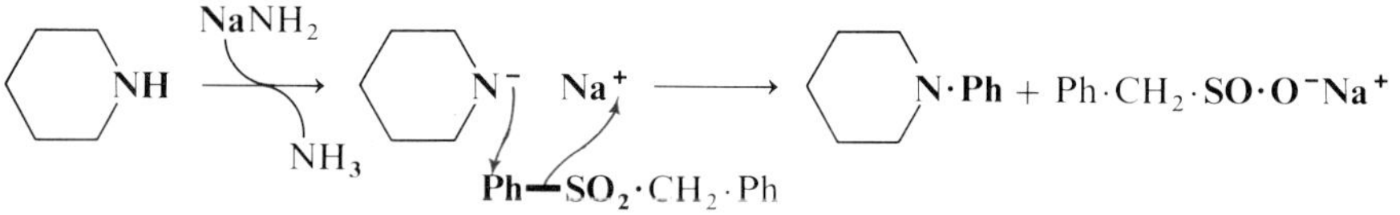

$$C_5H_{10}NH \xrightarrow[-NH_3]{NaNH_2} C_5H_{10}N^- \; Na^+ \; + \; Ph—SO_2 \cdot CH_2 \cdot Ph \longrightarrow C_5H_{10}N \cdot Ph + Ph \cdot CH_2 \cdot SO \cdot O^-Na^+$$

## SULPHONIC ACIDS

### General Properties

Alkane and arylsulphonic acids whilst exhibiting some differences in physical properties have many chemical properties in common. Alkanesulphonic acids, which are exemplified by methanesulphonic acid, are usually viscous highly water-soluble liquids. Typical aromatic sulphonic acids, such as benzenesulphonic and toluenesulphonic acids, are crystalline solids, though hygroscopic and readily soluble in water. All sulphonic acids are strong acids, comparable in strength to sulphuric acid, and react with metal hydroxides and carbonates, and with organic bases to form fully ionisable water-soluble salts. The solubilising effect of alkali metal salts of sulphonic acids has been widely used in the preparation of water-soluble dyes, such as *Indigo Carmine* (Chapter 23). The solubility of sulphonic acids also in many organic solvents makes them valuable reagents as acidic catalysts.

**Sulphonyl Halides**

Sulphonic acids resemble carboxylic acids in the formation of acid halides, sulphonyl chlorides and bromides, when treated with phosphorus pentahalides.

$$CH_3{\cdot}SO_2{\cdot}OH + PCl_5 \longrightarrow CH_3{\cdot}CH_2{\cdot}SO_2{\cdot}Cl + POCl_3 + HCl$$

Methanesulphonic acid — Methanesulphonyl chloride

$$CH_3{\cdot}C_6H_4{\cdot}SO_2{\cdot}OH + PCl_5 \longrightarrow CH_3{\cdot}C_6H_4{\cdot}SO_2Cl + POCl_3 + HCl$$

*p*-Toluene sulphonic acid — *p*-Toluenesulphonyl chloride

Sulphonyl halides differ from acyl halides (Chapter 11) in their greater stability to hydrolysis, which takes place only slowly in water, and also in their inability to react with diazomethane. They resemble acyl halides, however, in their reactivity towards ammonia, amines, alcohols and phenols. Sulphonyl chlorides react with ammonia, and both aliphatic and aromatic primary and secondary amines to form sulphonamides.

$$CH_3{\cdot}SO_2{\cdot}Cl + 2NH_3 \longrightarrow CH_3{\cdot}SO_2{\cdot}NH_2 + NH_4Cl$$

Methanesulphonamide

$$CH_3{\cdot}C_6H_4{\cdot}SO_2{\cdot}Cl + H_2N{\cdot}C_6H_5 \xrightarrow{-HCl} CH_3{\cdot}C_6H_4{\cdot}SO_2{\cdot}NH{\cdot}C_6H_5$$

*p*-Toluenesulphonamide

Sulphonamides derived from aromatic sulphonic acids in particular are usually easily crystallised, and are thus considered useful derivatives for the characterisation of primary and secondary amines.

Sulphonyl chlorides also react with alcohols and phenols, readily in the presence of tertiary bases, to form sulphonate esters.

$$CH_3{\cdot}SO_2{\cdot}Cl + CH_3{\cdot}CH_2{\cdot}OH \xrightarrow{-HCl} CH_3{\cdot}SO_2{\cdot}O{\cdot}CH_2{\cdot}CH_3$$

Ethyl methanesulphonate

$$CH_3{\cdot}C_6H_4{\cdot}SO_2{\cdot}Cl + CH_3{\cdot}OH \xrightarrow{-HCl} CH_3{\cdot}C_6H_4{\cdot}SO_2{\cdot}O{\cdot}CH_3$$

Methyl *p*-toluenesulphonate

**Sulphonate Esters**

The electron-attracting sulphonyl group of sulphonate esters promotes ready fission of the C—O bond. The sulphonate group is, therefore, an effective leaving group in nucleophilic displacement reactions.

$$CH_3\cdot SO_2\cdot O\!-\!CH_2\cdot CH_3 \ (Na^+,\ Br^-) \xrightarrow{\text{Ethylene glycol}} CH_3\cdot SO_2O^-Na^+ + CH_3\cdot CH_2\cdot Br$$

Similarly, tertiary amines are readily converted to quaternary ammonium salts by methyl *p*-toluenesulphonate and methyl methanesulphonate.

$$CH_3\cdot C_6H_4\cdot SO_2\cdot O\!-\!Me \ (R_3\ddot{N}) \longrightarrow R_3\cdot \overset{+}{N}\cdot Me \ \ CH_3\cdot C_6H_4\cdot SO_2O^-$$

Trialkylmethylammonium *p*-toluenesulphonate

*Busulphan* (Busulfan) is an effective di-functional biological alkylating agent and is used for its cytotoxic effect (Ross, 1962). Contrary to earlier reports (Timmis, Hudson, Marshall and Bierman, 1962), *Busulphan* is not hydrolysed in aqueous solution to 1,4-butanediol, since the monosulphonate (4-methane-sulphonyloxybutanol) is first formed cyclises rapidly ($\frac{1}{2}$-life, 12 min.; 37°C; pH 3.0 and pH 7.4) to form tetrahydrofuran (Feit and Rastrup-Anderson, 1973).

$$\underset{\textit{Busulphan}}{CH_3\cdot SO_2\cdot O(CH_2)_4O\cdot SO_2\cdot CH_3} \xrightarrow[-CH_3\cdot SO_2\cdot OH]{H_2O} CH_3\cdot SO_2\cdot O\cdot (CH_4)_4OH \xrightarrow[-CH_3\cdot SO_2\cdot OH]{} \text{tetrahydrofuran (O)}$$

**Desulphonation of Aromatic Sulphonic Acids**

Aliphatic and aromatic sulphonic acids are distinguishable by the ease with which the latter undergo desulphonation on treatment with superheated steam. This is the reverse of the sulphonation reaction whereby they are formed from the aromatic hydrocarbon and sulphuric, or better, fuming sulphuric acid. Aliphatic sulphonic acids are stable under these conditions.

$$C_6H_5\cdot SO_2\cdot OH + H_2O \rightleftharpoons C_6H_6 + H_2SO_4$$

**Displacement Reactions of Aromatic Sulphonic Acids**

Aromatic sulphonic acids similarly undergo a displacement of the $SO_2OH$ group when fused with sodium hydroxide, sodamide, sodium cyanide or potassium hydrogen sulphide, with formation of the corresponding amine, phenol, nitrile or thiophenol, and the metal sulphite. Aliphatic sulphonic acids are stable to these reagents.

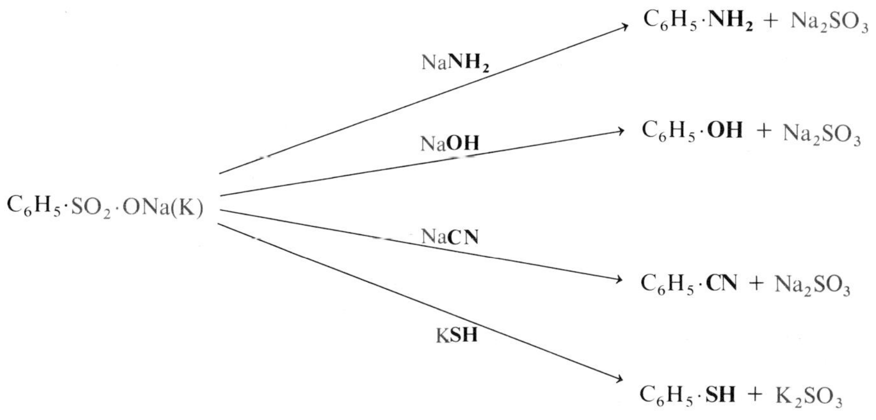

$$C_6H_5 \cdot SO_2 \cdot ONa(K) \xrightarrow{NaNH_2} C_6H_5 \cdot NH_2 + Na_2SO_3$$

$$C_6H_5 \cdot SO_2 \cdot ONa(K) \xrightarrow{NaOH} C_6H_5 \cdot OH + Na_2SO_3$$

$$C_6H_5 \cdot SO_2 \cdot ONa(K) \xrightarrow{NaCN} C_6H_5 \cdot CN + Na_2SO_3$$

$$C_6H_5 \cdot SO_2 \cdot ONa(K) \xrightarrow{KSH} C_6H_5 \cdot SH + K_2SO_3$$

## SULTONES

Sultones are cyclic sulphonic acid esters, and as such are structurally analogous to lactones (Chapter 13). Typical sultones, which include 1,8-naphthosultone and $\delta$-butyrosultone, are neutral crystalline compounds, insoluble in water and stable to cold aqueous sodium hydroxide. The sultone ring is cleaved by ammonia (0.880) or by hot aqueous alkali to give the hydroxysulphonamide and hydroxysulphonic acid respectively. In contrast to lactones, however, 5-membered ring sultones appear to be more reactive and less stable than the 6-membered ring compounds.

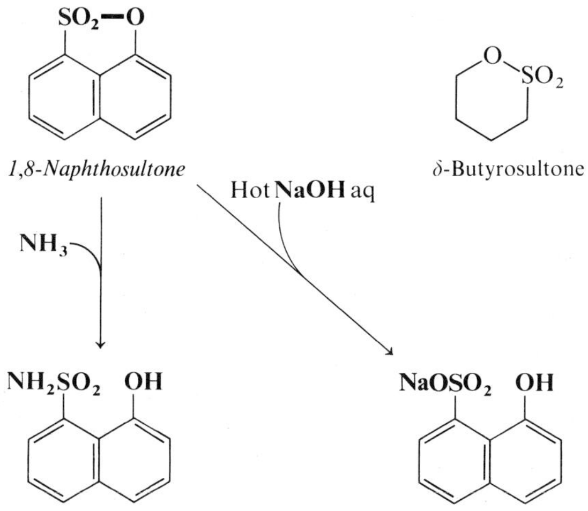

$SO_2$—O

*1,8-Naphthosultone*

O $SO_2$

$\delta$-Butyrosultone

$NH_3$      Hot NaOH aq

$NH_2SO_2$ OH      $NaOSO_2$ OH

### Sultones as Alkylating Agents

As cyclic sulphonic acid esters, sultones are powerful alkylating agents and react readily with alcohols, phenols, amines, metal alkoxides, and salts of organic acids (Helberger, Manecke and Heyden, 1949).

$$\text{(3-methyl-1,2-oxathiolane 2,2-dioxide)} \xrightarrow{C_6H_5ONa} CH_3\cdot CH(SO_2\cdot ONa)\cdot CH_2\cdot CH_2\cdot \mathbf{OC_6H_5}$$

$$\xrightarrow{R\cdot NH_2} CH_3\cdot CH(SO_2\cdot OH)\cdot CH_2\cdot CH_2\cdot \mathbf{NHR}$$

$$\xrightarrow{R\cdot CO\cdot ONa} CH_3\cdot CH(SO_2\cdot ONa)\cdot CH_2\cdot CH_2\cdot \mathbf{O\cdot CO\cdot R}$$

## SULPHONAMIDES

### General Properties

Sulphonamides are mostly stable, neutral crystalline compounds. Unsubstituted sulphonamides ($—SO_2\cdot NH_2$) and monosubstituted sulphonamides ($—SO_2\cdot NHR$) have acidic properties and are soluble in sodium hydroxide (but not sodium carbonate) forming water-soluble sodium salts.

$$C_6H_5\cdot SO_2\cdot NH\cdot CH_3 + \mathbf{NaOH} \xrightarrow{} C_6H_5\cdot SO_2\cdot N(Na)\cdot CH_3 + \mathbf{H_2O}$$

The ability to form these water-soluble salts provides the foundation for the Hinsberg separation of mixtures of primary, secondary and tertiary amines as toluenesulphonamides. Tertiary amines do not react with toluene sulphonyl chloride, and can be separated from the reaction mixture by distillation; secondary amines form *NN*-disubstituted toluenesulphonamides ($CH_3\cdot C_6H_4\cdot SO_2\cdot NR_2$), which are insoluble in alkali, whilst primary amines form alkali-soluble sulphonamides. Partition of the product between aqueous alkali and an organic solvent therefore gives an effective separation of the primary and secondary amine sulphonamides.

The highly water-soluble sodium salts, such as *Sulphadimidine Sodium* (Sulfamethazine; soluble 1 in 2.5) provide an excellent means of preparing solutions of antibacterial sulphonamides for injection. The injection solutions, like the parent substances (see below), are sensitive to light-catalysed air oxidation, and must be prepared with water free of dissolved air, using an antioxidant, and stored under nitrogen in the absence of light.

Sulphonamides are fairly stable substances, but amines can be released by hydrolysis with strong mineral acid (*10N* sulphuric or hydrochloric acids). The sulphonamide link is not cleaved by alkali.

Sulphonamides are susceptible to radical and light-catalysed oxidation. Radical oxidation of *Sulphanilamide* (Sulfanilamide) (Seikel, 1940) with potassium ferricyanide or hydrogen peroxide gives rise to the corresponding azobenzene-4,4′-disulphonamide and azoxybenzene-4,4′-disulphonamide. Light-catalysed oxidation gives analogous products. Medicinal sulphonamides must in consequence be stored in well-closed containers and be protected from light.

$H_2N \cdot C_6H_4 \cdot SO_2 \cdot NH_2$

$\xrightarrow[CH_3 \cdot CO \cdot OH]{H_2O_2} \; (-H_2O)$

$H_2N \cdot SO_2 \cdot C_6H_4 \cdot N{=}N \cdot C_6H_4 \cdot SO_2NH_2$

+

$H_2N \cdot SO_2 \cdot C_6H_4 \cdot \overset{+}{N}(O^-){=}N \cdot C_6H_4 \cdot SO_2 \cdot NH_2$

Azoxy compounds, so formed, are themselves light-sensitive (Badger and Buttery, 1954).

Azoxybenzene $\xrightarrow{h\nu}$ 2-Hydroxyazobenzene

## Chloramines

Treatment of *p*-toluenesulphonamide with sodium hypochlorite in the presence of sodium hydroxide gives rise to *Chloramine* (Chloramine T). If excess hypochlorite is used, further *N*-chlorination occurs with the formation of *Dichloramine*.

$CH_3 \cdot C_6H_4 \cdot SO_2 \cdot NH_2 \xrightarrow[-H_2O]{NaOCl} CH_3 \cdot C_6H_4 \cdot SO_2 \cdot N(Cl)Na \xrightarrow[-NaOH]{NaOCl + H_2O} CH_3 \cdot C_6H_4 \cdot SO_2 \cdot NCl_2$

*p*-Toluene sulphonamide — *Chloramine* — *Dichloramine*

Both compounds are stable solids, but provide a useful source of available chlorine, as hypochlorite in the presence of acid, and for this reason are useful as mild antiseptics. *Chloramine* is soluble in water, whereas *Dichloramine* is not; the latter, however, is soluble in organic solvents and oils.

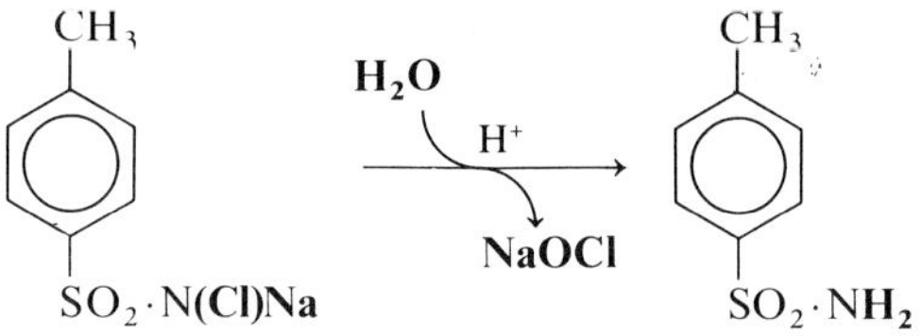

$CH_3$ | $H_2O$ | $CH_3$

$SO_2 \cdot N(Cl)Na$ —($H^+$, NaOCl)→ $SO_2 \cdot NH_2$

## Antibacterial Sulphonamides

### Discovery

The antibacterial sulphonamides (or sulpha drugs) stem from Domagk's discovery in the course of a survey of medicinal dyestuffs (1935) of the effects of *Prontosil*. *Prontosil*, a high melting brick-red powder, was found to be highly active *in vivo* against *Streptococci*, but suffered from the disadvantage of a low solubility in water (*ca* 1 in 400). This was overcome with the elaboration of *Prontosil Soluble*, which retained the antibacterial activity of *Prontosil*, whilst being much more soluble in water (1 in 25). As a result, these drugs were soon established as effective in the treatment of puerperal septicaemia (Colebrook and Kenny, 1936) and other streptococcal infections.

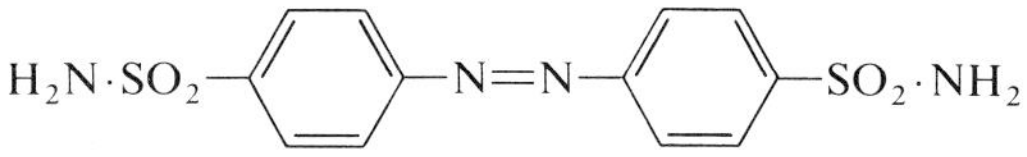

$H_2N \cdot SO_2-C_6H_4-N{=}N-C_6H_4-SO_2 \cdot NH_2$

*Prontosil*

$H_2N \cdot SO_2$ —N=N— OH, $NH \cdot CO \cdot CH_3$, $NaO \cdot O_2S$, $SO_2 \cdot ONa$

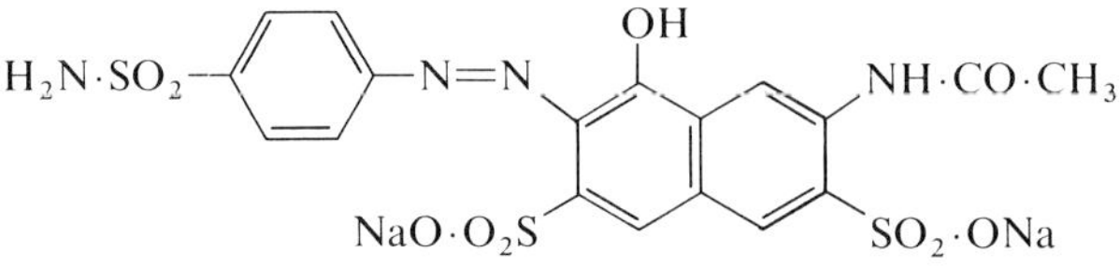

*Prontosil Soluble*

It was soon shown by Tréfoüel and his collaborators (1935) that although *Prontosil* was inactive *in vitro*, it became antibacterially active *in vitro* in the presence of reducing agents. It was also rapidly established that the *m*-phenylenediamine fragment of *Prontosil* was inactive, that the activity resided entirely in the *p*-aminobenzenesulphonamide (sulphanilamide) moiety, and that it was this compound which was formed *in vivo* by metabolic reduction.

*Sulphanilamide* had been known as a chemical entity since its discovery in 1908, and its manufacture was established by the following route from acetanilide.

$SO_2 \cdot Cl$ ... $SO_2 \cdot NH_2$

$\xrightarrow{Cl \cdot SO_2 \cdot OH}$ $\xrightarrow{NH_3}$

$NH \cdot CO \cdot CH_3$ — *Acetanilide*

$NH \cdot CO \cdot CH_3$ — *p*-Acetylaminobenzene sulphonyl chloride

$NH \cdot CO \cdot CH_3$

$\downarrow H^+$

$SO_2 \cdot \overset{1}{N}H_2$

$^4NH_2$

*Sulphanilamide*

The discovery of the clinical effectiveness of *Sulphanilamide* in the control of streptococcal infections was quickly followed by the introduction of *Sulphapyridine* (Sulfapyridine; M & B 693), which was especially effective against pneumonia, *Sulphathiazole* (Sulfathiazole) and various other sulpha drugs, all of which were capable of being manufactured from a common intermediate, *p*-acetylaminobenzenesulphonyl chloride.

**Essential Structural Parameters for Antibacterial Activity of Sulphonamides**

The synthesis of substituted sulphonamides rapidly led to the identification of certain structural features common to nearly all antibacterially active sulphonamides. The essential features appear to be:

(a) *Para relationship of amino and sulphonamido groups.* The corresponding *o*- and *m*-aminobenzenesulphonamides have only low levels of antibacterial activity. Additional substitution of nuclear hydrogens in the 2-, 3-, 5- and 6-positions by other groups also markedly reduces activity. Structural similarity to PABA, therefore, appears to be essential for activity.

(b) *Unsubstituted $N^4$-amino groups.* Substitution of nuclear amino group hydrogens by alkyl and aryl groups in general results in compounds with a low level of activity. A few appropriately substituted compounds capable of metabolic degradation to give a free amino group do possess considerable activity. *Prontosil* is a notable example of such a compound which is metabolically activated by an azoreductase in the liver. Other examples include *Succinylsulphathiazole* (Succinylsulfathiazole) and *Phthalylsulphathiazole* (Phthalysulfathiazole), which owe their activity to hydrolysis of the protecting acyl groups by digestive or bacterial proteases in the gut, with release of the active sulphonamide, *Sulphathiazole.* These acylamides have low water solubility and are poorly absorbed from the gut. Moreover, they are also extensively ionised in the alkaline conditions of the intestine, and in this state are incapable of being absorbed.

$$HO{\cdot}OC{\cdot}CH_2{\cdot}CH_2{\cdot}CO{\cdot}NH{-}C_6H_4{-}SO_2{\cdot}NH{-}(C_3H_2NS)$$

*Succinylsulphathiazole*

$$C_6H_4(CO{\cdot}OH){-}CO{\cdot}NH{-}C_6H_4{-}SO_2{\cdot}NH{-}(C_3H_2NS)$$

*Phthalylsulphathiazole*

$$H_2N{-}C_6H_4{-}SO_2{\cdot}NH{-}(C_5H_4N)$$

*Sulphapyridine*

$$H_2N{-}C_6H_4{-}SO_2{\cdot}NH{-}(C_3H_2NS)$$

*Sulphathiazole*

$$H_2N{-}C_6H_4{-}SO_2{\cdot}NH{-}(C_4H_3N_2)$$

*Sulphadiazine*

(c) *Substitution of the sulphonamide group.* $N^1$-substituents on the sulphonamide group, particularly heterocyclic substituents, as for example in *Sulphathiazole*, *Sulphapyridine* and *Sulphadiazine*, lead to valuable derivatives which are often more potent, and in some cases more specific in their antibacterial action than *Sulphanilamide* itself. Large heterocyclic $N^1$-substituents undoubtedly reduce the overall structural similarity to PABA, but several groups have drawn attention to the rôle of this and other related factors, notably ionisation, which are important in determining the level of activity. It is now generally agreed that sulphonamides penetrate the bacterial cell wall in an undissociated form. Inside the cell, the sulphonamide is dissociated, and it is the concentration of sulphonamide ion which determines the degree of antagonism.

**The Mode of Action of Sulphonamides**

In 1939, Stamp observed that some tissue extracts, pus, bacteria and especially yeast extract, contained a low molecular weight, thermostable, factor which was capable of inhibiting the antibacterial action of sulphonamides. Woods (1940) established by chemical tests that the inhibitory substance in yeast concentrates was an aromatic amino acid and that its antagonism was competitive, such that the ratio, effective concentration/sulphonamide concentration, was constant over a range of sulphonamide concentrations. Competitive antagonism between closely-related chemical substances as substrates for some enzyme systems was already known, and from this Woods concluded that the antagonist was structurally similar to *Sulphanilamide*. Woods, therefore, tested *p*-aminobenzoic acid (PABA) and established that it was a potent inhibitor of *Sulphanilamide*. Finally, PABA was isolated from yeast as its benzoyl derivative by Rubbo and Gillespie (1940) and shown to be an essential growth factor for *Clostridium acetobutylicum*.

It is now established that the antibacterial activity of sulphonamides, generally, is due to their chemical and structural similarity to *p*-aminobenzoic acid (PABA), which is a key substrate in bacterial synthesis of folic acid. All sulphonamides with structures essentially similar to *p*-aminobenzoic acid are antagonised by very much smaller concentrations of the latter both *in vitro* and *in vivo*. Thus, the inhibition of *Streptococcus haemolyticus* cultures by *Sulphanilamide* at a concentration of $3.03 \times 10^{-4}$ M is reversed by PABA at a concentration of only $1.2 \times 10^{-8}$ M. This 4000-fold difference in concentration between inhibitor and metabolite antagonist has been explained in terms of the much higher specificity which the enzyme has for its natural substrate (PABA).

The ratio of antagonising PABA concentration to inhibitory sulphonamide concentration remains constant irrespective of the concentration of sulphonamide. More active sulphonamides require correspondingly greater concentrations of PABA to effect the same degree of antagonism. Thus, *Sulphapyridine* is about five times as active as *Sulphanilamide*, and some five times as much PABA is required to antagonise the action of *Sulphapyridine* compared with that necessary to antagonise the action of *Sulphanilamide* against the same organism. However, as indicated below, Fox and Rose (1942) established that the amount of PABA required to antagonise the minimum effective concentration of the sulphonamide was constant, irrespective of its bacteriostatic activity (Table 45).

Strains of organisms resistant to *Sulphanilamide* are generally found to be resistant to other sulphonamides which are antagonised by *p*-aminobenzoic

Table 45. Ionisation and Bacteriostatic Activity of Sulphonamides (Fox and Rose, 1942)

| Sulphonamide | Min. Effective Conc. ($M \times 10^{-6}$) (MEC) | $pK_a$ (—$SO_2NR$) | Conc. of Sulphonamide Ion ($M \times 10^{-6}$) pH.7.0 | Conc. of PABA to prevent bacteriostasis | PABA/ Sulphonamide ion ratio |
|---|---|---|---|---|---|
| Sulphanilamide | 2500 | 10.5 | 0.71 | 0.5 | 1/1.4 |
| Sulphapyridine | 20 | 8.5 | 0.68 | 0.5 | 1/1.4 |
| Sulphathiazole | 4 | 6.8 | 2.46 | 0.5 | 1/4.9 |
| Sulphadiazine (Sulfamethazine) | 4 | 6.4 | 3.2 | 0.5 | 1/6.4 |

acid. These resistant strains, however, are susceptible to the action of sulphonamides which are not antagonised by PABA, such as *Marfanil* (Mafenide), in which the $N^4$-amino group is separated from the ring by a methylene group.

$H_2N{\cdot}CH_2-C_6H_4-SO_2{\cdot}NH_2$

Marfanil

## Ionisation of Sulphonamides

Antibacterially active sulphonamides have two ionisable groups, the $N^4$-primary amino group and the sulphonamido group. Modification of the primary amino group by substitution in general diminishes antibacterial activity, and all useful sulphonamides have either the unsubstituted or $N^1$-substituted *p*-sulphonamido group as the sole second substituent in the ring. It is not surprising, therefore, that all active sulphonamides antagonised by PABA have base dissociation constants within narrow limits (p$K_a$ 1.64–2.34; Tolstoouhov, 1955) close to that of the *p*-amino substituent of PABA, i.e. *ca* p$K_a$ 2.4. In contrast, antibacterially active sulphonamides have acidic sulphonamido groups with widely differing dissociation constants within the range p$K_{a_2}$ 2–11. Ionisation of the sulphonamido group, therefore, may reasonably provide an explanation of the range of activity of antibacterial sulphonamides which is actually observed.

Fox and Rose (1942) pointed out that although *Sulphathiazole* was about six hundred times more active than *Sulphanilamide*, the minimum bacteriostatic concentration of each is antagonised by the same amount of PABA. This suggested that the active fraction is similar in each case, and calculation showed that at the minimum effective bacteriostatic concentration, the concentration of sulphonamide **ion** ($-SO_2\bar{N}R$) present in solution at pH 7 remained relatively constant (Table 45). They concluded that the sulphonamide ion was the active antibacterial moiety and confirmed this by observations of the activity of *Sulphanilamide* in media at different pH. Between pH 6.8 and 7.8, there is a tenfold increase in ionisation, and an eightfold decrease in the molar minimum bacteriostatic concentration.

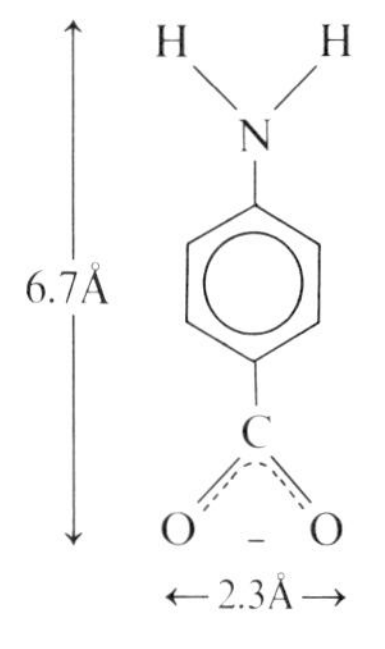


*p*-Aminobenzeate ion

H H
N
R·N
$S^{2+}$
O O
6.9Å
←2.4Å→

Sulphonamide ion

The results and conclusions of Fox and Rose were complemented by those of Bell and Roblin (1942). They suggested that if the bacteriostatic action of sulphonamides is due to competition with PABA, then the more closely the competitor resembles its metabolite, the greater will be its bacteriostatic activity. They drew attention to the structural and dimensional similarity between the *p*-aminobenzoate ion, in which form the acid exists almost exclusively at pH 7, and that of the sulphonamide ion.

The weakness of this simplistic comparison is, however, evident from the electron density distributions of *Sulphanilamide* (Foernzler and Martin, 1967) and the *p*-aminobenzoate ion (Pullman and Pullman, 1963).

$H_2N$ (+0.156)—C₆H₄—C (+0.265)(O$^{-0.377}$)(O$_{-0.377}$)

$H_2N$ (+0.152)—C₆H₄—S (+1.189)(O$^{-0.662}$)(O$_{-0.662}$)—$NH_2$ (+0.081)

Nevertheless, Bell and Roblin (1942) showed that when the antibacterial activities of a series of sulphonamides at physiological pH are plotted against their p$K_a$, a parabolic curve results (Fig. 21). Now, if ionic dissociation were the only factor involved, the activity of thc sulphonamides would be expected to increase with increasing acidic strength up to a maximum at a limiting p$K_a$ about 4–5, below which all would be 100% ionised. One explanation of this apparent anomaly suggested by Cowles (1942) and Brueckner (1943) is that the bacterial

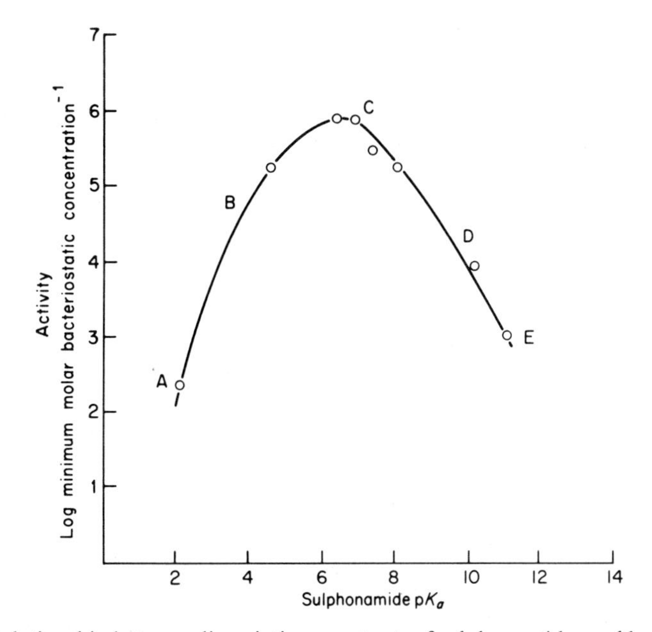


Fig. 21 Relationship between dissociation constants of sulphonamides and bacteriostatic activity

cell wall is impermeable to the sulphonamide ion, and that optimum activity is only attained by those substances which yield adequate concentrations of both ionised and unionised forms under physiological conditions. Thus, the part curve AB (Fig. 21) represents compounds which are completely unionised; they are able to penetrate the bacterial cell wall, but are then insufficiently ionised to be fully active. The part curve DE on the other hand encompasses compounds which are too highly ionised for efficient penetration of the cell wall, and therefore are inactive. The curve BCD represents the optimum conditions for activity, giving both cell penetration and ionisation inside the cell.

Despite the attractiveness of these developments of the Bell and Roblin hypothesis, it still possesses a number of weaknesses, as indicated by Seydel (1968). The first, acknowledged by Bell and Roblin, is the high level of antibacterial activity of sulphaguanidine, although it is too weak an acid to be measured in aqueous solution. Others include the unexpectedly low activity of sulphanilylurea (p$K_{a_2}$ 5.6) and the activity, almost as great as *Sulphanilamide*, of $N^1N^1$-dimethylsulphanilamide, which has no acidic hydrogen, and hence no second acid dissociation constant (p$K_{a_2}$).

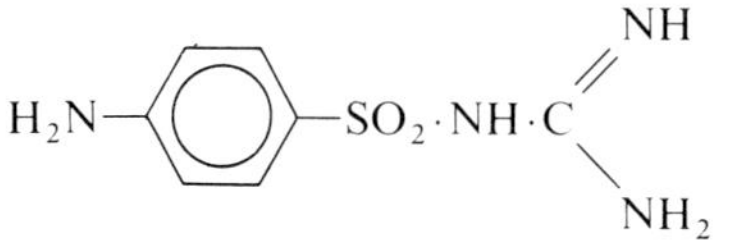

$H_2N-C_6H_4-SO_2\cdot NH\cdot C(=NH)NH_2$

*Sulphaguanidine*

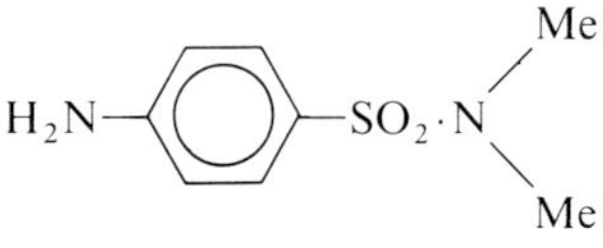

$H_2N-C_6H_4-SO_2\cdot N(Me)_2$

$N^1N^1$-Dimethylsulphanilamide

There is now a considerable body of evidence to show that the $N^4$-amino group of sulphonamides plays a prominent rôle in their mechanism of action. This stems from the suggestion by Tschesche (1947) that sulphonamides compete with PABA by forming a Schiff base with pteridinealdehyde. The importance of this concept was supported by experiments which showed that folate precursor pteridines can inhibit sulphonamide action (Tschesche, Korte and Korte, 1950) and the views of other research groups who had obtained similar results (O'Meara, McNally and Nelson, 1947; Forrest and Walker, 1948, 1949).

It is now accepted without question that the effectiveness of sulphonamides in the control of bacterial infections in human medicine lies in one major difference between bacterial and human metabolism of tetrahydrofolic acid. In mammals, it is not synthesised, but absorbed intact from the diet. In contrast, most bacteria and many pathogenic ones in particular are unable to absorb folic acid, and require to synthesise it *de novo*. The importance of tetrahydrofolic acid for the growth of micro-organisms lies in the vital rôle which it plays in one-carbon transfer processes. These form an important pathway to the biosynthesis of purine and pyrimidine nucleotides (Chapter 23) which in turn exert a controlling influence on DNA synthesis.

It is also established that the primary site of sulphonamide action in micro-organisms is the utilisation of *p*-aminobenzoic acid in tetrahydrofolic acid synthesis (Fig. 22). This has been demonstrated in a number of ways. Thus, Miller

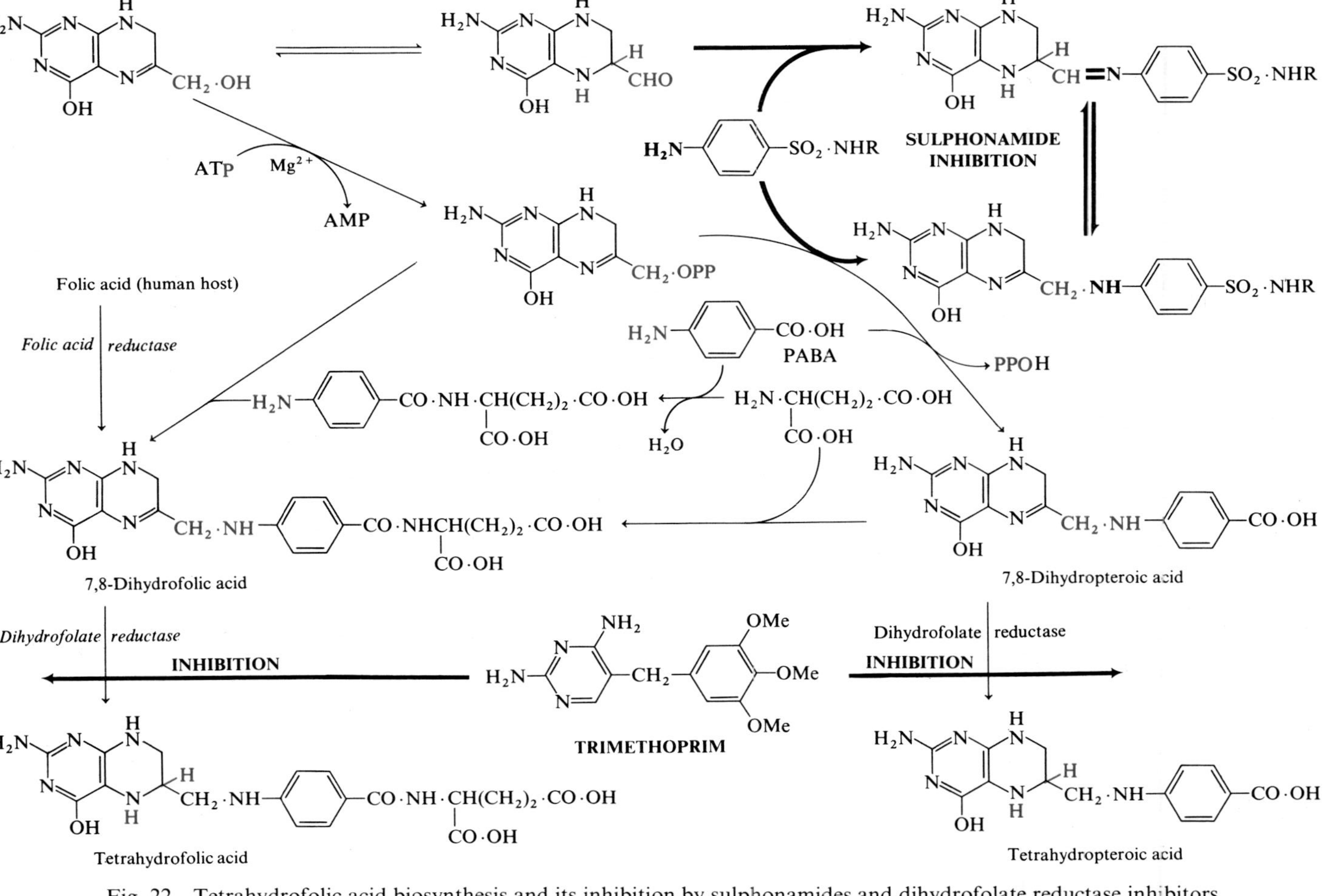


Fig. 22 Tetrahydrofolic acid biosynthesis and its inhibition by sulphonamides and dihydrofolate reductase inhibitors

(1944) has shown that bacteria grown in sub-inhibitory concentrations of sulphonamides synthesise only reduced amounts of folic acid. Similarly, *Lactobacillus casei* in the presence of folic acid is insensitive to sulphonamide inhibition (Lampen and Jones, 1946).

The biosynthesis of tetrahydrofolic acid (Fig. 22) in *E. coli* has been shown to proceed from 2-amino-4-hydroxy-6-hydroxymethyl-7,8-dihydropteridine, which is phosphorylated, possibly stepwise, by ATP to form the 6-diphosphate (Jaenicke and Chan, 1960; Weisman and Brown, 1964) before combining with *p*-aminobenzoylglutamic acid or *p*-aminobenzoic acid (PABA) to form dihydrofolic acid (DHFA) or dihydropteroic acid, respectively. The tautomeric 2-amino-4-hydroxy-6-formyltetrahydrofolic acid proved to be an equally effective precursor (Brown, Weisman and Molnar, 1961). The final step in the biosynthesis of tetrahydrofolic acid (THFA) is the reduction of dihydrofolic acid by dihydrofolate reductase. Studies on sulphonamide inhibition of tetrahydrofolic acid biosynthesis on cell-free extracts of *E. coli* have shown that inhibition occurs at the stage of dihydropteroic acid synthesis from 2-amino-4-hydroxy-6-hydroxymethyldihydropteridine and PABA (Brown, 1962).

If, as seems likely, inhibition results from nucleophilic attack of the sulphonamide $N^4$-amino group on either the pyrimidine-6-formyl group to form a Schiff's base or on the pyrimidine 6-pyrophosphoryloxymethyl group, this would be sensitive to the extent of dissociation of the amino group in the micro-environment of the biosynthetic reactions involved. In these circumstances, small variations in the dissociation constant of the amino group ($pK_{a_1}$) could well account for significant differences in bacterial effect. Since, also, the stronger the second acid dissociation, the weaker the first acid dissociation constant and hence the greater the availability of undissociated $N^4$-amino groups for attack, the correlations between $pK_{a_2}$ and activity may be capable of explanation in these terms.

It has become evident that the final step in the biosynthesis of tetrahydrofolic acid is also capable of inhibition by drugs which act as dihydrofolate reductase inhibitors. Substantial numbers of 2,4-diaminopyrimidines have been shown to have this capability, though they are often selective if not specific for different species. In particular, the antimalarial, *Pyrimethamine*, and the antibacterial, *Trimethoprim*, have found acceptance on account of this property. Because both sulphonamides and *Trimethoprim* act at different stages in the biosynthesis of tetrahydrofolic acid, *Trimethoprim* functions as a sulphonamide potentiator (Burchall and Hitchings, 1965; Bushby and Hitchings, 1968). It not only enhances the action of the sulphonamides, but also extends the spectrum of their activity to organisms such as *Proteus* against which activity is otherwise marginal.

### Classification of Antibacterial Sulphonamides

Antibacterial sulphonamides are broadly classified as long or short-acting (Struller, 1968) depending upon their dissociation, protein-binding, metabolism

| Sulphonamide $H_2N-C_6H_4-SO_2 \cdot NH \cdot R$ 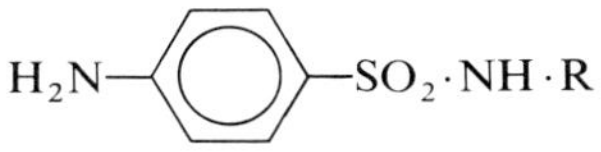 | Structure R | p$K_a$ | Protein binding % | Lipid solubility % | Half-life | Classification |
|---|---|---|---|---|---|---|
| Sulphafurazole | 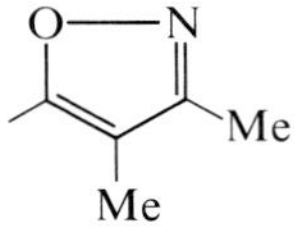 O—N, Me, Me | 4.9 | 86 | 5 | 6 | Short-acting; treatment of urinary tract infections |
| Sulphamethizole | 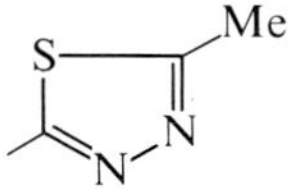 S, N, N, Me | 5.1 | 80 | 1 | 2.5 | |
| Sulphadiazine (Sulfamethazine) | 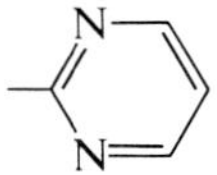 N, N | 6.4 | 45 | 27 | 17 | Short-acting; treatment of systemic infections |
| Sulphathiazole | 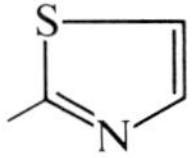 S, N | 7.1 | 77 | 15 | 4 | |
| Sulphadimidine | 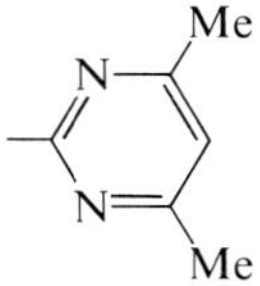 N, N, Me, Me | 7.4 | 80 | 83 | 7 | |
| Sulphadimethoxine | 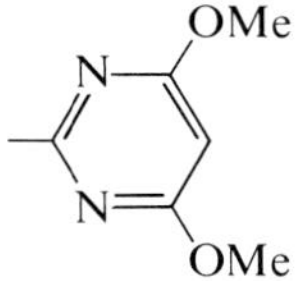 N, N, OMe, OMe | 6.1 | 99 | 79 | 40 | Long-acting; treatment of systemic infections |
| Sulphamethoxydiazine (Sulfameter) | 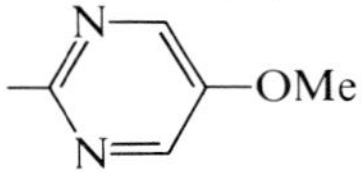 N, N, OMe | 7.0 | 87 | 64 | 37 | |
| Sulphapyridazine | 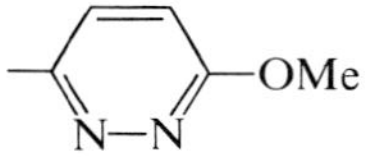 N—N, OMe | 7.2 | 90 | 70 | 37 | |

and hence excretion rates. Strongly acidic compounds, such as *Sulphafurazole* (Sulfisoxazole; p$K_a$ 4.9) and *Sulphamethizole* (Sulfamethizole; p$K_a$ 5.5), which are almost completely ionised at physiological pH, are excreted in the urine relatively rapidly, and hence find special favour in the treatment of urinary tract infections. Most other sulphonamides used in the treatment of general systemic infections have p$K_a$ between 6.1 and 7.4. Those which are less strongly bound to serum protein (<80% bound) and which also have limited lipid solubilities (Table 46) are rapidly excreted and only partially metabolised in man mainly to the $N^4$-acetylamino compound. Thus, 63% of the administered dose of *Sulphathiazole* appears in the urine unchanged, and only 29% is metabolised to $N^4$-acetylsulphathiazole. Like *Sulphadimidine* and, to a lesser extent, *Sulphadiazine*, it has only a relatively short duration of action.

Long-acting compounds such as *Sulphadimethoxine* (Sulfadimethoxine) have greatly extended half-lives, due primarily to protein-binding, but also to increased affinity for lipid tissue. These properties delay excretion and metabolism. In man, urinary excretion accounts for only 25% of the dose in 24 hr; of this 25%, only 2% is unchanged sulphonamide; 5% appears as the $N^4$-acetyl and 16% as $N^1$-glucuronide (Bridges, Kibby and Williams, 1965). It is worthy of

$H_2N-C_6H_4-SO_2\cdot N(Na)\cdot CO\cdot CH_3$

*Sulphacetamide Sodium*

$\downarrow H_2O$

$H_2N-C_6H_4-SO_2\cdot NH_2 + CH_3\cdot CO\cdot ONa$

$\cdot O-O\cdot \downarrow H_2O$

$H_2N\cdot SO_2-C_6H_4-N=N-C_6H_4-SO_2\cdot NH_2$

Azobenzene-4-4′-disulphonamide

+

$H_2N\cdot SO_2-C_6H_4-\overset{+}{N}(O^-)=N-C_6H_4-SO_2\cdot NH_2$

Azoxybenzene-4,4′-disulphonamide

comment that the stronger the acid, the weaker will be the $N^4$ base, discouraging acetylation in accord with the observed extent of metabolism.

The $N^4$-acylated sulphonamides, *Succinylsulphathiazole* and *Phthalylsulphathiazole*, have very low water solubilities; they are also ionised in the alkaline conditions of the human gut and hence are poorly absorbed. They are subject to hydrolysis by digestive and bacterial peptidases in the gut with release of the parent sulphonamide, and are therefore useful in the treatment of intestinal infections. *Sulphaguanidine* (Sulfaguanidine), too, has a very low water solubility; as a guanidine derivative, it is also an exceptionally strong base, a property which inhibits absorption in the stomach and upper regions of the intestinal tract where the pH is such that the base is fully ionised.

The $N^1$-acetylsulphonamide, *Sulphacetamide* (Sulfacetamide), in contrast, is much more readily water-soluble (1 in 150). It is used as its highly water-soluble (1 in 1.5) salt, *Sulphacetamide Sodium* (Sulfacetamide Sodium), mainly for the treatment of eye infections. *Sulphacetamide*, like most other sulphonamides, is sensitive to light-catalysed oxidation, and should be stored in tightly-sealed containers and in the absence of light. Antioxidants (*Sodium Metabisulphite*) are usually employed in the preparation of *Sulphacetamide Eye Drops* (Sulfacetamide Eye Drops), but in the absence of such precautions, hydrolysis and oxidation has been shown to occur (Clarke, 1965; see scheme page 424).

## Hypoglycaemic Sulphonamides

Certain types of diabetes, where some insulin-producing ($\beta$-cell) tissue still remains functional, may benefit by treatment with orally-administered hypoglycaemic sulphonamides in place of insulin. Important compounds of this type include the sulphonylureas *Carbutamide*, *Tolbutamide* and *Chlorpropamide*. They appear to act primarily by stimulating insulin secretion by $\beta$-cells of the pancreas. *Carbutamide*, additionally, has been shown to inhibit liver insulinase, which destroys insulin (Mirsky, Perisutti and Diengott, 1956). *Tolbutamide* is metabolised in man by oxidation of the ring methyl substituent to the alcohol (30%) and carboxylic acid (60%) (Thomas and Ikeda, 1966). In consequence, its action is relatively short-lived. *Chlorpropamide* is excreted largely unchanged and, therefore, has a much longer duration of action.

$$H_2N-C_6H_4-SO_2\cdot NH\cdot CO\cdot NH(CH_2)_3\cdot CH_3$$

*Carbutamide*

$$Cl-C_6H_4-SO_2\cdot NH\cdot CO\cdot NH(CH_2)_3\cdot CH_3$$

*Chlorpropamide*

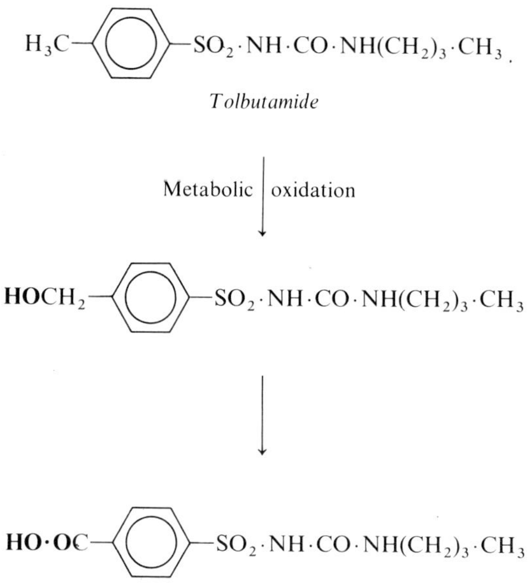

$H_3C-C_6H_4-SO_2\cdot NH\cdot CO\cdot NH(CH_2)_3\cdot CH_3$.

*Tolbutamide*

Metabolic oxidation ↓

$HOCH_2-C_6H_4-SO_2\cdot NH\cdot CO\cdot NH(CH_2)_3\cdot CH_3$

↓

$HO\cdot OC-C_6H_4-SO_2\cdot NH\cdot CO\cdot NH(CH_2)_3\cdot CH_3$

## Carbonic Anhydrase Inhibitors

*Sulphanilamide* therapy was found to lead to the excretion of large volumes of alkaline urine containing bicarbonate. Similar effects shown by other sulphonamides with unsubstituted sulphonamide groups ($-SO_2NH_2$) led to the development of the sulphonamide diuretics, of which *Acetazolamide* is a typical example. These compounds owe their diuretic action to inhibition of the enzyme carbonic anhydrase, which controls the equilibrium between carbon dioxide and carbonic acid in the proximal tubules of the kidney.

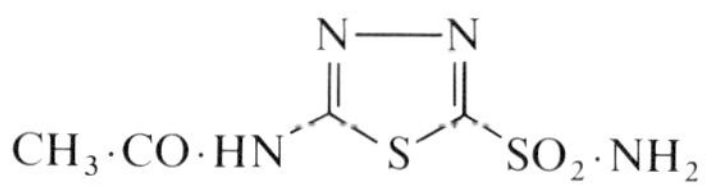

$CH_3\cdot CO\cdot HN-C_2N_2S-SO_2\cdot NH_2$ (1,3,4-thiadiazole ring: N—N, S)

*Acetazolamide*

$$H_2O + CO_2 \rightleftharpoons H_2CO_3 \rightleftharpoons H^+ + HCO_3^-$$

The electrochemical gradient between the tubular lumen (negative) and the interstitial fluid (positive) controls the flow of ions from the interstitial cells so that normally $HCO_3^-$ is totally reabsorbed, whilst $H^+$ is secreted into the glomerular filtrate (Fig. 23). The reabsorption of $HCO_3^-$, and also of $Cl^-$, is ionically balanced by the active reabsorption of $Na^+$ (against the electrochemical gradient), so that in effect $Na^+$ ions are largely exchanged for $H^+$ in urine. Water is also reabsorbed to maintain osmotic balance.

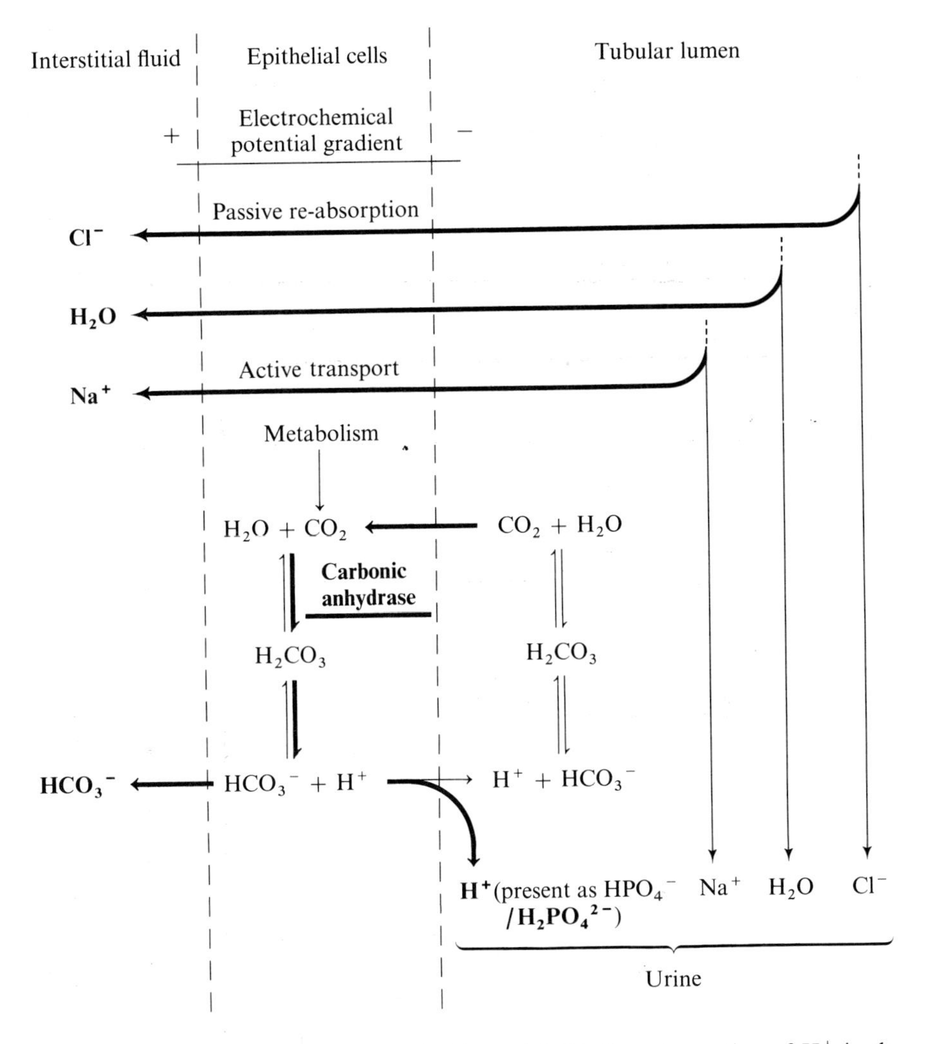


Fig. 23 Normal re-absorption of $Na^+$, $Cl^-$ and $HCO_3^-$, and secretion of $H^+$ in the promimal tubules of the kidney

Carbonic anhydrase inhibitors block the formation of $H_2CO_3$ in the epithelial cells. As a result, there is no reabsorption of $HCO_3^-$ or secretion of $H^+$ and diffusion of $CO_2$ into the tubular lumen increases (Fig. 24). Formation of $HCO_3^-$ in the lumen and absence of $H^-$ secretion promote excretion of $Na^+$ ions. $K^+$ ion excretion is also increased by the lack of $H^+$ for exchange in the distal tubule. Increased excretion of $Na^+$, $K^+$ and $HCO_3^-$ leads to increased retention of water in the tubule, and consequently a resulting diuresis.

Kinetic studies (Taylor, King and Burgen, 1970) have established that the sulphonamide inhibits the transfer of the enzyme co-ordinated metal ion which

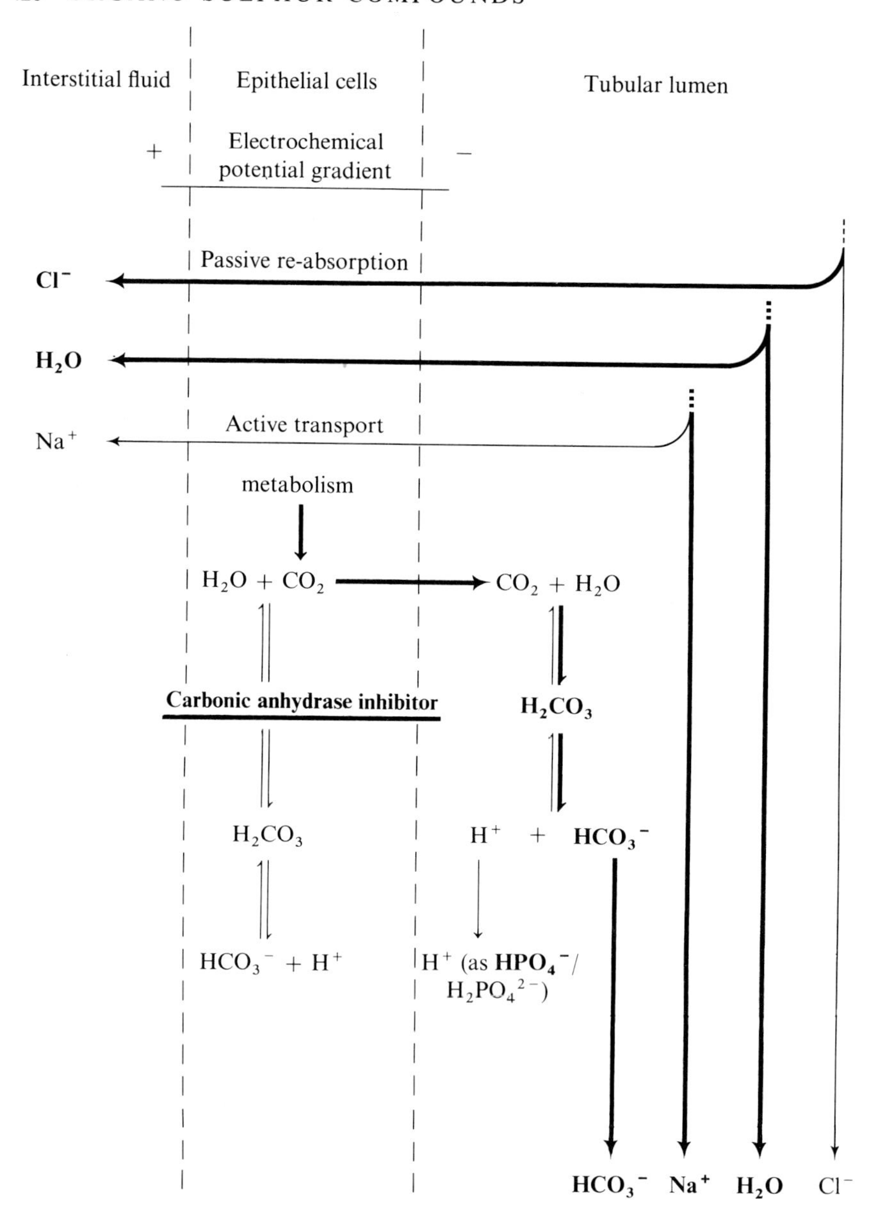


Fig. 24 Suppression of $H^+$ secretion and promotion of $Na^+$ and the $HCO_3^-$ secretion by carbonic anhydrase inhibitor in the proximal tubules of the kidney

catalyses the hydration of carbon dioxide. This process may be represented as follows.

Enz—$\ddot{N}H_2$ H H O $M^+$ O=C=O ⟶ Enz—$\overset{+}{N}H_3$ $M^{+-}$O—C(OH)=O ⟶ Enz—$\ddot{N}H_2$ $M^+$ HO—C(OH)=O

The sulphonamide—$SO_2NH_2$ is similarly able to supply a proton for the displacement of the metal ion. This is probably a direct metal proton displacement as there is no evidence for the existence of sulphonamide enolate ions.

Enz—$\ddot{N}H_2$ $M^+$ H H N $SO_2$ $NH_2$ ⟶ Enz—$\overset{+}{N}H_3$ $M^+$ H $^-N$ $SO_2$ $NH_2$

## Sultams

Sultams are cyclic sulphonamides, and as such are structural analogues of lactams (Chapter 18). Typical sultams, such as 1,8-naphthosultam, are weakly

$O_2S$—NH ⟶ (Cold aqueous **NaOH**) $O_2S$—NNa + $H_2O$

1,8-Naphthosultam

*Hot aqueous* **NaOH** ⟶ $SO_2 \cdot ONa$ $NH_2$

1-Naphthylamine-8-sodium sulphonate

acidic and soluble in cold aqueous sodium hydroxide to form water-soluble sodium salts; hot aqueous sodium hydroxide disrupts the sultam ring to form the aminosulphonate.

Alkali-metal derivatives of sultams react with alkyl and acyl halides to form *N*-alkyl and *N*-acylsultams.

### Thiazide Diuretics

The thiazide diuretics are alkali-soluble, but owe this solubility to the presence of both sultam and sulphonamido substituents. The anti-convulsant, *Sulthiame*, in which the sultam group is *N*-substituted and, therefore, incapable of forming a sodium salt, is also alkali-soluble by virtue of the $—SO_2NH_2$ group which is also present.

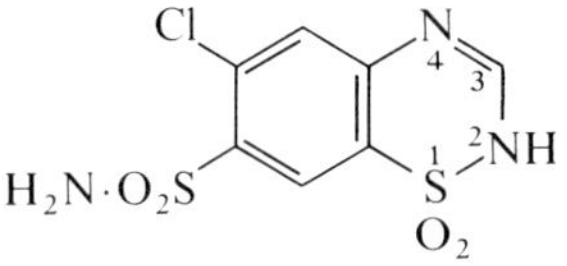

*Chlorothiazide*

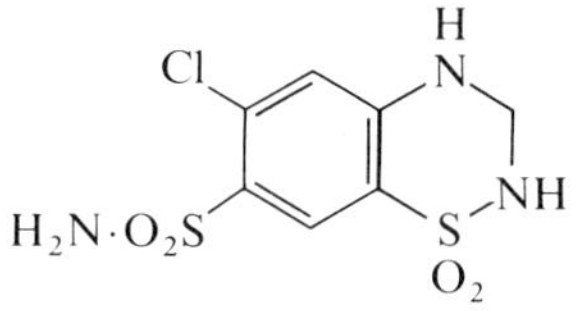

*Hydrochlorothiazide*

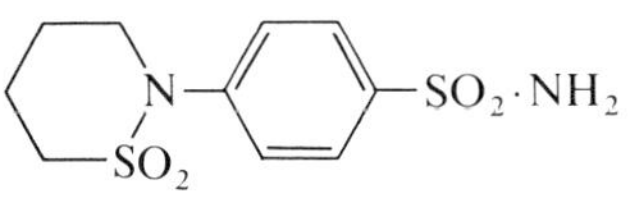

*Sulthiame*

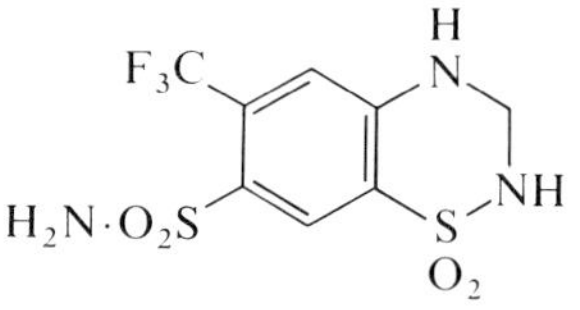

*Hydroflumethoazide*

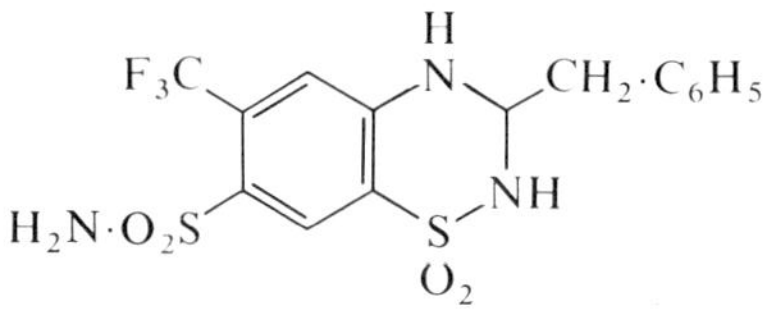

*Bendrofluazide*

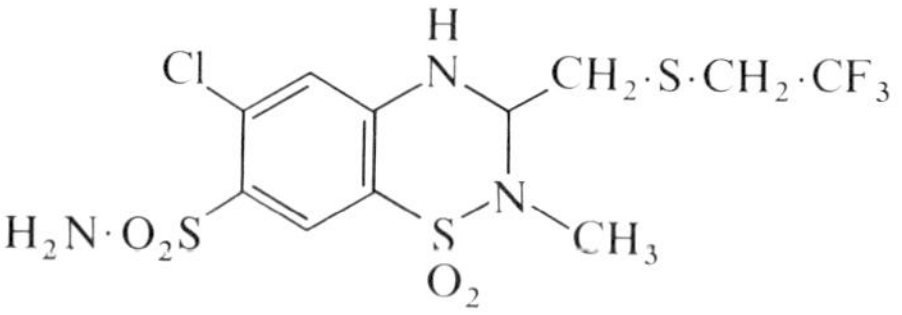

*Polythiazide*

Although the thiazide diuretics have some inhibitory effect on carbonic anhydrase, they cause diuresis mainly by a direct action on the tubules, thereby decreasing re-absorption of $Na^+$, $K^+$, $Cl^-$ and, to a more limited extent, $HCO_3^-$ ions. *Hydrochlorothiazide*, in which the 3,4-double bond of *Chlorothiazide* is saturated, is some 10–20 times more effective as a diuretic than the latter, although it is less effective as an inhibitor of carbonic anhydrase (De Stevens, Werner, Halamandani and Ricca, 1958). Substitution in the 3-position as in *Bendrofluazide* (Bendroflumethiazide) also increases diuretic potency.

*Chlorothiazide* and *Hydrochlorothiazide* are stable metabolically and are excreted unchanged in man. Thiazides, appropriately substituted in the 3-position, such as *Polythiazide* and *Bendrofluazide*, although also largely excreted unchanged, are metabolised by hydrolytic ring cleavage (Pinson, Schreiber, Wiseman, Chaini and Baumgartner, 1962).

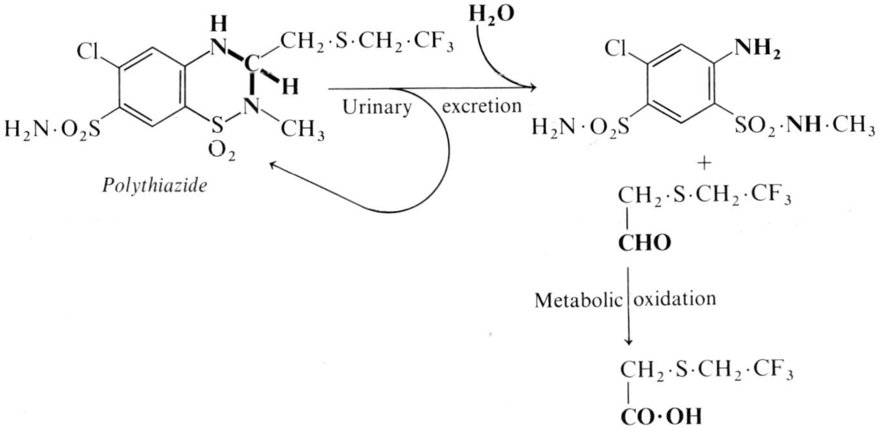

Cl, H, N, C, $CH_2\cdot S\cdot CH_2\cdot CF_3$, H, N, $CH_3$, $H_2N\cdot O_2S$, S, $O_2$

*Polythiazide*

$H_2O$

Urinary excretion

Cl, $NH_2$, $H_2N\cdot O_2S$, $SO_2\cdot NH\cdot CH_3$

+

$CH_2\cdot S\cdot CH_2\cdot CF_3$
|
CHO

Metabolic oxidation

$CH_2\cdot S\cdot CH_2\cdot CF_3$
|
$CO\cdot OH$

## Saccharin

*Saccharin*, which is formed by cyclodehydration of *o*-sulphamoylbenzoic acid, can be regarded as an unsubstituted acyl-sultam. The combination of the powerful electron-attracting aryl carbonyl and sulphonyl groups makes the NH group of saccharin strongly acidic. Saccharin is, therefore, readily soluble in solutions of quite weak bases, such as dilute ammonia or aqueous sodium bicarbonate. Saccharin is intensely sweet, being approximately 550 times as sweet as sucrose. It is manufactured from toluene-*o*-sulphonamide by oxidation. *Saccharin* is poorly soluble in water (1 in 290), but *Saccharin Sodium* is soluble 1 in 1.5 in water.

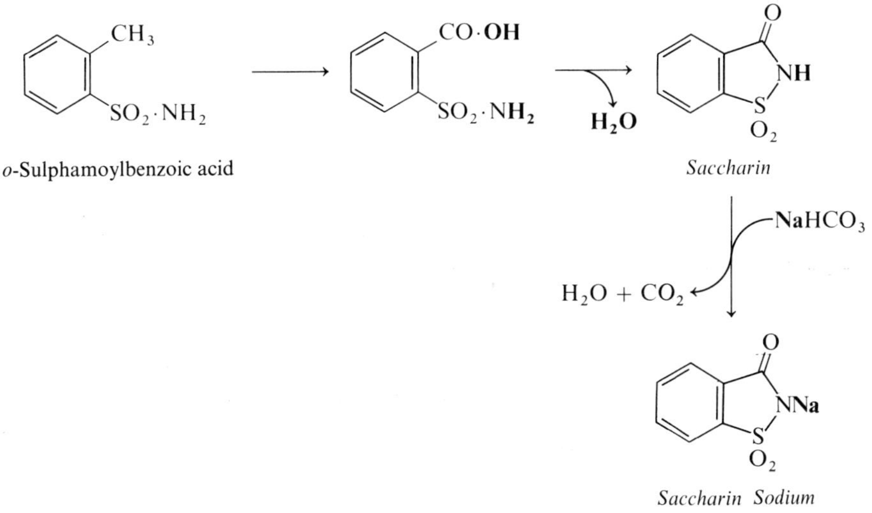

$CH_3$, $SO_2\cdot NH_2$ → $CO\cdot OH$, $SO_2\cdot NH_2$ → ($H_2O$) O, NH, S, $O_2$

*o*-Sulphamoylbenzoic acid

*Saccharin*

$NaHCO_3$

$H_2O + CO_2$

O, NNa, S, $O_2$

*Saccharin Sodium*

# THIOACIDS, THIOESTERS AND THIOAMIDES

## Thioacids

Thioacids exist as tautomers of the two possible, **thiolic** and **thionic** forms.

$$R\cdot C(=O)SH \rightleftharpoons R\cdot C(=S)OH$$

For the most part, they behave as thiolic acids, but the existence of tautomerism is evident from the formation of both *S*-alkyl and *O*-alkyl esters.

### General Properties

Thioformic acid is unstable, but thioacetic acid and its homologues exist as unpleasant-smelling liquids. Thioacetic acid (p$K_a$ 3.33) is a somewhat stronger acid than acetic acid (p$K_a$ 4.73). This is in agreement with the predominance of the thiol tautomer, since electron-withdrawal will be promoted to a greater extent by its carbonyl group than in the carboxylic acid group (Chapter 11).

Thioacids also form water-soluble alkali metal thiol salts, which are readily oxidised by iodine and other oxidants to form di-acyl disulphides.

$$CH_3\cdot CO\cdot SNa + I_2 \longrightarrow CH_3\cdot CO\cdot S\cdot S\cdot CO\cdot CH_3 + 2NaI$$

The thioacids add readily to olefines in a light-catalysed anti-Markownikoff radical reaction, to yield thiol-esters.

$$CH_3\cdot CO\cdot S{-}H \xrightarrow{-H\cdot} CH_3\cdot CO\cdot \dot{S} + Me_2\cdot C{=}CH\cdot Et \longrightarrow Me_2\cdot \dot{C}{-}CH(S\cdot CO\cdot CH_3)\cdot Et \xrightarrow{H\cdot} Me_2\cdot CH\cdot CH(S\cdot CO\cdot CH_3)\cdot Et$$

Thioacids and their esters are powerful acylating agents; thioacetic acid acetylates amines at room temperature.

$$CH_3\cdot C(=O)SH + R\cdot NH_2 \xrightarrow{-H^+} CH_3\cdot C(SH)(O^-)\cdot NH\cdot R \xrightarrow{-HS^-} CH_3\cdot C(=O)NH\cdot R$$

Thioacids are also extremely reactive towards aldehydes, to give relatively stable acyl hemi-thioacetals.

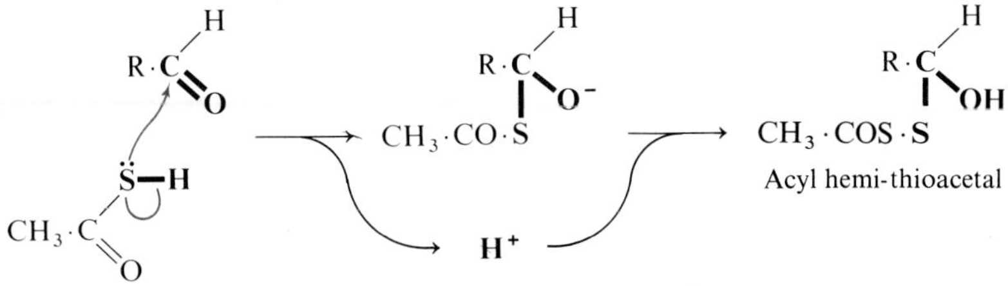

$$R{\cdot}CHO + CH_3{\cdot}COSH \longrightarrow R{\cdot}CH(O^-){\cdot}S{\cdot}CO{\cdot}CH_3 + H^+ \longrightarrow R{\cdot}CH(OH){\cdot}S{\cdot}COCH_3$$

Acyl hemi-thioacetal

## Thioesters

Thioacids form both *S*-alkyl and *O*-alkyl esters.

$$R{\cdot}C(=O){\cdot}SR' \qquad R{\cdot}C(=S){\cdot}OR'$$

### General Properties of Thiol Esters

Thiol esters resemble their oxygen analogues in many respects. They are, however, less stable, and hence show greater reactivity, due in part to the reduced mobility of unpaired electrons on sulphur. Resonance interactions are, therefore, less favourable than in the corresponding oxygen esters.

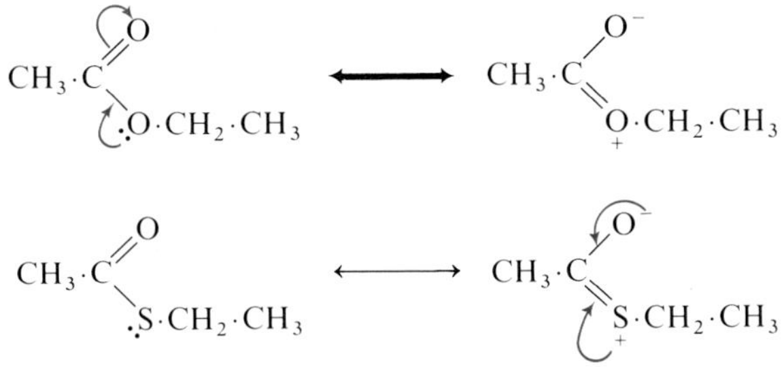

$$CH_3{\cdot}C(=O){\cdot}\ddot{O}{\cdot}CH_2{\cdot}CH_3 \longleftrightarrow CH_3{\cdot}C(O^-)={\overset{+}{O}}{\cdot}CH_2{\cdot}CH_3$$

$$CH_3{\cdot}C(=O){\cdot}\ddot{S}{\cdot}CH_2{\cdot}CH_3 \longleftrightarrow CH_3{\cdot}C(O^-)={\overset{+}{S}}{\cdot}CH_2{\cdot}CH_3$$

The inherent instability of thiol esters is exemplified by the antileprotic drug, *Ditophal*, which is the isophthalate di-ester of ethanethiol. Its action is due to the release of ethanethiol by metabolic hydrolysis (Ellard, Garrod, Scales and Snow, 1965). Incipient hydrolysis occurs in the ester, and since it is usually administered by inunction for rapid absorption through the skin, it is applied as a perfumed preparation (Etisul) to disguise the distinctive odour of ethanethiol.

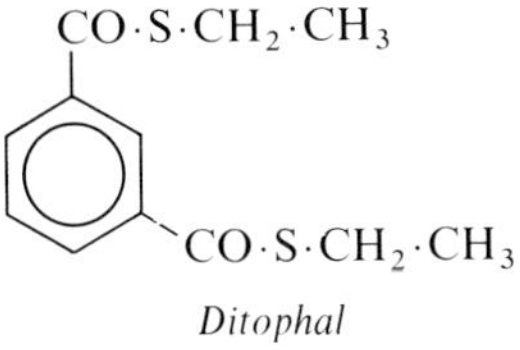

$$C_6H_4(CO{\cdot}S{\cdot}CH_2{\cdot}CH_3)_2$$

*Ditophal*

Thiol esters are hydrolysed in alkaline solution at rates similar to those of *O*-esters, but rather less readily in acidic solution (sulphur is less electronegative than oxygen and protonated less readily). They are, however, much more effective in the acylation of amines, due to the fact that thioalkoxides ($RS^-$) are weaker bases than alkoxides ($RO^-$) and hence make better leaving groups.

$$CH_3\cdot C(=O)SEt + R\cdot \ddot{N}H_2 \longrightarrow CH_3\cdot C(SEt)(O^-)\cdot \overset{+}{N}H_2R \longrightarrow CH_3\cdot CO\cdot \overset{+}{N}H_2R + Et\cdot S^- \longrightarrow CH_3\cdot CO\cdot NHR + EtSH$$

*S*-Acyl coenzyme A is a powerful biological acetylating agent and is responsible for the transfer of acetyl in a wide variety of systems, including the physiological biosynthesis of acetylcholine by choline acetylase, the metabolism of aromatic amines, such as *Sulphanilamide*, and the acetylation of *S*-alkyl- and *S*-aryl-cysteines in mercapturic acid conjugation in mammals.

$$\underset{\text{Choline}}{[HO\cdot CH_2\cdot CH_2\cdot \overset{+}{N}Me_3]X^-} \xrightarrow[\;CH_3\cdot CO\cdot S\cdot \overline{CoA}\;\to\; HS\cdot \overline{CoA}\;]{\text{Choline acetylase}} \underset{\text{Acetylcholine}}{[CH_3\cdot CO\cdot O\cdot CH_2\cdot CH_2\overset{+}{N}Me_3]X^-}$$

Sulphanilamide ($NH_2\cdot C_6H_4\cdot SO_2\cdot NH_2$) + $CH_3CO\cdot S\cdot \overline{CoA}$ → $N^4$-Acetylsulphanilamide ($CH_3\cdot CO\cdot NH\cdot C_6H_4\cdot SO_2\cdot NH_2$) + $HS\cdot \overline{CoA}$

Sulphanilamide + $CH_3\cdot CO\cdot S\cdot \overline{CoA}$ → $N^1$-Acetylsulphanilamide ($NH_2\cdot C_6H_4\cdot SO_2\cdot NH\cdot CO\cdot CH_3$) + $HS\cdot \overline{CoA}$

$N^4$-Acetylsulphanilamide / $N^1$-Acetylsulphanilamide → $N^1N^4$-Diacetylsulphanilamide ($CH_3CO\cdot NH\cdot C_6H_4\cdot SO_2\cdot NH\cdot COCH_3$)

$$\text{2,4-}(O_2N)_2C_6H_3\cdot S\cdot CH_2\cdot CH(NH_2)\cdot CO\cdot OH \xrightarrow[CH_3\cdot CO\cdot S\cdot \overline{CoA} \;\rightarrow\; HS\cdot \overline{CoA}]{} \text{2,4-}(O_2N)_2C_6H_3\cdot S\cdot CH_2\cdot CH(NH\cdot CO\cdot CH_3)\cdot CO\cdot OH$$

*S*-2,4-Dinitrophenylmercapturic acid

Thiolesters show greater acidity in their α C—H bonds, than in those of carboxylic acid oxy-esters. Accordingly, they readily form carbanions and show greater reactivity in Claisen condensations (Cronyn, Chang and Wall, 1955). Thus, *S*-acetyl coenzyme A plays a key rôle in the biogenesis and metabolism of fatty acids. It undergoes a Claisen-type condensation with the thiol ester carbanion malonyl-S·$\overline{CoA}$ to give a β-ketoacid, which decarboxylates to acetoacetylcoenzyme A.

$$CH_3\cdot CO\cdot S\cdot \overline{CoA} + {}^{-}CH(CO\cdot OH)\cdot CO\cdot S\cdot \overline{CoA} \rightleftharpoons CH_3\cdot CO\cdot CH(CO\cdot OH)\cdot CO\cdot S\cdot \overline{CoA} + {}^{-}S\cdot \overline{CoA}$$

## Thioamides

Thioamides are derived from the reputed thionic-form of the thioacids which according to Crouch (1952) is not seen in the free acid. Their properties, however, resemble those of *O*-amides. Thus, *Ethionamide*, the antitubercular drug, is readily hydrolysed by both alkali and acid.

$$\text{Ethionamide (2-Et-pyridine-4-}CS\cdot NH_2) \xrightarrow{NaOH} \text{2-Et-pyridine-4-}C(=O)S^{-}Na^{+} \rightarrow \text{2-Et-pyridine-4-}C(=O)O^{-}Na^{+}$$

$$\text{Ethionamide} \xrightarrow{NaOH} \text{2-Et-pyridine-4-}C(=O)NH_2 \rightarrow \text{2-Et-pyridine-4-}C(=O)O^{-}Na^{+}$$

*Ethionamide*

Alkaline hydrolysis gives a mixture of carboxamide, carboxylic acid, and according to Seydal (1966), the thionic-acid (—CS·OH), though no evidence is presented for the structure of the latter. Acid hydrolysis gives only the carboxylic acid.

Thioamides, unlike their *O*-analogues, react with ammonia and primary amines to form amidines. The equilibrium between amide and amidine hydrosulphide, initially established, is disturbed in favour of the amidine by addition of mercuric chloride.

$$R{\cdot}CS{\cdot}NH_2 + R'{\cdot}NH_2 \xrightarrow[HgCl_2 \ \rightarrow \ HgS + 2HCl]{} R{\cdot}C(=NR')NH_2$$

Pyrolysis of thioamides resembles that of amides and yields the corresponding nitrile.

$$R{\cdot}CS{\cdot}NH_2 \xrightarrow[-H_2S]{} R{\cdot}CN$$

Oxidation of thioamides with peroxide yields the *S*-oxide (Walter, Curts and Pawelzik, 1961).

$$\text{2-ethylpyridine-4-}CS{\cdot}NH_2 \xrightarrow[(H_2O_2 \text{ or metabolic})]{\text{Oxidation}} \text{2-ethylpyridine-4-}C(=\overset{+}{S}{-}O^-)NH_2$$

In this connection, it is interesting that *Ethionamide* is metabolised in humans to the *S*-oxide (Kane, 1962), which also appears to be an effective tuberculostatic *in vivo*. It is also established that *Ethionamide* inhibits catalase activity *in vitro* and that this enzyme is present in *Myco. tuberculosis* (Eilhauer and Kraemer, 1966). *N*-Ethylethionamide is inactive against tuberculosis and does not inhibit catalase.

# 15 Nitro Compounds

## STRUCTURE AND PROPERTIES

Nitro compounds are characterised by the $—NO_2$ group, attached to aliphatic (nitroparaffins) or aromatic hydrocarbons by a carbon—nitrogen single bond as in nitromethane ($H_3C—NO_2$). This distinguishes them from the corresponding nitrites, which are isomeric and have the structure R—O·NO. The nitro group is a resonance hybrid of two equivalent structures.

$$H_3C-\overset{+}{N}(-O^-)(=O) \longleftrightarrow H_3C-\overset{+}{N}(=O)(-O^-)$$

These may be represented, alternatively, as

$$H_3C-N(\rightarrow O)(=O) \longleftrightarrow H_3C-N(=O)(\rightarrow O)$$

### Nomenclature

Both aliphatic and aromatic nitro compounds are usuallly named as substitution products of the parent hydrocarbon (Table 47).

### General Properties

Nitroparaffins are colourless, volatile liquids, almost insoluble in water. Most aromatic nitro compounds are coloured, usually yellow, orange or red. They may be either liquids or crystalline solids at ambient temperatures. Nitroparaffins, and nitrobenzene which is also a liquid, have good solvent powers for a wide range of organic compounds. Nitroparaffins are also particularly good solvents for rubber, resins and certain polymers, notably cellulose esters. Nitrobenzene is an excellent solvent for electrophilic substitution reactions of aromatic compounds, e.g. Friedel–Crafts acylation, due to the powerful deactivating effect of the nitro group and the consequent low reactivity of the nitrobenzene ring toward electrophilic reagents. All nitro compounds are dangerously toxic and care should be taken in handling them to avoid contact with the skin or inhalation of vapour.

Table 47. Nomenclature of Nitro Compounds

| Structure | Systematic Name | b.p. (°C) |
|---|---|---|
| $CH_3 \cdot NO_2$ | Nitromethane | 102 |
| $CH_3 \cdot CH_2 \cdot NO_2$ | Nitroethane | 114 |
| $CH_3 \cdot CH_2 \cdot CH_2 \cdot NO_2$ | 1 Nitropropanc | 132 |
| $(CH_3)_2 \cdot CH \cdot NO_2$ | 2-Nitropropane | 120 |
| $CH_3 \cdot CH_2 \cdot CH_2 \cdot CH_2 \cdot NO_2$ | 1-Nitrobutane | 153 |
| $(CH_3)_3C \cdot NO_2$ | 2-Methyl-2-nitropropane | 127 |
| $C_6H_5 \cdot NO_2$ | Nitrobenzene | 211 |
| *o*-$CH_3 \cdot C_6H_4 \cdot NO_2$ | *o*-Nitrotoluene | 222 |
| *m*-$CH_3 \cdot C_6H_4 \cdot NO_2$ | *m*-Nitrotoluene | 227 |
| *p*-$CH_3 \cdot C_6H_4 \cdot NO_2$ | *p*-Nitrotoluene | 238 |

## Electron-withdrawing Properties

The nitro group is strongly electron-withdrawing. This effect is evident from the much greater acidic strength of the *m*- and *p*-nitrobenzoic acids compared to that of benzoic acid.

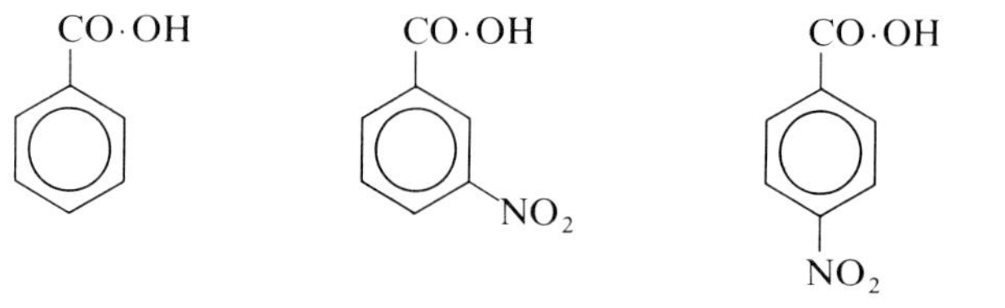

CO·OH | CO·OH, NO₂ | CO·OH, NO₂

| | Benzoic acid | *m*-Nitrobenzoic acid | *p*-Nitrobenzoic acid |
|---|---|---|---|
| $pK_a$ | 4.18 | 3.49 | 3.36 |

*o*-Nitrobenzoic acid is an even stronger acid than *m*- and *p*-nitrobenzoic acids due partly to steric effects of the *ortho* substituent and partly to the stabilisation of the resulting anion by charge interaction between the positively charged nitrogen and the negatively charged oxygen of the carboxylate ion.

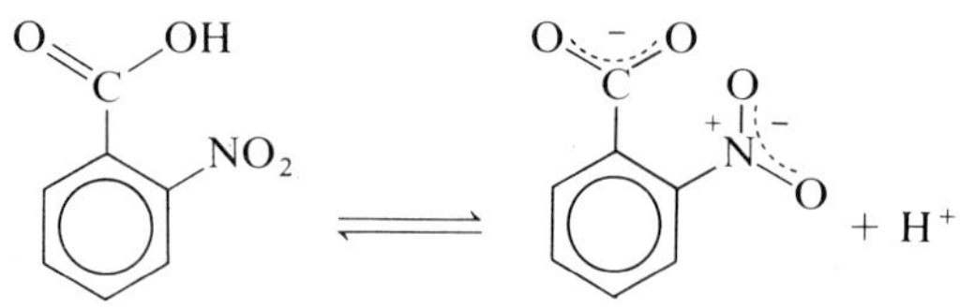

O, OH, C, $NO_2$ ⇌ O, −, O, C, O, +, N, −, O + $H^+$

Nitro substituents exert an even more powerful effect on the dissociation of aromatic amines. Here, the effect is base-weakening, so that their conjugate acids ($Ar\overset{+}{N}H_3$) are increased in strength.

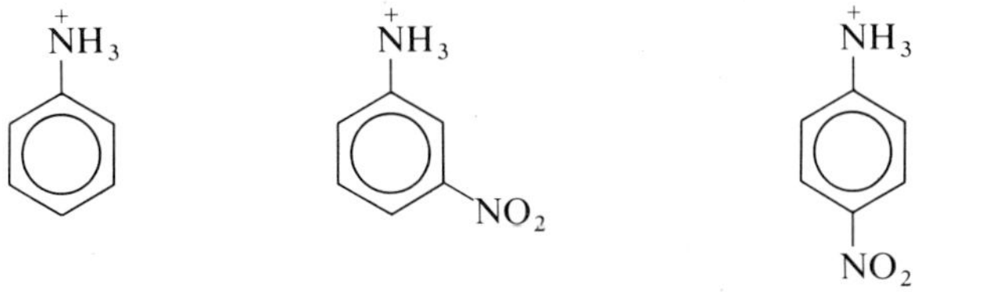

| | Anilinium ion | *m*-Nitroanilinium ion | *p*-Nitroanilinium ion |
|---|---|---|---|
| $pK_a$ | 4.66 | 1.6 | 2.04 |

### Charge-transfer Complexes

The powerful electron-withdrawing properties of the nitro group deactivate the ring of aromatic nitro compounds. Multiple nitro-substitution in such compounds as trinitrobenzene and picric acid promotes electron-withdrawal to such a degree that they are able to form relatively stable charge-transfer complexes by plane-to-plane alignment with other electron-releasing compounds. The trinitrobenzene–mesitylene complex is a good example of this type of interaction. The deepening of colour of polynitroaromatic compounds in the presence of ammonia or hydroxide ions is probably also due to complex formation of a similar sort.

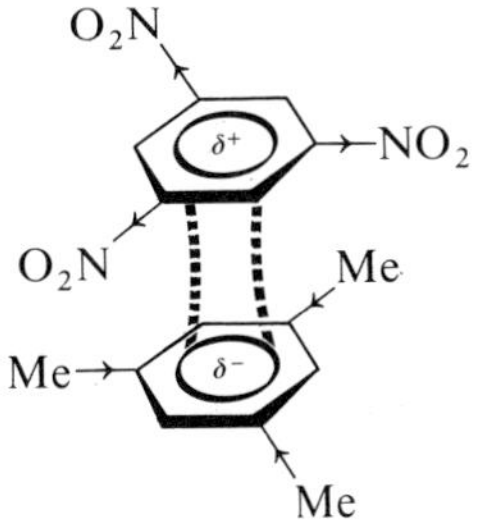

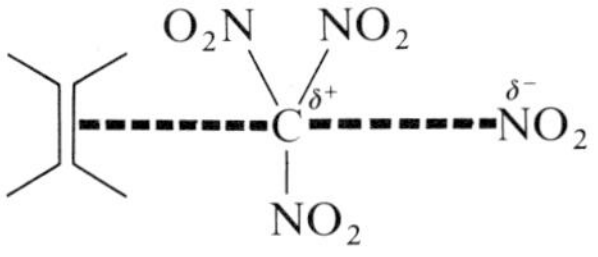

Charge-transfer complexes are also formed by tetranitromethane and ethylenic compounds (Heilbronner, 1953). The nature of these complexes is not known, but the ease with which one of the four nitro groups of tetranitromethane is displaced by base suggests a five-membered structure in which the central $C(NO_3)_3$ group is planar, analogous to the typical $S_N2$ transition state. The complexes, which are yellow or orange depending on the extent and nature of ethylenic substitution, provide a useful means of identifying unsaturated compounds.

### Acidic Properties of Primary and Secondary Nitroparaffins

The nitro group is strongly electron-attracting. Protons are easily lost from the α-carbon atoms of primary ($R \cdot CH_2 \cdot NO_2$) and secondary ($R_2 \cdot CH \cdot NO_2$) nitroalkanes, which are in tautomeric equilibrium with the corresponding *aci*-nitro compounds and hence are acidic in character.

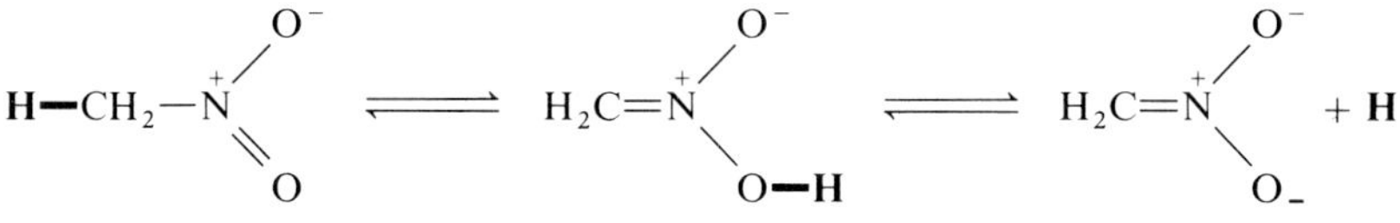

$$\mathrm{H{-}CH_2{-}\overset{+}{N}(=O)O^-} \rightleftharpoons \mathrm{H_2C{=}\overset{+}{N}(O^-)O{-}H} \rightleftharpoons \mathrm{H_2C{=}\overset{+}{N}(O^-)O^-} + \mathrm{H^+}$$

Tertiary nitroparaffins ($R_3C\cdot NO_2$) and aromatic nitro compounds ($Ar\cdot NO_2$) have no $\alpha$ C—H groups and, therefore, are neutral.

The equilibrium in primary and secondary nitroalkanes normally strongly favours the nitroparaffin rather than the *aci*-nitro compound, but they readily form water-soluble salts with strong bases, such as sodium hydroxide. The anion, so formed, is a resonance hybrid of two equivalent nitro carbanions and a third derived from the tautomeric *aci*-nitro compound or nitronic acid.

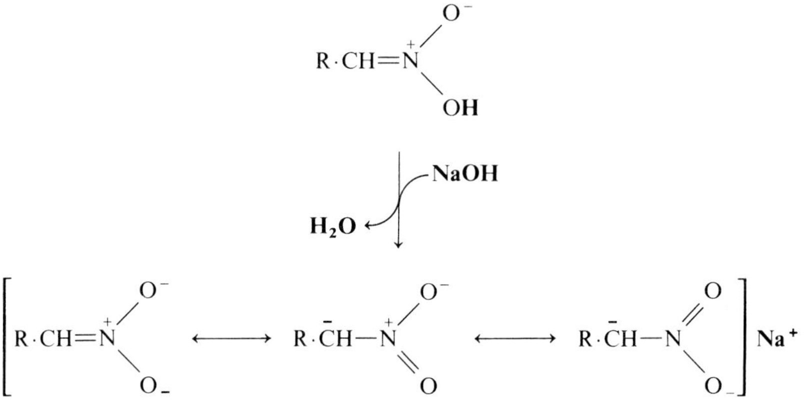

$$\mathrm{R\cdot CH{=}\overset{+}{N}(O^-)OH} \xrightarrow[\ H_2O\ \leftarrow\ ]{\ NaOH\ } \left[\mathrm{R\cdot CH{=}\overset{+}{N}(O^-)O^-} \longleftrightarrow \mathrm{R\cdot \bar{C}H{-}\overset{+}{N}(=O)O^-} \longleftrightarrow \mathrm{R\cdot \bar{C}H{-}N(=O)O^-}\right]\mathrm{Na^+}$$

Acidification of the salt solutions yields the *aci*-nitro compound which slowly reverts to the more stable nitro tautomer.

Careful acidification of the salt at low temperature releases the *aci*-form, which is usually water-soluble. The tautomeric equilibrium, however, favours the nitro tautomer of nitromethane, which slowly separates from solution as oily droplets. Some substituted-nitroparaffins have been isolated as stable *aci*-nitro compounds. All primary and secondary nitroparaffins, however, can be obtained as their *aci*-nitro esters, on treatment of either form with diazomethane.

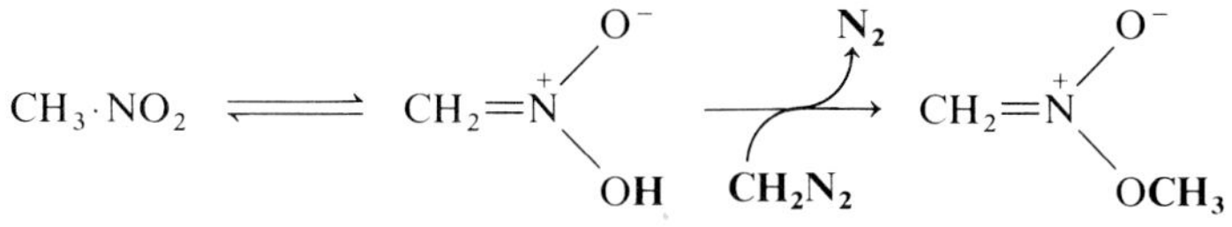

$$\mathrm{CH_3\cdot NO_2} \rightleftharpoons \mathrm{CH_2{=}\overset{+}{N}(O^-)OH} \xrightarrow[CH_2N_2]{\ \ N_2\ \uparrow\ \ } \mathrm{CH_2{=}\overset{+}{N}(O^-)OCH_3}$$

**Nef Reaction.** Decomposition of the sodium salts of ethyl and higher alkylnitronic acids with aqueous sulphuric acid gives good yields of carbonyl compounds (Johnson and Degering, 1943).

$$2CH_3\cdot CH{=}\overset{+}{N}(O^-)(ONa) + 2H_2SO_4 \longrightarrow 2CH_3\cdot CHO + 2NaHSO_4 + N_2O + H_2O$$

$$(CH_3)_2C{=}\overset{+}{N}(O^-)(ONa) + 2H_2SO_4 \longrightarrow (CH_3)_2\cdot CO + 2NaHSO_4 + N_2O + H_2O$$

**Reactions of Nitroparaffin and Nitrobenzyl Carbanions**

Formation of a carbanion by loss of a proton from primary and secondary nitroparaffins (p. 440) accounts for their reactivity in the presence of suitable proton acceptors, such as nitrous acid, halogens and carbonyl compounds.

*Ortho*- and *para*-nitrotoluenes also readily form benzylic anions in the presence of a suitable base. Anion formation is promoted by resonance stabilisation.

$$O_2N{-}C_6H_4{-}CH_3 \xrightarrow[H_2O]{HO^-} {}^-O(O{=})\overset{+}{N}{-}C_6H_4{-}\bar{C}H_2 \longleftrightarrow ({}^-O)_2\overset{+}{N}{=}C_6H_4{=}CH_2$$

**Nitrous acid.** The nitroparaffins are stronger acids than nitrous acid, and protonate the latter. This promotes attack by the carbanion so produced on the

$$R\cdot CH\cdot NO_2(H) + H{-}\ddot{O}{-}N{=}O \longrightarrow R\cdot\bar{C}H\cdot NO_2 + H_2\overset{+}{O}{-}N{=}O \xrightarrow{-H_2O} R(H)C(NO_2){-}N{=}O \xrightarrow{H^+,\ -H^+} \underset{\text{Nitrolic acid}}{R\cdot C(NO_2){=}N\cdot OH} \xrightarrow[H_2O]{KOH} \underset{\text{(red solution)}}{R\cdot C(NO_2){=}NO^-K^+}$$

$$R\cdot C(R')(H)\cdot NO_2 + H-\ddot{O}-N{=}O \longrightarrow R\cdot \bar{C}(R')\cdot NO_2 + H_2\overset{+}{O}-N{=}O \xrightarrow{-H_2O} RR'C(NO_2)N{=}O$$

Pseudonitrole
(blue solution in chloroform)

protonated nitrous acid with formation of either a nitrolic acid from a primary nitroparaffin or a pseudonitrole from a secondary nitroparaffin. Nitrolic acids form bright red alkali metal salts with alkali metal hydroxides. Pseudonitroles are insoluble in solutions of alkali metal hydroxides, but are readily soluble in chloroform to form deep blue solutions. The reaction with nitrous acid, therefore, forms a useful means of distinguishing between primary, secondary and tertiary nitroparaffins.

**Base-catalysed α-halogenation.** Primary and secondary nitroparaffins are readily α-halogenated in the presence of alkali.

$$R\cdot CH_2\cdot NO_2 \xrightarrow[-H_2O]{HO^-} R\cdot \bar{C}H\cdot NO_2 \xrightarrow[-Br^-]{Br-Br} R\cdot CH\mathbf{Br}\cdot NO_2$$

In contrast, free radical halogenation of nitroparaffins in the gas phase or in solution leads to halogen substitution both at the α- and other more distant carbon atoms.

**Addition to carbonyl compounds.** Primary and secondary nitroparaffins undergo base-catalysed aldol-type additions to aldehydes and ketones. Addition

$$H_2C{=}O + \bar{C}H_2\cdot NO_2 \xrightarrow{KHCO_3} O_2N\cdot CH_2\cdot \mathbf{CH_2\cdot O^-} \xrightarrow{H^+} O_2N\cdot CH_2\cdot \mathbf{CH_2OH} \dashrightarrow O_2N\cdot C\mathbf{(CH_2OH)_3}$$

$$C_6H_5\cdot \mathbf{CHO} + CH_3\cdot NO_2 \xrightarrow[-H_2O]{\text{Ethanolic KOH}} [C_6H_5\cdot \mathbf{CHOK}\cdot CH_2\cdot NO_2] \xrightarrow[-KX + H_2\mathbf{O}]{H^+X^-} C_6H_5\mathbf{CH{=}}CH\cdot NO_2$$

ω-Nitrostyrene

products formed from aromatic aldehydes often undergo spontaneous dehydration to give $\omega$-nitrostyrenes.

*o*- and *p*-Nitrotoluenes similarly undergo base-catalysed benzylic addition to carbonyl compounds, usually with spontaneous dehydration, so that a styrene derivative results.

$$O_2N\text{—}C_6H_4\text{—}CH_3 \xrightarrow[-H_2O]{HO^-,\ \text{Ethanolic KOH}} O_2N\text{—}C_6H_4\text{—}\bar{C}H_2$$

$$\xrightarrow{C_6H_5\cdot CHO} O_2N\text{—}C_6H_4\text{—}CH_2\cdot CH(OK)\text{—}C_6H_5$$

$$\xrightarrow[-H_2O]{H^+} O_2N\text{—}C_6H_4\text{—}CH{=}CH\text{—}C_6H_5$$

**Michael addition.** Acidic nitroparaffins also participate in base-catalysed Michael additions to suitably substituted unsaturated compounds.

$$O_2N\text{—}C(CH_3)_2\text{—}H + CH_2{=}CH\cdot CN \longrightarrow O_2N\cdot C(CH_3)_2\cdot CH_2\cdot CH_2\cdot CN$$

## Reduction of Nitro Compounds

Nitro compounds are in a high state of oxidation, and are readily reduced both in the laboratory and in a wide variety of living systems. They are fully reduced to the corresponding primary amine by catalytic hydrogenation (Pd and $H_2$), or nascent hydrogen produced by dissolving zinc, iron or tin in hydrochloric acid. Aromatic nitro compounds are also partially reduced by zinc with neutral or alkaline reagents to a number of products in intermediate oxidation states, depending on the pH of the solution (Table 48).

## Metabolic Reduction

Aromatic nitro compounds, such as *Chloramphenicol* (p. 449), are reduced to the corresponding primary amines by nitro reductases (**2**, 5) present in the microsomal and soluble fractions of the liver and kidney (Mueller and Miller, 1950).

Table 48. Reduction Products of Nitrobenzene

| State of Solution | Reagent | Principal Product | |
|---|---|---|---|
| Neutral | $Zn/NH_4Cl$aq | Ph·NHOH | β-Phenylhydroxylamine |
| Alkaline | Zn/MeOH | $Ph\cdot\overset{+}{N}(O^-)—N\cdot Ph$ | Azoxybenzene |
| Alkaline | Zn/MeOH—NaOH | Ph·N=N·Ph | Azobenzene |
| Alkaline | Zn/NaOHaq | Ph·NH·NH·Ph | Hydrazobenzene |
| Acidic | Zn/HCl | $Ph\cdot NH_2$ | Aniline |

These reductases are flavoproteins with flavine adenine dinucleotide (FAD) as the effective reducing agent. The enzyme systems are capable of using either NADH or NADPH as hydrogen donors, and are inhibited by atmospheric oxygen. There is some evidence that nitroso and hydroxylamino compounds are formed as intermediates in the reduction, since although nitro reductases reduce nitrosobenzene and β-phenylhydroxylamine to aniline more rapidly than they reduce nitrobenzene (Fouts and Brodie, 1957), metabolism of *m*-dinitrobenzene in the rabbit gave trace amounts of *m*-nitrosonitrobenzene and *m*-nitrophenylhydroxylamine in addition to the main product *m*-nitroaniline (Parke, 1961).

Nitro reductases are widely distributed in mammalian and other species, including bacteria (Adamson, Dixon, Francis and Rall, 1965). Reduction of *Nitrofurazone* (5-nitro-2-furfuraldehyde semicarbazone) by *Escherichia coli*, *Staphylococcus aureus* and *Aerobacter aerogenes* has been shown to give rise to inactive metabolites (Asnis, Cohen and Gots, 1952), which have been identified spectroscopically as the corresponding hydroxylamine and amine (Beckett and Robinson, 1956, 1957, 1959).

$O_2N$–(furan)–$CH{=}N\cdot NH\cdot CO\cdot NH_2$ → $HO\cdot HN$–(furan)–$CH{=}N\cdot NH\cdot CO\cdot NH_2$ → $H_2N$–(furan)–$CH{=}N\cdot NH\cdot CO\cdot NH_2$

*Nitrofurazone*

It is interesting to note, however, that whereas the amino metabolite of the schistosomicide, *Niridazole*, reaches far higher concentrations in the blood than does the unchanged nitro compound (Faigle and Keberle, 1966), it is the latter which is taken up by and exerts its killing effect on the parasites in infected patients (Faigle and Keberle, 1969). *Metronidazole* which acts similarly is also extensively metabolised to the amino compound, though the latter is not excreted as such (Mitchard, 1971).

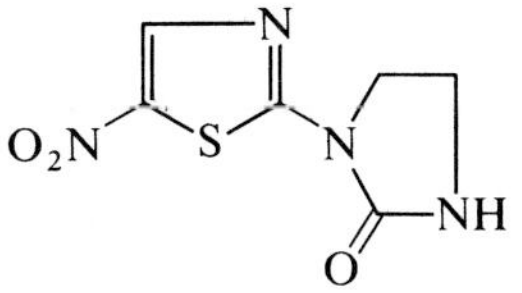

*Niridazole*

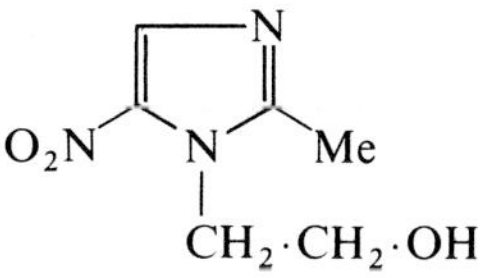

*Metronidazole*

**Intramolecular Reduction**

In certain compounds, the nitro group is capable of being reduced at the expense of an adjacent oxidisable group. Thus, *o*-nitrotoluene undergoes an intramolecular oxidation–reduction in ethanolic sodium hydroxide to form anthranilic acid.

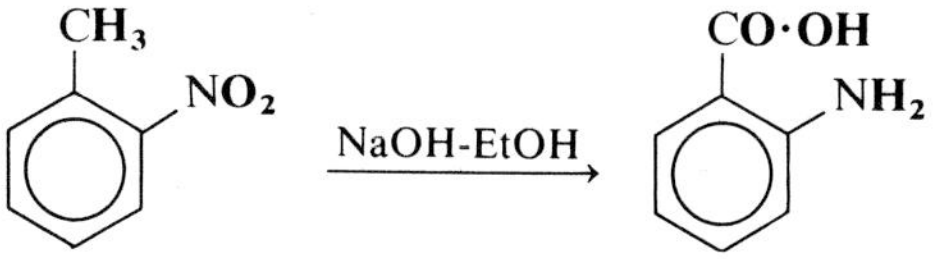

*o*-Nitrotoluene $\xrightarrow{\text{NaOH-EtOH}}$ Anthranilic acid

A similar type of neighbouring group oxidation at the α-carbon atom appears to be involved in the conversion of nitromethane to formic acid and hydroxylamine (Lippincott and Hass, 1939), which occurs in strongly acidic solutions (85% sulphuric acid). Formic hydroxamic acid is formed first, and then hydrolysed to formic acid and hydroxylamine, possibly by the following pathway.

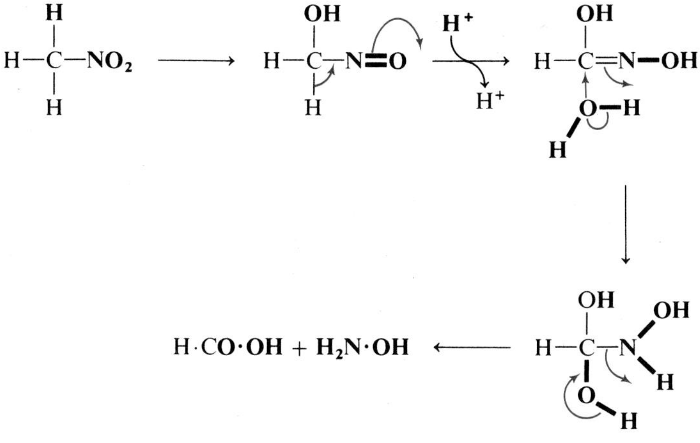

$$H \cdot CO \cdot OH + H_2N \cdot OH$$

**Ring-substitution of Aromatic Nitro Compounds**

Aromatic rings are deactivated by nitro substituents through electron withdrawal. This effect is greatest in the *ortho* and *para* positions, so that electrophilic substitutions which do occur give rise to *meta* substitution, as for example in the nitration of nitrobenzene (Chapter 5).

$$\text{C}_6\text{H}_5\text{NO}_2 \longrightarrow m\text{-C}_6\text{H}_4(\text{NO}_2)_2 \longrightarrow 1,3,5\text{-C}_6\text{H}_3(\text{NO}_2)_3$$

Nitrobenzene — *m*-Dinitrobenzene — 1,3,5-Trinitrobenzene

**Nucleophilic Displacement in Dinitroaromatic Halogen Compounds**

The powerful electron-withdrawing effect of the nitro group is able to promote nucleophilic displacement of halogens from the *ortho* and *para* positions of *m*-dinitro compounds under very mild conditions ($NaHCO_3$). The use of 2,4-dinitrofluorobenzene in the identification of *N*-terminal amino acids of peptides and proteins depends on such displacements to convert their free amino ($NH_2$) groups to the corresponding *N*-2,4-dinitrophenylamino derivatives. Hydrolysis

$$2,4\text{-}(\text{O}_2\text{N})_2\text{C}_6\text{H}_3\text{F} + \text{H}_2\text{N}\cdot\text{CH(CH}_3)\cdot\text{CO}\cdot\text{NH}\cdot\text{CH(R)}\cdot\text{CO}\cdot\text{NH}\text{---}\text{CO}\cdot\text{OH}$$

$$\downarrow \text{NaHCO}_3 \quad (\rightarrow \text{NaF} + \text{H}_2\text{O} + \text{CO}_2)$$

$$2,4\text{-}(\text{O}_2\text{N})_2\text{C}_6\text{H}_3\text{-NH}\cdot\text{CH(CH}_3)\cdot\text{CO}\cdot\text{NH}\cdot\text{CH(R)}\cdot\text{CO}\cdot\text{NH}\text{---}\text{CO}\cdot\text{OH}$$

$$\downarrow \text{H}_2\text{O},\ \text{H}^+$$

$$2,4\text{-}(\text{O}_2\text{N})_2\text{C}_6\text{H}_3\text{-NH}\cdot\text{CH(CH}_3)\cdot\text{CO}\cdot\text{OH}$$

$$+$$

$$\left.\begin{array}{c}\text{H}_2\text{N}\cdot\text{CH(R)}\cdot\text{CO}\cdot\text{OH} \\ + \\ \text{H}_2\text{N}\text{---}\text{CO}\cdot\text{OH}\end{array}\right\} \text{Constituent amino acids}$$

of the *N*-2,4-dinitrophenylpeptides gives *N*-2,4-dinitrophenylamino acids, which identify *N*-terminal and other amino acids with free amino groups (Sanger, 1945). Analogous displacement of halogen by the sulphydryl compound, glutathione, occurs in mammalian metabolism of 2,4-dinitrohalobenzenes (**1**, 14; **2**, 5).

## Nucleophilic Displacement of Aromatic Nitro Groups

One nitro group can activate another for nucleophilic displacement in *ortho* and *para* nitro compounds giving rise to substitution by hydroxyl, methoxyl, ammonia or amines.

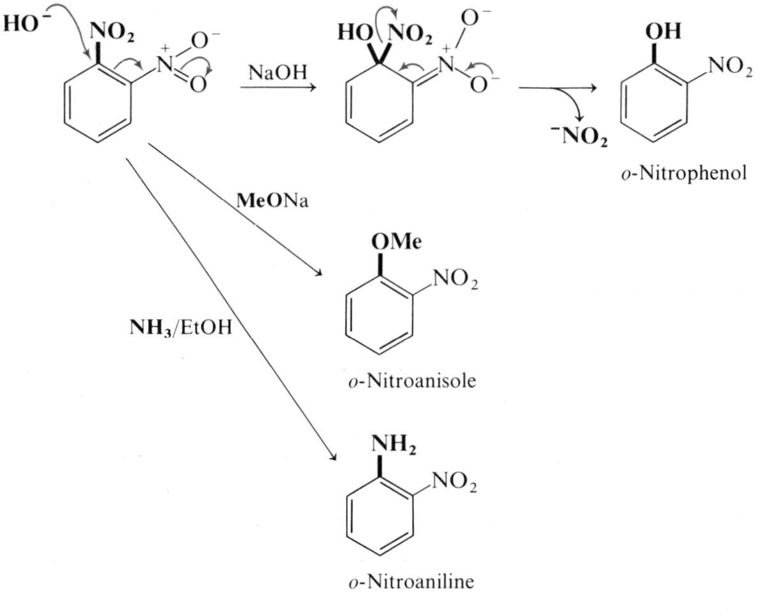

NaOH → $^{-}NO_2$ → *o*-Nitrophenol

MeONa → *o*-Nitroanisole

$NH_3$/EtOH → *o*-Nitroaniline

## Radical Oxidation of Aromatic Nitro Compounds

Although resistant to attack by electrophilic reagents, nitrobenzene reacts readily with Fenton's reagent (hydrogen peroxide and ferrous sulphate) and

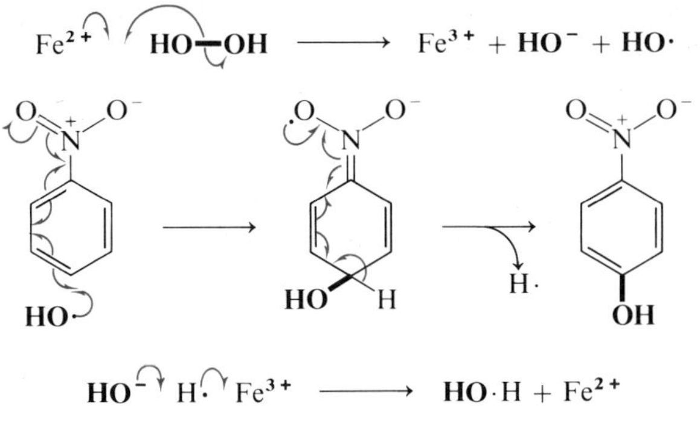

$$Fe^{2+} \quad HO{-}OH \longrightarrow Fe^{3+} + HO^{-} + HO\cdot$$

$$HO^{-} \quad H\cdot \quad Fe^{3+} \longrightarrow HO{\cdot}H + Fe^{2+}$$

alkaline potassium ferricyanide to form nitrophenols. These are radical oxidations and, predictably, nitrobenzene gives all three isomeric nitrophenols in yields of 22% *ortho*-, 22% *meta*- and 50% *para*-isomers respectively with Fenton's reagent (Loebl, Stein and Weiss, 1949).

Similarly, *m*-dinitrobenzene is oxidised to nitrophenols by alkaline potassium ferricyanide.

$$K_3^+[Fe(CN)_6]^{3-} + K^{+\,-}OH \longrightarrow K_4^+[Fe(CN)_6]^{4-} + HO\cdot$$

NO₂ NO₂ HO· → H· NO₂ NO₂ OH + NO₂ OH NO₂

Some nuclear hydroxylation also occurs as an alternative pathway to reduction in the mammalian metabolism of aromatic nitro compounds (Parke, 1968).

## CHLORAMPHENICOL

The antibiotic, *Chloramphenicol*, which was originally isolated from *Streptomyces venezuelae* (Ehrlich, Bartz, Smith, Joslyn and Burkholder, 1947), is now produced synthetically (Long and Troutman, 1949). Its use, however, has been severely restricted because of its effects on bone marrow which give rise to asplastic anaemia and agranulocytosis in some patients. Thus, although *Chloramphenicol* has a broad antibacterial spectrum (McLean, Schwab, Hillegas and Schlingman, 1949; Smith, Joslyn, Gruhzit, McLean, Penner and Ehrlich, 1948), it remains the antibiotic of choice only in the treatment of typhoid and paratyphoid infections.

### General Properties

Despite its rather low solubility in water (1 in 400), *Chloramphenicol* is well absorbed when administered orally (Ley, Smadel and Crocker, 1948), but because of its intensely bitter taste is usually administered in either capsule form or in paediatric suspensions as the tasteless *Chloramphenicol Palmitate* (Chapter 11). Aqueous solutions are stable. Moreover, its low aqueous solubility can be increased up to about 5% by propylene glycol and to about 1% by borax (0.6%) to give more concentrated preparations for use as ear and eye drops respectively. The latter solution, however, is alkaline, and this promotes hydrolysis of the important dichloroacetamido group (Higuchi and Bias, 1953). The most satisfactory formulation in aqueous solution, therefore, is one buffered to pH 7.0 with boric acid and borax.

$$O_2N-C_6H_4-CH(OH)-CH(NHCOCHCl_2)-CH_2OR$$

*Chloramphenicol* (R = H)
*Chloramphenicol Palmitate* (R = $CH_3(CH_2)_{14}\cdot CO-$)

$$\xrightarrow{HO^-} O_2N-C_6H_4-CH(OH)-CH(NH_2)-CH_2OH + Cl_2CH\cdot CO\cdot O^-$$

**Specificity and Mechanism of Action**

Large numbers of *Chloramphenicol* derivatives and analogues have been examined for antibacterial activity, but few are more effective than the parent antibiotic. The arrangement for the side-chain substituents is stereospecific, and the natural D-*threo*-isomer the only one of the four possible stereoisomers to show significant activity (Controulis, Rebstock and Crooks, 1949).

Similarly, modification of the side-chain by replacement of either hydroxyl function by hydrogen (Rebstock, 1951), or substitution of other halogen-substituted acetic acids for the *N*-dichloroacetyl group (Rebstock, 1950) lead to loss of activity.

The electron-withdrawing nitro group of *Chloramphenicol* is also essential for its antibacterial activity. Its replacement by other substituents leads to a fall in antibacterial activity. Analogues with other electron-attracting groups such as Cl, Br or I retain some activity, though at a significantly reduced level (Buu-Hoï, Hoán, Jacquignon and Khöi, 1950; Dann, Ulrich and Möller, 1950). The significance of these observations has not been precisely evaluated. It is known, however, that *Chloramphenicol* inhibits protein synthesis in bacterial ribosomes (Gale and Folkes, 1953; Hancock and Park, 1958; Rendi and Ochoa, 1962), and that this action is selective, since inhibition of protein synthesis only occurs in mammalian cells at much higher *Chloramphenicol* concentrations (Weisberger, 1968). Although *Chloramphenicol* inhibits protein synthesis, it does not affect the synthesis of nucleic acids (Wisseman, Smadel, Hahn and Hopps, 1954), so RNA accumulates in the cell.

It has been shown that *Chloramphenicol* binds only to the 50S sub-unit of 70S ribosomes, possibly by charge-transfer complexing (Cammarata, 1967), in sensitive organisms (Vazquez, 1964), and it appears certain that protein synthesis is inhibited due to the block in peptide bond formation which results therefrom (Julian, 1965). Experiments with $^{14}C$-uracil labelled mRNA show that it continues to enter poly-ribosomes, and also tRNA-containing ribosomal sub-units continue to move along mRNA in normal peptide synthesis, but peptide bond formation is inhibited due to distortion by *Chloramphenicol* binding in the 50S components (Gurgo, Apirion and Schlessinger, 1969).

It has been suggested (Hopkins, 1959) that the activity of *Chloramphenicol* may be ascribed to its structural relationship to uridine, which is characteristic of RNA (in place of thymine in DNA), and an important factor in the coding of amino acid sequences in protein synthesis (Speyer, Lengyel, Basilio and Ochoa, 1962). Both ring systems are strongly electron-deficient. It seems, therefore, that *Chloramphenicol* may compete with uridine for some $\pi$-excessive system associated with bacterial ribosomes. This concept is supported by the NMR studies of Jardetzky (1963) and the X-ray crystallographic studies of Dunitz (1952) which demonstrated the structural similarity of *Chloramphenicol* to uridine-5′-phosphate.

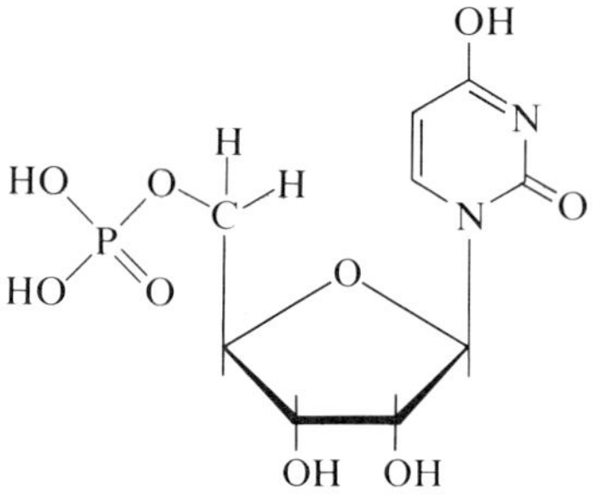

Uridine-5[1]-phosphate

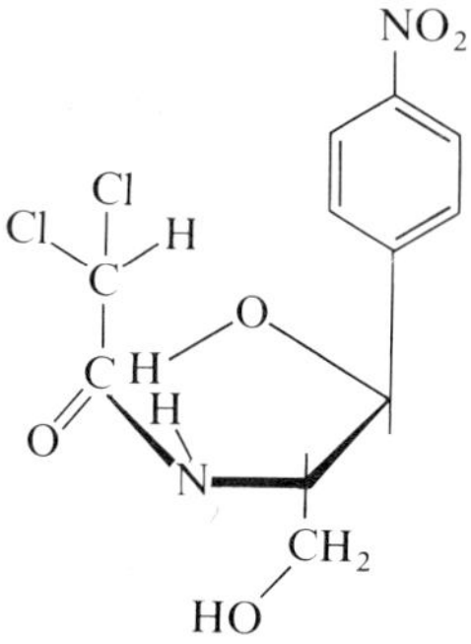

*Chloramphenicol*

Transferable bacterial resistance to *Chloramphenicol* is discussed elsewhere (**2**, 5).

# 16 Amines and Quaternary Ammonium Salts

## AMINES

Amines are basic compounds related to ammonia, with one or more alkyl or aryl groups attached to nitrogen in place of hydrogen. Amines are described as primary (1°), secondary (2°) or tertiary (3°) according to the number of *N*-alkyl or *N*-aryl substituents as follows.

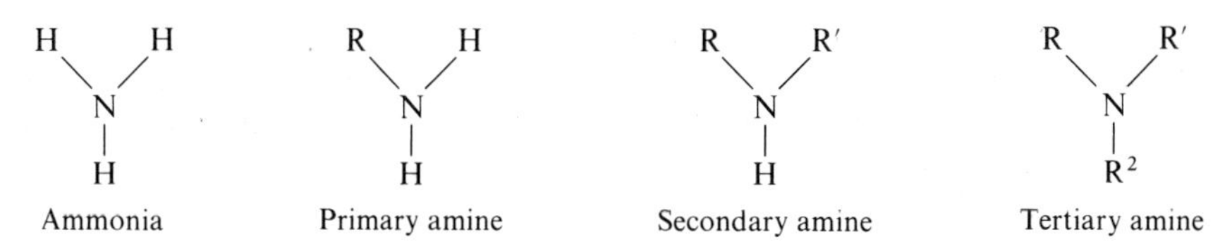


The nitrogen atom in amines is $sp^3$ hybridised. Hence the molecules are not planar as depicted above, but tetrahedral with the atoms disposed at the apices of the tetrahedron, and the lone pair of electrons occupying the fourth corner. As a result, asymmetric amines, in which all three *N*-substituents are different, form non-superposable mirror images. Although such compounds are, therefore, theoretically capable of resolution into optically active forms, simple amines of this type have not been resolved in practice. This is due to the low activation energy of racemisation and the extreme ease with which it occurs at normal temperatures. The nitrogen atom is, therefore, in oscillation with respect to the other three atoms, and is said to undergo rapid **inversion**.

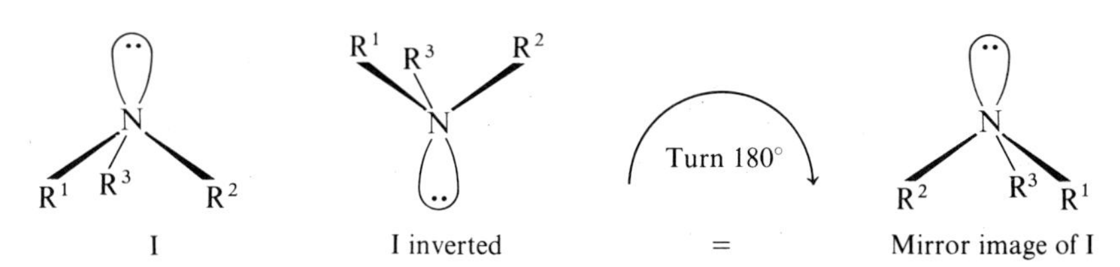


Certain asymmetric amines with structures sufficiently rigid to inhibit *N*-inversion have been resolved into their enantiomers.

### Nomenclature

Amines are named according to IUPAC rules by prefixing the hydrocarbon radicals present in alphabetical order to the suffix, **amine**. In complex amines, where ambiguity may arise over the position of the alkyl substituent, attachment to nitrogen may be indicated by the prefix ***N*-**.

| | $CH_3 \cdot NH_2$ | $(CH_3)_2NH$ | $(CH_3)_3N$ |
|---|---|---|---|
| | Methylamine | Dimethylamine | Trimethylamine |
| b.p. (°C) | −6.3 | 7 | 4 |

| | $CH_3 \cdot N(CH_2 \cdot CH_3) \cdot CH_2 \cdot CH_2 \cdot CH_2 \cdot NH_2$ | $C_6H_5 \cdot NH_2$ |
|---|---|---|
| | Ethylmethylpropylamine | Aniline |
| b.p. (°C) | 92 | 186 |

| | $C_6H_5 \cdot NH \cdot CH_3$ | $C_6H_5 \cdot N(CH_3)_2$ |
|---|---|---|
| | *N*-Methylaniline | *NN*-Dimethylaniline |
| b.p. (°C) | 196 | 194 |

## General Properties

The lower molecular weight aliphatic amines are gases or low boiling liquids with strong ammoniacal or fishy odours. They are generally soluble in water, but solubility in water and odour decreases with increasing molecular weight, so that higher members increasingly resemble hydrocarbons with low water solubility and little odour. A few more complex bases, such as amphetamine ($C_6H_5 \cdot CH_2 \cdot CH(CH_3) \cdot NH_2$), still possess sufficient volatility to be effective by inhalation in the relief of nasal congestion. Aromatic amines have low water solubility, and much higher boiling points than aliphatic amines. Both aromatic and aliphatic amines are soluble in a wide range of organic solvents.

The solubilities of aromatic amines are often significantly modified by ring substitution. Thus, halogen substitution in the nucleus of the arylalkylamines, *Chlorpromazine*, *Chlorphentermine* and *Chloramphetamine*, leads in each to increased lipid solubility compared with the corresponding unsubstituted compound. This enhanced lipid solubility has a marked effect on their rate of excretion, which is lower than that of the unsubstituted compounds (Beckett and Brookes, 1971).

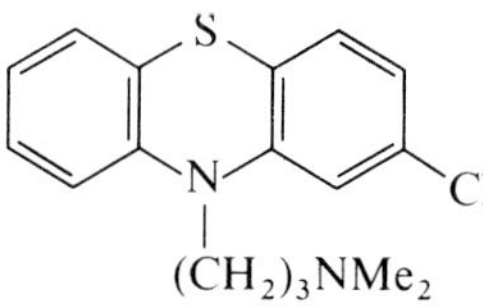

*Chlorpromazine*

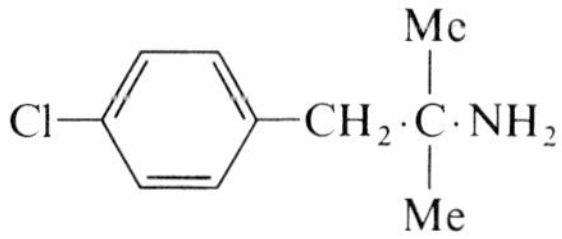

*Chlorphentermine*

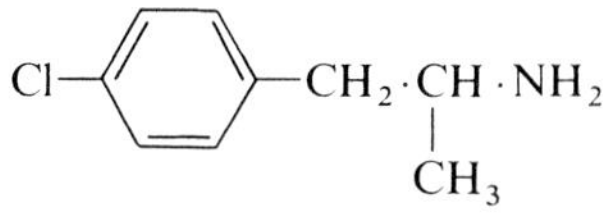

*Chloramphetamine*

Phenolic bases, including the analgesics, *Morphine* and *Phenazocine*, and the morphine antagonist, *Nalorphine*, are zwitterionic. Accordingly, not only are

Table 49. Solubilities of Morphine and Codeine

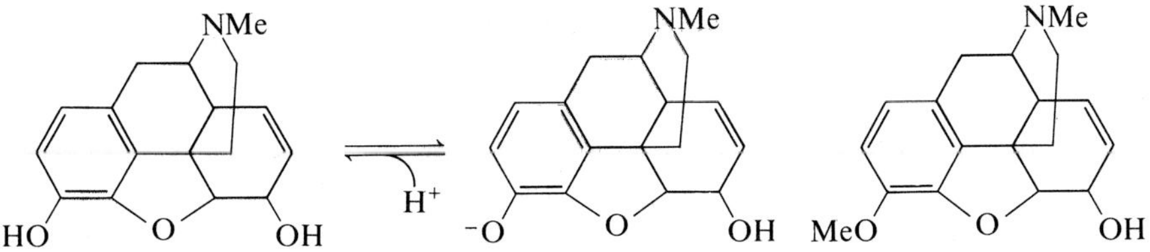

| Solvent | Morphine (p$K_a$ 7.87) | Codeine (p$K_a$ 8.95) |
|---|---|---|
| Water | 1 in 5000 | 1 in 120 |
| Alcohol | 1 in 250 | 1 in 2 |
| Benzene | 1 in 3250 | 1 in 13 |
| Chloroform | 1 in 1500 | 1 in 2 |

they somewhat weaker bases than the corresponding unsubstituted and methoxyl-substituted arylalkylamines (p. 456), but also have solubilities which are markedly different from the latter. Thus, they are not at all readily soluble in solvents such as chloroform or ether (Table 49), but being both acidic and basic form watersoluble sodium and amine salts.

**Basic Strength**

The characteristic properties of amines arise essentially from the lone pair of electrons on the nitrogen atom. Amines are electron donors and are able to accept a proton, as for example, from water to form an ammonium ion; hence, amines are bases.

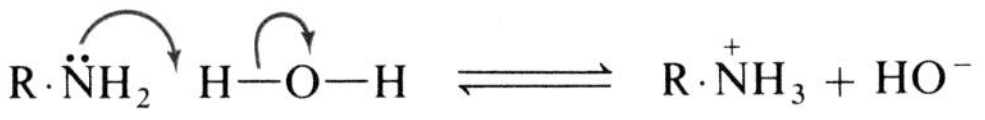

$$R\cdot\ddot{N}H_2 \quad H{-}O{-}H \rightleftharpoons R\cdot\overset{+}{N}H_3 + HO^-$$

The strength of the base is measured by the equilibrium or basicity constant $K_b$ for the reaction, which is given by the expression:

$$K_b = \frac{[R\cdot\overset{+}{N}H_3][HO^-]}{[R\cdot NH_2]}$$

Just as acidic strengths are conveniently indicated by the dissociation constant exponent p$K_a$, so base strengths are indicated by a base dissociation constant exponent p$K_b$($= -\log K_b$). Strong bases have low p$K_b$ and weak bases high p$K_b$. It is often more convenient, and more usual to express the strengths of bases in terms of the p$K_a$ of the base conjugate acid ($R\cdot\overset{+}{N}H_3$), or ammonium ion.

$$R\cdot\overset{+}{N}H_3 \rightleftharpoons R\cdot\ddot{N}H_2 + H^+$$

for which

$$K_a = \frac{[R \cdot NH_2][H^+]}{[R \cdot \overset{+}{N}H_3]}$$

where $pK_a = -\log K_a$.

The $pK_b$ of the base, and the $pK_a$ of the base conjugate acid, can be related one to the other, since:

$$K_a \cdot K_b = \frac{[R \cdot NH_2][H^+]}{[R \cdot \overset{+}{N}H_3]} \cdot \frac{[R \cdot \overset{+}{N}H_3][HO^-]}{[R \cdot NH_2]}$$

$$\therefore \quad K_a \cdot K_b = [H^+][HO^-]$$

$$\therefore \quad K_a \cdot K_b = K_w$$

and

$$pK_a + pK_b = pK_w \qquad (\text{where } pK_w = 14).$$

Thus, a strong base (low $pK_b$) will have a high $pK_a$ value, and a weak base (high $pK_b$) will have a low $pK_a$ value.

The basic strength of aliphatic amines is determined by the number and nature of alkyl or aryl substituents. Alkyl groups are electron-repelling. They increase the availability of electrons on nitrogen, and hence basic strength increases with increasing alkyl substitution. Thus, methylamine ($pK_a$ 10.64) is a much stronger base than ammonia ($pK_a$ 9.3), and dimethylamine slightly stronger again ($pK_a$ 10.71). Contrary to expectation, however, the corresponding tertiary amine, trimethylamine, is a weaker base ($pK_a$ 9.72). This is because basic strength is due not only to the extent of alkyl substitution, but also to the stabilisation by solvation of the cation formed on protonation. The influence of both effects is optimal in dialkylamines. In accordance with this view, basicity increases regularly from primary through secondary to tertiary amines in non-polar solvents.

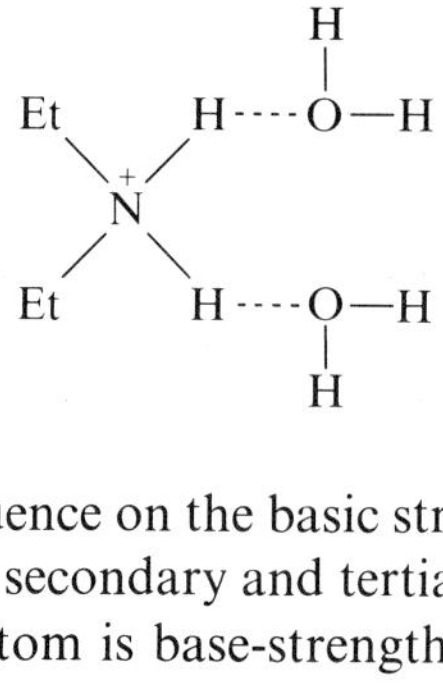

```
                 H
                 |
Et      H····O—H
  \  +  /
    N
  /     \
Et      H····O—H
                 |
                 H
```

Chain length has little influence on the basic strength of primary amines, but more pronounced effect with secondary and tertiary amines (Table 50). Chain-branching on the α-carbon atom is base-strengthening except where steric in-

Table 50. Dissociation Constants of Aliphatic Amines

| Primary | p$K_a$ | Secondary | p$K_a$ | Tertiary | p$K_a$ |
|---|---|---|---|---|---|
| $Me \cdot NH_2$ | 10.64 | $Me_2 \cdot NH$ | 10.71 | $Me_3N$ | 9.72 |
| $Et \cdot NH_2$ | 10.63 | $Et_2 \cdot NH$ | 10.93 | $Et_3N$ | 10.87 |
| $Pr \cdot NH_2$ | 10.58 | $Pr_2 \cdot NH$ | 10.98 | $Pr_3N$ | 10.74 |
| $Pr^i \cdot NH_2$ | 10.63 | $Pr^i_2 \cdot NH$ | 10.96 | | |
| n-Bu $\cdot NH_2$ | 10.61 | | | | |
| t-Bu $\cdot NH_2$ | 10.45 | | | | |
| Bz $\cdot NH_2$ | 9.33 | | | | |

teractions are appreciable (t-butylamine). The maximum effect is seen in the ganglion blocker, *Pempidine*, (1,2,2,6,6-pentamethylpiperidine), where the molecular conformation reduces steric interaction.

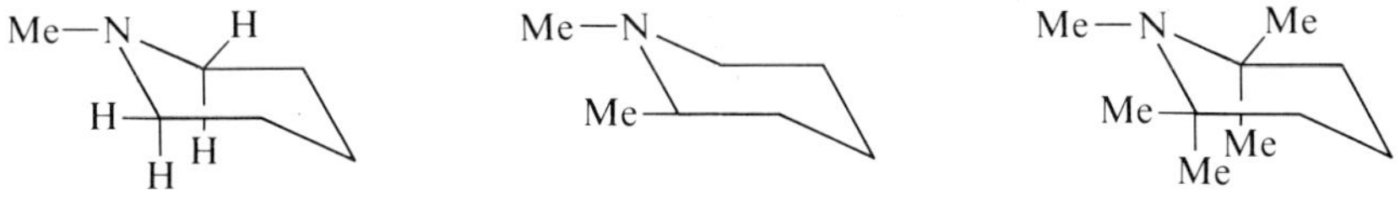

| | 1-Methylpiperidine | 1,2-Dimethylpiperidine | *Pempidine* |
|---|---|---|---|
| p$K_a$ | 10.08 | 10.22 | 11.25 |

In contrast to aliphatic amines, aromatic amines are weaker bases than ammonia. Thus, aniline has a p$K_a$ 4.63. This base-weakening effect of aromatic substituents is due to interaction of the nitrogen lone-pair of electrons with the delocalised $\pi$-orbitals of the aromatic ring in the base in a way which is not possible with the corresponding anilinium ion.

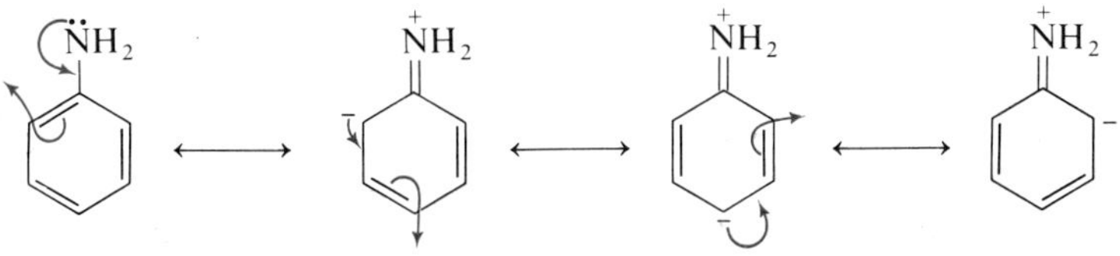

As might be anticipated, diphenylamine, $(C_6H_5)_2NH$, is an extremely weak base (p$K_a$ 0.8) and triphenylamine, $(C_6H_5)_3N$, is so weakly basic as not to be measurable in aqueous solution.

Resonance, however, is not necessarily base-weakening. Indeed, the opposite is true where resonance is greater in the ionised base than in the base itself. Albert and Goldacre (1943) have shown that the basic strength of acridine (p$K_a$ 4.8), a weak base comparable in strength to aniline, is little affected by amino substituents in the 1-, 2- or 4-positions.

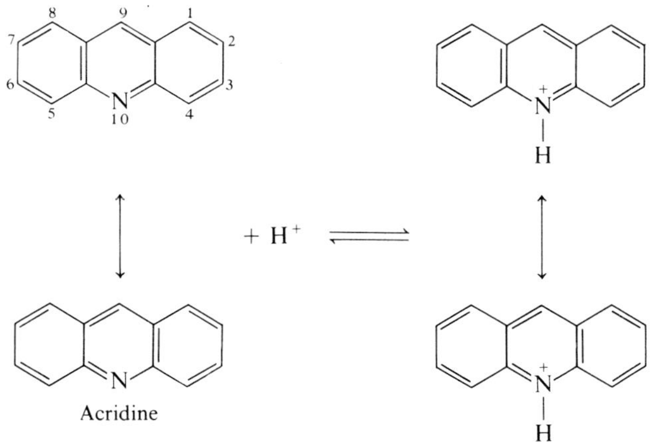

Acridine

Acridinium ion

In contrast, however, 3-aminoacridine (p$K_a$ 8.2) and 9-aminoacridine (p$K_a$ 9.9) are both strong bases as a result of the increased resonance which is possible in the corresponding acridinium ions.

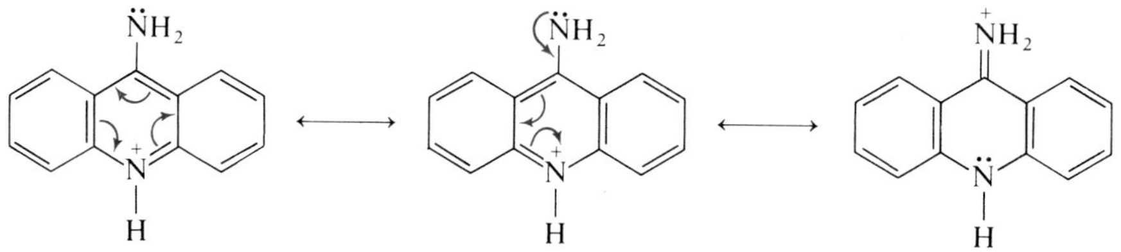

*N*-Alkylation of aniline (p$K_a$ 4.63) increases basic strength to p$K_a$ 4.83 in *N*-methylaniline and p$K_a$ 5.15 in *NN*-dimethylaniline by increasing electron availability on nitrogen. This is contrary to the expectation that the +I effect of the alkyl group should lead to increased resonance interaction between the lone pair on the nitrogen and ring, an effect which would decrease basic strength. Since *N*-ethyl substituents increase basic strength even more than methyl substituents and *NN*-dialkylanilines are stronger bases than *N*-alkylanilines, base-strengthening is considered to be due to steric inhibition of resonance. Thus, steric interactions between the *ortho*-hydrogen atoms of the ring and the *N*-alkyl-substituents forces these resonance-participating groups out-of-plane. The resulting inhibition of resonance leads to greater availability of the lone-apir of electrons on nitrogen for protonation, and hence increase in the basic strength (Brown, 1956).

Electron-repelling substituents (Me and MeO) in the *para* position are also base-strengthening, since they also oppose resonance of the $NH_2$ lone pair with the ring, though by a different mechanism. In contrast, *ortho* electron-repelling substituents are base-weakening, due to steric repulsion of approaching protons. The same substituent in the *meta* position, however, has only a marginal effect on p$K_a$.

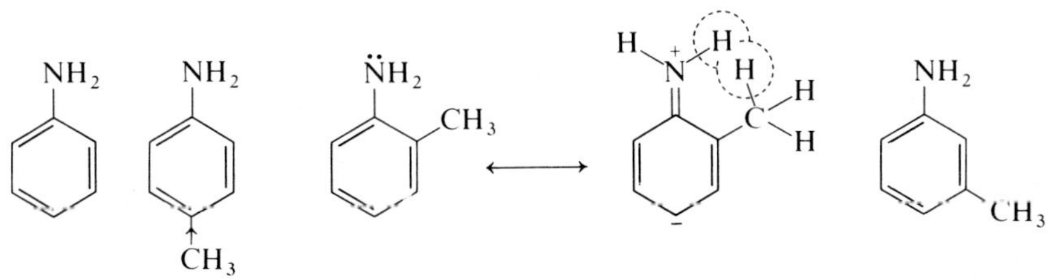

| | $NH_2$ | $NH_2$ / $CH_3$ | $\ddot{N}H_2$ / $CH_3$ ⟷ $H\overset{+}{N}H$ / $CH_3$ | $NH_2$ / $CH_3$ |
|---|---|---|---|---|
| $pK_a$ | 4.63 | 5.08 | 4.44 | 4.73 |

Electron-attracting substituents are base-weakening irrespective of their location in the ring, as a result of their $-I$ effects (Table 51). *Ortho* and *para* substituents assist resonance of the amino lone pair with the ring, and hence show greater base-weakening effects than *meta* substituents. *Ortho* substituents also cause steric effects, so that electron-attracting *ortho* substituents produce the greatest base-weakening effects.

Table 51. Dissociation Constants ($pK_a$) of Nuclear Substituted Anilines

| Substituent (R) | 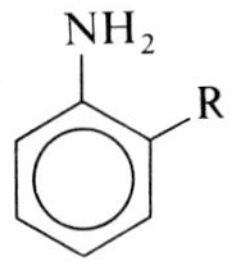 $NH_2$ / R | 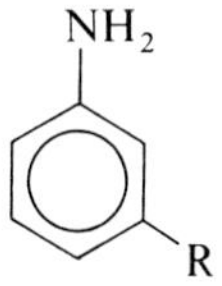 $NH_2$ / R | 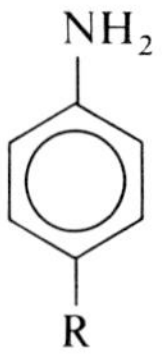 $NH_2$ / R |
|---|---|---|---|
| OH | 4.34 | | 5.84 |
| F | 3.20 | 3.50 | 4.65 |
| Cl | 2.65 | 3.46 | 4.15 |
| Br | 2.53 | 3.54 | 3.86 |
| $SO_2 \cdot OH$ | 2.46 | 3.74 | 3.2 |
| $SO_2 \cdot NH_2$ | | | 2.36 |
| $CO \cdot OH$ | 2.1 | | 2.5 |
| $NO_2$ | −0.2 | 2.6 | 2.04 |

The strength of ionised bases is important in determining the extent to which the base is ionised under physiological conditions. This influences transport, distribution and excretion patterns in humans (**2**, 4) and also determines the extent of ionic bond formation with carrier and receptor proteins (**2**, 2).

**Salt Formation**

Many important drugs are organic bases. Mineral acid salts, particularly hydrochlorides, hydrobromides and sulphates, and to a lesser extent phosphates and nitrates, provide a convenient form for medicinal use. The salts are almost invariably crystalline, more readily purified, and generally more stable to oxidative degradation than the corresponding bases. In contrast to their parent bases, the salts are for the most part appreciably water-soluble, and hence provide a convenient form for the preparation of sterile aqueous solutions for injection, and rapid dissolution from solid dosage forms such as tablets and capsules. A few notable exceptions include *Quinine Sulphate* (solubility in water *ca* 0.1%), *Ethopropazine Hydrochloride* (0.3%) and *Meclozine Hydrochloride* (Meclizine Hydrochloride; 0.1%). Water solubility, where it is not otherwise obtainable, is achieved by the use of acid salts of both dibasic inorganic acids (*Quinine Bisulphate*, solubility in water *ca* 12.5%) and di- or tri-basic organic acids [*Adrenaline Acid Tartrate* (Epinephrine Bitartrate), *Ergometrine Maleate* (Ergonovine Maleate, *Diethylcarbamazine Citrate*] or strong monobasic organic acids (*Benztropine Mesylate*, *Phentolamine Mesylate*).

Low solubility in water and organic solvents is a valuable property of salts required for isolation and characterisation of bases such as oxalates and picrates. Cyclohexylamine forms stable, crystalline salts with penicillins and is widely used in their isolation and purification. *Procaine Penicillin* (Penicillin G Procaine), which has a solubility of about 0.5% in water, is used for the preparation of aqueous suspensions to provide depôt medication by intramuscular injection, from which the penicillin is slowly leached by dissociation. Other water-insoluble salts, such as *Dextropropoxyphene Napsylate* and *Chlorpromazine Embonate* (Chlorpromazine Pamoate), are considerably less bitter than the corresponding hydrochlorides and are used in preparations where it is desirable to disguise the otherwise unpleasant taste.

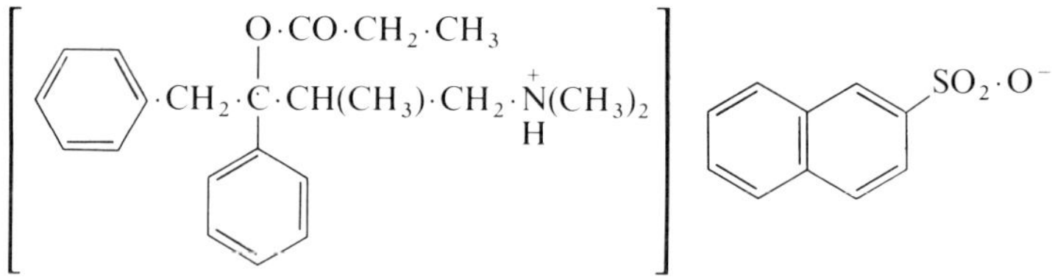

*Dextropropoxyphene Napsylate*

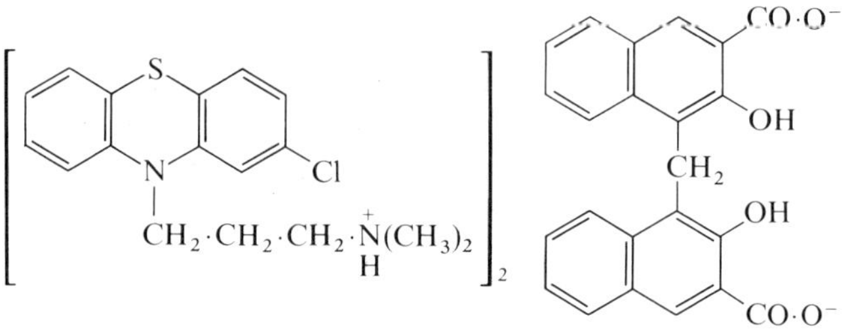

*Chlorpromazine Embonate*

A few rare examples of amine salts are known in which both base and acid components have therapeutic indications. One such is *Dimenhydrinate*, which is diphenhydramine 8-chlorotheophyllinate.

$[(C_6H_5)_2 \cdot CH_2 \cdot O \cdot CH_2 \cdot CH_2 \cdot N(CH_3)_2]$

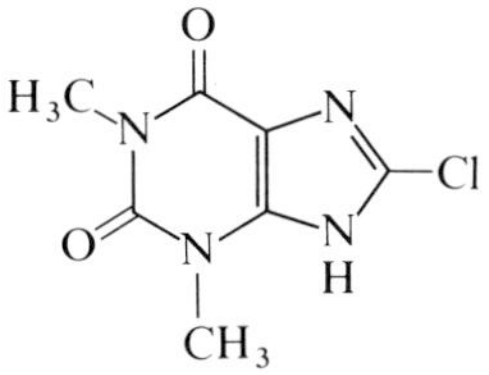

Dimenhydrinate

**Tertiary Amine Bicarbonates**

Hydrated tertiary amines, and anhydrous tertiary amines in the presence of moisture absorb carbon dioxide to form base bicarbonates, which are insoluble in organic solvents. Their formation may affect the extraction of tertiary bases from aqueous solution, if carbonates have been used for neutralisation. Absorption of carbon dioxide by tertiary bases may represent a storage hazard, but it is doubtful whether formation of base bicarbonates in pharmaceutical dosage forms affects the availability of the drug if they are for oral administration into the gastro-intestinal tract. Formation of base bicarbonates from the free base in products for direct inhalation, administration by aerosol or by buccal absorption, could lead to poor absorption characteristics (**2**, 4).

$$R_3N + H_2O + CO_2 \longrightarrow R_3\overset{+}{N}H \quad HCO_3^-$$

Primary and secondary amines react with carbon dioxide to form carbamates (see below).

***N*-Alkylation**

Amines readily release their unshared pair of electrons to form a new bond in nucleophilic attack on electron-deficient centres, as in protonation and salt formation. Nucleophilic attack on carbon in alkyl halides and sulphates leads to displacement of the halogen and results in *N*-alkylation.

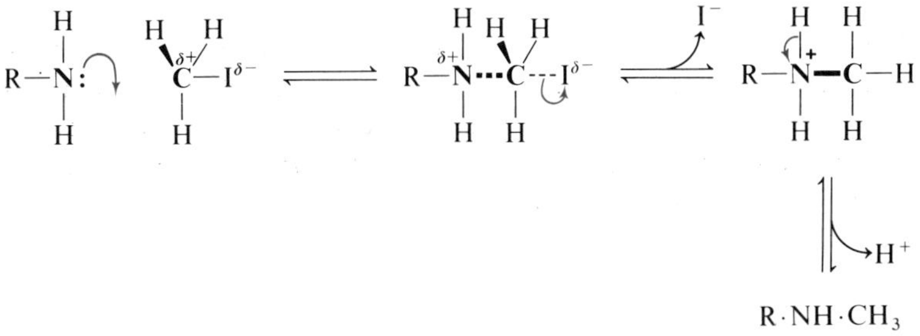

$$R{-}NH_2: + CH_3{-}I \rightleftharpoons R{-}NH_2\cdots CH_3\cdots I^{\delta-} \rightleftharpoons R{-}\overset{+}{N}H_2{-}CH_3 \; (+I^-) \xrightarrow{-H^+} R \cdot NH \cdot CH_3$$

Mixtures of 1°, 2° and 3° amines and quaternary ammonium salts are almost invariably formed when ammonia is heated under pressure in a sealed tube with an alkyl halide (Hofmann, 1850).

$$R{\cdot}X + NH_3 \longrightarrow R{\cdot}\overset{+}{N}H_3X^- \xrightarrow[\ NH_4CXl\ ]{NH_3} R{\cdot}NH_2 \quad \text{Primary (1°) amine}$$

$$R{\cdot}X + R{\cdot}NH_2 \longrightarrow R_2{\cdot}\overset{+}{N}H_2X^- \xrightarrow[\ NH_4CXl\ ]{NH_3} R_2{\cdot}NH \quad \text{Secondary (2°) amine}$$

$$R{\cdot}X + R_2{\cdot}NH \longrightarrow R_3{\cdot}\overset{+}{N}HX^- \xrightarrow[\ NH_4CXl\ ]{NH_3} R_3{\cdot}N \quad \text{Tertiary (3°) amine}$$

$$R{\cdot}X + R_3{\cdot}N \longrightarrow R_4{\cdot}\overset{+}{N}X^- \quad \text{Quaternary ammonium salt}$$

The proportion of each product formed depends on various factors, including the reactivity and relative proportions of amine and alkyl halide. It also depends on the relative basic strengths of the reacting base and products; the stronger the base, the greater the availability of the unshared electron pair, and hence the greater the reactivity of the base. Aromatic bases such as aniline are, therefore, much less readily alkylated than aliphatic amines.

The reactivity of primary alkyl halides in *N*-alkylation is in the following order:

$$RI > RBr > RCl$$

Secondary and tertiary halides ($R_2CHX$ and $R_3CX$) are much less efficient in *N*-alkylation reactions, because dehydrohalogenation to olefines occurs more readily with increasing substitution.

Although direct alkylations usually give rise to mixtures, reaction conditions can be manipulated to give good yields of particular products. Primary amines in large excess over alkyl halide give good yields of secondary amines. Also, secondary amines with the equivalent of alkyl halide give tertiary amines, though not always in good yield.

$$C_6H_5{\cdot}NH{\cdot}CH_3 + CH_3I \longrightarrow C_6H_5{\cdot}\overset{+}{N}H(CH_3)_2 \quad I^-$$

Quaternary ammonium salts are prepared by the reaction between a tertiary amine and an alkyl halide at ambient temperature. In some cases, it may be necessary to heat the reactants with or without solvent, under reflux.

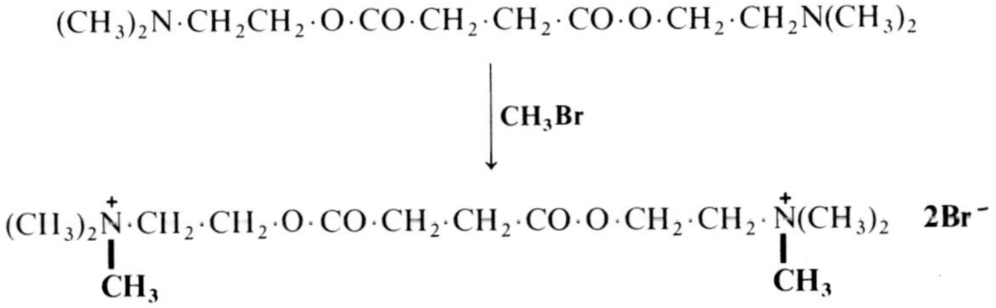

$$(CH_3)_2N\cdot CH_2CH_2\cdot O\cdot CO\cdot CH_2\cdot CH_2\cdot CO\cdot O\cdot CH_2\cdot CH_2N(CH_3)_2$$

$$\downarrow CH_3Br$$

$$(CH_3)_2\overset{+}{N}(CH_3)\cdot CH_2\cdot CH_2\cdot O\cdot CO\cdot CH_2\cdot CH_2\cdot CO\cdot O\cdot CH_2\cdot CH_2\cdot \overset{+}{N}(CH_3)(CH_3)_2 \quad 2Br^-$$

*Suxamethonium Bromide*

**Biological Alkylations**

*N*-Alkylation occurs in the biosynthesis of acetylcholine, according to the following pathway from serine (MacIntosh, 1959), *N*-methylation being effected by methyl transfer from *S*-adenosylmethionine.

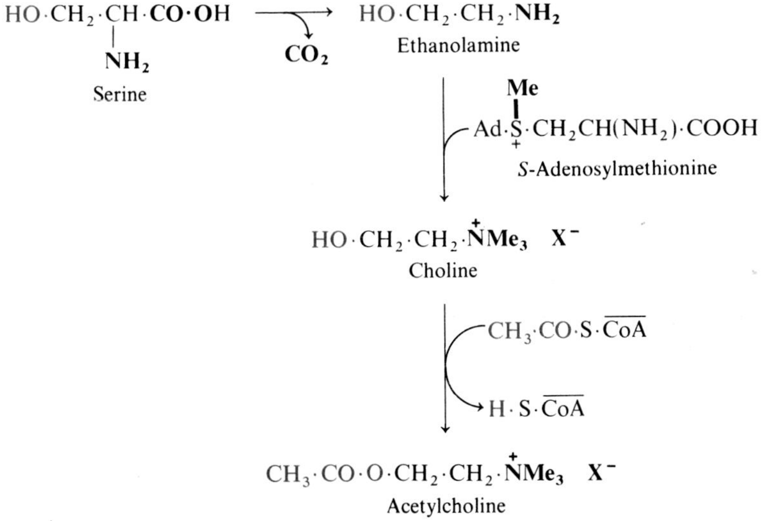

$$HO\cdot CH_2\cdot CH(NH_2)\cdot CO\cdot OH \xrightarrow{-CO_2} HO\cdot CH_2\cdot CH_2\cdot NH_2$$

Serine → Ethanolamine

$$\downarrow Ad\cdot \overset{+}{S}(Me)\cdot CH_2CH(NH_2)\cdot COOH \; (S\text{-Adenosylmethionine})$$

$$HO\cdot CH_2\cdot CH_2\cdot \overset{+}{N}Me_3 \quad X^-$$

Choline

$$\downarrow CH_3\cdot CO\cdot S\cdot \overline{CoA} \rightarrow H\cdot S\cdot \overline{CoA}$$

$$CH_3\cdot CO\cdot O\cdot CH_2\cdot CH_2\cdot \overset{+}{N}Me_3 \quad X^-$$

Acetylcholine

Biological alkylation also forms the basis of a large group of drugs used in cancer chemotherapy as cytotoxic agents. The simplest and best known of these are the nitrogen mustards, bis-$\beta$-chloroethylamines which are isosteric with mustard gas, $S(CH_2\cdot CH_2\cdot Cl)_2$.

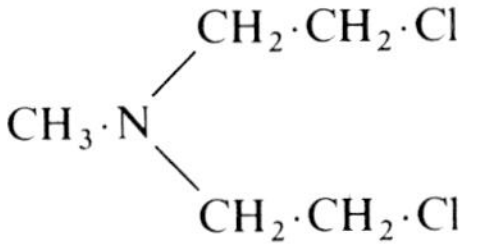

$$CH_3\cdot N(CH_2\cdot CH_2\cdot Cl)_2$$

*Mustine Hydrochloride*
(Mechorethamine Hydrochloride)

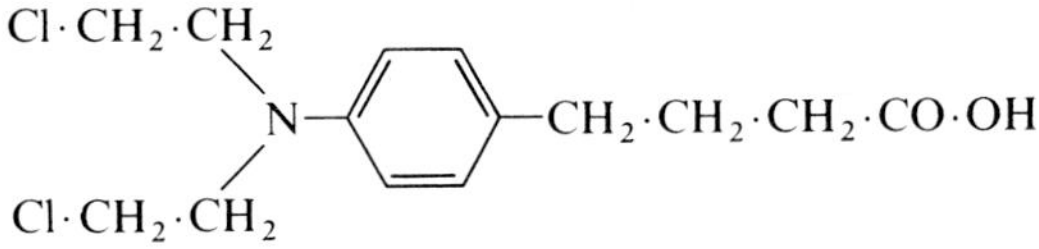

$$(Cl\cdot CH_2\cdot CH_2)_2N-C_6H_4-CH_2\cdot CH_2\cdot CH_2\cdot CO\cdot OH$$

*Chlorambucil*

$$\text{(cyclic } -CH_2CH_2CH_2-NH-P(=O)-O-\text{)}\; N(CH_2\cdot CH_2\cdot Cl)_2,\; H_2O$$

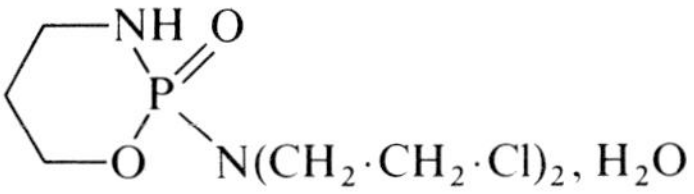

*Cyclophosphamide*

Alkylation proceeds by an $S_Ni$ mechanism via an intermediate imminium ion (**2**, 2).

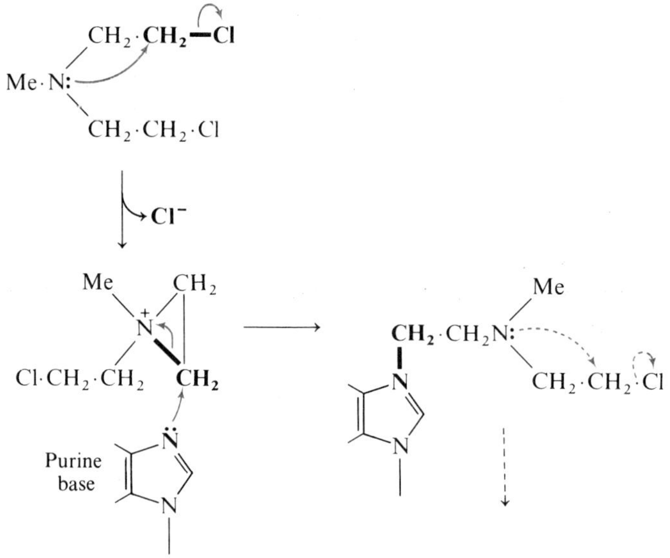

CH₂·CH₂—Cl
Me·N:
CH₂·CH₂·Cl
Cl⁻
Me
CH₂
N
Cl·CH₂·CH₂
CH₂
Purine
base
N
N
Me
CH₂·CH₂N:
CH₂·CH₂·Cl

### *N*-Dealkylation

Tertiary alkylamines can be dealkylated to the corresponding secondary amine in the laboratory, though not always in good yield, by reaction with cyanogen

$$NaCN + Br_2 \longrightarrow BrCN + NaBr$$

Br—CN
N—Me
HO
O
OH
Br⁻
Me
N—CN
HO
O
HO
MeBr
N·CN
HO
O
OH

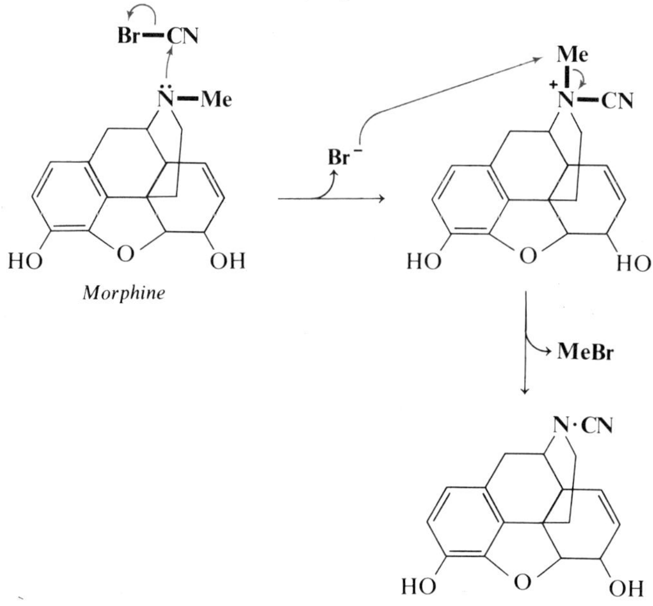

*Morphine*

Dialkylcyanamide

bromide (prepared from sodium cyanide and bromine water). The reaction, which proceeds through the formation of a dialkylcyanamide, has been used to prepare valuable intermediates, not otherwise readily available, as for example normorphine from morphine.

The cyanamide is hydrolysed by acid or alkali, though sometimes not at all readily, to a substituted carbamic acid which decarboxylates spontaneously to yield the secondary amine.

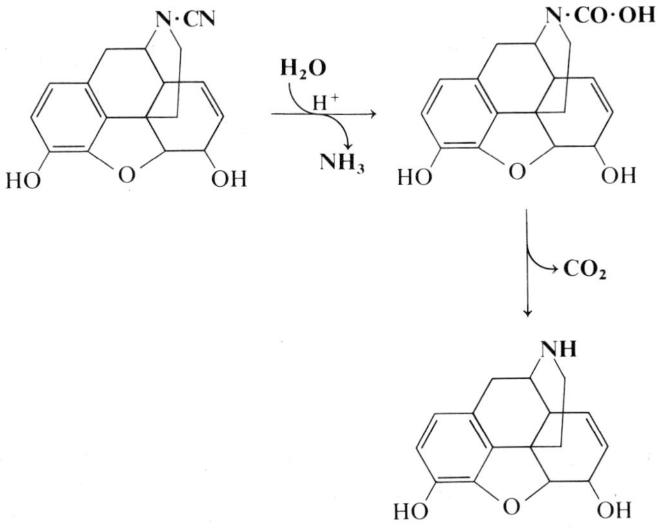

N·CN  HO  O  OH  $\xrightarrow[NH_3]{H_2O,\ H^+}$  N·CO·OH  HO  O  OH  $\xrightarrow{-CO_2}$  NH  HO  O  OH

Normorphine

Tertiary amine salts also dealkylate when strongly heated (*ca* 300°C).

$$R_2\cdot\overset{+}{N}\cdot\mathbf{MeH}\ \ \mathbf{I^-} \longrightarrow R_2\cdot NH + \mathbf{MeI}$$

*N*-Dealkylation is also a prominent feature of the metabolism of many tertiary bases. In the case of the antidepressant, *Imipramine*, the product secondary base, *Desipramine*, is a much more powerful drug which when administered as such is much more rapidly acting; with others, however, the process may represent a true detoxification. The process is an oxidative one (**2**, 5), leading to the separation of the alkyl group as an aldehyde (Mueller and Miller, 1953; La Du, Gaudette, Trousof and Brodie, 1955; Gaudette and Brodie, 1959; Dingell, Sulser and Gillette, 1964).

The precise mechanism of oxidative dealkylation is still not clear. Ziegler and Pettit (1964) have produced indirect evidence that the corresponding *N*-oxide is an intermediate in the oxidative dealkylation of *NN*-dimethylaniline, by showing that *N*-oxide formation occurs more readily than the overall *N*-demethylation in pig and rat liver microsomes. An alternative pathway involving a hydroxy alkyl intermediate, $>N\cdot CH_2OH$, (Keberle, Riesse, Schmidt and Hofmann, 1963) is based on the isolation of *N*-hydroxymethylglutethimide glucuronide as a metabolite of *N*-methylglutethimide. This seems more probable since it has also been observed that although *Dextropropoxyphene N*-oxide is appreciably more lipid-soluble than *Dextropropoxyphene*, the *N*-oxide is dealkylated more slowly than the parent amine.

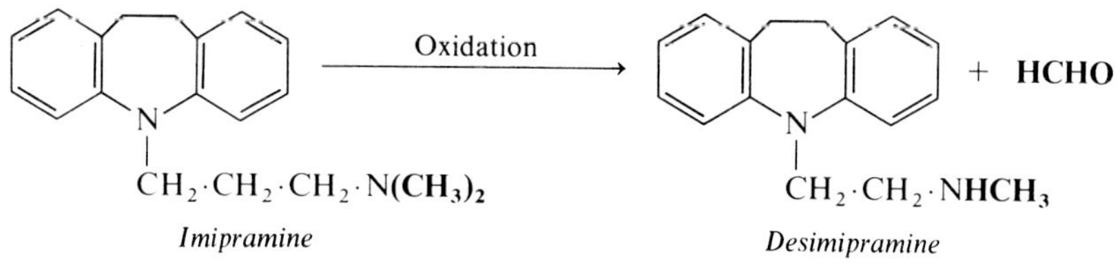

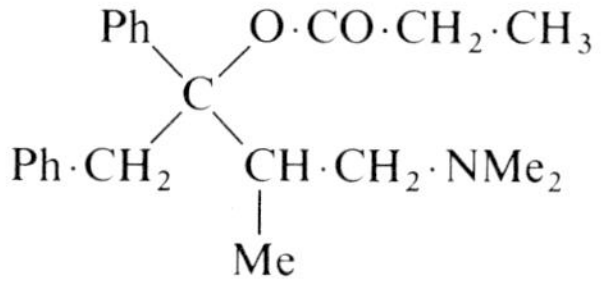

Oxidation

$CH_2 \cdot CH_2 \cdot CH_2 \cdot N(CH_3)_2$ → $CH_2 \cdot CH_2 \cdot NHCH_3$ + HCHO

*Imipramine* *Desimipramine*

Ph, $O \cdot CO \cdot CH_2 \cdot CH_3$, C, $Ph \cdot CH_2$, $CH \cdot CH_2 \cdot NMe_2$, Me

*Dextropropoxyphene*

## *N*-Acylation

Acylation of primary and secondary amines is analogous to *N*-alkylation in that it occurs by nucleophilic attack, at a centre of electron deficiency; in acylation, this is usually the carbonyl carbon of an acid chloride or acid anhydride. The product is an acid amide (Chapter 11).

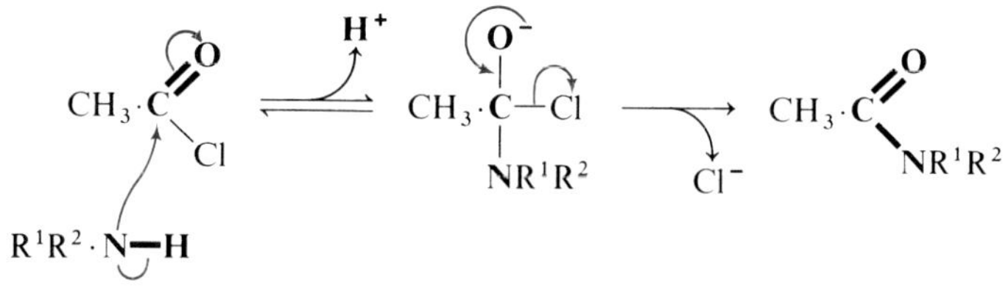

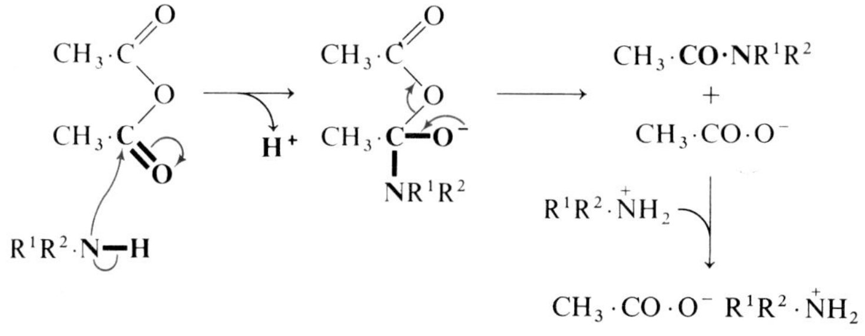

$CH_3 \cdot C(=O)Cl$ + $R^1R^2 \cdot N{-}H$ $\rightleftharpoons$ ($H^+$) $CH_3 \cdot C(O^-)(Cl)NR^1R^2$ → ($Cl^-$) $CH_3 \cdot C(=O)NR^1R^2$

$CH_3 \cdot C(=O) \cdot O \cdot C(=O) \cdot CH_3$ + $R^1R^2 \cdot N{-}H$ → ($H^+$) $CH_3 \cdot C(=O) \cdot O \cdot C(O^-)(CH_3)NR^1R^2$ → $CH_3 \cdot CO \cdot NR^1R^2$ + $CH_3 \cdot CO \cdot O^-$

$CH_3 \cdot CO \cdot O^-$ + $R^1R^2 \cdot \overset{+}{N}H_2$ → $CH_3 \cdot CO \cdot O^-\ R^1R^2 \cdot \overset{+}{N}H_2$

Acylation of aliphatic amines with carbonyl chloride (phosgene) yields symmetrical *NN*-tetra-substituted ureas.

$$COCl_2 + R^1R^2{\cdot}NH \longrightarrow R^1R^2N{\cdot}CO{\cdot}NR^1R^2 + 2HCl$$

*N*-Acetylation forms an important pathway for the metabolism of primary aromatic amines (*p*-aminobenzoic acid; sulphonamides) in most mammalian systems, except the dog (2, 5). Acetyl groups are transferred from $CH_3{\cdot}CO{\cdot}S{\cdot}\overline{CoA}$ by a specific enzyme, arylamine acetyltransferase, present in liver, kidney and in some species also in the reticuloendothelial cells of the lungs, spleen and gastro-intestinal mucosa (Govier, 1965; Hartiala and Terho, 1965).

$$H_2N{-}C_6H_4{-}SO_2{\cdot}NH_2 \xrightarrow[CH_3{\cdot}CO{\cdot}S{\cdot}\overline{CoA} \;\to\; HS{\cdot}\overline{CoA}]{\text{Arylamine acetyltransferase}} CH_3{\cdot}CO{\cdot}NH{-}C_6H_4{-}SO_2{\cdot}NH_2$$

Lack of amino-acetylation in the dog is thought to be due to the presence of an arylamine acetyltransferase inhibitor in the liver and kidney (Leibman and Anaclerio, 1961).

Drugs, such as *Paracetamol* (Acetaminophen), which are already *N*-acetylated are water-soluble (1 in 70 at 20°C) and readily excreted in the urine. $N^4$-Acetylation of sulphonamides however, often gives poorly soluble products, which can give rise to crystallurea. *N*-acetylation generally reduces toxicity.

$$HO{-}C_6H_4{-}NH{\cdot}CO{\cdot}CH_3$$

*Paracetamol* (Acetaminophen)

### Carbamylation

Primary and secondary amines tend to undergo carbamylation when exposed to carbon dioxide in a reaction which is the result of nucleophilic attack on the carbonyl carbon of carbon dioxide. The carbamic acids are unstable, but are stabilised in the presence of excess amine by salt formation.

$$R^1R^2N{-}H + O{=}C{=}O \rightleftharpoons R^1{\cdot}R^2{\cdot}N{\cdot}CO{\cdot}OH \xrightarrow{R^1{\cdot}R^2{\cdot}NH} R^1{\cdot}R^2{\cdot}N{\cdot}CO{\cdot}O^- \; R^1{\cdot}R^2{\cdot}\overset{+}{N}H_2$$

An analogous reaction catalysed by carbamyl phosphate synthetase is responsible for the incorporation of ammonia into urea and pyrimidines in microbiological and mammalian systems.

$$CO_2 + NH_3 + 2ATP + H_2O \rightleftharpoons H_2N{\cdot}CO{\cdot}O{\cdot}PO_3H_2 + 2ADP + H_3PO_4$$

**Sulphonylation**

Sulphonylation of primary and secondary amines with benzene-, toluene- or methane-sulphonyl chloride to yield sulphonamides is essentially analogous to acylation, and forms the basis of the Hinsberg method for the separation of primary, secondary and tertiary amines (Chapter 14).

**Azomethines (Imines) and Enamines**

Primary and secondary amines react readily with aldehydes, and less readily with ketones, by nucleophilic addition to the carbonyl carbon. Formaldehyde forms *N*-hydroxymethyl derivatives from primary amines, which often react with a second molecule of base to give a diaminomethane.

$$R\cdot NH_2 + H_2C{=}O \longrightarrow R\cdot NH\cdot CH_2\cdot OH \xrightarrow[-H_2O]{R\cdot NH_2} R\cdot NH\cdot CH_2\cdot NH\cdot R$$

The *N*-hydroxymethyl derivatives derived from higher aliphatic and aromatic aldehydes and primary amines are unstable and readily lose water to form azomethines or Schiff's bases (Chapter 10), which are the only isolable products of reaction in acidic solution.

$$R\cdot \overset{+}{N}H_3 \underset{H^+}{\rightleftharpoons} R\cdot \ddot{N}H_2 + Ph(H)C{=}\overset{+}{O}H \underset{H^+}{\rightleftharpoons} Ph(H)C{=}O$$

$$\downarrow\uparrow \text{ Slow}$$

$$R\cdot \overset{+}{N}H_2\text{—}CH(Ph)\text{—}OH \xrightarrow[-H_2O,\ -H^+]{\text{Fast}} R\cdot N{=}CH\cdot Ph$$

Schiff's Base

This reaction is of considerable significance in connection with extemporaneous additions of primary amino drug additives for convenient administration in large volume intravenous infusions containing *Dextrose* and *Laevulose*. During sterilisation by autoclaving, the sugars in these solutions tend to undergo some cyclodehydration to form 2-hydroxymethylfurfuraldehyde (Chapter 21), which on oxidation gives rise to laevulinic and formic acids. As a result, *Dextrose* and *Laevulose* (Fructose) *Infusions* become distinctly acid on autoclaving (*Dextrose Infusion* pH 3.5–5.5; *Laevulose Infusion* pH 3.0–5.5). This acidity promotes interaction of 2-hydroxymethylfurfuraldehyde with primary amine additives

(Blaug and Huang, 1972), such as sulphonamides, *Ampicillin*, and other amino antibiotics, leading to loss of effective activity (Jacobs, Kletter, Superstine, Hill, Lynn and Webb, 1973; Lynn, 1974).

The rate-determining step is a nucleophilic addition of the base itself to the carbonyl carbon of 2-hydroxymethylfurfural. This is acid-catalysed due to protonation of the carbonyl oxygen, but reaction rates are pH-dependent with the optimum pH such that maximum protonation of the carbonyl group is achieved with minimum protonation of the attacking base. For this reason, a knowledge of the base dissociation constant is important in order to predetermine the likelihood of reaction.

The ionisation of organic bases may be expressed as follows:

$$BH^+ \rightleftharpoons B + H^+$$

from which

$$pK_a = pH + \log \frac{[BH^+]}{[B]}$$

This relationship between pH, $pK_a$ and ionisation can be more effectively represented diagrammatically, as in Fig. 25.

Figure 25 shows that weak aromatic and heteroaromatic amines, such as the antibacterial sulphonamides, which have $pK_a(Ar\overset{+}{N}H_3 \rightleftharpoons ArNH_2 + H^+)$ between 1.6 and 2.4, will be almost completely ionised at the pH of *Dextrose*

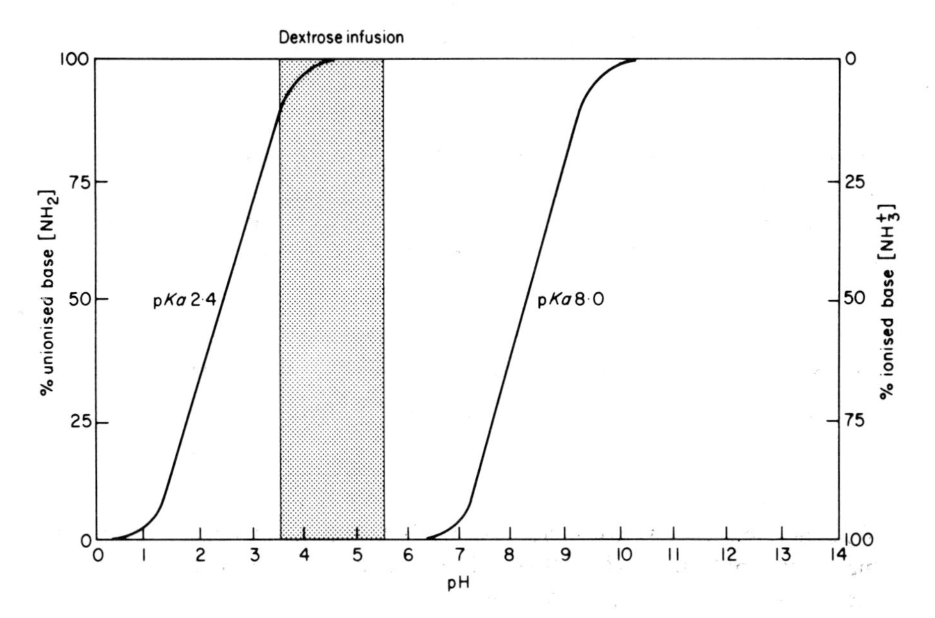


Fig. 25 Variation in ionisation of organic bases with pH

*Infusion* and hence readily capable of reaction, whereas stronger aliphatic bases of $pK_a$ 8 and above will be almost completely ionised and, hence, very much slower to react.

Schiff's base formation in such solutions is also catalysed by phosphate ions (Webb, 1935), presumably due to their buffering effect and its influence on dissociation. The presence of phosphate additives in conjunction with primary amino medicinal amines in infusion fluids should, therefore, be avoided. Similar catalytic effects have also been encountered with tartrate, citrate and acetate ions (Koshy, Duvall, Troup and Pyles, 1965), all of which are of common occurrence as anions in salts of medicinal amines. It would be best, therefore, if primary amino salts incorporating these and other similar anions such as maleate, succinate and, indeed, phosphate, did not find their way into *Dextrose* and *Laevulose Infusions*.

It should perhaps be noted that any Schiff's base condensation products of medicinal amines would almost certainly be metabolically degradable by hydrolysis to release the parent amine (**2**, 5). None-the-less, Schiff's base formation could well retard the onset of drug action, and impair not only excretion kinetics (**2**, 4), but also clinical efficacy of the drug.

Ketones similarly react with primary amines to form ketimines. The reaction, however, takes place less readily and only if catalysed by acid or metal salt catalysts ($ZnCl_2$). Primary amines also undergo reductive condensation with aldehydes and ketones in the presence of hydrogen and Raney nickel.

$$R^1R^2C{=}O + H_2N{\cdot}R^3 \xrightarrow[-H_2O]{H_2,\ Ni} R^1R^2CH{\cdot}NH{\cdot}R^3$$

Aromatic Schiff's bases are readily hydrolysed by aqueous acid to the component aldehyde and amine.

Secondary amines cannot form imines, but react with aldehydes and ketones with C—H bonds α to the carbonyl carbon to form enamines.

$$Et_2NH + H(CH_2R)C{=}\overset{+}{O}H \xrightarrow{-H^+} Et_2N{-}C(H)(OH){-}CH_2R \xrightarrow[-H_2O]{H^+} Et_2N{-}CH{=}CHR$$

Enamine

One such compound is formed in the metabolism of the analgesic, *Methadone*, which is metabolically *N*-demethylated to a secondary amine capable of undergoing intramolecular enamine formation.

$$Me_2N\cdot CH(Me)\cdot CH_2\cdot CPh_2\cdot \mathbf{CO}\cdot CH_2\cdot CH_3 \dashrightarrow MeNH\cdot CH(Me)\cdot CH_2\cdot CPh_2\cdot C(=\overset{+}{O}H)\cdot CH_2CH_3$$

$$\downarrow$$

Enamine (1,5-dimethyl-3,3-diphenyl-2-ethylidenepyrrolidine) $\xleftarrow{-H_2O}$ cyclic carbinolamine (2-hydroxy intermediate, loses $H^+$)

## OXIDATION OF AMINES

All amines are susceptible to oxidation and aromatic amines in particular discolour rapidly on exposure to air. Vigorous oxidation of primary and secondary amines with potassium permanganate gives rise primarily to aldimines and ketimines, but further oxidation results in cleavage of the C—N bond.

$$R\cdot CH_2\cdot NH_2 \xrightarrow{\text{Oxidation}} R\cdot CH{=}NH \xrightarrow{\text{Oxidation}} R\cdot \mathbf{CO\cdot OH}$$

$$R_2\cdot CH\cdot NH_2 \xrightarrow{\text{Oxidation}} R_2C{=}NH$$

Primary aliphatic amines lacking an $\alpha$ C—H bond are oxidised by Caro's acid (peroxymonosulphuric acid, $H_2SO_5$) to *N*-alkylhydroxylamines (Bamberger and Seligman, 1903), and by potassium permanganate to nitroalkanes (Kornblum and Cutter, 1954).

$$(CH_3)_3\cdot C\cdot \mathbf{NH_2} \longrightarrow (CH_3)_3\cdot C\cdot \mathbf{NHOH} \longrightarrow (CH_3)_3C\cdot \mathbf{NO}$$

$$\downarrow$$

$$(CH_3)_3C\cdot \mathbf{NO_2}$$

Oxidation of primary aromatic amines with Caro's acid gives a mixture of nitroso and nitrobenzenes. The latter are obtained as the sole product in excellent yield with $H_2O_2$ and $CF_3\cdot COOH$; chromic acid oxidises primary aromatic amines to *p*-benzoquinones.

**Autoxidation**

Aliphatic amines are readily susceptible to light-catalysed air oxidation. Although some indoles have been shown to undergo radical-induced hydroperoxidations (Witkop and Patrick, 1952), induced initially by radical cleavage of the N—H bond, it is clear because of the oxidative instability of tertiary amines that this mechanism is by no means essential. There is little doubt, therefore, that the primary point for hydroperoxidation is at an α C—H bond, i.e. adjacent to nitrogen (Schenck, 1957; Höft and Schultze, 1967; Hawkins, 1951, 1969). Subsequent decomposition is dependent upon the conditions and reagents present.

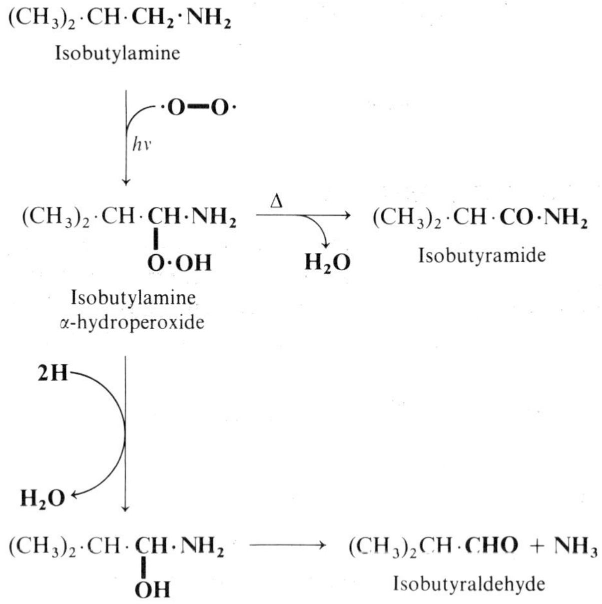
(CH3)2·CH·CH2·NH2
Isobutylamine
·O—O·
hν
(CH3)2·CH·CH·NH2
O·OH
Isobutylamine
α-hydroperoxide
Δ
H2O
(CH3)2·CH·CO·NH2
Isobutyramide
2H
H2O
(CH3)2·CH·CH·NH2
OH
(CH3)2CH·CHO + NH3
Isobutyraldehyde

Microsomal hydroxylation of tertiary amines often proceeds, similarly, by oxidation of an α C—H bond adjacent to nitrogen. The resulting carbinolamine is frequently unstable and subject to further metabolic transformation such as *N*-dealkylation. In certain appropriately-substituted compounds, such as *Diazepam*, a stable carbinolamine results (Sadèe, Garland and Castagnoli, 1971).

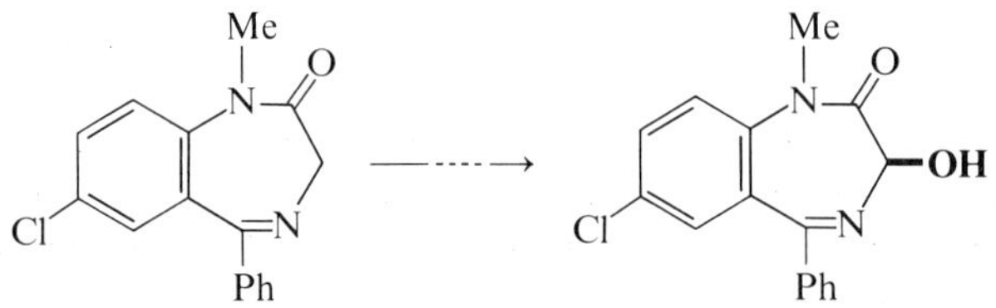
Me
O
N
Cl
N
Ph
Me
O
N
OH
Cl
N
Ph

*Diazepam*

Secondary and tertiary amines undergo autoxidation and subsequent degradation by similar pathways.

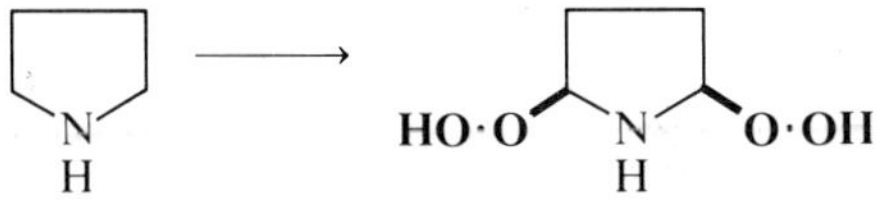

Pyrrolidine

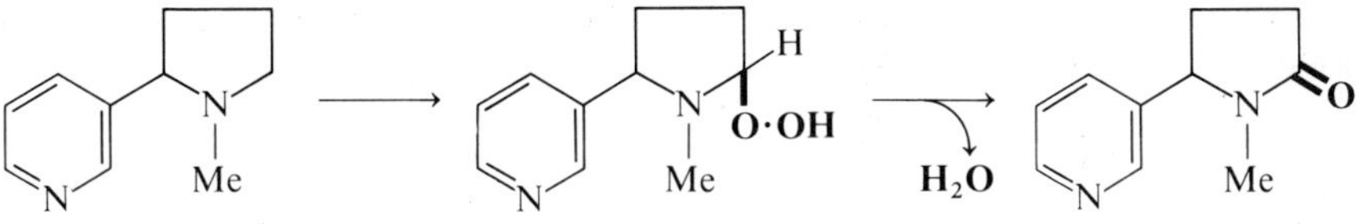

Nicotine — Nicotine-1′-hydroperoxide — $H_2O$ — Cotarnine

Primary aromatic amines are also susceptible to air oxidation forming phenazines, particularly in the presence of metal catalysts, as exemplified by cuprous chloride in pyridine (Terent'ev and Mogilyanskii, 1958).

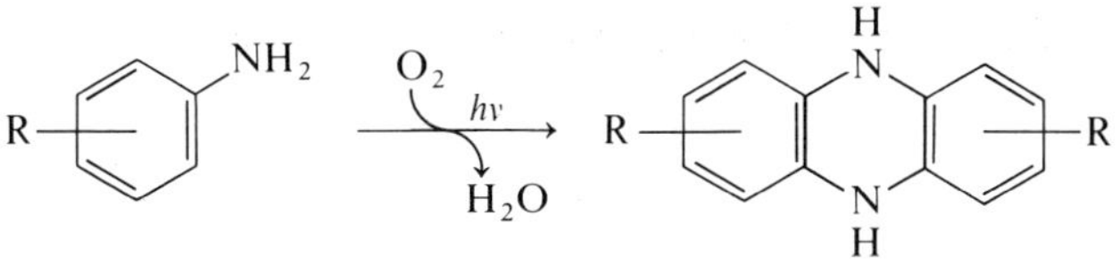

### *N*-Oxides and Nitrones

Tertiary amines are resistant to oxidation by potassium permanganate, but are oxidised by peroxymonosulphuric acid or hydrogen peroxide to *N*-oxides (Cope and Towle, 1949; Cope and Hiok-Huang Lee, 1957).

$$R_3{\cdot}N \longrightarrow R_3{\cdot}\overset{+}{N}{-}O^-$$

The *N*-oxides are hygroscopic, but readily form crystalline hydrates ($Me_3\overset{+}{N}{-}\overset{-}{O}, 2H_2O$). They are thermally unstable, but anhydrous amine oxides can usually be obtained by carefully heating *in vacuo*. They are weakly basic; methylamine *N*-oxide has $pK_a$ 4.65, but readily forms a stable crystalline picrate. They react with hydrogen halides and alkyl halides to form *N*-hydroxy- and *N*-alkoxy-ammonium salts.

$$(CH_3)_3{\cdot}\overset{+}{N}{-}O^- \quad H{-}X \longrightarrow (CH_3)_3\overset{+}{N}{-}OH \quad X^-$$

$$(CH_3)_3\overset{+}{N}{-}O^- \quad CH_3{-}I \longrightarrow (CH_3)_3\overset{+}{N}{-}OCH_3 \quad I^-$$

Thermal decomposition of *N*-oxides gives rise to either an olefine and hydroxylamine or, alternatively, if formation of an olefine is not feasible, to a secondary amine and an aldehyde, as in the case of trimethylamine-*N*-oxide.

$$CH_3\cdot CH_2\cdot CH_2\cdot \overset{+}{N}(Me)_2-O^- \longrightarrow CH_3\cdot CH{=}CH_2 + Me_2NOH$$

$$Me_3\overset{+}{N}-O \longrightarrow Me_2NH + HCHO$$

Such decomposition as this undoubtedly accounts for the apparent spontaneous decomposition of chlorpromazine-*N*-oxide to desmethylchlorpromazine which had been isolated as a metabolite of chlorpromazine from human urine.

$CH_2\cdot CH_2\cdot CH_2\cdot \overset{+}{N}(CH_3)_2-O^-$ ⟶ $CH_2\cdot CH_2\cdot CH_2NH\cdot CH_3$ + HCHO

Chlorpromazine-*N*-oxide — Desmethylchlorpromazine

Unsaturated *N*-oxides, known as nitrones, have structures as follows.

$$R^1R^2\overset{+}{C}-\ddot{N}(O^-)R^3 \longleftrightarrow R^1R^2C{=}\overset{+}{N}(O^-)R^3$$

They are light-sensitive and rearrange to oxiridines on exposure to light. The tranquilliser, *Chlordiazepoxide*, undergoes such a light-catalysed rearrangement. It is, however, reversible by heat (in isopropanol) or by the action of dilute mineral acid (Sternbach, Koechlin and Reeder, 1962). The oxiridine can be distinguished by its ability to release iodine from potassium iodide, a property not shown by the nitrone.

*Chlordiazepoxide* $\underset{\Delta}{\overset{h\nu}{\rightleftharpoons}}$ Oxaziridine

It is interesting to note that many oxiridines, not only revert thermally to the nitrone, but also undergo rearrangement to amides (Spence, Taylor and Buchardt, 1970).

$$R\cdot CH{=}\overset{+}{N}(O^-)-R^1 \underset{\Delta}{\overset{h\nu}{\rightleftharpoons}} (H)(R)C\overset{O}{-}N-R^1 \overset{\Delta}{\longrightarrow} O{=}C(H)-N(R^1)(R)$$

# BIOLOGICAL OXIDATION OF AMINES

## Oxidative Deamination by Monoamine Oxidase

Primary amines are deaminated by monoamine oxidase (MAO), an enzyme which is present in the liver, kidney, heart, lung, intestines and central nervous system of mammals. It is present in the mitochondrial fraction of liver homogenates alongside the cytochrome–cytochrome oxidase enzymes which promote further oxidation of the aldehydes formed by monoamine oxidase. Monoamine oxidase deaminates both endogenous amines such as tyramine, 5-hydroxytryptamine and catecholamines, and also amines of exogenous origin.

(HO)(HO)$C_6H_3$—CHOH·$CH_2$·$NH_2$ —Monoamine oxidase→ (HO)(HO)$C_6H_3$—CHOH·CHO + $NH_3$ (→ Urea cycle)

(HO)(HO)$C_6H_3$—CHOH·CHO —[Cytochrome(ox) / Cytochrome oxidase / Cytochrome(red)]→ (HO)(HO)$C_6H_3$—CHOH·CO·OH

Monoamine oxidase is also capable of oxidising secondary amines to the corresponding aldehyde; some 25% of *Adrenaline* (Epinephrine) is considered to be oxidised directly by this enzyme; the remainder undergoes 3-*O*-methylation to metanephrine. Monoamine oxidase shows some substrate specificity which, with aliphatic amines, is at a peak with a chain length of four carbon atoms. The enzyme is stable to acid, but pH-dependent (Hare, 1928) and inhibited in competition with hydrogen ions. It is destroyed by alkali at pH 9 and above. Chain-branching at the $\alpha$ carbon atom as in *Amphetamine* causes resistance to oxidation by monoamine oxidase. *Amphetamine*, however, is capable of being oxidatively deaminated in the rabbit (but not the dog or rat) by an amine oxidase, which differs from monoamine oxidase (Axelrod, 1955).

## MAO Inhibitors

The development of monoamine oxidase inhibitors stems from observations of marked central nervous anti-depressant activity in the antitubercular drug, *Iproniazid*. This was found to be due to its ability to inhibit monoamine oxidase. Its site of action, however, is not limited to the central nervous system, and inhibition of the enzyme in liver, spleen, kidney and other tissues is also marked.

As a result, it can cause acute and sometimes fatal liver damage, and for this reason has been replaced by other antidepressants with greater selectivity towards the inhibition of brain MAO only.

$(CH_3)_2\cdot CH\cdot NH\cdot NH\cdot CO\cdot C_5H_4N$

*Iproniazid*

$C_6H_5\cdot CH_2\cdot CH_2\cdot NH\cdot NH_2,\ H_2SO_4$

*Phenelzine Sulphate*

$[C_6H_5\cdot C_3H_4\cdot NH_2]_2\ H_2SO_4$

*Tranylcypromine Sulphate*

$C_5H_4N\cdot CO\cdot NH\cdot NH\cdot CH_2\cdot CH_2\cdot CO\cdot NH\cdot CH_2\cdot Ph$

*Nialamide*

$\text{Me-isoxazolyl}\cdot CO\cdot NH\cdot NH\cdot CH_2\cdot Ph$

*Isocarboxazid*

The majority of MAO inhibitors now used are hydrazide derivatives of the *Isoproniazid* type; these include *Nialamide*, *Isocarboxazid* and *Phenelzine*. Compounds of this type are non-competitive inhibitors and with actions which are only slowly reversible. None is free from effect on liver MAO, and hence all are potentially capable of causing hepatic damage.

In contrast, *Tranylcypromine* is a competitive inhibitor of MAO. It is much more rapid in both onset and cessation of action, and is widely used as an antidepressant.

Considerable care is essential to avoid hypertensive crises in patients undergoing treatment with MAO inhibitors which can occur if pressor amines (amphetamines, catecholamines, ephedrine), tryptophan or methyldopa are administered. Certain foodstuffs, particularly cheese and various meat and yeast extracts, with high contents of amines (tyramine) or amino acids (tyrosine) which are normally metabolised by MAO may also precipitate similar hypertensive crises.

## Oxidative Deamination by Diamine Oxidases

Diamine oxidases oxidatively deaminate $\alpha\Omega$-diamines such as putrescine, cadaverine and histamine. Diamine oxidases are distributed in the mitochondria of a wide variety of tissues, but particularly in the liver, kidney, intestinal mucosa and blood plasma, which contain a number of specific diamine oxidases, including histaminase. They are capable of oxidising only at one end of the diamine chain.

$$H_2N(CH_2)_n \cdot \mathbf{NH_2} \xrightarrow[\mathbf{NH_3}]{\text{Diamine oxidase}} H_2N(CH_2)_{n-1} \cdot CHO \longrightarrow H_2N(CH_2)_{n-1} \cdot CO \cdot OH$$

Diamines with nine or more carbon atoms are not deaminated by diamine oxidase, but are inhibited by monoamine oxidase. Diamine oxidases are specific for primary diamines only, but **histaminase** is able to metabolise histamine by oxidation of the terminal $NH_2$ group.

$CH_2 \cdot CH_2 \cdot \mathbf{NH_2}$ (HN, N imidazole ring) — Histamine $\xrightarrow[\mathbf{NH_3}]{\text{Histaminase}}$ $CH_2 \cdot CHO$ (HN, N imidazole ring) $\longrightarrow$ $CH_2 \cdot CO \cdot OH$ (HN, N imidazole ring)

Histamine

Imidazolylacetic acid

Substitution on the terminal amino nitrogen renders the compound incapable of being oxidised by this pathway and leads to active enzyme inhibitors.

Diamine oxidase activity in the blood is enormously increased in pregnancy due to its release from the placenta. This increase begins in the third month, rising to a maximum in the sixth month when it may be as much as a thousand times the normal level (Southren, Kobayashi, Sherman, Levine, Gordon and Weingold, 1964). The significance of this increase, however, is not known.

### *N*-Oxidation of Primary and Secondary Amines

Primary and secondary amines also undergo *N*-oxidation in the liver to nitroso and hydroxylamino compounds. Thus, Boyland and Manson (1966) observed the NADPH-dependent oxidation of 2-naphthylamine to 2-napthylhydroxylamine in both liver and lung.

2-Naphthyl-$\mathbf{NH_2}$ $\xrightarrow[\mathbf{H_2O} + \text{NADP}]{\mathbf{O_2} + \text{NADPH}}$ 2-Naphthyl-$\mathbf{NHOH}$

Such oxidation is characteristic of arylamines and aliphatic amines which do not posses $\alpha$ C—H bonds (Beckett, Van Dyk, Chissick and Gorrod, 1971; Beckett and Bélanger, 1974). These *N*-oxidations, like $\alpha C$-oxidation, are mediated by flavoprotein enzymes, though unlike *C*-oxidation, the enzyme cytochrome P-450 is not involved.

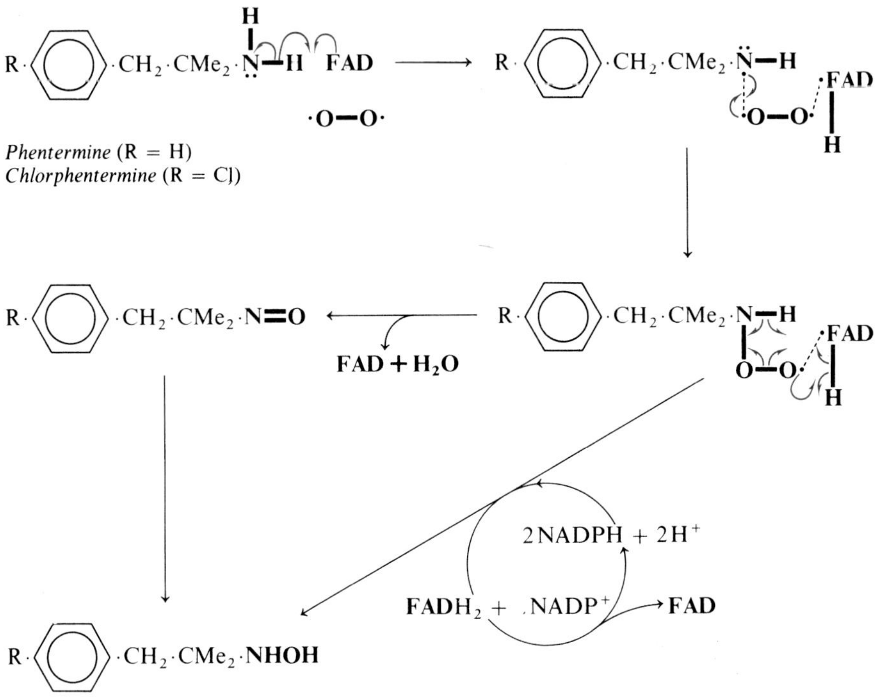

R·C₆H₄·CH₂·CMe₂·N̈—H FAD ·O—O· → R·C₆H₄·CH₂·CMe₂·N̈—H ·O—O·FAD—H

*Phentermine* (R = H)
*Chlorphentermine* (R = Cl)

R·C₆H₄·CH₂·CMe₂·N=O ← (FAD + $H_2O$) ← R·C₆H₄·CH₂·CMe₂·N—H, O—O·FAD—H

2NADPH + 2H⁺, FADH₂ + NADP⁺ → FAD

R·C₆H₄·CH₂·CMe₂·NHOH

There is some evidence that the nitroso and hydroxyamino compounds so formed are not only carcinogenic (Brill and Radomski, 1971), but also responsible for the development of other toxic side-effects (Nery, 1971).

### *N*-Oxidation of Tertiary Amines

A number of tertiary amines have also been shown to be metabolised in humans by *N*-oxidation, as for example in nicotine (Dagne and Castagnoli, 1972), which is converted to cotarnine-*N*-oxide, and in *Chloropromazine* (Beckett, Gorrod and Lazarus, 1971).

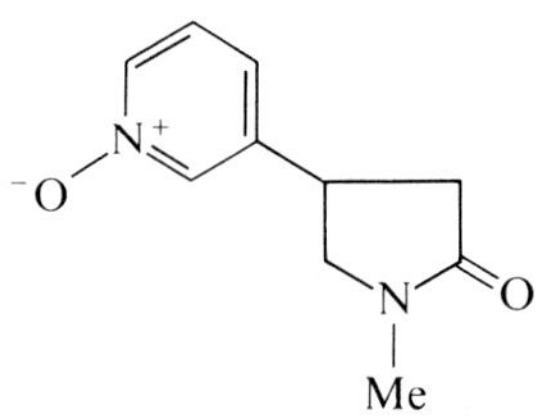

Cotarnine-*N*-oxide

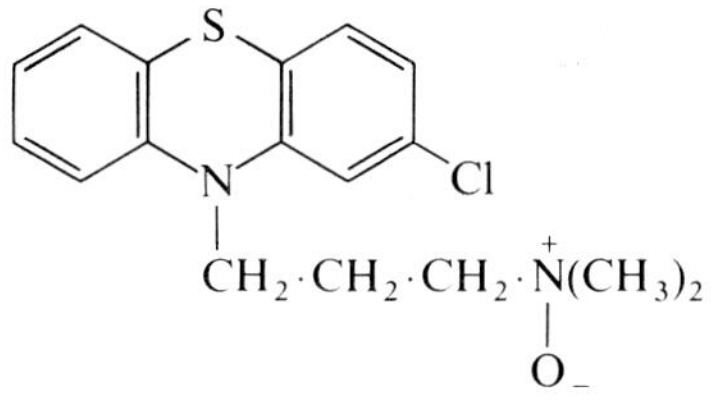

Chlorpromazine-*N*-oxide

## HISTAMINE AND ANTIHISTAMINES

### Histamine

Histamine is widely distributed, particularly in the connective tissues and capsular membranes of organs. In connective tissue, it is stored in subcellular components of **mast** cells in association with the sulphated polysaccharide, heparin (Chapter 21), with which it forms ionic linkages.

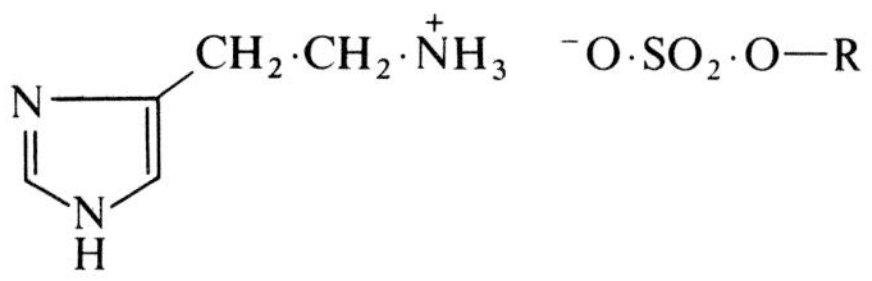

$CH_2{\cdot}CH_2{\cdot}\overset{+}{N}H_3 \quad {}^{-}O{\cdot}SO_2{\cdot}O{-}R$

N

N
H

It is also found in the blood in the basophil leucocytes in association with another sulphated polysaccharide, chondroitin sulphate, and in platelets linked to adenosine diphosphate and triphosphate. Histamine so stored is of endogenous origin and arises by decarboxylation of histidine, under the action of histidine decarboxylase. Biosynthesis is not limited to any one tissue, but large amounts are formed in the gut by bacterial metabolism of purines, a process which involves incorporation of N-1 and C-2 of the purine nucleus into the imidazole ring (Chapter 23).

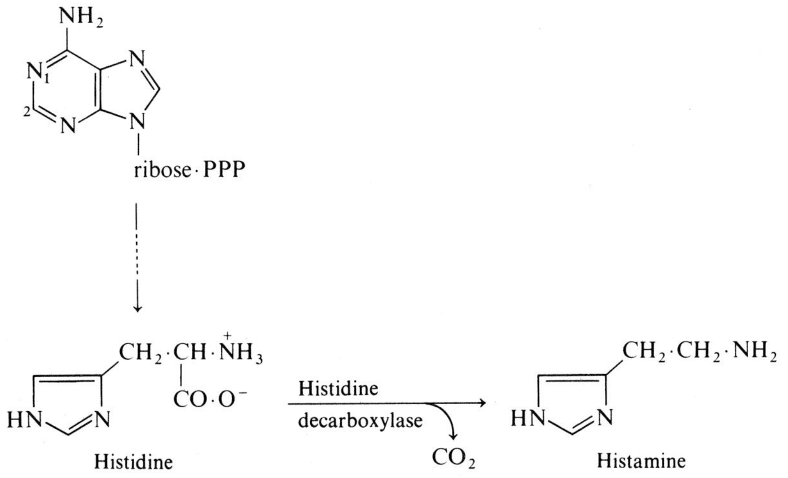

$NH_2$

$N_1$ N

2 N N

ribose·PPP

$CH_2{\cdot}CH{\cdot}\overset{+}{N}H_3$

$CO{\cdot}O^-$

HN N

Histidine

Histidine

decarboxylase

$CO_2$

$CH_2{\cdot}CH_2{\cdot}NH_2$

HN N

Histamine

Histidine decarboxylase is inhibited by the action of α-methylhistidine (Kahlson, Rosengren and Svensson, 1962) and other imidazole derivatives (Robinson and Shepherd, 1962). As already described, one of the main pathways for metabolism of histamine in man is by oxidation with histaminase to imidazolyacetic acid, which is excreted in the urine as a conjugate. Histamine is also metabolised to 4-(2′-aminoethyl)-1-methyl-imidazole, and *N*-acetylhistamine in man.

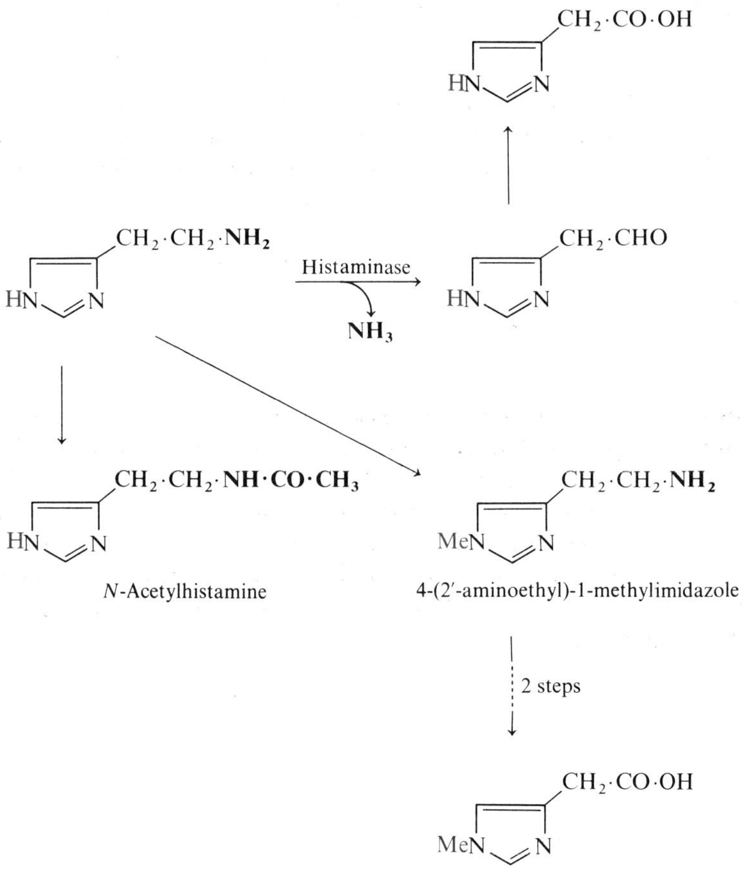

$CH_2 \cdot CO \cdot OH$

HN N

$CH_2 \cdot CH_2 \cdot \mathbf{NH_2}$

Histaminase

$\mathbf{NH_3}$

$CH_2 \cdot CHO$

HN N

HN N

$CH_2 \cdot CH_2 \cdot \mathbf{NH \cdot CO \cdot CH_3}$

HN N

*N*-Acetylhistamine

$CH_2 \cdot CH_2 \cdot \mathbf{NH_2}$

MeN N

4-(2′-aminoethyl)-1-methylimidazole

2 steps

$CH_2 \cdot CO \cdot OH$

MeN N

**Histamine Release**

Histamine is usually released in response to tissue damage. It causes contraction of smooth muscle in the gut, uterus and bronchi. Its release is considered to be the primary cause of constriction of air passages in asthmatic attacks. Histamine also increases permeability of capillary cell walls leading to oedema, when released as a result of skin damage due to burns and scalds. It also increases secretion of hydrochloric acid from acid-secreting glands in the stomach, and similarly increases secretion from the mucous-secreting glands in the nose.

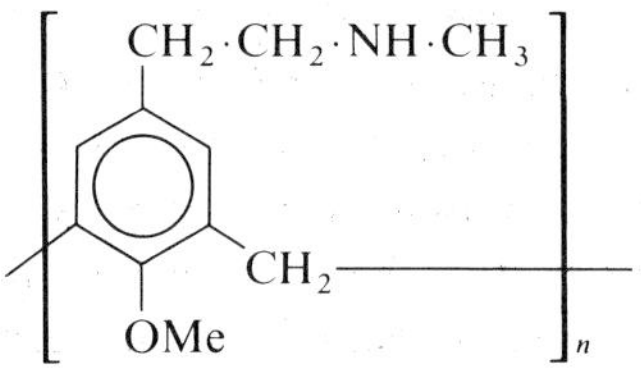

$CH_2 \cdot CH_2 \cdot NH \cdot CH_3$

$CH_2$

OMe

$n$

Polymer 48/80

Histamine release from mast cells is also brought about by a number of organic bases such as morphine. This is believed to be due to simple displacement of histamine by a stronger base from complexation with heparin. The polymer 48/80, which is also able to displace histamine, requires oxygen to be effective and for this reason is considered to depend on some more extensive metabolic effect.

**Antihistamines**

Antihistamines antagonise the actions of histamine at certain of their cellular sites of action, and find practical use principally in the treatment of allergic rhinitis and urticaria. The nature of the receptor sites is not understood, but the strongly basic character of both the histamine side-chain amine (p$K_a$ 9.70; Levy, 1935) and the terminal basic group of most potent antihistamines suggests that ionic bonding is involved through this group in its cationic form, which will be formed almost exclusively at physiological pH. This would explain the action of antihistamines as competitive antagonists of histamine (Bovet, 1947; Furchgott, 1964).

Since not all strong bases have either histamine-like or antihistamine activity, it is clear that other parts of the molecule are also involved. Lee and Jones (1949) concluded that histamine-like activity resided primarily in the moiety:

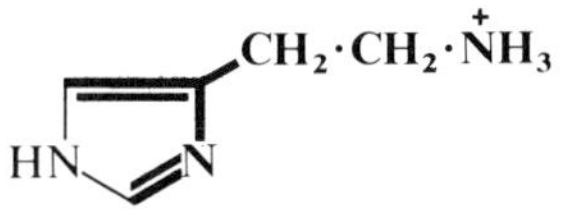

$CH_2{\cdot}CH_2{\cdot}\overset{+}{N}H_3$

HN N

This is supported by the study of a number of simple histamine analogues with analogous $\pi$-excessive five-membered heterocyclic rings.

Antihistamine activity varies greatly with the test preparation, depending whether this is measurement of the effect on gastric secretion, the ability to inhibit histamine-induced spasm of guinea pig ileum, inhibition of skin reactions, or protection against aerosol injection of histamine into the lung. Schild (1947) and his collaborators (Ash and Schild, 1966) distinguish two types of histamine receptor ($H_1$ and $H_2$). Most antihistamines are active only at $H_1$ receptors in smooth muscle (guinea pig ileum and bronchi) and in the skin, and show little or no activity in the stimulation of gastric secretion ($H_2$ receptors).

*N*-Substitution of the terminal amino group by even quite small alkyl groups markedly reduces histamine-like activity at $H_1$ receptors. Thus, histamine-like activity on cat blood pressure is halved in 4-($2^2$-methylaminoethyl)imidazole and reduced to 20% in 4-($2^1$-dimethylaminoethyl)imidazole. 4-Methylimidazole stimulates $H_2$ receptors selectively.

Both the imidazole ring, which is also basic (p$K_a$ 5.90), and the side-chain appear to be essential for activity, and some of the most potent antihistamines are structurally analogous *N′*-aromatic and *N′*-heteroaromatic derivatives of *NN*-dimethylaminoethylenediamine. The structure, p$K_a$ and p$A_2$ values for the more important antihistamines are given below (Table 52).

Table 52. Antihistamines

| Structure | Name | $pK_a$ | $^{1}pA_2$ (guinea-pig ileum) |
|---|---|---|---|
| 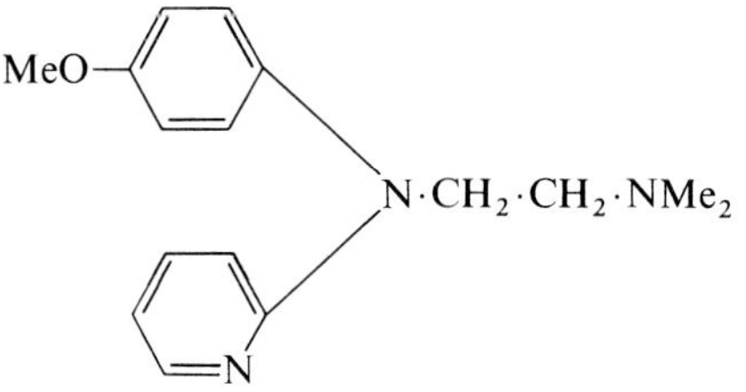 MeO— ; $N \cdot CH_2 \cdot CH_2 \cdot NMe_2$ ; =N | *Mepyramine* (Pyrilamine) | 8.85 | 9.36 |
| 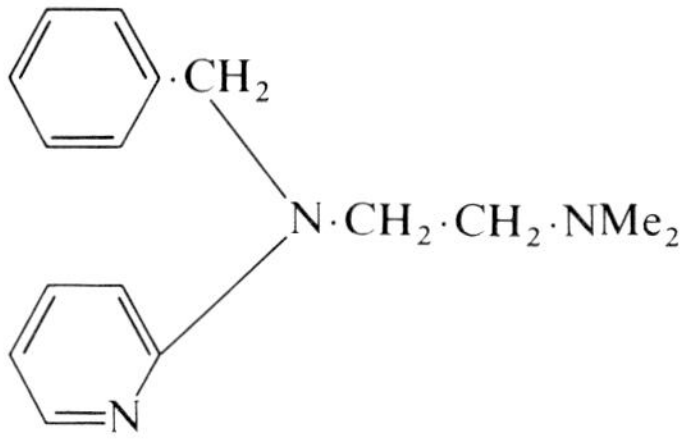 $\cdot CH_2$ ; $N \cdot CH_2 \cdot CH_2 \cdot NMe_2$ ; =N | *Tripelennamine* | 8.95 | 9.00 |
| 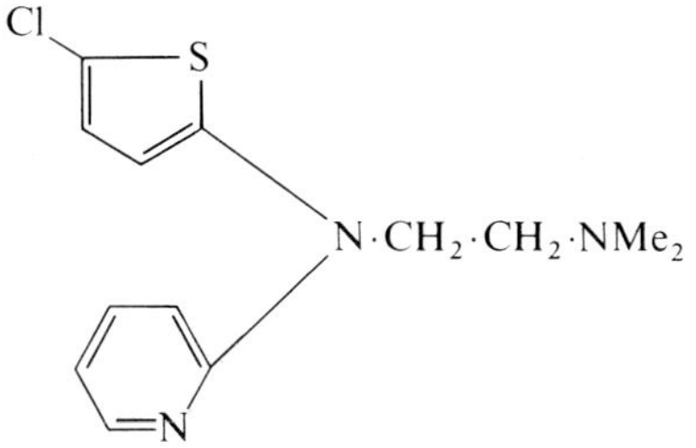 Cl ; S ; $N \cdot CH_2 \cdot CH_2 \cdot NMe_2$ ; =N | *Chloropyrilene* (Chlorothen) | 8.70 | 9.50 |
| 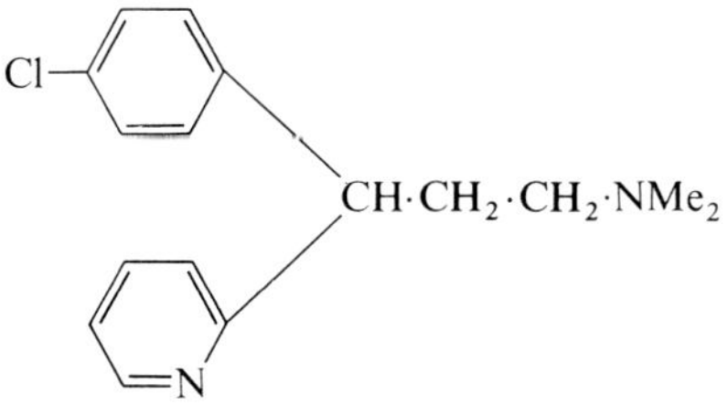 Cl— ; $CH \cdot CH_2 \cdot CH_2 \cdot NMe_2$ ; =N | *Chlorpheniramine* | 9.16 | 8.82 |
| 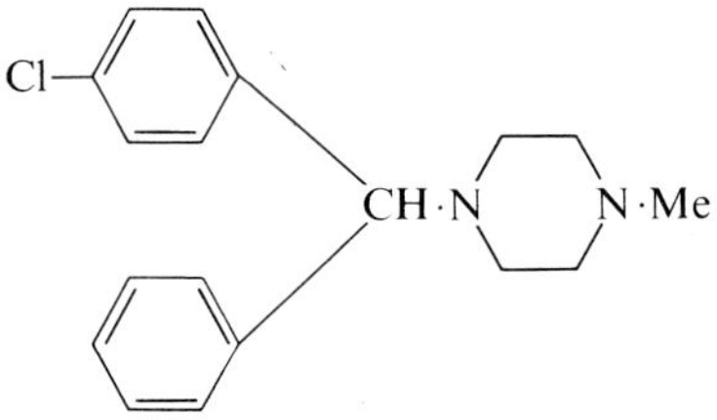 Cl— ; $CH \cdot N$ ; $N \cdot Me$ | *Chlorcyclizine* | 8.15 | 8.63 |

| | | | |
|---|---|---|---|
| Me–C₆H₄–C(C₅H₄N)=CH–$CH_2$·N(pyrrolidine) | *Triprolidine* | | 9.54 |
| $(C_6H_5)_2$CH·O·$CH_2$·$CH_2$·$NMe_2$ | *Diphenhydramine* | 8.98 | 8.14 |
| (phenothiazine) N·$CH_2$·CH(Me)·$NMe_2$ | *Promethazine* | 9.08 | 8.93 |
| (azaphenothiazine) N·$CH_2$·CH(Me)·$NMe_2$ | *Isothipendyl* | | 8.34 |
| $C_6H_5$·$CH_2$·N($C_6H_5$)·$CH_2$·(imidazolinyl, N, NH) | *Antazoline* | 10.00 | 7.67 |

---

[1] p$A_2$ is the negative logarithm of the antagonist concentration required to maintain a constant response when the agonist concentration is doubled.

Various attempts have been made to relate the three-dimensional structure of histamine, its derivatives and analogues, and certain antihistamines to the stereochemical requirements of $H_1$ and $H_2$ histamine receptors. The aminoethyl side-chain of histamine may adopt any one of three possible conformations about its carbon—carbon bond. Of these, one is trans and two are gauche, as represented by the Newman projections along the C—C bond which are shown below.

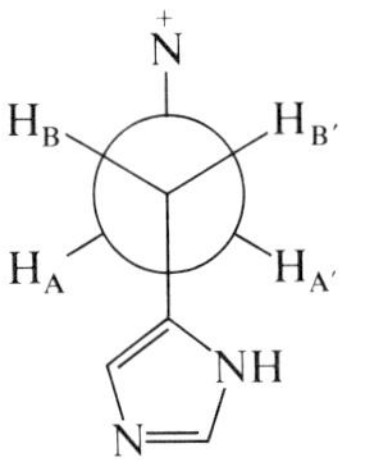

$\overset{+}{N}$ $H_B$ $H_{B'}$ $H_A$ $H_{A'}$ NH N

trans

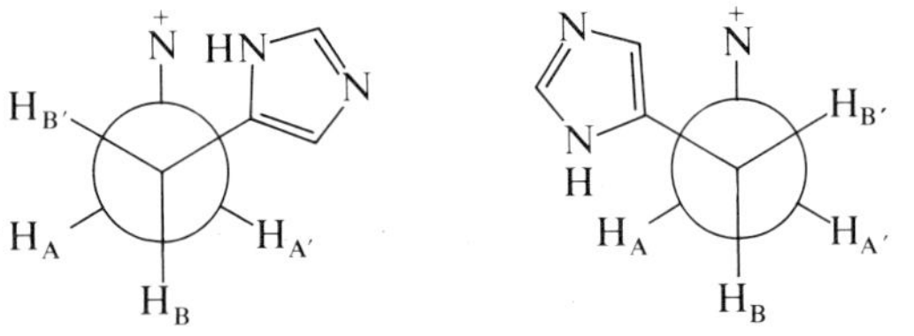

$\overset{+}{N}$ HN N $H_{B'}$ $H_A$ $H_{A'}$ $H_B$     N N H $\overset{+}{N}$ $H_{B'}$ $H_A$ $H_{A'}$ $H_B$

gauche

Kier (1968) found that the three conformations have almost equal total energies. X-ray crystallographic evidence (Veidis, Palenick, Schaffrin and Trotter, 1969) showed that histamine adopts the *trans* conformation in solid state from which it has been assumed, possibly erroneously, that the $H_1$ histamine receptor activity is mediated by the *trans* conformation. Similar conclusions have been reached based on studies of the configuration of a series of 4-amino-1,2-diarylbut-2-enes related to *Triprolidine* (Casy and Ison, 1970) and ethylenediamines (Ham, 1971). Thus, *Tripolidine* is the *cis*(H/pyrid-2-yl)-isomer, which is more potent than the *trans*-isomer (Adamson, Barrett, Billinghurst and Jones, 1957).

In contrast, more recent NMR solution studies (Ham, Casy and Ison, 1973) suggest the reverse conclusion, that $H_1$-receptor activity relates to the gauche conformation of histamine, and that the *trans*-rotamer is associated with gastric secretory ($H_2$-receptor) activity. Thus, histamine diphosphate, adopts the *trans* configuration in the solid state in conformity with the repulsive effects expected of the di-cation. Similarly, NMR studies confirm that the proportion of *trans*-rotamer, which is 47% at physiological pH, increases with decrease in pH. This, therefore, accords with the view that the *trans*-rotamer is important in the induction of gastric-secretory activity. Similarly, the gastric secretogogue, *NN*-dimethylhistamine, which has greatly reduced $H_1$-receptor activity (p. 479), has a higher proportion of the *trans*-rotamer than histamine, which likewise increases with increasing acidity (52% *trans* at pH 8.5; 60% at pH 6.5; 66% at pH 4.0).

The pyrazole analogue of histamine, *Betazole*, which exhibits mainly $H_2$-receptor activity, similarly shows a higher proportion (52%) of *trans*-rotamer than histamine at physiological pH which increases to 61% at pH 1.0. The evidence from *NNN*-trimethylhistamine, of which 76% is present in solution as

$CH_2 \cdot CH_2 \cdot \overset{+}{N}H \cdot Me_2$ HN N

*NN*-Dimethylhistamine

$CH_2 \cdot CH_2 \cdot \overset{+}{N}Me_3$ HN N

*NNN*-Trimethylhistamine

$CH_2 \cdot CH_2 \cdot \overset{+}{N}H_3$ N N H

*Betazole*

*trans*-rotamer is less convincing, since it is inactive both in guinea-pig ileum and gastric-secretory tests.

**$H_2$-Receptor Antagonists**

Some specific gastric secretory inhibitors have been reported. These include a number of phenoxypyridines and arylpyridines, notably 2,2′-bipyridine (Butler, Bass, Nordin, Hauck and L'Italien, 1971) and *Burimamide* (Black, Duncan, Durant, Ganellin and Parsons, 1972). The latter not only blocks histamine-induced gastric acid secretion, but also food- and pentagastrin-induced acid secretion. The response to *Burimamide* is indistinguishable irrespective of the method of stimulation, an observation which is considered to support the view that histamine is positively involved in gastric secretion (Code, 1965), rather than the reverse that it is not (Johnson, 1971). Further developments have led by isosteric replacement to *Metiamide* (Black, Durant, Emmett and Ganellin, 1974) which, because of evidence of its ability in common with other thioureas to cause agranulocytosis, was rapidly discarded in favour of *Cimetidine* (Brimblecombe, Duncan, Durant, Emmett, Ganellin and Parsons, 1975). *Cimetidine* has similar properties to *Burimamide* and *Metiamide* in blocking acid gastric secretion, however stimulated. It also promotes rapid healing of gastric ulcers, and in contrast to *Metiamide* does not cause agranulocytosis.

$(CH_2)_4 \cdot NH \cdot C(=S) \cdot NH_2$ HN N

*Burimamide*

Me $CH_2 \cdot S \cdot CH_2 \cdot CH_2 \cdot NH \cdot C(=S) \cdot NH \cdot Me$ HN N

*Metiamide*

Me $CH_2 \cdot S \cdot CH_2 \cdot CH_2 \cdot NH \cdot C(=N \cdot CN) \cdot NH \cdot Me$ HN N

*Cimetidine*

## REACTIONS OF AMINES WITH NITROUS ACID

### Primary Aliphatic Amines

Primary aliphatic amines react with nitrous acid (sodium nitrite and hydrochloric or sulphuric acid) with a quantitative evolution of nitrogen, a reaction which distinguishes primary from secondary and tertiary amines, and also forms the basis of the Van Slyke method for the determination of amino acids. It has often, and erroneously, been suggested that the principal product of the reaction is the corresponding alcohol. Austin (1960) has shown that whilst some methanol is formed from methylamine, the principal product is methyl nitrite. He has also shown that the primary reagent is not nitrous acid, but dinitrogen trioxide (nitrous anhydride), which is formed from it by protonation and loss of water.

$$HO{-}N{=}O \rightleftharpoons H^+ + O{=}N{-}O^-$$

$$HO{-}N{=}O \xrightarrow{H^+} H_2\overset{+}{O}{-}N{=}O$$

$$O{=}N{-}O^- + H_2\overset{+}{O}{-}N{=}O \longrightarrow O{=}N{-}O{-}N{=}O$$

Nitrous anhydride

$$CH_3{\cdot}\ddot{N}H_2 + O{=}N{-}O{-}N{=}O \xrightarrow{-\ {}^-O{-}N{=}O} CH_3{\cdot}\overset{+}{N}H_2{-}N{=}O \xrightarrow{-H^+} CH_3{\cdot}NH{-}N{=}O \xrightarrow{H^+} CH_3{\cdot}N{=}N{-}OH$$

$$CH_3{\cdot}N{=}N{-}OH \xrightarrow{H^+} CH_3{\cdot}\ddot{N}{=}N{-}\overset{+}{O}H_2 \longrightarrow CH_3{-}\overset{+}{N}{\equiv}N + H{-}O{-}H$$

$$CH_3{-}\overset{+}{N}{\equiv}N \xrightarrow{-N_2} \overset{+}{C}H_3$$

$$\overset{+}{C}H_3 + H{-}O{-}H \longrightarrow CH_3{\cdot}OH + H^+ \quad \text{Methanol}$$

$$\overset{+}{C}H_3 + {}^-O{-}N{=}O \longrightarrow CH_3{\cdot}O{\cdot}N{=}O \quad \text{Methyl nitrite}$$

### Secondary Aliphatic and Aromatic Amines

Secondary amines form nitrosamines with nitrous acid.

$$\underset{R}{\overset{CH_3}{}}\!\!>\ddot{N}H + O{=}N{-}O{-}N{=}O \xrightarrow{-NO_2^-} CH_3{-}\overset{+}{N}(H)(R){-}N{=}O \xrightarrow{-H^+} \underset{R}{\overset{CH_3}{}}\!\!>N{-}N{=}O$$

Nitrosamine

Nitrosamines are yellow, oily liquids, which are insoluble in water. They are decomposed by concentrated sulphuric acid, releasing nitrous acid and, as a result, react with phenol in the presence of sulphuric acid to give a red colour (Liebermann nitroso reaction; Chapter 8).

Secondary nitrosamines have been shown to be carcinogenic. It has been suggested that they are metabolised to diazoalkanes which then alkylate purine and pyrimidine bases in DNA (Magee and Barnes, 1967). Challis and Osborne (1972), however, favour the alternative hypothesis that *N*-nitroso compounds function as proximate carcinogens by transnitrosation, and cite the relative ease with which *N*-nitrosodiphenylamine and *N*-methylaniline are able to interact.

$$Ph_2 \cdot N \cdot \mathbf{NO} + Ph \cdot NHMe \rightleftharpoons Ph_2 \cdot \mathbf{NH} + Ph \cdot NMe \cdot \mathbf{NO}$$

Some concern has also been expressed on the possible carcinogenic effects of the use of sodium nitrite as a curing agent in the preparation of foodstuffs from animal sources (Anon, 1973).

### Tertiary Aliphatic and Aromatic Amines

In general, aliphatic tertiary amines do not react with nitrous acid. The latter may, therefore, be used to distinguish primary, secondary and tertiary aliphatic amines; primary amines evolve nitrogen, secondary amines give a yellow oily and water-insoluble nitrosamine, whilst tertiary amines do not react. Aromatic tertiary amines form nuclear substituted *p*-nitroso compounds (p. 493) provided always that the *p*-position is available for substitution.

## DIAZOTISATION

Primary aromatic amines form benzenediazonium salts in nitrous acid. The reaction is quantitative and titration with nitrous acid is used in the determination of sulphonamides.

$$Ar \cdot NH_2 + HNO_2 + HCl \longrightarrow Ar \cdot \overset{+}{N}_2Cl^- + 2H_2O$$

The diazonium salts have the structure, $Ar \cdot \overset{+}{N} \equiv N\ X^-$, and owe their stability relative to the much more unstable aliphatic diazonium salts to resonance.

$$C_6H_5-\ddot{N}=\overset{+}{N}: \longleftrightarrow C_6H_5-\overset{+}{N}\equiv N: \longleftrightarrow {}^{+}C_6H_5=\overset{+}{N}=\overset{-}{N} \longleftrightarrow {}^{+}C_6H_5=\overset{+}{N}=\overset{-}{N}$$

Aromatic diazonium salts can be isolated as colourless, crystalline solids. They are for the most part unstable, and may decompose violently when in the solid state. They are highly reactive in solution, and diazonium salts, prepared by treating a cooled solution of primary amine in dilute mineral acid with sodium nitrite, are generally used as reagents in further reactions without isolation of the diazonium salt.

Diazonium salts undergo two main types of reaction, those in which nitrogen is eliminated with displacement of the diazo ($-N_2X$) group by some other substituent, and those in which the two nitrogen atoms are retained.

### Displacement Reactions of Diazonium Salts

(a) **Replacement by halogen** is readily accomplished in the Sandmeyer reaction when the diazonium salt solution is heated with a solution of cuprous chloride or bromide (copper (I) halide) dissolved in the corresponding halogen acid.

The cuprous salt catalyses the formation of a complex diazonium salt ($Ph\cdot\overset{+}{N}\equiv NCuX_2^-$), which undergoes radical decomposition to yield monochlorobenzene and re-form copper(I) halide.

$$Ph-\overset{+}{N}\equiv N\ Cl^- \xrightarrow{CuCl} Ph\cdot + N\equiv\overset{+\cdot}{N};\quad Ph\cdot + Cl-Cu-Cl \longrightarrow Ph-Cl + \dot{C}u-Cl$$

Monochlorobenzene

$$N\equiv\overset{+\cdot}{N} + \dot{C}u-Cl \longrightarrow Cu-Cl + N_2$$

Replacement by chlorine or bromine can also be accomplished in the Gatterman reaction by treating the amine with sodium nitrite in the appropriate mineral acid (HCl or HBr) and heating with copper powder.

$$Ph-\overset{+}{N}\equiv N \xrightarrow[Cu^{2+}]{Cu^+} Ph-\overset{+}{N}\equiv N\cdot \xrightarrow[N_2]{} Ph\cdot \xrightarrow[Cu^+]{Cu^{2+}} Ph^+ \xrightarrow{Cl^-} Ph-Cl$$

Iodo-compounds are readily formed by boiling the diazonium salt solution with aqueous potassium iodide. No catalyst is necessary. Fluoro compounds are prepared via the diazonium fluoroborates. These form as insoluble salts when fluoroboric acid ($HBF_4$, prepared by dissolving $HBO_3$ in 50% HF) is added to the diazonium salt solution. The fluoroborate is collected by filtration and decomposed by heat.

$$Ph\cdot\overset{+}{N_2}Cl^- \xrightarrow[HCl]{HBF_4} Ph\cdot\overset{+}{N_2}BF_4^- \xrightarrow[N_2\ +\ BF_4^-]{\Delta} Ph^+ \xrightarrow{BF_3} Ph-F$$

(b) **Replacement by the cyano group** can also be effected with aqueous potassium cyanide either in the presence of copper (I) cyanide or copper powder.

$$\mathbf{Ph\cdot \overset{+}{N}_2HSO_4^-} \xrightarrow[\text{KHSO}_4 + \text{N}_2]{\text{KCN},\ \text{Cu}} \mathbf{Ph{-}CN}$$

(c) **Replacement by hydroxy** to form phenols occurs when diazonium sulphate solutions (obtained by diazotisation with sodium nitrite in sulphuric acid) are boiled, either under strongly acidic conditions (40–50% $H_2SO_4$) or with copper sulphate solution.

$$\mathbf{Ph\cdot \overset{+}{N}_2HSO_4^-} \xrightarrow[N_2 + HSO_4^-]{\Delta} \mathbf{Ph^+} \xrightarrow[\mathbf{H^+}]{\mathbf{H_2O}} \mathbf{Ph{-}OH}$$

(d) **Replacement by hydrogen** occurs in the decomposition of diazonium salts with hypophosphorous acid ($H_3PO_2$) in the presence of finely divided copper. The reaction involves a diazonium radical.

$$\mathbf{Ph}\cdot\overset{+}{N}{\equiv}N \xrightarrow[Cu^{2+}]{Cu^+} \mathbf{Ph}\cdot\overset{+}{N}{\equiv}N\cdot \xrightarrow[N_2]{} \mathbf{Ph}\cdot \xrightarrow[\cdot P(OH)_2]{H{-}P(OH)_2} \mathbf{Ph}\cdot H$$

$$\mathbf{Ph}{-}\overset{+}{N}{\equiv}N \quad \cdot P(OH)_2 \quad H{-}OH \longrightarrow \mathbf{Ph}\cdot + N_2 + H_3PO_3 + H^+$$

Ethanol may also be used as the reagent in this replacement reaction.

(e) **Other replacement reactions.** Under the appropriate conditions, the diazonium group may be replaced by numerous other groups including $—NO_2$, $—SH$, $—SCN$, $—NCO$, $—SO_2OH$, $—AsO_3H_2$, $—HgCl$, and aromatic groups.

### Coupling Reactions of Diazonium Salts with Retention of Nitrogen

The diazonium ion is capable of acting as an electrophile in aromatic substitution. It is, however, a relatively weak electrophile and only attacks suitably activated aromatic compounds (p. 129), notably phenols and amines. The reaction, which is generally known as diazo coupling, gives rise to azo dyes, and is useful not only in the manufacture of such dyes, but also as the basis of a wide range of colorimetric assays and limit tests for primary aromatic amines on the one hand, and, on the other hand, phenols and tertiary amines. Coupling with phenols takes place most readily in alkaline solution.

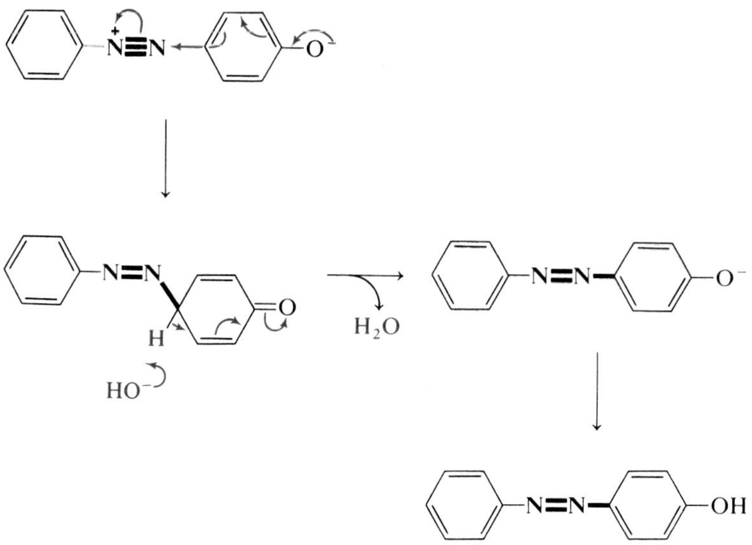

*p*-Hydroxyazobenzene

Coupling occurs in the *para* position to the phenolic hydroxyl. If the *para* position is blocked, the coupling occurs in the *ortho* position. $\beta$-Naphthol which is frequently used couples in the $\alpha$ position, and is often preferred since the benzeneazo-$\beta$-naphthol formed is insoluble in sodium hydroxide due to intramolecular hydrogen bonding.

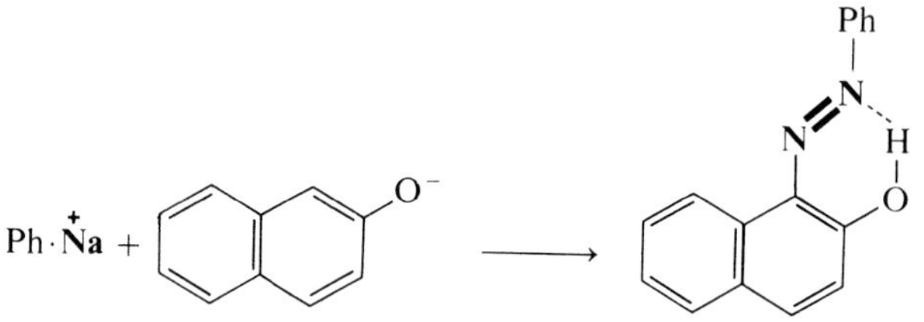

Coupling with aromatic tertiary amines, such as dimethylaniline, similarly results in *para*- or *ortho*-substitution. The reaction takes place most readily in weakly acidic solution, and is thought to involve the free base, and not its cation. The product resulting from the coupling of diazotised sulphanilic acid with dimethylaniline is the titration indicator, methyl orange.

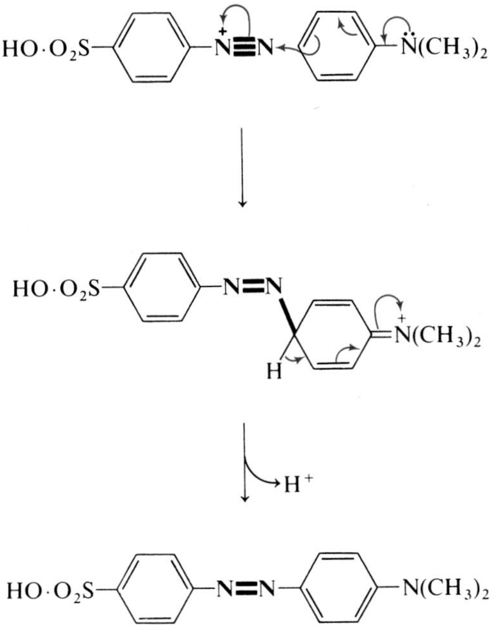

$$HO \cdot O_2S-C_6H_4-\overset{+}{N}\equiv N + C_6H_5-\ddot{N}(CH_3)_2$$

$$\downarrow$$

$$HO \cdot O_2S-C_6H_4-N=N-C_6H_5(H)=\overset{+}{N}(CH_3)_2$$

$$\downarrow -H^+$$

$$HO \cdot O_2S-C_6H_4-N=N-C_6H_4-N(CH_3)_2$$

Methyl Orange

Diazonium salts also couple with primary and secondary aromatic amines in neutral and alkaline solution to form diazoamino compounds. The latter rearrange on warming to *p*-aminoazobenzenes.

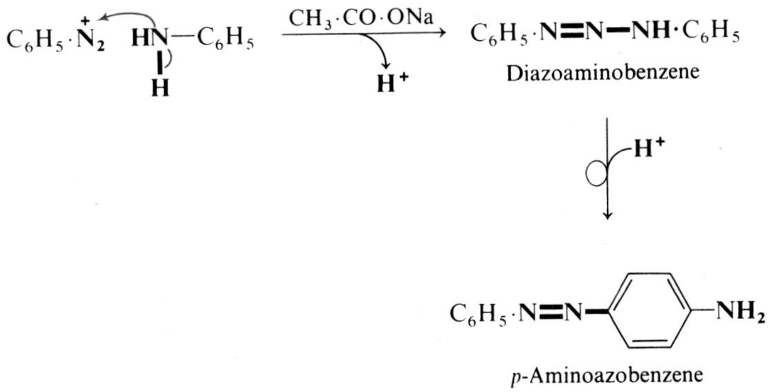

$$C_6H_5 \cdot \overset{+}{N}_2 \quad HN(H)-C_6H_5 \xrightarrow[\;H^+\;]{CH_3 \cdot CO \cdot ONa} C_6H_5 \cdot N=N-NH \cdot C_6H_5$$

Diazoaminobenzene

$$\downarrow H^+$$

$$C_6H_5 \cdot N=N-C_6H_4-NH_2$$

*p*-Aminoazobenzene

**Reduction of Diazonium Salts to Phenylhydrazines**

Diazonium salts are reduced with stannous chloride in hydrochloric acid to phenylhydrazines.

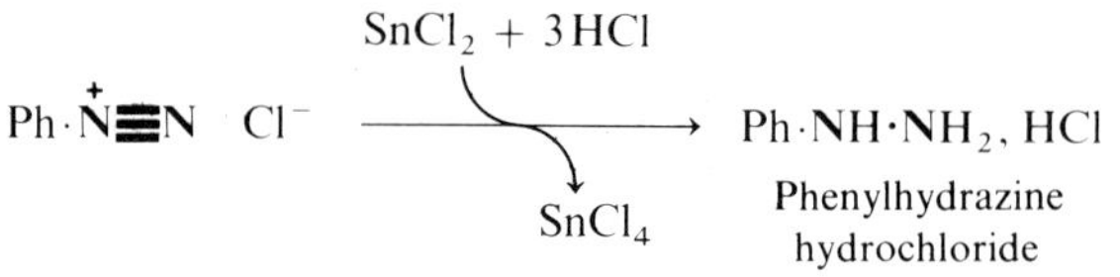

$$\mathrm{Ph{\cdot}\overset{+}{N}{\equiv}N\ \ Cl^-} \xrightarrow[\quad SnCl_4 \quad]{SnCl_2 + 3HCl} \mathrm{Ph{\cdot}NH{\cdot}NH_2,\ HCl}$$

Phenylhydrazine hydrochloride

**Azo Dyes**

A number of medicinal dyestuffs, including the colouring agents, *Amaranth* (FD & C Red No. 2), *Tartrazine* (FD & C Yellow No. 5) and *Sunset Yellow* (FD & C Yellow No. 6), and also *Azovan Blue* (Evans Blue), which is used for the measurement of blood volume, are azo compounds. The sodium sulphonate substituents confer water solubility.

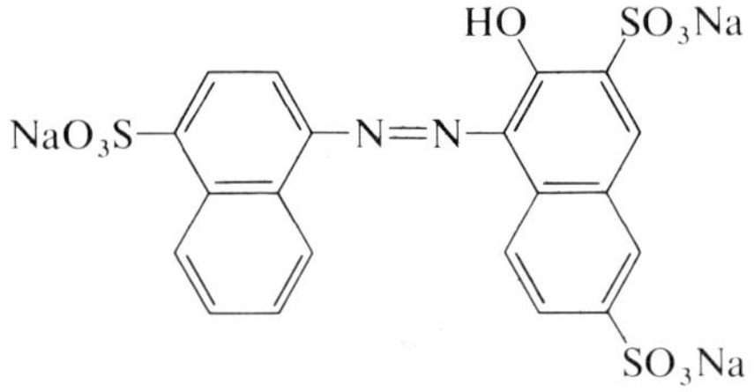

HO SO$_3$Na
NaO$_3$S— —N=N—
SO$_3$Na

*Amaranth*

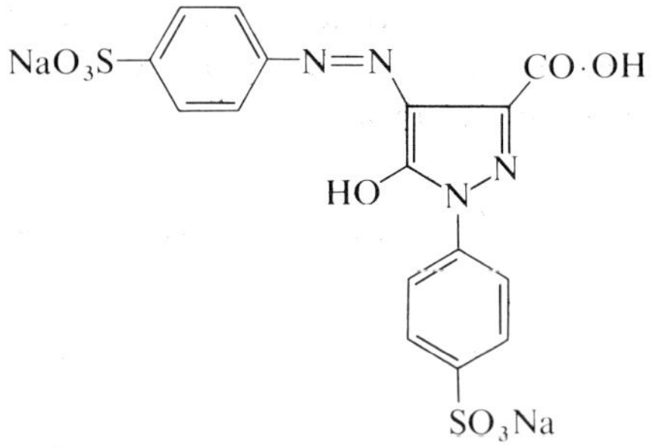

NaO$_3$S— —N=N CO·OH
HO N N
SO$_3$Na

*Tartrazine*

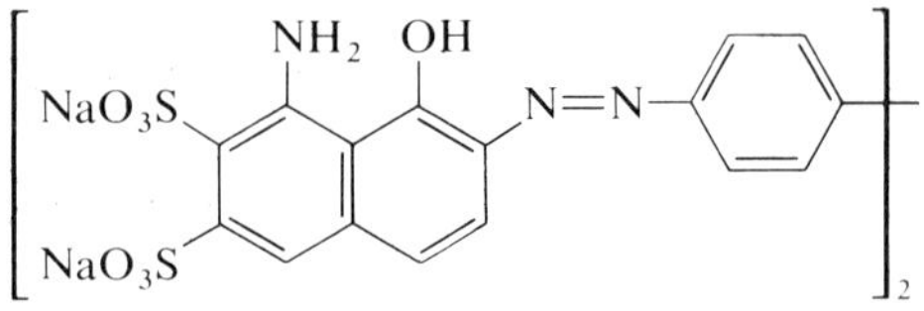

NH$_2$ OH
NaO$_3$S N=N—
NaO$_3$S ]$_2$

*Azovan Blue*

Fading can be a problem with phenolic azo dyes. It can, however, be overcome provided oxygen is excluded, since the dyes are unaffected by ultraviolet radiation alone, and only decompose when exposed to ultraviolet light in the presence of oxygen (Griffiths and Hawkins, 1972).

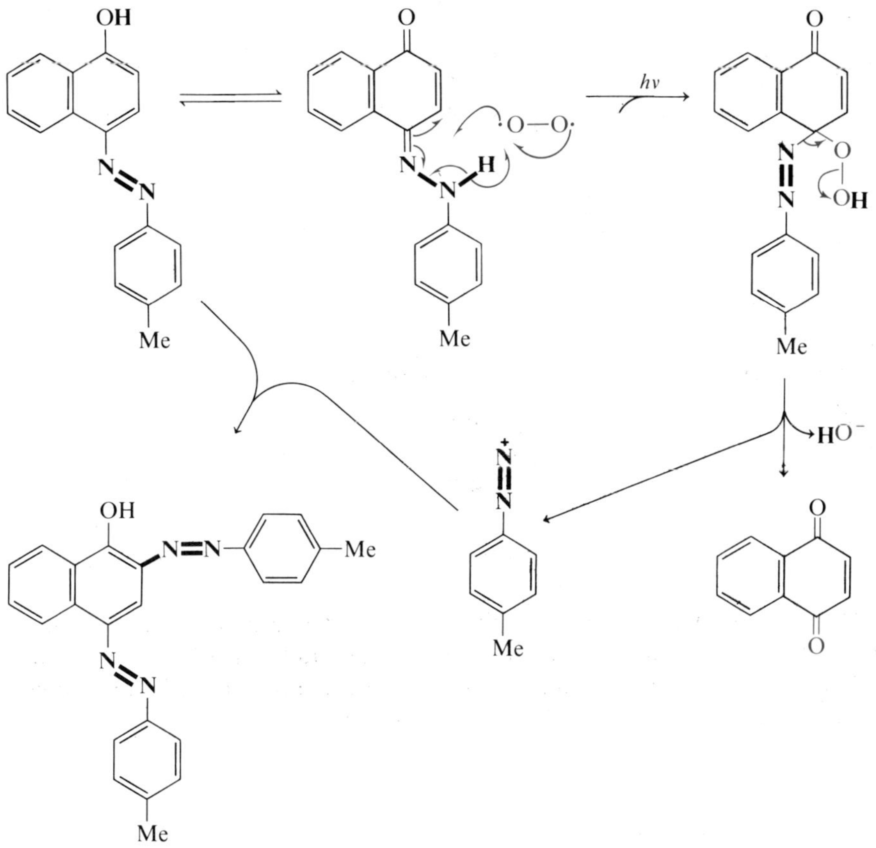

## NUCLEAR SUBSTITUTION OF AROMATIC AMINES

### Halogenation

Aromatic amines are activated for electrophilic substitution by resonance interaction of the nitrogen lone pair of electrons with the delocalised electrons of the aromatic ring. In consequence, aniline reacts instantaneously with chlorine- or bromine-water to give a precipitate of the 2,4,6-trihalogeno compound. In order to obtain ortho and para monosubstitution, the reactivity of the ring must be reduced by acetylation. The mono halogeno amines are obtained by acid hydrolysis of the product, which consists mainly of the *para*-isomer. *Ortho*- and *para*-bromoanilines can be separated by steam distillation, since only the former is appreciably volatile in steam.

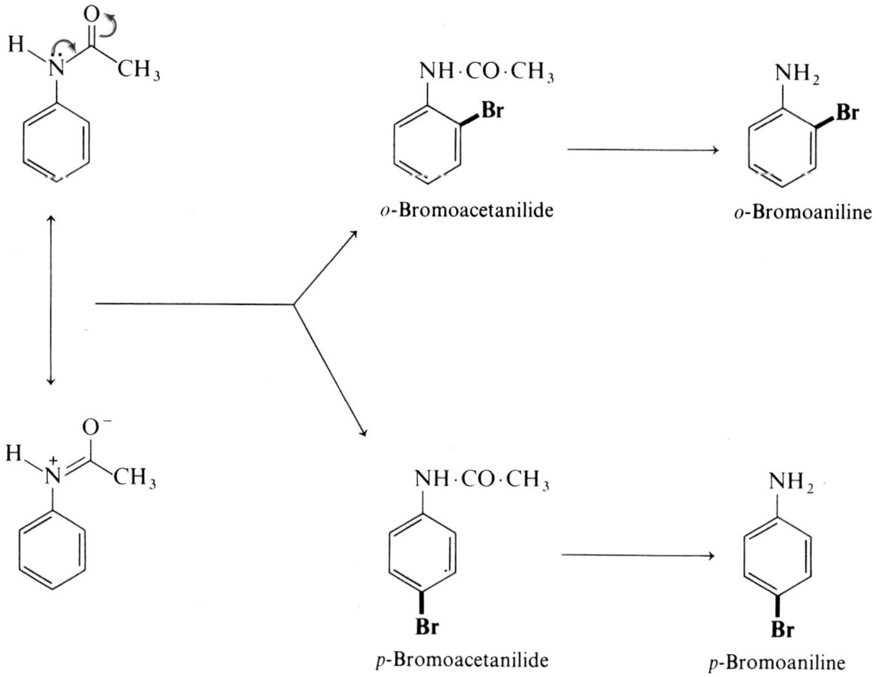

$NH \cdot CO \cdot CH_3$, Br

*o*-Bromoacetanilide

$NH_2$, Br

*o*-Bromoaniline

$NH \cdot CO \cdot CH_3$, Br

*p*-Bromoacetanilide

$NH_2$, Br

*p*-Bromoaniline

Halogen-substituted aromatic amines and their derivatives, such as *p*-chloroacetanilide, a trace impurity of *Phenacetin*, are generally regarded as toxic. The iodine-containing radio-opaque *Iopanoic Acid* is excreted in the bile, and is used as an X-ray contrast medium for outlining the gall bladder.

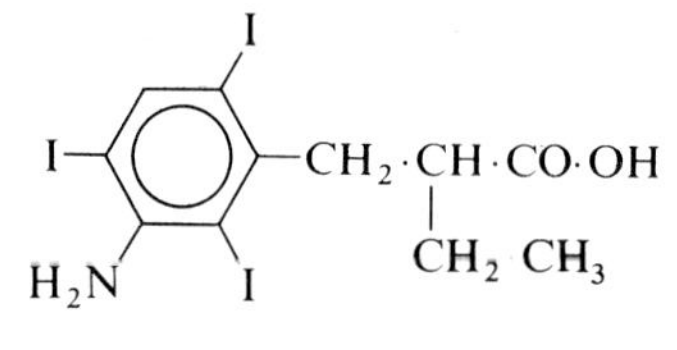

$CH_2 \cdot CH \cdot CO \cdot OH$ / $CH_2\ CH_3$ *Iopanoic Acid*

**Nitration and Nitrosation**

Direct nitration of aniline is not possible due to oxidation, but *p*-nitroaniline can be obtained by direct nitration of acetanilide and subsequent hydrolysis.

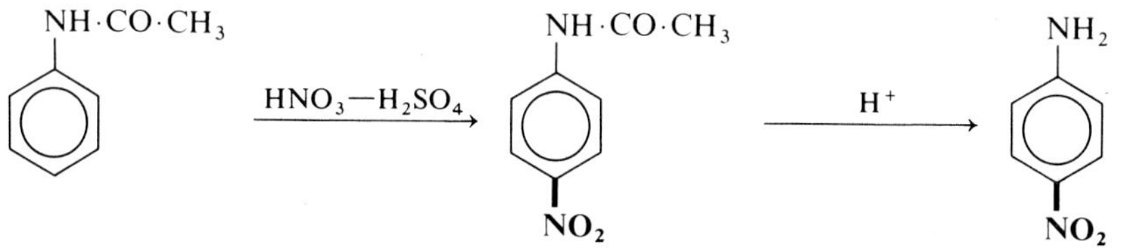

$$NH \cdot CO \cdot CH_3\text{-benzene} \xrightarrow{HNO_3-H_2SO_4} p\text{-}NO_2\text{-}C_6H_4\text{-}NH \cdot CO \cdot CH_3 \xrightarrow{H^+} p\text{-}NO_2\text{-}C_6H_4\text{-}NH_2$$

Direct nitration of tertiary aromatic amines is feasible in strongly acidic solution. The basic group, however, is protonated and as such is deactivating and *meta*-directing.

$$C_6H_5\overset{+}{N}HMe_2 \xrightarrow{HNO_3-H_2SO_4} m\text{-}O_2N{-}C_6H_4\overset{+}{N}HMe_2$$

Tertiary amines also react readily with nitrous acid to form *para*-nitroso compounds, or *ortho*-nitroso compounds if the *para* position is blocked.

$$C_6H_5NMe_2 \xrightarrow{HNO_2} p\text{-}ON{-}C_6H_4{-}NMe_2$$

The nitroso compounds are readily oxidised with dilute nitric acid to the corresponding nitro compound; they are also readily reduced to the corresponding primary amine.

### Basic Triphenylmethane Dyes

Dimethylaniline condenses with carbonyl chloride in the presence of aluminium chloride to form the substituted benzophenone known as Michler's ketone.

$$2Me_2N{-}C_6H_5 + COCl_2 \xrightarrow{AlCl_3} Me_2N{-}C_6H_4{-}CO{-}C_6H_4{-}NMe_2$$

Michler's Ketone

Michler's ketone is an important intermediate in the synthesis of the group of basic triphenylmethane dyes which includes *Crystal Violet*. This is formed when Michler's ketone is heated with dimethylaniline in the presence of a suitable acid catalyst ($AlCl_3$, HCl, $POCl_3$).

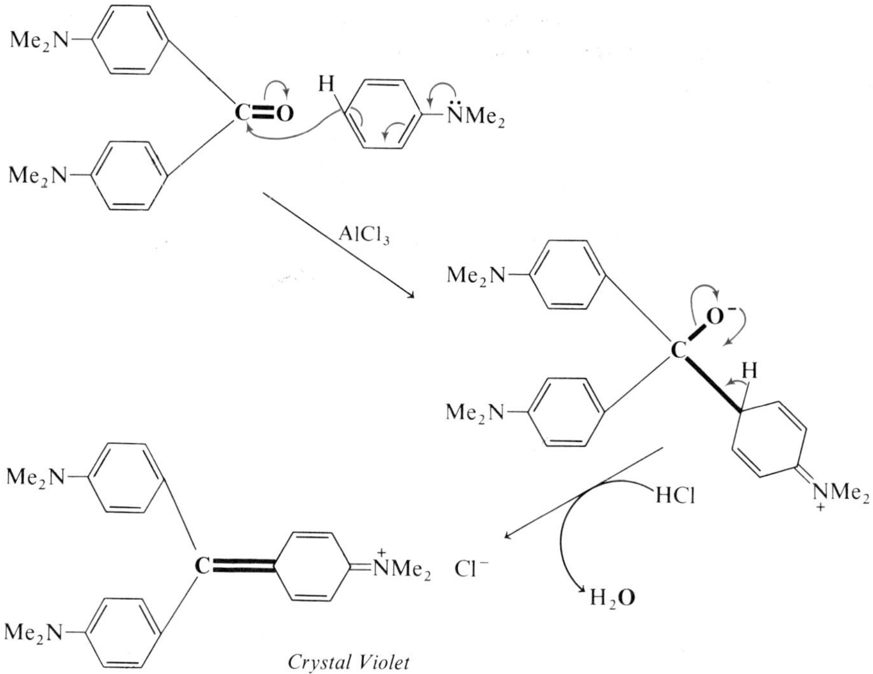

$Me_2N$ $Me_2N$ C=O H $\ddot{N}Me_2$

$AlCl_3$

$Me_2N$ $Me_2N$ C $O^-$ H $\overset{+}{N}Me_2$

HCl $H_2O$

$Me_2N$ $Me_2N$ C $=\overset{+}{N}Me_2$ $Cl^-$

*Crystal Violet*

Benzaldehyde similarly condenses with diethylaniline in hot concentrated sulphuric acid to form the triphenylmethane derivative, leuco brilliant green,

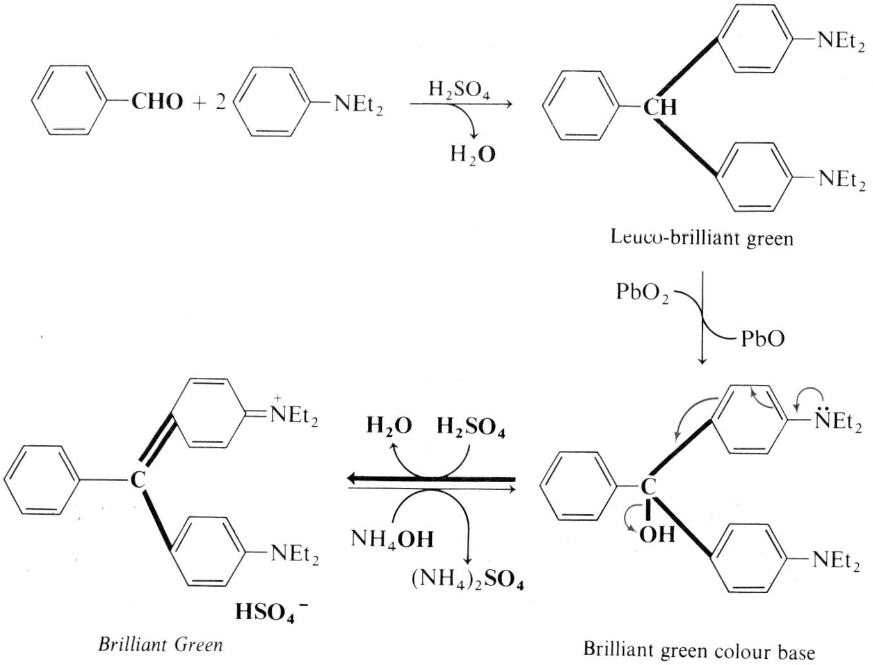

$$C_6H_5CHO + 2\,C_6H_5NEt_2 \xrightarrow[-H_2O]{H_2SO_4} C_6H_5CH(C_6H_4NEt_2)_2$$

Leuco-brilliant green

$PbO_2$ → $PbO$

$\overset{+}{N}Et_2$ C $NEt_2$ $HSO_4^-$ $H_2O$ $H_2SO_4$ $NH_4OH$ $(NH_4)_2SO_4$ C OH $\ddot{N}Et_2$ $NEt_2$

*Brilliant Green*

Brilliant green colour base

which on oxidation with lead peroxide or some other mild oxidising agent forms the dye, *Brilliant Green.*

Triphenylmethane dyes are readily soluble in water to form intensely coloured solutions. The intense colour is the result of resonance between equivalent structures.

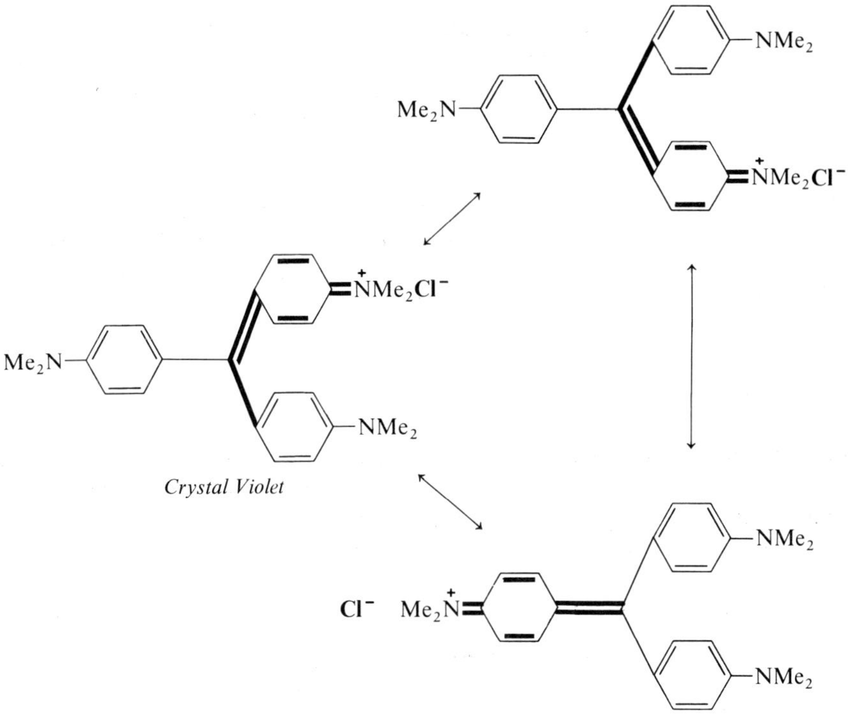
NMe$_2$
Me$_2$N
$\overset{+}{N}$Me$_2$Cl$^-$
$\overset{+}{N}$Me$_2$Cl$^-$
Me$_2$N
NMe$_2$
NMe$_2$
Cl$^-$ Me$_2\overset{+}{N}$
NMe$_2$

*Crystal Violet*

The purple colour of *Crystal Violet* changes to green in acidic solution due to the more limited resonance of the monoprotonated form (below). In this form, resonance in *Crystal Violet* is analogous to that in *Brilliant Green.*

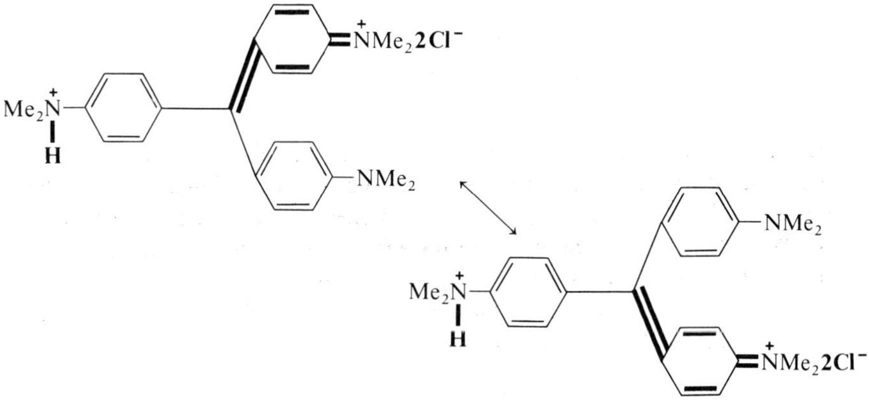
$\overset{+}{N}$Me$_2$2Cl$^-$
Me$_2\overset{+}{N}$
H
NMe$_2$
NMe$_2$
Me$_2\overset{+}{N}$
H
$\overset{+}{N}$Me$_2$2Cl$^-$

With further acidification, solutions of *Crystal Violet* change to pale-yellow with formation of the tri-cation in which resonance is largely suppressed.

$$Me_2\overset{+}{N}{=}C_6H_4{=}C(C_6H_4\text{-}\overset{+}{N}HMe_2)_2 \quad 3Cl^-$$

The dyes are precipitated as the colourless water-insoluble colour base on the addition of ammonia or sodium hydroxide solution. *Crystal Violet* and *Brilliant Green* are effective as antibacterial agents mainly against Gram-positive organisms and find limited use as skin antiseptics.

## ALKANOLAMINES

The term **alkanolamine**, although descriptive of hydroxyalkylamines generally, is usually reserved to $\beta$-hydroxyalkylamines. The simplest, ethanolamine, is formed biosynthetically in mammals by decarboxylation of serine, which in turn is derived from glucose via D-3-phosphoglyceric acid.

$$\underset{\text{D-3-Phosphoglyceric acid}}{HO\text{—}CH(CH_2\cdot O\cdot PO_3H_2)\text{—}CO\cdot O^-} \rightleftharpoons O{=}C(CH_2\cdot O\cdot PO_3H_2)\text{—}CO\cdot O^- \rightleftharpoons H\text{—}C(CH_2\cdot O\cdot PO_3H_2)(\overset{+}{N}H_3)\text{—}CO\cdot O^-$$

$$\rightleftharpoons H\text{—}C(CH_2OH)(\overset{+}{N}H_3)\text{—}CO\cdot O^- \xrightarrow{-CO_2} \underset{\text{Ethanolamine}}{HO\cdot CH_2\cdot CH_2NH_2}$$

Ethanolamine is an important intermediate in the biosynthesis of choline, acetylcholine, and various phospholipids (Chapter 11) including lecithin, cephalin and plasmalogens. The sympathetic or adrenergic transmitter is a mixture of two alkanolamines, adrenaline and noradrenaline, in which the latter predominates. They arise by biosynthesis from tyrosine (p. 497).

The sympathetic transmitter is stored in chromaffin cells (so named because they become coloured on oxidation) in association with adenosine phosphates.

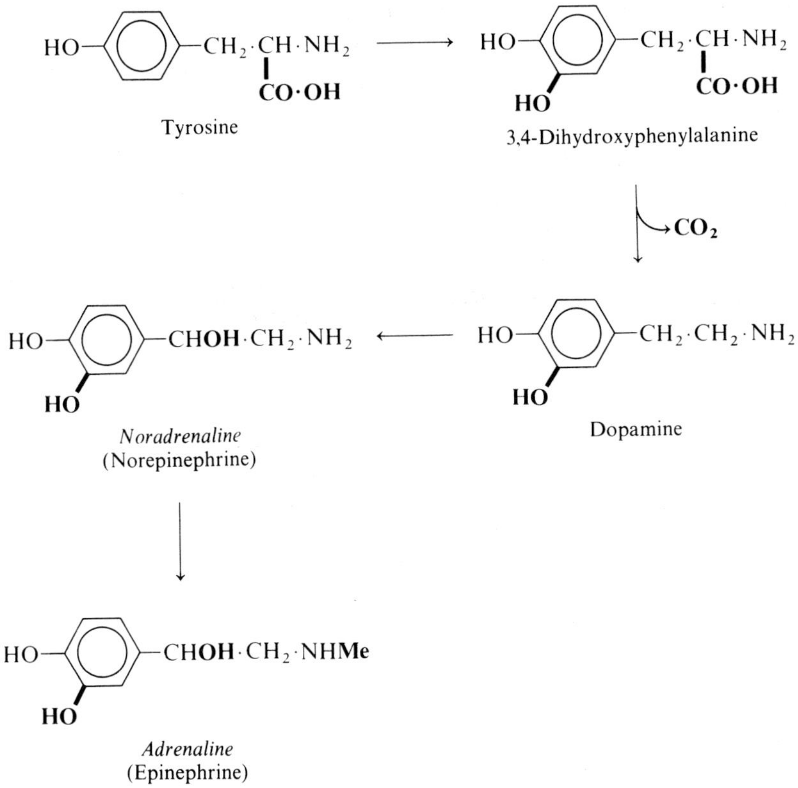

HO—C₆H₄—$CH_2 \cdot CH(CO \cdot OH) \cdot NH_2$ ⟶ HO, HO—C₆H₃—$CH_2 \cdot CH(CO \cdot OH) \cdot NH_2$

Tyrosine

3,4-Dihydroxyphenylalanine

↓ $CO_2$

HO, HO—C₆H₃—$CH_2 \cdot CH_2 \cdot NH_2$

Dopamine

HO, HO—C₆H₃—$CHOH \cdot CH_2 \cdot NH_2$

*Noradrenaline*
(Norepinephrine)

↓

HO, HO—C₆H₃—$CHOH \cdot CH_2 \cdot NHMe$

*Adrenaline*
(Epinephrine)

It is worthy of note that amongst its many actions, adrenaline stimulates adenylcyclase which affects the conversion of adenosine triphosphate to cyclic-3,5-adenosinemonophosphate in the liver and in muscle.

## General Properties

Simple alkanolamines, such as ethanolamine, are much more water-soluble than the corresponding alkylamines. The electron-withdrawing effect of the $\beta$-C—O bond is base-weakening, so that ethanolamine has p$K_a$ 9.5 compared with 10.63 for ethylamine. Ethanolamine, diethanolamine [$(HO \cdot CH_2 \cdot CH_2)_2NH$] and triethanolamine [$(HO \cdot CH_2 \cdot CH_2)_3N$] are nonetheless sufficiently basic to find considerable use in the preparation of water-soluble soaps with long-chain fatty acids, and also as buffering agents. Tris(hydroxymethyl)aminomethane [$H_2N \cdot C(CH_2OH)_3$; *Trometamol* (Tromethamine)] is also used for the preparation of soluble salts of poorly soluble fatty acids, such as for example certain prostaglandins.

Sympathomimetic alkanolamines, such as adrenaline and noradrenaline, are sufficiently ionised at physiological pH to function in the ionic form, and ionic bonding of agonists and blocking agents occurs at both $\alpha$- and $\beta$-adrenergic receptors, as classified by Ahlquist (1948). The responses to adrenaline and noradrenaline at these centres show a high degree of stereospecificity. Thus, (−)-adrenaline is some 20 times more active than its (+)-enantiomer in raising blood pressure in the cat (Tainter, 1930). Further aspects of these stereospecific interactions are discussed elsewhere (**2**, 2).

Alkanolamines for the most part exhibit the expected properties of alcohols and amines. A number of alkanolamines, and alkanolamine esters and amides have important pharmacological actions. The tropane alkaloids, which are esters of tropine and numerous structurally-related alkanolamine esters, such as *Dicyclomine* and *Benzhexol* (Trihexaphenidyl), are antagonists of acetylcholine at postganglionic cholinergic receptors, and find use as mydriatics and spasmolytics.

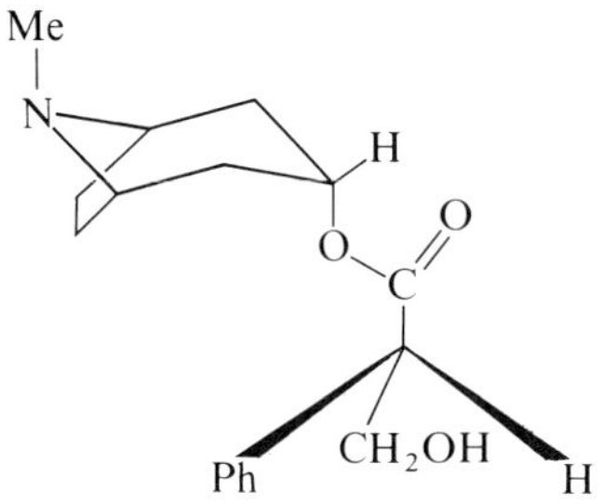

(S)-(−)-*Hyoscyamine*

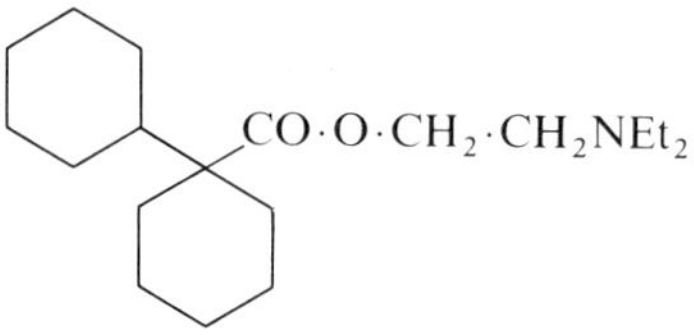

*Dicyclomine*

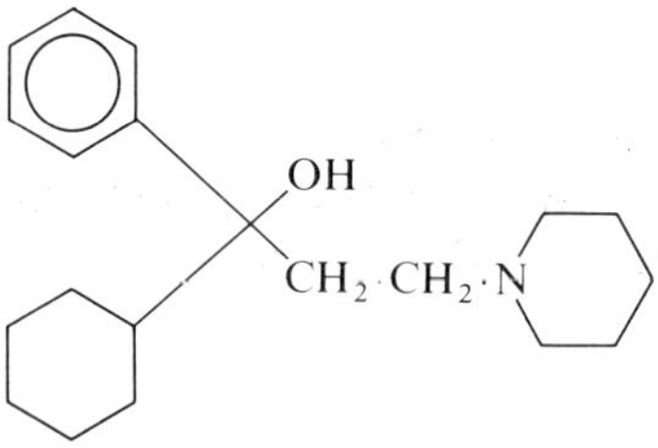

*Benzhexol*
(Trihexaphenidyl)

As might be anticipated, many of these compounds which are capable of existence in stereoisomeric forms show considerable stereospecificity in their pharmacological actions (**2**, 2).

*p*-Aminobenzoyl esters of diethylaminoethanol such as *Procaine* and related alkanolamines generally possess significant local anaesthetic activity. The ester link, however, appears to be merely a means of orientating the substituted aromatic ring and the basic side-chain in a suitable configuration for activity, since bioisosteric amides such as *Procainamide* have similar activities. The rôle of ionisation of the basic group, fat solubility, and substitution in the aromatic ring on local anaesthetic activity is discussed later (**2**, 2).

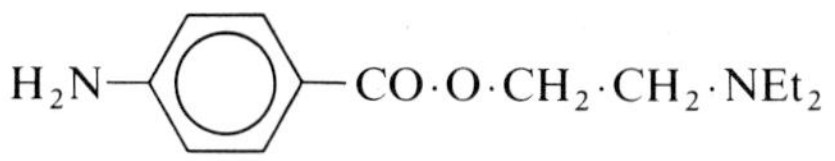

$$H_2N-C_6H_4-CO\cdot O\cdot CH_2\cdot CH_2\cdot NEt_2$$

*Procaine*

$$H_2N-C_6H_4-CO\cdot NH\cdot CH_2\cdot CH_2\cdot NEt_2$$

*Procainamide*

## QUATERNARY AMMONIUM SALTS

Quaternary ammonium salts are formed by the direct alkylation of tertiary amines with alkyl halides.

$$(CH_3)_3\cdot \ddot{N} \quad \mathbf{R{-}X} \longrightarrow (CH_3)_3\cdot \overset{+}{N}\cdot \mathbf{R} \quad X^-$$

The nitrogen atom is tetrahedral, all four N—C bonds being equivalent. This is demonstrated by the formation of one and the same product in both the following quaternisations.

$$(CH_3)\cdot N + \mathbf{EtI} \longrightarrow (CH_3)_3\overset{+}{N}\cdot \mathbf{Et}\mathbf{I}^- \longleftarrow (CH_3)_2\cdot N\cdot \mathbf{Et} + CH_3I$$

Acetylcholine, which plays a key rôle in autonomic nervous transmission at the ganglia, at postganglionic parasympathetic nerve endings, and at the neuromuscular junction, arises biosynthetically from ethanolamine by the following pathway.

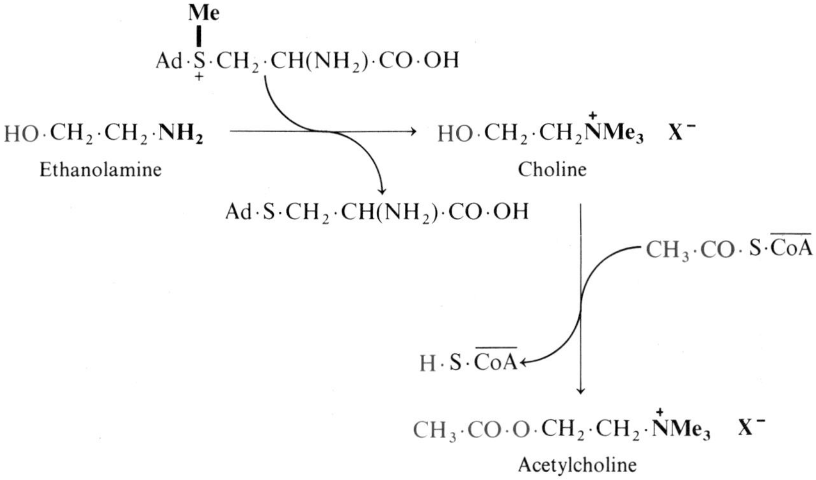

$$Ad\cdot \overset{+}{S}(\mathbf{Me})\cdot CH_2\cdot CH(NH_2)\cdot CO\cdot OH$$

$$HO\cdot CH_2\cdot CH_2\cdot \mathbf{NH_2} \longrightarrow HO\cdot CH_2\cdot CH_2\overset{+}{\mathbf{N}}\mathbf{Me_3} \quad X^-$$

Ethanolamine — Choline

$$Ad\cdot S\cdot CH_2\cdot CH(NH_2)\cdot CO\cdot OH$$

$$CH_3\cdot CO\cdot S\cdot \overline{CoA} \longrightarrow H\cdot S\cdot \overline{CoA}$$

$$CH_3\cdot CO\cdot O\cdot CH_2\cdot CH_2\cdot \overset{+}{\mathbf{N}}\mathbf{Me_3} \quad X^-$$

Acetylcholine

The rôle and stereospecificity of acetylcholine and acetylcholine-like molecules are discussed in later chapters (**2**, 2 and 3).

## General Properties

Quaternary ammonium salts are neutral compounds usually forming crystalline solids, often hygroscopic. They are usually readily soluble in water, soluble in ethanol, but generally with low solubilities in most other organic solvents, such as chloroform or ether, though quaternary iodides tend to be more soluble in these solvents than the corresponding chlorides and bromides. They are also insoluble in fats and oils. In consequence, only lower molecular weight, fat-soluble quaternary salts are readily transported across biological membranes (Table 53). Thus, the high molecular weight, lipid-insoluble quaternary salts, *Tubocurarine Chloride* and *Gallamine Triethiodide*, are not absorbed, and are only effective by injection. On the other hand, low molecular weight lipid-soluble salts, such as *Neostigmine Bromide* and *Pyridostigmine Bromide*, are moderately well, if somewhat variably, absorbed from the gastro-intestinal tract.

Table 53. Solubilities and Gastro-intestinal Absorption of Quaternary Salts

| Salt | Water solubility | Chloroform solubility | Absorption |
|---|---|---|---|
| Tubocurarine Chloride | 1 in 20 | Almost insoluble | – |
| Gallamine Triethiodide | 1 in 0.6 | 1 in 5000 | – |
| Neostigmine Bromide | 1 in 0.5 | Freely soluble | + |
| Pyridostigmine Bromide | 1 in $<1$ | 1 in 1 | + |

It has been suggested that lipid solubility and absorption of particular quaternary salts is due to ion-pairing in lipid solvents, an effect which can be demonstrated by NMR spectroscopy using shift reagents (Bladon and Stenlake, 1973).

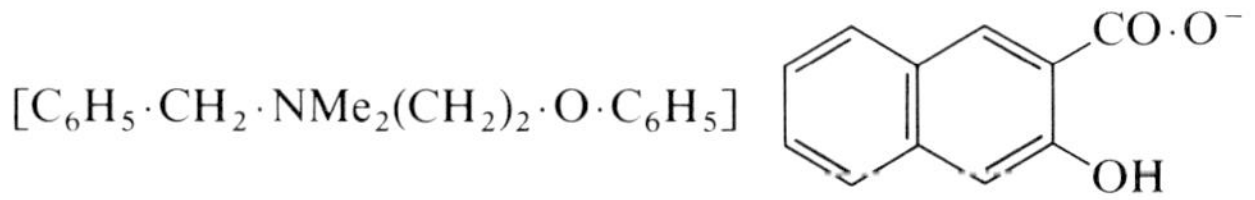

$[C_6H_5 \cdot CH_2 \cdot NMe_2(CH_2)_2 \cdot O \cdot C_6H_5]$ CO·O$^-$ OH

*Bephenium Hydroxynaphthoate*

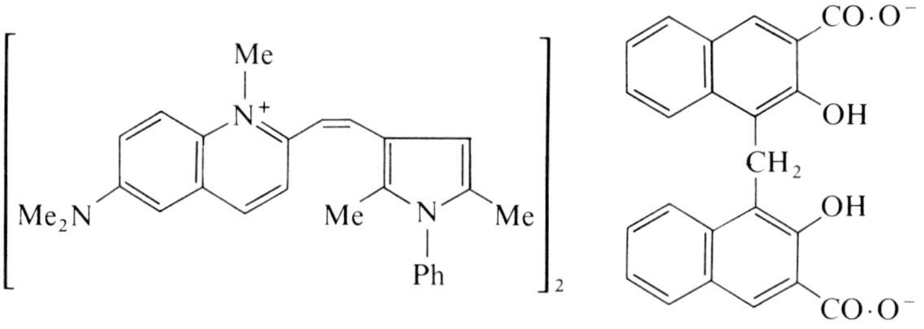

Me N$^+$ Me$_2$N Me N Me Ph ]$_2$ CO·O$^-$ OH CH$_2$ OH CO·O$^-$

*Viprynium Embonate*

Ion-pairing with lipid-soluble dyes has been used for the recovery and measurement of urinary excretion kinetics of quaternary ammonium compounds (Schill, 1965; Borg, Holgersson and Lagerström, 1970).

High molecular weight quaternary salts, such as *Bephenium Hydroxynaphthoate* and *Viprynium Embonate* (Pyrvinium Pamoate), which have very low lipid solubilities and also are almost completely insoluble in water, are not absorbed from the gastro-intestinal tract. In consequence, they pass down into the lower bowel where they are effective in the treatment of certain worm infections.

Long-chain aliphatic quaternary salts, such as *Cetylpyridinium Chloride* (Cetrimide) and *Benzalkonium Chloride*, are detergents because of their ability to concentrate at the oil–water interface. They may be used to increase the solubility of other compounds with low water solubility (Sjöblom, 1958), such as *Oestradiol* (**2**, 5).

$$(CH_3)_3\cdot\overset{+}{N}(CH_2)_{15-17}\cdot CH_3 \quad Cl^-$$

*Benzalkonium Chloride*

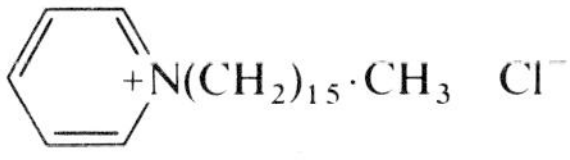

$$C_5H_5\overset{+}{N}(CH_2)_{15}\cdot CH_3 \quad Cl^-$$

*Cetylpyridinium Chloride*

These compounds, which are appreciably soluble in water, are effective as antibacterial agents, particularly against Gram-negative organisms. They are incompatible with soaps and anionic detergents due to the formation of insoluble salts.

## Chemical Properties

Quaternary ammonium salts decompose when heated strongly, usually with elimination of the smallest alkyl group.

$$Et_3\overset{+}{N}\cdot \mathbf{Me} \quad \mathbf{I}^- \longrightarrow Et_3N + \mathbf{MeI}$$

Treatment of quaternary ammonium salts with strong bases, such as potassium hydroxide, moist silver oxide or the hydroxyl form of an anionic exchange resin, gives the corresponding quaternary ammonium hydroxide. The hydroxides are highly water-soluble, deliquescent solids. They are strongly basic and re-form quaternary salts on neutralisation with acid.

Quaternary ammonium salts are decomposed by prolonged exposure to bases, such as potassium hydroxide, silver hydroxide or sodium ethoxide with the formation of olefines by $\beta$-elimination (Hofmann elimination), particularly at elevated temperatures. The direction of elimination is the reverse of that which applies in dehydrohalogenation of alkyl halides (Saytzeff rule; Chapter 9). Thus, in the Hofmann elimination, **hydrogen is eliminated from the β-carbon carrying the lesser number of alkyl groups**. There are two reasons for this. Firstly, the charged onium group is more effective than halogens in promoting the acidity of $\beta$-hydrogens by electron-withdrawal. In consequence, the acidity of respective $\beta$-hydrogens is more sensitive to the inductive effects of $\beta$-alkyl substituents. Thus, the action of base on dimethylethyl-n-propylammonium chloride yields ethylene rather than propylene, since the +I effect of the terminal methyl group lowers the acidity of the adjacent $\beta$-hydrogens ($b$) relative to those ($a$) on the ethyl group.

$$CH_3\cdot CH_2\cdot CH_2\cdot N(CH_3)_2 + \mathbf{H_2C{=}CH_2} + \mathbf{H_2O}$$

$$\overset{\beta(b)}{CH_3 \rightarrow CH_2}-CH_2\rightarrow \overset{+}{N}(CH_3)_2-CH_2-\overset{H^{\beta(a)}\ {}^{-}OH}{C}H_2$$

$$CH_3\cdot CH{=}CH_2 + (CH_3)_2N\cdot \mathbf{CH_2\cdot CH_3} + \mathbf{H_2O}$$

Secondly, steric factors, arising from the size of the onium group favour the formation of a transition state in which crowding is at a minimum; attack at the $\beta$-carbon carrying the lesser number of substituents is, therefore, again favoured. This steric effect is opposed, however, and overcome by the presence of electron-attracting substituents, such as the phenyl group in the $\beta$-position, and the Saytzeff rule then applies.

$$Ph\leftarrow CH_2-CH_2-\overset{+}{N}(CH_3)_2-CH_2-CH_3 \xrightarrow{HO^-} Ph\cdot \mathbf{CH{=}CH_2} + (CH_3)_2N\cdot CH_2\cdot CH_3 + \mathbf{H_2O}$$

Hofmann elimination of suitably-substituted sulphonium salts is known to be catalysed by certain enzyme systems in algae (Cantoni and Anderson, 1956).

$$Me_2\overset{+}{S}-\mathbf{CH_2}-\mathbf{CH_2}\cdot CO\cdot OH \longrightarrow Me_2S + \mathbf{H_2C{=}CH}\cdot CO\cdot OH$$

Dimethylpropiothelin

There are few well documented accounts of similar enzyme-catalysed Hofmann degradations of quaternary ammonium salts. It is possible, however, that the vinylic ether group of plasmalogens is derived from the corresponding choline ethers by elimination.

$$\begin{array}{l} CH_2\cdot O\cdot \mathbf{CH_2{-}CH}\cdot \overset{+}{N}(CH_3)_3 \\ \quad\quad\quad\quad\quad\quad\quad \mathbf{R} \\ R\cdot CO\cdot O\cdot CH \\ CH_2\cdot O\cdot \underset{O^-}{\overset{O}{\overset{\|}{P}}}\cdot O\cdot CH_2\cdot CH_2\cdot \overset{+}{N}(CH_3)_3 \end{array}$$

$$\xrightarrow[\quad \rightarrow \mathbf{H_2O} + (CH_3)_3N]{\mathbf{HO^-}}$$

$$\begin{array}{l} CH_2\cdot O\cdot \mathbf{CH{=}CH\cdot R} \\ R\cdot CO\cdot O\cdot CH \\ CH_2\cdot O\cdot \underset{O^-}{\overset{O}{\overset{\|}{P}}}\cdot O\cdot CH_2\cdot CH_2\cdot \overset{+}{N}(CH_3)_3 \end{array}$$

Plasmalogen

## Ylids

Treatment of quaternary ammonium salts with phenyl lithium gives rise to ylids.

$$Me_3\overset{+}{N}\cdot \mathbf{CH_3}Cl^- \xrightarrow[\mathbf{Ph{-}H}]{\mathbf{PhLi}} [Me_3\overset{+}{N}\cdot \mathbf{CH_2{-}Li}]Cl^- \xrightarrow[\mathbf{LiCl}]{} Me_3\overset{+}{N}\cdot \mathbf{CH_2}^-$$

Ylids are highly reactive with such compounds as methyl iodide and carbonyl compounds with the formation of substituted quaternary salts.

$$Me_3\overset{+}{N}\cdot \mathbf{CH_2}^- \ \mathbf{Me{-}I} \longrightarrow Me_3\overset{+}{N}\cdot \mathbf{CH_2\cdot Me} \quad \mathbf{I^-}$$

$$Me_3\overset{+}{N}\cdot \mathbf{CH_2}^- \ + \ \begin{array}{c} Me \\ \diagdown \\ \mathbf{C{=}O} \\ \diagup \\ Me \end{array} \longrightarrow Me_3\overset{+}{N}\cdot \mathbf{CH_2\cdot CO^-}\cdot Me_2$$

$$\downarrow HX$$

$$Me_3\overset{+}{N}\mathbf{CH_2\cdot C(OH)}\cdot Me_2 \quad X^-$$

# 17 Carbamic Acid Derivatives

## Carbonic Acid

Carbonic acid ($(HO)_2C{=}O$) is unknown in the free state, but occurs widely in the form of its salts (carbonates). It also forms water-soluble esters (alkyl carbonates) when silver carbonate is treated with an alkyl iodide.

$$Ag_2CO_3 + 2RI \longrightarrow (RO)_2CO + 2AgI$$

Carbonyl chloride (phosgene), the acid chloride of carbonic acid, is formed directly from chlorine and carbon monoxide either on exposure to light or, as in the manufacturing process, by the action of heat in the presence of activated charcoal.

$$Cl_2 + CO \longrightarrow COCl_2$$

Carbonyl chloride, a toxic gas (b.p. 8°C), behaves as a typical acid chloride and is reactive towards water, alcohols, ammonia, and primary and secondary amines. It is slowly decomposed by water forming carbon dioxide and hydrochloric acid.

$$COCl_2 + H_2O \longrightarrow CO_2 + 2HCl$$

The reaction with alcohols gives rise to either an alkyl chloroformate or, if the alcohol is present in excess, an alkyl carbonate.

$$COCl_2 + EtOH \longrightarrow \underset{\text{Ethyl chloroformate}}{Cl{\cdot}CO{\cdot}OEt} + HCl$$

$$COCl_2 + 2EtOH \longrightarrow \underset{\text{Ethyl carbonate}}{(EtO)_2{\cdot}CO} + 2HCl$$

Reaction with ammonia or amines (primary and secondary) yields either urea or *NN'*-disubstituted ureas.

$$COCl_2 + 2NH_3 \longrightarrow \underset{\text{Urea}}{H_2N{\cdot}CO{\cdot}NH_2} + 2HCl$$

$$COCl_2 + 2R{\cdot}NH_2 \longrightarrow RNH{\cdot}CO{\cdot}NHR + 2HCl$$

$$COCl_2 + 2R_2{\cdot}NH \longrightarrow R_2N{\cdot}CO{\cdot}NR_2 + 2HCl$$

Ethyl chloroformate reacts similarly with ammonia or amines to form urethanes.

## CARBAMATES (URETHANES)

Carbamic acid itself, $H_2N \cdot COOH$, does not exist, but it is known in the form of its salts, acid amides and esters. The ammonium salt is formed as a crystalline solid by reaction of dry ammonia with dry carbon dioxide.

$$CO_2 + 2NH_3 \longrightarrow H_2N \cdot CO \cdot O \cdot NH_4$$

When heated under pressure, ammonium carbamate loses water to form urea.

$$\underset{\text{Ammonium carbamate}}{H_2N \cdot CO \cdot O \cdot NH_4} \longrightarrow \underset{\text{Urea}}{H_2N \cdot CO \cdot NH_2} + H_2O$$

Calcium and barium carbamates, unlike the corresponding carbonates, are soluble in water.

Esters of carbamic acid, urethanes, are formed by the action of ammonia or amines on the corresponding chloroformates. They are also formed when urea is heated with an alcohol.

$$Cl \cdot CO \cdot OEt \xrightarrow[\text{HCl}]{\text{NH}_3} H_2N \cdot CO \cdot OEt \xleftarrow[\text{NH}_3]{\text{EtOH}} H_2N \cdot CO \cdot NH_2$$

**Properties**

Urethanes are generally moderately soluble in water, but unstable, undergoing hydrolysis readily in both acid and alkaline solution. Acid hydrolysis gives rise to a solution of ammonium chloride with release of carbon dioxide and the corresponding alcohol.

$$H_2N \cdot CO \cdot OEt + HCl + H_2O \longrightarrow NH_4Cl + EtOH + CO_2$$

In alkaline solution, ammonia and ethanol are released, and carbon dioxide is trapped as carbonate.

$$H_2N \cdot CO \cdot OEt + 2NaOH \longrightarrow Na_2CO_3 + EtOH + NH_3$$

A number of urethane derivatives which find important uses in medicine undergo hydrolysis at physiological pH (7.4) and either have their actions terminated thereby or, alternatively, promote the formation of otherwise active biotransformation products. Thus, the parasympathomimetic, *Carbachol*, and the anticholinesterases, *Neostigmine Bromide* and *Physostigmine* (Eserine) *Salicylate*, are all biodegraded either chemically or enzymatically to the corresponding alcohol or phenol with loss of their specific pharmacological activity. As quaternary salts, both *Carbachol* and *Neostigmine Bromide* are highly water-soluble forming neutral injection solutions, which are sufficiently stable to permit sterilisation by autoclaving without undergoing significant decomposition.

$H_2N \cdot CO \cdot O \cdot CH_2 \cdot CH_2 \cdot \overset{+}{N}(CH_3)_3Cl^-$

Carbachol

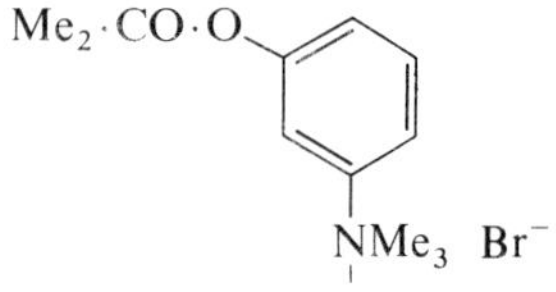

$Me_2 \cdot CO \cdot O$–$C_6H_4$–$\overset{+}{N}Me_3 \; Br^-$

Neostigmine Bromide

MeNH·CO·O– [tricyclic ring with N–Me, N–Me, Me] · [o-HO·C₆H₄·CO·OH]

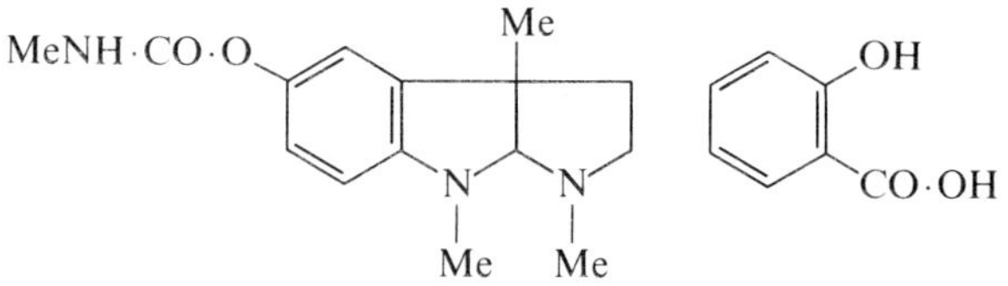

Physostigmine Salicylate

The antithyroid drug, *Carbimazole*, however, is hydrolysed and decarboxylated rapidly in human plasma to form *Methimazole*, which is itself a potent antithyroid agent.

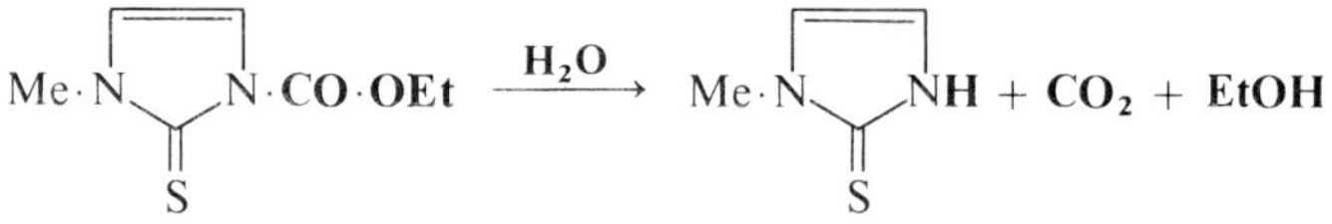

$$\text{Me}\cdot\text{N}\langle\text{imidazoline-2-thione}\rangle\text{N}\cdot\textbf{CO}\cdot\textbf{OEt} \xrightarrow{H_2O} \text{Me}\cdot\text{N}\langle\text{imidazoline-2-thione}\rangle\textbf{NH} + \textbf{CO}_2 + \textbf{EtOH}$$

Carbimazole — Methimazole

Urethane is a mild hypnotic, but is no longer used for this purpose since it has been shown to be carcinogenic in some animal species (Haddow, 1955). Such effects appear to be confined to urethane itself ($H_2N \cdot CO \cdot OEt$), and the urethane derivative, *Meprobamate*, is now widely used as a hypnotic in its place.

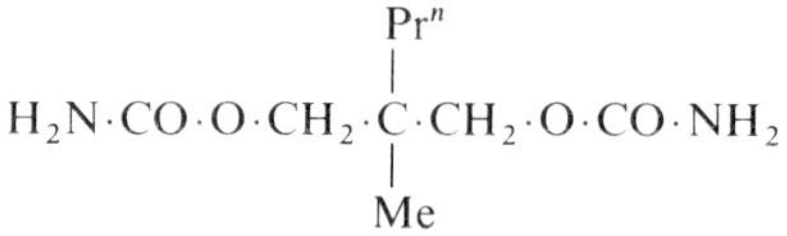

$$H_2N \cdot CO \cdot O \cdot CH_2 \cdot \underset{\text{Me}}{\overset{\text{Pr}^n}{C}} \cdot CH_2 \cdot O \cdot CO \cdot NH_2$$

Meprobamate

Urethane has been shown to be carcinogenic in the mouse (Nettleship, Henshaw and Meyer, 1943), in the rat (Tannenbaum, Vesselinovitch, Maltoni, and Stryzak Mitchell, 1962) and also in the hamster. Notwithstanding these carcinogenic effects, urethane inhibits the growth of Walker rat carcinoma 256 (Haddow and Sexton, 1946) and has been used in the treatment of leukaemia in man (Paterson, Ap Thomas, Haddow and Watkinson, 1946).

It has been suggested (Rogers, 1954) that urethane forms a conjugate with oxaloacetic acid or some related compound which interferes competitively with the formation of carbamoyl aspartate in the biosynthesis of orotic acid and hence in the formation of pyrimidines (Chapter 23). It is also established that urethane is metabolised to *N*-hydroxyurethane in the rat, the rabbit, and in man (Boyland, Nery, Peggie and Williams, 1963). Other metabolites include *S*-ethyl- and *S*-ethoxycarbonyl-*N*-acetylcysteine, but since urethane reacts only slowly with the sulphydryl group of glutathione in the presence of rat liver homogenate, whereas hydroxyurea reacts readily with glutathione in the absence of rat liver homogenate, it is suggested that *N*-hydroxyurethane could be acting as a biological alkylating agent.

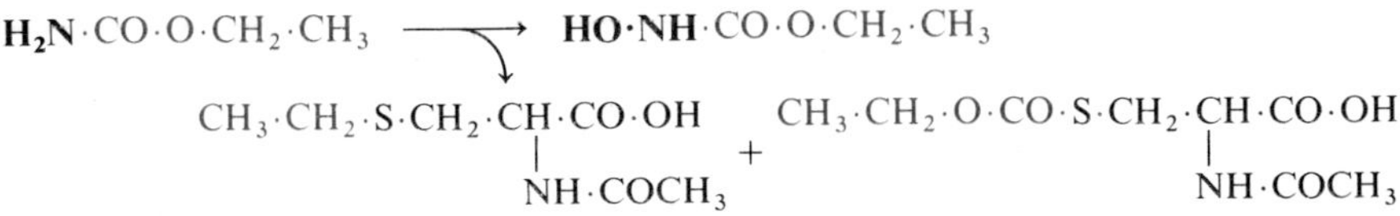

$$\mathbf{H_2N}\cdot CO\cdot O\cdot CH_2\cdot CH_3 \longrightarrow \mathbf{HO\cdot NH}\cdot CO\cdot O\cdot CH_2\cdot CH_3$$

$$CH_3\cdot CH_2\cdot S\cdot CH_2\cdot \underset{|}{CH}(NH\cdot COCH_3)\cdot CO\cdot OH + CH_3\cdot CH_2\cdot O\cdot CO\cdot S\cdot CH_2\cdot CH(NH\cdot COCH_3)\cdot CO\cdot OH$$

However, in view of the antitumour activity which has also been reported in *N*-hydroxyurea, HONH · CO · $NH_2$ (Stearns, Losee and Bernstein, 1963) and in *N*-hydroxyguanidine (Adamson, 1972), it seems more likely that in both cases this lies in the acylhydroxylamino group.

It is of interest that in the mouse, *N*-hydroxyurethane is only about one-third as carcinogenic as urethane itself (Berenblum, Ben-Ishai, Haran-Ghera, Lapidot, Simon and Trainin, 1959) and that significantly *N*-hydroxyurethane is rapidly reduced metabolically to urethane (Mirvish, 1964), in contrast to the oxidation of urethane which occurs in the rat, rabbit and man. Mirvish concludes that the carcinogenicity of *N*-hydroxyurethane in the mouse arises from this conversion to urethane.

Carbamoyl phosphate is a key intermediate in pyrimidine biosynthesis and for incorporation of ammonia into urea. Its synthesis in mammalian systems from carbon dioxide and ammonia is catalysed by **carbamoyl phosphate synthetase**, and requires *N*-acetylglutamate as coenzyme.

$$CO_2 + NH_3 + 2ATP \xrightarrow{N\text{-Acetylglutamate}} H_2N\cdot CO\cdot OPO_3H_2 + 2ADP + P$$

In pyrimidine biosynthesis, carbamoyl phosphate donates the carbamoyl group to the α-amino group of aspartate in a reaction catalysed by aspartate

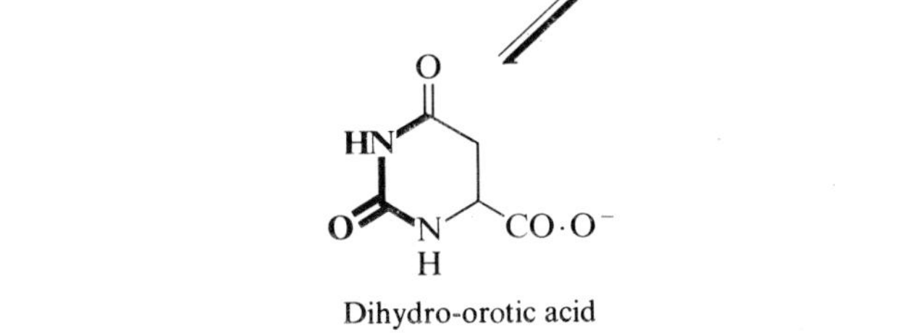

$$H_2N\cdot CO\cdot OPO_3H_2 + H_2N\cdot CH(CO\cdot O^-)\cdot CH_2\cdot CO\cdot O^- \rightleftharpoons \text{Carbamoyl aspartate} \rightleftharpoons \text{Dihydro-orotic acid}$$

Carbamoyl aspartate

Dihydro-orotic acid

**carbamate transferase**, the resulting carbamoyl aspartate then undergoing lactam formation (cyclo-dehydration) to give dihydro-orotic acid.

**Dithiocarbamates**

Dithiocarbamic acid ($H_2N \cdot CS \cdot SH$), like carbamic acid, is unstable and is known only in the form of its salts and derivatives. Stable amine salts of *N*-alkyldithiocarbamic acid are formed when primary and secondary amines are heated with carbon disulphide.

$$Et_2NH + CS_2 \longrightarrow Et_2N \cdot CS \cdot SH \xrightarrow{Et_2NH} Et_2N \cdot CS \cdot S^- \quad Et_2\overset{+}{N}H_2$$

The primary amine salts, however, decompose when heated to form *NN'*-dialkyl thioureas.

$$EtNH \cdot CS \cdot S^- \quad Et\overset{+}{N}H_3 \xrightarrow[H_2S]{} Et \cdot N{:}C{:}S + EtNH_2 \longrightarrow EtNH \cdot CS \cdot NHEt$$

Oxidation of *NN'*-diethyldithiocarbamic acid gives rise to tetraethylthiuram disulphide, *Disulfiram* (Antabuse), which is used in the control of alcoholism. This is reduced metabolically to *NN'*-diethyldithiocarbamic acid which inhibits the oxidation of acetaldehyde, itself formed by oxidation of ethanol. The build up of acetaldehyde rapidly leads to symptoms of flushing and nausea, the avoidance of which can only be achieved by restraint over the intake of alcohol.

$$\underset{\textit{Disulfiram}}{Et_2N \cdot CS \cdot S \cdot S \cdot CS \cdot NEt_2} \underset{O}{\overset{2H}{\rightleftharpoons}} 2Et_2N \cdot CS \cdot SH$$

Antabuse binds to serum albumin forming mixed disulphides with its one free protein sulphydryl group (Strömme, 1965).

$$\text{Albumin}—\mathbf{SH} + Et_2 \cdot CS \cdot S \cdot S \cdot CS \cdot NEt_2 \xrightarrow[Et_2N \cdot CS \cdot \mathbf{SH}]{} \text{Albumin}—S \cdot S \cdot CS \cdot NEt_2$$

## UREA

Urea is the amide of carbamic acid. It is isomeric with ammonium cyanate and is formed from the latter when an aqueous solution is slowly evaporated. This historic reaction, discovered by Wöhler in 1828, is the first recorded conversion of an inorganic into an organic compound.

$$\underset{\text{Ammonium cyanate}}{NH_4 \cdot O \cdot C{\equiv}N} \longrightarrow \underset{\text{Urea}}{H_2N \cdot CO \cdot NH_2}$$

Urea is formed biosynthetically in the liver in mammals as the end-product of nitrogen metabolism. Some 30 g is excreted in the urine daily in man. Ammonia, arising from the deamination of amino acids and from the hydrolysis of glutamine and asparagine, enters the urea cycle as carbamoyl phosphate and is transformed to urea.

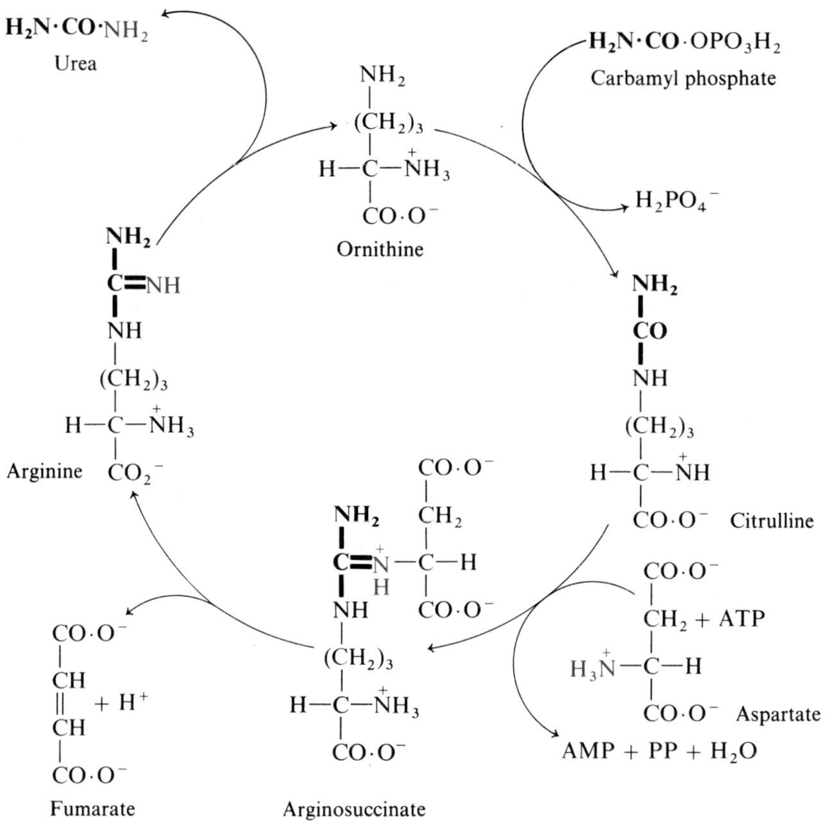

$H_2N \cdot CO \cdot NH_2$
Urea

$H_2N \cdot CO \cdot OPO_3H_2$
Carbamyl phosphate

$NH_2$–$(CH_2)_3$–$H$–$C$–$\overset{+}{N}H_3$–$CO \cdot O^-$
Ornithine

$H_2PO_4^-$

$NH_2$–$CO$–$NH$–$(CH_2)_3$–$H$–$C$–$\overset{+}{N}H$–$CO \cdot O^-$ Citrulline

$NH_2$–$C{=}NH$–$NH$–$(CH_2)_3$–$H$–$C$–$\overset{+}{N}H_3$–$CO_2^-$
Arginine

$CO \cdot O^-$–$CH_2$–$H_3\overset{+}{N}$–$C$–$H$–$CO \cdot O^-$ + ATP Aspartate

$NH_2$–$C{=}\overset{+}{N}H$–$C$–$H$ ($CH_2 \cdot CO \cdot O^-$, $CO \cdot O^-$)–$NH$–$(CH_2)_3$–$H$–$C$–$\overset{+}{N}H_3$–$CO \cdot O^-$
Arginosuccinate

$AMP + PP + H_2O$

$CO \cdot O^-$–$CH{=}CH$–$CO \cdot O^-$ + $H^+$
Fumarate

## General Properties

Urea is a crystalline solid, readily soluble in water to form neutral solutions from which it is not readily re-extracted by organic solvents. Its partial dipolar character, the result of resonance, accounts for its high water solubility (1 in 1) and insolubility in most organic solvents, except ethanol.

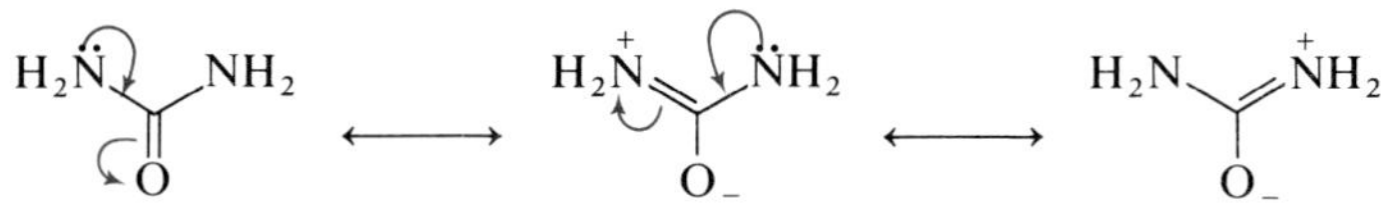

$$H_2N\text{–}C({=}O)\text{–}NH_2 \longleftrightarrow H_2\overset{+}{N}{=}C(O^-)\text{–}NH_2 \longleftrightarrow H_2N\text{–}C(O^-){=}\overset{+}{N}H_2$$

Urea at high concentration (6–8 M) is a powerful protein denaturant, and disrupts internal hydrogen bonds which maintain the integrity of tertiary and quaternary protein structures (Chapter 19).

Urea normally crystallises in a close-packed structure, but in the presence of certain long straight-chain hydrocarbons and related compounds, such as n-alcohols, n-acids and n-esters, urea crystallises in an alternative more loosely packed form which is channelled to include the hydrocarbon type molecules.

Branched-chain compounds are as a rule excluded, so that these inclusion complexes can be used as a means of separating straight and branched-chain hydrocarbons and related classes of molecule. The channels are of the order of 5 Å in diameter (7 Å in thiourea, which behaves similarly), and the complexes usually contain four or more molecules of urea to each molecule of hydrocarbon, the ratio increasing with chain length of the hydrocarbon.

## Chemical Properties

Urea is weakly basic and functions as a monoacidic base, the base-conjugate acid of which derives stability as a result of resonance.

$$H_2N{-}C({=}O){-}NH_2 \xrightarrow{H^+} H_2\overset{+}{N}{=}C(OH){-}\ddot{N}H_2 \longleftrightarrow H_2\ddot{N}{-}C(OH){=}\overset{+}{N}H_2 \longleftrightarrow H_2N{-}C({=}\overset{+}{O}H){-}NH_2$$

In consequence, urea forms a crystalline nitrate, $H_2N \cdot CO \cdot NH_2, HNO_3$, and oxalate $(H_2N \cdot CO \cdot NH_2)_2(COOH)_2$. The nitrate is dehydrated in concentrated sulphuric acid to form nitrourea.

$$H_2N \cdot CO \cdot NH_2, HNO_3 \xrightarrow{H_2SO_4} \underset{\text{Nitrourea}}{H_2N \cdot CO \cdot NH \cdot NO_2} + H_2O$$

Urea is readily hydrolysed to carbon dioxide and ammonia when heated in dilute mineral acid or in alkaline solution. The hydrolysis is therefore that of a typical amide giving ammonia and an acid (carbonic acid).

$$H_2N \cdot CO \cdot NH_2 + H_2O \longrightarrow CO_2 + 2NH_3$$
$$CO_2 + 2NH_3 \xrightarrow{2HCl} 2NH_4Cl + CO_2\uparrow$$
$$CO_2 + 2NH_3 \xrightarrow{2NaOH} Na_2CO_3 + H_2O + 2NH_3\uparrow$$

This same hydrolysis is also readily accomplished by the enzyme **urease** (urea amidohydrolase), which is present in soya bean and jack bean, and was the first enzyme to be obtained crystalline (Sumner, 1926).

Urea is also oxidatively decomposed by nitrous acid and by alkaline hypobromite (KBrO) to carbon dioxide and nitrogen. These reactions, like the urease hydrolysis, provide the basis for manometric determinations of urea in biological fluids.

$$H_2N \cdot CO \cdot NH_2 + 2HNO_2 \longrightarrow 2N_2\uparrow + CO_2\uparrow + 3H_2O$$

$$H_2N \cdot CO \cdot NH_2 + 3KBrO \longrightarrow N_2\uparrow + CO_2\uparrow + 2H_2O + 3KBr$$

Crystalline urea when gently heated above its melting point undergoes self-condensation with loss of ammonia to yield biuret (p. 526), which gives an intense violet colour with traces of copper sulphate solution, and aqueous sodium hydroxide.

### Biotin

Biotin is a bicyclic dialkylurea, present in yeast, egg-yolk, various micro-organisms, and all plant and animal tissues. It is essential for growth in mammals and certain bacteria. Nutritional requirements in man are supplied in the diet and from microbiological synthesis by the intestinal flora. Biotin plays a part in the incorporation of pantothenic acid

$$(HO \cdot CH_2 \cdot C(CH_3)_2 \cdot CHOH \cdot CO \cdot NH \cdot CH_2 \cdot CH_2 \cdot COOH)$$

into coenzyme A, and exerts a key rôle in carboxylation reactions.

Biotin is strongly bound to biotin-dependent enzymes through the ε-amino group of lysine as demonstrated by the isolation of biocytin (ε-*N*-biotinyl-lysine; Lane and Lynen, 1963).

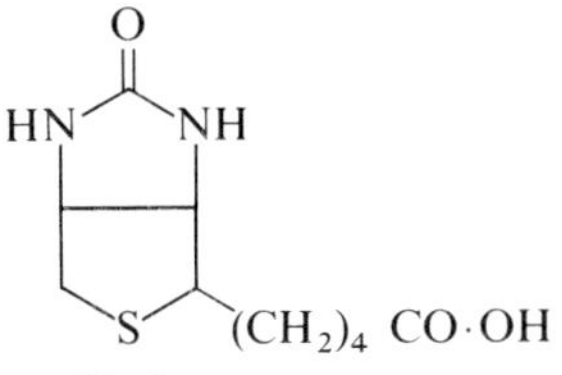

O, HN, NH, S, $(CH_2)_4$ CO·OH

Biotin

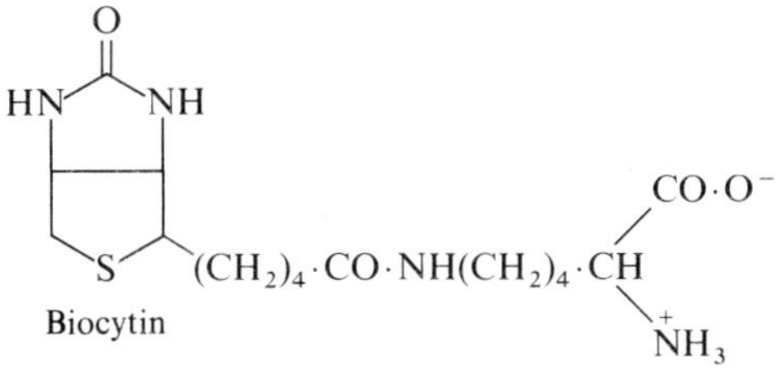

O, HN, NH, S, $(CH_2)_4 \cdot CO \cdot NH(CH_2)_4 \cdot CH$, $CO \cdot O^-$, $\overset{+}{N}H_3$

Biocytin

It is now established that carboxylation is achieved through *N*-carboxybiotin (Lynen, 1959; Knappe, Biederbick and Bruemmer, 1962) as follows.

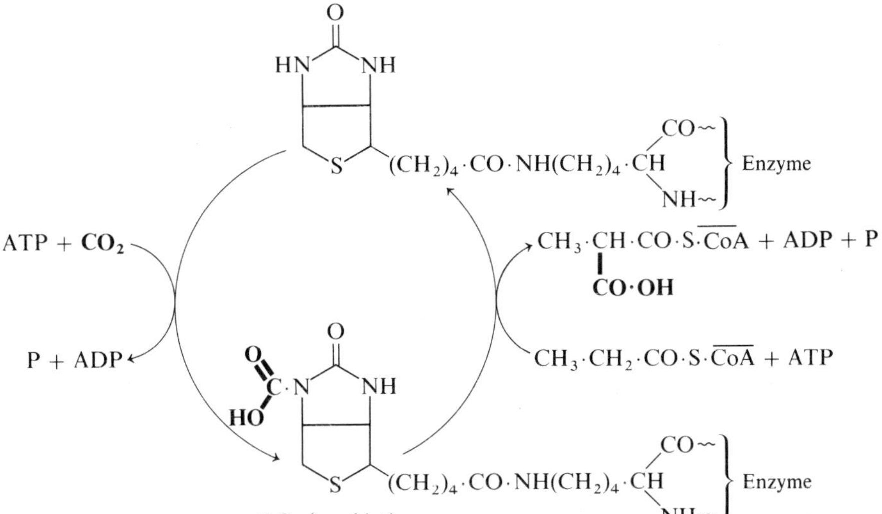

O, HN, NH, S, $(CH_2)_4 \cdot CO \cdot NH(CH_2)_4 \cdot CH$, CO~, NH~ } Enzyme

ATP + $CO_2$

P + ADP

$CH_3 \cdot CH \cdot CO \cdot S \cdot \overline{CoA}$ + ADP + P
with CO·OH on CH

$CH_3 \cdot CH_2 \cdot CO \cdot S \cdot \overline{CoA}$ + ATP

O, HO·C(=O)·N, NH, S, $(CH_2)_4 \cdot CO \cdot NH(CH_2)_4 \cdot CH$, CO~, NH~ } Enzyme

*N*-Carboxybiotin

## UREIDES

Urea reacts with acid chlorides, esters and anhydrides to form acyl ureas, or ureides. With acetyl chloride, it forms acetylurea (I, R = $CH_3$). Ureides may be of the open-chain type, as in (I) and (II) or cyclic as in (III) when the acylating acid is dibasic. Purine derivatives, including uric acid, caffeine and theophylline, may also be regarded as cyclic ureides (Chapter 23).

$R{\cdot}CO{\cdot}NH{\cdot}CO{\cdot}NH_2$ (I)

$R{\cdot}CO{\cdot}NH{\cdot}CO{\cdot}NH{\cdot}CO{\cdot}R$ (II)

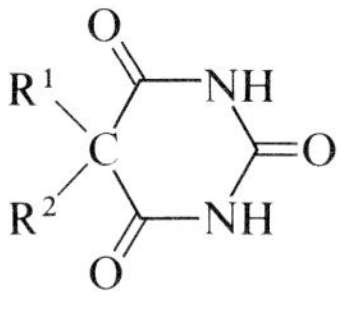

$R^1R^2C(CO{\cdot}NH)_2CO$ (III)

Ureides possess important sedative and hypnotic properties and these are most pronounced in the cyclic ureides or barbiturates. *Carbromal* [α-bromo-isovalerylurea, $(CH_3CH_2)_2C{\cdot}Br{\cdot}CO{\cdot}NH{\cdot}CO{\cdot}NH_2$] and *Bromvaletone* [Brom-isovalum; $(CH_3)_2CH{\cdot}CHBr{\cdot}CO{\cdot}NH{\cdot}CO{\cdot}NH_2$] are the only two examples of open-chain ureides of medical importance. They are less powerful hypnotics, but also less toxic than most barbiturates.

### Barbiturates

The parent substance of this group of compounds is barbituric acid (malonyl-urea), which may be prepared by condensing malonic acid with dry urea in ethanol in the presence of sodium ethoxide.

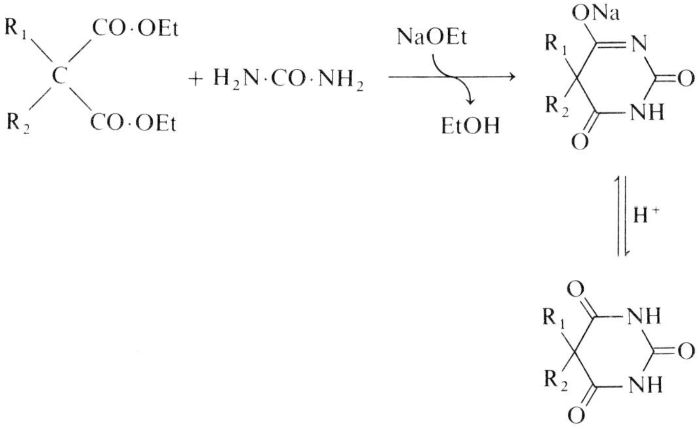

$$R_1R_2C(CO{\cdot}OEt)_2 + H_2N{\cdot}CO{\cdot}NH_2 \xrightarrow[EtOH]{NaOEt} \text{sodium salt (ONa, =N, =O, NH)} \underset{H^+}{\rightleftharpoons} R_1R_2C(CO{\cdot}NH)_2CO$$

All medicinally-important barbiturates are related to barbituric acid, 2-thiobarbituric acid or their *N*-alkyl derivatives by substitution in the 5-position. They are prepared by condensing the appropriately 2,2′-disubstituted diethyl malonate with either urea, thiourea or one of the *N*-alkylureas. The product is always isolated as the parent barbiturate, even when the sodium salt is required.

The reason for this lies in the fact that *Barbitone* (Barbital; 5,5-diethylbarbituric acid) for example, in contrast to *Barbitone Sodium* (Barbital Sodium), is soluble in organic solvents and readily crystallisable from toluene and ethanol. The purified *Barbitone* is reconverted to *Barbitone Sodium* by treating a solution in toluene with the theoretical amount of sodium ethoxide (in ethanol), warming and allowing the sodio-derivative to crystallise.

The general solubilities and chemical properties of *Barbitone* and *Barbitone Sodium* are typical of the series as a whole. The parent barbiturate is soluble in organic solvents, but only slightly soluble (1 in 160) in water. It is capable of existence in various tautomeric forms though the tri-keto form normally predominates.

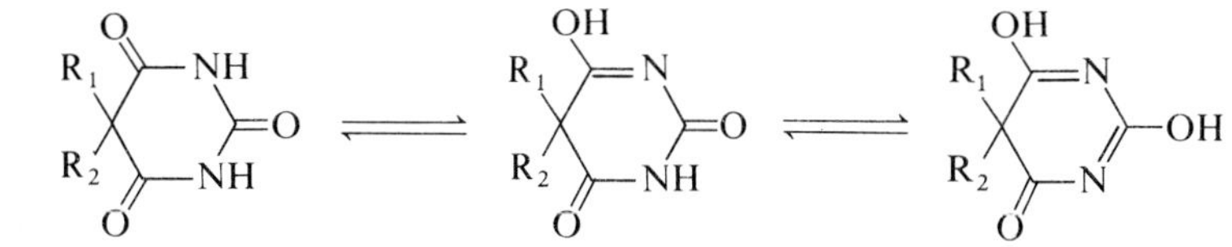

*Barbitone* ($R_1{=}R_2{=}Et$)

As a consequence of the existence of enolic forms, barbiturates function as weak acids and form water-soluble sodium salts with alkali and alkaline earth metals, as for example *Phenobarbitone* (Phenobarbital) *Sodium*. These sodium salts are, typically, hygroscopic solids, which are readily soluble in water to give quite distinctly alkaline solutions (pH 10–11), and are used in the preparation of injection solutions. The sodium salts, even in the solid state, decompose when exposed to carbon dioxide to form the parent water-insoluble barbiturate. Faulty storage may, therefore, render them insoluble in water or only partially so with the passage of time.

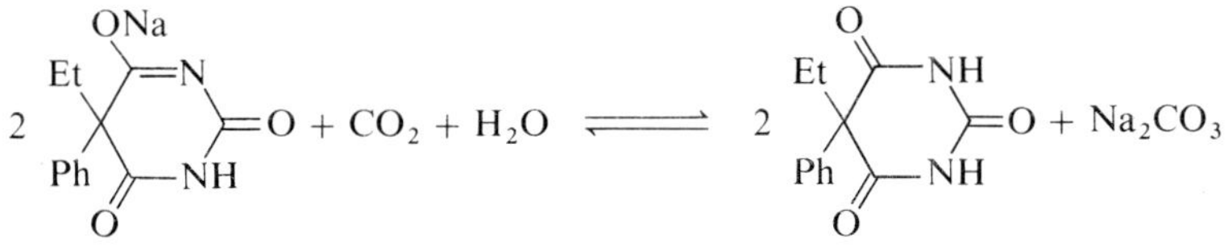

$2\ [\text{Phenobarbitone Sodium}] + CO_2 + H_2O \rightleftharpoons 2\ [\text{Phenobarbitone}] + Na_2CO_3$

*Phenobarbitone Sodium* *Phenobarbitone*

The sodium salts also undergo hydrolysis fairly easily in aqueous solution, especially at elevated temperatures, and solutions for injection may not be sterilised by autoclaving. They are prepared either by dissolving the contents of a sealed container in *Water for Injections* (sterile) which must be free from carbon dioxide (*Amylobarbitone Injection*; Amylobarbital Injection) or alternatively by sterilising a solution in aqueous propylene glycol (90%), which suppresses hydrolysis at a temperature of not more than 100°C (*Phenobarbitone Injection*; Phenobarbital Injection).

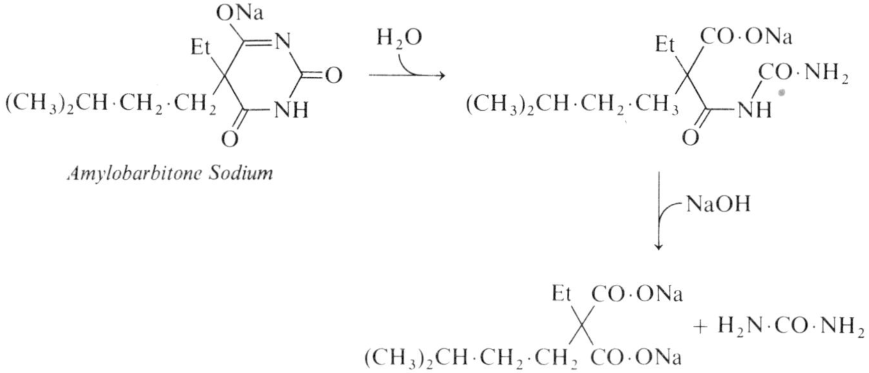

*Amylobarbitone Sodium*

The effect of varying the substituents on the 5-position of the barbiturate ring on absorption and transport in the body, including protein binding and fat deposition, and also on metabolic oxidation and hypnotic activity are discussed elsewhere (**2**, 4).

*Bemegride*, the structural analogue of barbiturates, acts as a barbiturate antagonist, because of a stimulant effect on the medulla.

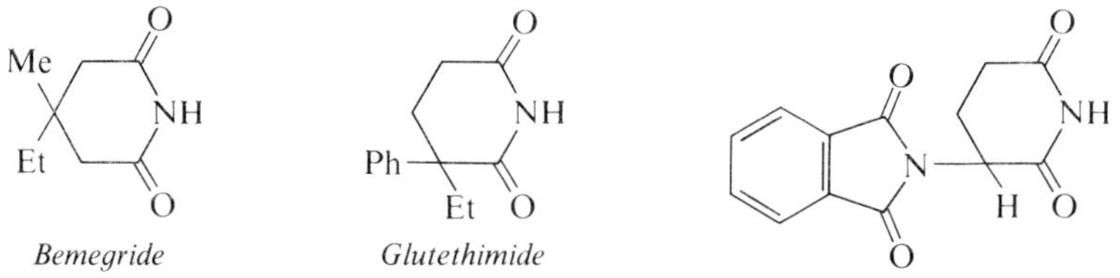

*Bemegride* *Glutethimide* *Thalidomide*

*Glutethimide* and *Thalidomide*, however, which are also glutarimide derivatives, are sedatives.

## THIOUREA

### General Properties

In contrast to urea, thiourea itself has no place in human metabolism. Certain of its derivatives, however, are important. It is formed when ammonium thiocyanate is heated to a temperature between 140° and 180°C when the two compounds exist in equilibrium. Thiourea, however, is present only to the extent of about 25% in this equilibrium mixture, but is obtained when the product is poured into water and the solution evaporated, due to its low solubility.

Thiourea has no definite melting point, and rapidly changes to ammonium thiocyanate when heated above 180°C. Like urea, it is capable of forming channel inclusion complexes. The diameter of the channel is about 7 Å, so that methyl-branched hydrocarbons, and even some alicyclic, aromatic and heterocyclic compounds can be accommodated within the crystal lattice (Schlenck, 1951).

### Oxidation

Thiourea is neutral in reaction, but is capable of functioning as a monoacidic base, forming comparatively stable salts, such as the hydrochloride. It is oxidised by permanganate in acidic solution to formamidine disulphide. Such oxidation is typical of thiols, and is evidence that thiourea exists as the thiol cation in acidic solution.

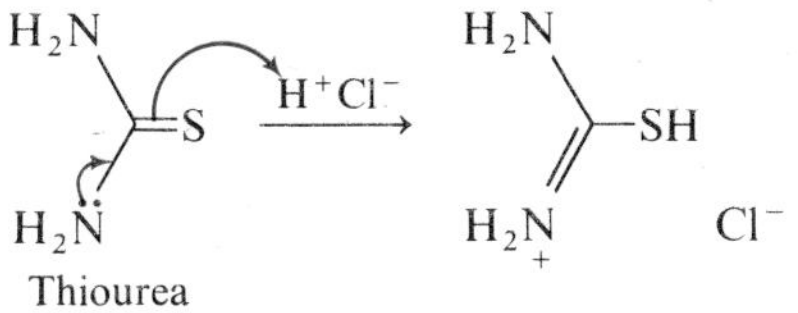

$$(H_2N)_2C{=}S \xrightarrow{H^+Cl^-} [H_2N{-}C(SH){=}\overset{+}{N}H_2]\ Cl^-$$

Thiourea

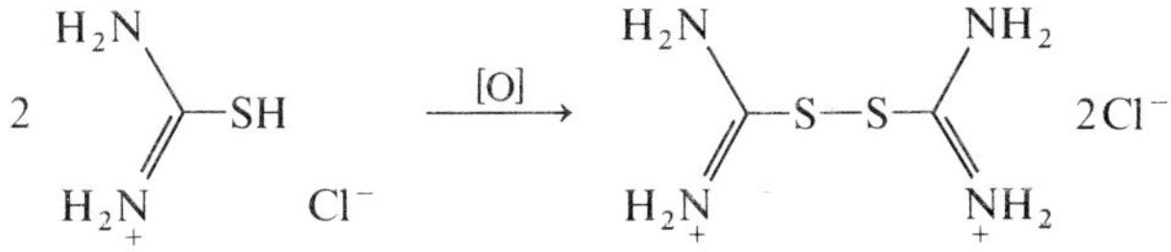

$$2\ [H_2N{-}C(SH){=}\overset{+}{N}H_2]\ Cl^- \xrightarrow{[O]} [H_2\overset{+}{N}{=}C(NH_2){-}S{-}S{-}C(NH_2){=}\overset{+}{N}H_2]\ 2Cl^-$$

Formamidine disulphide hydrochloride

Formamidine disulphide is also obtained by oxidation of thiourea with iodine. Thiourea is oxidised by lead and mercury oxides at room temperature to form cyanamide.

$$(H_2N)_2CS + HgO \longrightarrow H_2N{\cdot}CN + HgS + H_2O$$

Symmetrical *NN′*-substituted thioureas, however, are oxidised by the same reagents in low boiling organic solvents, such as ether, benzene or acetone to form carbodiimides (p. 517).

### Metabolic Oxidation, and Distribution of Thiourea Derivatives

Oxidative desulphurisation is characteristic of a number of thiourea derivatives in mammals. Thus, phenylthiourea is transformed to phenylurea (Scheline,

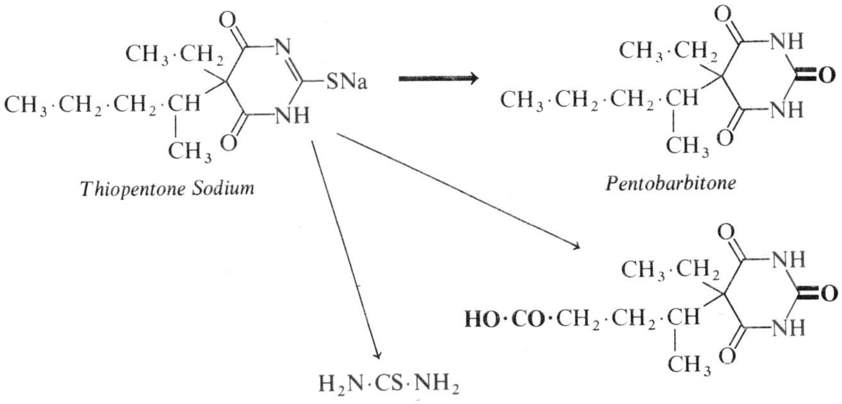

Thiopentone Sodium ($CH_3{\cdot}CH_2$ and $CH_3{\cdot}CH_2{\cdot}CH_2{\cdot}CH(CH_3)$ substituted 2-thio-barbiturate, SNa) $\longrightarrow$ Pentobarbitone ($CH_3{\cdot}CH_2$, $CH_3{\cdot}CH_2{\cdot}CH_2{\cdot}CH(CH_3)$ barbituric acid, C=O)

Thiopentone Sodium $\longrightarrow$ $HO{\cdot}CO{\cdot}CH_2{\cdot}CH_2{\cdot}CH(CH_3)$ / $CH_3{\cdot}CH_2$ barbituric acid

Thiopentone Sodium $\longrightarrow$ $H_2N{\cdot}CS{\cdot}NH_2$

*Thiopentone Sodium*   *Pentobarbitone*

Smith and Williams, 1961), and *Thiopentone* (Thiopental) is mainly metabolised to its oxygen analogue, *Pentobarbitone*. Some oxidation of the side-chain and ring cleavage to thiourea also occurs (Frey, Doenicke and Jaeger, 1961).

Studies with the antithyroid drugs $^{35}$S-*Propylthiouracil* (Alexander, Evans, MacAulay, Gallagher and Londono, 1966) and both $^{35}$S-*Methimazole* and $2^{14}$-C-*Methimazole* have shown that the sulphur is rapidly detached in man (Stenlake, Williams and Skellern, 1973).

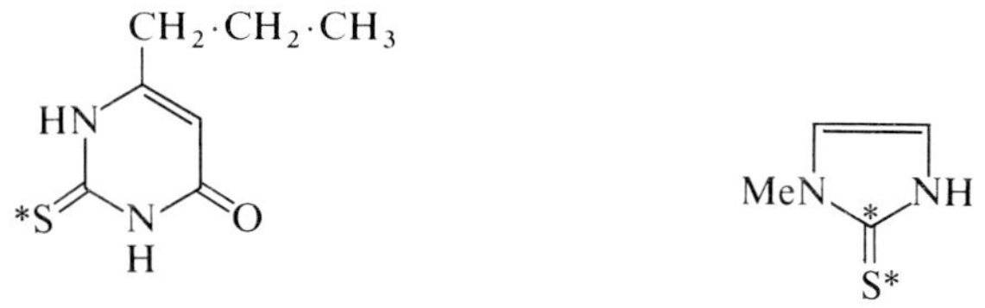

$CH_2{\cdot}CH_2{\cdot}CH_3$ HN *S N H O

*Propylthiouracil*

MeN NH * S*

*Methimazole*

The mechanism of this *S*-displacement is not clear, but it is evident that it only occurs in compounds which are capable of existing in a tautomeric sulphydryl form. Moreover, there is some evidence that one of the metabolites of *Methimazole* is its *S*-acetyl derivative, which should be susceptible to nucleophilic attack.

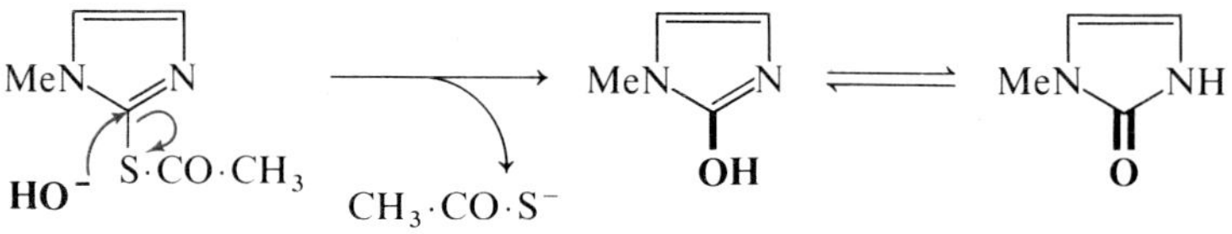

MeN N HO⁻ $S{\cdot}CO{\cdot}CH_3$ → $CH_3{\cdot}CO{\cdot}S^-$ MeN N OH ⇌ MeN NH O

It is equally possible that the metabolic displacement of sulphur may be oxidative, since thiourea is capable of similar oxidation with potassium permanganate in alkaline solution.

$$H_2N{\cdot}CS{\cdot}NH_2 \xrightarrow{[O]} H_2N{\cdot}CO{\cdot}NH_2 + S$$

These cyclic derivatives of thiourea are more lipid-soluble than their oxygen analogues and this accounts for their ready distribution into fatty and phospholipid based tissues in the body. They also form water-soluble sodium salts which assist the preparation of injection solutions. They are not, however, particularly stable and are susceptible to oxidation, giving the corresponding disulphide. The antithyroid drugs, *Propylthiouracil* and *Methimazole*, are also rapidly concentrated in the thyroid gland, where they specifically inhibit the enzyme responsible for the incorporation of iodine into thyroxine and triiodothyronine. Jiroüsek and Pritchard (1970) have suggested that the key intermediate in the iodination is a sulphenyl iodide, in accord with the views of Cunningham, 1964; Chapter 14).

### *S*-Alkylation

Treatment of thiourea with alkyl halides similarly leads to derivatives of isothiourea, a reaction which is characteristic of dipolar ion character in thiourea. The product is an *S*-alkylisothiouronium salt.

$$(H_2\ddot{N})(H_2N)C{=}S \longleftrightarrow (H_2\overset{+}{N}{=})(H_2N)C{-}S^- \xrightarrow{CH_3{-}I} \left[(H_2\overset{+}{N}{=})(H_2N)C{-}S{-}CH_3\right] I^-$$

*S*-Methylisothiouronium iodide

*S*-Benzylthiouronium chloride is used to characterise carboxylic acids with which it forms relatively insoluble crystalline salts.

$$Ph\cdot CH_2\cdot S\cdot C(=\overset{+}{N}H_2)(NH_2)\ Cl^- + \mathbf{R\cdot CO\cdot OH} \longrightarrow \left[Ph\cdot CH_2\cdot S\cdot C(=\overset{+}{N}H_2)(NH_2)\right]\mathbf{R\cdot CO\cdot O^-} + \mathbf{HCl}$$

*S*-Methylisothiourea sulphate, prepared from thiourea and dimethyl sulphate, also condenses readily with primary and secondary alkylamines, to form *N*-alkylguanidines, with displacement of methanethiol.

$$(H_2N)(HN{=})C{-}SMe + \mathbf{R^1R^2NH} \longrightarrow (H_2N)(HN{=})C{-}\mathbf{NR^1R^2} + MeSH$$

### Hydrolysis

Like urea, thiourea is unstable to alkali, undergoing complete decomposition.

$$H_2N\cdot CS\cdot NH_2 + 2NaOH + H_2O \longrightarrow Na_2CO_3 + H_2S + 2NH_3$$

## CARBODIIMIDES

Carbodiimides are formed by oxidation of symmetrical *NN'*-disubstituted thioureas in low boiling organic solvents by means of lead or mercury oxides (p. 515).

$$C_6H_{11}{-}NH\cdot CS\cdot NH{-}C_6H_{11} \xrightarrow[HgS\ +\ H_2O]{HgO} C_6H_{11}{-}N{=}C{=}N{-}C_6H_{11}$$

*NN'*-Dicyclohexylthiourea — *NN'*-Dicyclohexylcarbodiimide

### Addition Reactions

Carbodiimides are only weakly basic, but very reactive in addition reactions with water, hydrogen sulphide and carboxylic acids.

$$R\cdot N{=}C{=}N\cdot R \xrightarrow{H_2O} R\cdot NH\cdot CO\cdot NH\cdot R$$

$$R\cdot N{=}C{=}N\cdot R \xrightarrow{H_2S} R\cdot NH\cdot CS\cdot NHR$$

$$R\cdot N{=}C{=}N\cdot R \xrightarrow{R^1\cdot CO\cdot OH} R\cdot NH\cdot CO\cdot N(R)(CO\cdot R^1)$$

### Peptide Bond Formation

Under appropriate conditions, carbodiimides also react with carboxylic acids, sulphonic acids and phosphoric acids forming the corresponding anhydrides. One of their most valuable uses is in peptide bond synthesis (Chapter 19). The carbodiimide functions in this reaction through the formation of a reactive anhydride type intermediate with the carboxylic acid.

$$C_6H_{11}\text{—}N{=}C{=}N\text{—}C_6H_{11} + Z\cdot NH\cdot CH(R)\cdot CO\cdot O\text{—}H \longrightarrow C_6H_{11}\text{—}N{=}C(O\cdot CO\cdot CH(R)\cdot NH\cdot Z)\text{—}NH\text{—}C_6H_{11}$$

$$\xrightarrow{H_2N\cdot CH(R^1)\cdot CO\text{—}} C_6H_{11}\text{—}NH\cdot CO\cdot NH\text{—}C_6H_{11} + Z\cdot NH\cdot CH(R)\cdot CO\cdot NH\cdot CH(R^1)\cdot CO\text{—}$$

## GUANIDINES

Guanidine, the imide of urea, and its derivatives are of considerable metabolic interest. The guanidino-amino acid, arginine, is formed in the urea cycle, and is a constituent amino acid of many important peptides and proteins. In particular, it is established that salt and hydrogen bond formation between the guanidino groups of arginine and the $\beta$- and $\gamma$-carboxyl groups of aspartic and glutamic acids play an important rôle in the stabilisation of protein conformations (Kennard and Walker, 1963), as for example in sperm-whale myoglobin (Kendrew, 1961).

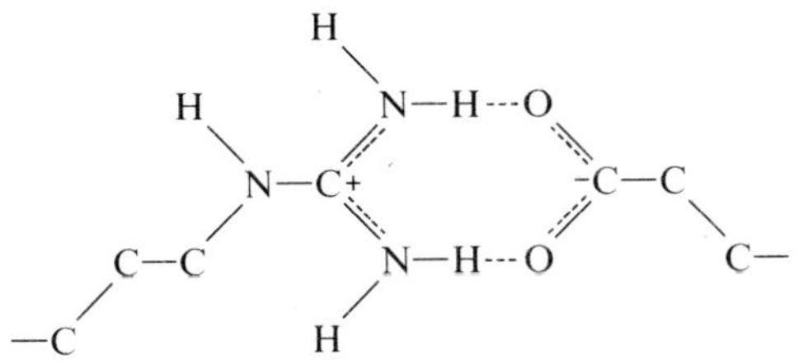

H N—H···O / H N—C+ -C—C / C—C N—H···O C— / —C H

For the same reason, most guanidino-compounds are strongly bound to protein, often with significant effect on their quaternary structure. Thus, guanidine

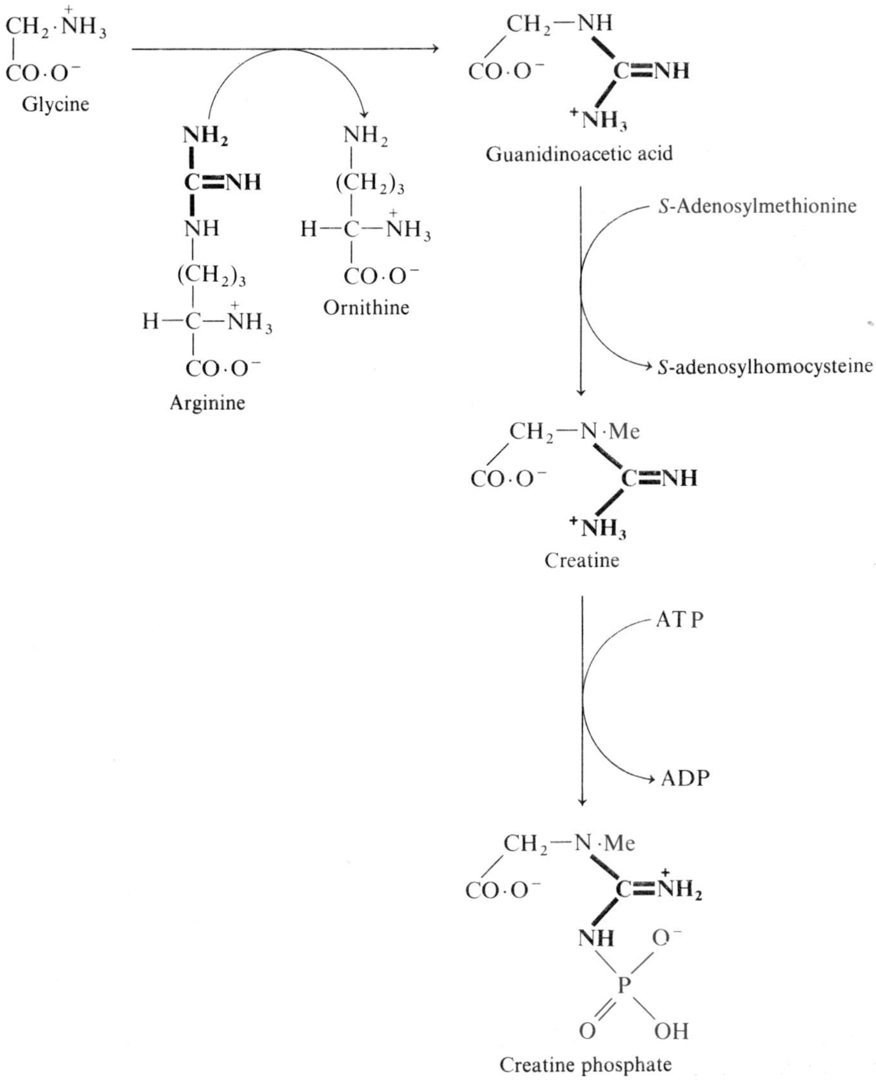

$CH_2 \cdot \overset{+}{N}H_3$ | $CO \cdot O^-$
Glycine

$NH_2$–C=NH–NH–$(CH_2)_3$–H–C–$\overset{+}{N}H_3$–$CO \cdot O^-$
Arginine

$NH_2$–$(CH_2)_3$–H–C–$\overset{+}{N}H_3$–$CO \cdot O^-$
Ornithine

$CH_2$—NH, $CO \cdot O^-$, C=NH, $^+NH_3$
Guanidinoacetic acid

*S*-Adenosylmethionine → *S*-adenosylhomocysteine

$CH_2$—N·Me, $CO \cdot O^-$, C=NH, $^+NH_3$
Creatine

ATP → ADP

$CH_2$—N·Me, $CO \cdot O^-$, C=$\overset{+}{N}H_2$, NH, P, $O^-$, O, OH
Creatine phosphate

hydrochloride is a powerful protein denaturant (Chapter 19) producing large changes in the physical and chemical properties of native proteins with changes in conformation from folded to random-coil structures (Tanford, Kawahara, Lapanje, Hooker, Zarlengo, Salahuddin, Aune and Takagi, 1967).

Arginine also plays a part in the biosynthesis of creatine (p. 519), which as creatine phosphate is an important form of energy storage in muscle.

### General Properties

Guanidine itself is a low melting deliquescent crystalline solid with caustic properties. It is readily soluble in water and ethanol. It melts at about 50°C, but decomposes with loss of ammonia at higher temperatures to form melamine (2,4,6-triamino-1,3,5-triazine).

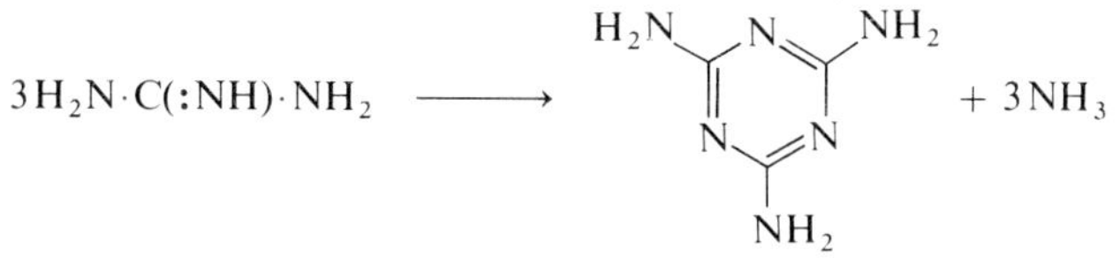

$$3H_2N{\cdot}C(:NH){\cdot}NH_2 \longrightarrow \text{Melamine} + 3NH_3$$

Melamine

### Basic Properties

Guanidine is a very strong base (p$K_a$ 13.65) and readily forms stable salts with even quite weak acids. Thus, not only does it form a stable hydrochloride and nitrate, but also reacts with picric acid and carbon dioxide to form the picrate, carbonate and bicarbonate respectively.

The strength of the base arises because of the symmetry in the guanidinium ion, which thus permits resonance between three equivalent structures (Pauling, 1960).

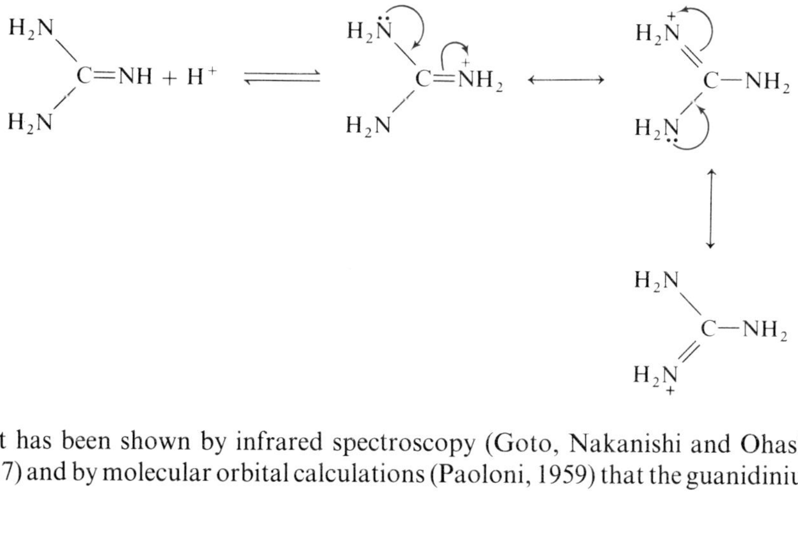

$$(H_2N)_2C{=}NH + H^+ \rightleftharpoons (H_2N)_2C{=}\overset{+}{N}H_2 \longleftrightarrow H_2\overset{+}{N}{=}C(NH_2)_2 \longleftrightarrow H_2N{-}C(NH_2){=}\overset{+}{N}H_2$$

It has been shown by infrared spectroscopy (Goto, Nakanishi and Ohashi, 1957) and by molecular orbital calculations (Paoloni, 1959) that the guanidinium ion behaves, as a result of resonance, as a triaminocarbonium ion, with the positive charge largely concentrated on the central carbon atom.

H
+0.086
N—H
H
+0.086 N—C +0.742
H
N—H
+0.086
H

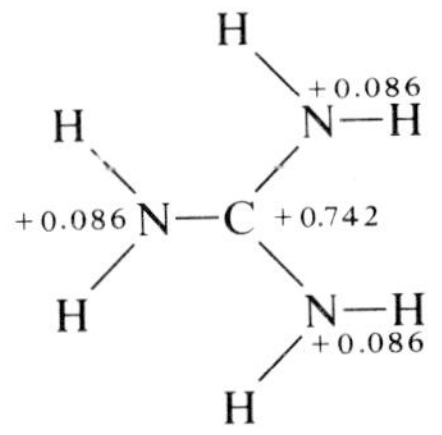

This concentration of positive charge on the central carbon atom no doubt explains the similarity of action of guanidinium compounds, such as *Guanethidine* and *Bethanidine*, which are fully ionised at physiological pH, with that of simple monoquaternary ammonium salts, as adrenergic neurone blocking agents. The similarity in ionic radius of the guanidine ion (*ca* 3 Å) and the tetramethylammonium ion 3.34 Å (Gill, 1965) is also important. The essential molecular requirements for adrenergic neurone blockade appear to be a strongly basic centre and a lipophilic group (Durant, Roe and Green, 1970). These requirements are met in the alkoxyalkylguanidines and alkoxyalkylaminoguanidines, which according to Augstein, Green, Monro, Wrigley, Katritzky and Tiddy (1967) adopt a conformation analogous to that of $\beta$-haloalkylamines (Belleau, 1958).

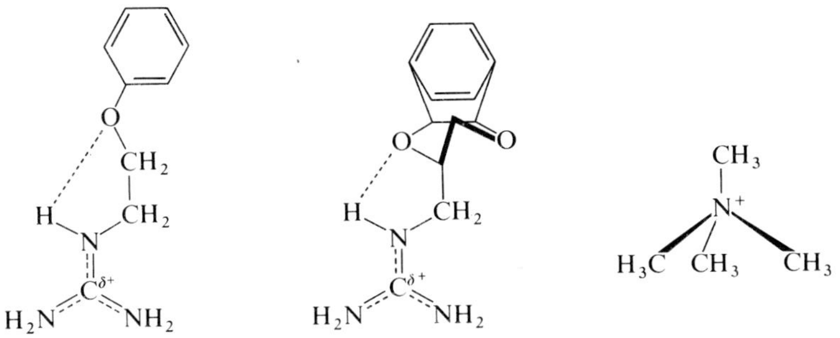

O CH$_2$ CH$_2$ H N C$^{\delta+}$ H$_2$N NH$_2$

O O H CH$_2$ N C$^{\delta+}$ H$_2$N NH$_2$

CH$_3$ N$^+$ H$_3$C CH$_3$ CH$_3$

Pressman and Park (1963) have shown that guanidine competes with magnesium ions for phosphate groups on mitochondrial binding sites. Hollunger (1955) and Pressman (1963) have also shown that guanidine and its derivatives are capable of inhibiting the formation of phosphate bonds in mitochondria.

**Basic Strength of Substituted Guanidines**

The strongly basic character of symmetrical guanidines is seen in *N,N′,N″*-trimethylguanidine (p$K_a$ 13.9). Mono-alkylguanidines are weaker bases (*N*-methylguanidine, p$K_a$ 13.4) despite the electron-repelling alkyl substituents, because of loss of symmetry in the ion. Electron-attracting substituents also reduce basicity, as in *N*-phenylguanidine (p$K_a$ 10.8) and *NN′*-diphenylguanidine (p$K_a$ 9.9). The basic strength of the guanidino group in arginine (p$K_a$ 12.48), however, is only slightly modified by the α-amino acid substituents, because of their mutual zwitterionic interaction.

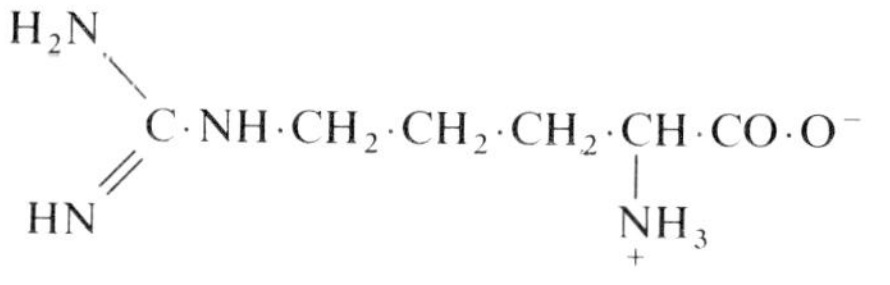

$$\begin{array}{l} H_2N \\ \quad \diagdown \\ \quad\; C\cdot NH\cdot CH_2\cdot CH_2\cdot CH_2\cdot CH\cdot CO\cdot O^- \\ \quad /\!/ \hspace{12em} | \\ HN \hspace{13em} \overset{+}{N}H_3 \end{array}$$

Arginine

$$\begin{array}{ccc} & Me & \\ & | & \\ & N & \\ CH_2 & & C{=}NH \\ | & & | \\ CO\cdot O^- & & {}^+NH_3 \end{array}$$

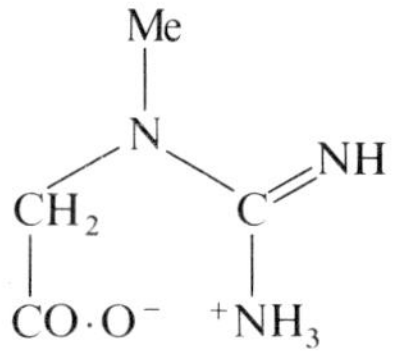

Creatine

Creatine, on the other hand, which lacks an α-amino substituent, is a much weaker base (p$K_a$ 11.02) due to zwitterionic interaction between the guanidino and carboxyl groups. Creatine is readily cyclised by dilute acid and alkali, even at physiological pH (7.4) and temperature, to the strongly basic creatinine, a proportion of which is normally excreted in the urine.

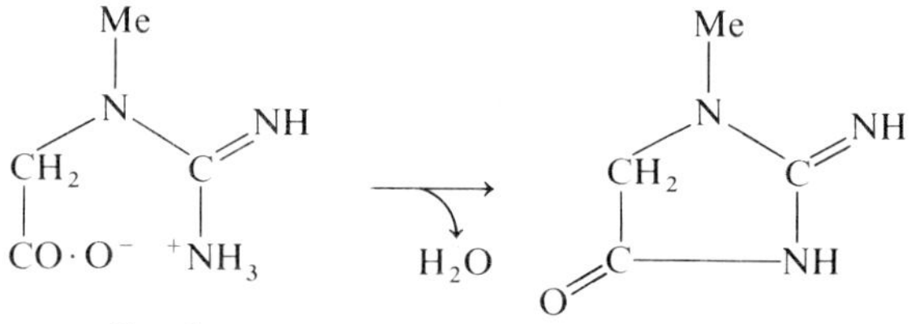

$$\begin{array}{ccc} & Me & \\ & | & \\ & N & \\ CH_2 & & C{=}NH \\ | & & | \\ CO\cdot O^- & & {}^+NH_3 \end{array} \xrightarrow[H_2O]{} \begin{array}{ccc} & Me & \\ & | & \\ & N & \\ CH_2 & & C{=}NH \\ | & & | \\ O{=}C & \text{———} & NH \end{array}$$

Creatine Creatinine

Phosphoguanidines, such as creatine phosphate (p. 519) and arginine phosphate, are important sources of energy in muscle and other tissues. The high energy content, which is released on hydrolysis of the N—P bond, arises as a result of inhibition of resonance through lack of symmetry in the phosphate substituted guanidinium ion. The inherent instability of these high energy derivatives favours a facile hydrolysis in which the driving force is the gain in resonance stabilisation due to the loss of the phosphate group (Chapter 2).

Since guanidines exist almost entirely in the ionic form under physiological conditions, hydrophilic guanidino compounds, such as *Streptomycin Sulphate*, are only very poorly absorbed from the gastro-intestinal tract, and hence for systemic effect require to be administered by injection. *Streptomycin*, so administered, is, however, rapidly excreted via the kidney (Spector, 1959; Weinstein and Ehrekranz, 1958; Garrod and O'Grady, 1968).

More lipophilic compounds, such as *Guanethidine* and *Bethanidine*, are absorbed from the gastro-intestinal tract, though somewhat erratically. Hence, they require to be administered in oral doses which are greater than might otherwise be necessary. The antibacterial sulphonamide, *Sulphaguanidine* (Chapter 14), which is both insoluble in water and strongly basic, and similarly is poorly

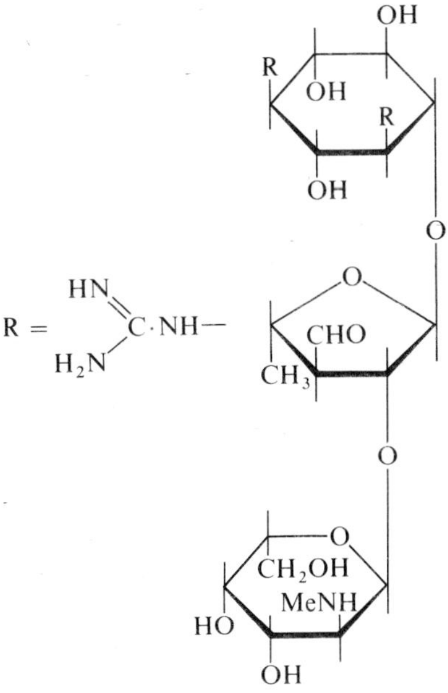

*Streptomycin*

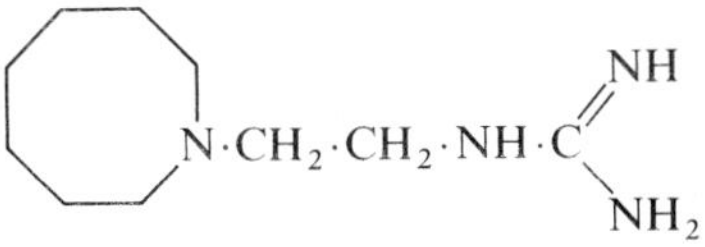

*Guanethidine*

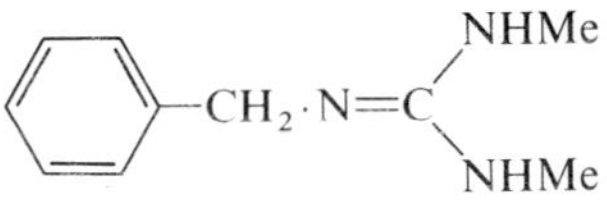

*Bethanidine*

absorbed from the gastro-intestinal tract is used to treat bacterial infections of the gut. The strongly basic character of guanidines also makes extraction from aqueous media difficult. Guanidino compounds can only be isolated from biological fluids, such as urine, with difficulty, and by extraction with polar solvents such as n-butanol.

**Acidic Properties and Complex Formation**

Guanidine possesses some acidic properties in that it readily forms insoluble salts with mono- and di-valent heavy metals such as silver and copper (II) (Krall, 1915). These appear to involve substitution of the amino hydrogens, since treatment of the salts with methyl iodide yields the corresponding *N*-methylguanidines.

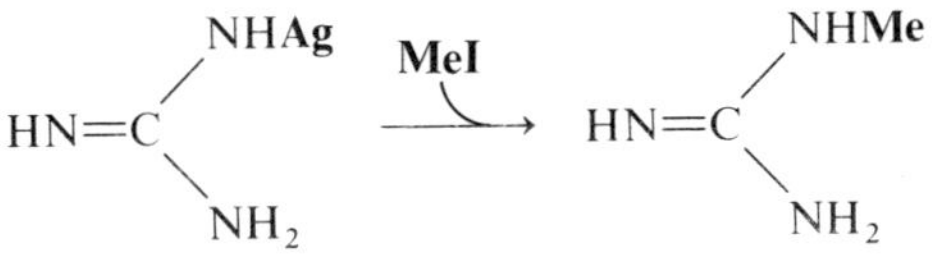

$$HN{=}C(NHAg)(NH_2) \xrightarrow{MeI} HN{=}C(NHMe)(NH_2)$$

*N*-Methylguanidine

Guanidine and even fully alkylated guanidines such as tetramethylguanidine also form coloured transition metal complexes with Co(II), Cu(II), Zn(II), Pd(II), Ni(II) and Co(III), which involve co-ordination via the lone pairs on the imino nitrogen (Longhi and Drago, 1965).

$$\left[ \mathrm{Co} \leftarrow \left( :\underset{\mathrm{H}}{\overset{|}{\mathrm{N}}}=\mathrm{C} \begin{matrix} \mathrm{NMe_2} \\ \mathrm{NMe_2} \end{matrix} \right)_4 \right]^{2+} (\mathrm{ClO_4})_2^-$$

**Nitroguanidine**

Guanidine nitrate resembles urea nitrate in that it is readily dehydrated by concentrated sulphuric acid. The product, nitroguanidine, is readily detonated and is used as an explosive propellant. It can be reduced to form first nitrosoguanidine, and finally aminoguanidine.

$$(\mathrm{H_2N})_2\mathrm{C}=\overset{+}{\mathrm{N}}\mathrm{H_2}\ \mathrm{NO_3^-} \xrightarrow{\mathrm{H_2SO_4}} (\mathrm{H_2N})_2\mathrm{C}=\mathrm{N\cdot NO_2}$$

Guanidine nitrate → Nitroguanidine

$$(\mathrm{H_2N})_2\mathrm{C}=\mathrm{N\cdot NO_2} \longrightarrow (\mathrm{H_2N})_2\mathrm{C}=\mathrm{NH\cdot NO}$$

Nitroguanidine → Nitrosoguanidine

$$\mathrm{HN}=\mathrm{C}(\mathrm{H_2N})-\mathrm{NH\cdot NH_2} \longleftarrow (\mathrm{H_2N})_2\mathrm{C}=\mathrm{NH\cdot NO}$$

Aminoguanidine ← Nitrosoguanidine

**Aminoguanidine**

Aminoguanidine, which is also formed by condensation of cyanamide and hydrazine hydrate, like guanidine, is a strong base (p$K_a$ 11.04).

$$\mathrm{H_2N\cdot C{\equiv}N} + \mathrm{H_2N\cdot NH_2\cdot H_2O} \xrightarrow{-\mathrm{H_2O}} \mathrm{HN}=\mathrm{C}(\mathrm{H_2N})-\mathrm{NH\cdot NH_2}$$

**Hydrolysis**

Guanidine, and substituted guanidines, are readily hydrolysed by barium hydroxide solution with release of ammonia and formation of the corresponding urea.

$$(\mathrm{H_2N})_2\mathrm{C}=\mathrm{NH} + \mathrm{H_2O} \xrightarrow{\mathrm{Ba(OH)_2}} (\mathrm{H_2N})_2\mathrm{C}=\mathrm{O} + \mathrm{NH_3}\uparrow$$

**Metabolism**

There is no evidence that guanidines are hydrolysed enzymically in mammalian systems, but transguanilation is well established in the conversion of arginine to ornithine in the urea cycle and in the conversion of glycine to guanidinoacetic acid. The metabolism of *Guanethidine Sulphate* has been studied extensively in both animals and man. Unlike many simpler non-endogenous compounds, such as methylguanidine which are rapidly excreted in the urine (Sperber, 1949), *Guanethidine* is only slowly excreted in humans (Dollery, Emslie-Smith and Milne, 1960). The guanidino group is not metabolised (Rahn and Dayton, 1969), and almost a quarter of the administered dose appears in the faeces in evidence of poor absorption. Almost 50% of the administered dose is excreted in the urine over a period of 3 days. The remainder is much more slowly excreted (McMartin, Rondel, Vinter, Allan, Humberstine, Leishman, Sandler and Thirkettle, 1970). The main pathways for metabolism appear to be oxidation to the *N*-oxide and ring scission (Rahn and Dayton, 1969; Abramson, Furst, McMartin and Wade, 1969; McMartin and Vintner, 1969).

$O^-$ / $^+N$ / NH / $CH_2 \cdot CH_2 \cdot NH \cdot C$ / $NH_2$

Guanethidine-*N*-oxide

Metabolism

NH / $N \cdot CH_2 \cdot CH_2 \cdot NH \cdot C$ / $NH_2$

*Guanethidine*

Urinary excretion

CO·OH / NH / NH·C / NH / $NH_2$

**Hydroxyguanidine**

The introduction of hydroxyguanidine as an anti-cancer agent (Adamson, 1972) was based on the observation of the presence of an RNA-dependent DNA polymerase (reverse transcriptase) in human leukaemic cells (Gallo, Yang and Ting, 1970), and reports of the concurrent presence of viral particles and reverse transcriptase in human milk of leukaemic subjects (Schlom, Spiegelman and Moore, 1971). The structure of hydroxyguanidine ($H_2N \cdot C(NH) \cdot NHOH$) combines the guanidino moiety which is known to have antiviral properties (Eggers, Ikegami and Tamm, 1965; Caliguiri and Tamm, 1970) with the hydroxylamino group which confers anti-tumour activity on hydroxyurea (Stearns,

Losee and Bernstein, 1963). Hydroxyguanidine shows antiviral and cytoxic activity *in vitro* and produces substantial increases in the survival time of mice bearing leukaemia and mast cell tumour P815, and rats bearing Walker 256 carcinoma.

## BIGUANIDES

Biguanide (guanylguanidine) bears the same relationship to guanidine as biuret does to urea, and is formed when guanidine hydrochloride is heated gently.

$$(H_2N)_2C{=}\overset{+}{N}H_2\,Cl^- \xrightarrow{-NH_4Cl} HN{=}C(NH_2){-}NH{-}C({=}NH)NH_2$$

Biguanide

$$H_2N{-}C({=}O){-}NH{-}C({=}O){-}NH_2$$

Biuret

### Biguanide Anti-diabetics

Biguanide, like guanidine, is strongly basic, and functions as a diacidic base ($pK_{a_1}$ 12.8, $pK_{a_2}$ 3.1). The biguanide oral anti-diabetics, *Phenformin Hydrochloride* (*N*-phenethylbiguanide) and *Metformin Hydrochloride*, are particularly useful in the treatment of maturity-onset diabetes, a type of diabetes in which insulin levels may be subnormal, but may often be normal or even above normal.

In contrast to the sulphonamide anti-diabetics (Chapter 14), these biguanides do not stimulate secretion of insulin from the pancreas. Further, they have no hypoglycaemic action in normal subjects, but are effective anti-hyperglycaemics in diabetics. Their mechanism of action is not fully understood. They are also useful in reducing blood urea levels in those diabetics where these are abnormally high, possibly by interfering with the urea cycle.

$$Ph{\cdot}CH_2{\cdot}CH_2{\cdot}NH{\cdot}C({=}NH){\cdot}NH{\cdot}C({=}NH)NH_2$$

*Phenformin*

$$HN{=}C(NMe_2){\cdot}NH{\cdot}C({=}NH)NH_2 \quad HCl$$

*Metformin Hydrochloride*

### Antimalarial Biguanides

The antimalarial, *Proguanil Hydrochloride* (Chloroguanide Hydrochloride), undergoes oxidative cyclisation in humans to give a 1,3,5-triazine derivative, which is the effective prophylactic agent (Crowther and Levi, 1953). The discovery of this biotransformation led to the development of *Pyrimethamine Hydrochloride* which is bio-isosteric with the active proguanil metabolite.

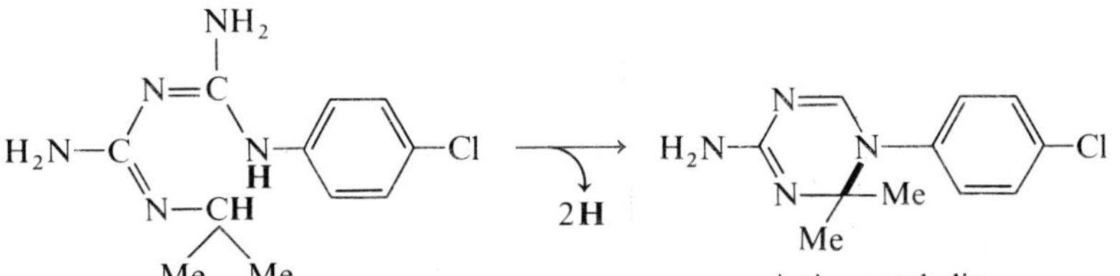

*Proguanil* → (2H) Active metabolite

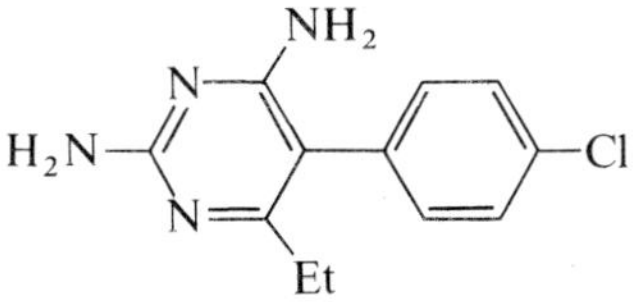

*Pyrimethamine*

**Antibacterial Biguanides**

The bis-biguanide, *Chlorhexidine Hydrochloride*, and the rather more water-soluble *Chlorhexidine Acetate* are powerful antibacterial agents. They are effective against both Gram-positive and Gram-negative organisms, relatively non-toxic, and useful in the cleansing of burns and wounds, and as skin disinfectants. Because of the relatively low water solubility of the salts, concentrated solutions are often prepared with gluconic acid or alternatively with the assistance of cationic detergents. *Chlorhexidine*, however, is incompatible with soaps and other anionic detergents due to the formation of insoluble salts; for the same reason, it is also incompatible with borates, carbonates, chlorides, phosphates, sulphates, and even citrates and tartrates.

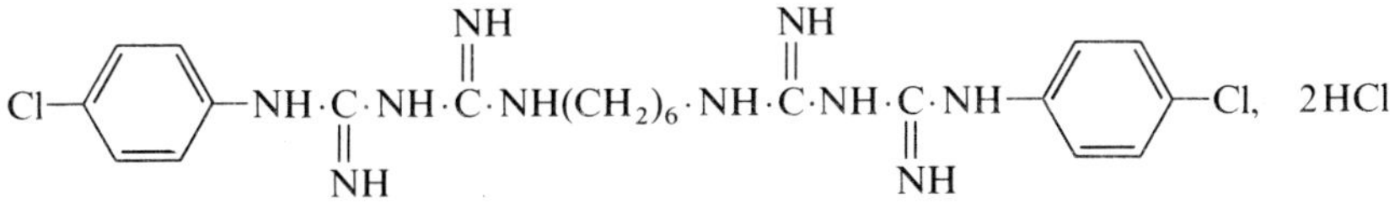

$$Cl-C_6H_4-NH\cdot C(=NH)\cdot NH\cdot C(=NH)\cdot NH(CH_2)_6\cdot NH\cdot C(=NH)\cdot NH\cdot C(=NH)\cdot NH-C_6H_4-Cl,\ 2HCl$$

*Chlorhexidine Hydrochloride*

# 18 Amino Acids

## α-AMINO ACIDS

### Introduction

α-Amino acids are of fundamental importance as the basic molecular units of the peptides and proteins, which are essential to the structure and function of all living matter. They are formed by enzymic digestion of proteins in the gastro-intestinal tract and absorbed by active transport mechanisms. They are normally present in human blood at a level of about 50 mg/100 ml; only tryptophan, however, is significantly protein-bound. They may be separated from the products of complete hydrolysis of peptides and proteins, and are formed in nature both by catabolism of proteins and by transamination from keto-acids.

Amino acids obtained by hydrolysis of natural peptides and proteins are almost exclusively α-amino acids, i.e. acids in which the amino group is attached to the carbon atom immediately adjacent to the carboxyl group. This close conjunction of basic and acidic groups influences both the physical and chemical properties of α-amino acids, accounting for their salt-like properties, which arise as the result of existence in a dipolar ion (zwitterion) form.

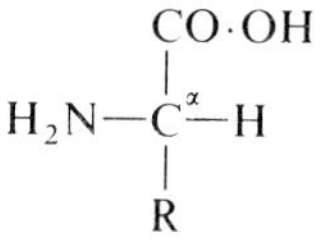

$$H_2N{-}\overset{\overset{\displaystyle CO{\cdot}OH}{|}}{\underset{\underset{\displaystyle R}{|}}{C^{\alpha}}}{-}H \qquad\qquad H_3\overset{+}{N}{-}\overset{\overset{\displaystyle CO{\cdot}O^-}{|}}{\underset{\underset{\displaystyle R}{|}}{C}}{-}H$$

α-Amino acid 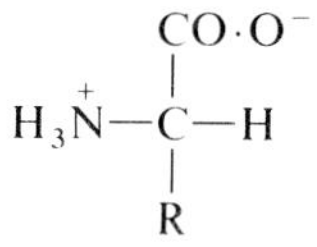 — Dipolar ion form

The simplest amino acid is glycine (R=H). Alternatively, R may be a simple alkyl group (alanine, valine, leucine), an aralkyl, substituted aralkyl or hetero-aralkyl group (Table 54). All such amino acids including hydroxy- and sulph-ydryl-substituted acids containing one amino and one carboxyl group are essentially neutral in character. In a few amino acids, the substituent R carries an additional basic (amino or guanidino) group or carboxyl group, conferring respectively basic or acidic properties on the molecule as a whole.

### Physical Properties

In accordance with their zwitterionic state, amino acids exhibit typical salt-like properties. They exist as colourless, crystalline solids, which melt only at high temperature (200–300°C), usually with decomposition. They are generally soluble in water, and insoluble in organic solvents, though solubility in water decreases rapidly with increasing molecular weight and molecular complexity (Table 54).

**Ionisation and Salt Formation**

Amino acids function as both acids and bases. Neutral amino acids (Table 54), e.g. glycine, have two ionisable groups, the first dissociation constant being associated with the carboxyl group, and the second with the amino group. Acidic and basic amino acids (Table 54), such as aspartic acid and lysine, have three ionisable groups.

$$H_3\overset{+}{N}\cdot CH_2\cdot CO\cdot OH \overset{pK_{a_1}}{\rightleftharpoons} H^+ + H_3\overset{+}{N}\cdot CH_2\cdot CO\cdot O^-$$

$$\Big\Updownarrow pK_{a_2}$$

$$H^+ + H_2N\cdot CH_2\cdot CO\cdot O^-$$

The carboxyl group is a stronger acid ($pK_{a_1}$ *ca* 2.0) than the corresponding alkanecarboxylic acids ($pK_a$ 4–5) due to the electron-attracting (-I) and proton-repelling properties of the ammonium group. The base-conjugate acid ($pK_{a_2}$ *ca* 9–10) is a weak acid on account of the electron-attracting (-I) properties of the carboxylate ion. It follows from these equilibria that there is a characteristic pH at which the molecule exists entirely in the dipolar ion form, and hence bears no net charge. At this pH, known as the **isoelectric point** and designated pI (Table 54), the molecule is electrically neutral, and is incapable of migration in an electric field. Solubility in water of amino acids is usually at a minimum at the isoelectric point.

Amino acids form salts with both acids and bases. Mineral acids form water-soluble amine salts, $X^-H_3\overset{+}{N}\cdot CHR\cdot COOH$, from which the amino acid can be regenerated by ion exchangers (Meyers and Miller, 1952). Most amino acids also form good crystalline salts with optically active acids such as tartaric acid and camphoric acid, which are, therefore, useful in the resolution of racemic amino acids. As carboxylic acids, amino acids also react with alkali metal hydroxides and carbonates to form water-soluble salts, as in the case of the thyroid hormone, 3,5,3′-tri-iodothyronine (*Liothyronine Sodium*).

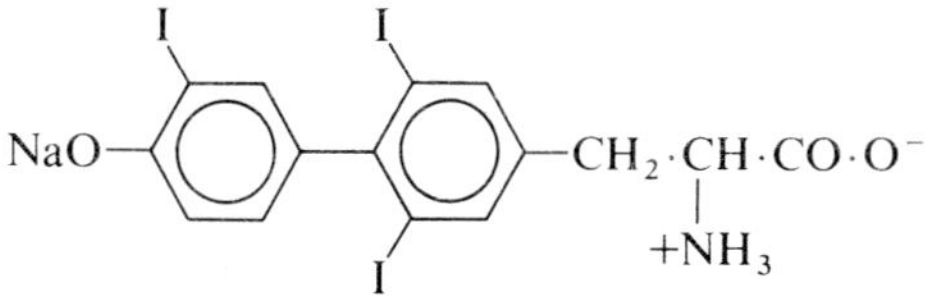

*Liothyronine Sodium*

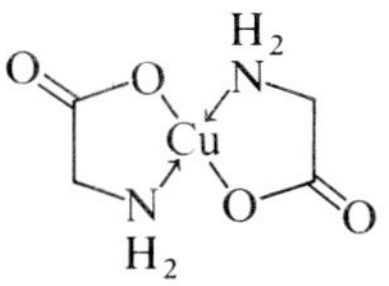

Glycine–Cu(II)-complex

Use has been made of the buffering capacity of the zwitterionic form of glycine in the formulation of tablets containing active ingredients, which are unduly susceptible to acid. The theory, underlying the use of glycine in this way, is that it stabilises pH at the tablet surface—gastric fluid interface.

Table 54. Nomenclature, Structure, and Physical Constants of α-Amino Acids

| Name | Abbreviation | Structure | Solubility (water) | p$K_a$ | Isoelectric point (pI) |
|---|---|---|---|---|---|
| **Neutral amino acids** | | | | | |
| Glycine | Gly | $H_2N \cdot CH_2 \cdot COOH$ | 1 in 4.5 | 2.34<br>9.60 | 5.97 |
| Alanine | Ala | $CH_3 \cdot CH(NH_2) \cdot COOH$ | 1 in 8 | 2.34<br>9.69 | 6.02 |
| Valine | Val | $(CH_3)_2 \cdot CH \cdot CH(NH_2) \cdot COOH$ | 1 in 11 | 2.32<br>9.62 | 5.97 |
| Leucine | Leu | $(CH_3)_2 \cdot CH \cdot CH_2 \cdot CH(NH_2) \cdot COOH$ | 1 in 40 | 2.36<br>9.60 | 5.98 |
| Isoleucine | Ile | $CH_3 \cdot CH_2 \cdot CH(CH_3) \cdot CH(NH_2) \cdot COOH$ | | 2.36<br>9.68 | 6.02 |
| Serine | Ser | $HO \cdot CH_2 \cdot CH(NH_2) \cdot COOH$ | | 2.21<br>9.15 | 5.68 |
| Homoserine | — | $HO(CH_2)_2 \cdot CH(NH_2) \cdot COOH$ | | | |
| Threonine | Thr | $CH_3 \cdot CH(OH) \cdot CH(NH_2) \cdot COOH$ | | 2.63<br>10.43 | 6.53 |
| Proline | Pro | (pyrrolidine ring, N–H) CO·OH | | 1.99<br>10.60 | 6.10 |
| Hydroxyproline | Hypro | HO–(pyrrolidine ring, N–H) CO·OH | | 1.92<br>9.73 | 5.83 |

| | | | | |
|---|---|---|---|---|
| Cysteine | Cys | $HS \cdot CH_2 \cdot CH(NH_2) \cdot COOH$ | 1.71<br>8.33(SH)<br>10.78($NH_2$) | 5.02 |
| Cystine | CyS-SCy | $-[S \cdot CH_2 \cdot CH(NH_2) \cdot COOH]_2$ | 1.65, 2.26(COOH)<br>7.85, 9.85($NH_2$) | 5.06 |
| Homocysteine | — | $HS(CH_2)_2 \cdot CH(NH_2) \cdot COOH$ | | |
| Methionine | Met | $CH_3S(CH_2)_2 \cdot CH(NH_2) \cdot COOH$ | 2.28<br>9.21 | 5.75 |
| Penicillamine | — | $HS \cdot C(CH_3)_2 \cdot CH(NH_2) \cdot COOH$ | | |
| Phenylalanine | Phe | $C_6H_5 \cdot CH_2 \cdot CH(NH_2) \cdot COOH$ | 1.83<br>9.13 | 5.98 |
| Tyrosine | Tyr | $HO-C_6H_4-CH_2 \cdot CH(NH_2) \cdot COOH$ | 2.20<br>9.11<br>10.97(OH) | 5.65 |
| Dopa | — | $(HO)_2C_6H_3-CH_2 \cdot CH(NH_2) \cdot COOH$ | | |
| Monoiodotyrosine | — | $HO(I)C_6H_3-CH_2 \cdot CH(NH_2) \cdot COOH$ | | |
| Di-iodotyrosine | — | $HO(I)_2C_6H_2-CH_2 \cdot CH(NH_2) \cdot COOH$ | | |

Table 54 (*continued*)

| Name | Abbreviation | Structure | Solubility (water) | p$K_a$ | Isoelectric point (pI) |
|---|---|---|---|---|---|
| Thyroxine | — | 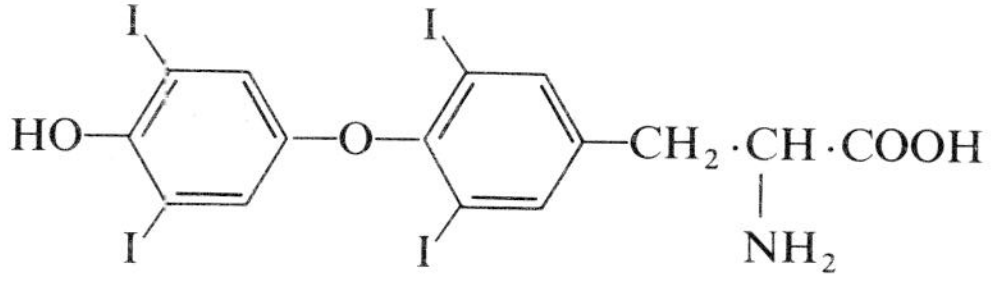 I, I, HO—, —O—, I, I, $—CH_2 \cdot CH(NH_2) \cdot COOH$ | | | |
| Tri-iodothyronine | — | 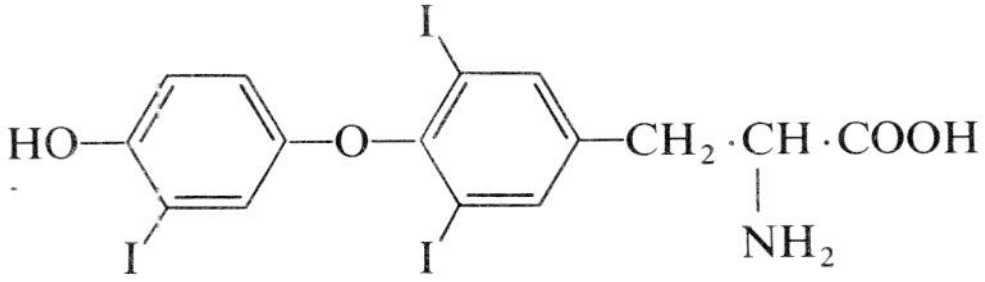 I, HO—, —O—, I, I, $—CH_2 \cdot CH(NH_2) \cdot COOH$ | | | |
| Trytophan | Trp | 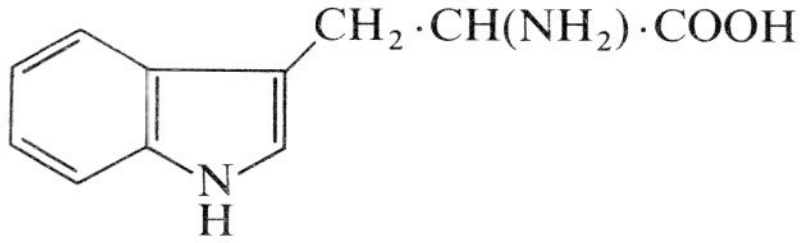 $CH_2 \cdot CH(NH_2) \cdot COOH$, N, H | | 2.38<br>9.39 | 5.88 |
| **Basic amino acids and their derivatives** | | | | | |
| Ornithine | Orn | $H_2N(CH_2)_3 \cdot CH(NH_2) \cdot COOH$ | | | |
| Arginine | Arg | 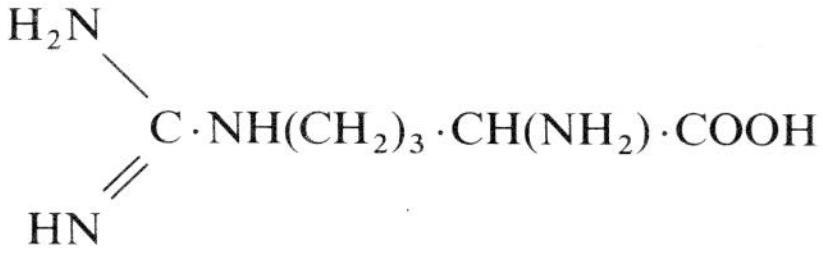 $H_2N$, $C \cdot NH(CH_2)_3 \cdot CH(NH_2) \cdot COOH$, HN | | 2.17<br>9.04($\alpha$-$NH_2$)<br>12.48(guanidino) | 10.76 |

| | | | | |
|---|---|---|---|---|
| Citrulline | Cit | $H_2N\text{—}C(\text{=}O)\cdot NH(CH_2)_3\cdot CH(NH_2)\cdot COOH$ | | |
| Lysine | Lys | $H_2N(CH_2)_4\cdot CH(NH_2)\cdot COOH$ | 2.18<br>8.95($\alpha$-$NH_2$)<br>10.53($\varepsilon$-$NH_2$) | 9.74 |
| Hydroxylysine | Hylys | $H_2N\cdot CH_2\cdot CH(OH)\cdot (CH_2)_2\cdot CH(NH_2)\cdot COOH$ | 2.13<br>8.62($\alpha$-$NH_2$)<br>9.67($\varepsilon$-$NH_2$) | 9.15 |
| Histidine | His | (imidazole ring: N, NH) $CH_2\cdot CH(NH_2)\cdot COOH$ | 1.82<br>6.00(imidazole)<br>9.17($NH_2$) | 7.58 |
| **Acidic amino acids and their derivatives** | | | | |
| Aspartic acid | Asp | $HOOC\cdot CH_2\cdot CH(NH_2)\cdot COOH$ | 2.09($\alpha$-COOH)<br>3.86($\beta$-COOH)<br>9.82 | 2.87 |
| Asparagine | $AspNH_2$ | $H_2N\cdot CO\cdot CH_2\cdot CH(NH_2)\cdot COOH$ | 2.02<br>8.80 | 5.41 |
| Glutamic acid | Glu | $HOOC\cdot (CH_2)_2\cdot CH(NH_2)\cdot COOH$ | 2.19($\alpha$-COOH)<br>4.25($\beta$-COOH)<br>9.67 | 3.22 |
| Glutamine | $GluNH_2$ | $H_2N\cdot CO(CH_2)_2\cdot CH(NH_2)\cdot COOH$ | 2.17<br>9.13 | 5.65 |

### Complex Formation

Because of the close proximity of carboxyl and amino groups, α-amino acids form metal chelate complexes with transition metals in which the metal ion is attached through the carboxyl and amino groups (Stosick, 1945). The overall stability constant of such complexes is given by the relationship $K = k_1 \cdot k_2$, where $k_1$ and $k_2$ are derived from the following equilibria, and $A^-$ represents the amino acid in its dipolar ion form.

$$M^{2+} + A^- \underset{}{\overset{k_1}{\rightleftharpoons}} M^+A + H^+$$

$$M^+A + A^- \underset{}{\overset{k_2}{\rightleftharpoons}} MA_2 + H^+$$

Some values for log $K$ determined for a number of amino acids in water at 25°C are given in Table 55 (Maley and Mellor, 1950). This shows that the stability constants for glycine, alanine, valine and leucine are all of the same order, and that the order of stability for the metals listed is similar to that for other ligands (Mellor and Maley, 1948). The much greater stability constant for histidine is consistent with a different attachment through its amino and imino groups (Burk, Hearon, Caroline and Schade, 1946).

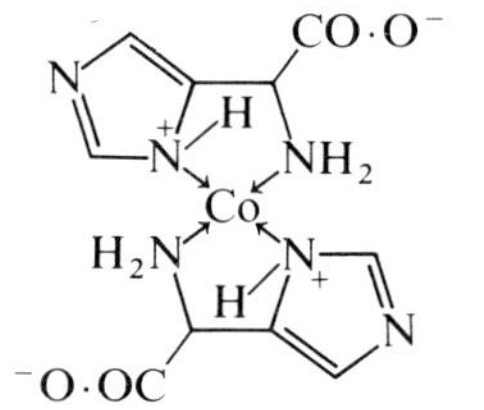

Histidine–Co(II)-complex

The enhanced stability of these histidine complexes would appear to explain why histamine is able to nullify the inhibitory effect of cobalt on various micro-organisms and tumours (Burk, Schade, Hesselbach, Hearon and Fischer, 1946; Hearon, Schade, Levy and Burk, 1947).

Zinc, which is known to be important for wound healing, is bound to low molecular weight components of human serum proteins (Prasad and Oberleas, 1970). Bradley and Sen (1974) have shown in experiments with $^{65}Zn$ that higher

Table 55. Stability Constant Exponents (log $K$) of Metal Amino Acid Complexes in Water at 25°C

| Metal | Glycine | Alanine | Valine | Leucine | Histidine |
|---|---|---|---|---|---|
| Cu | 15.42 | 14.83 | 14.45 | 14.34 | 18.33 |
| Zn | 9.72 | 9.50 | 9.10 | 8.93 | 12.88 |
| Co | 8.94 | 8.78 | 8.24 | 8.26 | 13.86 |
| Mn | 6.63 | 6.05 | 5.56 | 5.45 | 7.74 |

Table 56. Stability Constant Exponents (log $K$) for L-Cysteine and D-Penicillamine Metal Complexes at 20°C

| | L-Cysteine | | | D-Penicillamine | | |
|---|---|---|---|---|---|---|
| | log $K_1$ | log $K_2$ | log $K_3$ | log $K_1$ | log $K_2$ | log $K_3$ |
| $Mg^{2+}$ | 2.75 | — | — | 2.73 | — | — |
| $Ca^{2+}$ | 2.50 | — | — | 2.70 | — | — |
| $Mn^{2+}$ | 4.90 | 3.75 | — | 5.80 | 4.52 | — |
| $Fe^{2+}$ | 6.66 | 5.52 | — | 7.69 | 6.55 | — |
| $Co^{2+}$ | 8.46 | 7.67 | — | 9.35 | 8.09 | — |
| $Ni^{2+}$ | 9.83 | 10.38 | 2.87 | 10.80 | 12.65 | 3.22 |
| $Zn^{2+}$ | 9.67 | 9.04 | 2.93 | 10.33 | 9.64 | 3.02 |
| $Pb^{2+}$ | 12.75 | 4.17 | 2.59 | 13.75 | 4.32 | 3.14 |

levels of unbound serum zinc are obtained at low zinc concentrations with $^{65}Zn(L\text{-}His)_2, 2H_2O$ than with $^{65}ZnCl_2$ indicating the importance of low molecular weight complexes in achieving high free (unbound) zinc serum levels.

The cysteine derivative, *Penicillamine*, is used for internal administration as a chelating agent to increase urinary excretion of copper in the treatment of liver degeneration in Wilson's disease. It is also used in the treatment of lead and iron poisoning, and to aid the excretion of iron in haemosiderosis, a disease seen in the Bantu natives of Johannesburg resulting from dietary iron overload (Seftel, 1970). The precise structure of the chelate has not been established; several chelates are possible. Comparison of the stability constants of the $Zn^{2+}$—cysteine complex with model compounds suggests that *N*- and *S*-co-ordination (C) is favoured (Martin, 1964) rather than *S*- and *O*- (A) or *N*- and *O*-co-ordination (B). Similar five-membered ring *N*- and *S*-co-ordination complexes should also be favoured in the case of *Penicillamine*. This is confirmed by comparison of stability constants of L-cysteine and D-penicillamine with $Ni^{2+}$, $Zn^{2+}$ and $Pb^{2+}$ (Table 56; Doornbos and Faber, 1964; Doornbos, 1968).

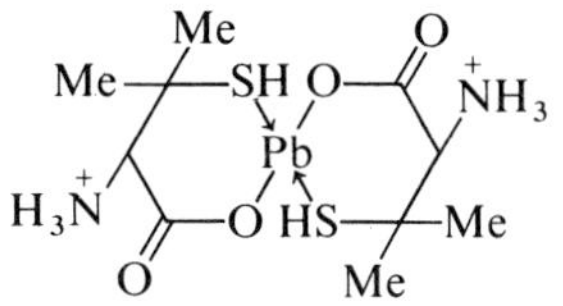

(A)

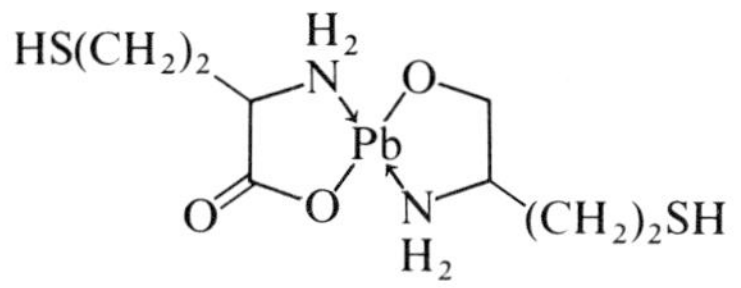

(B)

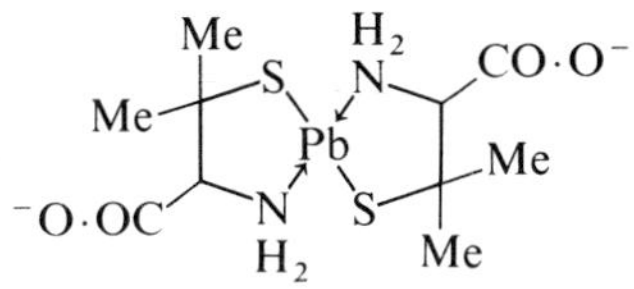

(C)

Ethylenediamine tetra-acetic acid (EDTA) reacts readily with most polyvalent metal ions to form stable water-soluble complexes, which cannot be extracted from aqueous solution into organic solvents. Chelating agents with properties such as these are known as **sequestering agents**. EDTA forms 1:1-complexes with most cations, irrespective of the valency of the ion, according to the following equations in which M is a metal and $[H_2X]^{2-}$ is the anion of disodium edetate.

$$M^{2+} + [H_2X]^{2-} \longrightarrow [MX]^{2-} + 2H^+$$

$$M^{3+} + [H_2X]^{2-} \longrightarrow [MX]^{-} + 2H^+$$

$$M^{4+} + [H_2X]^{2-} \longrightarrow [MX] + 2H^+$$

The structures of these complexes with di-, tri- and tetra-valent metals are as follows.

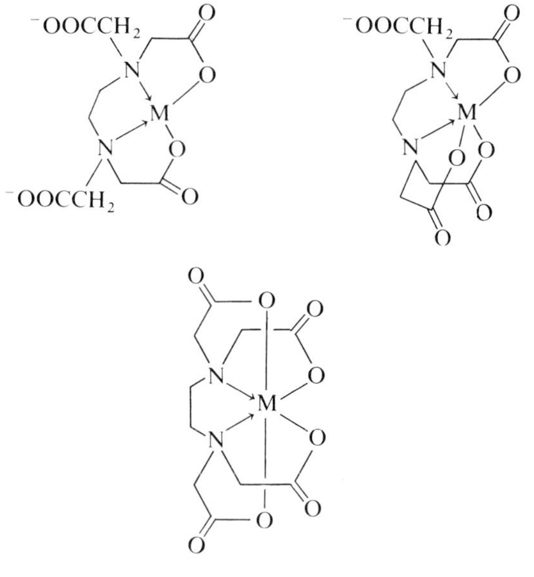

*Disodium Edetate* (Edetate Disodium) is used in the formulation of some injections and creams as a stabilising agent to remove traces of heavy metals which might otherwise catalyse decomposition of the product. It is also used as *Sodium Calcium Edetate Injection* (Edetate Calcium Disodium) in the treatment of lead and vanadium poisoning to assist the elimination of these metals from the body in a water-soluble form (**2**, 2).

## Optical Isomerism

All α-amino acids, with the exception of glycine, possess an asymmetric centre, and are capable of existing in optically active forms. The majority of natural amino acids found as constituents of proteins have the L-configuration. A number of peptide antibiotics, notably penicillins, gramicidins, bacitracin and actinomycin C, contain amino acids of the D-series. L-Serine, originally chosen as the reference amino acid for configurational purposes because of its structural resemblance to glyceraldehyde, has been shown to have the same absolute configuration as the **enantiomer** of D-(+)-glyceraldehyde, which is the configurational reference compound for carbohydrates.

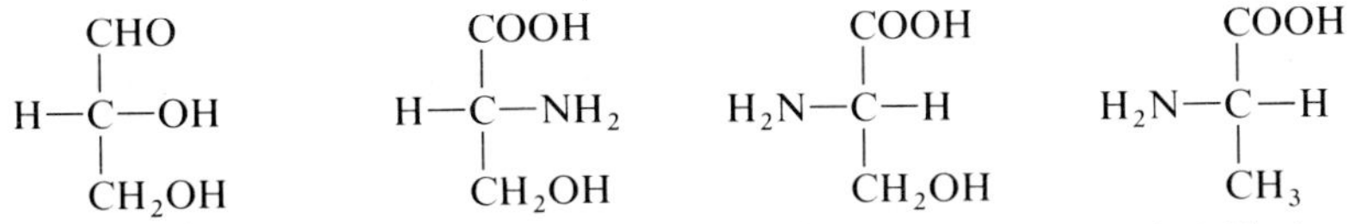

| CHO | COOH | COOH | COOH |
|---|---|---|---|
| H—C—OH | H—C—$NH_2$ | $H_2N$—C—H | $H_2N$—C—H |
| $CH_2OH$ | $CH_2OH$ | $CH_2OH$ | $CH_3$ |
| D-(+)-Glyceraldehyde | D-(+)-Serine | L-(−)-Serine | L-(+)-Alanine |

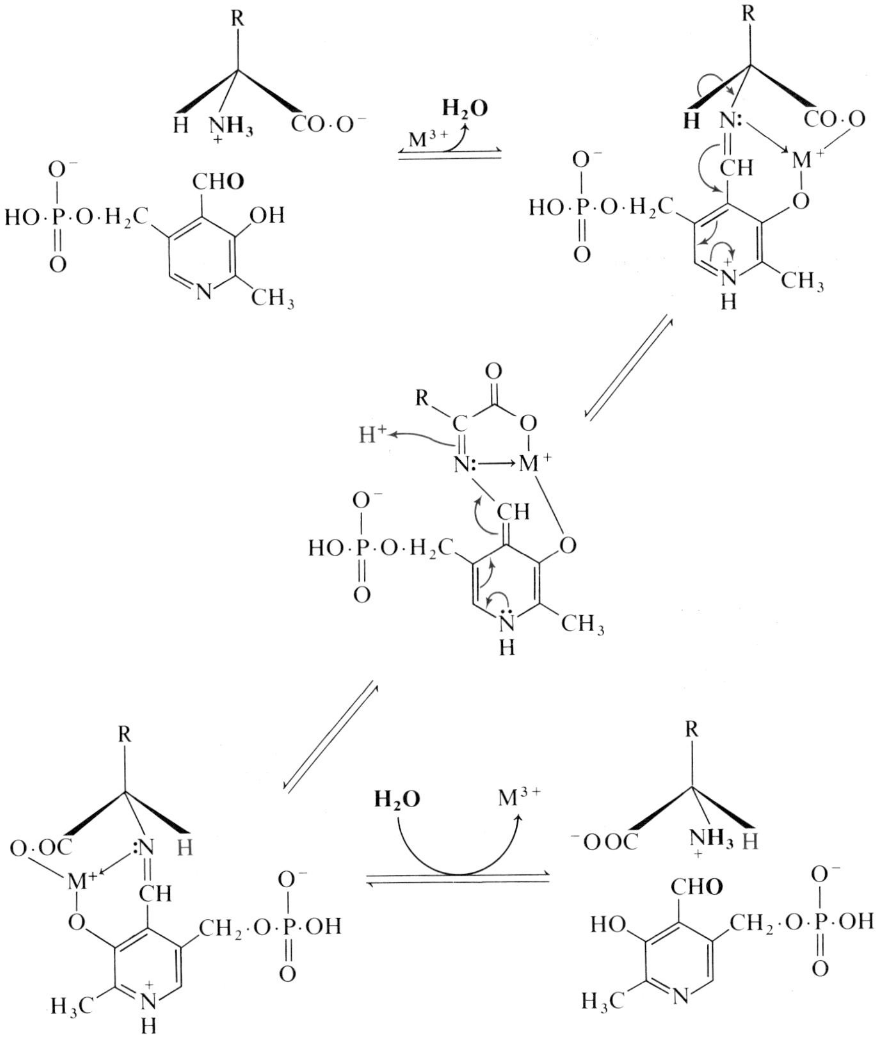

Amino acids of the L- and D-series can also be distinguished by taste, those of the L-series being generally bitter, and those of the D-series sweet. Optical rotation, however, varies with the extent of ionisation, and hence pH. The rotation of L-amino acids becomes more positive, and that of D-amino acids more negative in acidic solution (Clough, 1918; Lutz and Jirgensons, 1930, 1931).

Racemic amino acids can be resolved by crystallisation of their diastereoisomeric salts with (+)-camphorsulphonic acid and (+)-tartaric acid, and by crystallisation of alkaloid salts of *N*-acylamino acids (see below). Microbiological resolution, involving the metabolic transformation of one enantiomer, has been reported for a number of amino acids including the formation of D-leucine and D-glutamic acid (Schulze and Likiernik, 1893) and D-methionine (Stumpf and Green, 1944).

Racemisation of α-amino acids is readily promoted by pyridoxal phosphate. The Schiff's base, which is formed through the amino group, readily loses a proton from the α-carbon atom. In the reverse reaction, re-addition of a proton to the planar trigonal carbon can occur from either side, so that racemisation of the optical centre results (Olivard, Metzler and Snell, 1952; Metzler, Ikawa and Snell, 1954; Snell, Metzler and Ikawa, 1954).

Similar enzyme-catalysed processes, utilising pyridoxal phosphate as coenzyme, are probably also responsible for specific L- to D-configurational inversions of amino acids.

### Self-condensation

α-Amino acids undergo self-condensation to diketopiperazines when heated in high boiling solvents.

$$2H_2N\cdot CH(R)\cdot CO\cdot \mathbf{OH} \xrightarrow[2\mathbf{H_2O}]{} \text{(diketopiperazine: ring with O, NH, H, R; H, HN, R, O)}$$

They may also be made to condense with one another at ambient temperatures in presence of dicyclohexylcarbodiimide to form peptides (see p. 539).

### *N*-Acylamino Acids

Amino acids react with acid chlorides and acid anhydrides to form *N*-acylamino acids, which being no longer basic and zwitterionic function as simple carboxylic acids. They readily form crystalline salts with alkaloids, such as strychnine, brucine, quinine and quinidine, which have been widely used for optical resolution of racemic amino acids (Velluz, Amiard and Héymes, 1955).

$$C_6H_5\cdot \mathbf{CO\cdot Cl} + \mathbf{H_2N}\cdot CH_2\cdot CO\cdot OH \xrightarrow[\mathbf{HCl}]{} C_6H_5\cdot \mathbf{CO\cdot NH}\cdot CH_2\cdot CO\cdot OH$$

*N*-Benzoylglycine

The formation of *N*-acylamino acids, mediated by ATP and coenzyme A, is an important feature of the human metabolism and excretion of aromatic and heterocyclic acids, such as benzoic, salicylic and nicotinic acids. The amino acid normally involved in man is glycine, but glutamic acid conjugates are sometimes formed, as for example with phenylacetic acid. In birds and reptiles, conjugation usually occurs with ornithine.

$$C_6H_5\cdot CO\cdot \mathbf{OH} \xrightarrow{ATP \to PP} C_6H_5\cdot CO\cdot \overline{\mathbf{AMP}} \xrightarrow{HS\cdot CoA \to AMP} C_6H_5\cdot CO\cdot \mathbf{S}\cdot \overline{\mathbf{CoA}}$$

$$\xrightarrow{H_2NCH_2\cdot CO\cdot OH \to HS\cdot \overline{CoA}} C_6H_5\cdot CO\cdot \mathbf{NH}\cdot CH_2\cdot CO\cdot OH$$

Benzoylglycine (hippuric acid)

*N*-Benzyloxycarbonylamino acids ($C_6H_5\cdot CH_2\cdot O\cdot CO\cdot NH\cdot CH(R)\cdot COOH$) and t-butoxycarbonylamino acids (t-BOC-amino acids), formed by the action of benzyloxycarbonyl chloride and t-butyl azidoformate respectively on amino acids, are used as a means of protecting the terminal amino groups in peptide synthesis. The protected amino acids readily form acid chlorides for coupling with a second amino acid or, alternatively, may be coupled directly to another amino acid using dicyclohexylcarbodiimide. The protecting group is removable by catalytic hydrogenation in the case of the carbobenzoxy group (usually designated as Z in the formula), and by treatment with mineral acid (aqueous HCl or trifluoracetic acid) in the case of the t-butoxycarbonyl group (designated t-BOC).

$$\mathbf{H_2N}\cdot CH_2\cdot CO\cdot OH \xrightarrow{C_6H_5\cdot CH_2\cdot \mathbf{O\cdot CO\cdot Cl} \to \mathbf{HCl}} C_6H_5\cdot CH_2\cdot \mathbf{O\cdot CONH}\cdot CH_2\cdot CO\cdot OH$$

$$\xrightarrow{C_6H_{11}-N{=}C{=}N-C_6H_{11},\ H_2N\cdot CH(CH_3)\cdot CO\cdot OH \to C_6H_{11}-NH\cdot CO\cdot NH-C_6H_{11}} Z\cdot \mathbf{NH}\cdot CH_2\cdot CO\cdot NH\cdot CH(CH_3)\cdot CO\cdot OH$$

$$\xrightarrow{H_2/Pd{-}C \to CO_2 + C_6H_5\cdot CH_3} \mathbf{H_2N}\cdot CH_2\cdot CO\cdot NH\cdot CH(CH_3)\cdot CO\cdot OH$$

The formation of acylamino links in natural peptide and protein synthesis is a function of *N*-acylation of one aminoacyl-tRNA by a second aminoacyl-tRNA held in adjacence by an appropriately-coded tRNA. Peptide formation is assisted by $K^+$ ions, which presumably activate the ester carbonyl group of one molecule for attack by the amino group of the other.

$$H_2N\cdot CHR\cdot C(=O)\cdot O\cdot tRNA \;(K^+) + H{-}NH\cdot CHR^1\cdot CO\cdot O\cdot tRNA \xrightarrow{-H^+} H_2N\cdot CHR\cdot C(O{-}K)(O\cdot tRNA)\cdot NH\cdot CHR^1\cdot CO\cdot O\cdot tRNA \xrightarrow{-K^+,\ ^-O\cdot tRNA} H_2N\cdot CHR\cdot C(=O)\cdot NH\cdot CHR^1\cdot CO\cdot O\cdot tRNA$$

*N*-Benzoylamino acids readily condense with aldehydes and cyclize in the presence of acetic anhydride and sodium acetate to form azlactones (Erlenmeyer's azlactone synthesis; Carter, 1946). Reduction and hydrolysis of azlactones provides a useful route to the synthesis of higher amino acids.

$$R\cdot CHO + H_2C(NH\cdot CO\cdot C_6H_5)\cdot CO\cdot OH \xrightarrow[-2H_2O]{Ac_2O/CH_3\cdot CO\cdot ONa} R\cdot CH{=}C\text{(azlactone ring: }C(=O)\text{–O–}C(C_6H_5){=}N)$$

*N*-Benzoylglycine → Azlactone

$$\text{Azlactone} \xrightarrow[\to NaOH + Hg]{H_2O,\ Na/Hg} R\cdot CH_2\cdot CH(NH\cdot CO\cdot C_6H_5)\cdot CO\cdot OH \xrightarrow[\to C_6H_5\cdot CO\cdot OH]{H_2O,\ H^+} R\cdot CH_2\cdot CH(NH_2)\cdot CO\cdot OH$$

## *N*-Alkylation

Direct *N*-alkylation of amino acids is readily achieved by reaction with simple alkyl iodides or dialkyl sulphates, but often results in mixtures of mono- and di-alkyl derivatives with the corresponding betaine.

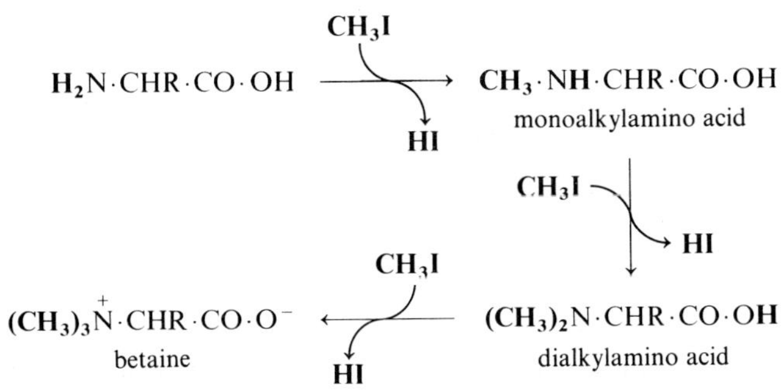

$$H_2N \cdot CHR \cdot CO \cdot OH \xrightarrow[HI]{CH_3I} CH_3 \cdot NH \cdot CHR \cdot CO \cdot OH$$

monoalkylamino acid

$$\downarrow CH_3I, \; -HI$$

$$\overset{+}{(CH_3)_3N} \cdot CHR \cdot CO \cdot O^- \xleftarrow[HI]{CH_3I} (CH_3)_2N \cdot CHR \cdot CO \cdot OH$$

betaine — dialkylamino acid

Betaines are formed almost exclusively with dimethyl sulphate in alkaline solution. Long chain (n = 11–17) substituted betaines function as neutral ionic detergents.

$$H_2N \cdot \underset{(CH_2)_n \cdot CH_3}{CH} \cdot CO \cdot OH \xrightarrow{(CH_3)_2SO_4/NaOH} (CH_3)_3\overset{+}{N} \cdot \underset{(CH_2)_n \cdot CH_3}{CH} \cdot CO \cdot O^-$$

*N*-Benzylation ($C_6H_5CH_2 \cdot NH—$) and *N*-tritylation[$(C_6H_5)_3CH \cdot NH—$] have been used to protect the amino group and so facilitate coupling through the

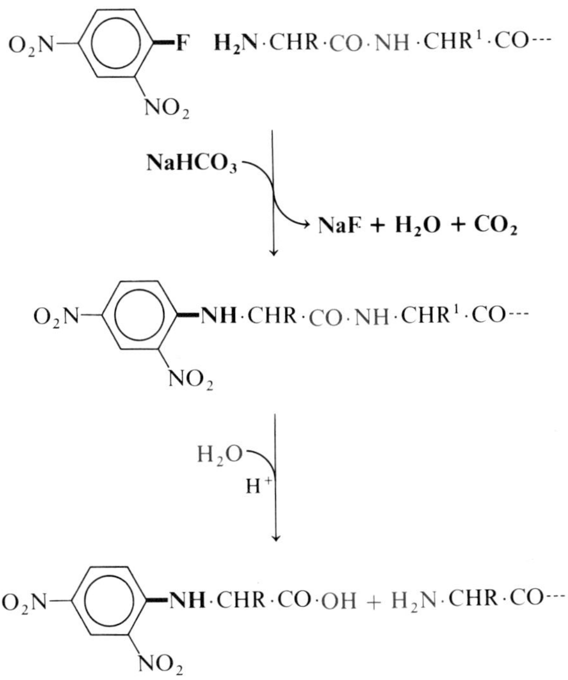

$$O_2N\text{—}C_6H_3(NO_2)\text{—}F \quad H_2N \cdot CHR \cdot CO \cdot NH \cdot CHR^1 \cdot CO\cdots$$

$$\downarrow NaHCO_3, \; \rightarrow NaF + H_2O + CO_2$$

$$O_2N\text{—}C_6H_3(NO_2)\text{—}NH \cdot CHR \cdot CO \cdot NH \cdot CHR^1 \cdot CO\cdots$$

$$\downarrow H_2O, \; H^+$$

$$O_2N\text{—}C_6H_3(NO_2)\text{—}NH \cdot CHR \cdot CO \cdot OH + H_2N \cdot CHR \cdot CO\cdots$$

2,4-Dinitrophenyl-*N*-terminal amino acid

carboxyl group with a second amino acid in peptide synthesis. The protecting groups can be removed by hydrogenolysis.

*N*-Arylation with 2,4-dinitrofluorobenzene is readily effected (p. 541) under mild conditions in the presence of sodium bicarbonate. The reagent has been used extensively as a marker for free *N*-terminal amino acid residues in peptides and proteins (Sanger, 1945, 1952). Hydrolysis of the *N*-2,4-dinitrophenyl-peptides yields *N*-2,4-dinitrophenylamino acids, which can be isolated by extraction and identified, thus establishing which units in the intact peptide have free reactive *N*-terminal amino groups. It should be noted that the terminal amino groups of diamino acids such as lysine also react with 2,4-dinitrofluorobenzene.

Dansyl chloride has also been used in end-group analysis of peptides (Hartley and Massey, 1956) because the intensely yellow fluorescence of the *N*-dansyl-amino acids formed on hydrolysis is easily detected in chromatographic analysis.

$SO_2 \cdot Cl$

$Me_2N$

Dansyl chloride

$+ H_2N \cdot CH_2 \cdot CO \cdot OH$

Glycine

↓

$SO_2 \cdot NH \cdot CH_2 \cdot COOH$

$Me_2N$

Dansylglycine

Dansylglycine shows enhanced fluorescence, accompanied by blue (hypsochromic) shifts in the wavelength of the fluorescence emission maximum in solvents of low dielectric constant, such as dioxan (Chen, 1967). For this reason, dansylglycine has been used as a fluorescent probe for investigating the binding of acidic drugs at hydrophobic binding sites in plasma proteins (Chignell, 1969). Dansylglycine is bound at such a site in human serum albumin with a large increase in the fluorescence quantum yield (0.051 → 0.443 with excitation at 350 nm), and a marked blue shift in its fluorescence emission spectrum (580 → 480 nm).

### Formol Titration

The reaction of amino acids with formaldehyde, which forms the basis of the so-called formol titration, is now known to be much more complex than formation of a simple aldimine. The formation of a cyclic compound, and *N*-mono- and di-hydroxymethyl compounds have been suggested (French and Edsall, 1945).

$HO\cdot CH_2\cdot NH\cdot CHR\cdot CO\cdot OH$

$H_2C{=}N\cdot CHR\cdot CO\cdot OH$

Aldimine

$(HO\cdot CH_2)_2N\cdot CHR\cdot CO\cdot OH$

Hydroxymethyl compounds

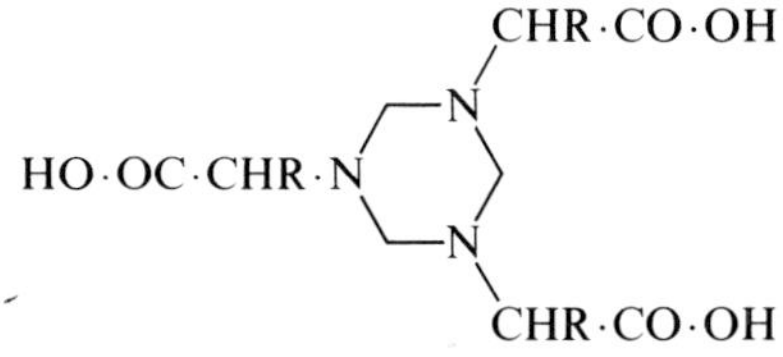

CHR·CO·OH

HO·OC·CHR·N (triazine ring) N

CHR·CO·OH

Whatever the nature of the product(s), blocking of the basic amino group by formaldehyde destroys the zwitterionic character of the amino acid, so permitting direct titration with standard alkali solution. The use of formaldehyde in the detoxification of diphtheria toxoid from diphtheria toxin is based on a similar reaction with free amino groups; toxicity is reduced without significant loss of antigenicity.

The condensation of sugars with aminoacids which can occur with discolouration in hyperalimentation solutions is a related reaction (Chapter 21).

**Oxidative Deamination**

Oxidising agents generally bring about oxidative deamination of amino acids with the formation of an aldehyde, carbon dioxide and ammonia (Johnson and McCaldin, 1958). Deamination with hydrogen peroxide in presence of ferrous ions ($Fe^{2+}$) occurs readily at ambient temperature (Dakin, 1906). Molecular

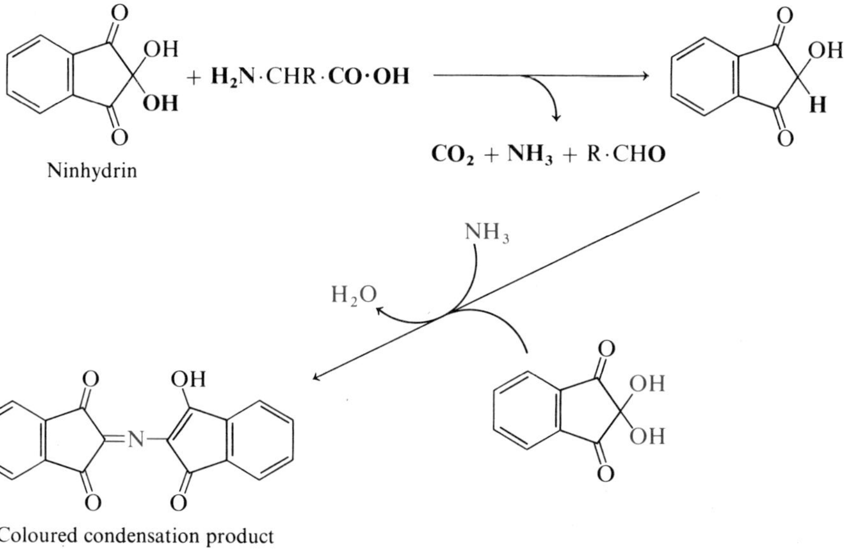

Ninhydrin + $H_2N\cdot CHR\cdot CO\cdot OH$ → (hydrindantin-type product) + $CO_2 + NH_3 + R\cdot CHO$

+ $NH_3$ + ninhydrin, − $H_2O$ →

Coloured condensation product

oxygen also effects oxidative deamination of α-amino acids in aqueous solution, in the presence of activated charcoal (Wieland and Bergel, 1924), which possibly exerts its catalytic effect as a result of adsorbed metal ions. The ninhydrin (triketohydrindene hydrate) reaction is of special importance in the detection and determination of amino acids after paper and thin-layer chromatography. The coloured product is formed by condensation of ninhydrin, its reduction product, and ammonia (scheme p. 543).

Oxidative deamination by ninhydrin, like that by *Phytomenadione* (Phytonadione; vitamin $K_1$) and *Menadione* (Chapter 22) are typical examples of the Strecker degradation of α-amino acids by α-diketones and their vinylogues of general structure, $-CO\left[-C=C-\right]_n CO$ in which $n = 0$ or an integer (Schönberg and Moubasher, 1952). Thus, glyoxal and methylglyoxal degrade α-amino acids rapidly (Schönberg, Moubasher and Mostafa, 1948), whilst deamination with *Menadione* and vitamin $K_1$ occurs under physiological conditions (Schönberg, Moubasher and Said, 1949).

$$R(R^1)C(=O)-C=O + H_2N\cdot CHR^2\cdot CO\cdot OH \longrightarrow R(R^1CO)C=N-CHR^2-C(=O)-O-H \xrightarrow{-CO_2} R-\bar{C}(R^1CO)-N=CH\cdot R^2 \xrightarrow{H_3\overset{+}{O}} R-CH(NH_2)-C(=O)R^1 + OCH\cdot R^2$$

This latter reaction simulates the enzyme-catalysed transaminations (Herbst, 1944; Snell, 1958) prominent in mammalian amino acid metabolism (Chapters 13 and 23), which are mediated by pyridoxal phosphate through the same Schiff's base intermediate as obtained in racemisation and inversion of the optical centre (p. 537).

Carbohydrates, such as glucose, only deaminate amino acids in the presence of oxygen, presumably due to initial oxidation to a 1,2-di-carbonyl compound (Schönberg and Moubasher, 1952). Care should, therefore, be exercised in any admixture of amino acid supplements in solution with *Dextrose* in intravenous infusions to ensure that air is excluded.

Oxidative deamination also occurs in the reaction of amino acids with nitrous acid. This reaction forms the basis of the Van Slyke method (1929) of analysis. The nitrogen which is evolved is measured manometrically.

$$H_2N{\cdot}CHR{\cdot}CO{\cdot}OH + HNO_2 \longrightarrow HO{\cdot}CHR{\cdot}CO{\cdot}OH + N_2\uparrow + H_2O$$

*N*-Alkylamino acids form *N*-nitrosoamino acids with nitrous acid, which on treatment with acetic anhydride yield cyclic meso-ionic compounds known as sydnones (Earl and Mackney, 1935; Baker and Ollis, 1957).

$$RNH{\cdot}CH_2{\cdot}CO{\cdot}OH \xrightarrow[-H_2O]{HNO_2} R{\cdot}N(N{=}O){\cdot}CH_2{\cdot}CO{\cdot}OH$$

$$\downarrow Ac_2O \quad (\rightarrow CH_3{\cdot}CO{\cdot}OH)$$

R·N⁺ ring (N–O–C(O⁻)=CH) 

Sydnone

## Decarboxylation

Thermal decarboxylation of amino acids requires fairly severe heat treatment at around 200°C (Kanoa, 1947). Enzyme-catalysed decarboxylation occurs much more readily, and provides the main pathway for the biosynthesis of a number of physiologically important amines, including γ-aminobutyric acid (from glutamic acid), histamine (from histidine) and dopamine (from *Levodopa*). The anti-hypertensive, *Methyldopa* (α-methyldopa), acts as a competitor for dopa decarboxylase which normally brings about the decarboxylation of *Levodopa*. Since the latter is a key step in the biosynthesis of noradrenaline, formation of noradrenaline is inhibited (scheme p. 546). The hypertensive effect of noradrenaline, however, is not completely suppressed, since α-methyldopa is itself slowly decarboxylated by dopa decarboxylase and the resulting α-methyldopamine is oxidised to α-methylnoradrenaline. Chain-branching further inhibits the metabolism of α-methylnoradrenaline by monoamine oxidase, so that is accumulates and is eventually released as a false transmitter from adrenergic neurones. α-Methylnoradrenaline, however, is less effective than the natural transmitter, and leads to a diminished transmitter response, with some reduction in the level of hypertensive activity.

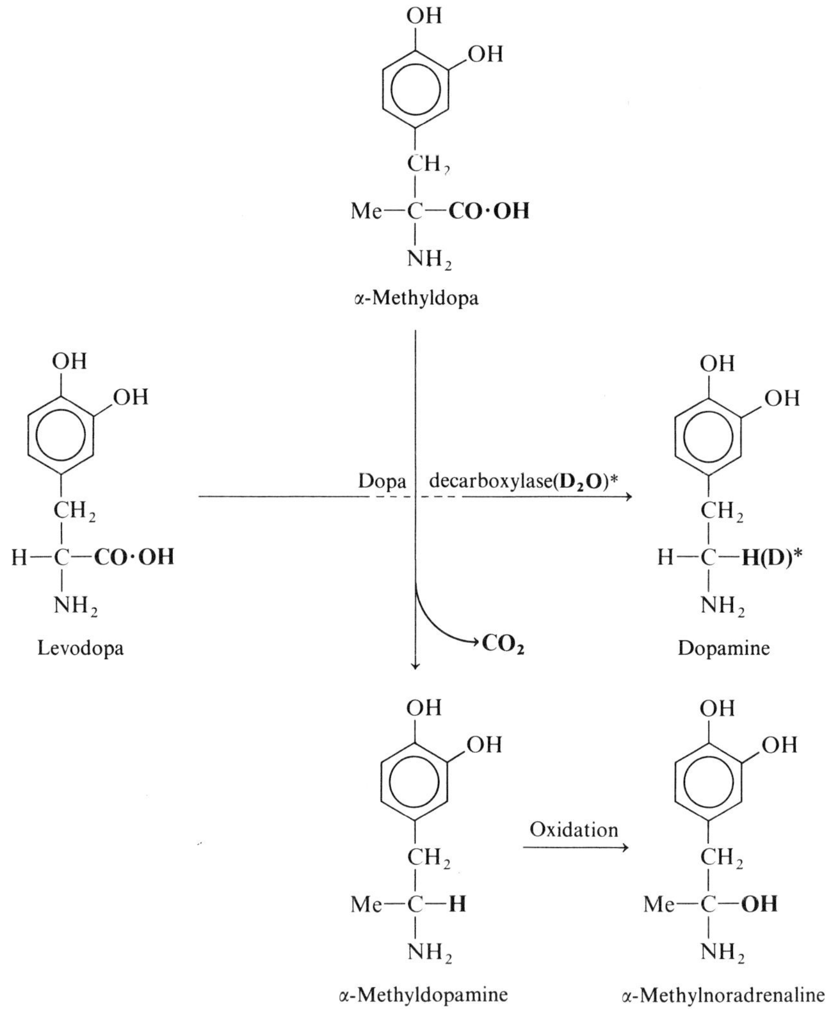

OH, OH; $CH_2$; Me—C—CO·OH; $NH_2$

α-Methyldopa

OH, OH; $CH_2$; H—C—CO·OH; $NH_2$

Levodopa

Dopa decarboxylase($D_2O$)* →

OH, OH; $CH_2$; H—C—H(D)*; $NH_2$

Dopamine

→ $CO_2$

OH, OH; $CH_2$; Me—C—H; $NH_2$

α-Methyldopamine

Oxidation →

OH, OH; $CH_2$; Me—C—OH; $NH_2$

α-Methylnoradrenaline

* Enzymic decarboxylation in presence of $D_2O$ leads to a monodeuteriodopamine

Enzymic decarboxylation requires pyridoxal phosphate as coenzyme (Gale, 1946; Braunshtein, 1960), which is also capable of catalysing non-enzymic decarboxylation in the presence of metal ions by the pathway shown on p. 547.

Dopa decarboxylase is also powerfully inhibited by *Carbidopa*. This compound does not cross the blood-brain barrier, and therefore only inhibits the decarboxylation of *Levodopa* in the gastro-intestinal tract and elsewhere outside the central nervous system. When co-administered with *Levodopa*, *Carbidopa* substantially increases the amount of *Levodopa* available for transport into the brain for conversion into dopamine. This enables large reductions to be made in the dosage of *Levodopa* necessary for relief of tremor and other symptoms in

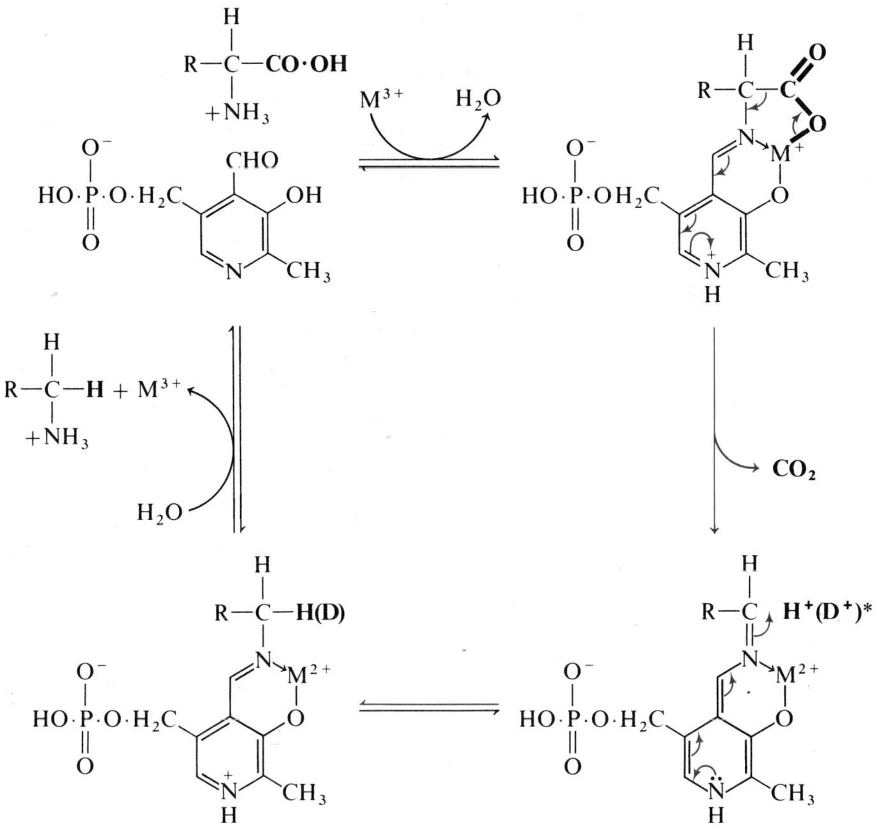

* Decarboxylation in presence of $D_2O$ leads to the formation of a α-deuteroamine

the treatment of Parkinson's disease. The lower dosage means a concomitant reduction in side-reactions.

*Carbidopa* is decomposed by heat in the solid state with loss of the hydrazino function.

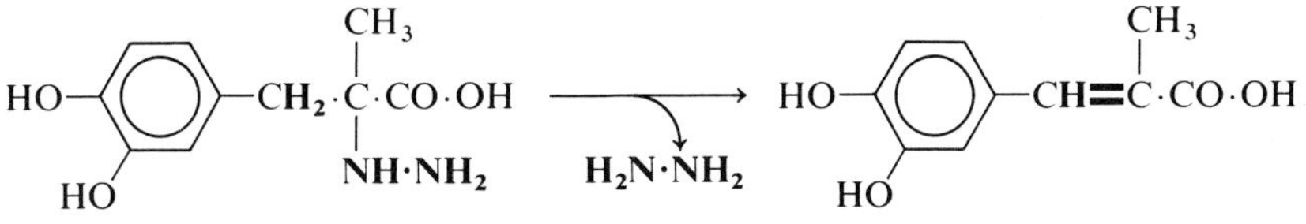

$$\text{HO}(\text{HO})\text{C}_6\text{H}_3\text{—CH}_2\cdot\overset{\text{CH}_3}{\underset{\text{NH}\cdot\text{NH}_2}{\text{C}}}\cdot\text{CO}\cdot\text{OH} \xrightarrow{-\text{H}_2\text{N}\cdot\text{NH}_2} \text{HO}(\text{HO})\text{C}_6\text{H}_3\text{—CH}{=}\overset{\text{CH}_3}{\text{C}}\cdot\text{CO}\cdot\text{OH}$$

*Carbidopa*

It is also unstable in neutral or alkaline solution, but rather more stable under acidic conditions. Decomposition in aqueous solution is catalysed by heavy metal ions ($Fe^{3+}$, $Cu^{2+}$). *Carbidopa* is also sensitive to oxidation. Its rapid decomposition in alkaline solution is accompanied by discolouration and is, therefore, probably also due in part to oxidation.

Other physiologically-important decarboxylations include the decarboxylations of aspartic acid to $\beta$-alanine, and of cysteine to $\beta$-mercaptoethylamine, which are both important intermediates in the biosynthesis of coenzyme A (Chapter 14).

Enzymic decarboxylation carried out in the presence of deuterium oxide ($D_2O$) leads to deuterated products. Similar decarboxylation of L-tyrosine in the presence of $D_2O$ establishes that the product amine is formed with overall retention of configuration (Belleau and Burba, 1960).

**Esterification**

$\alpha$-Amino acids are readily converted to the ester hydrochloride by refluxing with an alcohol in the presence of dry hydrogen chloride. In contrast to the parent amino acid, the ester base is no longer zwitterionic, and in consequence is readily soluble in organic solvents. The esters, however, readily undergo self-condensation when heated to form diketopiperazines.

$$2H_2N{\cdot}CHR{\cdot}CO{\cdot}OMe \xrightarrow{-MeOH} \text{(2,5-diketopiperazine: R, H substituted ring with two NH and two C=O)}$$

Amino acid esters in dilute hydrochloric acid react with sodium nitrite to form diazoesters. These yellow-coloured, volatile liquids, which are extremely reactive as nucleophiles, are used in the formation of carbenes (p. 552 and Chapter 2).

$$H_2N{\cdot}CH_2{\cdot}CO{\cdot}OEt \xrightarrow[-H_2O]{HCl-NaNO_2} N_2{\cdot}CH{\cdot}CO{\cdot}OEt$$

Aminoacyl phosphates are of considerable biosynthetic importance as intermediates in the formation of aminoacyl-tRNA and hence in protein synthesis.

$$CH_3S{\cdot}CH_2{\cdot}CH_2{\cdot}\underset{\underset{+NH_3}{|}}{CH}{\cdot}CO{\cdot}O^- + ATP$$

$$\Big\Updownarrow \; Mg^{2+}$$

$$CH_3S{\cdot}CH_2{\cdot}CH_2{\cdot}\underset{\underset{+NH_3}{|}}{CH}{\cdot}CO{\cdot}O{\cdot}\overset{\overset{O}{\|}}{\underset{\underset{O_-}{|}}{P}}{\cdot}O{\cdot}CH_2\text{-(ribose, OH OH)-Adenyl}$$

Aminoacyl-AMP

The intermediate aminoacyladenosine monophosphate (aminoacyl-AMP) is formed as an enzyme complex from the amino acid and adenosine triphosphate in the presence of amino acid specific synthetases (De Moss, Genuth and Novelli, 1956; Berg, 1958; Moldave, Castelfranco and Meister, 1959).

Interaction of the aminoacyl-AMP—enzyme complex with the appropriate amino acid specific tRNA gives rise to the corresponding aminoacyl-tRNA.

$$\text{Aminoacyl-AMP—Enz} + \text{tRNA} \rightleftharpoons \text{Aminoacyl-tRNA} + \text{Enz} + \text{AMP}$$

**Merrifield Solid-state Peptide Synthesis** (1965)

The Merrifield solid-state peptide synthesis is based essentially on the use of esterification to link the *C*-terminal amino acid through its carboxyl group to the supporting solid phase. The latter consists of a styrene–divinylbenzene copolymer (98:2) which is activated for coupling with the *C*-terminal amino acid

$$\text{t-BuO·CO·NH·CHR·}\mathbf{CO·O^-} + \mathbf{Cl·CH_2}\text{–C}_6\text{H}_4\text{–Polymer}$$

$$\xrightarrow{\text{Et}_3\overset{+}{\text{N}}\text{H}} \quad (\rightarrow \text{Et}_3\overset{+}{\text{N}}\text{HCl}^-)$$

$$\text{t-BuO·CO·NH·CHR·}\mathbf{CO·O·CH_2}\text{–C}_6\text{H}_4\text{–Polymer}$$

$$\xrightarrow{\text{1. } N \text{ HCl; 2. Et}_3\text{N}} \quad (\rightarrow \text{t-BuOH} + \text{CO}_2)$$

$$\mathbf{H_2N·}\text{CHR·}\mathbf{CO·O·CH_2}\text{–C}_6\text{H}_4\text{–Polymer}$$

$$+ \text{ t-BuO·CO·NH·CHR}^1\text{·}\mathbf{CO·OH} + \text{C}_6\text{H}_{11}\text{–N=C=N–C}_6\text{H}_{11} \quad (\rightarrow \text{C}_6\text{H}_{11}\text{–NH·CO·NH–C}_6\text{H}_{11})$$

$$\text{t-BuO·CO·NH·CHR}^1\text{·}\mathbf{CO·NH·}\text{CHR·}\mathbf{CO·O·CH_2}\text{–C}_6\text{H}_4\text{–Polymer}$$

$$\xrightarrow{\text{HBr/HAc}}$$

$$\text{H}_2\text{N·CHR}^1\text{·}\mathbf{CO·NH·}\text{CHR·}\mathbf{CO·OH}$$

by chloromethylation of the aromatic rings. Coupling is effected with the *N*-t-butoxycarbonyl-*C*-terminal amino acid of the desired peptide in the presence of triethylamine (stage 1). The protecting *N*-t-butoxycarbonyl group is readily removed by treatment with dilute acid (stage 2).

The first peptide bond is thus formed by reaction of the exposed amino group with the *N*-t-butoxycarbonyl derivative of the second amino acid (stage 3). Stages 1–3 are then repeated as often as necessary until the entire amino acid sequence is complete. In the final step (stage 4), the complete peptide is released from combination with its polymer carrier, with concomitant loss of the remaining *N*-terminal protecting butoxycarbonyl group.

## DIAZO ESTERS

Simple diazo esters, which are formed by reaction of amino acid esters with nitrous acid (p. 548), are yellow volatile liquids, which are only slightly soluble in water, but readily miscible with ethanol and ether. More complex diazo esters, such as ethyl diazoacetylglycinate, formed from glycylglycine, are crystalline solids.

### Reaction with Acids

Diazoacetic esters are extremely reactive as nucleophiles, combining with hydroxylic substances such as water, alcohols, phenols and carboxylic acids, with loss of nitrogen.

$$N{\equiv}\overset{+}{N}{-}\overset{-}{C}H{\cdot}CO{\cdot}OEt \xrightarrow[-HO^-]{H{-}OH} N{\equiv}\overset{+}{N}{-}CH_2{\cdot}CO{\cdot}OEt \xrightarrow[-N_2]{} \overset{+}{C}H_2{\cdot}CO{\cdot}OEt \xrightarrow{^-OH} HO{\cdot}CH_2{\cdot}CO{\cdot}OEt$$

Ethyl diazoacetate → Ethyl glycollate

$$N{\equiv}\overset{+}{N}{\cdot}\overset{-}{C}H{\cdot}CO{\cdot}OEt \xrightarrow[-N_2]{C_6H_5{\cdot}OH} C_6H_5{\cdot}O{\cdot}CH_2{\cdot}CO{\cdot}OEt$$

Ethyl phenoxyacetate

$$N{\equiv}\overset{+}{N}{\cdot}\overset{-}{C}H{\cdot}CO{\cdot}OEt \xrightarrow[-N_2]{CH_3{\cdot}CO{\cdot}OH} CH_3CO{\cdot}CH_2{\cdot}CO{\cdot}OEt$$

Ethyl acetoacetate

Ethyl diazoacetate also reacts readily with halogen acids and halogens to form α-halogeno- and α-dihalogeno-acetic esters respectively.

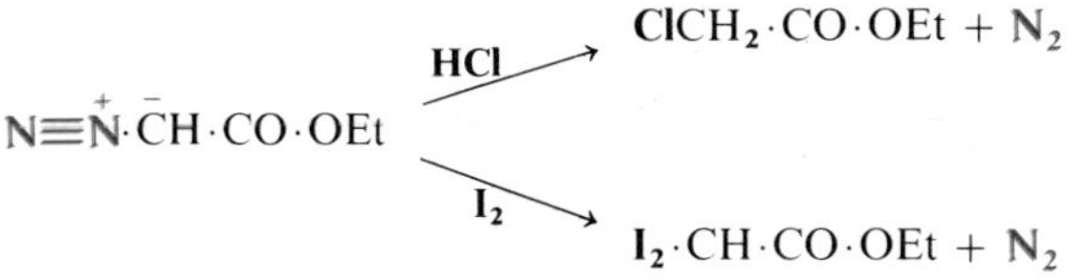

$$N{\equiv}\overset{+}{N}{\cdot}\overset{-}{C}H{\cdot}CO{\cdot}OEt \xrightarrow{HCl} ClCH_2{\cdot}CO{\cdot}OEt + N_2$$

$$N{\equiv}\overset{+}{N}{\cdot}\overset{-}{C}H{\cdot}CO{\cdot}OEt \xrightarrow{I_2} I_2{\cdot}CH{\cdot}CO{\cdot}OEt + N_2$$

**Addition to Aldehydes**

Ethyl diazoacetate undergoes nucleophilic addition to aldehydes.

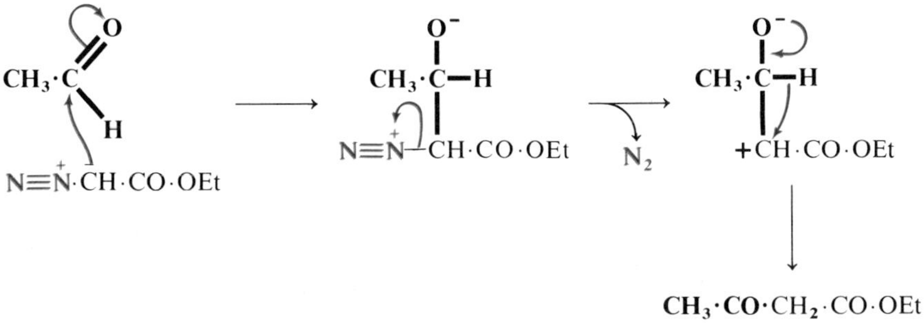

$$CH_3{\cdot}CHO + N{\equiv}\overset{+}{N}{\cdot}\overset{-}{C}H{\cdot}CO{\cdot}OEt \longrightarrow CH_3{\cdot}CH(O^-){-}CH(\overset{+}{N}{\equiv}N){\cdot}CO{\cdot}OEt \xrightarrow{-N_2} CH_3{\cdot}CH(O^-){-}\overset{+}{C}H{\cdot}CO{\cdot}OEt \longrightarrow CH_3{\cdot}CO{\cdot}CH_2{\cdot}CO{\cdot}OEt$$

**Addition to Olefines** (and Acetylenes)

Addition of diazo esters to olefines, suitably activated by electron-attracting groups, gives rise to pyrazolines. The latter frequently undergo pyrolytic decomposition with loss of nitrogen to form cyclopropane derivatives.

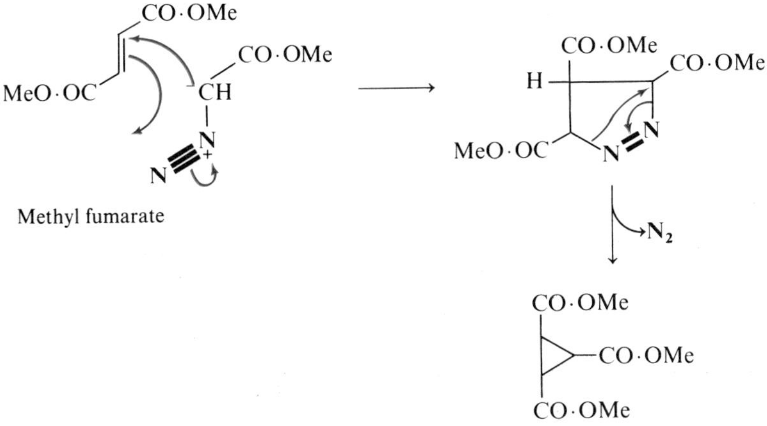

$$MeO{\cdot}OC{\cdot}CH{=}CH{\cdot}CO{\cdot}OMe + N{\equiv}\overset{+}{N}{\cdot}\overset{-}{C}H{\cdot}CO{\cdot}OMe \longrightarrow \text{pyrazoline} \xrightarrow{-N_2} \text{cyclopropane}$$

Methyl fumarate

Methyl cyclopropane-1,2,3 tricarboxylate

**Reaction as Carbenes**

Diazoesters undergo thermal and photochemical decomposition with loss of nitrogen to form carbenes, which are intensely reactive. Thermal decomposition of ethyl diazoacetate in benzene gives rise to a mixture of products, isolated after hydrolysis as carboxylic acids, which result from the formation of a carbene intermediate and its reaction with benzene (Buchner, 1885–98, 1908; Schenck and Ziegler, 1953). The principal product is cycloheptatrienecarboxylic acid.

$$N{\equiv}\overset{+}{N}{-}\overset{-}{C}H{\cdot}CO{\cdot}OEt \xrightarrow{-N_2} \overset{-}{C}H{\cdot}CO{\cdot}OEt \xrightarrow[\text{2. } H^+]{\text{1. Heat in benzene}}$$

H

CO·OH

CO·OH ⟵ CO·OH

Ethyl diazoacetate forms the same reactive carbene on photochemical decomposition in ultraviolet light. Combination with styrene provides a useful route to ethyl 2-phenyl-1-cyclopropanecarboxylate, an intermediate in the formation of *Tranylcypromine*.

$$N{\equiv}\overset{+}{N}{-}\overset{-}{C}H{\cdot}CO{\cdot}OEt \xrightarrow{-N_2} \overset{-}{C}H{\cdot}CO{\cdot}OEt + Ph{\cdot}CH{=}CH_2 \longrightarrow$$

H CO·OEt

Ph H

several stages

H $NH_2$

Ph H

*Tranylcypromine*

**L-Azaserine**

*Azaserine*, *O*-diazoacetyl-L-serine, is an antibiotic isolable from certain strains of streptomyces (Stock *et al.*, 1954; Bartz *et al.*, 1954), but available synthetically. It has antibacterial and antifungal action, and inhibits the Crocker mouse sarcoma 180 in rodents.

*Azaserine* is a specific inhibitor of purine synthesis, and thus blocks the incorporation of purines into nucleic acids, both in normal and tumour tissue. This azaserine-induced inhibition, which can be reversed by glutamine, results in the accumulation of formylglycinamidylribotide in the cells (Goldthwait,

Peabody and Greenberg, 1954; Hartman, Levenberg and Buchanan, 1955). The synthesis of inosinic acid is thus specifically inhibited by interference with the following reaction.

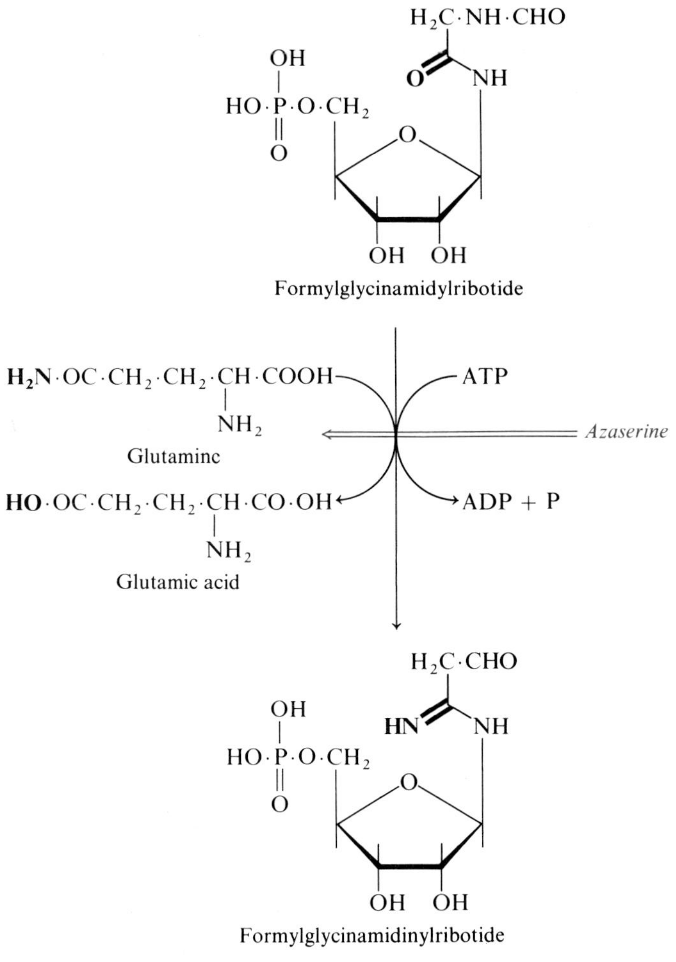

$H_2C{\cdot}NH{\cdot}CHO$

OH

$HO{\cdot}P{\cdot}O{\cdot}CH_2$

O

NH

O

OH OH

Formylglycinamidylribotide

$H_2N{\cdot}OC{\cdot}CH_2{\cdot}CH_2{\cdot}CH{\cdot}COOH$

$NH_2$

Glutaminc

ATP

*Azaserine*

$HO{\cdot}OC{\cdot}CH_2{\cdot}CH_2{\cdot}CH{\cdot}CO{\cdot}OH$

$NH_2$

Glutamic acid

ADP + P

$H_2C{\cdot}CHO$

OH

HN

NH

$HO{\cdot}P{\cdot}O{\cdot}CH_2$

O

O

OH OH

Formylglycinamidinylribotide

## β-AMINO ACIDS

The simplest $\beta$-amino acid, $\beta$-alanine, occurs widely in both animal and plant kingdoms, where it arises by decarboxylation of aspartic acid. It is found in various micro-organisms, and in muscle, liver and nervous tissue, and in particular as a component of pantothenic acid, one of the B-group of vitamins, and in coenzyme A. Incorporation of $\beta$-alanine into pantothenic acid is inhibited by the analogous sulphonic acid, taurine, in some micro-organisms (Sarett and Cheldelin, 1945).

$H_2N \cdot CH_2 \cdot CH_2 \cdot CO \cdot OH$

β-Alanine

$HO \cdot CH_2 \cdot C(Me)_2 \cdot CHOH \cdot CO \cdot NH \cdot CH_2 \cdot CH_2 \cdot CO \cdot OH$

Pantothenic acid

$H_2N \cdot CH_2 \cdot CH_2 \cdot SO_2OH$

Taurine

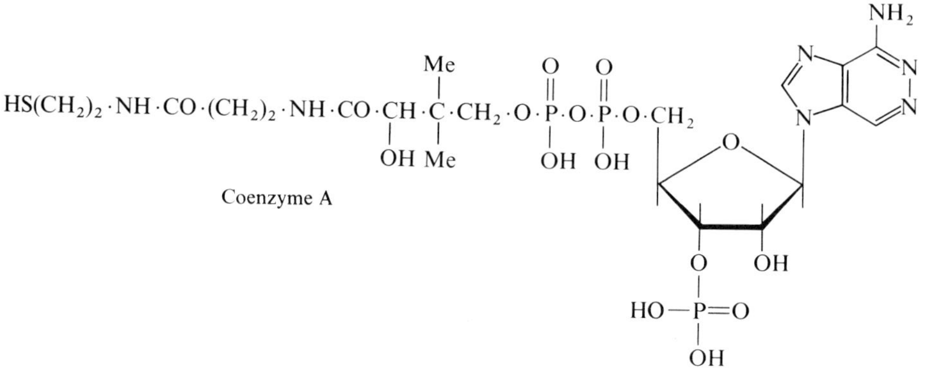

$HS(CH_2)_2 \cdot NH \cdot CO \cdot (CH_2)_2 \cdot NH \cdot CO \cdot CH(OH) \cdot C(Me)_2 \cdot CH_2 \cdot O \cdot P(O)(OH) \cdot O \cdot P(O)(OH) \cdot O \cdot CH_2$–[ribose, 3′-O–P(=O)(OH)–OH, 2′-OH]–adenine ($NH_2$)

Coenzyme A

The dipeptide, carnosine, β-alaninylhistidine is also present in muscle. t-BOC-β-alanine is also a component of the synthetic peptide, *Pentagastrin.*

### β-Lactams

β-Amino acids, their esters and acyl derivatives, are cyclised in the presence of base, though often with some difficulty, to four-membered cyclic amides, known as β-lactams.

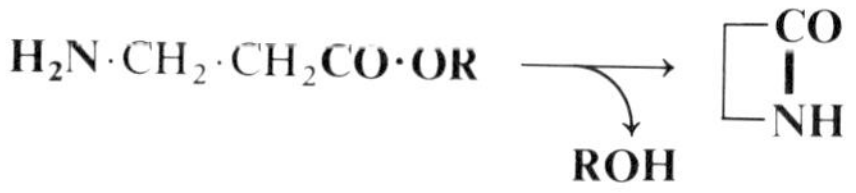

$$H_2N \cdot CH_2 \cdot CH_2CO \cdot OR \xrightarrow{-ROH} \text{[cyclic } CH_2CH_2\text{–CO–NH]}$$

β-Lactams are much more readily hydrolysed than simple amides due to ring strain. The penicillins and cephalosporins are typical β-lactams, and owe their antibiotic action to reactivity of the β-lactam ring in biological acylations (pp. 565 and 580, and **2**, 2).

## γ-AMINO ACIDS

γ-Aminobutyric acid (GABA) is widely distributed in micro-organism, plant, and animal tissue, including brain, liver and muscle, where it arises by decarboxylation of glutamic acid. It is known to be implicated in both transmitter and inhibitor functions in the human brain acting either *per se* or in combination as γ-amino-β-hydroxybutyric acid or γ-aminobutyrocholine.

$$\begin{array}{l} CO{\cdot}OH \\ | \\ CH_2 \\ | \\ CH_2 \\ | \\ CH{\cdot}\mathbf{CO{\cdot}O^-} \\ | \\ {}^+NH_3 \end{array} \xrightarrow{\;-\mathbf{CO_2}\;} \begin{array}{l} CO{\cdot}OH \\ | \\ CH_2 \\ | \\ CH_2 \\ | \\ CH_2{\cdot}NH_2 \end{array}$$

γ-aminobutyric acid

Oxidation →

$$\begin{array}{l} CO{\cdot}OH \\ | \\ CH_2 \\ | \\ CH\mathbf{OH} \\ | \\ CH_2{\cdot}NH_2 \end{array}$$

γ-amino-β-hydroxybutyric acid

→

$$\begin{array}{l} CO{\cdot}O{\cdot}CH_2{\cdot}CH_2{\cdot}\overset{+}{N}Me_3 \\ | \\ CH_2 \\ | \\ CH_2 \\ | \\ CH_2NH_2 \end{array}$$

γ-aminobutyrocholine

**γ- and δ-Lactams**

γ-Aminobutyric acid readily cyclises on heating to give the five-membered cyclic γ-lactam, pyrrolidone. Six-membered δ-lactams are similarly formed from δ-amino acids.

$$\begin{array}{l} CO{\cdot}\mathbf{OH} \\ | \\ CH_2 \\ | \\ CH_2 \\ | \\ CH_2{\cdot}\mathbf{NH_2} \end{array} \xrightarrow{\;-\mathbf{H_2O}\;}$$

Pyrrolidone(γ-butyrolactone)

On a commercial scale, pyrrolidone is prepared by electrolytic reduction of succinimide, and also from butyrolactone by treatment with ammonia.

Lactams, as cyclic amides, exhibit weakly acidic and basic properties, and undergo hydrolysis to the corresponding amino acids.

**Polyvinylpyrrolidone**

Pyrrolidone condenses with acetylene to form vinylpyrrolidone, which is readily polymerised in aqueous solution with hydrogen peroxide, and ammonia acting as an accelerator.

The pure polyvinylpyrrolidone polymers are glass-like solids or powders, which soften at temperatures above 100°C. The molecular weights of the polymers vary widely from 10 000 to 700 000 and careful fractionation is necessary to obtain products with uniform properties. The average molecular weight of polymer fractions can be determined by their viscosity in aqueous solution,

$$HC{\equiv}CH$$

Vinylpyrrolidone

$$H_2O_2$$

$-CH-CH_2-CH-CH_2-CH-$

*Polyvinylpyrrolidone* (Povidone)

which increases with increasing molecular weight. Viscosities are usually expressed as K values (Fikentscher, 1932).

*Povidone* is hygroscopic, with a significant moisture pick-up, and reaches an equilibrium moisture content of 40% at 70% relative humidity. It is readily soluble in water, though aqueous solubility decreases as the average molecular weight increases. *Povidone* is also soluble in ethanol and chloroform, but insoluble in ether. Addition of salts precipitates the polymer from aqueous solution, but solutions are stable to heat and can be sterilised by autoclaving.

Polymer fractions for parenteral administration should have a molecular weight of about 40 000 (K values 28–32), and may be used as a plasma expander. About 60–70% of administered *Povidone* is excreted unchanged in the urine within 24 hr. Higher molecular weight grades (K value ≈ 50) are used as suspending and dispersing agents, and as a binder in the preparation of tablet granules. The polymers are especially suitable for the granulation of moisture- and heat-sensitive materials, since they can be used in organic solvents which are both anhydrous and readily volatile. Solutions in organic solvents will dry to a film which is both hard and adhesive; polymer solutions, often with cellulose acetate phthalate and polyvinyl alcohol, are therefore widely used in the film-coating of tablets. The water solubility of *Povidone* aids solution of the polymer film following ingestion. Concentrated aqueous solutions (10–25%) are also useful as delayed-release vehicles for a variety of drugs including antibiotics and steroids.

## PENICILLINS AND CEPHALOSPORINS

### Natural Penicillins

#### Discovery

The discovery by Sir Alexander Fleming in 1929 that media in which *Penicillium notatum* and *P. chrysogenum* have been cultivated possess antibiotic properties, led to the isolation of not one, but a number of penicillins. These are all related

Table 57. Natural Penicillins

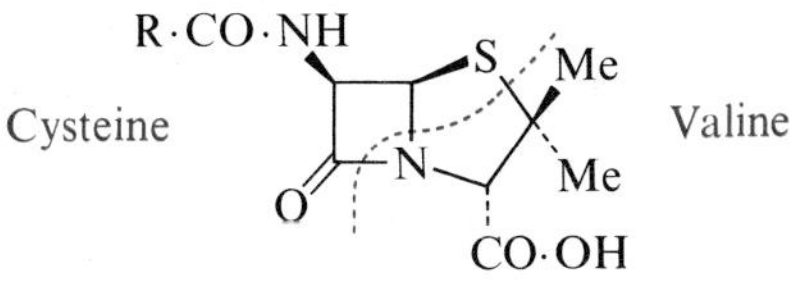

R·CO·NH, S, Me, Me, N, O, CO·OH — Cysteine, Valine

| | | Penicillin | R | Antibacterial Activity (International units/mg) |
|---|---|---|---|---|
| I | (F) | pentenyl | $CH_3 \cdot CH_2CH{=}CH \cdot CH_2$ | 1600 |
| II | (G) | benzyl | $C_6H_5 \cdot CH_2$ | 1666 (Na salt); 1600 (K salt) |
| III | (X) | *p*-hydroxybenzyl | *p*-$HO \cdot C_6H_4 \cdot CH_2$ | 900 |
| IV | (K) | heptyl | $CH_3(CH_2)_6$ | 2300 |
| V | | phenoxymethyl | $C_6H_5 \cdot O \cdot CH_2$ | 1600 |

to the parent substance, 6-aminopenicillanic acid, which is regarded as being derived biogenetically from the amino acids, cysteine and valine. Fleming was unable to isolate penicillin, and it was left to others to isolate the antibiotic and demonstrate its clinical usefulness (Florey *et al.*, 1940, 1941, 1949; Chain, 1948).

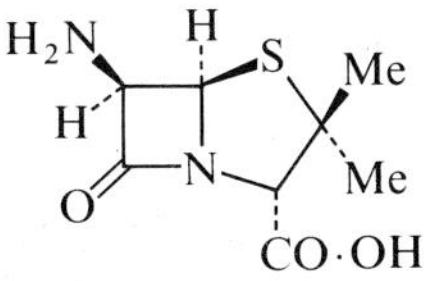

$H_2N$, H, S, Me, H, N, Me, O, CO·OH

6-Aminopenicillanic acid

Natural penicillium strains produce mainly Penicillin G when corn-steep liquor, which contains phenylacetic acid, is present in the medium (deep-culture fermentation). The formation of some penicillins, such as *Benzylpenicillin* (Penicillin G) and *Phenoxymethylpenicillin* (Penicillin V), can be controlled (Brandl and Margreiter, 1954) by addition of the appropriate chemical side-chain precursor to the culture medium (Table 57). This, however, is not generally applicable, and the number of penicillins which can be produced economically by direct fermentation on an industrial scale is strictly limited. On the other hand, 6-aminopenicillanic acid can be produced in reasonable yield by fermentation, and its production by this means now serves as the basis from which semi-synthetic penicillins can be manufactured (p. 566).

**Antibacterial Activity**

The natural penicillins are active mainly against Gram-positive organisms. They vary both in potency and in antibacterial spectrum. In recent years, a number of penicillin-resistant strains of staphylococci have emerged. This is not due to mutation, but to progressive suppression of sensitive organisms, and their gradual replacement by surviving resistant organisms from the original cultures (**2**, 5). Such strains will only succumb to attack by certain of the semi-synthetic penicillins such as *Cloxacillin* and *Flucloxacillin* (Floxacillin; p. 568).

## BENZYLPENICILLIN

### General Properties

*Benzylpenicillin* is generally available as its colourless crystalline sodium or potassium salt. The salts melt with decomposition (Na salt, 215°C; K salt, 214–217°C). They are optically active, [*Benzylpenicillin* (Penicillin G) *Sodium* $[\alpha]_D^{25}$ + 305° (*c*, 0.821 in $H_2O$), *Benzylpenicillin* (Penicillin G) *Potassium* $[\alpha]_D^{22}$ + 285–310° (in $H_2O$)], possessing three asymmetric centres. Fusion of the thiazolidine and β-lactam ring creates a molecule which is folded at the ring junction, and L-shaped.

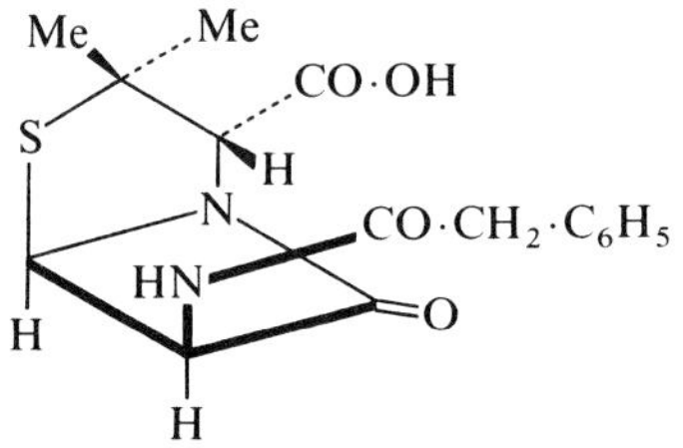

The alkali metal salts are readily soluble in water, but practically insoluble in organic solvents, fixed oils and *Liquid Paraffin*. The free acid, however, is only sparingly soluble in water, but readily soluble in organic solvents. Aqueous and ethanolic solutions of *Benzylpenicillin* show characteristic benzenoid absorption in the ultraviolet at 259 nm and 265 nm. Penicillins lacking an aromatic group do not show these characteristic maxima, but merely exhibit weak end-absorption.

### Thermal and Hydrolytic Stability

*Benzylpenicillin*, and its sodium and potassium salts, are stable in the solid state at room temperature when completely dry. They are, however, all hygroscopic and rapidly inactivated in the presence of moisture. The moisture content of the crystalline salts is strictly limited for this reason (*Benzylpenicillin* $\ngtr$ 1% loss on drying at 105°C), and a test for heat stability applied (Hodge, Senkus and Riddick, 1946), which consists of heating a sample for four days in an open vial at 105°C ($\ngtr$ 10% loss of activity).

In aqueous solution, the rate at which *Sodium Benzylpenicillin* is inactivated depends on both the pH of the solution and the temperature. In general, decomposition is retarded at low temperatures (0–10°C) and between pH 5 and 8 (Abraham, Chain, Fletcher, Florey, Gardner, Heatley and Jennings, 1941;

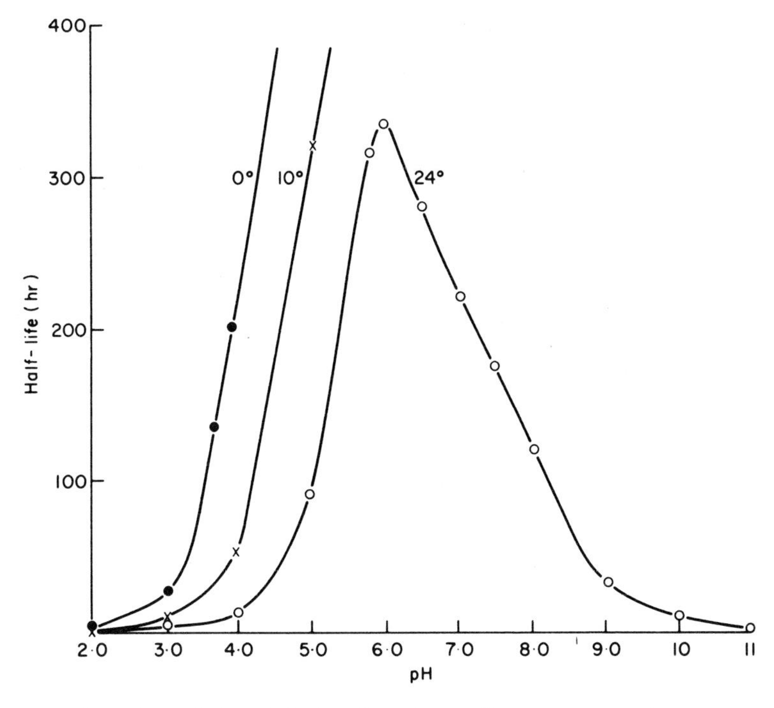


Fig. 26 Stability of *Sodium Benzylpenicillin* in aqueous solution

Abraham, Chain and Holiday, 1942). Decomposition is accelerated under more acidic and alkaline conditions at all temperatures (Benedict, Schmidt and Coghill, 1946). It is particularly marked, however, in acidic solution where the formation of benzylpenillic acid and benzylpenicilloic acid accelerates the reaction (Fig. 26).

Optimum stability is achieved by stabilisation in aqueous solutions at pH 6.0 by the use of phosphate and citrate buffers (Pulvertaft and Yudkin, 1946; Hahn, 1947; Pratt, 1947). (Table 58).

Table 58. Stability of Sodium Benzylpenicillin in Citrate and Phosphate Buffers at pH 6 and 37°C (Hahn, 1947)

| Reaction time (hr) | Diameter of Inhibition Zone (mm) due to Benzylpenicillin Na (4 units/ml) | | |
|---|---|---|---|
| | NaCl Solution | Phosphate Buffer | Citrate—NaCl Buffer |
| 0 | 19 | 18 | 18 |
| 12 | 7 | 14 | 18 |
| 24 | 6 | 14 | 18 |
| 48 | 0 | 12 | 18 |

**Inactivation by Alkali**

Hydrolysis of *Benzylpenicillin* occurs in alkaline solution at pH 10–12 to give benzylpenicilloic acid as sole product. Alkaline degradation is catalysed by metal ions, such as $Cu^{2+}$ (Cressman, Sugita, Doluisio and Nubergall, 1969).

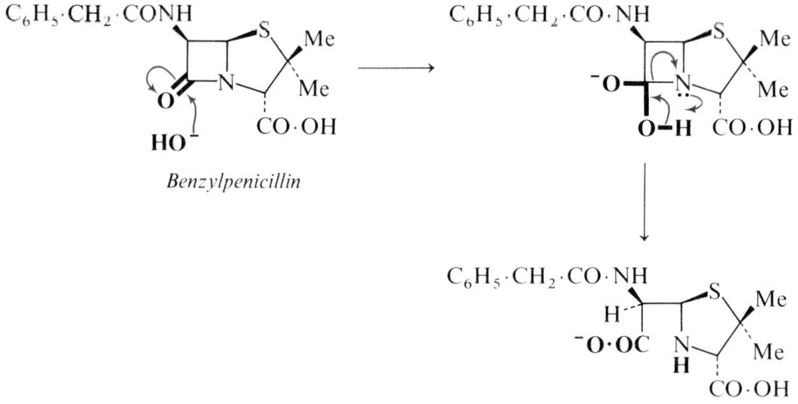

$C_6H_5 \cdot CH_2 \cdot CONH$ S Me Me O N $HO^-$ CO·OH

*Benzylpenicillin*

$C_6H_5 \cdot CH_2 \cdot CO \cdot NH$ S Me Me $^-O$ N O–H CO·OH

$C_6H_5 \cdot CH_2 \cdot CO \cdot NH$ H S Me Me $^-O \cdot OC$ N H CO·OH

Benzylpenicilloic acid

**Inactivation by Acid**

Hydrolysis in acidic solution is rather more complicated. Treatment of *Benzylpenicillin* or its salts with dilute mineral acid at pH 2.5 gives some benzylpenicilloic acid, but mainly an alternative degradation product, benzylpenillic acid. The formation of benzylpenicilloic acid has been explained as the straightforward hydrolysis of the strained $\beta$-lactam ring (Schwartz, 1965; compare amide hydrolysis, p. 302), but is more likely to be formed concurrently with benzylpenillic acid by hydrolysis of benzylpenicillenic acid, the initial product of rearrangement in acidic solution.

This rearrangement, which results in cleavage of the $\beta$-lactam ring, has been the subject of much discussion, since the four-membered $\beta$-lactam ring is subject to Baeyer strain and, hence, is intrinsically of low stability. It is however significantly less stable in acidic solution than the $\beta$-lactam ring in desthiobenzylpenicillin and 2-phenylacetylamino-*N*-phenylbutyrolactam. Hence, the thiazolidine ring is an important factor in its instability to acid.

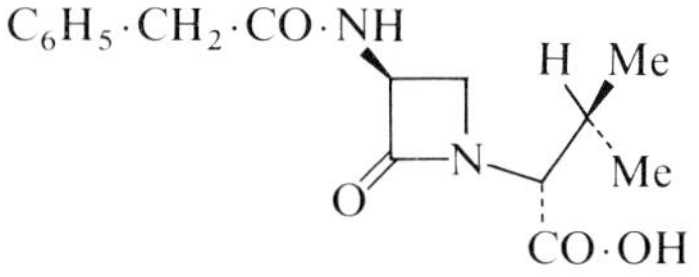

$C_6H_5 \cdot CH_2 \cdot CO \cdot NH$ H Me O N Me CO·OH

Desthiobenzylpenicillin

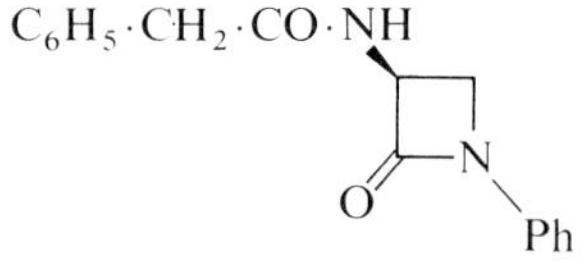

$C_6H_5 \cdot CH_2 \cdot CO \cdot NH$ O N Ph

2-Phenylacetylamino-*N*-phenylbutyrolactam

*Benzylpenicillin* is also less stable to acid than the synthetic fused thiazolidine-$\beta$-lactam, which lacks an acylamino side-chain.

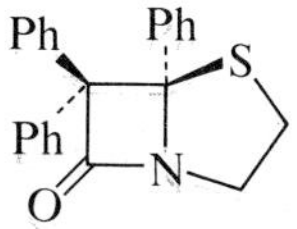

*Phenoxymethylpenicillin* (Penicillin V) and 6-aminopenicillanic acid both have enhanced acidic stability. Thus, the nature of the acylamino side-chain, the $\beta$-lactam ring and the presence of the thiazolidine ring all contribute to the acidic instability of *Benzylpenicillin*.

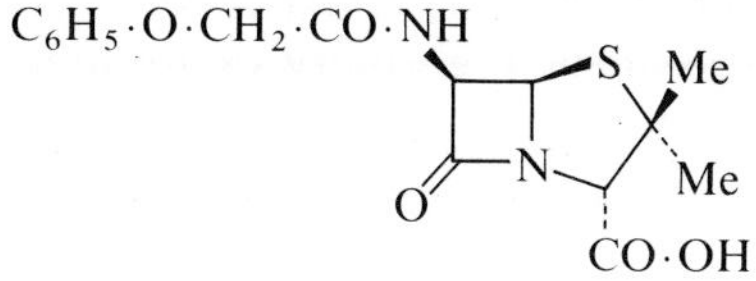

*Phenoxymethylpenicillin*

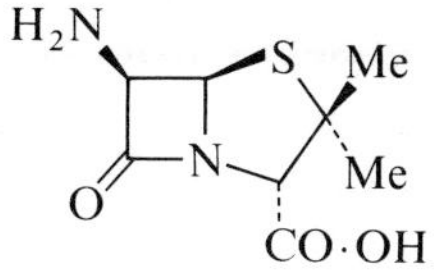

6-Aminopenicillanic acid

The mechanism of rearrangement (inactivation) of *Benzylpenicillin* in acidic solution is, therefore, probably a concerted one, involving both rings and the

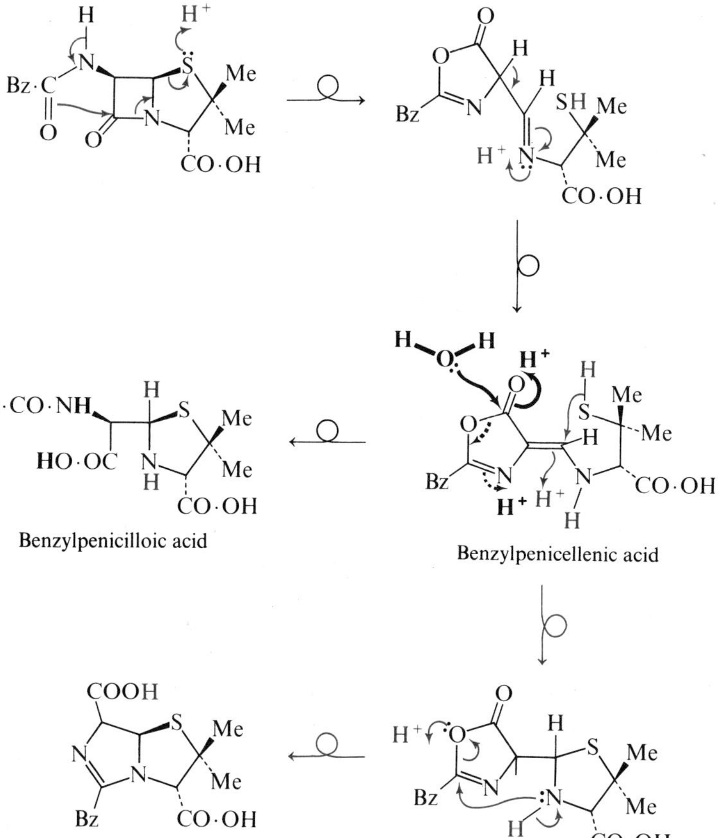

Benzylpenicilloic acid

Benzylpenicellenic acid

Benzylpenillic acid

acylamino side-chain. It is probably initiated by proton-acceptance at the electrons of the sulphur atom in the thiazolidine ring.

Evidence in support of this mechanism comes from studies on the acidic decomposition of *Methicillin*, which initially gives the penicillenic acid, and only forms the corresponding penicilloic and penillic acids on prolonged standing (Johnson and Panetta, 1964). An alternative mechanism, which does not involve opening of the thiazolidine ring in the first step has also been proposed (Bundgaard, 1971).

The acidic stability conferred by the phenoxymethyl group in *Phenoxymethylpenicillin* stems from the greater electron-attracting ability of the phenoxymethyl group compared with that of the benzyl group in *Benzylpenicillin*. This tends to oppose the electron drift which initiates the above rearrangement.

**Inactivation by Metal Ions**

Aqueous solutions of *Sodium Benzylpenicillin* are inactivated by $Zn^{2+}$, $Cd^{2+}$, $Cu^{2+}$ and $Pb^{2+}$. The reactions which are catalytic are thought to be induced by an attack of the metal ion on the $\beta$-lactam carbonyl group. The reaction results in the formation of benzylpenicilloic acid which is precipitated as the metal salt. It should be noted that there is, therefore, a need for precautions in the preparation and handling of penicillin injections to minimise contact with metal ions.

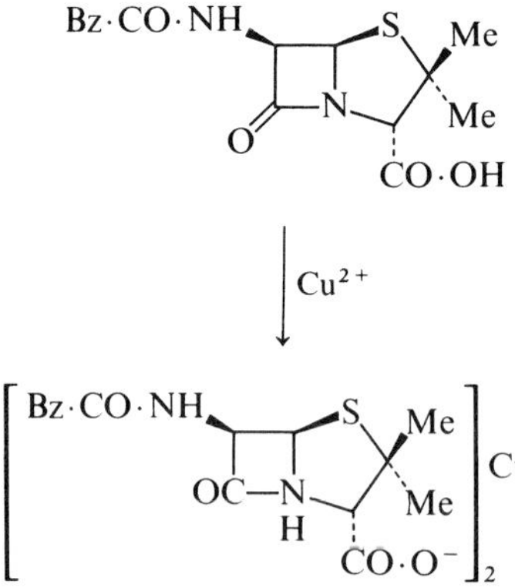

Bz·CO·NH, S, Me, Me, N, O, CO·OH

↓ $Cu^{2+}$

[Bz·CO·NH, S, Me, Me, OC—N, H, CO·O⁻]₂ $Cu^{2+}$

Mercuric chloride also reacts with *Sodium Benzylpenicillin* in aqueous solution bringing about a more profound decomposition, with evolution of $CO_2$ and precipitation of a complex formed from $HgCl_2$ and D-penicillamine, a decomposition product. The intact antibiotic reacts quite slowly with mercuric chloride, in contrast to benzylpenicilloic acid, which is readily decomposed by this reagent. It seems likely, therefore, that decomposition of *Sodium Benzylpenicillin* proceeds via an initial hydrolytic fission to sodium benzylpenicilloate.

D-Penicillamine, which is formed by this decomposition, has no antibiotic activity, but is used in medicine as a metal-chelating agent to aid the elimination of toxic metal ions from the body in copper, lead and mercury poisoning (p. 535 and **2**, 2).

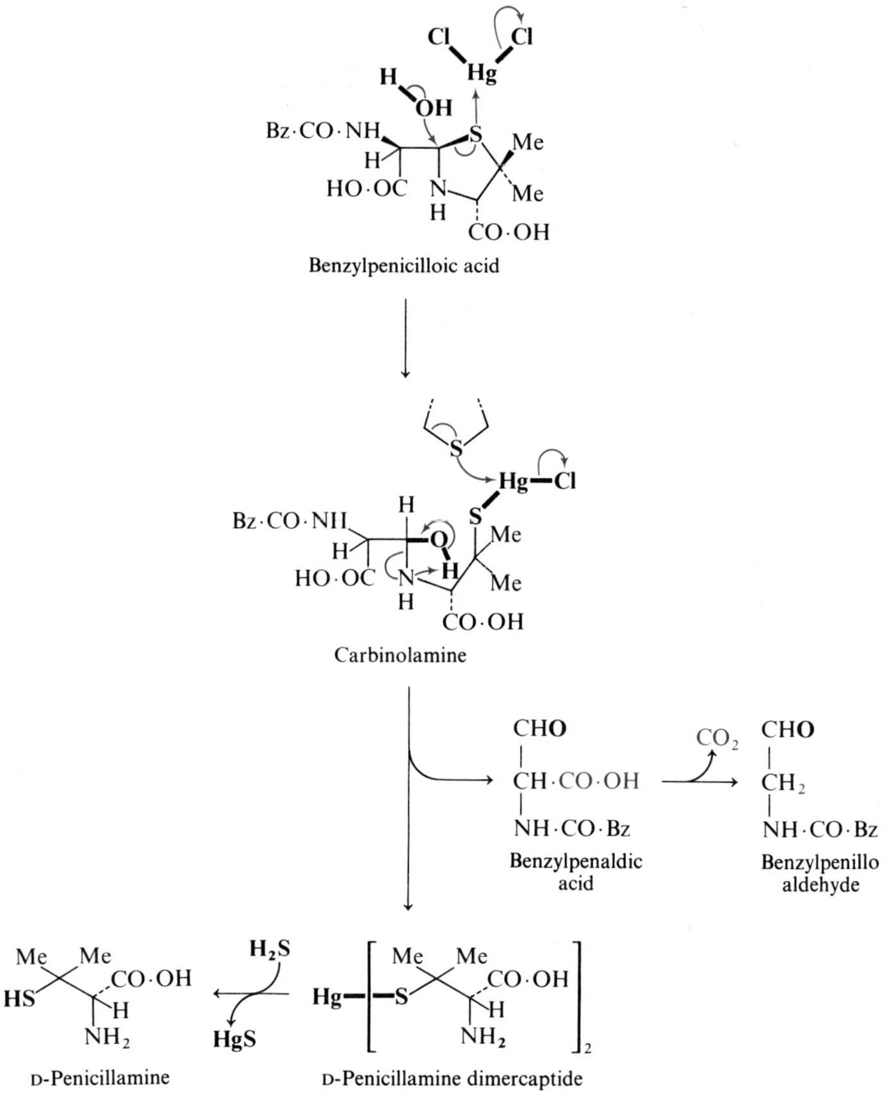

Benzylpenicilloic acid

Carbinolamine

Benzylpenaldic acid

Benzylpenillo aldehyde

D-Penicillamine

D-Penicillamine dimercaptide

**Alcoholysis and Aminolysis**

The $\beta$-lactam ring is also sensitive to primary alcohols, amines, hydroxylamines and amino acids (such as cysteine), and undergoes alcoholysis or aminolysis to give the corresponding $\alpha$-benzylpenicilloic esters and amides, with loss of antibiotic activity (Table 59).

Alcoholysis, like hydrolysis, is catalysed by metal ions (Chain, Philpot and Callow, 1948) including $Cu^{2+}$, $Zn^{2+}$ and $Sn^{2+}$. Desthiobenzylpenicillin is not susceptible to metal-catalysed alcoholysis, suggesting that breakdown occurs through a concerted mechanism as in acid hydrolysis.

Table 59. Metal-catalysed Methanolysis of Benzylpenicillin at 37°C (Chain *et al.*, 1948)

| | | Activity of Penicillin (units/ml) | | |
|---|---|---|---|---|
| | | Time (hr) | | |
| Cation | Concentration ($\mu$g/ml) | 0 | 6 | 24 |
| $Zn^{2+}$ | 0.1 | 540 | 415 | 0 |
| | 0.5 | 500 | 0 | 0 |
| $Sn^{2+}$ | 0.1 | 515 | 300 | 0 |
| | 0.5 | 490 | 0 | 0 |
| $Cu^{2+}$ | 0.1 | 430 | 445 | 345 |
| | 0.5 | 445 | 390 | 0 |

The partial inactivation of *Benzylpenicillin* by *Dextrose* and *Sucrose* (Schneider and de Week, 1967; Hem, Russo, Bahal and Levi, 1973) appears to be due to alcoholysis of the $\beta$-lactam ring with formation of an $\alpha$-ester of benzylpenicilloic acid.

Bz·CO·NH, S, Me, Me, N, O, CO·OH — R·OH → Bz·CO·NH, S, Me, RO·OC$^{\alpha}$, NH, Me, $\beta$, CO·OH

*Benzylpenicillin* — $\alpha$-Ester of benzylpenicilloic acid

*Benzylpenicillin* is also sensitive to aminolysis, the rate of decomposition being dependent on the p$K_a$ of the amine (Yamana, Tsuji, Miyamoto and Kiya, 1975). Amines with p$K_a$ about 7 are most reactive (half-life *ca* 9 min). Stronger bases are much less reactive, the reaction with bases of p$K_a$ 10–11 having a half-life of 100–400 hr. It seems, therefore, that $\varepsilon$-amino groups of lysine in plasma proteins are unlikely to be involved in the aminolysis of penicillin *in vivo*, an interaction which has been suggested as a source of penicillin allergy.

## SEMI-SYNTHETIC PENICILLINS

### 6-Aminopenicillanic Acid

All natural penicillins may be regarded as acyl derivatives of a parent amine, 6-aminopenicillanic acid, which Sakaguchi and Murao (1950) reported to be formed by the action of an amidase present in a strain of *Penicillium chrysogenum*. Kato (1953) also described a fermentation product of *Penicillium chrysogenum*, which was biologically inactive, but showed some of the chemical properties of *Benzylpenicillin.* This substance, identified as 6-aminopenicillanic acid, was first obtained synthetically by Sheehan and his co-workers (1953, 1958).

6-Aminopenicillanic acid was finally isolated from penicillin fermentation media by Batchelor, Doyle, Nayler and Rolinson in 1959, after observing discrepancies between the results of chemical and biological assays on fermentation products obtained in the absence of side-chain precursors. It is now prepared in this way by fermentation of *Penicillium chrysogenum* in culture media, which lack the usual penicillin side-chain precursors. 6-Aminopenicillanic acid remains in the medium after removal of the natural penicillins by extraction with organic solvents at low pH. Adjustment of the pH of the solution to the isoelectric point (*ca* 4.3) then permits extraction of the product.

A number of amidases are known which catalyse the formation of 6-aminopenicillanic acid from various penicillins by hydrolysis of the side-chain. Amidases present in certain fungi and actinomycetes de-acylate *Phenoxymethylpenicillin* more rapidly than *Benzylpenicillin*; others found in *Escherichia* and *Alcaligenes* destroy *Benzylpenicillin* more rapidly than *Phenoxymethylpenicillin.*

6-Aminopenicillanic acid possesses antibacterial properties, but of a much lower order than those of *Benzylpenicillin.* It is destroyed by penicillinase ($\beta$-lactamase; p. 570), which opens the lactam ring, though much more slowly than in *Benzylpenicillin.* It shows the same chemical instability to alkali as the penicillins and is rapidly decomposed with cleavage of the $\beta$-lactam ring to penicilloic acid. It is, however, comparatively stable to acids due to the lack of a 6-acylsubstituent and hence its structural inability to rearrange to a penillic acid (*cf Benzylpenicillin*, p. 561). It is only slowly decomposed, therefore, by acid. On the other hand, carbon dioxide readily attacks 6-aminopenicillanic acid to yield a biologically inactive dicarboxylic acid. The latter contains a labile carboxyl group and loses carbon dioxide (1 mol) when boiled in water for 2–3 hr, or

$H_2N$ S Me Me N O CO·OH —$CO_2$→ [HO·OC·NH S Me Me N O CO·OH]

HO·OC HN S Me Me N O CO·OH —100°, $-CO_2$→ HN S Me Me N O CO·OH treated with mercuric chloride solution, yielding a crystalline isomer of 6-aminopenicillanic acid.

**Discovery and Synthesis**

The discovery that 6-aminopenicillanic acid could be readily acylated with phenylacetyl chloride to give *Benzylpenicillin*, and phenoxyacetyl chloride to give *Phenoxymethylpenicillin*, opened the way to the preparation of a large number of semi-synthetic penicillins for screening against various micro-organisms. As a result, a whole range of new penicillin-type antibiotics with enhanced stability towards acid and β-lactamases (penicillinases) have been introduced into general use in recent years.

The general method of preparation is the direct condensation of the appropriate acid chloride with 6-aminopenicillanic acid either in water in the presence of sodium bicarbonate, or in an anhydrous solvent containing a tertiary base (Doyle, Hardy, Nayler, Soulal, Stove and Waddington, 1962).

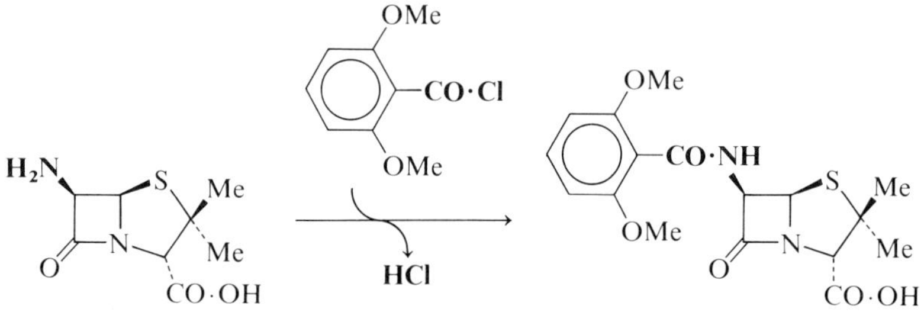

A modified synthetic route is needed in the case of *Ampicillin* because of the need to protect the amino group of the acylating acid (Doyle, Fosker, Nayler and Smith, 1962; reaction scheme page 567).

**Stability**

The stability of semi-synthetic penicillins in the solid state is determined mainly by the often significant amounts of water, and traces of alcohols and organic bases remaining from the manufacturing process. These are capable of triggering off hydrolysis, alcoholysis or aminolysis at rates, which are a function of the storage conditions, especially temperature and humidity. For the most part, potency losses at ambient temperatures or, where recommended, lower storage temperatures, remain within reasonable limits throughout the life of the product. A number, including *Ampicillin Sodium*, *Cloxacillin Sodium*, *Phenethicillin Potassium*, *Phenoxymethylpenicillin Potassium* (Penicillin V Potassium) and *Propicillin Potassium* are moderately stable to heat in the dry state at temperatures up to 100°C for limited periods of time (Lynn, 1970). *Carbenicillin Sodium* (Carbenicillin Disodium), however, shows 20% decomposition after 3 hr at 100°C. *Ampicillin Trihydrate* likewise showed 20, 30 and 40% decomposition at

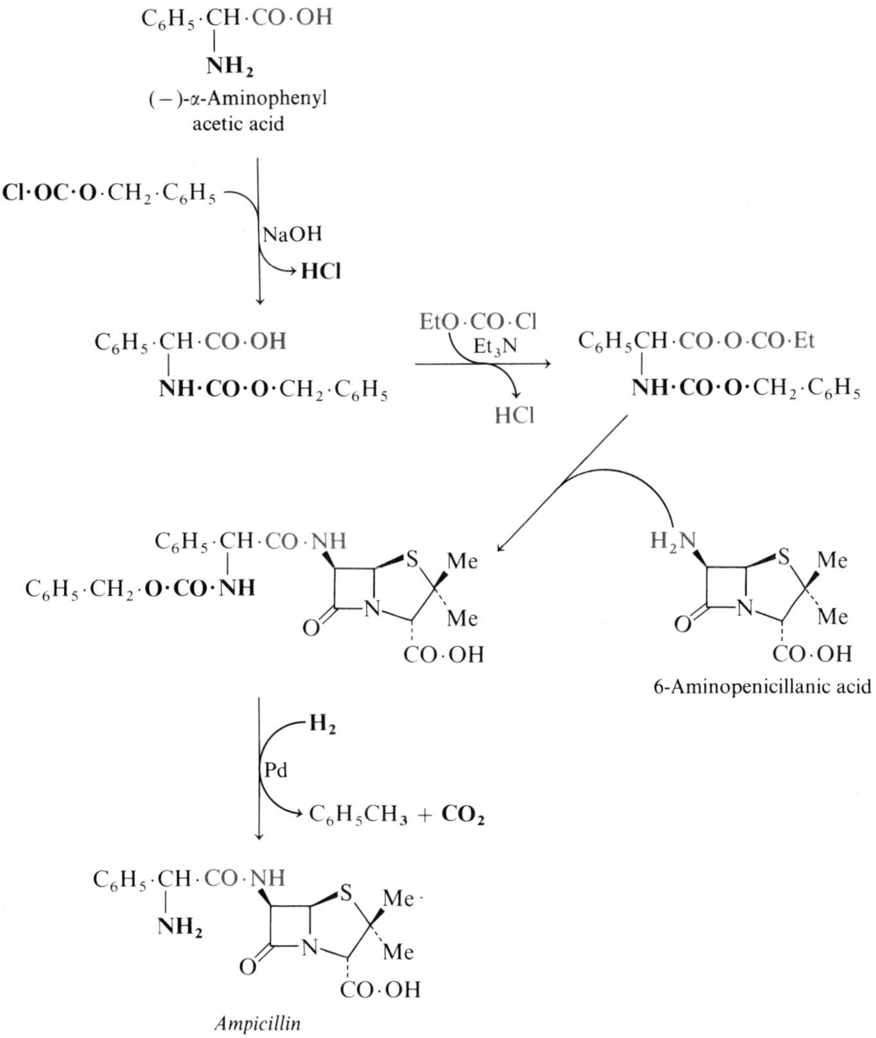

$C_6H_5 \cdot CH(NH_2) \cdot CO \cdot OH$

(−)-α-Aminophenyl acetic acid

$Cl \cdot OC \cdot O \cdot CH_2 \cdot C_6H_5$ → NaOH → $HCl$

$C_6H_5 \cdot CH(NH \cdot CO \cdot O \cdot CH_2 \cdot C_6H_5) \cdot CO \cdot OH$

$EtO \cdot CO \cdot Cl$, $Et_3N$ → $HCl$

$C_6H_5CH(NH \cdot CO \cdot O \cdot CH_2 \cdot C_6H_5) \cdot CO \cdot O \cdot CO \cdot Et$

$C_6H_5 \cdot CH(C_6H_5 \cdot CH_2 \cdot O \cdot CO \cdot NH) \cdot CO \cdot NH$–(penam: S, Me, Me, N, O, CO·OH)

$H_2N$–(penam: S, Me, Me, N, O, CO·OH)

6-Aminopenicillanic acid

$H_2$, Pd → $C_6H_5CH_3 + CO_2$

$C_6H_5 \cdot CH(NH_2) \cdot CO \cdot NH$–(penam: S, Me, Me, N, O, CO·OH)

*Ampicillin*

110°C after 1, 2 and 3 hr respectively. At higher temperatures (150°C), all are unstable.

In aqueous solution, stability of semi-synthetic penicillins, like that of *Benzylpenicillin* (p. 558), is a function of pH. Stability in acidic solution is unequivocally correlated with the structure and properties of the 6-acyl substituent. Electron-withdrawing substituents, as in *Phenoxymethylpenicillin*, *Oxacillin*, *Cloxacillin* and *Ampicillin*, inhibit the electron drift from the 6-acyl carbonyl group, which is an essential feature of the proposed mechanism of decomposition in acidic solution (p. 561). Electron-withdrawing substituents, therefore, increase acid stability, whilst electron-donating substituents, as for example in *Methicillin*,

| | | |
|---|---|---|
| Penicillin | 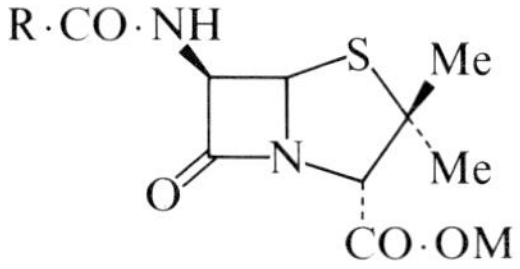 R·CO·NH–(β-lactam/thiazolidine ring: S, Me, Me, N, O, CO·OM) | M |
| | R | |
| *Cloxacillin* | 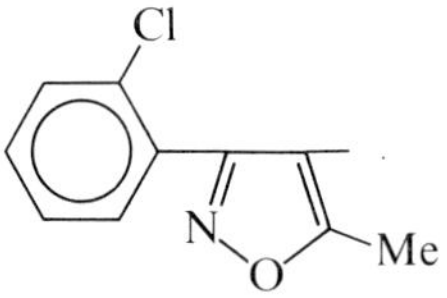 2-chlorophenyl-5-methylisoxazol-4-yl (Cl, N, O, Me) | Na |
| *Flucloxacillin* | 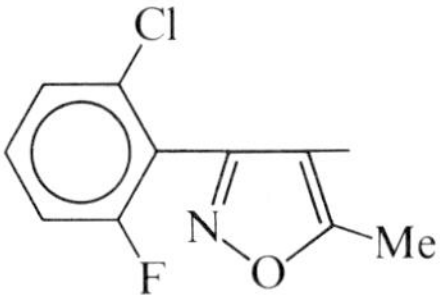 2-chloro-6-fluorophenyl-5-methylisoxazol-4-yl (Cl, F, N, O, Me) | Na |
| *Oxacillin* | 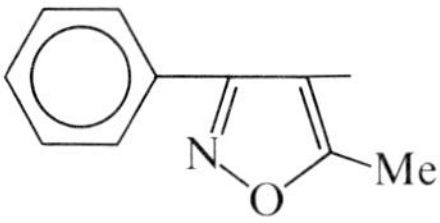 phenyl-5-methylisoxazol-4-yl (N, O, Me) | Na |
| *Phenethicillin* | 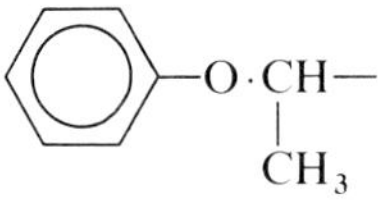 $C_6H_5$–O·CH(CH$_3$)– | K |
| *Phenoxymethylpenicillin* | 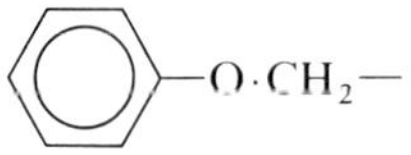 $C_6H_5$–O·CH$_2$– | K |
| *Propicillin* | 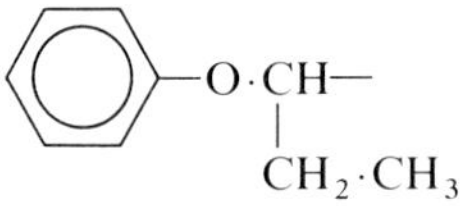 $C_6H_5$–O·CH(CH$_2$·CH$_3$)– | K |

decrease stability to acid (Doyle and Nayler, 1964). These conclusions are confirmed by other studies, which similarly show that the rate of inactivation in acid can be correlated quantitatively with both the Hammett constant for the side-chain substituents (Panarin, Solotovskii and Ekzemplyarov, 1967; Panarin and Solotovskii, 1970), and the p$K_a$ of the side-chain acids (Table 60).

Table 60. Solubility and Stability in Acid of Semi-Synthetic Penicillins

| Penicillin | Solubility in Water (20°C) | Half-life (min)[1] | p$K_a$ of Side-chain acid |
|---|---|---|---|
| Benzylpenicillin Potassium | <1 in 1 | 3.5[1] | 5.6 |
| α-Methoxybenzylpenicillin | | 77[1] | 4.6 |
| α-Chlorobenzylpenicillin | | 300[1] | 4.0 |
| Ampicillin | 1 in 170 (Anhydrous) 1 in 150 (Trihydrate) | 660[1] | 3.2 |
| Ampicillin Sodium | 1 in 2 | | |
| Carbenicillin Sodium | 1 in 1.2 | 30[2] | |
| Cloxacillin Sodium | 1 in 2.5 | 160[2,3] | |
| Methicillin Sodium | 1 in 0.6 | 2.3[3] | |
| Phenethicillin Potassium | 1 in 1.5 | *ca* 160[2] (68[4]) | |
| Phenoxymethylpenicillin Potassium | 1 in 1.5 | 160[3] | |
| Propicillin Potassium | 1 in 1.5 | *ca* 160[2] | |

[1] Half-life at 35°C in 50% aqueous ethanol at pH 1.3 (Doyle, Nayler, Smith and Stove, 1961)
[2] Lynn, 1970
[3] Doyle, Long, Nayler and Stove, 1962
[4] In aqueous solutions (Schwartz, Granatek and Buckwalter, 1962)

*Carbenicillin Sodium*, which incorporates the electron-withdrawing carboxyl group is acid-sensitive, and appears at first sight to be the exception to the above rule. Acidification, however, releases the free acid, which because it is a β-ketoacid, is acid-labile (Chapter 13) and readily decarboxylates to form the acid-labile *Benzylpenicillin*.

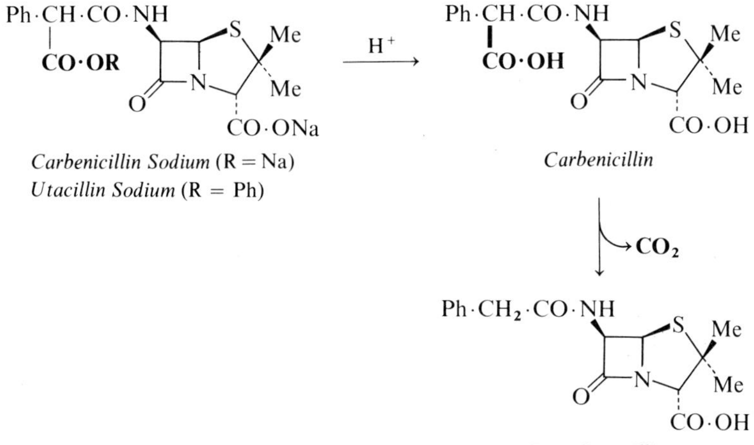


*Carbenicillin Sodium* (R = Na)
*Utacillin Sodium* (R = Ph)

*Carbenicillin*

*Benzylpenicillin*

Because of this acid instability, *Carbenicillin Sodium* is not suitable for oral administration. It is rapidly excreted by the kidney, serum half-life in normal subjects being as low as one hour (Kass, 1970; Hoffman, Cestero and Bullock, 1970). It is especially useful, therefore, in the treatment of urinary tract infections, particularly arising from *Pseudomonas aeruginosa* and *Proteus vulgaris*, which are sensitive to it. The problem of availability of *Carbenicillin* by oral administration is overcome by the use of its phenolic ester, *Utacillin*. As an ester, this is not subject to decarboxylation, whilst the electron-attracting properties of the phenyl carboxylate group ensure the acid stability of the β-lactam ring.

Although injection solutions of *Cloxacillin Sodium* and *Carbenicillin Sodium* are close to neutrality, and at worst only slightly alkaline, they are still subject to hydrolytic cleavage ($HO^-$-catalysed) of the β-lactam ring. *Cloxacillin Injection*, once re-constituted, is required to be used within 24 hr, but may be stored for up to four days if kept between 2 and 10°C. *Carbenicillin Injection*, which is somewhat more alkaline, if stored after reconstitution, must be maintained at $\ngtr$ 5°C. *Ampicillin Injection* is even more alkaline (a 10% solution of *Ampicillin Sodium* has a pH between 8 and 10). Moreover, its rate of decomposition and hence inactivation is concentration-dependent. Thus 1, 5, 10 and 25% solutions in water at pH 8.5 undergo 8, 32, 52 and 74% decomposition respectively in five days at 5°C (Lynn, 1970).

The hydrolytic instability of the penicillins poses serious problems when attempts are made to administer them in large-volume intravenous infusions, and substantial potency losses have been reported for solutions of *Ampicillin Sodium* in *Dextrose* and *Laevulose* (Fructose) *Infusions* (Lynn, 1970, 1971).

**Enzymic Inactivation of Penicillins**

Penicillinases catalyse the hydrolytic cleavage of the β-lactam ring in *Benzylpenicillin* to yield the biologically inactive penicilloic acid. For this reason, they are now generally known as β-lactamases. The first such enzyme was discovered by Abraham and Chain (1940) in the cells of *E. coli*. Similar enzymes are produced by many Gram-positive and Gram-negative bacteria. Many organisms only release the enzyme on autolysis (death and disruption), but release of β-lactamase from *Bacillus cereus* and *Staphylococcus aureus* (resistant strains) is actually stimulated in the presence of penicillins. β-Lactamases are heat-sensitive, and rapidly inactivated above 60°C.

*Benzylpenicillin* is sensitive to all known β-lactamases and rapidly inactivated. *Ampicillin* is even more unstable (Hou and Poole, 1973). On the other hand, 6-aminopenicillanic acid and penicillins with sterically-hindered side-chains,

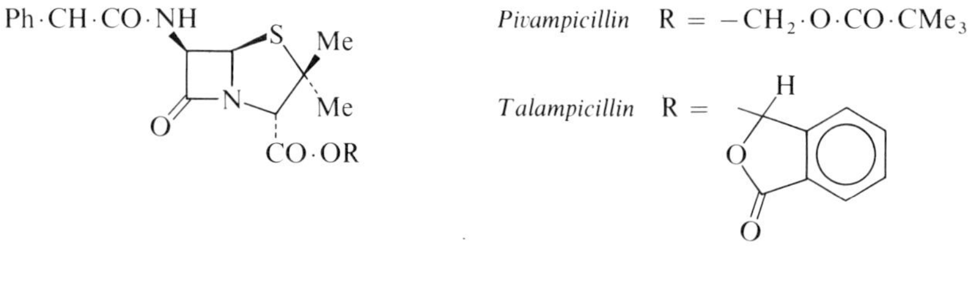

Ph·CH·CO·NH ... S, Me, Me, N, O, CO·OR

*Pivampicillin* R = $-CH_2·O·CO·CMe_3$

*Talampicillin* R = (H, O, O)

such as *Methicillin*, *Nafcillin* (Naficillin Sodium), *Oxacillin* and *Cloxacillin*, are resistant to most β-lactamases (Nayler, 1973). Another factor which reduces enzymic hydrolysis of the β-lactam ring, is esterification of the carboxyl group, as for example in *Pivampicillin* and *Talampicillin*.

**Chemical Incompatabilities**

Chemical incompatability of penicillin salts results from the thoughtless admixture of antibiotic-containing syrups with other syrups containing organic bases (Lynn, 1970). Gould and Brown (1974) reported an immediate precipitation in a prescribed mixture of *Orbenin Syrup* (*Cloxacillin Sodium*) with *Phensadyl Syrup (Promethazine Hydrochloride*, *Codeine Phosphate* and *Ephedrine Hydrochloride)* in equal parts. The precipitation is clearly double decomposition with the formation of insoluble salts, comparable with such products as *Benzathine Penicillin* (Penicillin G). The pH of the syrup mixture was 3.5, and precipitation was predictably accompanied by a 20% loss of antibiotic activity in 5 hr, and a 99% loss in 5 days.

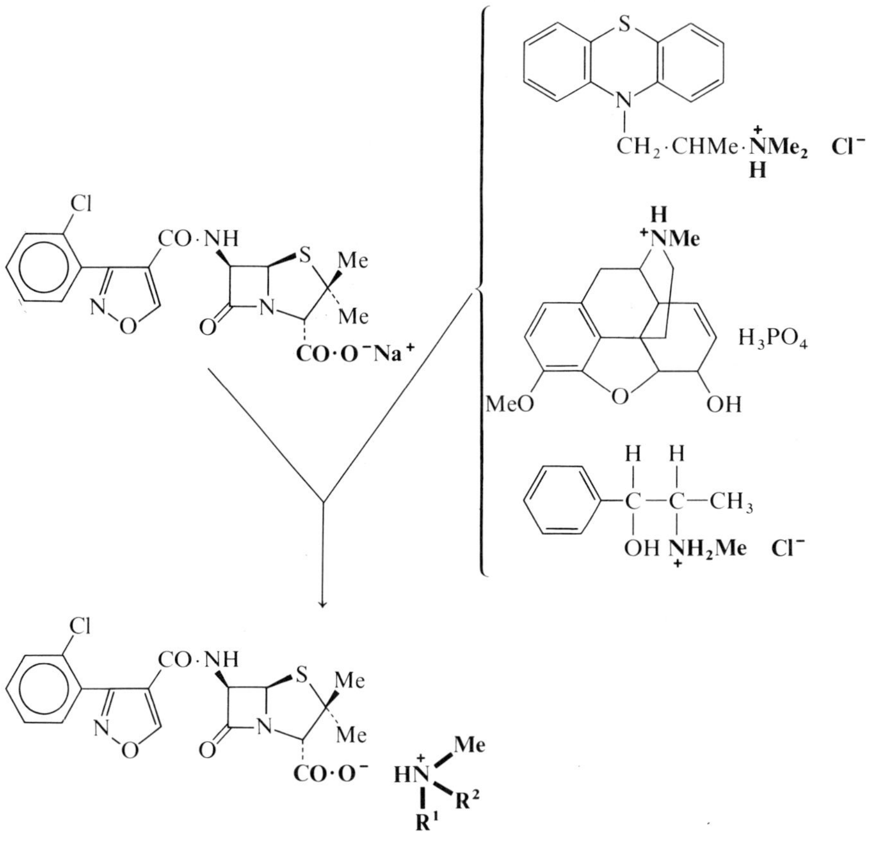

Cl
CO·NH
S
Me
N
O
N
Me
O
CO·O⁻Na⁺

S
N
CH₂·CHMe·N⁺HMe₂ Cl⁻

H
⁺NMe
H₃PO₄
MeO
O
OH

H H
C—C—CH₃
OH NH₂⁺Me Cl⁻

Cl
CO·NH
S
Me
N
O
N
Me
O
CO·O⁻ HN⁺ Me R² R¹

Similar incompatabilities resulting in the precipitation of insoluble salts have been reported for admixture of various penicillins with basic antibiotics including *Streptomycin*, *Kanamycin*, *Gentamicin*, *Erythromycin*, *Tetracycline* and *Oxytetracycline* (Lynn, 1970), and with local anaesthetics, such as *Lignocaine Hydrochloride* (Lidocaine Hydrochloride; Lynn, 1971).

The condensation of *Ampicillin* with 2-hydroxymethylfurfural in a so-called browning reaction, where the antibiotic is added to *Dextrose* and *Laevulose Infusions*, can also lead to discolouration and inactivation (Stenlake, 1975). An analogous condensation with acetone is involved in the formation of *Hetacillin* from *Ampicillin* (Jusko and Lewis, 1973). *Hetacillin*, which has little or no antibacterial activity *per se*, is stable in the solid state, but is rapidly converted to *Ampicillin* in aqueous media.

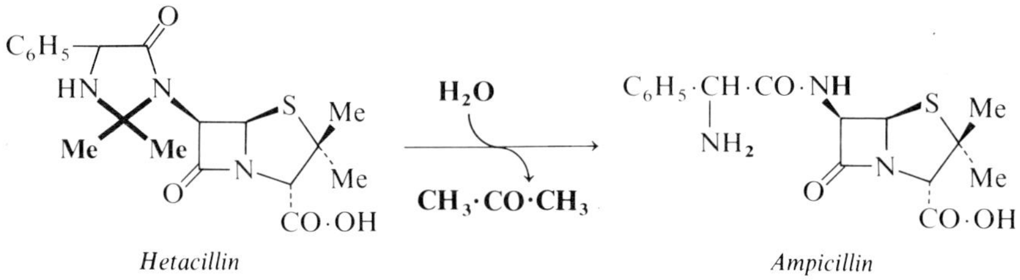


## Antibiotic Activity of Semi-synthetic Penicillins

*Phenethicillin*, the first of the semi-synthetic penicillins, has a similar range of antibacterial activity to *Benzylpenicillin*, and is about equi-active with the latter against sensitive strains of *Staphylococcus aureus*. *Phenethicillin*, however, is rather more effective against *Benzylpenicillin*-resistant, β-lactamase-producing strains (Garrod, 1960), and has the advantage that it is stable to acid.

*Phenethicillin* is also relatively better absorbed from the gastro-intestinal tract, and serum levels at least twice those obtainable with equivalent doses of *Phenoxymethylpenicillin* can be obtained by oral administration (Knudsen and Rolinson, 1959; Fairbrother and Taylor, 1960). Concentrations of *Phenethicillin* in blood serum obtained by oral administration are of the same order as those obtained by intramuscular injection of equivalent doses of *Benzylpenicillin* (Knudsen and Rolinson, 1959; McNeill, 1962).

*Propicillin* and *Phenbenicillin* (Rollo, Somers and Burley, 1962) are marginally more effective than *Phenethicillin* against β-lactamase-producing strains of *Staphyloccocus aureus*, but both also give enhanced serum levels, which in the case of *Propicillin* are said to be about four times those obtainable with *Phenoxymethylpenicillin* (Williamson, Morrison and Stevens, 1961). The enhanced efficacy of *Propicillin* and *Phenbenicillin* clearly derives from the increased steric hindrance to enzyme attack at the amide and lactam carbonyls, which is conferred by the bulkier α-substituents, though in the case of *Phenbenicillin*, this effect is also aided by the electron-withdrawing properties of the α-phenyl group, which oppose the hydrolytic mechanisms.

The conclusion that stability to β-lactamase was the result of steric hindrance in the region of the side-chain amide links was confirmed by the observation that the effect was enhanced by bulky substituents (Brain, Doyle, Hardy, Long, Mehta, Miller, Nayler, Soulal, Stove and Thomas, 1962). The preparation of *Methicillin* represented a logical extension of this deduction, the flanking methoxyl groups achieving the same measure of steric hindrance as seen in triphenylmethylpenicillin. *Methicillin*, however, is something like one hundred times less active than *Benzylpenicillin* against non-β-lactamase-producing strains of *Staphylococcus aureus* (*in vitro*), but because of its resistance to destruction by β-lactamase, it is about twenty times more potent against the β-lactamase-producing strains (Fairbrother and Taylor, 1960; Knox, 1960, 1961; Rolinson, Batchelor, Stevens, Cameron-Wood and Chain, 1960). Its value as a therapeutic agent, therefore, depends upon its ability to control infection due to *Benzylpenicillin*-resistant organisms. It is, however, rapidly excreted in the urine and frequent injections are essential to maintain adequate serum concentrations. It is also unstable to acid, losing 50% of its activity within 20 min at pH 2.0 and 25°C. For these reasons, it has been superceded in use by other, more stable penicillins.

*Oxacillin*, *Cloxacillin* and *Flucloxacillin* appear to be free from the major disadvantages of *Methicillin*. *Cloxacillin*, which is available as the sodium salt, is freely soluble in cold water, and solutions stored at 5°C are stable for one week. In 50% aqueous ethanol at pH 1.3 and 35°C, solutions of both *Oxacillin* and *Cloxacillin* have a half-life of 160 min (Table 60). *Cloxacillin* and *Flucloxacillin* are highly active against *Staphylococcus aureus* including β-lactamase-resistant strains, *Streptococcus pyogenes*, *Streptococcus viridans* and *Streptococcus pneumoniae*, but have little action against Gram-negative organisms.

The basic side-chain of *Ampicillin*, D-(−)-α-aminophenylpenicillin, confers activity against Gram-negative organisms. As might be anticipated, the −I effect of the α-amino side-chain substituent also confers acid stability on the molecule. In common with other electron-attracting substituents, the α-amino group opposes the electron drift proposed by Robinson and Woodward as the mechanism for the opening of the β-lactam ring, and, as already implied, a direct correlation can be demonstrated between the p$K_a$ of the acylating acid and the time for 50% loss of activity (Table 60). In accordance with this, acid stability among amino-substituted penicillins declines as the amino group is removed from proximity to the amide link, and is virtually lost in 5-aminopentylpenicillin. With the exception of the amino group, however, the presence of electron-attracting α-substituents in the penicilllin side-chain in general leads to loss of antibacterial potency against Gram-negative organisms.

The water solubility of the D-(−)-α-aminophenylpenicillin is greater than that of the epimeric L-(+)-form; the former also showed greater antibacterial activity against most organisms (Rolinson and Stevens, 1961; Brown and Acred, 1961). As a consequence of its acid stability, *Ampicillin* can be administered orally, and although it is absorbed somewhat erratically, more satisfactory and more sustained blood levels can be obtained than with either *Phenethicillin* or *Phenoxymethylpenicillin*.

The rate of urinary excretion in *Ampicillin* is significantly less than that of *Methicillin*, some 30% only being excreted in 6–8 hr (Knudsen, Rolinson and Stevens, 1961). Nevertheless, significant urine concentrations are achieved. Significant amounts are also eliminated unchanged in the bile in sufficient concentration to inhibit most strains of *Salmonella typhi* (Stewart and Harrison, 1961).

*Ampicillin* is sensitive to staphylococcal β-lactamases, and also to the β-lactamases of some Gram-negative organisms, particularly those of some *Proteus* species and *Aerobacter aerogenes*. It is, however, in general, more effective than either *Chloramphenicol* or *Tetracycline* against infection by Gram-negative organisms.

It is much easier to achieve antibiotically effective blood levels by oral administration of *Ampicillin* esters, such as *Pivampicillin* and *Talampicillin*, which are much more readily absorbed, and metabolised to the parent antibiotic by hydrolysis in the liver and plasma. Both absorption and hydrolysis of the *Ampicillin* esters is rapid and two to four-fold enhancement of *Ampicillin* bioavailability, as judged by the AUC (area under the curve; **2**, 4), can be obtained on an equivalent dose basis.

*Amoxycillin* (6-[D-(−)-α-amino-4-hydroxyphenylacetamido]penicillanic acid), the *p*-hydroxy derivative of *Ampicillin*, is also much better absorbed than the latter, less firmly bound to plasma proteins, and more readily excreted in the urine (Nayler, 1971; Sutherland, Croydon and Rolinson, 1972). *Amoxycillin*, like *Ampicillin*, is available as a trihydrate which is relatively insoluble in water (1 in 400), but reasonably soluble in phosphate buffer at pH 8.0.

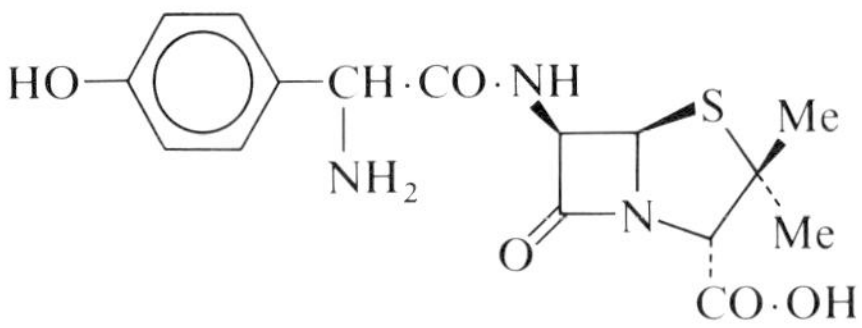

HO—C6H4—CH·CO·NH ; NH2 ; S ; Me ; Me ; N ; O ; CO·OH

*Amoxycillin*

*Carbenicillin Sodium* and its ester, *Utacillin Sodium*, which are also effective against Gram-negative organisms, are also sensitive to staphylococcal β-lactamases. They are particularly valuable because of their effectiveness in the treatment of *Pseudomonas* infections and infections due to ampicillin-resistant strains of *Proteus*.

## CEPHALOSPORINS

### Cephalosporin N

Cephalosporin N alone, of all the natural penicillins, possesses significant antibacterial activity against Gram-negative organisms. This antibiotic, which is D-4-amino-carboxybutylpenicillin, is produced by *Cephalosporium acremonium* isolated by Brotzu from a sewage effluent off the Sardinian coast in 1948 (Brotzu, 1948; Abraham, Newton, Olson, Schuurmans, Schenck, Hargie, Fisher and Fusari, 1955).

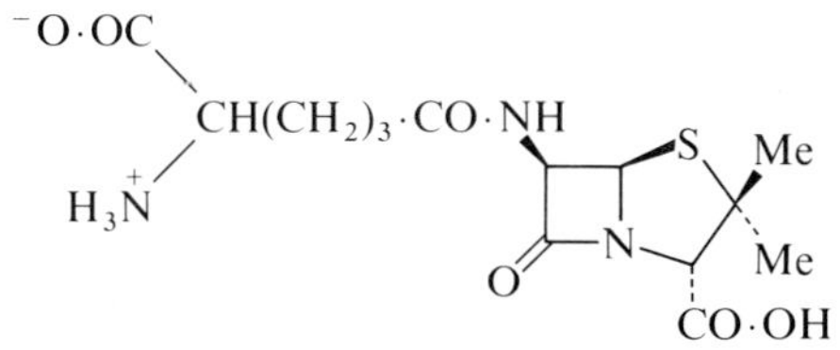

$^-$O·OC
CH(CH$_2$)$_3$·CO·NH
$H_3\overset{+}{N}$
S Me
N Me
O
CO·OH

*Cephalosporin N*

The zwitterionic side-chain endows the molecule with enhanced activity against such Gram-negative organisms as *Salmonella typhi*, but much reduced activity against *Staphylococcus aureus*. The spectrum of activity, however, is shifted by acylation of the amino group, in favour of activity against Gram-positive organisms, whilst activity against Gram-negative ones is much reduced (Newton and Abraham, 1956a). A similar shift in activity has also been observed with the acylation of 4-aminobenzylpenicillin (Tosoni, Glass and Goldsmith, 1958; Doyle and Nayler, 1960). The unusual solubility properties of the zwitterionic side-chain, which is strongly hydrophilic, also confer an absorption and excretion pattern markedly different from that of *Benzylpenicillin*. Thus, it is only slowly absorbed from the intestines, and this in conjunction with its greater effectiveness against Gram-negative organisms, made it, prior to the advent of *Ampicillin*, a potentially useful antibiotic for the treatment of bowel infections (Benavides, Olson, Varela and Holt, 1955).

**Cephalosporin C**

The antibiotic, *Cephalosporin C*, which is also produced by various *cephalosporium* species was discovered as a contaminant of Cephalosporin N, and isolated by Newton and Abraham (1956b) as its sodium salt. Its similarity to Cephalosporin N was suggested by its behaviour as a monoaminodicarboxylic acid, and by the presence of a band at 1780 cm$^{-1}$ in its infrared absorption spectrum characteristic of the penicillin $\beta$-lactam carbonyl group. *Cephalosporin C* differs, however, from Cephalosporin N in a number of important respects. Thus, its molecular formula is $C_{16}H_{21}O_8N_3S$ (Cephalosporin N is $C_{14}H_{21}O_6N_3S$). It does not yield D-penicillamine, and exhibits an ultraviolet absorption maximum at 260 nm.

*Cephalosporin C* has only about one-tenth of the antibacterial activity of Cephalosporin N, although it exhibits a similar spectrum of activity. It has only one-thousandth of the activity of *Benzylpenicillin* against both resistant and non-resistant strains of *Staphylococcus aureus*. It is, however, excreted readily by the kidneys after parenteral administration, and high concentrations are thus produced in the urine, giving good curative effect against urinary tract infections by some Gram-negative organisms. It is able to induce $\beta$-lactamase formation

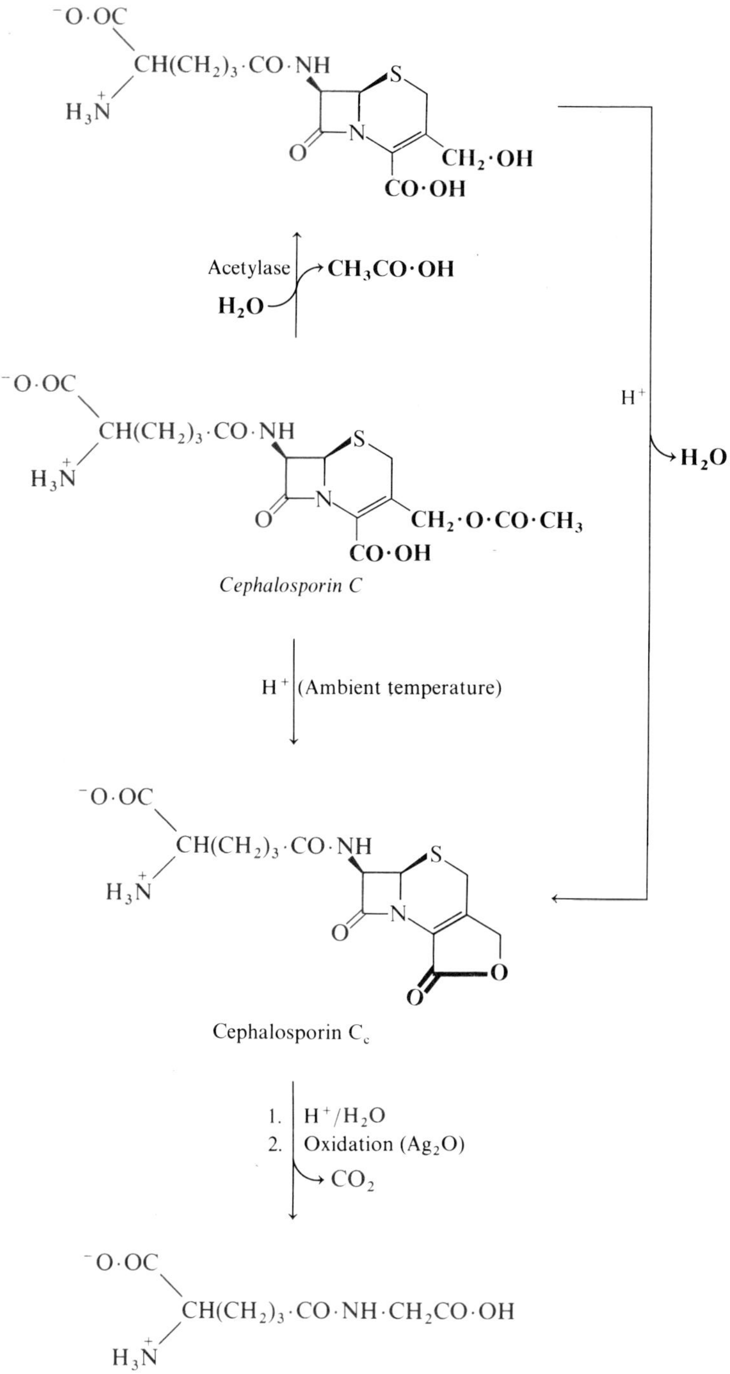

$^{-}O \cdot OC$
$CH(CH_2)_3 \cdot CO \cdot NH$
$H_3\overset{+}{N}$
S
N
O
$CH_2 \cdot OH$
$CO \cdot OH$

Acetylase
$H_2O$
$\rightarrow CH_3CO \cdot OH$

$H^+$
$\rightarrow H_2O$

$^{-}O \cdot OC$
$CH(CH_2)_3 \cdot CO \cdot NH$
$H_3\overset{+}{N}$
S
N
O
$CH_2 \cdot O \cdot CO \cdot CH_3$
$CO \cdot OH$

*Cephalosporin C*

$H^+$ (Ambient temperature)

$^{-}O \cdot OC$
$CH(CH_2)_3 \cdot CO \cdot NH$
$H_3\overset{+}{N}$
S
N
O
O
O

Cephalosporin $C_c$

1. $H^+/H_2O$
2. Oxidation ($Ag_2O$)
$\rightarrow CO_2$

$^{-}O \cdot OC$
$CH(CH_2)_3 \cdot CO \cdot NH \cdot CH_2CO \cdot OH$
$H_3\overset{+}{N}$

$\delta$-Amino-$\delta$-carboxyvalerylglycine

in *Bacillus cereus* and *B. subtilis* (Pollock, 1957), but it is also a competitive inhibitor of β-lactamase (Newton and Abraham, 1956b) and is remarkably stable to enzyme hydrolysis. The rate of inhibition by purified β-lactamase from *Bacillus cereus* is only one-fivethousandth of the rate at which *Benzylpenicillin* is inhibited.

The D-α-aminoadipic acid structure of the side-chain and β-lactam ring of *Cephalosporin C* follows from its behaviour on enzymic or acidic hydrolysis. Enzymic hydrolysis brings about de-acetylation (Jeffrey, Abraham and Newton, 1961) to yield de-acetylcephalosporin, which on treatment with dilute hydrochloric acid at room temperature gives a neutral substance cephalosporin $C_c$ (de-acetylcephalosporin lactone) with the release of acetic acid. More vigorous treatment with hot acid presumably cleaves the β-lactam ring, as with Cephalosporin N, since oxidation of the masked β-aldehydic acid, and decarboxylation of the product, leads to formation of δ-amino-δ-carboxyvalerylglycine. (Reaction scheme on page 576.)

Further evidence for the structure of *Cephalosporin C* was deduced from Raney nickel hydrogenolysis of the antibiotic and cephalosporin $C_c$. It was confirmed by crystallographic analysis (Crowfoot, Hodgkin and Maslen, 1961), and by chemical conversion of a phenoxymethylpenicillin derivative to the corresponding cephalosporin by a mechanism consistent with the accepted structure of both nuclei (Morin, Jackson, Mueller, Lavagnino, Scanlon and Andrews, 1963).

Ph·O·CH₂·CO·NH ... S, Me, Me, N, O, CO·OH

1. $HIO_4$
2. $CH_2N_2$

Ph·O·CH₂·CO·NH ... $H^+$, O, S, C—H, H, H, Me, N, O, CO·OMe

SO₂·OH

xylene

Ph·O·CH₂·CO·NH ... OH, S, $CH_2$, Me, N, O, H, OH, CO·OMe

$H_2O$

Ph·O·CH₂·CO·NH ... S, N, O, Me, CO·OMe

Studies on the chemical relationships between cephalosporins and penicillins (Barton and Sammes, 1971) are based on the ready oxidation of penicillins to the corresponding penicillin sulphoxides (Morin, Jackson, Mueller, Lavagnino, Scanlon and Andrews, 1963, 1969). The oxidation, which is stereospecific (Johnson and McCants, 1965), gives rise to the (*S*)-isomer because of the stabilising influence of hydrogen bonding to the side-chain amide proton (Henbest, 1963).

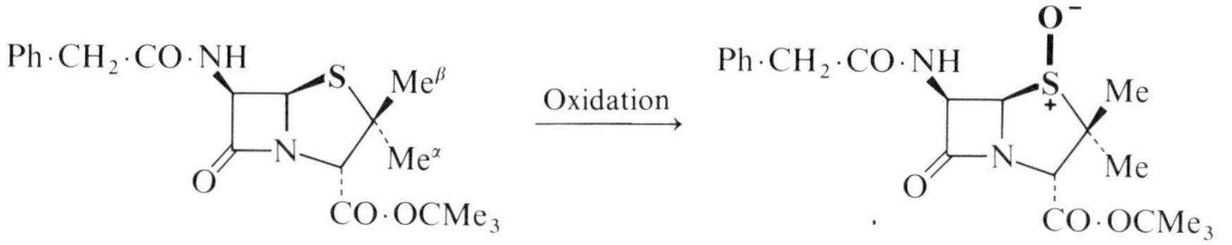

Oxidation

(*S*)-6-phenylacetamidopenam derivative

The corresponding (*R*)-sulphoxide is thermodynamically unstable (Barton, Comer, Greig, Sammes, Cooper, Hewitt and Underwood, 1971), and is rearranged in refluxing benzene to an acyclic sulphenic acid which is then ring-closed via the 2α-methyl group.

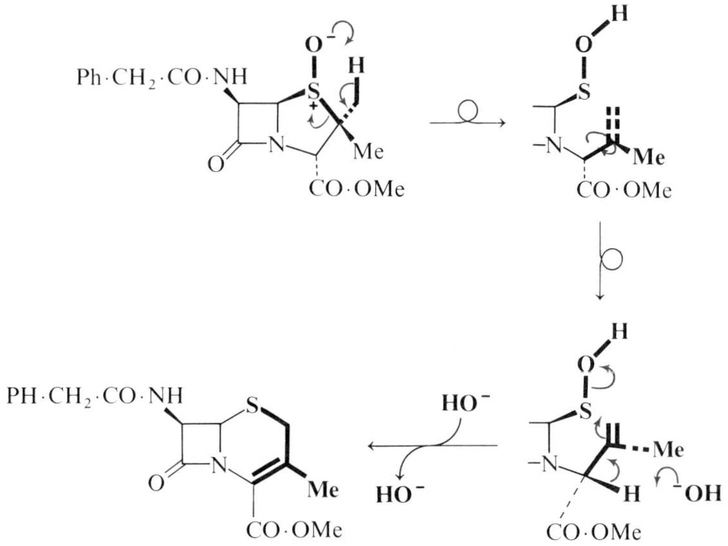

Treatment of the *p*-nitrobenzyl ester of the 6-phenylacetamidopenam-(*S*)-sulphide with acetic anhydride gave by rearrangement and substitution mainly an acetoxypenam derivative, whilst chloracetic anhydride gave mainly a chloracetoxycepham derivative. The rearrangement is considered to proceed through the sulphenic acid and a cyclic intermediate, of which the direction of ring opening is dependent upon the strength of the acid.

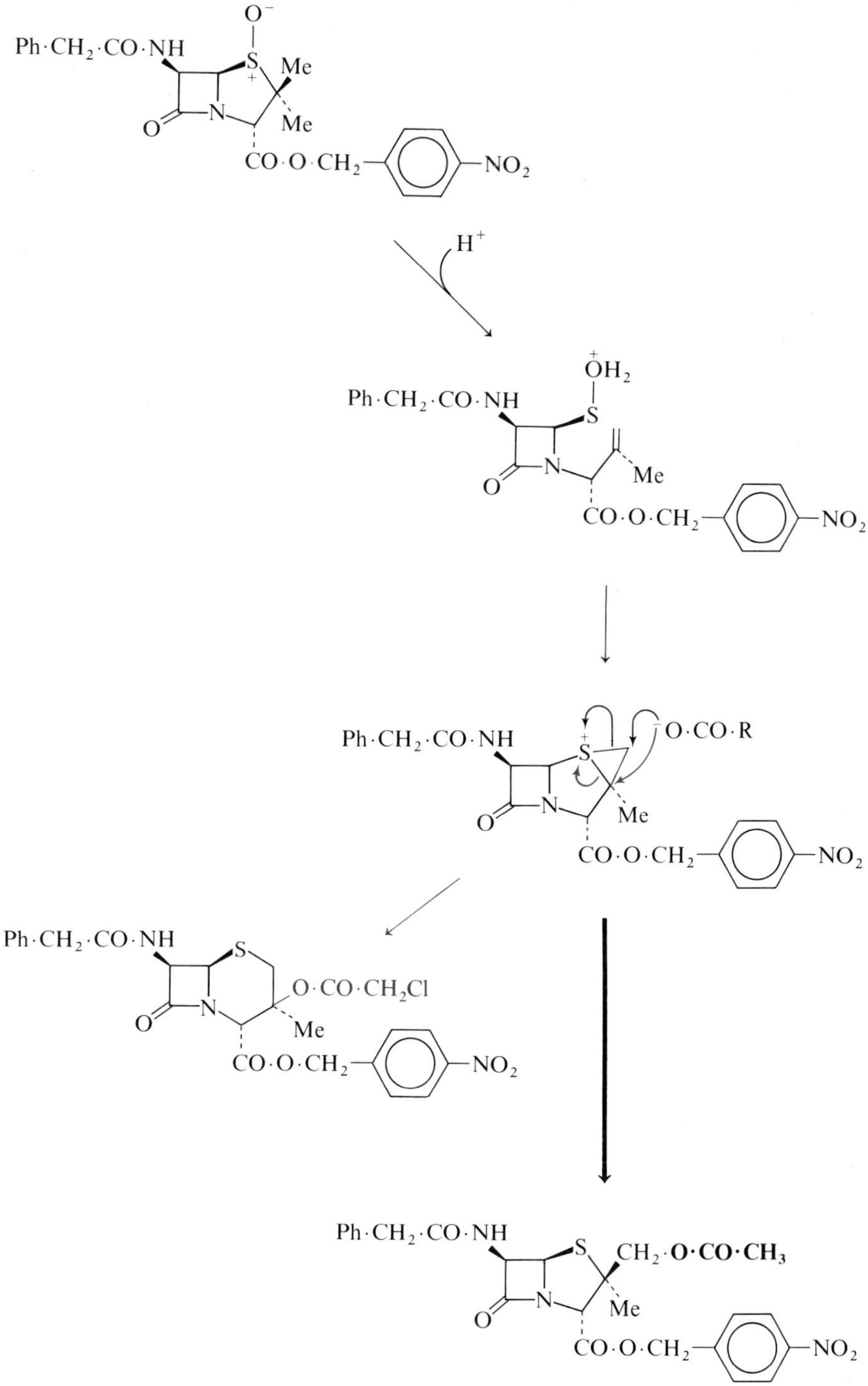

$O^-$

$Ph{\cdot}CH_2{\cdot}CO{\cdot}NH$ S$^+$ Me

O N Me

$CO{\cdot}O{\cdot}CH_2$—$NO_2$

$H^+$

$\overset{+}{O}H_2$

$Ph{\cdot}CH_2{\cdot}CO{\cdot}NH$ S

O N Me

$CO{\cdot}O{\cdot}CH_2$—$NO_2$

$Ph{\cdot}CH_2{\cdot}CO{\cdot}NH$ $\overset{+}{S}$ $^-O{\cdot}CO{\cdot}R$

O N Me

$CO{\cdot}O{\cdot}CH_2$—$NO_2$

$Ph{\cdot}CH_2{\cdot}CO{\cdot}NH$ S

$O{\cdot}CO{\cdot}CH_2Cl$

O N Me

$CO{\cdot}O{\cdot}CH_2$—$NO_2$

$Ph{\cdot}CH_2{\cdot}CO{\cdot}NH$ S $CH_2{\cdot}$**O·CO·CH$_3$**

O N Me

$CO{\cdot}O{\cdot}CH_2$—$NO_2$

## Cephalosporin C Derivatives

*Cephalosporin C* may be regarded as an acyl derivative of the parent 7-amino-cephalosporanic acid in the same way as Cephalosporin N and the penicillins are derived from 6-aminopenicillanic acid. *Cephalosporin C*, unlike the penicillins, fails to undergo a penicillic acid type of rearrangement owing to the much greater stability of its β-lactam ring. Careful treatment with dilute hydrochloric acid (20°C; 3 days) yields a mixture of cephalosporin $C_c$, 7-aminocephalosporanic acid, and the corresponding derivative of cephalosporin $C_c$, which can be separated by a combination of anion exchange and paper electrophoresis.

$^{-}O{\cdot}OC$ / $H_3\overset{+}{N}$ — $CH(CH_2)_3{\cdot}CO{\cdot}NH$ — [bicyclic ring with S, N, O] — $CH_2{\cdot}O{\cdot}CO{\cdot}CH_3$, $CO{\cdot}OH$

Cephalosporin C

$\downarrow H^+$

$H_2N$ — [bicyclic ring with S, N, O] — $CH_2{\cdot}O{\cdot}CO{\cdot}CH_3$, $CO{\cdot}OH$ + $H_2N$ — [tricyclic lactone with S, N, O, O, O]

7-Aminocephalosporanic acid

7-Aminocephalosporanic acid has been converted by acylation to semisynthetic cephalosporins. The benzylpenicillin analogue, *Cephaloram* (Cefaloram; phenylacetylcephalosporanic acid), obtained by treating 7-aminocephalosporanic acid with phenylacetyl chloride, and the *N*-phenoxyacetyl derivative, are about 100 times more active than *Cephalosporin C*, but only one-fifth as active as *Benzylpenicillin* against *Staphylococcus aureus* (*in vitro*). *Cephaloram*, however, is only slightly more effective *in vivo* than *Cephalosporin*, affording little protection to mice infected with *Staphylococcus aureus*, owing to metabolic de-acetylation by body esterases. Similar de-acetylation is brought about by liver de-acetylases in man. *Cephalothin* (Cefalotin; 7-(2-thienylacetyl)-cephalosporanic acid), is a more effective antibiotic, but also suffers from de-activation by de-acetylation. It is not, however, inactivated by β-lactamase, and is useful in the treatment of resistant, β-lactamase-producing staphylococci.

Replacement of a 3-acetoxymethyl group by a 3-methyl substituent in *Cephalexin* (Cefalexin) provides a means of avoiding rapid de-activation. The 7-D-

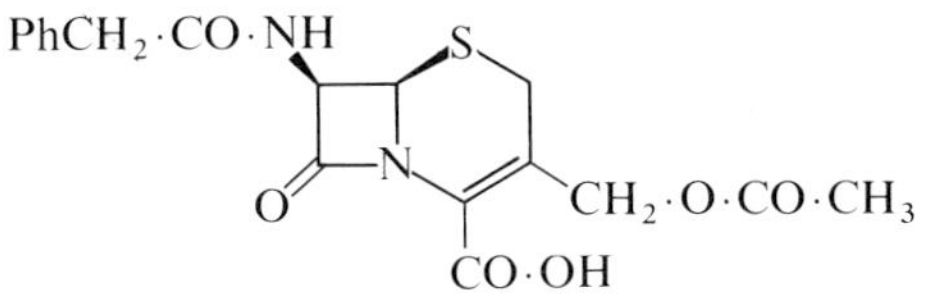

PhCH$_2$·CO·NH S N O CH$_2$·O·CO·CH$_3$ CO·OH

*Cephaloram*

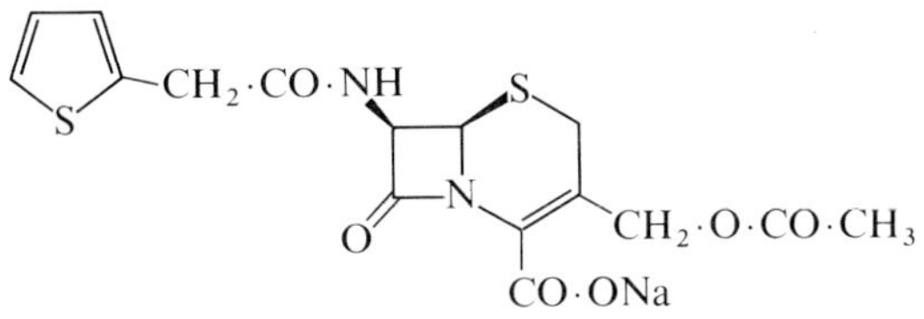

S CH$_2$·CO·NH S N O CH$_2$·O·CO·CH$_3$ CO·ONa

*Cephalothin sodium*

PhCH·CO·NH NH$_2$ S N O Me CO·OH

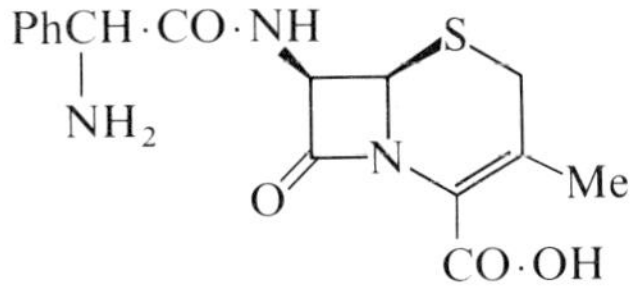

*Cephalexin*

aminophenylacetamido substituent makes the antibiotic effective against Gram-negative as well as Gram-positive organisms.

**Cephaloridine**

Treatment of *Cephalosporin C* with pyridine in acetate buffer yields cephalosporin C$_A$, in which the acetoxy group has been replaced by the pyridinium group (Hale, Newton & Abraham, 1961; Abraham & Newton, 1961; Loder, Newton & Abraham, 1961) which unlike the acetoxy group is not subject to enzymatic cleavage *in vivo*. Alkylation of the tertiary base occurs by virtue of the allylic position of the acetoxy group, and the reaction has been shown to be general for a series of tertiary bases, giving rise to a family of cephalosporin C$_A$ derivatives. Hydrolysis of *Cephalosporin C* with dilute acid cleaves the 7-acyl link to give the parent 7-amino compound, which can also be obtained by the action of aqueous pyridine on 7-aminocephalosporanic acid. *Cephaloridine* is prepared from this compound by direct *N*-acylation with 2-thienylacetic acid, as in the penicillin series (reaction scheme page 582).

*Cephaloridine* (Cefaloridine) is active against a wide range of organisms, including staphylococci, streptococci, coli, salmonella, proteus and shigella species both *in vitro* and *in vivo*. It is highly active against penicillin-resistant strains of *Staphylococcus aureus* on account of its stability to $\beta$-lactamase, which is about 4000 times that of *Benzylpenicillin*. *Cephaloridine* is also highly stable

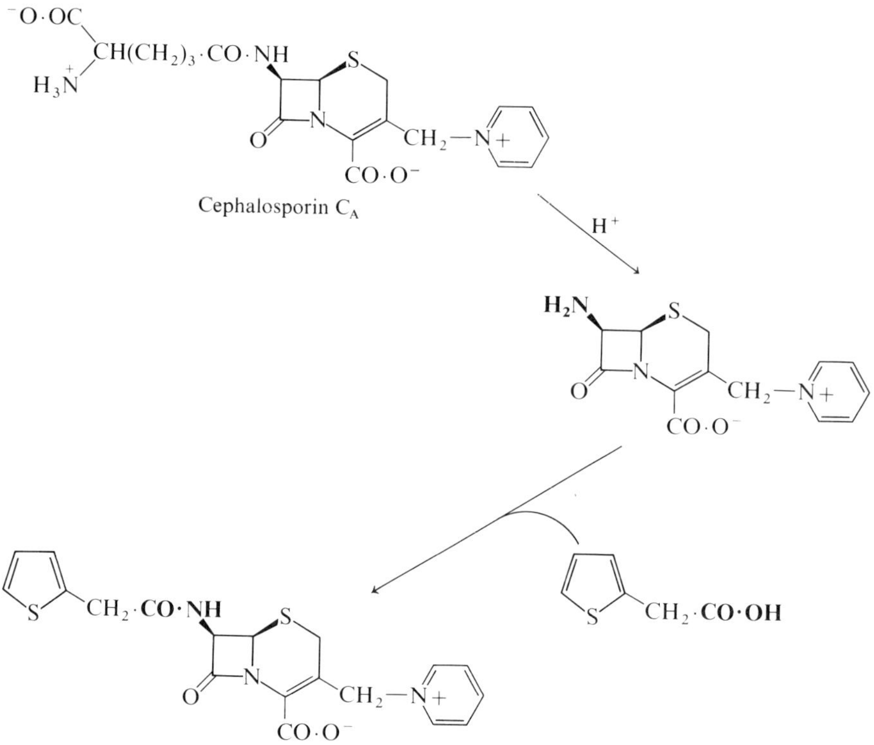

$^{-}O \cdot OC$
$CH(CH_2)_3 \cdot CO \cdot NH$
$H_3\overset{+}{N}$
S
N
O
$CH_2—N+$
$CO \cdot O^-$

Cephalosporin $C_A$

$H^+$

$H_2N$
S
N
O
$CH_2—N+$
$CO \cdot O^-$

S
$CH_2 \cdot$**CO·OH**

S
$CH_2 \cdot$**CO·NH**
S
N
O
$CH_2—N+$
$CO \cdot O^-$

*Cephaloridine*

to acid, and is unaffected in passing through the stomach. Its zwitterionic character leads to poor absorption from the gastro-intestinal tract, hence the need to administer it by injection if it is required for general systemic use.

*Cephazolin Sodium* (Cefazolin Sodium), another cephalosporin derivative, has useful activity against *E. coli* and *Klebsiella pneumoniae*, which are strongly resistant to *Cephalothin* (Cefalotin) and *Cephaloridine* (Kariyone, Harada, Kurita and Takano, 1970).

Hydrolysis of *Cephaloridine* by β-lactamases causes expulsion of pyridine (O'Callaghan, Kirby, Morris, Waller and Dunscombe, 1972). Application of this effect where the leaving group forms the antibacterial agent, 2-mercaptopyridine-*N*-oxide (omadine), has led to the development of the celphalosporin, MCO (O'Callaghan, Sykes and Staniforth, 1976; Greenwood and O'Grady, 1976). MCO has a dual mode of action. Organisms which do not produce an effective amount of β-lactamase are inhibited by the intact cephalosporin. β-Lactamase-producing organisms are also inhibited, though somewhat less effectively by the omadine, which is released in their presence according to the mechanism proposed by Hamilton-Miller, Newton and Abraham (1970). Additionally, because of the 2-mercaptopyridine-*N*-oxide component, MCO, in contrast to other cephalosporins, is active against yeasts and fungi.

Ph·CHOH·CO·NH

S

N

O

CH$_2$—S

N$^+$

O$_-$

CO·OH

Enz〰N̈ucleophile

MCO

Ph·CHOH·CO·NH

S

Enz〰N

N

O$_-$

CH$_2$—S

N$^+$

CO·OH

O$_-$

Ph·CHOH·CO·NH

S

Enz〰N

N

O

CH$_2$

CO·OH

+

$^-$S

N$^+$

O$_-$

# 19 Peptides and Proteins

## INTRODUCTION

Peptides and proteins are formed by multiple condensation of amino acids, each amino acid unit being joined through its carboxyl group with the α-amino group of the next. The simplest examples are dipeptides, consisting of only two amino acid units which may be the same or different, as in glycylglycine or glycylalanine.

$$H_2N{\cdot}CH_2{\cdot}CO{\cdot}NH{\cdot}CH_2{\cdot}CO{\cdot}OH$$

Glycylglycine

$$H_2N{\cdot}CH_2{\cdot}CO{\cdot}NH{\cdot}\underset{\displaystyle CH_3}{\underset{|}{CH}}{\cdot}CO{\cdot}OH$$

Glycylalanine

Although the peptide bond (—CONH—) may be considered to be derived from the constituent amino acids by the elimination of a molecule of water, it can only be prepared indirectly. Dipeptides consisting of the same two amino acid units can be obtained via the amino acid ester and the diketopiperazine which can be formed from it by cyclisation. Partial hydrolysis of the latter gives the dipeptide.

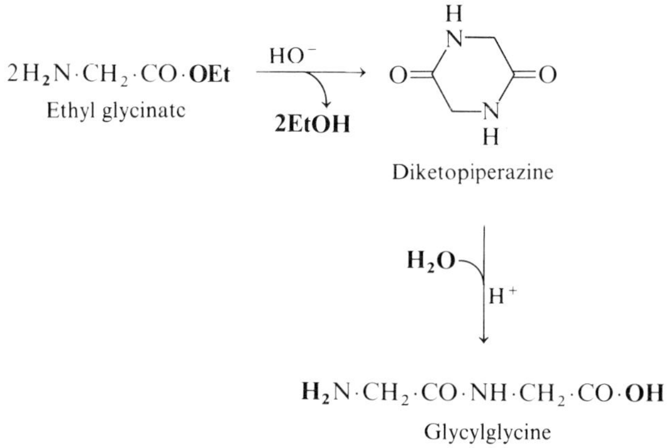

$$2H_2N{\cdot}CH_2{\cdot}CO{\cdot}\mathbf{OEt} \xrightarrow[2EtOH]{HO^-} \text{Diketopiperazine}$$

Ethyl glycinate

Diketopiperazine

$$\xrightarrow[H^+]{H_2O} H_2N{\cdot}CH_2{\cdot}CO{\cdot}NH{\cdot}CH_2{\cdot}CO{\cdot}\mathbf{OH}$$

Glycylglycine

Mixed dipeptides, tripeptides and higher peptides can only be synthesised by more complex processes, which involve protection of the amino group prior to the formation, if only transiently as with dicyclohexylcarbodiimide (Chapter 18), of a reactive derivative of the carboxyl group capable of condensing with the amino group of the next amino acid unit. Many condensation processes are known, but one of the most important for the synthesis of complex peptides and simple proteins is the Merrifield synthesis described in Chapter 18.

The essential distinction between polypeptides and proteins, both of which contain large numbers of amino acid units, is basically one of size. There is no sharp cut-off point between the two classes of compounds, but as a general rule, compounds of molecular weight up to around 6000 are classed as polypeptides. Larger compounds with molecular weights ranging from *ca* 6000 upwards, in some cases to as high as 2 m, are regarded as proteins.

Peptides and proteins are vital constituents of all living matter with important metabolic and structural functions. The tripeptide, glutathione (glutamyl-cysteinyl-glycine), which is present in most living cells, is an important component of many oxidation–reduction systems. Naturally-occurring peptides include such hormones as oxytocin, adrenocorticotrophic hormone (ACTH), and insulin, which contain 9, 39 and 51 amino acid units respectively. Other important hormones, include glucagon, which antagonises the hypoglycaemic effects of insulin; vasopressin, which influences blood pressure; thyrocalcitonin which lowers plasma calcium levels; the angiotensins, and many of the anterior pituitary hormones such as the gonadotrophins, follicle stimulating hormone, luteinising hormone and LHRH (luteinising hormone–releasing hormone; *Gonadorelin*). Certain micro-organisms produce peptide antibiotics, such as *Bacitracin*, *Gramicidin*, *Viomycin* and valinomycin, which inhibit the growth of other organisms.

The structure and properties of proteins is very much related to their function. Those with structural functions such as muscle proteins (myosins), the proteins of skin and connective tissue (collagens), and of hair and nails (keratins) are linear or sheet-like, fibrous, and insoluble in water and salt solutions. On heating with water, however, collagen is converted into gelatin, which is water-soluble. In contrast, the polypeptide chains of many enzymic proteins and the blood proteins, albumin and globulin, are folded and, therefore, globular in shape. Many enzymes are water-soluble, though some are membrane-dependent and hence insoluble. Blood plasma proteins, likewise, are differentiated mainly by their solubilities, though also by other properties. Thus, whereas albumin is soluble in water, the globulins are insoluble, but soluble in dilute salt solutions. Other water-soluble proteins include the histones of red blood corpuscles and leucocytes, and protamines of fish, which give alkaline solutions due to the high proportion of basic amino acids (diamino acids).

A number of important proteins occur in conjugation with other molecular species. These **conjugated proteins** include phosphoproteins, lipoproteins, nucleoproteins and glucoproteins. Phosphoproteins contain phosphoric acid. In consequence, they are insoluble in water, but soluble in alkali and re-precipitated on acidification. Casein of milk and vitelline of egg-yolk are typical phosphoproteins. Lipoproteins, which are the principal fat-carrying proteins of blood plasma are of two types, high-density ($\alpha$-lipoproteins) and low-density ($\beta$-lipoproteins). The two high-density $\alpha$-lipoproteins contain some 67 and 43% of lipids respectively, consisting mainly of phospholipids (*ca* 45%), free (10%) and esterified (30%) cholesterol, and about 15% of neutral fat. The low density $\beta$-lipoproteins, on the other hand, contain a much higher proportion of lipids (80–95%) consisting mainly of either neutral fat or cholesterol. Blood plasma at certain times, particularly after the absorption of fat from the intestines, also contains chylomicrons which consist of about 99% lipid, associated with a very small amount of protein. Glycoproteins, in which the protein is associated with carbohydrates, occur in blood plasma, and in mucins in saliva. The cell walls of bacteria (Chapter 21), and certain antibiotics such as *Vancomycin* (McCormick, Stark, Pittenger, Pittenger and McGuire, 1956; Marshall, 1965) may be considered as glycopeptides. Protein-bound carbohydrates in blood plasma consist mainly of galactose, mannose, glucosamine, and galactosamine. Nucleoproteins, in which protein is associated with nucleic acids, occur in the nuclei of cells. They are weakly acidic, and soluble in water and salt solutions.

Despite an enormous range in the properties and function of natural mammalian peptides and proteins, they are all composed, without exception, of no more than about twenty different L-amino acids. The cell walls of bacteria, which consist of peptide-linked carbohydrates (Chapter 21) and certain peptide antibiotics, such as *Gramicidin S*, valinomycin and *Actinomycin D* (Dactinomycin), contain both D- and L-amino acids, including some of somewhat unusual structure. The enormous range in size of these molecules, many consisting of large numbers of amino acid units, permits incorporation of the individual amino acids in a countless variety of ways. Their physical and chemical properties, their shape, the specificity of action of enzymes, hormones, and antibiotics, and the specific characteristics of the various structural proteins, stem in each case essentially from the specific ordering of the component amino acids.

The order in which the individual amino acids are linked is known as the **primary structure**. This not only determines the stability of the chain to chemical and enzymic degradation, but also pre-determines certain of the intra- and intermolecular constraints to which the resulting peptide chain is subjected. These give rise to structural features, known as **secondary structure**, which are characteristic of the peptide. The coiled $\alpha$-helix, and $\alpha$- and $\beta$-pleated sheets are typical examples of secondary structure. Chain-folding, stabilised by disulphide bridging, hydrogen bonding and other cross-linking devices, which is typical of the globular soluble proteins, is known as **tertiary structure**. Some globular protein molecules as, for example, haemoglobin, are further associated in multiple units, giving rise to an element of **quaternary structure**.

The complete characterisation of individual peptides and proteins requires the establishment of all elements of its primary and secondary structure, and where evident, tertiary and quaternary structure as well. Structural elucidation, however, is only possible if completely homogeneous materials are available for study. Consequently, isolation, purification and determination of homogeneity are essential preliminaries to any such study.

**Isolation and Purification**

Extraction and purification of peptides and proteins are essential preliminaries to any study of their basic amino acid composition. Primary structure is not readily disrupted chemically except by hydrolysis, but changes in secondary and tertiary structure are often readily effected by heat, pH shifts and addition of chemical reagents. The protein is then said to be **denatured**. Denaturation may or may not be reversible. Irreversible denaturation leads to permanent changes in the physical and biological properties of the protein (p. 611). Changes in solubility arising as a result of mild reversible denaturation, however, do no permanent damage, and are particularly useful in the isolation and purification of proteins.

If the protein is cellular in origin, then some preliminary disruption of cellular structure is almost certainly essential. This is usually best accomplished mechanically, by grinding, shaking with glass beads, homogenisation, or by ultrasonic vibration, since physical methods are least likely to degrade the peptide chain. Lipid material is next removed by extraction with organic solvents such as ether, petrol or hexane in which the protein is insoluble, and the protein then extracted into dilute salt solution. Low molecular weight impurities and electrolytes are removed by dialysis, by electrodialysis in an applied field, or by ultracentrifugation through semi-permeable membranes. Certain enzymic proteins, e.g. glucuronyltransferase, are attached to cellular membranes by phospholipid components, and only retain their specific enzymic activity whilst so attached. They may, however, be solubilised by the action of phospholipase, and reactivated by addition of phospholipid (**2**, 5). Fractionation of partially purified peptides and proteins is achieved by a variety of techniques which take advantage of differences in physical and chemical properties of naturally associated, but different fractions. The solubility of proteins is markedly influenced by pH, and by the ionic strength, dielectric constant and temperature of the solvent. Solubility is usually at a minimum at the isoelectric point, and is further depressed by increasing the ionic strength, and by reducing the dielectric constant of the solvent by addition of non-ionic solvents such as ethanol (Fevold, 1951). Repeated fractionation by manipulation of one or more of these factors, either separately or in combination, is usually capable of effecting a high degree of purification, though complete homogeneity is rarely achieved without resort to more refined physical separation techniques. On a manufacturing scale, however, reasonable homogeneity can often be achieved with general fractionation procedures, though in a few cases the formation of peptide or protein specific complexes, such as the crystallisation of insulin in the presence of zinc chloride, provides a basis for further purification of the product. So-called mono-component insulins are further purified by gel filtration (Brandenburg, 1969). A typical example of solvent/pH fractionation is set out in Fig. 34, which shows the fractionation of human blood plasma proteins by means of solvent ether and pH adjustment.

Small-scale isolation procedures, where a much higher degree of homogeneity is essential as a preliminary to analysis, make use of a variety of modern physical separation techniques, which include adsorption and ion-exchange chromatography, gel filtration, electrophoresis and ultracentrifugation. The last step in the isolation of purified peptide and protein fractions, the removal of the solvent, is usually effected by freeze-drying or by cold precipitation with absolute ethanol, and drying under vacuum (Taylor, 1953). Many criteria of protein purity have been proposed, but ion-exchange and zone electrophoretic characteristics are considered to be the most satisfactory (Lontie, 1962).

## PRIMARY STRUCTURE

Although many proteins are of extremely high molecular weight, all natural peptides and proteins are composed exclusively of no more than twenty different amino acids. Individual peptides and proteins differ, of course, in the particular amino acids present, in the numbers of each unit present, and in the order in which they are linked. Despite, therefore, the small number of different amino acids involved, the number of alternative combinations is almost infinite.

The properties of peptides and proteins are determined both by their amino acid composition, and by the order in which the amino acids are arranged. In particular, the presence of diamino acids, guanidino amino acids, and amino-dicarboxylic acids in the peptide chain, confers basic and acidic properties resulting from the presence of uncombined amino, guanidino and carboxylic groups. Histamine, with its imidazoline ring system, can also act as a proton acceptor. The balance between basic and acidic side-chains determines whether the protein is essentially acidic, basic or neutral in character, and hence whether its isoelectric point is above or below pH 7. The isoelectric point, the pH at which the protein is electrically-neutral, represents the pH at which the maximum number of basic and acidic groups are ionised. For this reason, peptides and proteins are usually most readily salted out of solution at their isoelectric point.

$$+NH_3 \quad CO{\cdot}OH \xleftarrow{H^+} +NH_3 \quad CO{\cdot}O^- \xrightarrow{HO^-} NH_2 \quad CO{\cdot}O^-$$

isoelectric point

The presence of acidic and basic amino acids also enables the compound to act as a proton donor or acceptor, thereby explaining the catalytic rôle of enzymes in acid- and base-catalysed reactions. Similarly, the hydroxyl group of serine is capable of activation by hydrogen bonding to serve as a powerful nucleophile as, for example, at the active site of acetylcholinesterase (**2**, 1), whilst cysteine provides a sulphydryl group, which is not only reactive as such, but a potential centre for intramolecular cross-linking by the formation of disulphide bonds. Hydrocarbon groups of neutral amino acids are important factors in determining the extent to which proteins containing them show affinity for lipid molecules by hydrophobic bonding (**2**, 4).

### Amino Acid Composition

The individual amino acids present in any peptide or protein are readily identified by hydrolysis followed by two-dimensional partition chromatography on paper or thin-layer. Each amino acid moves at a characteristic rate in relation to the solvent front, which is expressed as an $R_F$ value. Development in two directions consecutively at right angles spreads out the amino acids more than a single development in one direction only, and makes for greater accuracy in identification. Acid hydrolysis with hot *10N* mineral acid (HCl or $H_2SO_4$) for 18–24 hr is usually used. Some destruction of more sensitive amino acids occurs, particularly tryptophan, cystine, serine and threonine, and hydrolysis is accompanied by the formation of a black solid (humin). For this reason, the recovery of amino acids is neither quantitative nor precisely stoichiometric. The extent of decomposition of sensitive amino acids under normal hydrolytic conditions is well-established, so that the correct stoichiometry can be determined by the use of appropriate correction factors in the quantitative analysis of the separated amino acids. Automated methods of amino acid analysis are now frequently used in this type of work.

Tryptophan is more stable to alkaline hydrolysis with *6N* sodium or barium hydroxide, which is used occasionally as an alternative. Alkaline hydrolysis, however, leads to considerable decomposition of serine and threonine, also cysteine and arginine, and further is more likely to cause racemisation. The amino acid composition of a number of important peptide hormones and proteins is given in Table 61.

The molecular weight of peptides and proteins is often difficult to determine precisely. Various physical methods which can be used include sedimentation by ultracentrifugation, diffusion, osmotic pressure, viscosity and light-scattering measurements in solution. Approximate molecular weights, so obtained, give some indication of the total number of amino acid residues present, but may conceal association giving rise to quaternary structure, as in the case of insulin. Thus, some physical methods gave the molecular weight of insulin as 24 000 and 48 000 (Behrens and Bromer, 1958), but these were later shown to be aggregations of units of molecular weight 12 000. Still later, the true molecular weight of insulin was shown to be 6000 by counter-current fractionation of the dinitrophenylated hormone. The unit of molecular weight 12 000 is considered to be that of the zinc complex in which two molecules of insulin are associated with one atom of zinc (page 609).

### *N*-Terminal Amino Acids

Various techniques have been used in the all-important sequence analysis of peptide chains. These include the identification of *N*-terminal and *C*-terminal amino acids, sequential removal of *N*-terminal residues, and partial chemical or

Table 61. Amino Acid Composition (%) of some Peptides and Proteins

| | Insulin | ACTH | Albumin | γ-Globulin | Fibrinogen |
|---|---|---|---|---|---|
| Glycine | 4.3 | 8.0 | 1.6 | 4.2 | 5.6 |
| Alanine | 4.5 | — | — | — | 3.7 |
| Valine | 7.8 | 3.4 | 7.7 | 9.7 | 4.1 |
| Leucine | 13.2 | 7.8 | 11.0 | 9.3 | 7.1 |
| Isoleucine | 2.8 | 3.1 | 1.7 | 2.7 | 4.8 |
| Serine | 5.2 | 6.0 | 3.7 | 11.4 | 7.0 |
| Threonine | 2.1 | 3.2 | 5.0 | 8.4 | 6.1 |
| Proline | 2.5 | 8.2 | 5.1 | 8.1 | 5.7 |
| Cysteine | 12.5 | 7.2 | 0.7 | 0.7 | 0.4 |
| Cystine | — | — | 5.6 | 2.4 | 2.2 |
| Methionine | — | 1.9 | 1.3 | 1.1 | 2.5 |
| Phenylalanine | 8.1 | 4.0 | 7.8 | 4.6 | 4.6 |
| Tyrosine | 13.0 | 2.4 | 4.7 | 6.8 | 5.5 |
| Tryptophan | — | — | 0.2 | 2.9 | 3.3 |
| Arginine | 3.1 | 8.7 | 6.2 | 4.8 | 7.8 |
| Lysine | 2.5 | 5.0 | 12.3 | 8.1 | 9.2 |
| Histidine | 4.9 | 1.3 | 3.5 | 2.5 | 2.6 |
| Aspartic acid | 6.8 | 6.7 | 10.4 | 8.8 | 13.1 |
| Glutamic acid | 18.6 | 15.6 | 17.4 | 11.8 | 14.5 |

enzymic hydrolysis to smaller identifiable peptide units. The identification of *N*-terminal residues by reaction with 2,4-dinitrofluorobenzene (Sanger, 1945; 1952) has already been described (Chapter 18). In many respects, however, the Edman (1950) phenylthiocarbamyl (PTC) method in which the *N*-terminal amino acid is removed as a thiohydantoin is more useful, since *N*-terminal residues can be split off sequentially. The PTC-peptide generated by reaction with phenylisothiocyanate in aqueous solution at pH 9, is readily cleaved by dilute acid to yield the thiohydantoin of the *N*-terminal amino acid and the

$$C_6H_5\cdot\mathbf{N{=}C{=}S} + \mathbf{H_2N}\cdot CHMe\cdot CO\cdot NH\cdot CHR\cdot CO—$$

$$\downarrow \text{pH}_9$$

$$C_6H_5\cdot\mathbf{NH\cdot CS\cdot NH}\cdot CHMe\cdot CO\cdot NH\cdot CHR\cdot CO—$$

$$\downarrow H^+$$

$$\text{(thiohydantoin ring: } C_6H_5\cdot N\text{–CO–CH(Me)–NH–C(=S)–)} + H_2N\cdot CHR\cdot CO—$$

Substituted thiohydantoin

*des-N*-terminal peptide. The thiohydantoin is soluble in organic solvents, and is readily extracted and identified by paper chromatography in comparison with authentic material, whilst the *des-N*-terminal peptide is available for further degradation.

### *C*-Terminal Amino Acids

*C*-Terminal residues of peptides are identified by the use of carboxypeptidase, an enzyme present in beef pancreas. Carboxypeptidase is specific for the hydrolysis of peptide links adjacent to a free carboxyl group, and its peptidase activity is blocked by *C*-terminal amides and esters, and also by certain specific amino acids in the *C*-terminal position (lysine, proline and arginine).

$$H_2N\cdot CHR''\cdot CO\cdot NH\cdot CHR'\cdot \mathbf{CO\cdot NH}\cdot CHR\cdot CO\cdot OH$$

$$\downarrow \text{Carboxypeptidase}$$

$$H_2N\cdot CHR''\cdot CO\cdot NH\cdot CHR'\cdot \mathbf{CO\cdot OH} + \mathbf{H_2N}\cdot CHR\cdot CO\cdot OH$$

The enzyme is activated by $Zn^{2+}$ ions, which are thought to form a complex with the terminal carboxylate ion and the enzyme protein, the adjacent peptide link undergoing hydrolysis (Valee, Rupley, Coombs and Neurath, 1960). Comparison of the stability constant for the complexing of carboxypeptidase to $Zn^{2+}$ with similar stability constants for a number of model compounds indicates that both a cysteine sulphydryl and an amino group are implicated in the binding of the metal ion. Enzyme inhibition by terminal lysine and arginine residues is thus perhaps explained in terms of competition of their free amino groups with the enzyme amino group.

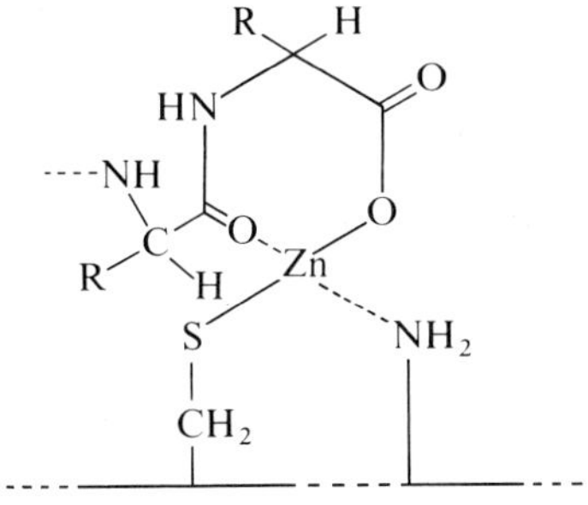

Non-enzymic methods for the detection of *C*-terminal residues include hydrazinolysis, and reduction of the terminal carboxylate group. In hydrazinolysis, all peptide links, but not the *C*-terminal carboxylic acid groups, are attacked by hydrazine hydrate to form aminoacyl hydrazines. These are readily separated from the *C*-terminal acid, which is thus readily identifiable.

$$H_2N\cdot CHR''\cdot\mathbf{CO\cdot NH}\cdot CHR'\cdot\mathbf{CO\cdot NH}\cdot CHR\cdot CO\cdot OH$$

$$\xrightarrow{H_2N\cdot NH_2}$$

$$H_2N\cdot CHR''\cdot\mathbf{CO\cdot NH\cdot NH_2} + \mathbf{H_2N}\cdot CHR'\cdot\mathbf{CO\cdot NH\cdot NH_2} + \mathbf{H_2N}\cdot CHR\cdot CO\cdot OH$$

Aminoacylhydrazines | *C*-Terminal amino acid

Reductive identification of the *C*-terminal amino acid requires esterification of the terminal carboxylic acid group, reduction with sodium or lithium borohydride to the corresponding primary alcohol, and hydrolysis of the modified peptide with subsequent identification of the *C*-terminal-derived amino alcohol (Chibnall and Rees, 1958).

$$H_2N\cdot CHR''\cdot CO\cdot NH\cdot CHR'\cdot CO\cdot NH\cdot CHR\cdot\mathbf{CO\cdot OH}$$

1. MeOH/HCl
2. $NaBH_4$
3. $H^+$

$$\downarrow$$

$$H_2N\cdot CHR''\cdot CO\cdot OH + H_2N\cdot CHR'\cdot CO\cdot OH + H_2N\cdot CHR\cdot\mathbf{CH_2OH}$$

**Sequential Analysis**

Complete sequential analysis requires partial rather than complete hydrolysis to cleave the polypeptide unit into smaller peptides with common amino acid residues. Partial hydrolysis can be achieved by treatment with dilute acid under conditions which are less severe than those used for complete breakdown of the peptide chain. Partial hydrolysis can also be achieved enzymatically by the use of specific peptidases which cleave the chain preferentially at peptide links associated with particular amino acids. Carboxypeptidase, already discussed (p. 591), is specific for the cleavage of *C*-terminal peptide links. For activation, it requires the formation of a metal ion ($Zn^{2+}$) complex involving the *C*-terminal peptide bond and the adjacent *C*-terminal carboxylic acid group. It is particularly efficient in the cleavage of *C*-terminal peptide bonds associated with aromatic amino acids. In a similar way, leucine aminopeptidase is specific for the cleavage of *N*-terminal peptide links, and for activation requires the formation of a divalent metal ion complex involving the peptide nitrogen and adjacent *N*-terminal amino group.

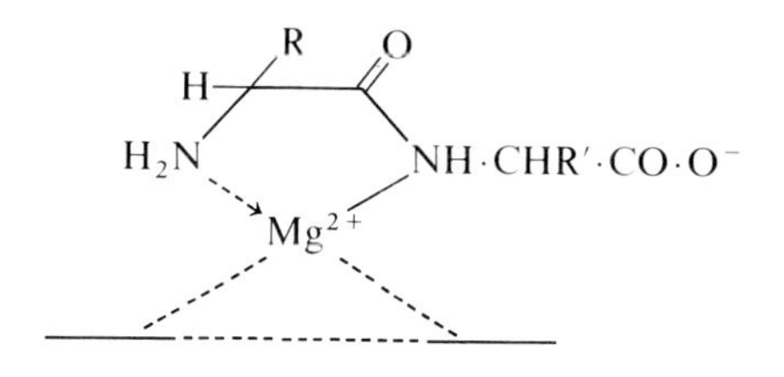

Other amino acid peptidases include **Pepsin**, which preferentially hydrolyses peptide bond linkages involving tyrosine, phenylalanine and tryptophan; **Chymotrypsin**, which similarly is specific for peptide bonds derived from aromatic amino acids; and **Trypsin**, which is specific for peptide bonds derived from the basic amino acids, lysine and arginine.

Treatment with ethyl trifluorothioacetate followed by trypsin provides a refinement which distinguishes between the peptide links of lysine and arginine. Of the two, only lysine reacts with this reagent through its ε-amino group. The resulting trifluoroacetamide is resistant to trypsin hydrolysis, so that cleavage occurs only at the arginine peptide bonds.

$$CF_3{\cdot}CO{\cdot}SEt + H_2N{\cdot}CH_2{\cdot}CH_2\text{---} \longrightarrow CF_3{\cdot}CO{\cdot}NH{\cdot}CH_2{\cdot}CH_2\text{---}$$

The residual trifluoracetyl-lysylpeptides can be degraded with alkali to remove the enzyme-blocking trifluoroacetyl group, and on subsequent trypsinisation then undergo specific cleavage at the lysine-peptide bonds.

Trypsin can also be made to cleave at cysteinyl residues which have been modified chemically to lysine-like residues. One such method of modification is the aminomethylation of cysteine to thialysine with ethylenimine (Cole, 1967).

H—S + $CH_2$—$CH_2$ (N–H ring), cysteinyl residue —NH—CH($CH_2$SH)—CO—NH— —Trypsin→ ✗ (no cleavage)

↓ pH 8.6

$CH_2{\cdot}NH_2$—$CH_2$—S—$CH_2$— on —NH—CH—CO┆NH— (Thialysylpeptide; Trypsinisation)

compare

$CH_2NH_2$—$CH_2$—$CH_2$—$CH_2$— on —NH—CH—CO┆NH— (Lysine; Trypsinisation)

An alternative method is the *S*-carboxamidomethylation of cysteine with iodoacetamide to give a *S*-carboxamidomethylcysteinyl peptide, which is also readily hydrolysed by trypsin.

$$\begin{array}{c} SH \\ | \\ CH_2 \\ | \\ -NH-CH-CO-NH- \end{array} \quad I\cdot CH_2\cdot CO\cdot NH_2 \longrightarrow \begin{array}{c} CO\cdot NH_2 \\ | \\ CH_2 \\ | \\ S \\ | \\ CH_2 \\ | \\ -NH-CH-CO \vdots NH- \end{array}$$

**Trypsinisation**

*S*-Carboxyamidomethylcysteinyl peptide

Similarly, *O*-glycylserines [$H_2N \cdot CH_2 \cdot CO \cdot OCH_2 \cdot CH(NH_2)CONH$---] also serve as substrates for trypsin.

Whatever methods of chain degradation are used, sequential analysis rests on the detection of overlapping peptide fragments relating to one and the same portion of the entire peptide chain. This is best illustrated by reference to the hydrolysis products of an imaginary hexapeptide found to contain tyrosine, leucine, phenylalanine, valine, histidine and glycine. If the *N*-terminal amino acid is identified as tyrosine, and partial hydrolysis gave the following fragments:

Val-Gly
Tyr-Leu
Val-Gly-Phe
Leu-His-Val

then it could be deduced that the correct peptide sequence is:

Tyr-Leu-His-Val-Gly-Phe

## SECONDARY STRUCTURE

The secondary structure of peptides and proteins defines the shape assumed by the polypeptide chain. Low molecular weight peptides are usually flexible, suffering only the constraints imposed by the fixed geometry of the peptide bond. Such conformations are usually described as **unordered** or **random coil**. More complex peptides and proteins usually adopt defined secondary structures as a result of various constraining factors. These include the *trans* geometry of the peptide bond, and hydrogen bonding between the amidic hydrogen of one peptide bond and the carbonyl group of another. Such hydrogen bonding may be either intramolecular or intermolecular. Intramolecular hydrogen bonding is the key factor in stabilising the spiral conformation of the **α-helix**, whilst intermolecular hydrogen bonding gives rise to the laminated structures of the **parallel** and **antiparallel β-pleated sheet** structures. Intramolecular cyclisation to form cyclic peptides, as in *Oxytoxin* and *Gramicidin S*, and cyclic depsipeptides (ester-linked peptides) also gives rise to products with defined secondary structures, which are also frequently stabilised by intramolecular hydrogen bonding.

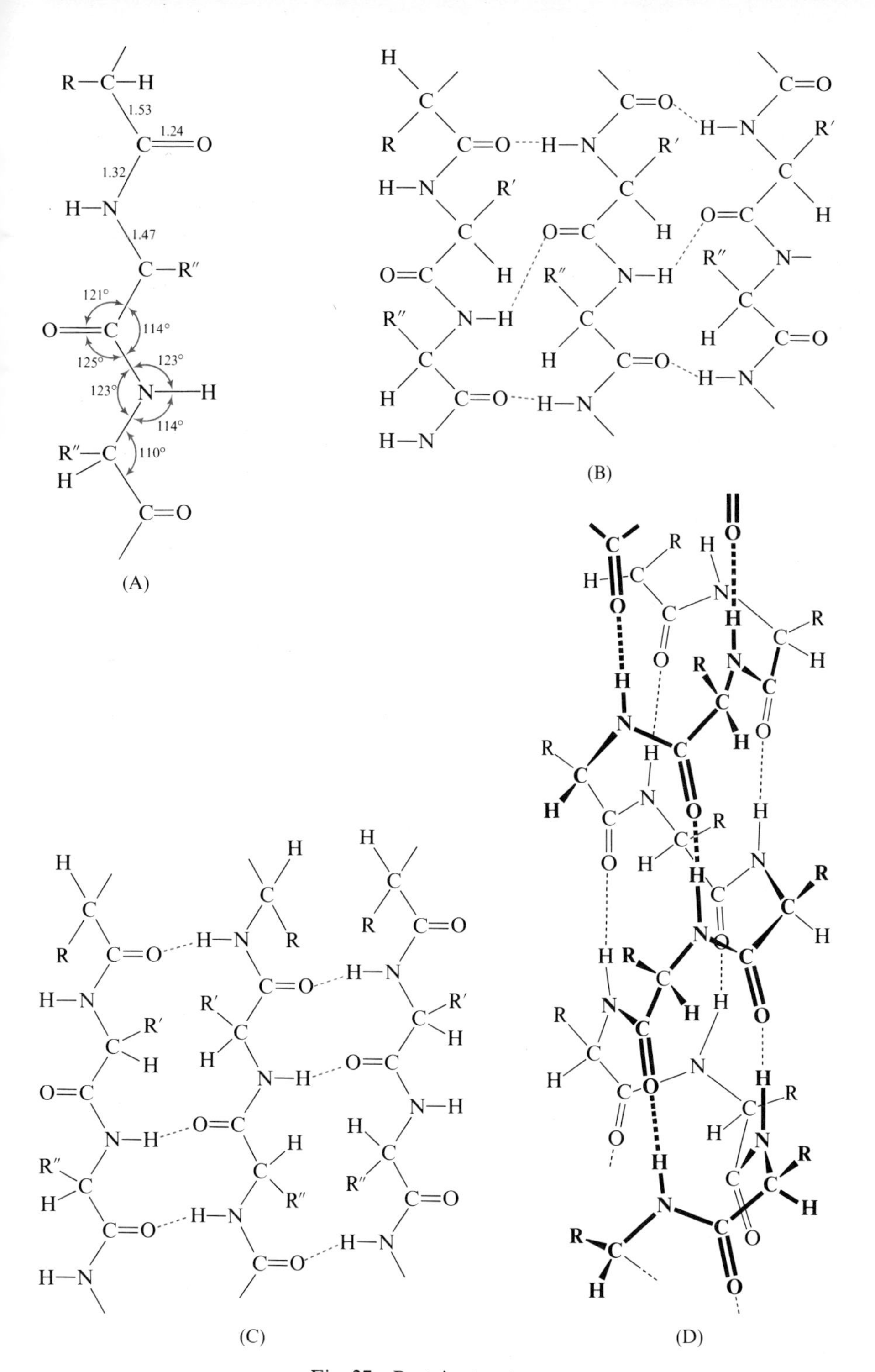


Fig. 27 Protein structures

## Random Coil Structures

Small peptides with less than five amino acid residues do not as a rule adopt helical or other organised conformations, probably because they are unable to generate a sufficient number of intramolecular hydrogen bonds to stabilise the conformation. This is evident from their amide carbonyl infrared absorption frequencies (p. 598), and is supported by their optical rotatory dispersion (Goodman, Schmitt and Yphantis, 1962) and circular dichroism spectra (Timasheff and Gorbunoff, 1967), which arise from light absorption by functional groups adjacent to asymmetric $\alpha$-carbon atoms.

There is evidence that small peptides do not necessarily adopt random coil structures, but show definite preferences for particular non-helix conformations. Thus, the tetrapeptide BOC-Val-Val-Ala-Gly-OEt shows secondary structures in both methanol and chloroform, which appears to be $\alpha$-helical and $\beta$-sheet respectively (Shields and McDowell, 1967).

## α-Helical Structures

The $\alpha$-helical structure of $\alpha$-keratin and soluble proteins is characterised by coiling of the peptide chain backbone in either a left-handed or right-handed spiral structure. The helix is stabilised by internal C—O···H—N bonds lying parallel to the axis of the coil and between individual turns of the coil (Pauling, 1960). As a result, each amino acid residue is twisted about its $C_\alpha$—C(O)—N(H)—$C_\alpha$ axis, so that the $\alpha$-carbon atoms lie at a radius of 2.3 Å from the axis of the coil, with an angular rotation of approximately 105° about the axis per amino acid residue. There are, therefore, about 3.6 amino acid units for each complete turn of the helix, and the pitch, the distance along the axis in which one complete turn of the helix is achieved, is 5.4 Å. The hydrogen bonds which stabilise the helix are formed between peptide carbonyl and amido hydrogens in adjacent turns of the spiral, i.e. between the first and fifth, second and sixth amino acid units, and so on. The right-handed $\alpha$-helix is theoretically more stable (Fig. 27D).

The precise helical structure of any peptide can be defined by the rotational angles about each of the two single bonds linking each pair of adjacent peptide groups to their intervening $\alpha$-carbon atom. Thus, in the dipeptide chain the

$$\mathrm{C_{(1)}—\underset{\underset{O}{\|}}{C_{(1')}}—\overset{\overset{H}{|}}{N_{(1)}}—C_{(2)}—\underset{\underset{O}{\|}}{C_{(2')}}—\overset{\overset{H}{|}}{N_{(2)}}—C_{(3)}}$$

twisting of the two peptide groups about the $N_{(1)}$—$C_{(2)}$ and $C_{(2)}$—$C_{(2')}$ bonds is denoted by two dihedral angles $\phi$ and $\phi'$ respectively. Each pair of dihedral angles $\phi$ and $\phi'$ expresses the extent of rotation with respect to a standard conformation ($\phi = \phi' = 0°$) when the planes of the two peptide groups and the $N_{(1)}$—$C_{(2)}$—$C_{(2')}$ group coincide, and the two peptide NH groups face each other (Ramachandran, Ramakrishnan and Sasisekharan, 1963).

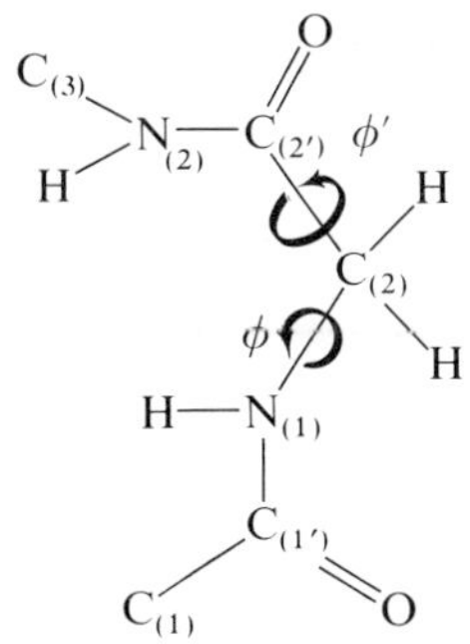

**β-Pleated Sheet Structures**

In the pleated sheet structures, the peptide chain is slightly contracted compared with the fully extended structure, the amino acid repeat distance being about 6.7 Å between successive $\alpha$-carbon atoms (7.23 Å in the fully extended structure). Two types of pleated sheet structure are possible (Pauling, 1960); one in which the peptide chains are parallel (Fig. 27B) with a separation of 6.5 Å, as in $\beta$-keratin, and the other with an antiparallel arrangement (Fig. 27C) of peptide chains with a separation of 7.0 Å as seen in tussah silk fibroin and $\beta$-poly-L-alanine.

**The *trans* Peptide Bond**

The constraints imposed on the completely free rotation about the

$$—C(O)—N—C_\alpha—C(O)—$$

bonds, which form the backbone of the peptide chain, are important factors in the determination of secondary peptide structure. These constraints arise primarily from the properties of the amide bond, and to a lesser extent, hydrogen bonding between spacially adjacent groups. Thus, although free rotation about the N—$C_\alpha$ bond is possible, even in the simplest peptides, the amide C—N bond has a considerable element of double bond character, due to steric hindrance and resonance contributions from charged forms. This is revealed by X-ray diffraction which shows considerable shortening of the C—N bond, coplanarity with adjacent elements, and normally a *trans* arrangement of the amidic >C=O and >N—H functions (Chapters 11 and **2**, 3). In practice, however, the amide bonds of complex proteins experience significant deformation from the *all-trans* state as a result of the protein adopting various stabilised conformations.

**Hydrogen Bonding**

A typical, fully extended peptide chain with *all-trans* amide link requirements is shown in Fig. 27A (Pauling and Corey, 1951). There is, however, considerable evidence from X-ray crystallographic studies and deuterium exchange in solution to suggest that hydrogen bonding between amide carbonyl and amide

Table 62. Amide I Infrared Frequencies and Circular Dichroism of Protein Conformations

| Conformation | Amide I infrared absorption frequency ($cm^{-1}$) | Circular dichroism: Peak (nm) | Circular dichroism: Molecular elipticity $[\theta] \times 10^{-3}$ deg. $cm^2$/decimol |
|---|---|---|---|
| α-Helix | 1650 (s) | 221–2 | −30 |
| | 1646 (w) | 207 | −28 to −29 |
| | | 190–2 | 52–55 |
| Unordered | 1658 (s) | 235–8 | *ca* −0.2 |
| | | 217 | *ca* 2 |
| | | 196–7 | −25 to −35 |
| Anti-parallel β-pleated sheet | 1685 (w) | 217 | −14 to −17 |
| | 1632 (s) | 195 | 21 |
| | 1668 (vw) | | |
| Parallel β-pleated sheet | 1648 (w) | 216 | — |
| | 1632 (s) | 181 | — |

hydrogen, >C=O···H—N<, plays an important part in the stabilisation of secondary protein structure. These hydrogen bonds, which are linear, with an almost constant nitrogen to oxygen bond distance of 2.8 ± 1 Å may be either intermolecular giving rise to the flat pleated sheet structures of the fibrous proteins, β-keratin and tussah silk fibroin, or, alternatively intramolecular, giving rise to the α-helical structure characteristic of α-keratin and many soluble proteins.

The double bond character of the amide carbonyl group is modified by hydrogen bonding, and its infrared stretching frequency is lowered. Interpretation of the amide I (1600 to 1700 $cm^{-1}$; C=O stretching) and amide II (1500 to 1550 $cm^{-1}$; N—H plane bending and C—N stretching) bands in the infrared spectra of peptides and proteins therefore provides a means of distinguishing between random coil and the various ordered conformations (Fig. 27) of α-helical, parallel β-pleated sheet and anti-parallel β-pleated sheet structures (Miyazawa, 1960; Krimm, 1962). Further valuable information on protein conformation can also be derived from the circular dichroism spectra. Table 62 shows some calculated amide I frequencies and the diagnostic circular dichroism peaks and intensities (Timasheff and Gorbunoff, 1967) for the principal types of protein conformation.

### Trans-annular Hydrogen Bonding

Hydrogen bonding is also responsible for stabilising particular conformations which many cyclic peptides and cyclic depsipeptides (ester-linked peptides)

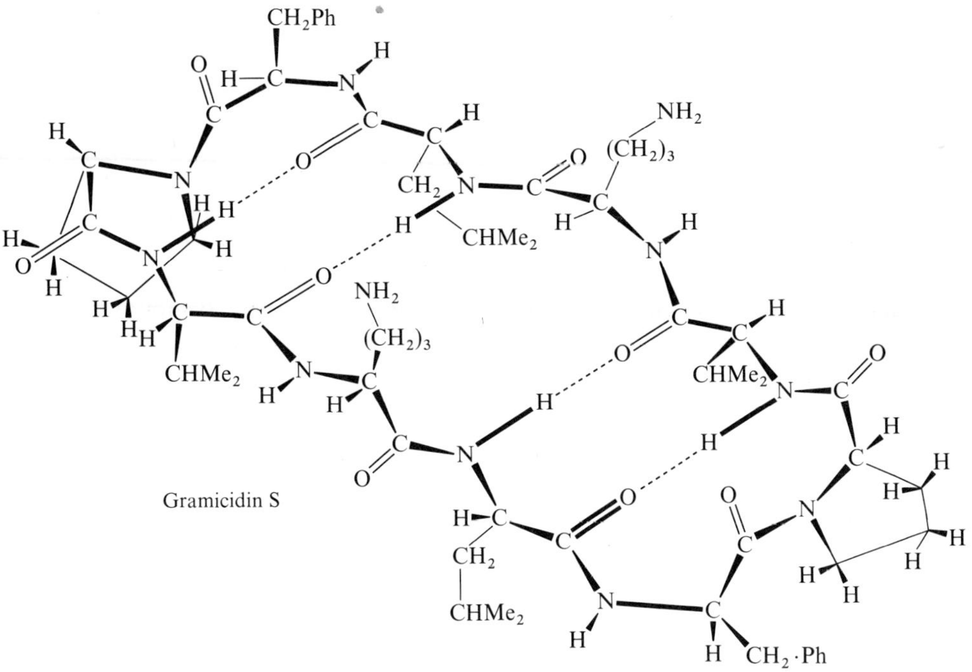

Gramicidin S

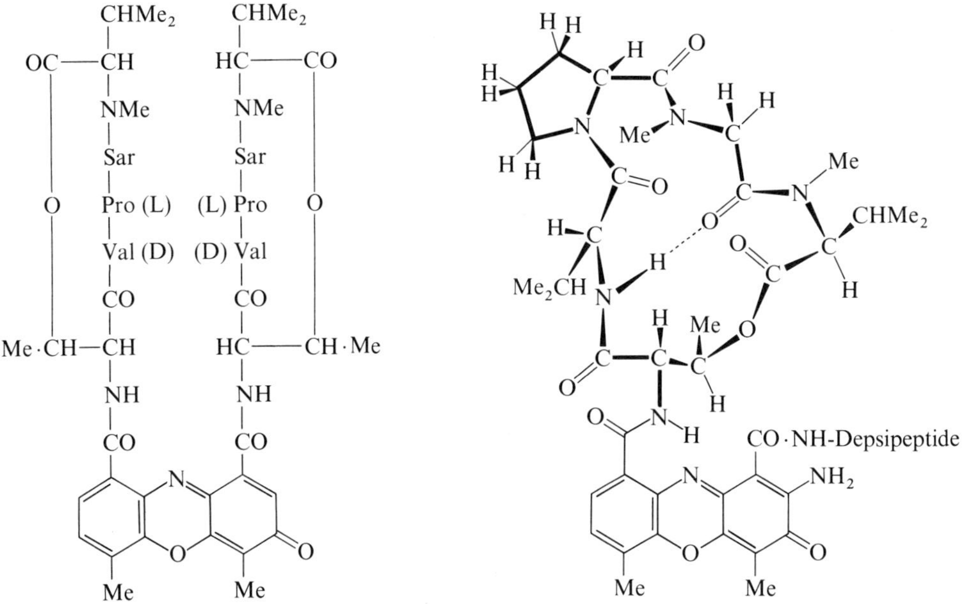

Actinomycin D

Fig. 28 Cyclic peptide and depsipeptide conformations

appear to adopt. A characteristic feature of many such compounds is the 10-membered rings arising from trans-annular hydrogen bonding, similar to those which are formed by intermolecular hydrogen bonding in $\beta$-antiparallel pleated sheet protein conformations, and also to those formed by intramolecular hydrogen bonding in certain helix conformations (Geddes, Parker, Atkins and Beighton, 1968). This particular feature has been described as the $\beta$-loop (Hassall and Thomas, 1971), and is responsible for stabilising the conformations of the peptide antibiotic, *Gramicidin S* (Fig. 28), and those of the depsipeptide antibiotics, *Actinomycin D* (Lackner, 1970), and valinomycin (Pinkerton, Steinrauf and Dawkins, 1969; Haynes, Kowalsky and Pressman, 1969; Ivanov, Laine, Abdullaev, Senyavina, Popov, Ovchinnikov and Shemyakin, 1969).

**Ester Links**

Ester linkages are important features of the primary structure of the depsipeptides, and contribute both to the stabilisation of their secondary structures, and particularly to the formation of alkali-metal complexes. Some depsipeptides such as *Enniatin-B* and valinomycin have cyclic structures which consist of alternating $\alpha$-amino and $\alpha$-hydroxy acids; others such as *Actinomycin D* have less regular structures with ester links arising from hydroxyl groups of hydroxy-amino acids. Both *Enniatin-B* and valinomycin form stable complexes with alkali metal ions which have interesting implications for the transport of these ions across biological membranes. Formation of these complexes is a reflection of the symmetry of the compounds. Thus, in the $K^+$-*Enniatin-B* complex, the cation is co-ordinated to the six carbonyl oxygen atoms, three above and three below the plane of the molecule (Fig. 29). Folding of the deca-depsipeptide valinomycin is such that a cylindrical $K^+$ complex results.

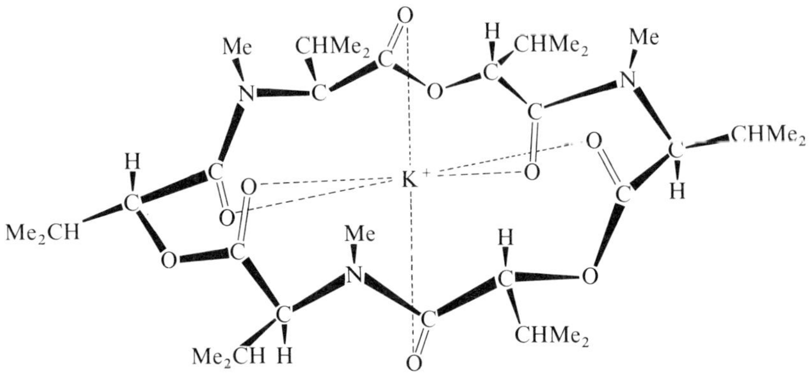

Fig. 29 $K^+$—Enniatin B complex

Schwyzer and his collaborators (1970) have prepared a synthetic disulphide cross-linked bicyclopeptide. This forms a cylindrical $K^+$ complex, which is readily transported across lipid membranes. The monocyclic peptide formed by reduction of the disulphide bridge does not complex with $K^+$ ions, so that ion transport is only feasible in the oxidised state.

Gly—Gly
Cys Pro + $K^+$ ⇌
Gly—Gly
SH

Gly—Gly
Cys O Pro
Gly—Gly
O O
S
$K^+$
S
O O
Gly—Gly
Cys O Pro
Gly—Gly

This type of complexing of $K^+$ ions may well prove important in ion transport, and could possibly provide the nucleus of an explanation for the penetration of such ions into the central nervous system and, for example, effectiveness of lithium in the treatment of manic depressive disorders.

**Non-bonded Interactions**

There is considerable evidence that both non-bonded interactions and van der Waals contacts also play an important part in the stabilisation of both secondary and tertiary structures of peptides and proteins. Thus, poly-L-proline exists in one of two helical forms, I (right-handed helix with *all-cis* peptide bonds, giving an amino acid residue translation along the helical axis of 1.85 Å) and II (left-handed helix with *all-trans* peptide bonds, giving an amino acid residue translation along the helical axis of 3.12 Å) despite the fact that intramolecular hydrogen bonding is impossible, since all amide nitrogens are fully alkylated (Traub and Shmueli, 1963; Cowan and McGavin, 1955). Its acyclic analogue, poly-*N*-methyl-L-alanine, which is similarly incapable of intramolecular hydrogen bonding, is also helical, indicating that this property is independent of the geometrical restrictions imposed by the pyrrolidine ring (Goodman and Fried, 1967).

Non-bonded backbone interactions also play a considerable part in limiting the number of stable conformers. Thus, in the series of small peptides, glycylglycine, glycylalanine and glycylvaline, increasing chain-branching at or about the α-carbon atom successively reduces the actual number of major conformers from 50%, through 16%, to 5% of those conceivably possible (Leach, Némethy and Scheraga, 1966). This effect is confirmed by conformational energy calculations for poly-*N*-methyl-L-alanine which show that the right-handed α-helical conformation found experimentally indeed represents the theoretically most probable low energy conformation (Mark and Goodman, 1967).

## TERTIARY STRUCTURE

Tertiary structure results from the characteristic folding which arises in globular proteins such as insulin, ribonuclease and serum albumin. For example, disulphide bridges in insulin create a number of conformation constraints, though there is crystallographic and other evidence that the B-chain is present as a right-handed α-helix, and that the A-chain is also helical, but left-handed from residues 1 to 9 and right-handed from residues 10 to 21 (Lindley and Rollett, 1955). None-the-less, the integrity of the insulin conformation both in solid state and in solution is obviously largely dependent on the disulphide bridges, since their destruction by oxidation or by mercaptolysis leads to complete loss of insulin activity (Acher, 1960).

### Disulphide Cross-linking

Cross-linking in protein structures occurs most frequently as a result of formation of disulphide bridges. Thus, insulin consists of two discrete peptide chains joined by two disulphide bridges. These not only link the two peptide chains, but impose a certain constraint on conformation both by the existence of the R—S—S—R′ bridge and by virtue of the characteristic 90° dihedral angle of the R—S and S—R′ bonds.

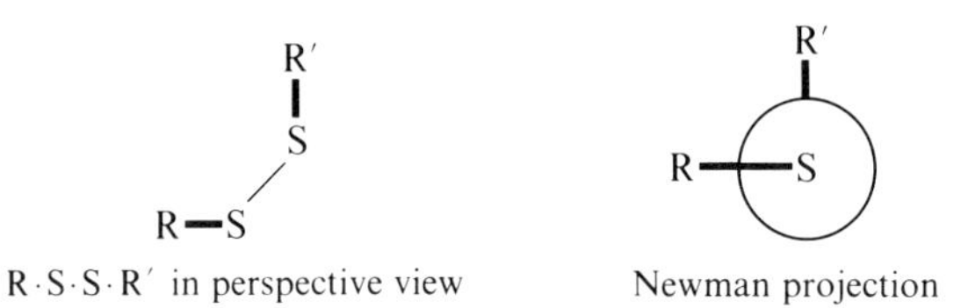

R·S·S·R′ in perspective view          Newman projection

Disulphide bonds are also responsible for the formation of the intrachain loops in the (A)-chain of insulin, and *Calcitonin* (salmon).

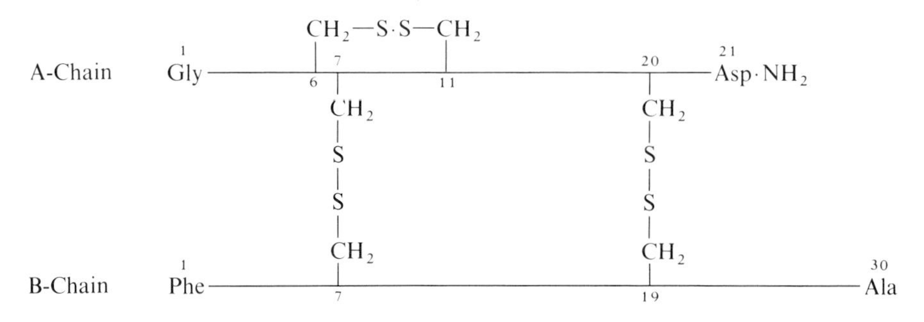


The characteristic folding and enzymatic activity of ribonuclease also arises as a result of intrachain cross-linking. The enzyme, which consists of a single peptide chain of 124 amino acids, derives its convoluted conformation and associated specific activity from four disulphide cross-links formed between cysteine residues at positions 26 and 84, 40 and 95, 58 and 110, and 65 and 72. Cleavage of these S—S bonds by reduction with mercaptoethanol in the presence of urea destroys its specific enzymic activity. Removal of urea and re-oxidation with molecular oxygen, however, not only substantially restores these characteristic disulphide bridges, in preference to alternative cross-links, but also largely restores the enzyme's specific activity, thus indicating that folding of the peptide chain is largely determined already by the primary peptide chain structure (Epstein, Goldberger and Anfinsen, 1963).

**Chain-folding**

Most globular proteins possess chains consisting of many amino acid residues characterised by a secondary structure which consists of a number of helical regions interspersed by random coil sections at which folding occurs. Myoglobin, the oxygen-carrying protein of mammalian muscle, is a typical globular protein, consisting of a single polypeptide chain of 153 amino acid units co-ordinatively bound through one of its histidine residues to a single iron-porphyrin (haem) group. Cysteine residues are absent. It has, therefore, proved an excellent subject for examination of the factors which lead to chain-folding. Its tertiary structure, determined by the elegant crystallographic studies of Kendrew and his collaborators (1961), shows eight straight right-handed α-helices of 24, 19, 9, 20, 7, 7, 16 and 16 amino acid residues in order from the *C*-terminal end of the molecule (Fig. 30). In accord with this structure, the infra-red spectrum of native myoglobin in $D_2O$ shows a sharp amide I band at 1650 $cm^{-1}$ characteristic of the α-helix (Timasheff and Susi, 1966).

The *C*-terminal tetrapeptide is random coil, and random coil sections consisting of up to eight residues separate each of the helical segments. A single alanine separates two of the helical segments, and an *N*-terminal proline separates two other helices. Because of the rigidity imposed by the pyrrolidine ring, proline

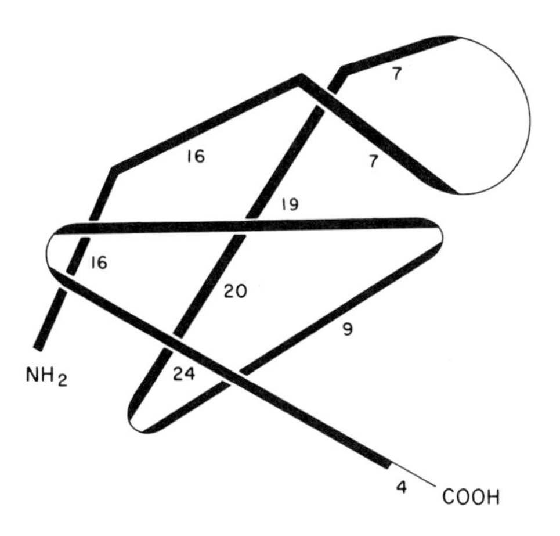


Fig. 30 Chain-folding of myoglobin at the proline residues

can only be accommodated at the *N*-terminal position of the α-helix. It is significant, therefore, that all three of the remaining proline residues in myoglobin are also situated in random coil segments and clearly contribute materially to the folding process.

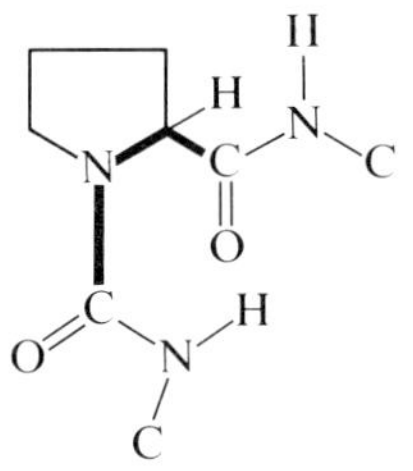

Chain-folding of myoglobin at the proline residues

Folding is such that hydrophilic polar amino acid groups are exposed at the surface of the protein, and hydrophobic non-polar aliphatic chains and aromatic groups are buried inside the globular conformation which, therefore, possesses a micellar-like structure. Surface-exposed polar groups create a strong affinity for water, and hence enhance water solubility, though extra-helical hydroxyls of serine, threonine and tyrosine and the imidazole $>$NH group of the iron-porphyrin-histidine complex in myoglobin (Fig. 31) are hydrogen-bonded to

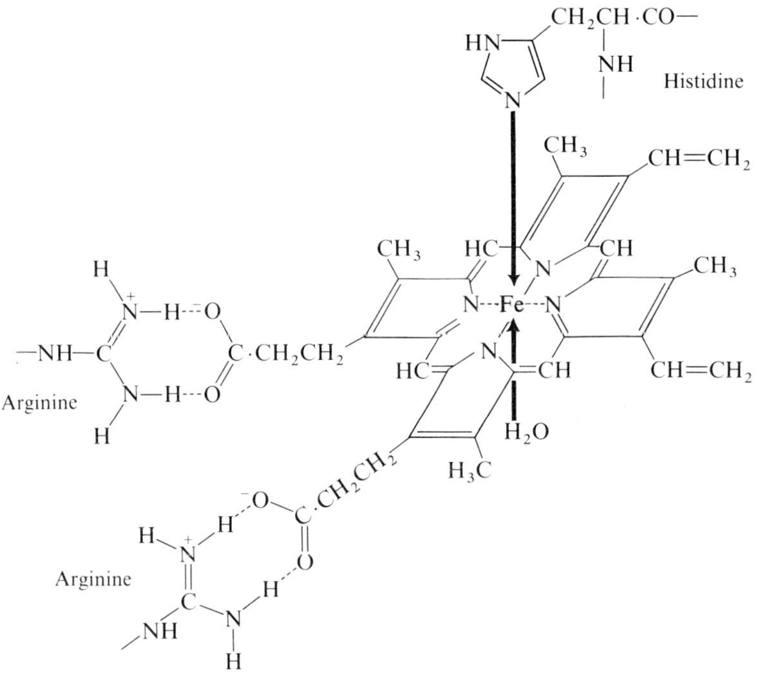

Fig. 31 Histidine co-ordination and hydrogen bonding of haem to arginine in myoglobin

peptide carbonyl groups. Arginine is hydrogen-bonded to the porphyrin carboxyl groups, which are also surface-orientated, whilst the vinyl side-chains of the porphyrin unit seek affinity with the other hydrocarbon side-chain residues towards the centre of the molecule.

### Van der Waals Contacts and Hydrophobic Interactions

Secondary, but more especially, tertiary structure of globular proteins derive considerable stability from van der Waals contacts and hydrophobic interactions between non-polar residues, which are more often than not buried collectively within the folds of the protein chain, as in the case of myoglobin. The globular structure of insulin in solution appears to be stabilised by van der Waals attraction of the non-polar hydrophobic residues, which tend as in myoglobin to be in micellar-like concentration towards the centre of the molecule. Optical rotatory properties of insulin solutions, which resemble those of other globular proteins rather than those of simple helical structures (Schellman, 1958), also support the concept of a globular structure for the hormone in solution.

Fluorescent probes such as dansylglycine (1-dimethylamino-5-naphthylsulphonylglycine) can be used to study the hydrophobic regions of proteins. The use of this reagent depends upon the blue-shift which the fluorescence absorption maximum undergoes in solvents of low dielectric constant, such as dioxan (Chen, 1967), and the observation that similar changes occur when dansylglycine binds to bovine serum albumin. The binding of acidic drugs, such as *Phenylbutazone* (Phenbutazone), *Flufenamic Acid* and *Dicoumarol* (Dicumarol), to serum albumin has been followed by their ability to displace dansylglycine competitively (Chignell, 1969).

### Hydrogen Bonding

Hydrogen bonding also plays a significant rôle in maintaining the shape of globular proteins in solution. This includes the intramolecular N—H···O=C— hydrogen bonding between individual turns of helical segments, intramolecular —O—H···O=C— hydrogen bonding between serine, threonine and tyrosine residues and amide carbonyl, as in myoglobin, and intermolecular hydrogen bonding to water at the surface of the molecule.

Hydrogen bonding plays an important part in determining the shape of the insulin molecule in solution, since the dimer of MW 12 000 dissociates in guanidine hydrochloride solution (Jirgensons, 1962). In the monomer (MW 6000), deuterium exchange of amide hydrogen occurs readily, though not in those residues forming part of the intra-A-chain 20-membered disulphide loop, which is associated with the break in the helix structure between residues 9 and 10 (Hvidt and Linderstrom-Lang, 1954; Hvidt, 1955).

### Surface-exposed Groups

The nature of surface-exposed groups plays an important part in determining not only the solubility of proteins, but also their reactivity in biological reactions.

The extent to which particular residues, such as histidine, tyrosine or tryptophan, are surface-exposed or buried within the folds of proteins can be determined by their reactivity to particular reagents. For example, both histidine residues in the B-chain of insulin are accessible to diazo-1-*H*-tetrazole (Horinishi, Hachimori, Kurihara and Shibata, 1964), which couples with them to form a coloured complex. They are also freely available for photo-oxidation (Weil, Seibles and Herskovits, 1965). Solvent perturbation spectroscopy (Laskowski, 1966), however, shows that the four tyrosine residues of insulin are of three types. Two residues of one type appear to be buried within the molecule of Zn insulin (Weil, Seibles and Herskovits, 1965), but photo-oxidation experiments reveal that one of these two buried tyrosyl residues is exposed when Zn is removed. Notwithstanding this effect, comparison of the acid titration curves of crystalline zinc insulin with those of zinc-free insulin suggest that zinc is complexed to two imidazole and to two appropriately-situated carboxyl groups in each of the two monomer insulin units of the complex (Tanford and Epstein, 1954). Marcker (1960), however, reaches the alternative conclusion based on experiments with *N*-phenylureido-insulin derivatives, that it is the *N*-terminal phenylalanine residues of the B-chain and not imidazole groups which form the point of attachment for the zinc ion.

Surface-exposed polar groups and some, though not all, chain-terminal residues seem to be unimportant in determining the hypoglycaemic activity of insulin. Thus, notwithstanding the conclusion (Marcker, 1960) that the *N*-terminal phenylalanine residue is essential for complex formation with zinc, removal of the entire phenylalanine-containing *N*-terminal hexapeptide [Phe-Val-Asp($NH_2$)-Glu($NH_2$)-His-Leu-] from the B-chain by hydrolysis with aminopeptidase does not reduce hypoglycaemic activity in the resulting des-hexapeptido-insulin (Smith, Hill and Borman, 1958). Similarly, removal of the B-chain *C*-terminal alanine with carboxypeptidase to form des-alanine insulin does not affect its potency (Slobin and Carpenter, 1963). Desalanine-insulin is also formed by trypsinisation of pro-insulin, the biosynthetic precursor of porcine insulin, in which the A and B-chains are linked by a loop of 33 amino acids (Chance, Ellis and Bromer, 1968).

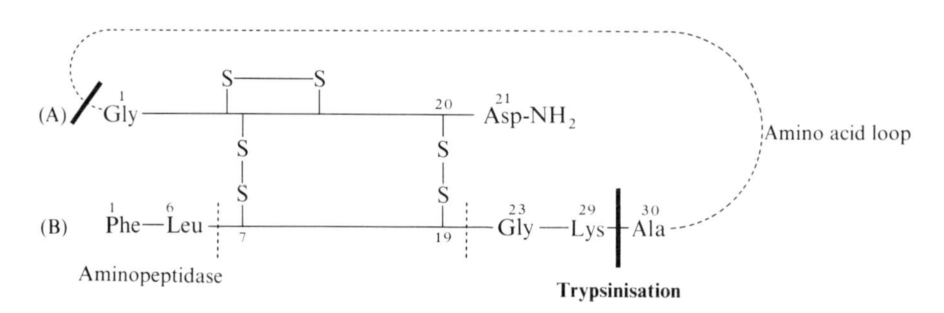


Des-alanine insulin

Complete removal of the *C*-terminal octapeptide (Gly-Phe-Phe-Tyr-Thr-Pro-Lys-Ala) from the B-chain does not destroy, but merely reduces activity in the resulting DHA-insulin to about 15% of that of insulin itself. This indicates that whilst this fragment plays some part in determining the level of potency, it does not represent the active centre (Sanger, 1960). Desamido-insulin, which merely lacks the A-chain *C*-terminal asparagine amido group, also has appreciable hypoglycaemic activity. Desamido-insulin can be separated from commercial samples of crystalline bovine insulin (Chrambach and Carpenter, 1960) and is formed by the action of very dilute acid on insulin (Slobin and Carpenter, 1963). In contrast to desamido-insulin, cleavage of the entire *C*-terminal asparagine group results in a complete loss of hypoglycaemic activity.

There is some evidence that the so-called monocomponent insulins, prepared by gel filtration, undergo spontaneous deamidation on formulation, due to the acidity of the injection solutions, which in the case of *Insulin Injection*, is of the order of pH 3.0 to 3.5. It is doubtful, therefore, whether monocomponent insulin formulations prepared in this way remain homogeneous on storage. Most other official insulin injection formulations are either neutral [*Neutral Insulin Injection* (pH 6.6 to 7.7)], or very slightly alkaline (*Protamine Zinc Insulin Injection*; Protamine Zinc Insulin Suspension, pH 6.9 to 7.4; *Insulin Zinc Suspension*, pH 7.0 to 7.5).

Although the inter-chain disulphide ring almost certainly houses the seat of biological activity, the A-chain intra-chain disulphide loop is, in contrast, relatively unimportant (Table 63), since it is here that the major structural distinctions between insulins from different species are found (Brown, Sanger and Kitai, 1955; Harris, Sanger and Naughton, 1956; Sanger, 1960).

That exposed polar groups are largely inessential for insulin activity is shown by blocking of serine hydroxyl groups and free amino groups. This has little effect on potency. Photo-oxidation experiments (Weil, Seibles and Herskovits, 1965) on the other hand clearly implicate the two histidine residues of the B-chain, which readily undergo photo-oxidation at pH 7.0. Below 10°C, only histidine residues are oxidised, but at somewhat higher temperatures (20–40°C), oxidation of tyrosine residues also occurs. In 8 M urea, however, unmasking and

Table 63. Species Variations in Insulin

| Source | A-Chain | B-Chain |
|---|---|---|
| | S—S (Cy⁶ to Cy¹¹) | |
| Beef | —$Cy^{6}$—$Cy^{7}$—$Ala^{8}$—$Ser^{9}$—$Val^{10}$—$Cy^{11}$ | $Ala^{30}$ |
| Pork | —Thr—Ser—Ileu— | $Ala^{30}$ |
| Human | —Thr—Ser—Ileu— | $Ser^{30}$ |
| Rabbit | —Thr—Ser—Ileu— | $Thr^{30}$ |
| Sheep | —Ala—Gly—Val— | $Ala^{30}$ |
| Horse | —Thr—Gly—Ileu— | $Ala^{30}$ |
| Whale | —Thr—Ser—Ileu— | $Ala^{30}$ |

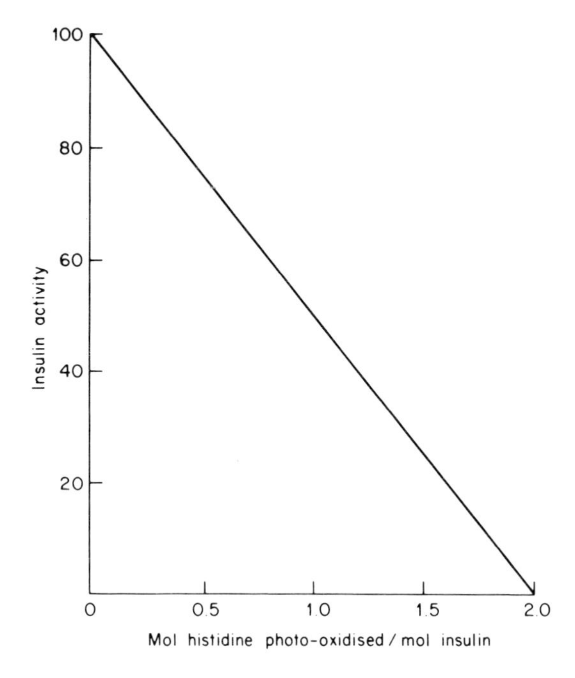


Fig. 32 Relationship between available histidine residues and activity of insulin

photo-oxidation of tyrosine residues occurs even at low temperatures, indicating that photo-oxidation is markedly influenced by the conformation of the molecule. The low temperature oxidation of histidine, however, in zinc-free insulin indicates a direct correlation between biological activity and available histidine residues (Fig. 32). Both photo-oxidation of histidine residues in insulin and destruction of insulin-activity follow identical first order kinetics. Since, however, photo-oxidation of histidine itself proceeds much faster under the same conditions, with zero order kinetics, the rate of photo-oxidation of histidine in insulin must be a function of the insulin conformation.

The photo-oxidation of zinc-insulin appears to be much slower than that of either zinc-free insulin or the histidine-zinc complex, in conformity with the increased molecular complexity and possibly, also, folding of zinc-insulin.

## QUATERNARY STRUCTURE

Quaternary structure of proteins represents a level of organisation, characteristic of certain globular proteins, arising from the association of two or more sub-units. The sub-units may be identical as, for example, in insulin and glutamic hydrogenase, but it is not uncommon for them to be of more than one type, as exemplified by the lactic dehydrogenases. Association is usually controlled by a number of factors. It may be different in solid and solution states, and in the latter can be determined by concentration, pH, ionic strength, and by the presence of metal ions or other quaternary structure stabilisers.

Native glutamic hydrogenase is a typical example of a protein showing association as a tetramer of MW $1.0 - 1.6 \times 10^6$. Dilution of the enzyme promotes reversible dissociation into the four enzymatically active sub-units of MW 250 000 to 350 000 (Frieden, 1963). Both NADPH and guanylic acid, which associate with these sub-units, prevent their re-association to the tetramer as a result of conformational changes which they induce. Treatment of the sub-units with 6 M urea promotes a further dissociation into a number of identical sub-units of MW 40 000, which is irreversible. Haemoglobin, the oxygen-carrying haem-protein of blood, is also a tetramer (MW 64 450), but consists of two pairs of identical symmetrically placed sub-units, each pair consisting of an A-chain and B-chain structure, arranged tetrahedrally. Interactions between the individual haem units in the tetrahedral structure account for the increasing ease with which each of the four oxygen atoms, which combine successively with haemoglobin, react with the protein.

Association of the insulin monomer (MW 5750) into dimers and hexamers is complex. In the absence of zinc ($Zn^{2+}$) ions, association is largely dependent on pH and concentration. In neutral solution, the average molecular weight ranges from dimer to hexamer depending upon concentration in the concentration range of 0.1 to 1%. Dilute solutions (0.1 to 0.25%) dissociate with increasing pH to monomers at pH 9.1 to 9.5, whilst dimerisation is predominant in acid solution, with the proportion of dimers increasing as pH and ionic strength decrease.

Association of insulin also occurs in solution above pH 3.5 in the presence of $Zn^{2+}$ and other suitable divalent cations ($Cd^{2+}$, $Co^{2+}$, $Ni^{2+}$, $Cu^{2+}$, $Fe^{2+}$, $Mn^{2+}$) to give both dimers and hexamers. Dimers are present in most such aqueous solutions, but hexamers predominate at neutral and moderately alkaline (7.5) pH. In this pH range, the proportion of hexamers increases with zinc concentration up to a content of *ca* 0.35, which is equivalent to two atoms of zinc per hexamer. At higher zinc concentrations, binding of zinc continues to at least a further seven atoms per hexamer at pH 8.0 (Cunningham, Fischer and Vestling, 1955).

Crystal symmetry in the hexamer requires that it be organised as three equivalent dimers about a three-fold axis (perpendicular to the plane of the paper in Fig. 33). Each of the dimers is situated asymmetrically about one of three non-crystallographic approximately two-fold axes of symmetry which lie at right angles to the three-fold axis (in the plane of the paper in Fig. 33). The two zinc ions are 17 Å apart, situated in the three-fold axis above and below the two-fold axes. The two monomer components of each dimer are arranged in different conformations with the C-terminous of the B chains antiparallel to give an antiparallel $\beta$-pleated sheet structure with four hydrogen bonds. The two-fold symmetry of dimer is further stabilised by hydrophobic interactions involving the B chain valine and phenylalanine units.

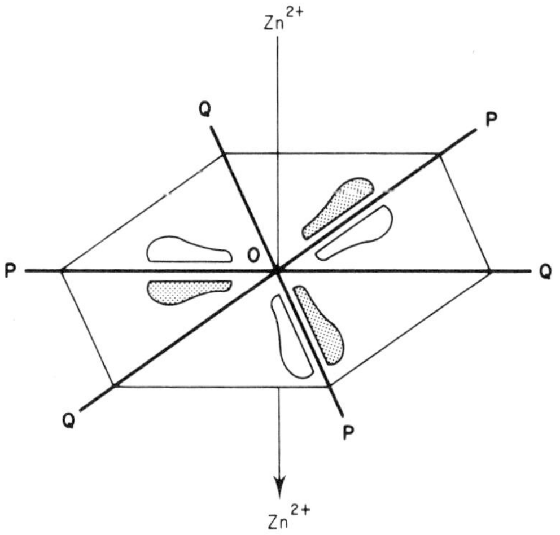


Fig. 33 Insulin hexamer (after Blundell, Dodson, Hodkin and Mercola, 1972)
POQ—two-fold axes of symmetry
$Zn^{2+}$—O—$Zn^{2+}$—three-fold axis of symmetry

Although dimeric insulin persists in most aqueous solutions, dissociation to the monomer is affected by a variety of agents capable of hydrogen bond breaking, including guanidine hydrochloride, urea and dimethylformamide.

Blundell (1975) has shown that the 29 amino acid units of glucagon (Bromer, Staub, Diller, Bird, Sinn and Behrens, 1957) adopt an α-helical conformation at bioactive concentrations in solution, which is extended into a non-helical structure at the *N*-terminal end. In the solid state, and presumably in its inactive depôt form in the pancreas, glucagon adopts a non-globular trimeric conformation, which is stabilised by hydrophobic interactions between the helical segments of the contributing manomers.

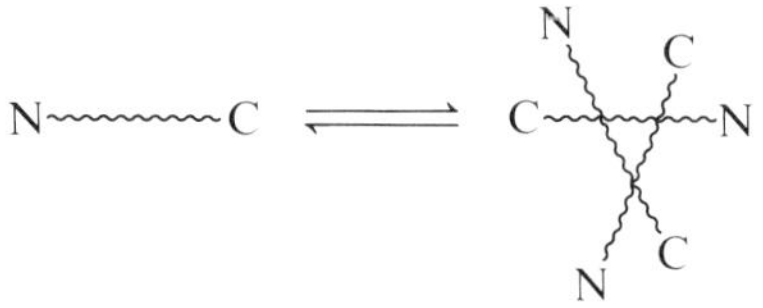


When glucagon is released from the pancreas in response to low circulating blood sugar levels, the trimer dissociates, and the monomer binds to a receptor protein in the liver (Rodbell, Krans, Pohl and Birnbaumer, 1971). This complex regulates the catalytic activity of adenylcyclase, the enzyme responsible for the conversion of ATP to cyclic AMP (Chapter 23). The enhanced levels of cyclic AMP then activate enzymes responsible for the breakdown of glycogen to glucose (Sutherland and Robison, 1966), thus restoring blood sugar levels to normal. Binding of the monomer to the receptor protein, like association into the trimer, appears to be determined mainly by hydrophobic interactions between the binding protein and the helical moiety of the peptide. The free, extended, non-helical, *N*-terminal end of the polypeptide, therefore, appears to be the centre of its catalytic activity.

The phenomenon of isoenzyme formation is related to the association of dissimilar sub-units (Cahn, Kaplan, Levine and Zwilling, 1962), which has been studied in depth with the lactic dehydrogenases. Two distinct lactic dehydrogenases are found in the chicken, one in breast muscle and the other in heart muscle. Three other lactic dehydrogenases are known from other species with properties intermediate between those of the chicken breast and heart enzymes. The relationship is established by the dissociation of each of these enzymes in guanidine hydrochloride and mercaptoethanol into four sub-units, of two types, H (heart) and M (muscle). Association of any four gives rise to five possible tetramers, $H_4$, $H_3M$, $H_2M_2$, $HM_3$ and $M_4$, consistent with the five known enzymes and the observed gradation of properties between them.

## PROTEIN DENATURATION

Denaturation occurs as a result of major changes in the native, secondary, tertiary or quaternary structure of proteins. The primary structure remains unchanged. Denaturation, which may or may not be reversible, therefore, results in a major departure from the ordered, coiled, sheet or folded structure of the native state generally either to an alternative ordered form or to random coil conformations. Helix–random coil transitions usually occur within a narrow temperature range, and even short peptide chains of around 7-amino acid units, which almost invariably possess random coil structures at room temperature, may favour helical forms at somewhat lower temperature. Helix–random coil transitions are also accompanied by an increase in hydrophilic properties as the molecule changes from an internally hydrogen-bonded to an externally hydrogen-bonded structure. Strong hydrogen bonding reagents, such as urea, guanidinium salts, formic, dichloracetic and trichloracetic acids, therefore, favour formation of random-coil structures, whilst solvents with poor, or little hydrogen bonding capability, such as chloroform, methylene dichloride and chloroethanol, promote helix formation.

Denaturation may also involve the breaking of disulphide bridges, but such changes are usually irreversible. Cleavage of S—S links is essential if a complete random coil conformation is to be achieved; this latter, however, is always inhibited whenever proline or hydroxyproline units are present because of the physical restraint imposed by the pyrrolidine ring (p. 604).

The conformational changes which accompany denaturation can be followed by changes in the physical properties of the protein, and may lead to solubilisation of insoluble proteins or precipitation of soluble proteins. In solution, changes in viscosity, infrared spectra, optical rotatory dispersion and circular dichroism spectra, and tritium exchange of otherwise **exchange-resistant (hard-to-exchange) protons** are readily measured and serve as useful, but sometimes divergent, criteria of protein change.

Globular proteins may sometimes acquire alternative conformations resembling those of their native state in their compactness, but are none-the-less entirely different in structure. Such structures are described as quasi-native states (Tanford, 1968). Thus, $\beta$-lactoglobulin undergoes a transition between pH 7 and 8, which despite the fact that the protein retains its compact globular form, is accompanied by significant changes in the ORD curve about 250 nm. It is also accompanied by exposure of a carboxylic acid group at the protein surface, which in its native state is buried within the folds of the protein chain (Tanford, Bunville and Nozaki, 1959; Tanford and Taggart, 1961).

### Salt Precipitation

In general, solubility of peptides and proteins decreases with increasing molecular weight and molecular asymmetry. Aqueous solubility of macromolecules increases with ionisation, and for this reason proteins are most readily precipitated at the isoelectric point, where the number of positively- and negatively-charged groups are in balance. Inorganic salts, such as ammonium sulphate which has a high water solubility, are effective protein precipitants as a result of solvent competition for the more soluble ions. It follows that high molecular weight proteins are precipitated more readily than low molecular weight proteins. Thus, high molecular weight globulins are salted out by half-saturated ammonium sulphate, whilst the much lower molecular weight albumins are only salted out from saturated ammonium sulphate. For similar reasons, ammonium and sodium sulphates and phosphate buffers are much more effective than alkali metal chlorides, because of the much greater ionic strengths achieved at saturation. Salt precipitation, however, is generally reversible with resolubilisation of the protein, on removal of the precipitant by dialysis.

Some indication of the nature of the changes which occur in reversible denaturation with inorganic salts has been deduced from studies of the effect of lithium bromide on various globular proteins such as serum albumin, lysozyme and ribonuclease in solution. Physical changes are much less pronounced than with more powerful denaturants such as guanidine hydrochloride, and the increase in viscosity of serum albumin, for example, is such as to suggest that the protein, disulphide bridges apart, is not in a fully randomised coil conformation. In contrast, denaturation of the fibrous protein, myosin, with lithium bromide is accompanied by a fall in intrinsic viscosity to such extremely low values as to be indicative of change to a compact globular structure.

### Solvent Precipitation

Hydrophilic organic solvents, such as ethanol, acetone and ether, promote precipitation of proteins, as a result of both solvent–protein hydrogen bonding, and repulsion of hydrophilic protein groups by solvent hydrocarbon chains. Ethanol

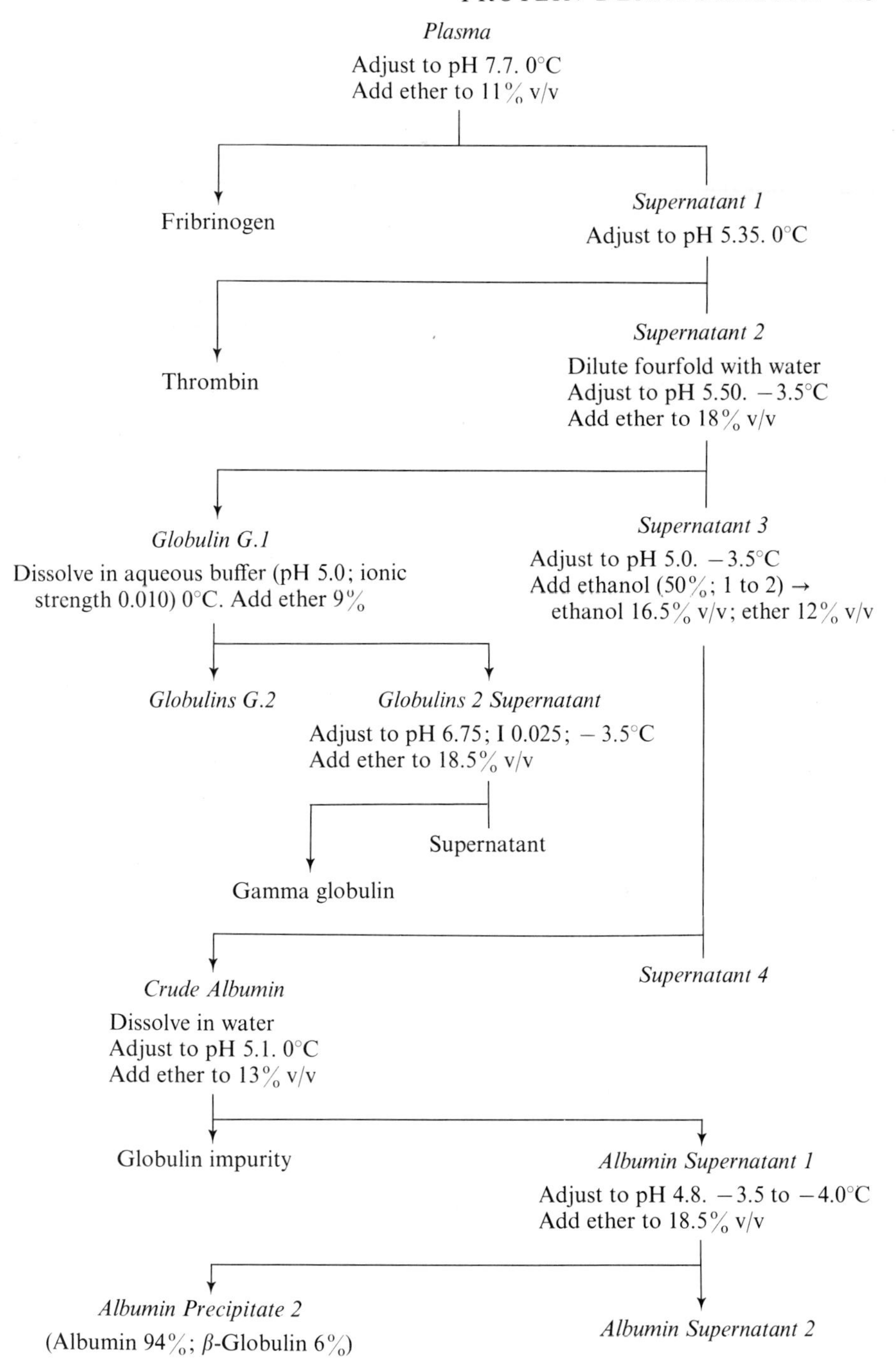


Fig. 34 Ether fractionation of plasma proteins

concentrations below about 20%, however, have little effect on protein structure, but presumably the protein denaturant action of ethanol must have some bearing on its use as a mild disinfectant, for which the most effective concentration is around 70%. Cold ethanol fractionation (Cohn, Strong, Hughes, Mulford, Ashworth, Melin and Taylor, 1946) has been widely used in the United States for fractional precipitation of blood proteins. A similar pH-ether precipitation process has been developed (Kekwick and MacKay, 1954) for use in the United Kingdom (Fig. 34). Crude insulin, likewise, is usually precipitated in the course of manufacture by ethanol concentrations of 90–95%.

The effects of organic solvents contrast with those of other denaturants in that the denatured protein, whilst adopting a conformation distinctly different from that of the native protein, retains a high degree of ordered structure with a significant α-helical content (Urnes and Doty, 1961; Jirgensons, 1967). The mechanism of denaturation is not clear, but the lower dielectric constant of alcohols compared with water favours aggregation and precipitation of proteins in the region of the isoelectric point. Studies of ribonuclease in aqueous chloroethanol, however, have established that the transition from native to denatured state takes place in two stages, first uncoiling of the native protein, and secondly coiling and folding into the denatured conformation. These changes bear some relationship to some enzyme-substrate and enzyme-inhibitor interactions, and to other allosteric effects. These and other drug protein interactions are further discussed elsewhere (**2**, 1 and **2**, 4).

Ethylene and propylene glycols, glycerol and sucrose are much less effective than aliphatic alcohols as denaturants; dimethyl sulphoxide (Hamaguchi, 1964) is also less effective than ethanol. Ethylene glycol has, however, been shown to induce changes in protein conformation (Nozaki and Tanford, 1965). Somewhat greater care is essential with the use of more reactive organic solvents, such as acetone, as protein precipitants. Thus, oxytocin, for example, is inactivated by acetone due to formation of a cyclic derivative at the *N*-terminal Cys-Tyr link (Hruby, Yamashiro and du Vigneaud, 1968; Yamashiro and du Vigneaud, 1968).

$$HS\cdot CH_2\cdot CH(NH_2)\cdot CO\cdot NH\cdot CH(CH_2C_6H_4OH)\cdot CO\text{—} \xrightarrow[\;-H_2O\;]{CH_3CO\cdot CH_3} HS\cdot CH_2\text{—}[\text{ring: }HN\text{—}C(Me)(Me)\text{—}N\cdot CH(CH_2C_6H_4OH)\cdot CO\text{—},\ C{=}O]$$

## pH Effects and Salt Formation

Even at quite low temperatures, some proteins undergo denaturation in acid and alkaline solution with substantial uncoiling of the helix structure. Thus, glyceraldehyde-3-phosphate dehydrogenase, which is stable at 2.4°C between pH 4.0 and 11.0, undergoes dissociation into sub-units, and suffers drastic conformational changes with complete loss of activity below pH 4.0 (Shibata and Kronman, 1967). Fumarase and aldolase behave similarly, also carbonic anhydrase and haemoglobin. All have buried, uncharged histidyl residues in the native state, accounting for conformational instability in acid solution, which arises from the drive to expose histidyl residues to the proton source.

The combined action of acid (HCl) and acetone brings about reversible dissociation of haemoglobins into a colourless apoprotein, globin, and its coloured haem prosthetic group. Under anaerobic conditions, the iron remains in the ferrous state, but in the presence of oxygen, it is oxidised to ferric iron as in haemoglobin. Provided this oxidation is prevented, re-constitution of the native protein can be effected by returning the pH to neutrality. Similarly, recombination of globin with iron-porphyrins other than the native material gives rise to *pseudohaemoglobins* possessing oxygen carrying capacity, but differing somewhat in their physical properties from native haemoglobin.

Whilst some proteins are equally sensitive to both acid and alkali, others including haemoglobin and myoglobin, are much more stable to alkali, since exposure of uncharged histidyl residues constitutes a driving force only at low pH. Similarly, the phenolic hydroxyl of tyrosyl residues and acidic groups of glutamic and aspartic acids can also act as a drive to denaturation by alkali, if similarly buried within the native protein structure. Thus, thin-film dialysis and ORD studies have shown that a remarkable increase in the size of the angiotensin II molecule occurs with ionisation of the tyrosine phenolic group. More generally, however, alkaline denaturation is often associated with disruption of disulphide bonds, and other oxidative changes.

Effective chain length of proteins, even in the denatured random coil state, is affected by pH and ionic strength. Thus, the imbalance of charge increases rapidly as the pH is moved away from the isoelectric point, and increasing repulsion of charges leads to some chain elongation. This effect, however, is largely eliminated by the addition of electrolytes. Nevertheless, pH changes provide a useful method of effecting precipitation of soluble proteins, and also of effecting solution of insoluble ones. Thus, glucagon, which has been shown by ORD measurements to exist largely as a random coil structure in aqueous solution, is precipitated in acid solution in a form which adopts the anti-parallel $\beta$-pleated sheet conformation (Gratzer, Beavan, Rattle and Bradbury, 1968). Insulin, too, undergoes reversible precipitation as fibrils at low pH, and solutions in *0.2N* acid may actually be boiled for short periods with retention of activity. At pH 2.0 and a temperature of 90–100°C, fibril formation occurs with the formation of a thixotropic gel (Elliott, Ambrose and Robinson, 1950; Ambrose and Elliott, 1951). The fibrils, which are often up to 100 Å in width and 16 000 Å in length, have a $\beta$-cross structure, in which the chain runs perpendicular to the fibre axis (Astbury, Dickinson and Bailey, 1935). Fibril formation is used in the purification of insulins, since they are also readily precipitated by seeding crude insulin solutions in acid with fibrils prepared from crystalline insulin; the precipitated hormone can then be crystallised. The fibrillar form is biologically inactive, but activity can be restored by neutralisation with alkali.

A number of organic acids, such as dichloracetic acid and trichloracetic acid, are extremely powerful denaturants, often more effective than even guanidine hydrochloride (Sage and Fasman, 1966). They appear to function, however, more by molecular association, such as ionic bonding and hydrogen bonding, than by protonation (Stewart, Mandelkern and Glick, 1967). Formation of specific insoluble crystalline salts with organic acids and bases, such as insulin picrate and oxytocin flavianate (Pierce, Gordon and du Vigneaud, 1952) provides a valuable method of purification for certain peptides and proteins. Interaction of insulin with basic proteins, such as protamine and globin, forms the basis for the preparation of insoluble long-acting insulin preparations, such as *Isophane Insulin Injection* (crystalline protamine zinc insulin) and *Globin Zinc Insulin Injection.* Unfortunately, these preparations may occasionally give rise to allergic reactions arising from the added protein. This difficulty, however, is overcome with the use of the much more insoluble acetate buffered *Insulin Zinc Suspensions* (Hallas-Møller, Petersen and Schlichtkrull, 1952), which unlike citrate- and phosphate-buffered material remain insoluble at physiological pH. Although amorphous when precipitated from acetate buffer, the product crystallises on standing within certain defined pH limits (4.7–5.6), and remains crystalline, provided there is no change in pH, and zinc sequestering ions such as citrate and phosphate are absent.

**Disruption of Hydrogen Bonds**

One of the most widely used and most effective denaturants, guanidine hydrochloride, almost certainly acts by disruption of hydrogen bonds. Most proteins are denatured by it in 6–8 M solution at room temperature to random coil structures (Tanford, 1968). Exceptionally, however, some more stable proteins, such as ribonuclease, may not be completely denatured even under these conditions, since both p*K* and ultraviolet absorption measurements reveal that some tyrosine residues remain in less polar environments than others. Denaturation of proteins is always accompanied by significant changes in intrinsic viscosity. Typical polypeptides, such as insulin, and globular proteins, such as serum albumin and thyroglobulin, which have low intrinsic viscosities in their native state, in general, show an increase in intrinsic viscosity on denaturation. In contrast, the very high intrinsic viscosities of native fibrous proteins, such as myosin, decrease as a result of transition to random coil structures. The rate of exchange of H for $^{3}H$ increases sharply with myosin in guanidine hydrochloride solution as restraining hydrogen bonds are broken, thus confirming denaturation to a random coil structure (Segal and Harrington, 1967). Both the anions and cations of guanidine hydrochloride are involved in the denaturation process, as, significantly, whilst guanidine thiocyanate is an even more powerful denaturant, guanidine sulphate has little effect (von Hippel and Wong, 1964, 1965).

Urea (6–16 M) is also widely used as a denaturing agent. Although many proteins are not fully denatured by it, serum albumin is over 50% denatured by 8–10 M urea at room temperature, and almost completely denatured in 9 M solution at elevated temperatures. Ribonuclease, similarly, is fully denatured by urea, whereas, in contrast, immunoglobulin and lysozyme are barely affected, and the quaternary structure only of haemoglobin merely undergoes part dissociation to two half-units. Like guanidine, urea is strongly hydrogen-bonded to proteins. As a small highly water-soluble molecule, restrictions on its close approach to macromolecules are minimal, and strong intermolecular hydrogen bonds, stronger than the intramolecular hydrogen bonds of the α-helix, are readily formed. The extent of urea denaturation of proteins is markedly influenced by pH and temperature. At elevated temperatures, isomerism of urea to ammonium cyanate is favoured with increased possibilities for attack at free amino groups and formation of carbamates.

**Cleavage of Disulphide Cross-linking**

The presence of disulphide bonds normally limits denaturation by hydrogen bond disruption. Thus, denaturation of serum albumin, lysozyme and ribonuclease with guanidine hydrochloride before and after cleavage of the disulphide bonds, which they all contain, shows that cross-linking decreases viscosity. This is due to the restraint which the disulphide bridges impose on the otherwise completely randomised coil conformation of the denatured protein (Tanford, Kawahara, Lapanje, Hooker, Zarlengo, Salahuddin, Aune and Takagi, 1967).

Irreversible denaturation, which occurs with urea at pH 7.0 and above, is probably also due to the chemical reactivity of sulphydryl and disulphide groups. Thus, urea denaturation of β-lactoglobulin (Kauzmann and Simpson, 1953) proceeds in two stages at pH 7.0; the first, fast, and the second much slower. The latter, however, becomes much more rapid at pH 8.7 and above. It is considered (Tanford, 1968) that the first step is reversible and the second irreversible.

Prolonged treatment of insulin with alkali at high pH results in irreversible changes, which appear to be associated with disruption of disulphide links. Insulin is also inactivated by thiol-containing reagents, such as glutathione, thioglycollic acid and dimercaprol, which bring about disulphide interchange. Disulphide interchange in the presence of base, such as triethylamine, occurs with oxytocin and leads to isomeric dimers probably with parallel and antiparallel structures. These dimers, which have low oxytocic activity (Yamashiro, Hope and du Vigneaud, 1968), are also formed as by-products in the production of synthetic oxytocin by oxidation of the intermediate oxytoceine.

**Thermal Denaturation**

Thermal denaturation arises out of increases in the bending, stretching and twisting vibrations of bonds which, if sufficiently great, lead to bond dissociation. Increased charge repulsions create additional instability, so that even at low temperatures, denaturation becomes favoured at low pH. Such denaturation, however, is frequently reversible. The products of thermal degradation usually have highly disordered conformations, but are not fully random coiled, containing some elements of ordered structure. This is evident from the further physical changes which occur after heat denaturation on addition of guanidine hydrochloride (Aune, Salahuddin, Zarlengo and Tanford, 1967), and the breaking of disulphide bonds (Ginsburg and Carroll, 1965).

Irreversible thermal denaturation is usually characteristic of exposure to heat at temperatures above ambient level. It is generally greatly accelerated by high protein concentrations and high pH. The nature of the process is complex and variable depending upon conditions, and includes formation of new disulphide bridges by disulphide interchange as in serum albumin (Warner and Levy, 1958), and attack on disulphide bonds by hydroxyl ion, in addition to helix-random coil and other analogous transitions. Another feature of thermal denaturation is that unfolding to random coil structures is not necessarily a prime requirement. Serum albumin, for example, appears to undergo direct irreversible thermal denaturation from its native state (Foster, 1960). This view is supported by the experiments of Clark and Gurd (1967) which showed that carboxamido- and carboxymethyl-myoglobin required a higher temperature for irreversible denaturation than unmodified myoglobin, and also gave a greater percentage reconversion to modified protein on cooling.

# 20 Glycols and Polyols

Dihydric alcohols, otherwise known as diols or glycols, contain two alcoholic hydroxyl groups in the molecule. Almost without exception stable compounds of this type have the hydroxyl groups attached to different carbon atoms. Where, however, a 1,1-dihydroxy compound (*gem*-diol) is formed in the course of a reaction, as a general rule water is immediately lost to form a carbonyl compound.

$$RR'C(OH)_2 \longrightarrow RR'C{=}O + H_2O$$

Stable *gem*-diols are formed where the diol group is immediately adjacent to a strongly electron-attracting group as in chloral hydrate, $Cl_3C \cdot CH(OH)_2$, and the hydrate of glyoxalic acid, $(HO)_2CH \cdot CO \cdot OH$. Stable derivatives of 1,1-dihydroxy compounds are also known, e.g. acetals and hemiacetals, which may be derived by condensation of aldehydes with alcohols under acidic conditions (Chapter 10).

Stable diols are formed when two hydroxyl groups are attached to different carbon atoms as in the 1,2-diols (*vic*-diols; $\alpha$-glycols), such as ethylene glycol and propylene glycol. Alternatively, the two hydroxyl groups may be linked at more remote intervals along the carbon chain, as in the 1,3-diols, such as trimethylene glycol, $(HO \cdot CH_2 \cdot CH_2 \cdot CH_2OH)$, and higher $\alpha,\omega$-diols, such as penta-, hexa- and deca-methylene diols, $[HO(CH_2)_nOH]$.

Glycerol (propane-1,2,3-triol) is widely distributed in nature in both animal and plant tissues. Phosphate esters of glycerol are found in the glycolytic pathway which is the principal route for the breakdown of glucose in some micro-organisms. Phosphate esters of glycerol also enter into the composition of phospholipids, and natural fats and oils (Chapter 11), from which glycerol is released on saponification with caustic alkali. Of the higher polyols, erythritol, arabitol, ribitol, sorbitol, iditol, galactitol and mannitol all occur naturally. Mannitol is the principal constituent of Manna, the dried exudate of the European flowering ash, *Fraximis ornus*, and is also found in high concentration in similar exudates of olive and plane trees; extensive amounts up to 20 % are also found in the brown seaweeds. Sorbitol is, similarly, widely distributed in seaweeds, particularly some of the red varieties, and in fruits of the natural order, Rosaceae, including apples, pears, cherries, plums, peaches and apricots. The cyclic polyol, *myo*inositol, is widely distributed in plants, but also forms an important constituent of mammalian inulin. Penta-erythritol is an entirely synthetic tetrahydric alcohol obtained by condensation of formaldehyde and acetaldehyde at room temperature in the presence of calcium hydroxide. The product is probably formed as a result of a mixed-aldol condensation followed by a Cannizzaro reaction.

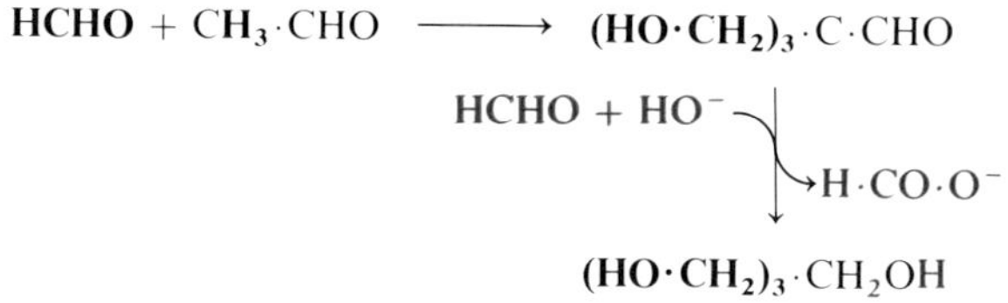

$$HCHO + CH_3{\cdot}CHO \longrightarrow (HO{\cdot}CH_2)_3{\cdot}C{\cdot}CHO$$

$$HCHO + HO^- \rightarrow H{\cdot}CO{\cdot}O^-$$

$$\downarrow$$

$$(HO{\cdot}CH_2)_3{\cdot}CH_2OH$$

## Nomenclature

α-Glycols are named as derivatives of the corresponding unsaturated hydrocarbon. The term **glycol** is used in trivial names, but the IUPAC system uses the suffix, **-diol**.

$HO{\cdot}CH_2{\cdot}CH_2{\cdot}OH$ Ethylene glycol

$CH_3{\cdot}CHOH{\cdot}CH_2OH$ Propylene glycol

$HO{\cdot}CH_2{\cdot}CH_2{\cdot}CH_2{\cdot}OH$ Propane-1,3-diol (Trimethylene glycol)

β- and γ-Glycols are described as **polymethylene** glycols, as for example, trimethylene glycol. Triols, tetritols, pentitols and hexitols are mostly known by trivial names, many of which relate to the corresponding sugars, from which they are readily obtained by reduction.

$HO{\cdot}CH_2{\cdot}CHOH{\cdot}CH_2{\cdot}OH$

Glycerol

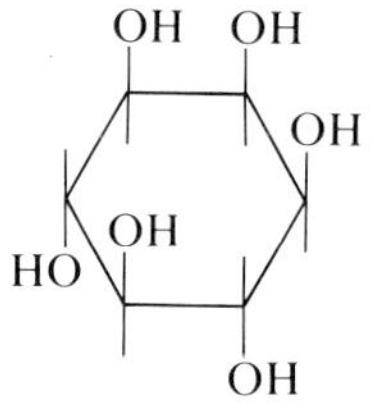

[cyclohexane ring bearing six OH groups]

*myo*-Inositol

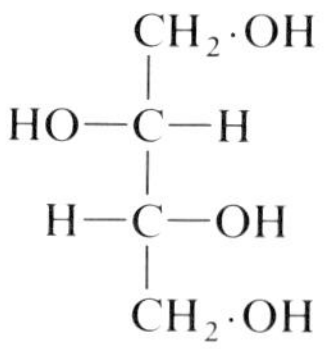

$CH_2{\cdot}OH$ / HO—C—H / H—C—OH / $CH_2{\cdot}OH$

D-Threitol

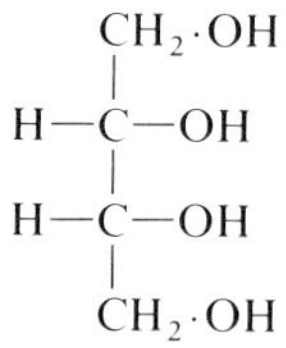

$CH_2{\cdot}OH$ / H—C—OH / H—C—OH / $CH_2{\cdot}OH$

Erythritol

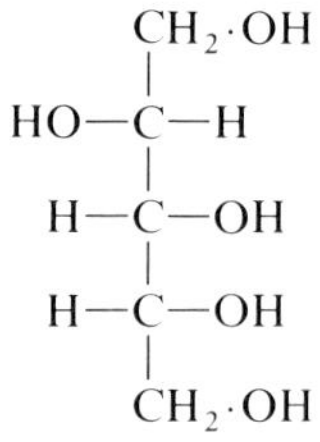

$CH_2{\cdot}OH$ / HO—C—H / H—C—OH / H—C—OH / $CH_2{\cdot}OH$

D-Arabitol

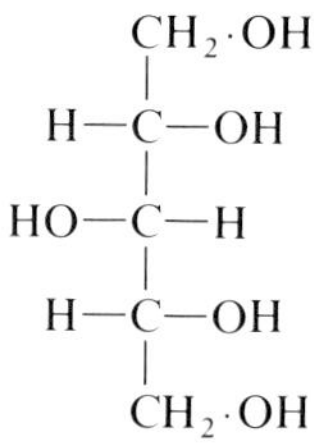

$CH_2{\cdot}OH$ / H—C—OH / HO—C—H / H—C—OH / $CH_2{\cdot}OH$

D-Xylitol

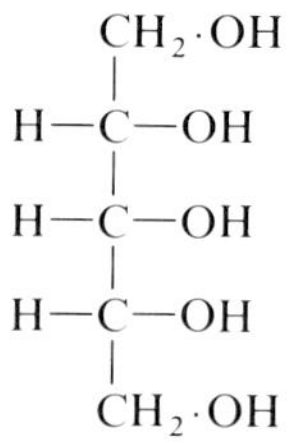

$CH_2{\cdot}OH$ / H—C—OH / H—C—OH / H—C—OH / $CH_2{\cdot}OH$

D-Ribitol

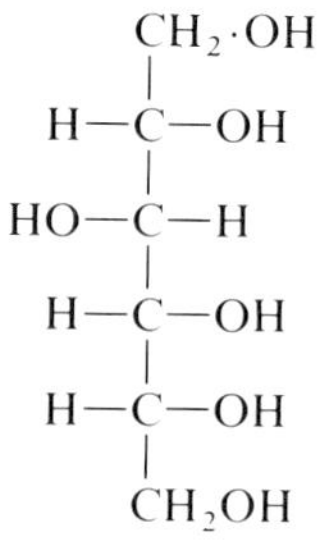

$CH_2{\cdot}OH$ / H—C—OH / HO—C—H / H—C—OH / H—C—OH / $CH_2OH$

D-Sorbitol

$CH_2{\cdot}OH$ / HO—C—H / HO—C—H / H—C—OH / H—C—OH / $CH_2{\cdot}OH$

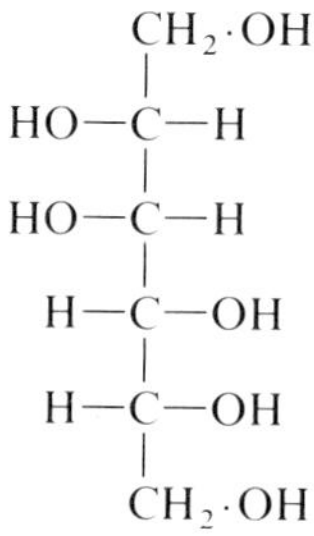

D-Mannitol

## PHYSICAL PROPERTIES

Ethylene and propylene glycols, and glycerol are all high-boiling, viscous hygroscopic, neutral liquids. Trimethylene glycol is also liquid, but the higher polyols are all crystalline solids. Pure dry glycerol is crystallisable, and the crystals melt at *ca* 25°C. They all have a sweet taste, the term **glycol** being derived from the Greek, *glukus*, meaning sweet. Both glycerol and sorbitol are used as sweetening agents in syrups and linctuses, whilst the sweetish taste of mannitol also contributes to its use as a bulk diluent in the manufacture of compressed tablets. The simple glycols and glycerol are all heavier than water, but completely miscible with it and also with ethanol in conformity with their polyhydroxylic character. Ethylene glycol and propylene glycol are also completely miscible with chloroform, but glycerol is not; the former two are also miscible with ether, whilst glycerol is not. All three are immiscible with petrol and fixed oils. Glycols and glycerol depress the freezing point of water, and ethylene glycol in particular is widely used as an antifreeze. The solid polyols are all highly water-soluble, a property which in the case of mannitol aids the disintegration of compressed tablets prepared with it.

Both ethylene and propylene glycols have wide solvent powers, but the use of ethylene glycol for this purpose is limited by its toxicity, which arises through metabolic oxidation to oxalic acid (**2**, 5). Propylene glycol, on the other hand, is metabolised to lactic acid, and for this reason is relatively non-toxic. Trimethylene glycol is more toxic than propylene glycol, it being metabolised to malonic acid, which forms an insoluble calcium salt and acts as an enzyme inhibitor. Glycerol is natural to mammalian metabolism, and in general presents no toxicity problems. It is, however, toxic in very large doses both orally and by injection.

Propylene glycol is occasionally used as an industrial solvent for the extraction of alkaloids, volatile oils, steroids and dyestuffs. Pharmaceutically, propylene glycol is valuable as a solvent for certain vitamins, sulpha drugs, barbiturates, steroids and other substances of low water solubility. It is used in this way in the preparation of *Chloramphenicol Ear Drops*, since chloramphenicol is soluble 1 in 7 in propylene glycol compared with about 1 in 400 in water. Propylene glycol is also used in the preparation of *Digoxin*, *Dimenhydrinate Melarsaprol* and *Phenobarbitone* (Phenobarbital) *Injections* to overcome problems of low solubility. It cannot, however, be used in preparations for instillation into the nose and eye because of the irritant effects which it produces (page 662). Glycerol shows the same disadvantage; none-the-less, considerable use is made of its wide solvent powers and viscosity as a liquid vehicle for the instillation of such widely differing compounds as alum, borax and phenol into the nose, mouth, pharynx and ear for local therapy.

Alkali metal salts of weak acids are more soluble and usually more stable in aqueous propylene glycol than in water, due to suppression of ionisation and hydrolysis (Greco, 1953). Thus, the solubility of *Secobarbitone Sodium* (Secobarbital Sodium) in aqueous systems increases in the presence of glycols in the order glycerol, propylene glycol, polyoxyethylene glycol 400, ethanol (Udani and Autian, 1960).

Glycols, including glycerol, are hygroscopic and are used as humectants, particularly in creams and ointments for topical application to the skin. The glycol forms a protective film on the epidermis which assists moisture retention, and by its emollient action hastens the healing of minor abrasions. Their hygroscopic action also accounts for their irritancy, and for the stinging sensation which is experienced when the anhydrous compounds come directly into contact with mucous surfaces and open wounds. This is due to the localised cellular dehydration which results. The use of glycerol as a laxative by rectal insertion in suppositories is likewise due to a similar localised dehydration, and consequent irritant action on the rectal mucosa as a stimulus to induce peristalsis in the pelvic and descending colons. Affinity for water also accounts for the mild diuretic properties of mannitol, which appears to act solely by an osmotic effect. The effectiveness of *Kaolin Poultice* when applied (hot) to boils and carbuncles is also at least in part ascribable to the hygroscopic properties of the glycerol which is present.

The glycols are all viscous liquids, and they are frequently used to increase the viscosity of aqueous solutions, such as *Insulin Injection*, and particularly to stabilise suspensions for injection. Glycerol also has very weak antiseptic powers, and is an effective preservative at concentrations over 50%. Sorbitol is used to increase the specific gravity of the liquid vehicle in aqueous suspensions for injection to match that of the suspended drug. This gives good dispersability and aids re-suspendability on re-constitution prior to injection. Mannitol is frequently used in the preparation of lozenges because of its **cooling taste**, which is due to its negative heat of solution (*ca* 120 kJ $g^{-1}$).

## CHEMICAL PROPERTIES

Chemically, 1,2-diols function as typical alcohols (Chapter 7). They are distinguished from other diols, however, by the ease with which they undergo oxidation. Like their monohydric counterparts, they are essentially neutral in character, but none-the-less display acidic function in the replacement of the hydroxylic hydrogens by alkali and alkaline earth metals. The first hydroxylic hydrogen is readily displaced by sodium at room temperature, but the second is only replaced at a much higher temperature (150°C), since the mono-alkoxide ion opposes the approach of the metal.

$$HO{\cdot}CH_2{\cdot}CH_2{\cdot}OH \xrightarrow[\downarrow H_2]{Na} HO{\cdot}CH_2{\cdot}CH_2{\cdot}O^-Na^+$$

$$\xrightarrow[\rightarrow H_2]{Na^+} Na^{+-}O{\cdot}CH_2{\cdot}CH_2{\cdot}O^-Na^+$$

Glycerol reacts similarly. Sodium readily displaces one of the α-hydroxylic hydrogens, and the second only with difficulty. No reaction occurs at the β-hydroxyl group.

$$HO\cdot CH_2\cdot CHOH\cdot CH_2\cdot OH \xrightarrow[-H_2]{Na} HO\cdot CH_2\cdot CHOH\cdot CH_2\cdot O^-Na^+ \xrightarrow[-H_2]{Na} Na^{+\,-}O\cdot CH_2\cdot CHOH\cdot CH_2O^-Na^+$$

**Esterification**

Hydrogen chloride similarly reacts with glycols in stages. With ethylene glycol, it first gives ethylene chlorhydrin (at 160°C), and then at a somewhat higher temperature (200°C), dichloroethane. Dichloroethane is also formed when ethylene glycol is treated with phosphorus trichloride or thionyl chloride.

$$HO\cdot CH_2\cdot CHOH \xrightarrow[-H_2O]{HCl,\ 160^\circ} \underset{\text{Ethylene chlorhydrin}}{HO\cdot CH_2\cdot CH_2\cdot \mathbf{Cl}} \xrightarrow[-H_2O]{HCl,\ 200^\circ} \underset{\text{Dichloroethane}}{\mathbf{Cl}\cdot CH_2\cdot CH_2\cdot \mathbf{Cl}}$$

Glycerol saturated with dry hydrogen chloride forms both possible (α and β) chlorhydrins, of which the α-form predominates.

$$HO\cdot CH_2\cdot CHOH\cdot CH_2\cdot \mathbf{Cl}$$

α-Monochlorhydrin
(1-Chloro-2,3-dihydroxypropane)

$$HO\cdot CH_2\cdot CH\cdot \mathbf{Cl}\cdot CH_2OH$$

β-Monochlorhydrin
(2-Chloro-1,3-dihydroxypropane)

With glacial acetic acid as solvent, however, hydrogen chloride gives mainly α-dichlorhydrin, which on treatment with sodium hydroxide solution yields epichlorhydrin.

$$HO\cdot CH_2\cdot CHOH\cdot CH_2OH \xrightarrow{HCl/HAc} \mathbf{Cl}\cdot CH_2\cdot CH\cdot OH\cdot CH_2\mathbf{Cl}$$

α-Dichlorhydrin
(1,3-Dichloro-2-hydroxypropane)

$$\xrightarrow[\to H_2O + \mathbf{Cl}^-]{HO^-} \underset{\text{(O bridging the first two carbons)}}{CH_2\cdot CH\cdot CH_2\cdot \mathbf{Cl}}$$

Epichlorhydrin

Glycerophosphates occur naturally in combination with choline as lecithin in soya-bean, egg-yolk, brain and nervous tissue, liver, and blood corpuscles. In mammalian systems, biosynthesis of triglycerides is initiated in the liver with the formation of L-glycerophosphate both from free glycerol, and also from dihydroxyacetone phosphate. The latter pathway is dominant in the intestinal mucosa and in adipose tissue which are virtually devoid of glycerokinase.

$$HO-CH(CH_2OH)-CH_2OH \xrightarrow[\text{ATP} \rightarrow \text{ADP}]{Mg^{2+}\ \text{Glycerokinase}} HO-CH(CH_2OH)-CH_2 \cdot O \cdot PO_3H^-$$

L-Glycerophosphate

$$HO-CH(CH_2OH)-CH_2 \cdot O \cdot PO_3H^- \xleftarrow[\text{NADH} \rightarrow \text{NAD}^+]{\text{L-Glycerophosphate-NAD-oxidoreductase}} CH_2OH \cdot C{=}O \cdot CH_2 \cdot O \cdot PO_3H^-$$

L-Glycerophosphate ← Dihydroxyacetone phosphate

The resulting L-glycerophosphate is then acylated by two moles of fatty acid-derived acyl—S·$\overline{\text{CoA}}$ to yield phosphatidic acids, which are key intermediates in the formation of triglycerides.

$$HO-CH(CH_2 \cdot OH)-CH_2 \cdot O \cdot PO_3H^- \xrightarrow[\text{HS}\cdot\overline{\text{CoA}} \quad \text{HS}\cdot\overline{\text{CoA}}]{\mathbf{R^1 \cdot CO} \cdot S \cdot \overline{\text{CoA}} \quad \mathbf{R^2 \cdot CO} \cdot S \cdot \overline{\text{CoA}}} \mathbf{R^2 \cdot CO} \cdot O-CH(CH_2 \cdot O \cdot \mathbf{CO \cdot R^1})-CH_2 \cdot O \cdot PO_3H^-$$

$$\xrightarrow[\rightarrow H_2PO_4^-]{-H_2O} \mathbf{R^2 \cdot CO \cdot O}-CH(CH_2 \cdot O \cdot \mathbf{CO \cdot R^1})-CH_2 \cdot OH$$

$$\mathbf{R^2 \cdot CO \cdot O}-CH(CH_2 \cdot O \cdot \mathbf{CO \cdot R^1})-CH_2 \cdot OH \xrightarrow[\text{HS}\cdot\overline{\text{CoA}}]{\mathbf{R^3 \cdot CO} \cdot S \cdot \overline{\text{CoA}}} \mathbf{R^2 \cdot CO} \cdot O-CH(CH_2 \cdot O \cdot \mathbf{CO \cdot R^1})-CH_2 \cdot O \cdot \mathbf{CO \cdot R^3}$$

The phosphatidic acids are also essential intermediates in the biosynthesis of phosphatidylserines, phosphatidylethanolamines, and the phosphatidylcholines

$$\begin{array}{l} CH_2{\cdot}O{\cdot}CO{\cdot}R^1 \\ R^2{\cdot}CO{\cdot}O{\cdot}CH \\ CH_2{\cdot}O{\cdot}PO_3H^- \end{array} \xrightarrow[\text{PP}]{\text{CTP}} \begin{array}{l} CH_2{\cdot}O{\cdot}CO{\cdot}R^1 \\ R^2{\cdot}CO{\cdot}O{\cdot}CH \\ CH_2{\cdot}O{\cdot}CDP \end{array}$$

Phosphatidic acid → CDP-Diglyceride

Phosphatidic acid —($H_2O$ in; $H_2PO_4^-$ out)→ αβ-Diglyceride

$$\begin{array}{l} CH_2{\cdot}O{\cdot}CO{\cdot}R^1 \\ R^2{\cdot}CO{\cdot}O{\cdot}CH \\ CH_2{\cdot}O{\cdot}H \end{array}$$

αβ-Diglyceride

CDP-Diglyceride —($HO{\cdot}CH_2{\cdot}CH(CO{\cdot}O^-){\cdot}\overset{+}{N}H_3$ in; CMP out)→ Phosphatidyl serine

$$\begin{array}{l} CH_2{\cdot}O{\cdot}CO{\cdot}R^1 \\ R^2{\cdot}CO{\cdot}O{\cdot}CH \\ CH_2{\cdot}O{\cdot}P(=O)(O^-){\cdot}O{\cdot}CH_2{\cdot}CH(CO{\cdot}O^-){\cdot}\overset{+}{N}H_3 \end{array}$$

Phosphatidyl serine

αβ-Diglyceride —($CDP{\cdot}O{\cdot}CH_2{\cdot}CH_2{\cdot}NH_2$ in; CMP out)→ Phosphatidylethanolamine

Phosphatidyl serine —($CO_2$ out)→ Phosphatidylethanolamine

$$\begin{array}{l} CH_2{\cdot}O{\cdot}CO{\cdot}R^1 \\ R^2{\cdot}CO{\cdot}O{\cdot}CH \\ CH_2{\cdot}O{\cdot}P(=O)(O^-){\cdot}O{\cdot}CH_2{\cdot}CH_2{\cdot}NH_2 \end{array}$$

Phosphatidylethanolamine

Phosphatidylethanolamine —(*S*-Adenosylmethionine in; *S*-Adenosylhomocysteine out)→ Phosphatidylcholine (lecithin)

αβ-Diglyceride —($CDP{-}O{\cdot}CH_2{\cdot}CH_2{\cdot}\overset{+}{N}(CH_3)_3$ in; CMP out)→ Phosphatidylcholine (lecithin)

$$\begin{array}{l} CH_2{\cdot}O{\cdot}CO{\cdot}R^1 \\ R^2{\cdot}CO{\cdot}O{\cdot}CH \\ CH_2{\cdot}O{\cdot}P(=O)(O^-){\cdot}O{\cdot}CH_2{\cdot}CH_2\overset{+}{N}(CH_3)_3HO^- \end{array}$$

Phosphatidylcholine (lecithin)

(lecithins). The latter are waxy, apparently crystalline solids, which swell in water to form translucent colloidal solutions, and are believed to have important functions in natural cell membrane structures.

The phosphatidylinositols, which are found in brain and other mammalian tissues, are formed from cytidine diphosphoryl-diglycerides (CDP-diglycerides) and inositol in a similar manner to the phosphatidylserines. The reaction is essentially a nucleophilic displacement involving inversion of configuration at the 1-position of inositol to yield 1-phosphatidyl-L-*myo*inositol.

$CH_2{\cdot}O{\cdot}CO{\cdot}R^1$
$R^2{\cdot}CO{\cdot}O{\cdot}CH$
$CH_2{\cdot}O{\cdot}$**CDP**

CDP-diglyceride

\+

HO OH OH OH 1 OH OH

Inositol

→ **CMP**

HO OH OH O
O·P·O·$CH_2$
OH OH O– HC·O·CO·$R^2$
$CH_2O{\cdot}CO{\cdot}R^1$

1-Phosphatidyl L *myo*-inositol

The glycerol- and ribitol-teichoic acids found as constituents of certain bacterial cell walls are polyphosphate polymers derived from L-glycerophosphate and ribitol-5′-phosphate respectively, *via* CDP-glycerol and CDP-ribitol.

*Glyceryl Tinitrate* [Nitroglycerin; $O_2NO{\cdot}CH_2{\cdot}CH(ONO_2){\cdot}CH_2{\cdot}ONO_2$], a synthetic ester prepared by adding glycerol to a cooled (10°C) mixture of concentrated sulphuric and fuming nitric acids, and pouring the product into water, is a colourless and odourless oil. Its vapour, however, is toxic and induces headaches. *Glyceryl Trinitrate* is used as a heart stimulant in the treatment of cardiac disorders, and is considered to act by reduction to the nitrite ester *in vivo*, and release of nitrite ions by hydrolysis. *Amyl* and *Octyl Nitrites*, which are sufficiently volatile to be administered by inhalation, are equally if not more efficient.

$$\begin{array}{l} CH_2\cdot OH \\ | \\ HO\cdot CH \\ | \\ CH_2\cdot O\cdot \mathbf{PO_3H^-} \end{array} \xrightarrow[\mathbf{PP}]{\mathbf{CTP}} \begin{array}{l} CH_2\cdot OH \\ | \\ HO\cdot CH \\ | \\ CH_2\cdot O\cdot \mathbf{CDP} \end{array}$$

L-Glycerophosphate

L-Glycerophosphate → **CMP**

$$\begin{array}{l} CH_2\cdot OH \\ | \\ HO\cdot CH \\ | \\ CH_2 - \end{array} O - \left[ \begin{array}{c} O \\ \| \\ P \\ | \\ O^- \end{array} - O\cdot CH_2 - HO\cdot CH - CH_2 - O \right]_n - \begin{array}{c} O \\ \| \\ P \\ | \\ O^- \end{array} - \begin{array}{l} O\cdot CH_2 \\ | \\ HO\cdot CH \\ | \\ CH_2\cdot O\cdot \mathbf{PO_3H^-} \end{array}$$

Nucleosidediphosphohexoside → Nucleotidediphosphate

$$\begin{array}{l} CH_2\cdot OH \\ | \\ Hexosido\cdot O\cdot CH \\ | \\ CH_2 - \end{array} O - \left[ \begin{array}{c} O \\ \| \\ P \\ | \\ O^- \end{array} - O\cdot CH_2 - Hexosido\cdot O\cdot CH - CH_2 - O \right]_n - \begin{array}{c} O \\ \| \\ P \\ | \\ O^- \end{array} - \begin{array}{l} O\cdot CH_2 \\ | \\ Hexosido\cdot O\cdot CH \\ | \\ CH_2\cdot O\cdot \mathbf{PO_3H^-} \end{array}$$

Glycerol teichoic acid

*Glyceryl Trinitrate* is inflammable, and burns harmlessly, but explodes if detonated, breaking down completely to nitrogen, carbon dioxide, oxygen and water vapour.

$$4C_3H_5(ONO_2)_3 \longrightarrow 6N_2 + 12CO_2 + O_2 + 10H_2O$$

Dynamite, which is manufactured by absorbing glyceryl trinitrate onto Kieselguhr, is also explosive, but can be handled more safely.

Polyhydric alcohols are equally readily esterified with organic acids. The process occurs stepwise, and the product is largely determined by the proportion of the reactants.

$$HO \cdot CH_2 \cdot CH_2 \cdot OH \xrightarrow{R \cdot CO \cdot OH/H^+} R \cdot CO \cdot O \cdot CH_2 \cdot CH_2 \cdot OH \xrightarrow{R \cdot CO \cdot OH/H^+} R \cdot CO \cdot O \cdot CH_2 \cdot CH_2 \cdot O \cdot COR$$

*Glyceryl Monostearate*, prepared from glycerol and stearic acid, is a mixture of mono-, di- and tri-stearates, containing between 30 and 40% of the monoester, together with free stearic acid and glycerol. It is a waxy solid, which is insoluble in cold water, but dispersable in hot water and at ambient temperatures in the presence of surfactants. It is used as an emulsifying agent, although it is not particularly efficient. It is useful, however, as an emulsion stabiliser, and is usually used in conjunction with a surfactant.

Reaction of diols with dibasic acids gives linear polyesters, as for example with terephthalic acid, which gives Terylene.

$$nHO \cdot CH_2 \cdot CH_2 \cdot OH + nHO \cdot OC \cdot C_6H_4 \cdot CO \cdot OH \longrightarrow -O \cdot CH_2 \cdot CH_2O \left( OC \cdot C_6H_4 \cdot CO \cdot O \cdot CH_2 \cdot CH_2 \cdot O \right)_n OC \cdot C_6H_4 \cdot CO-$$

1,2-Diols also form cyclic esters with boric acid in aqueous solution, which are strongly acidic, and which in the case of optically active diols show considerable exaltation of optical rotation.

$$R\text{-}CH(OH)\text{-}CH(OH)\text{-}R \xrightarrow{H_3BO_3} R\text{-}CH(O)\text{-}CH(O)\text{-}R\,\text{>}B\text{-}OH + \left[ \begin{array}{c} R \quad\quad\quad\quad R \\ H\text{-}C\text{-}O \quad O\text{-}C\text{-}H \\ B \\ H\text{-}C\text{-}O \quad O\text{-}C\text{-}H \\ R^1 \quad\quad\quad\quad R^1 \end{array} \right]^- H^+$$

Glycerol and mannitol react in this way with boric acid to give analogous complexes. These are strongly acidic, and their formation provides a basis for the alkalimetric determination of boric acid (Diehl, 1937).

Weser (1967) has shown that boric acid forms a 2:1-complex with nucleosides, and that borate interferes with RNA synthesis by complexing with the ribose-2′,3′-*cis*-diol group of nucleotides.

In a somewhat analogous way, 1,2-diols condense with aldehydes and ketones under anhydrous acid conditions to form cyclic acetals and ketals (1,3-dioxolans). Thus, glycerol reacts with formaldehyde to form glycerol formal, a useful water-miscible, neutral, non-toxic solvent for pharmacological studies of drugs. It may be susceptible to aqueous decomposition under acidic conditions.

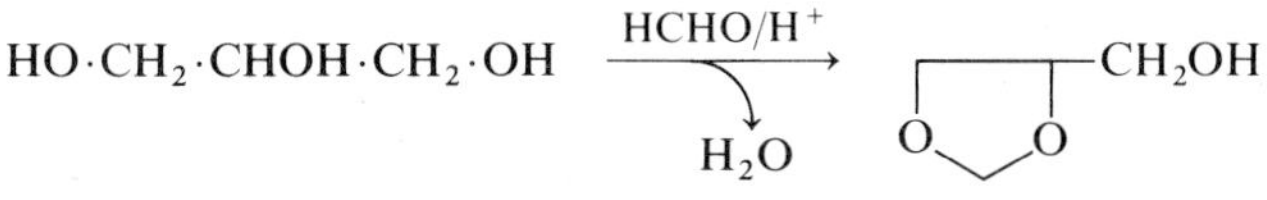

$$HO\cdot CH_2\cdot CHOH\cdot CH_2\cdot OH \xrightarrow[H_2O]{HCHO/H^+} \text{(1,3-dioxolan ring with } CH_2OH)$$

Glycerol formal

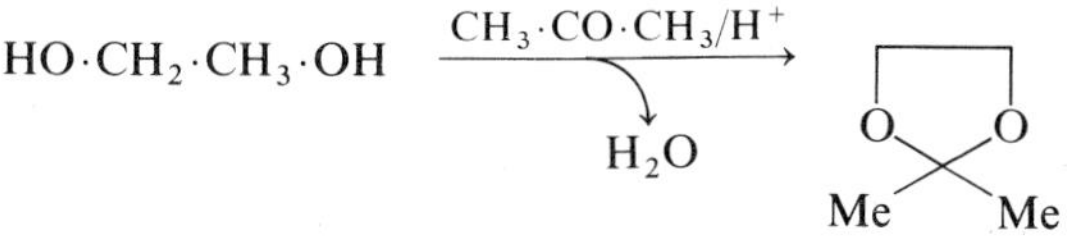

$$HO\cdot CH_2\cdot CH_3\cdot OH \xrightarrow[H_2O]{CH_3\cdot CO\cdot CH_3/H^+} \text{(1,3-dioxolan ring, 2-Me, 2-Me)}$$

2,2-Dimethyl-1,3-dioxolan

## Glycol Ethers

The formation and solvent properties of simple glycol ethers have been discussed elsewhere (Chapter 7). The plasmalogens are important vinyl ethers of phosphatidylethanolamines and cholines found in human plasma and spermatozoa (Chapter 16). The synthetic glycerol ethers, *Mephenesin*, *Chlorphenesin* and *Guaiphenesin* (Guaifenesin), have central muscle relaxant properties, and may be used as cough suppressants. *Mephenesin* is metabolically oxidised in man and excreted as its oxidation product, 3-*o*-tolyloxylactic acid.

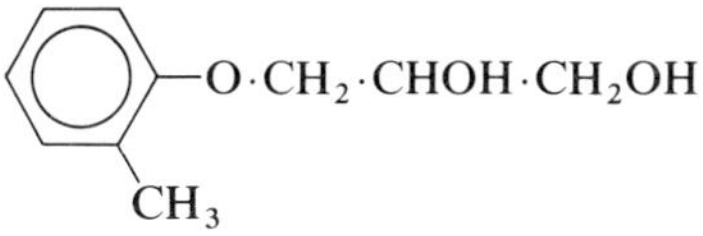

$$o\text{-}CH_3C_6H_4\text{—}O\cdot CH_2\cdot CHOH\cdot CH_2OH$$

*Mephenesin*

$$o\text{-}CH_3C_6H_4\text{—}O\cdot CH_2\cdot CHOH\cdot CO\cdot OH$$

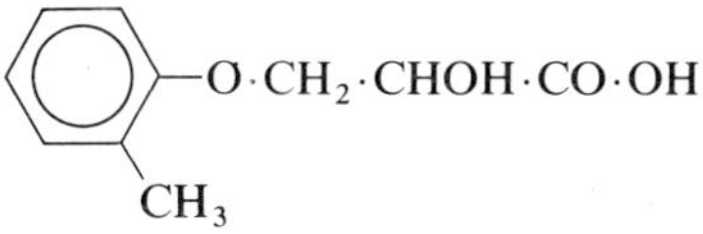

3-*o*-Tolyloxylatic acid

## Oxidation

Oxidation of ethylene glycol with nitric acid yields glycollic acid and/or oxalic acid. The intermediate products may be isolated, but cannot be obtained in appreciable yield by this method (reaction scheme page 630).

Metabolic oxidation of ethylene glycols in mammals proceeds by two pathways. The major pathway is to carbon dioxide and water, but oxidation also occurs by an alternative pathway to oxalic acid, to an extent which varies considerably with species, increasing in the order, rabbit, rat, cat (Gessner, Parke and Williams, 1961). This is also the species order of increasing toxicity due to ethylene glycol; oxalic acid formation also accounts for its toxicity in man. This

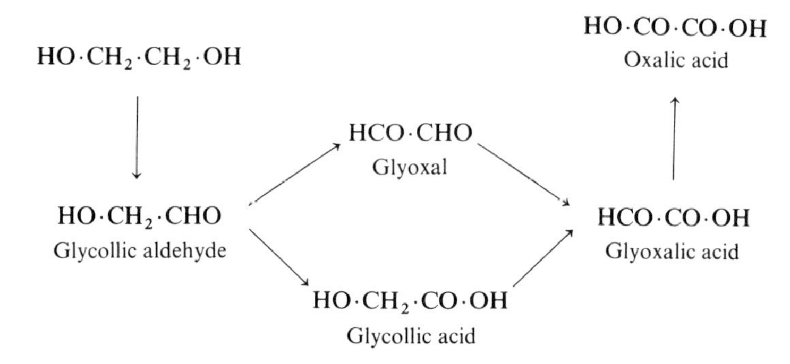


shows, though only when administered in large doses, as kidney damage due to deposition of oxalates in the renal tubules. Propylene glycol is oxidatively metabolised to lactic acid, which is innocuous, and further metabolised to formate and acetate. Its isomer, trimethylene glycol, is metabolised to malonic acid, which forms an insoluble calcium salt, and is also an inhibitor of succinic dehydrogenase.

$$\underset{\text{Trimethylene glycol}}{HO{\cdot}CH_2{\cdot}CH_2{\cdot}CH_2OH} \longrightarrow \underset{\beta\text{-Hydroxypropionic acid}}{HO{\cdot}CH_2{\cdot}CH_2{\cdot}CO{\cdot}OH}$$

$$\downarrow$$

$$\underset{\text{Malonic acid}}{HO{\cdot}CO{\cdot}CH_2{\cdot}CO{\cdot}OH}$$

Oxidation of glycerol with mild oxidising agents leads to the formation of a mixture of isomers, glyceraldehyde ($HO{\cdot}CH_2{\cdot}CHOH{\cdot}CHO$) and dihydroxyacetone ($HO{\cdot}CH_2{\cdot}CO{\cdot}CH_2OH$). The enzymically-controlled interconversions of L-glycerophosphate, dihydroxyacetone phosphate and D-glyceraldehyde-3-phosphate have important rôles in the Embden–Meyerhof–Parnas (EMP) glycolytic pathway for the breakdown of glucose and fructose to pyruvate (Chapter 21). The interconversion of dihydroxyacetone phosphate and D-glyceraldehyde-3-phosphate is controlled by triose-phosphate isomerase. It is found in most organisms, whilst that between L-glycerophosphate and dihydroxyacetone phosphate is confined to certain micro-organisms.

Both glyceraldehyde and dihydroxyacetone are formed when glycerol is oxidised with hydrogen peroxide, organic peroxides, ferrous salts, dilute nitric acid and other oxidising agents. Concentrated nitric acid converts glycerol to glyceric acid ($HO{\cdot}CH_2{\cdot}CHOH{\cdot}COOH$). Its derivatives, 1,3-diphosphoglycerate, 3-phosphoglycerate and 2-phosphoglycerate, are also important intermediates in the EMP-glycolysis pathway, being formed successively by oxidation of D-glyceraldehyde phosphate.

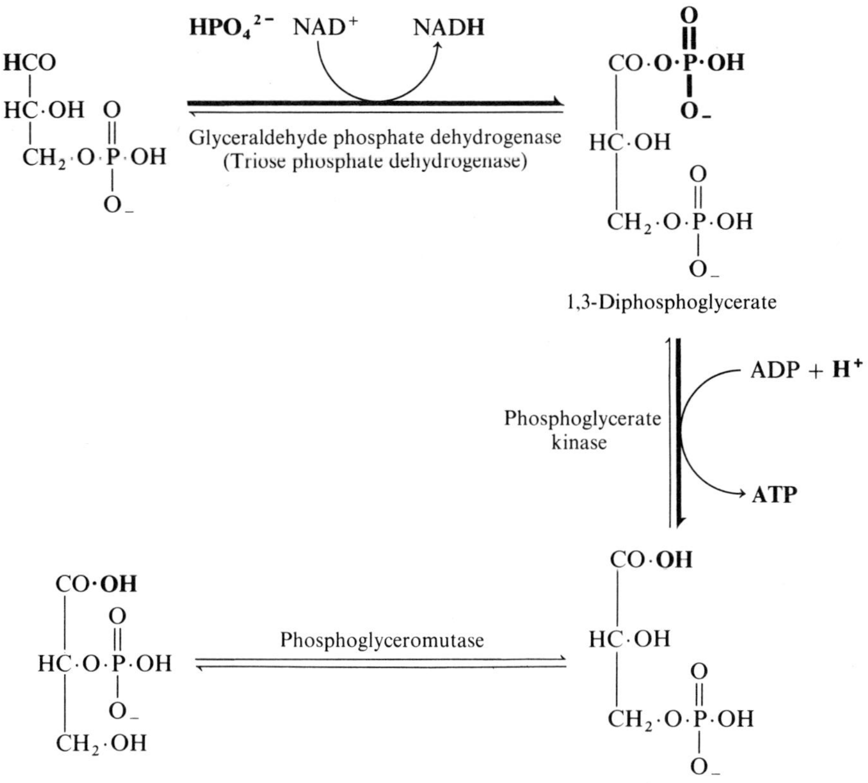

HCO · HC·OH · $CH_2·O·P(=O)(O^-)·OH$ + $HPO_4^{2-}$, $NAD^+$ → NADH ⇌ (Glyceraldehyde phosphate dehydrogenase (Triose phosphate dehydrogenase)) CO·O·P(=O)(O₋)·OH · HC·OH · $CH_2·O·P(=O)(O^-)·OH$

1,3-Diphosphoglycerate

⇅ Phosphoglycerate kinase (ADP + $H^+$ → ATP)

CO·OH · HC·OH · $CH_2·O·P(=O)(O^-)·OH$

D-3-Phosphoglycerate

⇌ Phosphoglyceromutase

CO·OH · HC·O·P(=O)(O₋)·OH · $CH_2·OH$

D-2-Phosphoglycerate

1,2-Diols are readily distinguished from other diols by the ease with which they undergo oxidation with periodic acid ($HIO_4$) with cleavage of the carbon–carbon bond.

$$HO·CH_2·CH_2·OH + HIO_4 \longrightarrow 2HCHO + H_2O + HIO_3$$

The reaction which is general to all 1,2-diols proceeds through the formation of a cyclic ester.

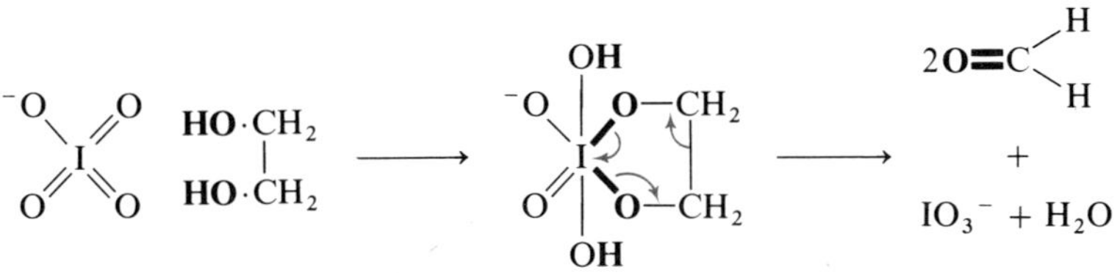

$$^-O-IO_3 + HO·CH_2–HO·CH_2 \longrightarrow \text{cyclic ester } (^-O)(O)I(OH)_2(O–CH_2)(O–CH_2) \longrightarrow 2O{=}CH_2 + IO_3^- + H_2O$$

Periodic acid oxidation cleaves the molecule of glycerol to formaldehyde (2 mols) and formic acid (1 mol).

$$HO{\cdot}CH_2{\cdot}CHOH{\cdot}CH_2OH + 2HIO_4 \longrightarrow 2HCHO + H{\cdot}CO{\cdot}OH + 2HIO_3 + H_2O$$

Periodate oxidation with C—C bond cleavage also extends to 1,2-diketones, to $\alpha$-ketols and 1,2-amino alcohols, and has been widely used in structural determinations.

Catalytic oxidation of cyclitols over Pt catalysts, and bacteriological oxidation is stereoselective for C—H equatorial bonds. Thus, *myo*-inositol is oxidised catalytically (Heyns and Paulsen, 1953) and also, similarly, by *Acetobacter suboxydans* (Kluyver and Boezaardt, 1939) to *scyllo*-inosose.

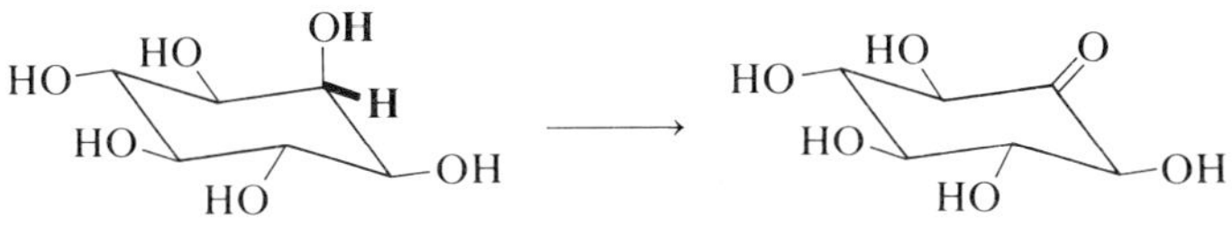


*myo*-Inositol    *Scyllo*-inosose

**Dehydration**

Ethylene glycol loses water to form acetaldehyde on heating with dehydrating agents such as zinc chloride.

$$HO{\cdot}CH_2{\cdot}CH_2{\cdot}OH \xrightarrow[H_2O]{ZnCl_2} [CH_2{=}CH{\cdot}OH] \longrightarrow \underset{\text{Acetaldehyde}}{CH_3{\cdot}CHO}$$

Dehydration of 1,2-dialkyl-1,2-diols occurs similarly and gives rise to a mixture of ketones.

Treatment with acidic dehydrating agents, such as sulphuric or phosphoric acids, leads to self-condensation with the formation of diethylene, triethylene and higher glycols or dioxan depending upon the actual conditions employed.

$$2HO{\cdot}CH_2{\cdot}CH_2{\cdot}OH \xrightarrow[H_2O]{H^+} \underset{\text{Diethylene glycol}}{HO{\cdot}CH_2{\cdot}CH_2{\cdot}O{\cdot}CH_2{\cdot}CH_2OH}$$

$$\downarrow H^+ \quad (\rightarrow H_2O)$$

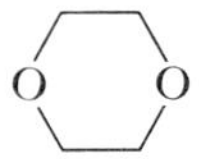


Dioxan

Glycerol is readily dehydrated when heated with anhydrous potassium sulphate or concentrated sulphuric acid to yield the unsaturated aldehyde, acrolein. The reaction appears to proceed by a combination of an acid-catalysed dehydration, tautomerism and a second acid-catalysed dehydration.

$$HO{\cdot}CH_2{\cdot}CH(\overset{+}{O}H_2){\cdot}CH_2{\cdot}OH \xrightarrow{-H_2O} HO{\cdot}CH_2{\cdot}\overset{+}{C}H{-}CH_2{-}OH \xrightarrow{-H^+} HO{\cdot}CH_2{\cdot}CH{=}CH{-}O{-}H$$

$$HO{\cdot}CH_2{\cdot}CH{=}CH{-}OH \longrightarrow HO{\cdot}CH_2{\cdot}CH_2{\cdot}CHO \xrightarrow{H^+} H\overset{+}{O}H{-}CH_2{-}CH_2{\cdot}CHO \xrightarrow{-(H_2O + H^+)} CH_2{=}CH{\cdot}CHO$$

Acrolein

Acid-catalysed dehydration of di-tertiary 1,2-glycols triggers off the pinacol-pinacolone rearrangement, which proceeds as follows.

$$\underset{\text{Pinacol}}{(CH_3)_2{\cdot}C(OH){-}C(OH)(CH_3)_2} \underset{}{\overset{H^+}{\rightleftharpoons}} (CH_3)_2{\cdot}C(\overset{+}{O}H_2){-}C(OH)(CH_3)_2 \xrightarrow{-H_2O} (CH_3)_2\overset{+}{C}{-}C(CH_3)(OH){-}CH_3$$

$$(CH_3)_2\overset{+}{C}{-}C(CH_3)(OH){-}CH_3 \longrightarrow (CH_3)_3C{-}\overset{+}{C}(OH){-}CH_3 \overset{-H^+}{\rightleftharpoons} \underset{\text{Pinacolone}}{(CH_3)_3C{-}CO{-}CH_3}$$

**Diethyleneglycol** (Di-2-hydroxyethyl ether)

Diethyleneglycol is less toxic than ethylene glycol, but also undergoes oxidative metabolism in mammals to oxalic acid. Its monomethyl ether (methyl carbitol, Chapter 7) has important solvent properties for organic compounds.

In contrast to ethylene glycol, diethylene glycol is subject to autoxidation (Lloyd, 1956). This is subject to a slow induction period of 1–200 hr depending upon the purity and history of the sample, during which only small amounts of oxygen are consumed. Subsequent autoxidation, which proceeds rapidly with oxygen consumption at rates of about 20 mmol $l^{-1}$ $hr^{-1}$, is markedly accelerated by acid between pH 1.8 and 5.2 and temperatures above 35°C. The principal products of autoxidation, ethylene glycol, formic acid and formaldehyde, which are formed in a 1 : 1 ratio, are consistent with the formation of the hydroperoxide, 1-hydroperoxy-2,2′-dihydroxydiethyl ether, and its subsequent acid-catalysed decomposition.

$$HO{\cdot}CH_2{\cdot}CH_2{\cdot}O{\cdot}CH_2{\cdot}CH_2{\cdot}OH$$

$$\downarrow O_2$$

$$HO{\cdot}CH_2{\cdot}CH_2{\cdot}O{\cdot}\underset{\displaystyle O{\cdot}OH}{CH}{\cdot}CH_2OH \xrightarrow{H^+} HO{\cdot}CH_2{\cdot}CH_2{\cdot}O{\cdot}\underset{\displaystyle H_2\overset{+}{O}{-}O}{CH}{-}\overset{H}{\underset{H}{C}}{-}O{-}H$$

$$\downarrow H_2O + H^+$$

$$HO{\cdot}CH_2{\cdot}CH_2{\cdot}OH + HCO{\cdot}OH \xleftarrow{H_2O/H^+} HO{\cdot}CH_2{\cdot}CH_2{\cdot}O{\cdot}C(H){=}O + HCHO$$

**Dioxan**

Dioxan is also a valuable solvent both for organic compounds and certain inorganic salts. It is largely used as an industrial solvent, and is particularly useful on account of its miscibility with water. Dioxan undergoes some oxidative decomposition on storage and may contain traces of acetaldehyde, as its acetal with ethylene glycol. Unless rigorously dried, it nearly always contains some water. Dioxan may be purified by treatment with dilute hydrochloric acid at its boiling point in a stream of nitrogen to remove acetaldehyde liberated from the contaminant by hydrolysis. This treatment is followed by distillation.

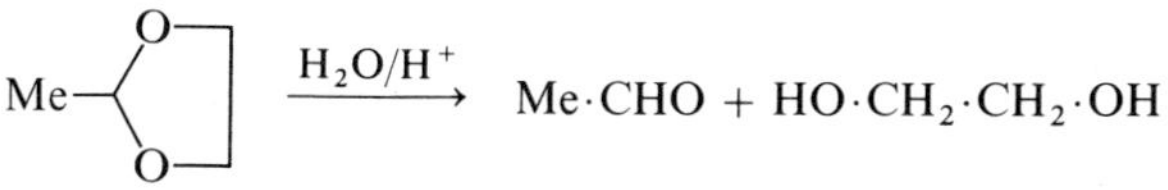

$$\text{Me}-\text{CH}\langle\begin{matrix}\text{O}-\\\text{O}-\end{matrix}\rangle \xrightarrow{H_2O/H^+} \text{Me}\cdot\text{CHO} + \text{HO}\cdot\text{CH}_2\cdot\text{CH}_2\cdot\text{OH}$$

Acetaldehyde ethyleneglycal — Acetaldehyde

## Sorbitan

Sorbitan, 1,4-anhydro-D-sorbitol, obtained by direct acid-catalysed dehydration of sorbitol, forms the base of the polysorbate esters used as solubilising agents (p. 640).

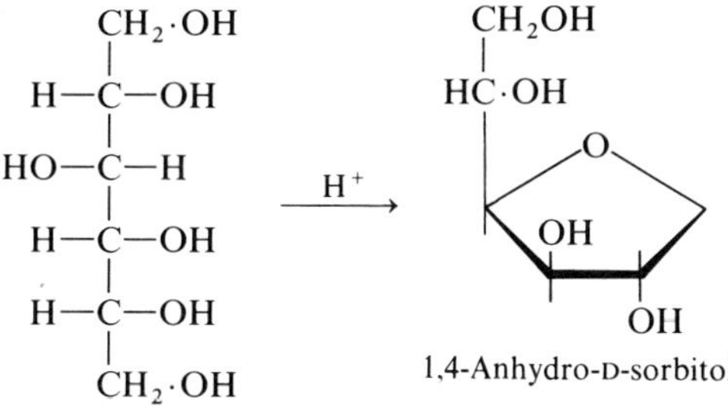

$$\begin{matrix}CH_2\cdot OH\\ H-C-OH\\ HO-C-H\\ H-C-OH\\ H-C-OH\\ CH_2\cdot OH\end{matrix} \xrightarrow{H^+} \begin{matrix}CH_2OH\\ HC\cdot OH\\ \text{(furanose ring with O, OH, OH)}\end{matrix}$$

D-Sorbitol — 1,4-Anhydro-D-sorbitol

# POLYOXYALKYLENE GLYCOLS

## Preparation

Polyoxyethylene glycols are formed by the condensation of ethylene oxide with either water, ethylene glycol or, preferably, diethylene glycol, at high temperature and pressure, and in the presence of a small quantity of sodium hydroxide.

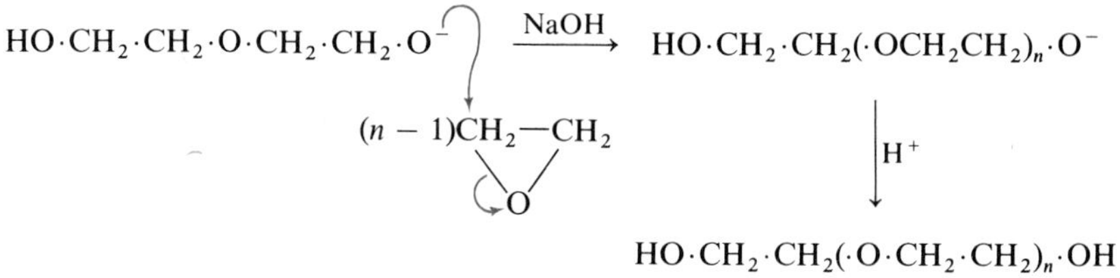

$$HO\cdot CH_2\cdot CH_2\cdot O\cdot CH_2\cdot CH_2\cdot O^- + (n-1)CH_2{-}CH_2(O) \xrightarrow{NaOH} HO\cdot CH_2\cdot CH_2(\cdot OCH_2CH_2)_n\cdot O^-$$

$$\xrightarrow{H^+} HO\cdot CH_2\cdot CH_2(\cdot O\cdot CH_2\cdot CH_2)_n\cdot OH$$

The resulting polyoxyethylene glycols can be formed up to any required molecular weight, the process being controlled by the supply of ethylene oxide. Polyoxypropylene glycols are prepared similarly from propylene glycol and propylene oxide. Mixed polyoxyethylene-polyoxypropylene glycols are also available.

## Grades

Polyoxyethylene glycols (Macrogols; Carbowaxes) are available in various grades of average molecular weight ranging from 200 to 6000, and polyoxypropylene glycols in average molecular weights from 150 to 2025. The various

Table 64. Properties of Polyoxyalkylene Glycols

| Grade | n | Average molecular weight | m.p. (°C) | Consistency at ambient temperatures |
|---|---|---|---|---|
| Polyoxyethylene glycols | | | | |
| 200 | 3–4 | 190–200 | −15 | liquid |
| 300 | 4–5 | 285–315 | −12 | liquid |
| 400 | 7–9 | 380–420 | 7 | liquid |
| 600 | 11–13 | 570–630 | 18 | liquid |
| 800 | 16–18 | 760–840 | 29 | grease |
| 1000 | 20–22 | 950–1050 | 36 | grease |
| 1540 | 31–34 | 1430–1570 | 45 | soft wax |
| 4000 | 70–85 | 3300–3600 | 51 | hard wax |
| 6000 | 105–120 | 6000–7500 | 55 | hard wax |
| Polyoxypropylene glycols | | | | |
| 150 | 2 | 140–160 | | liquid |
| 425 | 4–6 | 350–450 | −60 | liquid |
| 750 | 10–12 | 700–800 | | liquid |
| 1025 | 14–16 | 900–1050 | −50 | liquid |
| 1200 | 17–21 | 1100–1300 | | liquid |
| 2025 | 33–35 | 1900–2100 | −45 | liquid |

molecular weight grades show a wide range of physical properties (Table 64). Low molecular weight polyoxyethylene glycols are colourless, viscous liquids, miscible with water in all proportions; those of molecular weight between 700 and 1000 are soft greases, and higher molecular weight polymers are wax-like solids. Some grades are formed by admixture; thus, Carbowax 1500 is actually a mixture of the 1540 and 300 grades; it has an average molecular weight of 500–600, melts at 38°C, and is soft and waxy in character.

## Properties

All grades of polyoxyethylene glycols are soluble in water, but higher molecular weight fractions are only partially miscible, and show increasing solubility in organic solvents such as chloroform. All grades, however, are insoluble in ether. In general, polyoxypropylene glycols have lower melting points, lower water solubility and greater solubility in organic solvents than the corresponding grade of polyoxyethylene glycol. All are reasonably stable, though low molecular weight grades have been shown to exhibit oxidising effects due to the presence of peroxides (Coates, Pashley and Tattersall, 1961). Polyoxyalkylene glycols do not undergo hydrolysis, and will not support mould growth. They are reputed to be incompatible with tannic acid, phenol and salicylic acid.

Aqueous solutions of polyoxyethylene glycols assist the solution of drugs with poor water solubility. In this way, they increase the absorption giving higher plasma levels of orally-administered *Griseofulvin* (Marvel, Schlichting, Denton, Levy and Cahn, 1964). Like the simple polyhydric alcohols, they suppress ionisation and hydrolysis of weak acid salts. As a result, solutions of *Amylobarbitone* (Amylobarbital) *Sodium* and *Quinalbarbitone* (Quinalbarbital) *Sodium* in 50% aqueous polyoxyethylene glycol 200 are less irritant than simple aqueous solutions on intramuscular injection (Swanson, Anderson, Harris and Rose, 1953). Polyoxyethylene glycols are also useful in the preparation of aqueous solutions containing up to 30% of iodine, the iodine being freely available for antiseptic purposes on dilution with water (Hurst, 1963).

## POLYOXYALKYLENE GLYCOL ETHERS

Cetomacrogol 1000, a typical polyoxyethylene glycol ether, is prepared by condensing either cetyl or ceto-stearyl alcohol with ethylene oxide at a temperature of 140 to 180°C under pressure.

$$CH_3(CH_2)_m{\cdot}OH + n\underbrace{CH_2{-}CH_2}_{O} \longrightarrow CH_3(CH_2)_m({\cdot}OCH_2CH_2)_n{\cdot}OH$$

$$m = 15 - 17 \qquad n = 20 - 24$$

### Physical Properties

The product is soluble in water, acetone and ethanol, but insoluble in light-petroleum fractions. Molecular association to form micelles occurs in aqueous solution. The critical micellar concentration (CMC), the concentration at which micelles are formed, depends on molecular structure, and in particular the chain length of the polyoxyethylene moiety. As this increases in length, the monomer becomes more hydrophilic, and in consequence the CMC increases. The CMC of Cetomacrogol 1000 is about $4.0 \times 10^{-6}$ $mol\,l^{-1}$. It is used extensively in the preparation of oil-in-water emulsions, which are unaffected by high concentrations of electrolytes, and stable over a wide range of pH. Studies of the effect of varying the polyoxyethylene chain length on emulsion stability show that this increases as the number of polyoxyethylene units (n) increases from 3 to 9 (Elworthy and Florence, 1963).

### Solubilisation

Polyoxyalkylene glycol ethers at concentrations above the critical micelle concentration (CMC) are also widely used to enhance the aqueous solubility of drug substances with limited solubility. In some cases, however, such as that of *Sulphisoxazole* (Sulfisoxole), increased micellar solubilisation is medically disadvantageous, inhibiting rectal absorption of the unionised sulphonamide (Kakemi, Arita and Muranishi, 1965). By suitable choice of particular polymers with appropriate hydrocarbon and oxyethylene chains, solubilisation of a wide range of chemical types can be achieved. In general, the shorter the hydrophilic oxyethylene chain relative to the hydrophobic moiety, the greater the solubilising effect for non-polar molecules (Saito and Shinoda, 1967).

Solubilities of pure hydrocarbons (decane and benzene), *Sulphadiazine* (Sulfamethazine), *p*-hydroxybenzoic acid, and ethyl and butyl *p*-hydroxybenzoates increase linearly with surfactant concentration above the CMC. Because of their relatively low solubility in both water and hydrocarbon solvents, both *Sulphadiazine* (Corby and Elworthy, 1971) and *Griseofulvin* (Elworthy and Lipscombe, 1968) show only small increases in solubility in cetomacrogol solutions. For many compounds, however, micellar solubility in cetomacrogol solutions runs parallel with solubility in water and contrary to solubility in hydrocarbon solutions. Ultraviolet and NMR studies show that whereas decane and benzene are solubilised mainly in the hydrocarbon core of the micelle, *p*-hydroxybenzoic acid is solubilised in the polyoxyethylene layer, whilst ethyl *p*-hydroxybenzoate is solubilised partly in the hydrocarbon core, but mostly by the oxyethylene chains of the micelle (Corby and Elworthy, 1971).

The solubilisation of bactericidal agents with Cetomacrogol 1000 and other polyoxyalkylene ethers in the preparation of disinfectant solutions has been studied extensively. Phenolic compounds, such as *Chloroxylenol*, form hydrogen bonds between the phenolic hydroxyls and ether oxygens of the polyoxyalkylene chain, and solubility increases with polymer concentration above the CMC (Mulley and Metcalf, 1956).

The antibacterial activity of the *Chloroxylenol*, however, is solely related to the amount of phenol free in aqueous solutions, irrespective of the quantity present in the micelles (Bean and Berry, 1951; Mitchell, 1964). Solubilisation of antiseptic phenols with polyoxyalkylene glycols, therefore, provides a means of preparing concentrated solutions for sale and distribution. To be fully effective, however, they must be diluted before use.

Iodine is similarly solubilised by polyoxyalkylene ethers such as *Cetomacrogol* (Terry and Shelanski, 1952), due it is said to direct intermolecular interaction between the ether oxygens and molecular iodine by charge-transfer (Henderson and Newton, 1966), but this has been disputed (Elworthy, Florence and Macfarlane, 1968). The solutions, like their phenolic counterparts, provide high concentrations of iodine in a non-volatile, water-soluble form to buffer losses from the system in contact with bacteria. The surfactant, further, aids penetration of hair follicles, which are impervious to aqueous iodine-KI solutions (Allawala and Riegelman, 1953), and also inhibits metal corrosion.

**Stability and Incompatabilities**

Certain organic bases and their salts have been shown to deteriorate in the presence of *Cetomacrogol*. Thus, aqueous solutions of *Benzocaine Hydrochloride* (0.25 M) containing *Cetomacrogol* (3%) become yellow within a week at room temperature (Azaz, Donbrow and Hamburger, 1973). This deterioration is due to autoxidation of *Cetomacrogol* leading to formation of peroxides, and can be suppressed by the use of such antioxidants as *Butylated Hydroxytoluene* (BHT), *Propyl Gallate* and *Sodium Metabisulphite* (Sodium Metabisulfite).

Peroxidation of *Cetomacrogol* in aqueous solution is retarded by cooling, but accelerated by mineral acid to the extent that solutions (3%) in *N* hydrochloric acid give rise to a colourless precipitate after six months at ambient temperatures. These effects are consistent with the heterolytic peroxide cleavage shown by diethylene glycol (p. 634).

$$HO{\cdot}CH_2{\cdot}CH_2({\cdot}O{\cdot}CH_2{\cdot}CH_2)_n{\cdot}OH$$

$$\downarrow O_2$$

$$HO{\cdot}CH_2{\cdot}CH_2({\cdot}O{\cdot}CH_2{\cdot}CH_2)_{n-1}{\cdot}O{\cdot}\underset{\displaystyle O{\cdot}OH}{\underset{|}{C}H}{\cdot}CH_2{\cdot}OH$$

$$\downarrow H^+$$

$$HO{\cdot}CH_2{\cdot}CH_2({\cdot}O{\cdot}CH_2{\cdot}CH_2)_{n-1}{\cdot}O{\cdot}C(H){=}O + HCHO$$

$$\downarrow H^+ \quad \rightarrow H{\cdot}CO{\cdot}OH$$

$$HO{\cdot}CH_2{\cdot}CH_2(O{\cdot}CH_2{\cdot}CH_2)_{n-1}{\cdot}OH + (HCHO)_n\downarrow$$

## POLYOXYALKYLENE GLYCOL ESTERS

### Preparation

Polyoxyalkylene glycol esters are usually prepared by heating a particular grade of polyoxyethylene glycol with a long-chain fatty acid at 120–130°C.

$$R(CH_2)_m{\cdot}CO{\cdot}OH + HO{\cdot}CH_2{\cdot}CH_2({\cdot}OCH_2{\cdot}CH_2)_nOH$$

$$\downarrow$$

$$R(CH_2)_m{\cdot}CO{\cdot}O{\cdot}CH_2{\cdot}CH_2({\cdot}O{\cdot}CH_2{\cdot}CH_2)_nOH$$

Alternatively, they can also be made by heating fatty acids with ethylene oxide at 140 to 180°C under pressure.

The parent anhydro-hexitol, sorbitan, from which the polyoxyethylene sorbitan esters are derived is obtained by acid-catalysed dehydration of sorbitol (Chapter 21). Sorbitan esters with lauric, palmitic, stearic and oleic acids known as **Spans** (20, 40, 60 and 80 respectively), which are formed on heating sorbitan with the required acid, are mixtures of undefined constitution. They conform, however, approximately to monosorbitan monoesters.

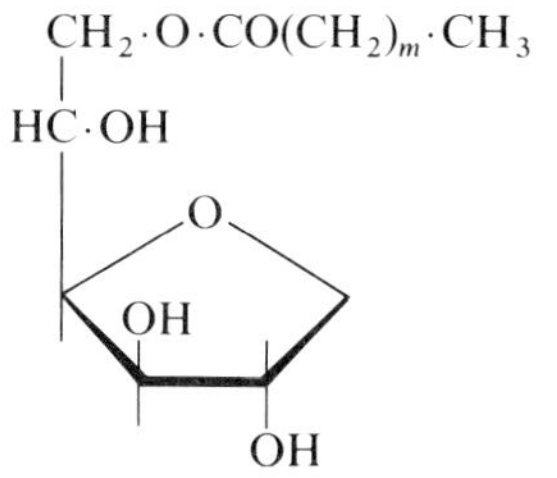

$CH_2 \cdot O \cdot CO(CH_2)_m \cdot CH_3$
$HC \cdot OH$
O
OH
OH

Spans

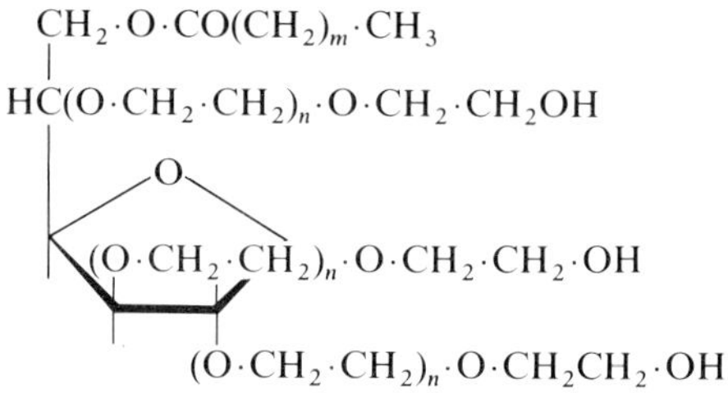

$CH_2 \cdot O \cdot CO(CH_2)_m \cdot CH_3$
$HC(O \cdot CH_2 \cdot CH_2)_n \cdot O \cdot CH_2 \cdot CH_2OH$
O
$(O \cdot CH_2 \cdot CH_2)_n \cdot O \cdot CH_2 \cdot CH_2 \cdot OH$
$(O \cdot CH_2 \cdot CH_2)_n \cdot O \cdot CH_2CH_2 \cdot OH$

Tweens (Polysorbates)

Treatment of Spans with ethylene oxide gives the corresponding **Tweens** or polysorbates, which are polyoxyethylene glycol ethers of the sorbitan esters containing some 10 to 30 oxyethylene units.

## Properties

The properties of the products depend both on the grade of polyalkylene glycol used, and on the nature of the fatty acid component. They often contain small quantities of free fatty acid, which is advantageous in that it improves their emulsifying properties. In general, polyoxyalkylene glycols of molecular weight less than 400 form esters which are soluble in organic solvents, but only form colloidal dispersions in water. Higher molecular weight polymer esters are water-soluble, but show decreased oil solubility. They are used mainly as emulsifying and solubilising agents and in the preparation of ointments, creams and suspensions, as also are the corresponding polysorbates, which are polyoxyethylene sorbitan esters.

Both Spans and Tweens have surfactant properties, which combined with their low toxicity and bland taste, make them extremely useful as commercial solubilisers and emulsifiers, particularly in products, both medicinal and food, for internal administration. The Spans, although dispersable in water, are not soluble. Tweens, however, being much more hydrophilic in structure, are water-soluble. They are also neutral in reaction, and largely unaffected by electrolytes. The grades of polysorbate correspond to those of the Spans (Table 65).

Polysorbates, like their counterparts, the polyoxyalkylene glycols, are widely used to assist solubilisation and suspendability of poorly soluble compounds in the preparation of compounded medicines, including *Vitamin A Alcohol*

Table 65. Nomenclature of Spans and Tweens

| | |
|---|---|
| Span 20 | Sorbitan monolaurate |
| Span 40 | Sorbitan monopalmitate |
| Span 60 | Sorbitan monostearate |
| Span 80 | Sorbitan mono-oleate |
| Span 85 | Sorbitan tri-oleate |
| Tween 20 (Polysorbate 20) | Polyoxyethylene sorbitan monolaurate |
| Tween 40 (Polysorbate 40) | Polyoxyethylene sorbitan monopalmitate |
| Tween 60 (Polysorbate 60) | Polyoxyethylene sorbitan monostearate |
| Tween 80 (Polysorbate 80) | Polyoxyethylene sorbitan mono-oleate |

(retinol) and *Palmitate*, *Phytomenadione* (Phytonadione; Vitamin $K_1$), *Cholecalciferol* (Gstirner and Tata, 1958; Nakagawa and Muneyuki, 1954), anti-inflammatory steroids for ophthalmic use (Güttman, Hamlin, Shell and Wagner, 1961), synthetic oestrogens (Nakagawa, 1954, 1956) and *Chloramphenicol* (Matsumura *et al.*, 1958). The use of *Polysorbate 80* as a deflocculating agent in the preparation of injectable aqueous suspensions of water-insoluble compounds is especially valuable, since it gives rise to products with excellent solid dispersability and prevents **claying**.

There is also evidence that combinations of polysorbate with glycerol or sorbitol are more effective solubilisers of *Vitamin A Palmitate* than polysorbate itself (Huettenrauch and Klotz, 1963). This is due to interchain hydrogen bonding of the hydroxylic polyol which is sandwiched in molecular association between the polyoxyethylene chains. As a result, hydration of the polymer is inhibited, and its effective hydrophilic character increased, so that it more closely resembles pure glycol.

Retinol is undoubtedly more stable in aqueous *Polysorbate 20* than in oily solution (Kern and Antoshkiw, 1950). Occasionally, however, clouding of polysorbate solutions causes problems with some compounds such as *Retinol* and *Cholecalciferol*, and separation into layers or actual precipitation may occur (Mima, 1958). This may be due to the susceptibility of such active ingredients to autoxidation, which clearly must promote autoxidation of the *Polysorbate* in the same way as for *Cetomacrogol* (p. 638).

## GLYCAMINES

Glycamines are derivatives of polyols in which one or other of the hydroxyl groups have been replaced by a primary or secondary amine function. *Riboflavine* (Chapter 23) is a naturally-occurring derivative of D-ribamine (1-amino-1-deoxyribitol). Other important natural glycamines are sphingosine and dihydrosphingosine and the related sphingomyelins and cerebrosides which are important constituents of nerve and brain tissue.

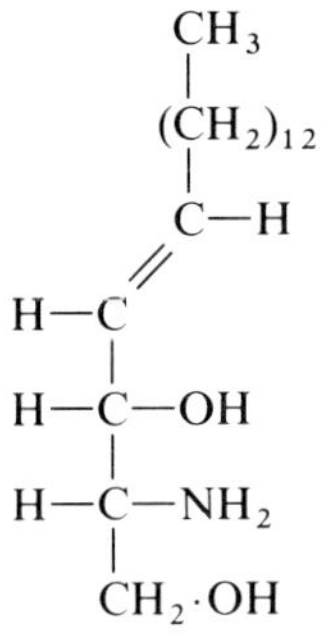

$$
\begin{array}{l}
\quad CH_3 \\
\quad | \\
\quad (CH_2)_{12} \\
\quad | \\
\quad C{-}H \\
\quad /\!/ \\
H{-}C \\
\quad | \\
H{-}C{-}OH \\
\quad | \\
H{-}C{-}NH_2 \\
\quad | \\
\quad CH_2{\cdot}OH
\end{array}
$$

Sphingosine

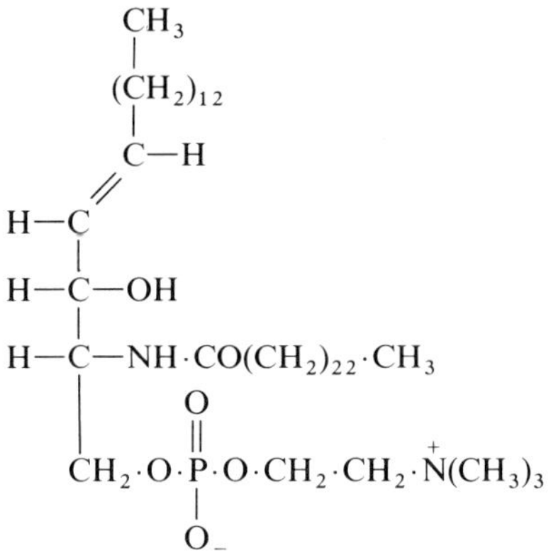

$$
\begin{array}{l}
\quad CH_3 \\
\quad | \\
\quad (CH_2)_{12} \\
\quad | \\
\quad C{-}H \\
\quad /\!/ \\
H{-}C \\
\quad | \\
H{-}C{-}OH \\
\quad | \\
H{-}C{-}NH{\cdot}CO(CH_2)_{22}{\cdot}CH_3 \\
\quad | \qquad\quad O \\
\quad | \qquad\quad \| \\
\quad CH_2{\cdot}O{\cdot}P{\cdot}O{\cdot}CH_2{\cdot}CH_2{\cdot}\overset{+}{N}(CH_3)_3 \\
\qquad\qquad\;\; | \\
\qquad\qquad\;\; O^- 
\end{array}
$$

Sphingomyelin

$$
\begin{array}{l}
\quad CH_3 \\
\quad | \\
\quad (CH_2)_{12} \\
\quad | \\
\quad C{-}H \\
\quad /\!/ \\
H{-}C \\
\quad | \\
H{-}C{-}OH \\
\quad | \\
H{-}C{-}NH{\cdot}\mathbf{CO{\cdot}R} \\
\quad | \\
\quad CH_2 \\
\quad | \\
\quad O \\
\quad | \\
\text{(galactose ring: } CH_2{\cdot}OH,\ HO,\ HO,\ OH)
\end{array}
$$

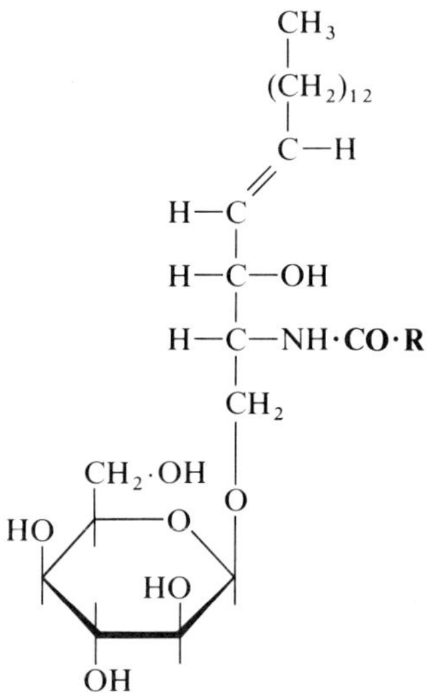

| Cerebrosides | **R·CO—** | |
|---|---|---|
| Kerasin | Lignoceric acid | $CH_3(CH_2)_{22}{\cdot}CO{\cdot}OH$ |
| Phrenesin | Cerebronic acid | $CH_3(CH_2)_{21}{\cdot}CHOH{\cdot}CO{\cdot}OH$ |
| Nervone | Nervonic acid | $CH_3(CH_2)_7{\cdot}CH{=}CH(CH_2)_{13}{\cdot}CO{\cdot}OH$ |
| Hydroxynervone | Hydroxynervonic acid | $CH_3(CH_2)_7{\cdot}CH{=}CH(CH_2)_{12}{\cdot}CHOH{\cdot}CO{\cdot}OH$ |

Gangliosides are complex sphingolipids derived from sphingosine, acetyl-neuraminic acid, hexoses and amino hexoses, and stearic acid, which are widely distributed in brain and parenchymatous tissue such as that of the erythrocytes.

Glucamine is obtained synthetically by reduction of glucose-oxime. Its *N*-methyl derivative, *Meglumine*, is a highly water-soluble base, which is used to

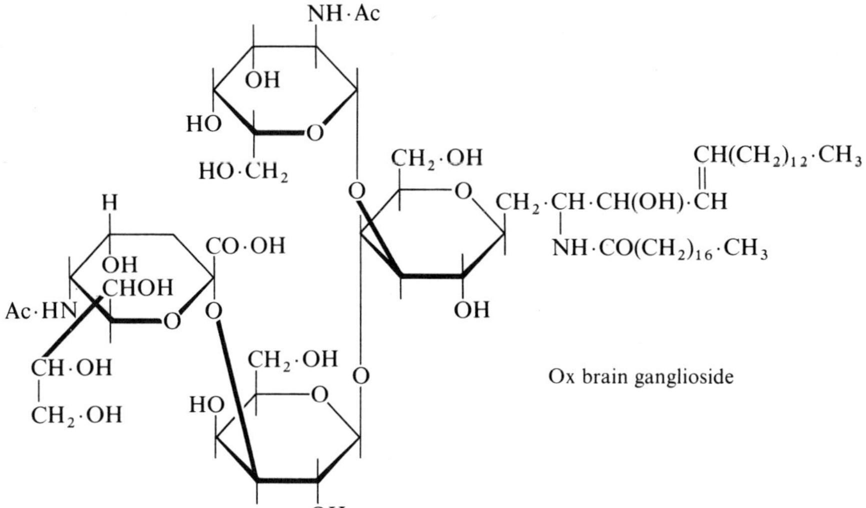

NH·Ac OH HO O HO·CH₂ CH₂·OH CH(CH₂)₁₂·CH₃ O O CH₂·CH·CH(OH)·CH H CO·OH NH·CO(CH₂)₁₆·CH₃ OH CHOH Ac·HN O O OH CH·OH CH₂·OH O CH₂·OH HO O OH

Ox brain ganglioside

H₂N O OH CHOH OH CO·OH CHOH CH₂·OH

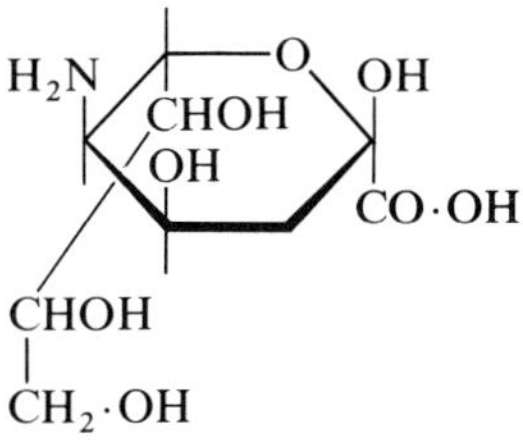

Neuraminic acid

produce water-soluble salts of insoluble iodine-containing aromatic acids such as *Iothalamic Acid* for injection as X-ray contrast media.

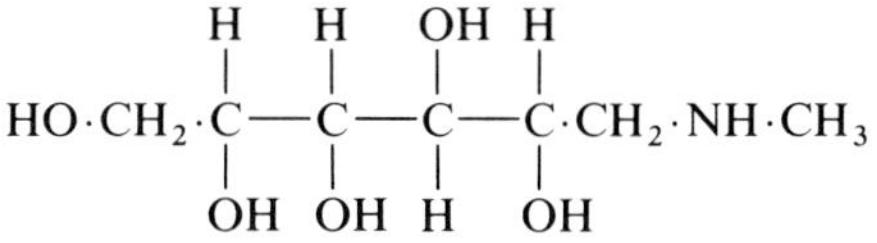

$$HO{\cdot}CH_2{\cdot}\underset{OH}{\overset{H}{C}}{-}\underset{OH}{\overset{H}{C}}{-}\underset{H}{\overset{OH}{C}}{-}\underset{OH}{\overset{H}{C}}{\cdot}CH_2{\cdot}NH{\cdot}CH_3$$

*Meglumine*

CO·OH I I CH₃·CO·HN CO·NH·CH₃ I

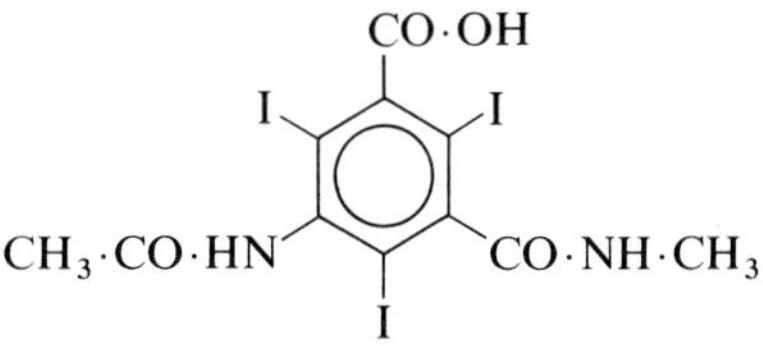

*Iothalamic Acid*

A number of important aminocyclitols, notably streptamine and its derivatives, occur naturally in the polysaccharide antibiotics, *Streptomycin*, *Neomycin*, *Kanamycin*, *Paramomycin* and *Gentamicin* (Chapter 21).

## 1,6-DIBROMO-1,6-DIDEOXYHEXOSES

*Mitobronitol* and *Mannomustine* are used as cytotoxic agents in the control of leukaemia and other forms of neoplastic disease (Ramanan and Israëls, 1969).

The mechanism of action of *Mannomustine* is similar to that of *Mustine* (Mechlorethamine Hydrochloride; **2**, 2), but it is much less toxic. According to Harper (1962), the polyol chain plays a vital rôle in transportation of the drug *in vivo*, since the analogous non-hydroxylic 1,6-di(2′-chloroethylamino)-n-hexane is inactive (Vargha, Toldy, Feher and Lendvai, 1957).

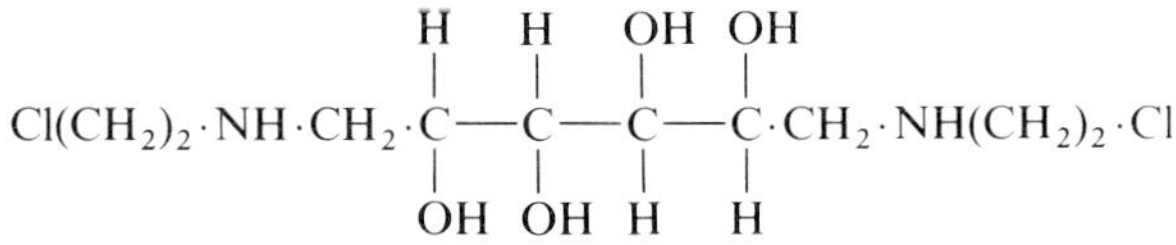

$$Cl(CH_2)_2\cdot NH\cdot CH_2\cdot \underset{OH}{\overset{H}{C}}-\underset{OH}{\overset{H}{C}}-\underset{H}{\overset{OH}{C}}-\underset{H}{\overset{OH}{C}}\cdot CH_2\cdot NH(CH_2)_2\cdot Cl$$

*Mannomustine*

$$Cl(CH_2)_2\cdot NH(CH_2)_6\cdot NH(CH_2)_2\cdot Cl$$

1,6-di(2′-chloroethylamino)-n-hexane

The stereochemical configuration of the polyol moiety is also important. The D-dulcitol (D-galactitol) analogue is inactive; and the D-sorbitol, L-mannitol and L-iditol derivatives, although active as growth inhibitors of certain transplanted tumours, are less effective than *Mannomustine*.

Mannitol-1,6-dimethanesulphonate (mannitol myleran) is also an effective tumour inhibitor (Haddow, Timmis and Brown, 1958).

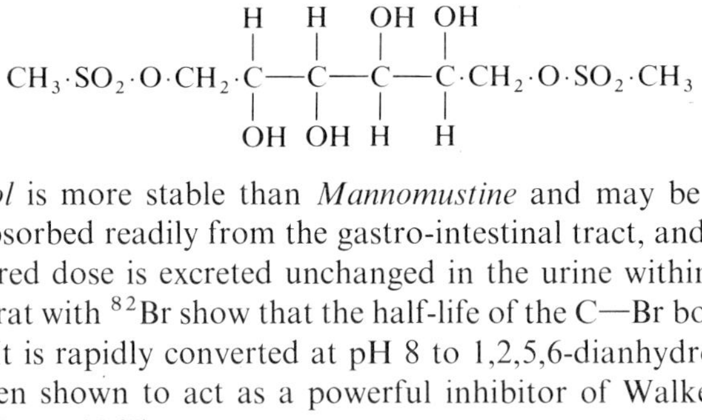

$$CH_3\cdot SO_2\cdot O\cdot CH_2\cdot \underset{OH}{\overset{H}{C}}-\underset{OH}{\overset{H}{C}}-\underset{H}{\overset{OH}{C}}-\underset{H}{\overset{OH}{C}}\cdot CH_2\cdot O\cdot SO_2\cdot CH_3$$

*Mitobronitol* is more stable than *Mannomustine* and may be administered orally. It is absorbed readily from the gastro-intestinal tract, and some 20% of the administered dose is excreted unchanged in the urine within 6 hr. Tracer studies in the rat with $^{82}Br$ show that the half-life of the C—Br bond is between 5 and 10 hr. It is rapidly converted at pH 8 to 1,2,5,6-dianhydro-D-mannitol, which has been shown to act as a powerful inhibitor of Walker rat tumour (Jarman and Ross, 1967).

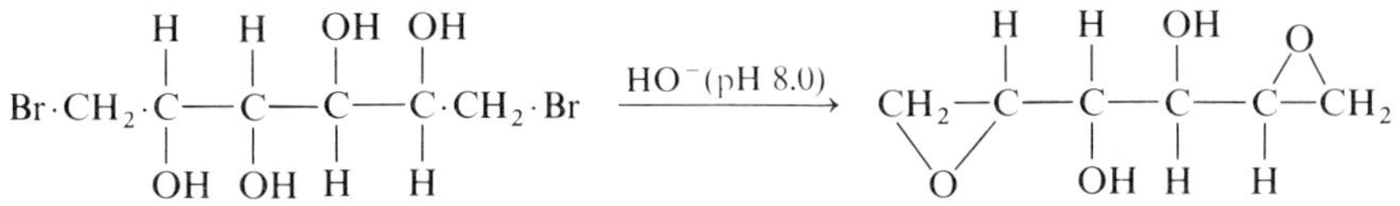

$$Br\cdot CH_2\cdot \underset{OH}{\overset{H}{C}}-\underset{OH}{\overset{H}{C}}-\underset{H}{\overset{OH}{C}}-\underset{H}{\overset{OH}{C}}\cdot CH_2\cdot Br \xrightarrow{HO^-(pH\ 8.0)} \underbrace{CH_2-\overset{H}{C}}_{O}-\underset{OH}{\overset{H}{C}}-\underset{H}{\overset{OH}{C}}-\underbrace{\underset{H}{C}-CH_2}_{O}$$

Mitobronitol — 1,2,5,6-Dianhydromannitol

# 21 Carbohydrates

### Introduction

Monosaccharides are the simplest carbohydrates, a term which embraces almost all carbon compounds of general formula $C_n(H_2O)_m$. Oligosaccharides are carbohydrates in which two, three, four or occasionally more monosaccharide units (up to at most nine or ten) are linked glycosidically. Monosaccharides and oligosaccharides are usually crystalline and have a sweet taste, and are therefore also known as sugars. In contrast, polysaccharides, which contain anything from ten to several hundred monosaccharide units, are amorphous and are not sweet in taste.

Many typical monosaccharides containing from three to eight carbon atoms occur naturally, but pentoses and hexoses containing five or six carbon atoms respectively are most widely distributed. They represent important sources of energy in most living systems. Normal human blood levels of glucose are about 100 mg/100 ml. Small amounts of pentoses, *ca* 2.5 mg/100 ml, are also present. Pentoses, particularly ribose and desoxyribose, are essential components of nucleotides and the nucleic acids, which have a central rôle in cell replication. Monosaccharides have molecular formulae which conform to $C_n(H_2O)_n$.

## MONOSACCHARIDES

### Nomenclature and Stereochemistry

Both monosaccharides and oligosaccharides are characterised by the use of names which carry the suffix **-ose**, as in glucose and fructose. Monosaccharides with 3, 4, 5, 6 and 7 carbon atoms are described generically as trioses, tetroses, pentoses, hexoses and heptoses respectively. According to whether they possess aldehydic or ketonic properties in their open chain form, they are also classified as aldose or ketose sugars of general formula:

$$HO{\cdot}CH_2(CHOH)_n{\cdot}CHO \qquad HO{\cdot}CH_2(CHOH)_n{\cdot}CO{\cdot}CH_2OH$$

Aldose — Ketose

Although glycollic aldehyde is strictly an aldose ($n = 0$), the triose glyceraldehyde ($n = 1$) is usually considered the simplest aldose. In contrast to its isomer, the ketose, dihydroxyacetone ($n = 0$), glyceraldehyde possesses an asymmetric carbon atom and exists in optically active D- and L-forms. The addition of each (CHOH) group as both aldose and ketose series are ascended similarly gives rise to a further homologous pair of isomers, related stereochemically to either D- or L-glyceraldehyde.

$$\begin{array}{rcl} & \mathrm{CHO} & \\ & | & \\ \mathrm{H}- & \mathrm{C} & -\mathrm{OH} \\ & | & \\ & \mathrm{CH_2OH} & \end{array}$$

D-Glyceraldehyde

$$\begin{array}{rcl} & \mathrm{CHO} & \\ & | & \\ \mathrm{H}- & \mathrm{C} & -\mathrm{OH} \\ & | & \\ \mathrm{H}- & \mathrm{C} & -\mathrm{OH} \\ & | & \\ & \mathrm{CH_2OH} & \end{array}$$

D-Erythrose

$$\begin{array}{rcl} & \mathrm{CHO} & \\ & | & \\ \mathrm{HO}- & \mathrm{C} & -\mathrm{H} \\ & | & \\ \mathrm{H}- & \mathrm{C} & -\mathrm{OH} \\ & | & \\ & \mathrm{CH_2OH} & \end{array}$$

D-Threose

$$\begin{array}{rcl} & \mathrm{CHO} & \\ & | & \\ \mathrm{H}- & \mathrm{C} & -\mathrm{OH} \\ & | & \\ \mathrm{H}- & \mathrm{C} & -\mathrm{OH} \\ & | & \\ \mathrm{H}- & \mathrm{C} & -\mathrm{OH} \\ & | & \\ & \mathrm{CH_2OH} & \end{array}$$

D-Ribose

$$\begin{array}{rcl} & \mathrm{CHO} & \\ & | & \\ \mathrm{HO}- & \mathrm{C} & -\mathrm{H} \\ & | & \\ \mathrm{H}- & \mathrm{C} & -\mathrm{OH} \\ & | & \\ \mathrm{H}- & \mathrm{C} & -\mathrm{OH} \\ & | & \\ & \mathrm{CH_2OH} & \end{array}$$

D-Arabinose

$$\begin{array}{rcl} & \mathrm{CHO} & \\ & | & \\ \mathrm{H}- & \mathrm{C} & -\mathrm{OH} \\ & | & \\ \mathrm{HO}- & \mathrm{C} & -\mathrm{H} \\ & | & \\ \mathrm{H}- & \mathrm{C} & -\mathrm{OH} \\ & | & \\ & \mathrm{CH_2OH} & \end{array}$$

D-Xylose

$$\begin{array}{rcl} & \mathrm{CHO} & \\ & | & \\ \mathrm{HO}- & \mathrm{C} & -\mathrm{H} \\ & | & \\ \mathrm{HO}- & \mathrm{C} & -\mathrm{H} \\ & | & \\ \mathrm{H}- & \mathrm{C} & -\mathrm{OH} \\ & | & \\ & \mathrm{CH_2OH} & \end{array}$$

D-Lyxose

$$\begin{array}{rcl} & \mathrm{CHO} & \\ & | & \\ \mathrm{H}- & \mathrm{C} & -\mathrm{OH} \\ & | & \\ \mathrm{H}- & \mathrm{C} & -\mathrm{OH} \\ & | & \\ \mathrm{H}- & \mathrm{C} & -\mathrm{OH} \\ & | & \\ \mathrm{H}- & \mathrm{C} & -\mathrm{OH} \\ & | & \\ & \mathrm{CH_2OH} & \end{array}$$

D-Allose

$$\begin{array}{rcl} & \mathrm{CHO} & \\ & | & \\ \mathrm{HO}- & \mathrm{C} & -\mathrm{H} \\ & | & \\ \mathrm{H}- & \mathrm{C} & -\mathrm{OH} \\ & | & \\ \mathrm{H}- & \mathrm{C} & -\mathrm{OH} \\ & | & \\ \mathrm{H}- & \mathrm{C} & -\mathrm{OH} \\ & | & \\ & \mathrm{CH_2OH} & \end{array}$$

D-Altrose

$$\begin{array}{rcl} & \mathrm{CHO} & \\ & | & \\ \mathrm{H}- & \mathrm{C} & -\mathrm{OH} \\ & | & \\ \mathrm{HO}- & \mathrm{C} & -\mathrm{H} \\ & | & \\ \mathrm{H}- & \mathrm{C} & -\mathrm{OH} \\ & | & \\ \mathrm{H}- & \mathrm{C} & -\mathrm{OH} \\ & | & \\ & \mathrm{CH_2OH} & \end{array}$$

D-Glucose

$$\begin{array}{rcl} & \mathrm{CHO} & \\ & | & \\ \mathrm{HO}- & \mathrm{C} & -\mathrm{H} \\ & | & \\ \mathrm{HO}- & \mathrm{C} & -\mathrm{H} \\ & | & \\ \mathrm{H}- & \mathrm{C} & -\mathrm{OH} \\ & | & \\ \mathrm{H}- & \mathrm{C} & -\mathrm{OH} \\ & | & \\ & \mathrm{CH_2OH} & \end{array}$$

D-Mannose

$$\begin{array}{rcl} & \mathrm{CHO} & \\ & | & \\ \mathrm{H}- & \mathrm{C} & -\mathrm{OH} \\ & | & \\ \mathrm{H}- & \mathrm{C} & -\mathrm{OH} \\ & | & \\ \mathrm{HO}- & \mathrm{C} & -\mathrm{H} \\ & | & \\ \mathrm{H}- & \mathrm{C} & -\mathrm{OH} \\ & | & \\ & \mathrm{CH_2OH} & \end{array}$$

D-Gulose

$$\begin{array}{rcl} & \mathrm{CHO} & \\ & | & \\ \mathrm{HO}- & \mathrm{C} & -\mathrm{H} \\ & | & \\ \mathrm{H}- & \mathrm{C} & -\mathrm{OH} \\ & | & \\ \mathrm{HO}- & \mathrm{C} & -\mathrm{H} \\ & | & \\ \mathrm{H}- & \mathrm{C} & -\mathrm{OH} \\ & | & \\ & \mathrm{CH_2OH} & \end{array}$$

D-Idose

$$\begin{array}{rcl} & \mathrm{CHO} & \\ & | & \\ \mathrm{H}- & \mathrm{C} & -\mathrm{OH} \\ & | & \\ \mathrm{HO}- & \mathrm{C} & -\mathrm{H} \\ & | & \\ \mathrm{HO}- & \mathrm{C} & -\mathrm{H} \\ & | & \\ \mathrm{H}- & \mathrm{C} & -\mathrm{OH} \\ & | & \\ & \mathrm{CH_2OH} & \end{array}$$

D-Galactose

$$\begin{array}{rcl} & \mathrm{CHO} & \\ & | & \\ \mathrm{HO}- & \mathrm{C} & -\mathrm{H} \\ & | & \\ \mathrm{HO}- & \mathrm{C} & -\mathrm{H} \\ & | & \\ \mathrm{HO}- & \mathrm{C} & -\mathrm{H} \\ & | & \\ \mathrm{H}- & \mathrm{C} & -\mathrm{OH} \\ & | & \\ & \mathrm{CH_2OH} & \end{array}$$

D-Talose

D-Aldoses derived from D-glyceraldehyde

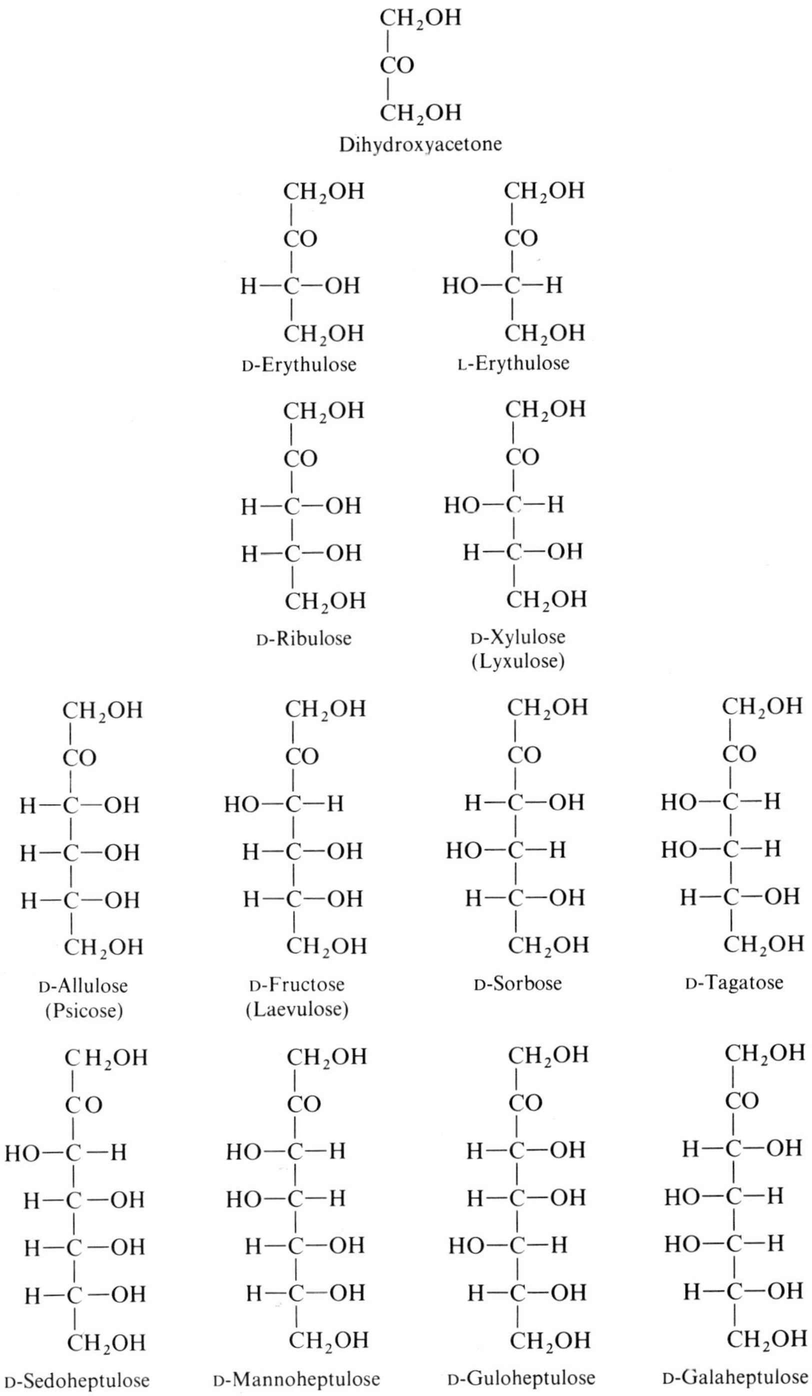

$$\begin{array}{c} CH_2OH \\ | \\ CO \\ | \\ CH_2OH \end{array}$$

Dihydroxyacetone

$$\begin{array}{c} CH_2OH \\ | \\ CO \\ | \\ H-C-OH \\ | \\ CH_2OH \end{array}$$

D-Erythulose

$$\begin{array}{c} CH_2OH \\ | \\ CO \\ | \\ HO-C-H \\ | \\ CH_2OH \end{array}$$

L-Erythulose

$$\begin{array}{c} CH_2OH \\ | \\ CO \\ | \\ H-C-OH \\ | \\ H-C-OH \\ | \\ CH_2OH \end{array}$$

D-Ribulose

$$\begin{array}{c} CH_2OH \\ | \\ CO \\ | \\ HO-C-H \\ | \\ H-C-OH \\ | \\ CH_2OH \end{array}$$

D-Xylulose
(Lyxulose)

$$\begin{array}{c} CH_2OH \\ | \\ CO \\ | \\ H-C-OH \\ | \\ H-C-OH \\ | \\ H-C-OH \\ | \\ CH_2OH \end{array}$$

D-Allulose
(Psicose)

$$\begin{array}{c} CH_2OH \\ | \\ CO \\ | \\ HO-C-H \\ | \\ H-C-OH \\ | \\ H-C-OH \\ | \\ CH_2OH \end{array}$$

D-Fructose
(Laevulose)

$$\begin{array}{c} CH_2OH \\ | \\ CO \\ | \\ H-C-OH \\ | \\ HO-C-H \\ | \\ H-C-OH \\ | \\ CH_2OH \end{array}$$

D-Sorbose

$$\begin{array}{c} CH_2OH \\ | \\ CO \\ | \\ HO-C-H \\ | \\ HO-C-H \\ | \\ H-C-OH \\ | \\ CH_2OH \end{array}$$

D-Tagatose

$$\begin{array}{c} CH_2OH \\ | \\ CO \\ | \\ HO-C-H \\ | \\ H-C-OH \\ | \\ H-C-OH \\ | \\ H-C-OH \\ | \\ CH_2OH \end{array}$$

D-Sedoheptulose

$$\begin{array}{c} CH_2OH \\ | \\ CO \\ | \\ HO-C-H \\ | \\ HO-C-H \\ | \\ H-C-OH \\ | \\ H-C-OH \\ | \\ CH_2OH \end{array}$$

D-Mannoheptulose

$$\begin{array}{c} CH_2OH \\ | \\ CO \\ | \\ H-C-OH \\ | \\ H-C-OH \\ | \\ HO-C-H \\ | \\ H-C-OH \\ | \\ CH_2OH \end{array}$$

D-Guloheptulose

$$\begin{array}{c} CH_2OH \\ | \\ CO \\ | \\ H-C-OH \\ | \\ HO-C-H \\ | \\ HO-C-H \\ | \\ H-C-OH \\ | \\ CH_2OH \end{array}$$

D-Galaheptulose

Some ketose sugars derived from dihydroxyacetone

**Ring Structure and Conformation**

The trioses and erythulose are obligate open-chain compounds. All other aldoses and ketoses are usually stabilised as cyclic hemiacetals and hemiketals respectively, which exist in equilibrium with the corresponding open-chain form only in solution. These cyclic structures are the result of intramolecular interaction of a $\gamma$- or $\delta$-hydroxyl with the aldehydic or ketonic carbonyl of the open-chain structure, giving rise to five- or six-membered rings related to furan and pyran.

Furan

Pyran

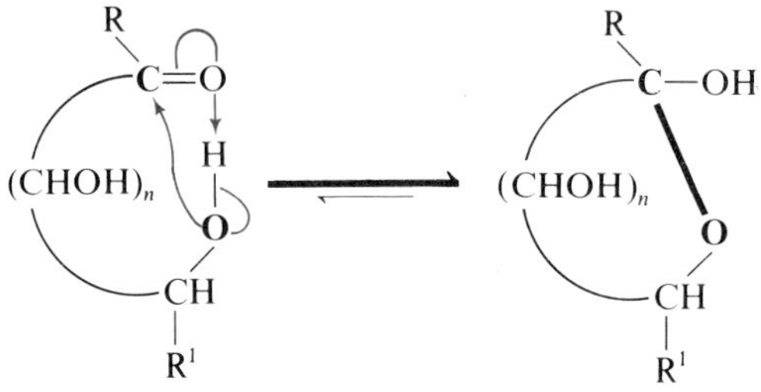

R = H or $—CH_2OH$
$R^1$ = H, $—CH_2OH$ or $—CHOH·CH_2OH$
$n$ = 2 or 3

Sugars with five-membered ring structures are known as **furanoses** and those with six-membered rings, **pyranoses**. Some sugars may exist in both furanose and pyranose forms, though one is usually preferred. Glucose exists almost exclusively in the pyranose form. Fructose similarly crystallises as the pyranose, but fructofuranose is formed in solution, and is present in some of its derivatives. Cyclisation results in the creation of an additional asymmetric centre, and in consequence each D- or L-furanose and pyranose gives rise to two further isomers, designated $\alpha$- and $\beta$-, as for example $\alpha$-D-glucose and $\beta$-D-glucose. $\alpha$-D-Glucose is obtained by crystallisation from ethanol-water, and has $[\alpha]_D + 113°$; crystallisation of $\alpha$-D-glucose from concentrated aqueous solution at elevated temperatures gives $\beta$-D-glucose, $[\alpha]_D + 19°$. Neither of the two crystalline forms is stable in aqueous solution, and the optical rotation changes slowly to an equilibrium value, $[\alpha]_D + 52°$. The isomers are said to undergo **mutarotation**, which is the attainment of an equilibrium between $\alpha$- and $\beta$-forms via the common open-chain aldose structure. The aldose form is present only in minute concentration, *ca* 0.01% at equilibrium, but its presence accounts for the characteristic aldehydic properties of glucose described on page 654. Ketonic properties of ketose sugars are similarly explained. The attainment of equilibrium in mutarotation is greatly accelerated by acid or base catalysis.

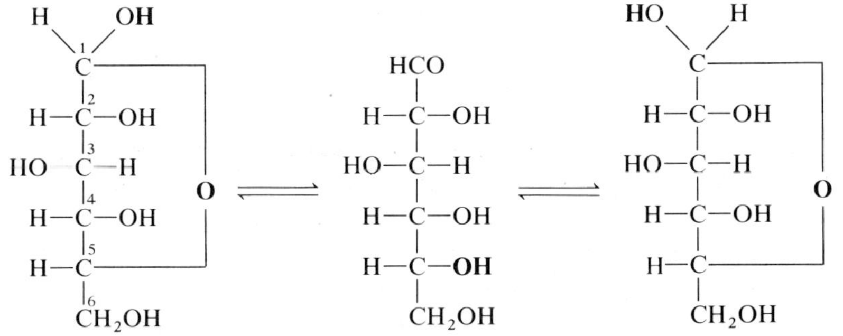

α-D-Glucose $[\alpha]_D + 113°$

Aldehydic structure

β-D-Glucose $[\alpha]_D + 19°$

The specific term **anomer** is used to describe carbohydrate stereoisomers at the hemiacetal carbon atom (C-1 in aldoses and C-2 in ketoses. By convention, the isomer in the D-series which is more strongly dextrorotatory, is always designated the α-anomer, whilst in the L-series the isomer with the higher laevo-rotation is α. Also by convention, in the Fischer projection formulae shown above, the $C_1$—OH group is written on the right of the asymmetric carbon atom in the more dextrorotatory isomer of each pair (α-D- or β-L-configuration) and on the left in the more laevorotatory isomer of each pair (β-D- or α-L-configuration).

**Projection Formulae**

Fischer projection formulae for pyranoses and furanoses, however, suffer from one serious disadvantage. They do not correctly depict the actual configuration of $C_{(4)}$—O and $C_{(5)}$—O bonds forming the oxygen bridge of the furanose and pyranose structures. This disadvantage is overcome in the alternative ring formulae due to Haworth, which provide a pictorial representation of the relative orientation of all substituents.

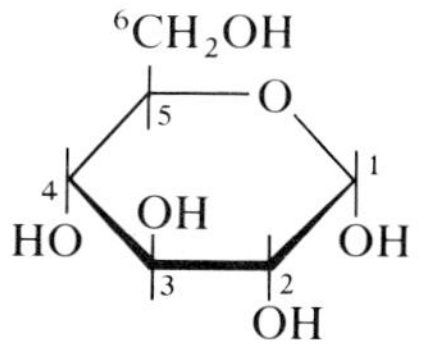

α-D-Glucopyranose

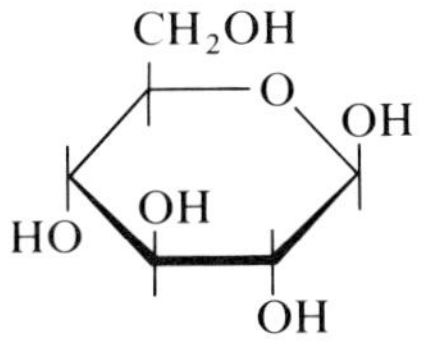

β-D-Glucopyranose

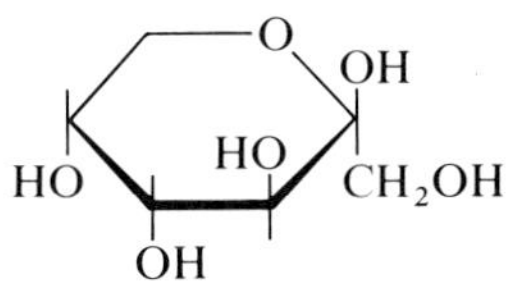

β-D-Fructopyranose

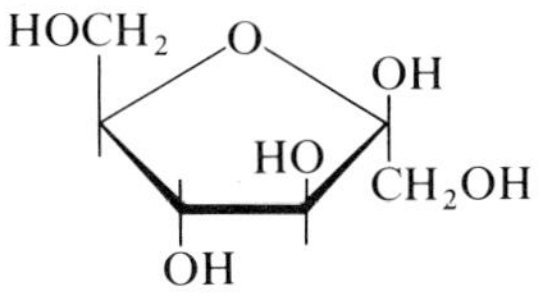

β-D-Fructofuranose

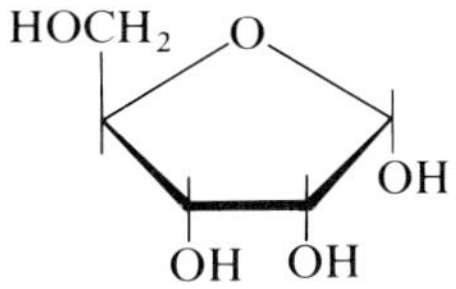

α-D-Ribofuranose

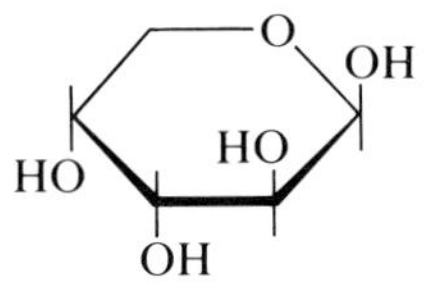

β-D-Arabinose

Conversion of Fischer projection formulae to Haworth structures is exemplified by the following procedure for α-D-glucose. Interchange of the hydrogen on $C_5$ in the Fischer projection with the $C_5$—$CH_2OH$ (to give representation A), followed by the further interchange of H on $C_5$ with the $C_5$—oxygen bridge function to give representation B, is the equivalent of two successive Walden inversions. Thus B has the same configuration at $C_5$ as the original Fischer projection. Bonds drawn horizontally in B attach to groups lying above the plane of the paper, whilst the oxygen bridge lies below the plane of the paper. If B is now turned clockwise so that the oxide ring is at right angles to the plane of the paper as in the Haworth projection, the attached groups automatically fall into their correct orientation above and below the plane of the paper.

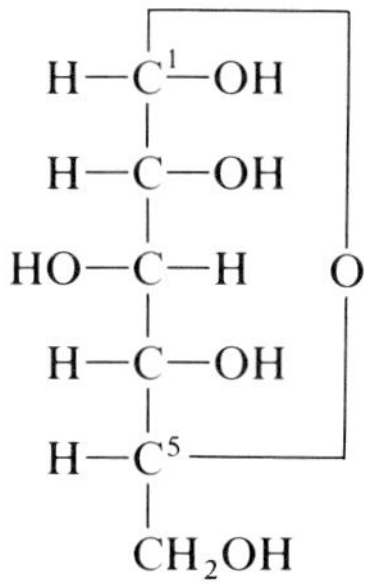

Fischer projection

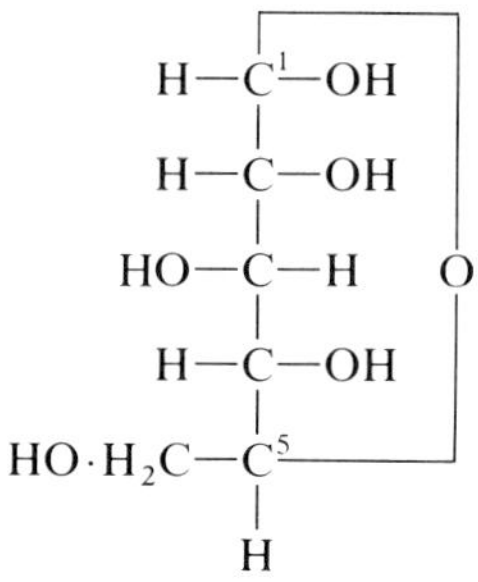

Representation A

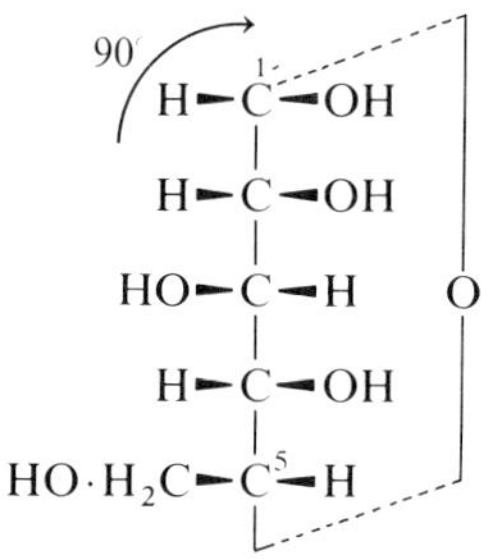

Representation B

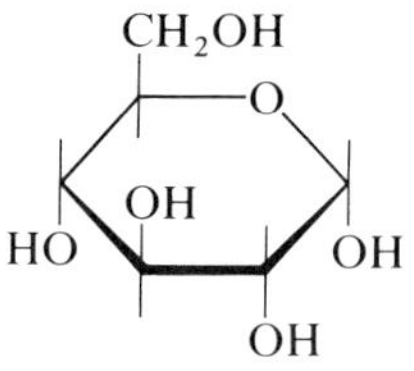

Haworth projection

The furanose ring is planar or almost so, but it is established that the pyranose ring is not actually planar as represented in the Haworth structures, but puckered so that it adopts a chair conformation analogous to that of cyclohexane (Chapters 3 and **2**, 3). Two forms of chair are possible, the Cl as in α-D-glucopyranose, and the 1C as in β-D-fructopyranose.

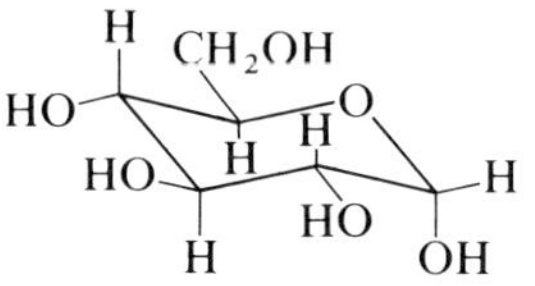

α-D-Glucopyranose
(Cl conformation)

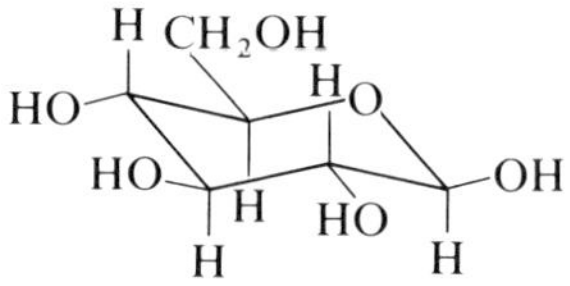

β-D-Glucopyranose
(Cl conformation)

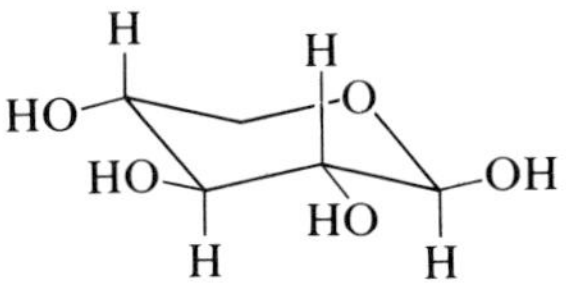

H H HO -O HO OH HO H H

β-D-Xylose
(Cl conformation)

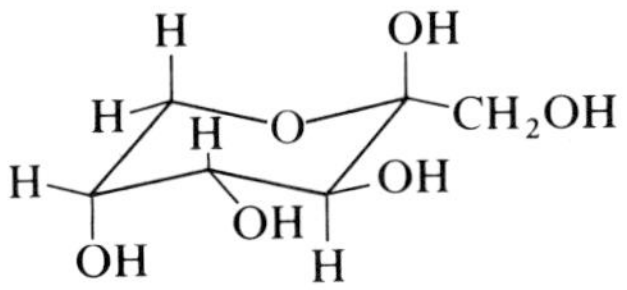

H OH H H O $CH_2OH$ H OH OH OH H

β-D-Fructopyranose
(lC conformation)

### Conformational Factors Affecting the Absorption and Transport of Sugars

The absorption of simple sugars across biological membranes is mediated by special carrier mechanisms. Thus, glucose penetrates into human red cells much faster from lower than from higher sugar concentrations (Ege, Gottlieb and Rakestraw, 1925), and xylose transport from the small intestine in the rat occurs by a similar mobile carrier process (Alvarado, 1966). It is now established that for aldoses there is a definite correlation between conformational stability factors (Reeves, 1950, 1951) and affinity for the red blood cells monosaccharide transfer system (Le Fevre and Marshall, 1958; Le Fevre, 1961). β-D-Glucose, which lacks axial substituents (other than hydrogen) when in the Cl conformation, has the highest affinity, whilst the sterically-analogous β-D-xylose is similarly the most favoured pentose. Interestingly, their mirror image enantiomorphs, L-glucose and L-xylose, although comparably stable, exist in the alternative 1C conformation, and in consequence show extremely low affinity for the carrier. Similarly, α-methyl-D-glucoside, α-methyl-D-mannoside and other derivatives with axial methoxy- and methyl-substituents are not transported, whereas their corresponding equatorial β-epimers, β-methyl D-glucoside and β-methyl D-mannoside, have high affinities for the carrier mechanism.

### General Properties

Monosaccharides are sweet-tasting, crystalline solids, often hygroscopic, and, being polyhydroxylic, highly water-soluble and insoluble in most general organic solvents. Aqueous solutions of glucose and fructose (laevulose) containing 5.25% w/v are isotonic with human serum and are used for intravenous drip administration. The solutions can be sterilised by autoclaving, but become acidic, pH 3.5 to 6.5, depending upon concentration and extent of heat treatment, as a result of decomposition via 2-hydroxylmethylfurfural. Thermal decomposition of glucose to 2-hydroxymethylfurfural is acid-catalysed, but proceeds in the absence of acid at temperatures of 100 to 150°C, probably because the reaction is catalysed at glass surfaces at a rate proportional to cation availability (Orsi, 1967). The latter promotes formation of the ene-diol from the aldehydic form of glucose which is considered to be the rate-determining step in the decomposition (Haworth and Jones, 1944). Cyclo-dehydration of the ene-diol gives the anhydrodiol, which on further dehydration gives 2-hydroxymethylfurfural.

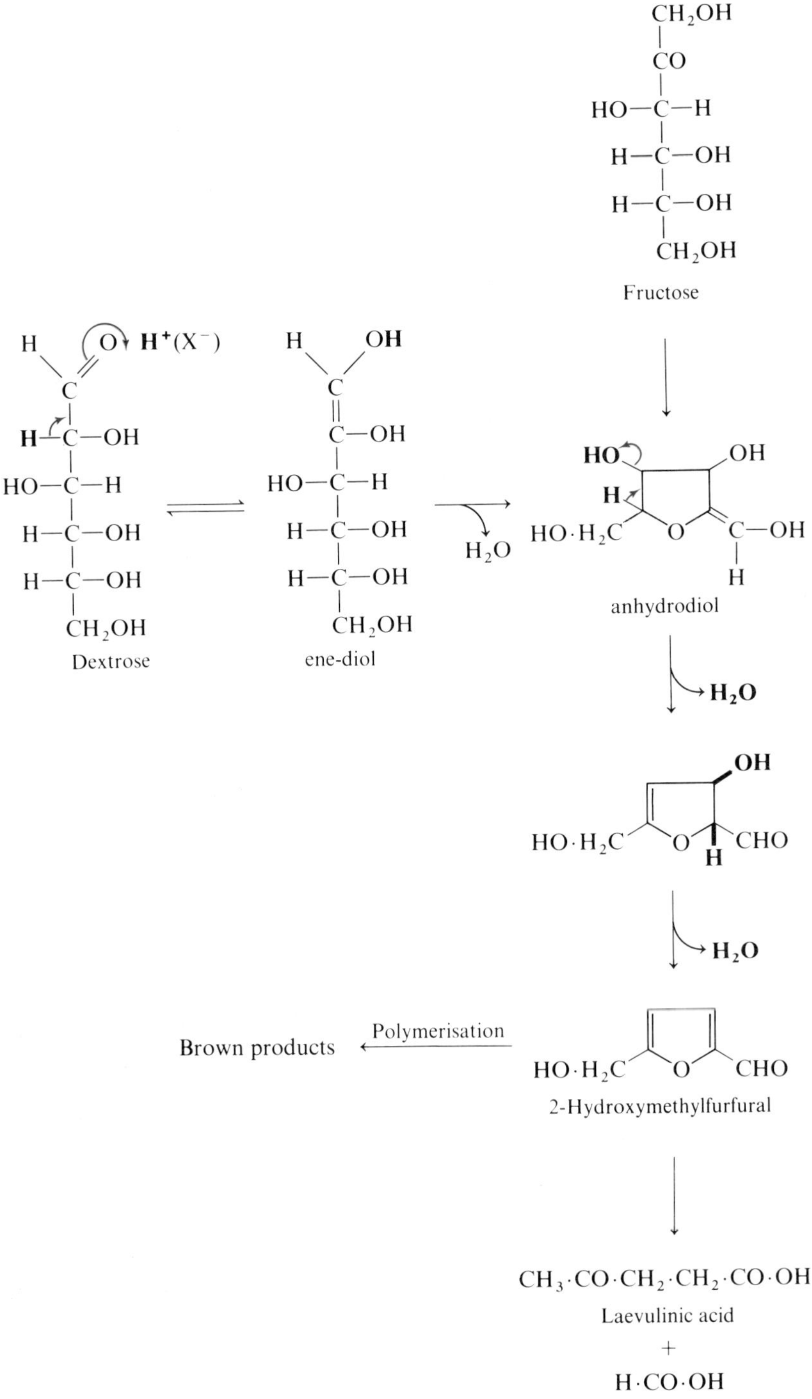

$CH_2OH$
|
CO
|
HO—C—H
|
H—C—OH
|
H—C—OH
|
$CH_2OH$

Fructose

↓

Dextrose: H, O, $H^+(X^-)$; C; H—C—OH; HO—C—H; H—C—OH; H—C—OH; $CH_2OH$

⇌

ene-diol: H, OH; C; C—OH; HO—C—H; H—C—OH; H—C—OH; $CH_2OH$

→ $H_2O$

anhydrodiol: HO, OH, H, $HO{\cdot}H_2C$, O, C—OH, H

↓ $H_2O$

OH; $HO{\cdot}H_2C$, O, CHO, H

↓ $H_2O$

Brown products ← Polymerisation — $HO{\cdot}H_2C$, O, CHO

2-Hydroxymethylfurfural

↓

$CH_3{\cdot}CO{\cdot}CH_2{\cdot}CH_2{\cdot}CO{\cdot}OH$

Laevulinic acid

+

$H{\cdot}CO{\cdot}OH$

Table 66. pH and Degradation of Antibiotic Solutions in Intravenous Dextrose Infusions

| Antibiotic | pH | Loss of Antibiotic Activity (%) | Time (hr) |
|---|---|---|---|
| Nafcillin Sodium | 4.6 | 47 | 24 |
| Oxytetracycline | 3.5 | 25 | 24 |
| Ampicillin | 8.2 | 24 | 8 |
| Methicillin Sodium | 3.6 | 50 | 4.6 |

Fructose is also decomposed thermally to 2-hydroxymethylfurfural, but much more rapidly than glucose, as a result of direct cyclodehydration to the anhydrodiol. Hydroxymethylfurfural readily undergoes polymerisation with the formation of the brown products, which are responsible for the discolouration of glucose and fructose solutions when sterilised by autoclaving. The acidity of these solutions arises from the alternative hydrolytic decomposition to formic and laevulinic acids (van Ekenstein and Blanksma, 1909, 1910).

The addition of antibiotics to intravenous infusion fluids containing dextrose can lead to significant pH changes with serious medical consequences, and also incompatability resulting in rapid degradation of the antibiotic (Engel and Dodson, 1971). This is especially evident with *Nafcillin Sodium*, *Oxytetracycline*, *Ampicillin* and *Methicillin Sodium* (Table 66).

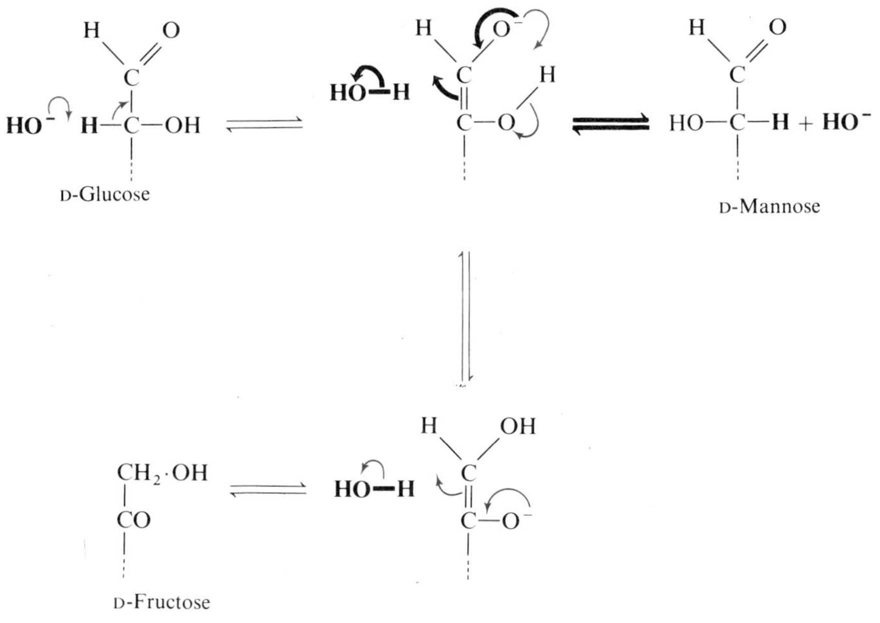

$HO^- + H{-}C(OH)(CHO)$ ⇌ $HO{-}H +$ enediolate ⇌ $HO{-}C{-}H\,(CHO) + HO^-$

D-Glucose

D-Mannose

$CH_2{\cdot}OH{-}CO$ ⇌ $HO{-}H +$ enediolate

D-Fructose

Most sugars are sensitive to alkali. Strongly alkaline solutions discolour rapidly, becoming first yellow, then brown, and finally forming resinous products. In contrast, mild alkali or amines merely cause rearrangement, and D-glucose for example is converted rapidly by dilute aqueous sodium hydroxide to an equilibrium mixture of D-glucose, D-fructose and D-mannose. More than one mechanism has been proposed to explain the rearrangement, including the reaction scheme on page 653.

An analogous mechanism has also been suggested for the interconversion of D-glucose-6-phosphate and D-fructose-6-phosphate catalysed by phosphoglucoisomerase (Rose, 1962).

## Aldehydic and Ketonic Properties

Although it is clearly established that the majority of aldoses and pentoses adopt cyclic hemi-acetal and hemi-ketal structures, they are capable of entering into reaction in their aldehydic and ketonic forms. Thus aldoses are oxidised to aldonic acids (p. 660), and both aldoses and ketoses are readily reduced to the corresponding polyols. Both aldoses and ketoses also condense readily with hydroxylamine to form oximes, and with phenylhydrazine to form phenylhydrazones and characteristic crystalline osazones. Formation of osazones occurs with excess of reagent, and several mechanisms have been proposed to explain the reaction (Fischer, 1887; Weygand and Reckhaus, 1949; Bloink and Pausacker, 1952), none of which is entirely satisfactory. Both the Weygand–Reckhaus and Bloink–Pausacker mechanisms however, rest on an Amadori rearrangement of the α-hydroxy-phenylhydrazone to the corresponding β-ketohydrazine. (See reaction scheme on page 655.)

It will be apparent that glucosazone and fructosazone are identical, since both C-1 and C-2 are involved in the reaction in each case.

Acid hydrolysis of osazones gives rise to osones. These are reactive and have the properties of ene-diols, similar to those of ascorbic acid (p. 669).

HC=N·NH·Ph
|
C=N·NH·Ph
|
HO—C—H
|
H—C—OH
|
H—C—OH
|
$CH_2OH$

Glucosazone

$\xrightarrow{H_2O/H^+}$

HC=O
|
C=O
|
HO—C—H
|
H—C—OH
|
H—C—OH
|
$CH_2OH$

⇌

HO H
\ /
C (ring O to C-5)
|
HO—C
||
HO—C
|
H—C—OH
|
H—C (ring O)
|
$CH_2OH$

Glucosone

$$\begin{array}{c} CHO \\ H-C-OH \\ HO-C-H \\ H-C-OH \\ H-C-OH \\ CH_2OH \end{array} \xrightarrow[-H_2O]{PhNH\cdot NH_2} \begin{array}{c} H-C=N\cdot NH\cdot Ph \\ H-C-OH \\ HO-C-H \\ H-C-OH \\ H-C-OH \\ CH_2OH \end{array} \xrightarrow{H^+} \begin{array}{c} H-C=\overset{+}{N}H\cdot NH\cdot Ph \\ H-C-OH \\ \vdots \end{array}$$

D-Glucose — D-Glucose phenylhydrazone

$$\xrightarrow{H^+} \begin{array}{c} H-C-NH\cdot NH\cdot Ph \\ \| \\ C-O-H \\ \vdots \end{array} + H^+ \xrightarrow{H^+} \begin{array}{c} H_2C-NH\cdot NH\cdot Ph \\ C=O \\ \vdots \end{array}$$

$$\xrightarrow[-H_2O]{Ph\cdot NH\cdot NH_2} \begin{array}{c} H_2C-NH\cdot NH\cdot Ph \\ C=N\cdot NHPh \\ \vdots \end{array} \xrightarrow[H^+]{Ph\cdot NH\cdot NH_2} \begin{array}{c} H_3\overset{+}{N}\cdots NH\cdot Ph \\ \vdots \quad \vdots \\ H \quad H \\ \vdots \quad \vdots \\ H-C-N\cdot NH\cdot Ph \\ C=N\cdot NH\cdot Ph \\ \vdots \end{array}$$

$$\longrightarrow \overset{+}{N}H_4 + Ph\cdot NH_2 + \begin{array}{c} H-C=N\cdot NH\cdot Ph \\ C=N\cdot NH\cdot Ph \\ HO-C-H \\ H-C-OH \\ H-C-OH \\ CH_2OH \end{array}$$

Glucosazone

## Shikimic Acid

The condensation of phospho-enol pyruvate with the carbonyl group of D-erythrose-4-phosphate forms the basis of the biosynthesis of aromatic compounds in bacteria and plants (but not animals) by the shikimic acid pathway (Gibson and Pittard, 1968).

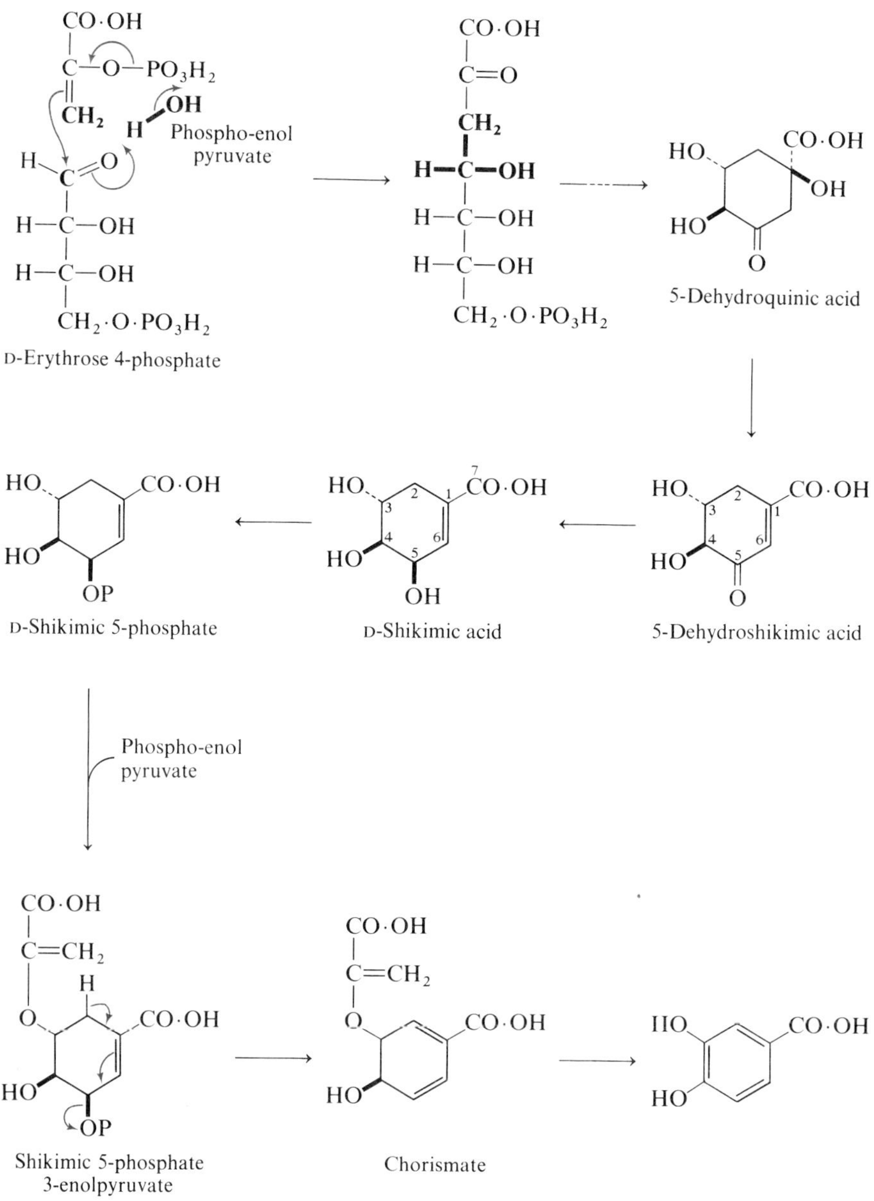

CO·OH
C—O—$PO_3H_2$
$CH_2$ H OH
Phospho-enol pyruvate
H—C=O
H—C—OH
H—C—OH
$CH_2·O·PO_3H_2$
D-Erythrose 4-phosphate

CO·OH
C=O
$CH_2$
**H—C—OH**
H—C—OH
H—C—OH
$CH_2·O·PO_3H_2$

HO CO·OH OH HO O
5-Dehydroquinic acid

HO CO·OH HO O
5-Dehydroshikimic acid

HO CO·OH HO OH
D-Shikimic acid

HO CO·OH HO OP
D-Shikimic 5-phosphate

Phospho-enol pyruvate

CO·OH
C=$CH_2$
H
O CO·OH HO OP
Shikimic 5-phosphate 3-enolpyruvate

CO·OH
C=$CH_2$
O CO·OH HO
Chorismate

HO CO·OH HO

### Esters

The polyhydroxylic properties of monosaccharides are readily apparent in their reaction with acetic anhydride which, for example, converts glucose to its penta-acetate.

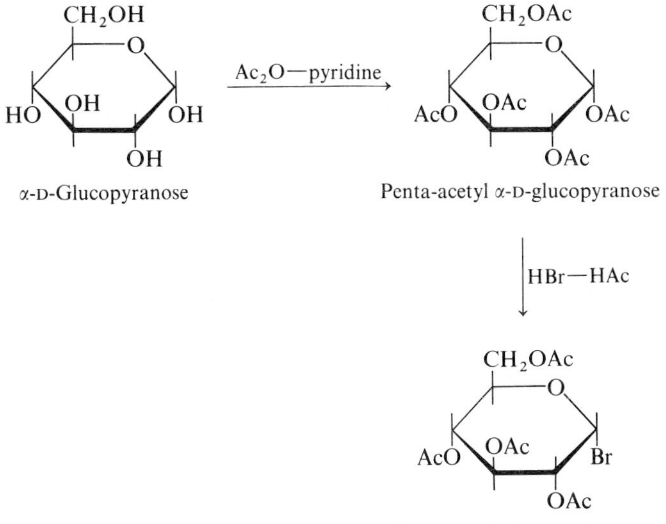

α-D-Glucopyranose $\xrightarrow{Ac_2O\text{—pyridine}}$ Penta-acetyl α-D-glucopyranose

$\downarrow$ HBr—HAc

Tetra-acetyl α-bromoglucose

The product is insoluble in water, and does not show mutarotation, since there is no hydroxyl group on C-1. The acetyl groups are readily removed by mild alkaline hydrolysis (ethanolic ammonia). The C-1 *O*-acetyl group is readily displaced on treatment with a solution of hydrogen bromide in acetic acid to yield tetra-acetyl α-bromoglucose. Substitution of 1-acetoxy-substituents in this reaction, and also the acid hydrolysis of aldose-1-phosphates, which occurs with considerable facility, is due to the formation of an intermediate mesomeric ion which is stabilised by charge delocalisation (see glycosides, p. 662).

Phosphate esters are important in the biosynthesis and breakdown of polysaccharides, and also as key intermediates in the Embden–Meyerhof–Parnas glycolytic pathway for carbohydrate metabolism. In man, glucose-1-phosphate is released on demand from glycogen stores by the action of α-1,4-glucan phosphorylase. The key intermediate in the glycolytic pathway, however, is glucose-6-phosphate, which arises either by isomerism of glucose-1-phosphate with phosphoglucomutase or by direct phosphorylation of D-glucose with ATP. Phosphoglucose isomerase converts glucose-6-phosphate to fructose-6-phosphate, which is further phosphorylated by phosphofructokinase to fructose-1,6-diphosphate. The latter is then degraded to dihydroxyacetone and glyceraldehyde and thence by pathways already described (Chapter 20) to glycerol and pyruvate.

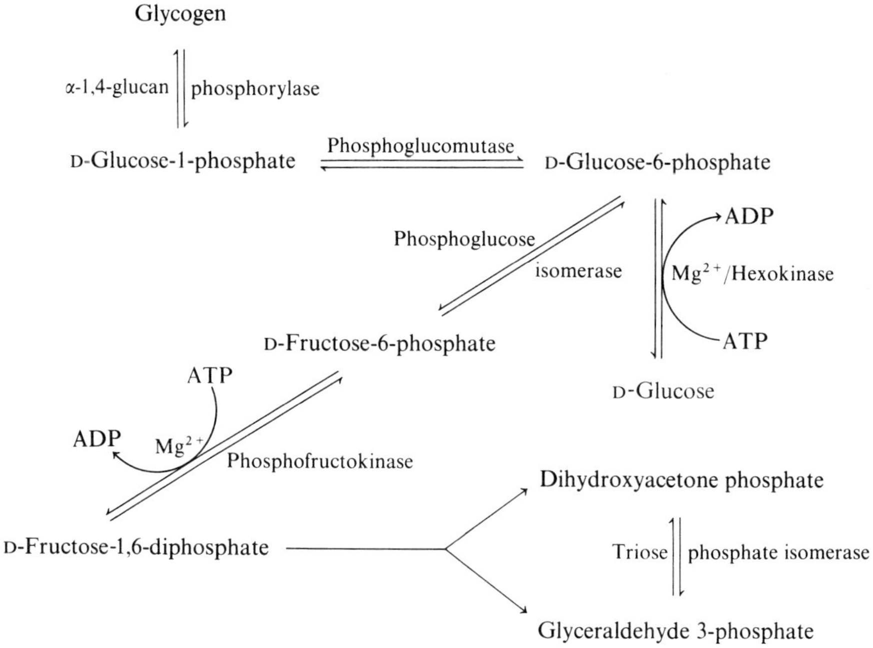

Glycogen

α-1,4-glucan phosphorylase

D-Glucose-1-phosphate ⇌ (Phosphoglucomutase) D-Glucose-6-phosphate

Phosphoglucose isomerase

$Mg^{2+}$/Hexokinase (ATP → ADP)

D-Fructose-6-phosphate

D-Glucose

Phosphofructokinase ($Mg^{2+}$; ATP → ADP)

D-Fructose-1,6-diphosphate

Dihydroxyacetone phosphate

Triose phosphate isomerase

Glyceraldehyde 3-phosphate

Sugar phosphate esters are strongly acidic (p$K_{a_1}$ *ca* 1.0; p$K_{a_2}$ *ca* 6.1) and readily form crystalline metal salts. They also readily undergo rearrangement in acidic solution (Levene and Raymond, 1934) with phosphate migration to proximate hydroxyl groups, via a cyclic 3′,5′-intermediate.

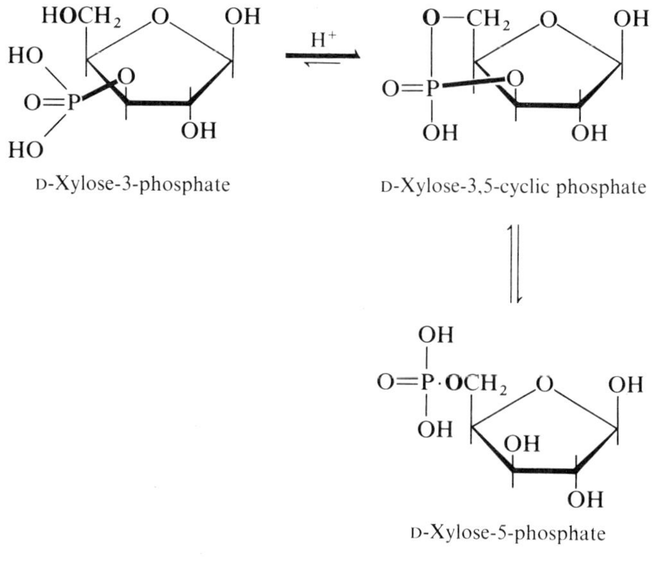

D-Xylose-3-phosphate ⇌ ($H^+$) D-Xylose-3,5-cyclic phosphate

⇌

D-Xylose-5-phosphate

Cyclic phosphates can be obtained chemically from the appropriate monophosphate in aqueous solution by the action of dicyclohexyldicarbodiimide (DCC) (Dekker and Khorana, 1954) and are readily reconverted to a mixture of monophosphates on treatment with dilute acid or alkali.

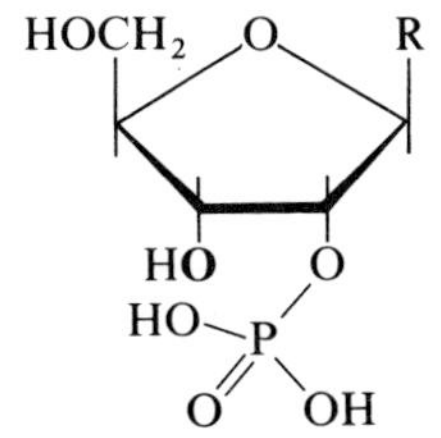

Adenosine-2′-phosphate

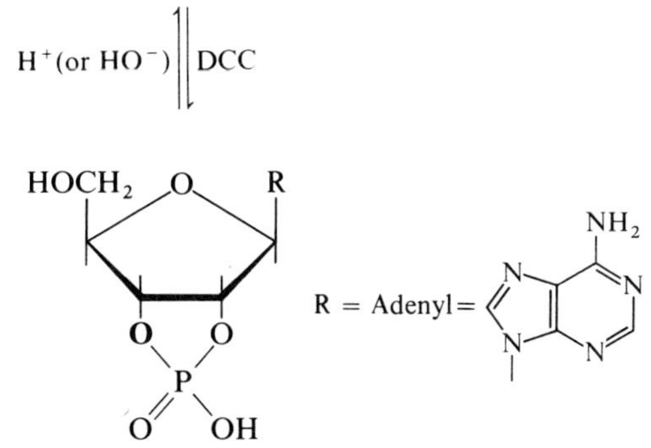

$H^+$(or $HO^-$) ⇅ DCC

Adenosine-2′,3′-cyclic phosphate

R = Adenyl

The formation of cyclic-AMP (adenosine-3′,5′-phosphate) from ATP by adenyl cyclase is an important factor in the breakdown of glycogen to glucose-1-phosphate, and is influenced by the action of adrenaline (**2**, 3). Cyclic-AMP is required to activate phosphorylase kinase, which in turn converts an inactive phosphorylase *b* to phosphorylase *a*, which is active in the breakdown of glycogen (see reaction scheme page 660).

Acid hydrolysis of sugar phosphates is generally slower than acid-catalysed phosphate migration. Sugar-1-phosphates are readily hydrolysed, but esters of primary hydroxyl groups as in glucose-6-phosphate are more resistant to hydrolysis (Leloir, 1951). This difference in reactivity is also evident in the enzymatic hydrolysis of glucose-1,6-diphosphate formed as intermediate in the interconversion of glucose-1-phosphate and glucose-6-phosphate. Acid hydrolysis occurs with C—O bond cleavage (Cohn, 1949), and in keeping with this, hydrolysis of the less hindered equatorial ester link in *β*-D-glucose-1-phosphate occurs faster than hydrolysis of the corresponding axial bond in *α*-D-glucose-1-phosphate (Wolfram, Smith, Fletcher and Brown, 1942).

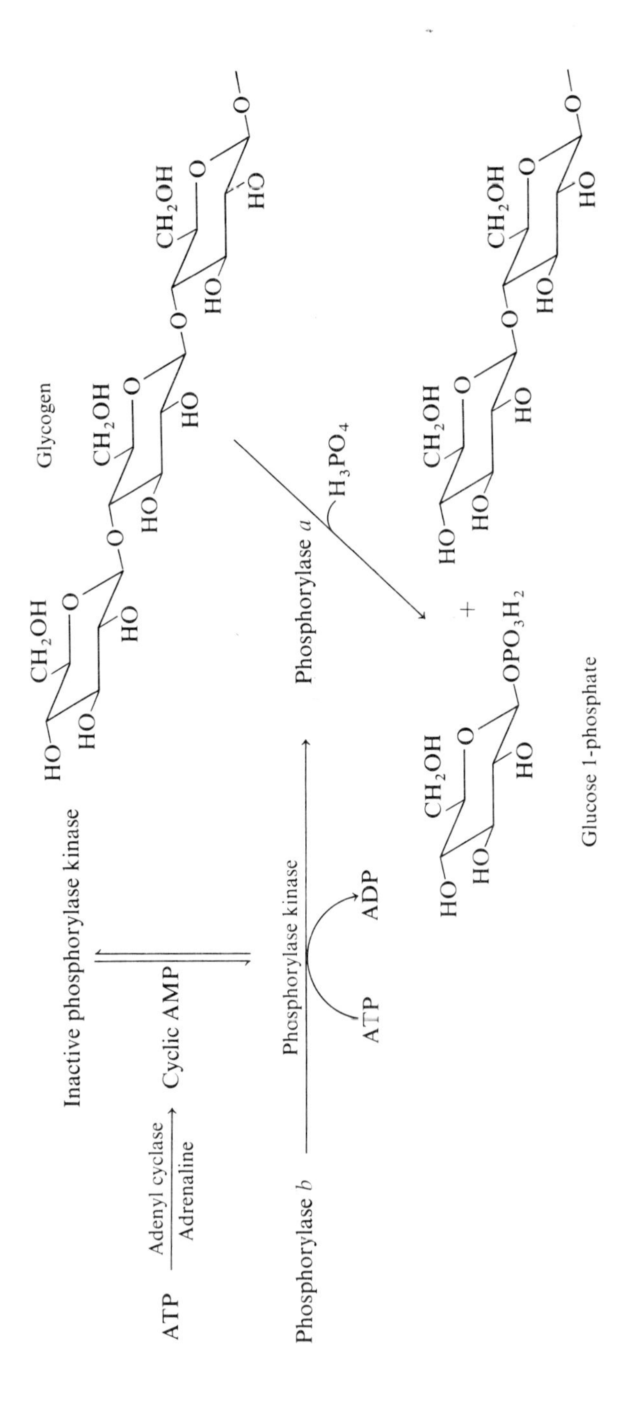

ATP
Adenyl cyclase
Adrenaline
Cyclic AMP
Inactive phosphorylase kinase
Phosphorylase b
Phosphorylase kinase
ATP
ADP
Phosphorylase a
$H_3PO_4$
Glycogen
$CH_2OH$
HO
+
$OPO_3H_2$
Glucose 1-phosphate

**Ethers and Glycosides**

Treatment of sugars with dimethyl sulphate in the presence of sodium hydroxide leads to complete methylation of all free hydroxyl groups. Glucose is converted to a mixture of α- and β-pentamethylglucose.

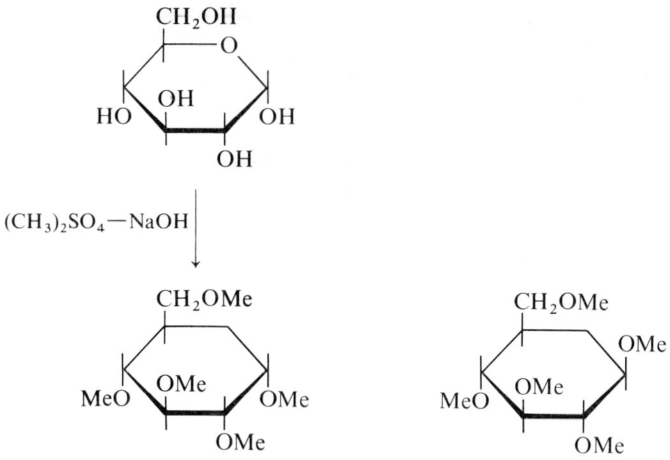

$CH_2OH$

O

HO OH OH

OH

$(CH_3)_2SO_4$—NaOH

$CH_2OMe$ $CH_2OMe$

OMe

MeO OMe OMe MeO OMe

OMe OMe

α- β-

Pentamethyl-D-glucose

Methylation of hydroxyl groups protects the ring structure from oxidative degradation. The methoxy groups are also resistant to hydrolysis with the exception of the glycosidic methoxy link ($C_1$ in glucose) which is readily hydrolysed in dilute acid.

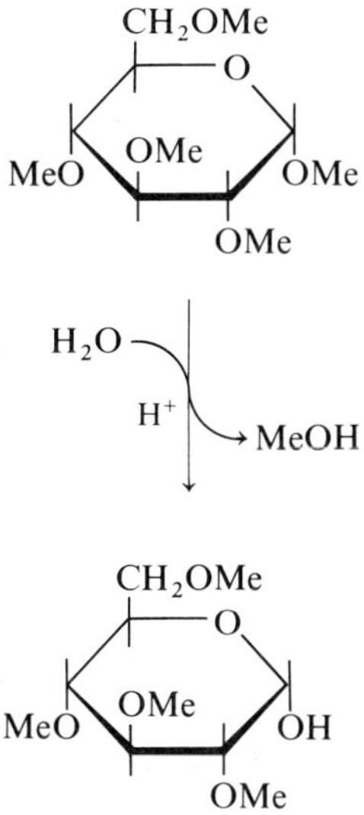

$CH_2OMe$

O

MeO OMe OMe

OMe

$H_2O$

$H^+$ MeOH

$CH_2OMe$

O

MeO OMe OH

OMe

2,3,4,6-Tetramethyl α-D-glucose

The ease of hydrolysis is due to the formation of an intermediate mesomeric ion, which is stabilised by the charge delocalisation thus achieved.

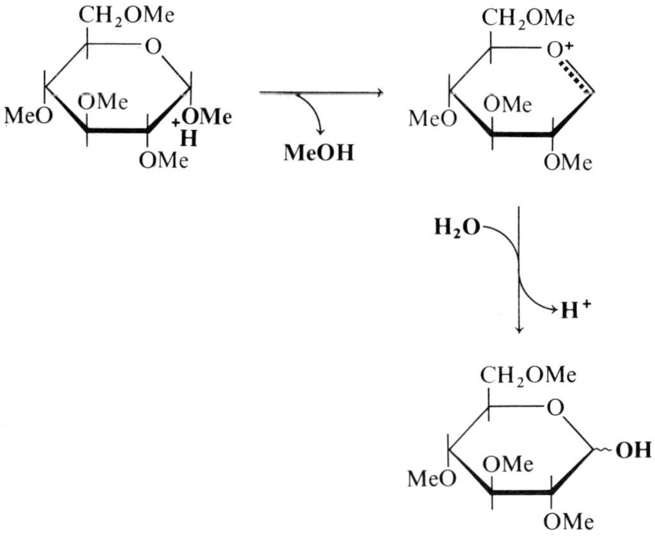

$CH_2OMe$ O MeO OMe $\overset{+}{O}Me$ H OMe → (MeOH) $CH_2OMe$ $O^+$ MeO OMe OMe

$H_2O$ ↓ $H^+$

$CH_2OMe$ O MeO OMe OH OMe

The susceptibility of unprotected alcoholic hydroxyl groups and the stability of methoxyl functions to oxidation is also evident in the formation of 2,3,4-trimethoxyglutaric acid from 2,3,4,6-tetramethyl-α-D-glucose.

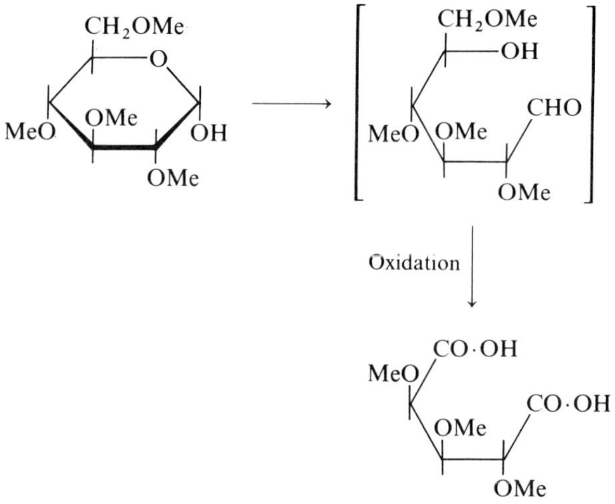

$CH_2OMe$ O MeO OMe OH OMe → [ $CH_2OMe$ OH CHO MeO OMe OMe ]

Oxidation ↓

CO·OH MeO CO·OH OMe OMe

2,3,4-Trimethoxyglutaric acid

Methylation of sugars with methanolic hydrogen chloride gives mainly the monomethylated α-derivative, α-methyl-glucoside, together with a smaller amount of the corresponding β-isomer.

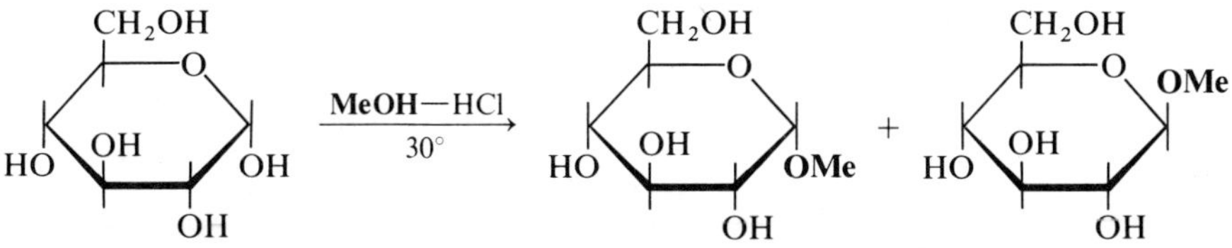

CH$_2$OH O HO OH OH OH — MeOH—HCl, 30° → CH$_2$OH O HO OH OMe OH + CH$_2$OH O OMe HO OH OH

α-D-Glucose α-Methyl-D-glucoside β-Methyl-D-glucoside

The products no longer exhibit aldehydic properties and do not mutarotate. Like the pentamethyl D-glucose described above, both α- and β-methylglucosides, which in effect are typical acetals (Chapter 10), are readily hydrolysed in aqueous solution by dilute acid.

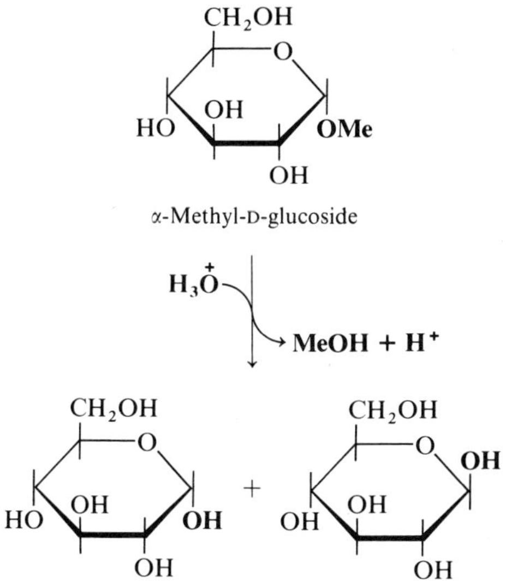

CH$_2$OH O HO OH OMe OH

α-Methyl-D-glucoside

$H_3\overset{+}{O}$ → MeOH + $H^+$

CH$_2$OH O HO OH OH OH + CH$_2$OH O OH OH OH OH

α-D-Glucose β-D-Glucose

The methyl glucosides are characteristic of a wide range of related compounds, known generally as **glycosides**. They are mostly, though not all, of natural origin. Specific glycosides are designated as glucosides, galactosides, glucuronides, fructosides and so on, according to the constituent sugar; the non-sugar moiety which may be an alcohol or phenol is termed the **aglycone**. Most naturally occurring glycosides are β-glycosides, as for example, the β-glucuronide conjugates formed in the metabolism of drugs in man (**2**, 5). These result from the stereospecificity of the conjugating enzyme. Similarly, enzymes capable of hydrolysing glycosides (i.e. **glycosidases**) are usually stereospecific. Thus, maltase hydrolyses only α-aldopyranosides, whilst emulsin which is present in almonds is specific for β-aldopyranosides; glucofuranosides on the other hand are not hydrolysed by either. Likewise, β-glucuronidase, which is present in animal tissues, but notably in the liver, kidney, spleen and intestinal tract, is specific for the hydrolysis of β-glucuronides.

The rate of hydrolysis of glycosides by acid is affected by the nature of the aglycone, its configuration (α- or β-), ring size, and ring substitution. In general, glycosides derived from aromatic aglycones are more susceptible to hydrolysis than their aliphatic counterparts; the relative rates of hydrolysis of D-glucopyranosides are in the order Ph > $PhCH_2$ > $CH_3$ (Heidt and Purves, 1944). Most β-methyl-glycosides are more readily hydrolysed than their α-anomers, but α-phenyl-glycosides are usually more readily hydrolysed than their β-anomers (Wolfram and Thompson, 1957). Ring size has a major influence. Aldofuranosides are hydrolysed some 50 to 200 times as fast as aldopyranosides (Haworth, 1932). Fructofuranosides, however, are hydrolysed at similar rates to fructopyranosides (Purves and Hudson, 1937). The effect of adjacent substituents is apparent in the much more rapid hydrolysis of 2-deoxypentosides and 2-deoxyhexosides by acid than of their corresponding 2-hydroxy-counterparts (Butler, Laland, Overend and Stacey, 1950).

Glycosides in general are stable to alkali. Phenolic glycosides, and glycosides derived from aliphatic aglycones with electron-attracting substituents are alkali-sensitive. Alkaline hydrolysis of phenyl β-D-glucopyranoside at elevated temperatures (100°C) is accompanied by dehydration with formation of the glucosan, 1,6-anhydro-β-D-glucopyranose (Lemieux, 1954).

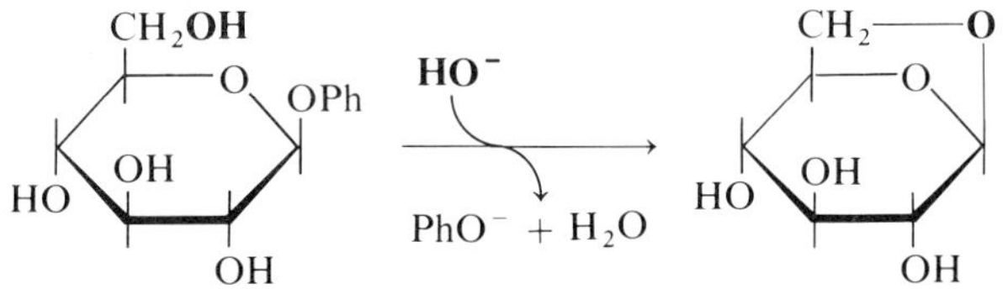

$CH_2OH$ O OPh HO OH OH — $HO^-$ → $PhO^- + H_2O$ — $CH_2$—O O HO OH OH

Glycosides are susceptible to metal-catalysed air oxidation. The attack is selective for equatorial C—H bonds; thus, methyl β-D-ribopyranoside in its preferred C1 conformation is attacked at the sole equatorial C—H bond at C-3 to give the predicted methyl β-D-*erythro*pentopyranosyl-3-ulose (Brimacombe, Brimacombe and Lindberg, 1960).

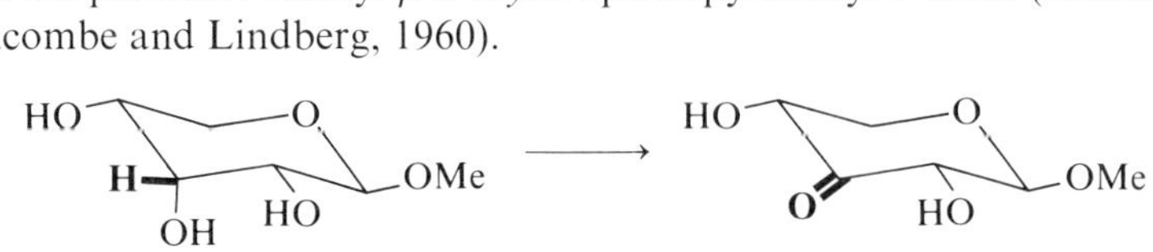

HO O H OMe OH HO → HO O O OMe HO

### *N*- and *S*-Glycosides

The nucleotides which form the basic units of nucleic acid polymers, RNA and DNA, are typical *N*-ribosides and *N*-deoxyriboside phosphates. *N*-Glucuronides, such as sulphanilamide-$N^4$-glucuronide, have also been identified as products of drug metabolism in man and other species. Such compounds, however, are considered to be formed as artefacts by direct chemical interaction of the amine with glucuronic acid, and not to be metabolites. Such *N*-glycosides are more acid-labile than the corresponding *O*-glycoside (Butler, Laland, Overend and Stacey, 1950), and are thus readily distinguished. *N*-Glycosides of

2-deoxy-sugars, like the analogous *O*-glycosides, are also hydrolysed more readily than the corresponding 2-hydroxy-sugar *N*-glycosides. Amidic *N*-glucuronides, such as Meprobamate-*N*-glucuronide, are much more stable than amine-*N*-glucuronides in acid solution (Parke, 1968).

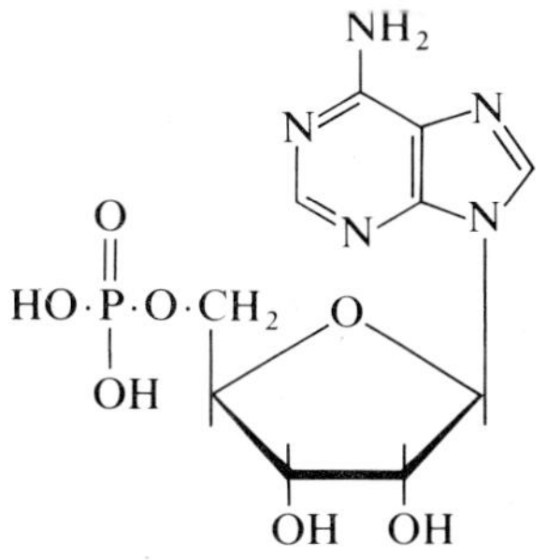

Adenylic acid

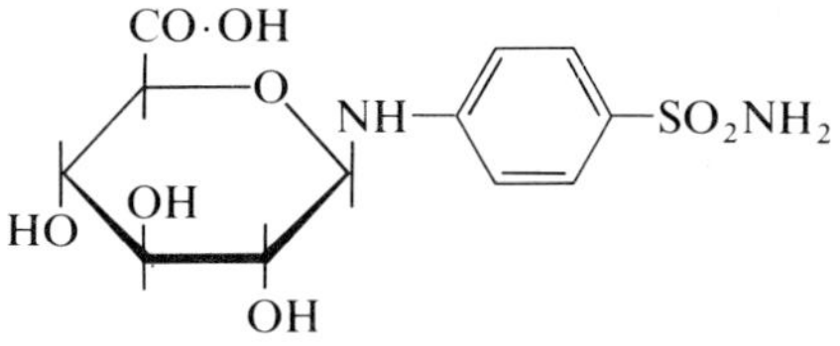

Sulphanilamide-$N^4$-glucuronide

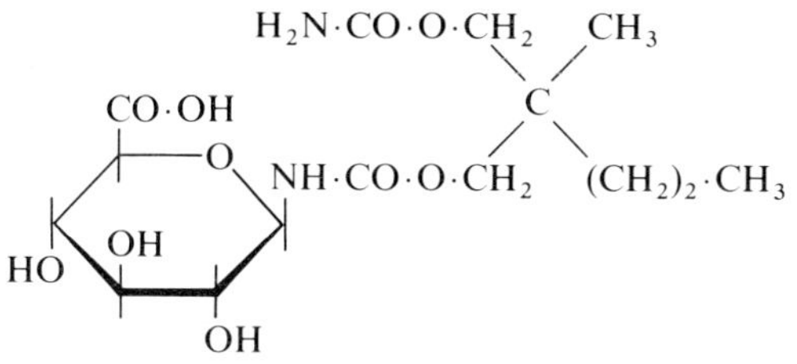

Meprobamate-*N*-glucuronide

A number of sulphur compounds, such as thiophenol and *Antabuse*, are metabolised to *S*-glucuronides (Kaslander, 1963), which like *N*-glucuronides are relatively unstable to acid and readily hydrolysed.

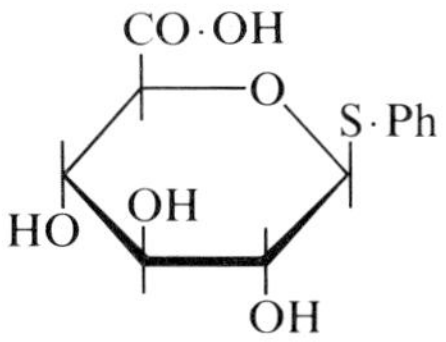

Thiophenyl-*S*-glucuronide

*Antabuse*

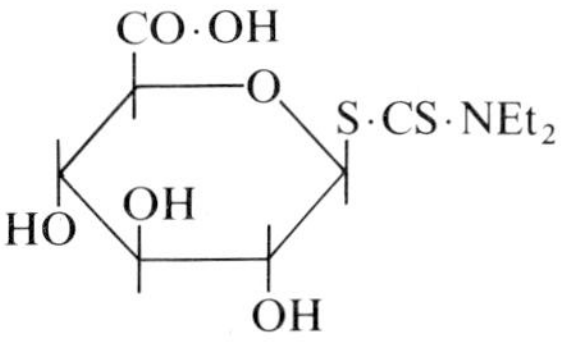

Diethyldithiocarbamyl-*S*-glucuronide

**Oxidation**

Both aldoses and ketoses are readily oxidised, and hence are powerful reducing agents. Fehling's solution (alkaline copper sulphate) and ammoniacal silver nitrate are both rapidly reduced when heated with aldoses and ketoses. Both reagents give complex mixtures of products; oxidation of glucose with Fehling's solution gives rise to a mixture of aldohexonic and lower molecular weight hydroxy acids.

Controlled oxidation of aldoses with bromine in aqueous solution provides high yields of the corresponding aldonic acid. Ketoses, however, resist oxidation by bromine. The effective oxidant is probably hypobromous acid, but at less acidic pH (*ca* 5.4) in the presence of barium carbonate, direct oxidation by the free halogen occurs, and D-glucopyranose, for example, is oxidised directly to D-gluconolactone (Isbell, 1932; Isbell and Hudson, 1932). In general, β-anomers of aldohexoses are oxidised much more rapidly than the corresponding α-anomer; β-D-glucose is oxidised about forty times more rapidly than α-D-glucose (Isbell and Pigman, 1937).

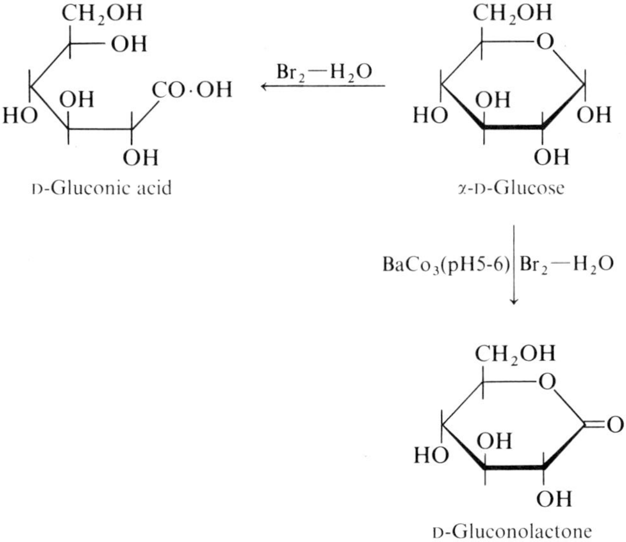


D-Gluconic acid — α-D-Glucose — D-Gluconolactone

Sugars are readily oxidised by molecular oxygen in alkaline solution. Aldoses undergo oxidation at C-1 and C-2 with loss of formic acid to give the corresponding lower aldonic acid. Thus, D-glucose is converted to *D*-arabonic acid. The product is stable, but further degradation occurs in the presence of metal ions, including copper, iron, cobalt and nickel. L-Sorbose is converted first to 2-keto-L-gulonic acid and then to L-xylonic acid.

Photo-oxidation of sugars also occurs readily in neutral aqueous solutions on exposure to ultraviolet light of wavelength between 230 and 250 nm. Experiments based on D-sorbitol which is readily photo-oxidised in the presence of oxygen to D-glucose show that the latter is further degraded to D-gluconic acid, and thence to D-arabinose with loss of carbon dioxide (Phillips and Criddle, 1963).

Oxidation also proceeds at the other end of the molecule to give L-gulose, which in turn is degraded through L-gulonic acid to L-xylose. It is further established that the reaction is the result of direct photolysis of the carbohydrate, and does not proceed as a result of a prior photolysis of water to hydrogen and

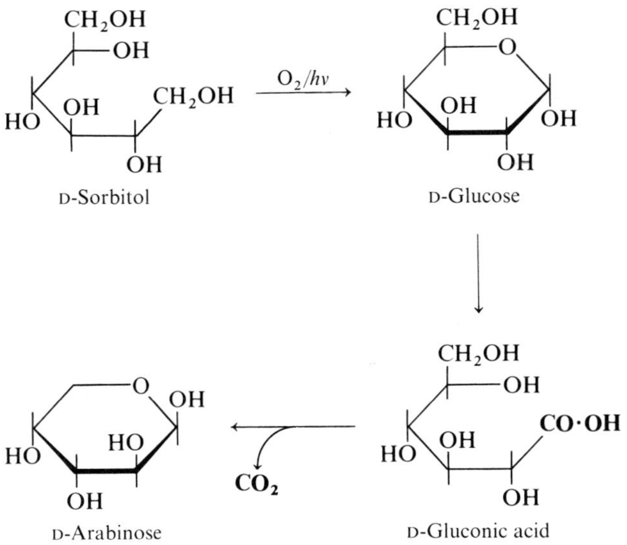

$O_2/h\nu$

D-Sorbitol → D-Glucose

D-Gluconic acid → D-Arabinose ($CO_2$)

hydroxyl radicals. The primary process for D-sorbitol is considered to be excitation of non-bonded orbitals on the oxygen atom to anti-bonding orbitals (n—$\sigma^*$ transition). Light of wavelength 230–254 nm corresponding to an energy of 124–113 kJ mol$^{-1}$, which is absorbed in the process, is sufficiently energetic to rupture a C—C, C—O or C—H bond. Photolysis is completely inhibited in hard glass vessels which cut off light of wavelength less than 294 nm (Phillips and Barber, 1963).

**Gluconic Acid**

Gluconic acid is prepared either by chemical or bacterial oxidation of glucose. It readily forms a water-soluble calcium salt, which is used in the treatment of calcium deficiency. The calcium salt is only slowly soluble in water, but can be used to prepare supersaturated solutions for injection. The stability of these injection solutions depends on complete freedom from solid particles, otherwise crystallisation can result. Inclusion of small amounts, up to 5% of calcium saccharate, provides a means of increasing the stability of the solution.

6-Phosphogluconic acid is an intermediate in the pentose-phosphate pathway for the metabolism of glucose.

**Glucuronic Acid**

Glucuronic acid occurs widely distributed in nature. In man and other mammalian species, it performs an important function in the metabolism of drugs and other foreign compounds, many of which are converted in the liver to glucuronides to aid biliary excretion and subsequent elimination from the body via the urine and faeces. The formation of glucuronides is particularly important in the metabolism of alcohols and can also be important for the excretion of certain phenols, acids, thiols and amines. Glucuronide formation also occurs in the

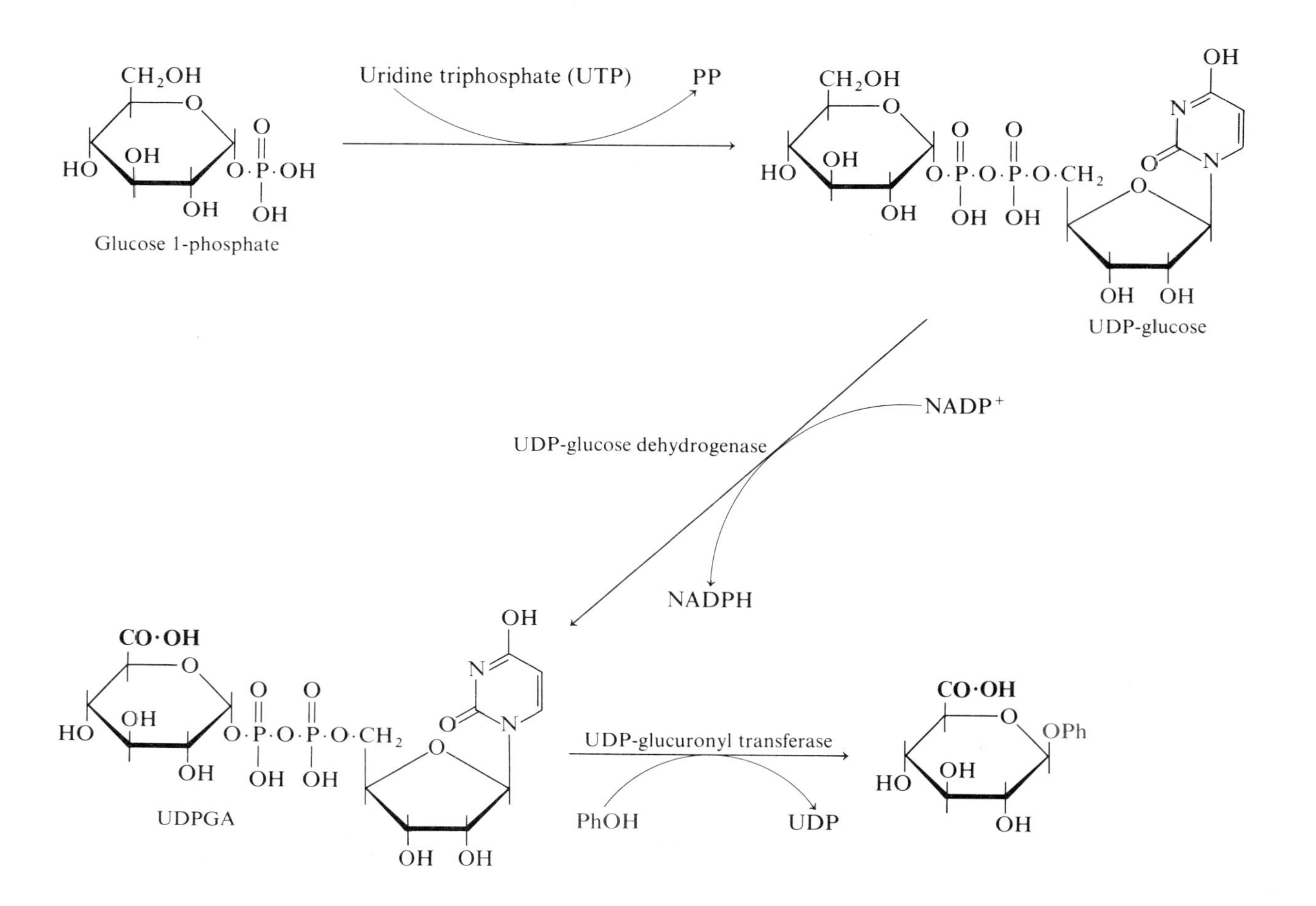
Glucose 1-phosphate
Uridine triphosphate (UTP)
PP
UDP-glucose
UDP-glucose dehydrogenase
NADP⁺
NADPH
UDPGA
UDP-glucuronyl transferase
PhOH
UDP

kidney, in the gastro-intestinal tract and the skin. The key intermediate in glucuronide formation is uridine diphosphoglucuronic acid (UDPGA). This is formed biosynthetically from glucose-1-phosphate and converted to glucuronide in the reaction scheme on page 668.

### Ascorbic Acid

Glucuronic acid is converted to ascorbic acid (vitamin C) in both plant and animal kingdoms. Most mammals are able to complete the biosynthesis of ascorbic acid from glucuronic acid. Man, primates and the guinea pig, however, are unable to bring about the conversion of L-gulonolactone to 2-keto-L-gulonolactone. Ascorbic acid is, therefore, a vitamin, and for man, green vegetables, citrus fruit and potatoes are valuable dietary sources. The complete biosynthetic pathway in those species capable of achieving it is as follows.

D-Glucuronic acid → (NADPH → $NADP^+$) → L-Gulonic acid

L-Gulonic acid → ($-H_2O$) → L-Gulonolactone

L-Gulonolactone → (Blocked in man) → 2-Keto-L-gulonolactone

2-Keto-L-gulonolactone → Ascorbic acid

Pure ascorbic acid is a freely water-soluble crystalline compound. It is distinctly acidic and has a pleasant acidic taste. The acid readily forms a sodium enolate even with sodium bicarbonate. The acid is stabilised in the ene-diol form, but is capable of reacting in solution as the tautomeric 3-keto-L-gulonolactone, which has been obtained as its hydrazone.

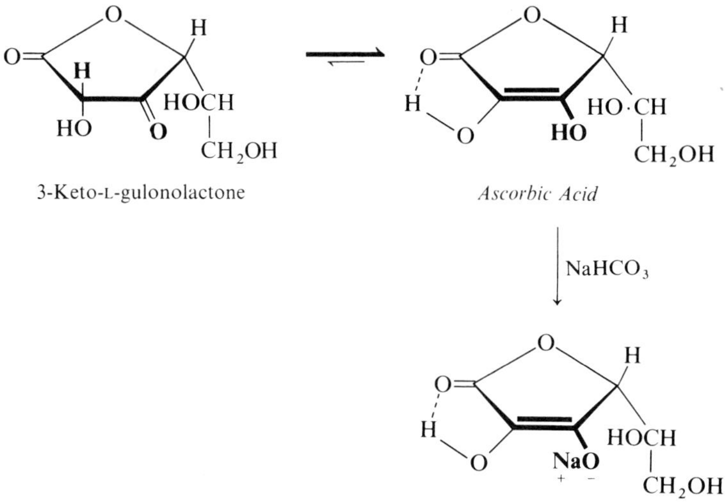

3-Keto-L-gulonolactone

*Ascorbic Acid*

$NaHCO_3$

Ascorbic acid is a powerful reducing agent and readily oxidised in solution, so that the natural vitamin is often destroyed in the cooking of fruits and vegetables.

Oxidative decomposition is accelerated under alkaline conditions, when air oxidation occurs readily. Oxidation, which is catalysed by light and heat, results in the formation of dehydroascorbic acid. This despite its name is a neutral lactone. Oxidation seems likely to occur by a radical mechanism.

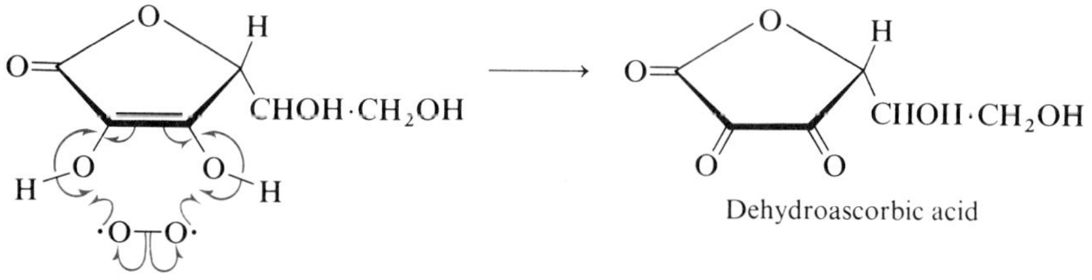

Dehydroascorbic acid

This same oxidation forms the basis of a number of methods for the determination of vitamin C, such as that employing the pink dye 2,6-dichlorophenolindophenol.

In man, ascorbic acid and dehydroascorbic acid are associated in a number of oxidation–reduction systems involving thiols, notably cysteine and glutathione. Some further oxidation of dehydroascorbic acid also occurs with the formation of oxalic acid, and in certain metabolic disorders this pathway can become a major one for the metabolism of ascorbic acid.

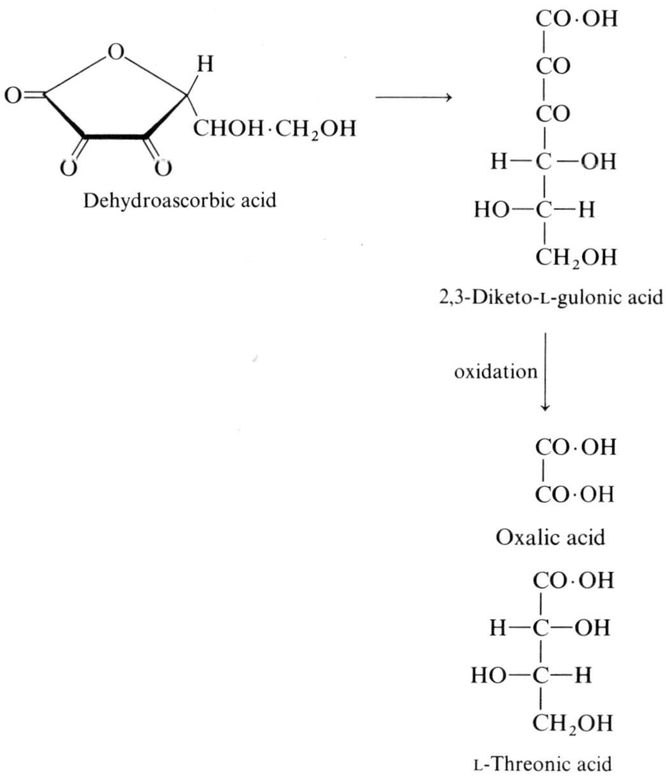

Dehydroascorbic acid $\longrightarrow$

$$\begin{array}{c} CO\cdot OH \\ | \\ CO \\ | \\ CO \\ | \\ H-C-OH \\ | \\ HO-C-H \\ | \\ CH_2OH \end{array}$$

2,3-Diketo-L-gulonic acid

$\downarrow$ oxidation

$$\begin{array}{c} CO\cdot OH \\ | \\ CO\cdot OH \end{array}$$

Oxalic acid

$$\begin{array}{c} CO\cdot OH \\ | \\ H-C-OH \\ | \\ HO-C-H \\ | \\ CH_2OH \end{array}$$

L-Threonic acid

In siderosis, the iron overload disease seen in the Johannesburg Bantu, the reversible reaction between L-ascorbic acid and dehydro-L-ascorbic acid, which normally favours the former, is reversed by the abnormal amounts of ferric iron. As a result, dehydro-L-ascorbic acid accumulates and is then irreversibly oxidised to 2,3-diketogulonic acid and then to oxalic and threonic acids (Seftel, 1970).

Ascorbic acid is involved in the hydroxylation of proline containing peptides in collagen biosynthesis, so that ascorbic acid deficiencies can also lead to defective collagen biosynthesis.

## AMINO-DEOXY SUGARS

Amino-deoxy sugars are widely distributed as constituents of structural polysaccharides such as chitin, which occurs principally in the exoskeleton of crustaceans, the glycoproteins or mureins of bacterial cell walls, and the chondroitin of cartilage. Amino-deoxy sugars are also important constituents of the mucopolysaccharides, such as hyaluronic acid, heparin, and blood specific polysaccharides, and a number of oligosaccharide antibiotics, such as *Streptomycin*, *Neomycin*, *Kanamycin* and *Gentamicin*.

Most amino-deoxy sugars occurring in structural polysaccharides are 2-amino-2-deoxyaldoses, notably glucosamine (2-amino-2-deoxy-D-glucose) and galactosamine (2-amino-2-deoxy-D-galactose) or their *N*-acetyl derivatives. 2-Amino-2-deoxy sugars exhibit typical properties of bases, such as salt formation, and many of the typical properties of the parent sugars. Thus, glucosamine retains the reducing properties of glucose, and, consequently, is readily oxidised to glucosaminic acid. The amino group, however, inhibits lactone formation in the resultant acid as a result of zwitterion formation.

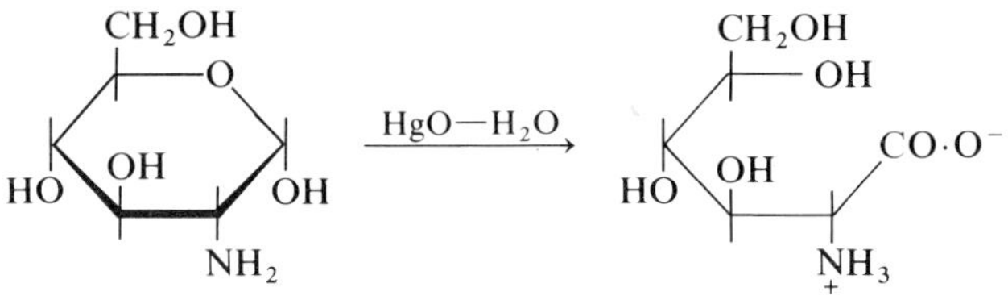

$CH_2OH$ O HO OH OH $NH_2$ $\xrightarrow{HgO—H_2O}$ $CH_2OH$ OH HO OH $CO{\cdot}O^-$ $\overset{+}{N}H_3$

D-Glucosamine D-Glucosaminic acid

The amino substituent shows such typical reactions as Schiff's base formation with aldehydes, and readily acylates with acid chlorides and anhydrides. The resultant *N*-acylamino sugars are predictably more stable to acid hydrolysis than the corresponding *O*-acetyl derivatives. Treatment of *N*-acetylglucosamine or *N*-acetylgalactosamine with alkali causes cyclodehydration to an oxazoline which on further treatment with Ehrlich's reagent (*p*-dimethylaminobenzaldehyde in HCl) gives a purplish red colour. Similar cyclisations of glucosamine with acetylacetone to form a pyrrole derivative and reaction with Ehrlich's reagent form the basis of quantitative methods for its determination (Elson and Morgan, 1933).

2-Amino-substituents also inhibit the formation of glycosides; protonation of the amino group by the acidic catalyst inhibits protonation of the adjacent anomeric hydroxyl group as a result of charge repulsion, thereby blocking the reaction. For the same reason, 2-amino-deoxyglycosides are generally highly resistant to hydrolysis. Thus, the *N*-methylglucosaminide, streptobiosamine, which is formed by mild acid hydrolysis of *Streptomycin*, is only cleaved on prolonged hydrolysis with strong acid (*ca 6N*), under conditions which lead to the complete disruption of the streptose moiety (p. 676). 3- and 6-Amino-deoxysugar glycosides are more readily susceptible to hydrolysis.

## OLIGOSACCHARIDES

Oligo- and poly-saccharides are characterised essentially by the linking of two or more monosaccharides, notably but not exclusively aldohexoses. The constituent units are linked glycosidically, i.e. through the anomeric hydroxyl group at C-1. Disaccharides are known in which the link involves the glycosidic hydroxyl of both sugars as in sucrose (1 → 2 glycosidic link). Such compounds have no reducing properties, and in this respect differ from others, such as mannose or lactose, in which the glycosidic hydroxyl of one monosaccharide component is linked to the hydroxyl at C-4 in the second sugar component (1 → 4 glycosidic link).

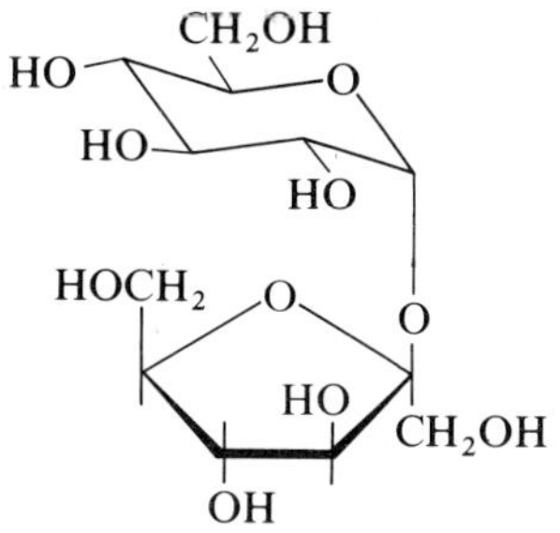

Sucrose
(α-D-glucopyranosyl-β-D-fructofuranoside

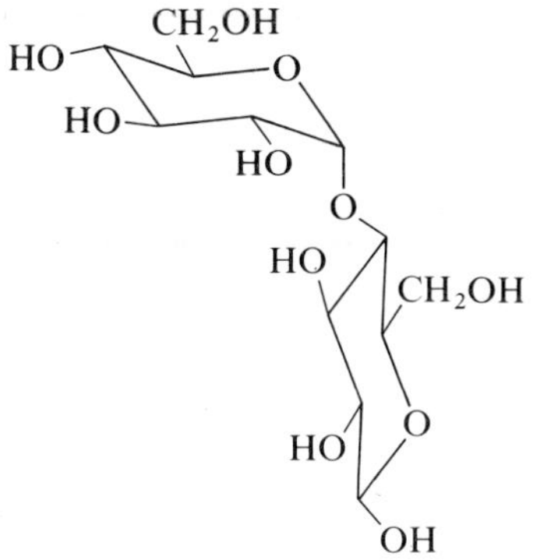

Maltose
4′-[α-D-glucopyranosido]-β-D-glucopyranose

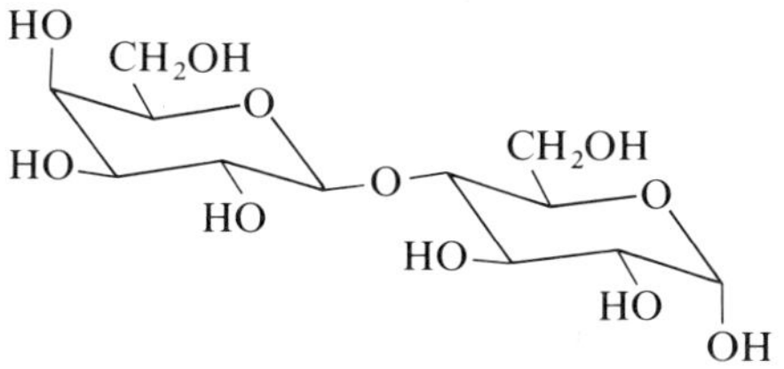

Lactose
4′-[β-D-galactopyranosido]-α-D-glucopyranose

Such (1 → 4)-linked disaccharides have one free reducing group and hence exhibit reducing properties, though these are not so pronounced as in the corresponding monosaccharides.

In other respects, most oligosaccharides resemble monosaccharides in many of their physical and chemical properties. Thus, they are generally highly water-soluble, colourless crystalline solids. Sucrose is usually obtained anhydrous, and only absorbs moisture at high relative humidities. Lactose, however, occurs as a monohydrate. Both sugars are readily soluble in water (sucrose in less than 1 part of water; lactose about 1 in 6). Both are sweet to the taste, but sucrose is significantly sweeter than lactose. A concentrated sugar solution (*Syrup*) containing 67.5% w/w of sucrose, provides a useful sweet-tasting liquid vehicle for the administration of unpleasant tasting drugs. Such solutions containing over 65% w/w sucrose have a sufficiently high osmotic pressure to inhibit the growth of moulds and other microbiological contaminants, and may be stored for considerable periods without deterioration. Syrup is also widely used in the sugar coating of tablets. Sugar, ingested orally, is digested rapidly, and marked elevations of blood sugar have been reported within minutes of ingestion. It is rapidly degraded by **sucrase** from the intestinal mucosa to glucose and fructose, which are then absorbed. As a fructofuranoside, sucrose is also extremely labile under acidic conditions and readily hydrolyses to glucose and fructose.

Sucrose

$H_2O$
Sucrase (Invertase)

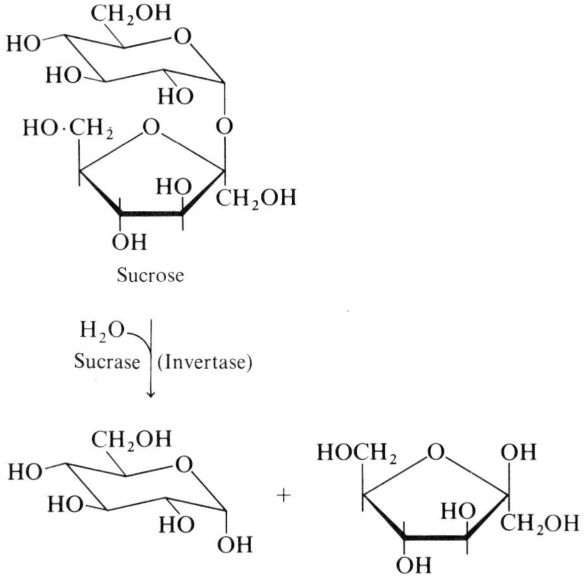

α-D-Glucose + β-D-Fructose

Lactose has certain advantages as a sweetening agent, particularly as an additive to milk foods for use in infant feeding, since it is less liable to cause intestinal fermentation and gas production. It does, however, promote the growth of acid producing micro-organisms in the intestine, notably *Lactobacillus acidophilus*, which converts lactose to lactic acid. Lactose supplements in infant milk foods therefore promote lactic acid formation and consequent lowering of faecal pH to around pH 5.0. Lactose also finds considerable use, pharmaceutically, as a solid diluent to give bulk to low dose medicines prior to tabletting or encapsulation, and to visualise the contents of ampoules which contain very small amounts of solid medicament. Lactose monohydrate can occasionally cause or accelerate decomposition of admixed medicaments, and may require to be used in the anhydrous form, which is readily obtained on drying at 105°C. Spray-dried lactose can, however, contain appreciable amounts (7–8%) of 2-hydroxymethylfurfural, which can be the cause of manufacturing losses in the tabletting of certain compounds (Janicki and Almond, 1974). As a reducing sugar, lactose is also more prone than sucrose to oxidation by appropriate medicaments.

# OLIGOSACCHARIDE AND RELATED ANTIBIOTICS

## Streptomycin

*Streptomycin* (Schatz, Bugie and Waksman, 1944) is a water-soluble non-crystalline, tri-acidic base, which readily forms stable salts such as the sulphate [$(Base)_2$, $3H_2SO_4$], hydrochloride [Base, 3HCl], and a double salt of the latter with calcium chloride [$(Base, 3HCl)_2$ $CaCl_2$]. The salts, which are colourless crystalline solids, are hygroscopic and deliquesce in moist air. They are all readily soluble in water to give solutions which are only slightly acidic in reaction. In aqueous solution, the salts are stable between pH 4 and 7, but alkaline solutions are unstable, *0.1N* sodium hydroxide causing fairly rapid hydrolysis

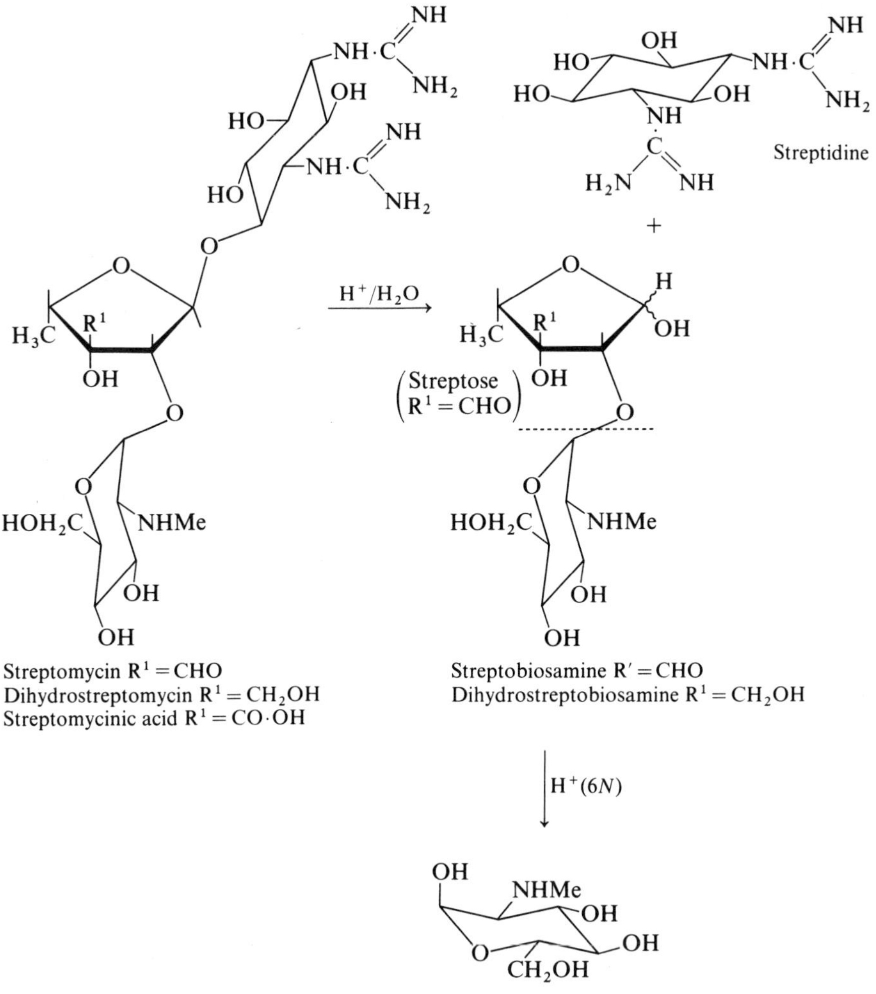

$H^+/H_2O$ →

Streptidine

+

(Streptose $R^1 = CHO$)

Streptomycin $R^1 = CHO$
Dihydrostreptomycin $R^1 = CH_2OH$
Streptomycinic acid $R^1 = CO{\cdot}OH$

Streptobiosamine $R' = CHO$
Dihydrostreptobiosamine $R^1 = CH_2OH$

↓ $H^+(6N)$

*N*-Methyl-L-glucosamine

with loss of antibiotic activity and formation of maltol. Below pH 4, the molecule is disrupted by hydrolysis yielding two antibiotically inactive bases, streptidine and streptobiosamine. Streptidine separates as the rather insoluble streptidine sulphate on treatment with sulphuric acid and, like *Streptomycin*, gives colour reactions characteristic of their guanidino substituents. Aqueous solutions of *Streptomycin Sulphate* are acidic (pH 5.0–6.5) and are insufficiently stable to permit sterilisation for injection by autoclaving.

The pyranoside link of streptobiosamine is stabilised towards acid hydrolysis by protonation of the neighbouring methylamino group (p. 672). The link is broken by prolonged hydrolysis with strong acid (6*N*) with the formation of *N*-methyl-L-glucosamine, but the streptose moiety is destroyed in the process.

*Streptomycin* exhibits a number of properties characteristic of a free aldehyde group in the molecule. It readily forms an oxime, hydrazone and semicarbazone, all of which are biologically inactive (Donovick, Rake and Fried, 1946). On the other hand, some antitubercular activity has been demonstrated in streptomycylidene isonicotinylhydrazine, which was obtained by condensation of *Streptomycin* and *Isoniazid* (Yale, Losee, Martins, Holsing, Perry and Bernstein, 1953).

The antibiotic can also be reduced catalytically in aqueous solution with palladium and hydrogen without loss of antibacterial activity to yield *Dihydrostreptomycin*. Streptomycinic acid, the corresponding carboxylic acid (p*K* 2.35) which is formed by oxidation of *Streptomycin* with bromine water, however, is inactive.

*Streptomycin* is unstable in alkaline solution and decomposes to form the γ-pyrone derivative, maltol. On the other hand, neither *Dihydrostreptomycin* nor streptomycinic acid rearrange in the presence of alkali, indicating that maltol is derived from the streptose fragment of the molecule in a reaction involving the aldehyde group. The carbon skeleton of maltol differs from that of streptose and it has been deduced that it is formed by a carbon–carbon rearrangement similar to that of α-hydroxyisobutyraldehyde to acetoin (Schenck and Spielman, 1945; Lemieux and Wolfrom, 1948), as set out below and on page 677.

$(H_3C)_2C(OH){-}CH{=}O \xrightarrow{HO^-} (H_3C)_2C(O^-)(OH\cdots){-}CH(OH){-}O^- \xrightarrow{HO^-} CH_3{\cdot}CO{\cdot}CHOH{\cdot}CH_3$

Acetoin

*Streptomycin* is widely used in the treatment of tuberculosis. Studies on *E. coli* have shown that the antibacterial activity of *Streptomycin* is reduced in the presence of cysteine. The latter, however, does not inactivate *Dihydrostreptomycin*, which is almost as active antibiotically as *Streptomycin* itself. It appears, therefore, that the streptose-aldehydic function is a key centre for activity. This is supported by the inactivity of streptomycinic acid and the inhibitory effect of aldehyde-blocking reagents (Donovick, Rake and Fried, 1946).

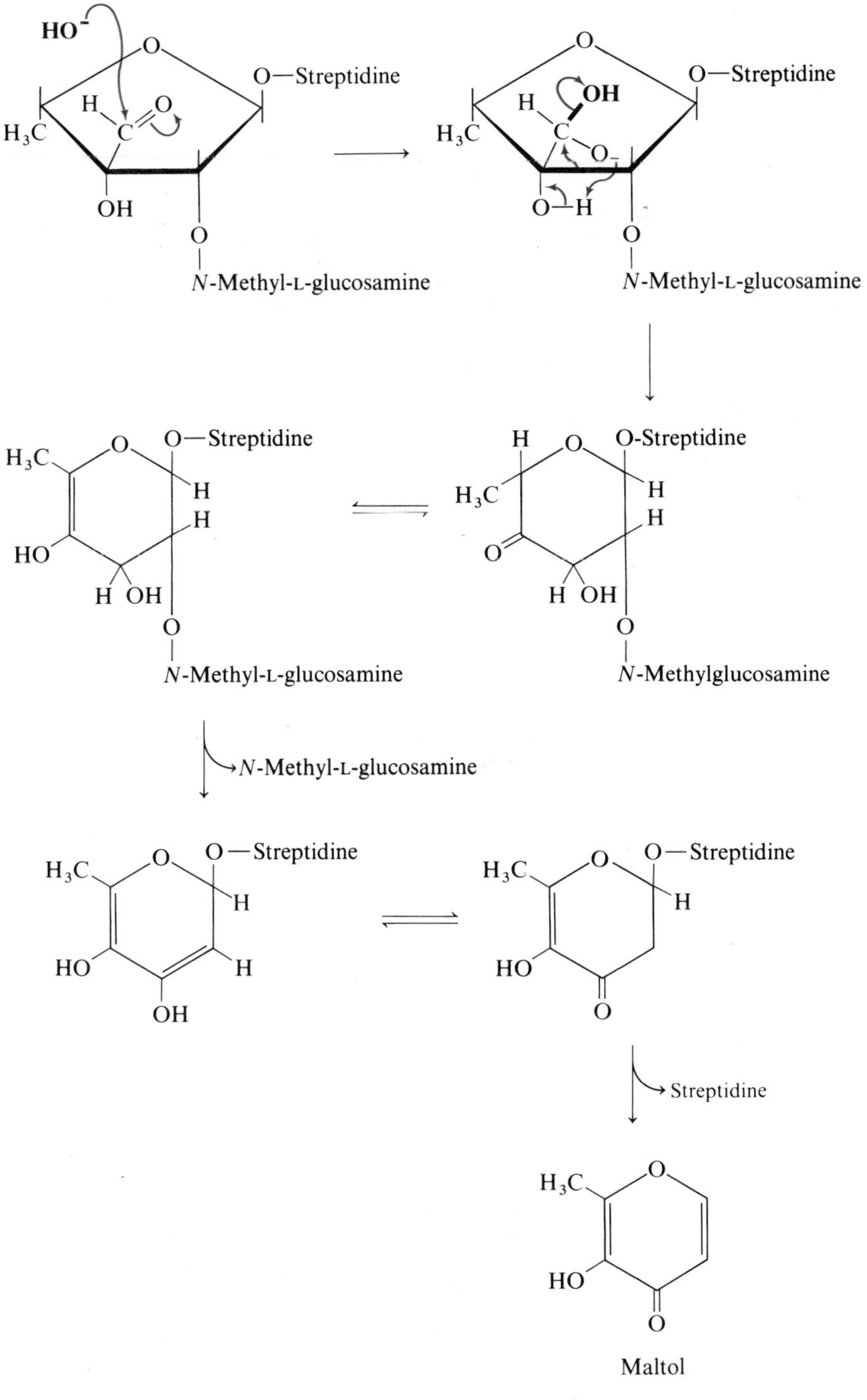

HO⁻

O—Streptidine

*N*-Methyl-L-glucosamine

O—Streptidine

*N*-Methyl-L-glucosamine

O—Streptidine

*N*-Methyl-L-glucosamine

O-Streptidine

*N*-Methylglucosamine

→ *N*-Methyl-L-glucosamine

O—Streptidine

O—Streptidine

→ Streptidine

Maltol

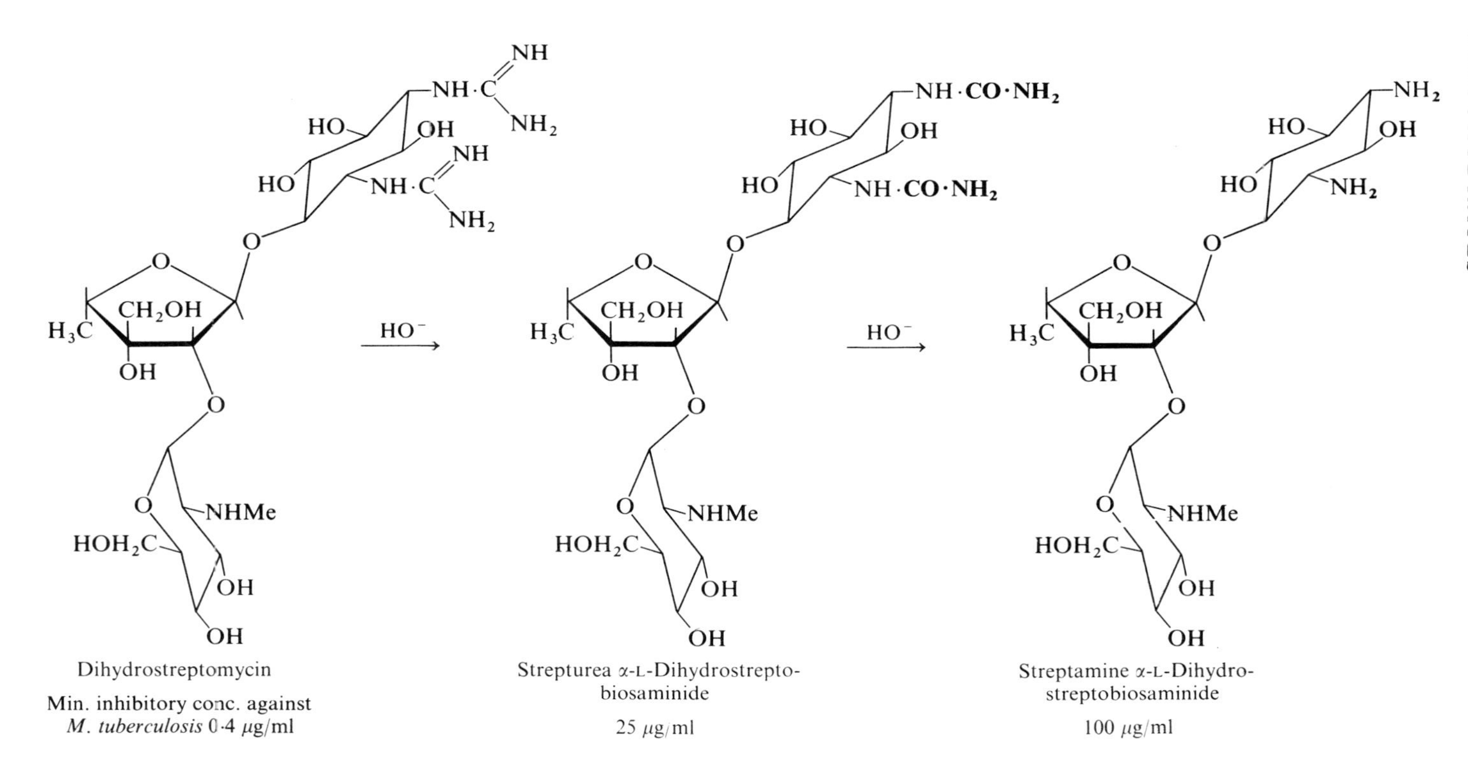


Dihydrostreptomycin

Min. inhibitory conc. against *M. tuberculosis* 0·4 μg/ml

Strepturea α-L-Dihydrostreptobiosaminide

25 μg/ml

Streptamine α-L-Dihydrostreptobiosaminide

100 μg/ml

Bailey and Cavallito (1947) have ascribed the effective tuberculostatic activity of *Dihydrostreptomycin* to the availability of the streptose —$CH_2OH$ hydroxyl for re-oxidation *in vivo* to *Streptomycin*. This view is supported by the inactivity of dihydrostreptomycin derivatives in which this group is blocked in the form of metabolically-stable trityl ($Ph_3C$—) ethers (Comrie, Mital and Stenlake, 1960).

The strongly basic guanidino groups together with the general hydrophilic character of the molecule inhibit the gastro-intestinal absorption of *Streptomycin*, which in consequence requires to be administered by injection. The guanidino groups are also important for effective antibiotic action against *M. tuberculosis*. Thus, undeca-acetyldihydrostreptomycin and dodeca-acetyldihydrostreptomycin are inactive, whilst strepturea dihydrostreptobiosaminide and streptamine dihydrostreptobiosaminide obtained by progressive alkaline degradation of dihydrostreptomycin have only low levels of activity against *M. tuberculosis* (Comrie, Mital and Stenlake, 1960). (See scheme on page 678.)

Abraham and Duthie (1946) have also shown that the inhibitory action of *Streptomycin* against Gram-positive bacteria is depressed with decreasing pH of the growth medium, and have suggested that the antibiotic functions in its cationic form, competing with hydrogen ions for the same sites on the bacterial cells.

Transferable drug resistance to *Streptomycin* has been shown to be due to development of specific drug-metabolising enzymes in bacteria capable of converting the antibiotic to inactive derivatives (**2**, 5).

### Neomycin

*Neomycin* (Waksman and Lechevalier, 1949) is primarily neomycin B admixed with varying amounts, usually about 10%, of its more toxic and antibiotically less potent epimer, neomycin C. Most samples of *Neomycin Sulphate* also contain small amounts of an antibiotically inactive hydrolytic decomposition product neamine (neomycin A), which is formed in the course of fermentation and extraction of the antibiotic from *Streptomyces fradiae* (Swart, Hutchinson and Waksman, 1949, 1951). According to Sebak (1955), certain species of coccobacterium, which are inhibited by neomycin B at concentrations of 15–30 $\mu$g/ml, are unaffected by either neomycin C or neamine at 50 000 $\mu$g/ml; these organisms are used to assay *Neomycin Sulphate* for neomycin B. *Framycetin* (Neomycin; Tremblay, Destouches and Karatchenzeff, 1953) consists almost exclusively of neomycin B. (See structure on page 680.)

*Neomycin Sulphate* is a hygroscopic powder, which is readily soluble in water (1 in 3) to give slightly acidic solutions. Both solid and solutions discolour on prolonged storage, on exposure to light or ionising radiation, due to oxidative radical-induced decomposition. Neomycins B and C are stable to alkali, and although stable at pH 2.0 for at least 24 hr at room temperature, undergo hydrolysis on heating with *N* mineral acid to yield the antibiotically inactive neamine (Leach and Teeters, 1951).

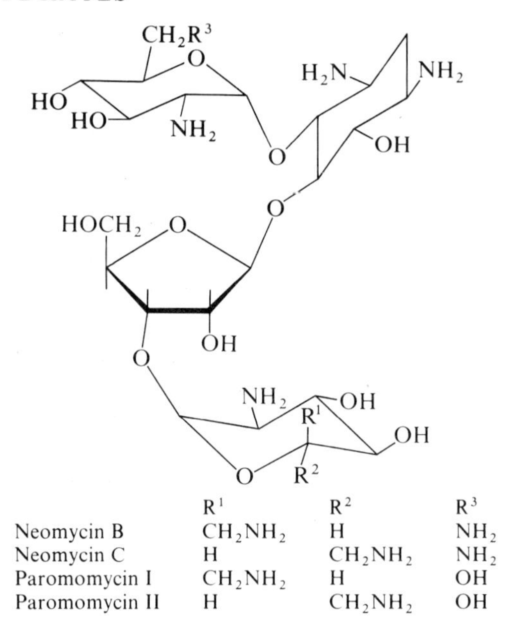


| | $R^1$ | $R^2$ | $R^3$ |
|---|---|---|---|
| Neomycin B | $CH_2NH_2$ | H | $NH_2$ |
| Neomycin C | H | $CH_2NH_2$ | $NH_2$ |
| Paromomycin I | $CH_2NH_2$ | H | OH |
| Paromomycin II | H | $CH_2NH_2$ | OH |

$\downarrow H^+$

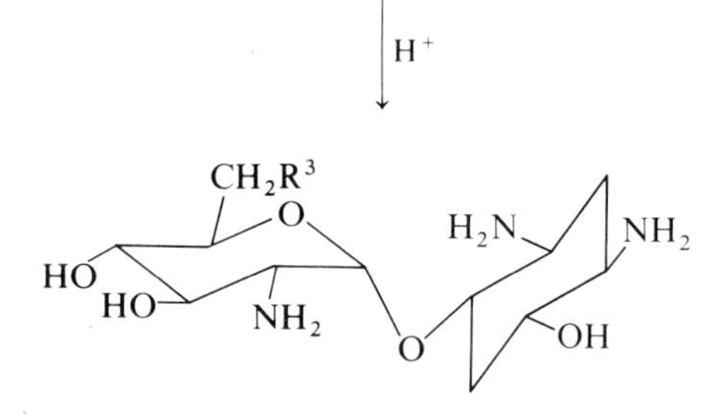


Neamine $R^3 = NH_2$
Paromamine $R^3 = OH$

+

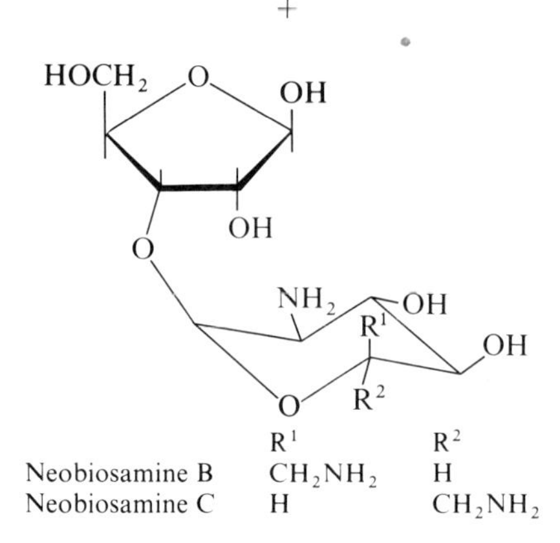


| | $R^1$ | $R^2$ |
|---|---|---|
| Neobiosamine B | $CH_2NH_2$ | H |
| Neobiosamine C | H | $CH_2NH_2$ |

*Neomycin* is an effective antibiotic against Gram-positive and Gram-negative organisms, including streptomycin-resistant strains; it is also effective against *Myco. tuberculosis* (Leach, DeVries, Nelson, Jackson and Evans, 1951). It is used as the sulphate, a water-soluble salt in the treatment of topical infections. It is also administered orally in the treatment of bowel infections, on account of its comparative stability to acidic, alkaline and enzymatic hydrolysis, and its poor absorption from the gastro-intestinal tract, which is a function of its strongly hydrophilic character. Systemic use is limited by the nephrotoxicity and ototoxicity, which it produces on prolonged administration.

*Neomycin* is a strong base and is incompatible with anionic substances. It precipitates with *Sodium Lauryl Sulphate*, with *Sodium Alginate* and *Sodium Carboxymethylcellulose*, *Amaranth* (FD & C Red No. 2) and other basic dyes. It is also firmly bound to *Bentonite*, but is compatible with *Acacia*, *Tragacanth*, *Methylcellulose* and *Povidone* (Dale and Rundman, 1957).

The paromomycins (Coffrey *et al.*, 1959) are closely related to the neomycins. *Paromomycin I*, the major component of the antibiotic, yields paromamine and neobiosamine B on acid hydrolysis, whilst *Paromomycin II* is similarly related to neomycin C. Paramomycin has a similar antibacterial spectrum to neomycin and, like the latter, is poorly absorbed from the alimentary canal. It is used as *Paromomycin Sulphate*, which is a hygroscopic, water-soluble (1 in 20) solid by oral administration and is particularly effective against *E. histolytica*, shigella and salmonella infections. It is of value in the treatment of amoebic and bacterial dysentery, and in gastroenteritis due to shigella and typhoid (Pratt, 1962).

**Kanamycin**

*Kanamycin* (Takeuchi, Hikiji, Nitta, Yamazaki, Abe, Takayama and Umezawa, 1957), is also obtained as a mixture of closely-related components A, B and C, the major one being kanamycin A. It is effective, but less so than *Streptomycin* and *Neomycin* against *Myco. tuberculosis* (Pratt, 1962). It is also bactericidal for a wide range of Gram-positive and Gram-negative organisms, and finds special use in the treatment of persistent urinary tract infections with penicillin-resistant strains, which fortunately do not as a rule show cross-resistance. Like *Neomycin*, *Kanamycin* is poorly absorbed from the alimentary tract. It is, therefore, usually administered as the sulphate by intramuscular injection.

Both glycosidic links at the 4- and 6-positions of 2-deoxystreptamine are adjacent to amino substituents. The antibiotic is, therefore, considerably more stable to acid hydrolysis than either *Neomycin* or *Streptomycin*, and injection solutions of *Kanamycin Sulphate*, which have a pH of about 6.0, can be sterilised by autoclaving without decomposition.

*Kanamycin-*, *Neomycin-* and *Paromomycin*-resistant strains of *E. coli* have been shown to inactivate the antibiotic by specific phosphorylation at C-3 (*Kanamycins A* and *C*, *Neomycin* and *Paromomycin*) and *N*-acetylation of the 6-amino-6-deoxy-D-glucose fragment (*Kanamycin C* only) respectively (Umezawa *et al.*, 1967a and b). The antibiotics, *Gentamicin* (Copper, Daniels, Yudis,

Marigliano, Guthrie and Bukhari, 1971) and *Tobramycin*, are similar in structure to *Kanamycin*. *Tobramycin*, however, lacks the C-3 hydroxyl group in the 6-amino-6-deoxy-D-glucose moiety, and is not inactivated by phosphorylation (**2**, 5).

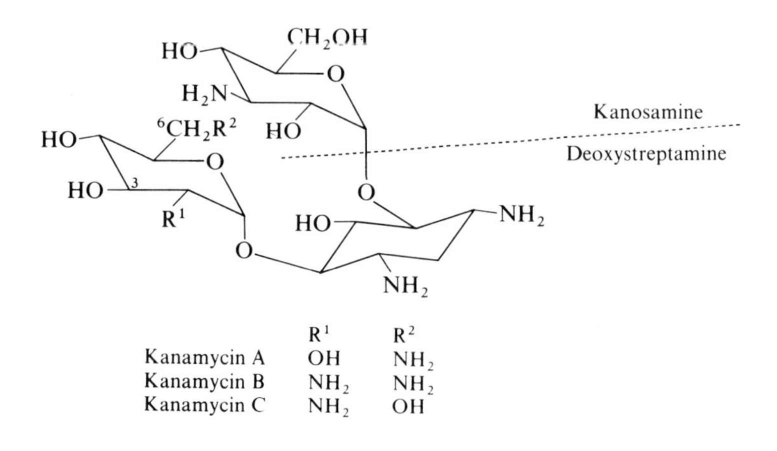


| | $R^1$ | $R^2$ |
|---|---|---|
| Kanamycin A | OH | $NH_2$ |
| Kanamycin B | $NH_2$ | $NH_2$ |
| Kanamycin C | $NH_2$ | OH |

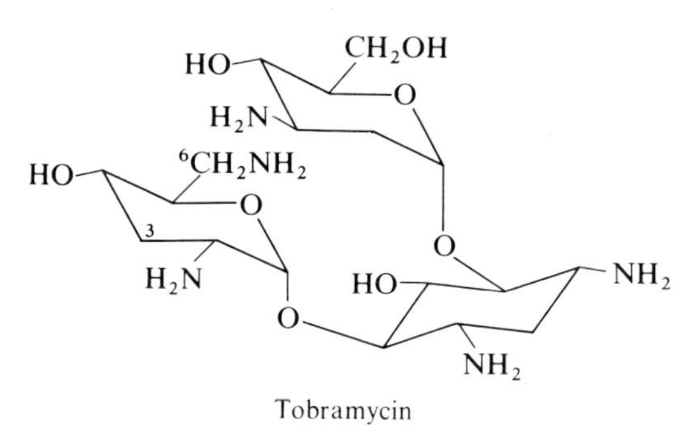

Tobramycin

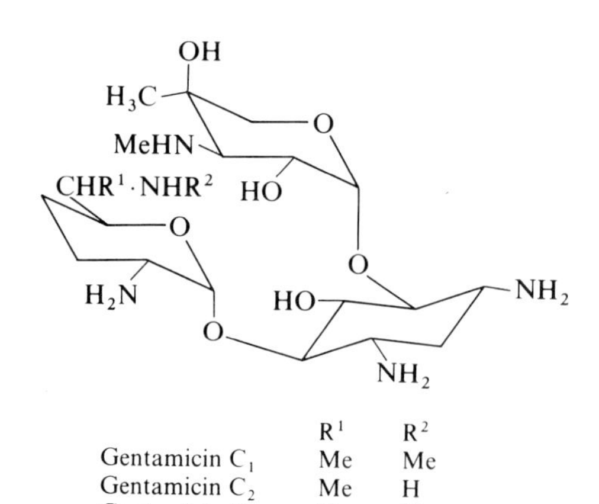

| | $R^1$ | $R^2$ |
|---|---|---|
| Gentamicin $C_1$ | Me | Me |
| Gentamicin $C_2$ | Me | H |
| Gentamicin $C_{1A}$ | H | H |

**Lincomycin**

Lincomycin (Mason, Dietz and DeBoer, 1962; Herr and Bergy, 1962) is a water-soluble, monobasic antibiotic obtained as the hydrochloride by fermentation of *Streptomyces lincolnensis*. The free base is soluble in water, also in most organic solvents, thus accounting for its ready absorption from the gastro-intestinal tract, and wide tissue distribution. The hydrochloride, in which form it is usually administered, is water-soluble (1 in 2) to give distinctly acidic solutions, about pH 5.0. The antibiotic is unstable on prolonged exposure to acid, undergoing hydrolysis primarily with hydrolysis of the thioether link to yield methanethiol (Hoeksema *et al.*, 1964; Schroeder, Bannister and Hoeksema, 1967). Strong acid cleaves the amide link. For this reason, injection solutions must be sterilised by filtration (and not autoclaving) and must be stored at less than 20°C. The antibiotic is also sensitive to light, probably undergoing radical oxidation to the sulphoxide and/or bond cleavage also at the thio-glycoside link.

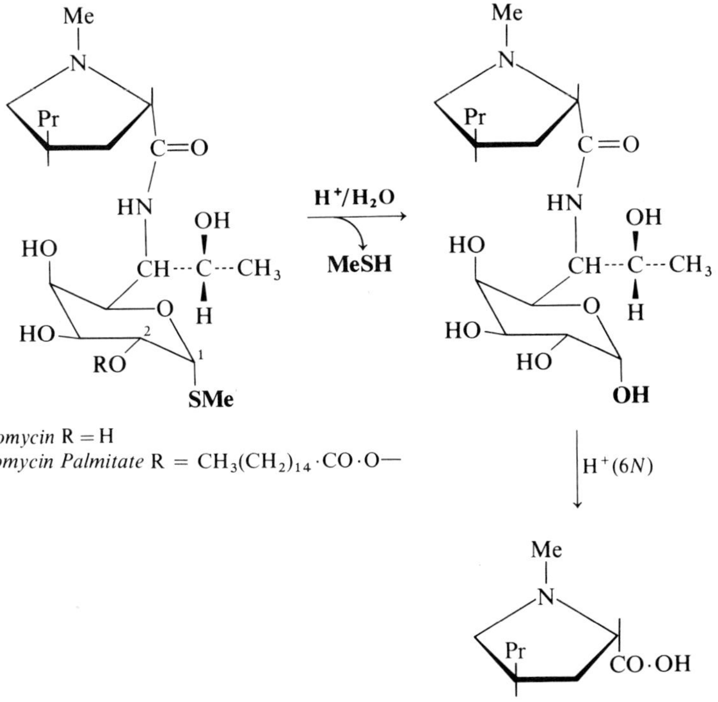

$H^+/H_2O$ → MeSH

*Lincomycin* R = H
*Lincomycin Palmitate* R = $CH_3(CH_2)_{14} \cdot CO \cdot O-$

$H^+(6N)$

*trans*-L-4-n-propylhygric acid

The antibiotic is effective mainly against Gram-positive organisms and is used in the treatment of staphylococcal, streptococcal and pneumococcal infections. It may be administered orally or parenterally for more rapid effect, though 30–60% of the parenterally-administered dose is excreted within 24 hr (Kaplan

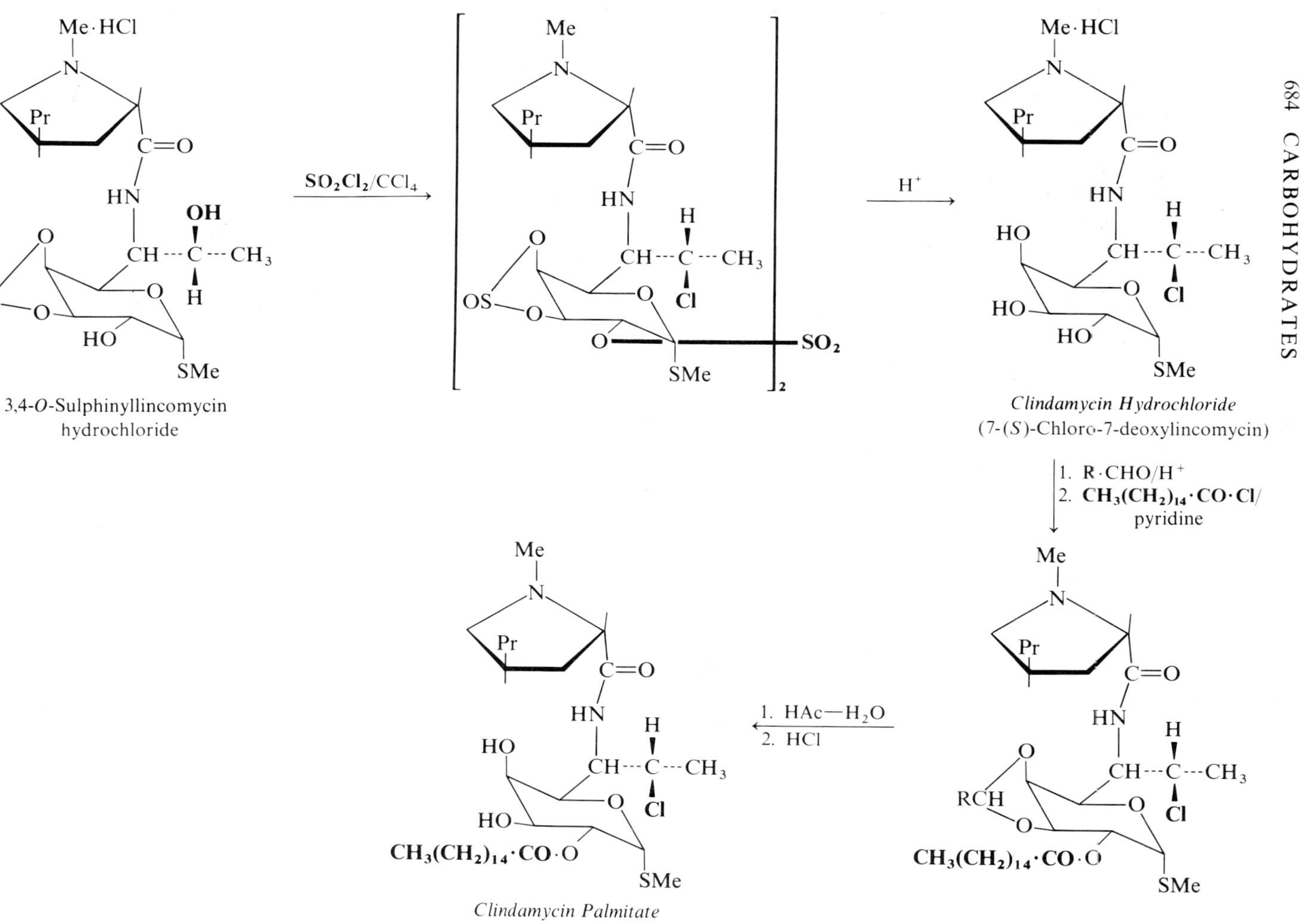
Me·HCl
Pr
N
C=O
HN
CH
OH
H
CH₃
O
OS
HO
SMe
3,4-O-Sulphinyllincomycin hydrochloride
SO₂Cl₂/CCl₄
Me
Cl
SO₂
2
H⁺
Clindamycin Hydrochloride
(7-(S)-Chloro-7-deoxylincomycin)
1. R·CHO/H⁺
2. CH₃(CH₂)₁₄·CO·Cl/ pyridine
RCH
CH₃(CH₂)₁₄·CO·O
1. HAc—H₂O
2. HCl
Clindamycin Palmitate

and Weinstein, 1965), despite the fact that some 70% of plasma *Lincomycin* is protein-bound (Gordon, Regamy and Kirby, 1973). It interferes with the utilisation of $^{14}C$-lysine in the log phase of growing *Staphylococcus aureus* cultures, due apparently to inhibition of DNA synthesis (Josten and Allen, 1964).

*Lincomycin B*, a minor contaminant of most commercial samples of lincomycin, is the lower homologue derived from *trans*-L-4-n-ethylhygric acid. It has a similar microbiological spectrum, but has only about 25% of the potency of *Lincomycin* (Mason and Lewis, 1964).

2-Monoesters of *Lincomycin* formed from long-chain fatty acids, such as *Lincomycin Palmitate* are tasteless, effective antibiotics (Morozowich, Sinkula, MacKeller and Lewis, 1973). The ester grouping reduces water solubility, but they are nonetheless suitable for formulation in paediatric preparations on account of taste.

**Clindamycin**

A number of chemical modifications of *Lincomycin* have been prepared and examined (Magerlein, Birkenmeyer and Kagan, 1966), of which the most important is *Clindamycin* and its palmitate (see reaction scheme page 684). *Clindamycin*, 7-(*S*)-chloro-7-deoxylincomycin, has about four times the antibacterial activity of *Lincomycin*, but is more strongly bound (94%) to plasma proteins (Gordon, Regemy and Kirby, 1973); the corresponding 7-(*R*)-chloro-7-deoxy-derivative is only about 1.6 times as active.

*Clindamycin Palmitate Hydrochloride* has the advantage over *Clindamycin Hydrochloride* in that it is slightly sweet, instead of bitter, to the taste. The palmitate hydrochloride is water-soluble and gives a weakly acidic solution. Clindamycin palmitate base is precipitated as an oil above pH 6.0.

**Spectinomycin**

*Spectinomycin* (Hoeksema, Argoudelis and Wiley, 1962) is a basic antibiotic, derived from the amino-cyclitol, actinamine (Slomp and MacKellar, 1962), which is used in the treatment of gonorrhea. It is used as the hydrochloride which is freely soluble in water to give weakly acidic solutions (*ca* pH 4.0) suitable for intramuscular injection. Prolonged exposure to acid, particularly at low pH causes decomposition to the antibiotically inactive actinamine (Chapman, Autrey, Gourlay, Johnson, Souto and Tarbell, 1962; Bergy, Eble and Herr, 1961). For this reason, injections are prepared as a sterile freeze-dried powder for reconstitution as an injectable suspension. The antibiotic is also rapidly hydrolysed to spectinomycinic acid in alkaline solution [*0.1N* $Ba(OH)_2$] at ambient temperature with concomitant loss of antibiotic activity (Wiley, 1962; Wiley, Argoudelis and Hoeksema, 1963). In contrast, dihydrospectinomycin, which is also antibiotically active, is stable in alkaline solution. Garrett and Umbreit (1962) have examined the stability of *Spectinomycin* to buffer anions, including borate and phosphate, and shown solutions to have a half-life of about 1000 days at pH 7.0, and about 100 days at pH 8.0.

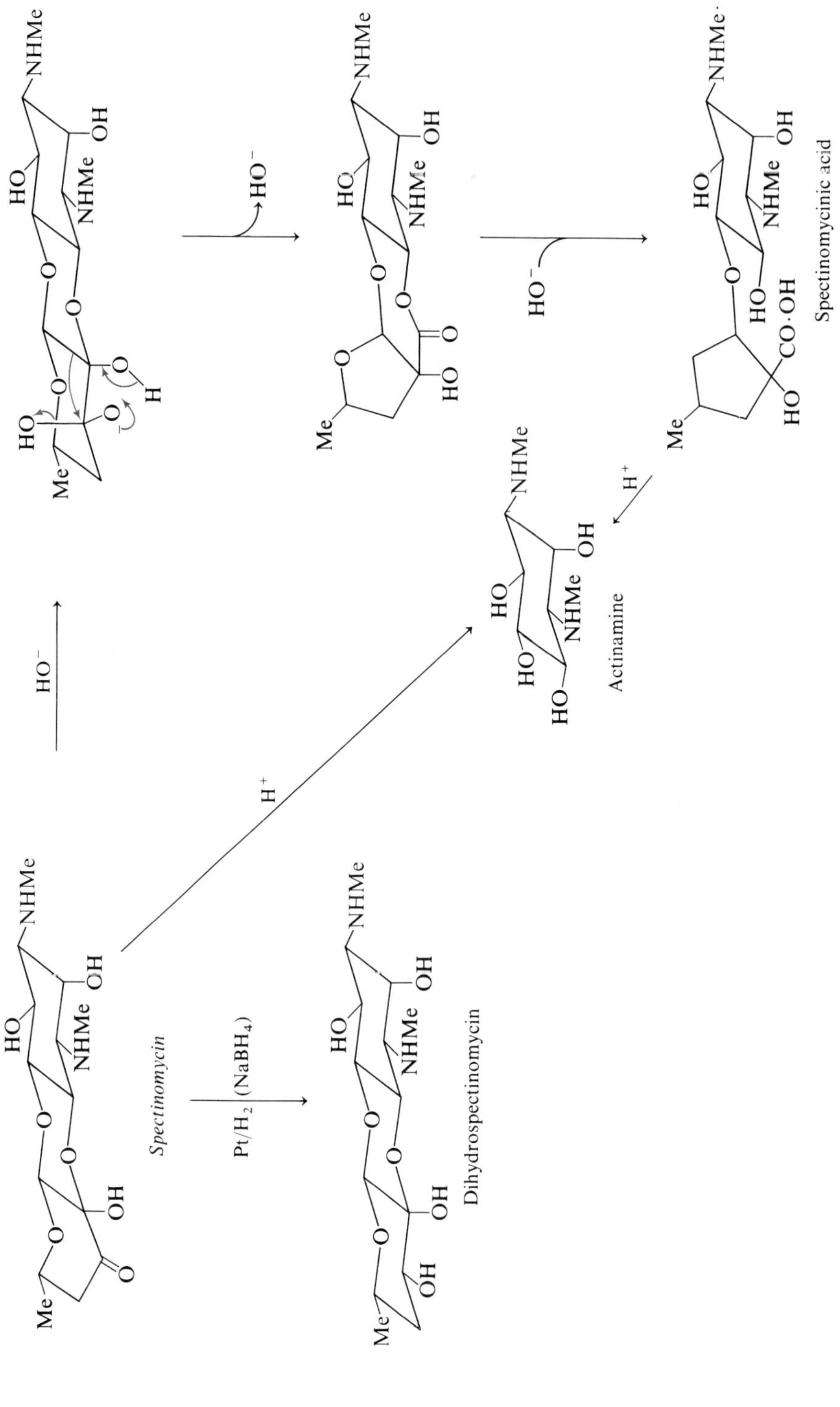

*Spectinomycin*

$Pt/H_2$ ($NaBH_4$)

Dihydrospectinomycin

$H^+$

$HO^-$

Actinamine

Spectinomycinic acid

## POLYSACCHARIDES (GLYCANS)

Polysaccharides are high molecular weight condensation polymers consisting of large numbers of monosaccharide units linked glycosidically by the elimination of water. Many polysaccharides have old and well-established trivial names, as in cellulose, glycogen, chitin, and so on. Others, more recently identified, have names with the suffix **-an**, as in dextran, to identify them as polysaccharides, to which the general name, **glycan**, is now attributed.

The majority of polysaccharides are of natural origin. They are of special importance in the structural organisation of plants, and may compose as much as 80% of the dried substance. Polysaccharides are also important constituents of the cell wall in micro-organisms and of the exo-skeleton of arthropods. Their function in animals, however, is somewhat different. Mucopolysaccharides, such as hyaluronic acid and chondroitin sulphate, are widely distributed in the ground cement substance of interstitial tissue and collagen, and as such make a structural contribution. Others, however, have more specialised functions, such as that of glycogen in energy storage, mucin as a protective of mucous membranes, and heparin as an anticoagulant. In a like manner starches, gums, mucilages and other polysaccharides also have energy-storage and other specialised functions in plants.

### General Properties

The vast majority of natural polysaccharides have either (1 → 4) or (1 → 6) glycosidic links or both. The position and configuration of these glycosidic links largely determines the shape and flexibility of the polymer, and these are probably the most important determinants of their physico-chemical properties (Rees, 1969). Thus, although polysaccharides are polyhydroxylic, and therefore in principle hydrophilic, high molecular weight and multiple inter- and intramolecular hydrogen bonds lead to low water solubility in the more compact polymer structures.

Computer model building (Rees and Scott, 1969) shows that for homopolysaccharides in which the glycosidic C—O bond is equatorial to both sugar rings, β(1 → 4) linkages lead to stiff ribbon-like chains, and β(1 → 3) linkages give rise to stiff helices capable of intertwining into multiple-stranded ropes. In contrast β(1 → 6) polymer links form jointed chains with high flexibility and solubility, whilst β(1 → 2) linkages lead to stiff contorted chains unlikely to form stable tertiary structures.

### β(1 → 4)-Linked Polysaccharides

The native purified α-cellulose of cotton, a β(1 → 4) D-glucose polymer (Fig. 35A) of average molecular weight in excess of 1 million forms long ribbon-like chains, which are subject to considerable restraint. The chains are, however, capable of close lateral association and cohesion as a result of interchain hydrogen bonding. This explains the polymer's complete water insolubility and

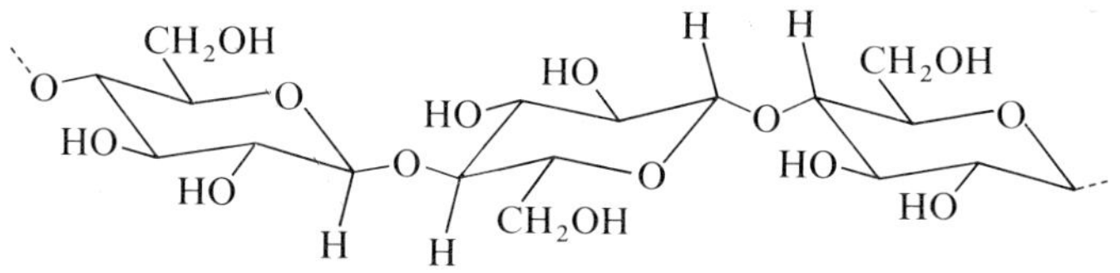

(A) Bent $\beta(1 \rightarrow 4)$ chain conformation of cellulose (Marchessault and Sarko, 1967)

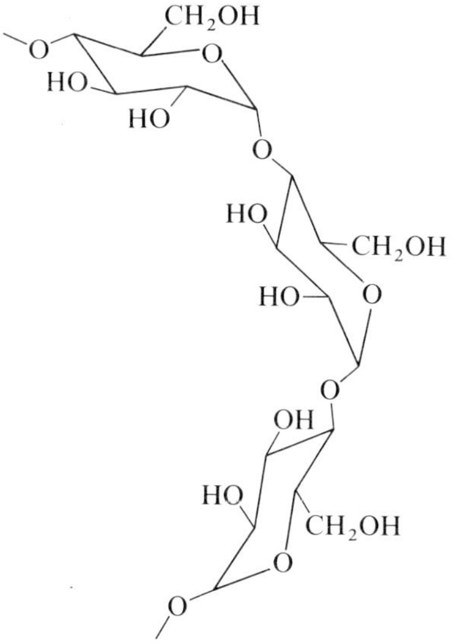

(B) $\alpha(1 \rightarrow 4)$ linked helix of amylose

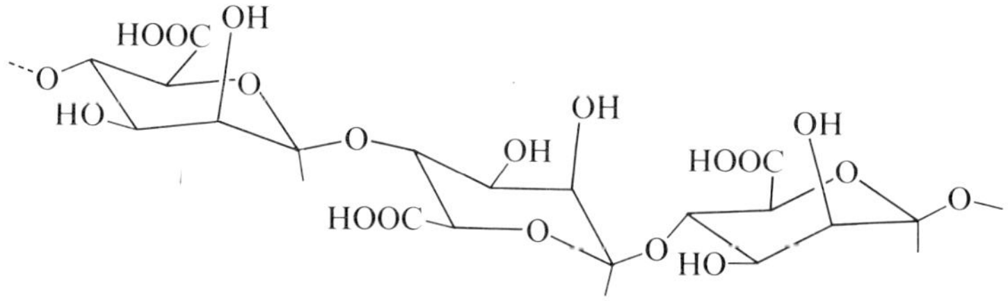

(C) $\beta(1 \rightarrow 4)$ linked chain of alginic acid

Fig. 35 Some polysaccharide conformations

accounts for its strength and utility as a structural material. Purified native cellulose ($\alpha$-cellulose) is, however, capable of absorbing water due to its lack of physical homogeneity. Thus, it consists of molecular chains in densely packed **crystalline** areas intermingled with more loosely packed **amorphous** areas, which thus permit some penetration of water into the structural interstices.

Rayon, a form of artificial silk prepared by dissolving native cellulose in ammoniacal copper hydroxide solution and forcing the solution through a spinneret into dilute sulphuric acid, has a similar structure, though the degree of polymerisation (average number of units per polymer chain) is reduced from around 10 000 to between 350 and 450 in the process. This reduction in polymer chain length together with its more randomised packing increases the number of free (non hydrogen-bonded) hydroxyl groups for association with, and hence, absorption of water. In consequence, the tensile strength of rayon is much lower than that of natural cellulose, especially when it is wet. All forms of cellulose are susceptible to oxidative degradation in air on exposure to ultraviolet light. Oxidative scission of the polymer chain occurs with some loss of tensile strength and a corresponding decrease in the intrinsic viscosity (Grassie, 1956).

## Microcrystalline Cellulose

Microcrystalline cellulose (Battista and Smith, 1961; Battista, 1962) is also an artificially produced form of cellulose, but with completely different physical properties (Battista and Smith, 1962). It is formed by treating natural α-cellulose with acid, which being able to penetrate the more randomly packed **amorphous** areas of the polymer fibre, causes partial hydrolysis, but without disruption of the chains in the closely packed crystalline areas. This results in the separation of crystalline fragments, which are then disintegrated mechanically to form an aqueous dispersion. Spray-drying of this dispersion gives microcrystalline cellulose as a spongy, porous, powder of low bulk density 270–350 kg $m^{-3}$.

Microcrystalline cellulose consists of polymer chains with a degree of polymerisation between about 185 and 300 (average molecular weight 30 000 to 50 000). These polymer chains in the microcrystals associate by intermolecular hydrogen bonding of hydroxyl groups. Its spongy consistency and the loose packing of the polymer chains permits the absorption of water and oils. Water disrupts interchain hydrogen bonds, so that microcrystalline cellulose forms stable colloidal suspensions, creams and spreadable gels, depending upon concentration; 5% forms a suspension, 10% a pourable cream, and 20% or more a mouldable gel. Viscosity of colloidal dispersions increases with pH reaching a maximum around pH 9 to 10. Addition of salts increases the apparent viscosity of dispersion whilst addition of emulsified oils and fats, and also sucrose, permits repeated freezing and thawing of the product without breakdown of the dispersion. Aqueous dispersions can also be autoclaved at 115°C without breakdown, though some increase of viscosity is not unusual.

For similar reasons, microcrystalline cellulose now finds widespread use as a binder disintegrant in tabletting, forming strong hard tablets when compressed, which readily disintegrate in water (Battista and Smith, 1962). The tablets owe their strength to hydrogen bonding, but in water immediately soften, swell and disintegrate as absorbed water disrupts the compacted hydrogen bonds (Fox, Richman and Shangraw, 1965). By contrast, microcrystalline cellulose tablets do not disintegrate in solvents of low dielectric constant which are less able to hydrogen-bond with the cellulose polymer (Reier and Shangraw, 1966).

Because of its high porosity, microcrystalline cellulose is an excellent absorbent of fixed and volatile oils, and is therefore of considerable value for the incorporation of such compounds in tablets. It is also capable of admixture with plastic solids, such as hydrogenated fats and molten chocolate, and with viscous solutions, such as molasses, honey and maple syrup to form free-flowing powders, a property which is of considerable value in the food and confectionery industries and of potential use in the manufacture of medicines in solid dosage form. The high absorbency of microcrystalline cellulose makes it an excellent support material for column partition chromatography and also for thin-layer systems.

### Sodium Alginate

Alginic acid, which is obtained from brown algae also consists of $\beta(1 \rightarrow 4)$-linked chains (Fig. 35C) of alternating D-mannuronic and L-guluronic acid residues. The acid is insoluble in water, as might be anticipated from its structure. It readily forms a neutral sodium salt, *Sodium Alginate*, which swells and dissolves in water to form viscous solutions, which are useful for stabilising pharmaceutical suspensions.

### $\alpha(1 \rightarrow 4)$-Linked Polysaccharides

In contrast to the stiff ribbon-like structures of $\beta(1 \rightarrow 4)$-linked polysaccharides, the $\alpha(1 \rightarrow 4)$-linked D-glucose polymer chain of amylose (Fig. 35B) is helical with six glucose units to each complete turn of the helix (Haworth, 1946). This much more open structure permits penetration and ready hydration, and in consequence amylose is water-soluble. It can, however, be precipitated from aqueous solution by the addition of butanol. Amylose is one of the principal constituents of starch along with amylopectin, which is usually though not always present as the major component (72–80%). Amylose forms a blue complex with iodine, in which the latter is trapped, approximately one molecule within each turn of the helix.

Amylopectin is much less soluble in water, and merely swells in contact with it to form a gel. This is due to its much more complex polymer mesh-like structure, arising from a combination of $\alpha(1 \rightarrow 4)$ and $\alpha(1 \rightarrow 6)$ glucosidic linkages, which permits absorption of water, but in insufficient amount for solution. Thus, chains of 20–25 D-glucose units linked by $\alpha(1 \rightarrow 4)$ bonds with branch points formed by $\alpha(1 \rightarrow 6)$ bonds give rise to structures (Fig. 36) in which only a proportion of the chain can be disrupted by $\beta$-amylase to yield maltose. The non-hydrolysable polymer fragment is termed **limit dextrin**. It is usually more readily water-soluble, but also gives rise to mobile solutions of much lower viscosity.

*Starch* granules behave in water more like amylopectin than amylose. They swell rapidly; hence starch is widely used as a tablet disintegrant. When aqueous suspensions are heated, the granules swell to form a translucent viscous gel, and only disrupt on boiling. Treatment of starch with cold dilute hydrochloric acid causes partial hydrolysis of glucosidic links with a reduction in molecular weight.

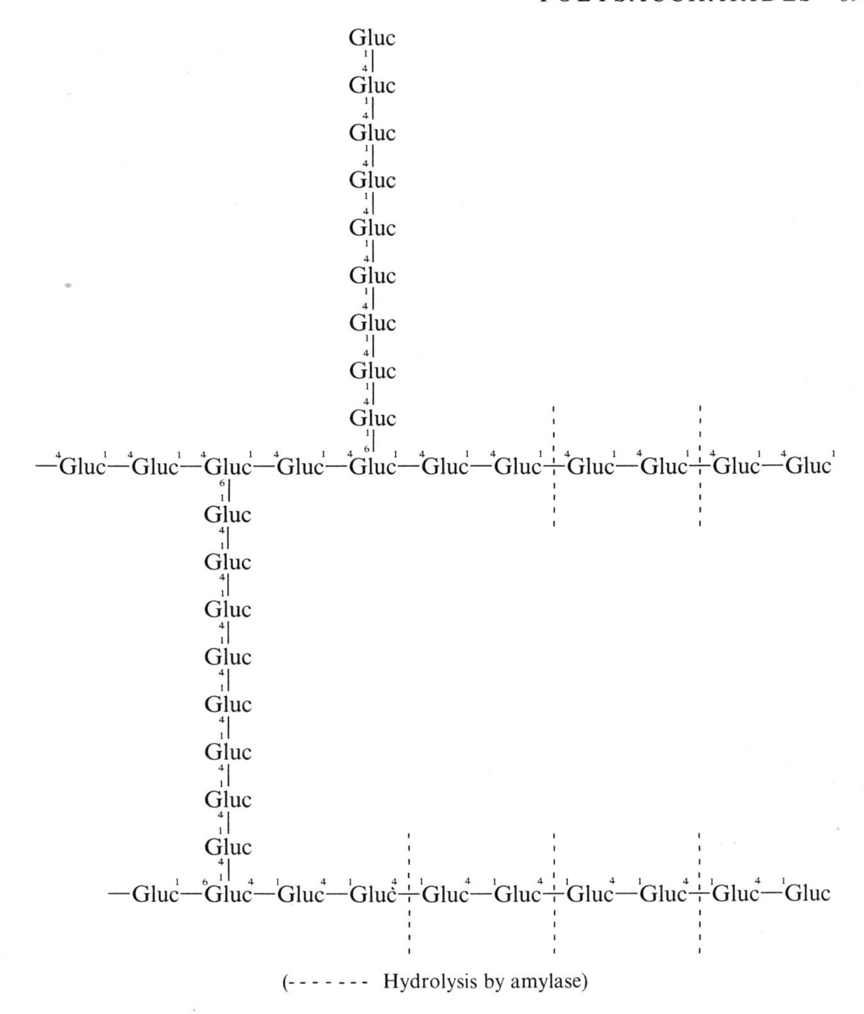


Fig. 36 Branched chain of amylopectin

The product, soluble starch, is still insoluble in cold water, but dissolves in hot water to form a non-gelatinous, transparent mobile solution, still capable of complexing with iodine to form a blue colour. It is used as an indicator in iodine titrations.

Glycogen, the principal carbohydrate storage material in mammals, is a D-glucose polymer, somewhat similar in structure to amylopectin. The chains of $\alpha(1 \rightarrow 4)$-linked glucose residues are much shorter (12–14 units), and the external chains, capable of hydrolysis by amylase, are also considerably shorter, being some 5–8 units compared with 14–19 in amylopectin. As a result, the molecules are more spherical, absorb water more readily, and are water-soluble, giving solutions of low viscosity.

### $(1 \rightarrow 6)$- and $(1 \rightarrow 3)$-Linked Polysaccharides

The outstanding characteristic of the $(1 \rightarrow 6)$-glycosidic link is its flexibility (Rees and Scott, 1969). The large increase in the number of possible conformations arising from bonding to the 6-position provides for a large entropy change in favour of the solution state rather than the solid state. In general, therefore, $(1 \rightarrow 6)$ and $(3 \rightarrow 6)$-linked glycosidic polymers have good water solubility, and high viscosity in solution, particularly when the chain is branched as in the dextrans and in plant gums such as acacia (Aspinall, 1969).

*Dextrans*, which are of bacterial origin, are D-glucose polymers containing a high proportion of $\alpha(1 \rightarrow 6)$ links with some $(1 \rightarrow 4)$ or $(1 \rightarrow 3)$ bonds to provide weight 40 000 or 100 000 is used in injection solutions as a plasma expander for the temporary restoration of blood volume. *Dextran Injection* is of much for the temporary restoration of blood volume. *Dextran 40* injection is of much lower viscosity than *Dextran 100*, and is useful when it is required to reduce overall blood viscosity and assist blood flow. The solutions are sufficiently stable to be sterilised by autoclaving, but solid deposits of unknown composition may form on storage, particularly if there are significant temperature fluctuations.

*Acacia Gum* consists of neutral salts of a highly branched acidic polysaccharide incorporating $\alpha(1 \rightarrow 4)$-L-rhamnopyranosylglucuronic acid, $\beta(1 \rightarrow 6)$-D-glucopyranosyl-D-galactose, and $\alpha(1 \rightarrow 3)$-D-galactopyranosyl-L-arabinose links, appropriately joined to an inner poly-$\alpha(1 \rightarrow 3)$-D-galactopyranosyl-D-galactose chain (Aspinall, 1969). It is highly water-soluble, and is almost entirely soluble in an equal weight of cold water to yield a translucent viscous solution, which is used pharmaceutically as a suspending and emulsifying agent, and as an emulsion stabiliser.

The alternating $\beta(1 \rightarrow 3)$ and $\beta(1 \rightarrow 4)$ links between glucuronic acid and *N*-acetylglucosamine in hyaluronic acid, and between 3,6-anhydrogalactose and galactose in agarose, give readily hydrateable structures, which account for the ability of these substances to retain water in gel form. Hyaluronic acid which is widely distributed in the interstitial spaces of animal tissue, in the vitreous and aqueous humour of the eye, and the synovial fluid, therefore assists the retention of water, gives elasticity to the tissue and acts as a natural lubricant in the eye and in joints.

### Bacterial Cell-wall Polysaccharides

The cell walls, which impart the characteristic shape to the bacterial cell and prevent its disruption, owe their strength and rigidity to a characteristic group of glycoproteins known as mureins. These substances, which vary in detailed structure with the bacterial species, are composed of alternative $\beta(1 \rightarrow 4)$-glycosidically-linked units of *N*-acetylglucosamine (NAG) and *N*-acetylmuramic acid (NAM). Each *N*-acetylmuramic acid unit is linked through its carboxyl group to a tetrapeptide consisting of both D- and L-amino acids, usually D- and L-alanine, D-glutamic acid and either L-lysine or LL-$\alpha\varepsilon$-diaminopimelic acid, which forms cross links between the polysaccharide chains either directly as in *E. coli*, or indirectly via pentaglycyl bridges, as in *Staphylococcus aureus* (Fig. 37).

The $\beta(1 \rightarrow 4)$-linked glycosidic chain gives the same degree of structural stability as in cellulose. Peptide cross-links assist formation of a lattice-like framework which provides for the integrity and mechanical strength of the cell wall. It is these $\beta(1 \rightarrow 4)$ glycosidic bonds which are cleaved by the protective enzyme lysozyme, with breakdown of the polysaccharide chain, so leading to loss of cell-wall integrity and lysis of the underlying cytoplasm (p. 696).

In addition to mureins, bacterial cell walls contain a number of other polymers, including both antigenic and protective substances, such as the lipopolysaccharide-*O*-antigens of the *Enterobacteriaceae*, and the teichoic acids (p. 626) which account for up to 60% of the dry weight of the cell wall in some Gram-positive organisms (i.e. *Staphylococcus aureus*). The relative proportions of mureins and teichoic acids are very variable even amongst Gram-positive organisms. In contrast, both lipopolysaccharide and murein development are limited in Gram-negative organisms, and the cell wall usually includes a substantial lipid and lipoprotein component. A more detailed discussion of bacterial and fungal polysaccharides is given in the Chemical Society's specialist periodical report on Carbohydrate Chemistry No. 1 (1968).

**Inhibition of Bacterial Cell-wall Synthesis by Antibiotics**

Antibiotics, such as penicillins, which exert their action mainly against actively dividing bacteria (Hobbey, Meyer and Chaffee, 1942; Chain, Duthie and Callow, 1945) interfere with cell wall formation, so that the observed cell lysis is a consequence of the deficiency of cell-wall material (Park and Strominger, 1957). Park (1952, 1957), Strominger (1959) and their collaborators have shown that a number of uridine nucleotides accumulate in the cells of *Staphylococcus aureus* grown in the presence of penicillin. The ratio, *N*-acetylmuramic acid : D-glutamic acid : L-lysine : alanine which is 1 : 1 : 1 : 3 in the uridine-5′-pyrophosphate-*N*-acetylmuramic acid pentapeptide, known as Park Nucleotide I, is the same in the intact bacterial cell wall.

The stages in cell wall synthesis are illustrated in Fig. 37. *N*-Acetylglucosamine is first converted via UDP-*N*-acetylglucosamine (Park Nucleotide II) by lactoylation into UDP-*N*-acetylmuramic acid. The acidic lactoyl function is then involved in peptide formation, first with L-alanine and then by serial addition of D-glutamic acid, L-lysine and the dipeptide, D-Ala—D-Ala, to form the UDP-*N*-acetylmuramylpentapeptide known as Park Nucleotide I. The antibiotic, *Cycloserine* (Chapter 18), inhibits the formation of Park Nucleotide I, not only by blocking the racemisation of L-alanine, and hence the source of D-alanine, but also by inhibiting the action of D-Ala—D-Ala synthetase, as a result of more favourable binding to the enzymes (Roze and Strominger, 1966).

Cell-wall synthesis continues with disaccharide formation from Park Nucleotides I and II, following attachment of Park Nucleotide I to a membrane-bound $C_{55}$-isoprenoid pyrophosphate carrier. The resulting disaccharide is further

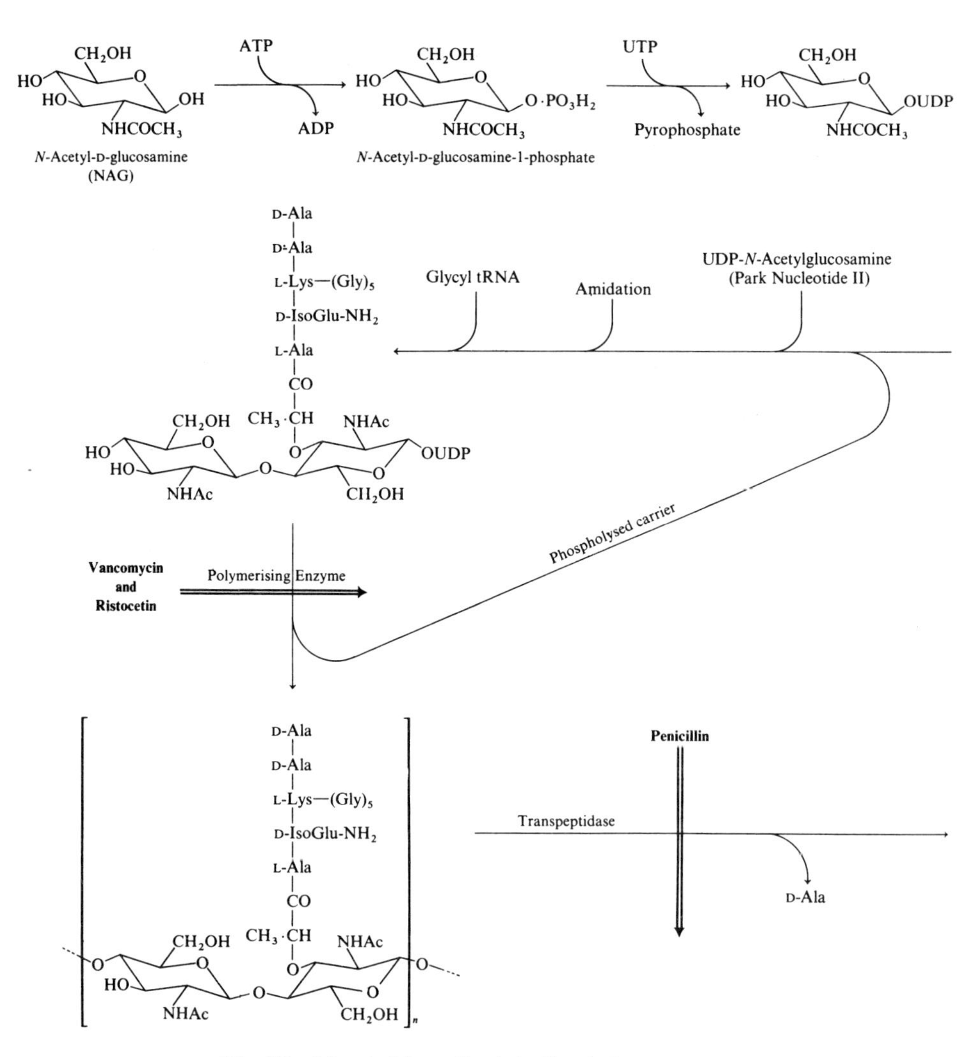


Fig. 37 Murein biosynthesis in *Staphylococcus aureus*

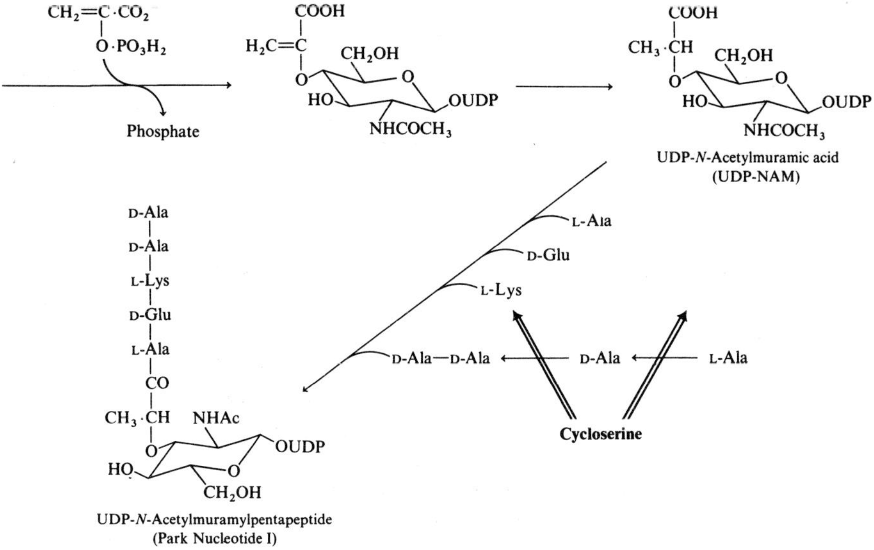

$CH_2{=}C{\cdot}CO_2^-$ / $O{\cdot}PO_3H_2$ — Phosphate

UDP-*N*-Acetylmuramic acid (UDP-NAM)

L-Ala, D-Glu, L-Lys, D-Ala—D-Ala ← D-Ala ← L-Ala

Cycloserine

UDP-*N*-Acetylmuramylpentapeptide (Park Nucleotide I)

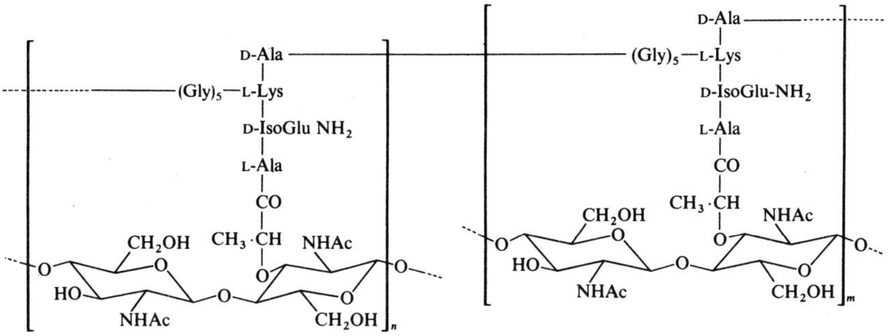

D-Ala — (Gly)$_5$—L-Lys — D-IsoGlu NH$_2$ — L-Ala — CO — $CH_3{\cdot}CH$

$CH_2OH$, NHAc, HO, $]_n$, $]_m$

modified by amidation of the D-glutamic acid unit to D-isoglutamine and the linking of a pentaglycyl unit derived from glycyl-tRNA (Matsuhashi, Dietrich and Strominger, 1965). The antibiotics, *Ristocetin* and *Vancomycin*, inhibit this phospholipid carrier mechanism.

The final steps in cell-wall synthesis consists of polyglycan formation from the disaccharide units with release of the phospholipid carrier, and the formation of cross-links by transpeptidisation. The latter occurs with displacement of the terminal D-alanine residue from the pentapeptide chains by a terminal glycyl residue from an adjacent chain. Chain-termination is brought about by D-alaninecarboxypeptidase, which splits off D-alanine and hence inhibits transpeptidation (Strominger, Izaki, Matsuhashi and Tipper, 1967). It is this phase which is inhibited by penicillin.

Martin (1963) has established that the structurally-weakened cell walls of penicillin-induced spheroplasts of *Proteus murabilis* can be dissolved by lysozyme which specifically attacks the glycosidic links of the cell-wall polysaccharide. It, therefore, follows that penicillin inhibits the transpeptidase responsible for formation of the peptide cross-linkages (Martin, 1963, 1964), and murein formed in the presence of penicillin has been shown to have an unusually high content of alanine and of terminal glycine amino groups (Wise and Park, 1965). Fractionation procedures have also demonstrated the presence of abnormally high concentrations of non cross-linked polysaccharide, and correspondingly low concentrations of cross-linked material after growth in contact with penicillin (Tipper and Strominger, 1965; Tipper, 1966).

Several penicillin-binding sites have been identified, of which two are penicillin-resistant in penicillin-resistant organisms. One such resistant binding site is probably on D-Ala carboxypeptidase.

The low murein content of the cell in Gram-negative organisms accounts for the general ineffectiveness of most penicillins against this class of organism (Chapter 18). *Ampicillin* is an exception, but its mode of action is not clearly established.

## Bacterial Cell-wall Lysis

The enzyme lysozyme, which is present in mucous, lachrymal secretion and spleen is capable of hydrolysing the glycosidic links of bacterial mureins, and hence has a defensive rôle in combating bacterial infection. The enzyme has a molecular weight of about 14 600, and consists of a single polypeptide chain of 129 units held in a folded conformation by four disulphide bridges (Jollès, Jauregui-Adell and Jollès, 1964; Canfield and Liu, 1965). The tertiary structure of hen eggwhite lysozyme has been identified by crystallographic analyses (Blake *et al.*, 1965, 1967a and b; Phillips, 1967). These show that the enzyme is ellipsoidal in shape with a large surface cleft, which has been identified as the active site of the enzyme.

X-ray studies and model building of enzyme–substrate complexes, such as the enzyme–*N*-acetylglucamine hexamer complex show that the entire $(NAG)_6$

hexamer just fills the active site cleft (Blake *et al.*, 1967b; Phillips, 1967; Harte and Rupley, 1968). This observation considered in the light of the resistance of the enzyme–$(NAG)_3$ complex to hydrolysis suggests that the polysaccharide chain of the hexamer, $(NAG)_6$, is cleaved by the enzymic hydrolysis between residues four and five. Cleavage occurs at the $C_{(1)}$—O bond with retention of configuration at C-1. A number of proposals for the mechanism of hydrolysis have been put forward (Hollaway, 1968 reviews). Retention of configuration, however, is considered to exclude single displacement reactions (Raftery and Rand-Meir, 1968). A favoured mechanism involves the intervention of both

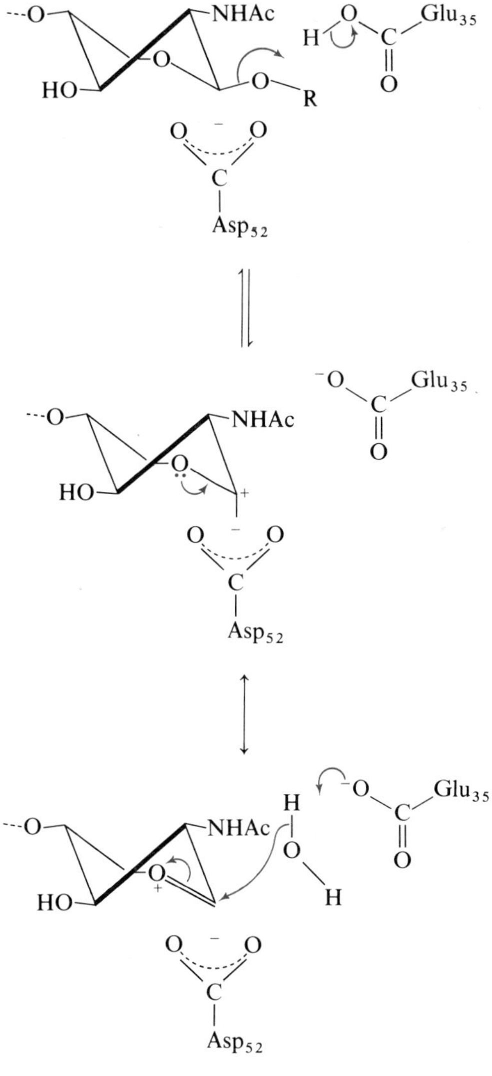

NHAc
H
O
C
$Glu_{35}$
O
O
R
HO
O
O
C
$Asp_{52}$

$^{-}O$
$Glu_{35}$
C
O
NHAc
HO
O
+
O
O
C
$Asp_{52}$

H
$^{-}O$
$Glu_{35}$
NHAc
O
C
O
HO
+
H
O
O
C
$Asp_{52}$

aspartyl-52 and glutamyl-35 residues, which are situated on either side of the fourth and fifth glycosidic link in the polysaccharide chain, provided ring D adopts a half-chair conformation. It is suggested that protonated Glu-35 provides general acid catalysis, whilst the ionised aspartyl residue stabilises the intermediate carbonium ion (Blake *et al.*, 1967b; Phillips, 1967; Vernon, 1967).

An alternative mechanism (Lowe *et al.*, 1967a and b; 1968) has been proposed in which the 2-acetamido group provides anchimeric assistance, whilst Glu-35 acts as a general acid catalyst. Anchimeric assistance, however, is not essential (Raftery and Rand-Meir, 1968) although its intervention cannot be excluded.

## SULPHATED POLYSACCHARIDES

### Heparin

*Heparin* is a natural, sulphated aminopolysaccharide of molecular weight about 17 000 consisting mainly of alternating glucuronic acid and 2-amino-2-deoxy glucose units in α(1 → 4)-glucosaminylglucuronic acid, and both α(1 → 4)- and α(1 → 6)-glucuronosylglucosamine links (Foster, Harrison, Inch, Stacey and

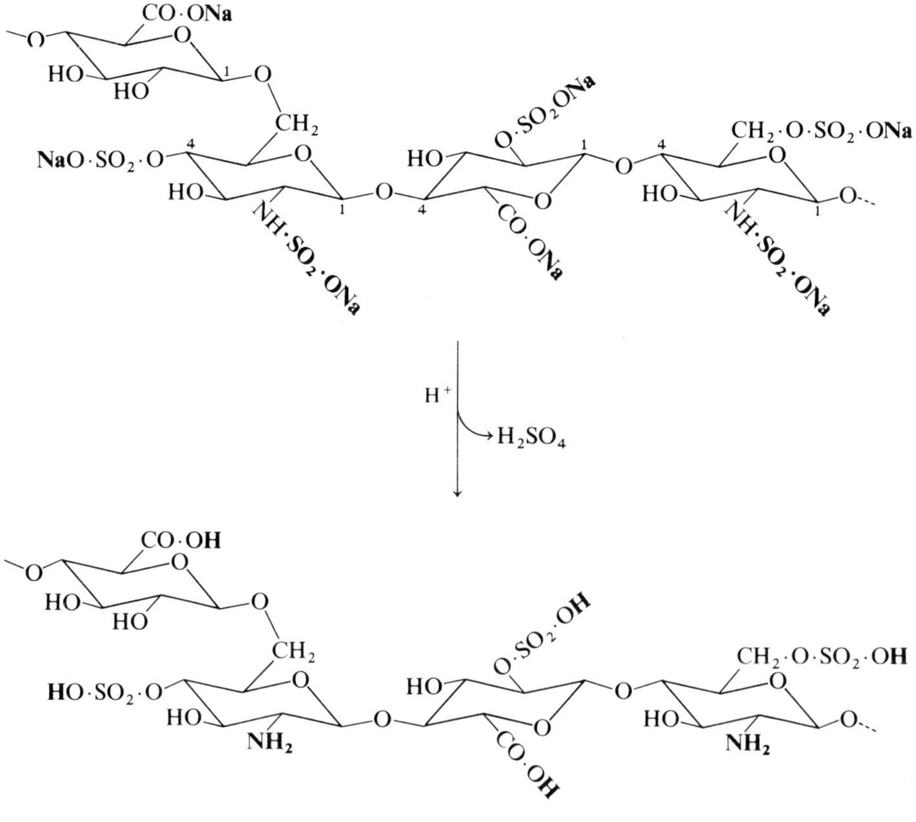

Webber, 1963). *Heparin* is the natural anticoagulant of the blood, acting primarily by inhibiting the action of thrombin in the conversion of fibrinogen to fibrin. It is present in most cells in the walls of blood vessels, the liver and lung tissue and in the intestinal mucosa.

The polysaccharide chain is both *N*- and *O*-sulphated, the sulphate ester content being approximately 5.2 groups per tetrasaccharide unit. In consequence, *Heparin*, which is the sodium salt, is readily soluble in water. The sulphamate groups are readily hydrolysed by acid under mild conditions; the sulphate ester groups, however, are only hydrolysed under more strongly acidic conditions. The presence of the sulphate groups are essential for its anticoagulent action since the desulphated polysaccharide is devoid of activity. In consequence, *Heparin* shows a sharp fall in potency to about 50% of its theoretical anticoagulant activity if added to intravenous infusion fluids containing *Dextrose* and *Laevulose* (Fructose), due to their pronounced acidity (p. 651).

The anticoagulant action of *Heparin* is also reversed by salt formation with the strongly basic protein, protamine, which forms an insoluble complex. Mono-, di- and oligo-saccharide sulphates inhibit pepsin and show anti-ulcerogenic properties similar to those of sulphated polysaccharides (Namekata *et al.*, 1967).

*Dextran Sulphate*, formed by sulphating a low molecular weight dextran hydrolysate (MW *ca* 7500), is also used as a blood anticoagulant.

## CHEMICAL MODIFICATION OF POLYSACCHARIDES

### Cellulose Esters

The free hydroxyl groups of cellulose are capable of chemical modification leading to the formation of ethers, esters and oxidation products with modified properties. Cellulose is readily nitrated with nitric-sulphuric acid mixtures to give cellulose nitrates of varying nitrogen content. The products, such as *Pyroxylin*, are highly inflammable and soluble in organic solvents, such as acetone, ethanol and ether. A solution in ethanol–ether containing resin (colophony) and castor oil dries on exposure to air to give a flexible, adhesive and transparent film which is useful as a skin protective.

*Cellacephate* (Cellacefate) obtained by partial acetylation of cellulose with acetic anhydride, and further reaction with phthalic anhydride, is a hygroscopic powder, which is water-insoluble and barely soluble in ethanol and chloroform. It is, however, reasonably soluble in acetone and in ethyl acetate–isopropanol, and solutions are used as spray-coatings for the enteric coating of tablets and capsules. *Cellacephate* films are stable in acid media, but soften and swell in aqueous alkali; hence their intrinsic suitability for use as enteric coatings. Pure *Cellacephate* films, however, are somewhat brittle, and coating solutions generally incorporate diethyl phthalate or glyceryl triacetate as plasticiser to render the coat more pliable and less liable to chip. Fats or waxes are also usually added to aid water-resistance in acid media.

**Cellulose Ethers**

Methylcelluloses, containing about 29% of methoxy groups, are prepared by the action of methyl chloride on cellulose in the presence of sodium hydroxide. They are available in a number of grades with different degrees of polymerisation which are designated by numerals indicative of the viscosity of their aqueous solutions. *Methylcellulose 450* (viscosity of 2% solution, 400–500 centistokes at 20°C) is a colourless, hygroscopic powder, which slowly swells and dissolves in cold water to produce a neutral viscous colloidal solution. It is insoluble in hot water, but wetted by it, and clear aqueous solutions are best prepared by pre-wetting with boiling water prior to the addition of cold water or ice. Methylcellulose solutions are stable over a wide range of pH (2 to 12), but sensitive to the addition of electrolytes, which tend to salt out the polysaccharide and so increase viscosity.

*Ethylcellulose*, a cellulose ether containing about 48% of ethoxy groups, is being used increasingly in applications of the National Cash Register micro-encapsulation process (US Patent 3 155 590) for micro-encapsulation of drugs, such as aspirin, to control release rates and hence blood levels (British Patent 1 016 839; Bell, Berdick and Holliday, 1966). The process has also been adapted for the coating of micronised benzylpenicillin with a mixture of ethylcellulose and an appropriate wax, such as spermaceti, to protect the antibiotic from acid decomposition in the stomach when administered orally (Granatek, De Murio, Nunning, Athanas, Dana, Granatek and Daoust, 1969).

Hypromelloses are closely related derivatives, prepared by alkylating cellulose under alkaline conditions with a mixture of methyl chloride and propylene oxide. They are similarly graded according to degree of polymerisation and viscosity of their solutions. Their properties are closely similar to those of the corresponding methylcelluloses, though gel temperatures are usually significantly higher, and solutions are less prone to exhibit opalescence. They are, therefore, preferable to the methylcelluloses for increasing the viscosity of solutions for use in ophthalmology.

*Carboxymethylcellulose* is similarly prepared by alkylation of cellulose with chloracetic acid in the presence of sodium hydroxide. Its sodium salt, *Sodium Carboxymethylcellulose*, is also available in various grades with different degrees of substitution and polymerisation and differing viscosity in solution. *Sodium Carboxymethylcellulose* is a colourless water-soluble powder, which forms almost neutral solutions. The dry powder is sufficiently heat-stable to permit sterilisation at 160°C for 1 hr, although some depolymerisation and reduction in intrinsic viscosity occurs. Viscosity is also markedly reduced in acid solution below pH 5 due to formation and precipitation of the free acid. For similar reasons, it is incompatible with heavy metal ions which deposit insoluble salts.

Medium viscosity grades of *Sodium Carboxymethylcellulose* are used at low concentration (1%) as suspending agents. Higher concentrations of medium viscosity grades and high viscosity grades are useful in the preparation of gels and pastes. Both *Methylcellulose* and *Sodium Carboxymethylcellulose* are also

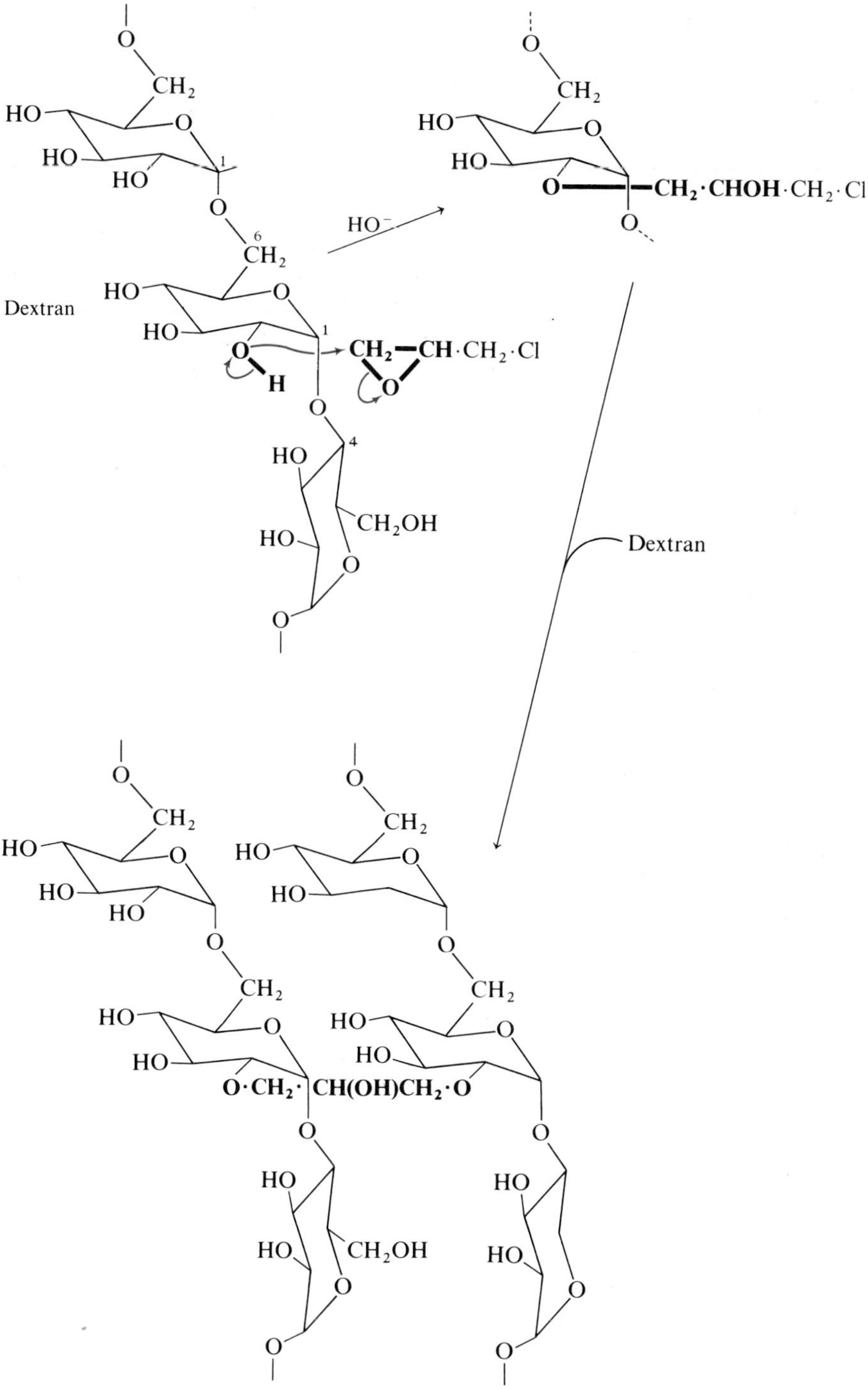

Dextran

HO⁻

$-CH_2\cdot CHOH\cdot CH_2\cdot Cl$

$CH_2-CH\cdot CH_2\cdot Cl$

Dextran

$O\cdot CH_2\cdot CH(OH)CH_2\cdot O$

used for their medicinal value as bulk laxatives. For this purpose, they require to be administered with copious volumes of fluid to ensure passage into the intestine.

*Diethylaminoethylcellulose* (DEAE-cellulose) and carboxymethylcellulose are also valuable ion-exchange materials.

**Oxidised Cellulose**

Oxidation of the primary alcoholic hydroxyl groups of cellulose with oxides of nitrogen to the corresponding carboxylic acids gives the corresponding anhydroglucuronic acid polymer, known as *Oxidised Cellulose*. Oxidation is usually carried out directly on cotton or viscose rayon gauze dressings, as the individual fibres show considerable loss of tensile strength. The oxidised cellulose dressing is used directly as an absorbable haemostat which interacts directly with fresh blood to form a coagulum. It is capable of inactivating acid-sensitive medicaments such as *Benzylpenicillin* (Penicillin G).

**Polyglycerylenedextrans**

Dextrans of molecular weight about 40 000 can be cross-linked by treatment with epichlorhydrin to give a range of polyglycerylenedextran polymers (Porath and Flodin, 1959). (See reaction scheme on page 701.)

The polymers exhibit polar character as a result of their polyhydroxylic structure and swell in water, although they are insoluble. The gel network is permeable to small molecules, which tend therefore to be preferentially retained whilst larger molecules are excluded. This property has given rise to the term **molecular sieves** and their widespread use as *Sephadex* in chromatographic procedures for the separation and purification of proteins, enzymes and other high molecular weight materials. They are also useful in the study of drug binding to natural macromolecules (Barlow, Firemark and Roth, 1962; Cooper and Wood, 1968; Wood and Cooper, 1970). The polymers have also been used to promote wound healing as a result of the absorption of water.

*Secholex*, a particular grade of 2-diethylaminoethylpolyglycerylenedextran (DEAE–Sephadex), is used as a cholesterol-lowering agent. It binds bile acids present in the intestine, thus promoting their excretion rather than reabsorption. As a result, the liver draws on plasma cholesterol and cholesterol depôts for breakdown and excretion as bile acids.

# 22 Fused-ring Hydrocarbons

## AROMATIC COMPOUNDS

### Introduction

Typical fused-ring aromatic hydrocarbons are those in which two adjacent carbon atoms of a benzene ring form part of a second carbocyclic structure. The majority of such compounds consist entirely of six-membered rings or combinations of six- and five- or six- and seven-membered rings.

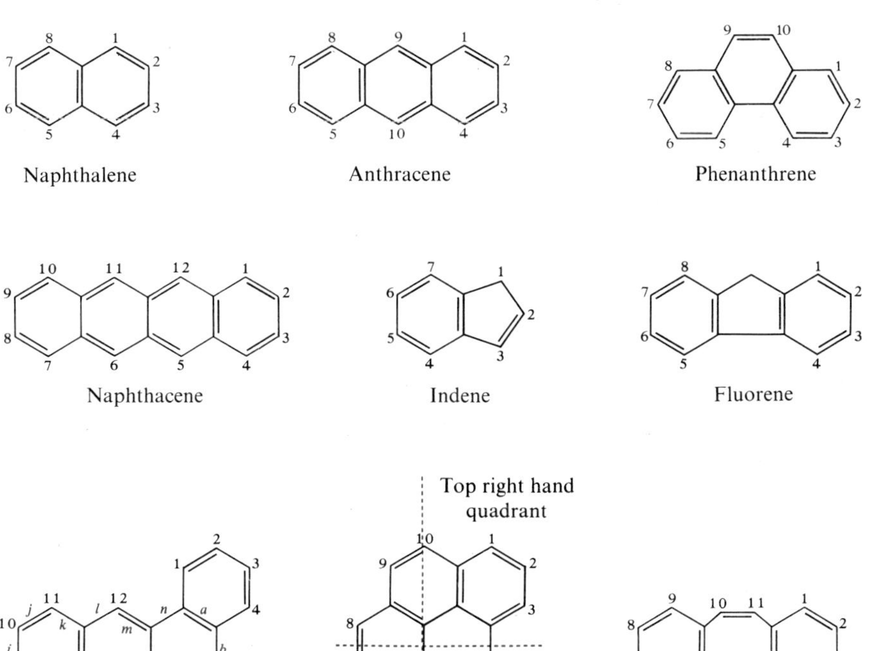


### Nomenclature

Most common polycyclic hydrocarbons have trivial names. IUPAC nomenclature rules are based on these trivial names, and with the exception of anthracene and phcnanthrene, structures are numbered according to an agreed convention. By this, they are first drawn with the maximum number of rings in a horizontal line and maximum number in the top right-hand quadrant.

Substituent positions are then numbered clockwise from the first carbon atom not engaged in ring fusion in the right-hand ring of the uppermost row.

Higher polycyclic hydrocarbons without trivial names are named as derivatives of the largest unit possessing a trivial name. The peripheral faces of this unit are lettered consecutively, *a*, *b*, *c* and so on, in a clockwise direction starting from the C-1–C-2 face. The name of the hydrocarbon is constructed with the name of the substituent, and its face-position given by the appropriate letter (in square brackets) prefixing that of the basic trivially-named unit, as in benzo[*a*]-anthracene and dibenzo[*a*,*d*]cycloheptatriene. For numbering, the entire structure must, if necessary, be re-orientated and re-numbered in accordance with the basic numbering rule stated above.

**Structure**

Fused aromatic six-membered ring structures are resonance-stabilised. Linearly-fused hydrocarbons containing n benzene rings are hybrids of $(n + 1)$ unexcited resonance forms (Kekulé structures) as in naphthalene and anthracene.

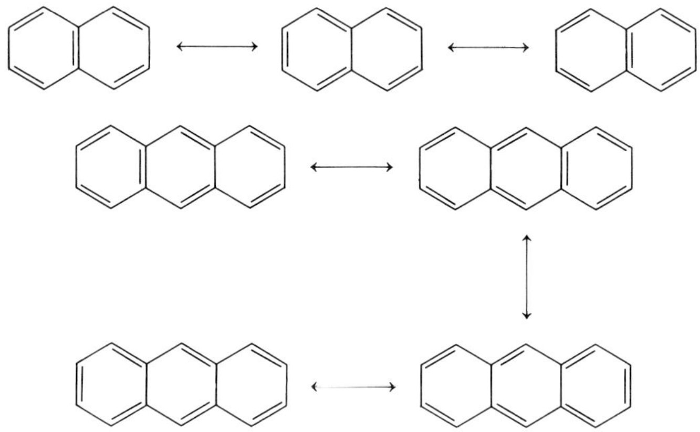

Angular compounds, such as phenanthrene, formed from $n$ benzene rings have $(n + 2)$ unexcited resonance forms contributing to the resonance hybrid.

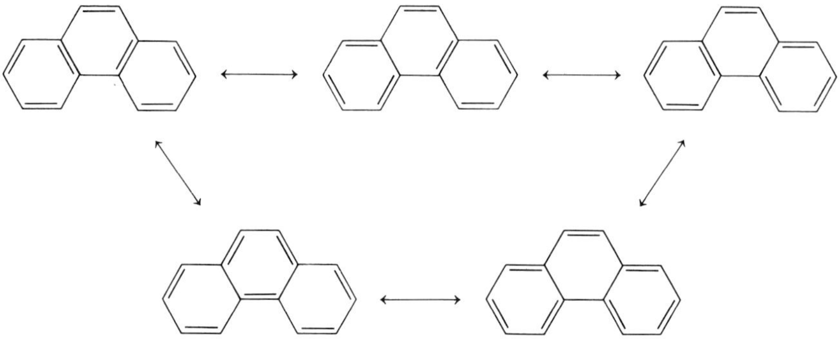

Polycyclic aromatic hydrocarbons consisting entirely of six-membered ring structures are planar, like benzene itself. In a few compounds, appropriately placed substituents can lead to steric repulsion, which is accommodated by distortion of the otherwise planar multi-ring structure into a non-planar configuration. This occurs in 1,12-dimethylbenzo[*c*]phenanthrene-5-acetic acid, which has been prepared in optically active forms (Newman and Wise, 1956).

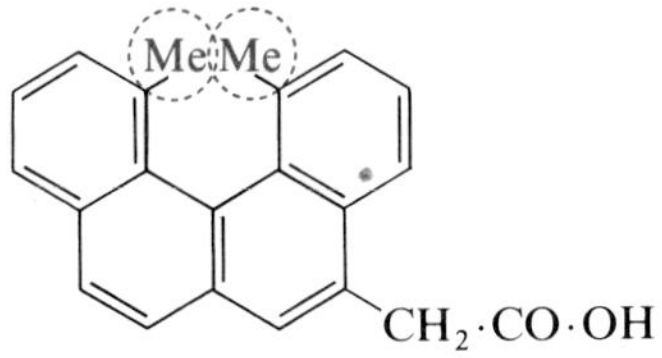

1,12-Dimethylbenzo[*c*]-phenanthrene-5-acetic acid

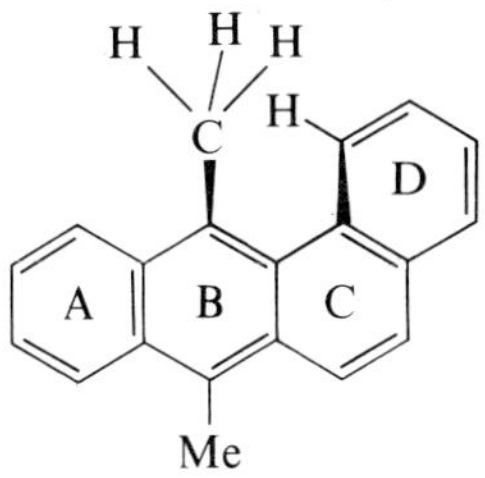

7,12-Dimethylbenz[*a*]anthracene

Another interesting exception is 7,12-dimethylbenz[*a*]anthracene, a potent carcinogen in which ring D has been shown to be inclined at an angle of about 20° downwards out of the general plane of the molecule to release strain (Sayre and Friedlander, 1960). Compounds with odd numbers of carbon atoms, such as indene, fluorene and dibenzocycloheptatriene, do not have fully conjugated systems. They, therefore, contain one or more methylene groups and in consequence are also non-planar.

**Double-bond Character**

The chemical reactivity of individual aromatic fused-ring compounds is linked to calculations of bond order, which derive from considerations of resonance (Pauling and Wheland, 1933). More precise calculations of double bond character by molecular orbital methods correlate well with experimentally-determined bond lengths for a number of fused-ring polycyclic hydrocarbons (Coulson and Longuet-Higgins, 1947; Berthier, Coulson, Greenwood and Pullman, 1948; Baldock, Berthier and Pullman, 1949), and thus establish the main centres of reactivity.

This evidence of double bond character explains the ease with which anthracene and phenanthrene form stable addition compounds with osmium tetroxide, ozone and aliphatic diazo compounds, compared with benzene and naphthalene. Addition of osmium tetroxide is slow, but, in contrast to the other reagents, it adds almost exclusively to the most reactive bonds (Cook and Schoental, 1948), i.e. to those with the highest bond order. Thus, it adds to the 1,2-bond of anthracene, to the 9,10-bond of phenanthrene, to the 3,4-bond of 1,2-benzathracene, and to the 6,7-bond of 3,4-benzpyrene (see also p. 715). These centres appear to be especially important in relation to carcinogenicity.

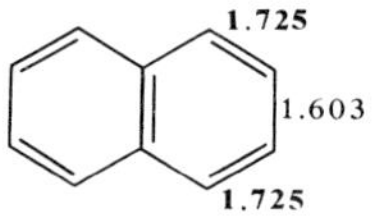

1.725 1.603 1.725

Naphthalene

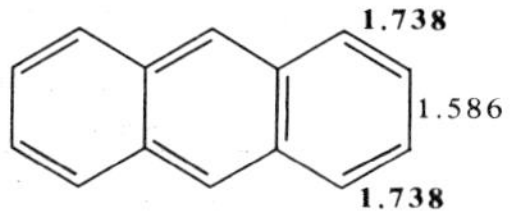

1.738 1.586 1.738

Anthracene

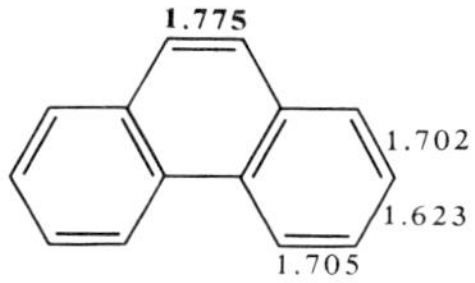

1.775 1.702 1.623 1.705

Phenanthrene

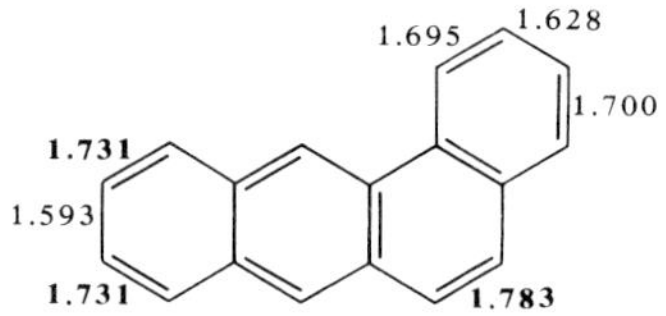

1.695 1.628 1.700 1.731 1.593 1.731 1.783

Benz[*a*]anthracene

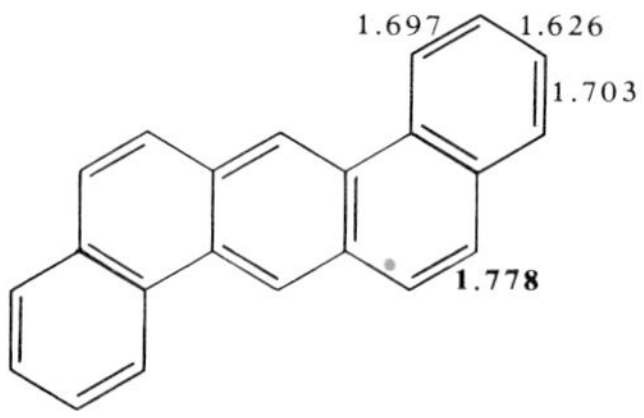

1.697 1.626 1.703 1.778

Dibenz[*a,h*]anthracene

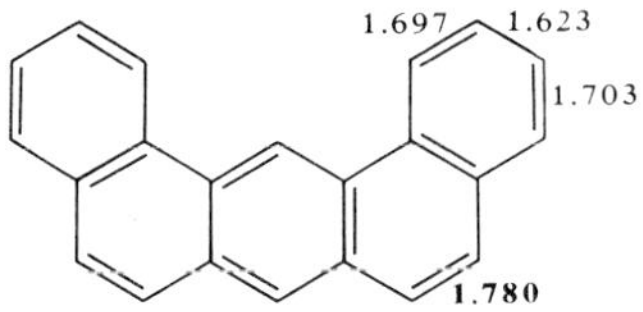

1.697 1.623 1.703 1.780

Dibenz[*a, j*]anthracene

## Carcinogenic Hydrocarbons

It has long been known that the presence of meso (9,10)-phenanthrenic regions of high $\pi$-electron density (K-regions) correlated with the carcinogenic activity of polycyclic aromatic hydrocarbons (Schmidt, 1938, 1939, 1941), and that the rate of reaction with osmium tetroxide increases in parallel with carcinogenic activity (Table 67, Badger, 1949).

This concept of the central rôle of K-regions in the carcinogenic activity of polycyclic aromatic hydrocarbons has been developed by Pullman and Pullman (1946), and extended by Chalvet, Daudel and Moser (1958). The latter assume that interaction with cellular macromolecules may involve either addition at the K-region, or alternatively a Diels–Alder diene addition at the meso(9,10)-anthracenic region (L-region). The bio-receptor, similarly, has been the object of considerable discussion and speculation (Arcos and Arcos, 1962; Jones and Matthews, 1974), and it is by no means certain whether it is protein or nucleic acid.

The interaction with the bio-receptor is considered by some to be a form of charge-transfer complexing, a view which stemmed from the known ability of polycyclic aromatic hydrocarbons to form such complexes with iodine and picric acid. Their planar structure and ability to form stable monomolecular films with similarly shaped sterols (Davis, Krahl and Clowes, 1940) are not inconsistent with this concept, which is also supported by the observed correlation

Table 67. Reaction Rates with Osmium Tetroxide and Carcinogenicity of Polycyclic Aromatic Hydrocarbons (Badger, 1949)

| Hydrocarbon | $10^3$ k | Carcinogenic activity |
|---|---|---|
| 5,6,9,10-Tetramethylbenz[*a*]anthracene | 2.9 | +++ |
| 9,10-Dimethylbenz[*a*]anthracene | 2.7 | ++++ |
| 20-Methylcholanthrene | 1.1 | +++ |
| Acenaphthanthracene | 1.1 | + |
| Cholanthrene | 1.0 | ++++ |
| Benzo[*a*]pyrene | 0.94 | ++++ |
| 10-Methylbenz[*a*]anthracene | 0.91 | +++ |
| 5,6-Dimethylbenz[*a*]anthracene | 0.64 | ++ |
| Dibenz[*a,h*]anthracene | 0.64 | ++ |

The nomenclature used in this table is in accord with the rules described on p. 703. The original text uses the notation described by Arcos and Arcos (1962).

between the carcinogenicity index (Pullman and Pullman, 1955) and the energy $E_1$ of the first excited $\pi$-singlet state of these compounds (Birks, 1961). Of 63 compounds examined by Birks, 22 out of 23 carcinogens, but only 12 out of 46 non-carcinogens, were found to have $E_1$ between 3.04 and 3.20 eV. A plot of wavelengths corresponding to the mean energies of the excited state against carcinogenic activity of these compounds showed that the optimum wavelength for maximum carcinogenicity corresponds to the peak wavelength of the tryptophan fluorescence emission spectrum. Calculation of the overlap integrals for the first singlet absorption bands ($J_1$) as a measure of dipole–dipole interaction with tryptophan shows a high degree of correlation between $J_1$ and carcinogenic index (Table 68).

Birks is careful to point out that other cellular components, such as the guanine and adenine residues of DNA and RNA, which have $\pi$-excitation states around 4 eV and fluoresce weakly in the region of 350 nm, may also be involved. In contrast, Mason (1958) avers that the dipole–dipole interaction between hydrocarbon and protein is best considered with the protein in an acceptor rather than a donor rôle, with the absorbed energy being dissipated in keto-enol tautomerism of the peptide chain, leading to concomitant changes in tertiary structure. Serious objections have, however, been made to Mason's proposed model and in particular to the idea that matching of energy levels is essential for charge-transfer complexing, which in fact depends more on the ionisation potential of the donor and electron affinity of the acceptor (Pullman and Pullman, 1962). Similar objections also appear to attach to alternative proposals that the carcinogenic activity of polycyclic aromatic hydrocarbons involves analogous interactions with purine and pyrimidine bases in nucleic acids (Pullman and Pullman, 1959; Hoffman and Ladik, 1961).

Two other factors, molecular size and shape, are also important determinants of the carcinogenic activity of polycyclic hydrocarbons. Arcos and Arcos (1955)

Table 68. Tryptophan Overlap Resonance Integrals and Carcinogenic Index of Polycyclic Aromatic Hydrocarbons (Birks, 1961)

| Hydrocarbon | Overlap integral | Carcinogenic activity |
|---|---|---|
| Benzene | 0 | 0 |
| Naphthalene | 0 | 0 |
| Phenanthrene | 0 | 0 |
| Pentacene | 0 | 0 |
| Pentaphene | 100 | 0 |
| Benzo[*e*]pyrene | 200 | 0 |
| Picene | 260 | 0 |
| Benzo[*c*]phenanthrene | 400 | + |
| Chrysene | 500 | 0 |
| Dibenz[*a,c*]anthracene | 1000 | 0 |
| Benz[*a*]anthracene | 1350 | ? |
| Anthracene | 1350 | ? |
| Dibenz[*a,j*]anthracene | 1900 | + |
| Dibenz[*a,h*]anthracene | 2400 | ++ |
| Dibenzo[*a,h*]pyrene | 2500 | ++++ |
| Dibenzo[*b,c*]pyrene | 3300 | +++ |
| Benzo[*a*]pyrene | 6000 | ++++ |

The nomenclature used in this table is in accord with the rules described on p. 703. The original text uses the notation described by Arcos and Arcos (1962).

have shown that the area of the minimum rectangular envelope (incumbrance area) lies within about 95 and 135 Å for the majority of carcinogenic hydrocarbons. A substantial element of co-planarity is also important. Molecular distortion as in 9,10-dimethylbenz[*a*]anthracene, or arising from methylene bridges, as in cholanthrene, is unimportant, so long as the substituents do not protrude excessively from the general plane of the molecule (Fieser, 1945). This requirement has been expressed in terms of a maximum molecular thickness for carcinogenicity of about 4 Å (Arcos and Arcos, 1956), which if exceeded by the presence of substituents, as in 15,20- and 16,20-dimethylcholanthrene, giving a molecular thickness of the order of 5.11 Å, leads to a fall in activity. Reduction of benzanthracenes to dihydrobenzanthracenes similarly leads to a substantial loss of co-planarity (Ferrier and Iball, 1954; Beckett and Mulley, 1955) and of carcinogenic activity.

Attention has also been drawn to the importance of K-region epoxides, which are formed metabolically in cell microsomes (Sims, 1967) and are both mutagenic and carcinogenic (Kuroki, Huberman, Marquardt, Selkirk, Heidelberger, Grover and Sims, 1972). Their importance, however, has been partly discounted by Dipple, Lawley and Brookes, (1968), who consider that M-region epoxides degrade to excretable hydroxylic compounds through a carbonium ion intermediate, and that K-region epoxides, although reactive as carcinogenic intermediates, exert this function through similar carbonium ion intermediates.

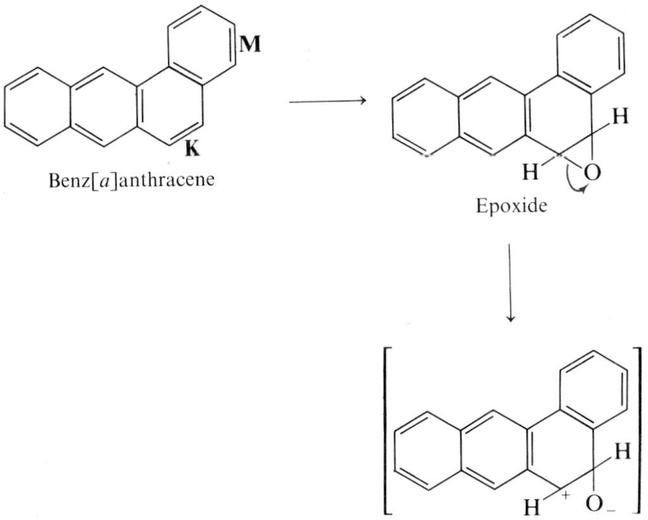

M

K

Benz[$a$]anthracene

H

H O

Epoxide

H

H + O−

Carbonium ion intermediate

**Substitution Reactions**

Polycyclic aromatic hydrocarbons undergo electrophilic substitution more readily than benzene. The point of substitution is largely governed by the stability of the intermediate carbonium ion. Thus, although naphthalene can be expected to substitute in either the $\alpha$-(1,4,5 or 8)-position or the $\beta$-(2,3,6 or 7)-position, halogenation, nitration and sulphonation (at low temperature; 40°C) give predominantly the $\alpha$-substituted compounds, since the carbonium ion intermediate formed by substitution in the $\alpha$-position has six resonance forms, four of which retain one benzenoid ring, whereas $\beta$-substitution gives a carbonium ion intermediate with fewer contributing forms, only two of which are aromatic.

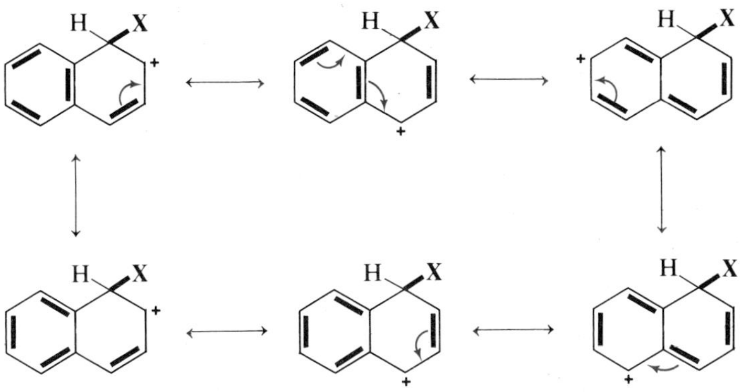

H X

H X

H X

H X

H X

H X

$\alpha$-Substitution carbonium ion intermediate

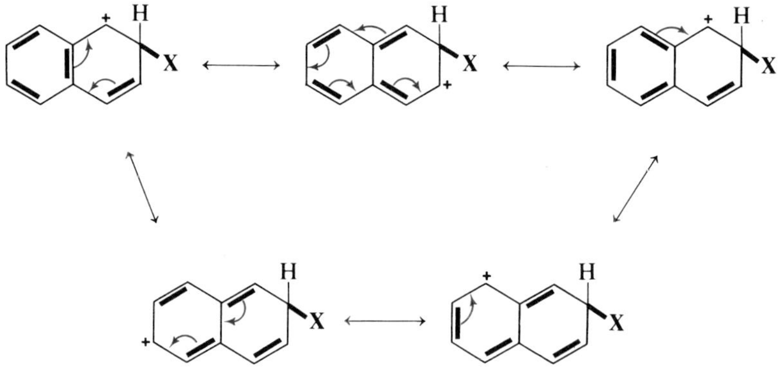

$\beta$-Substitution carbonium ion intermediate

Formation of the $\beta$-sulphonic acid at high temperature (*ca* 160°C) is due to an equilibrium between $\alpha$- and $\beta$-naphthalenesulphonic acids and unchanged naphthalene which is rapidly attained at higher temperatures (Euwes, 1909; Lantz, 1935). The acids may be separated by crystallisation of their calcium or lead salts, those of the $\alpha$-acid being more soluble than those of the $\beta$-acid. Naphthalene-2-sulphonic acid (napsylic acid) is used in the preparation of the water-insoluble *Dextropropoxyphene Napsylate*, which is less bitter and less irritant to mucous surfaces than the corresponding hydrochloride. The naphthalenesulphonic acids are used in the production of $\alpha$- and $\beta$-naphthols by fusion with caustic alkali.

Friedel–Crafts alkylation and acylation with anhydrous aluminium chloride can also be anomalous and either $\alpha$- or $\beta$-substituted naphthalenes can be obtained (Baddeley, 1949). The reaction is sterically-controlled, conditions which give rise to bulky complexes favouring $\beta$-substitution.

The position of entry of a second substituent depends not only on the substituent already present, but also on its position ($\alpha$- or $\beta$-) and the nature of the entering group. In general, substituents which are ortho/para-directing promote substitution in the same ring; groups which deactivate (meta-directing) promote substitution in the other ring. Ortho/para-directing substituents in the 1-position promote mainly 4-substitution (with some 2-substitution). More strongly electronegative o/p-directing groups in the 2-position (halogen or $NH_2$) promote hetero-annular substitution (i.e. in the other ring) in the 5- or 8-position, whilst less strongly electronegative o/p-directing groups ($CH_3$, HO, $CH_3O$, NHR and $NHCOCH_3$) lead to homo-annular substitution (i.e. in the same ring) in the 1-position. Meta-directing groups ($NO_2$ and $SO_2OH$) in either the 1- or 2-position favour substitution in positions 5 and 8.

Embonic acid, which is formed by the condensation of 2-hydroxynaphthalene-3-carboxylic acid with formaldehyde, illustrates the influence of two competing groups on further substitution. The acid is particularly suitable for the preparation of salts, which have low water solubility, from organic bases. Such salts may be used to delay release of the base. Low solubility also reduces the unpleasant taste associated with more soluble salts of the same base, as for example with the antitussive, *Benproperine Embonate*, which is much more suitable than the corresponding phosphate for use in oral preparations (i.e. linctus).

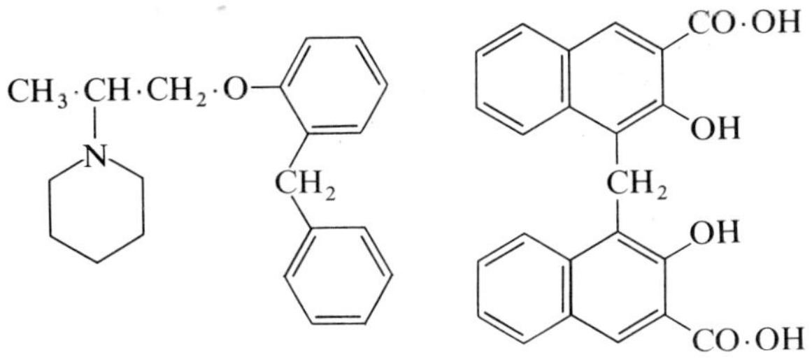

*Benproperine Embonate*

Electrophilic substitution of phenanthrene and anthracene is more complex. The 9,10-double bond of phenanthrene is reactive to addition of most reagents which add to the ethylenic bond. Both 9- and 10-double bonds are equally reactive in anthracene though the 9,10-dihydro-9,10-substituted products frequently undergo a ready elimination to give the 9-substituted anthracene.

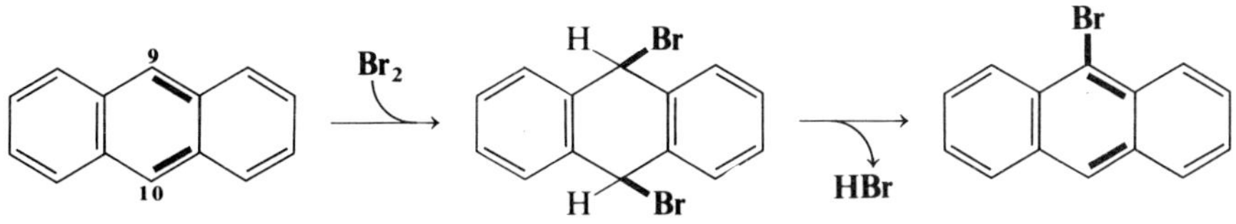

The 9,10-bond in phenanthrene is only slightly less reactive, and both 9,10-addition and 9-substitution occur concurrently.

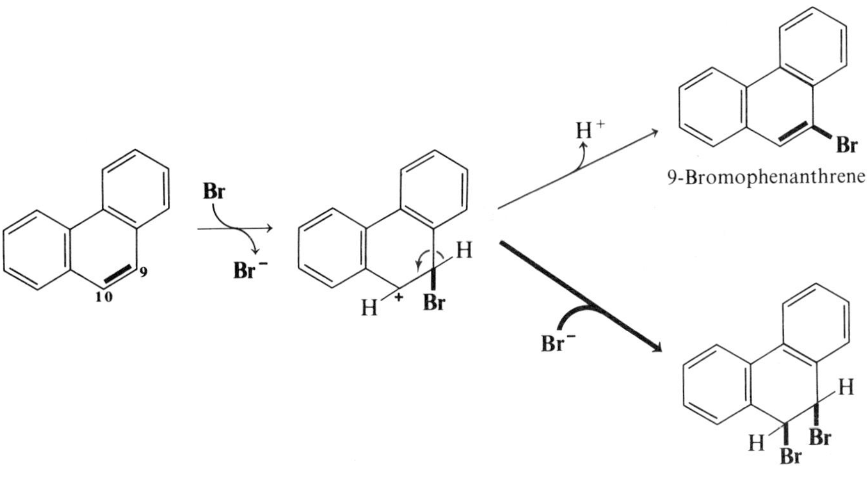

9-Bromophenanthrene

9,10-Dibromophenanthrene

**Reduction**

Naphthalene can be reduced with sodium and ethanol to form 1,4-dihydronaphthalene (1,4-dialin), which is unstable and rearranges when heated with ethanolic sodium ethoxide to 1,2-dialin, in which the double bond is conjugated with the aromatic ring. Hydrogenation of naphthalene over a nickel catalyst or reduction with sodium in boiling amyl alcohol gives tetralin (1,2,3,4-tetrahydronaphthalene). The latter is used extensively as a solvent (b.p. 207°C). Complete hydrogenation of tetralin gives decalin (decahydronaphthalene).

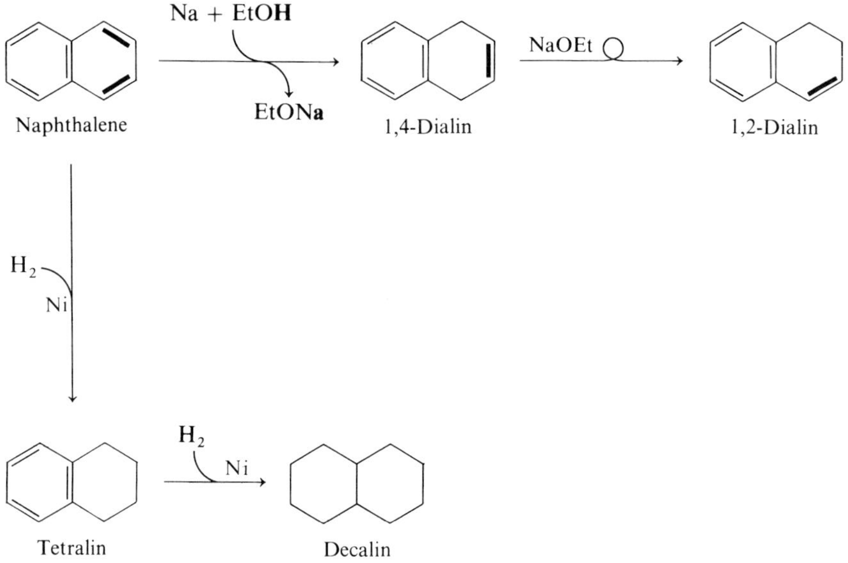

Na + EtOH

EtONa

NaOEt

Naphthalene 1,4-Dialin 1,2-Dialin

$H_2$ Ni

$H_2$ Ni

Tetralin Decalin

Halogens under radical conditions (in presence of light) substitute in the alicyclic ring to give mono- and di-substituted compounds; in the absence of light and in the presence of a halogen carrier (iron), substitution occurs in the aromatic ring of tetralin in the 5- and 6-positions. Tetralin also readily undergoes autoxidation in air to yield a hydroperoxide, which decomposes with sodium hydroxide to yield α-tetralone.

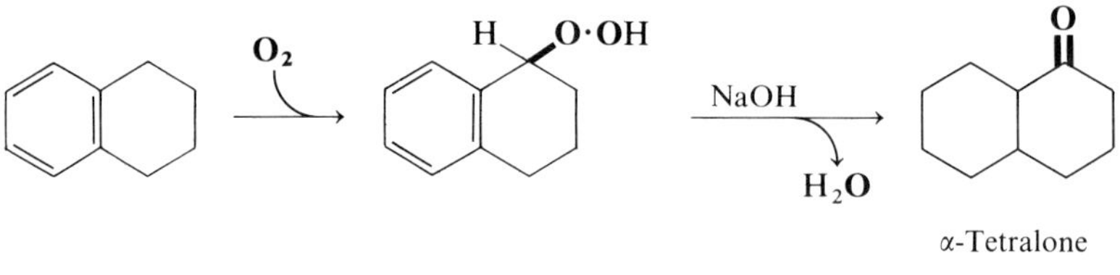

$O_2$

H O·OH

NaOH

$H_2O$

O

α-Tetralone

Reduction of naphthalene to tetralin destroys the planarity of the molecule giving rise to a half-chair structure (Raphael and Stenlake, 1953) of the alicyclic ring, as in cyclohexene. In this, C-1 and C-4 are perforce co-planar with the remaining aromatic ring. The tetrahydro-ring, however, still retains some element of flexibility so that two conformations are possible in which C-2 and C-3 are either above and below the plane of the aromatic ring, or the reverse thereof.

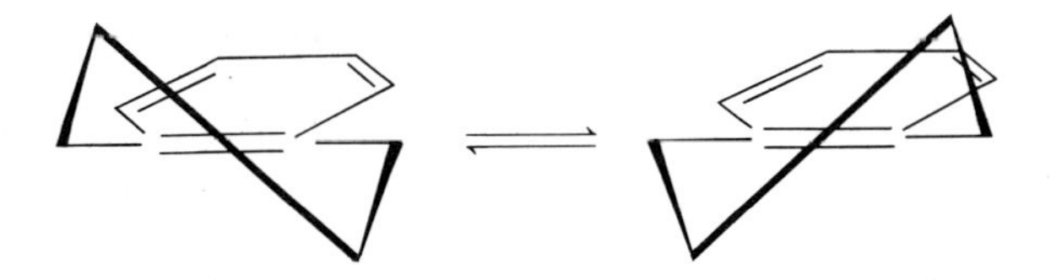

Decalin exists in the *cis*- and *trans*-forms. Examination of models shows that *trans*-fusion of two saturated six-membered rings (Chapter 3) is possible only in one conformation in which both carbon—hydrogen bonds at the ring junction are axial. In contrast, two alternative conformations are possible for *cis*-decalin, depending upon which of the two carbon—hydrogen bonds at the ring junction is equatorial and which is axial.

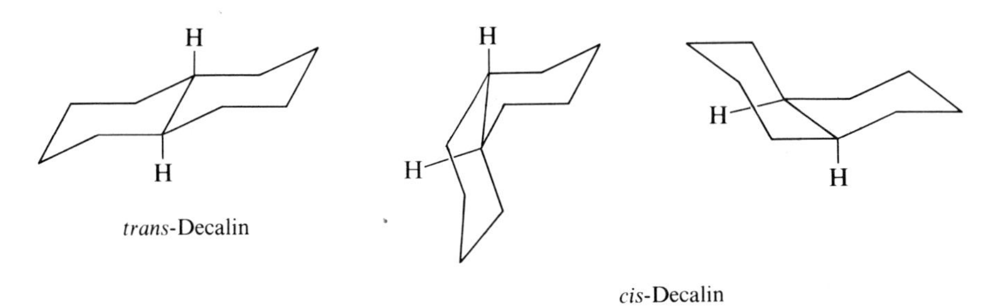


*trans*-Decalin

*cis*-Decalin

Anthracene adds hydrogen much more readily than naphthalene, to form the stable 9,10-dihydroanthracene, which is no longer planar, but butterfly-shaped with a dihedral angle of 145° about the 9,10-axis (Ferrier and Iball, 1954). The molecule is flexible and this has implications for the conformation of 9,10-disubstituted-9,10-dihydroanthracenes which may adopt either the folded or reverse-folded conformation according to the configuration of the 9- and 10-substituents (Beckett and Mulley, 1955).

### Oxidation

Drastic oxidation of polycyclic aromatic hydrocarbons leads to ring cleavage. Thus, naphthalene is oxidised with chromic-sulphuric acid, or catalytically with vanadium pentoxide in air at high temperatures to form phthalic acid. Under somewhat milder conditions (catalytically in air in presence of chromium), oxidation of polycyclic hydrocarbons stops short of ring cleavage with the formation of either a 1,4-quinone (naphthalene and anthracene) or a 1,2-quinone (phenanthrene).

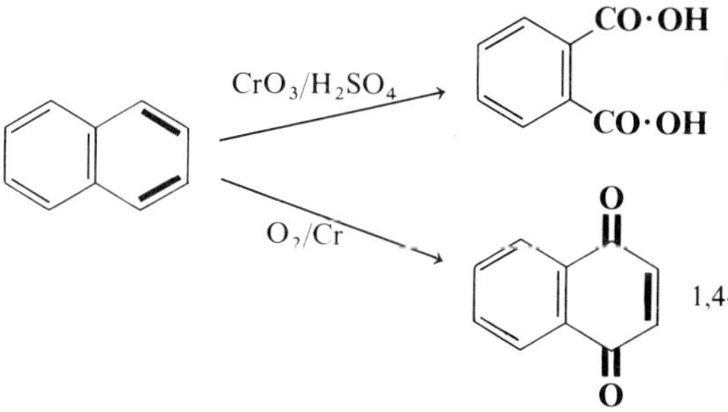

$CrO_3/H_2SO_4$ → CO·OH, CO·OH — Phthalic acid

$O_2/Cr$ → 1,4-Naphthoquinone

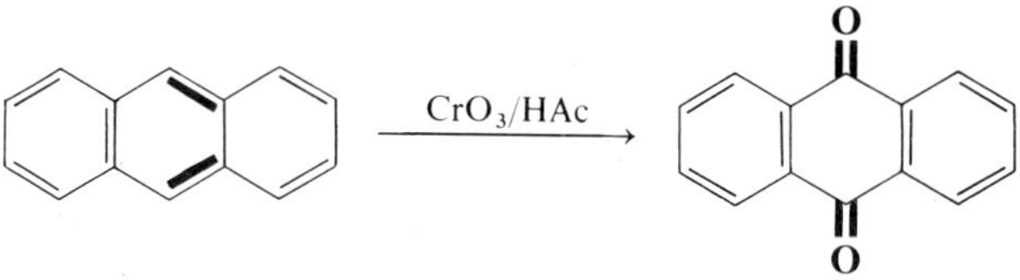

$CrO_3/HAc$ →

Anthraquinone

$CrO_3/HAc$ →

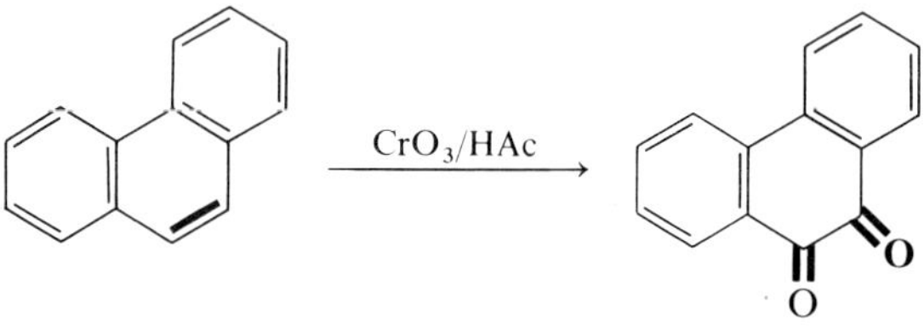

Phenanthraquinone

Anthracene is particularly susceptible to oxidation, and undergoes direct oxidation in the presence of light to give anthracene photoxide, a typical peroxide formed by addition of molecular oxygen in the meso(9,10)-position, which releases iodine from potassium iodide.

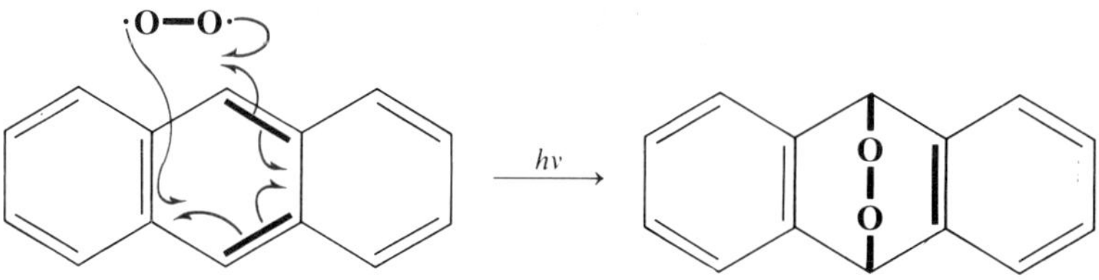

·O—O· $h\nu$ →

Anthraquinone photoxide

Metabolic oxidation of polycyclic aromatic hydrocarbons in mammalian liver follows a pathway similar to that of benzene, proceeding via an epoxide (Boyland and Williams, 1965) to form the corresponding 1,2-dihydrodiols, 1,2-dihydro-1-hydroxy-2-mercapturic acids and monohydric phenols. Epoxide formation occurs at those bonds with the greater double bond character, that is at the 1,2-bond in naphthalene, the 4,5-bond in pyrene (Boyland and Sims, 1964a) and at the 3,4-, 5,6-, 8,9- and 10,11-bonds of benzanthracene (Boyland and Sims, 1964b).

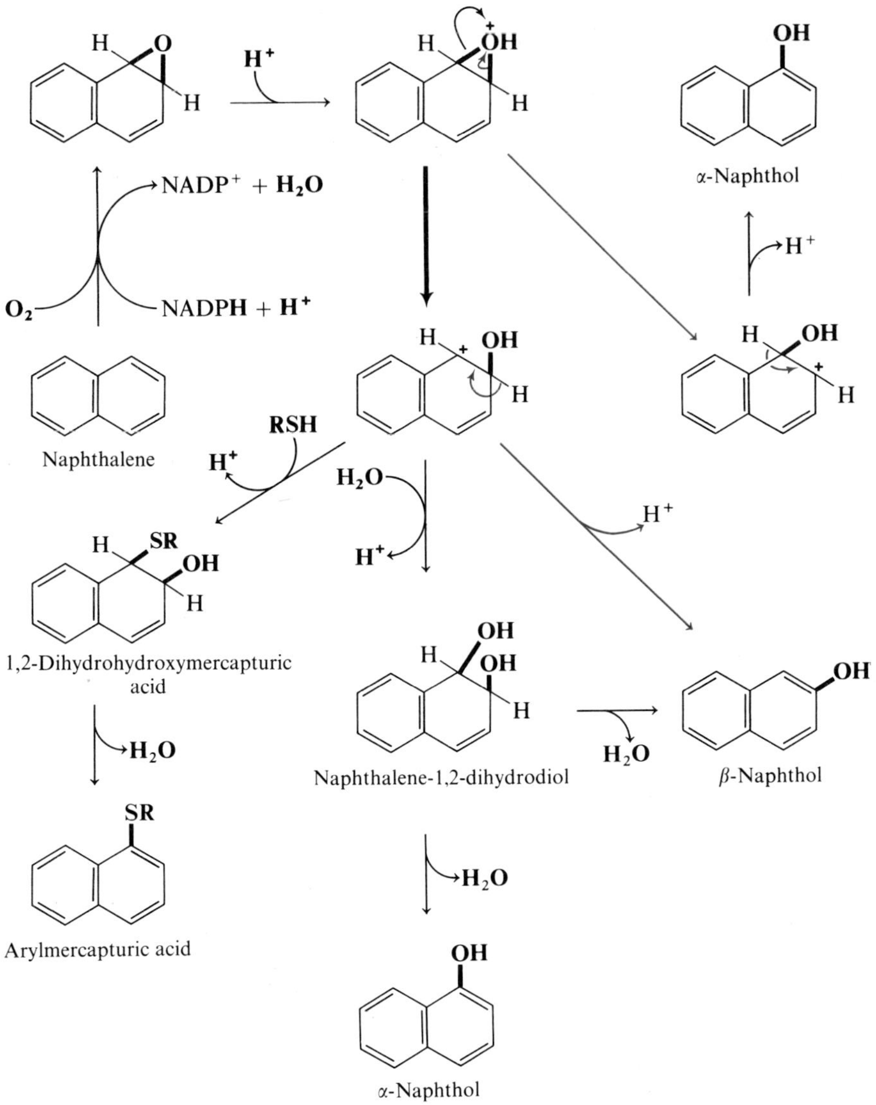

In most cases, the major metabolite is the 1,2-dihydrohydroxymercapturic acid. This is formed by interaction of glutathione with the epoxide and subsequent degradative metabolism of the glutathione conjugate (Chapter 14 and **2**, 5). Loss of water to give the arylmercapturic acid is a non-enzymic process (acid-catalysed dehydration), but often occurs in the isolation of such metabolites from urine.

## VITAMINS K

### Nomenclature

The term vitamin K was first applied by Dam (1935) to a fat-soluble dietary factor essential for blood coagulation (**K**oagulations vitamin). A number of naturally occurring vitamins K, all 1,4-naphthoquinones, have been identified. That from green plants, vegetables, and algae, 2-methyl-3-phytyl-1,4-naphthoquinone, is designated *Vitamin* $K_1$, and known as *Phytomenadione* (Phytonadione) or alternatively phylloquinone (IUPAC-IUB Commission on Biochemical Nomenclature). The vitamins $K_2$ with polyunsaturated isoprenoid side-chains, known collectively as menaquinones, are found in bacteria ($n = 1$ to 13), fish ($n = 6$) and animals ($n = 4$). Synthetic vitamin K, 2-methyl-1,4-naphthoquinone, is designated *Vitamin* $K_3$, and known as *Menaphthone* (Menadione).

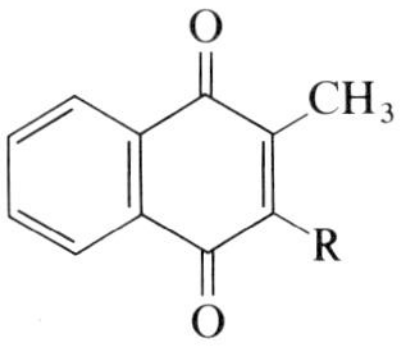

$R = -CH_2 \cdot CH{=}C(CH_3) \cdot CH_2(CH_2 \cdot CH_2 \cdot CH(CH_3))_2 \cdot CH_2 \cdot CH_2 \cdot CH(CH_3)_2$

*Phytomenadione* (Vitamin $K_1$)

R = H *Menaphthone* (Menadione)

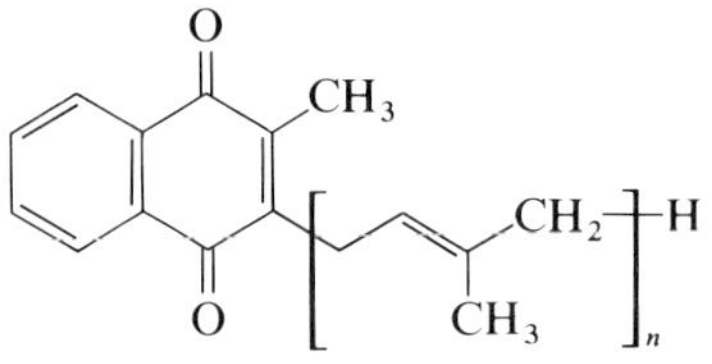

Menaquinones (Vitamins $K_2$)
Bacterial $n = 1$ to 13
Piscine $n = 6$
Animal $n = 4$

### Biosynthesis

There is considerable evidence that biosynthesis of vitamins K, at least in bacteria, is based on the shikimic acid pathway (Threfall, 1971).

Studies with radioactive shikimic acid have shown that all seven carbon atoms are incorporated into the naphthalenic skeleton of the naphthoquines, the ring carbons into ring A (Cox and Gibson, 1966) and the carboxylic acid carbon (C-7) into ring B at either $C_1$ or $C_4$ (Leistner, Schmitt and Zenk, 1967). There is now considerable evidence that the key intermediate is not shikimic acid itself, but its derivative, chorismate, (Dansette and Azerad, 1970) and that the remaining three-carbon unit of ring B is derived from L-glutamate (Campbell, 1969; Robins, Campbell and Bentley, 1970) via a succinylsemialdehyde thiamine pyrophosphate anion.

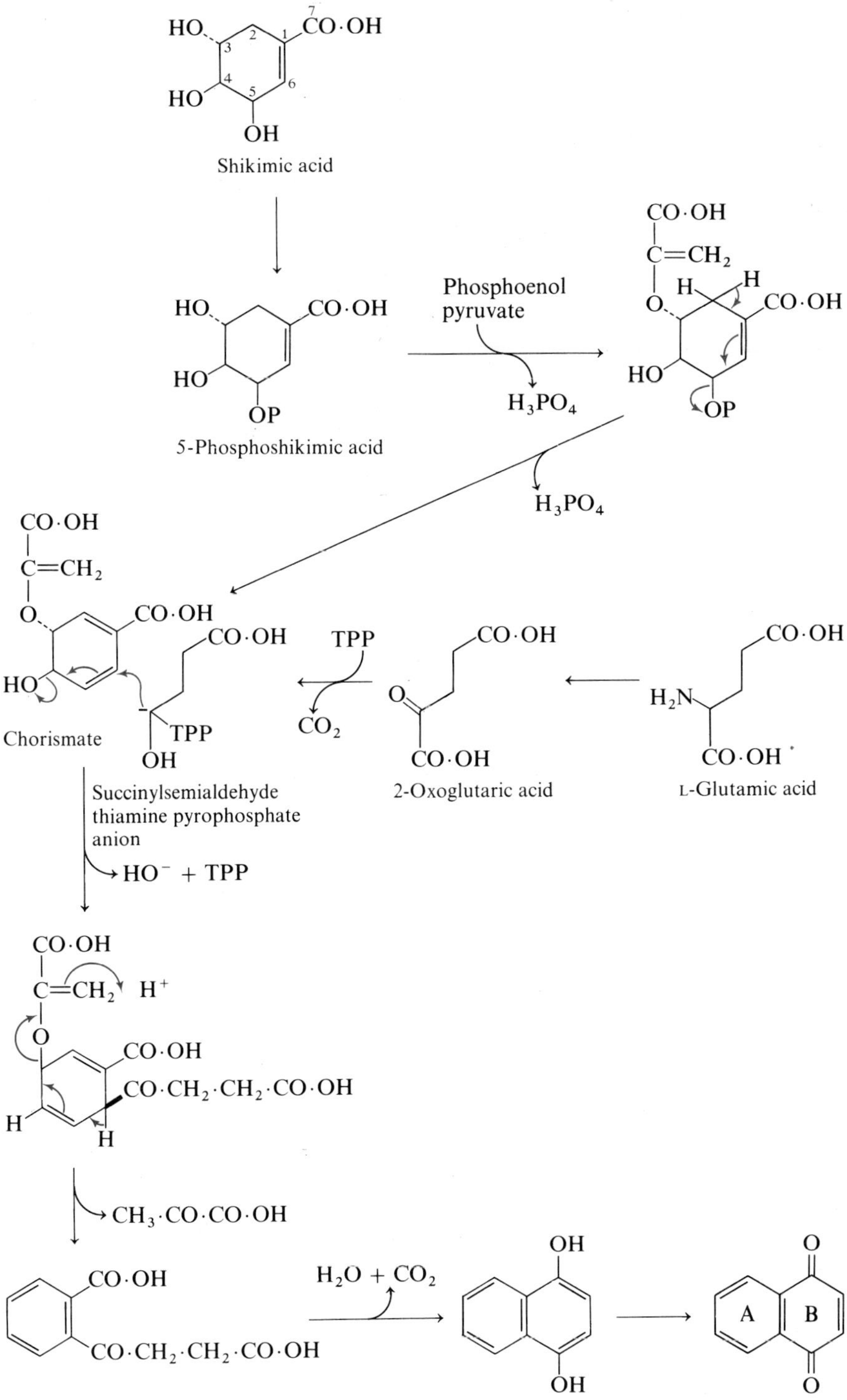

HO 3 2 1 7 CO·OH
4 6
HO 5
OH

Shikimic acid

HO CO·OH
HO
OP

5-Phosphoshikimic acid

Phosphoenol pyruvate

$H_3PO_4$

CO·OH
$C{=}CH_2$
O H H CO·OH
HO
OP

$H_3PO_4$

CO·OH
$C{=}CH_2$
O CO·OH
HO

Chorismate

CO·OH
TPP
OH

Succinylsemialdehyde thiamine pyrophosphate anion

TPP

$CO_2$

CO·OH
O
CO·OH

2-Oxoglutaric acid

CO·OH
$H_2N$
CO·OH

L-Glutamic acid

$HO^- + TPP$

CO·OH
$C{=}CH_2$ $H^+$
O
CO·OH
$CO\cdot CH_2\cdot CH_2\cdot CO\cdot OH$
H H

$CH_3\cdot CO\cdot CO\cdot OH$

CO·OH
$CO\cdot CH_2\cdot CH_2\cdot CO\cdot OH$

$H_2O + CO_2$

OH
OH

O
A B
O

**Function**

The natural vitamins K regulate the formation of thrombin from its precursor, prothrombin, in the liver. They also affect the clotting mechanism by controlling biosynthesis of preconvertin (factor VIII), Stuart factor (factor X) and Christmas factor (factor IX), which are all related to prothrombin. In healthy adults, supplies of the vitamins are derived from the diet and by the biosynthetic activities of intestinal bacteria (Udall, 1965). Dietary deficiency, faulty absorption, or liver disease can result in low blood prothrombin levels (hypoprothrombinaemia) or deficiencies of the other vitamin K-dependent factors, with prolongation of clotting time and consequent danger in the event of haemorrhage. The natural vitamins are water-insoluble oils, and absorption from the gastro-intestinal tract is dependent on complex formation with bile salts (Cilento, 1960), so that liver and gall bladder diseases resulting in a deficiency of bile can also deplete circulating blood levels of vitamins K. Similarly, excessive use of antibiotics which destroy intestinal bacteria, can lead to prolongation of normal clotting time, due to deficiency in vitamin K of bacterial origin. Haemorrhage of the newborn, when the intestinal tract is virtually sterile and supply of human milk (as opposed to cows' milk) a poor source of the vitamin, can be readily counteracted by administration of vitamin K directly to the newborn child.

Both natural and synthetic vitamins K are readily soluble in organic solvents and fixed oils. *Phytomenadione* (Phytonadione) is administered orally in capsules and tablets, but may also be given by intramuscular injection as an aqueous dispersion, which is heat-stable and can be heat-sterilised. The vitamins are acid-stable, but readily decomposed by alkali, and highly sensitive to light with loss of vitamin activity (MacCorquodale, Cheney, Binkley, Holcomb, McKee, Thayer and Doisy, 1939) probably as a result of dimerisation.

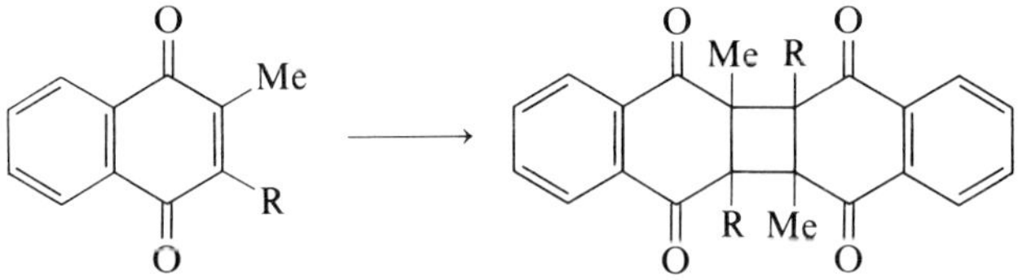

O Me R O → O Me R O R Me O O

**Mechanism of Action**

The mechanism whereby vitamins K regulate prothrombin synthesis is still not clear. The dependence of the body on external supply is, however, well established. Thus, the anticoagulant, *Dicoumarol* (Dicumarol), eliminates the effect of vitamin K almost instantaneously, although the vitamin K-dependent anticoagulant proteins disappear only at their individual turnover rates (Campbell and Link, 1941). This effect of *Dicoumarol*, however, is reversed rapidly by administration of *Phytomenodione* (Phytonadione), but not by *Menaphthone* (Smith, Fradkin and Lackey, 1946). It is now established that *Menaphthone* is a pro-vitamin K which is transformed metabolically after absorption from the gastro-intestinal tract to vitamin $K_2$ (Martius and Esser, 1959), most probably in the liver.

*Menaphthone* itself is now seldom used, and then is only administered as the much more stable *Acetomenaphthone* (Vitamin $K_4$) in the prophylaxis of neonatal haemorrhage. *Acetomenaphthone* is deacetylated by serum esterases *in vivo* and the 1,4-diol then oxidised to menaphthone prior to its conversion to vitamin $K_2$. This oxidation is reversible, as the main pathway for metabolism and excretion of vitamins K is via the corresponding 1,4-diols and their glucuronide and sulphate conjugates.

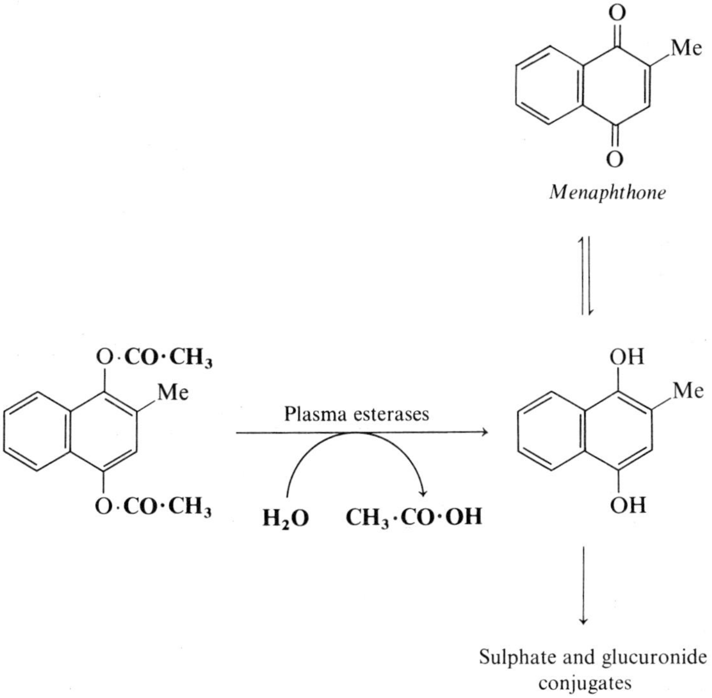

O
Me
O
*Menaphthone*

O·CO·CH₃
Me
O·CO·CH₃

Plasma esterases
$H_2O$ $CH_3 \cdot CO \cdot OH$

OH
Me
OH

Sulphate and glucuronide conjugates

Vitamin K in the liver has been reported to be concentrated predominantly in the mitochondria (Green, Søndergaard and Dam, 1956), but is clearly also present in microsomal tissue (Hill, Gaetani, Paolucci, Ramarao, Alden, Ranhotra, Shah, Shah and Johnson, 1968). It functions not as suggested by Olson (1964) in the stimulation of messenger RNA formation, but in the actual synthesis of prothrombin by mRNA. Thus, whilst vitamin K activity is inhibited by *Actinomycin D* (Dactinomycin), vitamin K deficiency does not inhibit protein synthesis generally, and the effect of blocking doses of *Actinomycin D* on prothrombin synthesis is reversed by subsequent treatment of *Phytomenadione*. The rôle of vitamin K at a late stage in protein synthesis is indicated by its effect in reversing blockade by cycloheximide which blocks protein synthesis at the transcription stage (Wettstein, Noll and Penman, 1964; Felicetti, Colombo and Baglioni, 1966). This conclusion is supported by the inability of vitamin K to reverse block of prothrombin biosynthesis by *Puromycin* which also acts at the translation stage of protein biosynthesis (**2**, 3).

Vitamin K has also been reported to effect uncoupling of oxidative phosphorylation, though conclusive support for this conclusion has not been forthcoming (Lederer and Vilkas, 1966). Nonetheless, Martius (1966) has alluded to the widespread distribution of vitamin K in all organs of the body (Wiss and Gloor, 1966) with high concentrations in the heart as strong evidence in favour of some general cellular function, such as cell respiration or oxidative phosphorylation. The ease and ready reversibility of the oxidation–reduction between the quinone and 1,4-diol have suggested its involvement as an electron transport carrier, or in the reversible oxidation–reduction of thiol–disulphide structures (Fieser, 1951).

The anticoagulant action of coumarin and indandione derivatives is only poorly understood (Green, 1966). Although a number of these compounds, such as *Ethyl Biscoumacetate*, are bi-functional (*bis*), this particular feature of the molecule is not essential for activity. Structural similarities to the quinonoid structure of the vitamins K is presumptive evidence for competitive inhibition at essential enzyme sites. Prydz (1965) has shown that *Warfarin Sodium*, like *Puromycin*, inhibits the biosynthesis of certain clotting factors. Factor VII in rat liver, which competes with vitamin K (Hill *et al.*, 1968) as a co-factor is thought to be essential for the creation of the folded quaternary structures of prothrombin and related clotting factors which are stabilised by multiple S—S bonds.

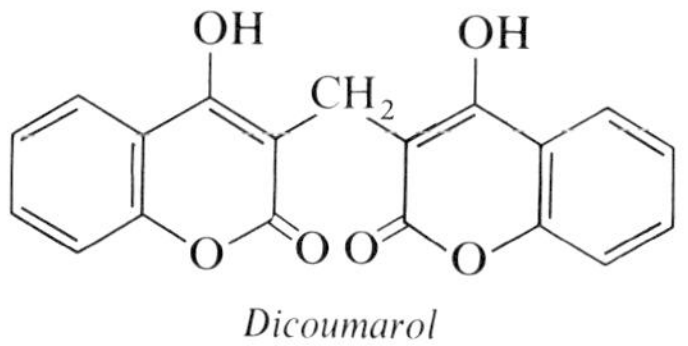

*Dicoumarol*

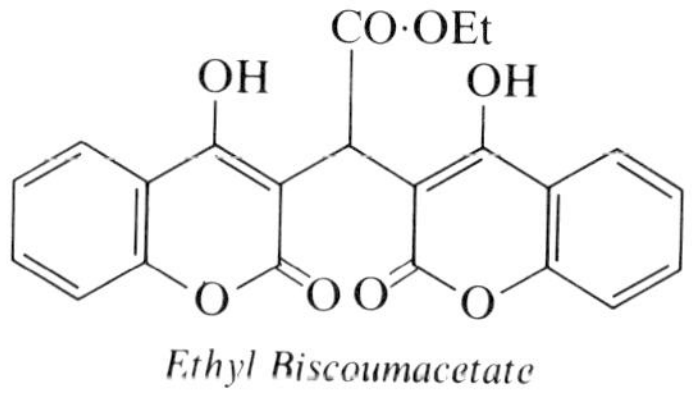

*Ethyl Biscoumacetate*

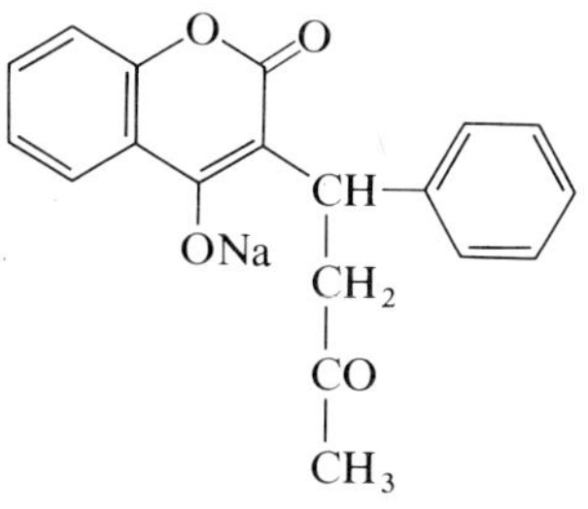

*Warfarin Sodium*

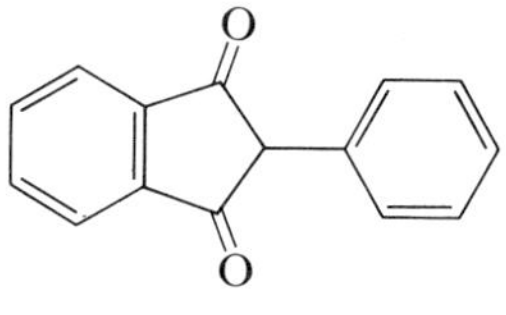

*Phenindione*

## ANTHRAQUINONES

Anthraquinones are resistant to further oxidation by mild oxidising agents, but like the parent 1,4-benzoquinones readily undergo stepwise reduction with metals, such as zinc, in appropriate solvents. The first product is that of a one-electron reduction to the semiquinone ion, which then readily picks up a proton in acidic solution to form the comparatively stable semiquinone (Gupta, 1952; Coffey, 1953).

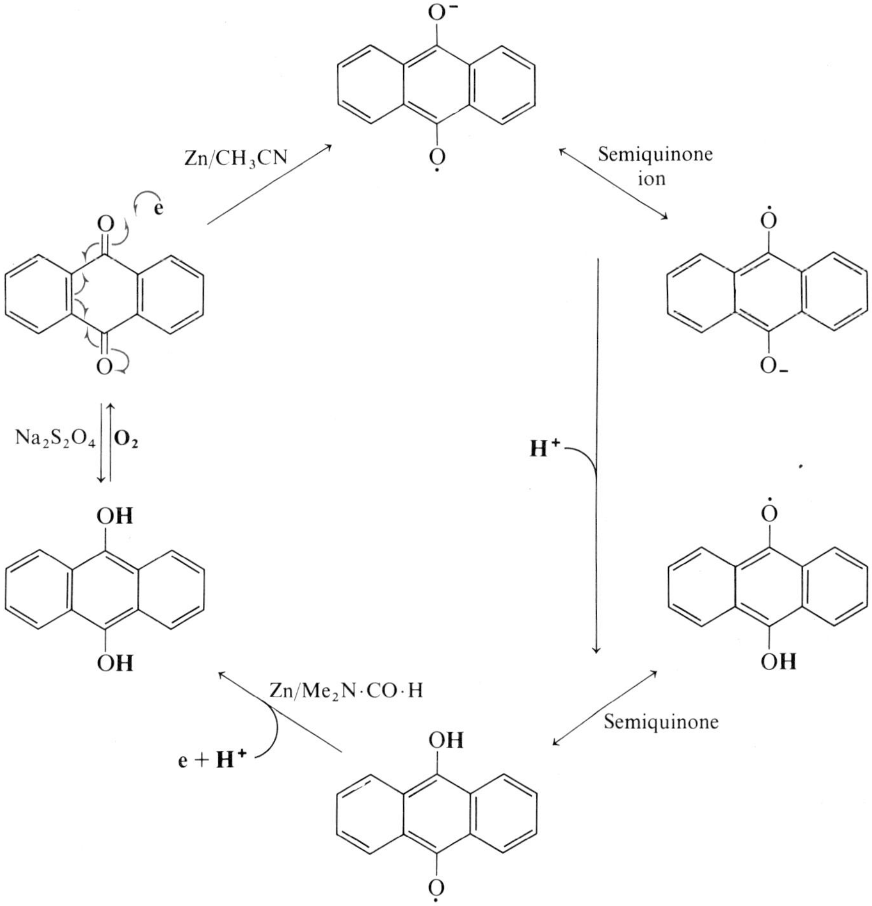

Zn/$CH_3CN$

Semiquinone ion

$Na_2S_2O_4$ $O_2$

$H^+$

Zn/$Me_2N \cdot CO \cdot H$

Semiquinone

e + $H^+$

Further reduction of the semiquinone gives 9,10-dihydroxyanthracene (anthrahydroquinone) which is formed in the normal two-electron reduction with zinc and sodium hydroxide or sodium dithionite ($Na_2S_2O_4$). The reduction product is pale-yellow, but becomes orange and red with successive ionisation of the 9- and 10-hydroxy groups in alkaline solution. The reduction, which is readily reversed with mild oxidising agents (shaking in air), forms the basis of the so-called vat dyeing process. Complex substituted anthraquinones are often highly coloured, but for lack of solubility do not impregnate the fabric to be dyed. Reduction to the dihydroxy-form gives an alkali-soluble form which, when impregnated into the fibres of the cloth, can be re-oxidised to form the stable anthraquinone dye *in situ*.

9,10-Dihydroxyanthracenes normally exist in tautomeric equilibrium with the corresponding oxanthrone, though in this case the equilibrium strongly favours the dihydroxyanthracene. The presence of substituents, however, markedly influences the equilibrium, and in some cases significantly favours the oxanthrone (Coffey, 1953).

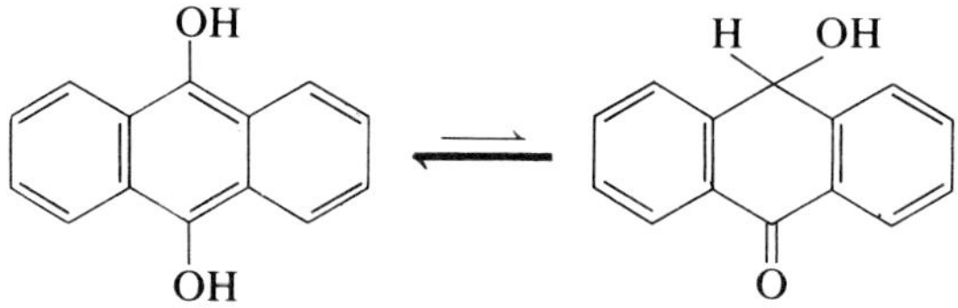

OH OH H OH O

9,10-Dihydroxyanthracene Oxanthrone

A similar equilibrium occurs between anthranol and anthrone, formed by reduction of anthraquinone in acidic solution (tin and acetic acid), though in this case it is the keto-form which is favoured.

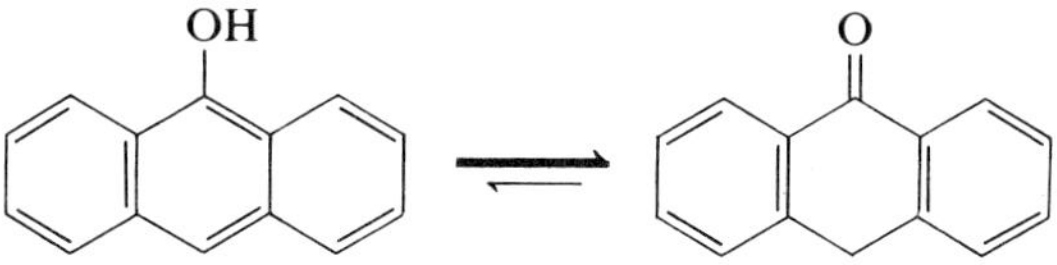

OH O

Anthranol Anthrone

## Hydroxyanthraquinones

A number of hydroxy-substituted anthraquinones and their glycosides are found as important constituents of laxative and purgative drugs, including senna, rhubarb, cascara and aloes. The anthraquinone aglycones, such as chrysophanol, present in rhubarb, and aloe-emodin, present in rhubarb, senna and cascara, are readily identified. The most active compounds as laxatives, however, appear to be the corresponding anthrol- and oxanthrone-glycosides in which the molecule is protected by the glycosidic link from oxidation to the comparatively inactive anthraquinone (Fairbairn, 1949).

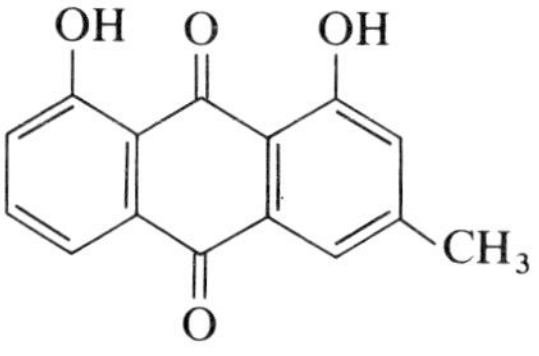

OH O OH $CH_3$ O OH O OH $CH_2OH$ O

Chrysophanol 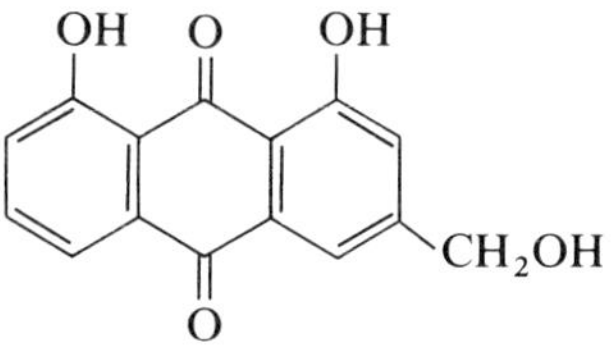 Aloe-emodin

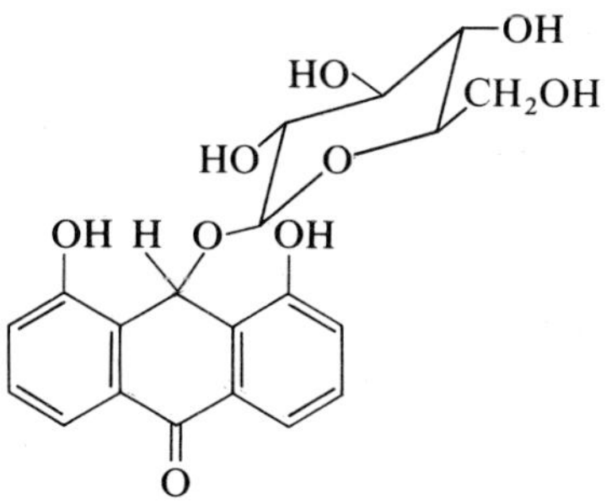

Hydroxyanthraquinone glycoside

The cytotoxic antibiotic, *Daunorubicin* (Daunomycin; Rubidomycin) can also be considered as a hydroxyanthraquinone derivative.

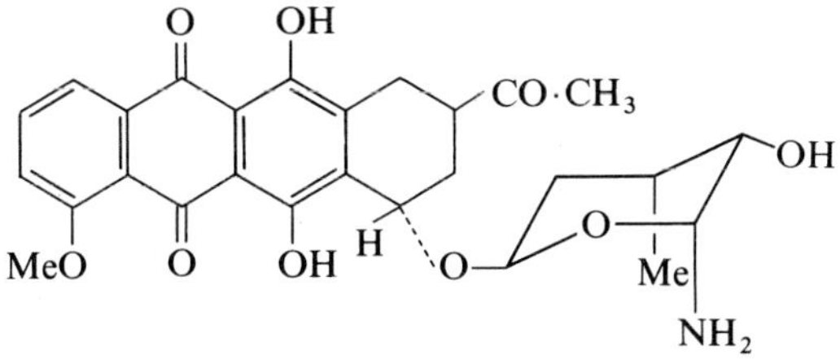

*Daunorubicin*

**Phenanthraquinones**

In contrast to anthraquinones, phenanthraquinones are readily cleaved by oxidation to the corresponding diphenic acid. They are also sensitive to alkali and readily undergo the benzilic acid rearrangement to form 9-hydroxyfluorene-9-carboxylic acid.

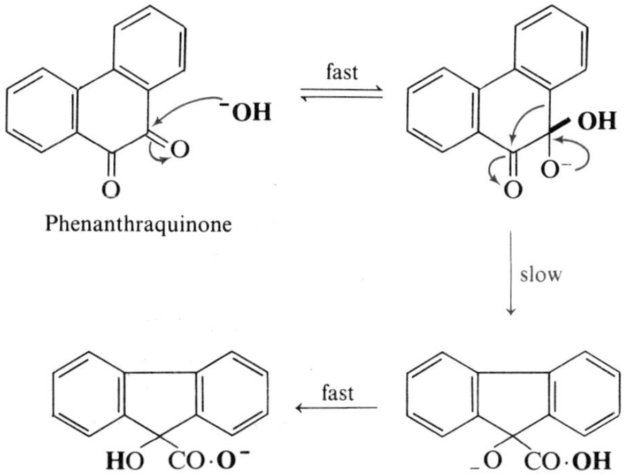

Phenanthraquinone

## TETRACYCLINES

The tetracyclines form a group of naturally occurring antibiotics with chemical structures derived from the tetracyclic hydrocarbon, naphthacene. *Chlortetracycline* (Aureomycin), the first member of this group to be isolated, was obtained from culture filtrates of *Streptomyces aureofaciens* (Duggar, 1948). Catalytic dehydrohalogenation of *Chlortetracycline* with hydrogen and a palladium-charcoal catalyst gives the parent antibiotic, *Tetracycline*, though this is now manufactured by fermentation of a mutant strain of *Streptomyces aureofaciens* in a chloride-free medium.

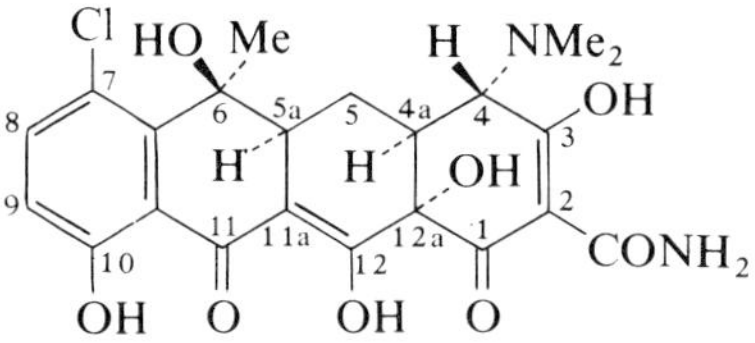

*Chlortetracycline* 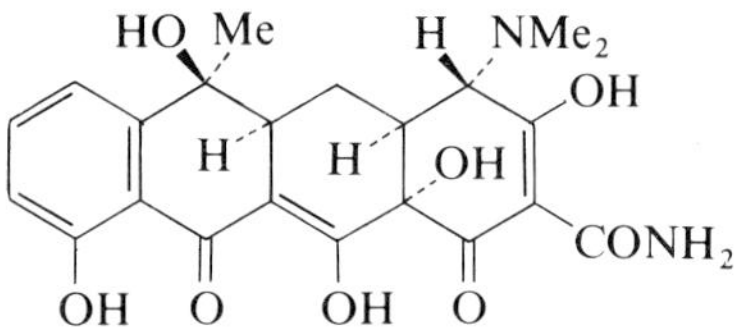 *Tetracycline*

*Oxytetracycline* (Terramycin), which is obtained by fermentation of *Streptomyces rimosus*, is generally isolated as a so-called dihydrate, the structure of which has been the matter of considerable debate. There is, however, now a considerable body of evidence based on infrared spectroscopy, X-ray diffraction and electron microscopy, which indicates that the compound is not a dihydrate, but a separate entity (in the crystalline state), the 1,1,3,5,6,10,11,11,12,12a-decahydroxy compound.

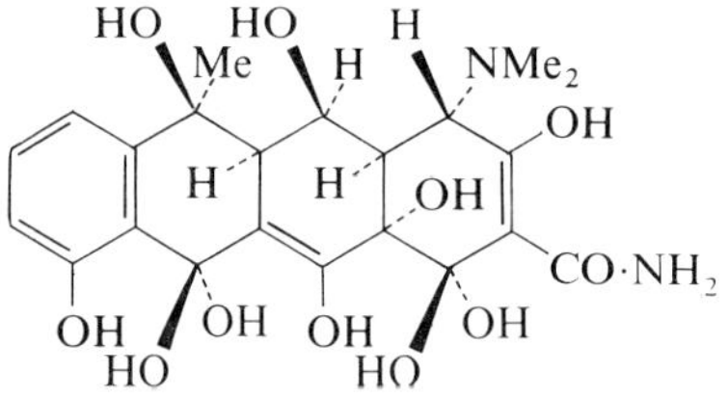

*Oxytetracycline Dihydrate* 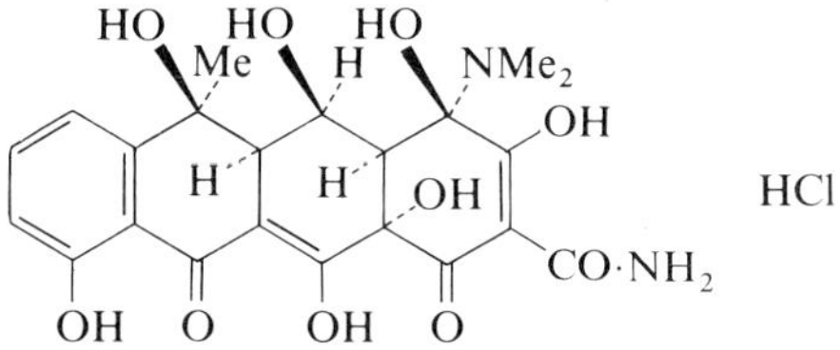 *Oxytetracycline Hydrochloride*

**General Properties**

The tetracycline antibiotics are yellow, bitter-tasting, amphoteric compounds. They are effective as antibacterial agents against a wide range of Gram-positive and Gram-negative organisms. Potentiometric titration reveals the presence of three dissociating groups, which in *Chlortetracycline* have p$K_a$ 3.30, 7.44 and 9.27. These constants are associated with the ionisation of the carboxamido-1,3-enolone system, the 4-dimethylamino group and the 10-phenolic-11,12-enolone system respectively (Stephens, Murai, Brunings and Woodward, 1956). The antibiotics also form coloured complexes with heavy metals, the stability of which decrease in the order $Fe^{3+} > Cu^{2+} > Fe^{2+} > Co^{2+} > Zn^{2+} > Mn^{2+}$

(Albert, 1953). An interesting correlation has been observed between the stability of these metal chelates and the ability of metal cations to reverse the action of tetracycline antibiotics. The greater the stability of the complex, the more readily is antibiotic action reversed by the corresponding metallic cation (Soncin, 1953; Price, Zolli, Atkinson and Luther, 1957). Complexation can, therefore, affect the antibiotic action of the tetracyclines *in vivo* (**2**, 2).

Tetracyclines are generally used as their water-soluble hydrochlorides since the free bases, including *Oxytetracycline Dihydrate*, are only slightly soluble in water. Indeed, solutions of *Oxytetracycline Hydrochloride*, which is soluble in 2 parts of water, deposit the crystalline dihydrate, unless the solution is acidified below pH 1.5. Both bases and hydrochlorides are stable in the dry state and *Chlortetracycline Hydrochloride*, for example, can be sterilised by dry heat at 120°C without loss of potency. *Chlortetracycline* also darkens in colour on exposure to light for prolonged periods, though without detectable loss of potency.

Stability in aqueous solution is a function of pH and at an optimum about pH 2.5. Solutions of *Chlortetracycline Hydrochloride* (1% aqueous solution has pH *ca* 2.6) retain their potency for long periods at 4°C, but at room temperature may lose up to 50% of the activity within 14 days. Under mildly alkaline conditions (pH 8.5), however, 40% of the activity is lost within 2 hr. *Oxytetracycline Hydrochloride* is similarly affected (Regna and Solomons, 1952; Table 69), though it is somewhat more stable in alkaline solution and rather less stable in acidic solution.

Decomposition in alkaline solution is due to the lability of the enolone at C-11, 11a and 12 when attacked by hydroxyl ions. The overall electron-withdrawing effect of the 7-chloro group and the consequent strengthening of hydrogen bonding between the 10-phenoxy and 11-carbonyl oxygen probably helps to promote attack at the carbonyl carbon in the more labile chlortetracycline. (See reaction scheme on page 726.)

Instability of tetracyclines in acidic solution is the result of epimerisation at C-4 to yield the corresponding quatrimycins, which have substantially reduced antibiotic activity, and/or acid-catalysed dehydration at C-6 to form the *epi*-anhydro- and anhydro-tetracyclines respectively (Doerschuk, Bitler and McCormick, 1955). Epimerisation at C-4 occurs in aqueous solution between pH 2 and 6. The reaction, which is pH-dependent and catalysed by phosphate, acetate

Table 69. Stability of Oxytetracycline in Aqueous Solution at varying pH (20°C)

| pH | Half-life (hr) | pH | Half-life (hr) |
|---|---|---|---|
| 1.0 | 114 | 7.0 | 26 |
| 2.5 | 134 | 8.5 | 33 |
| 4.6 | 45 | 10.0 | 14 |
| 5.5 | 45 | | |

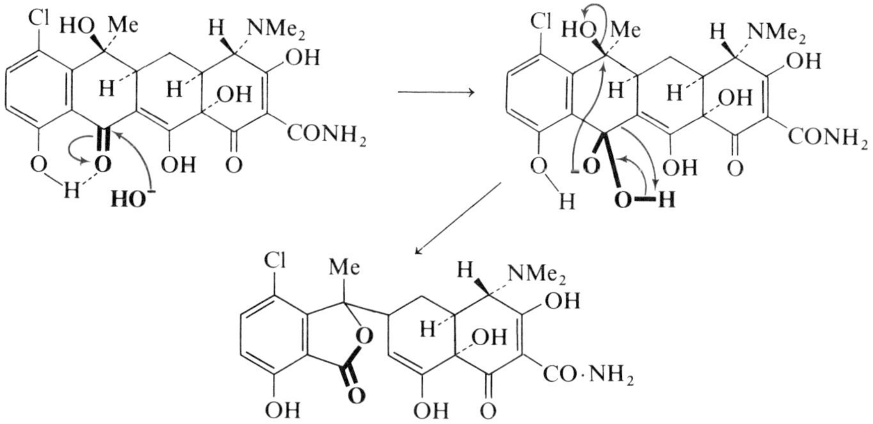

and citrate ions, reaches an equilibrium when some 20–25% of epimer is produced (McCormick *et al.*, 1957). More strongly acidic conditions cause dehydration at C-6. The somewhat greater stability of *Chlortetracycline* to acid compared with that of *Tetracycline* possibly derives from steric hindrance to protonation of the 6-hydroxyl due to the adjacent C-7 chloro substituent, though the 5-hydroxy group in *Oxytetracycline* is also somewhat similarly placed. (See reaction scheme on page 727.)

The enhanced acid stability of the 6-demethyltetracyclines, which are produced by mutant strains of *Streptomyces aureofaciens*, follows from the fact that the secondary hydroxylic group at C-6 will be less susceptible to acid-catalysed dehydration than the tertiary hydroxyl of the corresponding tetracyclines. *Demeclocycline* (Demecycline; chlor-6-demethyltetracycline) and 6-demethyltetracycline have the same level and spectrum of antibacterial activity and are also more stable in alkali as well as in acid than the corresponding tetracyclines (McCormick, Sjølander, Hirsch, Jensen and Doerschuk, 1957; Table 70).

Table 70. Relative Stability and Antibacterial Activity of Tetracyclines

| Antibiotic | Half-life (min at 100°C) in | | Activity against *Staph. aureus* |
|---|---|---|---|
| | $H_2SO_4$(*N*) | NaOH(*0.1N*) | |
| Tetracycline | 1.0 | 6.8 | 25 |
| 4-*epi*-Tetracycline | 0.9 | 7.2 | 1.6 |
| 6-Demethyltetracycline | 24.8 | 31.5 | 24 |
| 6-Demethyl-4-*epi*-tetracycline | 25.8 | 46 | 3 |
| 7-Chlortetracycline | 2.1 | $<0.3$ | 100 |
| 7-Chloro-4-*epi*-tetracycline | 5.8 | $<0.6$ | 4.2 |
| Demeclocycline | 44.5 | 40 | 75 |
| 7-Chloro-6-demethyl-4-*epi*-tetracycline | 32.2 | 35.6 | 7 |
| Oxytetracycline | 4.5 | 2.2 | 24 |
| 4-*epi*-Oxytetracycline | 2.6 | 3.3 | 1.1 |

(McCormick, Sjølander, Hirsch, Jensen and Doerschuk, 1957)

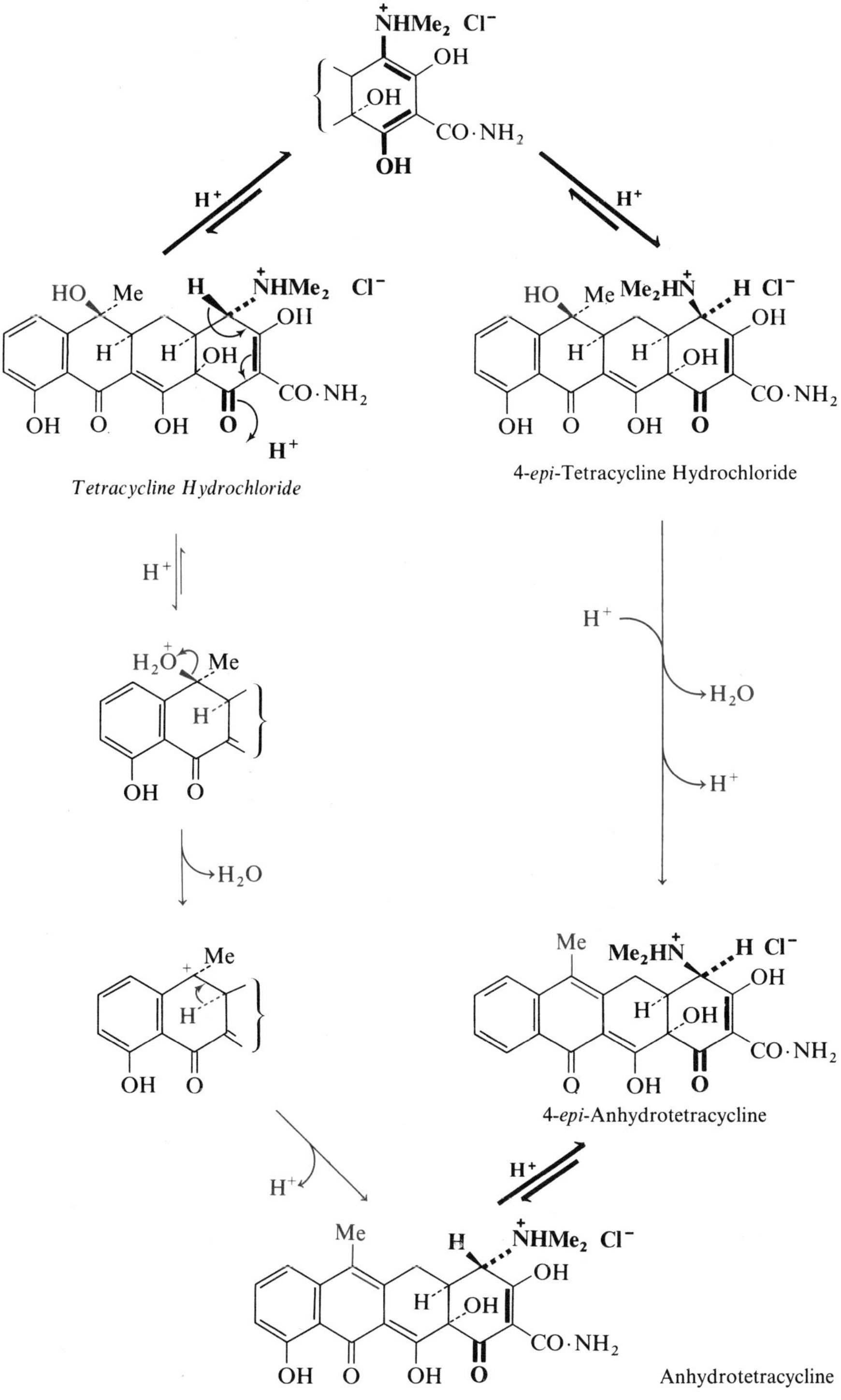

Tetracycline Hydrochloride

4-*epi*-Tetracycline Hydrochloride

4-*epi*-Anhydrotetracycline

Anhydrotetracycline

**Semi-synthetic Tetracyclines**

Hydrogenolysis of tetracyclines at a palladium charcoal catalyst under acidic conditions removes the 6-hydroxy-group to give 6-deoxytetracyclines (Stephens, Murai, Rennhard, Conover and Brunings, 1958). Both 6-deoxytetracycline and 6-deoxy-6-demethyltetracycline have antibacterial spectra *in vitro* very similar to those of their respective parent compounds. 6-Demethyl-6-deoxytetracycline has about twice the activity of *Tetracycline* against *Staphylococcus aureus* (McCormick, Jensen, Miller and Doerschuk, 1960; Table 71). The most important property of the 6-deoxytetracyclines, however, is the enhanced stability to acid and alkali, which loss of the C-6 hydroxyl confers on the molecule (Table 71).

Table 71. Relative Stability and Antibacterial Activity of 6-Deoxytetracyclines

| Antibiotic | Half-life (min at 100°C) in HCl(*3N*) | Half-life (min at 100°C) in NaOH(*0.1N*) | Activity against *Staph. aureus* |
|---|---|---|---|
| Tetracycline | <1 | 6.8 | 100 |
| 6-Deoxytetracycline | 1600 | 570 | 70 |
| 6-Demethyltetracycline | 1.4 | 32 | 100 |
| 6-Deoxy-6-demethyltetracycline | 1600 | 200 | 200 |
| Oxytetracycline | <1 | 2.2 | 80 |
| Doxycycline (6-Deoxy-5-hydroxytetracycline) | 2700 | 630 | 50 |

Their chemical stability has permitted the study of a series of electrophilic substitution reactions with 6-deoxytetracycline and 6-demethyl-6-deoxytetracycline under strongly acidic conditions (Beereboom, Ursprung, Rennhard and Stephens, 1960; Boothe, Hlavka, Petisi and Spencer, 1960) to give a series of 7- and 9-halogeno-, nitro-, and amino-tetracyclines. Few of these compounds proved more useful than natural tetracyclines; 7-iodo-6-deoxytetracycline, however, is localised in some tumour tissues and is of special value in their detection by X-ray (Hlavka and Buyske, 1960). Nitration of 6-demethyl-6-deoxytetracycline and reductive methylation of the resulting 7-nitro compound gives *Minocycline* (Church, Schaub and Weiss, 1971). It is reported to be active against other tetracycline-resistant strains of *Staphylococcus aureus* (Redin, 1967; Martell and Boothe, 1967), and has been used in the treatment of gonorrhoea.

A series of so-called carboxamido derivatives prepared from tetracyclines, formaldehyde, and various bases have been described (Gottstein, Minor and Cheney, 1959) reputedly with high activity, high water solubility and good absorption characteristics giving higher serum concentrations than with *Tetracycline Hydrochloride* (Kaplan, Albright and Buckwalter, 1960). *Lymecycline* is probably similarly constituted as a soluble tetracycline derivative incorporating tetracycline, formaldehyde and lysine.

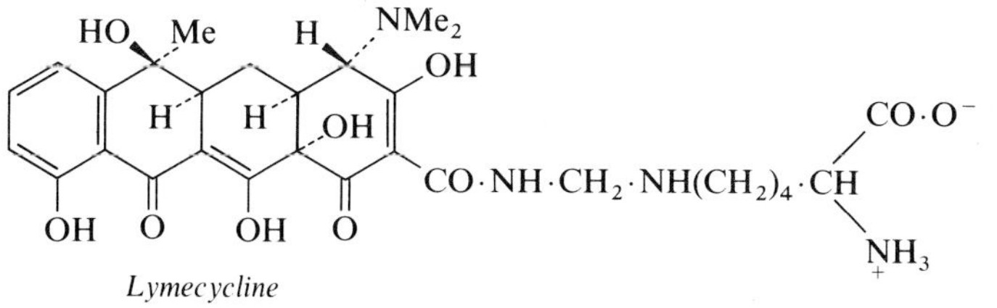

*Lymecycline*

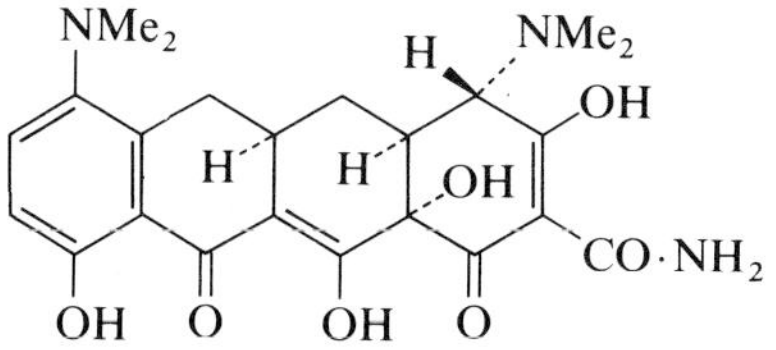

*Minocycline*

## STEROIDS

### Natural Distribution

Steroids form a chemically-related group of compounds based on the perhydrocyclopentanophenanthrene nucleus, which are widely distributed amongst living organisms of both plant and animal kingdoms, including fungi and protozoa, though significantly not in bacteria. In man and higher animals, steroids have important hormonal functions in sex differentiation, sexual activity, growth, and in the control of carbohydrate and mineral metabolism. In contrast, the rôle of plant steroids is by no means clear despite the fact that yeasts, moulds and oil-producing seeds and fruits are important sources of steroids. Thus, ecdysone, a steroid hormone, which controls moulting and growth in insects has also been found as a constituent of bracken and other plants. Some plant steroid derivatives, such as the cardiac glycosides, have important uses in medicine.

Cholesterol, which is particularly widely distributed in mammals, has been known since the eighteenth century as the principal constituent of gall stones which consists of 70–80% pure cholesterol. It is present in the blood, in brain and nervous tissue, is an important constituent of egg-yolk, and also found in wool fat which contains a high percentage both free, and esterified as oleate, palmitate and stearate. Commercially, cholesterol is obtained from the non-saponifiable fraction of wool fat.

Table 72. Mean Human Plasma Cholesterol Levels (Böhle, Böttcher, Piekarski and Biegler, 1956)

| Adult males | Free cholesterol (mg/100 ml) | Total cholesterol (mg/100 ml) |
|---|---|---|
| 18–30 yr | 62 | 177 |
| 31–45 yr | 76 | 199 |
| 46–60 yr | 78 | 205 |
| Adult females | | |
| 18–30 yr | 67 | 180 |
| 31–45 yr | 72 | 192 |
| 46–60 yr | 79 | 204 |

Cholesterol occurs in the blood in both the free and combined states. Free cholesterol is distributed between the plasma and erythrocytes, but cholesterol esters with long-chain fatty acids are present only in the plasma. Blood cholesterol levels, both free and combined, increase with age in both sexes (Table 72).

### Structure and Stereochemistry

All steroids are derived from the perhydrocyclopentanophenanthrene nucleus, a tetracyclic-fused ring system in which the rings are lettered and numbered as follows.

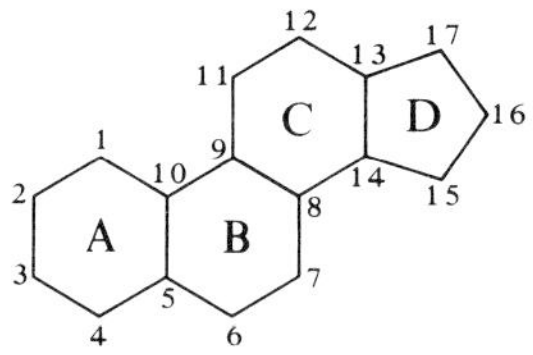

The majority of steroids have angular methyl groups at C-10 and C-13 and a substituent at C-17. Oestrogenic steroids, which form the principal exception, have no angular methyl group at C-10 since ring A is aromatic. These apart, the typical steroid has seven asymmetric centres at C-5, C-8, C-9, C-10, C-13, C-14 and C-17, giving rise to a theoretical $2^7$ possible stereoisomers. With the exception of C-5 and to a much more limited extent C-14, most natural steroids have the same absolute configuration at these centres. By convention, their stereochemistry is illustrated as set out on pages 732 and 733, substituents projecting above the plane of the rings ($\beta$-substituents) being drawn with bonds in full, heavy lines, whilst $\alpha$-substituents, which project below the plane of the rings, are drawn with bonds in light or broken lines. Unless specifically designated as hydrogen, angular substituents are assumed to be methyl; only the latter are numbered. These principles are illustrated in the structures of the more important fundamental steroid hydrocarbons also shown on pages 732–3.

The configuration at C-5 defines the stereochemistry at the A/B ring junction; where the C-5 hydrogen is $\alpha$, rings A and B are *trans*-fused; conversely, where the C-5 hydrogen is $\beta$, rings A and B are *cis*-fused. Similarly, the configuration of the C-14 hydrogen defines the fusion of rings C and D. In older texts, steroids with rings A/B *trans*-fused are sometimes described as **allo**-steroids, as in allopregnane, whilst steroids with rings A/B *cis*-fused are said to be **normal** steroids. This form of nomenclature is now obsolete.

The three-dimensional structure of the steroids and its implications for biological activity is only fully appreciated by considering steroid conformations. The *trans*-decalin-like fusion of rings B and C gives an essential rigidity to all steroids, so that flexibility is restricted to the possibility of the alternative chair, boat or intermediate conformations in ring A, and the minor conformational changes possible in the five-membered ring D constrained by fusion to ring C. Boat structures for ring A are the exception rather than the rule, so that ignoring conformational alternatives for ring D, steroids in general preferentially adopt conformations in which all the six-membered rings have chair conformations. Depending upon whether rings A/B are *cis*- or *trans*-fused, the entire steroid molecule has either a roughly flattened shape (*trans* A/B) or is bent at the A/B junction (*cis* A/B).

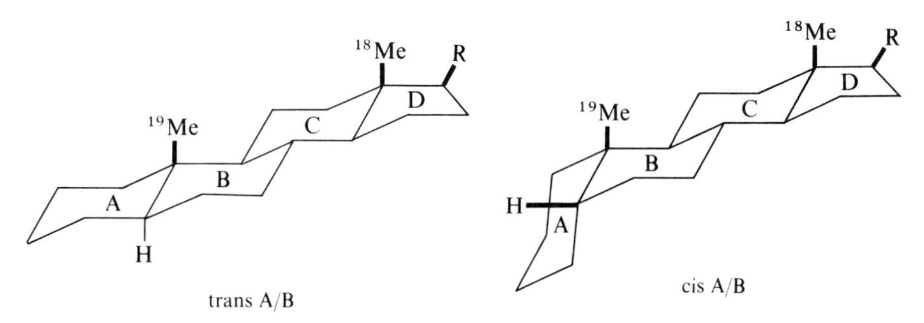


The conformation of the steroid skeleton to some extent determines the stability and reactivity of particular substituents, depending amongst other factors on whether or not they are equatorially or axially disposed. Thus, compounds with equatorially-bonded substituents are usually more stable than the corresponding axial epimers, since the effect of 1,3-non-bonded interactions is greater in the latter. This explains why, for example, cholestanol which has A/B trans and a 3$\beta$-hydroxy group is more stable than *epi*-cholestanol (A/B trans, 3$\alpha$-hydroxy group), whilst in contrast *epi*-coprostanol (A/B cis,3$\alpha$-hydroxy group) is more stable than its epimer, coprostanol (A/B cis,3$\beta$-hydroxy group). In both cases, the more stable epimer has the 3-OH and 5-H groups in a trans configuration, with the 3-OH equatorial to ring A. Also, equilibration with sodium amylate in boiling amyl alcohol, which proceeds by air oxidation to the ketone and subsequent reduction, favours the more stable isomer of each epimeric pair, with the equatorial 3-OH group. (See reaction scheme on page 734.)

STEROID HYDROCARBONS

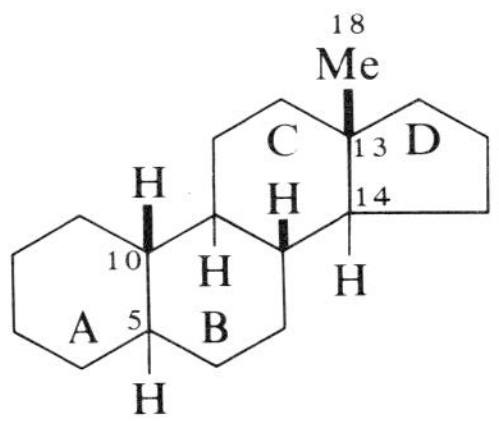

5-$\alpha$-Oestrane

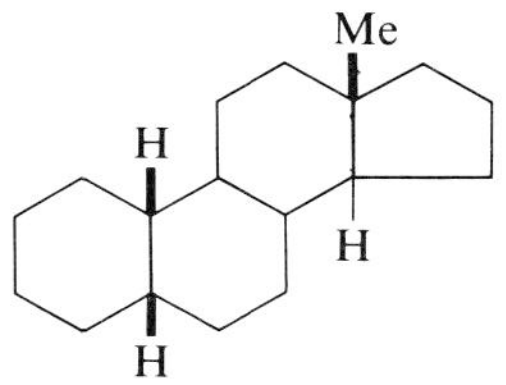

5-$\beta$-Oestrane

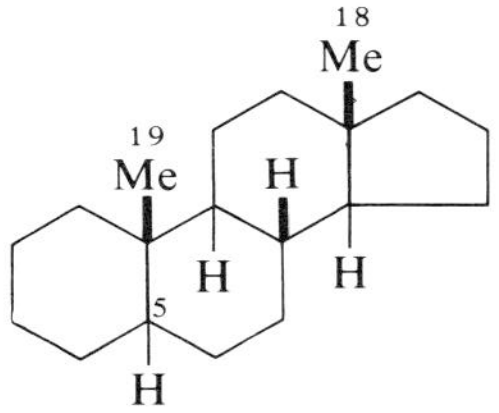

5-$\alpha$-Androstane

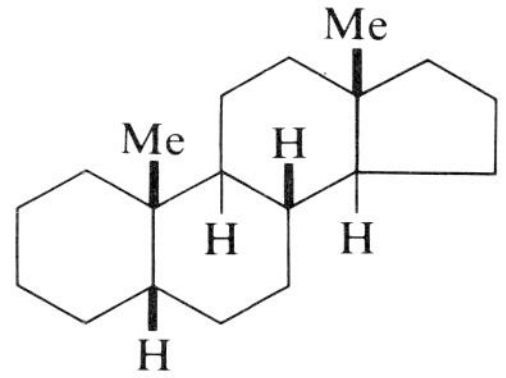

5-$\beta$-Androstane

TYPICAL STEROIDS

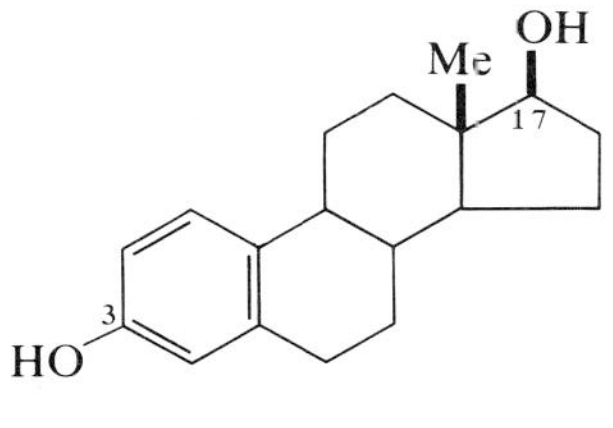

*Oestradiol*

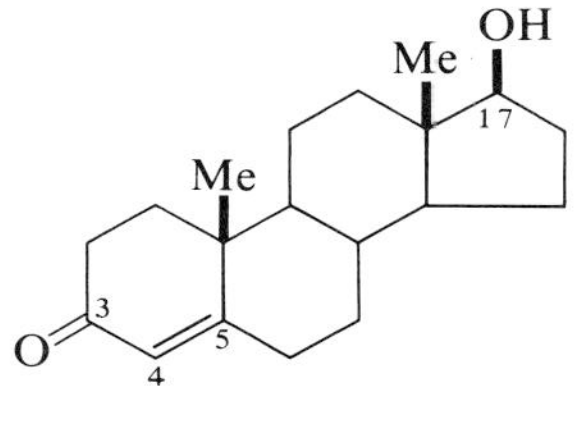

*Testosterone*

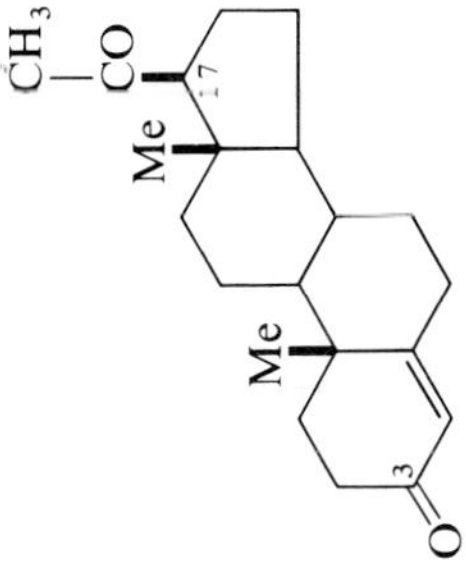

5-α-Pregnane

5β-Pregnane

*Progesterone*

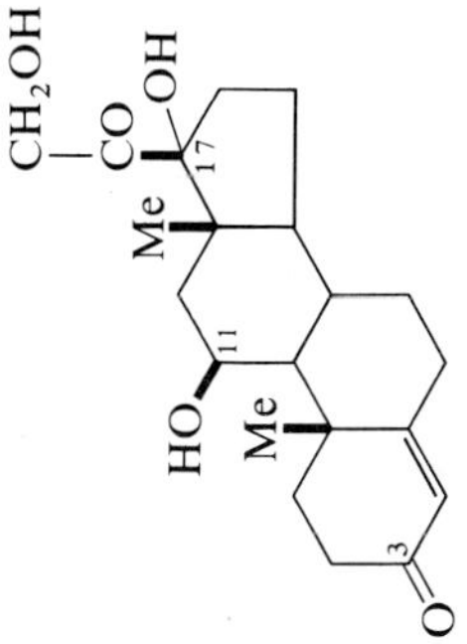

*Hydrocortisone*

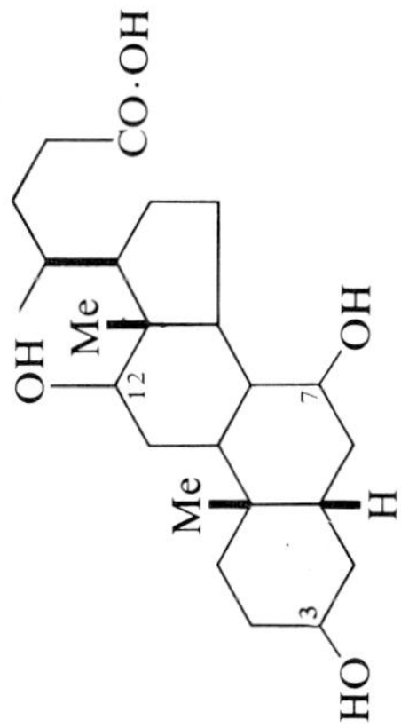

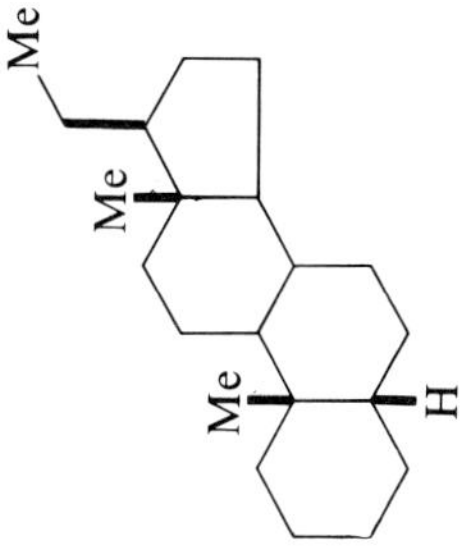

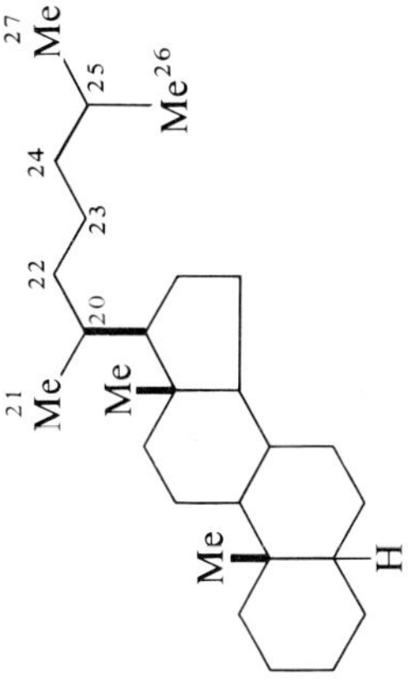

4-α-Cholestane

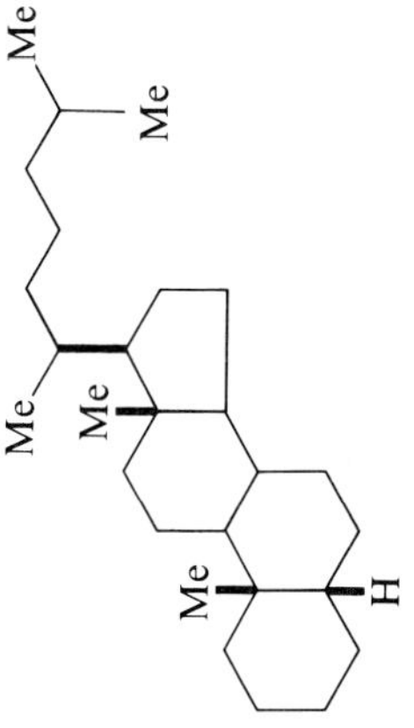

5-β-Cholestane
(Coprostane)

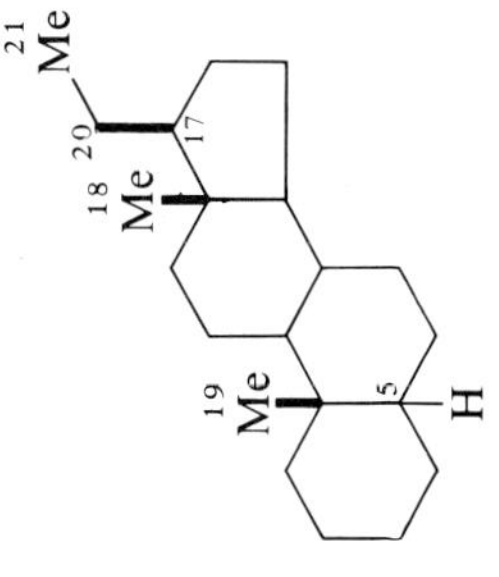

Cholic acid

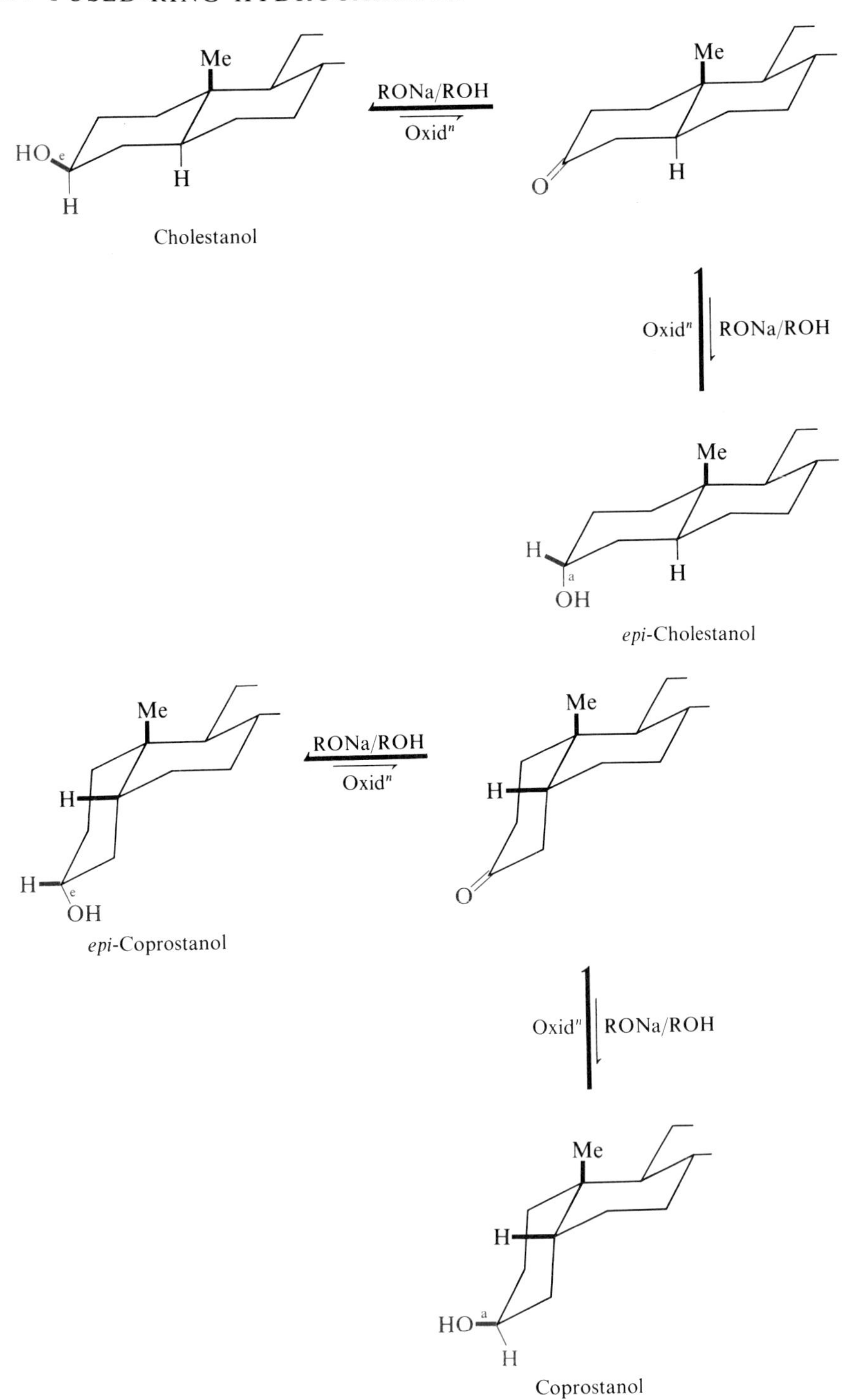


Cholestanol

*epi*-Cholestanol

*epi*-Coprostanol

Coprostanol

The ease of esterification of the 11-hydroxyl and the rate of hydrolysis of 11-acetoxy groups is similarly determined by conformational considerations. Thus, the equatorial 11α-hydroxyl of 11-*epi*-hydrocortisone is capable of acetylation with acetic anhydride (Long and Gallagher, 1946), whereas acetylation of the axial 11β-hydroxyl of *Hydrocortisone* is impeded by 1,3-non-bonded interactions with the C-18 and C-19 angular methyl groups. In consequence, acetylation of *Hydrocortisone* with acetic anhydride gives the 21-monoacetate (*Hydrocortisone Acetate*), whereas acetylation of 11-*epi*-hydrocortisone gives the 11,21-diacetate.

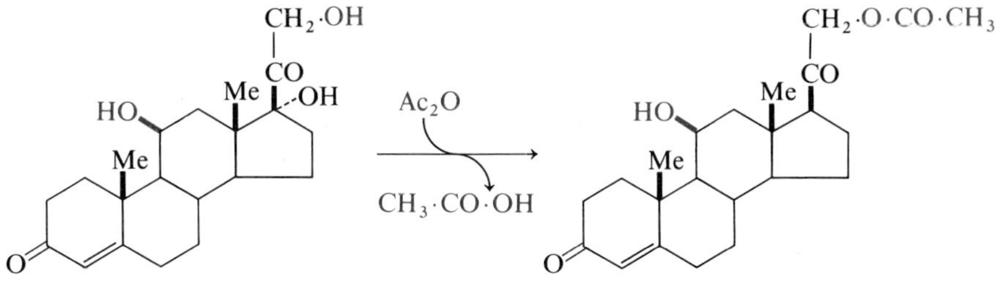

$CH_2{\cdot}OH$, CO, Me, OH, HO, Me, O — $Ac_2O$ → $CH_3{\cdot}CO{\cdot}OH$ — $CH_2{\cdot}O{\cdot}CO{\cdot}CH_3$, CO, Me, HO, Me, O

*Hydrocortisone* | *Hydrocortisone Acetate*

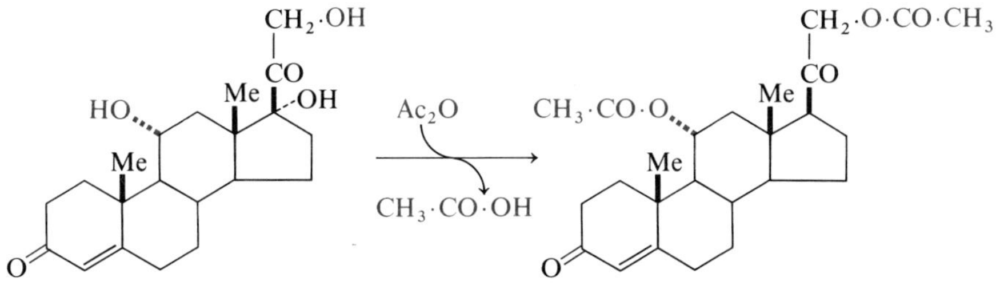

$CH_2{\cdot}OH$, CO, Me, OH, HO, Me, O — $Ac_2O$ → $CH_3{\cdot}CO{\cdot}OH$ — $CH_2{\cdot}O{\cdot}CO{\cdot}CH_3$, CO, Me, $CH_3{\cdot}CO{\cdot}O$, Me, O

11-*epi*-Hydrocortisone | 11-*epi*-Hydrocortisone 11,21-Diacetate

Inversion of the stereochemistry at the A/B and B/C ring junctions in the 9β,10α(*retro*)-steroids leads to a major alteration in the shape of the molecule (Braun, Hornstra and Leenhouts, 1969).

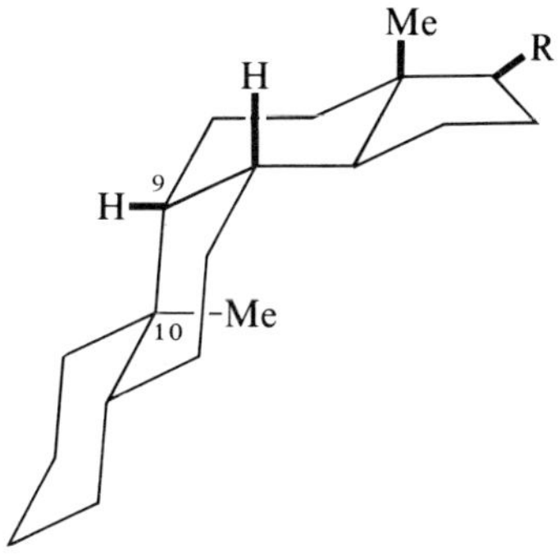

Me, R, H, 9, H, 10, -Me

## Structural and Stereochemical Specificity in the Biological Action of Steroids

Molecular shape is an important factor in determining the nature and level of activity of hormonal steroids (Bush, 1962). Oxygen functions at C-3, C-17, C-20 and C-21 play a major rôle in such action, so whatever the mechanism of action at molecular level, other important structural features of the steroid nucleus must either assist or at least should not impede reaction with these centres.

### Androgens

Ringold (1961) in an extensive review of compounds with male hormone activity concludes that it involves reaction at the lower, unhindered α-face of the molecule. Thus, the saturated 17β-hydroxyandrostan-3-one shows a similar level of androgenic activity to testosterone. In both compounds, the major ring substituents (C-18 and C-19 methyl groups, and C-17 hydroxyl) are all β-orientated, so that approach to the lower α-face of the molecule is relatively unhindered. Also, both molecules have substantially the same orientation of C-3 and C-17 oxygen functions, despite the difference in shape of ring A giving a long flattish structure.

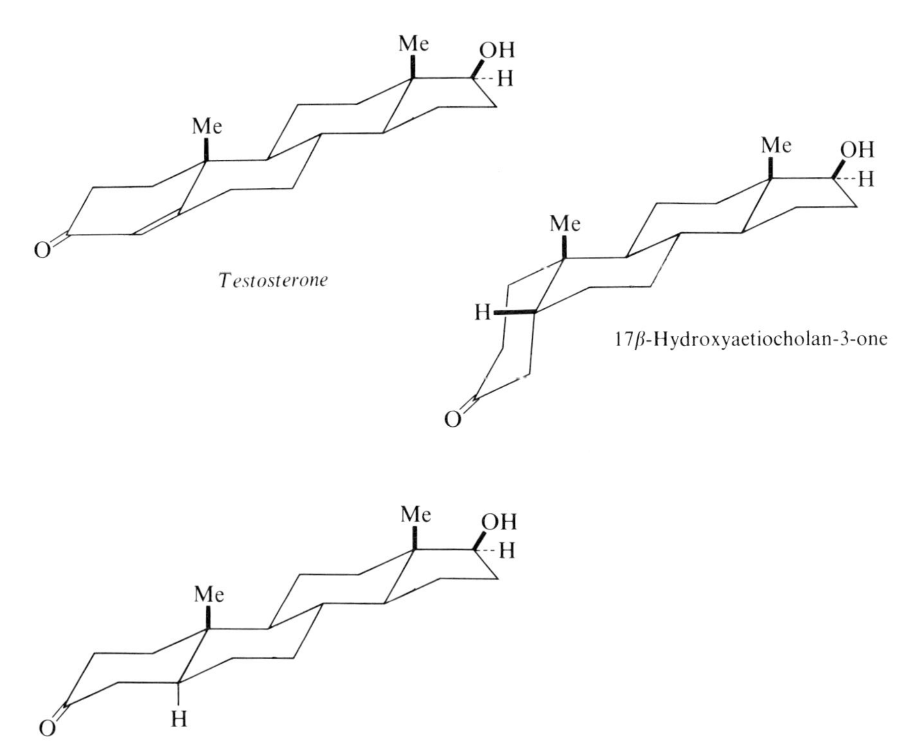


*Testosterone*

17β-Hydroxyaetiocholan-3-one

17β-Hydroxyandrostan-3-one

In contrast, 17$\beta$-hydroxyaetiocholan-3-one in which rings A/B are *cis*, so that ring A is folded downwards on the $\alpha$-face, has no androgenic activity. Similarly, whilst 17$\alpha$-methyltestosterone and 17$\beta$-hydroxy-17$\alpha$-methylandrostan-3-one are potent androgens, the corresponding 17$\alpha$-ethyl and 17$\alpha$-vinyl compounds have only a low level of male hormone activity. Further evidence in support of essential interactions at the $\alpha$-face of androgenic steroids is the retention of hormonal activity, though with some loss in potency in 19-nortestosterone and 17$\beta$-hydroxy-19-norandrostan-3-one (Hershberger, Shipley and Meyer, 1953), which both lack the $\beta$-orientated C-19 methyl group, and more significantly the complete lack of androgenic activity in 17$\beta$-hydroxy-1$\alpha$-methyl-19-norandrostan-3-one with its downward axial ($\alpha$-) projecting 1-methyl substituent (Bowers, Ringold and Dorfman, 1957; Bowers, Ringold and Denot, 1958).

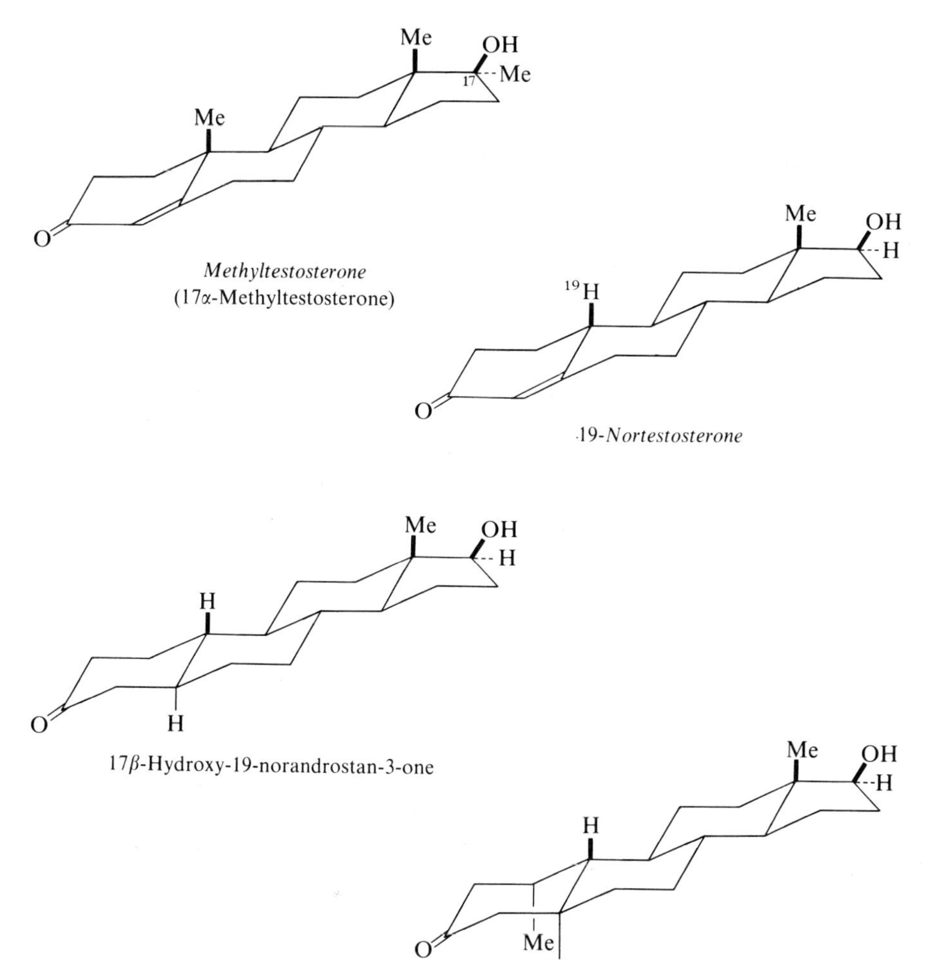


*Methyltestosterone*
(17$\alpha$-Methyltestosterone)

19-*Nortestosterone*

17$\beta$-Hydroxy-19-norandrostan-3-one

17$\beta$-Hydroxy-1$\alpha$-methyl-19-norandrostan-3-one

## Anabolic Steroids

Androgenic hormones also exhibit anabolic activity. Thus, testosterone has a marked effect on the incorporation of amino acids into protein, apparently due to stimulation of RNA-polymerase (Williams-Ashman, Liao, Hancock, Jurkowitz and Silverman, 1964). The value of testosterone either alone or in combination with oestrogen in the treatment of osteoporosis with acute hypercalcaemia, which is often seen in geriatrics (Weddon, 1956; Mason, 1957), led to attempts to divorce anabolic from male hormone activity.

Useful compounds have been found in the nortestosterone series. Thus, 19-nortestosterone (Birch, 1950) has only about one-fifth of the androgenic effect of testosterone, but the same level of anabolic potency (Hershberger, Shipley and Meyer, 1953). Moreover, its anabolic action can be prolonged by esterification with, for example, benzoic or phenylpropionic acids (Barnes, Stafford, Guild and Olson, 1954; Overbeek and de Visser, 1957). *Norethandrolone* (17α-ethyl-19-nortestosterone) similarly shows a favourable ratio of anabolic/androgenic potencies (Drill and Saunders, 1956), but significantly also possesses progestational properties. A more interesting development, however, is the development of 17β-hydroxy-17α-methylandrostano[3,2-*c*]pyrazole, which shows anabolic activity with little or no virilising effect.

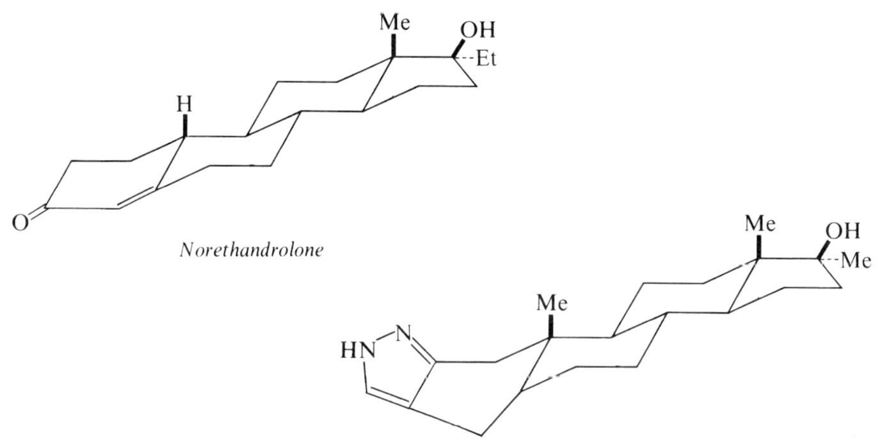


17β-Hydroxy-17α-methylandrostano[3,2,c]pyrazole

## Oestrogens

The absence of a β-orientated C-19 methyl group and distortion of ring B resulting from aromatisation of ring A in the natural oestrogen, 17β-oestradiol, together with the lack of activity in the epimeric 17α-oestradiol, and the potency of *Ethinyloestradiol* (Ethinyl Estradiol) and *Mestranol* all point to reaction at the β-face as an essential feature of the oestrogenic response (Ringold, 1961). This generalisation must, however, be treated with some reserve, since although the aromatic ring itself is twisted upwards, β-wise, compared with androgenic and other steroids, 6β-methyl substituents mitigate against activity.

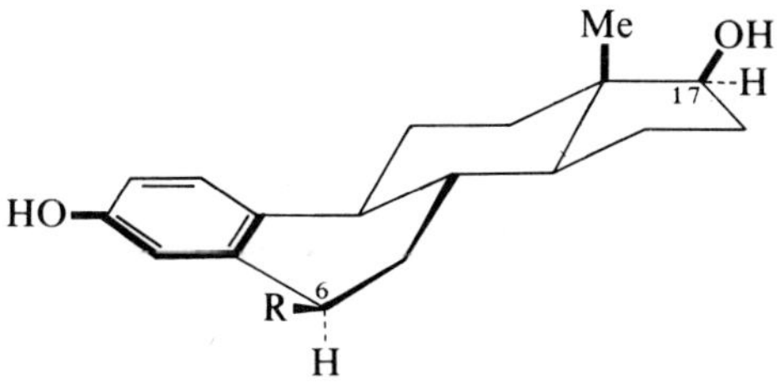

Oestradiol (R = H)

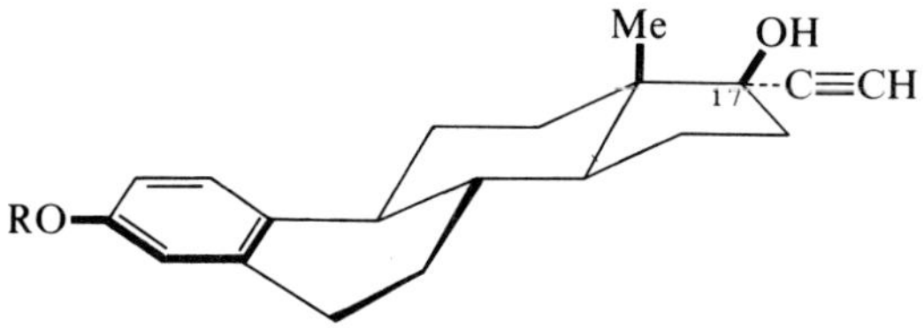

Ethinyloestradiol (R = H)
Mestranol (R = Me)

Powerful oestrogenic activity is also seen in analogous compounds [*Stilboestrol* (Diethylstilbestrol), *Dienoestrol* (Dienestrol), *Hexoestrol* (Hexestrol) and *Chlorotrianisene*] which are comparable in shape, giving similar spacing between two phenolic or related oxygen functions, as in 17$\beta$-oestradiol (**2**, 3).

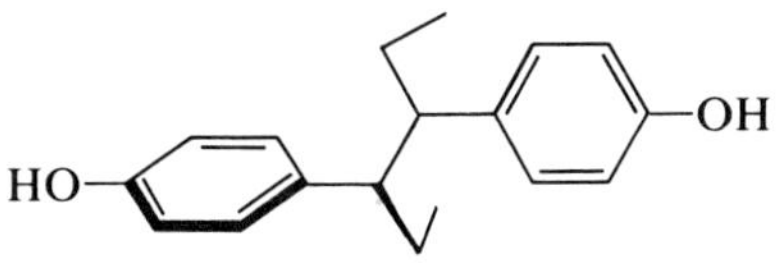

*meso*-Hexoestrol

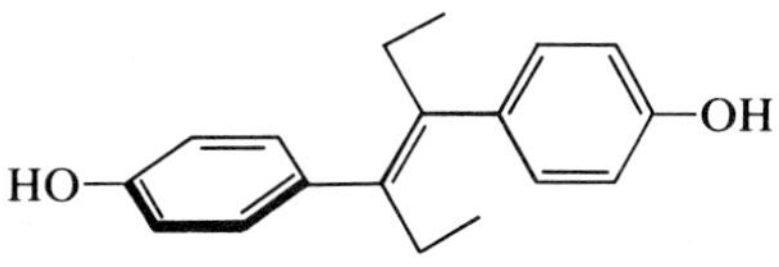

*Stilboestrol*

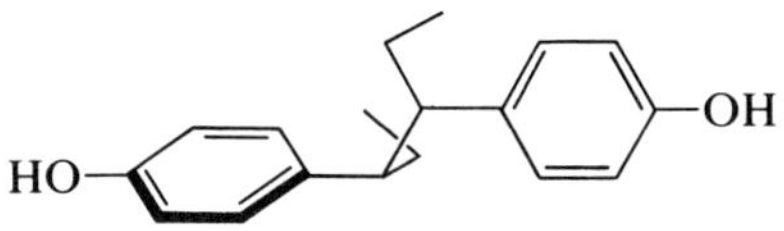

(A)

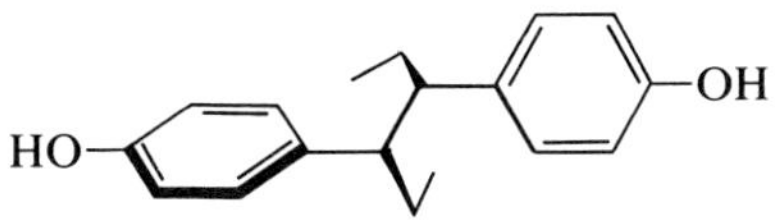

(B)

Hexoestrol optical isomers

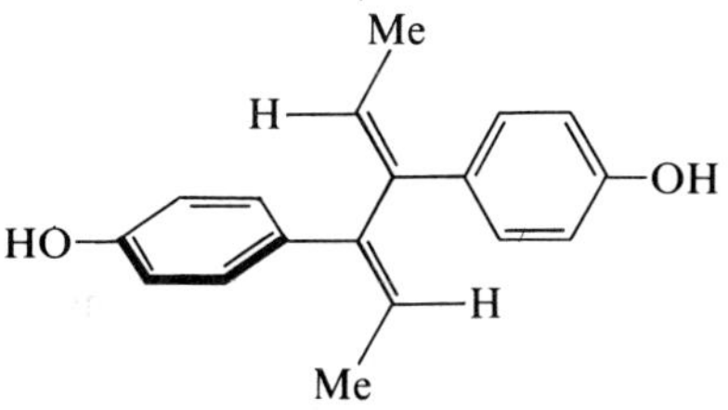

*Dienoestrol*

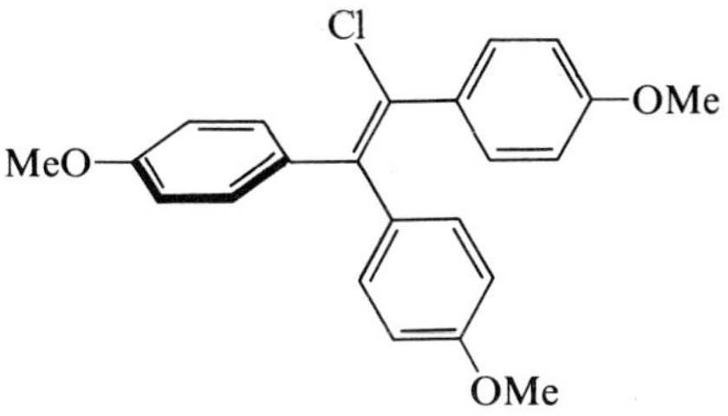

*Chlortrianisene*

It is clear from the potency of *Chlorotrianisene* and *Mestranol* (17α-ethinyl-oestradiol-3-methyl ether) that a free phenolic group is not essential for oestrogenic activity. The rôle of the aromatic ring, therefore, could well be one of providing a centre of high electron density for protein-binding. Thus, it is established that all oestrogens are bound to specific proteins in the uterus (Puca and Bresciani, 1969). It is not clear whether these are enzymic proteins, though it is established that oestradiol increases protein synthesis in rat uterus, by increasing the uptake of [$^3$H]-uridine into ribosomal RNA, thus stimulating RNA synthesis (Aizawa and Mueller, 1961), which in turn stimulates protein synthesis.

The suppression of activity by methyl substitution in ring A of oestradiol, which is greatest when these substituents are in either the 2- or 4-positions (Ringold, 1961), appears to be steric rather than electronic, since the *ortho*-methyl substituents oppose the electron drift from oxygen to the ring. The same substituents would, however, impede the approach of a reactant molecule towards the β-face, because of the dihedral angle between rings A and C.

**Progestational Agents**

Progesterone, the natural gestational hormone secreted by the corpus luteum after ovulation, is responsible for preparing the uterus to receive the fertilised ovum, and for the maintenance of pregnancy by suppression of the ovulatory cycle. It is this latter property which has provided the basis for the use of progestogens as oral contraceptives. Not only do they exert feedback suppression over the secretion of luteinising hormone-releasing factor (LH–RF) from the hypothalamus (p. 742), but they cause considerable thickening of the cervical mucous. The coiling of mucoid particles into micellar structures leads to the formation of a dense mucoid network which largely prevents the entry of spermatozoa into the uterine cavity.

Progestational steroids fall into two main groups, progesterone derivatives and C-17α-alkyl-, alkenyl- or alkynyl-testosterones. Whilst these two groups have distinctly different substituents at C-17, with radically different steric consequences, the balance of evidence suggests that reaction with receptor molecules is at the β-face of the molecule. Thus, replacement of the C-10 angular methyl group by hydrogen leads to a substantial increase in potency, as in 19-norprogesterone (Djerassi, Miramontes and Rosenkranz, 1953; Tullner and Hertz, 1953) and *Norethisterone* (Norethindrone Acetate; Djerassi, Miramontes, Rosenkranz and Sondheimer, 1954; Hertz, Tullner and Raffelt, 1954).

In contrast, bulky 17α-substituents, as in *Ethisterone*, *Norethisterone*, and 16α-methyl-17α-hydroxyprogesterone acetate (Babcock, Gutsell, Herr, Hogg, Stucki, Barnes and Dulin, 1958) do not depress activity. Also, the buckling of rings A and B in *Norethynodrel* distorts the molecule towards the α-face, in keeping with its retention of progestational activity. Similarly, a marked potentiation of activity occurs with the insertion of appropriately-orientated substituents at C-6 of the steroid nucleus. It has been suggested that substitution at

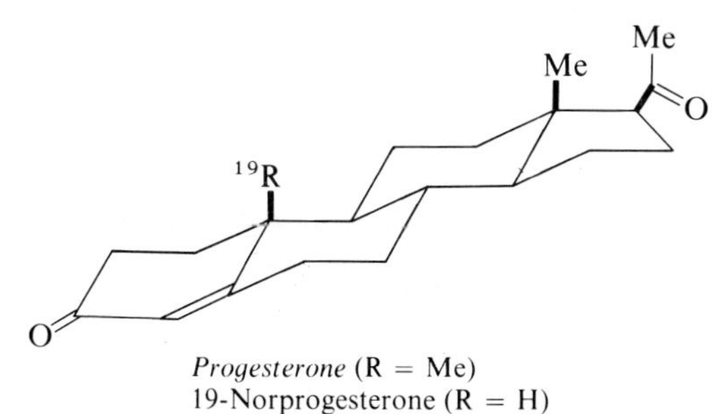


*Progesterone* (R = Me)
19-Norprogesterone (R = H)

*Ethisterone* ($R^1$ – Mc; $R^2$ = $R^3$ = H)
*Norethisterone* ($R^1$ = $R^2$ = $R^3$ = H)
6α, 21-Dimethylethisterone ($R^1$ = $R^2$ = $R^3$ = Me)

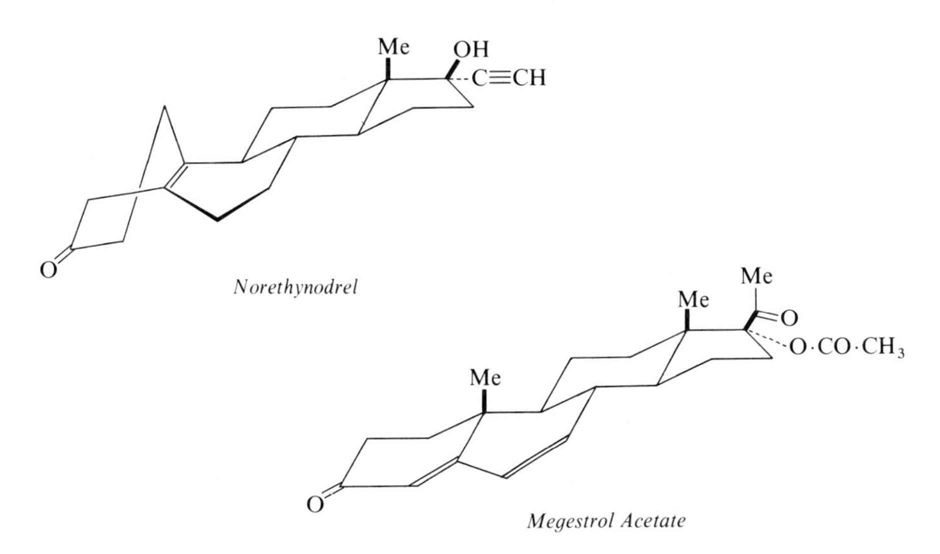


*Norethynodrel*

*Megestrol Acetate*

C-6 inhibits metabolic oxidation which is often favoured at this position. Compounds, such as 6α-methyl- and 6α,21-dimethylethisterone, show seven and twelve times the potency of *Ethisterone*; on the other hand, 6β-methylethisterone shows only one-third of the activity of *Ethisterone* (David, Hartley, Millson and Petrow, 1957). Similarly, *Megestrol Acetate* is about 50 times as active as progesterone.

**Control of Gonadal Steroids**

Secretion of gonadal steroids is under the control of gonadotrophic hormones of the anterior pituitary gland. Follicle stimulating hormone (FSH) maintains gametogenesis in both sexes, and in particular influences the secretion of oestrogens. Luteinising hormone (LH) is responsible for ovulation, the suppression of oestrogen secretion, and secretion of progesterone. The release of both gonadotrophins is under the control of gonadotrophin-releasing factors (GRF) in the hypothalamus, whilst the hormonal steroids themselves are able to depress the biosynthesis of luteinising hormone-releasing factor (LH–RF) by a negative feedback mechanism, when present at high blood levels. It is for this reason that small doses of oestrogen are included along with the progestagen in the majority of oral contraceptives (Bingel and Benoit, 1973). The interplay of these factors in oral contraception is illustrated in Fig. 38.

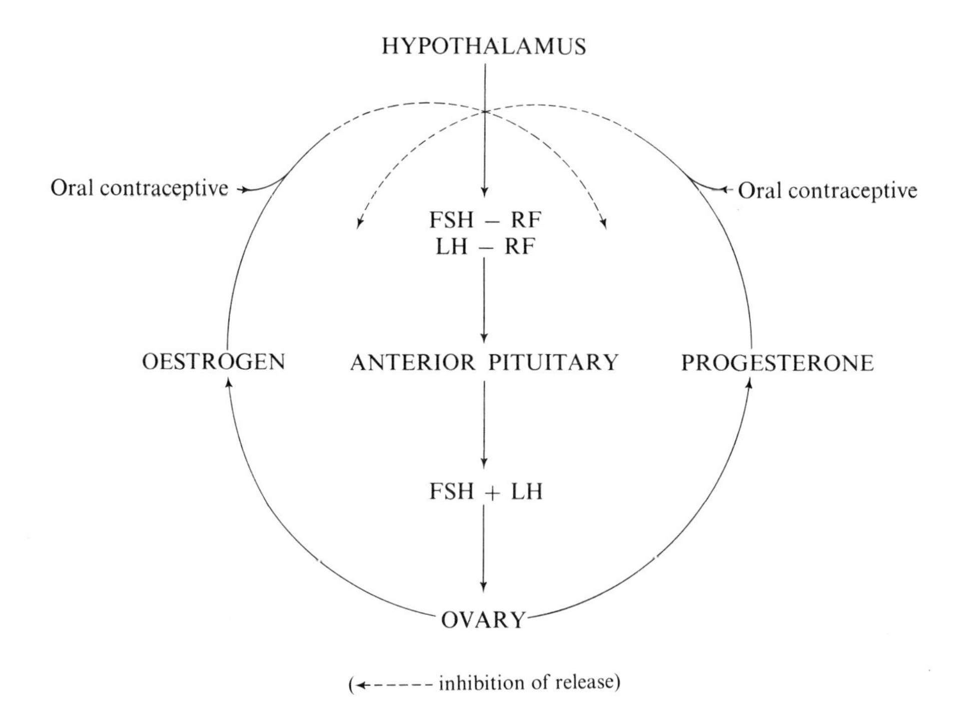


Fig. 38 Release and feedback control of gonadal steroids

The oestrogen content of the contraceptive pill has been shown to carry enhanced risk of venous thrombosis and pulmonary embolism (Inman and Vessey, 1968). This may be due to platelet adhesiveness (Elkeles, Hampton and Mitchell, 1968).

**Adrenal Corticosteroids**

The characteristic structural feature of the natural adrenal corticosteroids, irrespective of whether they influence carbohydrate metabolism (gluco-corticoids) or mineral metabolism (mineralo-corticoids), is the 17$\beta$-ketol side-chain. This, as shown by Cole and Williams (1968), is orientated such that the C-20 carbonyl group is directed back towards ring D with the C-21 hydroxyl weakly hydrogen-bonded to the C-20 carbonyl group. The most potent mineralo-corticoid hormone, *Aldosterone*, however, is also characterised by the presence of the hemi-acetal structure which bridges the C-11, C-12 and C-13 positions.

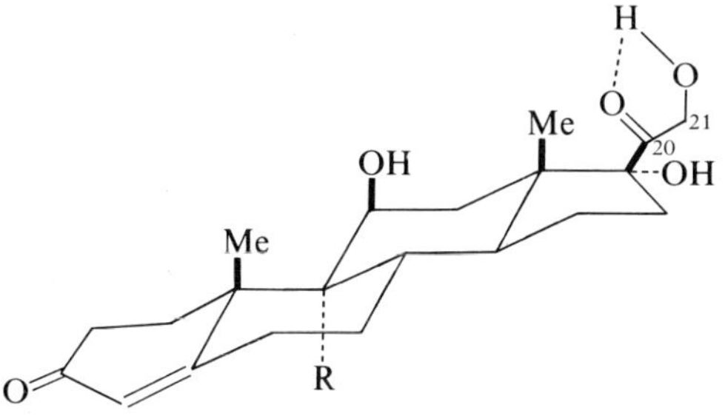

*Hydrocortisone* (R = H)
*Fludrocortisone* (R = F)

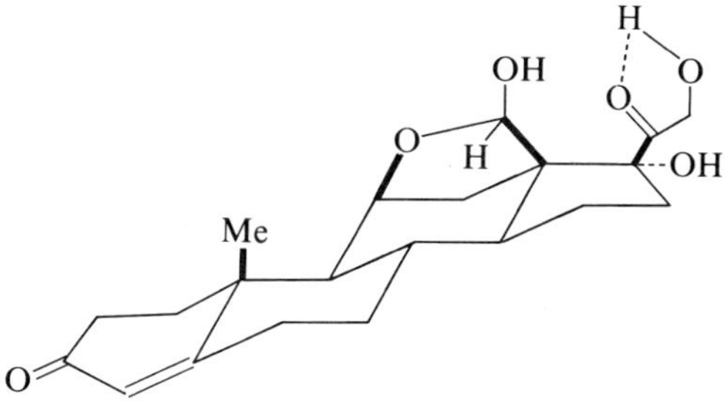

*Aldosterone*

The important anti-inflammatory properties of *Hydrocortisone* have been developed at the expense of gluco- and mineralo-corticoid properties mainly by introduction of 9$\alpha$-halogen substituents, additional double bonds in the 1,2-position, and 16$\alpha$-methyl and 16$\alpha$-hydroxyl groups in such compounds as *Fludrocortisone*, *Prednisolone*, *Dexamethasone* and *Triamcinolone*.

Fat solubility is increased and absorption and distribution facilitated in the C-21 esters, such as *Hydrocortisone Acetate*. They are, however, rapidly hydro-lysed *in vivo* and owe their action to metabolic release of the parent C-21 alcohols. The function of the 9$\alpha$-fluoro substituents in many of the more effective anti-inflammatory agents is not established, though it is clear that it enhances chloroform solubility and hence lipid solubility. Thus, *Fludrocortisone Acetate* (solubility 1 in 50) is some three times more soluble in chloroform than *Hydro-cortisone Acetate* (solubility 1 in 150). A similar difference is also seen in the

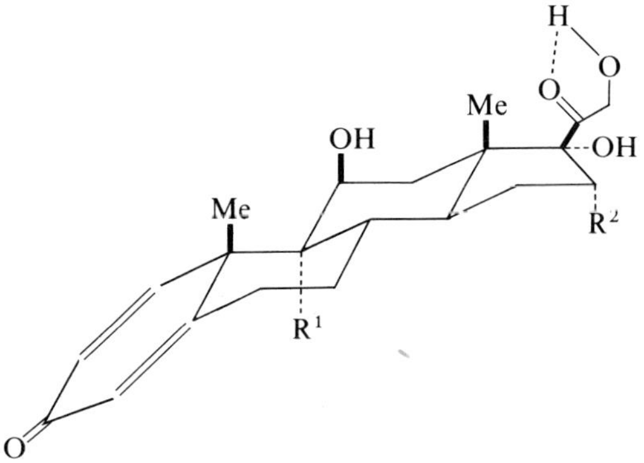

H O O Me OH OH Me R[2] R[1] O

*Prednisolone* ($R^1 = R^2 = H$)
*Dexamethasone* ($R^1 = F$; $R^2 = Me$)
*Triamcinolone* ($R^1 = F$; $R^2 = OH$)

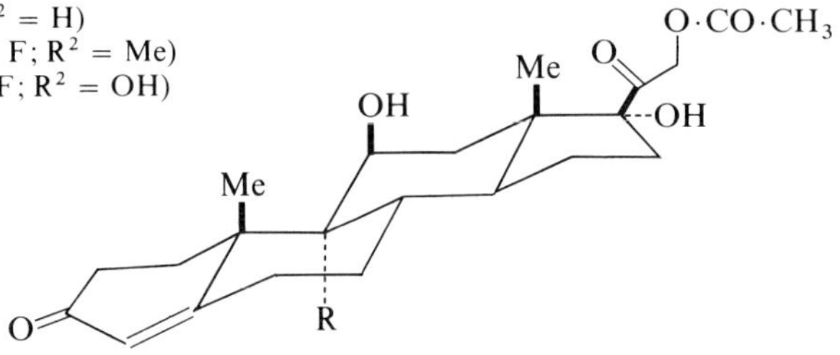

O·CO·$CH_3$ Me O OH OH Me R O

*Hydrocortisone Acetate* (R = H)
*Fludrocortisone Acetate* (R = F)

chloroform solubilities of *Fluocinolone Acetonide* (1 in 15) and *Triamcinolone Acetonide* (1 in 40).

Enhancement of anti-inflammatory activity also occurs in the C-17 esters, such as *Betamethasone Valerate*. Unfortunately, these compounds suffer from the disadvantage that a markedly de-activating C-17/C-21 trans-esterification occurs readily at pH 7.0 and above. This, however, is prevented by the C-21 chloro group in the structurally-related compound, *Clobetasol* 17-*Propionate*, which also has valuable anti-inflammatory properties.

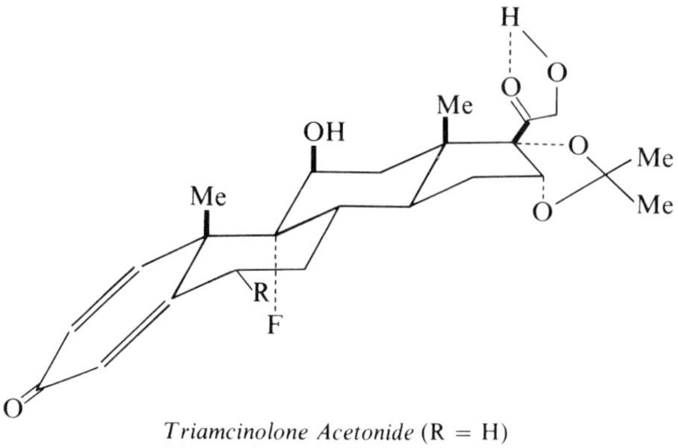

H O O Me OH O Me Me Me O R F O

*Triamcinolone Acetonide* (R = H)
*Fluocinolone Acetonide* (R = F)

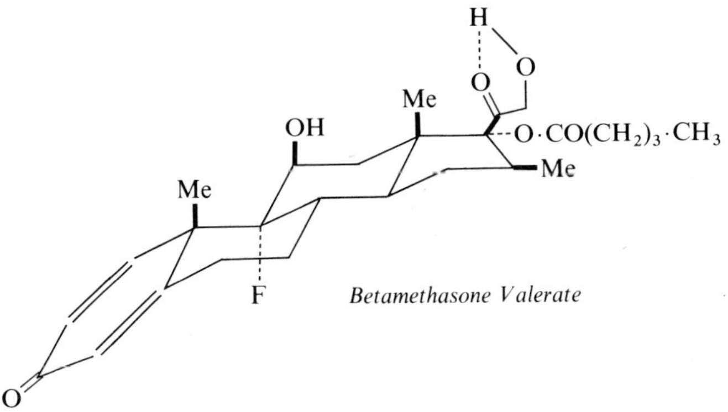

*Betamethasone Valerate*

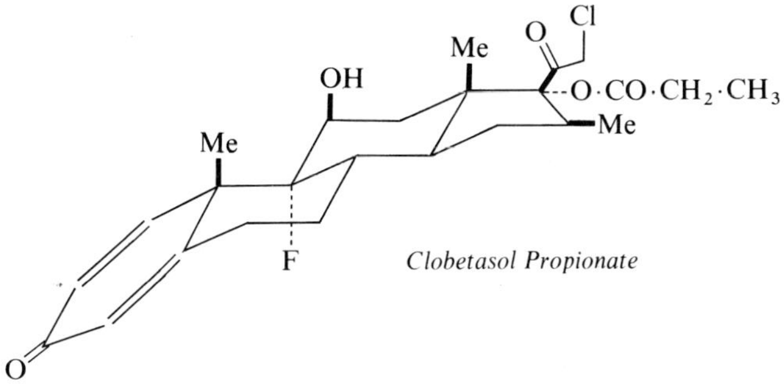

*Clobetasol Propionate*

## Cardiac Glycosides

The main group of cardiac glycosides are steroidal glycosides from plants of the orders *Apocynaceae*, *Liliaceae*, *Ranunculaceae* and *Scrophulariaceae*. The two most important sources are digitalis and strophanthus. The separation of individual glycosides is complicated by the ease with which the primary glycosides

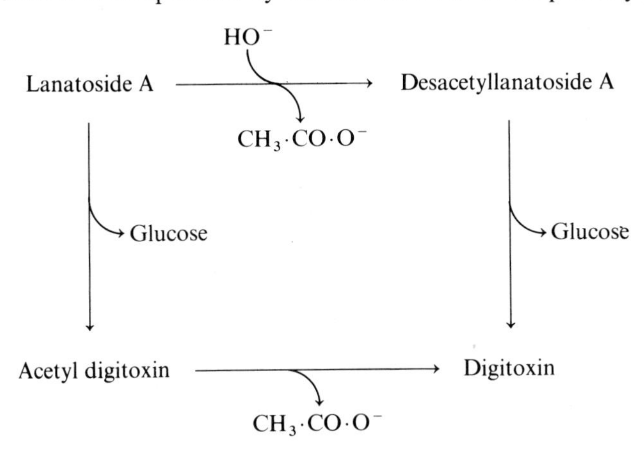

are hydrolysed with the loss of glucose and acetic acid residues. Three primary glycosides, *Lanatosides A*, *B* and *C*, have been isolated from *Digitalis lanata*. These can be hydrolysed as follows to give the secondary glycosides, digitoxin (from *Lanatoside A*), gitoxin (from *Lanatoside B*) and digoxin (from *Lanatoside C*) together with glucose (1 mol) and acetic acid in each case.

These glycosides are stable to heat in 70% ethanol, but aqueous solutions are liable to hydrolysis. Acidic conditions promote hydrolysis to the genin and digitoxose, whilst alkaline conditions disrupt the lactone ring.

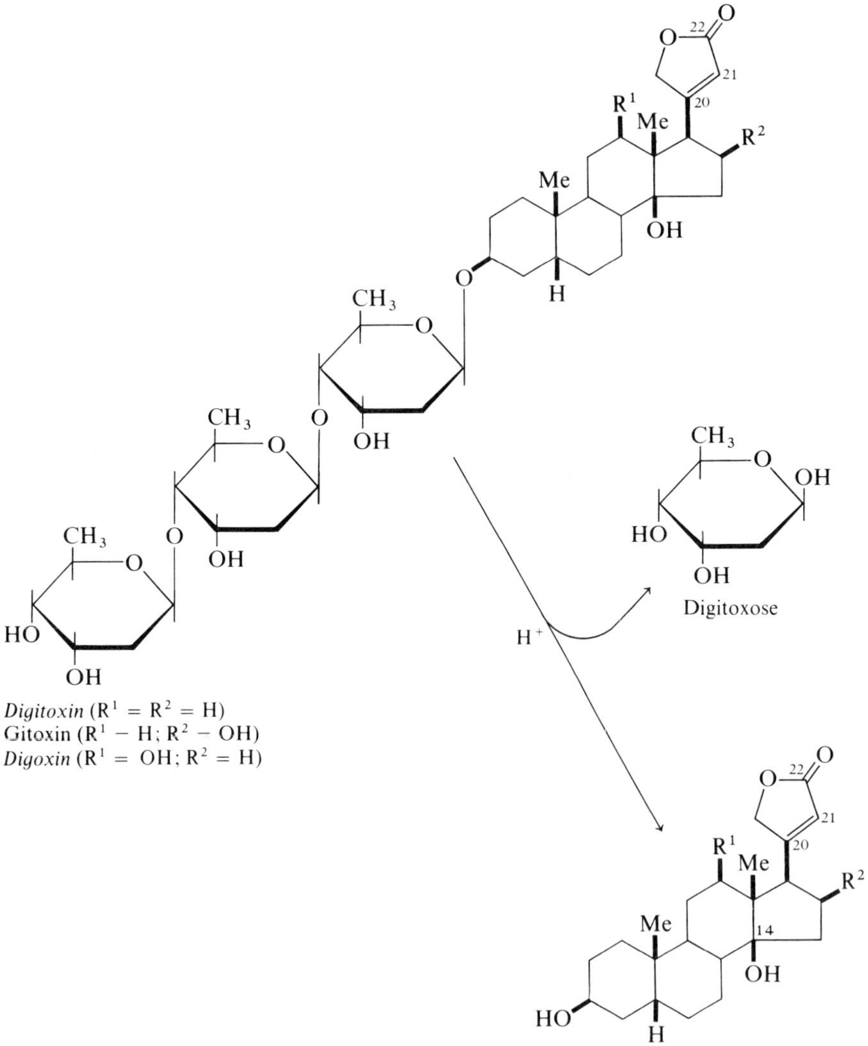

*Digitoxin* ($R^1 = R^2 = H$)
Gitoxin ($R^1 = H$; $R^2 = OH$)
*Digoxin* ($R^1 = OH$; $R^2 = H$)

Digitoxose

Digitoxigenin ($R^1 = R^2 = H$)
Gitoxigenin ($R^1 = H$; $R^2 = OH$)
Digoxigenin ($R^1 = OH$; $R^2 = H$)

Digitoxin, gitoxin and digoxin break down on further hydrolysis to give the sugar, digitoxose (3 mols) and the appropriate genin (digitoxigen, gitoxigenin and digoxigenin respectively). The genins are characterised by a steroidal nucleus with the A/B and C/D ring junctions both *cis*, a hydroxyl substituent at C-14 and the $\alpha\beta$-unsaturated $\gamma$-lactone ring derived from the C-17$\beta$ cholesterol side-chain.

The secondary glycosides, *Digitoxin* and *Digoxin*, also *Lanatoside C* and *Deslanoside* (deacetyl-lanatoside C) are used as myocardial stimulants. Their action on heart muscle is not fully elucidated, but it is evident that it is complex (Thomas, Boutagy and Gelbart, 1974) and primarily related to the active transport of sodium and potassium ions (Bowman, Rand and West, 1968).

The essential stereochemical features of the steroidal nucleus for cardiotonic activity appear to be the *cis* C/D ring junction and the 17$\beta$-orientation of the $\alpha\beta$-unsaturated $\gamma$-lactone function. Thus, 14$\alpha$-digitoxigin, which has the *trans* C/D ring junction (Zürcher, Weiss-Berg and Tamm, 1969), and both 17$\alpha$-cardenolides (Tham, 1963) and 14$\alpha$,17$\alpha$-cardenolides (Nambara, Shimada, Goto and Goya, 1971) are all inactive. The *cis* A/B ring junction favours, but is not essential to activity, since uzarigenin (5$\alpha$-digitoxigenin) is about half as potent as digitoxigenin.

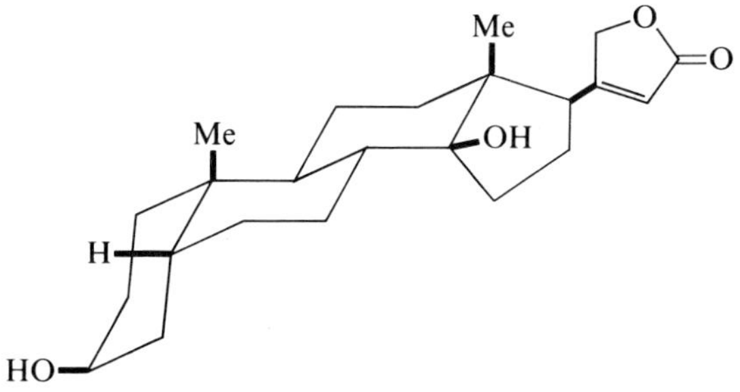

Digitoxigenin

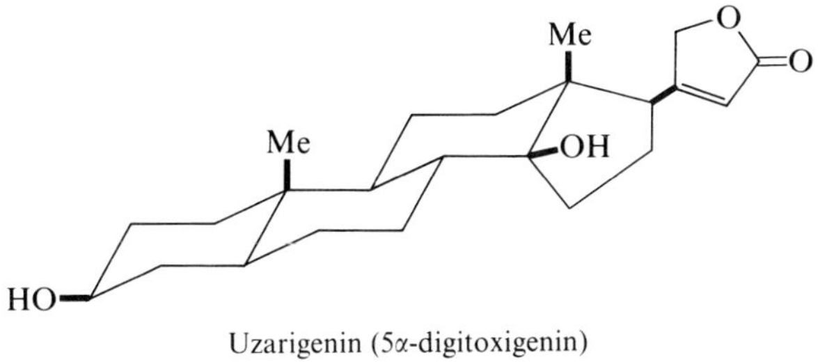

Uzarigenin (5$\alpha$-digitoxigenin)

The 3$\beta$-oxygen function gives rise to much more potent cardiotonic activity when the hydroxyl function is substituted as glycosides or esterified (Brown, Stafford and Wright, 1962). The 3$\beta$-oxy-function, however, is not essential for activity, since, whilst the 3$\alpha$-oxy compounds are virtually inactive (Repke, 1963), 3-deoxy-derivatives, such as 3-deoxydigitoxigenin (Zürcher, Weiss-Berg and

Tamm, 1969), exhibit the same level of cardiotonic activity as digitoxigenin. Similarly, the 14$\beta$-hydroxy group is not essential for activity.

The mechanism whereby the cardiotonic glycosides produce their inotropic (force-increasing) effect on heart muscle is undoubtedly complex. There is strong evidence, however, that their primary effect is inhibition of $Na^+$, $K^+$—ATPase and that this leads to an increase in activator—$Ca^{2+}$ and thence to the required inotropic effect. Thus, Repke (1965) has shown that similar stereochemical parameters apply to the binding and inhibitory activity of cardiotonic glycosides on $Na^+$, $K^+$—ATPase as apply to their cardiotonic activity. He visualised location and attachment to the enzyme surface as being primarily effected through a hydrogen bond to the carbonyl group of the $\alpha\beta$-unsaturated lactone ring. The main source of binding energy is supplied by van der Waals and/or hydrophobic interactions, between the steroid nucleus and the component amino acids of the enzyme protein.

Studies on the rôle of the lactone ring (Boutagy, Gelbart and Thomas, 1973) show that it can be replaced by open-chain substituents with a co-planar arrangement of atoms, —CH=CH—C—hetero-atom without loss of either $Na^+$, $K^+$—ATPase affinity or inotropic activity. It is considered that binding occurs between such side-chains and the enzyme within a cleft on the enzyme surface, involving hydrogen bonding at the electron-rich hetero atom and charge-transfer complexing with the electron-deficient C-20 atom (Fig. 39). This conclusion is compatible with the loss of potency on replacement of H by a methyl or methoxyl substituent at C-21, but retention of potency on replacement of H by fluorine at C-21 (Eberlein, Nickl, Heider, Dahms and Machleidt, 1972).

Because of its low water solubility, tablets containing *Digoxin*, the glycoside most widely used as a heart stimulant, can show wide variations in dissolution profile and bioavailability (Fraser, Leach, Poston, Bold, Culank and Lipede, 1973; Stenlake and Watt, 1975). The problem is largely overcome by reduction in particle size (Shaw, Carless, Howard and Raymond, 1973), though it is apparent that this is accompanied by changes from crystalline to amorphous state (Florence, Salole and Stenlake, 1974).

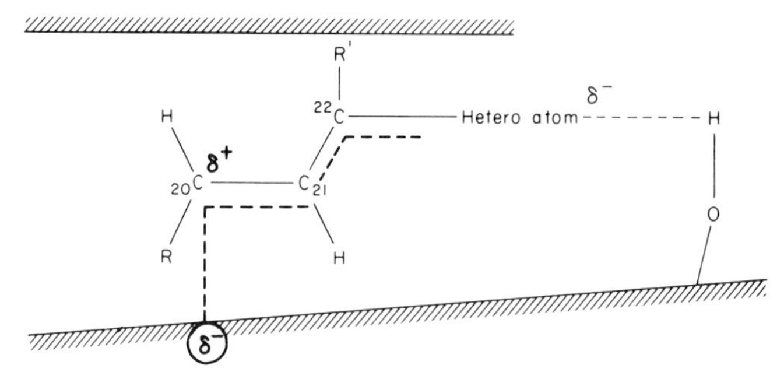


Fig. 39 Binding of cardiotonic $\alpha\beta$-unsaturated lactones and analogues to $Na^+$, $K^+$, ATPase (Adapted from Thomas, Boutagy and Gelbart, 1974)

Table 73. Solubilities and Gastro-intestinal Absorption of Cardiotonic Glycosides

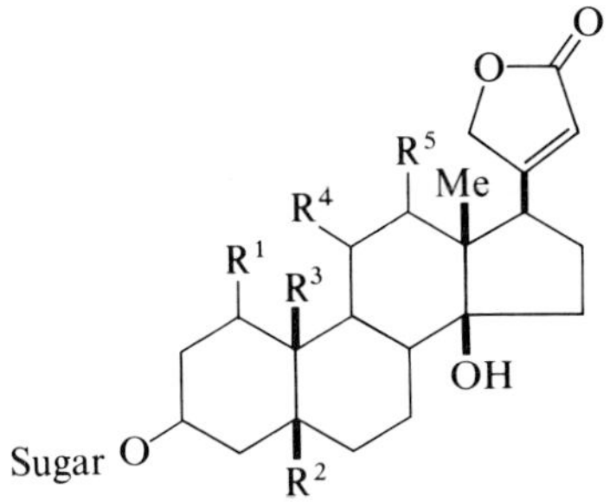

| Glycoside | $R^1$ | $R^2$ | $R^3$ | $R^4$ | $R^5$ | Solubility in Water | Solubility in $CHCl_3$ | Absorption |
|---|---|---|---|---|---|---|---|---|
| Digitoxin | H | H | $CH_3$ | H | H | Insoluble | 1 in 40 | +++ |
| Digoxin | H | H | $CH_3$ | H | OH | Slightly soluble | Slightly soluble | ++ |
| Ouabain | OH | OH | $CH_2OH$ | OH | H | 1 in 75 | Insoluble | ± |

The extent of nuclear hydroxylation materially affects both aqueous and lipid solubilities of cardiotonic glycosides, and hence the rate and extent of absorption from the gastro-intestinal tract. The closely-related glycosides, *Digitoxin*, *Digoxin* and *Ouabain*, which differ essentially (but not solely) in the increasing extent of hydroxylation of the steroid component (Table 73) all have comparatively low water solubilities, and in consequence only reach blood levels in the nanogram range. However, *Digitoxin*, the least hydroxylated with the lowest water solubility, but with the highest lipid solubility is the most readily absorbed from the gastro-intestinal tract and gives the highest blood levels of the three (Caldwell, Martin, Dutta and Greenberg, 1969).

In contrast, *Ouabain*, which has the greatest water solubility, is virtually insoluble in lipid solvents, and in consequence is poorly absorbed. The extent of absorption is, therefore, determined by fat solubility. The rate of absorption, however, depends upon the amount available in aqueous solution for partition across the lipid-absorbing membrane. Water solubility is, therefore, the rate-limiting property for gastro-intestinal absorption of these compounds.

Once absorbed, they show marked differences in distribution characteristics, depending on their lipid character. Thus, whilst *Digitoxin* is some 85% bound to plasma proteins, *Digoxin* is only about 35% bound. Digoxin, however, is much more extensively bound to skeletal and heart muscle. Thus, mean concentrations for plasma, skeletal muscle and heart muscle in patients on steady-state *Digoxin* dosage have been reported as being about 1 ng/ml, 11 ng/g and 77 ng/g (Coltart, Howard and Chamberlain, 1972). This affinity for muscle protein leads to extensive tissue accumulation, an effect which accounts for the toxicity of the cardiac glycosides.

Penta-acetylgitoxin, unlike the parent glycoside, gitoxin, is lipid-soluble and much more readily absorbed from the gastro-intestinal tract than the latter (Megges and Repke, 1963). Once absorbed, it is rapidly de-acetylated to yield the active glycoside, gitoxin. β-Methyldigoxin (Kaiser, 1971) and β-acetyldigoxin (Ruiz-Torres and Burmeister, 1972), similarly have been found to have more favourable absorption properties than *Digoxin*.

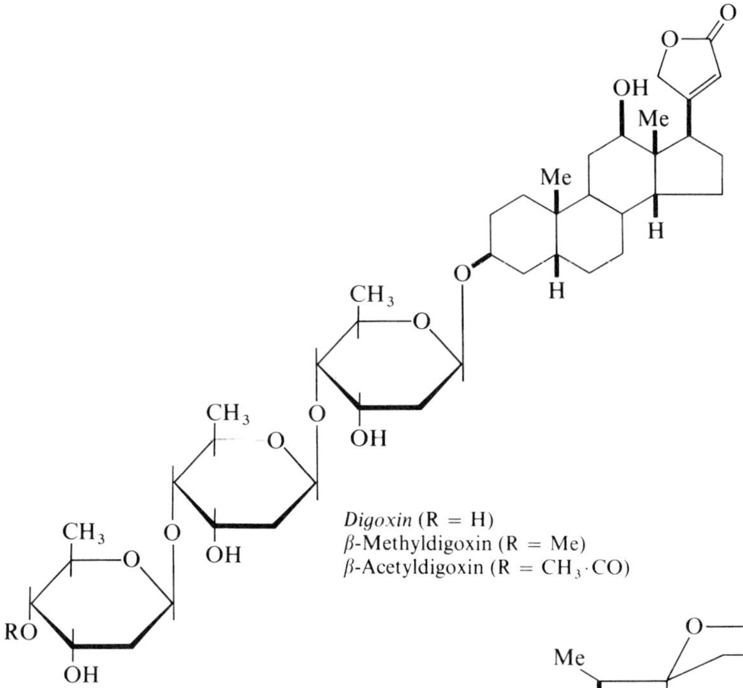

*Digoxin* (R = H)
β-Methyldigoxin (R = Me)
β-Acetyldigoxin (R = $CH_3 \cdot CO$)

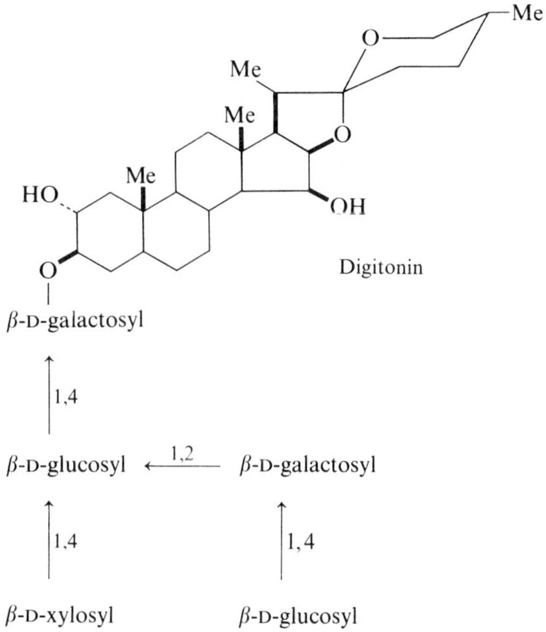

Digitonin

β-D-galactosyl ↑ 1,4 β-D-glucosyl ← 1,2 β-D-galactosyl

β-D-glucosyl ↑ 1,4 β-D-xylosyl

β-D-galactosyl ↑ 1,4 β-D-glucosyl

**Steroidal Saponins**

A large number of steroidal saponins of plant origin are known. They are polyglycosides, with high water solubility derived from the hydrophilic properties of the polysaccharide chain, high detergent activity, and foaming capability. Acidic hydrolysis cleaves the polysaccharide chain, with release of the steroidal sapogenin, which is characterised by the possession of a di-oxaspirostane group bridging the C-16 and C-17 positions of the steroidal nucleus. Steroidal sapogenins include digitogenin from digitalis species, diosgenin obtained from the Mexican yam, and hecogenin obtained from sisal. Digitonin is strongly haemolytic, but forms an insoluble 1:1 complex with cholesterol and 3$\beta$-hydroxysterols which is devoid of haemolytic activity. Diosgenin and hecogenin are used in the synthesis of *Progesterone* and *Cortisone*.

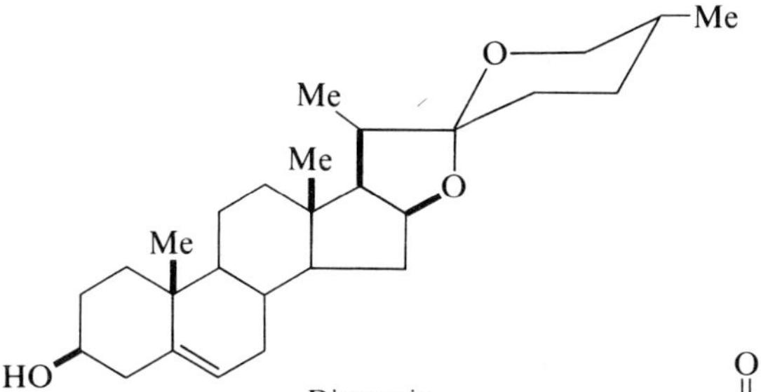

Diosgenin

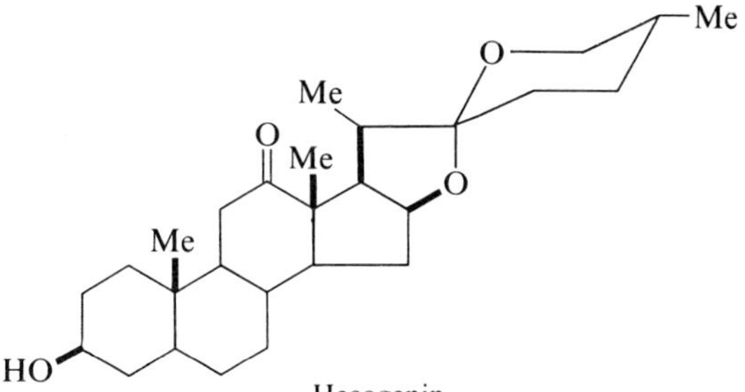

Hecogenin

## Steroid Biosynthesis

### Biosynthesis of Cholesterol

With the exception of insects, which require a dietary supply of preformed cholesterol, plants and animals obtain the bulk of their steroids by biosynthesis. In humans, small amounts of cholesterol are taken up from the diet, but cholesterol levels are mainly determined by the level of biosynthesis. Biosynthesis, and hence plasma cholesterol levels, are depressed when the subject is on a diet rich in linoleic, arachidonic and other poly-unsaturated fatty acids (Wagener, Mai and Schettler, 1963).

Biosynthesis of cholesterol is centred in the liver and follows a common pathway from mevalonic acid which is derived either from the acetate pool (Wright, 1961; Brodie, Wasson and Porter, 1963) via acetoacetyl-$S\cdot\overline{\text{CoA}}$ (Chapter 11) or the amino acid, leucine, as in the scheme on page 752.

Bloch and his collaborators established that $^{14}$C-labelled acetate is incorporated in cholesterol *in vivo*, and also in experiments with liver tissue slices (Bloch, Borek and Rittenberg, 1946), that both acetate carbons are incorporated,

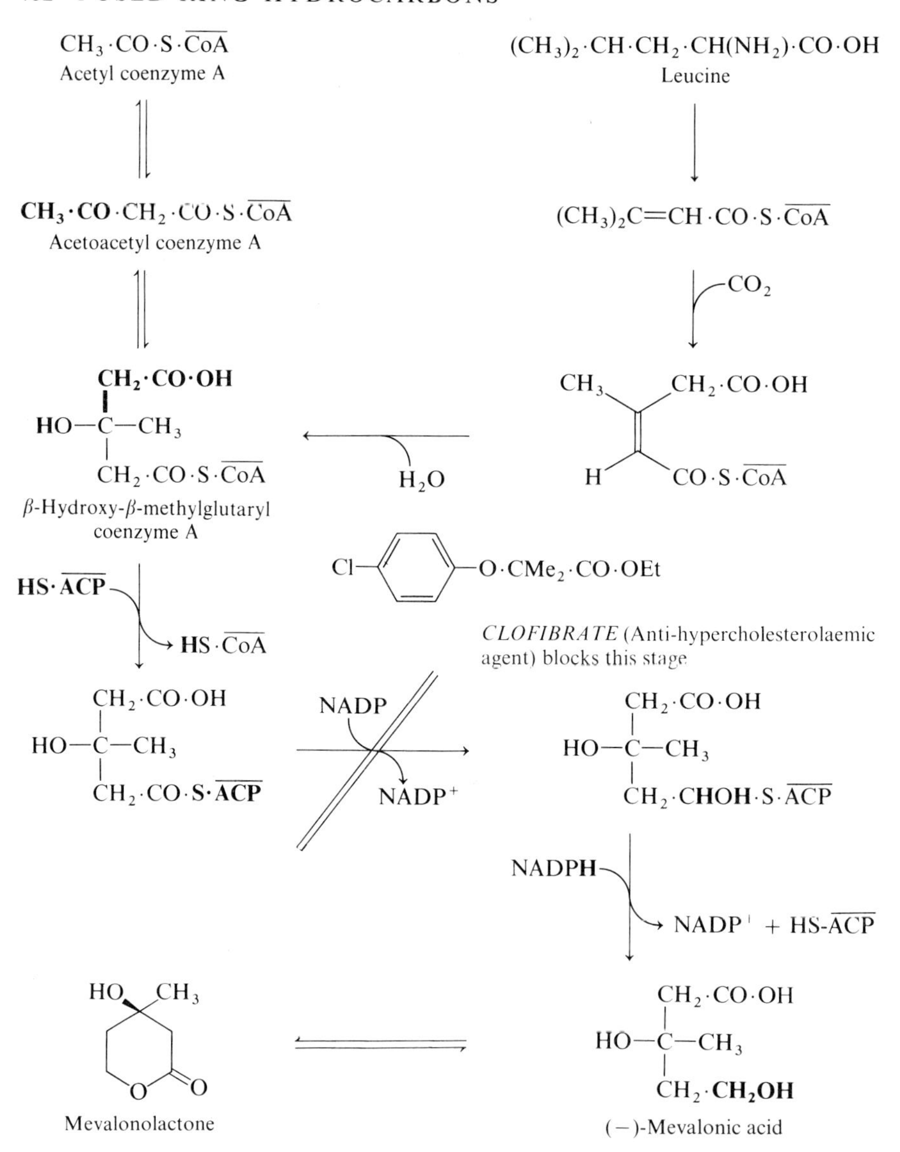

ACP = Acyl carrier protein

and that the entire cholesterol molecule is derived from acetate (Little and Bloch, 1950). It is now established that the route to cholesterol in mammalian tissue is via the triterpene hydrocarbon, squalene (Langdon and Bloch, 1953), and its 2,3-epoxide by a pattern of cyclisation, which accounts for the derivation of both lanosterol and cholesterol (Woodward and Bloch, 1953). Thus, it has been shown that lanosterol, biosynthesised in rat, is converted to cholesterol (Schneider, Clayton and Bloch, 1957).

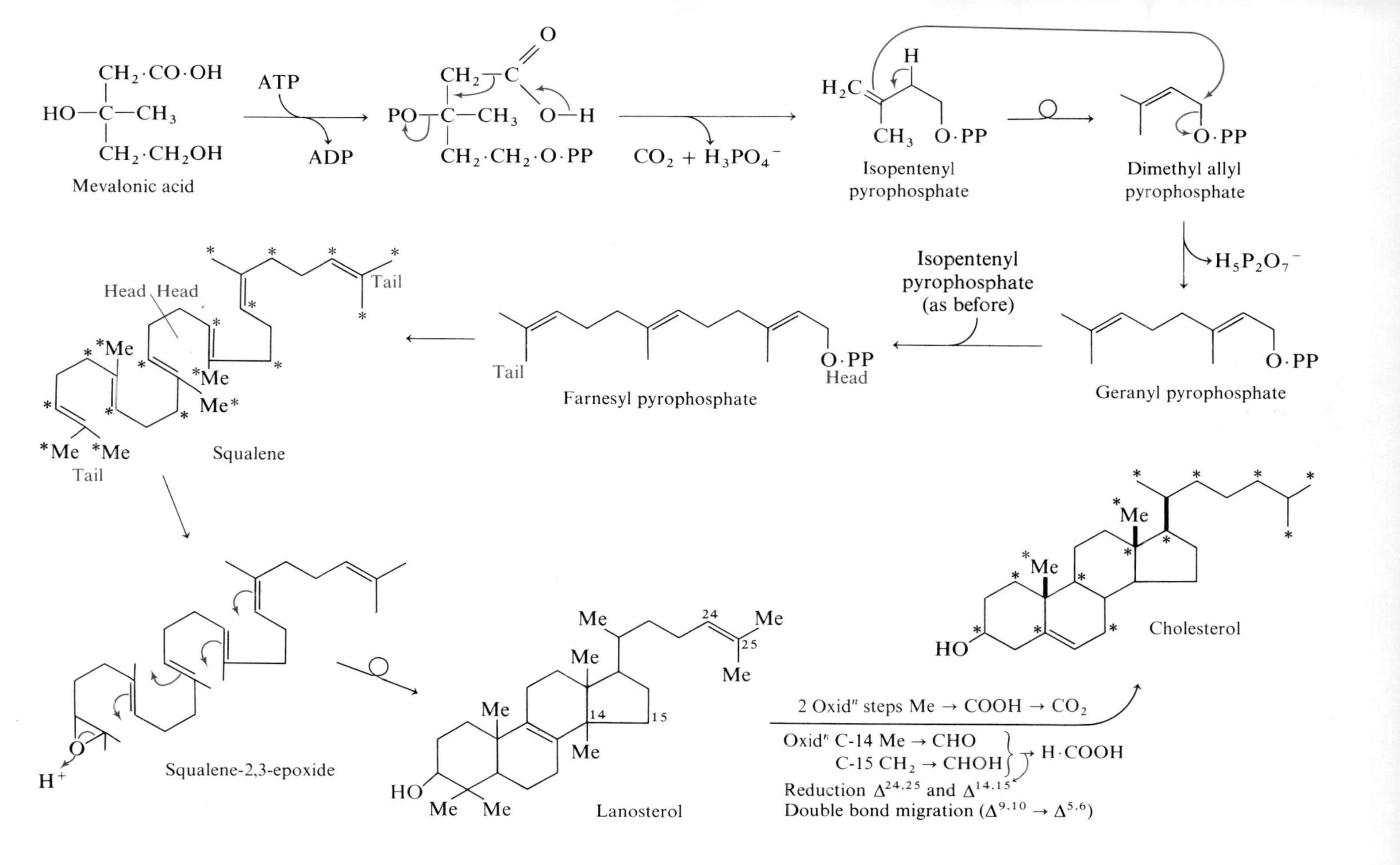

Mevalonic acid
ATP
ADP
$CO_2 + H_3PO_4^-$
Isopentenyl pyrophosphate
Dimethyl allyl pyrophosphate
$H_5P_2O_7^-$
Geranyl pyrophosphate
Isopentenyl pyrophosphate (as before)
Farnesyl pyrophosphate
Tail
Head
Squalene
Squalene-2,3-epoxide
Lanosterol
Cholesterol
2 Oxid$^n$ steps Me → COOH → $CO_2$
Oxid$^n$ C-14 Me → CHO
C-15 $CH_2$ → CHOH
H·COOH
Reduction $\Delta^{24,25}$ and $\Delta^{14,15}$
Double bond migration ($\Delta^{9,10} \rightarrow \Delta^{5,6}$)

Degradation of both squalene (Cornforth and Popják, 1954) and cholesterol (Cornforth, Hunter and Popják, 1953; Cornforth, Gore and Popják, 1957) obtained biosynthetically from 2[$^{14}$C]-labelled acetate ($CH_3 \cdot COOH$) has confirmed that cyclisation occurs with the labelling pattern shown.

### Biosynthesis of Natural Steroid Hormones

The biosynthesis of cholesterol is central to the formation of all other steroids, a process which is centred mainly, but not exclusively, in the adrenal cortex.

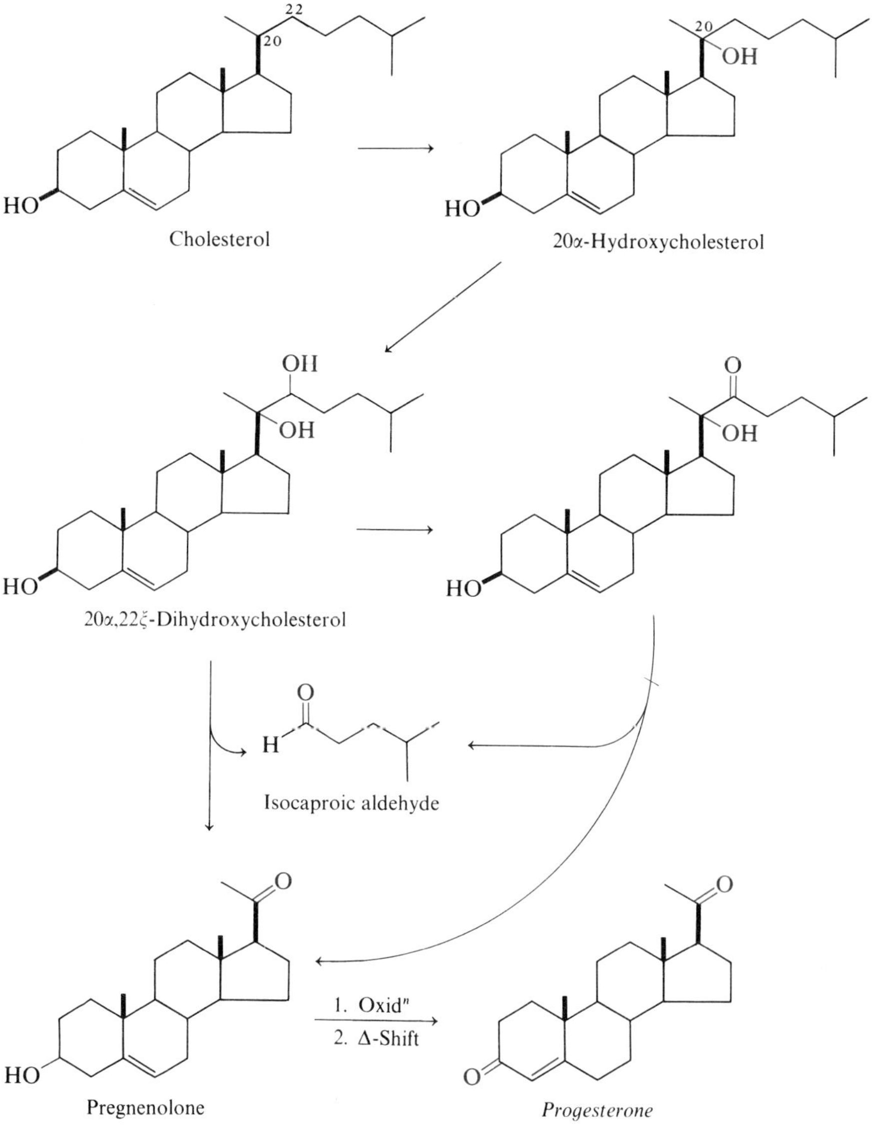

22
20
HO
Cholesterol

20
OH
HO
20α-Hydroxycholesterol

OH
OH
HO
20α,22ξ-Dihydroxycholesterol

O
OH
HO

O
H
Isocaproic aldehyde

O
HO
Pregnenolone

1. Oxid$^n$
2. Δ-Shift

O
O
*Progesterone*

The essential step in the biosynthesis of hormonal steroids from cholesterol is the oxidative cleavage of the C-17 side-chain to give pregnenolone and progesterone. Steroid biosynthesis in the adrenal cortex requires molecular oxygen, and utilisation of either NADH or NADPH by the oxidase enzymes involved. The output of adrenal steroids is controlled from the pituitary gland by corticotrophin (ACTH). Cholesterol side-chain degradation proceeds by hydroxylation at C-20 and thence by the pathways shown on page 754.

Progesterone and 17α-hydroxyprogesterone are key intermediates in the biosynthesis of the corticosteroids, *Aldosterone*, *Deoxycortone* (Desoxycorticosterone) *Acetate* and *Cortisol* (Hydrocortisone), as indicated in the reaction scheme on pages 756 and 757. 17α-Hydroxyprogesterone is also the precursor of $\Delta^4$-androstene-3,17-dione which affords both oestrogenic and androgenic hormones.

Biosynthesis of oestrogenic hormones from $\Delta^4$-androstenedione proceeds by the following pathway, via its 19-hydroxy derivative.

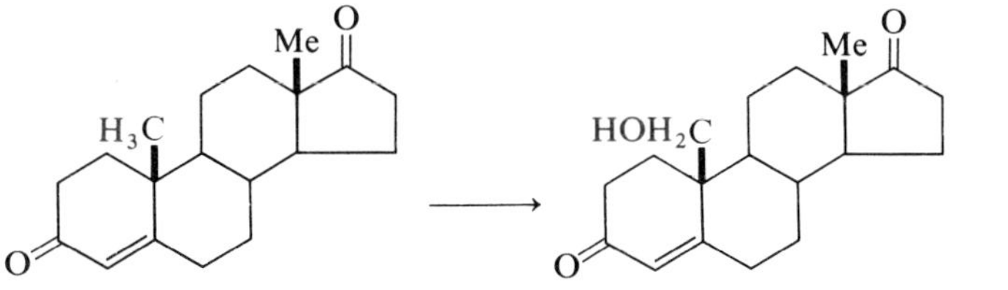

Me O H$_3$C O

$\Delta^{-4}$-Androstene-3,17-dione

Me O HOH$_2$C O

19-Hydroxy-$\Delta^4$-androstene-3,17-dione

Me O HO

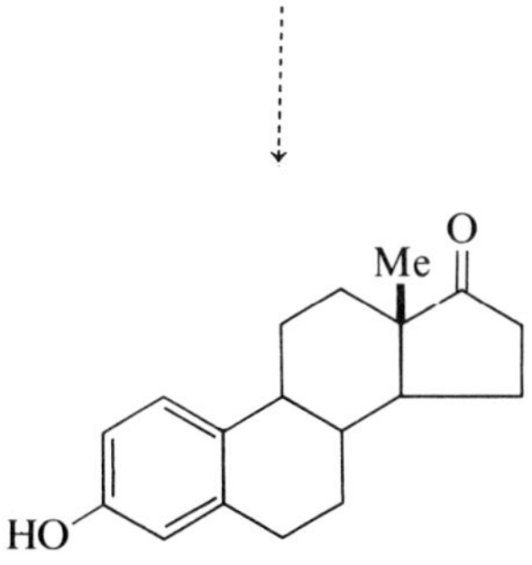

*Oestrone*

The actual mechanism for elimination of C-19 is not precisely established, but 19-hydroxy-$\Delta^4$-androstene-3,17-dione is a vinylogous aldol, and in the presence of base can undergo retro-aldolisation with formation of formaldehyde and a 3,4,5,6-dienol which can be readily dehydrogenated to the corresponding phenol, as shown on page 758.

It is not surprising in view of the centrol rôlc of cholesterol in steroid metabolism that certain metabolic disorders lead to abnormally high cholesterol levels in the body. Blood cholesterol levels in healthy adults range from about

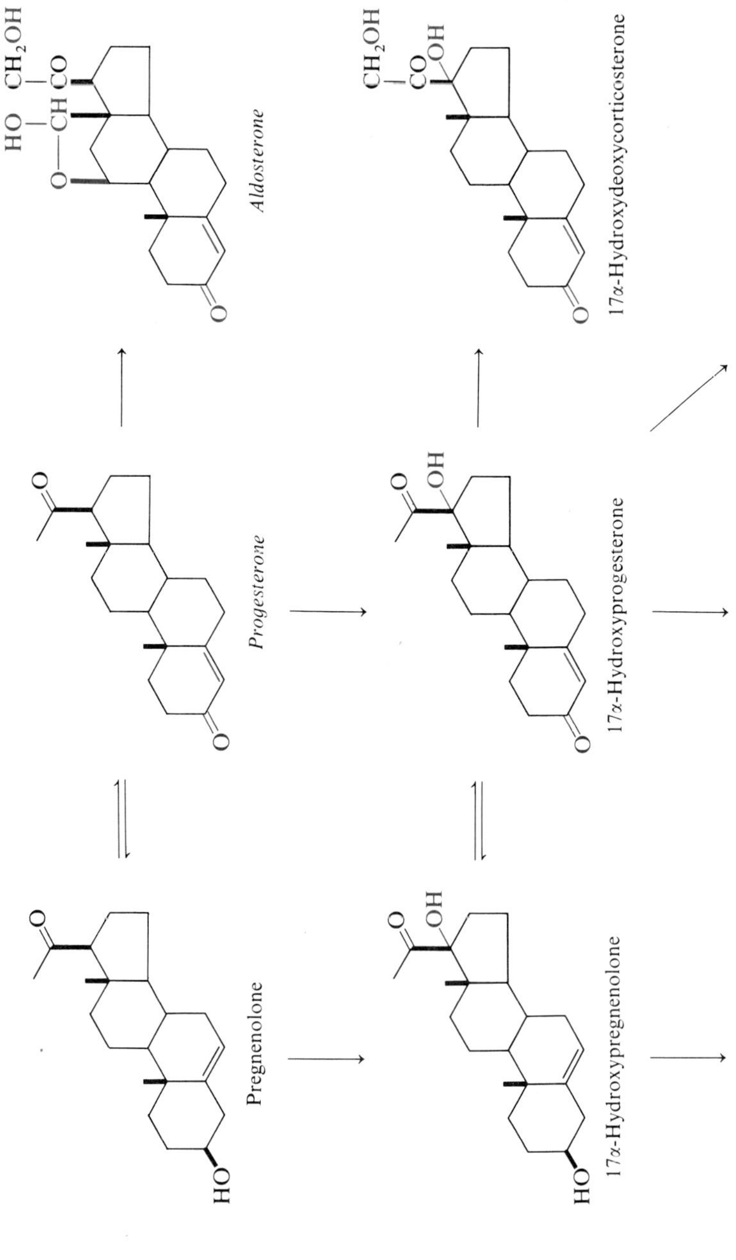

Pregnenolone

*Progesterone*

*Aldosterone*

17$\alpha$-Hydroxypregnenolone

17$\alpha$-Hydroxyprogesterone

17$\alpha$-Hydroxydeoxycorticosterone

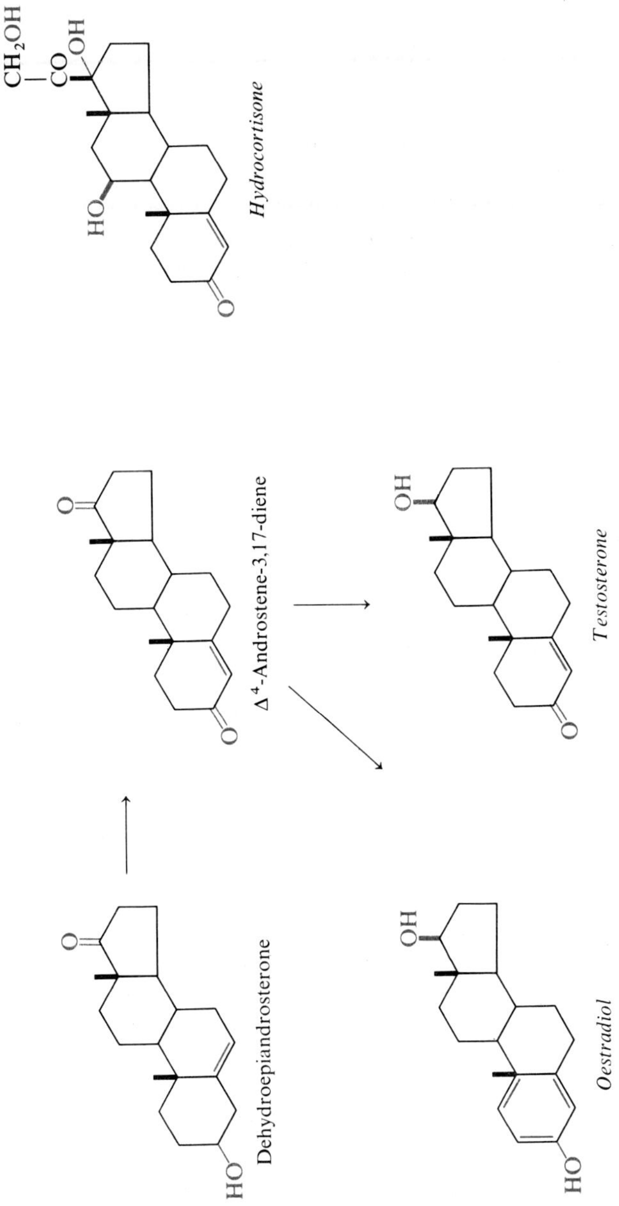

Dehydroepiandrosterone

$\Delta^4$-Androstene-3,17-diene

*Hydrocortisone*

*Oestradiol*

*Testosterone*

HO

H O CH$_2$

O

$H_2O + HCHO$

O

HO

175 to 225 mg/100 ml, but in certain diseases, notably atherosclerosis, the levels are raised significantly and actual deposition of cholesterol occurs. Blood cholesterol, which may be free or esterified with fatty acids, is significantly protein-bound; the sterol, but not its esters is also present in the erythrocytes. Blood cholesterol levels are markedly influenced by the level of saturated fatty acids in the diet, but much less by dietary cholesterol than is popularly supposed. Diets rich in unsaturated fatty acids aid the reduction of high serum cholesterol levels. The drug, *Clofibrate* (Atromid S), is also effective in the reduction of high serum cholesterol levels, by its action at an early stage of cholesterol biosynthesis in blocking the reduction of $\beta$-hydroxy-$\beta$-methylglutarate. This is the penultimate step in the biosynthesis of mevalonic acid, a key intermediate in cholesterol formation (p. 752). *Clofibrate* is activated by enzymic hydrolysis by plasma esterases to give the corresponding carboxylic acid, which is the effective blocking agent.

$$Cl-C_6H_4-O\cdot C(CH_3)_2\cdot CO\cdot O\cdot CH_2\cdot CH_3 \xrightarrow[\text{Esterase}]{H_2O} Cl-C_6H_4-O\cdot C(CH_3)_2\cdot CO\cdot OH + CH_3\cdot CH_2\cdot OH$$

*Clofibrate*

## Bile Acids

The bile acids are the second most abundant group of steroids in the human body. They are formed in the liver by hydroxylation of cholesterol, oxidation of the 3-hydroxyl, reduction of the 5,6-double bond to give a cis A/B ring junction, and side-chain oxidation. (See reaction pathway on page 759.)

The bile acids form amide conjugates with glycine and taurine. The alkali metal salts of these conjugates, e.g. sodium taurocholate and sodium glycocholate, are the bile salts, and are excreted as such in bile in which they are present at a concentration of about 6%. (See page 760.)

Bile salts are anionic surface-active agents consisting in physico-chemical terms of a large hydrocarbon fragment with polar and water-soluble groups at one side. Accordingly, they have powerful solubilising and emulsifying properties, which assist the absorption of fatty acids and other lipid materials (e.g. vitamins A and D) from the small intestine into the blood stream. They also promote the hydrolysis of fats, by emulsification which exposes the substrate (i.e. fat) to the action of lipase.

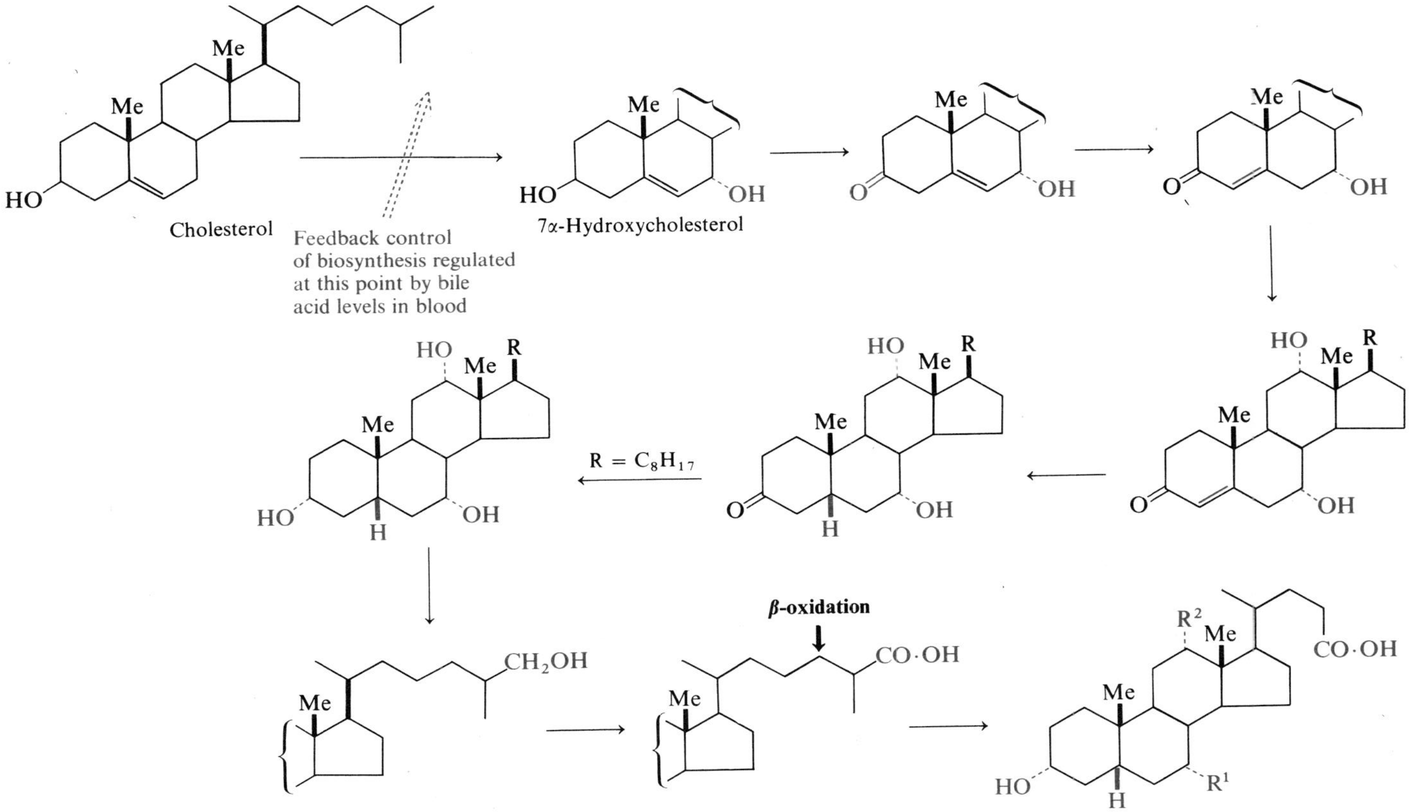

Cholesterol

Feedback control of biosynthesis regulated at this point by bile acid levels in blood

7α-Hydroxycholesterol

$R = C_8H_{17}$

**β-oxidation**

Lithocholic acid ($R^1 = R^2 = H$)
Desoxycholic acid ($R^1 = H$; $R^2 = OH$)

Chenodesoxycholic acid ($R^1 = OH$; $R^2 = H$)
Cholic acid ($R^1 = R^2 = OH$)

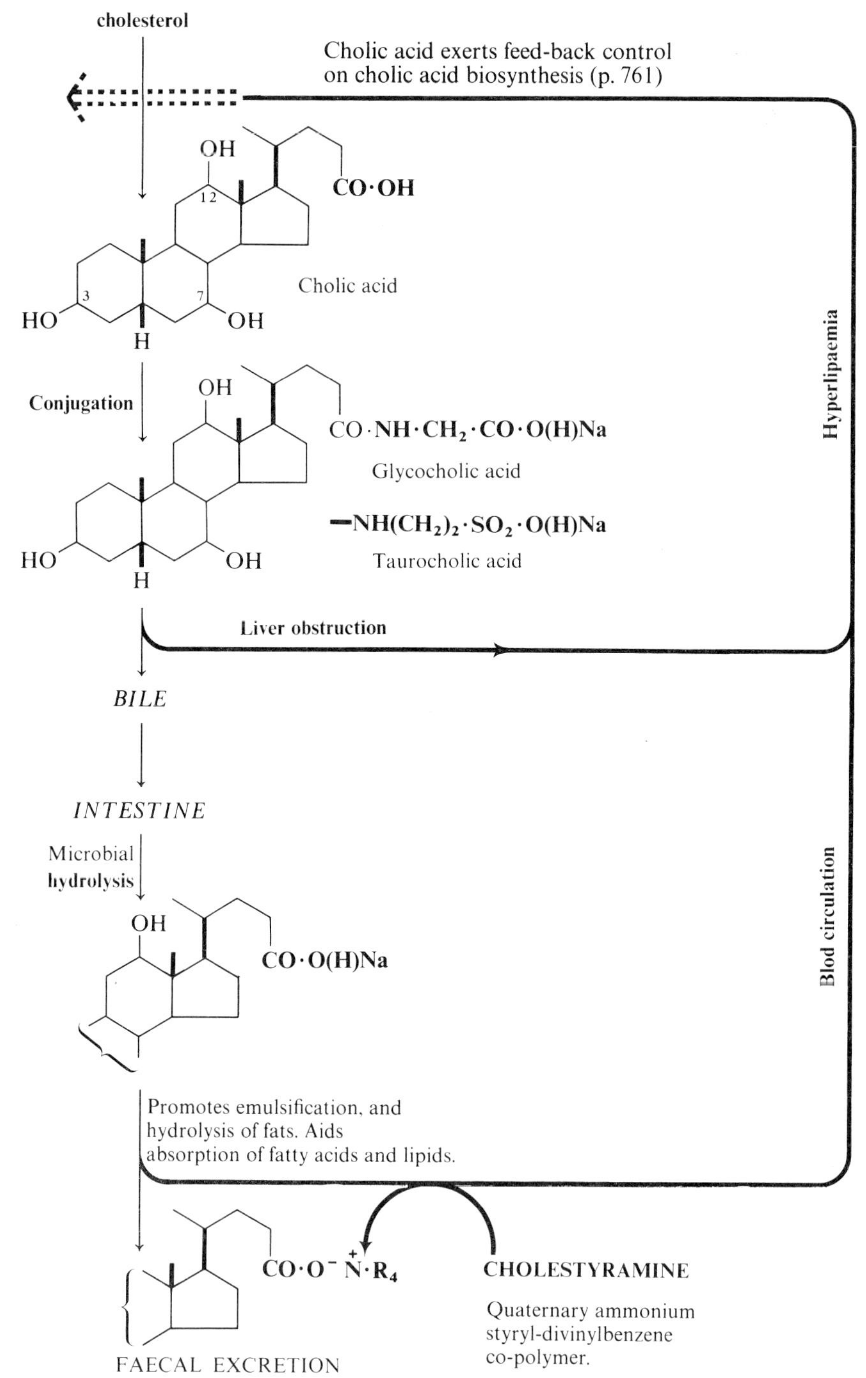
LIVER
cholesterol
Cholic acid exerts feed-back control
on cholic acid biosynthesis (p. 761)
OH
12
CO·OH
3
7
HO
OH
H
Cholic acid
Hyperlipaemia
Conjugation
OH
CO·NH·CH2·CO·O(H)Na
Glycocholic acid
—NH(CH2)2·SO2·O(H)Na
Taurocholic acid
HO
OH
H
Liver obstruction
BILE
INTESTINE
Microbial
hydrolysis
OH
CO·O(H)Na
Blod circulation
Promotes emulsification, and
hydrolysis of fats. Aids
absorption of fatty acids and lipids.
CO·O⁻ N·R4
CHOLESTYRAMINE
Quaternary ammonium
styryl-divinylbenzene
co-polymer.
FAECAL EXCRETION

The bile acid conjugates are also broken down in the intestine by microbial hydrolysis. The re-formed bile acids are normally re-absorbed, and exert control by a feed-back mechanism on cholic acid biosynthesis in the liver when plasma concentration reaches a high level (Mosbach, 1969).

Liver obstruction causes retention of bile salts in the blood. This results in jaundice (yellowing of the skin due to bile pigments). This is accompanied by accumulation of bile acids in the surface tissues, with irritation of the nerve endings and, in consequence, intense itching, known as pruritus. These effects may be alleviated, by oral administration of *Cholestyramine*, a quaternary ammonium cross-linked polystyryl-divinylbenzene co-polymer. Bile acids exchange for chloride, and are excreted ion-paired with the resin, thereby displacing bile acids from the blood and tissues. The overall effect is to enhance cholesterol catabolism, and so function indirectly as an anti-lipaemic agent.

## Physical Properties

### Solubility and Solubilisation

The substantially hydrocarbon skeleton, which characterises the typical steroid molecule, leads to compounds with low water solubility. This is rarely appreciable even in the presence of the hydrophilic oxygen functions which are typically present at C-3, C-11, C-17 and C-21. The presence of such substituents also usually limits the solubility of steroids in oils or fats, and appreciable solubility for administration in such media can usually only be achieved by conversion to an appropriate ester. Thus, oestradiol which has little oestrogenic activity when administered orally, can be given as its 3-monobenzoate, *Oestradiol* (Estradiol) *Benzoate* by intramuscular injection in solution in *Ethyl Oleate*. Likewise, *Hydrocortisone Acetate*, which is soluble in chloroform, is more suitable for formulation in ointment bases, such as *Wool Fat—Soft Paraffin*, 1:7 (Anhydrous Lanolin—White Petrolatum), than the parent alcohol, *Hydrocortisone*, which has very poor lipid solubility. More generally, the effect of substituents in the steroidal nucleus on partition between water and ether has been studied systematically (Flynn, 1971).

*Propyleneglycol*, polyethyleneglycols and other surfactants can be used to enhance the aqueous solubility of sterols (Güttman, Hamlin, Shell and Wagner, 1961; Sjöblom, 1958), but the most effective method of preparing aqueous solutions for injection is the formation of phosphate and hemisuccinate esters. Both *Betamethasone Sodium Phosphate* and *Hydrocortisone Sodium Succinate* are highly water-soluble, hygroscopic salts; both give rise to alkaline solutions, due to hydrolysis. Their use depends upon enzymic breakdown *in vivo* to the parent sterol. *Hydrocortisone Sodium Succinate* is readily hydrolysed by non-specific serum esterases, but the phosphate esters are much more stable chemically and only release the parent sterol in contact with the appropriate phosphatase enzymes (acid and alkaline phosphatase), though these are also present in the blood.

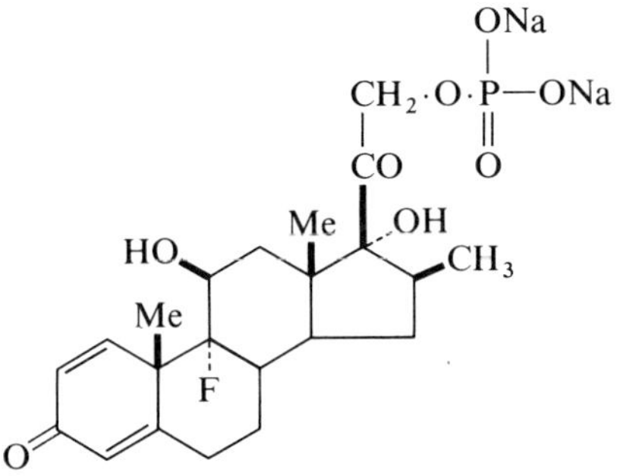

*Betamethasone Sodium Phosphate*

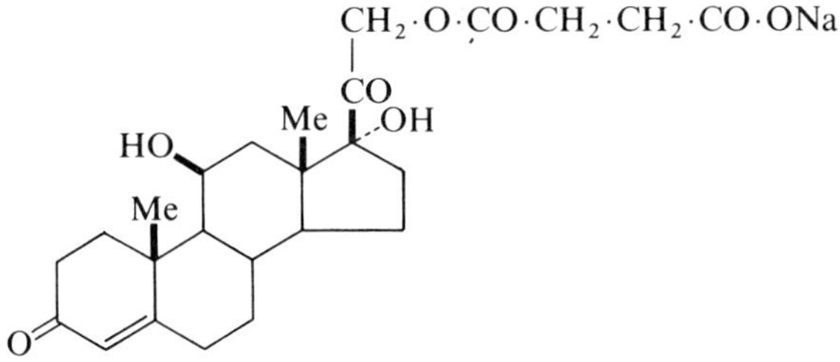

*Hydrocortisone Sodium Succinate*

A water-soluble, slow release oestradiol has also been produced in the form of a polyphosphate of high molecular weight.

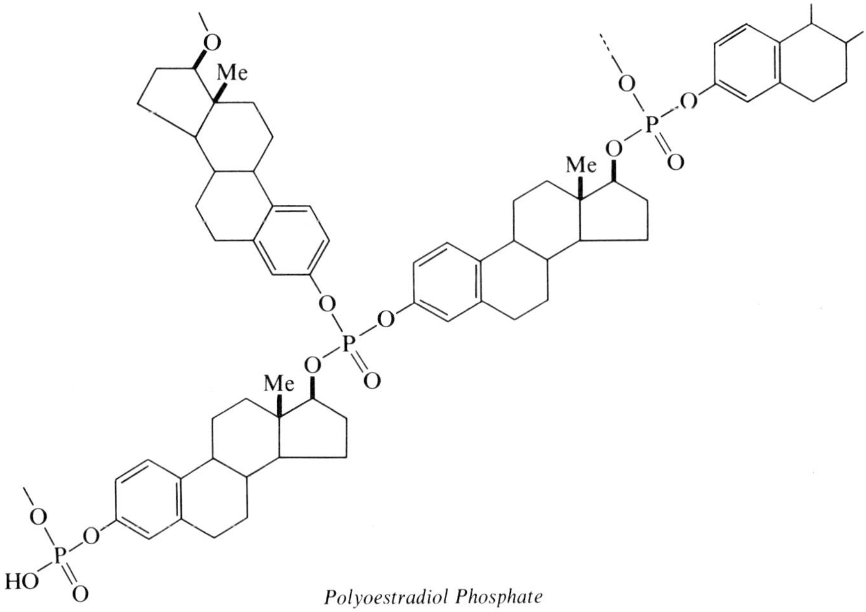

*Polyoestradiol Phosphate*

The physical state of steroids, particularly corticosteroids, can give rise to formulation problems. Thus, *Cortisone Acetate* is capable of existing in five different crystalline forms (Macek, 1954; Callow and Kennard, 1961) of which four of the forms present in the dry state are unstable and converted to an alternative form in contact with water. Unless suitable precautions are taken, microfine powder with a mean particle size of less than 5$\mu$ will convert rapidly in water to needle-shaped crystals up to 100$\mu$m in length in under 24 hr (Collard, 1961). This problem is now largely overcome in the manufacture of microfine *Cortisone Acetate*, either by milling in contact with water, or by crystallisation under carefully controlled conditions of temperature and concentration from a mixed (aqueous) solvent.

In some cases, advantage accrues from the use of low solubility crystal forms in depôt medication. Thus, certain microcrystalline forms of *Megestrol Acetate* give prolonged activity of up to six months after a single depôt injection (Helmreich and Huseby, 1965).

Biologically, the low water solubility of most steroids is unimportant, as most hormones are effective at extremely low concentration. Typical normal human plasma levels are given in Table 74. Turnover rates are, however, fairly high. The half-life of plasma oestrogens, for example, is estimated as 2–4 min. Medication in replacement therapy, in steroid treatment of other disorders, and in contraceptive pills, therefore requires administration at dose levels, which although still small in actual amount, are large by comparison with the tissue levels normally attained. Purines (Scott and Engel, 1957; Munck, Scott and Engel, 1957) and amino acids are known to increase the solubility of testosterone, progesterone and deoxycorticosterone, and may play a substantial rôle in their transport, and as receptor molecules for steroids *in vivo*.

Table 74. Human Plasma Levels of Steroid Hormones ($\mu$g/100 ml)

| Steroid | Men | Women (Normal) | Women (Pregnancy) 10th–42nd week |
|---|---|---|---|
| Aldosterone | 0.005–0.02 | 0.005–0.02 | 40–60 |
| Cortisol[1] | 15 | 18 | 1.29– 2.9 |
| Oestradiol | 0.015 | 0.01 –0.026 | 4.3 –17.5 |
| Oestriol | — | 0.025–0.037 | 2.7 –10.3 |
| Oestrone | 0.042 | 0.02 –0.70 | 6.4 –58.0 |
| Progesterone | 0.03 | 0.05 –1.52 | — |
| Testosterone | 0.7 | 0.03 –0.04 | — |

[1] Levels subject to diurnal variation. Figures quoted are morning values (minimal)

**Function in Phospholipid Membranes**

Cholesterol plays an important part in the structure of phospholipid membranes (Finean, 1953). It has been suggested, however, that only phospholipids containing double bonds interact with cholesterol and other sterols (Pethica, 1965).

This hypothesis finds support in the possible alignment in such membranes (Fig. 40) of fatty acid unsaturation with reactive steroid functional groups in, for example, the C-11 and C-21 positions, and is in accord both with the ideas of Willmer (1961) on the rôle of steroids in membrane structure and differentiation, and on the action of polyene antibiotics.

Willmer (1961) considers that differences in the action of major groups of physiologically active steroids may be explained by differences in the way they are packed into particular phospholipid membranes, and that cell-wall permeability is dependent on this factor. In support of this view, there is clear evidence that the extent of steroid–phospholipid interactions is markedly dependent on steroid structure. Gale and Saunders (1971) have shown that whereas introduction of a 17α-ethinyl group into testosterone causes a five-fold decrease in solubilisation by phosphatidylcholine, the same group introduced into oestradiol causes a five-fold increase in solubilisation. They relate the extent of solubilisation to the balance of polar characters between both ends of the molecule, which they express as coupled dipole moments derived by vector addition of calculated moments for rings A and D. This shows (Table 75) that solubility in water and solubilisation in lysophosphatidocholine decreases when the coupled dipole moment is reduced (*Testosterone* → *Ethisterone*), and that conversely, solubility and solubilisation is increased when the coupled dipole moment is increased (Oestradiol → *Ethinyloestradiol* (Ethinyl Estradiol).

Table 75. Calculated Dipole Moments and Solubilities of Steroids

| Steroid | Coupled Dipole Moment | | Solubility in Water ($\mu$mol $l^{-1}$ at 20°C) | Solubilisation in 4% Lysophosphatido-choline (mol/mol$^{-1}$) |
|---|---|---|---|---|
| Testosterone | (−)1.37 | $\Delta = -0.39$ | 84.6 | 0.113 |
| Ethisterone | ( )0.98 | | 1.92 | 0.0235 |
| Oestradiol | 1.26 | $\Delta = +0.54$ | 16.52 | 0.0636 |
| Ethinyloestradiol | 1.80 | | 34.41 | 0.262 |

**Interaction with Polyene Antifungal Antibiotics**

Polyene antibiotics, such as *Amphotericin B* and *Nystatin*, are active against yeasts and fungi, but not against bacteria. They are bound (reversibly) only to the cells of organisms sensitive to them (Lampen and Arnow, 1959), and then almost exclusively to the cell membrane. It is significant, however, that this group of antibiotics is only effective against organisms in which sterols are incorporated in the cell membrane, i.e. yeasts, fungi and protozoa, but not bacteria (Lampen, Arnow, Borowska and Laskin, 1962). This is illustrated particularly well in *Mycoplasma laidlawii* which is capable of growth in both the presence and absence of cholesterol. When cholesterol is present in the culture medium, it is incoporated into the cell membrane, and only then is the organism

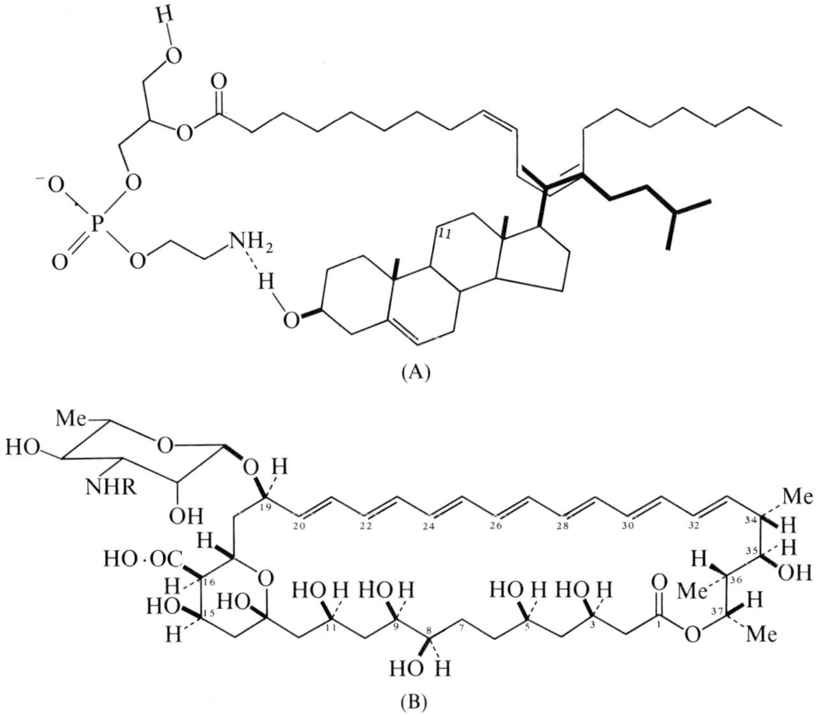

Fig. 40 A Phospholipid—steroid interactions in a typical phospholipid membrane
B Amphotericin B

sensitive to *Amphotericin B*. In sterol-free media, *Mycoplasma laidlawii* is resistant to the antibiotic (Feingold, 1965).

Polyene antibiotics bring about haemolysis of human erythrocytes in normal saline (Kinsky, 1963), an effect which fortunately is inhibited in whole blood and limited to pitting of the erythrocyte membrane. It is not established whether or not these effects are due to complexation with steroid components of the erythrocyte cell wall, but it undoubtedly accounts for the high toxicity and low margin of safety in the treatment of systemic fungal infections with such compounds as *Amphotericin B*. It is established, however, that polyene antibiotics interact preferentially with cholesterol in micromolecular lipid layers (Demel, Kinsky and van Deenen, 1965). *Amphotericin B* also causes marked swelling of egg–lethicin–sterol vesicles, which is dependent on the presence of the steroid (Bittman, Chen and Anderson, 1974).

## Chemical Properties

The steroidal skeleton in general exhibits a high degree of chemical and biochemical stability as befits a hydrocarbon structure. Thus, only minute amounts of $^{14}CO_2$ are expired by animals fed with skeletally-substituted $^{14}C$-steroids. Chemical reactivity, therefore, rests essentially in the substituents, which in general have the expected properties characteristic of the particular functional group(s). In some special cases, however, the geometry of the steroidal skeleton is such that the electronic and steric effects of one substituent can influence the properties of a second substituent at some apparently remote part of the molecule (Ringold, 1961).

**Oxidation–reduction**

All important natural and synthetic steroids possess oxygen functions, including alcoholic, phenolic and carbonyl groups. These show the expected properties. Thus, the phenolic (oestrogenic) steroids are acidic and soluble in alkali, whilst alcoholic hydroxyls are neutral, readily esterified unless sterically hindered, as for example in *Hydrocortisone Acetate*, and readily oxidised to the corresponding carbonyl compound. The oxidation of 11$\beta$-hydroxy to 11-ketosteroids in the solid state is both interesting and pharmaceutically important (Lewbart, 1969). Oxidation under these conditions is highly sensitive to the nature of the 17-hydroxy substituent, implying crystal lattice control of oxidation, since the same oxidation in solution shows no such sensitivity. Some 35% of the sterol is oxidised in light at room temperature in the course of 10 months.

The function of steroid hormones demands rapid turnover, so that reversible oxidation provides a useful means of shunting compounds between highly active alcoholic hydroxyl and less active ketonic forms (Grant, 1969).

$$\text{Cortisol} \rightleftharpoons \text{Cortisone}$$
$$\text{Testosterone} \rightleftharpoons \text{Androstenedione}$$
$$\text{Oestradiol} \rightleftharpoons \text{Oestrone}$$

The oestradiol–oestrone oxidation–reduction is now established as a co-enzyme function in placental transhydrogenase activity (Tomkins and Maxwell, 1963), in which hydrogen is transferred from NADPH to NADH.

NADPH → NADP⁺ ; NAD⁺ → NADH (ring D: Me, C=O ⇌ Me, OH, H)

This oestrogen-dependent transhydrogenase, also possesses dehydrogenase activity and is thermostable (Karavolas and Engel, 1966). It is also associated in placental tissue with a second thermolabile transhydrogenase which is similarly activated by oestradiol, but without concomitant oxidation to oestrone. Also, as might be anticipated, non-steroidal oestrogens have no effect on transhydrogenation, since they are entirely phenolic and hence incapable of undergoing comparable oxidation.

In contrast, reversible oxidation–reduction at C-17 cannot be an important requirement for androgenic activity, since both 17α-methyltestosterone and 17β-hydroxy-17α-methylandrostan-3-one are potent androgens, and being tertiary alcohols would undergo ring cleavage if oxidised. On the other hand, oxidation–reduction capability at C-3 could be significant in androgens, since both 3α,17β-androstanediol and its 3β-epimer are hormonally active, and equally significantly, 3β-methyl-3α,17β-androstanediol and 3α-methyl-3β,17β-androstanediol are inactive (Ringold, 1961).

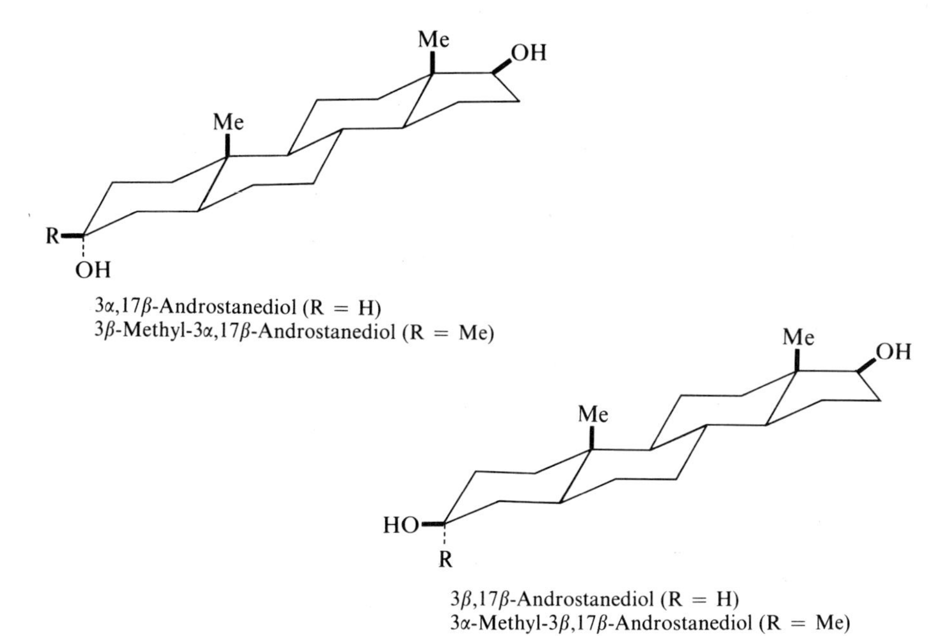


3α,17β-Androstanediol (R = H)
3β-Methyl-3α,17β-Androstanediol (R = Me)

3β,17β-Androstanediol (R = H)
3α-Methyl-3β,17β-Androstanediol (R = Me)

The greater hormone activity of 3α,17β-androstanediol compared to that of 3β,17β-androstanediol implies that if oxidation–reduction capability is essential for activity, this cannot be associated directly with the action of dehydrogenase enzyme activity solely at the α-face of the molecule (p. 736), since elimination of the 3β-hydrogen from the more active isomer demands enzymic approach to the west end of the molecule (Ringold, 1961). However, the known rôle of $NAD^+$ in such oxidations, and its molecular geometry, is not at all inconsistent with the concept of enzymic alignment at the α-face as shown on p. 770.

The rapid turnover of $\Delta^4$-3-ketosteroids in general is due to irreversible reductions of both the 3-carbonyl and 4-ethylenic group leading to hormonally-inactive products which are eliminated by urinary excretion, mainly as glucuronides. In the corticosteroids, reduction of the C-20 group also occurs following by oxidative cleavage of the entire C-17 side-chain (pp. 768–9).

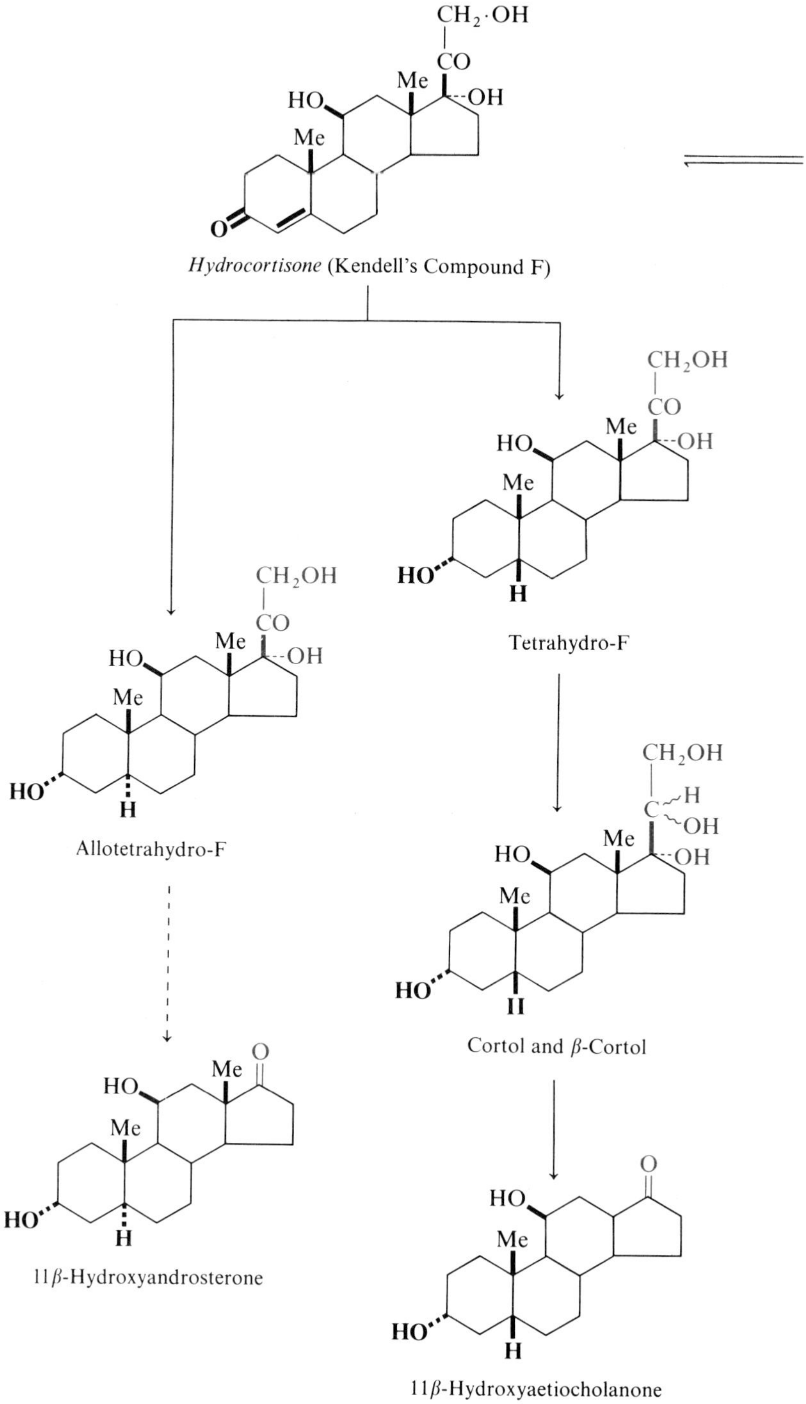

*Hydrocortisone* (Kendell's Compound F)

Tetrahydro-F

Allotetrahydro-F

Cortol and β-Cortol

11β-Hydroxyandrosterone

11β-Hydroxyaetiocholanone

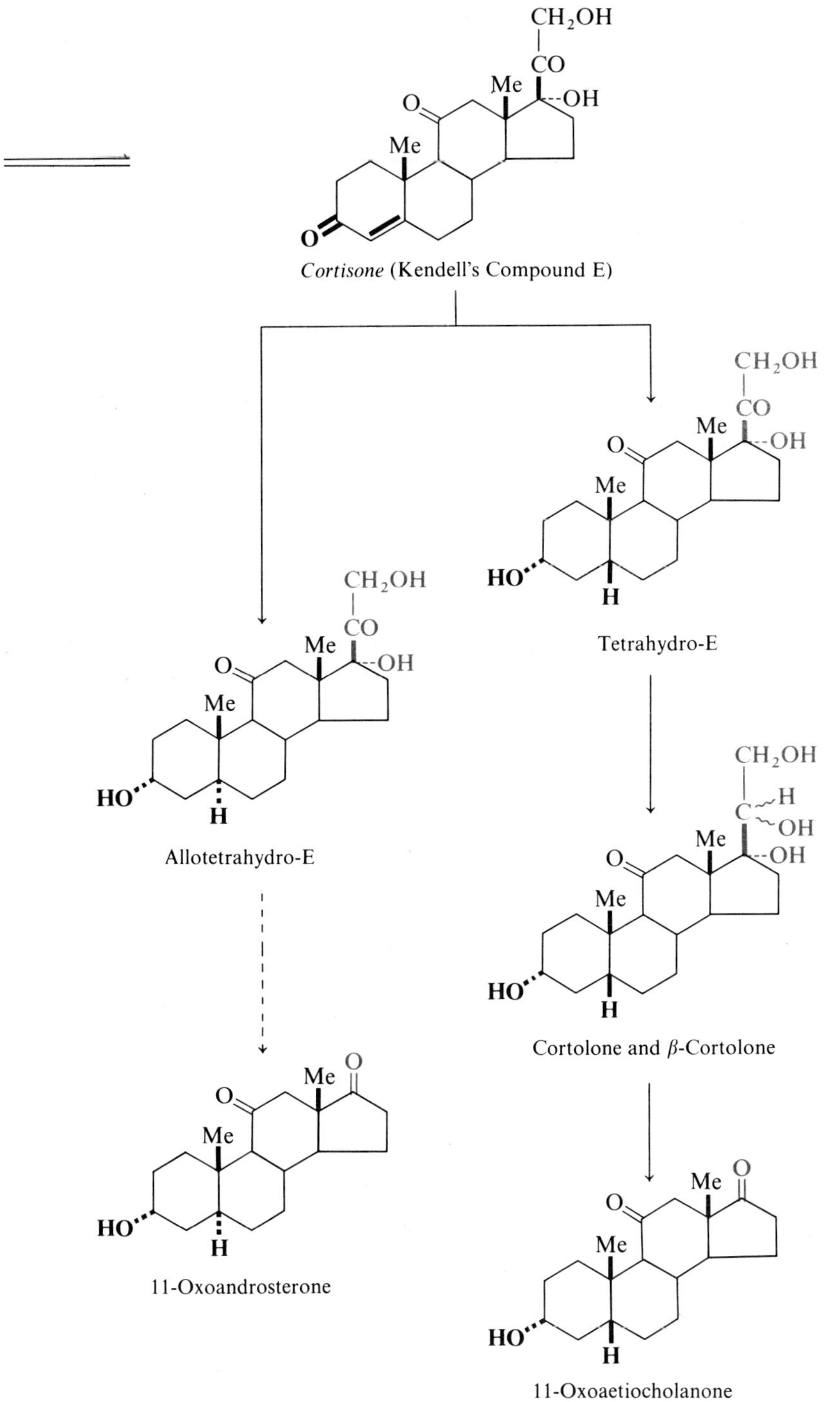

CH2OH
CO
Me
O
OH
Me
O

*Cortisone* (Kendell's Compound E)

CH2OH
CO
Me
O
OH
Me
HO
H

Tetrahydro-E

CH2OH
CO
Me
O
OH
Me
HO
H

Allotetrahydro-E

CH2OH
C
H
OH
Me
O
OH
Me
HO
H

Cortolone and β-Cortolone

O
Me
O
Me
HO
H

11-Oxoandrosterone

O
Me
O
Me
HO
H

11-Oxoaetiocholanone

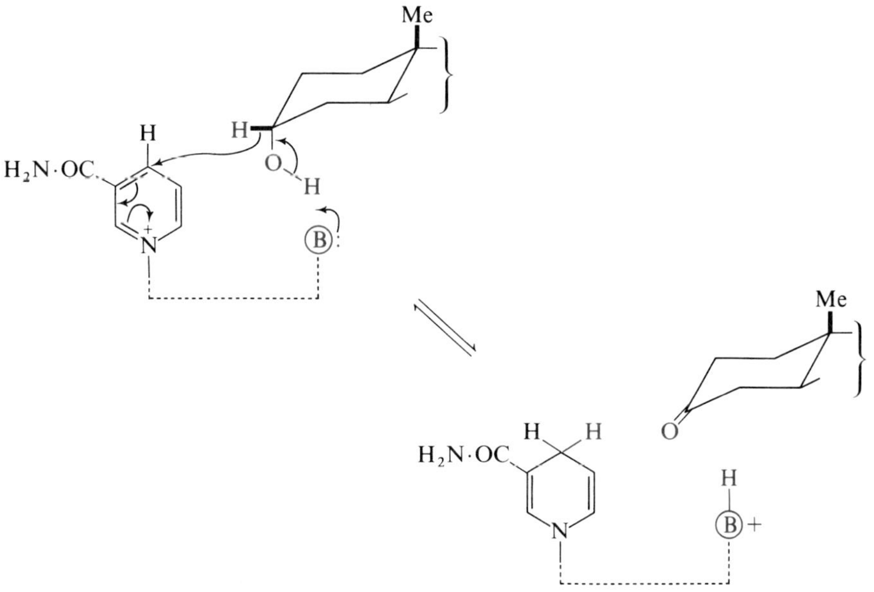

The oxidative instability of the C-20, 21 ketol of corticosteroid hormones, particularly under alkaline conditions, is an important factor in their handling, storage, and formulation, and has been studied in some detail (Velluz, Petit,

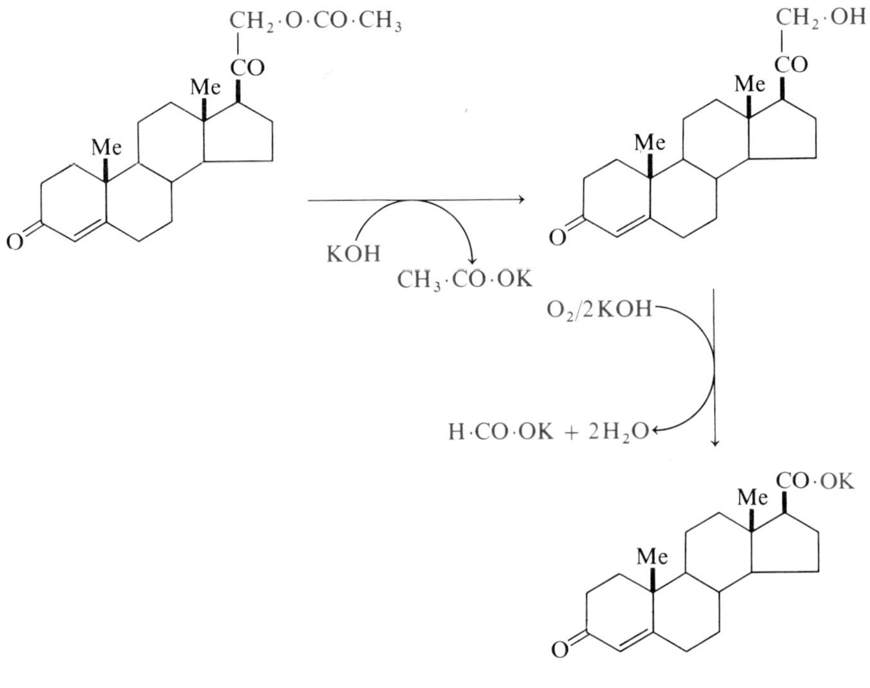

Pesez and Berret, 1947; Herzig and Ehrenstein, 1951). In the presence of oxygen, 20-keto-21-acetates are first hydrolysed to the ketol, which is then readily oxidised by the uptake of 1 mole of oxygen to the $C_{20}$ carboxylic acid.

Wendler and Graber (1956), however, have shown that even in the absence of oxygen, degradation of the 17-ketol side-chain of corticosteroids occurs readily with alkali to give two principal products. These are formed by retro-aldolisation of the equilibrium aldotriose, and by a saccharinic acid-type transformation of the intermediate keto-aldehyde.

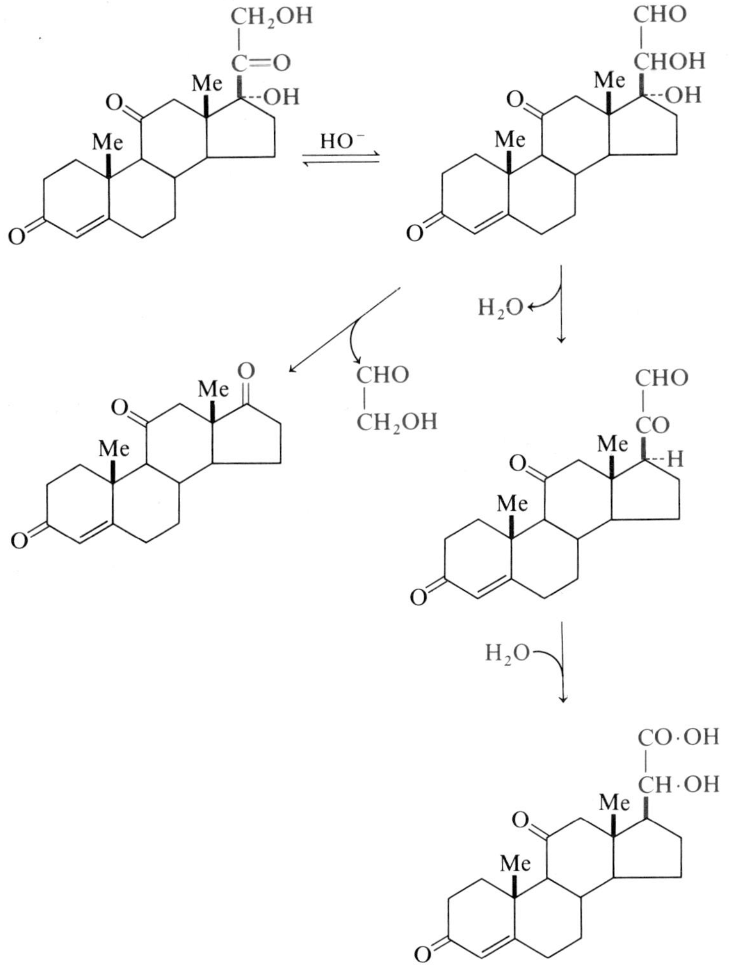

$CH_2OH$ | C=O; Me; --OH; O; Me; O; $HO^-$; CHO | CHOH; Me; --OH; O; Me; O

$H_2O$

O; Me; O; Me; O; CHO | $CH_2OH$; CHO | CO; Me; --H; O; Me; O

$H_2O$

CO·OH | CH·OH; Me; O; Me; O

These degradations are appreciable in dilute methanolic potassium hydroxide under nitrogen at room temperature within 30 min. Esterification of the C-21 hydroxyl group inhibits decomposition, provided hydrolysis of the ester group does not occur. They are relevant to the stability of corticosteroids when formulated with antacids, such as magnesium oxide, in preparations designed to inhibit peptic ulceration resulting from prolonged administration of corticosteroids. The problem, however, is largely overcome when the antacid is compressed around a steroid tablet core (Collard, 1961).

The corticosteroid side-chain undergoes catalytic dehydrogenation in the presence of traces of $Cu^{2+}$ ions, giving rise to the corresponding 21-dehydro-compounds (Lewbart and Mattox, 1959; Sunaga and Koide, 1968). Steroid carbonyl groups also exhibit typical condensation reactions, of which the stabilisation of the *Aldosterone* in the hemiacetal rather than the hydroxy-aldo form is of special interest.

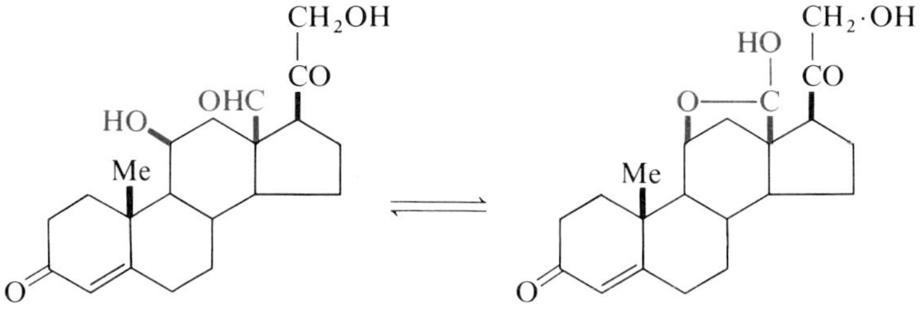

*Aldosterone*

*Aldosterone* is the natural steroid hormone responsible for the maintenance of mineral balance, and is about 50 times as active as *Deoxycortone* (11-deoxycorticosterone; Desoxycorticosterone) in the sodium retention test. Its mechanism of action is not fully understood, but it acts on the renal tubules to increase re-absorption of sodium. The ability of *Aldosterone* to react in the hydroxy-aldehydic form appears to be important, since the 11,21-diacetate has little sodium-retaining activity. The action of *Aldosterone* is antagonised by *Spironolactone*.

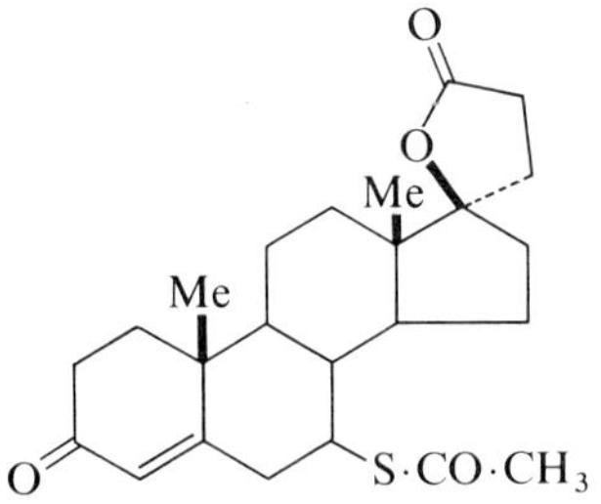

*Spironolactone*

**Oxidative Instability of Unsaturated Steroids**

On exposure to air, unsaturated steroids, such as cholesterol, are susceptible to oxidative addition at the double bond or, alternatively, at the allylic carbon

(C-7). Oxidative attack at the allylic position in cholesterol is used as the basis of a number of routes to 7,8-dehydrocholesterol, the key intermediate in the synthesis of natural vitamin D (Vitamin $D_3$).

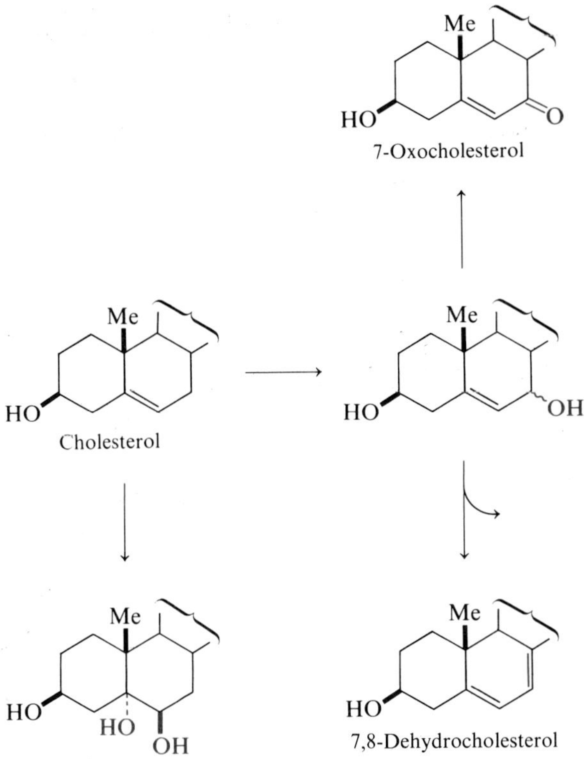

Me
HO
O
7-Oxocholesterol

Me
HO
Cholesterol

Me
HO
OH

Me
HO
HO
OH

Me
HO
7,8-Dehydrocholesterol

## Photochemical Reactions

The saturated hydrocarbon skeleton of steroids has the same degree of stability towards photochemically-activated attack as other similar and simpler hydrocarbons. Photochemical rearrangement and decomposition is only important when particularly reactive groupings are present. The ring A $\Delta^{1,4}$-dienone system, which is characteristic of the synthetic corticosteroids, *Prednisone*, *Prednisolone*, *Triamcinolone*, *Betamethasone* and *Dexamethasone*, is particularly sensitive to ultraviolet light (Barton and Taylor, 1958). Thus, *Prednisone Acetate* is readily converted to the inactive lumiprednisolone acetate on exposure to light in neutral ethanolic solution. In acidic solution (aqueous acetic acid), photodecomposition takes a different course resulting in an alternative rearrangement of rings A and B. (See reaction scheme on page 773.)

Photochemically-induced rearrangements are also important in the synthesis of vitamins D and aldosterone.

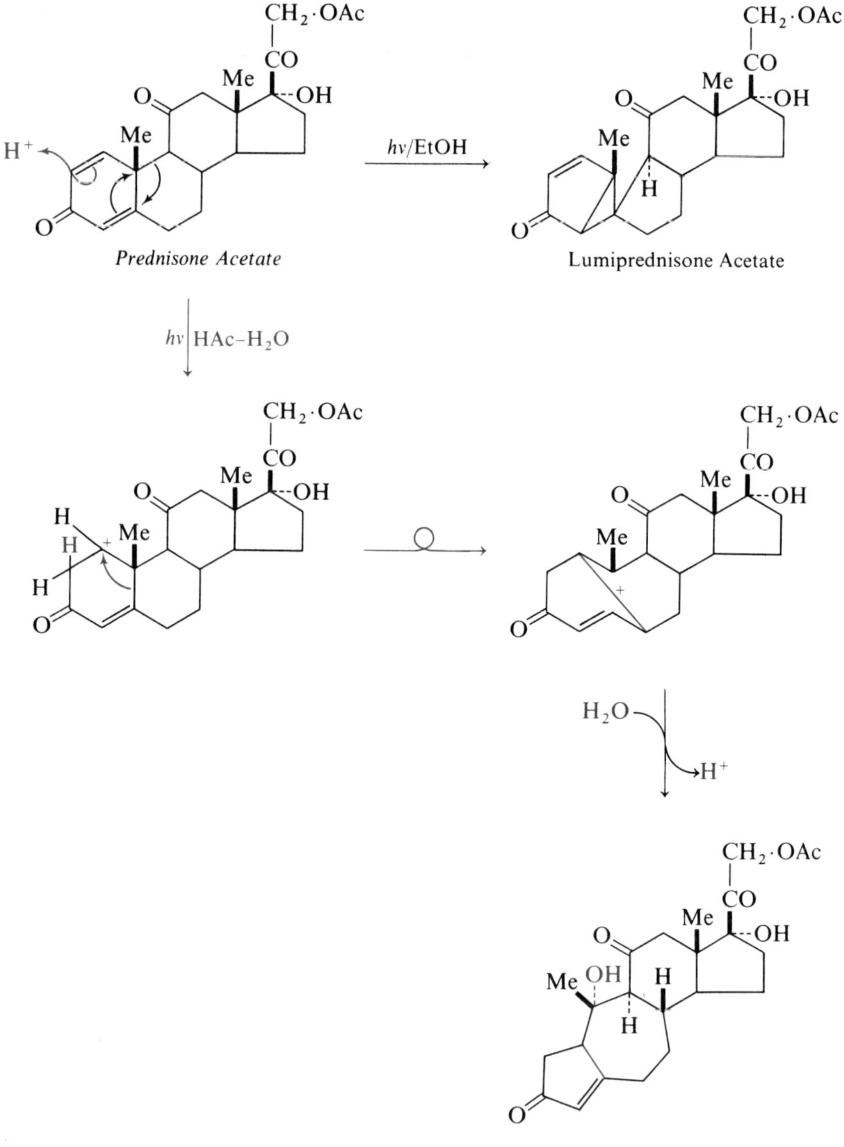

*Prednisone Acetate*

$h\nu$/EtOH

Lumiprednisone Acetate

$h\nu$ HAc–$H_2O$

$H_2O$

$H^+$

**Aldosterone Synthesis**

The generation of a free radical by light-catalysed decomposition of the labile corticosterone acetate 11-nitrite has been used as a means of promoting intramolecular attack on the adjacent C-18 methyl group leading to the synthesis of *Aldosterone Acetate* by the following route (Barton and Beaton, 1960; Barton, Beaton, Geller and Pechet, 1960). The C-19 methyl group, which is also adjacent, leads to some formation of a second product.

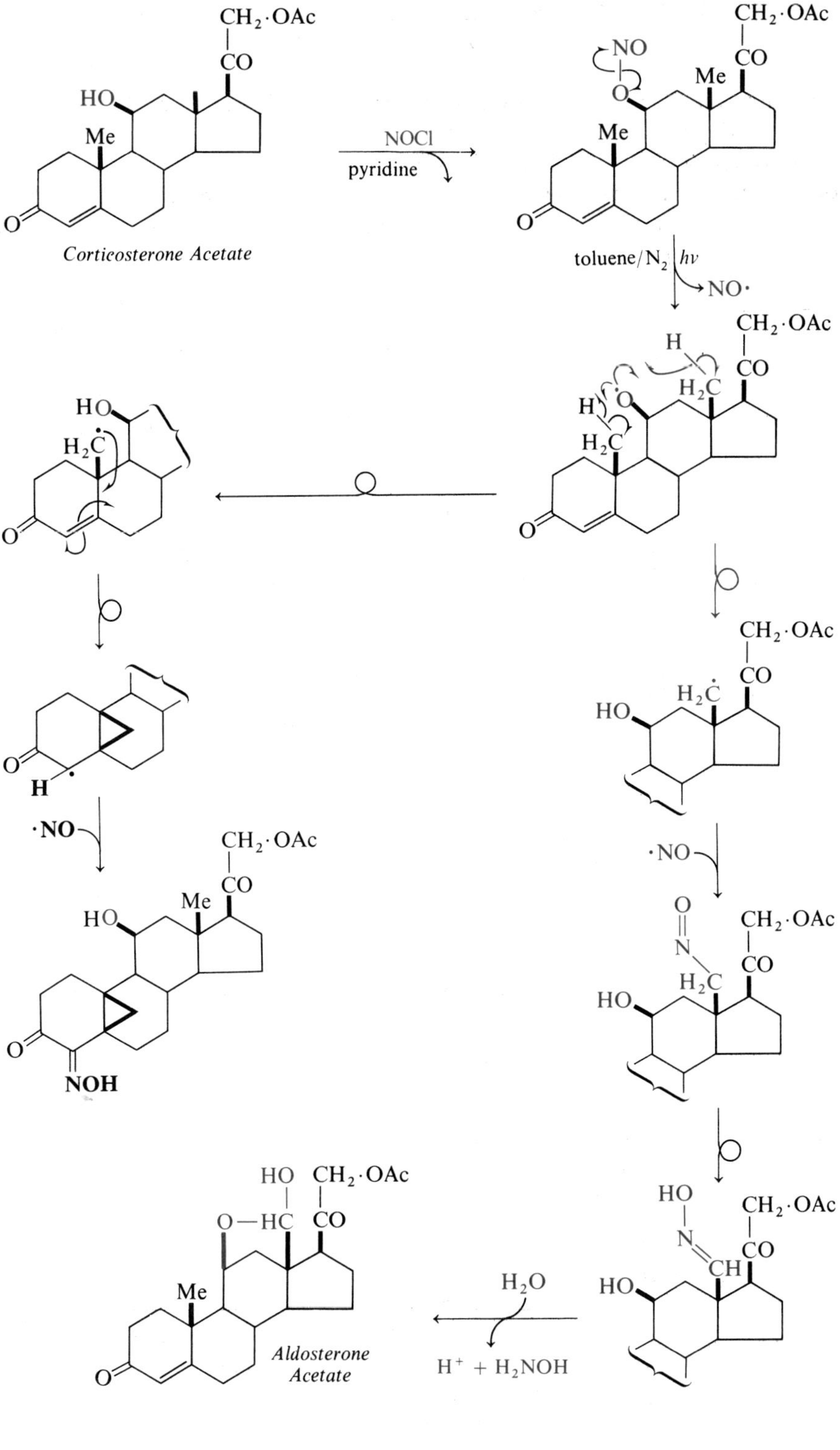
CH2·OAc
CO
HO
Me
O
Corticosterone Acetate
NOCl
pyridine
NO
O
toluene/N2 hν
NO·
H
H2C
·NO
NOH
HO CH2·OAc
O—HC CO
Aldosterone Acetate
H2O
H+ + H2NOH
CH

## VITAMINS D

The vitamins D consist of a closely-related group of 9,10-secosteroid-5,7,10(19)-trienes derived by photosynthesis from appropriate steroid precursors in man (*Cholecalciferol*; Vitamin $D_3$), in yeast (*Ergocalciferol*; Vitamin $D_2$) and other organisms. All are 5,6-*cis*-compounds and only those compounds with this stereochemistry have antirachitic (Vitamin D) activity.

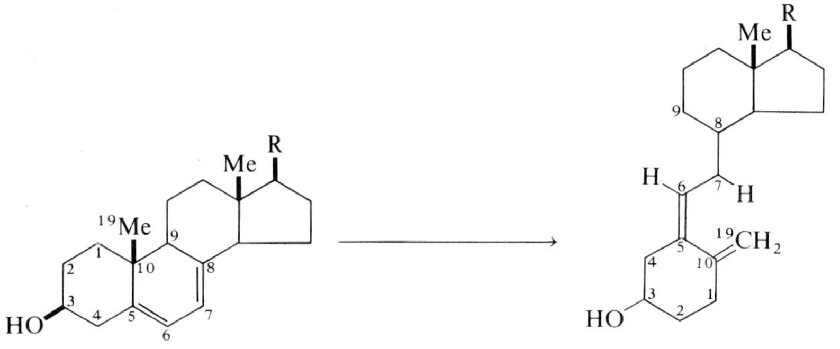

R =

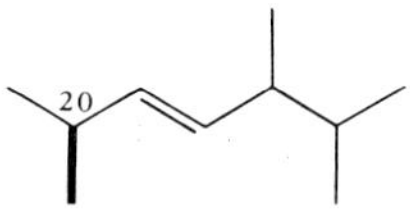

Ergosterol — Ergocalciferol (Vitamin $D_2$)

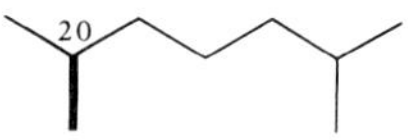

7-Dehydrocholesterol — Cholecalciferol (Vitamin $D_3$)

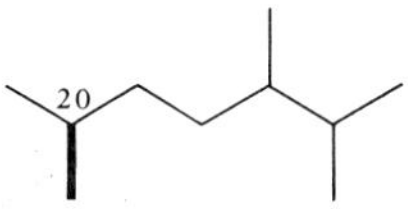

22-Dihydroergosterol — Dihydroergocalciferol (Vitamin $D_4$)

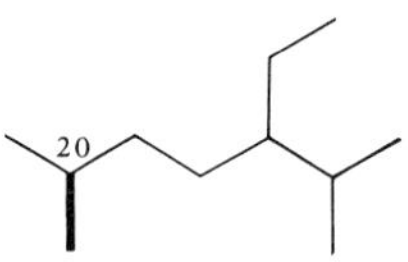

Vitamin $D_5$

The human body obtains its normal requirements of vitamins D either by direct absorption from food or endogenously by biosynthesis from steroid precursors. Biosynthesis from cholesterol occurs in two phases, of which the first consists of dehydrogenation in the intestinal mucosa to 7,8-dehydrocholesterol (Glover, Glover and Morton, 1952). The diene system so formed in ring B is sensitive to ultraviolet irradiation from sunlight, and in the second phase, 7,8-dehydrocholesterol undergoes photosynthetic conversion in the skin to *Cholecalciferol* (Vitamin $D_3$).

Irradiation of 7,8-dehydrocholesterol raises the molecule with its doubly allylic 9,10-bond, first to the excited singlet state with spin conservation, and thence by spin inversion and loss of energy to the excited triplet state. The latter then undergoes transformation to pre-calciferol via dihydronorlumisterol and the excited triplet state of *cis*-dihydronortachysterol. (See scheme pp. 778–9.)

It appears from detailed studies of the corresponding conversion of ergosterol to *Ergocalciferol* (vitamin $D_2$), which is carried out synthetically on a manufacturing scale, that the final step in the conversion of pre-calciferol to calciferol is a thermal process. It is not clear, however, how this step is achieved biosynthetically.

*Ergocalciferol* is sensitive to light, and over-irradiation in manufacture or unnecessary exposure to light on storage destroys vitamin activity as a result of its conversion to suprasterol-II (Dauben, Bell, Hutton, Laws, Rheiner and

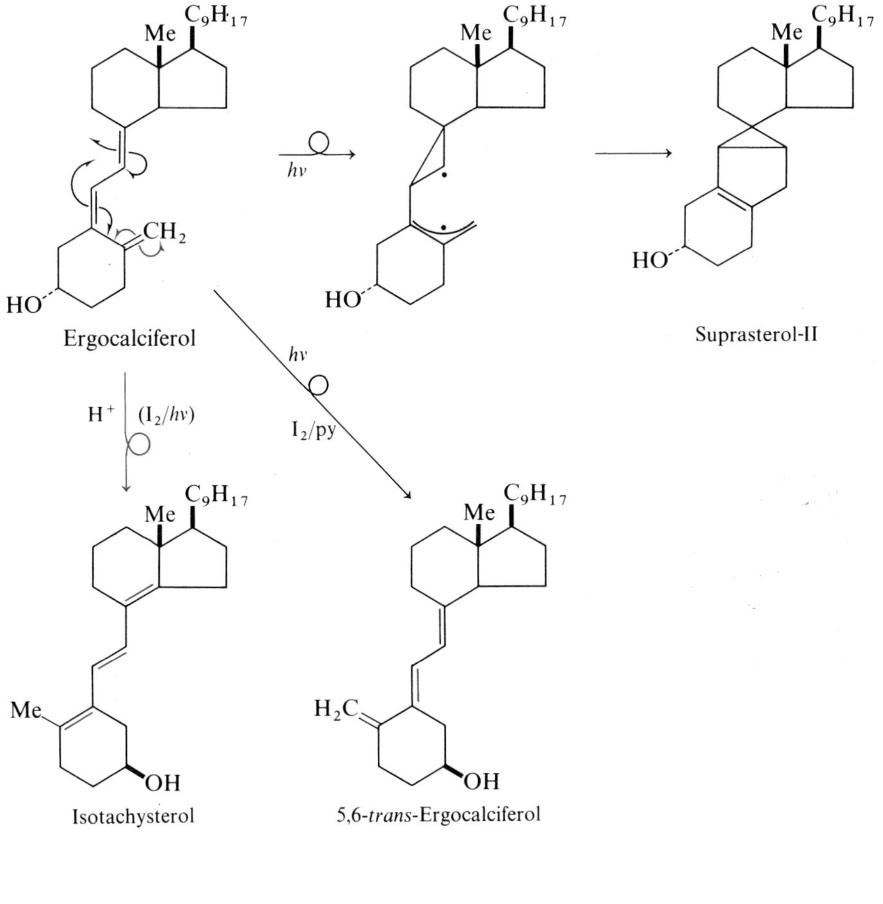

Me $C_9H_{17}$ HO $CH_2$ Ergocalciferol

$h\nu$ → Me $C_9H_{17}$ HO

→ Me $C_9H_{17}$ HO Suprasterol-II

$H^+$ ($I_2/h\nu$)

$h\nu$ $I_2$/py

Me $C_9H_{17}$ Me OH Isotachysterol

Me $C_9H_{17}$ $H_2C$ OH 5,6-*trans*-Ergocalciferol

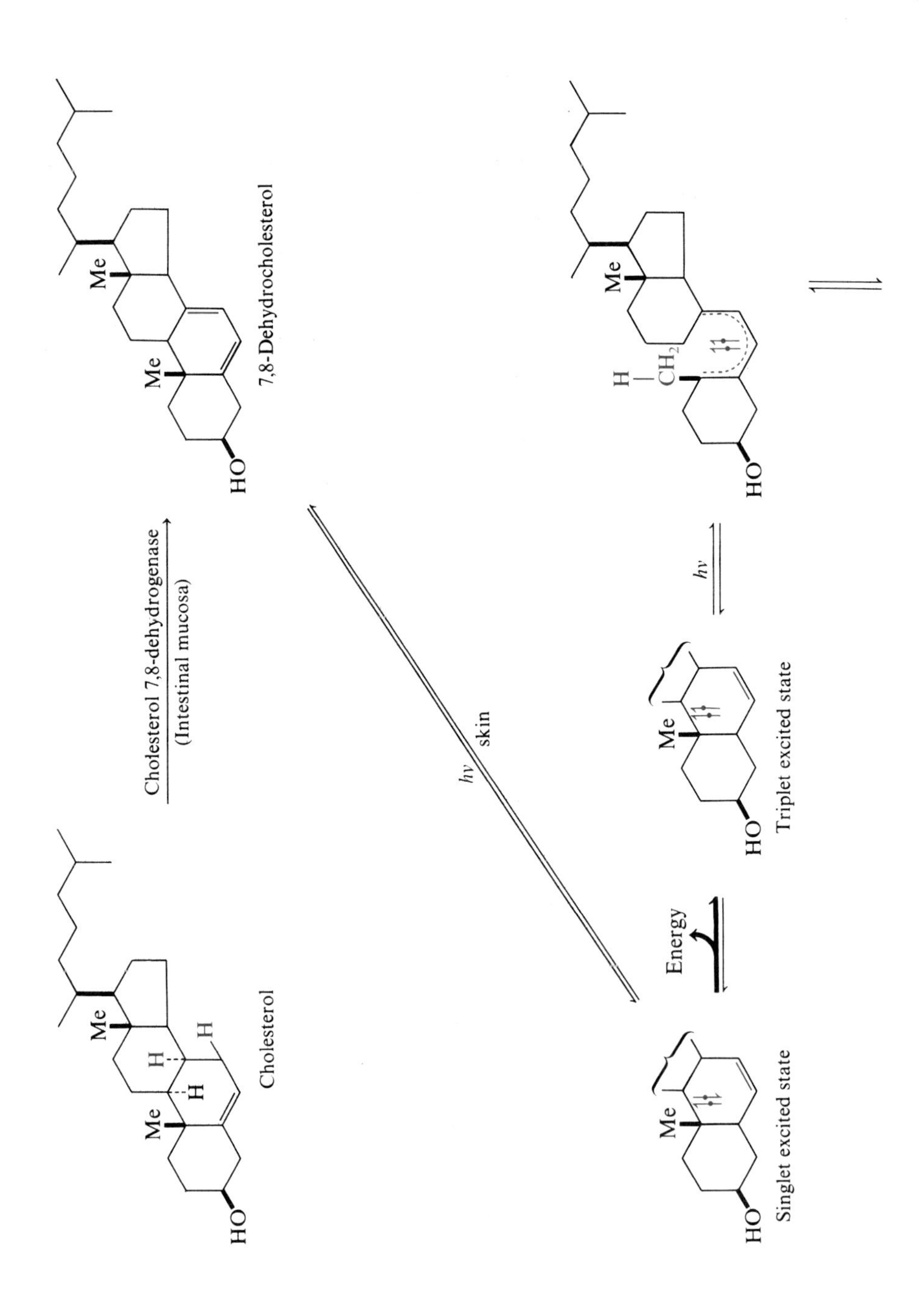
Cholesterol
Cholesterol 7,8-dehydrogenase
(Intestinal mucosa)
7,8-Dehydrocholesterol
hν
skin
Singlet excited state
Energy
Triplet excited state
hν
Me
HO
H
$CH_2$

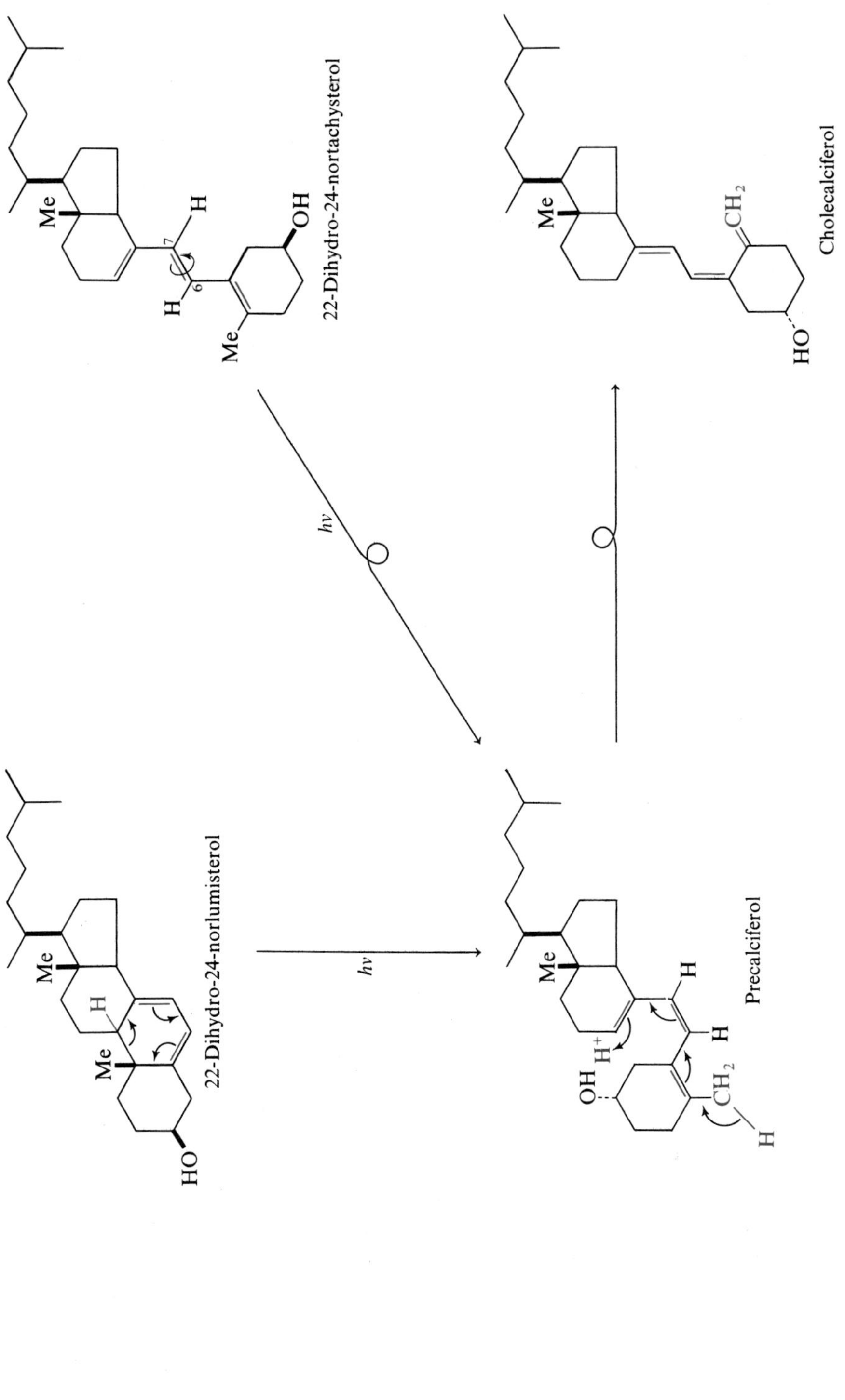

22-Dihydro-24-nortachysterol

$h\nu$

Cholecalciferol

22-Dihydro-24-norlumisterol

$h\nu$

Precalciferol

Urscheler, 1958; De Mayo and Reid, 1961). The vitamin is also sensitive to acid (Inhoffen, Brückner and Gründel, 1954), being converted to isotachysterol, and to iodine in daylight in the absence of acid (pyridine present) due to formation of 5,6-*trans*-ergocalciferol (Koevoet, Verloop and Havinga, 1955). Pre-ergocalciferol undergoes a similar transformation under the latter conditions with a comparable cis-trans rearrangement of the 6,7-double bond.

**Metabolic Activation of Vitamin D**

The action of vitamin D in the treatment of rickets arises primarily from its ability to increase the intestinal absorption of calcium (Nicolaysen, 1937, 1943),

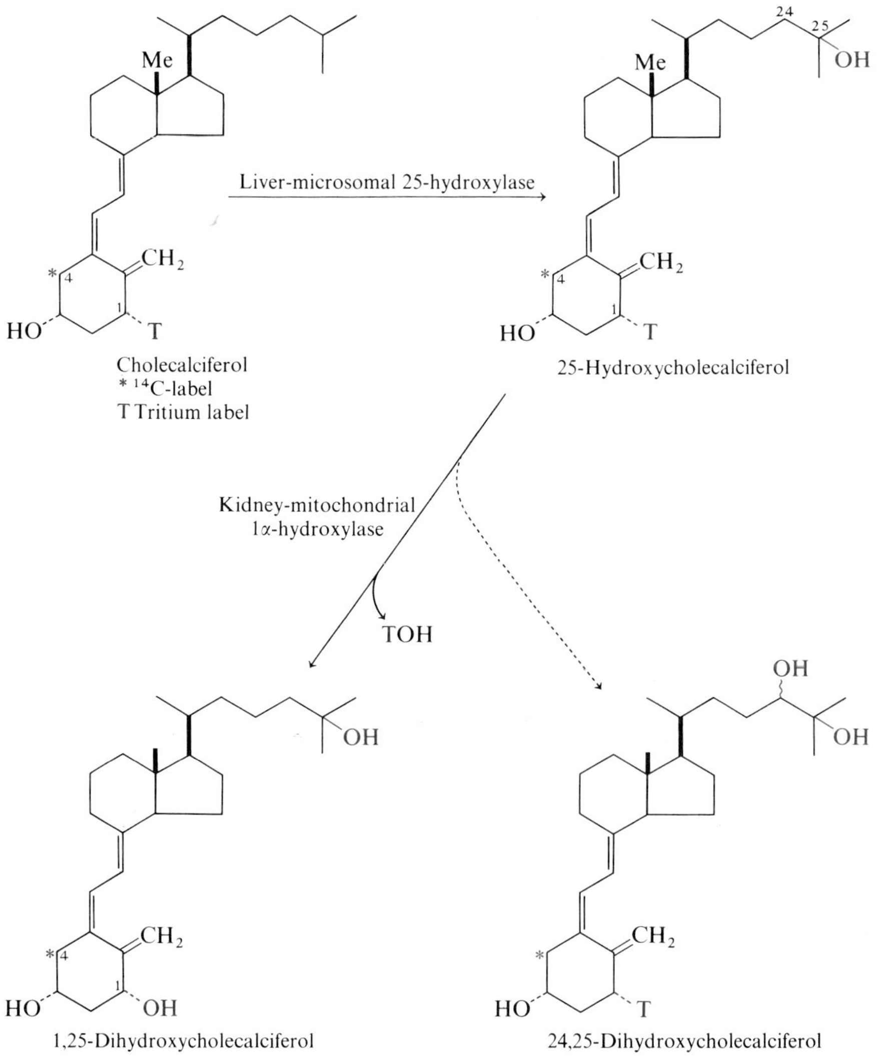

Cholecalciferol
* $^{14}C$-label
T Tritium label

Liver-microsomal 25-hydroxylase

25-Hydroxycholecalciferol

Kidney-mitochondrial
1α-hydroxylase

TOH

1,25-Dihydroxycholecalciferol

24,25-Dihydroxycholecalciferol

though it also acts directly on the calcium metabolism of bone. Metabolism in the liver and in kidney is essential for the activation of the vitamins (Kodicek, 1974). This is evident in the time lag of some 8 hr between administration of vitamin D and onset of activity. Metabolic activation consists of two distinct hydroxylation steps in liver and kidney respectively, with the vitamin and metabolites transported from the skin and from organ to organ in association with an $\alpha_2$-globulin.

The first step in the liver is the hydroxylation of *Cholecalciferol* to 25-hydroxycholecalciferol (Blunt, DeLuca and Schnoes, 1968; DeLuca, 1969). This has been shown to have about 1.4 times the antirachitic activity of *Cholecalciferol* in the rat.

Studies with *Cholecalciferol*, doubly labelled with $^{14}C$ at C-4 and $^{3}H$ at C-1 in rachitic chicks have shown that the kidney also plays a vital rôle in the activation of vitamin D (Lawson, Wilson and Kodicek, 1969; Kodicek, 1974). Loss of the C-1 tritium label with concurrent retention of the C-4 carbon-14 label in the resulting dihydroxy metabolite establishes the product as 1$\alpha$,25-dihydroxycholecalciferol.

Radiochromatograms of lipids, extracted from key tissues in the doubly-labelled *Cholecalciferol* studies, show the sharp decrease in the ratio of $^{3}H$ to $^{14}C$ radioactivity in liver, kidney and intestinal tissue homogenate in the fraction corresponding to the formation of 1$\alpha$,25-dihydroxycholecalciferol. They also show that an increasingly significant proportion of the total radioactivity is concentrated in the 1$\alpha$,25-dihydroxycholecalciferol fraction from liver (27% total radioactivity as 1$\alpha$,25-dihydroxycholecalciferol), kidney (36%), 800g-supernatant from intestinal tissue homogenate (68%) and cell nuclei from intestinal tissue supernatant (97%). The vital rôle of the kidney in the activation process is clearly established in studies involving nephrectomy in which the second hydroxylation step is inhibited.

1$\alpha$,25-Dihydroxycholecalciferol has 3–5 times the activity of 25-hydroxycholecalciferol in the transfer of $^{45}Ca$ across the intestinal wall. It is, however,

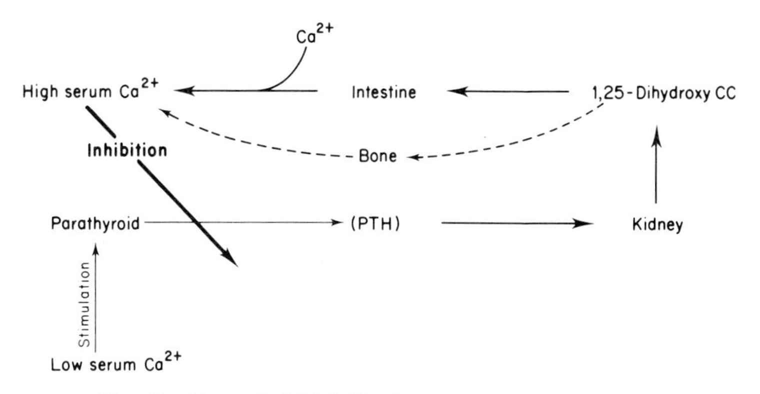


Fig. 41 Control of 1,25-dihydroxycalciferol biosynthesis

comparatively short-lived with a biological half-life of about 6–8 hr. Biosynthesis is under hormonal control from parathyroid hormone, since parathyroidectomy significantly reduces synthesis, and subsequent administration of parathyroid extract restores it. In normal animals, control of biosynthesis is effected through the release of parathyroid hormone, this being enhanced when serum calcium levels are low, and inhibited when serum calcium levels are high (Fig. 41). Synthetic material is available as *Calcitriol-1,25*.

### The Rôle of 1α,25-Dihydroxycholecalciferol in Calcium-binding Carrier Protein Synthesis

1α,25-Dihydroxycholecalciferol stimulates the synthesis of a calcium-binding carrier protein (CaBP) in the target organ (intestine). The evidence for this is now substantial. Thus, administration of vitamin D leads to an increase in the uptake of orotic acid by nuclear RNA as uradylic and cytidylic acids (Wasserman and Corradino, 1971). *Actinomycin D*, which is known to inhibit nuclear functions, also inhibits the ability of vitamin D to promote the synthesis of calcium-binding carrier protein (Zull, Czarnowska-Misztal and De Luca, 1966).

The evidence that vitamin D stimulates synthesis of mRNA for the calcium-binding protein of intestinal cells comes from experiments concerned with the formation of immuno-precipitable protein in a rabbit reticulocyte lysate, following the addition of RNA preparations from intestinal cell polysomes (Emtage, Lawson and Kodicek, 1973). Protein synthesis, as measured by the level of radioactivity precipitated by a specific antiserum to the calcium-binding protein, increases when RNA preparations from vitamin $D_3$-dosed chicks are added. No such effect is seen with RNA preparations from rachitic chicks (Table 76).

According to Kodicek (1974), the mechanism of calcium absorption depends initially upon the binding of 1α,25-dihydroxycholecalciferol in intestinal cells at a specific (acidic) cytoplasmic receptor protein, which sediments at about 5*S* (Fig. 42). This promotes release of mRNA, which appears to become attached to a polysomal array of 10–11 ribosomes for translation into CaBP. Synthesis of CaBP stimulates a series of extra-nuclear effects at the intestinal brush border,

Table 76. Synthesis of Immuno-precipitable Calcium-binding Carrier Protein (CaBP) in a Rabbit Reticulocyto Lysate Programmed with Intestinal Polysomal RNA

| RNA added (μg) | Total radioactivity incorporated (cpm per 10 μl) | $^{14}C$-precipitated by anti-CaBP (cpm) |
|---|---|---|
| None (control) | 14200 | 149 |
| 150 ($-D_3$) | 13700 | 155 |
| 108 ($+D_3$) | 14300 | 361 |
| 216 ($+D_3$) | 14000 | 551 |

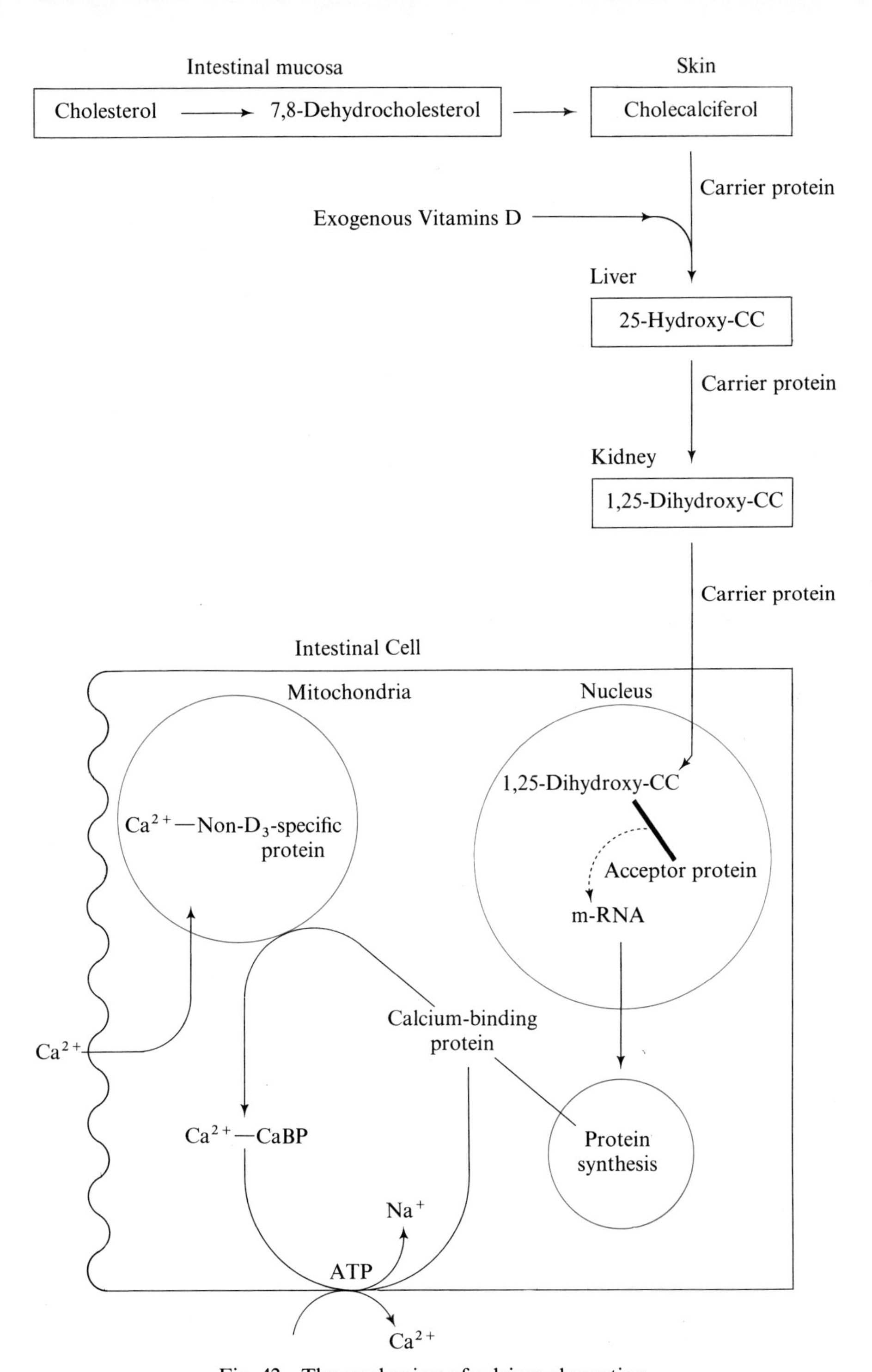


Fig. 42 The mechanism of calcium absorption

which results in increased absorption of $Ca^{2+}$, with a simultaneous increase in alkaline phosphatase.

$Ca^{2+}$, so translocated, is mopped up first by a non-$D_3$-dependent CaBP in mitochondria (Hamilton and Holdsworth, 1970) before being released to the $D_3$-specific CaBP which has a much higher affinity for the ion. Translocation outwards from the cell is considered to be mediated by an active energy-linked pump involving ATP, which may possibly also be linked with $Na^+$ ion influx.

### 1α-Hydroxycholecalciferol

1α,25-Dihydroxycholecalciferol has already been shown to be effective in raising serum $Ca^{2+}$ in patients with renal failure (Brickman, Coburn and Norman, 1972), and its use in other calcium deficiency states associated with metabolic bone disease (Catto, Macleod, Pelc and Kodicek, 1975) clearly has considerable potential. 1α-Hydroxycholecalciferol, which has been prepared synthetically (Barton, Hesse, Pechet and Rizzardo, 1973; Harrison, Lythgoe and Wright, 1973; Holick, Semmler, Schnoes and De Luca, 1973), has the advantage that since it by-passes the 1-hydroxylation stage in the kidney, it can be used in treatments where kidney failure is also a complication.

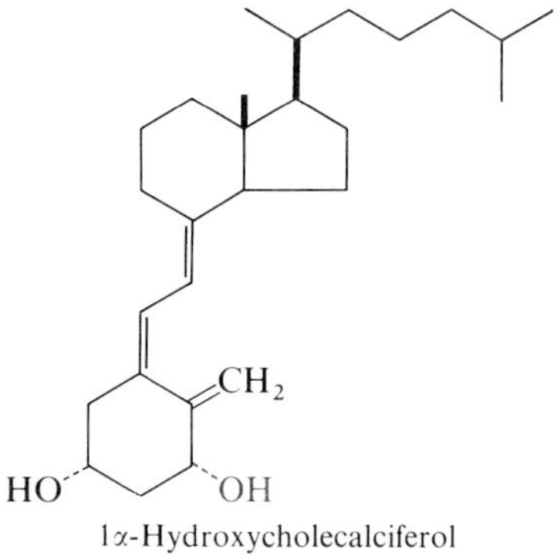

$CH_2$

HO OH

1α-Hydroxycholecalciferol

1α Hydroxycholecalciferol has the same antirachitic acitivity as *Cholecalciferol* in normal rats, and is also effective in nephrectomised animals. It has been shown to act like *Cholecalciferol* by increasing $Ca^{2+}$ absorption from the intestine and $Ca^{2+}$ mobilisation from bone.

## BRIDGED RING SYSTEMS

Decalin is at one and the same time both a fused-ring system consisting of two (fused) six-membered rings and a bridged cyclodecane. Such compounds are systematically defined by the total number of carbon atoms and the number of carbon atoms in each of the three linkages joining the two carbon atoms at the ring junction. In decalin, which has 10 carbon atoms, the linkages are 4, 4 and zero carbon atoms respectively; decalin is therefore systematically defined as bicyclo[4.4.0]decane.

*cis*-Decalin
bicyclo[4,4,0]decane

bicyclo[2,2,1]heptane

Adamantane

Systematic numbering of bridged rings requires one of the junctional carbons to be numbered C-1. The numbering then follows the longest chain to the next junction; numbering then continues along the next longest chain, and so on until it is completed along the shortest chain. On this basis, bicyclo[2,2,1] heptane and the tricyclic adamantane are numbered as shown.

A number of bridged ring compounds have been reported to be active anti-viral agents (Swallow, 1971). *Amantadine* (1-Adamantanamine) and 1(1′-adamantanyl)ethylamine have been reported to have useful prophylactic and therapeutic effects against various strains of influenza. *Amantadine* is also of value in the treatment of Parkinson's disease (Swab, England, Poskanzer and Young, 1969).

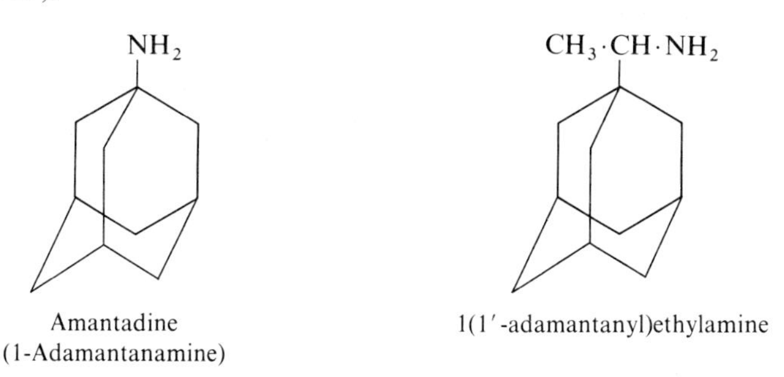


Amantadine
(1-Adamantanamine)

1(1′-adamantanyl)ethylamine

Various *endo*-2-aminobornanes and some *trans*-decalin derivatives are also considered to be of potential interest in the treatment of viral infections.

# 23 Heterocyclic Compounds

Heterocyclic compounds have cyclic structures in which the ring contains one or more atoms other than carbon. They are analogous, but not necessarily isosteric with the corresponding carbocyclic ring compounds and may be hetero-aromatic in character, or alternatively only partially unsaturated (heteroethyl-enic) or fully saturated (heteroparaffinic).

## Nomenclature

A number of common heterocyclic ring systems have trivial names. These include pyridine, piperidine, quinoline, isoquinoline, acridine, phenanthridine, pyrrole, indole, carbazole, pyrimidine, purine, furan, pyran, thiophen, imidezole (iminazole), oxazole and thiazole. Systematic names consist of a prefix denoting the hetero-atom(s) and a suffix denoting the ring size and type (i.e. whether nitrogenous or otherwise) in its most unsaturated form. Partially and fully saturated heterocyclics are usually named as derivatives of the most unsaturated form by appropriate additional prefixes such as dihydro- and tetrahydro-, though specific suffixes are used to denote the extent of unsaturation in all 3-, 4- and 5-membered rings and in some larger rings (Table 77). The principal hetero-atoms, oxygen, sulphur, nitrogen and phosphorus, are denoted by the prefixes **oxa-**, **thia-**, **aza-** and **phospha-** respectively, from which the terminal

Table 77. Suffixes used in Systematic Nomenclature of Heterocyclics

| | | Unsaturated Rings (No. of double bonds) | | | |
|---|---|---|---|---|---|
| Ring Size | Hetero-atom(s) | One | Two | Three | Saturated Rings |
| 3 | O ; S | -irene | — | — | -irane |
| 3 | N | -irine | — | — | -iridine |
| 4 | O ; S | -etene | -ete | — | -etane |
| 4 | N | -etine | -ete | — | -etidine |
| 5 | O ; S | -olene | -ole | — | -olane |
| 5 | N | -oline | -ole | — | -olidine |
| 6 | O ; S | — | — | -in | -ane |
| 6 | N | — | — | -ine | — |
| 7 | O ; S | — | — | -epin | -epane |
| 7 | N | — | — | -epine | — |
| 8 | O ; S | — | — | -ocin | -ocane |
| 8 | N | — | — | -ocine | — |

letter **a** may be elided if followed immediately by a vowel. If two or more different hetero-atoms are present, the order of prefix citation is that of descending Periodic Table Group number, and of increasing atomic number in the Group, i.e. in order oxa-, thia-, aza-, phospha-.

Heterocyclic rings containing a single hetero-atom are numbered from that atom. If more than one hetero-atom is present, numbering is such as to give the lowest locants to the hetero-atoms. If two or more different hetero-atoms are present, numbering starts with the hetero-atom which has citation precedence, and proceeds around the ring in the direction which gives the lowest number to the remaining hetero-atoms irrespective of precedence. These principles of nomenclature and numbering are illustrated by the following selection of the most commonly encountered heterocyclic rings.

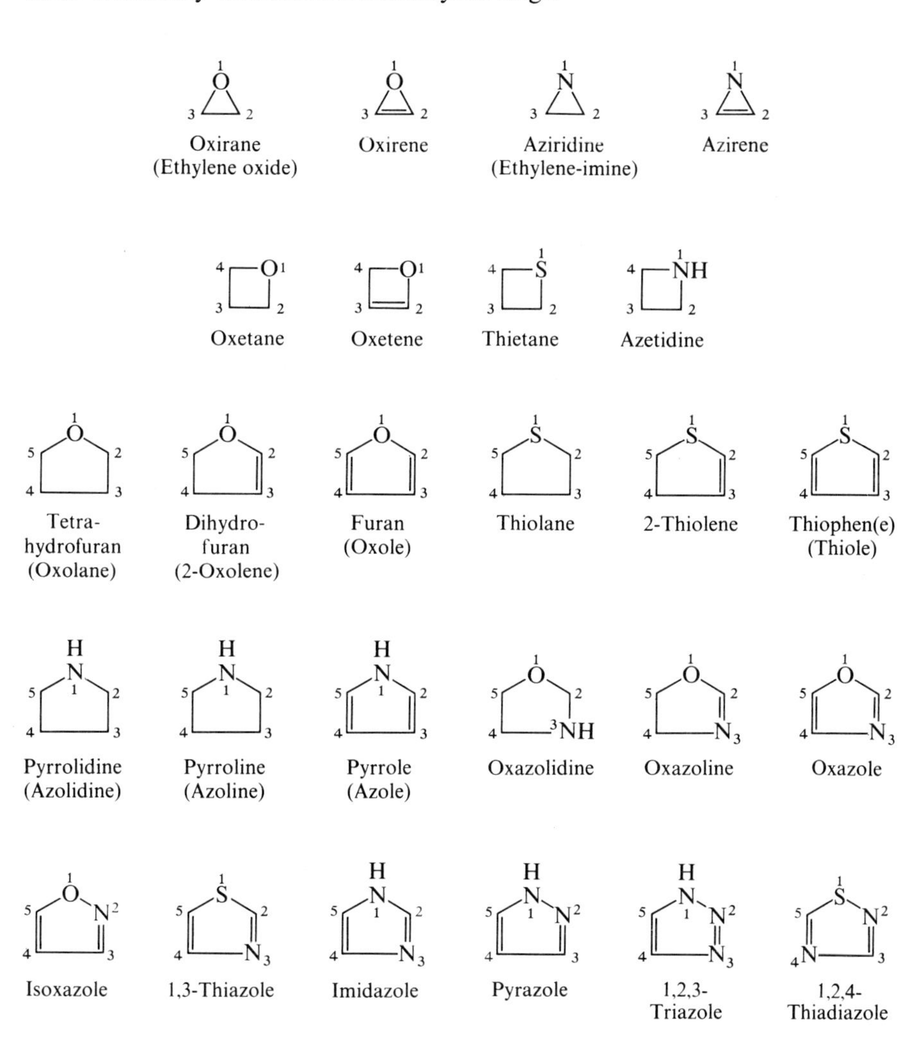

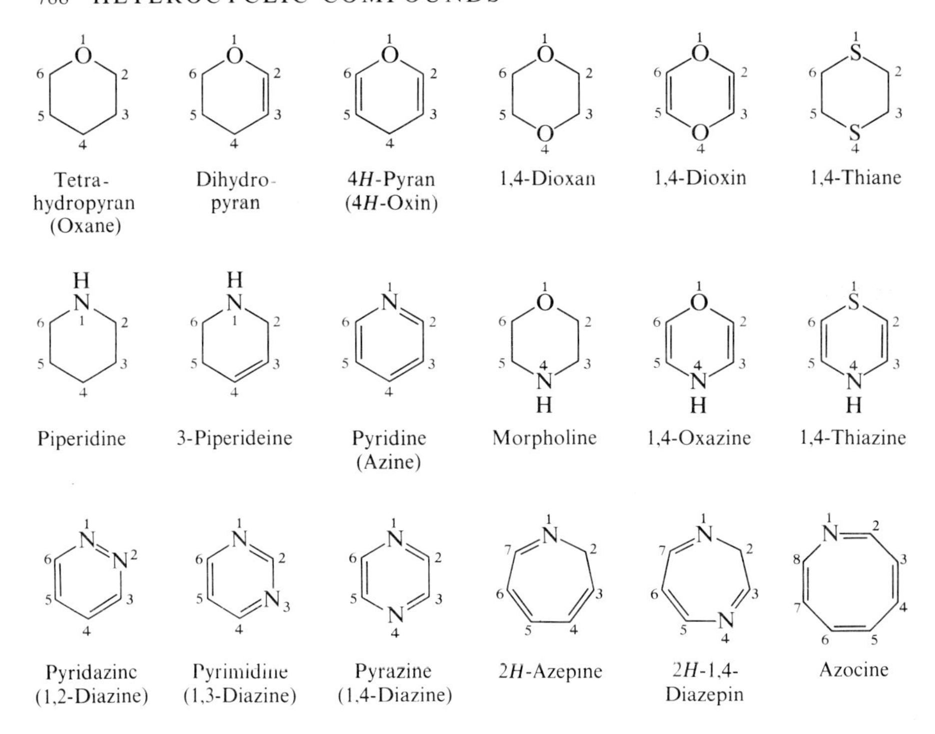


## CLASSIFICATION OF HETEROCYCLIC COMPOUNDS

Some unsaturated heterocyclic compounds have resonance energies of the same order as that of benzene. These are classed as hetero-aromatics.

Heteroaromaticity is found principally in the five- and six-membered heterocyclics, and is characterised not only by reactions which are typical of aromatic compounds in general, but also by modification of the normal properties of the carbon–hetero-atom bond as seen in open-chain compounds. In contrast, the carbon–hetero-atom bond in saturated heterocyclics is essentially the same as in corresponding aliphatic compounds, though it may be modified by steric and other constraints imposed by the ring structure. Thus, piperidine is essentially a secondary amine, with properties and basicity (p$K_a$ 11.1) closely resembling those of diethylamine (p$K_a$ 11.0), whilst pyridine, although a tertiary base, is a much weaker base (p$K_a$ 5.3) because of its aromatic character.

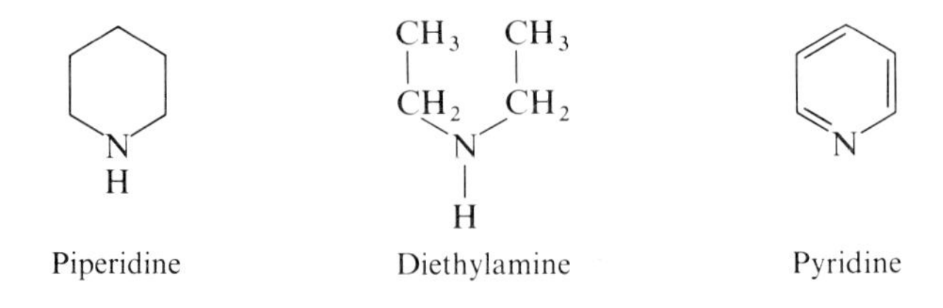

Similarly, tetrahydrofuran, dioxan and ethylene oxide are typical ethers. Tetrahydrofuran and dioxan show the typical stability, solvent properties and affinity for water of aliphatic ethers (Chapter 7). The carbon–oxygen bond, also, is sensitive to strong acids, as in diethyl ether, and both tetrahydrofuran and dioxan show the typical autoxidation of ethers with the formation of explosive peroxides. Whilst, however, ethylene oxide shows similar properties, it has greatly enhanced reactivity due to ring strain and the resulting energy gain on ring cleavage (Chapter 7).

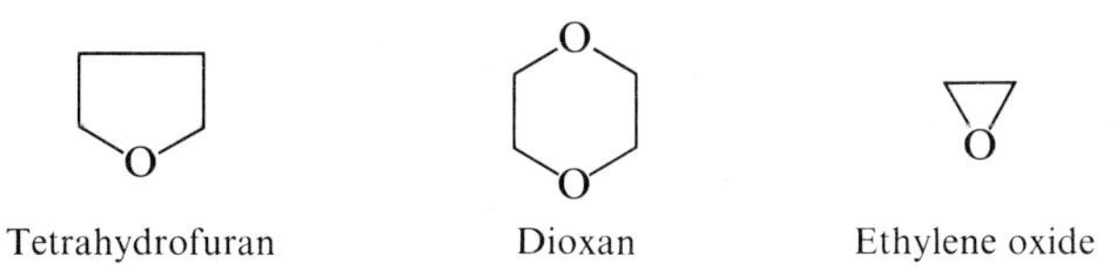


Tetrahydrofuran Dioxan Ethylene oxide

Saturated heterocyclics are generally more soluble in water than their aliphatic counterparts, due it is suggested to constraint imposed by the ring on the alkyl chain moieties, thereby permitting closer approach of water molecules to the heterocyclic atom (Ferguson, 1955). The rate of reaction of heterocyclic secondary bases, such as piperidine, with 1-fluoro-2,4-dinitrobenzene is similarly much greater than that of corresponding aliphatic bases, due also to greater ease of access to the heterocyclic atom (Brady and Cropper, 1950).

Hetero-paraffinics and hetero-ethylenics also resemble their carbocyclic counterparts in that they are generally isosteric and adopt similar conformations. Hetero-aromatics, resemble benzene, and are essentially planar.

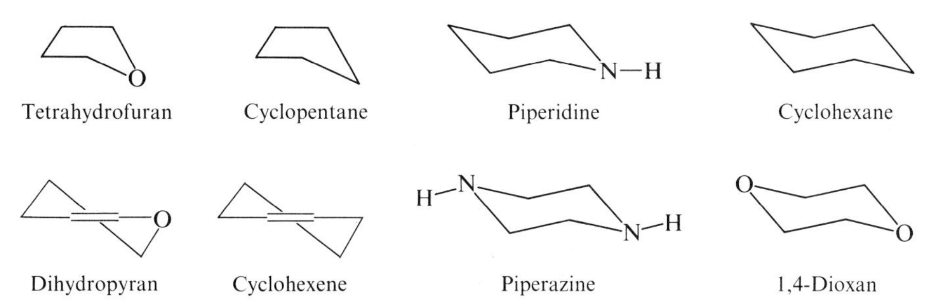


Tetrahydrofuran Cyclopentane Piperidine Cyclohexane

Dihydropyran Cyclohexene Piperazine 1,4-Dioxan

Hetero-aromatics are of two types, those in which the carbon atoms have a deficiency of $\pi$-electrons (**$\pi$-deficient**), and those in which the carbon atoms have an excess of $\pi$-electrons (**$\pi$-excessive**). These two types are illustrated by reference to the six-membered hetero-aromatic, pyridine, and to the five-membered compounds pyrrole, furan and thiophen respectively.

In pyridine, the aromatic sextet arises as a result of delocalisation of six $\pi$-electrons, one from each carbon atom and one from nitrogen. The resultant structure, like that of benzene, is planar hexagonal, though not quite regular; all C—C bonds are of equal length 1.39 Å, though bond order calculations show that they are not necessarily equally reactive in addition reactions. Also, whilst the two C—N bonds are similarly of equal length, they are very slightly shorter

(1.37 Å) than the C—C bonds. Moreover, they are shorter than the normal C—N bond (1.47 Å) and longer than the C=N bond (1.28 Å).

The electronegativity of the hetero-atom gives rise to a number of dipolar resonance forms which contribute significantly to the hybrid.

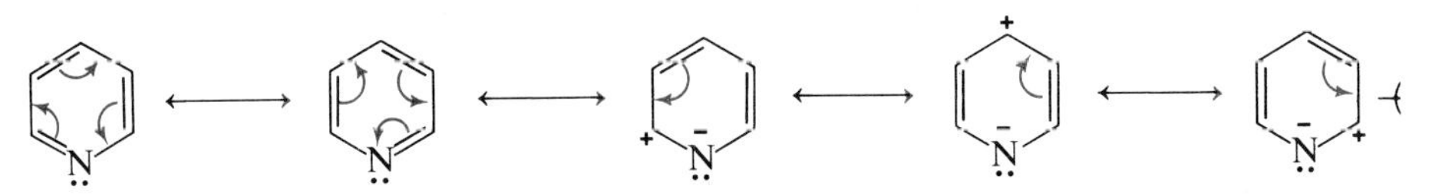

As a result, the hybrid shows a deficiency of electrons in the 2- and 4-positions. Hence the designation as $\pi$-deficient.

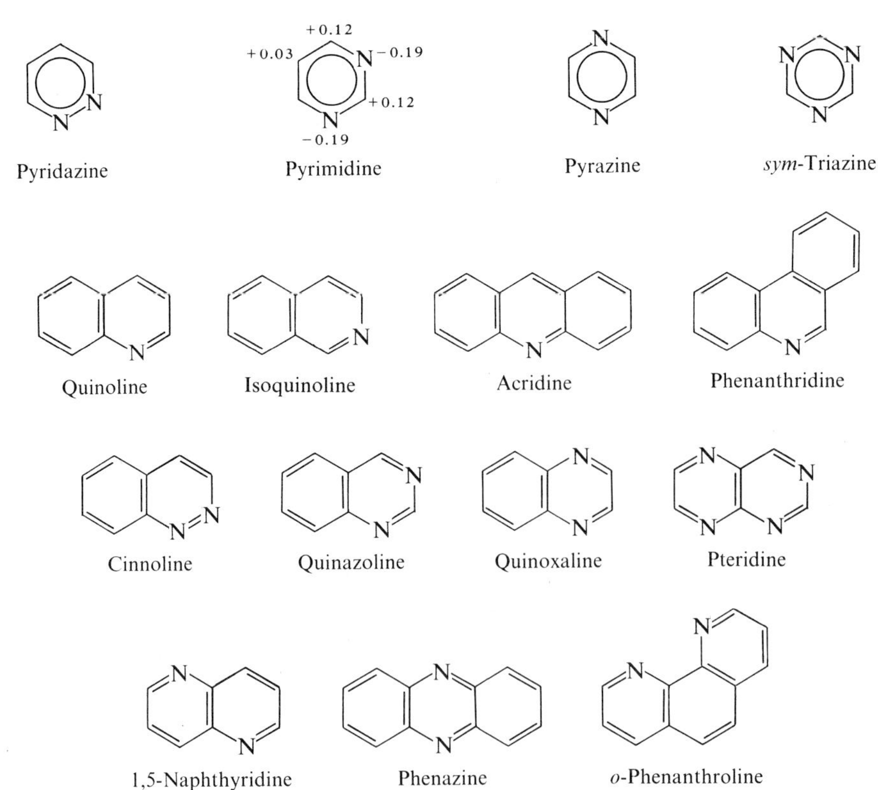

In the five-membered $\pi$-excessive hetero-aromatics, four of the six $\pi$-electrons comprising the aromatic sextet are derived one from each carbon atom, and the remaining two from the lone pair(s) of the heterocyclic atom (N,O,S). As a result, the hetero-atom is deficient in charge, and the excess is distributed, though not entirely evenly, over the four carbon atoms. Again, this particular charge distribution is reflected in calculations of bond order (Longuet-Higgins and Coulson, 1947).

−0.06
−0.10
N +0.32
H

Calculated net charge distributions

0.59
0.77
0.45
N
H

Calculated bond orders

Pyrrole

Other $\pi$-excessive heterocyclics include the following:

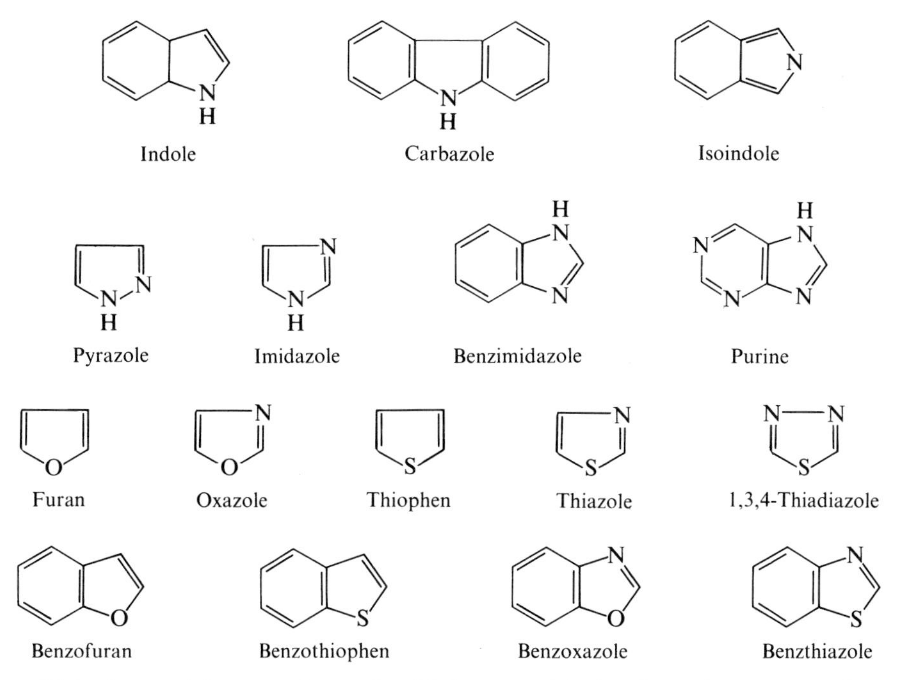


Indole　Carbazole　Isoindole

Pyrazole　Imidazole　Benzimidazole　Purine

Furan　Oxazole　Thiophen　Thiazole　1,3,4-Thiadiazole

Benzofuran　Benzothiophen　Benzoxazole　Benzthiazole

Thiophen and furan are isosteric with benzene, i.e. they have the same number and arrangement of electrons. The term **isosterism** was coined by Langmuir (1919) to indicate the relationship between pairs of molecules such as $N_2$ and CO, and $N_2O$ and $CO_2$, which have the same number and arrangement of electrons, and considerable similarity in physical properties. The extension of the concept of isosterism to the thiophen–benzene relationship was demonstrated by Erlenmeyer and Leo (1933), who showed the close similarity in physical properties for these two compounds and their derivatives.

## π-DEFICIENT N-HETERO-AROMATICS

### General Properties

Molecular planarity, which is a consequence of hetero-aromaticity, assists association and interaction with other aromatics in a number of important physiological and pharmacological reactions. Thus, it accounts for the ease with which strong hydrogen bonds are formed between substituents of purine and pyrimidine base-pairs in the double helices of the nucleic acids (p. 845); also, for the facility with which quinoline and acridine antimalarials are able to intercalate between one base pair and the next (**2**, 3). Similarly, molecular co-planarity assists charge-transfer interaction and stabilisation of the thiamine pyrophosphate—local anaesthetic complexes, which rae reputed to interact with membrane macromolecules and so modify membrane diffusion in local anaesthesia (Thyrum, Luchi and Thyrum, 1969; **2**, 2).

The presence of the hetero-atom lowers volatility. More importantly, the nitrogen lone pair of electrons confers hydrogen bonding capacity and favours water solubility of π-deficient-N-heterocyclics. Thus, pyridine, in contrast to benzene, is completely miscible with water. Pyridazine, pyrimidine and pyrazine likewise are also readily soluble in water. Fused hetero-aromatics incorporating benzenoid hydrocarbon rings (quinoline, isoquinoline, acridine and phenanthridene) have on the other hand only limited water solubility, though both quinoline and isoquinoline are hygroscopic. Fusion of two or more π-deficient N-hetero aromatics as in pteridine, however, leads as might be expected to appreciable water solubility (*ca* 1 in 7). This hydrophilic effect of the nitrogen atom in π-deficient-hetero-aromatics, and indeed with basic drugs in general, could be of considerable significance in determining the high level of biological activity found in such compounds. Some interesting examples are found among the pyridine-based antihistamines, such as *Mepyramine Maleate* (Pyrilamine Maleate) and *Chlorpheniramine Maleate*, which are more effective than their benzenoid analogues in competing with histamine for its receptors, though other factors such as electron-withdrawal are also prominent (Chapter 16).

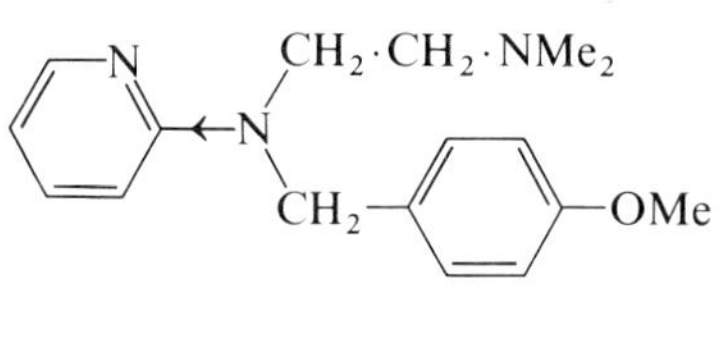

N
$CH_2{\cdot}CH_2{\cdot}NMe_2$
N
$CH_2$
OMe

*Mepyramine Maleate*

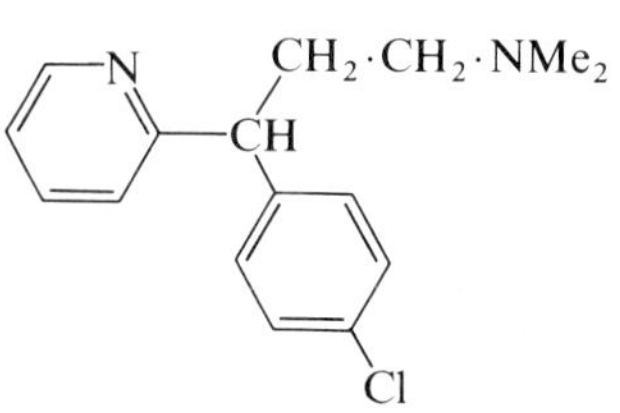

N
$CH_2{\cdot}CH_2{\cdot}NMe_2$
CH
Cl

*Chlorpheniramine Maleate*

The effect of many substituents on water solubility is largely predictable from the properties of comparably substituted aliphatic compounds. Hydroxyl, carboxyl, primary and secondary amino substituents, however, markedly reduce water solubility due to intermolecular hydrogen bonding with the ring nitrogen in the solid state. This is also reflected in their high melting point. Whether or not such compounds form zwitterionic or other tautomers in the solid state is uncertain. There is, however, spectroscopic evidence of zwitterion formation in solution; the extent to which they are formed, however, will be determined by the distance separating acidic and basic groups, and the p$K_a$ of the acidic group. Thus, whereas 3-hydroxypyridine is very soluble in water, nicotinic acid, which exists in aqueous solution almost exclusively as the zwitterion, is only slightly soluble.

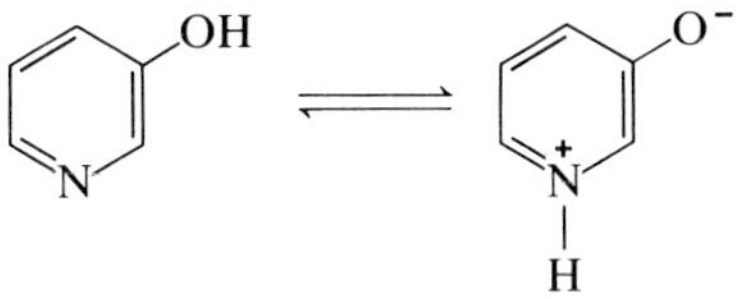

3-Hydroxypyridine (aq)

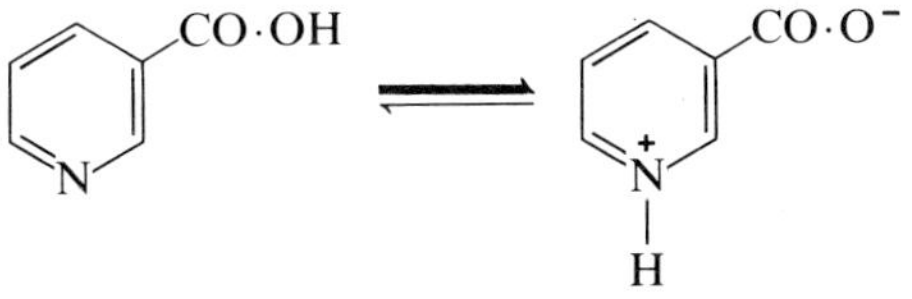

Nicotinic acid (aq)

## Basic Properties

π-Deficient-N-hetero-aromatics are basic, because the lone pair of electrons on the nitrogen is capable of accepting a proton. Heteroaromatic bases like pyridine, however, are much weaker bases than the corresponding heteroparaffinic base (piperidine), since the unshared electron pair is in an $sp^2$ planar trigonal orbital in the former and an $sp^3$ tetrahedral orbital in the latter. The higher *s* component of the $sp^2$ orbital means that the orbital is more symmetrical and that, in consequence, the electrons are held more closely to the nucleus of the nitrogen atom. Hence, they are less available for protonation, and the hetero-aromatic compound is significantly less basic.

Additional nitrogen atoms are still further base-weakening (Table 78). Thus, pyrimidine and pyrazine have p$K_a$ values of only 1.3 and 0.6. Quinazoline and pteridine are somewhat more strongly basic with p$K_a$ values of 3.5 and 4.1 due it is thought to covalent hydration of the cation (Albert, 1959).

Alkyl, halogen and amino substituents all have predictable effects on the ionisation of π-deficient hetero-aromatics. Electron-donating groups (methyl; amino) increase basic strength, whilst electron-attracting substituents (halogen; methoxycarbonyl) decrease basic strength with the greatest effect for both in the 2- and 4-positions in pyridine and pyrimidine (Table 78).

The increased basicity due to methoxyl substitution almost certainly accounts for the high level of serum-protein binding in the long-acting sulphonamides, *Sulphamethoxydiazine* (Sulfamethoxydiazine), *Sulphadimethoxine* (Sulfadimethoxine) and *Sulphamethoxypyridazine* (Sulfamethoxypyridazine); it also

Table 78. Ionisation Constants of Substituted $\pi$-Deficient N-Hetero-aromatics

| Base | $pK_{a_1}$ | $pK_{a_2}$ |
|---|---|---|
| Pyridine | 5.2 | |
| 2-amino- | 6.9 | |
| 3-amino- | 6.0 | |
| 4-amino- | 9.2 | |
| 2-carboxylic acid | 1.01 | 5.32 |
| 3-carboxylic acid | 2.07 | 4.81 |
| 4-carboxylic acid | 1.84 | 4.86 |
| 2-chloro- | 0.7 | |
| 3-chloro- | 2.8 | |
| 2-hydroxy- | 0.75 | 11.62 |
| 3-hydroxy- | 4.86 | 8.72 |
| 4-hydroxy- | 3.27 | 11.09 |
| 2-methoxy- | 3.28 | |
| 3-methoxy- | 4.88 | |
| 4-methoxy- | 6.62 | |
| 2-methoxycarbonyl- | 2.21 | |
| 3-methoxycarbonyl- | 3.13 | |
| 4-methoxycarbonyl- | 3.26 | |
| 2-methyl- | 6.0 | |
| 3-methyl- | 5.7 | |
| 4-methyl- | 6.0 | |
| Quinoline | 4.9 | |
| 2-amino- | 7.3 | |
| 3-amino- | 4.9 | |
| 4-amino- | 9.2 | |
| 2-hydroxy- | −0.31 | 11.74 |
| 3-hydroxy- | 4.3 | 8.06 |
| 4-hydroxy- | 2.27 | 11.25 |
| 8-hydroxy- | 5.13 | 9.89 |
| 2-methoxy- | 3.17 | |
| 4-methoxy- | 6.65 | |
| 2-methyl- | 5.4 | |
| 3-methyl- | 5.1 | |
| 4-methyl- | 5.2 | |
| Isoquinoline | 5.4 | |
| 1-amino- | 7.6 | |
| 3-amino- | 5.0 | |
| 4-amino- | 6.3 | |
| 1-hydroxy- | −1.2 | >12 |
| 1-methoxy- | 3.05 | |
| Acridine | 5.6 | |
| 1-amino | 4.4 | |
| 2-amino | 8.0 | |
| 3-amino | 5.9 | |
| 4-amino | 6.0 | |
| 5-amino | 10.0 | |
| 2,5-diamino- | 11.5 | |
| 2,8-diamino- | 9.6 | |
| 5-hydroxy- | −0.3 | |
| 5-methoxy- | 7.0 | |
| Pyrazine | 0.6 | |
| 2-amino- | 3.1 | |
| 2-methoxy- | 0.7 | |
| 2-methyl- | 1.4 | |
| Pyridazine | 2.3 | |
| Pyrimidine | 1.3 | |
| 2-amino- | 3.5 | |
| 2-amino-4-hydroxy- | 4.1 | |
| 4-amino- | 2.0 | |
| 4-amino-2-hydroxy- (cytosine) | 4.6 | |
| 5-amino- | 2.8 | |
| 2-hydroxy- | [1] | |
| 4-hydroxy- | [1] | |
| 2-methoxy- | <1 | |
| 4-methoxy- | 2.5 | |
| 4-methyl- | 2.0 | |
| Quinazoline | 3.5 | |
| 2-amino- | 4.8 | |
| 4-amino- | 5.8 | |
| 5-amino- | 3.6 | |
| 2-methoxy- | 1.3 | |
| 4-methoxy- | 3.1 | |
| 4-methyl- | 2.5 | |
| Pteridine | 4.1 | |
| 2-amino- | 4.3 | |
| 4-amino- | 3.6 | |
| 2-methoxy- | 2.1 | |
| 4-methoxy- | 1.0 | |
| 2-methyl- | 4.9 | |
| 4-methyl- | 2.9 | |

[1] Form stable salts with inorganic and organic acids; dissociation constants not recorded.

accounts for their relatively high solubility in urine (*ca* pH 6.0) which is so useful in the treatment of urinary tract infections.

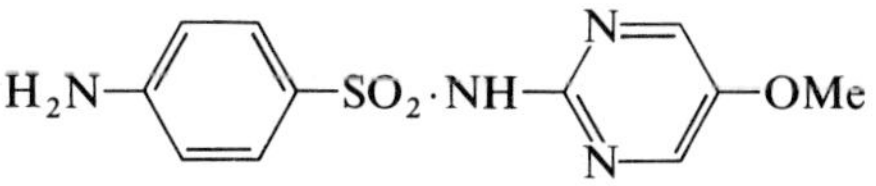

*Sulphamethoxydiazine*

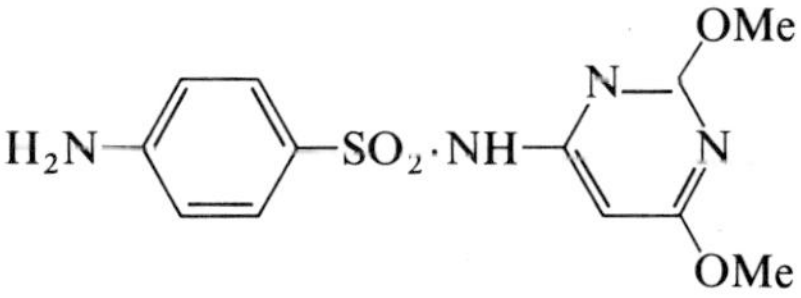

*Sulphadimethoxine*

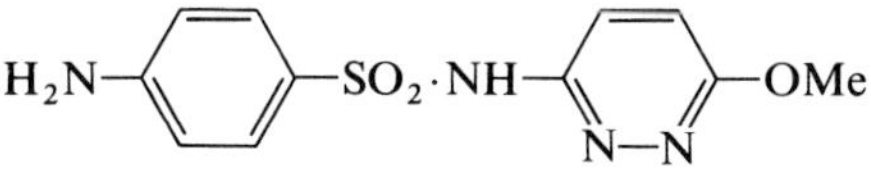

*Sulphamethoxypyridazine*

Similar effects are evident in the fused ring $\pi$-deficient hetero-aromatics quinoline, isoquinoline, acridine and pteridine. Steric effects, however, may dominate electronic effects as, for example, over the otherwise base-strengthening effect of the methyl group in 8-methylquinoline (p$K_a$ 4.6) compared with quinoline (p$K_a$ 4.9), and 4-methylpteridine (p$K_a$ 2.9) compared with pteridine (p$K_a$ 4.1), in which the methyl group adjacent to the hetero-atom tends to prevent the approach of a proton. The methoxy-group in 2-methoxypyridine (p$K_a$ 3.28) compared with 4-methoxypyridine (p$K_a$ 6.62), 1-methoxyisoquinoline (p$K_a$ 3.0), 2-methoxyquinazoline (p$K_a$ 1.3) and 4-methoxypteridine (p$K_a$ 1.0) shows a similar steric exclusion of protons.

$\alpha$-Amino and $\gamma$-amino-$\pi$-deficient hetero-aromatics usually show an unusually high enhancement of basicity. Such compounds are cyclic amidines ($\alpha$-amino)

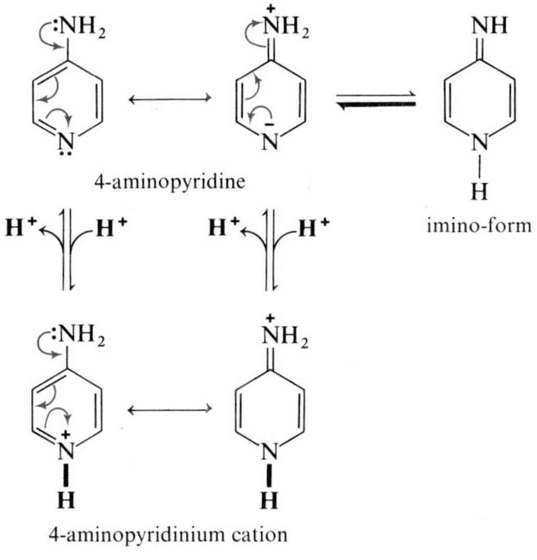

4-aminopyridine

imino-form

4-aminopyridinium cation

or vinologous cyclic amidines ($\gamma$-amino) capable of resonance primarily with a zwitterionic structure to the virtual exclusion of the alternative imino-form. The corresponding cations are similarly resonance-stabilised, hence the marked enhancement of basicity.

A similar base-strengthening effect of resonance is observed in 2- and 4-aminoquinolines, 2 and 5-aminoacridines and in 2,5- and 2,8-diaminoacridines.

Strongly basic aminoacridines, such as *Aminacrine* (5-aminoacridine) and *Proflavine* (2,8-diaminoacridine), are almost completely ionised at physiological pH, and since their antibacterial activity is due to the compound in the ionised state, these compounds are much more potent bacteriostats than other less strongly basic aminoacridines (**2**, 2).

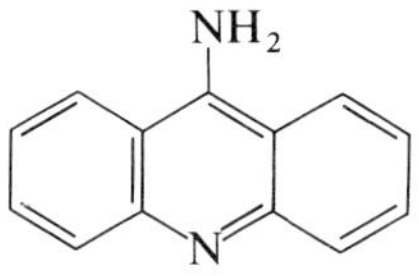

$NH_2$

*Aminacrine*

$H_2N$ N $NH_2$

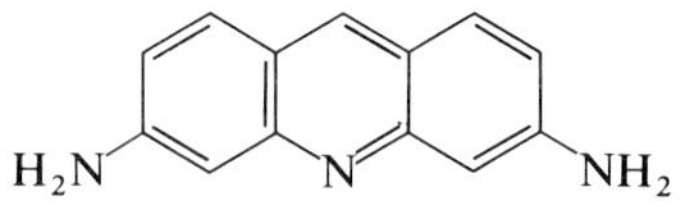

*Proflavine*

Similarly, the antimalarials, *Chloroquine* and *Mepacrine* (Quinacrine), which are diethylamino-derivatives of 4-aminoquinoline and 5-aminoacridine respectively, inhibit plasmodial DNA synthesis (Schellenberg and Coatney, 1961) by stabilising the DNA double helix, and thus preventing DNA replication and reproduction of the parasites. Stabilisation of the DNA double helix is due to

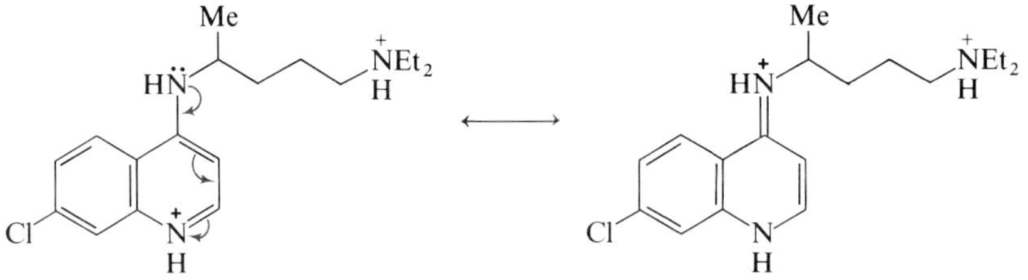

*Chloroquine*

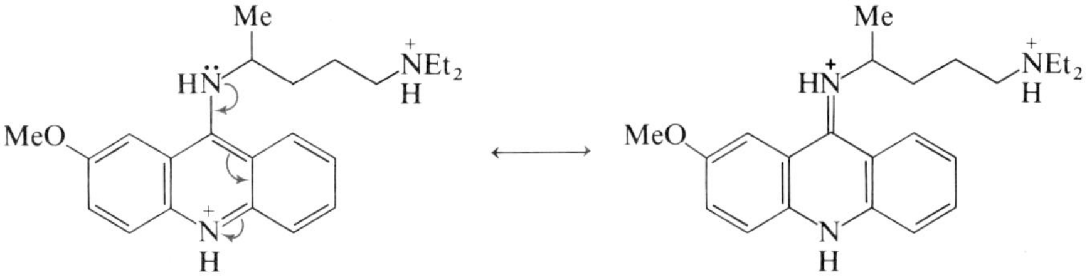

*Mepacrine*

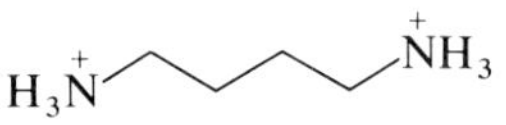

$H_3\overset{+}{N}$ $\overset{+}{N}H_3$

1,4-Diaminobutane (cation)

formation of a drug–DNA complex (Allison, O'Brien and Hahn, 1965), formed by intercalation of the heterocyclic ring between DNA base pairs, and by electrostatic interaction of the protonated diethylaminoalkylamino chain with negatively-charged DNA phosphate residues, so that it spans the minor groove between complementary strands of the DNA helix (O'Brien and Hahn, 1965) (**2**, 2).

The significance of resonance in the electrostatic binding of these drugs is apparent in that stabilisation of DNA temperature-dependent 'melting' (separation of the double helix strands) by aliphatic diamines is dependent on aliphatic chain length. The most effective diamines are 1,4-diaminobutane and 1,5-diaminopentane, which structurally resemble the di-cationic resonance form of the *Chloroquine* and *Mepacrine* side-chains (Mahler and Mehrotra, 1963). It is significant, too, that *Primaquine*, which is an 8-aminoquinoline derivative, is unable to compete with *Chloroquine* for DNA binding sites (Morris, Andrew, Wichard and Holbrook, 1970). The failure of quinine to do no more than suppress malaria, rather than eliminate the infecting parasite by disruption of its reproductive processes, is also apparent in its structure as a 4-hydroxyalkylquinoline derivative incapable of the apparently essential resonance phenomena.

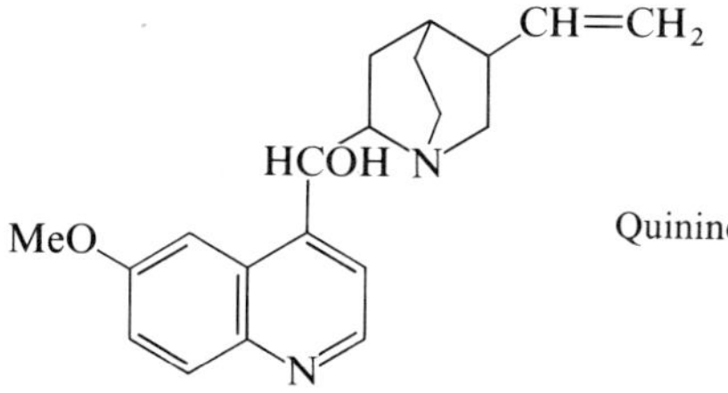

$CH{=}CH_2$

HCOH N

MeO

N

Quinine

The base-weakening effect of the electron-attracting methoxycarbonyl group is clearly evident in the methyl esters of all three pyridine carboxylic acids. Again, the steric effect of substituents adjacent to the hetero-atom shows in the significantly lower p$K_a$ for methyl pyridine-2-carboxylate compared with that

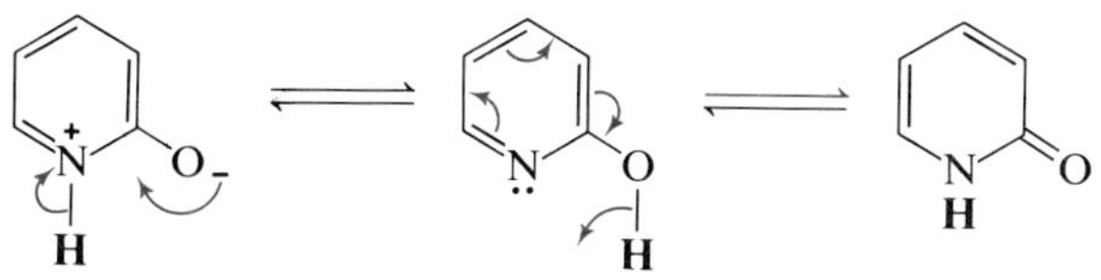

2-Hydroxypyridine α-pyridone

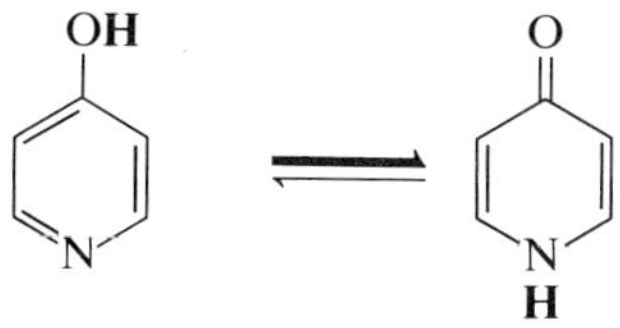

4-Hydroxypyridine γ-pyridone

of the 4-isomer. The lower $pK_a$ of the parent carboxylic acids reflects the substantial contribution of zwitterionic forms. Similarly, zwitterion formation in 3-hydroxypyridine is base-weakening. Zwitterions, however, are relatively unimportant in $\alpha$- and $\gamma$-hydroxy-compounds, such as 2- and 4-hydroxypyridines, 4,6-dihydroxypyrimidine (uracil), 2-hydroxyquinoline, 1-hydroxyquinoline and 5-hydroxyacridine, all of which are exceedingly weak bases due to the almost total predominance of the amide tautomer in aqueous solution. *N*-Substitution, as in the hypnotic, *Methaqualone*, and the X-ray contrast agent, *Iodoxyl* (Iodomethamate Sodium), fixes the structure in the pyridone form.

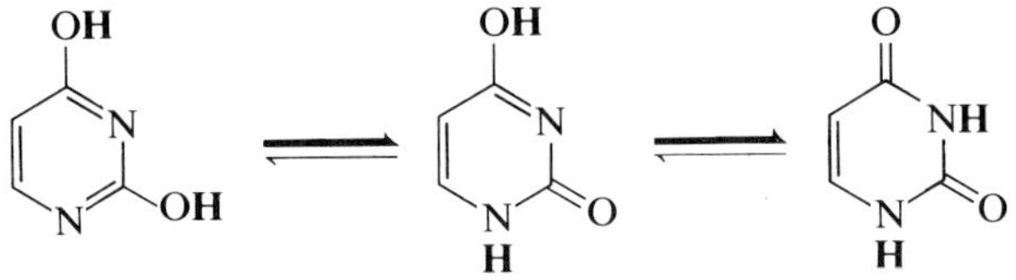

2,4-Dihydroxypyrimidine
(uracil)

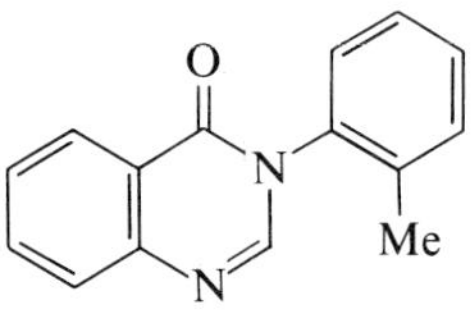

*Methaqualone*

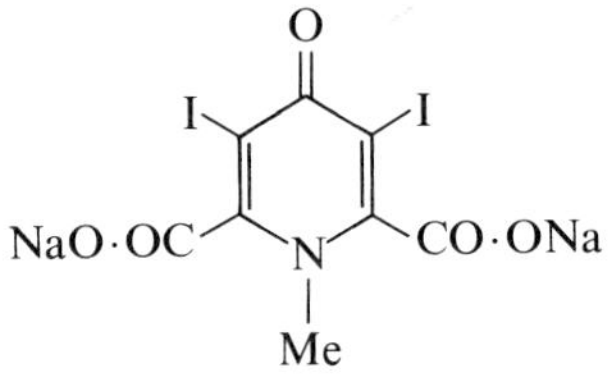

*Iodoxyl*

2- and 4-Mercapto-$\pi$-deficient hetero-aromatics, such as the important antithyroid agent, *Propylthiouracil*, also exists almost entirely in the thioamido-form in aqueous solution (Albert and Barlin, 1959).

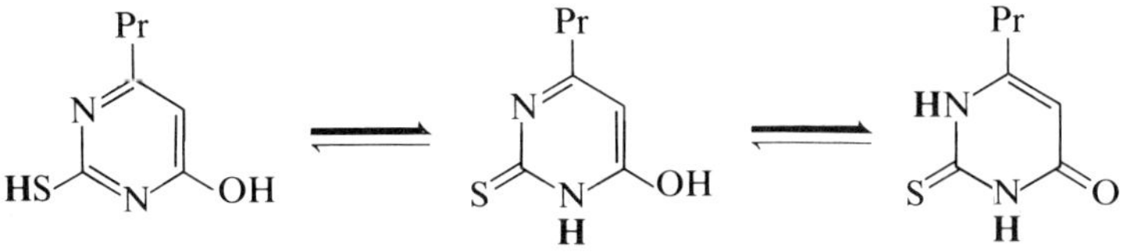

*Propylthiouracil*

The thiouracils and the purine derivative, *Theobromine* (3,7-dimethylxanthine), owe their solubility in aqueous alkali hydroxides to the existence of the hydroxylic tautomers, which are sufficiently acidic to form alkali metal salts. *Caffeine* (1,3,7-trimethylxanthine) in which the 1-position is blocked by a methyl substituent is insoluble in solutions of alkali hydroxides. (See p. 799.)

Uric acid (2,6,8-trihydroxypurine) also exhibits imido–imidole tautomerism. Absorption spectroscopy indicates that the amido-form normally predominates,

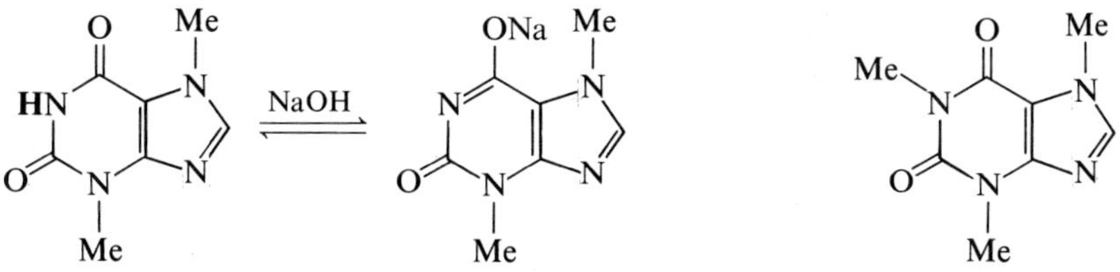

*Theobromine*

*Caffeine*

but alkali metal salts of the 2,6-dihydroxy-tautomer are readily formed. The lithium salts are the most soluble, hence the traditional use of lithium citrate for the relief of gout, an affliction which is accompanied by the separation of uric acid in the joints. The rationale for such treatment is weak, as the soluble lithium salt is unlikely to exist in the presence of excess sodium and potassium ions.

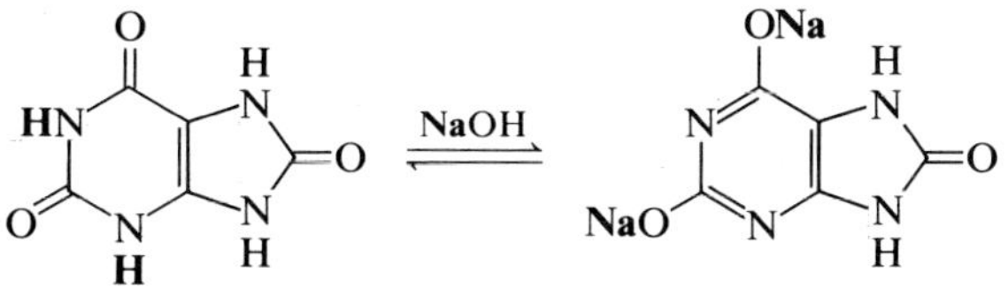

Uric acid

**Electrophilic Substitution**

Electrophilic substitution of π-deficient N-hetero-aromatics is hindered by the reduction of electron density on the ring carbon atoms. It occurs much less readily than in benzene, and takes place only with difficulty in pyridine and not at all in unsubstituted compounds with more than one hetero-atom (pyridazine, pyrimidine and pyrazine). Electrophilic substitution of pyridine with reagents, such as the nitronium and bromonium ions, which are only formed in an acidic environment, is also hindered by the positive charge of the protonated ring nitrogens and the ensuing enhanced electron-withdrawal.

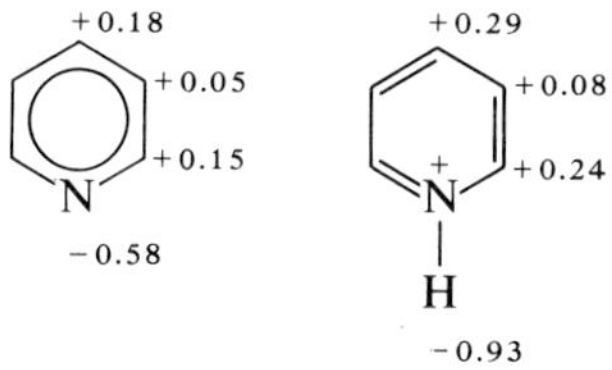

In consequence, substitution only occurs under somewhat drastic reaction conditions. Thus, nitration is only effective at 300°C, sulphonation with fuming sulphuric acid at 220°C, and bromination at 160°C. Substitution occurs at the point of minimum electron depletion, that is in the 3-position, and yields are generally poor (5% for nitration).

Electron-releasing substituents (amino; hydroxy; methoxy; sulphydryl) assist the reaction, and in pyridine direct the entering group into the ortho- and para-positions relative to the substituent, though the 4-position is avoided (Schofield, 1950).

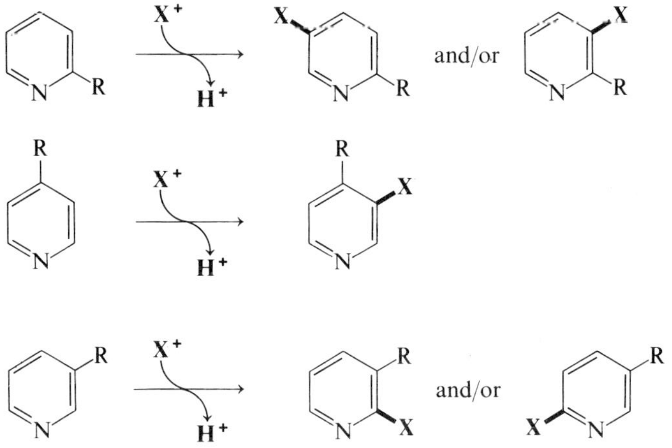

$R = —OH, —NH_2$ or $—OR$ $\quad X^+ = \overset{+}{N}O_2$, $Br^+$ or $Ph\overset{+}{N}_2$

Alkyl substituents are somewhat less effective than hydroxy, alkoxy, and amino substituents as electron-releasing groups, so that monomethyl pyridines fail to nitrate; 2,4,6-trimethylpyridine, however, undergoes nitration at 100°C.

Pyrimidines if suitably substituted with electron-releasing groups (amino; hydroxy; methoxy; sulphydryl) will also undergo electrophilic substitution, but

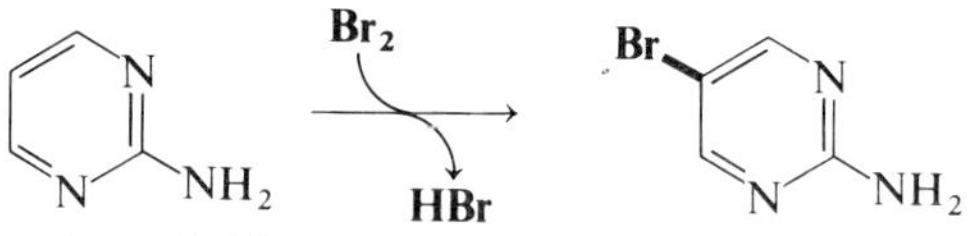

2-aminopyrimidine

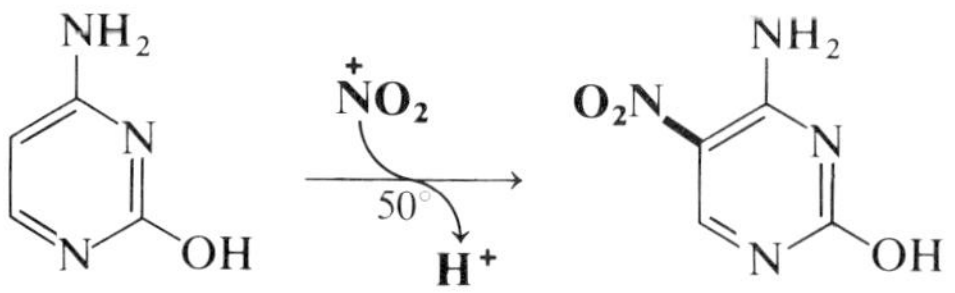

4-amino-2-hydroxypyrimidine (cytosine)

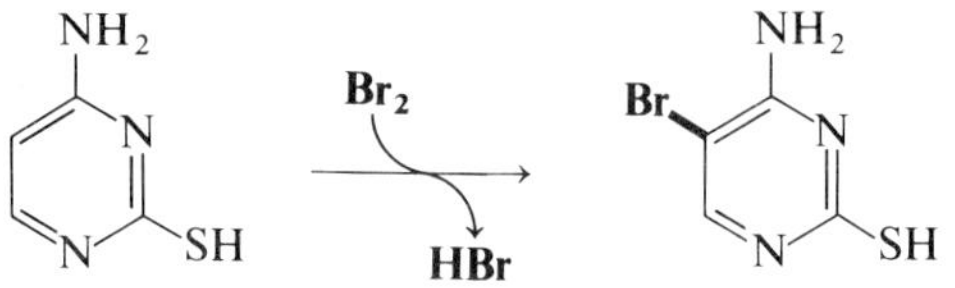

4-amino-2-mercaptopyrimidine

substitution occurs only at C-5, the point of minimum carbon electron depletion (p. 790). Nitration, nitrosation and coupling with diazonium salts all require double activation, i.e. at least two electron-releasing groups already present. A single activating group, however, will suffice for halogenation.

The fused-ring heterocyclic, quinoline, halogenates in the 3-position in neutral solution, roughly in accord with the known charge density distribution (Longuet-Higgins and Coulson, 1947). The latter, however, is modified in acidic solution (compare with pyridine, p. 799), so that halogenation in acidic solution occurs preferentially in the 5- and 8-positions. Nitration and sulphonation similarly occur in the benzene ring giving rise to the 5- and 8-nitroquinolines and quinolinesulphonic acids.

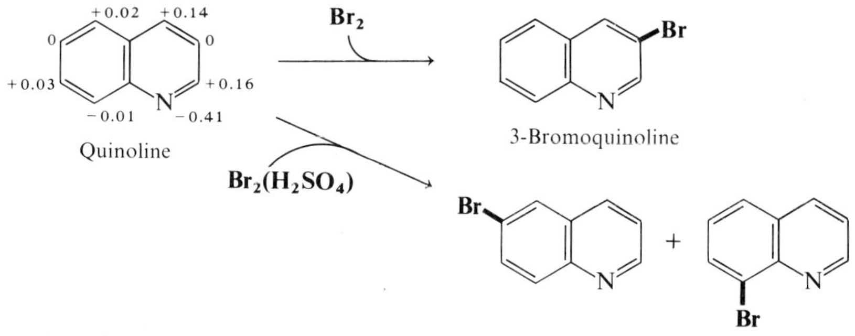

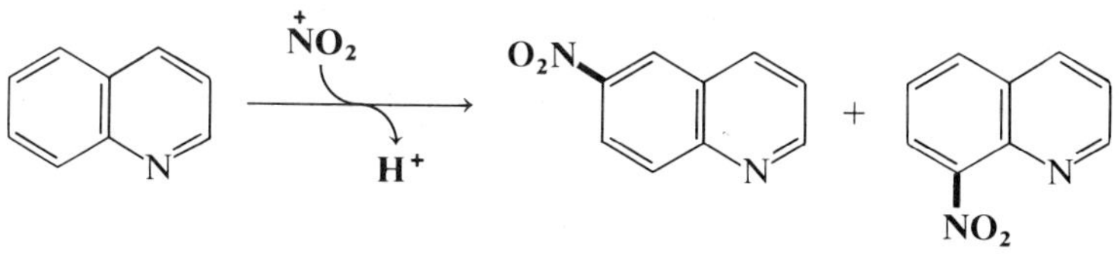

+0.02 +0.14 0 0 +0.03 +0.16 N −0.01 −0.41

Quinoline

$Br_2$

Br N

3-Bromoquinoline

$Br_2(H_2SO_4)$

Br N + N Br

$\overset{+}{N}O_2$ N $H^+$ $O_2N$ N + N $NO_2$

The presence of electron-releasing substituents in the benzenoid ring of quinoline causes even more ready substitution, and, as with pyridine, is generally ortho or para to the substituent. Other fused benzenoid hetero-aromatics, such as isoquinoline, acridine, and phenanthridine, also undergo electrophilic substitution principally in the benzene ring in accord with the charge distribution.

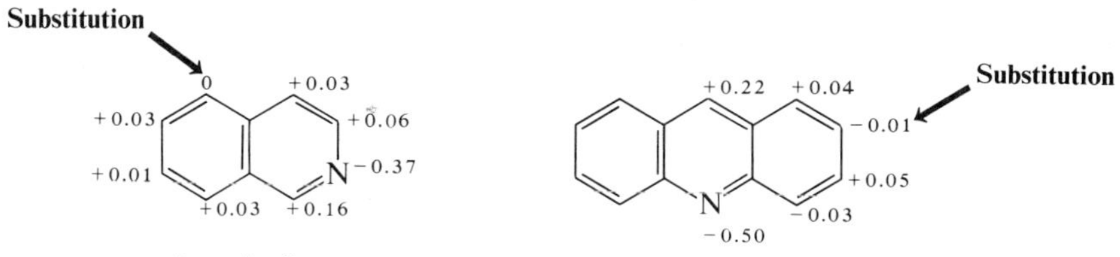

Substitution

0 +0.03 +0.03 +0.06 +0.01 N −0.37 +0.03 +0.16

Isoquinoline

+0.22 +0.04 Substitution −0.01 +0.05 N −0.03 −0.50

Acridine

## Nucleophilic Substitution

In contrast to electrophilic substitution, nucleophilic substitution is facilitated in $\pi$-deficient N-hetero-aromatics by the deficit of electrons on the ring carbon atoms. Thus, pyridine undergoes direct amination with sodamide in liquid ammonia (nucleophilic attack by $NH_2$) to form 2-aminopyridine. The latter reacts as it is formed with excess sodium hydride, to give the sodio-derivative; this is decomposed with mineral acid at the end of the reaction to yield 2-aminopyridine.

Na$^+$ $^-NH_2$ → [N(Na), H, $NH_2$ adduct] → (−NaH) 2-$NH_2$ pyridine; + NaH → (−$H_2$) 2-NHNa pyridine

Quinoline, isoquinoline, acridine and phenanthridine are similarly aminated with sodamide to 2- and 4-aminoquinoline, 1-aminoisoquinoline, 5-aminoacridine and 9-aminophenanthridine respectively.

Nucleophilic displacement of hydrogen by hydroxyl in $\pi$-deficient N-hetero-aromatics, such as pyridine and quinoline, only occurs at high temperature (250–300°C) in dry fused potassium hydroxide, to yield the 2-hydroxy-compounds.

K$^+$ $^-OH$ → [N(K), H, OH adduct] → 2-OH pyridine

On the other hand, the carbanions formed from lithium alkyls attack pyridine readily at room temperature, though the subsequent displacement of lithium hydride requires a somewhat higher temperature.

Li$^+$ Bu$^-$ → [N(Li), H, Bu adduct] → (−LiH) 2-Bu pyridine

The enhanced electron-withdrawal in π-deficient multi-N-hetero-aromatics promotes nucleophilic attack as demonstrated by their heightened hydrolytic instability. Thus, whereas pyridine is stable to hot aqueous alkali, pyrimidine is unstable, whilst pteridine (Albert, Brown and Wood, 1956) and *sym*-triazine (Grundmann and Kreutzberger, 1954) are rapidly hydrolysed within minutes by cold aqueous sodium carbonate and cold water respectively.

Protonation, which enhances electron-withdrawal from the ring carbons, promotes nucleophilic attack, so that whilst *sym*-triazine is hydrolysed within minutes in cold water, decomposition is almost instantaneous in acidic solution. Ammonium formate is the sole product. Pteridine likewise undergoes rapid hydrolysis in *N*-acid, with disruption of the pyrimidine ring.

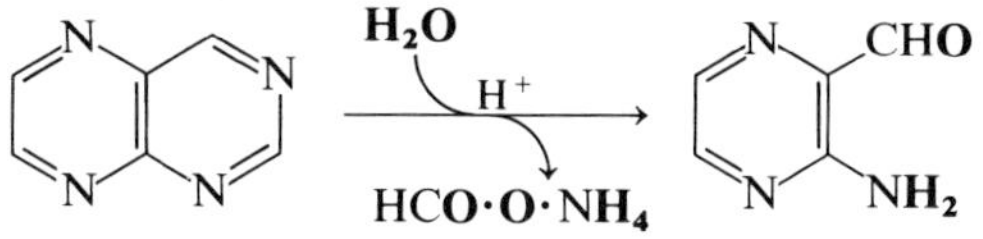


## Transmission of Electron-withdrawal to α- and γ-Substituents

The electron-attracting properties of the hetero-atom are transmitted primarily to the α- and γ-positions. Electron deficiency is greatest at these points, and halogen substituents so placed readily undergo nucleophilic displacement by alkoxyl ($RO^-$), phenoxyl ($PhO^-$) and hydroxyl ($HO^-$) ions, and also by amines and thiols. Similarly, whereas π-deficiency leads to all three pyridine carboxylic acids being stronger acids than benzoic acid, pyridine-2-carboxylic acid and pyridine-4-carboxylic acid are much stronger acids than nicotinic acid (pyridine-3-carboxylic acid) (Table 78). The ionisation of heteroaromatic *N*-substituted sulphonamides, which is relevant to their antibacterial action (Bell and Roblin, 1942; Chapter 14) is also directly related to the electron-attracting ability of the ring (Table 79).

Electron-withdrawal is also transmitted to α- and γ-methyl substituents, so that carbanion formation is readily promoted in the presence of reagents such as sodamide. The resulting carbanions are powerful nucleophiles and react readily with aldehydes. (See page 804.)

Activation of α- and γ-methyl groups is enhanced as the number of hetero-atoms increases. Thus, 2- and 4-methylpyrimidines react much more readily than the corresponding pyridines.

Table 79. Ionisation Constants of some π-Deficient N-Hetero-aromatic Sulphonamides

| | $pK_a$ | | $pK_a$ |
|---|---|---|---|
| Sulphanilamide | 10.43 | Sulphapyridazine | 7.06 |
| Sulphapyridine | 8.43 | Sulphadiazine | 6.48 |
| Sulphadimidine | 7.48 | Sulphapyrazine | 6.04 |

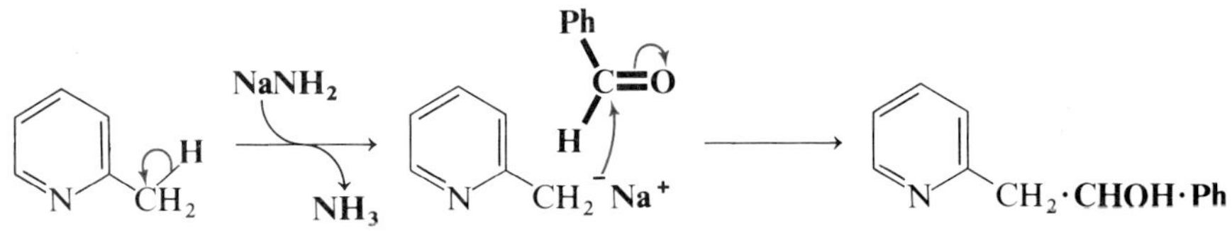

The condensation of amino acids with *Pyridoxal*, which entails formation of an azomethine by nucleophilic attack on the carbonyl group, is similarly facilitated by electron-withdrawal from the $\gamma$-position of the pyridine nucleus (p. 790). The reaction occurs in neutral solution at room temperature and forms the basis of the transamination reactions of amino acids (Chapter 18).

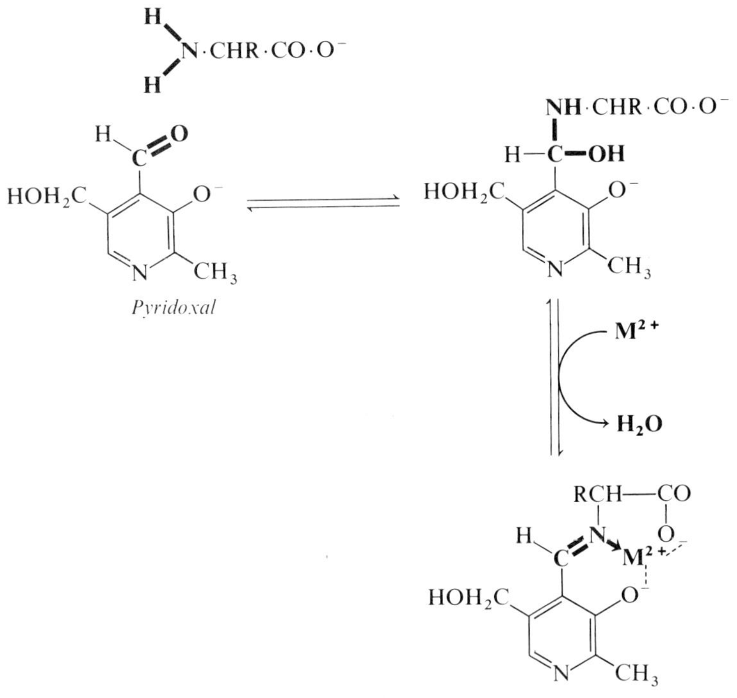

*Pyridoxal*

Activation of $\alpha$- (and $\gamma$-) substituents, promoting the formation of an electron-deficient centre in diphenylmethanes has been postulated by Perkow (1966) as

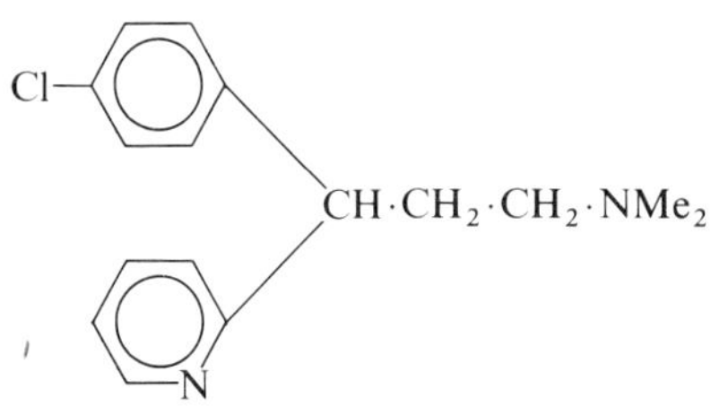

*Chlorpheniramine Maleate*

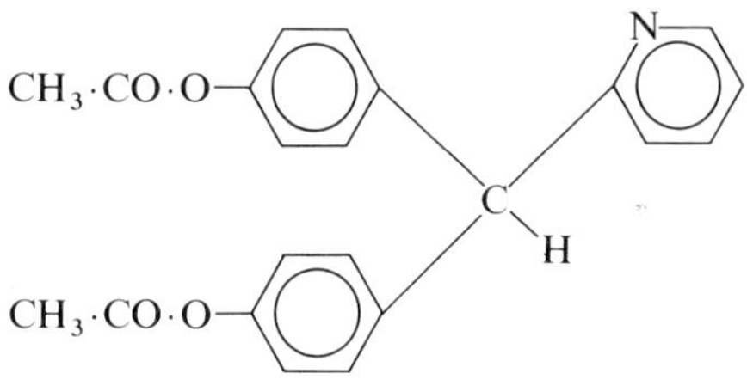

*Bisacodyl*

the mechanism whereby bio-receptor interaction of a wide range of medicinal agents is initiated. These include various antihistamines (Chapter 16), such as *Chlorpheniramine Maleate* and the laxative, *Bisacodyl*.

Radical formation on the $\alpha$-carbon atom of the cinchona alkaloids, which are 4-hydroxymethylquinoline derivatives, is considered by some to explain, at least in part, their sensitivity to light and air-oxidation, due to the ability of the hetero-aromatic ring to stabilise the radical so formed. It is clear, however, that the vinylic double bond is also sensitive to light-catalysed oxidation.

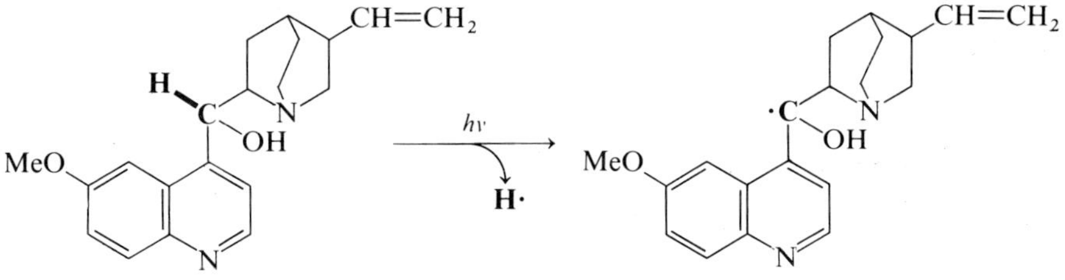

Quinine

It has also been suggested (Hansch, Kutter and Leo, 1969) that the 4-hydroxyalkyl group may interfere with radical reactions.

## Addition Reactions

Activation of substituents in the $\alpha$- and $\gamma$-positions by electron-withdrawal in $\pi$-deficient-N-hetero-aromatics is a manifestation of a degree of double bond fixation, in which certain bonds have a greater degree of double bond character than others (*cf* naphthalene, Chapter 22). This is particularly evident in the addition of weak acids, such as HCN, $NaHSO_3$, $H_2O$ and $NH_2OH$, which occurs readily in multi-N-hetero-aromatics, but only in the pyridine series after quaternisation.

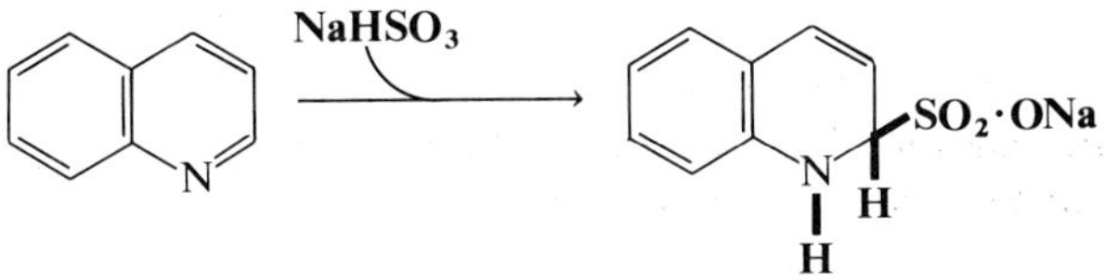

Double bond fixation and addition reactions are even more characteristic of hydroxy-amide tautomers. The addition of hydroxylamine and methoxyamine to cytosine, methylcytosine and hydroxymethylcytosine residues in DNA is an important feature of the mutagenic effect of these reagents (Freese and Freese, 1965; Lawley, 1967; Brown and Coe, 1970). The reaction follows two pathways, both involving an addition to the 5,6-double bond.

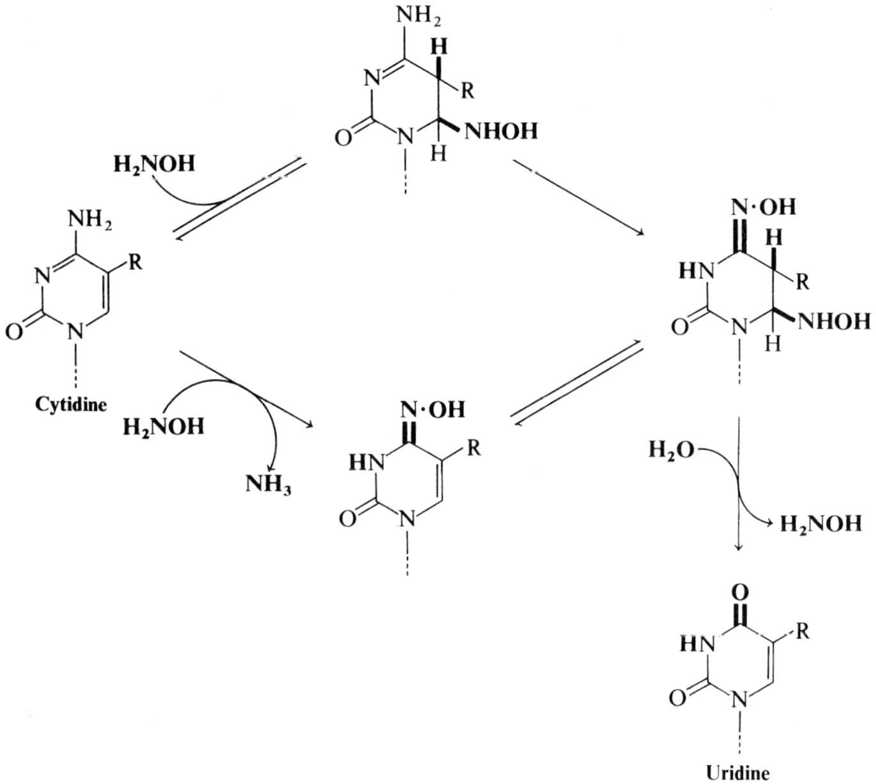

$NH_2$ H R NHOH H
$H_2NOH$
$NH_2$ R N O N
Cytidine
$H_2NOH$ $NH_3$
N·OH H HN R O N NHOH H
N·OH HN R O N
$H_2O$ $H_2NOH$
O HN R O N
Uridine

Subsequent elimination of the added hydroxylamine and hydrolysis of the oximo group results in the conversion of cytosine, methylcytosine or hydroxymethylcytosine to the corresponding uracil. Mutagenic effects can, therefore, result from transitions of base-pairs in the DNA helix (p. 847).

$$G\cdots C \longrightarrow G\cdots U$$

Dimerisation on exposure to ultraviolet light is also a typical manifestation of double bond fixation. This accounts for the mutagenic effects of ultraviolet light which is due to the formation of cross-links in the DNA double helix.

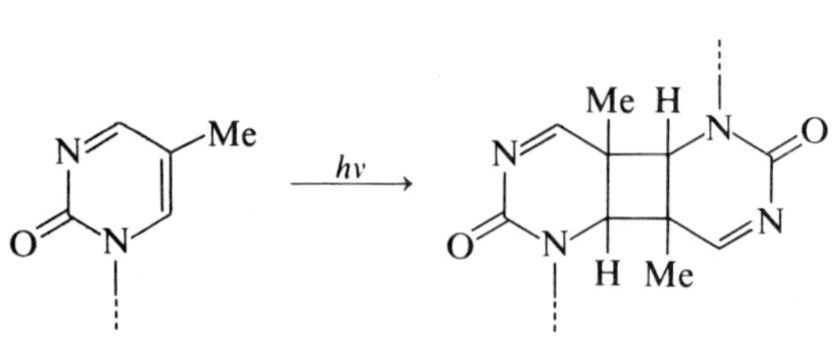

N Me O N $h\nu$ Me H N O N O N H Me

Thymine residue dimer

## Quaternary Salts

Despite the fact that pyridine, quinoline, isoquinoline, acridine and phenanthridine are all relatively weak bases, the high electron density on nitrogen facilitates reaction with alkyl halides to form quaternary salts.

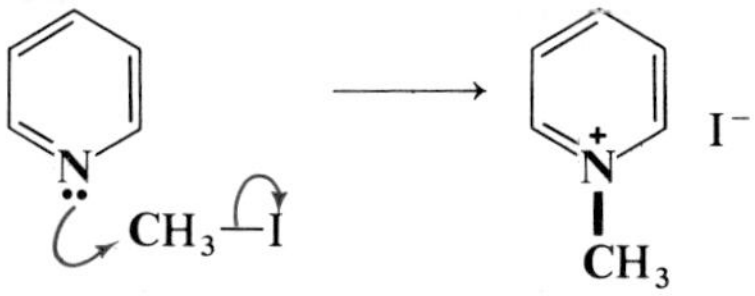

$CH_3$—I  ⟶  $N^+$—$CH_3$  $I^-$

The formation of similar salts in biological systems is uncommon, but a non-specific *N*-methyltransferase present in rabbit lung tissue is capable of converting pyridine and quinoline to their methoquaternary salts (Axelrod, 1962).

Nicotinic acid ribonucleotide, a naturally-occurring pyridinium salt, is formed biosynthetically in human erythrocytes by quaternisation of nicotinic acid with 5-phosphoribosyl-1-pyrophosphate and in the liver by an analogous pathway (from tryptophan) involving quinolinic acid. Nicotinic acid ribonucleotide is the precursor of NAD (nicotinamide adenine dinucleotide) which is formed from it via desamido-NAD. (See page 808.)

The quaternary salts are typically water-soluble compounds. Long-chain compounds such as *Cetylpyridinium Chloride* are somewhat less soluble (1 in 20), but surface-active and hence useful as cationic detergents. *Cetylpyridinium Chloride*, *Dequalinium Acetate* and *Dequalinium Chloride* have antibacterial and antifungal properties, and are used as surface antiseptics. They are incompatible with and inactivated by anionic compounds such as soaps, phenols, and anionic detergents. The two dequalinium salts differ markedly in water solubility, the acetate being very soluble and the chloride almost insoluble. *Viprynium Embonate* (Pyrvinium Pamoate), which is used in the treatment of threadworm infestation, combines the typical poor gastro-intestinal absorption of quaternary salts with low water solubility to ensure its passage and concentration near the seat of infection in the lower bowel.

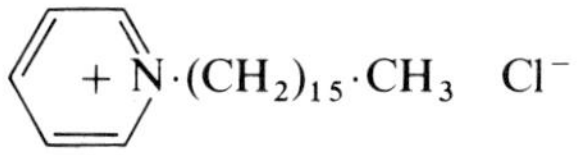

$+\,N\cdot(CH_2)_{15}\cdot CH_3\quad Cl^-$

*Cetylpyridinium Chloride*

$H_2N$— $^+N\cdot(CH_2)_{10}\cdot N^+$ —$NH_2$ ($CH_3$, $CH_3$)  $2CH_3\cdot CO\cdot O^-$

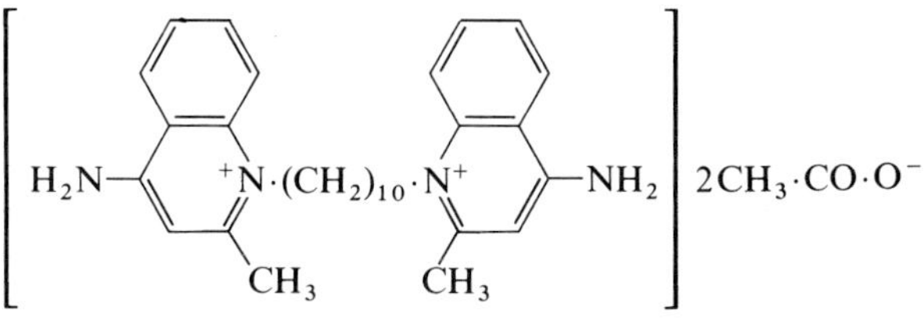

*Dequalinium Acetate*

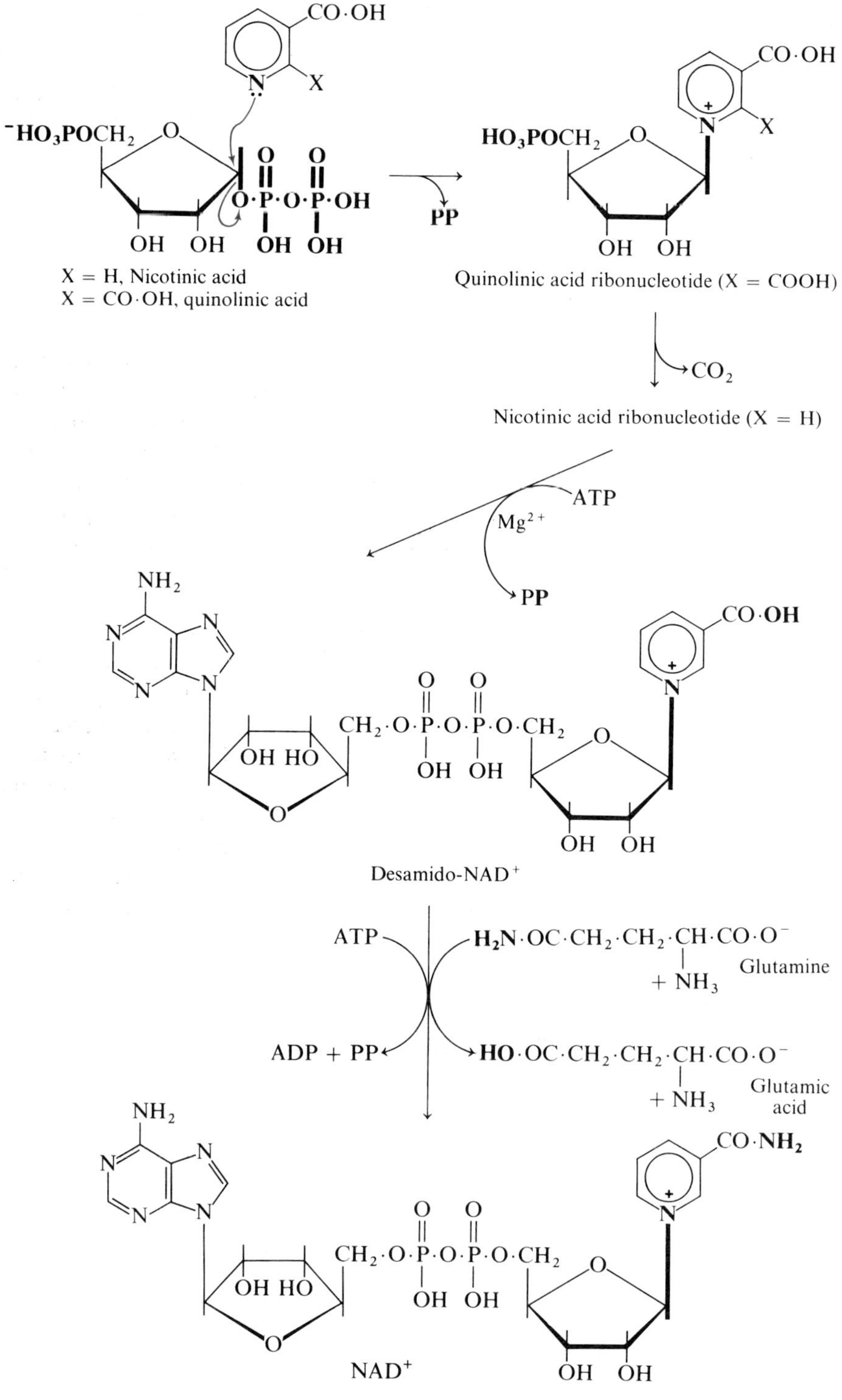

$^{-}HO_3POCH_2$ ... $O \cdot P \cdot O \cdot P \cdot OH$ (OH OH, OH OH) + nicotinic/quinolinic acid → $HO_3POCH_2$ ribonucleotide + PP

X = H, Nicotinic acid
X = CO·OH, quinolinic acid

Quinolinic acid ribonucleotide (X = COOH)

↓ → $CO_2$

Nicotinic acid ribonucleotide (X = H)

ATP, $Mg^{2+}$ → PP

$CH_2 \cdot O \cdot P \cdot O \cdot P \cdot O \cdot CH_2$ (CO·OH)

Desamido-NAD$^+$

ATP → ADP + PP

$H_2N \cdot OC \cdot CH_2 \cdot CH_2 \cdot CH(NH_3^+) \cdot CO \cdot O^-$ Glutamine

→ $HO \cdot OC \cdot CH_2 \cdot CH_2 \cdot CH(NH_3^+) \cdot CO \cdot O^-$ Glutamic acid

$CH_2 \cdot O \cdot P \cdot O \cdot P \cdot O \cdot CH_2$ (CO·NH$_2$)

NAD$^+$

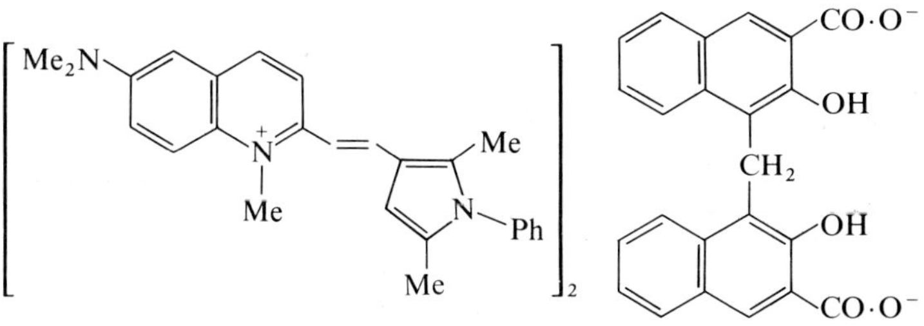

[ $Me_2N$ … $\overset{+}{N}$ … Me … Me … N … Ph … Me ]$_2$   $CO \cdot O^-$ … OH … $CH_2$ … OH … $CO \cdot O^-$

*Viprynium Embonate*

*Pyridostigmine*, an anticholinesterase used in the treatment of myasthenia gravis, owes its activity to structural resemblance to acetylcholine (Chapter 16). It is less potent than *Neostigmine*, which bears a somewhat closer structural analogy, but has a longer duration of action.

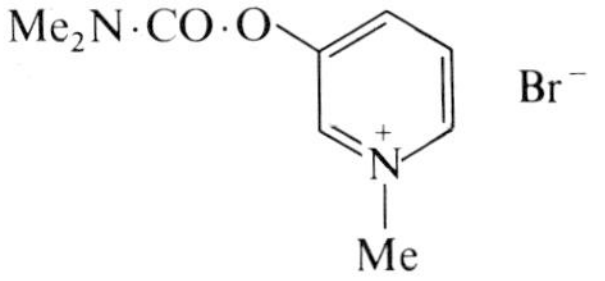

$Me_2N \cdot CO \cdot O$ … $\overset{+}{N}$ … Me   $Br^-$

$Me_2N \cdot CO \cdot O$ … $\overset{+}{N}Me_3$   $Br^-$

*Pyridostigmine Bromide*

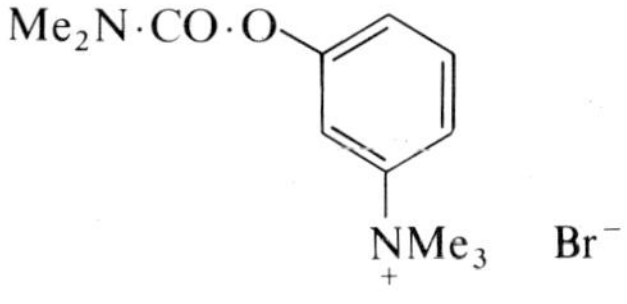

*Neostigmine Bromide*

In the presence of moist silver oxide or other strong alkali, the quaternary halides form quaternary hydroxides, which exist in solution in equilibrium with the corresponding pseudo bases (carbinol bases).

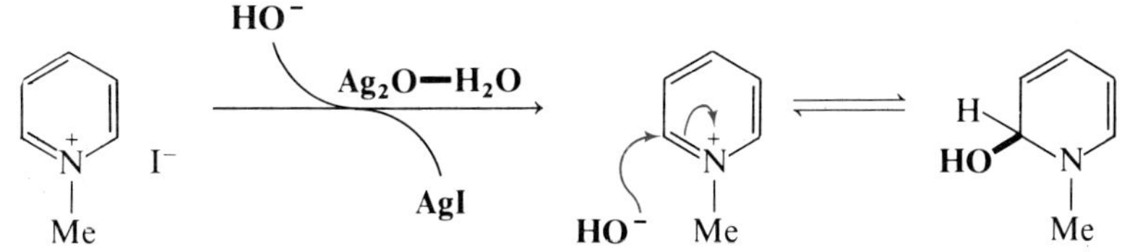

**HO⁻** … $\overset{+}{N}$ … Me   $I^-$   **$Ag_2O$—$H_2O$**, **AgI**   → **HO⁻** $\overset{+}{N}$ Me   ⇌   H, **HO** … N … Me

Pyridinium methiodide

Pseudo bases of some fused-ring hetero-aromatics have been isolated.

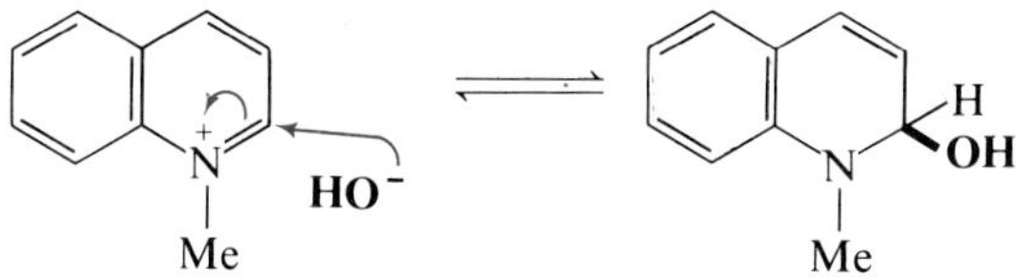

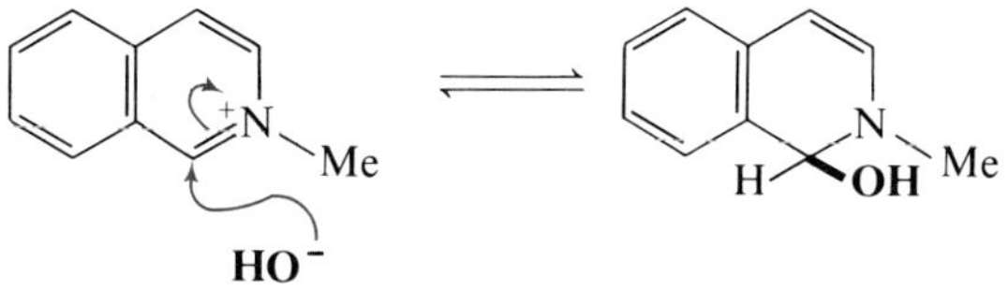

$\overset{+}{N}$ … Me   **HO⁻**   ⇌   N … Me … H … **OH**

$\overset{+}{N}$ … Me   **HO⁻**   ⇌   N … Me … H … **OH**

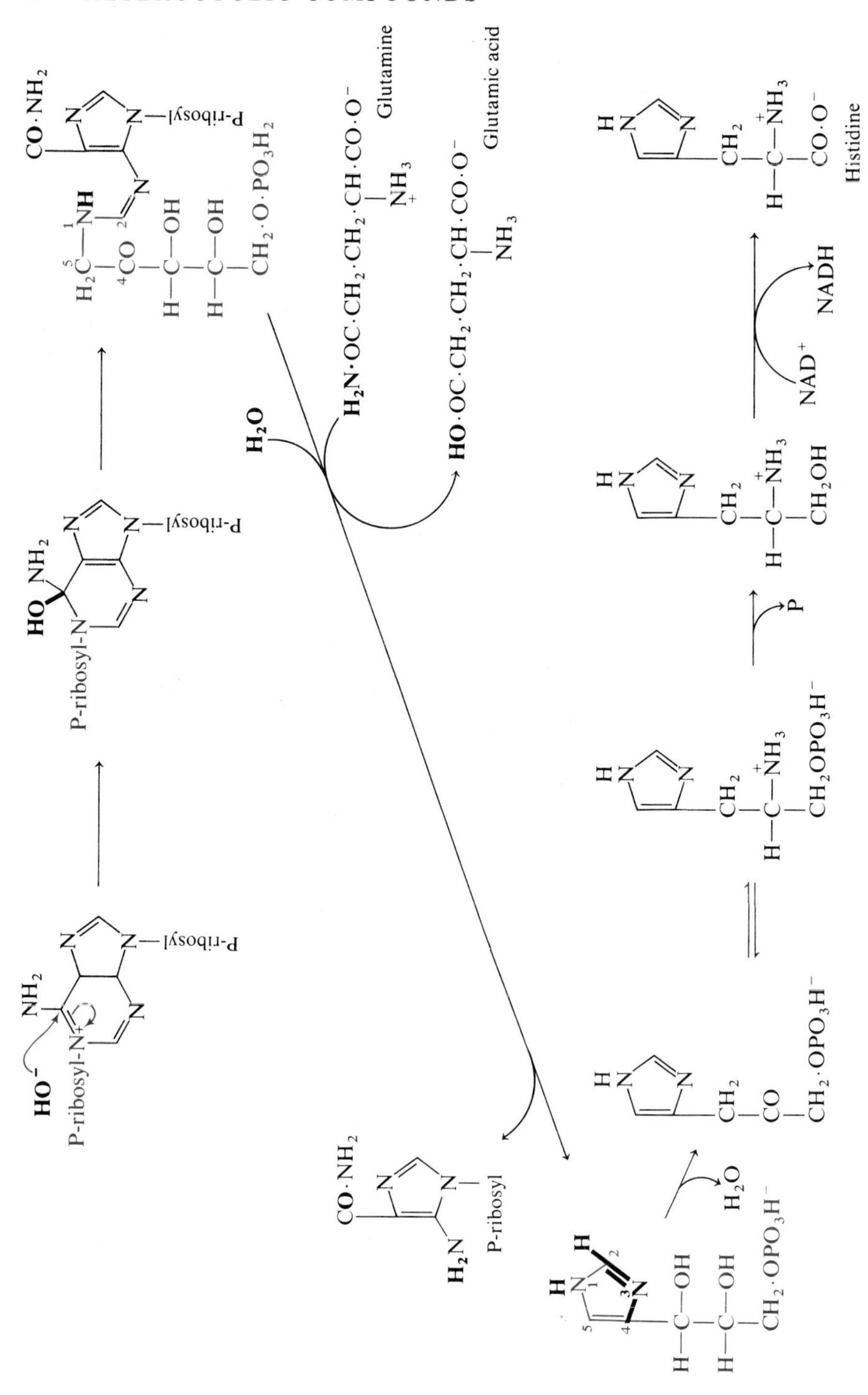
HO⁻
P-ribosyl
P-ribosyl
H₂O
H₂N·OC·CH₂·CH₂·CH·CO·O⁻
Glutamine
HO·OC·CH₂·CH₂·CH·CO·O⁻
Glutamic acid
NAD⁺
NADH
Histidine
P-ribosyl
H₂O
P

A similar carbinolamine intermediate could be involved in the disruption of the pyrimidine ring of *N*′-(5′-phosphoribosyl)-AMP leading to the biosynthesis of histidine in micro-organisms (Neidle and Waelsch, 1959; Moyed and Magasanik, 1957). (See page 810.)

The most important feature of the quaternary salts of $\pi$-deficient N-hetero-aromatics is the increased reactivity which results from the enhanced electron-withdrawal towards the positively-charged nitrogen. Nucleophilic substitution occurs much more readily than in the parent tertiary bases, and it is this enhanced reactivity which accounts for the ease with which $NAD^+$ undergoes hydride ion attack in the course of reduction to NADH by liver-alcohol dehydrogenase and various other enzymic systems.

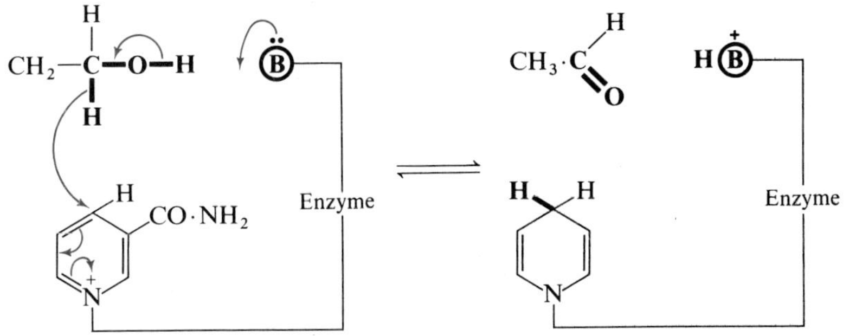

$CH_2$—C(H)(H)—O—H  B̈ ⇌ $CH_3 \cdot$C(H)=O  H$\overset{+}{\text{B}}$

CO·$NH_2$  Enzyme  H H  Enzyme

## Oxidation

$\pi$-Deficient N-hetero-aromatics are readily oxidised with peracids (perbenzoic acid), and with hydrogen peroxide in acetic acid to the corresponding *N*-oxides (den Hertog and Coombé, 1951). The smaller dipole moment of pyridine-*N*-oxide accords with the view that it is a resonance hybrid. This accounts for its enhanced reactivity in both electrophilic and nucleophilic substitution.

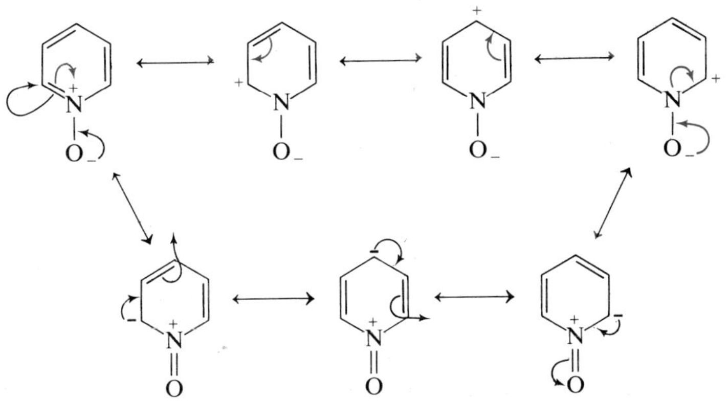

Thus, pyridine- and quinoline-*N*-oxides are nitrated at 100°C with potassium nitrate and fuming sulphuric acid to give the 4-nitro compounds.

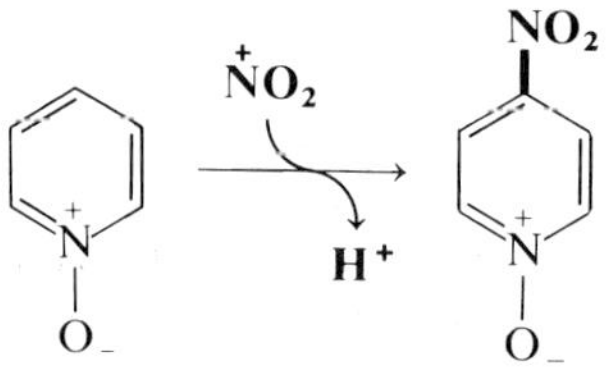

$\overset{+}{N}O_2$ $H^+$ $NO_2$ $\overset{+}{N}$ $O_-$

Whilst a number of heteroparaffinic bases are oxidised metabolically to *N*-oxides, metabolic oxidation of $\pi$-deficient-N-hetero-aromatics typically results in hydroxylation. Thus, pyridine hydroxylates in the 3-position, whilst quinoline yields 3-, 6- and 8-hydroxyquinolines together with 6-hydroxycarbostyril (Parke, 1968). The antihypertensive, *Hydrallazine* (Hydralazine Hydrochloride), is also metabolically hydroxylated, and then conjugated, in the benzenoid rather than in the heterocyclic ring.

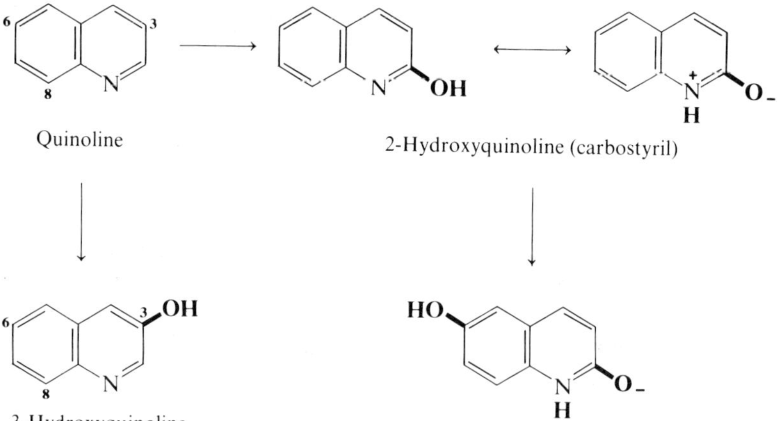

Quinoline

2-Hydroxyquinoline (carbostyril)

3-Hydroxyquinoline

6-Hydroxcarbostyril

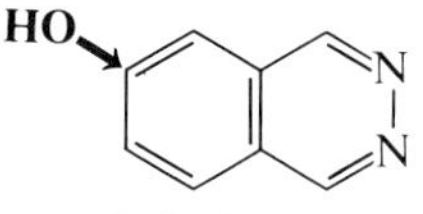

Hydrallazine

More complex $\pi$-deficient-N-hetero-aromatics are similarly hydroxylated. Thus, xanthine oxidase (xanthine:$O_2$-oxidoreductase) brings about hydroxylation of the purine, hypoxanthine, first in the pyrimidine ring to give xanthine, and subsequently in the imidazole ring to form uric acid. The xanthine oxidase inhibitor, *Allopurinol*, which is used in the treatment of gout, is similarly oxidised metabolically to alloxanthine (Elion, Kovensky, Hitchings, Metz and Rundles,

1966). The effect of *Allopurinol* as a xanthine oxidase inhibitor is to promote the renal clearance of hypoxanthine and xanthine and prevent the formation of excessive amounts of the much less soluble uric acid.

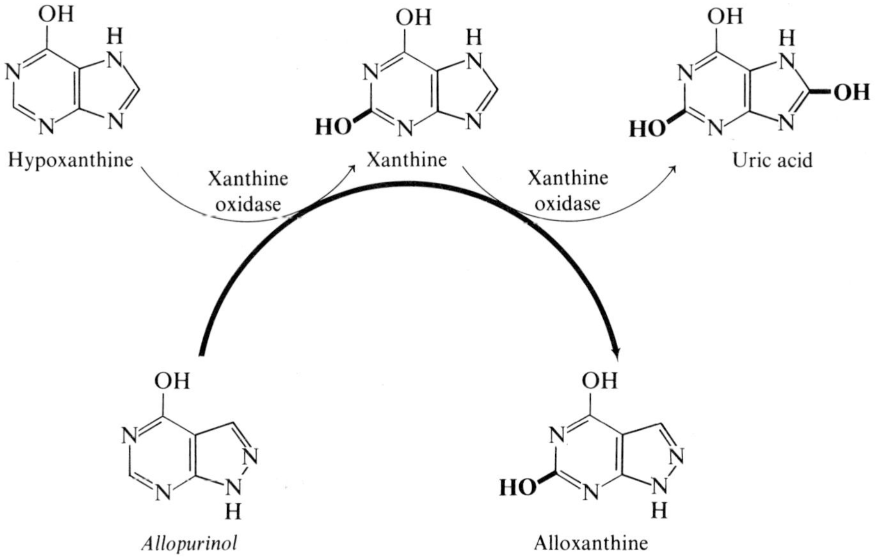

Hypoxanthine — Xanthine oxidase → Xanthine — Xanthine oxidase → Uric acid

*Allopurinol* → Alloxanthine

The antineoplastic agent, *Methotrexate*, similarly undergoes oxidation, and also oxidative deamination to 4,7-dihydroxy-4-desaminomethotrexate (Redetzski, Redetzski and Elias, 1966).

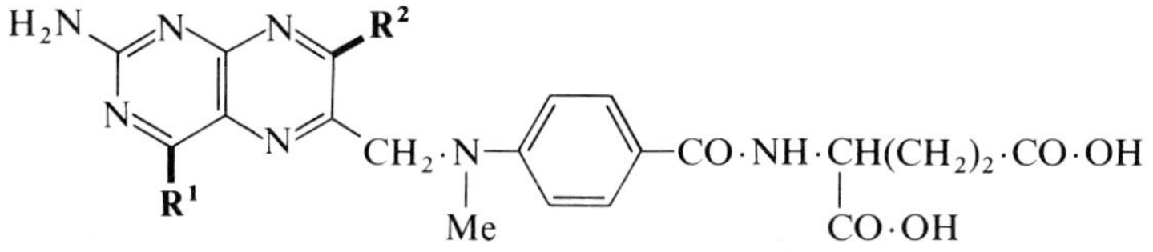

Methotrexate ($R^1 = NH_2$; $R^2 = H$)
4,7-Dihydroxy-4-desaminomethotrexate ($R^1 = R^2 = OH$)

## π-EXCESSIVE HETERO-AROMATICS

### General Properties

The resonance structure of π-excessive hetero-aromatics, which results from the contribution of the hetero-atom lone pair of electrons to the aromatic sextet, together with the characteristic five-membered ring leads to a planar structure. This assists face-to-face alignment and charge-transfer complex formation, with planar electron-deficient species, and accounts at least in part for the rôle of histidine and tryptophan in the binding of appropriately-substituted drug molecules to proteins.

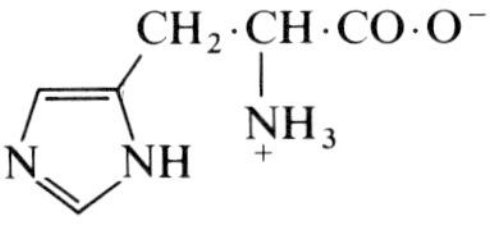

Histidine

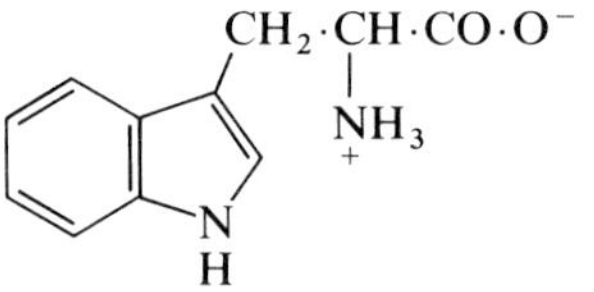

Tryptophan

The planarity of pyrrole also leads to the highly stable planar macrocyclic structure of the porphin ring system, from which the iron-protoporphyrin IX, haem, is derived.

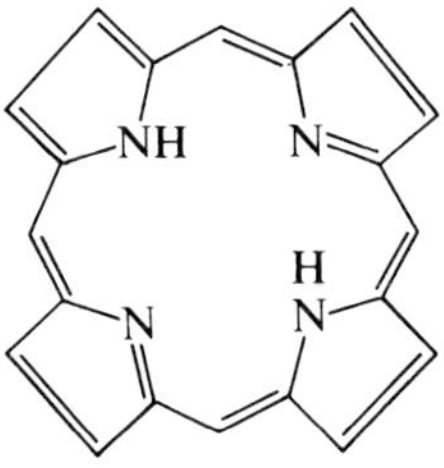

Porphin

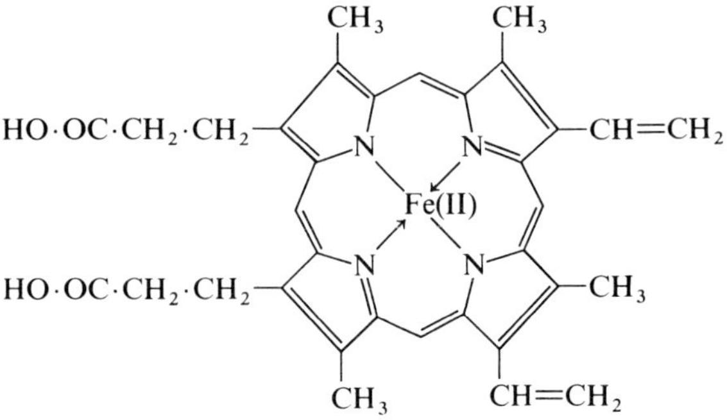

Haem

**Physical Properties**

Pyrrole, furan and thiophen are all volatile liquids, with boiling points which reflect the degree of charge distribution between the hetero-atom and the ring.

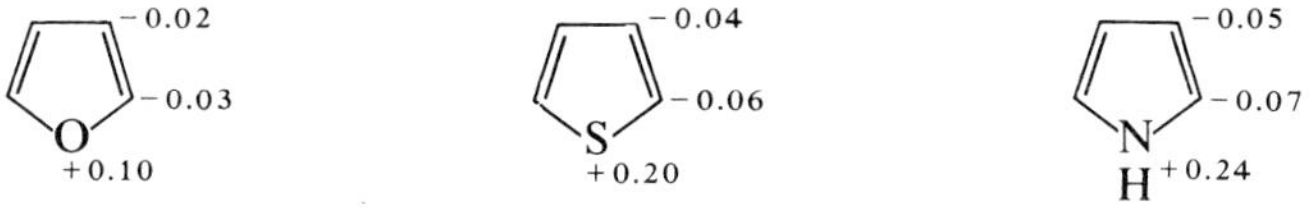

| | Furan | Thiophen | Pyrrole |
|---|---|---|---|
| b.p. | 31°C | 84°C | 131°C |

In contrast to pyridine, which is completely miscible with water, pyrrole is almost completely insoluble in water (*ca* 1 in 800). This is due to participation of the lone pair of electrons from nitrogen in the hetero-aromatic resonance, so that they are not readily available for hydrogen bonding to water.

Furan and thiophen are similarly almost completely insoluble in water despite the fact that the hetero-atom in both cases has an additional lone pair of electrons, which should theoretically be available for hydrogen bonding to the solvent.

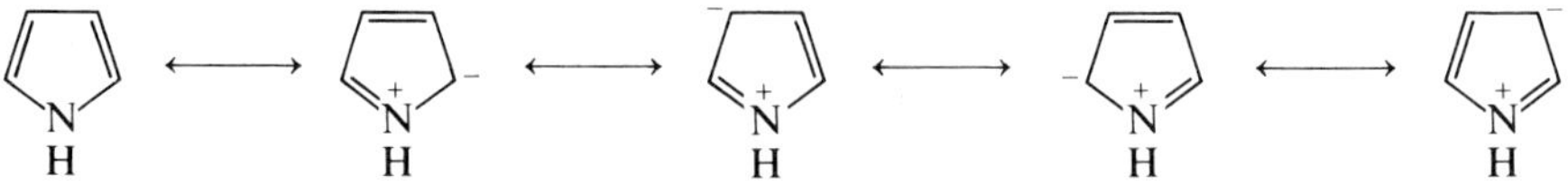

Additional nitrogenous hetero-atoms enhance water solubility, since the additional hetero-atom is necessarily in a C—N=C environment, with $\pi$-deficient characteristics and hence a lone pair available for hydrogen bonding, as in pyridine. Pyrazole, imidazole, oxazole, furazole and the thiadiazoles are, therefore, all much more readily water-soluble than the parent heterocycle.

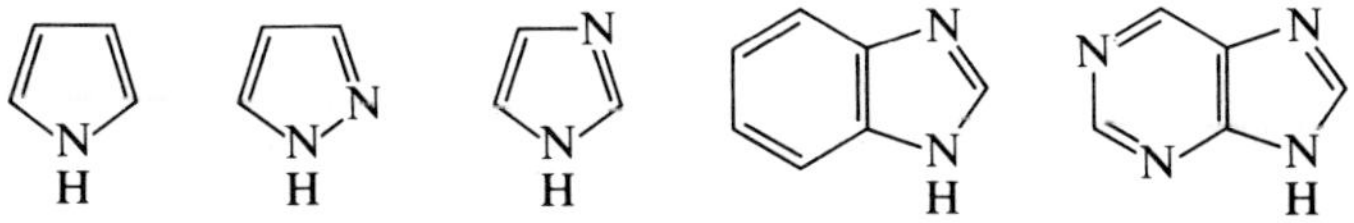

| | | Pyrrole | Pyrazole | Imidazole | Benzimidazole | Purine |
|---|---|---|---|---|---|---|
| Solubility in water | 1 in | 800 | 1 | < 1 | slightly soluble | 2 |

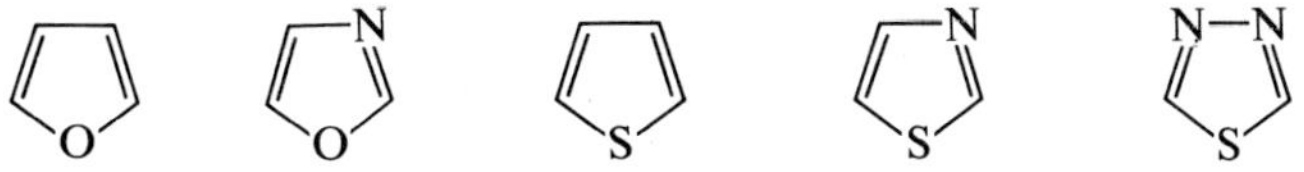

| | | Furan | Oxazole | Thiophen | Thiazole | 1,2,4-Thiadiazole |
|---|---|---|---|---|---|---|
| Solubility in water | 1 in | 100 | 1 | almost soluble | slightly soluble | |

Fusion with a benzenoid ring as in indole, benzofuran, benzimidazole and benzothiophen in general reduces solubility in water, but this effect on solubility is minimised in fused-ring compounds where the second ring is $\pi$-deficient-N-heteroaromatic, as in purine.

**Basic Properties**

In contrast to pyridine, pyrrole and indole have only very weakly basic properties (p$K_a$ of base-conjugated acid, $BH^+$, *ca* 0.4), since nitrogen has no unshared electrons which are completely free to accept a proton. In consequence, the equilibrium between pyrrole and a mineral acid favours dissociation.

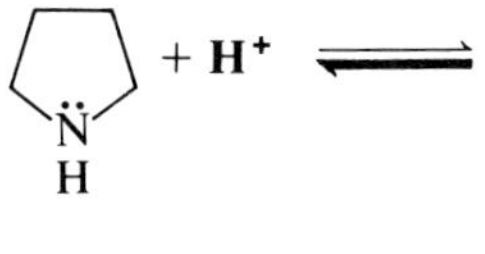

+ $H^+$

Such protonation as is favoured is more likely to occur on carbon, and this probably accounts for the polymerisation products which are formed from pyrrole (and furan) in acidic solution.

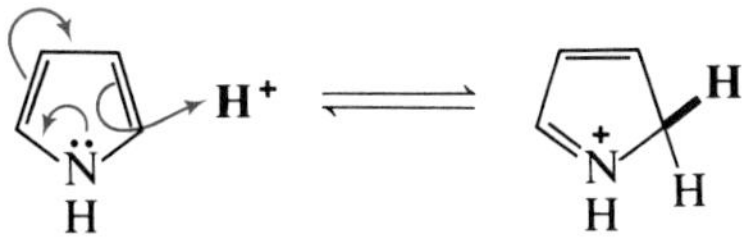

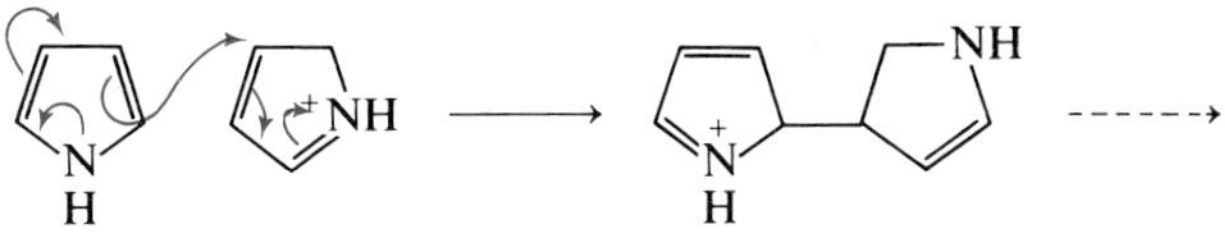

$H^+$

Indole similarly undergoes dimerisation in acidic solution, but carbazole and dibenzofuran are stable, since the reactive 2- and 3-positions are blocked by fusion with the adjoining benzene rings.

Furan is even more weakly basic than pyrrole, but similarly polymerises in acidic solution. 2,5-Disubstituted furans, in which both α-positions are blocked, undergo ring cleavage in acidic solution (Benson, 1936).

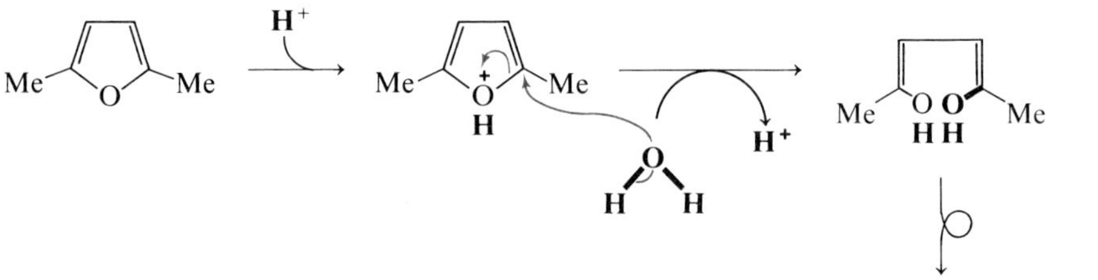

Me O Me $\xrightarrow{H^+}$ Me $\overset{+}{O}$ Me, H; $H_2O$ → $H^+$; Me O O Me, H H

↓

$CH_3 \cdot CO \cdot CH_2 \cdot CO \cdot CH_3$
Acetylacetone

Furfuryl alcohol also undergoes ring fission by a more complex mechanism to form laevulinic acid.

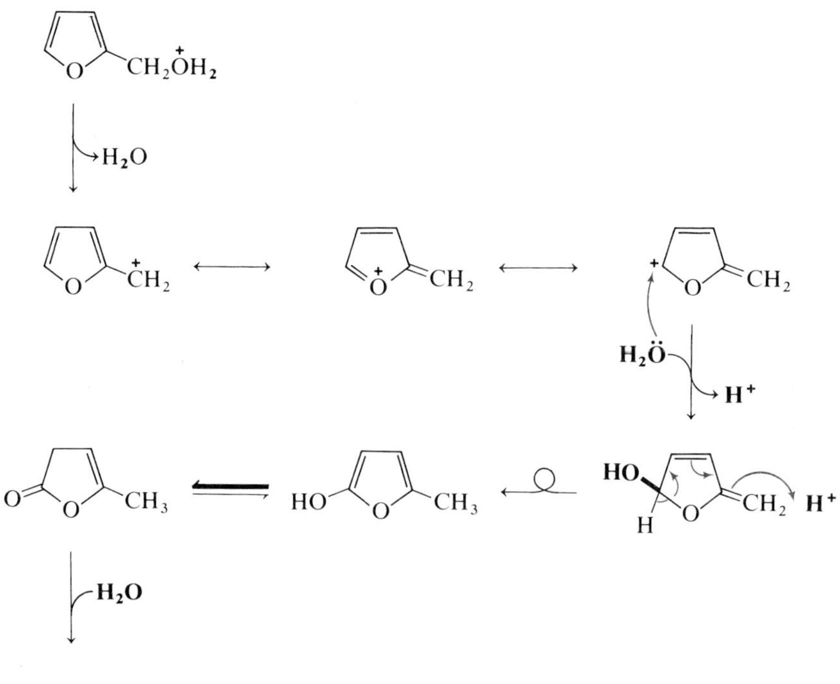

$CH_2\overset{+}{O}H_2$

↓ → $H_2O$

$\overset{+}{C}H_2$ ⟷ $\overset{+}{O}$ $CH_2$ ⟷ + O $CH_2$

$H_2\ddot{O}$ ↓ → $H^+$

O O $CH_3$ ⇌ HO O $CH_3$ ← HO H O $CH_2$ $H^+$

↓ $H_2O$

$CH_3 \cdot CO \cdot CH_2 \cdot CH_2 \cdot CO \cdot OH$
Laevulinic acid

Thiophen is virtually non-basic. Basicity of pyrrole is increased by the electron donating effect of 2- and 3-alkyl substituents sufficiently to permit the formation under anhydrous conditions of crystalline hydrobromides. Base-strengthening also occurs with the introduction of a second nitrogenous substituent in pyrazole (p$K_a$ 2.53) and imidazole (p$K_a$ 7.16). This is partly accounted for by the gain in resonance which is possible in the ion when the two nitrogen atoms become equivalent.

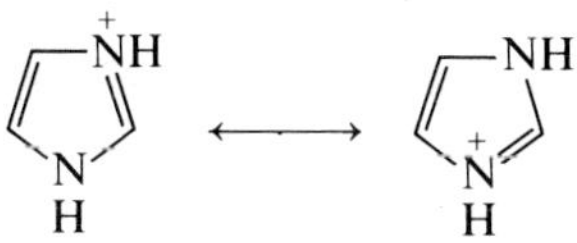

$\overset{+}{N}H$ ⟷ NH ... $\overset{+}{N}$ H

### Quaternary Salts

For similar reasons, whereas pyrrole and indole are too weakly basic to form quaternary ammonium salts, iminazole, oxazole and thiazole are able to form quaternary salts utilising the lone pair of the —C=N—C group. The formation of *Thiamine Chloride* (vitamin $B_1$), synthetically from 2-methyl-4-amino-5-hydroxymethylpyrimidyl bromide and biosynthetically from the corresponding phosphate is typical.

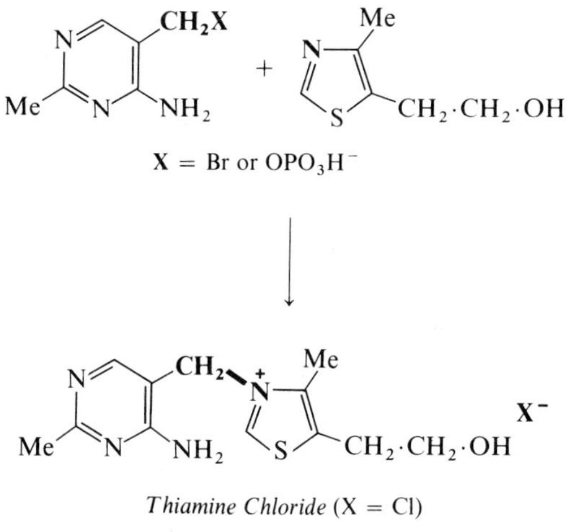

$CH_2X$ Me N N Me N $NH_2$ + S $CH_2 \cdot CH_2 \cdot OH$

X = Br or $OPO_3H^-$

↓

$CH_2$ $\overset{+}{N}$ Me N Me N $NH_2$ S $CH_2 \cdot CH_2 \cdot OH$ $X^-$

*Thiamine Chloride* (X = Cl)

*Thiamine Chloride* is readily water-soluble, stable to acid, but susceptible to nucleophilic attack by $HO^-$ in the 2-position of the thiazolium group in alkaline solution. The resulting 2-hydroxy-4-thiazoline is a thio-hemiacetal and undergoes spontaneous ring opening to yield the corresponding sulphydryl formylamine. The latter is readily oxidised spontaneously in air to the corresponding disulphide, and via the intermediate thioacetal to thiochrome.

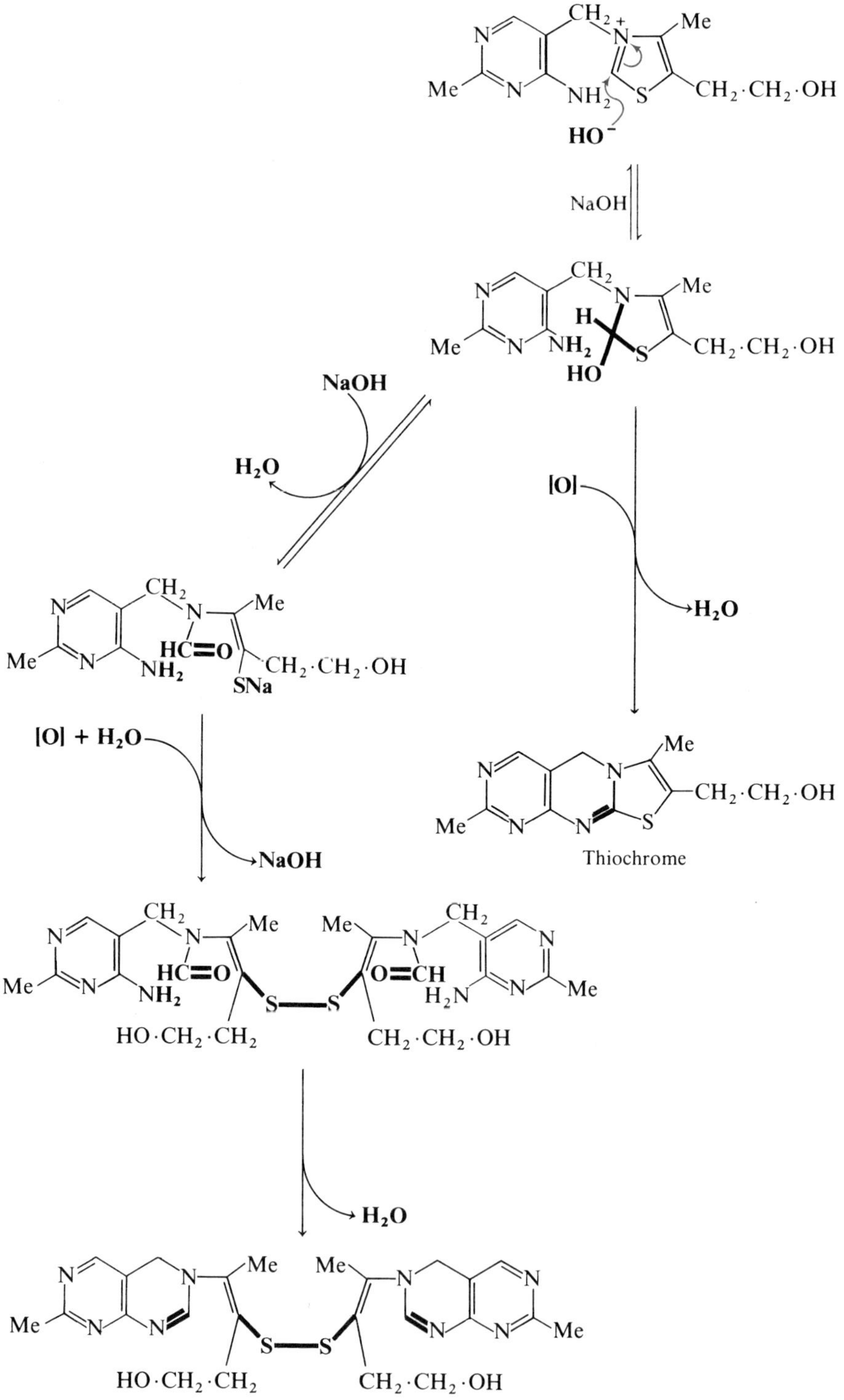

$HO^-$

NaOH

NaOH

$H_2O$

[O]

$H_2O$

[O] + $H_2O$

NaOH

Thiochrome

$H_2O$

The electron-attracting properties of the pyrimidylmethyl group destabilise the thiazolium group in *Thiamine Chloride*. Aqueous sodium sulphite cleaves the molecule by attack at the methylene group. The use of sulphur dioxide as a preservative or sodium metabisulphite as an antioxidant in preparations of the vitamin should, therefore, be avoided.

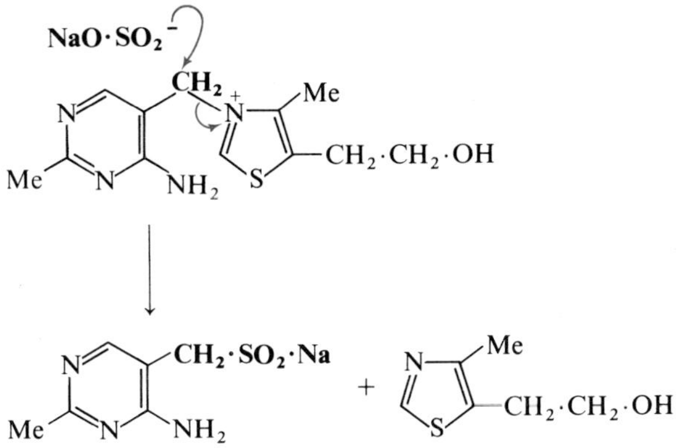

NaO·SO$_2^-$ CH$_2$ Me N N Me N NH$_2$ S CH$_2$·CH$_2$·OH

CH$_2$·SO$_2$·Na N Me N NH$_2$ + N Me S CH$_2$·CH$_2$·OH

Quaternary guaninium compounds formed as a result of biological alkylation of guanine nucleotides in DNA are similarly susceptible to nucleophilic attack with decomposition of the quaternary salt (**2**, 2).

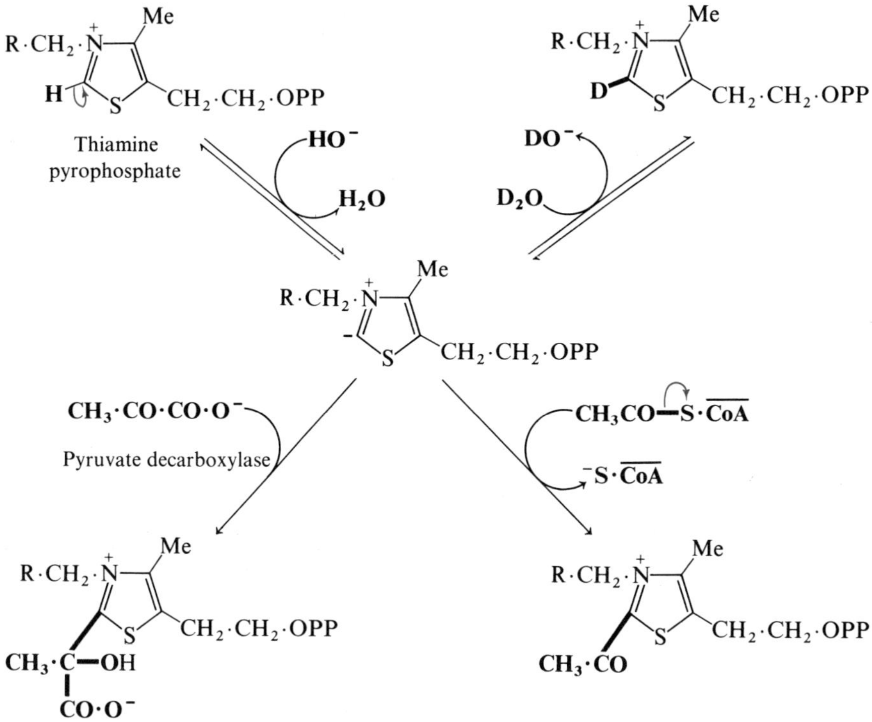

R·CH$_2$·N Me H S CH$_2$·CH$_2$·OPP
Thiamine pyrophosphate
HO$^-$ H$_2$O
DO$^-$ D$_2$O
R·CH$_2$·N Me D S CH$_2$·CH$_2$·OPP

R·CH$_2$·N Me S CH$_2$·CH$_2$·OPP

CH$_3$·CO·CO·O$^-$
Pyruvate decarboxylase
CH$_3$CO—S·CoA
$^-$S·CoA

R·CH$_2$·N Me S CH$_2$·CH$_2$·OPP CH$_3$·C—OH CO·O$^-$

R·CH$_2$·N Me S CH$_2$·CH$_2$·OPP CH$_3$·CO

The intervention of thiamine pyrophosphate in certain important biochemical reactions results from the lability of the C—H bond at C-2 in the thiazolinium ring. Hydrogen at C-2 readily exchanges with deuterium in deuterium oxide (Breslow, 1957), through the ylid intermediate, which is also the key reactant in acetoin biosynthesis and pyruvate decarboxylation (page 819 and Chapter 13).

Crystallographic studies confirm that the molecule of *Thiamine Chloride* is non-planar and show that the dihedral angle between the two rings is 76°, in such a way that the amino group lies close to the C-2 hydrogen of the thiazole ring (Kraut and Reed, 1962).

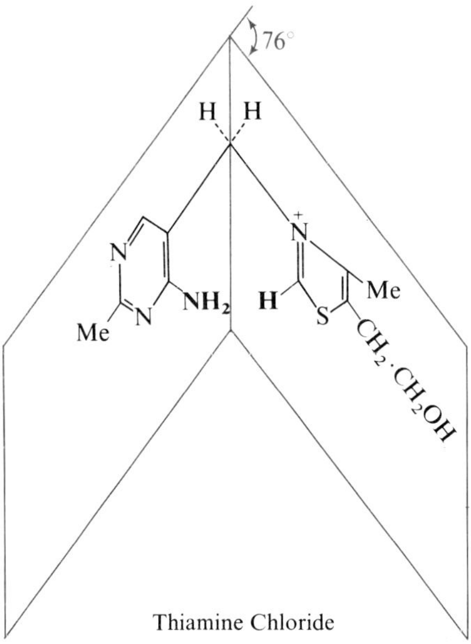

Thiamine Chloride

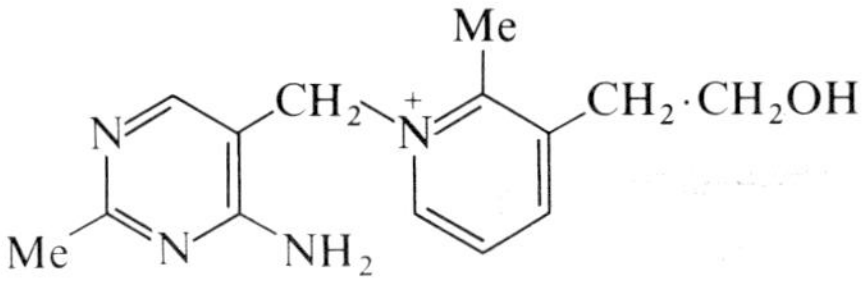

*Pyrithiamine*

Thiamine is an essential metabolite for bacteria, fungi, protozoa and higher animals, and has been identified as a component of a number of enzyme systems, particularly those associated with pyruvate metabolism (Chapter 13). Diphosphothiamine (co-carboxylase) has been identified as the enzyme responsible for the anaerobic decarboxylation of pyruvate in yeast (Lohmann and Schuster, 1937). *Pyrithiamine*, which retains the essential stereochemistry and most of the structural features of thiamine, inhibits the growth of micro-organisms unable to synthesise thiamine (Tracy and Elderfield, 1940). It is, however, unsuitable for use in human medicine owing to the appearance of toxic effects at blood concentrations which are bacteriostatic (Wyss, 1943).

**Acidic Properties**

Pyrrole and indole are also able to function as weak acids (p$K_a$ *ca* 15). This is evident in their ability to form alkali metal and Grignard salts under anhydrous conditions, which react with alkyl halides to give *N*-alkylpyrroles and *N*-alkylindoles.

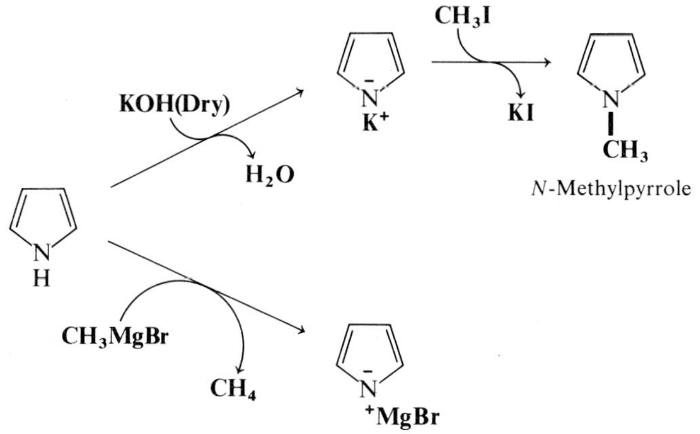

KOH(Dry)
$H_2O$
N
$K^+$
$CH_3I$
KI
N
$CH_3$
*N*-Methylpyrrole
N
H
$CH_3MgBr$
$CH_4$
N
$^+MgBr$

Loss of a proton is facilitated by the enhanced stability of the anion, which unlike pyrrole, has no **dipolar ion character** (p. 814).

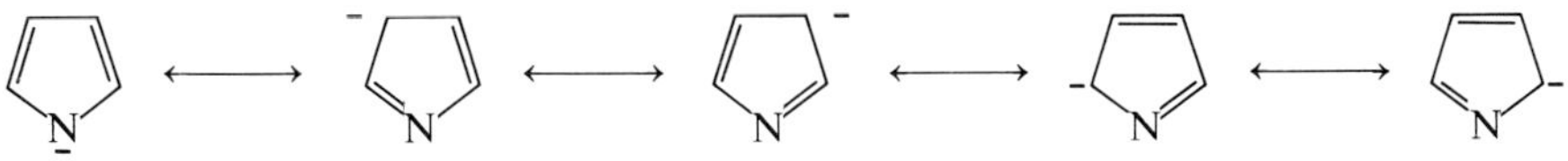

N ⟷ N ⟷ N ⟷ N ⟷ N

The acidic strength is enhanced by additional nitrogenous hetero-atoms which function as electron-attracting groups either in the same ring as in imidazole ($pK_a$ 14.5), 1,2,4-triazole ($pK_a$ 10.1) and benzimidazole ($pK_a$ 12.3), or in an adjoining ring as in purine (3,4,6-triaza-indole; $pK_a$ 8.93). Thus, the purine base, theophylline, is sufficiently acidic ($pK_a$ 8.77) to be soluble in ammonia, and to

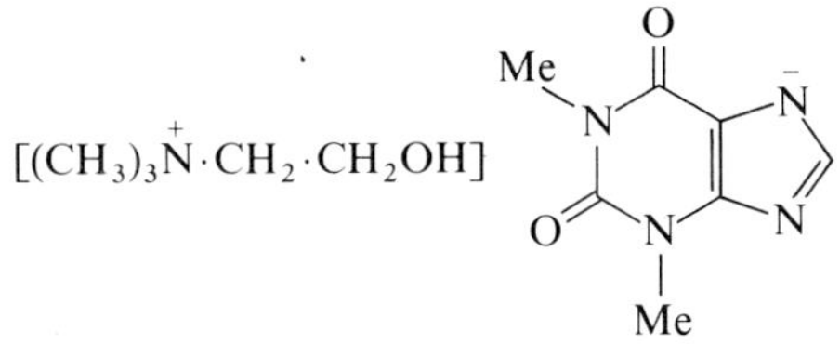

$[(CH_3)_3\overset{+}{N}\cdot CH_2\cdot CH_2OH]$ Me N O N N O N Me

*Choline Theophyllinate*

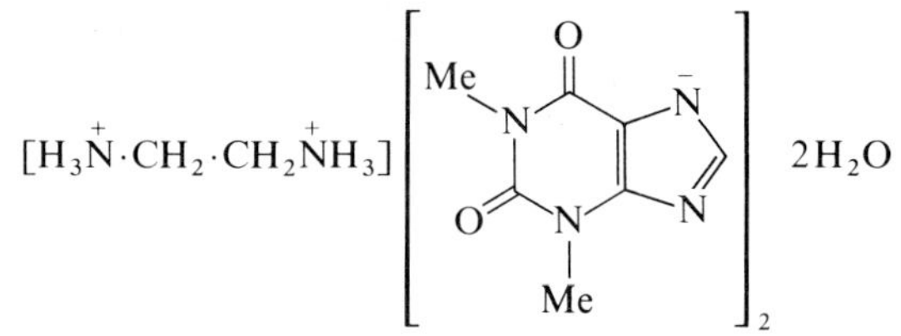

$[H_3\overset{+}{N}\cdot CH_2\cdot CH_2\overset{+}{N}H_3]$ [Me N O N N O N Me]$_2$ $2H_2O$

*Aminophylline*

form stable water-soluble salts with choline (*Choline Theophyllinate*) and ethylenediamine (Aminophylline) whilst *Theobromine*, *Mercaptopurine* and *Azathioprine* are soluble in solutions of alkali metal hydroxides.

Other electron-attracting groups also enhance the acidic strength, as in 8-chlorotheophylline, which is sufficiently acidic to form stable water- and lipid-soluble salts, such as *Dimenhydrinate* and *Promethazine Theoclate* (Promethazine 8-chlorotheophyllinate), for use in the control of motion sickness.

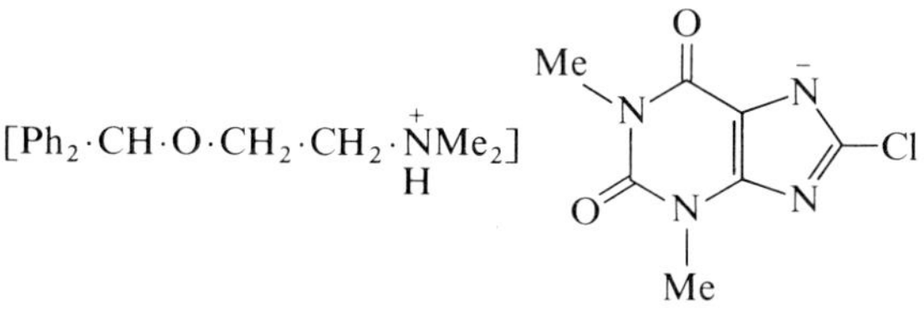

$[Ph_2 \cdot CH \cdot O \cdot CH_2 \cdot CH_2 \cdot \overset{+}{N}HMe_2]$

*Dimenhydrinate*

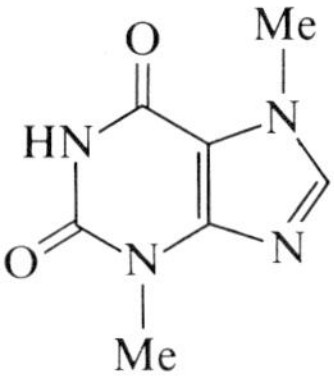

*Theobromine*

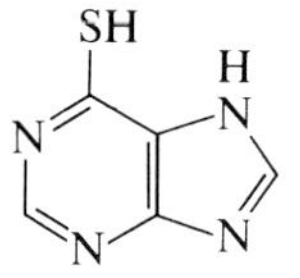

*Mercaptopurine*

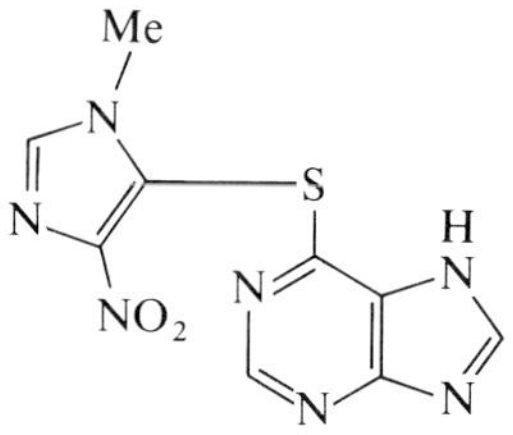

*Azathioprine*

**Porphyrins and Corins**

The ability to form stable metal salts reaches a peak in the cyclic tetrapyrrolic structure of the porphyrins, which are stable planar-4-co-ordinate and octahedral complexes with divalent metal ions. The iron–protoporphyrin IX complex, haem, is a square-planar chelate with ferrous iron linked by four resonance-hybridised N—Fe bonds. The stability of the complex is such that no exchange occurs with radio-labelled Fe(II), and iron is only displaced by treatment with *10N* hydrochloric acid.

Haem is capable of forming stable octahedral ($d^2sp^3$) complexes known as **haemochromes** by combination with two additional ligands. In both haemoglobin and myoglobin (Chapter 19), four haem units are linked to each molecule of carrier protein (globin) by one of the additional ligands via a histidine residue; the remaining ligand in each case is water. The latter, however, forms only a relatively weak ligand and is readily displaced by stronger field ligands such as $O_2$, CO, $CN^-$, NO and pyridine, which are capable of forming $\pi$-bonds. The oxygen complex, oxyhaemoglobin, the oxygen-carrier pigment present in fresh blood of all vertebrates, dissociates with release of oxygen in tissues under low oxygen tension, dissociation being enhanced by high $CO_2$ concentrations. Oxygen is readily displaced by carbon monoxide and cyanide ion to form the more stable complexes, carboxyhaemoglobin and cyanhaemoglobin.

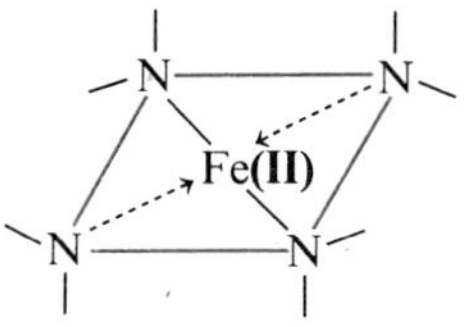

Fe(II)

Haem

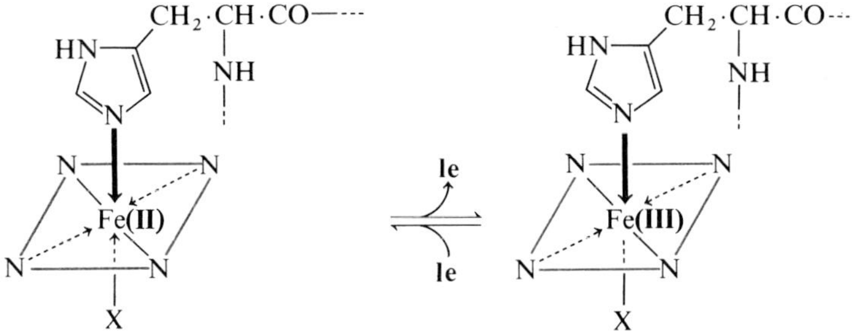

$X = H_2O$ Haemoglobin
$O_2$ Oxyhaemoglobin
CO Carboxyhaemoglobin
$CN^-$ Cyanhaemoglobin

$X = H_2O$ Hemiglobin
$HO^-$ Methaemoglobin

Ferrous iron in haem complexes can be oxidised to the ferric state to give penta-co-ordinate square pyramidal complexes such as haematin, which are readily converted to the corresponding octahedral ferric iron complexes known as **hemichromes**.

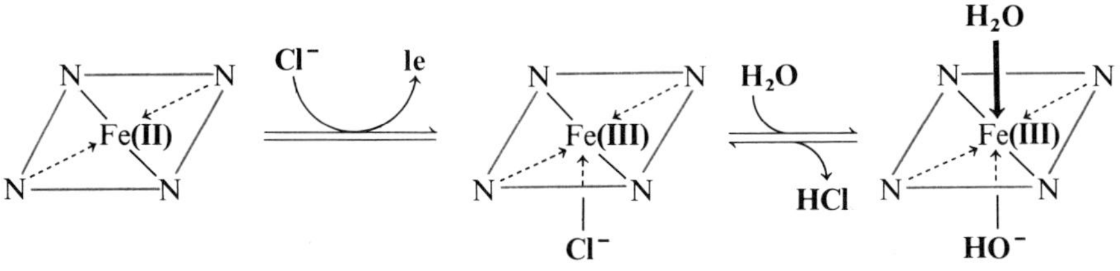

Haem

Haemin chloride

Haematin chloride

Metabolic breakdown of haemoglobin occurs by oxidative cleavage of an α-methine link with loss of one carbon atom to form thc open-chain tetrapyrrolic bile pigments, biliverdin and bilirubin. These are reduced by bacterial enzymes to stercobilinogen, which is the main excretory product.

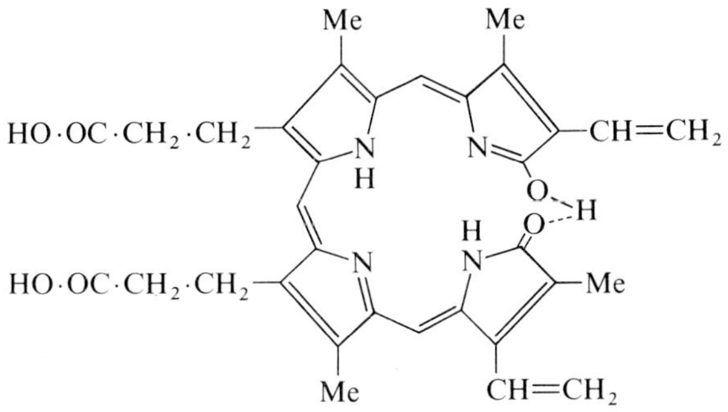

Biliverdin

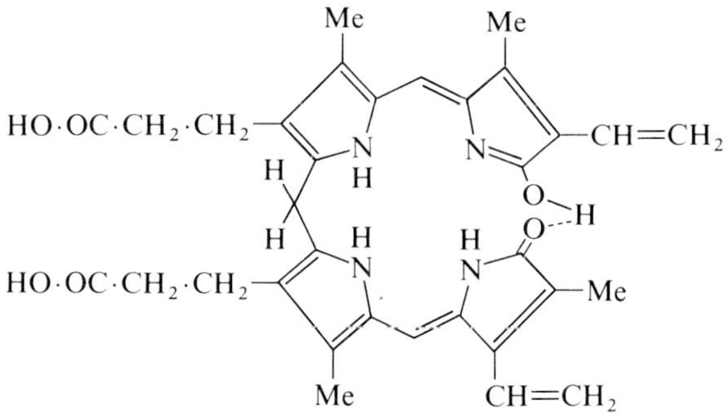

Bilirubin

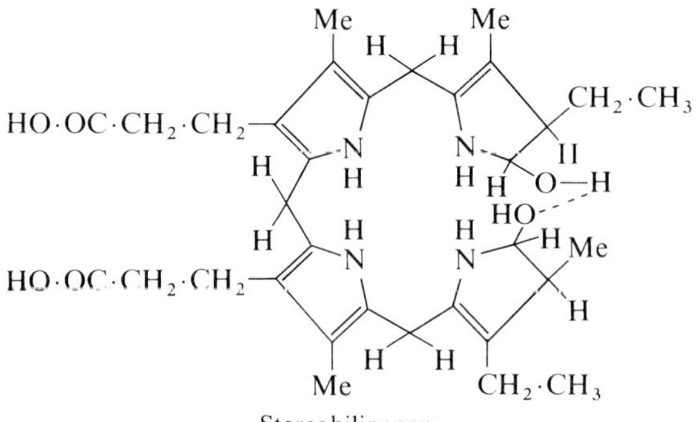

Stercobilinogen

The cytochromes form a group of haem-protein enzymes which function as oxidation–reduction systems in the mitochondrial electron transport chain. This effects the transfer of hydrogen to molecular oxygen to form water by a graded series of oxidation–reduction steps involving the oxidation state of the iron complex.

In cytochrome *c*, the haem group is covalently bound by sulphide bonds to the enzyme protein (Theorell, 1938, 1941), but in the other cytochromes, the link is non-covalent.

| 2NADH / $2NAD^+$ | $FlavH_2$ / Flav | 2Cyt *c* ($Fe^{2+}$ / $Fe^{3+}$) | 2Cyt $c_1$ ($Fe^{2+}$ / $Fe^{3+}$) | 2Cyt *a* ($Fe^{2+}$ / $Fe^{3+}$) | $H_2O$ / $\frac{1}{2}O_2$ |
|---|---|---|---|---|---|
| $E'_0$ (Volts) | −0.19 | +0.25 | +0.22 | +0.29 | |

*Cyanocobalamin* (vitamin $B_{12}$), the cobalt-containing vitamin which is present in liver and used in the treatment of pernicious anaemia, is an octahedral tetrapyrrole (**corin**) cobalt complex with cyanide ion ($CN^-$) and 5,6-dimethyl-1-($\alpha$-ribofuranosyl) benzimidazole-3′-phosphate ligands.

*Cyanocobalamin*

## Electrophilic Substitution

$\pi$-Excessive hetero-aromatics are activated for electrophilic attack due to their asymmetrical charge distribution, which leaves the ring carbons negatively-charged. As a result, they are more susceptible to electrophilic attack than benzene, with pyrrole being the most reactive and even more so than phenol. The order of reactivity is pyrrole > indole > furan > thiophen > benzene, with substitution occurring preferentially in the 2-position of the heterocyclic ring and in the 3-position if the latter is blocked. This substitution pattern not only reflects the higher charge density in the 2-position, but also the fact that substitution in this position gives the more stable transition state with three, as opposed to two, contributing resonance forms.

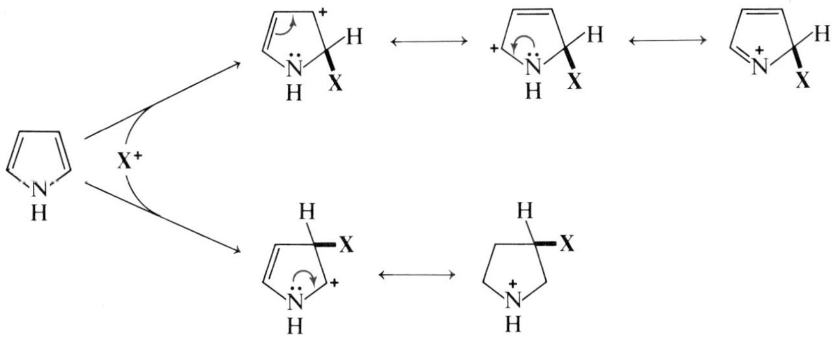

The tendency of pyrrole and furan to polymerise under strongly acidic conditions limits the choice of reagent and conditions, but pyrrole, furan and thiophene can all be nitrated, sulphonated, acylated and halogenated under appropriate conditions, and all undergo diazo coupling with diazonium salts. Typical reactions and products are shown in the following schemes for pyrrole.

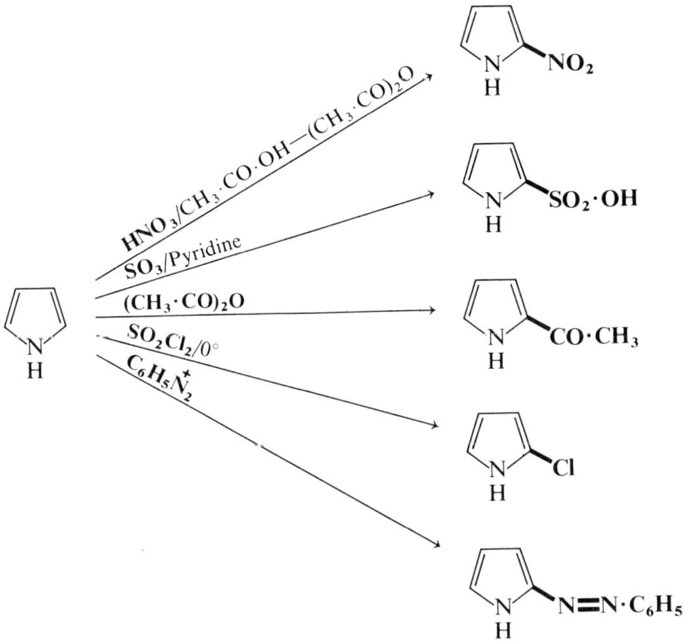

Similar substitutions can be carried out with furan and thiophen, and the same or similar reagents with minor modifications to the reaction conditions. In most cases, a second substituent usually enters the 5-position, irrespective of the nature of the 2-substituent. Thus, bromination of furoic acid gives 5-bromofuran-2-carboxylic acid, indicating that the directional influences which are decisive in the disubstitution of benzene are less important than the influence of the heterocyclic atom.

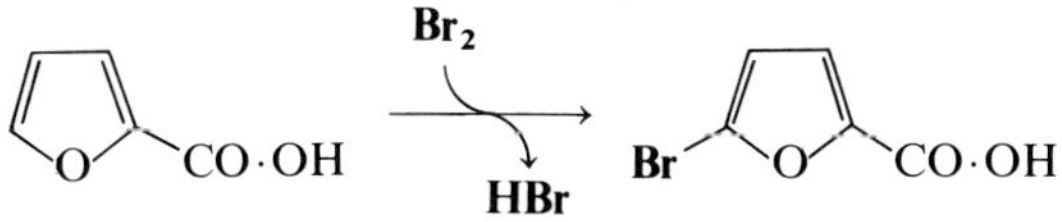

Br$_2$ → HBr
CO·OH — Br — CO·OH

Indole undergoes similar electrophilic substitution to pyrrole except that substitution almost invariably occurs in the 3-position in agreement with the electron density distribution. Sulphonation is the only exception, giving rise to indole-2-sulphonic acid.

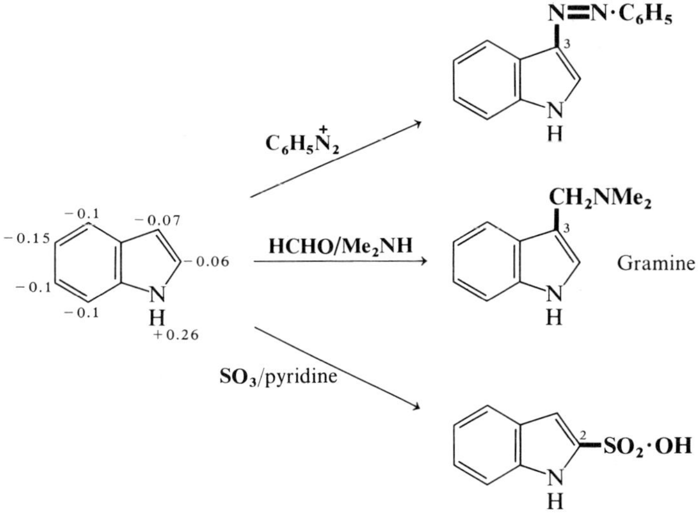

N=N·$C_6H_5$

$C_6H_5\overset{+}{N}_2$

$CH_2NMe_2$

−0.1 −0.07 −0.15 −0.06 −0.1 −0.1 H +0.26

HCHO/$Me_2NH$

Gramine

$SO_3$/pyridine

$SO_2$·OH

Gramine is an important intermediate in the synthesis of tryptophan, gramine methosulphate being condensed with acetamidomalonic ester.

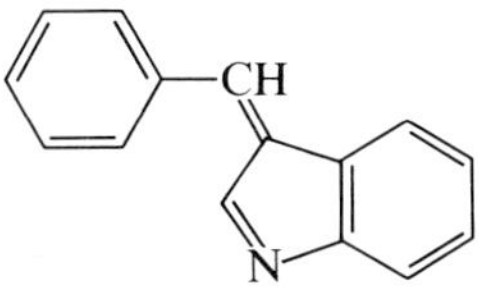

CH

$Me_2N$

Benzylidene indolenine

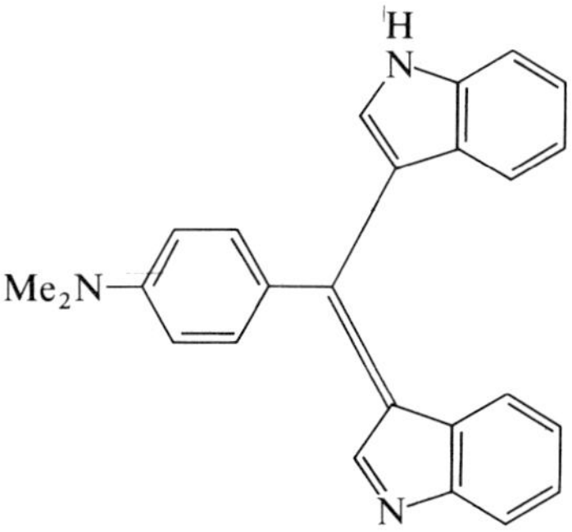

Di-indolylphenylmethane dye

Indoles condense readily with aromatic aldehydes to give benzylidene indolenines, and compounds related to the triphenylmethane dyes. The deep blue colour formed with *p*-dimethylaminobenzaldehyde in this way is widely used as an identification test, and for the quantitative colorimetric determination of ergot and other indolic alkaloids. (See page 827.)

### Substitution Products

Substituted $\pi$-excessive hetero-aromatics in general exhibit properties which resemble those of the corresponding benzenoid aromatics. Thus, furoic acid (furan-2-carboxylic acid; p$K_a$ 3.17) and thiophen-2-carboxylic acid (p$K_a$ 3.50) are typical aromatic carboxylic acids and, in keeping with the enhanced charge-density of the ring, are marginally stronger acids than benzoic acid (p$K_a$ 5.20). This effect is even more evident in the pyrrole-2-carboxylic acids which are readily decarboxylated even in boiling water. The acids form typical esters, such as *Diloxanide Furoate*. This is used in the treatment of chronic amoebiasis, since it has a specific action on the cysts of *E. histolytica.*

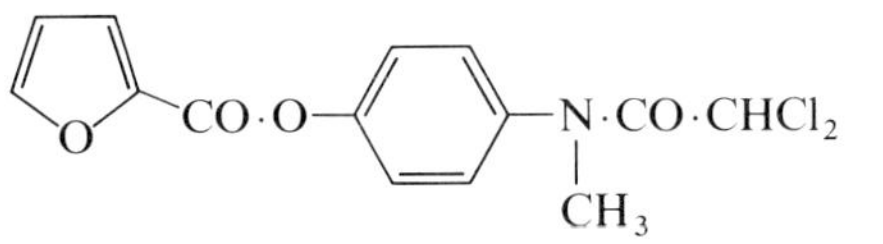

CO·O—N·CO·CHCl$_2$ (N–CH$_3$)

*Diloxanide Furoate*

Furfural and substituted furfurals function as typical aromatic aldehydes, undergoing a Cannizzaro reaction in the presence of aqueous sodium hydroxide.

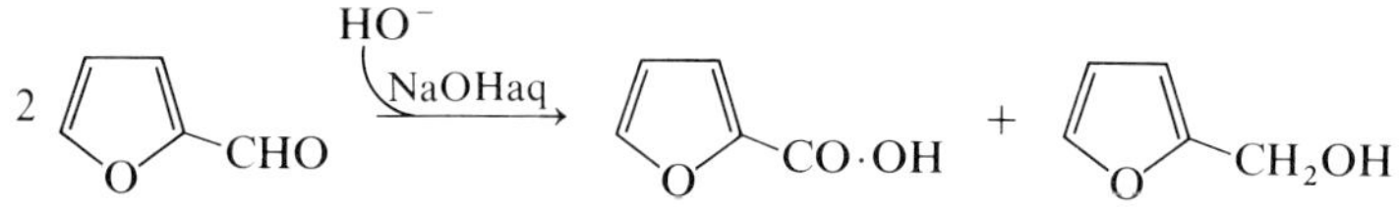

2 CHO —(HO$^-$, NaOHaq)→ CO·OH + CH$_2$OH

Furfural — 2-Furoic acid — Furfuryl alcohol

They also condense readily with hydroxylamine, hydrazine, phenylhydrazine, semicarbazide and also with amino compounds and active methylene compounds. A number of 5-nitrofurfurylidene compounds including *Nitrofurazone* and *Nitrofurantoin* find use as urinary antiseptics.

O$_2$N— —CH=N·NH·CO·NH$_2$

*Nitrofurazone*

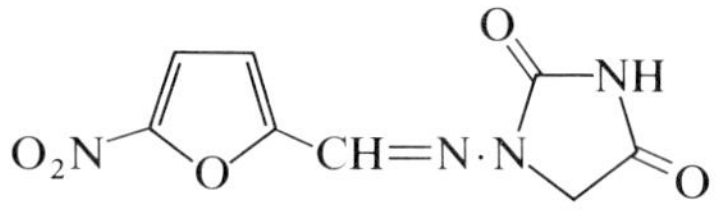

O$_2$N— —CH=N·N (O, NH, O)

*Nitrofurantoin*

Some compounds, notably amino- and hydroxy-derivatives, show evidence of anomalous behaviour. 2- and 3-Aminopyrroles are unknown, but substituted aminopyrroles and aminoindoles form diazonium salts on treatment with nitrous acid. Spectroscopic evidence (Kebrle and Hoffmann, 1956), however, shows that whereas 2-amino-1-methylindole favours the 2-amino structure rather than its 2-imino-tautomer, 2-aminoindole favours the 2-aminoindolenine structure, whilst both compounds form the 2-aminoindoleninium chloride in aqueous hydrochloric acid.

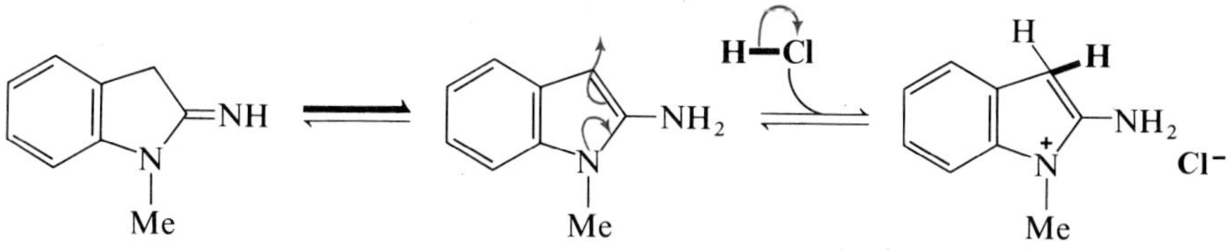

2-amino-1-methylindole

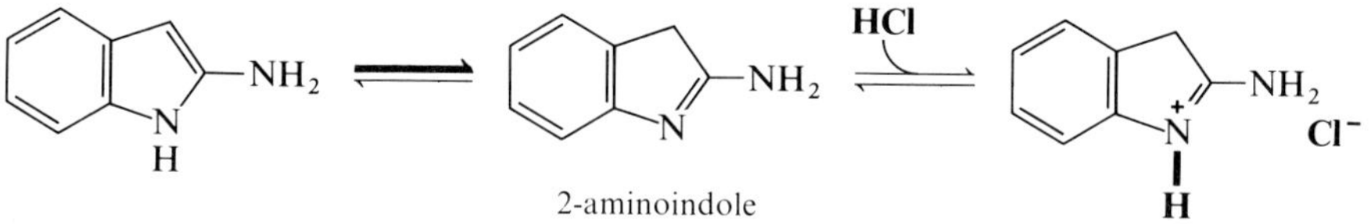

2-aminoindole

The corresponding aminofuran is believed to be formed in the metabolic reduction of *Nitrofurazone* (Beckett and Robinson, 1956). The product, however, appears to be unstable and could not be distinguished from the alternative 2-hydroxylamino compound, and like other 2-aminofurans is probably hydrolysed with loss of ammonia.

Hydroxypyrroles also show evidence of tautomerism, and in the absence of other substituents do not behave as phenols. However, there is considerable evidence to show that whilst such compounds as 2-hydroxy-4-carbethoxy-pyrrole (Grob and Ankli, 1949) show hydroxylic properties, they exist largely as the tautomeric 2-oxo-$\Delta^4$-pyrrolines, which are stabilised by amide type resonance.

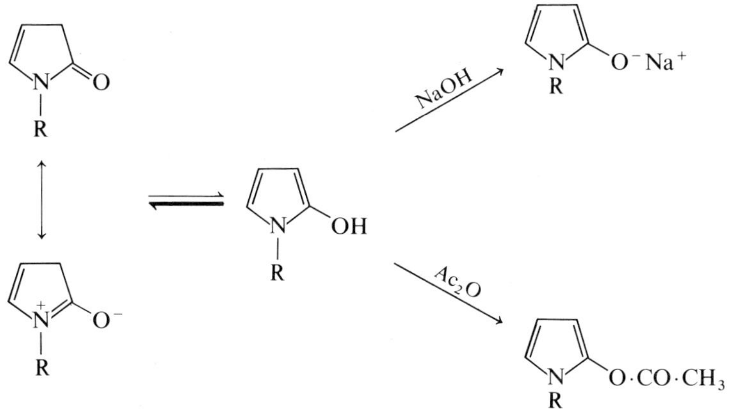

Similarly, 2,5-dihydroxypyrrole exists solely in the di-imido form, succinimide, 2,4-dihydroxyimidazole as hydantoin, and 2-hydroxyindole as oxindole. The 1-methyl-2-amino-4-hydroxyimidazole, creatinine, which is the lactam of creatine (Chapter 17), also exists in the amide form.

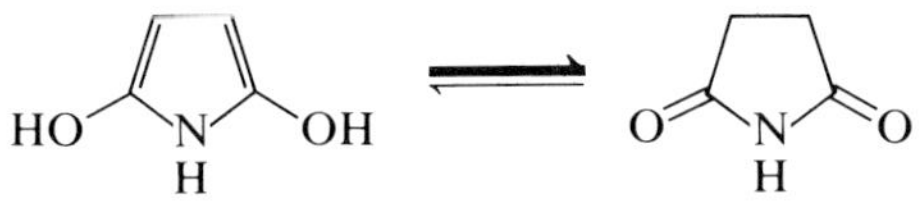

2,5-Dihydroxypyrrole ⇌ Succinimide

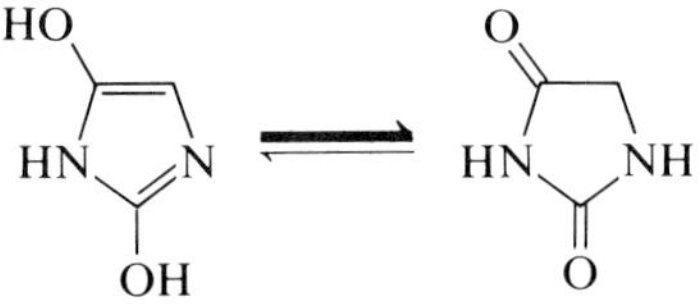

2,4-Dihydroxy-imidazole ⇌ Hydantoin

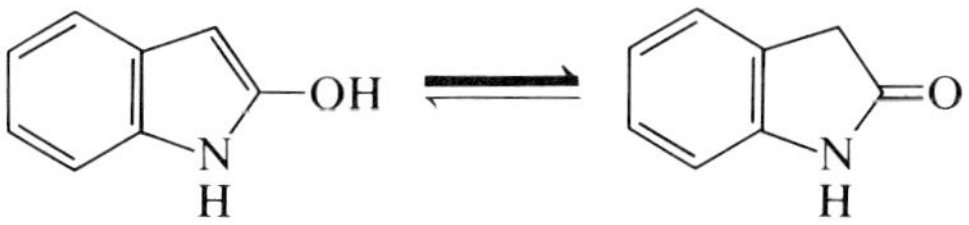

2-Hydroxyindole ⇌ Oxindole

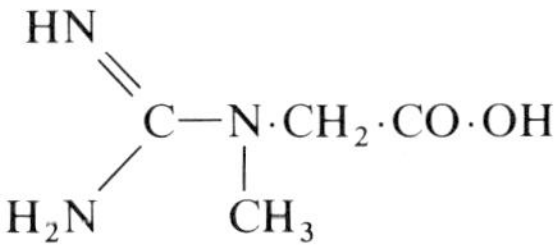

Creatine

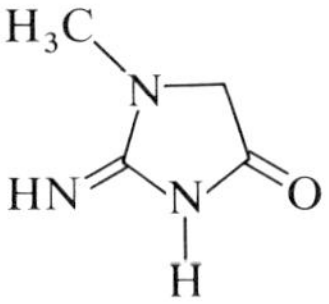

Creatinine

The amide structure of these compounds is further evidenced by the ease with which the active methylene group in such compounds as hydantoin and oxindole will condense with aromatic aldehydes to give benzylidene derivatives. Compounds of this type, nevertheless, still retain their hydroxylic properties as evidenced by the formation of stable water-soluble sodium salts, as in *Phenytoin Sodium*.

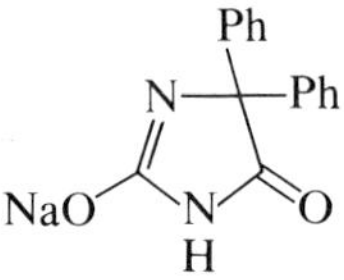

*Phenytoin Sodium*

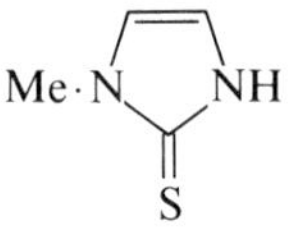

*Methimazole*

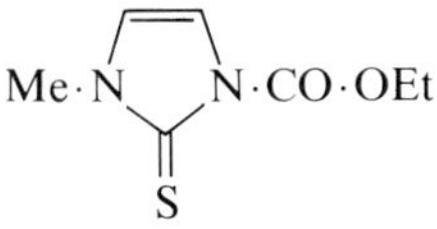

*Carbimazole*

2-Hydroxyfurans and 2-hydroxythiophens similarly exist primarily as the corresponding keto-tautomers, but also show enolic characters (Thomson, 1956). Sulphydryl compounds, such as the antithyroid compound, *Methimazole*, the parent of *Carbimazole*, on the other hand, exist predominantly in the thioamide form.

2-Hydroxyfuran ⇌ $\Delta^2$-Butenolide

2-Hydroxythiophen ⇌ $\Delta^2$-Thiobutenolide

**α-Pyrrolidone and Polyvinylpyrrolidones**

α-Hydroxypyrrolidine also exists in the amide form, α-pyrrolidone. This is the precursor of vinylpyrrolidone, which is readily polymerised in ammoniacal solution using a peroxide catalyst to form polyvinylpyrrolidone (*Povidone*).

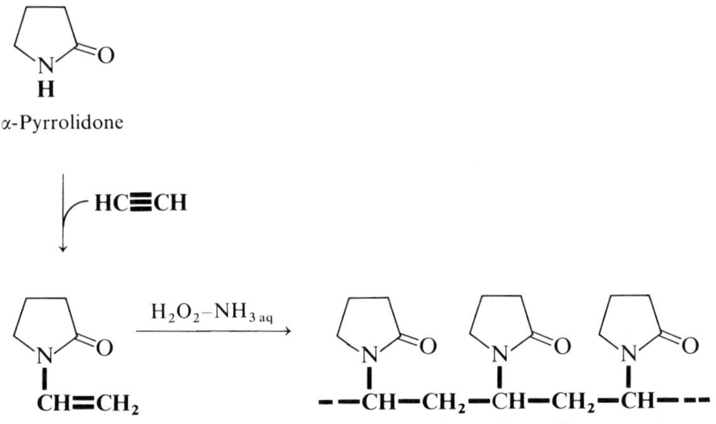

α-Pyrrolidone

↓ HC≡CH

Vinylpyrrolidone $\xrightarrow{H_2O_2-NH_{3\,aq}}$ Polyvinylpyrrolidone

$CH{=}CH_2$

$-CH-CH_2-CH-CH_2-CH-$

The pure polymers are hygroscopic, glass-like solids or powders, and are used as plasma expanders, as suspending and dispersing agents, and as a binder in tablet manufacture (Chapter 18).

**Addition Reactions**

Furan, alone, of the simple hetero-aromatics also functions as a simple diene undergoing 1,4-addition in the Diels–Alder reaction.

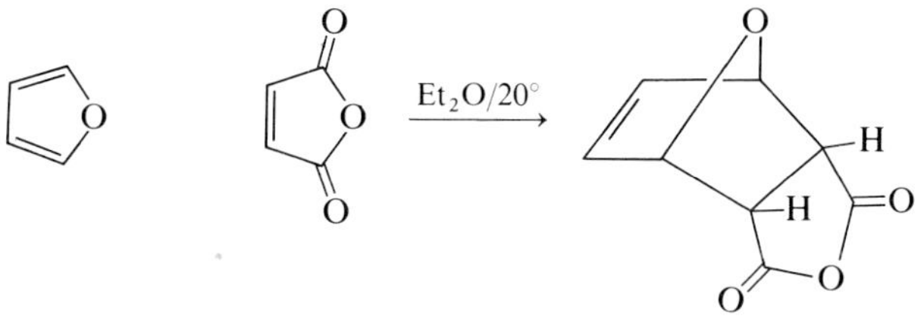

$Et_2O/20°$

## Reduction

Catalytic reduction in general gives the fully reduced heteroparaffinic compounds, pyrrolidine (from pyrrole), tetrahydrofuran (from furan) and less readily tetrahydrothiophen (from thiophen).

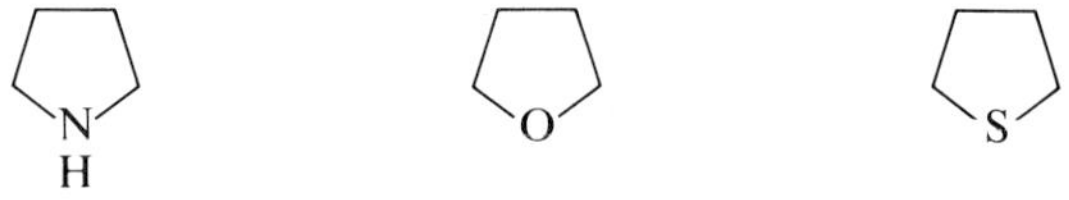

Pyrrolidine    Tetrahydrofuran    Tetrahydrothiophen

Reduction of pyrrole, indole and benzofuran under mild conditions by dissolving metals ($Zn/CH_3COOH$; Sn/HCl) gives the intermediate dihydro-compounds, 2,3-dihydropyrroline (3-pyrroline), 2,3-dihydroindole (indoline) and 2,3-dibenzofuran. Some dihydrothiophens are formed in the sodium liquid ammonia–methanol reduction of thiophen, but ring fission also occurs.

The colourless 2,3,5-triphenyl-2,1,3,4-tetrazolium chloride is readily reduced with ring cleavage to the corresponding 1,3,5-triphenylformazan, which is a bright-red dye. The triphenyltetrazolium chloride, which has an O/R potential of about $-0.08$ V independent of pH, is used as a sensitive detector of reducing agents, such as ascorbic acid in biological systems. It is also used extensively for

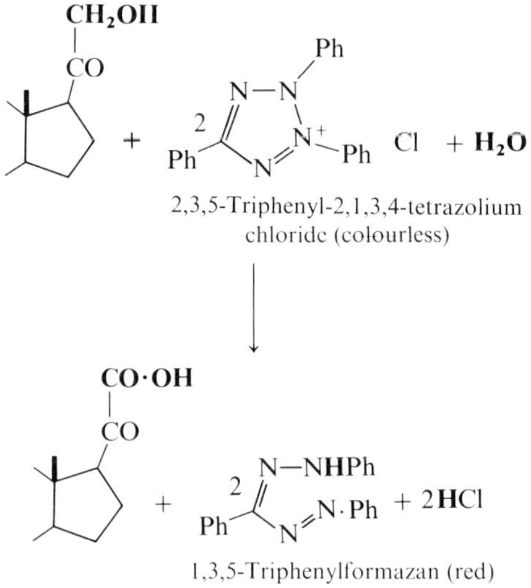

$CH_2OH$ CO + 2 Ph–C(=N–N(Ph))–N=N⁺–Ph Cl + $H_2O$

2,3,5-Triphenyl-2,1,3,4-tetrazolium chloride (colourless)

↓

CO·OH CO + 2 Ph–C(=N–NHPh)–N=N·Ph + 2HCl

1,3,5-Triphenylformazan (red)

the titration of readily oxidisable functions (Kuhn and Jerchel, 1941; Nineham, 1955), such as the C-20, 21 ketol group of corticosteroids in tablets and injections.

Triphenyltetrazolium is stable to acid and stable to oxidation. It is, however, sensitive to light, becoming pale-yellow on exposure to light in the solid state. Aqueous solutions are rapidly oxidised when exposed to ultraviolet light giving a mixture of the corresponding formazan, and a diphenylene.

Ph N—N N+ Ph Ph N $\xrightarrow{O_2/h\nu}$ N—NHPh Ph N N·Ph + 

## Oxidation

Pyrrole darkens on exposure to air undergoing a light-catalysed oxidation and giving rise to a mixture of products containing succinimide (2,5-dihydroxypyrrole), but consisting mainly of a substance known as pyrrole black. The constitution of the latter is unknown, but on further oxidation with chromic acid it yields maleimide.

Indole similarly undergoes light-catalysed oxidation in air to yield first indoxyl (3-hydroxyindole) and on further oxidation, indigo.

2 (indole) $\xrightarrow{O_2,\ h\nu}$ 2 Indoxyl $\xrightarrow[H_2O]{O_2}$ Indigo

Indigo $\xrightarrow{1.\ H_2SO_4\quad 2.\ NaOH}$ Indigo Carmine (HO·O$_2$S … SO$_2$·ONa)

Indigo $\underset{O_2}{\overset{Na_2S_2O_4}{\rightleftharpoons}}$ Leucindigo

*Indigo Carmine*

Leucindigo

Indigo, one of the oldest vegetable dyes, owes its importance to the quality of the colour which it produces, which is fast to light, acid and alkali. The dye is insoluble, and dyeing is achieved by a process known as vat-dyeing in which the fabric is immersed in a solution of its phenolic colourless alkali-soluble reduction product, leucindigo, and then exposed to air for reformation and precipitation of the dye within the fabric. Indigo is now used as a food colour (FD & C. Blue No. 2). It is only poorly absorbed from the gastro-intestinal tract. Its water-soluble derivative, *Indigo Carmine* (Indigotindisulfonate Sodium),

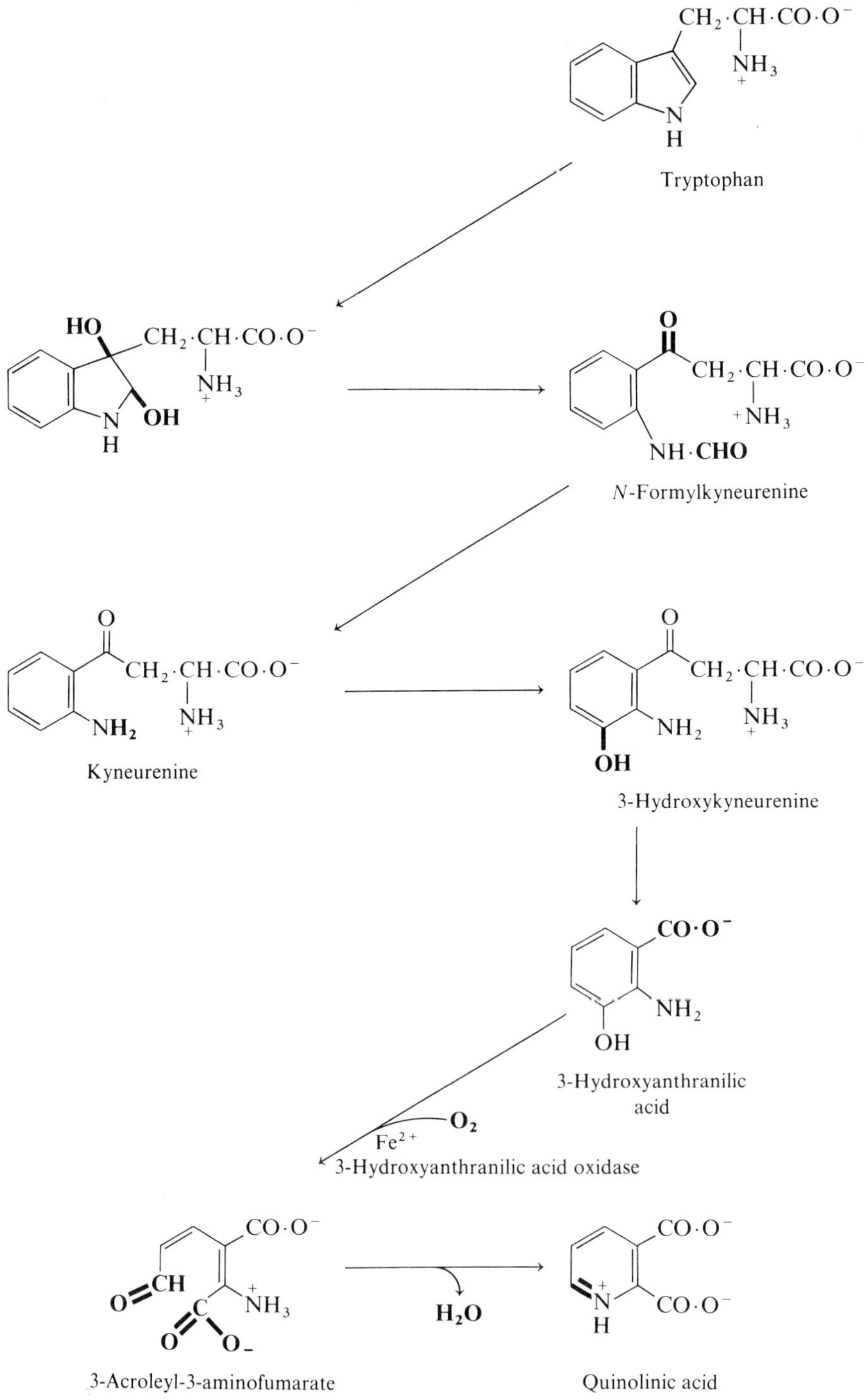

$CH_2 \cdot CH \cdot CO \cdot O^-$, $\overset{+}{N}H_3$

N H

Tryptophan

HO $CH_2 \cdot CH \cdot CO \cdot O^-$ $\overset{+}{N}H_3$

N OH H

O

$CH_2 \cdot CH \cdot CO \cdot O^-$ $^+NH_3$

NH·CHO

*N*-Formylkyneurenine

O

$CH_2 \cdot CH \cdot CO \cdot O^-$

$NH_2$ $\overset{+}{N}H_3$

Kyneurenine

O

$CH_2 \cdot CH \cdot CO \cdot O^-$

$NH_2$ $\overset{+}{N}H_3$

OH

3-Hydroxykyneurenine

$CO \cdot O^-$

$NH_2$

OH

3-Hydroxyanthranilic acid

$O_2$

$Fe^{2+}$

3-Hydroxyanthranilic acid oxidase

$CO \cdot O^-$

O=CH $\overset{+}{N}H_3$

C

O O−

3-Acroleyl-3-aminofumarate

$H_2O$

$CO \cdot O^-$

$\overset{+}{N}$ $CO \cdot O^-$

H

Quinolinic acid

which is prepared from it by sulphonation, is used as the sodium salt in renal function tests.

Tryptophan undergoes metabolic oxidation at both the 2- and 3-positions to form 2,3-dihydroxytryptophan as the first step in the ring cleavage to kyneurenine, an important intermediate in the biosynthesis of quinolinic and nicotinic acids. (See reaction scheme on page 834.)

Pyrazoles, iminazoles and triazoles are all much more stable than pyrrole to metabolic oxidation, but the presence of electron-releasing substituents, such as

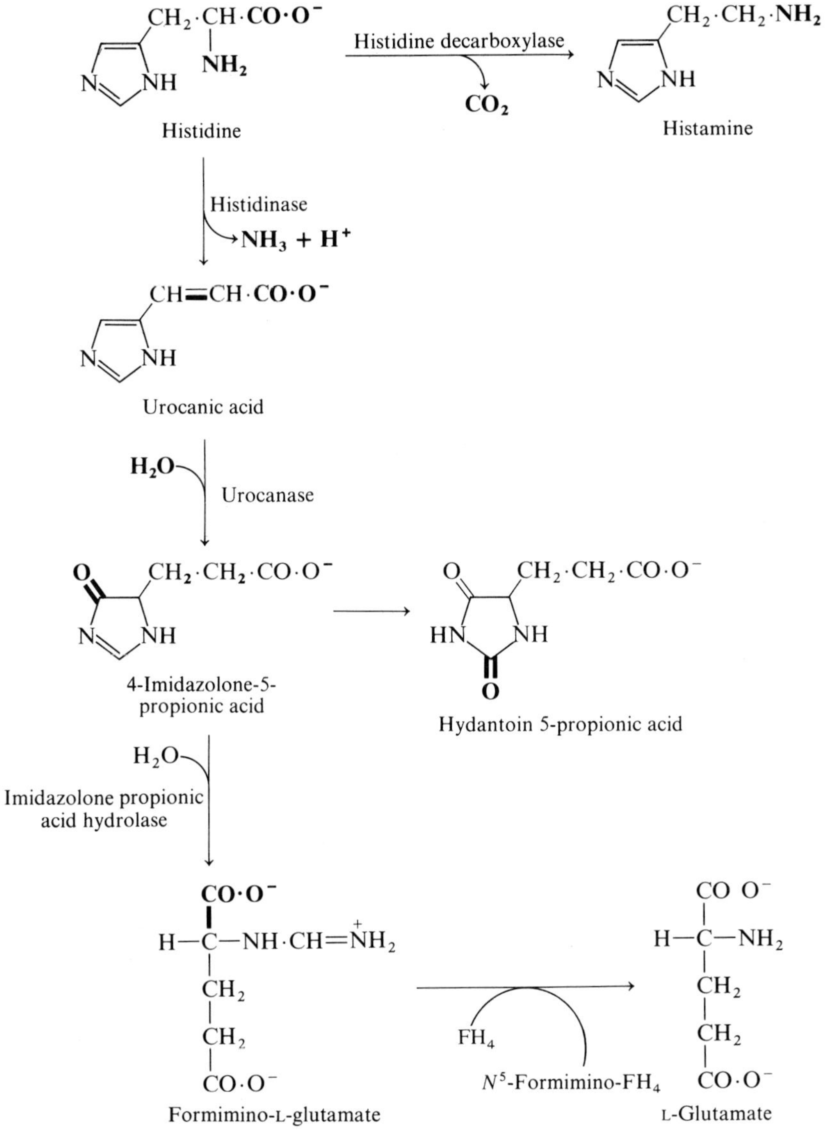

$CH_2 \cdot CH(NH_2) \cdot CO \cdot O^-$ (imidazole ring, N, NH)
Histidine

Histidine decarboxylase → ($CO_2$)

$CH_2 \cdot CH_2 \cdot NH_2$ (imidazole ring, N, NH)
Histamine

Histidine → Histidinase → $NH_3 + H^+$

$CH{=}CH \cdot CO \cdot O^-$ (imidazole ring, N, NH)
Urocanic acid

$H_2O$ → Urocanase

O, $CH_2 \cdot CH_2 \cdot CO \cdot O^-$ (ring, N, NH)
4-Imidazolone-5-propionic acid

→

O, $CH_2 \cdot CH_2 \cdot CO \cdot O^-$ (ring, HN, NH, O)
Hydantoin 5-propionic acid

$H_2O$ → Imidazolone propionic acid hydrolase

$$H{-}\underset{\displaystyle \underset{\displaystyle \underset{\displaystyle CO \cdot O^-}{|}}{\overset{|}{CH_2}} \atop |}{\overset{\displaystyle CO \cdot O^- \atop |}{C}}{-}NH \cdot CH{=}\overset{+}{N}H_2$$

Formimino-L-glutamate: $^-O \cdot OC{-}CH(NH \cdot CH{=}\overset{+}{N}H_2){-}CH_2{-}CH_2{-}CO \cdot O^-$

→ ($FH_4$ → $N^5$-Formimino-$FH_4$)

L-Glutamate: $^-O\,OC{-}CH(NH_2){-}CH_2{-}CH_2{-}CO \cdot O^-$

hydroxyl or amino, sensitise the rings to oxidation. Histidine, however, is either deaminated to urocanic acid and further oxidised by urocanase to 4-iminazolone-5-propionic acid, or decarboxylated to histamine. 4-Iminazolone-5-propionic acid is oxidised enzymatically in mammalian liver to hydantoin-5-propionic acid which is excreted without further metabolism; it also undergoes ring cleavage enzymatically in the presence of tetrahydrofolate ($FH_4$), and by two non-enzymatic pathways. (See reaction scheme p. 835).

Oxidation of thiophen with hydrogen peroxide gives rise to a tricyclic product which appears to be formed via the sulphoxide and sulphone by a Diels–Alder addition (Davies and James, 1954).

S → +S—O⁻ → SO / $SO_2$
$SO_2$

Benzothiophen also undergoes *S*-oxidation, but with the formation of benzothiophen-1,1-dioxide as the main product. Thiazoles are generally more stable to oxidation, though 4-methylthiazole is reported to be oxidised by peroxide to the *N*-oxide.

Furans are considerably less stable and undergo ring fission on oxidation in air or with Fenton's reagent, probably as a result of oxidative attack at the two points. Oxazoles, like thiazoles, are stable under mild oxidising conditions.

$$\text{Furfuraldehyde (O, CHO)} \xrightarrow{[O]\ (air)} H{\cdot}CO{\cdot}CH{=}CH{\cdot}CO{\cdot}OH + CO_2$$

Furfuraldehyde

## PYRIMIDINE AND PURINE NUCLEOTIDES

Pyrimidine and purine nucleotides form the basic units from which the nucleic acids are derived. Each nucleotide consists of a heterocyclic base linked to a pentose sugar phosphate, the sugar being either ribose or 2-deoxyribose. They are formed biosynthetically, but pyrimidine and purine nucleotides are formed by distinctly different pathways. The key intermediate in pyrimidine nucleotide biosynthesis is the preformed pyrimidine, orotic acid (Chapter 17). This is linked first to 5-phosphoribosyl-1-pyrophosphate (PRPP) to give orotidine-5′-phosphate, which is then decarboxylated to uridylic acid (uridine monophosphate; UMP). The latter is converted via UDP and UTP to cytidine triphosphate (CTP).

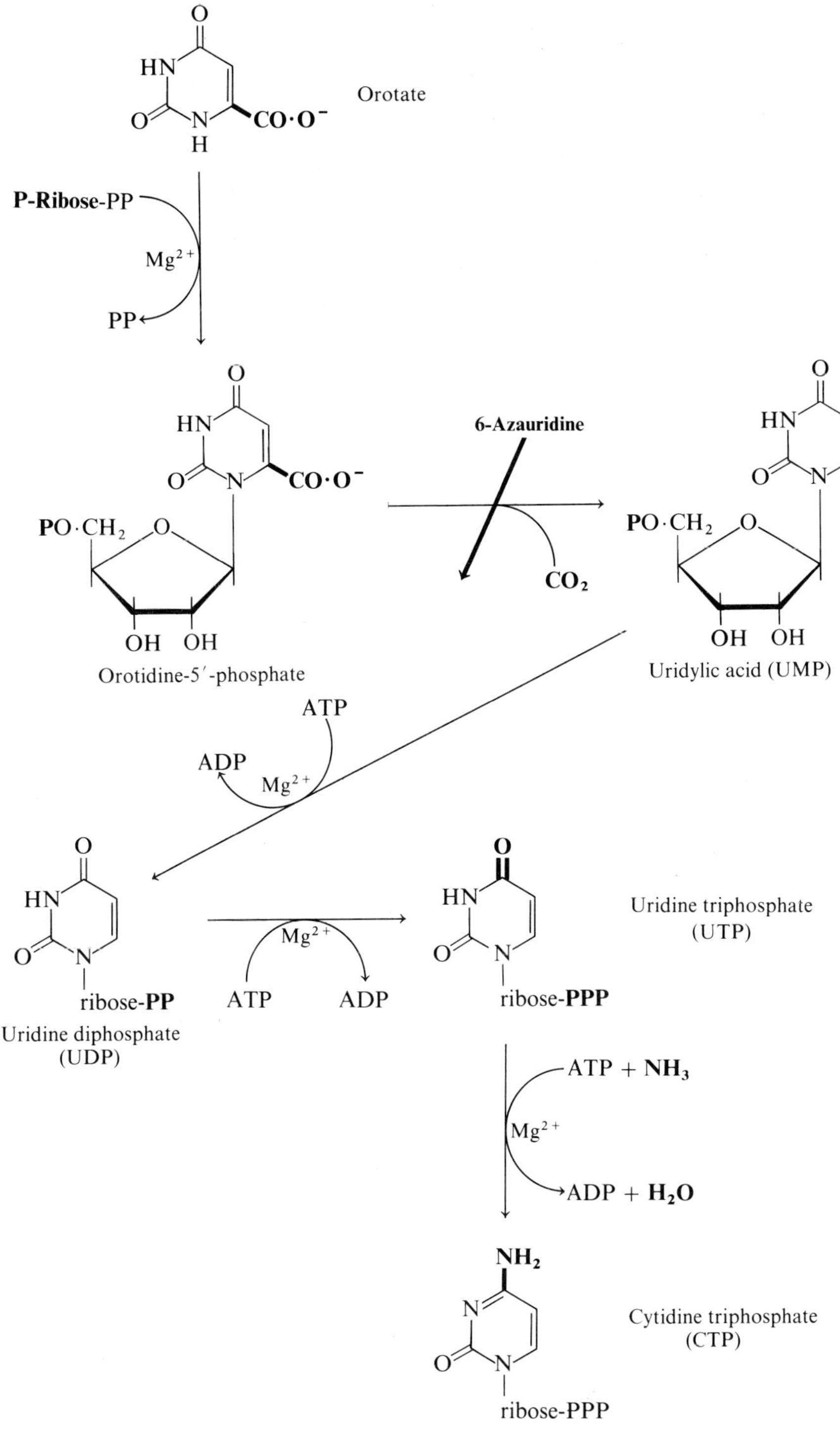

Orotate

P-Ribose-PP

$Mg^{2+}$

PP

6-Azauridine

$CO_2$

Orotidine-5′-phosphate

Uridylic acid (UMP)

ATP

ADP

$Mg^{2+}$

ribose-**PP**

Uridine diphosphate (UDP)

$Mg^{2+}$

ATP

ADP

ribose-**PPP**

Uridine triphosphate (UTP)

ATP + $NH_3$

$Mg^{2+}$

ADP + $H_2O$

ribose-PPP

Cytidine triphosphate (CTP)

The cytostatic agent, *Azauridine* (Skoda, Hess and Šorm, 1957; Prystas, Gut and Šorm, 1961; Elves, Buttoo, Israels and Wilkinson, 1963), and its triacetyl derivative (*Azarabine*), which is used in the treatment of psoriasis (Creasey, Fink, Handschumacher and Calabresi, 1963) bring about reversible inhibition of the decarboxylation of orotidylic acid to uridylic acid, thereby blocking protein synthesis. *Azarabine* undergoes metabolic de-acetylation to *Azauridine* which is converted by uridine kinase to the active metabolite, 6-azauridine-5′-phosphate. The latter is excreted via the kidney.

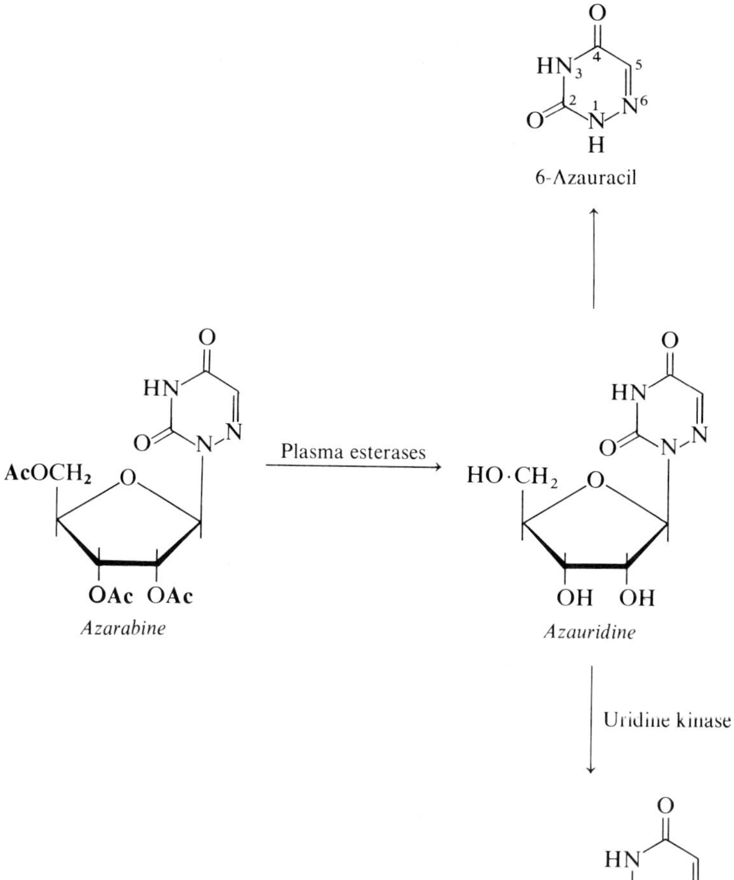

6-Azauracil

Plasma esterases

*Azarabine*

*Azauridine*

Uridine kinase

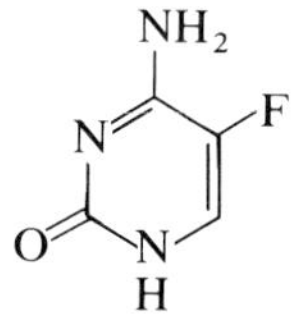

5-Fluorouracil

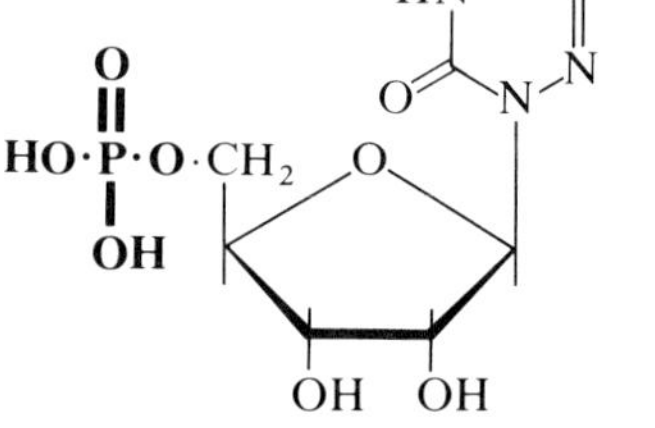

6-Azauridine-5′-phosphate

5-Fluorouracil is similarly converted to 5-fluorouracil deoxyribotide, which inhibits the synthesis of thymidylic acid (Chaudhuri, Montag and Heidelberger, 1958), forming the basis for its admittedly limited clinical effort in the control of carcinoma.

In contrast to the biosynthesis of pyrimidine nucleotides, purine nucleotides are built up directly from the sugar phosphate moiety and glycine, first to a 5-amino ribotide and thence by construction of the six-membered ring to inosinic acid. This is converted to adenylic and guanylic acids. The rôle of adenylic acid (adenosine monophosphate; AMP) and adenosine di- and tri-phosphates (ADP and ATP) in energy transfer and in phosphorylation is described in Chapters 2 and 7. (See pages 840 and 841.)

The antineoplastic agent, *Mercaptopurine*, is metabolised to 6-thioinosine monophosphate, which blocks the conversion of inosinic acid to xanthilic acid (Atkinson, Morton and Murray, 1963). The immunosuppressive agent, *Azathioprine*, functions similarly. 6-Thioinosine triphosphate, which is also formed, inhibits the transfer of adenylic acid from ATP as, for example, in the synthesis of NAD (Atkinson, Jackson and Morton, 1961).

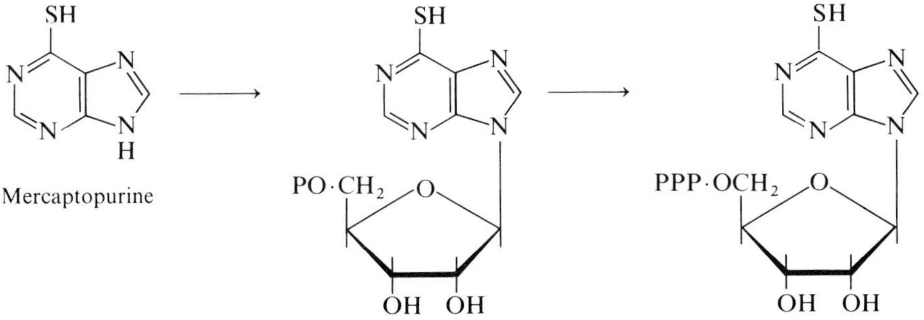

Mercaptopurine → 6-Thioinosine monophosphate → 6-Thioinosine triphosphate

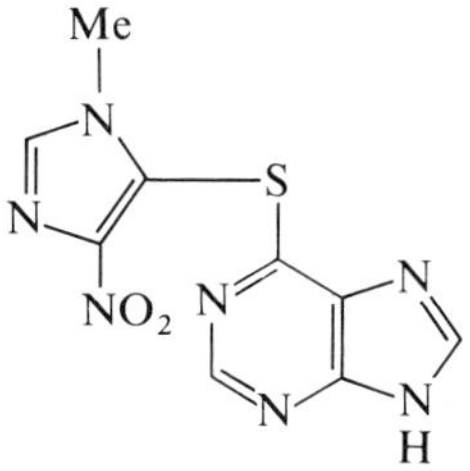

*Azathioprine*

Deoxyribonucleotides appear to be formed by reduction of the ribonucleotide, rather than biosynthetically from deoxyribose. It has been established in *E. coli*, for example, that the reduction takes place with nucleotide diphosphates at the expense of oxidation of a sulphydryl containing protein (Laurent, Moore and Reichard, 1964).

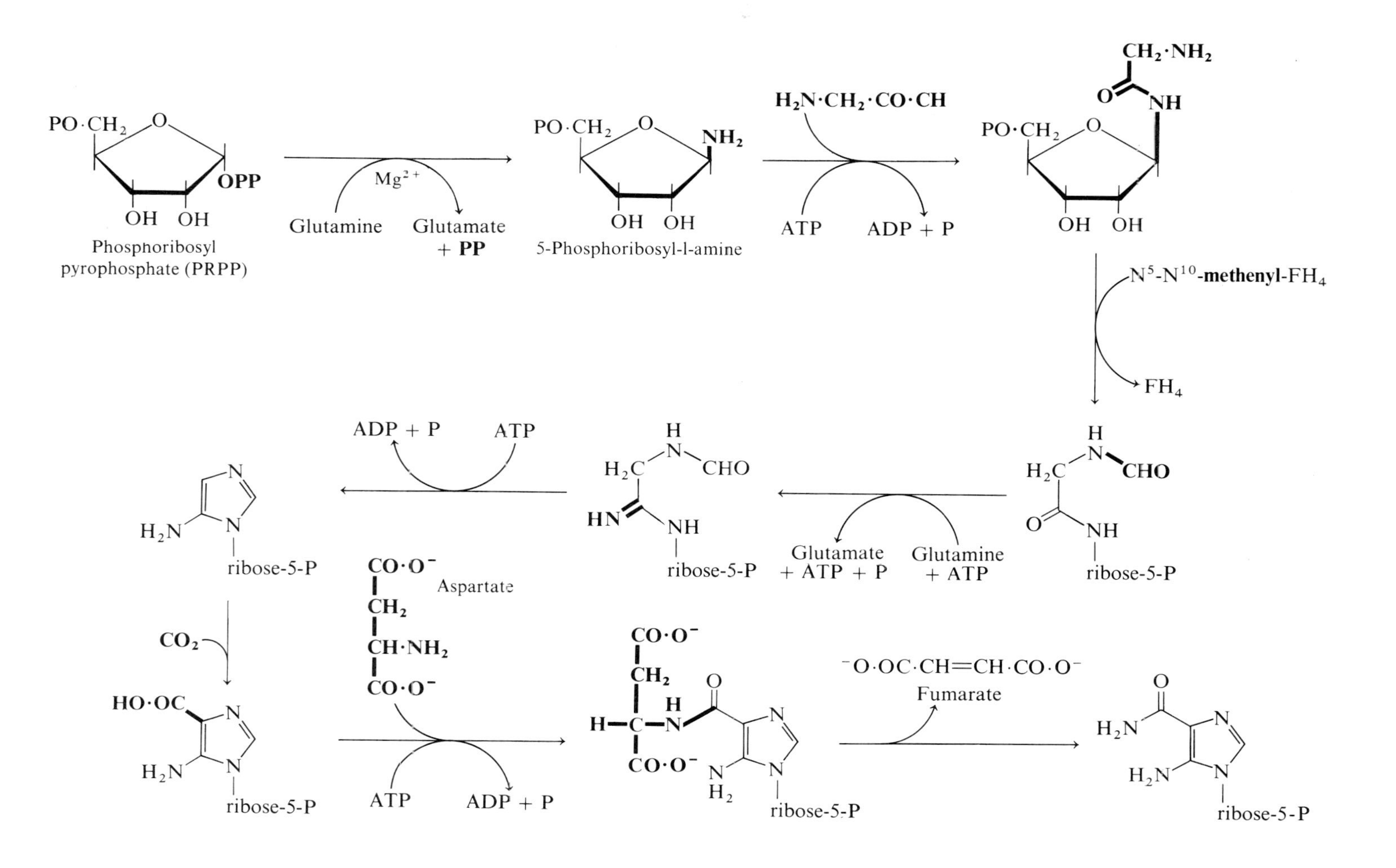

Phosphoribosyl pyrophosphate (PRPP)
Mg2+
Glutamine
Glutamate + PP
5-Phosphoribosyl-l-amine
H2N·CH2·CO·CH
ATP
ADP + P
N5-N10-methenyl-FH4
FH4
Glutamate + ATP + P
Glutamine + ATP
ribose-5-P
Aspartate
CO2
Fumarate

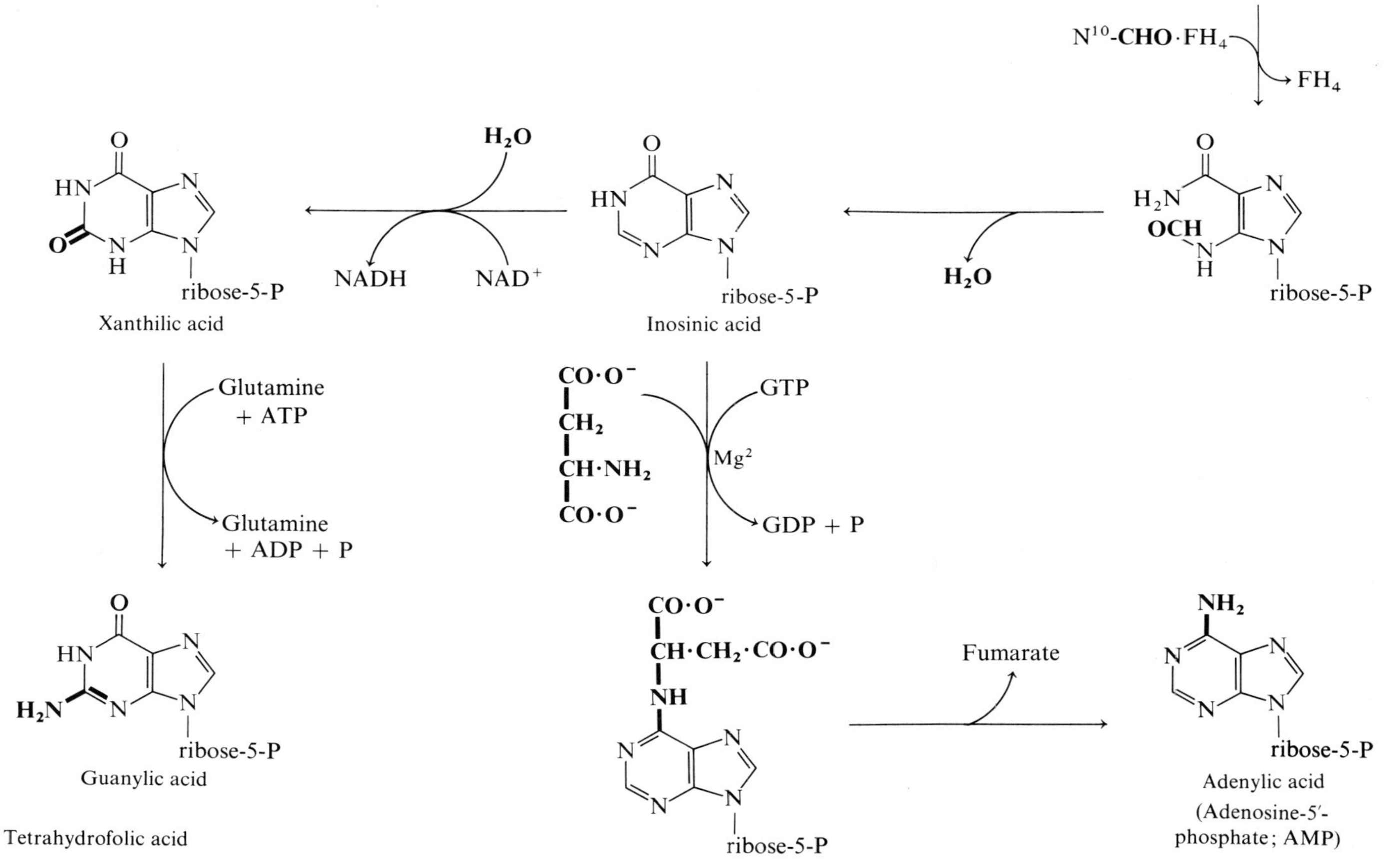

$N^{10}$-**CHO**·$FH_4$ → $FH_4$

$H_2N$ OCH NH ribose-5-P

→ **$H_2O$**

HN N N N ribose-5-P

Inosinic acid

**$H_2O$**, $NAD^+$ → NADH

HN **O** N H N N ribose-5-P

Xanthilic acid

Glutamine + ATP → Glutamine + ADP + P

**$H_2N$** HN N N N ribose-5-P

Guanylic acid

**CO·O⁻ | CH₂ | CH·NH₂ | CO·O⁻** + GTP → GDP + P ($Mg^2$)

**CO·O⁻ | CH·CH₂·CO·O⁻ | NH** N N N N ribose-5-P

→ Fumarate

**$NH_2$** N N N N ribose-5-P

Adenylic acid
(Adenosine-5′-phosphate; AMP)

$FH_4$ = Tetrahydrofolic acid

## NUCLEIC ACIDS

### Structure

The nucleic acids are so-called because they are present in the nuclei of all living cells. They are responsible for the storage and transmission of genetic information, and govern the synthesis of cellular proteins. They are high molecular weight nucleotide polymers of two types, ribonucleic acid (RNA) and deoxyribonucleic acid (DNA), according to whether they are derived from ribose or 2-deoxyribose. The polymers are linear structures in which the sugar moieties of individual monomeric nucleotide units are linked by 3′,5′-phosphodiester groups. The polymer chain which may terminate with either a free hydroxyl or a phosphate group in the terminal 5′-(head) and 3′-(tail) positions may be represented as follows.

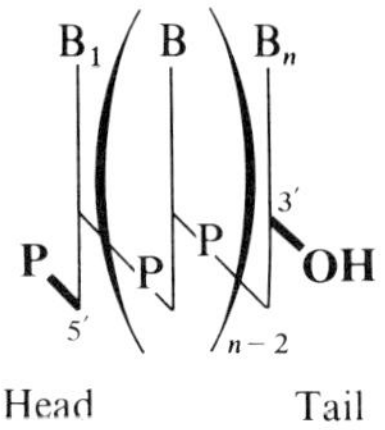

Head Tail

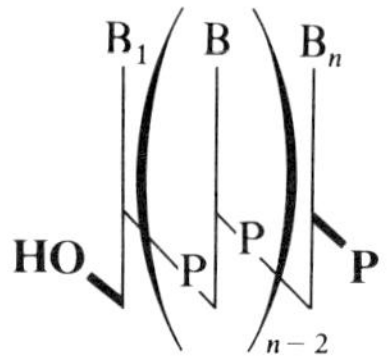

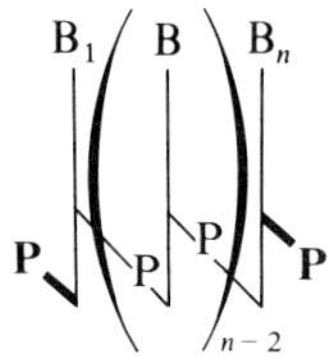

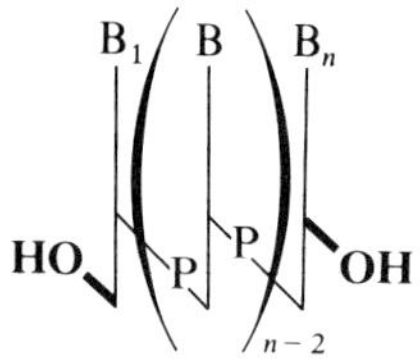

Each phosphate residue contributes an ionised hydroxyl group, so that the polymer is poly-anionic and strongly acidic. Nucleic acids are, therefore, usually found in association with various natural bases, which may be simple polyamines such as spermine, spermidine, putrescine or cadavarine, or alternatively basic proteins such as histones, and additionally $Mg^{2+}$ and alkaline earth cations.

Ribonucleic acid polymers are degraded by dilute alkali at room temperature into a mixture of 2′- and 3′-mononucleotides, which are formed via 2,3-cyclic phosphate intermediates.

A similar mechanism is involved in the breakdown of ribonucleic acid by pancreatic ribonuclease, in which the nucleophilic attack comes almost certainly from a histidine residue. Some ribonucleases are specific for cleavage of internucleotide phosphate links to 3′-purine nucleotides, whilst others, even more specific, attack exclusively 3′-guanylate or 3′-adenylate-linked phosphate bonds. Whatever the degree of nucleotide specificity, cleavage of the internuclear phosphate bond results in transfer of a phosphate group, which arises biosynthetically from a 5′-nucleotide, to the 2′ (or 3′)-position of the adjoining nucleotide unit.

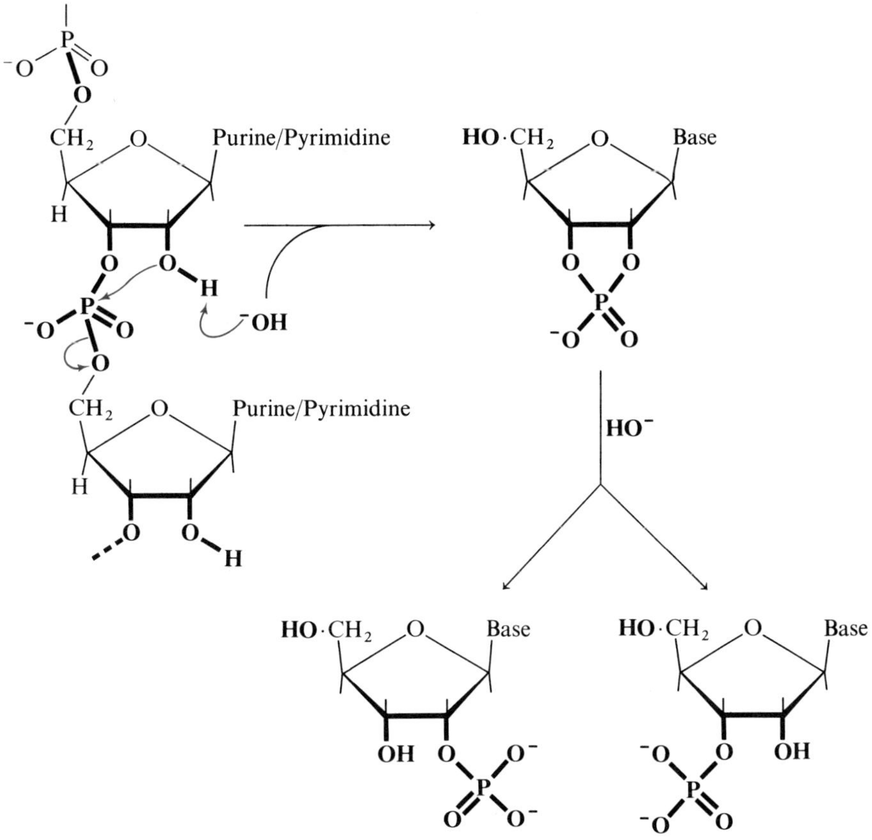

## DNA

Deoxyribonucleic acids, in contrast to ribonucleic acids, are devoid of 2′-hydroxy-groups. Neighbouring group participation in the hydrolysis is, therefore, not possible, and in consequence DNA's show much greater stability to alkali. Deoxyribonucleic acid can, however, be partially degraded enzymatically to oligodeoxyribonucleotides by, for example, pancrease deoxyribonuclease (DNase I), which brings about random internucleotide phosphate ester cleavage, usually between pyrimidine and purine nucleotide units. DNase I requires $Mg^{2+}$, is active in neutral solution, and forms oligodeoxyribonucleotides with a terminal 3′-hydroxy group at the tail and a terminal 5′-phosphate group at the head. In contrast, DNase II from mammalian spleen and thymus does not require $Mg^{2+}$, is active in acidic media, and forms oligodeoxyribonucleotides with a terminal 3′-phosphate group at the tail and a terminal 5′-hydroxy group at the head.

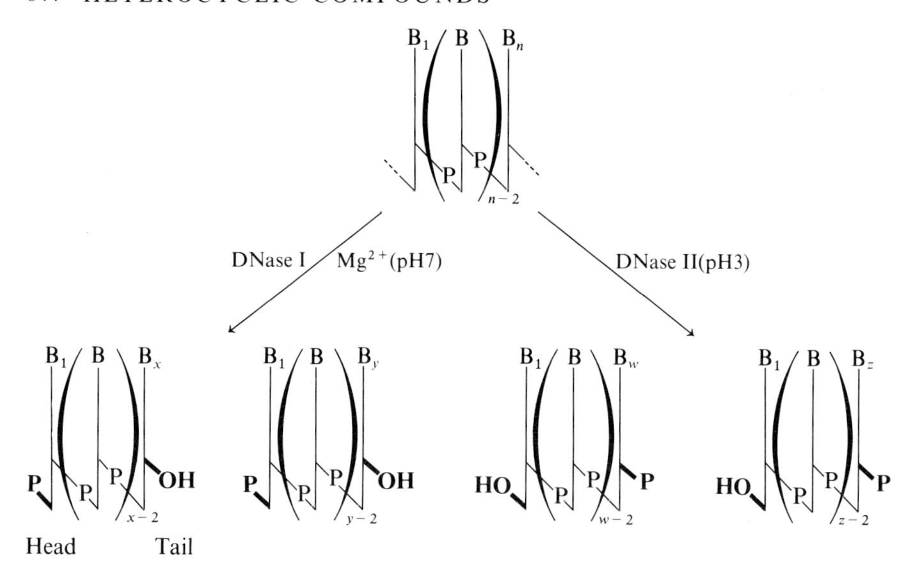


Deoxyribonucleotides are also degraded by heat (*ca* 60°C) under acidic conditions ($pH_3$ or less) with cleavage of deoxyribose purine bonds to give **apurinic** acids (Greer and Zamenhof, 1962). Degradation with hydrazine, similarly, gives **apyrimidinic** acids (Habermann, 1962).

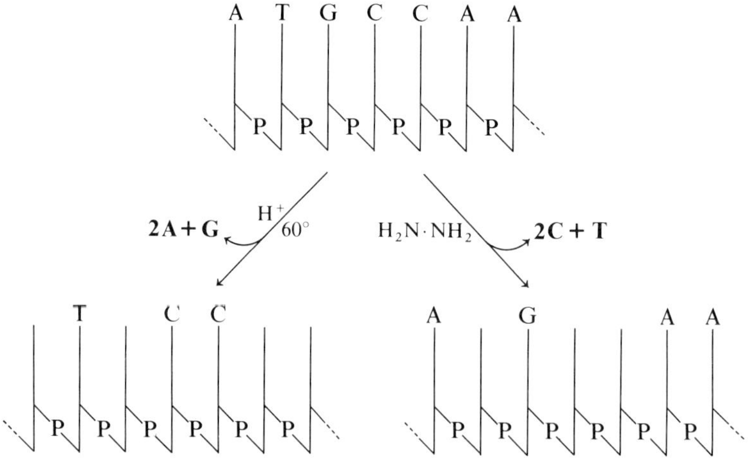

A T G C C A A

P P P P P P

**2A + G** ← $H^+$, 60° | $H_2N{\cdot}NH_2$ → **2C + T**

T C C — P P P P P P

A G A A — P P P P P P

DNA is clearly established as the seat of genetic information in all living cells. The precise composition of each DNA is characteristic of the organism, but all DNA's have certain common structural features. The characteristic feature is a double-stranded helix composed of two right-handed antiparallel chains, coiled in such a way that they can only be separated by complete unwinding. The constituent nucleotides are characterised by the presence of one or other of the purine and pyrimidine bases, adenine (A), guanine (G), cytosine (C) or thymine (T). They are arranged in purine—pyrimidine pairs which bridge the two strands of the helix in such a way that the base-pairs are stacked regularly 3.4 Å apart, one above the other with their planar heteroaromatic rings perpendicular to the main helical axis. Only certain particular base-pairs can be accommodated, the permitted combinations being A and T, and G and C. Base-pairing is achieved and stabilised by the formation of multiple hydrogen bonds.

Adenine-thymine pairing

Guanine-cytosine pairing

There are ten base-pairs to each turn of the helix, and one complete turn in each 34 Å along the length of the helix axis. The double helix, therefore, has two so-called grooves, a minor one *ca* 12 Å across, and a major one *ca* 22 Å in depth (Fig. 43).

Base-pairing and hence cohesion of the two strands of the helix is disturbed by a variety of reagents. Protonation of amino groups occurs first on adenine (p$K_a$ 4.2) and cytosine, and only on guanine (p$K_a$ 3.3) under more strongly acidic conditions. The effect of monoprotonation is to some extent stabilising, although it disrupts a hydrogen bond, since it actually strengthens the remaining adjacent hydrogen bonds. Under more acidic conditions, however, diprotonation of the guanosine–cytosine base-pair gives rise to charge repulsion, which destabilises

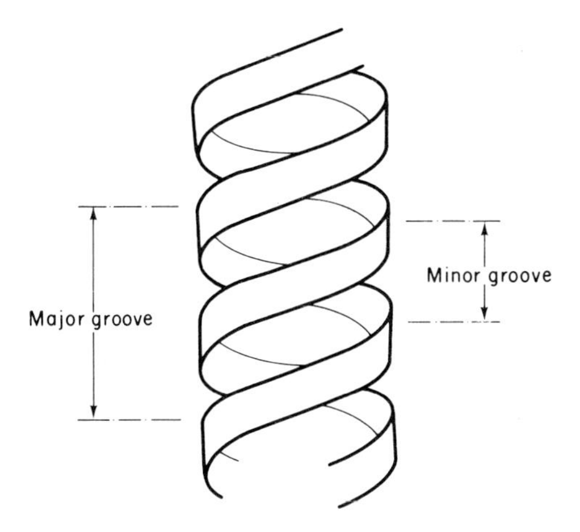


Fig. 43 Coiling of DNA double helix showing major and minor grooves

the helix. Strongly alkaline conditions above pH 12 give rise to the formation of enolate ions, which are formed with breaking of hydrogen bonds, thus increasing the prospect of helix disruption.

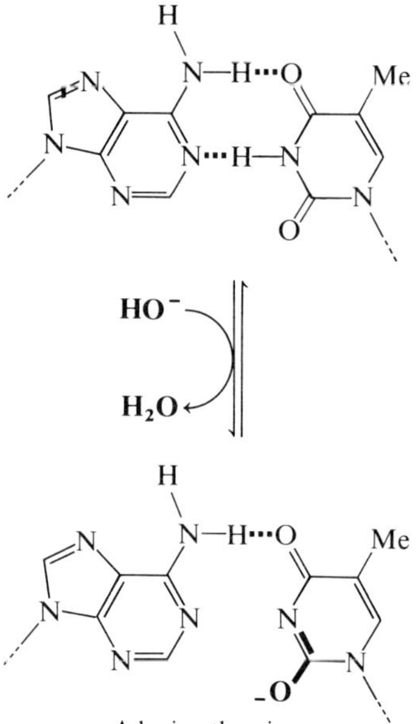

$H^+$

$HO^-$

$H_2O$

Adenine-thymine

$H^+$

$HO^-$

$H_2O$

$H^+$

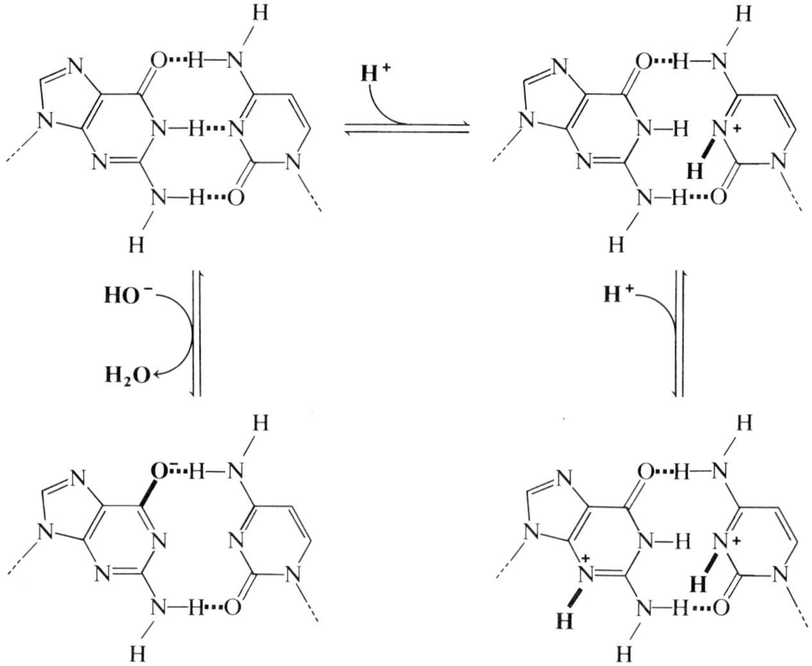

The amino substituents of adenine, cytosine and guanine residues in DNA are typically aromatic in character, and are attacked by nitrous acid in acetate buffer with the formation of diazonium salts, leading to the replacement of the amino substituent by hydroxyl. Adenine is thus converted to hypoxanthine, guanine to xanthine and cytosine to uracil.

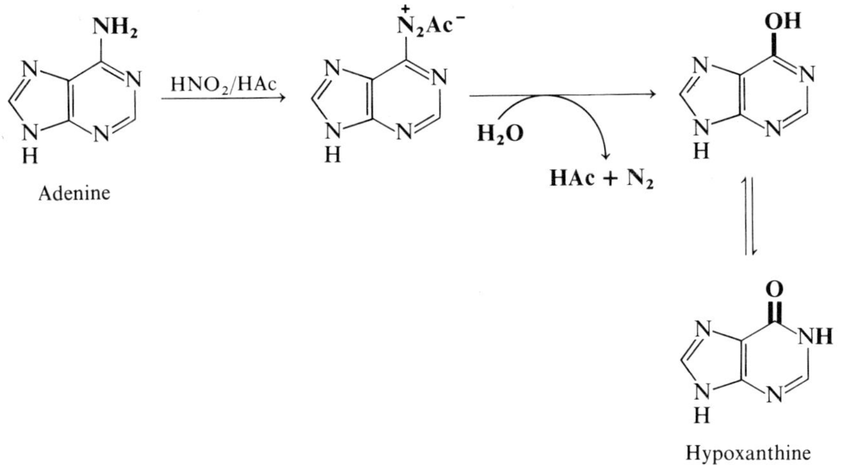

Adenine $\xrightarrow{HNO_2/HAc}$ ($\overset{+}{N}_2Ac^-$ derivative) $\xrightarrow[-\,HAc\,+\,N_2]{H_2O}$ (OH derivative) $\rightleftharpoons$ Hypoxanthine

Adenine

Hypoxanthine

## Biosynthesis of DNA and RNA

The biosynthesis of DNA from its constituent nucleotides is controlled by the enzyme DNA polymerase. The substrates are the nucleotide triphosphates, which are ordered in their correct chain sequence by alignment with the usual complementary base (A—T or G—C) on a single strand of DNA, which functions as a template.

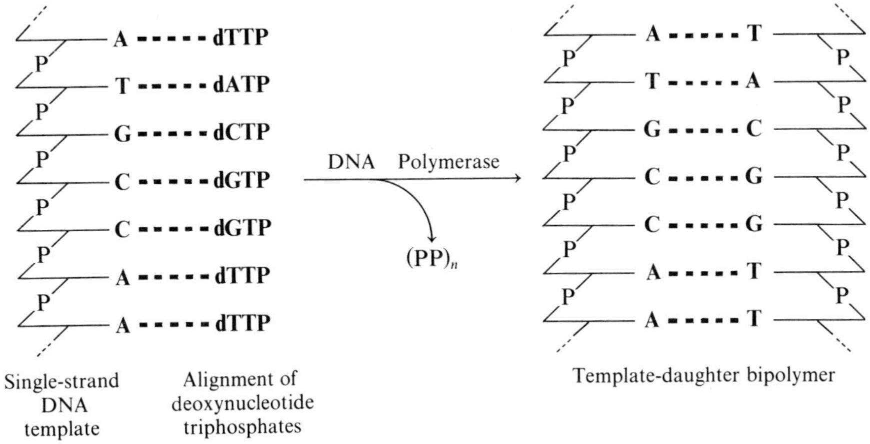

A ----- dTTP
T ----- dATP
G ----- dCTP
C ----- dGTP
C ----- dGTP
A ----- dTTP
A ----- dTTP

DNA Polymerase → $(PP)_n$

A ----- T
T ----- A
G ----- C
C ----- G
C ----- G
A ----- T
A ----- T

Single-strand DNA template

Alignment of deoxynucleotide triphosphates

Template-daughter bipolymer

The replication process requires the unwinding of the bi-polar DNA double helix to give the single-stranded polymer template. Factors, which inhibit the unwinding by cross-linking or other means, inhibit both DNA replication and protein synthesis (**2**, 2). Experiments by Meselson and Stahl (1958) have established that, as predicted by Watson and Crick (1953), both the parent template strand and the newly-synthesised daughter strand act as templates in the second and later stages of DNA biosynthesis.

The antiviral agent, *Idoxuridine*, which is used in the treatment of ocular herpes simplex (Kaufman, 1962; Perkins, Wood, Sears, Prusoff and Welch, 1962) is incorporated in place of thymidine in mammalian bacterial and viral DNA.

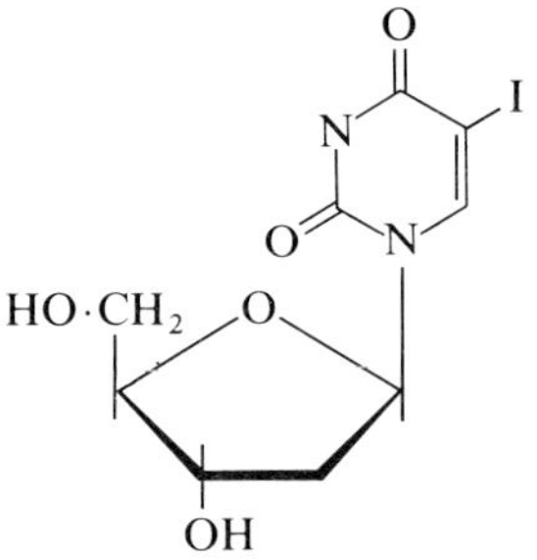

*Idoxuridine*

The halogen substituent in *Idoxuridine* is not particularly stable. For this reason, it should be protected from light, which catalyses its decomposition.

RNA biosynthesis occurs by a similar process involving the enzyme RNA polymerase (transcryptase) also with a single-stranded DNA acting as template. The purine bases in RNA are the same as those of DNA, i.e. adenine (A) and guanine (G); the characteristic pyrimidine bases, however, are cytosine (C), and uracil (U) (in place of thymine). The sequence of these nucleotides in the RNA-polymer will be determined, as in DNA synthesis, by the sequence of complementary bases (A—U or G—C) in the DNA template.

The inhibition of protein synthesis by *Chloramphenicol* is due at least in part to its interference with RNA-synthesis, with accumulation of what appears to be one of the natural ribosomal precursors. 8-Azaguanine is incorporated into RNA in tobacco mosaic virus, thereby preventing cell replication (Matthews, 1954). It also suppresses protein synthesis in *B. cereus* due to incorporation into tRNA.

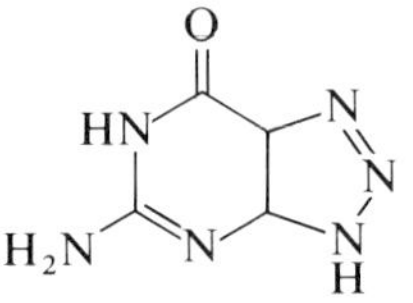

8-Azaguanine

**Protein Synthesis**

RNA carries the coded information for protein synthesis in the sequence of nucleotides, which is determined by the DNA template on which it was formed.

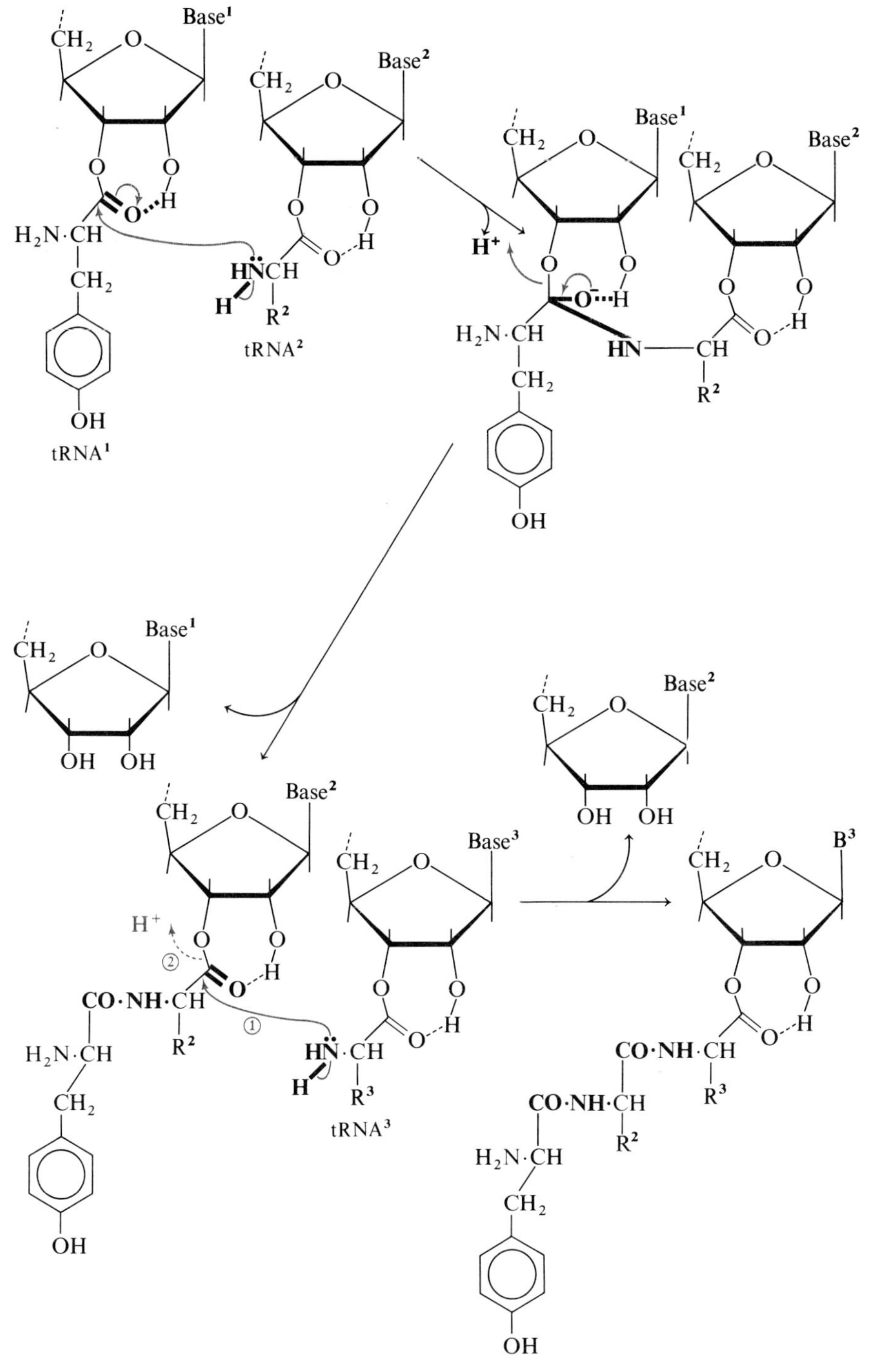

In this sense, RNA acts as messenger, and is known as messenger RNA (mRNA). This in turn functions as a template for the alignment of a second group of RNA's known as transfer RNA (tRNA), each one being specific for a particular amino acid, which is attached by an ester link in a terminal position.

Protein synthesis occurs in the ribosomes. Messenger RNA attaches itself to a specific site in the ribosomes. The first step in protein synthesis is attachment of the tRNA carrying the amino-terminal amino acid of the protein about to be

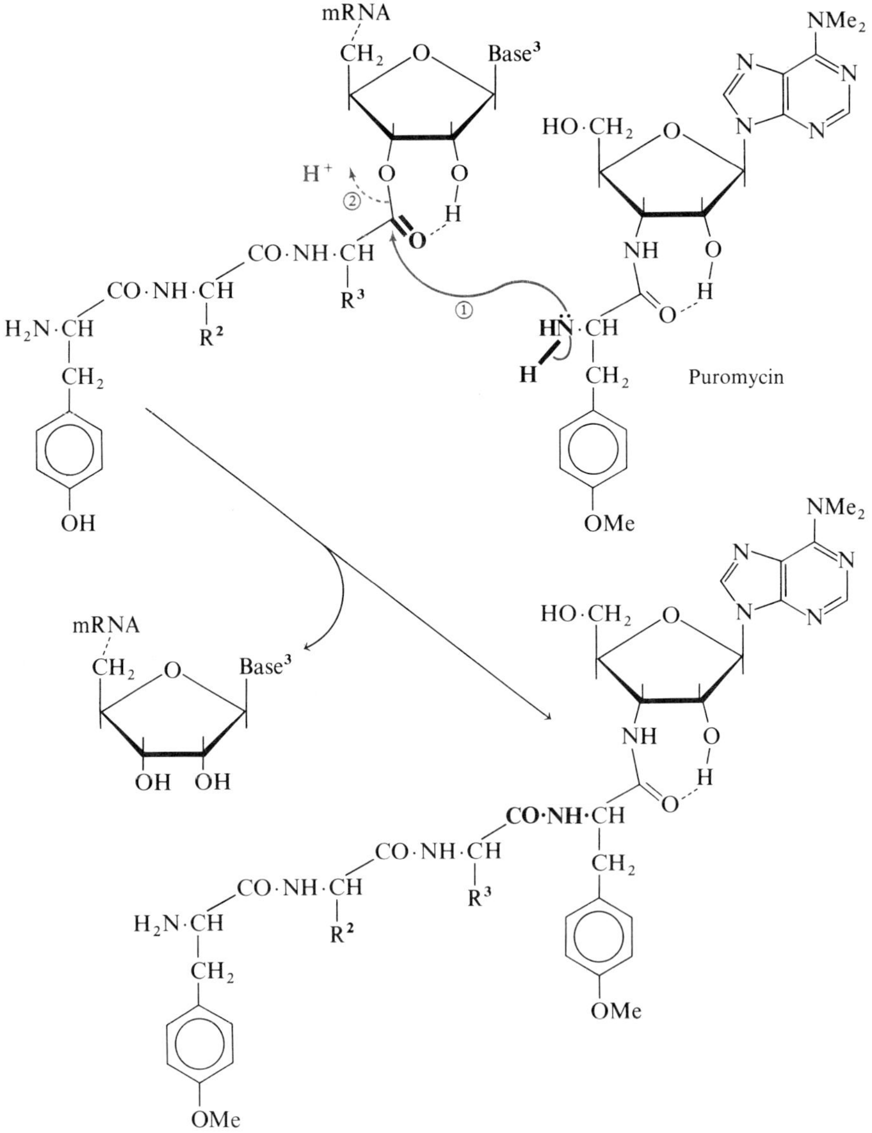

mRNA
$CH_2$ O Base[3]
$H^+$ O O H ② O
CO·NH·CH
$R^3$
CO·NH·CH
$R^2$
$H_2N$·CH
$CH_2$
OH
①
$NMe_2$
N N N N
HO·$CH_2$ O
NH O H O
HN·CH
H $CH_2$
OMe
Puromycin

mRNA
$CH_2$ O Base[3]
OH OH

$NMe_2$
N N N N
HO·$CH_2$ O
NH O H O
**CO·NH**·CH
$CH_2$
OMe
CO·NH·CH
$R^3$
CO·NH·CH
$R^2$
$H_2N$·CH
$CH_2$
OMe

Puromycyltripeptide

synthesised at its specific binding site on the mRNA molecule. Protein synthesis proceeds by the attachment of a second tRNA molecule at an adjacent binding site on the mRNA molecule. The close juxtaposition of the two tRNA terminal amino acid esters facilitates peptide bond formation by a typical nucleophilic attack of the amino group of the second amino ester on the ester carbonyl group of the first. The adjacent 2′-hydroxyl group may be expected to enhance electron deficiency on the carbonyl carbon and so assist formation of the new bond. Formation of the peptide bond necessitates severance of the link between the terminal amino acid and its tRNA, and this is followed immediately by detachment of this tRNA from mRNA. The latter is now free to proceed with the attachment of the third tRNA bringing the next amino acid, and for the cycle of reactions leading to formation of the second peptide bond to proceed (p. 849).

The nucleotide, *Puromycin*, inhibits ribosomal synthesis of proteins by competing with tRNA's in the formation of the growing amino acid chain. *Puromycin* is an analogue of the terminal tyrosinyladenosine unit of tyrosinyl-tRNA (Yarmolinsky and De La Haba, 1959). Its intervention results in complete severance of the growing peptide chain from attachment to its mRNA and hence the formation of a series of peptides, with the correct sequence of amino acids starting from the *N*-terminal end of the molecule and all terminating at the *C*-terminal end with a puromycin residue (Nathans, 1964; Williamson and Schweet, 1965). *Puromycin* has been used to treat amoebiasis (Hutchings, 1957).

## HETEROETHYLENIC COMPOUNDS

Heteroethylenic compounds are broadly of two types, being either partially saturated derivatives of 5- and 6-membered hetero-aromatics and other larger ring hetero-aromatics, or fused ring compounds in which unsaturation is contributed by the $\pi$-electrons of one or more fused benzenoid rings. Compounds in the first category have properties which in many ways resemble quite closely those of the corresponding open-chain ethylenic compounds, and are typically characterised by addition reactions to the ethylenic system. One or two such compounds not directly related to the 5- and 6-membered heteroaromatic compounds already discussed and of natural biochemical or special pharmaceutical interest have been singled out for discussion below.

### Stereochemical Considerations

Unsaturation in the second group of benzoheterocyclics is typically aromatic in character, but the compounds are of particular interest because of the steric restriction which the aromatic ring(s) places on the conformation of the hetero-aromatic ring, and in some cases this is an important factor in their biological action. This group of compounds includes the tetrahydroisoquinolines and the benzodioxans in which the heterocyclic ring adopts a typical half-chair conformation (Raphael and Stenlake, 1953), and the phenothiazines and dibenzthiones in which the heterocyclic ring adopts a boat conformation imparting a butterfly wing shape to the molecule as a whole (Stach and Pöldinger, 1966). The conformation of these and analogous tricyclic compounds is an important factor in determining whether such compounds possess either tranquillising or antidepressant properties (**2**, 3).

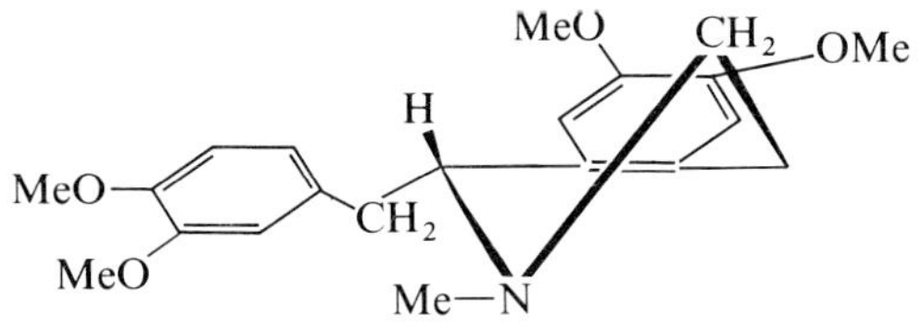

D-Laudanosine

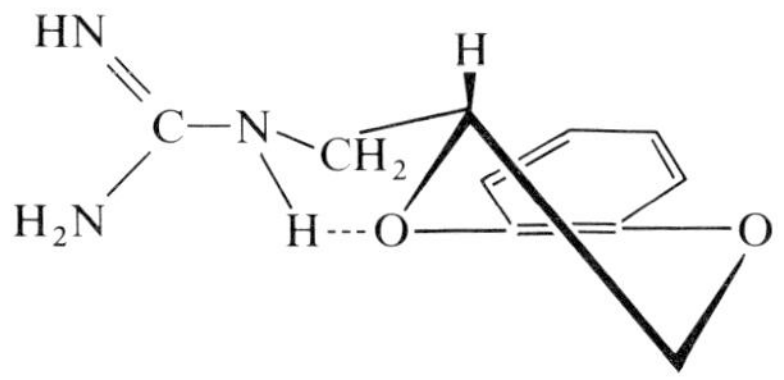

*Guanoxan*

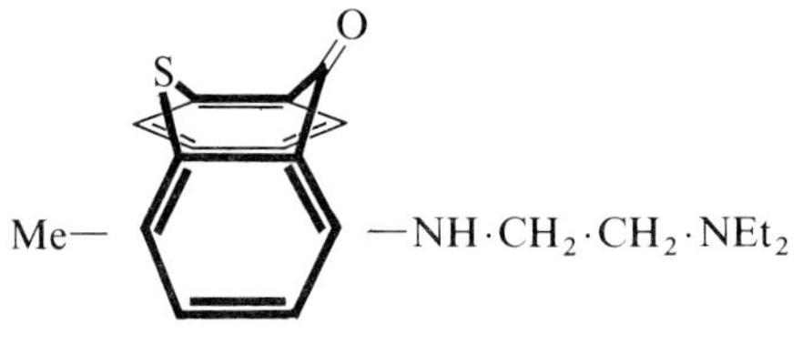

*Lucanthone*

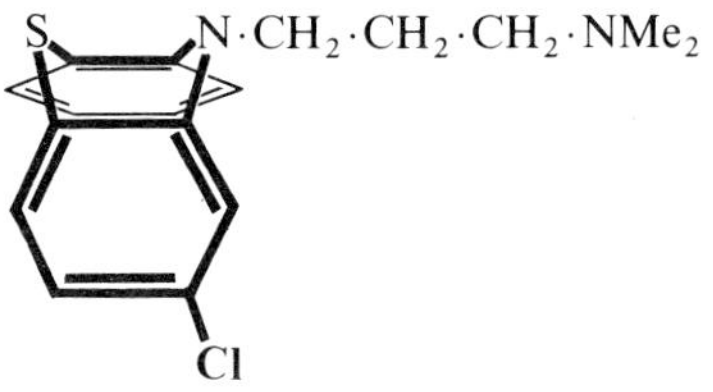

*Chlorpromazine*

**Riboflavine** (Vitamin $B_2$)

Riboflavine is a naturally occurring dihydrobenzpteridine derivative which is widely distributed throughout the entire animal and vegetable kingdom. It is formed biosynthetically from purine derivatives via a 2,4-dihydroxypteridine intermediate (Masuda, Sawa and Asai, 1955) and is essential to normal growth and development.

Riboflavine is only slightly soluble in water (*ca* 1 in 8000) forming yellow solutions which are intensely fluorescent. It is, however, amphoteric ($pK_a$ 4.7 and 11.2) and hence is much more readily soluble in aqueous acid or alkali, though its fluorescence is lost. Aqueous solutions are unstable on exposure to light both in the presence and absence of air (Kuhn and Wagner-Jauregg, 1933), and undergo severance of the ribitol side-chain, the point of cleavage being determined according to the pH of the solution.

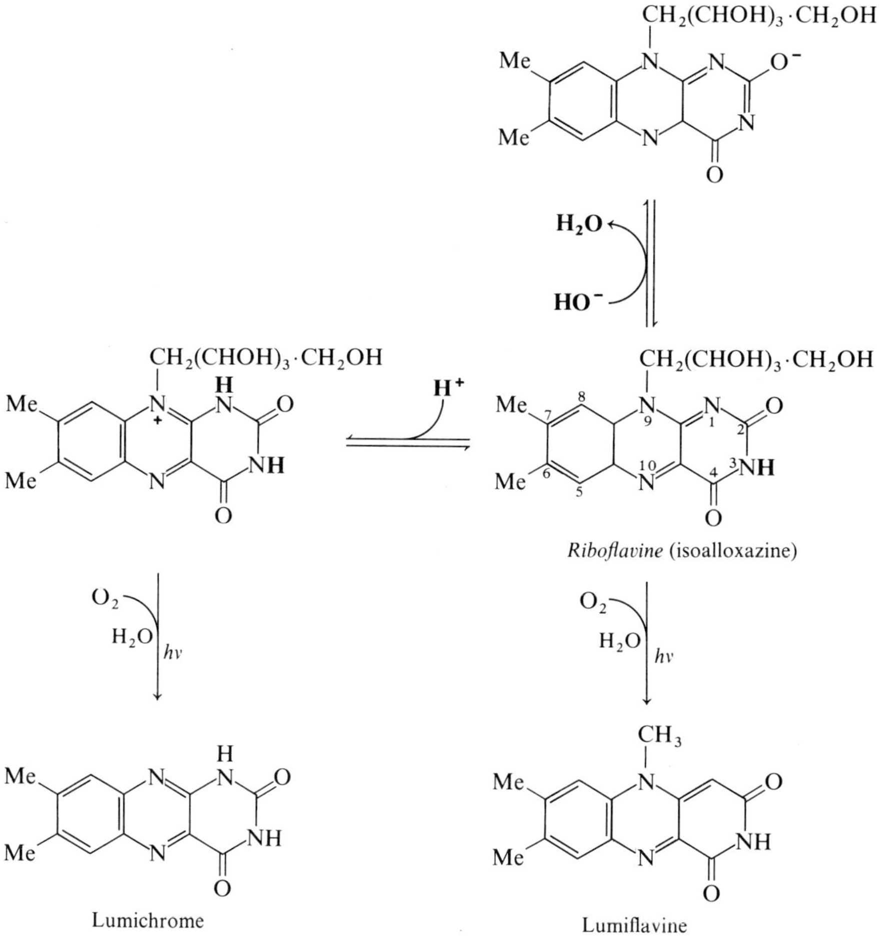

$CH_2(CHOH)_3 \cdot CH_2OH$

$H_2O$ / $HO^-$

$H^+$

*Riboflavine* (isoalloxazine)

$O_2$, $H_2O$, $h\nu$

Lumichrome

Lumiflavine

The monophosphate ester of riboflavine, isoalloxazine mononucleotide, is the prosthetic group of a class of respiratory enzymes known as the flavoprotein

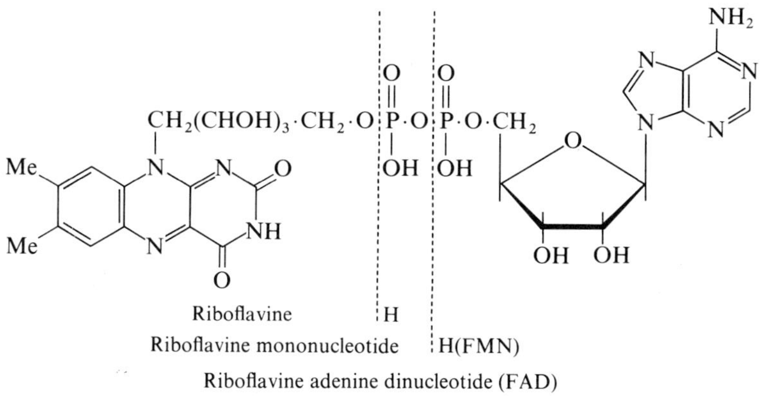

$CH_2(CHOH)_3 \cdot CH_2 \cdot O \cdot P(O)(OH) \cdot O \cdot P(O)(OH) \cdot O \cdot CH_2$

Riboflavine H

Riboflavine mononucleotide H(FMN)

Riboflavine adenine dinucleotide (FAD)

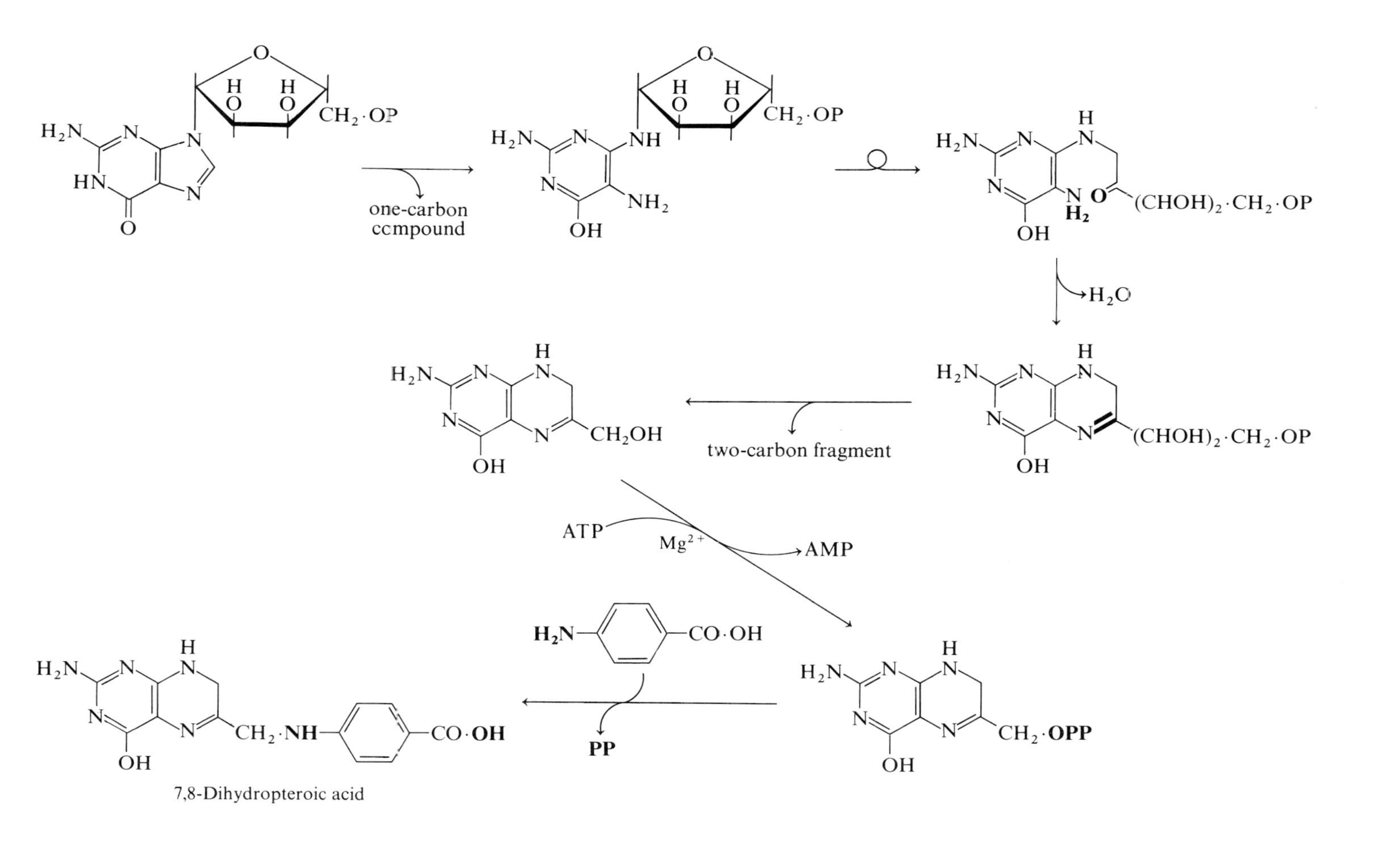
one-carbon compound
two-carbon fragment
ATP
Mg2+
AMP
PP
7,8-Dihydropteroic acid

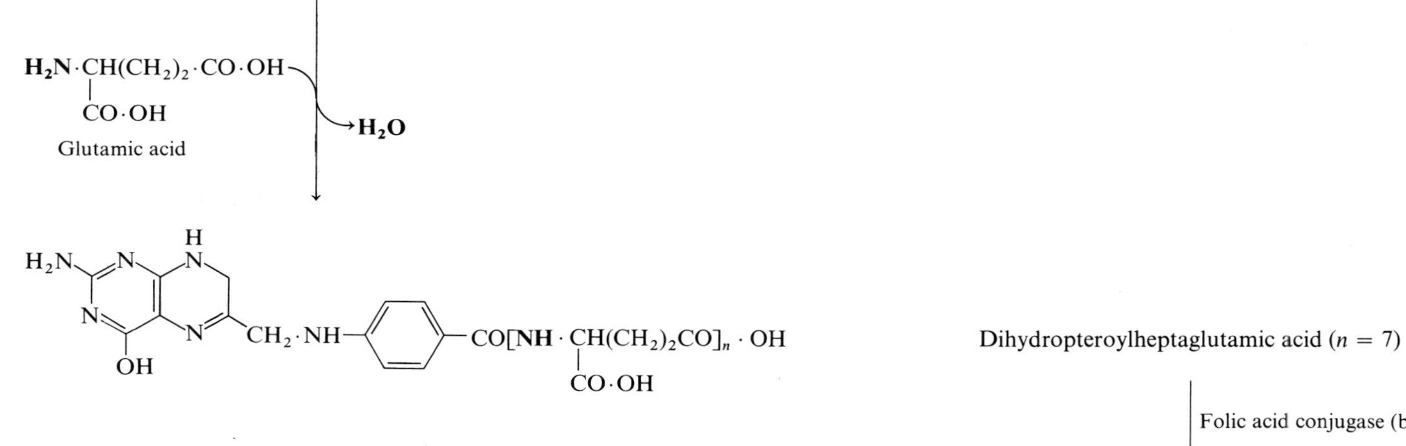


Dihydropteroylglutamic acid ($n$ = 1)

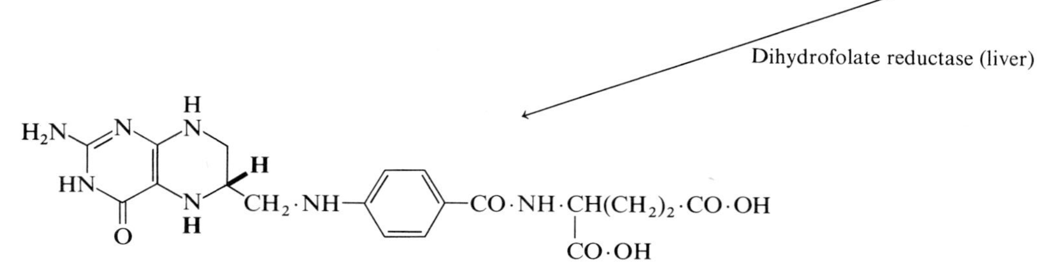

enzymes. Flavoprotein mononucleotide (FMN) and flavoprotein dinucleotide (riboflavine adenine dinucleotide FAD) catalyse the oxidation of pyridine nucleotides, α-amino acids and α-hydroxy acids.

These enzymes are able to effect oxidation as a result of the reduction of the isoalloxazine ring. The reduction is reversible and takes place in two distinct one-electron steps via a semiquinone intermediate (Chapters 2 and 10).

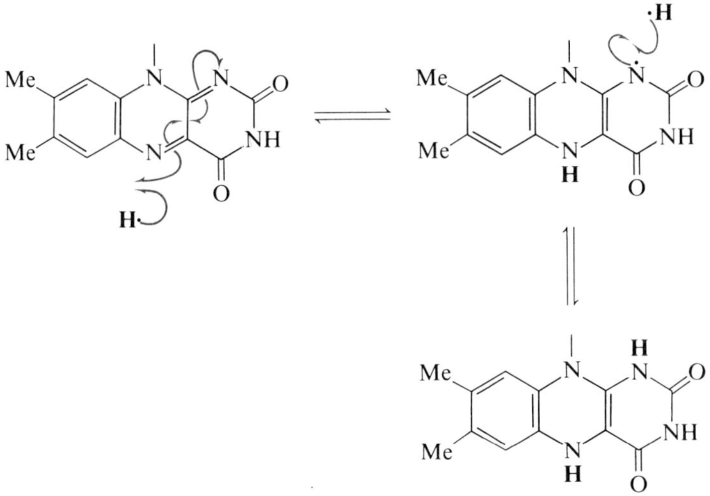

Together with the cytochrome enzymes which are responsible for their re-oxidation, the flavoprotein enzymes form part of the electron transport chain which is responsible for the transfer of hydrogen to molecular oxygen to form water.

### **Tetrahydrofolic Acid** ($FH_4$)

Tetrahydrofolic acid is an important factor in growth and reproduction, acting through its control of pyrimidine and purine biosynthesis. It is formed biosynthetically from dihydrofolic acid (dihydropteroylglutamic acid), which in man is derived largely from the intestinal flora in the form of dihydropteroyltri- and dihydropteroylhepta-glutamic acids. Formation of the pteridine nucleus of these folic acid derivatives starts in guanosine monophosphate (Reynolds and Brown, 1964) with loss of a one-carbon fragment and proceeds by an Amadori rearrangement (Chapter 21) to an intermediate capable of cyclising directly to a dihydropteridine (Griffin and Brown, 1964). (See pages 854 and 855.)

The reduction of folic acid to $FH_4$ is blocked by the antitumour agents, aminopterin and *Methotrexate*, which are strongly bound to dihydrofolate reductase (Brockman and Anderson, 1963). Although binding is reversible, it is not inhibited by folic acid. *Methotrexate* alone is, however, of limited value as a cytotoxic agent in the treatment of neoplastic disease, owing to the development of resistance to its effects. It is now used, however, with considerable success in children to maintain remission in acute lymphoblastic leukaemia which has been induced by combined treatment with *Prednisolone* and *Vincristine Sulphate*.

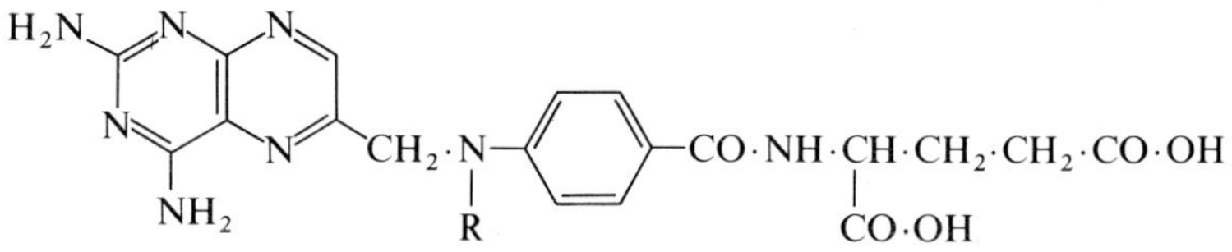

Aminopterin R = H
*Methotrexate* R = Me

The biosynthesis of tetrahydrofolic acid is also inhibited by *Trimethoprim* at the dihydrofolate reductase steps, and by sulphonamides at the point of incorporation of PABA (Chapter 14).

Tetrahydrofolic acid mediates the transfer of one-carbon fragments at various levels of oxidation in the biosynthesis of purines and pyrimidines (p. 836) and in the conversion of homocysteine to methionine. The formation and rôle of

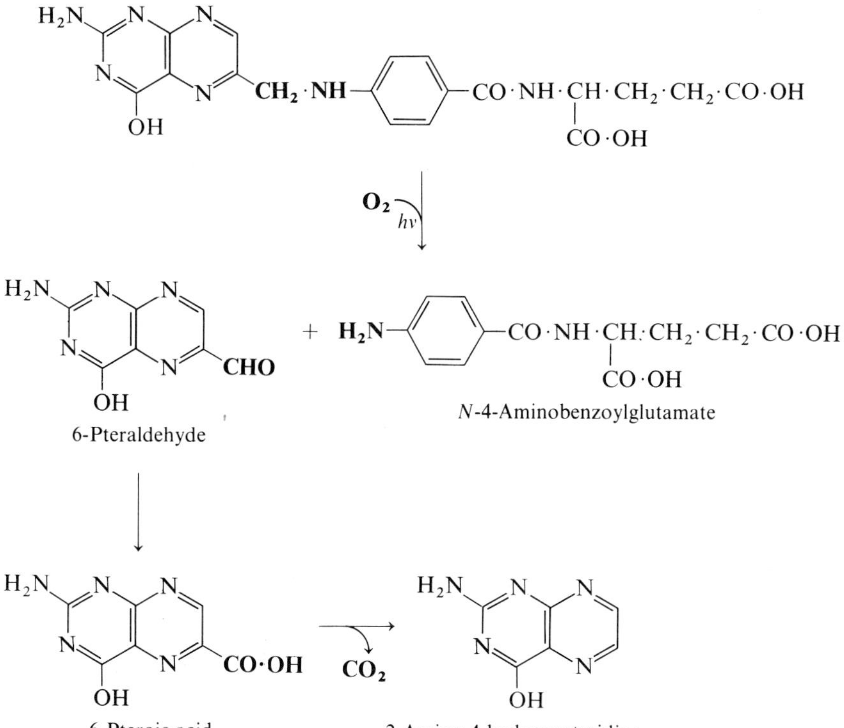

6-Pteraldehyde

*N*-4-Aminobenzoylglutamate

6-Pteroic acid

2-Amino-4-hydroxypteridine

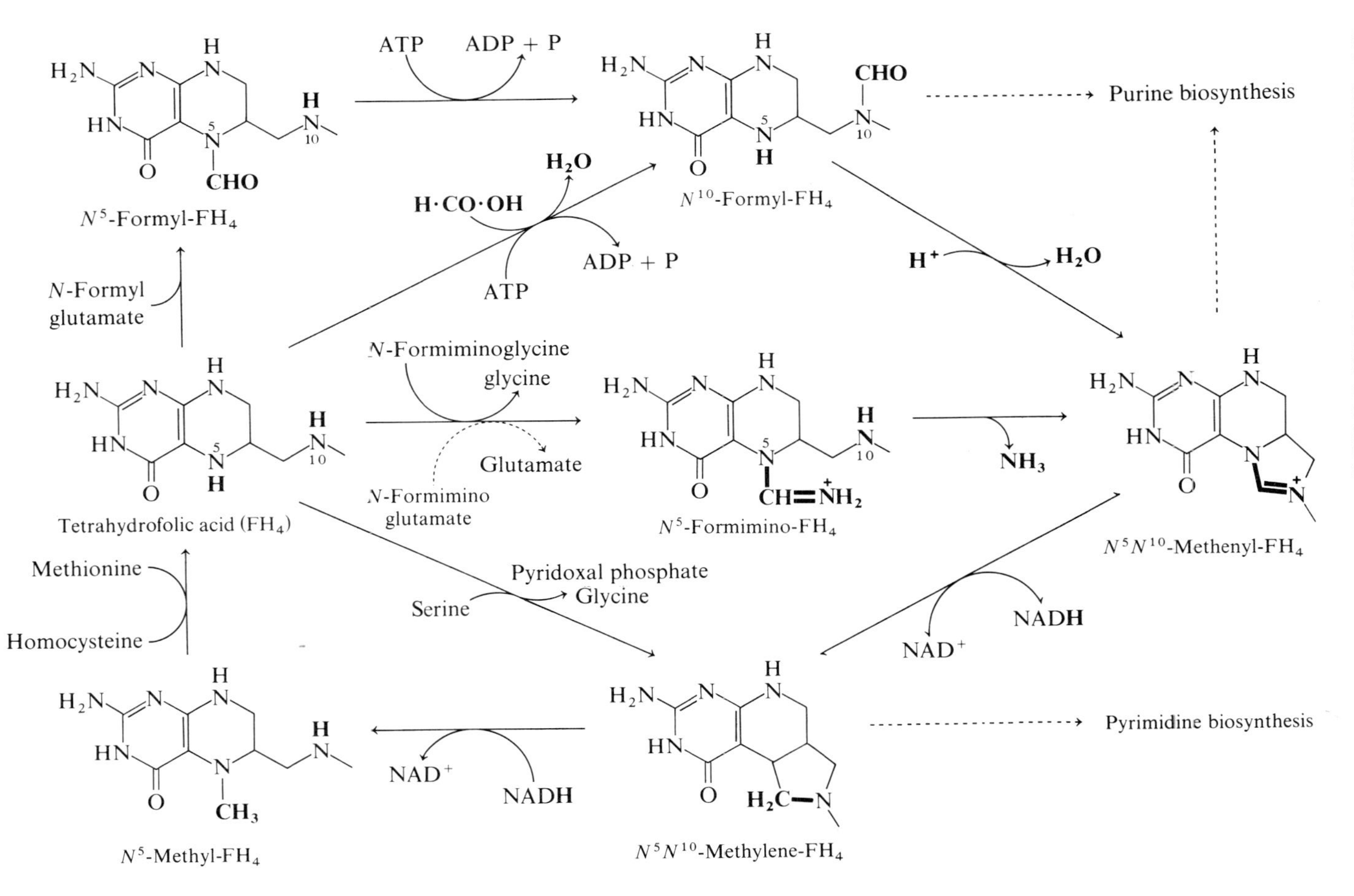

ATP
ADP + P
Purine biosynthesis
$N^5$-Formyl-$FH_4$
H·CO·OH
$H_2O$
ADP + P
ATP
$N^{10}$-Formyl-$FH_4$
$H^+$
$H_2O$
N-Formyl glutamate
N-Formiminoglycine
glycine
Glutamate
N-Formimino glutamate
$NH_3$
Tetrahydrofolic acid ($FH_4$)
$N^5$-Formimino-$FH_4$
$N^5N^{10}$-Methenyl-$FH_4$
Methionine
Homocysteine
Pyridoxal phosphate
Glycine
Serine
$NAD^+$
NADH
Pyrimidine biosynthesis
$N^5$-Methyl-$FH_4$
$N^5N^{10}$-Methylene-$FH_4$

the principal intermediates, $N^5$-formylFH$_4$, $N^5$-formiminoFH$_4$, $N^{10}$-formyl-FH$_4$, $N^5N^{10}$-methenylFH$_4$, $N^5N^{10}$-methyleneFH$_4$ and $N^5$-methylFH$_4$ is set out on the opposite page.

Folic acid deficiency in megaloblastic anaemia is treated by oral administration of folic acid. The latter, although of low solubility in water and lipid solvents, forms a readily water-soluble sodium salt, which is stable in neutral and alkaline solution. Acidic solutions are unstable due to hydrolytic cleavage of the pteridine ring (p. 803). Folic acid is also light-sensitive (Lowry, Bessey and Crawford, 1949) decomposing with oxidative disruption of the side-chain at the 6α-methylene group due to its activation by electron-withdrawal (p. 857).

**Benzazepines**

A number of compounds with valuable actions on the CNS are dihydroazepin and dihydrodiazepin derivatives. The azepines are seven-membered nitrogenous heterocyclics, of which the most stable is 3*H*-azepine. The corresponding 1*H*-azepine is unstable and unless substituted on nitrogen rearranges rapidly to the more stable tautomeric 3*H*-azepine.

1*H*-Azepine

3*H*-Azepine

The antidepressant, *Imipramine*, is a dibenzo-3,4-dihydro-1*H*-azepine. The dihydroazepine ring is constrained in a boat-like conformation resembling that of the phenothiazines, but leading to a different alignment of the two benzene rings relative to each other. The resultant twisted butterfly conformation is almost identical with that in the analogous non-heterocyclic dihydrodibenzcycloheptatriene antidepressant, *Amitriptyline* (**2**, 3).

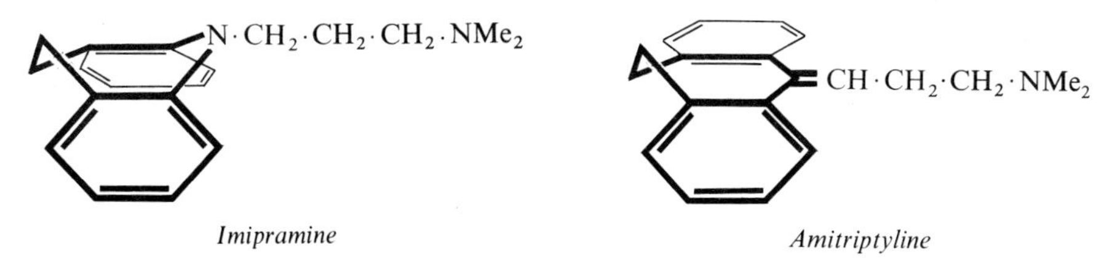


*Imipramine* *Amitriptyline*

**Benzodiazepines**

The tranquillizer, *Chlordiazepoxide*, is a benz-3*H*-diazepine and adopts a puckered cycloheptatriene-like conformation, somewhat like that of *Cyproheptidine*. The molecular symmetry of the latter and the dihedral angle between its two benzene rings is very similar to that of the phenothiazines in accord with its use as an antihistamine. In contrast, the benz-3*H*-diazepines, *Diazepam* and *Chlordiazepoxide*, merely exhibit sedative properties.

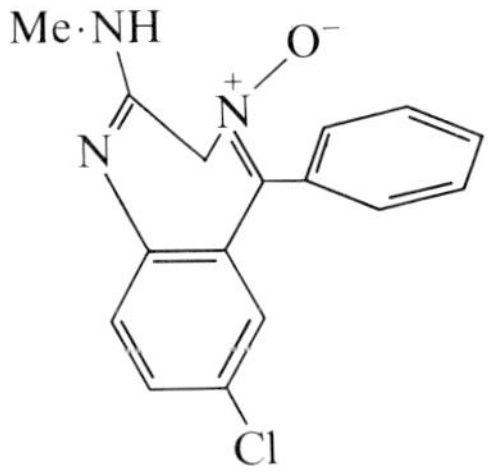

Me·NH O⁻ N N Cl

*Chlordiazepoxide*

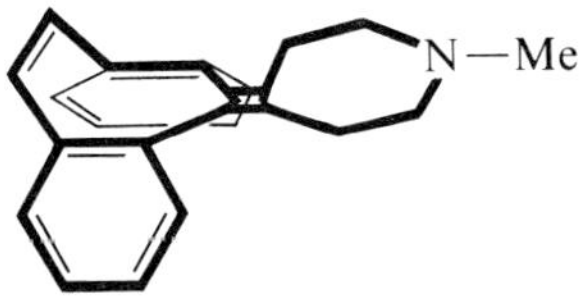

N—Me

*Cyproheptidine*

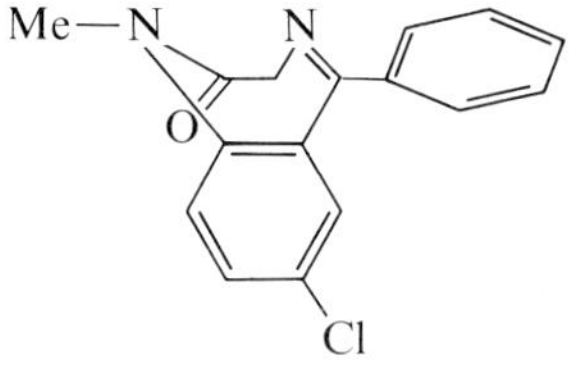

Me—N N O Cl

*Diazepam*

Benzodiazepines are reasonably stable under neutral conditions, but ring scission occurs in strongly acidic solution (Oelschläger, Volke and Kurek, 1964).

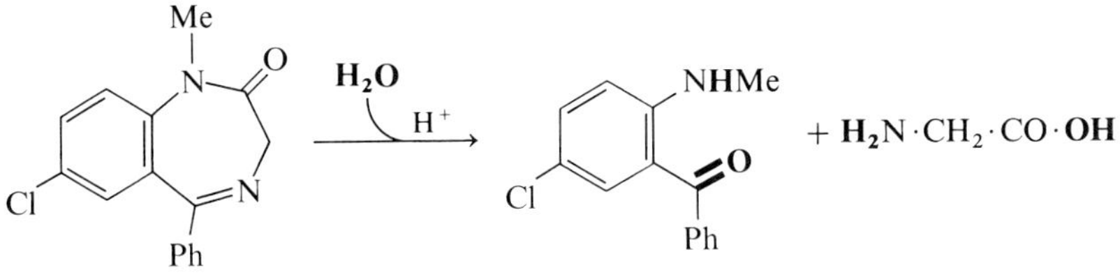

Me N O Cl N Ph $\xrightarrow[H^+]{H_2O}$ NHMe Cl O Ph + $H_2N \cdot CH_2 \cdot CO \cdot OH$

*Diazepam*

As an *N*-oxide, *Chlordiazepoxide* is light-sensitive and rearranges to the corresponding 4,5-oxide.

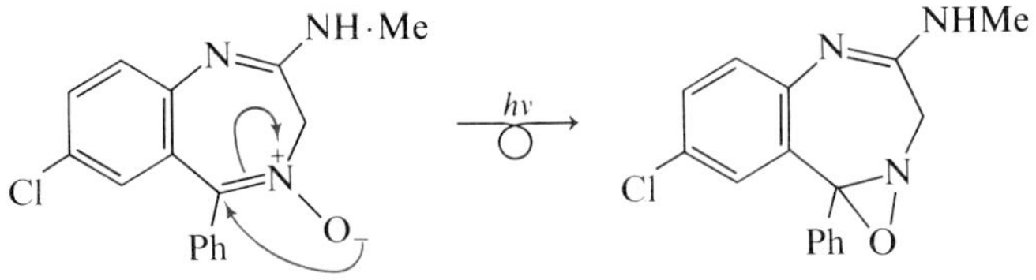

NH·Me N Cl N O⁻ Ph $\xrightarrow{h\nu}$ NHMe N Cl N Ph O

In human metabolism, however, oxidative *N*-demethylation predominates. The resultant cyclic amidine is hydrolysed and recyclised to the lactam, before undergoing rearrangement to *Oxazepam*, one of the products of the metabolic oxidation of *Diazepam*. *Desmethyldiazepam*, the other principal metabolite of *Diazepam*, is also metabolised to *Oxazepam*. *Desmethyldiazepam* is also a metabolite of *Clorazepate* and *Medazepam*.

*Chlordiazepoxide*

HCHO

*Oxazepam*

HCHO

*Diazepam*

HCHO

*Medazepam*

*Desmethyldiazepam*

$CO_2 + H_2O$

*Chlorazepate*

# References


Abraham, E. P. and Chain, E. B. (1940). *Nature, Lond.*, **146**, 837
Abraham, E. P., Chain, E. B., Fletcher, C. M. Florey, H. W., Gardner, A. D., Heatley, N. G. and Jennings, M. A. (1941). *Lancet*, **ii**, 177.
Abraham, E. P., Chain, E. B. and Holiday, E. R. (1942). *Br. J. exp. Path.*, **23**, 103.
Abraham, E. P. and Duthie, E. S. (1946). *Lancet*, **i**, 455.
Abraham, E. P. and Newton, G. G. F. (1961). *Biochem. J.*, **79**, 377.
Abraham, E. P., Newton, G. G. F., Olson, B. H., Schuurmans, D. M., Schenck, J. R., Hargie, M. P., Fisher, M. W. and Fusari, S. A. (1955). *Nature, Lond.*, **176**, 551.
Abramson, F. B., Furst, C. I., McMartin, C. and Wade, R. (1969). *Biochem. J.*, **113**, 143.
Acher, R. (1960). *A. Rev. Biochem.*, **29**, 547.
Adam, J., Gosselain, P. A. and Goldfinger, P. (1953). *Nature, Lond.*, **171**, 704.
Adamson, D. W., Barrett, P. A., Billinghurst, J. W. and Jones, T. S. G. (1957). *J. chem. Soc.*, 2315.
Adamson, R. H. (1972). *Nature, Lond.*, **236**, 400.
Adamson, R. H., Dixon, R. L., Francis, F. L. and Rall, D. P. (1965). *Proc. natn. Acad. Sci. U.S.A.*, **54**, 1386.
Adler, T. K. (1963). *J. Pharmac. exp. Ther.*, **140**, 155.
Agin, D. (1965). *Nature, Lond.*, **205**, 805.
Ahlquist, R. P. (1948). *Am. J. Physiol.*, **153**, 586.
Aizawa, Y. and Mueller, G. C. (1961). *J. biol. Chem.*, **236**, 381.
Akhtar, M., Munday, K. A., Rahimhula, A. D., Watkinson, I. A. and Wilton, D. C. (1969). *Chem. Commun.*, 1287.
Albert, A. (1953). *Biochem. J.*, **54**, 646.
—(1956). *Nature, Lond.*, **177**, 525.
—(1968). *Heterocyclic Chemistry*, 2nd edn. Athlone Press, London.
Albert, A. and Barlin, G. B. (1959). *J. chem. Soc.*, 2384.
Albert, A., Brown, D. J. and Wood, H. C. S. (1956). Ibid., 2066.
Albert, A. and Goldacre, R. J. (1943). Ibid., 454.
Albert, A., Rubbo, S. D., Goldacre, R. J. and Balfour, B. G. (1947). *Br. J. exp. Path.*, **28**, 69.
Alberts, A. W., Majerus, P. W., Talamo, B. and Vagelos, P. R. (1964). *Biochemistry, N. Y.*, **3**, 1563.
Alexander, W. D., Evans, V., MacAulay, A., Gallagher, T. F. and Londono, J. (1966). *Br. med. J.*, **2**, 290.
Allawala, N. A. and Riegelman, S. (1953). *J. Am. pharm. Ass. Sci. Edn.*, **42**, 396.
Allen, G. and Caldin, E. F. (1953). *Quart. Rev.*, **7**, 255.
Allison, J. L., O'Brien, R. L. and Hahn, F. E. (1965). *Science, N. Y.*, **149**, 1111.
Alpar, O., Deer, J. J., Hersey, J. A. and Shotton, E. (1969). *J. Pharm. Pharmac.*, **21**, (*suppl.*), 6S.
Alvarado, F. (1966). *Biochim. biophys. Acta*, **112**, 292.
Ambrose, E. J. and Elliott, A. (1951). *Proc. R. Soc. Ser. A.*, **208**, 75.
*American Industrial Hygiene Journal* (1972). **33**, 381.
*Anon.* (1973). *Pharm. J.*, **211**, 297.
Ansell, M. F., Hickinbottom, W. J. and Holton, P. J. (1955). *J. chem. Soc.*, 349.
Arcos, J. C. and Arcos, M. (1955). *Nuturwissenschaften*, **42**, 608, 651.
—(1956). *Bull. Soc. Chim. biol., Belges*, **65**, 5.
—(1962). *Progress in Drug Research*, vol. 4, p. 407, ed. Jucker, E. Birkhäuser, Basel.
Arens, J. F. and van Dorp, D. A. (1946). *Nature, Lond.*, **158**, 622.
Asano, A., Brodie, A. F., Wagner, A. F., Wittreich, P. E., and Folkers, K. (1962). *J. biol. Chem.*, **237**, PC2411.
Ash, A. S. F. and Schild, H. O. (1966). *Brit. J. Pharmac. Chemother.*, **27**, 427.
Asnis, R. E., Cohen, F. B. and Gots, J. S. (1952). *Antiobiotics Chemother.*, **2**, 123.
Aspinall, G. O. (1969). *Adv. Carbohydrate Chem. and Biochem.*, **24**, 333.

Astbury, W. T., Dickinson, S. and Bailey, K. (1935). *Biochem. J.* **29**, 2351.
Astill, B. D., Fassett, D. W. and Roudabush, R. L. (1960). Ibid., **75**, 543.
Astill, B. D., Mills, J., Fassett, D. W., Roudabush, R. L. and Terhaar, C. J. (1962). *J. agric. Fd Chem.*, **10**, 315.
Atherden, L. M. (1959). *Biochem. J.*, **71**, 411.
Atkinson, M. R., Jackson, J. F. and Morton, R. K. (1961). *Nature, Lond.*, **192**, 946.
Atkinson, M. R., Morton, R. K. and Murray, A. W. (1963). *Biochem. J.*, **89**, 167.
Aubort, J. D. and Hudson, R. F. (1970). *Chem. Commun.*, 938.
Augstein, J., Green, S. M., Monro, A. M., Wrigley, T. I., Katritzky, A. R. and Tiddy, G. J. T. (1967). *J. med. Chem.*, **10**, 391.
Aune, K. C., Salahuddin, A., Zarlengo, M. H. and Tanford, C. (1967). *J. biol. Chem.*, **242**, 4486.
Austin, A. T. (1960). *Nature, Lond.*, **188**, 1086.
Axelrod, J. (1955). *J. biol. Chem.*, **214**, 753.
—(1956). *Biochem. J.*, **63**, 634.
—(1962). *J. Pharmac. exp. Ther.*, **138**, 28.
Axelrod, J., Senoh, S. and Witkop, B. (1958). *J. biol. Chem.*, **233**, 697.
Azaz, E., Donbrow, M. and Hamburger, R. (1973). *Pharm. J.*, **211**, 15; *Analyst*, **98**, 663.
Babcock, J. C., Gutsell, E. S., Herr, M. E., Hogg, J. A., Stucki, J. C., Barnes, L. E. and Dulin, W. E. (1958). *J. Am. chem. Soc.*, **80**, 2904.
Baddeley, G. (1949). *J. chem. Soc.*, S99.
Badger, G. M. (1949). Ibid., 456.
Badger, G. M. and Buttery, R. G. (1954). Ibid., 2243.
Badger, G. M. and Rubalcava, H. (1954). *Proc. natn. Acad. Sci. U.S.A.*, **40**, 12.
Baeyer, von A. (1885). *Ber. dt. chem. Ges.*, **18**, 2269.
Bailey, J. H. and Cavallito, C. J. (1947). *J. Bact.*, **54**, 7.
Baker, W. and Ollis, W. D. (1957). *Quart. Rev.*, **11**, 15.
Baldock, G., Berthier, G. and Pullman, A. (1949). *C.r. hebd. Séanc. Acad. Sci., Paris*, **228**, 931.
Ball, S., Goodwin, T. W. and Morton, R. A. (1948). *Biochem. J.*, **42**, 516.
Bamberger, E. and Seligman, R. (1903). *Ber. dt. chem. Ges.*, **36**, 685.
Barbour, A. K. (1969). *Chemy Brit.*, **5**, 250.
Barlow, C. F., Firemark, H. and Roth, L. J. (1962). *J. Pharm. Pharmac.*, **14**, 550.
Barnes, L. E., Stafford, R. O., Guild, M. E. and Olson, K. J. (1954). *Proc. Soc. exp. Biol. Med.*, **87**, 35.
Barnsley, E. A. (1964). *Biochem. J.*, **90**, 9P.
—(1964). Ibid., **93**, 15P.
Bartlett, P. D. and Small, G. (1950). *J. Am. chem. Soc.*, **72**, 4867.
Barton, D. H. R. (1950). *Experientia*, **6**, 316.
Barton, D. H. R. and Beaton, J. M. (1960). *J. Am. chem. Soc.*, **82**, 2641.
Barton, D. H. R., Beaton, J. M., Geller, L. E. and Pechet, M. M. (1960). Ibid., **82**, 2640.
Barton, D. H. R., Comer, F., Greig, D. G. T., Sammes, P. G., Cooper, C. M., Hewitt, G. and Underwood, W. G. E. (1971). *J. chem. Soc.*, **C**, 3540.
Barton, D. H. R. and Cookson, R. C. (1965) *Quart. Rev.*, **10**, 44.
Barton, D. H. R., Hesse, R. H., Pechet, M. M. and Rizzardo, E. (1973). *J. Am. chem. Soc.*, **95**, 2748.
Barton, D. H. R. and Sammes, P. G. (1971). *Proc. R. Soc.*, **B.**, **179**, 345.
Barton, D. H. R. and Taylor, W. C. (1958). *J. chem. Soc.*, 2500.
Bartz, Q. R., Elder, C. C., Frohardt, R. P., Fusari, S. A., Haskell, T. H., Johannessen, D. W. and Ryder, A. (1954). *Nature, Lond.*, **173**, 72.
Batchelor, F. R., Doyle, F. P., Nayler, J. H. C. and Rolinson, G. N. (1959). Ibid., **183**, 257.
Battista, O. A. (1962). *U.S. Patent 3,023,104.*
Battista, O. A. and Smith, P. A. (1961). *U.S. Patent 2,978,446.*
—(1962). *Ind. Engng Chem.*, **54**, No. 9, 20.
Bean, H. S. and Berry, H. (1951). *J. Pharm. Pharmac.*, **3**, 639.
Beckett, A. H. (1956). Ibid., **8**, 848.
—(1962). *Enzymes and Drug Action*, pp. 15 and 238, ed. Mongar, J. L., Churchill, London.
Beckett, A. H. and Bélanger, P. M. (1974). *J. Pharm. Pharmac.*, **26**, 205, 558.
—(1974). *Xenobiotica*, **4**, 509.
Beckett, A. H. and Brookes, L. G. (1971). *J. Pharm. Pharmac.*, **23**, 288.
Beckett, A. H., Gorrod, J. W. and Lazarus, C. R. (1971). *Xenobiotica*, **1**, 535.
Beckett, A. H., Harper, N. J., Clitherow, J. W. and Lesser, E. (1961). *Nature, Lond.*, **189**, 671.

Beckett, A. H. and Hewick, D. S. (1967). *J. Pharm. Pharmac.*, **19**, 134.
Beckett, A. H. and Mulley, B. A. (1955). *Chemy Ind.*, 146.
Beckett, A. H. and Robinson, A. E. (1956). *J. Pharm. Pharmac.*, **8**, 1072.
—(1957). *Chemy Ind.*, 523.
—(1959). *J. med. Chem.*, **1**, 135.
Beckett, A. H. and Stenlake, J. B. (1975). *Practical Pharmaceutical Chemistry*, 3rd edn., Part 1, p. 229, Athlone Press, London.
Beckett, A. H., Van Dyk, J. M., Chissick, H. H. and Gorrod, J. W. (1971). *J. Pharm. Pharmac.*, **23**, 809.
Beereboom, J. J., Ursprung, J. J., Rennhard, H. H. and Stephens, C. R. (1960). *J. Am. chem. Soc.*, **82**, 1003.
Behnisch, R., Mietzsch, F. and Schmidt, H. (1950). *Am. Rev. Tuberc. pulm. Dis.*, **61**, 1.
Behrens, O. K. and Bromer, W. W. (1958). *A. Rev. Biochem.*, **27**, 57.
Bell, P. H. and Roblin, R. O. (1942). *J. Am. chem. Soc.*, **64**, 2905.
Bell, S. A., Berdick, M. and Holliday, W. M. (1966). *J. New Drugs*, **6**, 284.
Belleau, B. (1958). *Can. J. Biochem. Physiol.*, **36**, 731.
—(1960). *Adrenergic Mechanisms*, ed. Wolstenholme, G. E. W., Vane, J. R. and O'Connor, M., Churchill, London.
—(1967). *Ann. N. Y. Acad. Sci.*, **139**, 580.
Belleau, B. and Burba, J. (1960). *J. Am. chem. Soc.*, **82**, 5751.
Benavides, L., Olson, B. H., Varela, G. and Holt, S. H. (1955). *J. Am. med. Ass.*, **157**, 987.
Bender, M. L. and Kézdy, F. J. (1965). *A. Rev. Biochem.*, **34**, 49.
Benedict, R. G., Schmidt, W. H. and Coghill, R. D. (1946). *J. Bact.*, **51**, 291.
Benson, G. (1936). *Org. Synth.*, **16**, 26.
Berenblum, I., Ben-Ishai, D., Haran-Ghera, N., Lapidot, A., Simon, E. and Trainin, E. (1959). *Biochem. Pharmac.*, 2, 168.
Berg, P. (1958). *J. biol. Chem.*, **233**, 608.
Bergström, S., Carlson, L. A. and Weeks, J. R. (1968). *Pharmac. Rev.*, **20**, 1.
Bergy, M. E., Eble, T. E. and Herr, R. R. (1961). *Antibiotics Chemother.*, **11**, 661.
Berthier, G., Coulson, C. A., Greenwood, H. H. and Pullman, A. (1948). *C. r. hebd. Séanc. Acad. Sci., Paris*, **226**, 1906.
Bijvoet, J. M., Peerdeman, A. F. and van Bommel, A. J. (1951). *Nature, Lond.*, **168**, 271.
Bingel, A. S. and Benoit, P. S. (1973). *J. Pharm. Sci.*, **62**, 179.
Birch, A. J. (1950). *J. chem. Soc.*, 367.
Birch, A. J. and Donovan, F. W. (1953). *Aust. J. Chem.*, **6**, 360.
Birch, A. J., Massy-Westropp, R. A. and Moye, C. J. (1955). *Chemy Ind.*, 683.
Birks, J. B. (1961). *Nature, Lond.*, **190**, 232.
Bittman, R., Chen, W. C. and Anderson, O. R. (1974). *Biochemistry, N. Y.*, **13**, 1364.
Black, J. W., Duncan, W. A. M., Durant, C. J., Ganellin, C. R. and Parsons, E. M. (1972). *Nature, Lond.*, **236**, 385.
Black, J. W., Durant, G. J., Emmett, J. C. and Ganellin, C. R. (1974). ibid., **248**, 65.
Bladon, P. and Stenlake, J. B. (1973). Unpublished information.
Blake, C. C. F., Johnson, L. N., Mair, G. A., North, A. C. T., Phillips, D. C. and Sarma, V. R. (1967b). *Proc. R. Soc.*, **B**, **167**, 378.
Blake, C. C. F., Koenig, D. F., Mair, G. A., North, A. C. T., Phillips, D. C. and Sarma, V. R. (1965). *Nature, Lond.*, **206**, 757.
Blake, C. C. F., Mair, G. A., North, A. C. T., Phillips, D. C. and Sarma, V. R. (1967a). *Proc. R. Soc.*, **B**, **167**, 365.
Blanchard, H. S. (1960). *J. org. Chem.*, **25**, 264.
Blaug, S. M. and Huang, W. T. (1972). *J. Pharm. Sci.*, **61**, 1770.
Bloch, K., Borek, E. and Rittenberg, D. (1946). *J. biol. Chem.*, **162**, 441.
Bloch, K., Chaykin, S., Phillips, A. H. and De Waard, A. (1959). Ibid., **234**, 2595.
Bloink, G. J. and Pausacker, K. H. (1952). *J. chem. Soc.*, 661.
Bloom, B. M., Goldman, I. M. and Belleau, B. (1969). *Pharmac. Rev.*, **21**, 131.
Blundell, T. (1975). *New Scientist*, **67**, 667.
Blundell, T. L., Dodson, G. G., Hodgkin, D. C. and Mercola, D. A. (1972). *Adv. Protein Chem.*, **26**, 279.
Blunt, J. W., De Luca, H. F. and Schnoes, H. K. (1968). *Biochemistry, N. Y.*, **7**, 3317.
Böhle, E., Böttcher, K., Piekarski, H. G. and Biegler, R. (1956). *Dt. Arch. klin. Med.*, **203**, 29.

Boothe, J. H., Hlavka, J. J., Petisi, J. P. and Spencer, J. L. (1960). *J. Am. chem. Soc.*, **82**, 1253.
Borg, K. O., Holgersson, H. and Lagerström, P. O. (1970). *J. Pharm. Pharmac.*, **22**, 507.
Boutagy, J., Gelbart, A. and Thomas, R. (1973). *Aust. J. Sci.*, **[NS]2**, 41.
Bovet, D. (1947). *Rendiconto Instituto superiori di Sanita, Rome*, **10**, 1161.
Bowers, A., Ringold, H. J. and Denot, E. (1958). *J. Am. chem. Soc.*, **80**, 6115.
Bowers, A., Ringold, H. J. and Dorfman, R. I. (1957). Ibid., **79**, 4556.
Bowman, W. C., Rand, M. J. and West, G. B. (1968). *Textbook of Pharmacology*, p. 819, Blackwell, Oxford.
Boyland, E. and Manson, D. (1966). *Biochem. J.*, **101**, 84.
Boyland, E., Nery, R., Peggie, K. S. and Williams, K. (1963). Ibid., **89**, 113P.
Boyland, E. and Sims, P. (1964a). Ibid., **90**, 391.
—(1964b). Ibid., **91**, 493.
Boyland, E. and Williams, K. (1965). Ibid., **94**, 190.
Bradley, T. J. and Sen, H. (1974). *J. Pharm. Pharmac.*, **26**, (*suppl.*). 93P.
Brady, O. L. and Cropper, F. R. (1950). *J. chem. Soc.*, 507.
Brady, R. O., Bradley, R. M. and Trams, E. G. (1960). *J. biol. Chem.*, **235**, 3093.
Brain, E. G., Doyle, F. P., Hardy, K., Long, A. A. W., Mehta, M. D., Miller, D., Nayler, J. H. C., Soulal, M. J., Stove, E. R. and Thomas, G. R. (1962). *J. chem. Soc.*, 1445.
Brand, J. C. D., Jarvie, A. W. P. and Horning, W. C. (1959). Ibid., 3844.
Brandenburg, D. (1969). *Hoppe-Seyler's Z. physiol. Chem.*, **350**, 741.
Brandl, E. and Margreiter, H. (1954). *Öst. chem. Z.*, **55**, 11.
Braun, P. B., Hornstra, J. and Leenhouts, J. I. (1969). *Phillips Res. Reports*, **24**, 427.
Braunshtein, A. E. (1960). *The Enzymes*, p. 113, eds. Boyer, P. D., Lardy, H. and Myrbäck, K., Academic Press, New York.
Braunshtein, A. E. and Shemyakin, M. M. (1953). *Biokhimiya*, **18**, 393.
Breslow, R. (1957). *J. Am. chem. Soc.*, **79**, 1762.
—(1958). Ibid., **80**, 3719.
Breslow, R. and McNelis, E. (1959). Ibid., **81**, 3080.
Brestkin, A. P. and Rozengart, E. V. (1965). *Nature, Lond.*, **205**, 388.
Brickman, A. S., Coburn, J. W. and Norman, A. W. (1972). *New Engl. J. Med.*, **287**, 891.
Bridges, J. W., Kibby, M. R. and Williams, R. T. (1965). *Biochem. J.*, **96**, 829.
Brill, E. and Radomski, J. L. (1971). *Proceedings of the Symposium on the Biological Oxidation of Nitrogen in Organic Molecules, Xenobiotica*, p. 35, eds. Bridges, J. W., Gorrod, J. W., and Parke, D. V., Taylor and Francis, London.
Brimacombe, E., Brimacombe, J. S. and Lindberg, B. (1960). *Acta chem scand.*, **14**, 2236.
Brimblecombe, R. W., Duncan, W. A. M., Durant, G. J., Emmett, J. C., Ganellin, C. R. and Parsons, M. E. (1975). *J. int. med. Res.*, **3**, 86.
Brockman, R. W. and Anderson, E. P. (1963). *A. Rev. Biochem.*, **32**, 463.
Brodie, J. D., Wasson, G. and Porter, J. W. (1963). *J. biol. Chem.*, **238**, 1294.
Bromer, W. W., Staub, A., Diller, E. R., Bird, H. L., Sinn, L. G. and Behrens, O. K. (1957). *J. Am. chem. Soc.*, **79**, 2794.
Brotzu, G. (1948). *Lavori Inst. Ig., Cagliari.*
Brown, B. T., Stafford, A. and Wright, S. E. (1962). *Br. J. Pharmac. Chemother.*, **18**, 131.
Brown, D. M. and Acred, P. (1961). *Br. med. J.*, **ii**, 197.
Brown, D. M. and Coe, P. F. (1970). *Chem. Commun.*, 568.
Brown, G. M. (1962). *J. biol. Chem.*, **237**, 536.
Brown, G. M., Weisman, R. A. and Molnar, D. A. (1961). Ibid., **236**, 2534.
Brown, H., Sanger, F. and Kitai, R. (1955). *Biochem. J.*, **60**, 556.
Brown, H. C. (1956). *J. chem. Soc.*, 1248.
Brueckner, A. H. (1943). *Yale J. Biol. Med.*, **15**, 813.
Buchner, E. (1896). *Ber. dt. chem. Ges.*, **29**, 106.
—(1897). Ibid., **30**, 632.
—(1898). Ibid., **31**, 2241.
Buchner, E. and Curtius, T. (1885). Ibid., **18**, 2377.
Budziewicz, H., Djerassi, C. and Williams, D. H. (1964). *Structure Elucidation of Natural Products by Mass Spectrometry*, Holden-Day Inc., San Francisco.
Bundgaard, H. (1971). *J. Pharm. Sci.*, **60**, 1273.
Bunton, C. A., Llewellyn, D. R., Oldham, K. G. and Vernon, C. A. (1958). *J. chem. Soc.*, 3574, 3588.

Bunton, C. A., Mhala, M. M., Oldham, K. G. and Vernon, C. A. (1960). Ibid., 3293.
Bunton, C. A., Silver, B. L. and Vernon, C. A. (1957) *Proc. chem. Soc.*, 348.
Burchall, J. and Hitchings, G. H. (1965). *Molec. Pharmac.*, **1**, 126.
Burk, D., Hearon, J., Caroline, L. and Schade, A. L. (1946). *J. biol. Chem.*, **165**, 723.
Burk, D., Schade, A. L., Hesselbach, M. L., Hearon, J. and Fischer, C. E. (1946). *Cancer Res.*, **6**, 497.
Bush, I. E. (1962). *Pharmac Rev.*, **14**, 317.
Bushby, S. R. M. and Hitchings, G. H. (1968). *Brit. J. Pharmac. Chemother.*, **33**, 72.
Butler, D. E., Bass, P., Nordin, I. C., Hauck, F. P. and L'Italien, Y. J. (1971). *J. med. Chem.*, **14**, 575.
Butler, K., Laland, S., Overend, W. G. and Stacey, M. (1950). *J. chem. Soc.*, 1433.
Buu-Höi, N. P., Hoán, N., Jacquignon, P. and Khoi, N. H. (1950). Ibid., 2766.
Cahn, R. D., Kaplan, N. O., Levine, L. and Zwilling, E. (1962). *Science, N. Y.*, **136**, 962.
Cahn, R. S. (1964). *J. chem. Educ.*, **41**, 116.
Cahn, R. S., Ingold, C. K. and Prelog, V. (1956). *Experientia*, **12**, 81.
Caldwell, J. H., Martin, J. F., Dutta, S. and Greenberger, N. J. (1969). *Am. J. Physiol.*, **217**, 1747.
Caliguiri, L. A. and Tamm, I. (1970). *Ann. N. Y. Acad. Sci.*, **173**, 420.
Callow, R. K. and Kennard, O. (1961). *J. Pharm. Pharmac.*, **13**, 723.
Calvin, M. and Lemmon, R. M. (1947). *J. Am. chem. Soc.*, **69**, 1232.
Cammarata, A. (1967). *J. med. Chem.*, **10**, 525.
Campbell, D., Landgrebe, F. W. and Morgan, T. N. (1944). *Lancet*, **i**, 630.
Campbell, H. A. and Link, K. P. (1941). *J. biol. Chem.*, **138**, 21.
Campbell, I. M. (1969). *Tetrahedron Lett.*, **54**, 4777.
Canfield, R. E. and Liu, A. K. (1965). *J. biol. Chem.*, **240**, 1997.
Cantoni, G. L. and Anderson, D. G. (1956). Ibid., **222**, 171.
Carson, R. (1963). *Silent Spring*, Hamish Hamilton, London.
Carter, H. E. (1946). *Org. Reactions*, **3**, 198.
Casy, A. F. and Ison, R. R. (1970). *J. Pharm. Pharmac.*, **22**, 270.
Catto, G. R. D., Macleod, M., Pelc, B. and Kodicek, E. (1975). *Br. med. J.*, **i**, 12.
Cavill, G. W. K., Cole, E. R., Gilham, P. T. and McHugh, D. J. (1954). *J. chem. Soc.*, 2785.
Cecil, R. and McPhee, J. R. (1959). *Adv. Protein Chem.*, **14**, 255.
Chain, E. B. (1948). *A. Rev. Biochem.*, **17**, 657.
Chain, E. B., Duthie, E. S. and Callow, D. (1945). *Lancet*, **i**, 652.
Chain, E. B., Philpot, F. J. and Callow, D. S. (1948). *Archs Biochem.* **18**, 171.
Challis, B. C. and Osborne, M. R. (1972). *Chem. Commun.*, 518.
Chalvet, O., Daudel, R. and Moser, C. M. (1958). *Cancer Res.*, **18**, 1033.
Chance, R. E., Ellis, R. M. and Bromer, W. M. (1968). *Science, N. Y.*, **161**, 165.
Chapman, A. T. (1935). *J. Am. chem. Soc.*, **57**, 419.
Chapman, D. D., Autrey, R. L., Gourlay, R. H., Johnson, A. L., Souto, J. and Tarbell, D. S. (1962). *Proc. natn. Acad. Sci. U.S.A.*, **48**, 1108.
Chaudhuri, N. K., Montag, B. J., Heidelberger, C. (1958). *Cancer Res.*, **18**, 318.
Chen, R. F. (1967). *Archs Biochem. Biophys.*, **120**, 609.
Chibnall, A. C. and Rees, M. W. (1958). *Biochem. J.*, **68**, 105.
Chignell, C. F. (1969). *Molec. Pharmac.*, **5**, 244, 455.
Chrambach, A. and Carpenter, F. H. (1960). *J. biol. Chem.*, **235**, 3478.
Church, J. M. and Blumberg, R. (1951). *Ind. Engng Chem.* **43**, 1780.
Church, P. F. R., Schaub, R. E. and Weiss, M. J. (1971). *J. org. Chem.*, **36**, 723.
Cilento, G. (1960). *J. Med. Pharm. Chem.*, **2**, 241.
Clark, J. F. and Gurd, F. R. N. (1967). *J. biol. Chem.*, **242**, 3257.
Clark, P. A. (1965). *Pharm. J.*, **194**, 375.
Clayton, R. B. (1965). *Quart. Rev.*, **19**, 201.
Clough, G. W. (1918). *J. chem. Soc.*, 526.
Coates, L. V., Pashley, M. M. and Tattersall, K. (1961). *J. Pharm. Pharmac.*, **13**, 620.
Code, C. F. (1965). *Fedn Proc. Fedn Am. Socs exp. Biol.*, **24**, 1311.
Coffey, G. L., Anderson, L. E., Fisher, M. W., Galbraith, M. M., Hillegas, A. B., Kohberger, D. L., Thompson, P. E., Weston, K. S. and Ehrlich, J. (1959). *Antibiotics Chemother.*, **9**, 730.
Coffey, S. (1953) *Chemy Ind.*, 1068.
Cohen, J. A., Oosterbaan, R. A., Jansz, H. S. and Berends, F. (1959). *J. cell. comp. Physiol.*, **54**, 231.

Cohn, E. J., Strong, L. E., Hughes, W. L., Mulford, D. J., Ashworth, J. N., Melin, M. and Taylor, H. L. (1946). *J. Am. chem. Soc.*, **68**, 459.
Cohn, M. (1949). *J. biol. Chem.*, **180**, 771.
—(1956). *Biochim. biophys. Acta*, **20**, 92.
Cole, R. D. (1967). *Meth. Enzym.*, **11**, 315.
Cole, W. G. and Williams, D. H. (1968). *J. chem. Soc.*, **C**, 1849.
Colebrook, L. and Kenny, M. (1936). *Lancet*, **i**, 1279.
Collard, R. E. (1961). *Pharm. J.*, **186**, 113.
Coltart, J., Howard, M. and Chamberlain, D. (1972). *Br. med. J.*, **ii**, 318.
Comrie, A. M., Mital, H. C. and Stenlake, J. B. (1960). *J. med. Chem.*, **2**, 1, 153.
Controulis, J., Rebstock, M. C. and Crooks, H. M. (1949). *J. Am. chem. Soc.*, **71**, 2463.
Cook, C. D., Nash, N. G. and Flanagan, H. R. (1955). Ibid., **77**, 1783.
Cook, J. W. and Schoental, R. (1948). *J. chem. Soc.*, 170.
Cooper, J. R. and Brodie, B. B. (1955a). *J. Pharmac. exp. Ther.*, **120**, 75.
—(1955b). Ibid., **114**, 409.
Cooper, P. F. and Wood, G. C. (1968). *J. Pharm. Pharmac.*, **20**, 150S.
Cope, A. C. and Hiok-Huang Lee (1957). *J. Am. chem. Soc.*, **79**, 964.
Cope, A. C. and Towle, P. H. (1949). Ibid., **71**, 3423.
Copper, D. J., Daniels, P. J. L., Yudis, M. D., Marigliano, H. M. and Bukhari, S. T. K. (1971). *J. chem. Soc.*, *C*, 3126.
Corby, T. C. and Elworthy, P. H. (1971). *J. Pharm. Pharmac.*, **23**, 39S, 49S.
Cornforth, J. W., Gore, I. Y. and Popják, G. (1957). *Biochem. J.*, **65**, 94.
Cornforth, J. W., Hunter, G. D. and Popják, G. (1953). Ibid., **54**, 590, 597.
Cornforth, J. W. and Popják, G. (1954). Ibid., **58**, 403.
Cottrell, T. L. (1958). *The Strengths of Chemical Bonds*, 2nd edn., Butterworths, London.
Coulson, C. A. and Longuet-Higgins, H. C. (1947). *Rev. Sci.*, **85**, 929.
Cowan, P. M. and McGavin, S. (1955). *Nature, Lond.*, **176**, 501.
Cowles, P. B. (1942). *Yale J. Biol. Med.*, **14**, 599.
Cox, G. B. and Gibson, F. (1966). *Biochem. J.*, **100**, 1.
Cox, J. S. G., Beach, J. E., Blair, A. M. J. N., Clarke, A. J., King, J., Lee, T. B., Loveday, D. E. E., Moss, G. F., Orr, T. S. C., Ritchie, J. T. and Sheard, P. (1970). *Advances in Drug Research*, vol. 5, p. 115, eds. Harper, N. J. and Simmonds, A. B., Butterworths, London.
Craig, L. C., Shedlovsky, T., Gould, R. G. and Jacobs, W. A. (1938). *J. biol. Chem.*, **125**, 289.
Crawhall, J. C. and Watts, R. W. E. (1962). *Biochem. J.*, **85**, 163.
Creasey, W. A., Fink, M. E., Handschumacher, R. E. and Calabresi, P. (1963). *Cancer Res.*, **23**, 444.
Cressman, W. A., Sugita, E. T., Doluisio, J. T., Nubergall, P. J. (1969). *J. Pharm. Sci.*, **58**, 1471.
Cronyn, M. W., Chang, M. P. and Wall, R. A. (1955). *J. Am. chem. Soc.*, **77**, 3031.
Cross, A. D. (1964). *An Introduction to Practical Infrared Spectroscopy*, Butterworths, London.
Crouch, W. W. (1952). *J. Am. chem. Soc.*, **74**, 2926.
Crowfoot, D., Bunn, C. W., Rogers-Low, B. W. and Turner-Jones, A. (1949). *The Chemistry of Penicillin*, pp. 310–66, Princeton University Press.
Crowfoot, D., Hodgkin, D. and Maslen, E. N. (1961). *Biochem. J.*, **79**, 393.
Crowther, A. F. and Levi, A. A. (1953). *Br. J. Pharmac. Chemother.*, **8**, 93.
Cullen, S. C. and Gross, E. G. (1951). *Science, N. Y.*, **113**, 580.
Cunningham, L. W. (1964). *Biochemistry, N. Y.*, **3**, 1629.
Cunningham, L. W., Fischer, R. L. and Vestling, C. S. (1955). *J. Am. chem. Soc.*, **77**, 5703.
Cymerman-Craig, J., Rubbo, S. D., Willis, D. and Edgar, J. (1955). *Nature, Lond.*, **176**, 34.
Daemen, F. J. M. and Bonting, S. L. (1969). Ibid., **222**, 879.
Dagne, E. and Castagnoli, N. (Jr.). (1972). *J. med. Chem.*, **15**, 840.
Dakin, H. D. (1906). *J. biol. Chem.*, **1**, 171.
Dale, J. K. and Rundman, S. J. (1957). *J. Am. pharm. Ass. Pract. Pharm. edn.*, **18**, 421.
Dam, H. (1935). *Nature, Lond.*, **135**, 652; *Biochem. J.*, **29**, 1273.
Danielli, J. F., Danielli, M., Fraser, J. B., Mitchell, P. D., Owen, L. N. and Shaw, G. (1947). Ibid., **41**, 325.
Dann, O., Ulrich, H. and Möller, E. F. (1950). *Z. Naturf.*, **5b**, 446.
Dansette, P. and Azerad, R. (1970). *Biochem. biophys. Res. Commun.*, **40**, 1090.
Dauben, W. G., Bell, I., Hutton, T. W., Laws, G. F., Rheiner, A. and Urscheler, H. (1958). *J. Am. chem. Soc.*, **80**, 4116.

David, A., Hartley, F., Millson, D. R. and Petrow, V. (1957). *J. Pharm. Pharmac.*, **9**, 929.
Davies, W. and James, F. C. (1954). *J. chem. Soc.*, 15.
Davies, W. W., Krahl, M. E. and Clowes, G. H. A. (1940). *J. Am. chem. Soc.*, **62**, 3080.
Decker, F. and Lynen, F. (1955). *3 éme Congrès internationale de Biochemie, Résumés des communications*, 56.
Dekker, C. A. and Khorana, H. G. (1954). *J. Am. chem. Soc.*, **76**, 3522.
DeLuca, H. F. (1969). *Fedn. Proc. Fedn Am. Socs exp. Biol.*, **28**, 1678.
De Mayo, P. and Reid, S. T. (1961). *Quart. Rev.*, **15**, 393.
Demel, R. A., Kinsky, S. C. and van Deenen, L. L. M. (1965). *J. biol. Chem.*, **240**, 2749.
De Moss, J. A., Genuth, S. M. and Novelli, G. D. (1956). *Proc. natn. Acad. Sci. U.S.A.*, **42**, 325.
den Hertog, H. J. and Combé, W. P. (1951). *Recl Trav. chim. Pays-Bas Belg.*, **70**, 581.
Dermer, O. C. and Edmison, M. T. (1957). *Chem. Rev.*, **57**, 77.
De Stevens, G., Werner, L. H., Halamandani, A. and Ricca, S. (1958). *Experientia*, **14**, 463.
Dickens, F. (1964). *Br. med. Bull.*, **20**, 96.
Dickens, F. and Jones, H. E. H. (1961). *Br. J. Cancer*, **15**, 85.
—(1963). Ibid., **17**, 100.
Dickman, S. R. (1961). *The Enzymes*, vol. 5, p. 495, eds. Boyer, P. D., Lardy, H. and Mÿrback, K., Academic Press, New York.
Diehl, H. (1937). *Chem. Rev.*, **21**, 39.
Dingell, J. V., Sulser, F. and Gillette, J. R. (1964). *J. Pharmac. exp. Ther.*, **143**, 14.
Dipple, A., Lawley, P. D. and Brookes, P. (1968). *Eur. J. Cancer*, **4**, 493.
Djerassi, C., Miramontes, L. and Rosenkranz, G. (1953). *J. Am. chem. Soc.*, **75**, 4440.
Djerassi, C., Miramontes, L., Rosenkranz, G. and Sondheimer, F. (1954). Ibid., **76**, 4092.
Doerschuk, A. P., Bitler, B. A. and McCormick, J. R. D. (1955). Ibid., **77**, 4687.
Dollery, C. T., Emslie-Smith, D. and Milne, M. D. (1960). *Lancet*, **ii**, 381.
Domagk, G. (1935). *Dt. med. Wschr.*, **61**, 250.
Domagk, G., Behnische, R., Mietzsch, F. and Schmidt, H. (1946). *Naturwissenschaften*, **33**, 315.
Donovick, R., Rake, G. and Fried, J. (1946). *J. biol. Chem.*, **164**, 173.
Doornbos, D. A. (1968). *Pharm. Weekblad.*, **103**, 1213.
Doornbos, D. A. and Faber, J. S. (1964). ibid., **99**, 289.
Doudoroff, M., Barker, H. A. and Hassid, W. Z. (1947). *J. biol. Chem.*, **168**, 725.
Doyle, F. P., Fosker, G. R., Nayler, J. H. C. and Smith, H. (1962). *J. chem. Soc.*, 1440.
Doyle, F. P., Hardy, K., Nayler, J. H. C., Soulal, M. J., Stove, E. R. and Waddington, H. R. J. (1962). Ibid., 1453.
Doyle, F. P., Long, A. A. W., Nayler, J. H. C. and Stove, E. R. (1962). *Nature, Lond.*, **192**, 1183.
Doyle, F. P. and Nayler, J. H. C. (1960). *Brit. Pat.* 838,974.
—(1964). *Advances in Drug Research*, vol. 1, p. 1, Eds, Harper, N. J. and Simmonds, A. B., Academic Press, London.
Doyle, F. P., Nayler, J. H. C., Smith, H. and Stove, E. R. (1961). *Nature, Lond.*, **191**, 1091.
Drill, V. A. and Saunders, F. J. (1956). *Hormones and the Aging Process*, pp. 99–113, eds. Engel, L. L. and Pincus, G., Academic Press, New York.
Druckrey, H. and Raabe, S. (1952). *Klin. Wschr.*, **30**. 882.
Duggar, B. M. (1948). *Ann. N. Y. Acad. Sci.*, **51**, 177.
Dunitz, J. D. (1952). *J. Am. chem. Soc.*, **74**, 995.
Durant, G. J., Roe, A. M. and Green, A. L. (1970). *Progress in Medicinal Chemistry*, vol. 7, (I), p. 124, eds. Ellis, G. P. and West, G. B., Butterworths, London.
Earl, J. C. and Mackney, A. W. (1935). *J. chem. Soc.*, 899.
*East African/MRC Thiacetazone/Diphenylthiourea Investigation* (1959). *Tubercle, Lond.*, **40**, 399.
*East African/MRC Second Thiacetazone Investigation* (1963). Ibid., **44**, 301.
Eberlein, W., Nickl, J., Heider, J., Dahms, G. and Machleidt, H. (1972). *Chem. Ber.*, **105**, 3686.
Edman, P. (1950). *Acta chem. scand.*, **4**, 283.
Ege, R., Gottlieb, E. and Rakestraw, N. W. (1925). *Am. J. Physiol.*, **72**, 76.
Eggerer, H., Remberger, U. and Grüenewaelder, C. (1964). *Biochem. Z.*, **339**, 436.
Eggers, H. J. Ikegami, N. and Tamm, I. (1965). *Ann. N. Y. Acad Sci.*, **130**, 267.
Ehrlich, J., Bartz, Q. R., Smith, R. M., Joslyn, D. A. and Burkholder, P. R. (1947). *Science, N. Y.*, **106**, 417.
Eilhauer, H. D. and Kraemer, C. (1966). *Naturwissenschaften*, **53**, 107.
Elion, G. B., Kovensky, A., Hitchings, G. H., Metz, E. and Rundles, R. W. (1966). *Biochem. Pharmac.*, **15**, 863.

Elkeles, R. S., Hampton, J. R. and Mitchell, J. R. A. (1968). *Lancet*, **ii**, 315.
Ellard, G. A., Garrod, J. M. B., Scales, B. and Snow, G. A. (1965). *Biochem. Pharmac.*, **14**, 129.
Elliott, A., Ambrose, E. J. and Robinson, C. (1950). *Nature, Lond.*, **166**, 194.
Elliott, T. H., Robertson, J. S. and Williams, R. T. (1966). *Biochem. J.*, **100**, 403.
Elliott, T. H., Tao, R. C. C. and Williams, R. T. (1965). Ibid., **95**, 70.
Elson, L. A. and Morgan, W. T. J. (1933). Ibid., **27**, 1824.
Elves, M. W., Buttoo, A. S., Israels, M. C. G. and Wilkinson, J. F. (1963). *Br. med. J.*, **i**, 156.
Elworthy, P. H. (1963) *J. Pharm. Pharmac.*, **15**, 137T.
Elworthy, P. H. and Florence, A. T. (1969). Ibid., **21**, 70S.
Elworthy, P. H., Florence, A. T. and Macfarlane, C. B. (1968). *Solubilization by Surface-active Agents*, Chapman and Hall, London.
Elworthy, P. H. and Lipscomb, F. J. (1968). *J. Pharm. Pharmac.*, **20**, 817.
Embree, N. D. and Shantz. E. M. (1943). *J. Am. chem. Soc.*, **65**, 910.
Emtage, J. S., Lawson, D. E. M. and Kodicek, E. (1973). *Nature, Lond.*, **246**, 100.
Engel, G. B. and Dodson, L. F. (1971). *W. H. O. Pharmaceutical Bulletin*, no. 467.
Englard, S. and Colowick, S. P. (1957). *J. biol. Chem.*, **226**, 1047.
Epstein, C. J., Goldberger, R. F. and Anfinsen, C. B. (1963). *Cold Spring Harb. Symp. quant. Biol.*, **28**, 439.
Erlenmeyer, H. and Leo, M. (1933). *Helv. chim. Acta*, **16**, 1381.
Euwes, P. C. J. (1909). *Recl Trav. chim. Pays-Bas Belg.*, **28**, 298.
Exner, O. and Kakac, B. (1963). *Colln Czech. chem. Commun. Engl. Edn.*, **28**, 1656.
Fabro, S., Schumacher, H., Smith, R. L. and Williams, R. T. (1964). *Life Sci.*, **3**, 987.
Fabro, S., Smith, R. L. and Williams, R. T. (1965). *Nature, Lond.*, **208**, 1208.
Faigle, J. W. and Keberle, H. (1966). *Acta tropica separatum suppl.*, **9**, 8.
—(1969). *Ann. N. Y. Acad. Sci.*, **160**, 544.
Fairbairn, J. W. (1949). *J. Pharm. Pharmac.*, **1**, 683.
Fairbrother, R. W. and Taylor, G. (1960). *Lancet*, **ii**, 400.
Feingold, D. S. (1965). *Biophys. Res. Commun.*, **33**, 477.
Feit, P. W. and Rastrup-Anderson, N. (1973). *J. Pharm. Sci.*, **62**, 1007.
Felicetti, L., Colombo, B. and Baglioni, C. (1966). *Biochim. biophys Acta*, **119**, 120.
Ferguson, J. (1939). *Proc. R. Soc.*, **B**, **127**, 387.
Ferguson, L. N. (1955). *J. Am. chem. Soc.*, **77**, 5288.
Ferrier, W. G. and Iball, J. (1954). *Chemy Ind.*, 1296.
Fetizon, M. and Jurion, M. (1972). *Chem. Commun.*, 382.
Fevold, H. L. (1951). *Classification, Purification and Isolation of Proteins in Amino Acids and Proteins*, p. 256, eds. Greenberg, D. M. and Thomas, Charles C. Springfield, Illinois.
Field, L. and White, J. E. (1973). *Proc. natn. Acad. Sci. U.S.A.*, **70**, 328.
Fieser, L. (1951). *Ann. intern. Med.*, **15**, 648.
Fieser, L. F. (1945). *1944 Rep. A. A. A. S. Conf. Cancer, Washington*, 108.
Fikentscher, H. (1932). *Cellulosechemie*, **13**, 58.
Filippini, F. and Hudson, R. F. (1972). *Chem. Commun.*, 522.
Finean, J. B. (1953). *Experientia*, **9**, 17.
Fischer, E. (1887). *Ber. dt. chem. Ges.*, **20**, 821.
—(1891). Ibid., **24**, 2683.
Fleming, A. (1929). *Br. J. exp. Path.*, **10**, 226.
Flocker, R. H., Marberger, H., Begley, B. J. and Prendergist, L. J. (1955). *J. Urol.*, **74**, 549.
Florence, A. T., Salole, E. G. and Stenlake, J. B. (1974). *J. Pharm. Pharmac.*, **26**, 479.
Florey, H. W., Abraham, E. P., Chain, E., Fletcher, C. M., Gardner, A. D., Heatley, N. G., Jennings, M. A., Orr-Ewing, J. and Sanders, A. G. (1940). *Lancet*, **ii**, 226.
—(1941). Ibid., **ii**, 177.
Florey, H. W., Chain, E., Heatley, N. G., Jennings, M. A., Sanders, A. G., Abraham, E. P. and Florey, M. (1949). *Antibiotics*, Oxford University Press.
Flower, R. L. (1975). *Nature, Lond.*, **253**, 88.
Flynn, G. L. (1971). *J. Pharm. Sci.*, **60**, 345.
Foernzler, E. C. and Martin, A. N. (1967). Ibid., **56**, 608.
Forrest, H. S. and Walker, J. (1948). *Nature, Lond.*, **161**, 721.
—(1949). *J. chem. Soc.*, 2002.
Foss, O. and Tjomsland, O. (1958). *Acta chem. scand.*, **12**, 1810.
Foster, A. B., Harrison, R., Inch, T. D., Stacey, M. and Webber, J. M. (1963). *J. chem. Soc.*, 2279.

Foster, J. F. (1960). *The Plasma Proteins*, p. 179, ed. Putnam, F. W., Academic Press, New York.
Fouts, J. R. and Brodie, B. B. (1957). *J. Pharmac. exp. Ther.*, **119**, 197.
Fox, C. D., Richman, M. D., and Shangraw, R. F. (1965). *J. Pharm. Sci.*, **54**, 447.
Fox, C. L. and Rose, H. M. (1942). *Proc. Soc. exp. Biol. Med.*, **50**, 142.
Fraser, E. J., Leach, R. H., Poston, J. W., Bold, A. M., Culank, L. S. and Lipede, A. B. (1973). *J. Pharm. Pharmac.*, **25**, 968.
Freese, E. and Freese, E. B. (1965). *Biochemistry, N.Y.*, **4**, 2419.
French, D. and Edsall, J. T. (1945). *Adv. Protein Chem.*, **2**, 277.
Frey, H. H., Doenicke, A. and Jäeger, G. (1961). *Med. exp.* (*Basel*), **4**, 243.
Fridovich, I. and Westheimer, F. H. (1962). *J. Am. chem. Soc.*, **84**, 3208.
Frieden, C. (1963). *J. biol. Chem.*, **238**, 3286.
Furchgott, R. F. (1964). *A. Rev. Pharmac.*, **4**, 21, 24.
Futterman, S. (1963). *J. biol. Chem.*, **238**, 1145.
Gale, E. F. (1946). *Adv. Enzymol.*, **6**, 1.
Gale, E. F. and Folkes, J. P. (1953). *Biochem. J.*, **55**, 721, 730.
Gale, E. F. and Taylor, E. S. (1947). *J. gen. Microbiol.*, **1**, 314.
Gale, M. M. and Saunders, L. (1971). *Biochim. biophys. Acta*, **248**, 466.
Gallo, R. C., Yang, S. S. and Ting, R. C. (1970). *Nature, Lond.*, **228**, 927.
Gans, E. H. and Higuchi, T. (1957). *J. Am. pharm. Ass. Sci. Edn.*, **46**, 587.
Garrett, E. R. and Umbreit, G. R. (1962). *J. Pharm. Sci.*, **51**, 436.
Garrod, L. P. (1960). *Brit. med. J.*, **i**, 527.
Garrod, L. P. and O'Grady, F. (1968). *Antibiotics and Chemotherapy*, E. & S. Livingstone, Edinburgh.
Gaudette, L. E. and Brodie, B. H. (1959). *Biochem. Pharmac.*, **2**, 89.
Geddes, A. J., Parker, K. D., Atkins, E. D. T. and Beighton, E. (1968). *J. molec. Biol.*, **32**, 343.
Gelles, E. and Hay, R. W. (1958). *J. chem. Soc.*, 3673.
Gelles, E. and Salema, A. (1958). Ibid., 3683, 3689.
Gemmell, D. H. O. and Morrison, J. C. (1957). *J. Pharm. Pharmac.*, **9**, 641.
Gessner, P. K., Parke, D. V. and Williams, R. T. (1961). *Biochem. J.*, **79**, 482.
Gibson, F. and Pittard, J. (1968). *Bact. Rev.*, **32**, 465.
Gill, E. W. (1965). *Progress in Medicinal Chemistry*, vol. 4, p. 39, eds. Ellis, G. P. and West, G. B., Butterworths, London.
Ginsburg, A. and Carroll, W. R. (1965). *Biochemistry. N. Y.*, **4**, 2159.
Glover, J., Goodwin, T. W. and Morton, R. A. (1948). *Biochem. J.*, **43**, 512.
Glover, M., Glover, J. and Morton, R. A. (1952). Ibid., **51**, 1.
Goldman, P., Alberts, A. W. and Vagelos, P. R. (1963). *J. biol. Chem.*, **238**, 3579.
Goldthwait, D. A., Peabody, R. A. and Greenberg, G. R. (1954). *J. Am. chem. Soc.*, **76**, 5258.
Goodman, D. S. and Huang, H. S. (1965). *Science, N. Y.*, **149**, 879.
Goodman, M. and Fried, M. (1967). *J. Am. chem. Soc.*, **89**, 1264.
Goodman, M., Schmitt, E. E. and Yphantis, D. A. (1962). Ibid., **84**, 1288.
Gordon, R. C., Regamey, C. and Kirby, W. M. M. (1973). *J. Pharm. Sci.*, **62**, 1074.
Goto, T., Nakanishi, K. and Ohashi, M. (1957). *Bull. Chem. Soc., Japan*, **30**, 723.
Gottstein, W. J., Minor, W. F. and Cheney, L. C. (1959). *J. Am. chem. Soc.*, **81**, 1198.
Gould, L. and Brown, M. W. (1974). *Pharm. J.*, **212**, 276.
Govier, W. C. (1965). *J. Pharmac. exp. Ther.*, **150**, 305.
Granatek, A. P., DeMurio, M. P., Nunning, B. C., Athanas, N. G., Dana, R. L., Granatek, E. S. and Daoust, R. G. (1969). *S. African Patent* 6805,560.
Grant, J. K. (1969). *Essays in Biochemistry*, vol. 5, p. 1, eds. Campbell, P. N. and Greville, G. D., Academic Press, London.
Grassie, N. (1956). *Chemistry of High Polymer Degradation Processes*, p. 140, Butterworths, London
Gratzer, W. B., Beavan, C. H., Rattle, H. W. E. and Bradbury, E. M. (1968). *Eur. J. Biochem.*, **3**, 276.
Greco, S. J. (1953). *J. Am. pharm. Ass. Pract. Edn.*, **14**, 424.
Green, J. (1966). *Vitamins and Hormones*, **24**, 619.
Green, J. P., Søndergaard, E. and Dam, H. (1956). *Biochem. biophys. Acta*, **19**, 182.
Greenwood, D. and O'Grady, F. (1976). *Antimicrobial Agents and Chemotherapy*, **10**, 249.
Greer, S. and Zamenhof, S. (1962). *J. molec. Biol.*, **4**, 123.
Griffin, G. W., Basinski, J. E. and Vallturo, A. F. (1960). *Tetrahedron Lett.*, no. 3, p. 13.
Griffin, M. J. and Brown, G. M. (1964). *J. biol. Chem.*, **239**, 310.

Griffiths, J. and Hawkins, C. (1972). *Chem. Commun.*, 463.
Grob, C. A. and Ankli, P. (1949). *Helv. chim. Acta*, **32**, 2010.
Grundmann, C. and Kreutzberger, A. (1954). *J. Am. chem. Soc.*, **76**, 5646.
Gstirner, F. and Tata, P. S. (1958). *Mitt. Dt. pharm. Gesell.*, **28**, 191.
Gunstone, F. D. and Hilditch, T. P. (1946). *J. chem. Soc.*, 1022.
Gupta, A. K. (1952). Ibid., 3479.
Gurgo, C., Apirion, D. and Schlessinger, D. (1969). *J. molec. Biol.*, **45**, 205.
Güttman, D. E., Hamlin, W. E., Shell, J. W. and Wagner, J. G. (1961). *J. Pharm. Sci.*, **50**, 305.
Haber, F. and Weiss, J. (1934). *Proc. R. Soc.*, **A**, **147**, 332.
Habermann, V. (1962). *Biochim. biophys. Acta*, **55**, 999.
Haddow, A. (1955). *A. Rev. Biochem.*, **24**, 689.
Haddow, A. and Sexton, W. A. (1946). *Nature, Lond.*, **157**, 500.
Haddow, A., Timmis, G. M. and Brown, S. S. (1958). Ibid., **182**, 1164.
Hahn, L. A. (1947). *Lancet*, **i**, 408.
Hale, C. W., Newton, G. G. F. and Abraham, E. P. (1961). *Biochem. J.*, **79**, 403.
Halkerston, I. D. K., Eichhorn, J. and Hechter, O. (1961). *J. biol. Chem.*, **236**, 374.
Hallas-Møller, K., Petersen, K. and Schlichtkrull, J. (1952). *Science, N. Y.*, **116**, 394.
Ham, N. S. (1971). *J. Pharm. Sci.*, **60**, 1764.
Ham, N. S., Casy, A. F. and Ison, R. R. (1973). *J. med. Chem.*, **16**, 470.
Hamaguchi, K. (1964). *J. Biochem., Tokyo*, **56**, 441.
Hamilton, J. W. and Holdsworth, E. S. (1970). *Biochim. biophys. Res. Commun.*, **40**, 1325.
Hamilton-Miller, J. M. T., Newton, G. G. F. and Abraham, E. P. (1970). *Biochem. J.*, **116**, 371.
Hancock, R. and Park, J. T. (1958). *Nature, Lond.*, **181**, 1050.
Hansch, C., Kutter, E. and Leo, A. (1969). *J. med. Chem.*, **12**, 746.
Hare, M. L. C. (1928). *Biochem. J.*, **22**, 968.
Harper, N. J. (1962). *Progress in Drug Research*, vol. 4, pp. 221–96, ed. Jücker, E., Birkhäuser, Basel.
Harris, J. I., Sanger, F. and Naughton M. A. (1956). *Archs Biochem. Biophys.*, **65**, 427.
Harrison, R. G., Lythgoe, B. and Wright, P. W. (1973). *Tetrahedron Lett.*, No. 37, 3649.
Harte, R. A. and Rupley, J. A. (1968). *J. biol. Chem.*, **243**, 1663.
Hartiala, K. J. V. and Terho, T. (1965). *Nature, Lond.*, **205**, 809.
Hartley, B. S. and Massey, V. (1956). *Biochim. biophys. Acta.*, **21**, 58.
Hartman, S. C., Levenberg, B. and Buchanan, J. M. (1955). *J. Am. chem. Soc.*, **77**, 501.
Hassall, C. H. and Thomas, W. A. (1971). *Chemy Brit.*, **7**, 145.
Hassel, O. (1953). *Quart Rev.*, **7**, 221.
Hassel, O. and Viervoll, H. (1943). *Tidsskr. Kjemi, Bergvesen Met.*, **3**, 35.
Hawkins, D., Pinckard, R. N. and Farr, R. S. (1968). *Science, N. Y.*, **160**, 780.
Hawkins, E. G. E. (1951). *Organic Peroxides*, Spon, London.
—(1969). *J. chem. Soc.*, (C)., 2686.
Hawkins, E. G. E. and Hunter, R. F. (1944). *Biochem. J.*, **38**, 34.
Haworth, W. N. (1932). *Ber. dt. chem. Ges.*, **65A**, 43.
—(1946). *J. chem. Soc.*, 543.
Haworth, W. N. and Jones, W. G. M. (1944). Ibid., 667.
Hayano, M., Gut, M., Dorfman, R. I., Sebek, O. K. and Peterson, D. H. (1958). *J. Am. chem. Soc.*, **80**, 2336.
Haynes, D. H., Kowalsky, A. and Pressman, B. C. (1969). *J. biol. Chem.*, **244**, 502.
Hearon, J. Z., Schade, A. L., Levy, H. B. and Burk, D. (1947). *Cancer Res.*, **7**, 713.
Heidt, L. J. and Purves, C. B. (1944). *J. Am. chem. Soc.*, **66**, 1385.
Heilbronner, E. (1953). *Helv. chim. Acta*, **36**, 1121.
Helberger, J. H., Manecke, G., Heyden, R. (1949). *Annalen*, **565**, 22.
Helmreich, M. L. and Huseby, R. A. (1965). *Steroids*, (*suppl.*), **II**, 79.
Hem, S. L., Russo, E. J., Bahal, S. M. and Levi, R. S. (1973). *J. Pharm. Sci.*, **62**, 267.
Henbest, H. B. (1963). *Proc. chem. Soc.*, 159.
Henderson, G. and Newton, J. M. (1966). *Pharm. Acta helv.*, **41**, 228.
Herbst, R. M. (1944). *Adv. Enzymol.*, **4**, 75.
Herr, R. R. and Bergy, M. E. (1962). *Antimicrobial Agents and Chemotherapy*, 560.
Hershberger, L. G., Shipley, E. G. and Meyer, R. K. (1953). *Proc. Soc. exp. Biol. Med.*, **83**, 175.
Hertz, R., Tullner, W. W. and Raffelt, E. (1954). *Endocrinology*, **54**, 228.
Herzig, P. T. and Ehrenstein, M. (1951). *J. org. Chem.*, **16**, 1050.

Heyns, K. and Paulsen, H. (1953). *Chem. Ber.*, **86**, 833.
Higuchi, T. and Bias, C. D. (1953). *J. Amer. pharm. Ass. Sci. Edn.*, **42**, 707.
Higuchi, T., Gupta, M. and Busse, L. W. (1953). Ibid., **42**, 157.
Hill, R. B., Gaetani, S., Paolucci, A. M., RamaRao, P. B., Alden, R., Ranhotra, G. S., Shah, D. V., Shah, V. K. and Johnson, B. C., (1968). *J. biol. Chem.*, **243**, 3930.
Hlavka, J. J. and Buyske, D. A. (1960). *Nature, Lond.*, **186**, 1064.
Hobby, G. L., Meyer, K. and Chaffee, E. (1942). *Proc. Soc. exp. Biol. Med.*, **50**, 277.
Hodge, E. B., Senkus, M. and Riddick, J. A. (1946). *Chem. engng. News*, **24**, 2177.
Hoeksema, H., Argoudelis, A. D. and Wiley, P. F. (1962). *J. Am. chem. Soc.*, **84**, 3212.
Hoeksema, H., Bannister, B., Birkenmeyer, R. D., Kagan, F., Magerlein, B. J., MacKellar, F. A., Schroeder, W., Slomp, G. and Herr, R. R. (1964). Ibid., **86**, 4223.
Hoffman, T. A., Cestero, R. and Bullock, W. E. (1970). *J. infect. Dis.*, **122**, (*suppl.*), S75.
Hoffmann, T. A. and Ladik, J. (1961). *Cancer Res.*, **21**, 474.
Hoft, Von E. and Schultze, H. (1967). *Z. Chem.*, **7**, 137.
Holick, M. F., Semmler, E. J., Schnoes, H. K. and De Luca, H. F. (1973). *Science, N. Y.*, **180**, 190.
Hollaway, M. R. (1968). *Ann. Rep. Progr. Chem.*, **65**, 601.
Hollunger, G. (1955). *Acta pharmac. tox.*, **11**, (*Suppl.* 1), 84.
Hopkins, J. W. (1959). *Proc. natn. Acad. Sci. U.S.A.*, **45**, 1461.
Horinishi, H., Hachimori, Y., Kurihara, K. and Shibata, K. (1964). *Biochim. biophys. Acta*, **86**, 477.
Horton, E. W. (1969). *Physiol. Rev.*, **49**, 122.
Hotko, E. A. (1967). *U.S. Patent*, 3 340 152.
Hou, J. P. and Poole, J. W. (1973). *J. Pharm. Sci.*, **62**, 783.
Howard, L. C., Brum, D. R. and Blake, D. A. (1973). Ibid., **62**, 1021.
Hruby, V. J., Yamashiro, D. and du Vigneaud, V. (1968). *J. Am. chem. Soc.*, **90**, 7106.
Hucker, H. B., Ahmad, P. M. and Miller, E. A. (1966). *J. Pharmac. exp. Ther.*, **154**, 176.
Hüettenrauch, R. and Klotz, L. (1963). *Arch. Pharm.*, **296**, 145.
Hugo, W. B. (1957). *J. Pharm. Pharmac.*, **9**, 145.
Hurd, C. D. and Raterink, H. R. (1934). *J. Am. chem. Soc.*, **56**, 1348.
Hurst, E. D. (1963). *Chem. Prod.*, **26**, 48.
Hutchings, B. L. (1957). *Ciba Foundation Symposium on Chemistry and Biology of Purines*, pp. 177–88, Churchill, London.
Hvidt, A. and Linderstrom-Lang, K. (1954). *Biochim. biophys. Acta*, **14**, 574.
—(1955). Ibid., **16**, 168.
—(1955). Ibid., **18**, 306.
Inhoffen, H. H., Brückner, K. and Gründel, R. (1954). *Chem. Ber.*, **87**, 1.
Inman, W. H. W. and Mushin, W. W. (1974). *Br. med. J.*, **i**, 5.
Inman, W. H. W. and Vessey, M. P. (1968). Ibid., **ii**, 193.
Isbell, H. S. (1932). *Bur. Standards. J. Research*, **8**, 615.
Isbell, H. S. and Hudson, C. S. (1932). Ibid., **8**, 327.
Isbell, H. S. and Pigman, W. W. (1937). *J. Research Natl. Bur. Standards*, **18**, 141.
Isler, Von O., Huber, W., Ronco, A. and Kofler, M. (1947). *Helv. chim. Acta*, **30**, 1911.
Ivanov, V. T., Laine, I. A., Abdullaev, N. D., Senyavina, L. B., Popov, E. M., Ovchinnikov, Y. A. and Shemyakin, M. M. (1969). *Biochem. biophys. Res. Commun.*, **34**, 803.
Jacobs, J., Kletter, D., Superstine, E., Hill, K. R., Lynn, B. and Webb, R. A. (1973). *J. clin. Path.*, **26**, 742.
Jaenicke, L. and Chan, P. H. C. (1960). *Angew. Chem.*, **72**, 752.
Jaffé, H. H. (1953). *Chem. Rev.*, **53**, 191.
James, S. P. and Jeffery, D. J. (1964). *Biochem. J.*, **93**, 16P.
Janicki, C. A. and Almond, H. R. (1974). *J. Pharm. Sci.*, **63**, 41.
Jardetzky, O. (1963). *J. biol. Chem.*, **238**, 2498.
Jarman, M. and Ross, W. C. J. (1967). *Chemy Ind.*, 1789.
Jeffery, J. D'A., Abraham, E. P. and Newton, G. G. F. (1961). *Biochem. J.*, **81**, 591.
Jencks, W. P. (1962). *The Enzymes*, vol. 6, p. 339, eds. Boyer, P. D., Lardy, H. and Mÿrback, K., Academic Press, New York.
Jirgensons, B. (1962). *Natural Organic Macromolecules*, p. 258, Pergamon Press, Oxford.
—(1967). *J. biol. Chem.*, **242**, 912.
Jiroüsek, L. and Pritchard, E. T. (1970). *Biochim. biophys. Acta*, **208**, 275.
Johnson, A. W. and McCaldin, D. J. (1958). *J. chem. Soc.*, 817.
Johnson, C. R. and McCants, D. (1965). *J. Am. chem. Soc.*, **87**, 1109.

Johnson, D. A. and Panetta, C. A. (1964). *J. org. Chem.*, **29**, 1826.
Johnson, K. and Degering, E. F. (1943). Ibid., **8**, 10.
Johnson, L. R. (1971). *Gastroenterology*, **61**, 106.
Johnson, M. K. (1963). *Biochem. J.*, **87**, 9P.
Jollès, P., Jauregui-Adell, J. and Jollès, J. (1964). *C.r. hebd. Séanc. Acad. Sci., Paris*, **258**, 3926.
Jolliffe, N. and Goodhart, R. S. (1960). *A. Rev. Med.*, **11**, 257.
Jones, D. W. and Matthews, R. S. (1974). *Progress in Medicinal Chemistry*, vol. 10, p. 159, eds. Ellis, G. P. and West, G. B., Butterworths, London.
Josten, J. J. and Allen, P. M. (1964). *Biochem. biophys. Res. Commun.*, **14**, 241.
Julian, G. R. (1965). *J. molec. Biol.*, **12**, 9.
Jusko, W. J. and Lewis, G. P. (1973). *J. Pharm. Soc.*, **62**, 69.
Kahlson, G., Rosengren, E. and Svensson, S. E. (1962). *Nature, Lond.*, **194**, 876.
Kaiser, Von F. (1971). *Planta Med.* (*suppl.*), **4**, 52.
Kakemi, K., Arita, T. and Muranishi, S. (1965). *Chem. Pharm. Bull.*, **13**, 965, 966.
Kalow, W. (1952). *J. Pharmac. exp. Ther.*, **104**, 122.
Kane, P. O. (1962). *Nature, Lond.*, **195**, 495.
Kanoa, S. (1947). *J. Pharm. Soc., Japan*, **67**, 243.
Kaplan, K. and Weinstein, L. (1965). *Practitioner*, **194**, 834.
Kaplan, M. A., Albright, H. and Buckwalter, F. H. (1960). *Antibiotics A.*, 1959–60, 365.
Karavolas, H. J. and Engel, L. L. (1966). *J. biol. Chem.*, **241**, 3454.
Kariyone, K., Harada, H., Kurita, M. and Takano, T. (1970). *J. Antibiot.*, **23**, 131.
Kaslander, J. (1963). *Biochim. biophys. Acta*, **71**, 730.
Kass, E. H. (1970). *J. infect. Dis.*, **122**, (*suppl.*), S115.
Kato, K. (1953). *J. Antibiot. Japan*, **A6**, **130**, 184.
Kaufman, H. E. (1962). *Proc. Soc. exp. Biol. Med.*, **109**, 251.
Kauzmann, W. and Simpson, R. B. (1953). *J. Am. chem. Soc.*, **75**, 5154.
Keberle, H., Riess, W. and Hoffmann, K. (1963). *Archs int. Pharmacodyn. Thér.*, **142**, 117.
Keberle, H., Riess, W., Schmidt, K. and Hofmann, K. (1963). Ibid., **142**, 125.
Kebrle, J. and Hoffmann, K. (1956). *Helv. chim. Acta*, **39**, 116.
Keene, D. R. T. (1973). *Chemy Ind.*, 423.
Kekwick, R. A. and MacKay, M. E. (1954). *The Separation of Protein Fractions from Human Plasma*, H.M.S.O., London.
Kekwick, R. G. O., Archer, B. L., Barnard, D., Higgins, G. M. C., McSweeney, G. P. and Moore, C. G. (1959). *Nature, Lond.*, **184**, 268.
Kendrew, J. C. (1961). *Scient. Am.*, **205**, No. 6, 96.
Kendrew, J. C., Watson, H. C. Strandberg, B. E., Dickerson, R. E., Phillips, D. C. and Shore, V. C. (1961). *Nature, Lond.*, **190**, 666.
Kennard, O. and Walker, J. (1963). *J. chem. Soc.*, 5513.
Kern, C. J. and Antoshkiw, T. (1950). *Ind. Engng Chem.*, **42**, 709.
Kharasch, M. S. and Gladstone, M. T. (1943). *J. Am. chem. Soc.*, **65**, 15.
Kharasch, M. S. and Joshi, B. S. (1957). *J. org. Chem.*, **22**, 1439.
Kharasch, M. S., McBay, H. C. and Urry, W. H. (1945). Ibid., **10**, 394.
Kharasch, M. S. and Mayo, F. R. (1933). *J. Am. chem. Soc.*, **55**, 2468.
Kièr, L. B. (1968a). *Molec. Pharmac.*, **4**, 70.
—(1968b). *J. med. Chem.*, **11**, 441.
Kindler, K. (1926). *Annalen*, **450**, 1.
—(1927). Ibid., **452**, 90.
—(1936). *Ber. dt. chem. Ges.*, **69**, 2792.
Kinsky, S. C. (1963). *Archs. Biochem. Biophys.*, **102**, 180.
Kluyver, A. J. and Boezaardt, A. G. J. (1939). *Recl Trav. chim. Pays-Bas Belg.*, **58**, 956.
Knappe, J., Biederbick, K. and Bruemmer, W. (1962). *Angew. Chem.*, **74**, 432.
Knoop, F. (1905). *Beit. chem. Physiol. u. Path.*, **6**, 150.
Knox, R. (1960). *Br. med. J.*, **ii**, 690.
—(1961). *Nature, Lond.*, **192**, 492.
Knudsen, E. T. and Rolinson, G. N. (1959). *Lancet*, **ii**, 1105.
Knudsen, E. T., Rolinson, G. N. and Stevens, S. (1961). *Br. med. J.*, **ii**, 198.
Kodicek, E. (1974). *Lancet*, **i**, 325.
Koevoet, A. L., Verloop, A. and Havinga, E. (1955). *Recl Trav. chim. Pays-Bas Belg.*, **74**, 1125.
Koike, M., Reed, L. J. and Carroll, W. R. (1963). *J. biol. Chem.*, **238**, 30.

Kornblum, N. and Cutter, R. J. (1954). *J. Am. chem. Soc.*, **76**, 4494.
Koshy, K. T., Duvall, R. N., Troup, A. E. and Pyles, J. W. (1965). *J. Pharm. Sci.*, **54**, 549.
Kosower, E. M. (1962). *Molecular Biochemistry*, McGraw-Hill Book Company Inc.
Kossel, W. (1916). *Ann. Physik.*, **49**, 229.
Krall, H. (1915). *J. chem. Soc.*, 1396.
Krantz, J. C., Carr, C. J. and Knapp, M. J. (1951). *J. Pharmac. exp. Ther.*, **102**, 258.
Kraut, J. and Reed, H. J. (1962). *Acta cryst.*, **15**, 747.
Krimm, S. (1962). *J. molec. Biol.*, **4**, 528.
Krupa, R. M. (1966a). *Biochemistry, N. Y.*, **5**, 1983.
—(1966b). Ibid., **5**, 1988.
—(1967). Ibid., **6**, 1183.
Kuhn, R. and Jerchel, D. (1941). *Ber. dt. chem. Ges.*, **74**, 941.
Kuhn, R., Rudy, H. and Wagner-Jauregg, T. (1933). Ibid., **66**, 1950.
Kuhnle, J. A., Lunden, R. E. and Waiss, A. C. (1972). *Chem. Commun.*, 287.
Kumamoto, J., Cox, J. R. and Westheimer, F. H. (1956). *J. Am. chem. Soc.*, **78**, 4858.
Kuroki, T., Huberman, E., Marquardt, H., Selkirk, J. K., Heidelberger, C., Grover, P. L. and Sims, P. (1972). *Chem-Biol. Interactions*, **4**, 389.
Lack, L. (1961). *J. biol. Chem.*, **236**, 2835.
Lackner, H. (1970). *Tetrahedron Lett.*, 3189.
Ladomery, L. G., Ryan, A. J. and Wright, S. E. (1967). *J. Pharm. Pharmac.*, **19**, 383, 388.
La Du, B. N., Gaudette, L., Trousof, N. and Brodie, B. B. (1955). *J. biol. Chem.*, **214**, 741.
Lamfrom, H. and Nielsen, S. O. (1957). *J. Am. chem. Soc.*, **79**, 1966.
Lampen, J. O. and Arnow, P. M. (1959). *Proc. Soc. exp. Biol. Med.*, **101**, 792.
Lampen, J. O., Arnow, P. M., Borowska, Z. and Laskin, A. I. (1962). *J. Bact.*, **84**, 1152.
Lampen, J. O. and Jones, M. J. (1946). *J. biol. Chem.*, **166**, 435.
Lane, M. D. and Lynen, F. (1963). *Proc. natn. Acad. Sci. U.S.A.*, **49**, 379.
Langdon, R. G. and Block, K. (1953). *J. biol. Chem.*, **200**, 129.
Langmuir, I. (1919). *J. Am. chem. Soc.*, **41**, 868, 1543.
Lantz, R. (1935). *Bull. Soc. chim. Fr.* [V], 2, 1913.
Laskowski, M. (1966). *Fedn Proc. Fedn Am. Socs exp Biol.*, **25**, 20.
Launchbury, A. P. (1970). *Progress in Medicinal Chemistry*, vol. 7, p. 1, eds. Ellis, G. P. and West, G. B., Butterworth, London.
Laurent, T. C., Moore, E. C. and Reichard, P. (1964). *J. biol. Chem.*, **239**, 3436.
Lawley, P. D. (1967). *J. molec. Biol.*, **24**, 75.
Lawrence, J. H., Loomis, W. F., Tobias, C. A. and Turpin, F. H. (1946). *J. Physiol., Lond.*, **105**, 197.
Lawson, D. E. M., Wilson, P. W. and Kodicek, E. (1969). *Nature, Lond.*, **222**, 171.
Leach, B. E., DeVries, W. H., Nelson, H. A., Jackson, W. G. and Evans, J. S. (1951). *J. Am. chem. Soc.*, **73**, 2797.
Leach, B. E. and Teeters, C. M. (1951). Ibid., **73**, 2794.
Leach, S. J., Némethy, G. and Scheraga, H. A. (1966). *Biopolymers*, **4**, 369.
Lederer, E. and Vilkas, M. (1966). *Vitamins and Hormones*, **24**, 409.
Lee, H. M. and Jones, R. G. (1949). *J. Pharmac. exp. Ther.*, **95**, 71.
Leermakers, P. A., Warren, P. C. and Vesley, G. F. (1964). *J. Am. chem. Soc.*, **86**, 1768.
Le Fevre, P. G. (1961). *Pharmac. Rev.*, **13**, 39.
Le Fevre, P. G. and Marshall, J. K. (1958). *Am. J. Physiol.*, **194**, 333.
Leibman, K. C. and Anaclerio, A. M. (1961). *1st Int. Pharmac. Meet.*, **6**, 91.
Leistner, E., Schmitt, J. H. and Zenk, M. H. (1967). *Biochem. biophys. Res. Commun.*, **28**, 845.
Leloir, L. F. (1951). *Progress in the Chemistry of Organic Natural Products*, vol. 8, p. 47, ed. Zechmeister, L., Springer, Vienna.
Lemieux, R. U. (1954). *Adv. Carbohydrate Chem.*, **9**, 1.
Lemieux, R. U. and Wolfrom, M. L. (1948). Ibid., **3**, 337.
Leonard, N. J. and Hauck, F. P. (1957). *J. Am. chem. Soc.*, **79**, 5279.
Leonard, N. J., Oki, M. and Chiavarelli, S. (1955). Ibid., **77**, 6234.
Lesser, G. T., Blumberg, A. G., Steele, J. M., Reiter, H. and Porosowska, Y. (1952). *Am. J. Physiol.*, **169**, 545.
Levene, P. A. and Raymond, A. L. (1934). *J. biol. Chem.*, **107**, 75.
Levine, R. M. and Clark, B. B. (1955). *J. Pharmac. exp. Ther.*, **113**, 272.
Levy, M. (1935). *J. biol. Chem.*, **109**, 361.
Lewbart, M. L. (1969). *Nature, Lond.*, **222**, 663.

Lewbart, M. L. and Mattox, V. R. (1959). Ibid., **183**, 820.
Lewis, G. N. (1916). *J. Am. chem. Soc.*, **38**, 762.
Ley, H. L., Smadel, J. E. and Crocker, T. T. (1948). *Proc. Soc. exp. Biol. Med.*, **68**, 9.
Ley, K., Müller, E., Mayer, R. and Scheffler, K. (1958). *Ber. dt. chem. Ges.*, **91**, 2670.
Lindley, H. and Rollett, J. S. (1955). *Biochim. biophys Acta*, **18**, 183.
Lindsay-Smith, J. R. and Norman, R. O. C. (1963). *J. chem. Soc.*, 2897.
Lippincott, S. B. and Hass, H. B. (1939). *Ind. Engng Chem.* **31**, 118.
Little, H. N. and Bloch, K. (1950). *J. biol. Chem.*, **183**, 33.
Lloyd, W. G. (1956). *J. Am. chem. Soc.*, **78**, 72.
Loder, B., Newton, G. G. F. and Abraham, E. P. (1961). *Biochem. J.*, **79**, 408.
Loebl, H., Stein, G. and Weiss, J. (1949). *J. chem. Soc.*, 2074.
Loening, K. L., Garrett, A. B. and Newman, M. S. (1952). *J. Am. chem. Soc.*, **74**, 3929.
Lohmann, K. and Schuster, P. (1937). *Biochem. Z.*, **294**, 188.
Long, L. M. and Troutman, H. D. (1949). *J. Am. chem. Soc.*, **71**, 2469.
Long, W. P. and Gallagher, T. F. (1946). *J. biol. Chem.*, **162**, 511.
Longchampt, J. E., Gual, C., Ehrenstein, M. R. and Dorfman, R. I. (1960). *Endocrinology*, **66**, 416.
Longhi, R. and Drago, R. S. (1965). *Inorg. Chem.*, **4**, 11.
Longuet-Higgins, H. C. and Coulson, C. A. (1947). *Trans. Faraday Soc.*, **43**, 87.
Lontie, R. (1962). *Criteria of Purity of Proteins in Protides of Biological Fluids, Proceedings of the 9th Colloquium, Bruges* (1961). p. 1, ed. Peters, H., Elsevier, Amsterdam.
Louis, L. H., Fajans, S. S., Conn, J. W., Struck, W. A., Wright, J. B. and Johnson, J. L. (1956). *J. Am. chem. Soc.*, **78**, 5701.
Lowe, G. (1967a). *Proc. R. Soc.*, **B**, 167, 431.
Lowe, G. and Sheppard, G. (1968). *Chem. Commun.*, 529.
Lowe, G., Sheppard, G., Sinnott, M. L. and Williams, A. (1967b). *Biochem. J.*, **104**, 893.
Lowry, O. H., Bessey, O. A. and Crawford, E. J. (1949). *J. biol. Chem.*, **180**, 389.
Lutz, O. and Jirgensons, B. (1930). *Ber. dt. chem. Ges.*, **63**, 448.
—(1931). Ibid., **64**, 1221.
Lynen, F. (1959). *J. cell. comp. Physiol.*, **54**, (*suppl.*), **1**, 33.
Lynn, B. (1970). *J. Hosp. Pharm.*, **28**, 71.
—(1971). Ibid., **29**, 183.
—(1974). *Pharm. J.*, **213**, 399.
McCance, R. A. and Widdowson, E. M. (1946). *Nature, Lond.*, **157**, 837.
McCarthy, R. D. (1964). *Biochim. biophysics Acta*, **84**, 74.
McCormick, J. R. D., Fox, S. M., Smith, L. L., Bitler, B. A., Reichenthal, J., Origoni, V. E., Muller, W. H., Winterbottom, R. and Doerschuk, A. P. (1957). *J. Am. chem. Soc.*, **79**, 2849.
McCormick, J. R. D., Jensen, E. R., Miller, P. A. and Doerschuk, A. P. (1960). Ibid., **82**, 3381.
McCormick, J. R. D., Sjølander, N. O., Hirsch, U., Jensen, E. R. and Doerschuk, A. P. (1957). Ibid., **79**, 4561.
McCormick, M. H., Stark, W. M., Pittenger, G. E., Pittenger, R. C. and McGuire, J. M. *Antibiotics Annual* (1955–56), Medical Encyclopaedia Inc., New York, 1956, 606.
Thayer, S. A. and Doisy, E. A. (1939). *J. biol. Chem.*, **131**, 357.
MacCorquodale, D. W., Cheney, L. C., Binkley, S. B., Holcomb, W. F., McKee, R. W., Thayer, S. A. and Doisy, E. A. (1939). *J. biol. Chem.*, **131**, 357.
McGrath, B. P. and Tedder, J. M. (1961). *Proc. chem. Soc.*, **80**, 199.
McGuire, J. M., Bunch, R. L., Anderson, R. C., Boaz, H. E., Flynn, E. H., Powell, H. M. and Smith, J. W. (1952). *Antibiotics Chemother.*, **2**, 281.
MacIntosh, F. C. (1959). *Can. J. Biochem. Physiol.*, **37**, 343.
MacKay, E. M., Wick, A. N. and Barnum, C. P. (1940). *J. biol. Chem.*, **136**, 503.
McLean, I. W., Schwab, J. L., Hillegas, A. B. and Schlingman, A. S. (1949). *J. clin. Invest.*, **28**, 953.
McMartin, C., Rondel, R. K., Vinter, J., Allan, B. R., Humberstine, P. M., Leishman, A. W. D., Sanlder, G. and Thirkettle, J. L. (1970). *Clin. pharmac. Therap.*, **11**, 423.
McMartin, C. and Vinter, J. (1969). *J. Chromat.*, **41**, 188.
McNeill, R. A. (1962). *Br. med. J.*, **i**, 360.
Macek, T. J. (1954). *U.S. Patent* 2671750.
Magee, P. N. and Barnes, J. M. (1967). *Adv. Cancer Res.*, **10**, 163.
Magerlein, B. J., Birkenmeyer, R. D. and Kagan, F. (1966). *Antimicrobial Agents and Chemotherapy*, 727.
Mahler, H. R. and Mehrotra, B. D. (1963). *Biochim. biophys. Acta*, **68**, 211.

Maley, L. E. and Mellor, D. P. (1950). *Nature, Lond.*, **165**, 453.
Maloof, F. and Soodak, M. (1963). *Pharmac. Rev.*, **15**, 43.
Marchant, B. and Alexander, W. D. (1972). *Endocrinology*, **91**, 747.
Marchessault, R. H. and Sarko, A. (1967). *Adv. Carbohydrate Chem.*, **22**, 421.
Marcker, K. (1960). *Acta chem. scand.*, **14**, 2071.
Mark, J. E. and Goodman, M. (1967). *J. Am. chem. Soc.*, **89**, 1267.
Marshall, F. J. (1965). *J. med. Chem.*, **8**, 18.
Marshall, I. G., Murray, J. B., Smail, G. A. and Stenlake, J. B. (1967). *J. Pharm. Pharmac.*, **19**, 53.
Martell, M. J. and Boothe, J. H. (1967). *J. med. Chem.*, **10**, 44.
Martin, H. H. (1963). *Zentbl. Bakt. Parasit Kde Abt. I Orig.*, **191**, 409.
—(1964). *Proc. Sixth Int. Congr. Biochem., N. Y.*, 518.
Martin, R. B. (1964). *Introduction to Biophysical Chemistry*, McGraw-Hill, New York.
Martius, C. (1966). *Biochem. Z.*, **327**, 2169; *Vitamins and Hormones*, **24**, 441.
Martius, C. and Esser, H. O. (1959). *Biochem. Z.*, **331**, 1.
Marvel, J. R., Schlichting, D. A., Denton, C., Levy, E. J. and Cahn, M. M. (1964). *J. invest. Derm.*, **42**, 197.
—(1964). Ibid., **42**, 203.
Mason, A. S. (1957). *Lancet*, **i**, 911.
Mason, D. J., Dietz, A. and DeBoer, C. (1962). *Antimicrobial Agents and Chemotherapy*, 554.
Mason, D. J. and Lewis, C. (1964). Ibid., 7.
Mason, R. (1958). *Br. J. Cancer*, **12**, 469.
—(1958). *Nature, Lond.*, **181**, 820.
Masuda, T., Sawa, Y. and Asai, M. (1955). *Pharm. Bull.*, **3**, 375.
Matsuhashi, M., Dietrich, C. P. and Strominger, J. L. (1965). *Proc. natn. Acad. Sci. U.S.A.*, **54**, 587.
Matsumura, H. *et al.* (1958). *Yakuzaigaka*, **18**, 124 (From *Chemical Abstracts*, 1959, **53**, 6535f).
Matthews, R. E. F. (1954). *J. gen. Microbiol.*, **10**, 521.
Mazel, P., Henderson, J. F. and Axelrod, J. (1964). *J. Pharmac. exp. Ther.*, **143**, 1.
Megges, R. and Repke, K. (1963). *Proceedings of the First International Pharmacological Meeting, Stockholm, 1961*. Vol. 3, p. 271, eds. Wilbrandt, W. and Lindgren, P., Pergamon Press, Oxford.
Meister, A., Sober, H. A., Tice, S. V. and Fraser, P. E. (1952). *J. biol. Chem.*, **197**, 319.
Mellor, D. P. and Maley, L. E. (1948). *Nature, Lond.*, **161**, 436.
Merrifield, R. B. (1965). *Science, N. Y.*, **150**, 178.
Meselson, M. and Stahl, F. W. (1958). *Proc. natn. Acad. Sci. U.S.A.*, **44**, 671.
Metzler, D. E., Ikawa, M. and Snell, E. E. (1954). *J. Am. chem. Soc.*, **76**, 648.
Meyer, H. (1899). *Arch. exp. Path. Pharmak.*, **42**, 109, 119.
Meyer, K. H. (1911). *Annalen*, **380**, 212.
Meyers, C. Y. and Miller, L. E. (1952). *Organic Syntheses*, **32**, 13.
Miller, A. K. (1944). *Proc. Soc. exp. Biol. Med.*, **57**, 151.
Miller, W. H., Roblin, R. O. and Astwood, E. B. (1945). *J. Am. chem. Soc.*, **67**, 2201.
Mima, H. (1958). *J. pharm. Soc., Japan*, **78**, 381.
Mirsky, I. A., Perisutti, G. and Diengott, D. (1956). *Metabolism*, **5**, 156.
Mirvish, S. S. (1964). *Biochim. biophys. Acta*, **93**, 673.
Mitchard, M. (1971). *Xenobiotica*, **1**, 469.
Mitchell, A. G. (1964). *J. Pharm. Pharmac.*, **16**, 533.
Miyazawa, T. (1960). *J. chem. Phys.*, **32**, 1647.
Mohr, E. (1918). *J. prakt. Chem.*, **98**, 315.
Moldave, K., Castelfranco, P. and Meister, A. (1959). *J. biol. Chem.*, **234**, 841.
Moncada, S., Gryglewski, R., Bunting, S. and Vane, J. R. (1976). *Nature, Lond.*, **263**, 663.
Moore, T. (1930). *Biochem. J.*, **24**, 692.
Morato, T., Hayano, M., Dorfman, R. I. and Axelrod, L. R. (1961). *Biochem. biophys. Res. Commun.*, **6**, 334.
Morin, R. B., Jackson, B. G., Mueller, R. A., Lavagnino, E. R., Scanlon, W. B. and Andrews, S. L. (1963). *J. Am. chem. Soc.*, **85**, 1896.
—(1969). Ibid., **91**, 1401.
Morozowich, W., Sinkula, A. A., MacKeller, F. A. and Lewis, C. (1973). *J. Pharm. Sci.*, **62**, 1102.
Morris, A. and Russell, A. D. (1971). *Progress in Medicinal Chemistry*, vol. 8, p. 39, ed. Ellis, G. P. and West, G. B., Butterworths, London.
Morris, C. R., Andrew, L. V., Whichard, L. P. and Holbrook, D. J. (1970). *Molec. Pharmac.*, **6**, 240.

Mosbach, E. H. (1969). *Drugs Affecting Lipid Metabolism in Advances in Experimental Medicine* (1968). vol. 4, p. 421, eds. Holmes, W. L., Carlson, L. A. and Paoletti, R., Plenum Press, London.
Moyed, H. S. and Magasanik, B. (1957). *J. Am. chem. Soc.*, **79**, 4812.
Mueller, G. C. and Miller, J. A. (1950). *J. biol. Chem.*, **185**, 145.
—(1953). Ibid., **202**, 579.
Müller, N. and Mulliken, R. S. (1958). *J. Am. chem. Soc.*, **80**, 3489.
Mulley, B. A. and Metcalf, A. D. (1956). *J. Pharm. Pharmac.*, **8**, 774.
Mullins, J. D. and Macek, T. J. (1960). *J. Am. pharm. Ass. Sci. Edn.*, **49**, 245.
Munck, A., Scott, J. F. and Engel, L. L. (1957). *Biochim. biophys. Acta*, **26**, 397.
Musgrave, W. K. R. (1964). *Rodd's Chemistry of Carbon Compounds*, 2nd edn, vol. 1A, p. 485, ed. Coffey, S.
Nakada, H. I. and Weinhouse, S. (1953). *J. biol. Chem.*, **204**, 831.
Nakagawa, T. (1954). *J. Pharm. Soc., Japan*, **74**, 1116.
—(1956). Ibid., **76**, 1113.
Nakagawa, T. and Muneyuki, R. (1954). Ibid., **74**, 858.
Nakatsugawa, T., Ishida, M. and Dahm, P. A. (1965). *Biochem. Pharmac.*, **14**, 1853.
Nambara, T., Shimada, K., Goto, J. and Goya, S. (1971). *Chem. Pharm. Bull.*, **19**, 21.
Namekata, M., Matsuo, A., Momose, A. and Takagi, M. (1967). *J. Pharm. Soc., Japan*, **87**, 376.
Namekata, M., Sakamoto, N., Yokoyama, Y. and Takagi, M. (1967). Ibid., **87**, 778.
Nathans, D. (1964). *Proc. natn. Acad. Sci. U.S.A.*, **51**, 585.
Nayler, J. H. C. (1971). *Proc. R. Soc.*, **B 179**, 357.
—(1973). *Advances in Drug Research*, vol. 7, p. 43, eds. Harper, N. J. and Simmonds, A. B., Academic Press, London.
Needleman, P. and Krantz, J. C. (1965). *Biochem. Pharmac.*, **14**, 1225.
Needleman, P., Moncada, P., Bunting, S., Vane, J. R., Hamberg, M. and Samuelsson, B. (1976). *Nature, Lond.*, **261**, 550.
Neidle, A. and Waelsch, H. (1959). *J. biol. Chem.*, **234**, 586.
Nery, R. (1971a). *The Biological Oxidation of Nitrogen in Organic Molecules, Xenobiotica*, p. 27, eds. Bridges, J. W., Gorrod, J. W. and Parke, D. V., Taylor and Francis, London.
—(1971b). *Biochem. J.*, **122**, 311.
Nettleship, A., Henshaw, P. S. and Meyer, H. L. (1943). *J. natn. Cancer Inst.*, **4**, 309.
Newbould, B. B. and Kilpatrick, R. (1960). *Lancet*, **i**, 887.
Newman, M. S. and Wise, R. M. (1956). *J. Am. chem. Soc.*, **78**, 450.
Newton, G. G. F. and Abraham, E. P. (1956a). *Biochem. J.*, **63**, 628.
—(1956b). Ibid., **62**, 651.
Nicolaysen, R. (1937). *Biochem. J.*, **31**, 122.
—(1943). *Acta physiol. scand.*, **5**, 200.
Nineham, A. W. (1955). *Chem. Rev.*, **55**, 355.
Nozaki, Y. and Tanford, C. (1965). *J. biol. Chem.*, **240**, 3568.
O'Brien, R. L. and Hahn, F. E. (1965). *Antimicrobial Agents and Chemotherapy*, 315.
O'Callaghan, C. H., Kirby, S. M., Morris, A., Waller, R. E. and Dunscombe, R. E. (1972). *J. Bact.*, **110**, 988.
O'Callaghan, C. H., Sykes, R. B. and Staniforth, S. E. (1976). *Antimicrobial Agents and Chemotherapy*, **10**, 245.
Oelschläger, H., Volke, J. and Kurek, E. (1964). *Archs Pharm.*, **297**, 431.
Olivard, J., Metzler, D. E. and Snell, E. E. (1952). *J. biol. Chem.*, **199**, 669.
Olson, R. E. (1964). *Science, N. Y.*, **145**, 926.
O'Meara, R. A. Q., McNally, P. A. and Nelson, H. G. (1947). *Lancet*, **ii**, 747.
Orgel, L. E. (1958). *Biochem. Soc. Symp.*, vol. 15, p. 8, Cambridge University Press.
Orsi, F. (1967). *Magyar Kem. Folyóirat.*, **73**, 1.
Overbeek, G. A. and de Visser, J. (1957). *Acta endocr., Copnh.*, **24**, 209.
Overton, E. (1901). *Studien über die Narkose*, Jena, G. Fischer.
Oyakawa, E. K. and Levedahl, B. H. (1958). *Archs Biochem. Biophys.*, **74**, 17.
Palomaa, M. H. (1938). *Ber. dt. chem. Ges.*, **71**, **B**, 480.
Panarin, E. F. and Solotovskii, M. V. (1970). *Antibiotiki*, **15**, 426.
Panarin, E. F., Solotovskii, M. V. and Ekzemplyarov, O. N. (1967). Ibid., **12**, 643.
Paneth, F. A. (1929). *Z. angew. Chem.*, **42**, 189.
Paoloni, L. (1959). *Gazz. chim. ital.*, **89**, 957.
Park, J. T. (1952). *J. biol. Chem.*, **194**, 877, 885, 897.

Park, J. T. and Strominger, J. L. (1957). *Science, N. Y.*, **125**, 99.
Park, R. B. and Bonner, J. (1958). *J. biol. Chem.*, **233**, 340.
Parke, D. V. (1961). *Biochem. J.*, **78**, 262.
—(1968). *The Biochemistry of Foreign Compounds*, Pergamon, Oxford.
Paterson, E., Ap Thomas, I., Haddow, A. and Watkinson, J. M. (1946). *Lancet*, **i**, 677.
Pauling, L. (1961). *Science, N. Y.*, **134**, 15.
Pauling, L. and Corey, R. B. (1951). *Proc. natn. Acad. Sci. U.S.A.*, **37**, 251, 729.
Pauling, L. and Wheland, G. W. (1933). *J. chem. Phys.*, **1**, 362.
Pauling, L. C. (1960). *The Nature of the Chemical Bond*, 3rd edn, Cornell University Press, Ithaca, New York.
Pearson, R. G. (1949). *J. Am. chem. Soc.*, **71**, 2212.
Pedersen, K. J. (1938). Ibid., **60**, 595.
Perkins, E. S., Wood, R. M., Sears, M. L., Prusoff, W. H. and Welch, A. D. (1962). *Nature, Lond.*, **194**, 985.
Perkow, W. (1966). *Arzneimittel-Forsch.*, **16**, 429.
Peterson, J. E. and Robinson, W. H. (1964). *Toxicol. Appl. Pharmac.*, **6**, 321.
Pethica, B. A. (1965). *Symposium on Surface Activity and Microbial Cells*, vol. 19, p. 85, Society for Chemistry and Industry, London.
Phillips, B., Frostick, F. C. and Starcher, P. S. (1957). *J. Am. chem. Soc.*, **79**, 5982.
Phillips, D. C. (1967). *Proc. natn. Acad. Sci. U.S.A.*, **57**, 484.
Phillips, G. O. and Barber, P. (1963). *J. chem. Soc.*, 3990.
Phillips, G. O. and Criddle, W. J. (1963). Ibid., 3984.
Pickles, V. R. (1969). *Nature, Lond.*, **224**, 221.
Pierce, J. G., Gordon, S. and du Vigneaud, V. (1952). *J. biol. Chem.*, **199**, 929.
Pinckard, R. N., Hawkins, D. and Farr, R. S. (1968). *Nature, Lond.*, **219**, 68.
Pinkerton, M., Steinrauf, L. K. and Dawkins, P. (1969). *Biochem. biophys. Res. Commun.*, **35**, 512.
Pinson, E. R., Schreiber, E. C., Wiseman, E. H., Chiaini, J. A. and Baumgartner, D. (1962). *J. med. Pharm. Chem.*, **5**, 491.
Pitzer, K. S. (1937). *J. Am. chem. Soc.*, **59**, 2365.
—(1948). Ibid., **70**, 2140.
Pollock, M. R. (1957). *Biochem. J.*, **66**, 419.
Porath, P. and Flodin, P. (1959). *Nature, Lond.*, **183**, 1657.
Prasad, A. S. and Oberleas, D. (1970). *J. Lab. clin. Med.*, **76**, 416.
Pratt, R. (1947). *Nature, Lond.*, **159**, 233.
—(1962). *J. Pharm. Sci.*, **51**, 1.
Pressman, B. C. (1963). *J. biol. Chem.*, **238**, 401.
Pressman, D. and Park, J. K. (1963). *Biochem. biophys. Res. Commun.*, **11**, 182.
Price, C. C. (1948). *J. Polymer Sci.*, **3**, 772.
Price, K. E., Zolli, Z., Atkinson, J. C. and Luther, H. G. (1957). *Antibiotics Chemother.*, **7**, 689.
Prydz, H. (1965). *Scand. J. clin. Lab. Invest.*, **17**, 143.
Prystas, M., Gut, J. and Sŏrm, F. (1961). *Chemy Ind.*, 947.
Puca, G. A. and Bresciani, F. (1969). *Nature, Lond.*, **223**, 745.
Pullman, A. (1946). Thesis, University of Paris.
Pullman, A. and Pullman, B. (1946). *Experientia*, **2**, 364; *Rev. Sci.*, **84**, 145.
—(1955). *Cancérisation par les Substances Chimiques et Structure Moléculaire*, Masson et Cie, Paris.
—(1962). *Nature, Lond.*, **196**, 228.
Pullman, B. and Pullman, A. (1959). *Biochim. biophys. Acta*, **36**, 343.
—(1963). *Quantum Biochemistry*, Interscience, New York.
Pulvertaft, R. J. V. and Yudkin, J. (1946). *Lancet*, **ii**, 265.
Purves, C. B. and Hudson, C. S. (1937). *J. Am. chem. Soc.*, **59**, 1170.
Raftery, M. A. and Rand-Meir, T. (1968). *Biochemistry, N. Y.*, **7**, 3281.
Rahn, K. H. and Dayton, P. G. (1969). *Biochem. Pharmac.*, **18**, 1809.
Raley, J. H., Mullineaux, R. D. and Bittner, C. W. (1963). *J. Am. chem. Soc.*, **85**, 3174.
Rall, T. W. and Sutherland, E. W. (1959). *Pharmac. Rev.*, **11**, 464.
Ramachandran, G. N., Ramakrishnan, C. and Sasísekharan, J. (1963). *J. molec. Biol.*, **7**, 95.
Ramanan, C. V. and Israëls, M. C. G. (1969). *Lancet*, **ii**, 125.
Raphael, R. A. and Stenlake, J. B. (1953). *Chemy Ind.*, 1286.
Rapport, M. M. and Franzl, R. E. (1957). *J. biol. Chem.*, **225**, 851.

Rapport, M. M., Lerner, B., Alonzo, N. and Franzl, R. E. (1957). Ibid., **225**, 859.
Raschig, F. (1915). *Ber. dt. chem. Ges.*, **48**, 2088.
Rebstock, M. C. (1950). *J. Am. chem. Soc.*, **72**, 4800.
—(1951). Ibid., **73**, 3671.
Redetzki, H. M., Redetzki, J. E. and Elias, A. L. (1966). *Biochem. Pharmac.*, **15**, 425.
Redin, G. S. (1967). *Antimicrobial Agents and Chemotherapy*, 1966, 371.
Rees, D. A. (1969). *Adv. in Carbohydrate Chem. and Biochem.*, **24**, 267.
Rees, D. A. and Scott, W. E. (1969). *Chem. Commun.*, 1037.
Reeves, R. E. (1950). *J. Am. chem. Soc.*, **72**, 1499.
—(1951). *Adv. in Carbohydrate Chem.*, **6**, 107.
Regna, P. P. and Solomons, I. A. (1952). *Ann. N. Y. Acad. Sci.*, **15**, 12.
Reier, G. E. and Shangraw, R. F. (1966). *J. Pharm. Sci.*, **55**, 510.
Rendi, R. and Ochoa, S. (1962). *J. biol. Chem.*, **237**, 3711.
Renson, J., Weissbach, H., Udenfriend, S. (1965). *Molec. Pharmac.*, **1**, 145.
Repke, K. (1963). *Proceedings of the First International Pharmacological Meeting*, Stockholm, 1961, vol. 3, p. 47, eds. Wilbrandt, W. and Lindgren, P., Pergamon Press, Oxford.
—(1965). *Proceedings of the Second International Pharmacological Meeting*, Prague, 1963, vol. 4, p. 65, eds. Brodie, B. B. and Gillette, J. R., Pergamon Press, Oxford.
Repta, A. J., Baltezor, M. J. and Bansal, P. C. (1976). *J. Pharm. Sci.*, **65**, 238.
Reynolds, J. J. and Brown, G. M. (1964). *J. biol. Chem.*, **239**, 317.
Richardson, K. E. and Tolbert, N. E. (1961). Ibid., **236**, 1280.
Rieche, A. and Meister, R. (1936). *Angew. Chem.*, **49**, 101.
Ringold, H. J. (1961). *Mechanisms of Action of Steroid Hormones*, p. 1, eds. Villee, C. A. and Engel, L. L., Pergamon Press, Oxford.
Roberts, J. J. and Warwick, G. P. (1962). *40th Rep. Br. Emp. Cancer Campgn.*, 16.
Robins, D. J., Campbell, I. M. and Bentley, R. (1970). *Biochem Biophys. Res. Commun.*, **39**, 1081.
Robinson, B. and Shepherd, D. M. (1962). *J. Pharm. Pharmac.*, **14**, 9.
Rodbell, M., Krans, H. M. J., Pohl, S. L. and Birnbaumer, L. (1971). *J. biol. Chem.*, **246**, 1872.
Rogers, S. (1954). *Fedn Proc. Fedn Am. Socs exp. Biol.*, **13**, 442.
Rolinson, G. N., Batchelor, F. R., Stevens, S., Cameron-Wood, J. and Chain, E. B. (1960). *Lancet*, **ii**, 564.
Rolinson, G. N. and Stevens, S. (1961). *Br. med. J.*, **ii**, 191.
Rollo, I. M., Somers, G. F. and Burley, D. M. (1962). Ibid., **i**, 76.
Rose, I. A. (1962). *Brookhaven Symp. Biol.*, **15**, 293.
Rose, I. A. and Rieder, S. V. (1958). *J. biol. Chem.*, **231**, 315.
Ross, W. C. J. (1962). *Biological Alkylating Agents*, Butterworths, London.
Roze, U. and Strominger, J. L. (1966). *Molec. Pharmacol.*, **2**, 92.
Rubbo, S. D. and Gillespie, J. M. (1940). *Nature, Lond.*, **146**, 838.
Ruiz-Torres, A. and Burmeister, H. (1972). *Klin. Wschr.*, **50**, 191.
Sachse, H. (1890). *Ber. dt. chem. Ges.*, **23**, 1363.
Sadèe, W., Garland, Wm. and Castagnoli, N. (Jr.). (1971). *J. med. Chem.*, **14**, 643.
Sage, H. J. and Fasman, G. D. (1966). *Biochemistry, N. Y.*, **5**, 286.
Saito, H. and Shinoda, K. (1967). *J. Colloid Sci.*, **24**, 10.
Sakaguchi, K. and Murao, S. (1950). *J. agric. chem. Soc., Japan*, **23**, 411.
Sanger, F. (1945). *Biochem. J.*, **39**, 507.
—(1952). *Adv. Protein Chem.*, **7**, 1.
—(1960). *Br. med. Bull.*, **16**, 183.
—(1963). *Proc. chem. Soc.*, 76.
Sarett, H. P. and Cheldelin, V. H. (1945). *J. biol. Chem.*, **159**, 311.
Sargeant, K., Sheridan, A., O'Kelly, J. and Carnaghan, R. B. A. (1961). *Nature, Lond.*, **192**, 1096.
Sato, T., Fukuyama, T., Suzuki, T. and Yoshikawa, H. (1963). *J. Biochem., Tokyo*, **53**, 23.
Sayre, D. and Friedlander, P. H. (1960). *Nature, Lond.*, **187**, 139.
Schaffer, N. K., May, S. C. and Summerson, W. H. (1954). *J. biol. Chem.*, **206**, 201.
Schatz, A., Bugie, E. and Waksman, S. A. (1944). *Proc. Soc. exp. Biol. Med.*, **55**, 66.
Scheline, R. R., Smith, R. L. and Williams, R. T. (1961). *J. med. Chem.*, **4**, 109.
Schellenberg, K. A. and Coatney, G. R. (1961). *Biochem. Pharmac.*, **6**, 143.
Schellman, J. A. (1958). *C.r. Trav. Lab. Carlsberg*, **30**, 415.
Schenck, G. O. (1957). *Angew. Chem.*, **69**, 579.
Schenck, G. O. and Ziegler, H. (1953). *Annalen*, **584**, 221.

Schenck, J. R. and Spielman, M. A. (1945). *J. Am. chem. Soc.*, **67**, 2276.
Schild, H. O. (1947). *Br. J. Pharmac. Chemother.*, **2**, 189.
Schill, G. (1965). *Acta pharm. suecica*, **2**, 13.
Schlenk, W. (1951). *Annalen*, **573**, 142.
Schlom, J., Spiegelman, S. and Moore, D. (1971). *Nature, Lond.*, **231**, 97.
Schmidt, O. (1938). *Z. phys. Chem.*, **B39**, 59.
—(1939). Ibid., **B42**, 83; **B44**, 185, 194.
—(1941). *Naturwissenschaften*, **29**, 146.
Schneider, C. H. and de Week, A. L. (1967). *Immunochemistry*, **4**, 331.
Schneider, P. B., Clayton, R. B. and Bloch, K. (1957). *J. biol. Chem.*, **224**, 175.
Schofield, K. (1950). *Quart. Rev.*, **4**, 382.
Schönberg, A. and Moubasher, R. (1952). *Chem. Rev.*, **50**, 261.
Schönberg, A., Moubasher, R. and Mostafa, A. (1948). *J. chem. Soc.*, 176.
Schönberg, A., Moubasher, R. and Said, A. (1949). *Nature, Lond.*, **164**, 140.
Schroeder, W., Bannister, B. and Hoeksema, H. (1967). *J. Am. Chem. Soc.*, **89**, 2448.
Schulze, E. and Likiernik, A. (1893). *Z. phys. Chem.*, **17**, 513.
Schumacher, H., Smith, R. L. and Williams, R. T. (1965). *Br. J. Pharmac. Chemother.*, **25**, 324, 338.
Schwartz, M. A. (1965). *J. Pharm. Sci.*, **54**, 472.
Schwartz, M. A., Granatek, A. P. and Buckwalter, F. H. (1962). Ibid., **51**, 523.
Schwyzer, R. (1970). *Experientia*, **26**, 577.
Schwyzer, R., Aung Tun-Kyi, Caviezel, M. and Moser, P. (1970). *Helv. chim. Acta*, **53**, 15.
Scott, J. F. and Engel, L. L. (1957). *Biochim. biophys. Acta*, **23**, 665.
Scrutton, M. C., Keech, D. B. and Utter, M. F. (1965). *J. biol. Chem.*, **240**, 574.
Sebak, O. K. (1955). *Bact. Proc.*, 78.
Seftel, H. C. (1970). *Glaxo Volume*, **33**, 31.
—(1970). Ibid., **33**, 40.
Segal, D. M. and Harrington, W. F. (1967). *Biochemistry, N. Y.*, **6**, 768.
Seikel, M. K. (1940). *J. Am. chem. Soc.*, **62**, 1214.
Sekera, A., Sova, J. and Vrba, C. (1955). *Experientia*, **11**, 275.
Seltzer, S., Hamilton, G. A. and Westheimer, F. H. (1959). *J. Am. chem. Soc.*, **81**, 4018.
Senoh, S., Daly, J., Axelrod, J. and Witkop, B. (1959). Ibid., **81**, 6240.
Seydal, J. (1966). *Tetrahedron Lett.*, no. **11**, 1145.
Seydel, J. K. (1968). *J. Pharm. Sci.*, **57**, 1455.
Shami, E., Dudzinski, J., Lachman, L. and Tingstad, J. (1973). Ibid., **62**, 1283.
Shaw, T. R. D., Carless, J. E., Howard, M. R. and Raymond, K. (1973). *Lancet*, **ii**, 209.
Sheehan, J. C. (1958). *Amino Acids and Peptides with Antimetabolite Activity. Ciba Foundation Symposium*, p. 258, eds. Wolstenholme, G. E. W. and O'Connor, C. M. Churchill, London.
Sheehan, J. C., Henery-Logan, K. R. and Johnson, D. A. (1953). *J. Am. chem. Soc.*, **75**, 3292.
Shelton, J. R. and Davies, K. E. (1967). Ibid., **89**, 718.
Shibata, Y. and Kronman, M. J. (1967). *Archs Biochem. Biophys.*, **118**, 410.
Shields, J. E. and McDowell, S. T. (1967). *J. Am. chem. Soc.*, **89**, 2499.
Sidgwick, N. V. (1925). *J. chem. Soc.*, **127**, 907.
Sims, P. (1967). *Int. J. Cancer*, **2**, 505.
Sinclair, H. M. (1964). *Lipid Pharmacology*, p. 237, ed. Pauletti, R., Academic Press, London.
Sjöblom, L. (1958). *Acta Acad. Aboensis. Math. et Phys.*, **21**, no. 7.
Skellon, J. H. and Wharry, D. M. (1963). *Chemy Ind.*, 929.
Skipski, V. P., Barclay, M., Barclay, R. K., Fetzner, V. A., Good, J. J. and Archibald, F. M. (1967). *Biochem. J.*, **104**, 340.
Skoda, J., Hess, V. F. and Sŏrm, F. (1957). *Experientia*, **13**, 150.
Slobin, L. I. and Carpenter, F. H. (1963). *Biochemistry, N. Y.*, **2**, 16, 22.
Slomp, G. and MacKellar, F. A. (1962). *Tetrahedron, Lett.*, no. **12**, 521.
Smith, C. C., Fradkin, R. and Lackey, M. D. (1946). *Proc. Soc. exp. Biol. Med.*, **61**, 398.
Smith, E. L., Hill, R. L. and Borman, A. (1958). *Biochim. biophys. Acta*, **29**, 207.
Smith, J. B. and Willis, A. L. (1971). *Nature New Biology*, **231**, 235.
Smith, J. W., Dyke, R. W. and Griffith, R. S. (1953). *J. Am. med. Ass.*, **151**, 805.
Smith, R. M., Joslyn, D. A., Gruhzit, O. M., McLean, I. W., Penner, M. A. and Ehrlich, J. (1948). *J. Bact.*, **55**, 425.
Snell, E. E. (1958). *Vitamins and Hormones*, **16**, 77.
Snell, E. E., Metzler, D. E. and Ikawa, M. (1954). *J. Am. chem. Soc.*, **76**, 648.

Soncin, E. (1953). *Archs int. Pharmacodyn. Thér.*, **94**, 346.
Southren, A. L., Kobayashi, Y., Sherman, D. H., Levine, L., Gordon, G. and Weingold, A. B. (1964). *Am. J. Obstet. Gynec.*, **89**, 199.
Spector, W. S. (1959). (ed.) *Handbook of Toxicology*, W. B. Saunders Co., Philadelphia and London.
Spence, G. G., Taylor, E. C. and Buchardt, O. (1970). *Chem. Rev.*, **70**, 231.
Sperber, I. (1949). *K. LantbrHögsk. Annlr.*, **16**, 49.
Speyer, J. F., Lengyel, P., Basilio, C. and Ochoa, S. (1962). *Proc. natn. Acad. Sci. U.S.A.*, **48**, 282.
Stach, V. K. and Pöldinger, W. (1966). *Progress in Drug Research.* vol. 9, p. 129, ed. Jücker, E. Birkhäuser, Basel.
Stam, C. H. and MacGillavry, C. H. (1963). *Acta cryst.*, **16**, 62.
Stamp, T. C. (1939). *Lancet*, **ii**, 10.
Stearns, B., Losee, K. A. and Bernstein, J. (1963). *J. med. Chem.*, **6**, 201.
Stein, G. and Weiss, J. (1950). *Nature, Lond.*, **166**, 1104.
Stenlake, J. B. (1953). *Chemy Ind.*, 1089.
—(1955). *J. chem. Soc.*, 1626.
—(1963). *Progress in Medicinal Chemistry*, vol. 3, p. 1, eds. Ellis, G. P. and West, G. B., Butterworths, London.
—(1968a). *Ann. Pharm. Franç.*, **26**, 185.
—(1968b). Unpublished information.
—(1975). *Harrison Memorial Lecture, Pharm. J.*, **215**, 533.
Stenlake, J. B., Taylor, A. J. and Templeton, R. (1971). *J. Pharm. Pharmac.*, **23**, 221S.
Stenlake, J. B., Templeton, R. and Taylor, D. C. (1968). Ibid., **20**, 142S.
Stenlake, J. B. and Watt, D. (1975). *Pharm. J.*, **214**, 56.
Stenlake, J. B., Williams, W. D. and Skellern, G. G. (1973). *Xenobiotica*, **3**, 121.
Stephens, C. R., Murai, K., Brunings, K. J. and Woodward, R. B. (1956). *J. Am. chem. Soc.*, **78**, 4155.
Stephens, C. R., Murai, K., Rennhard, H. H., Conover, L. H. and Brunings, K. J. (1958). Ibid., **80**, 5342.
Sternbach, L. H., Koechlin, B. A. and Reeder, E. (1962). *J. org. Chem.*, **27**, 4671.
Stewart, G. T. and Harrison, P. M. (1961). *Br. J. Pharmac. Chemother.*, **17**, 414.
Stewart, W. E., Mandelkern, L. and Glick, R. E. (1967). *Biochemistry, N. Y.*, **6**, 143, 150.
Stock, C. C., Reilly, H. C., Buckley, S. M., Clarke, D. A. and Rhoads, C. P. (1954). *Nature, Lond.*, **173**, 71.
Stocken, L. A. and Thompson, R. H. S. (1946). *Biochem. J.*, **40**, 536.
Stosick, A. J. (1945). *J. Am. chem. Soc.*, **67**, 365.
Strauss, J. S., Pochi, P. E. and Whitman, E. N. (1967). *J. invest. Derm.*, **48**, 492.
Streitwieser, A. (1956). *Chem. Rev.*, **56**, 571.
Strittmatter, P. and Ball, E. G. (1955). *J. biol. Chem.*, **213**, 445.
Strominger, J. L., Izaki, K., Matsuhashi, M. and Tipper, D. J. (1967). *Fedn Proc. Fedn Am. Socs exp.* Biol., **26**, 9.
Strominger, J. L., Park, J. T. and Thompson, R. E. (1959). *J. biol. Chem.*, **234**, 3263.
Strömme, J. H. (1965). *Biochem. Pharmac.*, **14**, 381.
Struller, T. (1968). *Progress in Drug Research*, vol. 12, p. 389, ed. Jücker, E., Birkhäuser, Basel.
Stumpf, P. K. and Green, D. E. (1944). *J. biol. Chem.*, **153**, 387.
Sumner, J. B. (1926). Ibid., **69**, 435.
Sunaga, K. and Koide, S. S. (1968). *J. Pharm. Sci.*, **57**, 2116.
Suter, C. M. (1941). *Chem. Rev.*, **28**, 269.
Sutherland, E. W. and Robison, G. A. (1966). *Pharmac. Rev.*, **18**, 145.
Sutherland, R., Croydon, E. A. P. and Rolinson, G. N. (1972). *Br. med. J.*, **iii**, 13.
Swab, R. S., England, A. C., Poskanzer, D. C. and Young, R. R. (1969). *J. Am. med. Ass.*, **208**, 1168.
Swallow, D. L. (1971). *Progress in Medicinal Chemistry*, vol. 8, p. 119, eds. Ellis, G. P. and West, G. B., Butterworths, London.
Swanson, E. E., Anderson, R. C., Harris, P. N. and Rose, C. L. (1953). *J. Am. pharm. Ass. Sci. Edn.*, **42**, 571.
Swart, E. A., Hutchinson, D. J. and Waksman, S. A. (1949). *Archs Biochem.*, **24**, 92.
—(1951). *J. Am. chem. Soc.*, **73**, 3253.
Swern, D. (1961). *Fatty Acids*, 2nd edn., Part 2, p. 1387, ed. Markley, K. S., Interscience, New York.
Tainter, M. L. (1930). *J. Pharmac. exp. Ther.*, **40**, 43.

Takeuchi, T., Hikiji, T., Nitta, K., Yamazaki, S., Abe, M., Takayama, H. and Umezawa, H. (1957). *J. Antibiot. Japan*, **Ser. A. 10**, 228.
Tanford, C. (1968). *Adv. Protein Chem.*, **23**, 121.
Tanford, C., Bunville, L. G. and Nozaki, Y. (1959). *J. Am. chem. Soc.*, **81**, 4032.
Tanford, C. and Epstein, J. (1954). Ibid., **76**, 2170.
Tanford, C., Kawahara, K., Lapanje, S., Hooker, T. M., Zarlengo, M. H., Salahuddin, A., Aune, K. C. and Takagi, T. (1967). Ibid., **89**, 5023.
Tanford, C. and Taggart, V. G. (1961). Ibid., **83**, 1634.
Tannenbaum, A., Vesselinovitch, S. D., Maltoni, C. and Stryzak Mitchell, D. (1962). *Cancer Res.*, **22**, 1362.
Tarbell, D. S. and Harnish, D. P. (1951). *Chem. Rev.*, **49**, 1.
Taylor, J. F. (1953). *The Isolation of Proteins in the Proteins*, vol. 1A, p. 1, eds. Neurath, H. and Bailey, K., Academic Press, New York.
Taylor, P. W., King, R. W. and Burgen, A. S. V. (1970). *Biochemistry, N. Y.*, **9**, 3894.
Terent'ev, A. P. and Mogilyanskii, Ya. D. (1958). *Zhur. Obshchei Khim*, **28**, 1959.
Terry, D. H. and Shelanski, H. A. (1952). *Modern Sanitation*, **4**, 61.
Tham, C. (1963). *Proceedings of the First International Pharmacological Meeting, Stockholm 1961*, vol. 3, p. 11, eds. Wilbrandt, W. and Lindgren, P., Pergamon Press, Oxford.
Theorell, H. (1938). *Biochem. Z.*, **298**, 242.
—(1941). *J. Am. chem. Soc.*, **63**, 1820.
Thomas, R., Boutagy, J. and Gelbart, A. (1974). *J. Pharm. Sci.*, **63**, 1649.
Thomas, R. C. and Ikeda, G. J. (1966). *J. med. Pharm. Chem.*, **9**, 507.
Thomson, R. H. (1956). *Quart. Rev.*, **10**, 33.
Threfall, D. R. (1971). *Vitamins and Hormones*, **29**, 153.
Thyrum, P. T., Luchi, R. J. and Thyrum, E. M. (1969). *Nature, Lond.*, **223**, 747.
Timasheff, S. N. and Gorbunoff, M. J. (1967). *A. Rev. Biochem.*, **36**, Part 1, 13.
Timasheff, S. N. and Susi, H. (1966). *J. biol. Chem.*, **241**, 249.
Timmis, G. M., Hudson, R. F., Marshall, R. D. and Bierman, H. R. (1962). *U. S. Patent* 3041241.
Tipper, D. J. (1966). *Fedn Proc. Fedn Am. Socs exp. Biol.*, **25**, 344.
Tipper, D. J. and Strominger, J. L. (1965). *Proc. natn. Acad. Sci. U.S.A.*, **54**, 1133.
Tolstoouhov, A. V. (1955). *Ionic Interpretation of Drug Action in Chemotherapeutic Research*, Chemical Publishing Company, New York.
Tomkins, G. M. and Maxwell, E. S. (1963). *A. Rev. Biochem.*, **32**, 677.
Tommila, E. and Hinshelwood, C. N. (1938). *J. chem. Soc.*, 1801.
Tosoni, A. L., Glass, D. G. and Goldsmith, L. (1958). *Biochem. J.*, **69**, 476.
Tracy, A. H. and Elderfield, R. C. (1940). *Science, N. Y.*, **92**, 180.
Traub, W. and Shmueli, U. (1963). *Nature, Lond.*, **198**, 1165.
Tréfouël, J., Tréfouël, Mme. J., Nitti, F. and Bovet, D. (1935). *C.r. Séanc. Soc. Biol.*, **120**, 756.
Tremblay, E.-C., Destouches, P. and Karatchenzeff, N. (1953). *Presse méd.*, **61**, 292.
Tschesche, R. (1947). *Z. Naturf.*, **2b**, 10.
Tschesche, R., Korte, F. and Korte, I. (1950). Ibid., **5b**, 312.
Tsuji, F. I. and Foldes, F. F. (1953). *Fedn Proc. Fedn Am. Socs exp. Biol.*, **12**, 321, 374.
Tullner, W. W. and Hertz, R. (1953). *Endocrinology*, **52**, 359.
Udall, J. A. (1965). *J. Am. med. Ass.*, **194**, 127.
Udani, J. H. and Autian, J. (1960). *J. Am. pharm. Ass. Sci. Edn.*, **49**, 376.
Umezawa, H., Okanishi, M., Kondo, S., Hamana, K., Utahara, R., Maeda, K. and Mitsuhashi, S. (1967a). *Science, N. Y.*, **157**, 1559.
Umezawa, H., Okanishi, M., Utahara, R., Maeda, K. and Kondo, S. (1967b). *J. Antibiot., Japan, Ser, A.*, **20**, 136.
Urnes, P. J. and Doty, P. (1961). *Adv. Protein Chem.*, **16**, 401.
Vallee, B. L., Rupley, J. A., Coombs, T. L. and Neurath, H. (1960). *J. biol. Chem.*, **235**, 64.
Vane, J. R. (1971). *Nature New Biology*, **231**, 232.
van Ekenstein, W. A. and Blanksma, J. J. (1909). *Chem. Weekblad*, **6**, 717.
—(1910). Ibid., **7**, 387.
Van Slyke, D. D. (1929). *J. biol. Chem.*, **83**, 425.
Vargha, L., Toldy, L., Feher, O. and Lendvai, S. (1957). *J. chem. Soc.*, 805.
Vazquez, D. (1964). *Biochem. biophys. Res. Commun.*, **15**, 464.
—(1964). *Nature, Lond.*, **203**, 257.
Veidis, M. V., Palenick, G. J., Schaffrin, R. and Trotter, J. (1969). *J. chem. Soc., A*, 2659.

Velluz, L., Amiard, G. and Heymés, R. (1955). *Bull. Soc. Chim. Fr.*, **22**, 201.
Velluz, L., Petit, A., Pesez, M. and Berret, R. (1947). Ibid., **14**, 123.
Vernon, C. A. (1959). *Chemical Society Special Publication*, no. 8, p. 17.
—(1967). *Proc. R. Soc.*, **B 167**, 389.
Vigliani, E. C. and Forni, A. (1966). *Minerva Med.*, **57**, 3952.
Vilkas, M. and Lederer, E. (1962). *Experientia*, **18**, 546.
von Hippel, P. H. and Wong, K-Y. (1964). *Science, N. Y.*, **145**, 577.
—(1965). *J. biol. Chem.*, **240**, 3909.
Von Schmeissner, H. (1970). *Dtsch. zahnärzl. Zeitschr.*, **25**, 907.
Wagener, H., Mai, H. and Schettler, G. (1963). *Effects of Drugs on Mobilisation and Synthesis of Lipids*, vol. 2, p. 163, eds. Horning, E. C. and Lindgren, P., Pergamon, Oxford.
Waksman, S. A. and Lechevalier, H. A. (1949). *Science, N. Y.*, **109**, 305.
Walling, C. and Helmreich, W. (1959). *J. Am. chem. Soc.*, **81**, 1144.
Walpole, A. L., Roberts, D. C., Rose, F. L., Hendry, J. A. and Homer, R. F. (1954). *Br. J. Pharmac. Chemother.*, **9**, 306.
Walter, W., Curts, J. and Pawelzik, H. (1961). *Annalen*, **643**, 29.
Warner, R. C. and Levy, M. (1958). *J. Am. chem. Soc.*, **80**, 5735.
Wasserman, R. H. and Corradino, R. A. (1971). *A. Rev. Biochem.*, **40**, 501.
Watson, J. D. and Crick, F. H. C. (1953). *Nature, Lond.*, **171**, 737.
Weatherall, M. (1949). *J. Pharm. Pharmac.*, **1**, 576.
Webb, B. H. (1935). *J. Dairy Sci.*, **18**, 81.
Weddon, G. D. (1956). *Hormones and the Aging Process*, p. 221, eds. Engel, L. L. and Pincus, G., New York.
Weil, L., Seibles, T. S. and Herskovits, T. T. (1965). *Archs Biochem. Biophys.*, **111**, 308.
Weinstein, L. and Ehrenkranz, N. J. (1958). *Streptomycin and Dihydrostreptomycin*, Medical Encyclopoedia Inc., New York.
Weisberger, A. S. (1968). *The Interactions of Drugs and Subcellular Components in Animal Cells*, p. 133, ed. Campbell, P. N., Churchill, London.
Weisman, R. A. and Brown, G. M. (1964). *J. biol. Chem.*, **239**, 326.
Weiss, J. (1951). *Experientia*, **7**, 135.
Wendler, N. L. and Graber, R. P. (1956). *Chemy Ind.*, 549.
Weser, U. (1967). *Z. Naturf. B.*, **22**, 457.
Westheimer, F. H. and Nicolaides, N. (1949). *J. Am. chem. Soc.*, **71**, 25.
Wettstein, F. O., Noll, H. and Penman, S. (1964). *Biochim. biophys. Acta*, **87**, 525.
Wettstein, von A., Neher, R. and Urech, H. J. (1959). *Helv. chim. Acta*, **42**, 956.
Weygand, F. and Reckhaus, M. (1949). *Ber. dt. chem. Ges.*, **82**, 438.
Whittaker, V. P. and Wijesundera, S. (1952). *Biochem. J.*, **52**, 475.
Wieland, H. and Bergel, F. (1924). *Annalen*, **439**, 196.
Wiley, P. F. (1962). *J. Am. chem. Soc.*, **84**, 1514.
Wiley, P. F., Argoudelis, A. D. and Hoeksema, H. (1963). Ibid., **85**, 2652.
Williams-Ashman, H. G., Liao, S., Hancock, R. L., Jurkowitz, L. and Silverman, D. A. (1964). *Recent Prog. Horm. Res.*, **20**, 247.
Williamson, A. R. and Schweet, R. S. (1965). *J. molec. Biol.*, **11**, 358.
Williamson, G. M., Morrison, J. K. and Stevens, K. J. (1961). *Lancet*, **i**, 847.
Willmer, E. N. (1961). *Pathol. et biol., Semaine hôp.*, **9**, 885.
Wise, E. M. and Park, J. T. (1965). *Proc. natn. Acad. Sci. U.S.A.*, **54**, 75.
Wiss, O. and Gloor, H. (1966). *Vitamins and Hormones*, **24**, 575.
Wisseman, C. L., Smadel, J. E., Hahn, F. E. and Hopps, H. E. (1954). *J. Bact.*, **67**, 662.
Witkop, B. and Patrick, J. B. (1952). *J. Am. chem. Soc.*, **74**, 3855.
Wohl, A. and Momber, F. (1917). *Ber. dt. chem. Ges.*, **50**, 455.
Wolf, D. E., Hoffman, C. H., Aldrich, P. E., Skeggs, H. R., Wright, L. D. and Folkers, K. (1956). *J. Am. chem. Soc.*, **78**, 4499.
—(1957). Ibid., **79**, 1486.
Wolfrom, M. L., Smith, C. S., Pletcher, D. E. and Brown, A. E. (1942). Ibid., **64**, 23.
Wolfrom, M. L. and Thompson, A. (1957). *The Carbohydrates*, ed. Pigman, W., Academic Press, New York.
Wood, G. C. and Cooper, P. F. (1970). *Chromatogr. Revs.*, **12**, 88.
Woodward, R. B. and Bloch, K. (1953). *J. Am. chem. Soc.*, **75**, 2023.
Wright, L. D. (1961). *A Rev. Biochem.*, **30**, 525.

Wyss, O. (1943). *J. Bact.*, **46**, 483.
Yale, H. L., Losee, K., Martins, J., Holsing, M., Perry, F. M. and Bernstein, J. (1953). *J. Am. chem. Soc.*, **75**, 1933.
Yamamoto, T. and Otsu, T. (1967). *Chemy Ind.*, 787.
Yamana, T., Tsuji, A., Miyamoto, E. and Kiya, E. (1975). *J. Pharm. Pharmac.*, **27**, 56.
Yamashiro, D. and du Vigneaud, V. (1968). *J. Am. chem. Soc.*, **90**, 487.
Yamashiro, D., Hope, D. B. and du Vigneaud, V. (1968). Ibid., **90**, 3857.
Yarmolinsky, M. B. and De La Haba, G. L. (1959). *Proc. natn. Acad. Sci. U.S.A.*, **45**, 1721.
Zaugg, H. E. (1954). *Organic Reactions*, vol. 8, p. 305, ed. Adams, R., John Wiley and Sons, New York.
Zechmeister, L. (1960). *Fortschr. Chem. org. Naturst.*, **18**, 223.
—(1962). *Cis-Trans Isomeric Carotenoids, Vitamins A and Arylpolyenes*, Springer, Berlin.
Zechmeister, L. and Polygár, A. (1943). *J. Am. chem. Soc.*, **65**, 1522.
Ziegler, D. M. and Pettit, F. H. (1964). *Biochem. biophys. Res. Commun.*, **15**, 188.
Ziegler, K., Späth, A., Schaaf, E., Schumann, W. and Winkelmann, E. (1942). *Annalen*, **551**, 80.
Zull, J. E., Czarnowska-Misztal, E. and De Luca, H. F. (1966). *Proc. natn. Acad Sci. U.S.A.*, **55**, 177.
Zürcher, W., Weiss-Berg, E. and Tamm, C. (1969). *Helv. chim. Acta*, **52**, 2449.

# Index